INTERNATIONAL BACCALAUREATE

MATHEMATICS STANDARD LEVEL

3RD EDITION
3rd Imprint

Series editor: **Fabio Cirrito**

Contributing author
Patrick Tobin

Sean's Book

First published in 1997 by IBID Press
2nd Edition published in 1999 by IBID Press,
3rd Edition published in 2004 (2nd Imprint published in 2005) by IBID Press.
Reprinted 2007

Published by IBID Press, Victoria.
Library Catalogue:

Cirrito Fabio Editor., Tobin
1. Mathematics, 2. International Baccalaureate. Series Title: International Baccalaureate in Detail

ISBN: 1 876659 17 3 (10-digit)
978 1 876659 17 2 (13-digit)

We also wish to thank the Incorporated Association of Registered Teachers of Victoria for granting us permission to reproduced questions from their October Examination Papers.

Cover design by Adcore Creative.

Published by IBID Press, at www.ibid.com.au
For further information contact fabio@ibid.com.au

Printed in China through Trojan Press Book Printers
Preston, Victoria, Australia

PREFACE TO 3RD EDITION

The 3rd edition of the Mathematics Standard Level text has been completely revised and updated. Sections of the previous two editions are still present, but much has happened to improve the text both in content and accuracy.

In response to the many requests and suggestions from teachers worldwide the text was extensively revised. There are more examples for students to refer to when learning the subject matter for the first time or during their revision period. There is an abundance of well-graded exercises for students to hone their skills. Of course, it is not expected that any one student work through every question in this text - such a task would be quite a feat. It is hoped then that teachers will guide the students as to which questions to attempt. The questions serve to develop routine skills, reinforce concepts introduced in the topic and develop skills in making appropriate use of the graphics calculator.

The text has been written in a conversational style so that students will find that they are not simply making reference to an encyclopedia filled with mathematical facts, but rather find that they are in some way participating in or listening in on a discussion of the subject matter.

Throughout the text the subject matter is presented using graphical, numerical, algebraic and verbal means whenever appropriate. Classical approaches have been judiciously combined with modern approaches reflecting new technology - in particular the use of the graphics calculator.

The book has been specifically written to meet the demands of the Mathematics Standard Level course and has been pitched at a level that is appropriate for students studying this subject. The book presents an extensive coverage of the syllabus and in some areas goes beyond what is required of the student. Again, this is for the teacher to decide how best to use these sections.

Sets of revision exercises are included in the text. Many of the questions in these sets have been aimed at a level that is on par with what a student could expect on an examination paper. However, some of the questions do go beyond the level that students may expect to find on an examination paper. Success in examinations will depend on an individual's preparation and they will find that making use of a selection of questions from a number of sources will be very beneficial.

Chapter 23 is dedicated to portfolio work. It includes many topics that students can use, either as they stand or as a source of ideas from which to then produce their own investigation. Each suggestion has been extensively trialled and they have been included in the book because they will address the assessment criteria required for the successful completion of the profolio work.

I hope that most of the suggestions and recommendations that were brought forward have been addressed in this edition. However, there is always room for improvement. As always, I welcome and encourage teachers and students to contact me with feedback, not only on their likes and dislikes but suggestions on how the book can be improved as well as where errors and misprints occur. There will be updates on the IBID Press website in relation to errors that have been located in the book – so we suggest that you visit the IBID website at www.ibid.com.au. If you believe you have located an error or misprint please email me at fabio@ibid.com.au.

Fabio Cirrito, July 2004

PREFACE TO 2ND EDITION

The principles behind this second edition are very much the same as those of the first. Where possible we have made amendments based on recommendations and requests by teachers worldwide. Corrections to answers have been made as well as to the text. The support shown by the many teachers worldwide has been very encouraging, and for this reason we have this second edition as well as other support material relating to The Portfolio work and Examination practice papers for students as well as teachers support material.

Fabio Cirrito, 1998

PREFACE TO 1ST EDITION

This text has been produced independently as a resource to support the teaching of the Mathematical Methods Standard Level Course of the International Baccalaureate. The examples and questions do not necessarily reflect the views of the official senior examining team appointed by the International Baccalaureate Organisation.

The **notation** used is, as far as possible, that specified in the appropriate syllabus guidelines of the IB.

The **units** of physical measurements are in S.I.

The **language** and **spelling** are U.K. English.

Currency quantities are specified in dollars, though these could be read as any currency that is decimalised, such as Swiss francs, Lire etc.

The graphic calculators covered directly in the text are the Texas TI/82 and 83. Supplementary material is available from the publisher for students using some other makes and models of calculators. As it is important that students learn to interpret graphic calculator output, the text and answers present a mixture of graphic calculator screens and conventional diagrams when discussing graphs.

The text has been presented in the order in which the topics appear in the syllabus. This does not mean that the topics have to be treated in this order, though it is generally the case that the more fundamental topics appear at the front of the book. Students are reminded that it is the IB Syllabus that specifies the contents of the course and not this text. One of the keys to success in this course is to be thoroughly familiar with the course contents and the styles of questions that have been used in past examinations.

Fabio Cirrito, August 1997.

CONTENTS

NOTATION

The list below represents the signs and symbols which are recommended by the International Organization for Standardization as well as other symbols that are used in the text.

$\mathbb{N}$	the set of positive integers and zero, $\{0, 1, 2, 3,...\}$
$\mathbb{Z}$	the set of integers, $\{0, \pm 1, \pm 2, \pm 3...\}$
$\mathbb{Z}^+$	the set of positive integers, $\{1, 2, 3,...\}$
$\mathbb{Q}$	the set of rational numbers $\left\{x \mid x = \frac{a}{b}, b \neq 0, a, b \in \mathbb{Z}\right\}$
$\mathbb{Q}^+$	the set of positive rational numbers, $\{x \mid x \in \mathbb{Q}, x > 0\}$
$\mathbb{R}$	the set of real numbers
$\mathbb{R}^+$	the set of positive real numbers $\{x \mid x \in \mathbb{R}, x > 0\}$
$\mathbb{C}$	the set of complex numbers, $\{a + bi \mid a, b \in \mathbb{R}\}$
$\{x_1, x_2 \ldots\}$	the set with elements $x_1, x_2 \ldots$
$n(A)$	the number of elements in the finite set A
$\{x \mid \ \}$	the set of all x such that...
$\in$	is an element of
$\notin$	is not an element of
$\varnothing$	the empty (null) set
U	the universal set
$\cup$	union
$\cap$	intersection
$\subset$	is a proper subset of
$\subseteq$	is a proper subset of
A'	the complement of set A
$A \times B$	the Cartesian product of sets A & B, $(A \times B = \{(a, b) \mid a \in A, b \in B\})$
$a \mid b$	a divides b
$a^{1/n}, \sqrt[n]{a}$	a to the power $\frac{1}{n}$ or the nth root of a
$a^{1/2}, \sqrt{a}$	a to the power $\frac{1}{2}$ or the square root of $a \geq 0$
$\lvert x \rvert$	the modulus or absolute value of x $\begin{cases} x, & x \geq 0 \\ -x, & x < 0 \end{cases}$
$\equiv$	identity
$\approx$	is approximately equal to

$>$	is greater than
$\geq$	is greater than or equal to
$<$	is less than
$\leq$	is less than or equal to
$\ngtr$	is not greater than
$\nless$	is not less than
$[a, b]$	the closed interval $a \leq x \leq b$
$]a, b[$	the open interval $a < x < b$
u_n	the nth term of a sequence or series
d	the common difference of an arithmetic sequence
r	the common ratio of an geometric sequence
S_n	the sum of the first n terms of a sequence $u_1 + u_2 + u_3 + \ldots + u_n$
S_∞	the sum to infinity of a sequence $u_1 + u_2 + u_3 + \ldots$
$\sum_{i=1}^{n} u_i$	$u_1 + u_2 + \ldots + u_n$
$\prod_{i=1}^{n} u_i$	$u_1 \times u_2 \times \ldots \times u_n$
$\binom{n}{r}$	$\frac{n!}{r!(n-r)!}$ the rth binimial coefficient, r = 0, 1, 2, . . . in the expansion of $(a+b)^n$
$f : A \rightarrow B$	f is a function under which each element of set A has an image in set B
$f : x \mapsto y$	f is a function under which x is mapped to y
$f(x)$	the image of x under the function f
$f^{-1}(x)$	the inverse function of the function f
$f \circ g$	the composite function of f and g
$\lim_{x \to a} f(x)$	the limit of $f(x)$ as x tends to a
$\frac{dy}{dx}$	the derivative of y with respect to x
$f'(x)$	the derivative of $f(x)$ with respect to x
$\frac{d^2y}{dx^2}$	the second derivative of y with respect to x
$f''(x)$	the second derivative of $f(x)$ with respect to x
$\frac{d^ny}{dx^n}$	the nth derivative of y with respect to x

$f^{(n)}(x)$	the nth derivative of $f(x)$ with respect to x
$\int y dx$	the indefinite integral of y with respect to x
$\int_a^b y dx$	the indefinite integral of y with respect to x between the limits $x = a$ and $x = b$
e^x	the exponential function of x
$\log_a x$	logarithm to the base a of x
$\ln x$	the natural logarithm of x, $\log_e x$
sin,cos,tan	the circular functions
csc,sec,cot	the reciprocal circular functions
$A(x, y)$	the point A in the plane with Cartesian coordinates x and y
[AB]	the line segment with endpoints A and B
AB	the length of [AB]
(AB)	the line containing points A and B
$\hat{A}$	the angle at A
$C\hat{A}B$	the angle between the lines [CA] and [AB]
ΔABC	the triangle whose vertices are A, B and C
$\boldsymbol{v}$	the vector $\boldsymbol{v}$
$\overrightarrow{AB}$	the vector represented in magnitude and direction by the directed line segment from A to B
$\boldsymbol{a}$	the position vector $\overrightarrow{OA}$
$\boldsymbol{i}, \boldsymbol{j}, \boldsymbol{k}$	unit vectors in the directions of the Cartesian coordinate axes
$\lvert\boldsymbol{a}\rvert$	the magnitude of $\boldsymbol{a}$
$\lvert\overrightarrow{AB}\rvert$	the magnitude of $\overrightarrow{AB}$
$\boldsymbol{v} \cdot \boldsymbol{w}$	the scalar product of $\boldsymbol{v}$ and $\boldsymbol{w}$
A^{-1}	the inverse of the non-singular matrix A
A^T	the transpose of the matrix A
$\det A$	the determinant of the square matrix A
I	the identity matrix
$P(A)$	probability of event A
$P(A')$	probability of the event "not A"
$P(A \mid B)$	probability of the event A given B
$x_1, x_2, \ldots$	observations
$f_1, f_2, \ldots$	frequencies with which the observations $x_1, x_2, \ldots$ occur
P_x	probability distribution function $P(X = x)$ of the discrete random variable X

$\mathrm{E}(X)$	the expected value of the random variable X
$\mathrm{Var}(X)$	the variance of the random variable X
μ	population mean
σ^2	population variance, $\sigma^2 = \dfrac{\sum_{i=1}^{k} f_i(x_i - \mu)^2}{n}$ where $n = \sum_{i=1}^{k} f_i$
σ	population standard deviation
$\bar{x}$	sample mean
s_n^2	sample variance, $\sigma^2 = \dfrac{\sum_{i=1}^{k} f_i(x_i - \bar{x})^2}{n}$ where $n = \sum_{i=1}^{k} f_i$
s_n	standard deviation of the sample
s_{n-1}^2	unbiased estimate of the population variance $s_{n-1}^2 = \dfrac{n}{n-1} s_n^2$ or $\dfrac{\sum_{i=1}^{k} f_i(x_i - \bar{x})^2}{n-1}$
$\mathrm{B}(n,p)$	binomial distribution with parameters n and p
$\mathrm{N}(\mu, \sigma^2)$	normal distribution with mean μ and variance σ^2
$X \sim \mathrm{B}(n,p)$	the random variable X has a binomial distribution with parameters n and p
$X \sim \mathrm{N}(\mu, \sigma^2)$	the random variable X has a normal distribution with mean μ and variance σ^2
Φ	cumulative distribution function of the standardised normal variable: N(0,1)

1.1 PURE AND APPLIED MATHEMATICS

Mathematics has clearly played a significant part in the development of many past and present civilisations.

There is good evidence that mathematical, and probably astronomical techniques, were used to build the many stone circles of Europe which are thought to be at least three thousand years old (Thom). It is likely that the Egyptian pyramids and constructions on Aztec and Mayan sites in South America were also built by mathematically sophisticated architects. Similarly, cultures in China, India and throughout the Middle East developed mathematics a very long time ago. It is also the case that there have been very successful cultures that have found little use for mathematics. Ancient Rome, handicapped, as it was, by a non-place value number system did not develop a mathematical tradition at anything like the same level as that of Ancient Greece. Also, the Australian Aborigines who have one of the most long lasting and successful cultures in human history did not find much need for mathematical methods. The same is true of the many aboriginal cultures of Africa, Asia and the Americas. This may well be because these aboriginal cultures did not value ownership in the way that western culture does and had no need to count their possessions. Instead, to aboriginal cultures, a responsible and sustainable relationship with the environment is more important than acquisition and exploitation. Maybe we should learn from this before it is too late!

Mathematics has developed two distinct branches. Pure mathematics, which is studied for its own sake, and applied mathematics which is studied for its usefulness. This is not to say that the two branches have not cross-fertilised each other, for there have been many examples in which they have.

The pure mathematician Pierre de Fermat (1601-1665) guessed that the equation $x^n + y^n = z^n$ has whole numbered solutions for $n = 2$ only. To the pure mathematician, this type of problem is interesting for its own sake. To study it is to look for an essential truth, the 'majestic clockwork' of the universe. Pure mathematicians see 'beauty' and 'elegance' in a neat proof. To pure mathematicians, their subject is an art.

Applied mathematics seeks to develop mathematical objects such as equations and computer algorithms that can be used to predict what will happen if we follow a particular course of action. This is a very valuable capability. We no longer build bridges without making careful calculations as to whether or not they will stand. Airline pilots are able to experience serious failures in commercial jets without either risking lives or the airline's valuable aeroplanes or, indeed, without even leaving the ground.

1.2 AXIOMS

Mathematics is based on axioms. These are 'facts' that are assumed to be true. An axiom is a statement that is accepted without proof. Early sets of axioms contained statements that appeared to be obviously true. Euclid postulated a number of these 'obvious' axioms.

Example:

'Things equal to the same thing are equal to each other';

That is, if $y = a$ and $x = a$ then $y = x$.

Euclid was mainly interested in geometry and we still call plane geometry 'Euclidean'. In Euclidean space, the shortest distance between two points is a straight line. We will see later that it is possible to develop a useful, consistent mathematics that does not accept this axiom.

Most axiom systems have been based on the notion of a 'set', meaning a collection of objects. An example of a set axiom is the 'axiom of specification'. In crude terms, this says that if we have a set of objects and are looking at placing some condition or specification on this set, then the set thus specified must exist. We consider some examples of this axiom.

Example:

Assume that the set of citizens of China is defined. If we impose the condition that the members of this set must be female, then this new set (of Chinese females) is defined.

As a more mathematical example, if we assume that the set of whole numbers exists, then the set of even numbers (multiples of 2) must also exist.

A second example of a set axiom is the 'axiom of powers':

Example:

For each set, there exists a collection of sets that contains amongst its elements all the subsets of the original set. If we look at the set of cats in Bogota, then there must be a set that contains all the female cats in Bogota, another that contains all the cats with green eyes in Bogota, another that contains all the Bogota cats with black tails, etc. A good, but theoretical, account of axiomatic set theory can be found in Halmos, 1960.

Mathematics has, in some sense, been a search for the smallest possible set of consistent axioms. In the section on paradox, we will look further at the notion of axioms and the search for a set of assumptions that does not lead to contradictions. There is a very strong sense in which mathematics is an unusual pursuit in this respect. Pure mathematics is concerned with absolute truth only in the sense of creating a self-consistent structure of thinking.

As an example of some axioms that may not seem to be sensible, consider a geometry in which the shortest path between two points is the arc of a circle and all parallel lines meet. These 'axioms' do not seem to make sense in 'normal' geometry. The first mathematicians to investigate non-Euclidean geometry were the Russian, Nicolai Lobachevsky (1793-1856) and the Hungarian, Janos Bolyai (1802-1860). Independently, they developed self consistent geometries that did not include the so called parallel postulate which states that for every line AB and point C outside AB there is only one line through C that does not meet AB.

Since both lines extend to infinity in both directions, this seems to be 'obvious'. Non-Euclidean geometries do not include this postulate and assume either that there are no lines through C that do not meet AB or that there is more than one such line. It was the great achievement of Lobachevsky and Bolyai that they proved that these assumptions lead to geometries that are self consistent and thus acceptable as 'true' to pure mathematicians. In case you are thinking that this sort of activity is completely useless, one of the two non-Euclidean geometries discussed above has actually proved to be useful; the geometry of shapes drawn on a sphere. This is useful because it is the geometry used by the navigators of aeroplanes and ships.

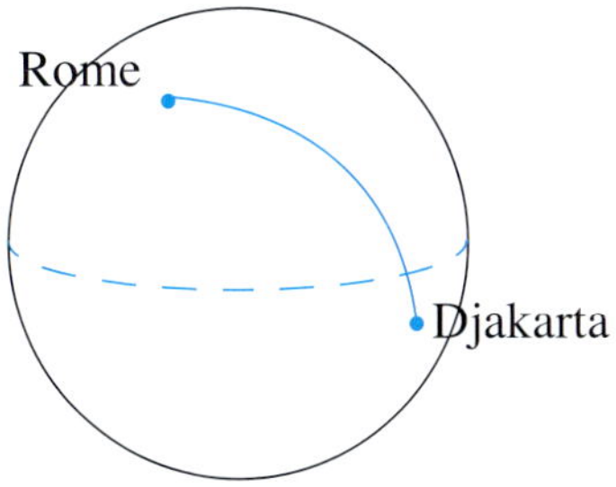

The first point about this geometry is that it is impossible to travel in straight lines. On the surface of a sphere, the shortest distance between two points is an arc of a circle centred at the centre of the sphere (a great circle). The shortest path from Rome to Djakarta is circular. If you want to see this path on a geographer's globe, take a length of sewing cotton and stretch it tightly between the two cities. The cotton will follow the approximate great circle route between the two cities.

If we now think of the arcs of great circles as our 'straight lines', what kind of geometry will we get? You can see some of these results without going into any complex calculations. For example, what would a triangle look like?

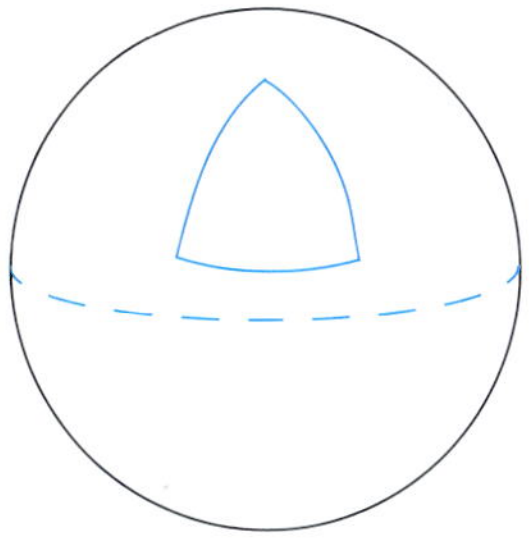

The first point is that the angles of this triangle add up to more than 180°. There are many other 'odd' features of this geometry. However, fortunately for the international airline trade, the geometry is self consistent and allows us to navigate safely around the surface of the globe. Thus non-Euclidean geometry is an acceptable pure mathematical structure.

While you are thinking about unusual geometries, what are the main features of the geometry of shapes drawn on the 'saddle surface'?

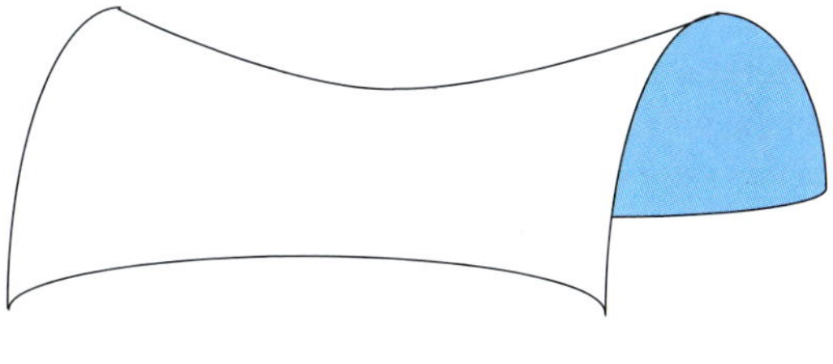

One final point on the subject of non-Euclidean geometries; it seems to be the case that our three dimensional universe is also curved. This was one of the great insights of Albert Einstein (1879-1955). We do not yet know if our universe is bent back on itself rather like a sphere or whether another model is appropriate. A short account of non-Euclidean Geometries can be found in Cameron (pp31-40).

By contrast, applied mathematics is judged more by its ability to predict the future, than by its self-consistency. Applied mathematics is also based on axioms, but these are judged more on their ability to lead to calculations that can predict eclipses, cyclones, whether or not a suspension bridge will be able to support traffic loads, etc. In some cases such mathematical models can be very complex and may not give very accurate predictions. Applied mathematics is about getting a prediction, evaluating it (seeing how well it predicts the future) and then improving the model.

In summary, both branches of mathematics are based on axioms. These may or may not be designed to be 'realistic'. What matters to the pure mathematician is that an axiom set should not lead to contradictions. The applied mathematician is looking for an axiom set and a mathematical structure built on these axioms that can be used to model the phenomena that we observe in nature. As we have seen, useful axiom sets need not start out being 'sensible'.

The system of deduction that we use to build the other truths of mathematics is known as **proof**.

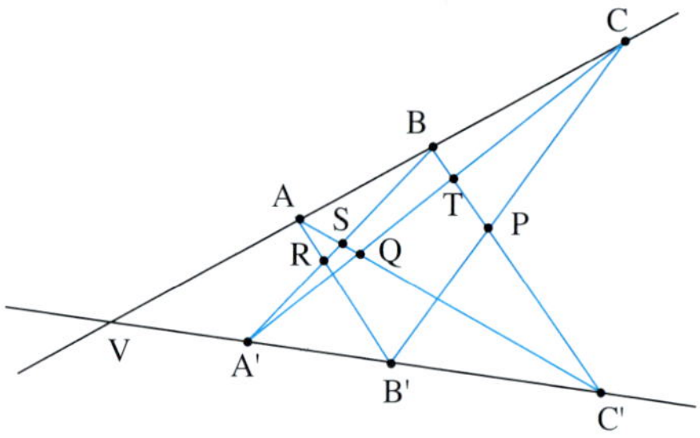

$\Rightarrow P, Q, R$ are collinear

1.3 PROOF

Proof has a very special meaning in mathematics. We use the word generally to mean 'proof beyond reasonable doubt' in situations such as law courts when we accept some doubt in a verdict. For mathematicians, proof is an argument that has *no* doubt at all. When a new proof is published, it is scrutinised and criticised by other mathematicians and is accepted when it is established that every step in the argument is legitimate. Only when this has happened does a proof become accepted.

Technically, every step in a proof rests on the axioms of the mathematics that is being used. As we have seen, there is more than one set of axioms that could be chosen. The statements that we prove from the axioms are known as **theorems**. Once we have a theorem, it becomes a statement that we accept as true and which can be used in the proof of other theorems. In this way we build up a structure that constitutes a 'mathematics'. The axioms are the foundations and the theorems are the superstructure. In the previous section we made use of the idea of consistency. This means that it must not be possible to use our axiom set to prove two theorems that are contradictory.

There are a variety of methods of proof available. This section will look at three of these in detail. We will mention others.

1.3.1 RULES OF INFERENCE

All proofs depend on rules of inference. Fundamental to these rules is the idea of 'implication'.

As an example, we can say that $2x = 4$ (which is known as a **proposition**) implies that $x = 2$ (provided that x is a normal real number and that we are talking about normal arithmetic).

In mathematical shorthand we would write this statement as $2x = 4 \Rightarrow x = 2$.

This implication works both ways because $x = 2$ implies that $2x = 4$ also.

This is written as $x = 2 \Rightarrow 2x = 4$ or the fact that the implication is both ways can be written as

$$x = 2 \Leftrightarrow 2x = 4 .$$

The $\Leftrightarrow$ symbol is read as '**If and only if**' or simply as '**Iff**', i.e., If with two fs.

Not every implication works both ways in this manner:

If $x = 2$ then we can conclude that $x^2 = 4$.
However, we cannot conclude the reverse:
i.e., $x^2 = 4$ implies that $x = 2$ is false because x might be –2.

Sothat $x = 2 \Rightarrow x^2 = 4$ is all that can be said in this case.

There are four main rules of inference:

1. The rule of detachment: from a is true and $a \Rightarrow b$ is true we can infer that b is true. a and b are propositions.

Example: If the following propositions are true:

It is raining.
If it is raining, I will take an umbrella.

We can infer that I will take an umbrella.

2. The rule of syllogism: from $a \Rightarrow b$ is true and $b \Rightarrow c$ is true, we can conclude that $a \Rightarrow c$ is true. a, b & c are propositions.

Example: If we accept as true that:
if x is an odd number then x is not divisible by 4 ($a \Rightarrow b$)and,
if x is not divisible by 4 then x is not divisible by 16 ($b \Rightarrow c$)

We can infer that the proposition;

if x is an odd number then x is not divisible by 16 ($a \Rightarrow c$) is true.

3. The rule of equivalence: at any stage in an argument we can replace any statement by an equivalent statement.

Example: If x is a whole number, the statement x is even could be replaced by the statement x is divisible by 2.

4. The rule of substitution: If we have a true statement about all the elements of a set, then that statement is true about any individual member of the set.

Example: If we accept that all lions have sharp teeth then Benji, who is a lion, must have sharp teeth.

Now that we have our rules of inference, we can look at some of the most commonly used methods of proof

1.3.2 PROOF BY EXHAUSTION

This method can be, as its name implies, exhausting! It depends on testing every possible case of a theorem.

Example:

Consider the theorem: Every year must contain at least one 'Friday the thirteenth'.

There are a limited number of possibilities as the first day of every year must be a Monday or a Tuesday or a Wednesday.... or a Sunday (7 possibilities). Taking the fact that the year might or might not be a leap year (with 366 days) means that there are going to be fourteen possibilities.

Once we have established all the possibilities, we would look at the calendar associated with each and establish whether or not it has a 'Friday the thirteenth'. If, for example, we are looking at a non-leap year in which January 1st is a Saturday, there will be a 'Friday the thirteenth' in May. Take a look at all the possibilities (an electronic organiser helps!). Is the theorem true?

1.3.3 DIRECT PROOF

The following diagrams represent a proof of the theorem of Pythagoras described in 'The Ascent of Man' (Bronowski pp 158-161). The theorem states that the area of a square drawn on the hypotenuse of a right angled triangle is equal to the sum of the areas of the squares drawn on the two shorter sides. The method is direct in the sense that it makes no assumptions at the start. Can you follow the steps of this proof and draw the appropriate conclusion?

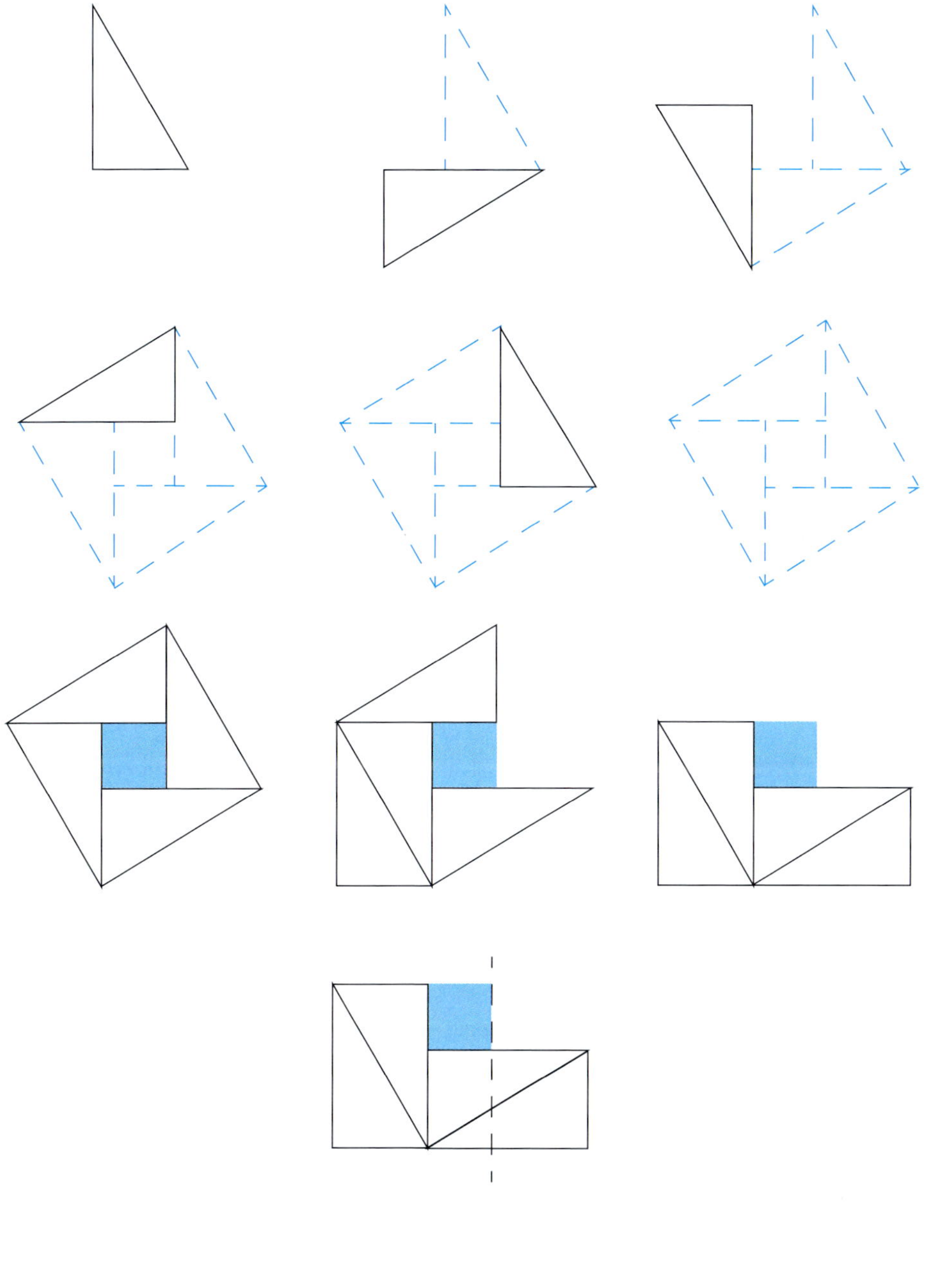

1.3.4 PROOF BY CONTRADICTION

This method works by assuming that the proposition is false and then proving that this assumption leads to a contradiction.

Example:

The number $\sqrt{2}$ greatly interested classical Greek mathematicians who were unable to find a number that, when it was squared, gave exactly 2.

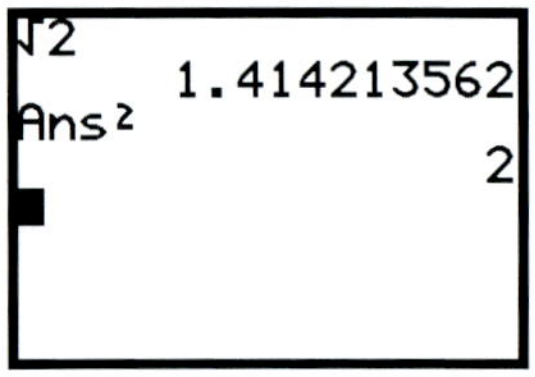

Modern students are often fooled into thinking that their calculators give an exact square root for 2 as when 2 is entered and the square root button is pressed, a result (depending on the model of calculator) of 1.414213562 is produced. When this is squared, exactly 2 results. This is not because we have an exact square root. It results from the way in which the calculator is designed to calculate with more figures than it actually displays. The first answer is stored to more figures than are shown, the result is rounded and then displayed. The same is true of the second result which only rounds to 2. Try squaring 1.414213562, the answer is not 2.

The theorem we shall prove is that there is *no* fraction that when squared gives 2. This also implies that there is no terminating or recurring decimal that, when squared, gives exactly 2, but this further theorem requires more argument.

The method begins by assuming that there *is* a fraction $\frac{p}{q}$ (p & q are integers) which has been cancelled to its lowest terms, such that $\frac{p}{q} = \sqrt{2}$. From the assumption, the argument proceeds:

$$\frac{p}{q} = \sqrt{2} \Rightarrow \frac{p^2}{q^2} = 2 \Rightarrow p^2 = 2q^2 \Rightarrow p^2 \text{ is even} \Rightarrow p \text{ is even}$$

As with most mathematical proofs, we have used simple axioms and theorems of arithmetic.

The most complex theorem used is that if p^2 is even, then p is even. Can you prove this?

The main proof continues with the deduction that if p is even there must be another *integer, r,* that is half p. $p = 2r \Rightarrow p^2 = 4r^2 \Rightarrow 2q^2 = 4r^2 \Rightarrow q^2 = 2r^2 \Rightarrow q^2$ is even $\Rightarrow q$ is even

We now have our contradiction as we assumed that $\frac{p}{q}$ was in its lowest terms so p & q cannot both be even. This proves the result, because we have a contradiction.

This theorem is a very strong statement of *impossibility*.

There are very few other areas of knowledge in which we can make similar statements. We might be virtually certain that we will never travel faster than the speed of light but it would be a brave physicist who would state with certainty that it is *impossible*. Other methods of proof include proof by induction which is mainly used to prove theorems involving sequences of statements. Whilst on the subject of proof, it is worth noting that it is much easier to disprove a statement than to prove it. When we succeed in disproving a statement, we have succeeded in proving its negation or reverse. To disprove a statement, all we need is a single example of a case in which the theorem does not hold. Such a case is known as a **counter-example**.

Example:

The theorem 'all prime numbers are odd' is false. This can be established by noting that 2 is an even prime and, therefore, is the only counter-example we need to give. By this method we have proved the theorem that 'not every prime number is odd'.

This is another example of the way in which pure mathematicians think in a slightly different way from other disciplines. Zoo-keepers (and indeed the rest of us) might be happy with the statement that 'all giraffes have long necks' and would not be very impressed with a pure mathematician who said that the statement was false because there was one giraffe (with a birth defect) who has a very short neck. This goes back to the slightly different standards of proof that are required in mathematics.

Counter-examples and proofs in mathematics may be difficult to find.

Consider the theorem that every odd positive integer is the sum of a prime number and twice the square of an integer. Examples of this theorem that do work are:

$$5 = 3 + 2 \times 1^2,\ 15 = 13 + 2 \times 1^2,\ 35 = 17 + 2 \times 3^2.$$

The theorem remains true for a very large number of cases and we do not arrive at a counter-example until 5777.

Another similar 'theorem' is known as the Goldbach Conjecture. Christian Goldbach (1690-1764) stated that every even number larger than 2 can be written as the sum of two primes. For example, $4 = 2 + 2$, $10 = 3 + 7$, $48 = 19 + 29$ etc. No-one has every found a counter-example to this 'simple' conjecture and yet no accepted proof has ever been produced, despite the fact that the conjecture is not exactly recent!

Finally, whilst considering proof, it would be a mistake to think that mathematics is a complete set of truths that has nothing which needs to be added. We have already seen that there are unproved theorems that we suspect to be true. It is also the case that new branches of mathematics are emerging with a fair degree of regularity. During this course you will study linear programming which was developed in the 1940s to help solve the problems associated with the distribution of limited resources. Recently, both pure and applied mathematics have been enriched by the development of 'Chaos Theory'. This has produced items of beauty such as the Mandelbrot set and insights into the workings of nature. It seems, for example, that the results of Chaos Theory indicate that accurate long term weather forecasts will never be possible (Mandelbrot).

1.4 PARADOX

1.4.1 WHAT IS A PARADOX?

Pure mathematics is a quest for a structure that does not contain internal contradictions. A satisfactory mathematics will contain no 'nonsense'.

Consider the following proof:

Let $x = 1$	
Then $x^2 - 1 = x - 1$	Try substituting $x = 1$ to check this line.
$(x+1)(x-1) = x - 1$	Factorising using the difference of two squares.
$x + 1 = 1$	Dividing both sides by $x - 1$.
$2 = 1$	Substituting $x = 1$.

There is obviously something wrong here as this is the sort of inconsistency that we have discussed earlier in this chapter, but what is wrong? To discover this, we must check each line of the argument for errors or faulty reasoning.

Line 1	must be acceptable as we are entitled to assign a numerical value to a pronumeral.
Line 2	is true because the left hand and right hand sides are the same if we substitute the given value of the pronumeral.
Line 3	is a simple factorisation of the left hand side.
Line 4	is obtained from line 3 by dividing both sides of the equation by $x - 1$ and should be acceptable as we have 'done the same thing' to both sides of the equation.
Line 5	is obtained from line 4 by substituting $x = 1$ and so should give the correct answer.

Obviously we have an unacceptable conclusion from a seemingly watertight argument. There must be something there that needs to be removed as an acceptable operation in mathematics.

The unacceptable operation is dividing both sides by $x - 1$ and then using a value of 1 for x. What we have effectively done is divide by a quantity that is zero. It is this operation that has allowed us to prove that $2 = 1$, an unacceptable result. When a **paradox** of this sort arises, we need to look at the steps of the proof to see if there is a faulty step. If there is, then the faulty step must be removed. In this case, we must add this rule to the allowed operations of mathematics:

Never divide by a quantity that is, or will become, zero.

This rule, often ignored by students, has important implications for Algebra and Calculus.

Some paradoxes are arguments that seem to be sound but contain a hidden error and thus do not contain serious implications for the structure of mathematical logic. An amusing compilation of simple paradoxes can be found in Gardner (1982). An example is the 'elevator paradox'.

Why does it always seem that when we are waiting for an elevator near the bottom of a tall building and wanting to go up, the first elevator to arrive is always going down? Also, when we want to go back down, why is the first elevator to arrive always going up? Is this a real phenomenon or is it just a subjective result of our impatience for the elevator to arrive? Or is it another example of Murphy's Law; whatever can go wrong will go wrong?

This is quite a complex question, but a simple explanation might run as follows:

If we are waiting near the bottom of a tall building, there are a small number of floors below us from which elevators that are going up might come and then pass our floor.

By contrast, there are more floors above us from which elevators might come and then pass our floor going down.

On the basis of this and assuming that the elevators are randomly distributed amongst the floors, it is more likely that the next elevator to pass will come from above and will, therefore, be going down.

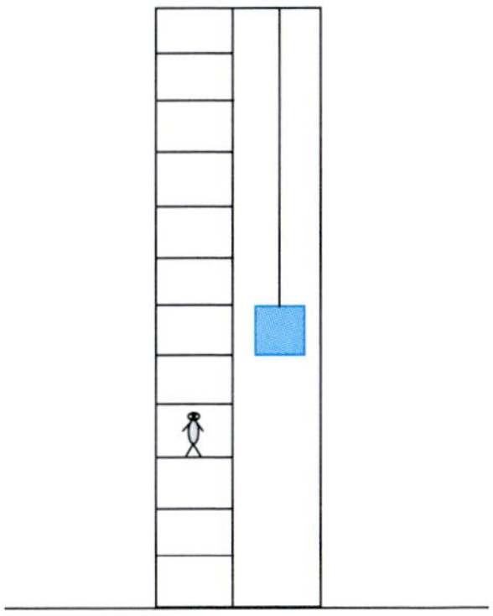

By contrast, if we are waiting near the top of a tall building, there are a small number of floors above us from which elevators that are going down might come and then pass our floor.

Also, there are more floors below us from which elevators might come and then pass our floor going up.

It is more likely that the next elevator to pass will come from below and will, therefore, be going up.

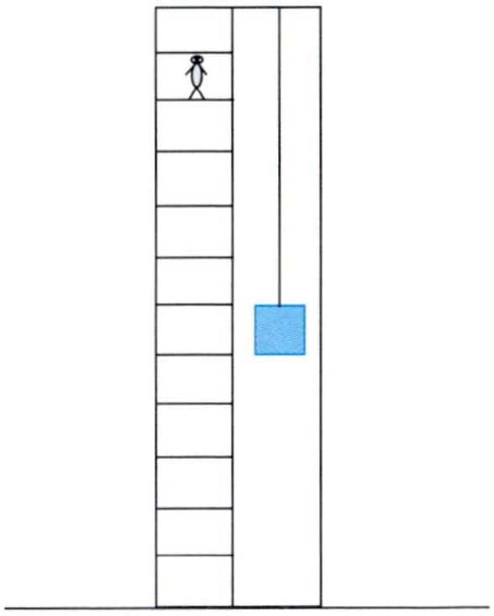

A fuller analysis of this paradox can be found in Gardner (pp96-97).

The elevator paradox does not contain serious implication for the structure of mathematics like our first example. We will conclude this section with a look at a modern paradox that did cause a re-evaluation of one of the basic ideas of mathematics, the set.

1.4.2 RUSSELL'S PARADOX

Bertrand Russell (1872-1970) looked in detail at the basic set axioms of mathematics. We do regard the existence of sets as axiomatic in all mathematical structures. Does this mean that we can make a set that contains 'everything'? There would seem to be no difficulty with this as we just move around the universe and sweep everything that we meet into our set, numbers, words, whales, motorcycles etc. and the result is the set that contains everything.

Russell posed the following question which we will relate in the context of library catalogues. Every library has a catalogue. There are various forms that this catalogue might take; a book, a set of cards, a computer disc etc. Whatever form the catalogue in your local library takes, there is a sense in which this catalogue is a book (or publication) owned by the library and, as such, should appear as an entry in the catalogue:

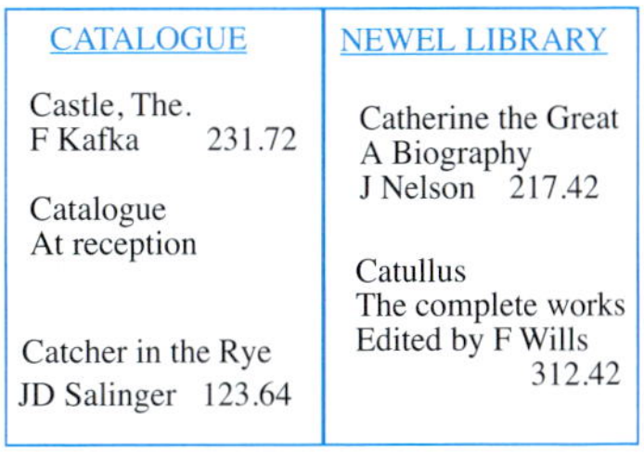

Of course, many librarians will decide that it is silly to include the catalogue as an entry in the catalogue because people who are already looking at the catalogue know where to find it in the library! It follows that library catalogues can be divided into two distinct groups:

Catalogues that do contain an entry describing themselves.
Catalogues that do not contain an entry describing themselves.

Next, let us make a catalogue of *all* the catalogues of type two, those that do not contain themselves.

This gives us a problem. Should we include an entry describing our new catalogue? If we do, then our catalogue ceases to be a catalogue of all those catalogues that do not contain themselves. If we do not, then our catalogue is no longer a complete catalogue of all those catalogues that do not contain themselves.

The conclusion is that making such a catalogue is impossible. This does not mean that the library catalogues themselves cannot exist. We have, however, defined an impossible catalogue.

In set terms, Russell's paradox says that sets are of two types:

Type 1 Sets that do contain themselves.
Type 2 Sets that do not contain themselves.

The set of all sets of type 2 cannot be properly defined without reaching a contradiction.

The most commonly accepted result of Russell's paradox is the conclusion that we have to be very careful when we talk about sets of everything. The most usual way out is to work within a carefully defined universal set, chosen to be appropriate to the mathematics that we are undertaking. If we are doing normal arithmetic, the universal set is the set of real numbers.

1.5 MATHEMATICS AND OTHER DISCIPLINES

When writing Theory of Knowledge essays, students are required to develop their arguments in a cross disciplinary way. For more details on this, you are strongly advised to read the task specifications and the assessment criteria that accompany the essay title. You are reminded that it is these statements that define what is expected of a good essay, not the contents of this Chapter which have been provided as a background resource. A good essay will only result if you develop your own ideas and examples in a clear and connected manner. Part of this process may include comparing the 'mathematical method' described earlier with the methods that are appropriate to other systems of knowledge.

As we have seen, mathematics rests on sets of axioms. This is true of many other disciplines. There is a sense in which many ethical systems also have their axioms such as 'Thou shalt not kill'.

The Ancient Greeks believed that beauty and harmony are based, almost axiomatically, on mathematical proportions. The golden mean is found by dividing a line in the following ratio:

The ratio of the length AB to the length BC is the same as the ratio of the length BC to the whole length AC. The actual ratio is $1:\frac{1}{2}(1+\sqrt{5})$ or about 1:1.618. The Greek idea was that if this line is converted into a rectangle, then the shape produced would be in perfect proportion:

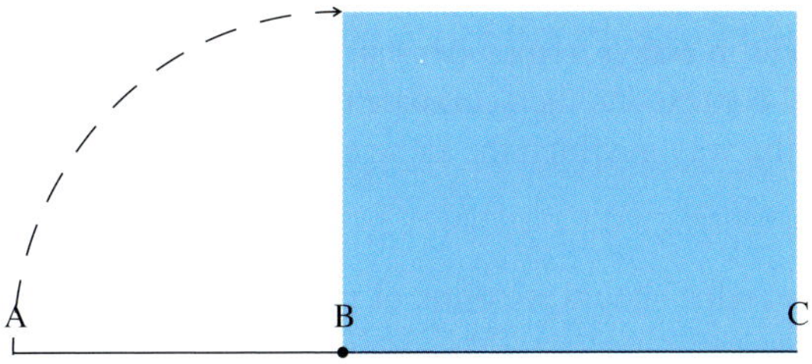

Likewise, the correct place to put the centre of interest in a picture is placed at the golden mean position between the sides and also at the golden mean between top and bottom. Take a look at the way in which television pictures are composed to see if we still use this idea:

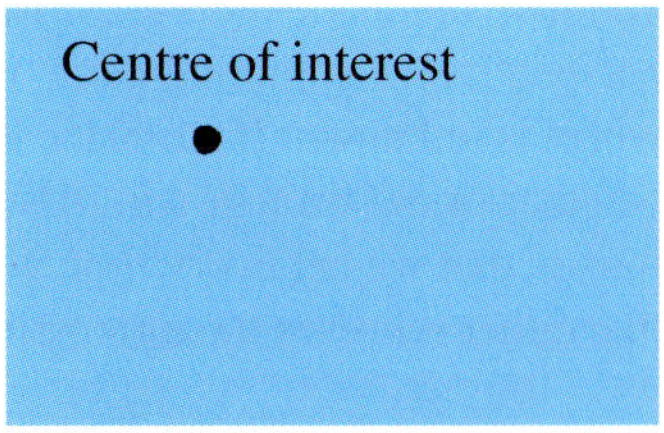

In a similar way, the Ancient Greeks believed that ratio determined harmony in music. If two similar strings whose lengths bear a simple ratio such as 1:2 or 2:3 are plucked together the resulting sound will be pleasant (harmonious). If the ratio of string lengths is 'awkward', such as 17:19, then the notes will be discordant. The same principle of simple ratios is used in tuning musical instruments (in most cultures) today.

The most common connection between mathematics and other disciplines is the use of mathematics as a tool. Examples are: the use of statistics by insurance actuaries, probability by quality control officers and of almost all branches of mathematics by engineers. Every time mathematics is used in this way, there is an assumption that the calculations will be done using techniques that produce consistent and correct answers. It is here that pure mathematical techniques, applied mathematical modelling and other disciplines interface.

In some of these examples, we apply very precise criteria to our calculations and are prepared to accept only very low levels of error. Navigation satellite systems work by measuring the position of a point on or above the Earth relative to the positions of satellites orbiting the Earth.

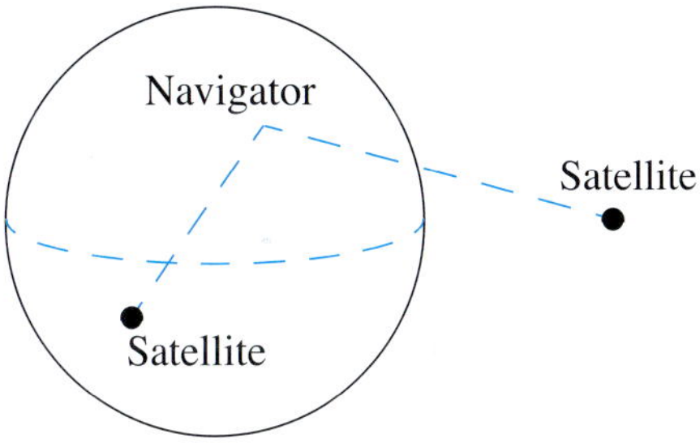

This system will only work if the positions of the satellites are known with very great precision.

By contrast, when calculations are made to forecast the weather, whilst they are done with as much precision as necessary, because the data is incomplete and the atmospheric models used are approximate, the results of the calculations are, at best, only an indication of what might happen. Fortunately, most of us expect this and are much more tolerant of errors in weather forecasting than we would be if airlines regularly failed to find their destinations!

There are, therefore a large number of ways in which mathematics complements other disciplines. In fact, because computers are essentially mathematical devices and we are increasingly dependent on them, it could be argued that mathematics and its methods underpin the modern world.

That is not to say that mathematics is 'everywhere'. Many very successful people have managed to avoid the subject altogether. Great art, music and poetry has been produced by people for whom mathematical ideas held little interest.

In using mathematical ideas in essays, remember that you should produce original examples, look at them in a mathematical context and then compare the ways in which the example might appear to a mathematician with the way in which the same example might appear to a thinker from another discipline.

As a very simple example, what should we think of gambling?

To the mathematician (Pascal was one of the first to look at this activity from the mathematical perspective), a gambling game is a probability event. The outcome of a single spin of a roulette wheel is unknown. If we place a single bet, we can only know the chances of winning, not whether or not we *will* win. Also, in the long run, we can expect to lose one thirtyseventh of any money that we bet every time we play. To the mathematician, (or at least to this mathematician) this rather removes the interest from the game!

Other people look at gambling from a different standpoint. To the politician, a casino is a source of revenue and possibly a focus of some social problems. To a social scientist, the major concern might be problem gamblers and the effect that gambling has on the fabric of society. A theologian might look at the ethical issues as being paramount. Is it ethical to take money for a service such as is provided by a casino? Many of these people might use mathematics in their investigations, but they are all bringing a slightly different view to the discussion.

As we can see, there are many sides to this question as there are many sides to most questions. Mathematics can often illuminate these, but will seldom provide all the answers. When you choose an essay title, you do not have to use mathematical ideas or a mathematical method to develop your analysis. However, we hope that if you do choose to do this, you will find the brief sketch of the mathematical method described in this Chapter helpful.

We will finish with one observation.

Mathematics and mathematicians are sometimes viewed as dry and unimaginative. This may be true in some cases, but definitely not all.

We conclude with some remarks by the mathematician Charles Dodgson (1832-1898), otherwise known as Lewis Carroll:

'The time has come', the Walrus said,
'To talk of many things:
Of shoes and ships and sealing wax,
Of cabbages and kings,
Of why the sea is boiling hot
And whether pigs have wings'.
Through the Looking Glass

References:
Megalithic Sites in Britain. Thom, A. (1967). U.K. Oxford University Press.
Heritage Mathematics. Cameron, M. (1984).U.K. E.J. Arnold.
The Ascent of Man, Bronowski, J. (1973).U.K. BBC.
The Fractal Geometry of Nature, Mandelbrot, B (1977), U.S. W.H. Freeman & Co. 1977.
Gotcha!, Gardner, M. (1977). U.S.A. W.H. Freeman & Co.

2.1 THE REAL NUMBER LINE

2.1.1 THE REAL NUMBER LINE

A visual method to represent the set of **real numbers**, $\mathbb{R}$, is to use a straight line. This geometrical representation is known as the **real number line**. It is usually drawn as a horizontal straight line extending out indefinitely in two directions with a point of reference known as the origin, usually denoted by the letter O. Corresponding to every real number x there is a point P, on the line, representing this value. If $x > 0$, the point P lies to the right of O. If $x < 0$ the point P lies to the left of O. If $x = 0$, the point P is at the origin, O.

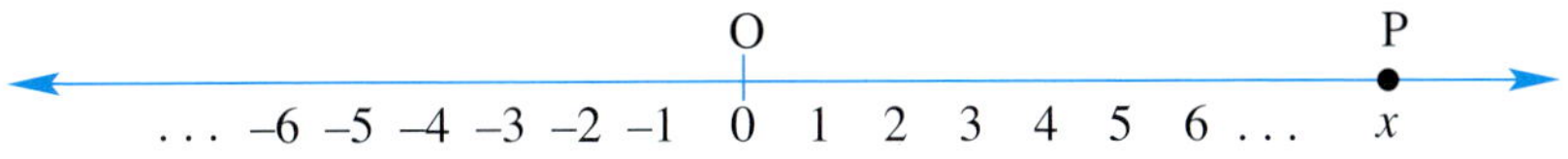

2.1.2 SET BUILDER NOTATION

The set of points on the real number line can also be written in an algebraic form:

$$\mathbb{R} = \{x : -\infty < x < \infty\}$$

This means that any real number set can be expressed algebraically. For example, the set of

positive real numbers $= \mathbb{R}^+ = \{x : x > 0\}$,
negative real numbers $= \mathbb{R}^- = \{x : x < 0\}$

Note that $\mathbb{R} = \mathbb{R}^- \cup \{0\} \cup \mathbb{R}^+$.

Similarly we can construct any subset of the real number line. The great thing about using set notation is that we can quickly identify if a point on the number line is included or excluded in the set. How do we represent these inclusions and exclusions on the real number line?

If the number is **included** in the set, you 'black-out' a circle at that point on the number line – this is called a **closed circle**.
For example, the set $\{x : x \geq 3\}$ has the following representation:

If the number is **excluded** from the set, you place a circle at that point on the number line – this is called an **open circle**.
For example, the set $\{x : x > 3\}$ has the following representation:

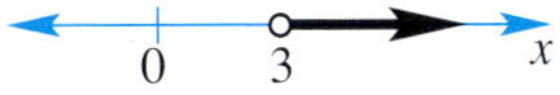

2.1.3 INTERVAL NOTATION

Another notation that is used to describe subsets of the real numbers is **interval notation**. This form of notation makes use of only 'square brackets' or 'square brackets' and 'round brackets' to indicate if a number is included or excluded. For the examples above we have:

$$\{x : x \geq 3\} = [3, \infty) \text{ or } [3, \infty[$$

and for $\{x : x > 3\} = (3, \infty) \text{ or }]3, \infty[$

Notice that '∞' is never included!

It should be noted that $\{x : x \geq 3\}$ can also be expressed as $\{x : 3 \leq x < \infty\}$. Hence the reason for having '∞' in the interval notation representation.

A summary of the different possible intervals is given in the table below:

Real Number Line	Set Notation	Interval Notation	Example
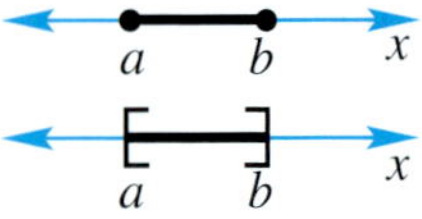	$\{x : a \leq x \leq b\}$	$x \in [a, b]$	 $\{x : 3 \leq x \leq 8\}$ or $x \in [3, 8]$
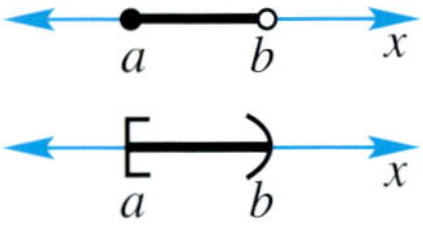	$\{x : a \leq x < b\}$	$x \in [a, b)$ or $x \in [a, b[$	 $\{x : 3 \leq x < 8\}$ or $x \in [3, 8)$ $x \in [3, 8[$
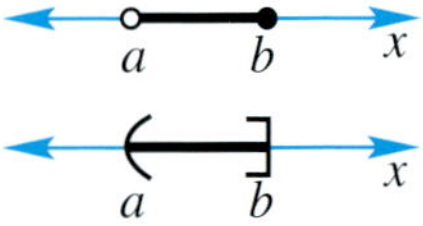	$\{x : a < x \leq b\}$	$x \in (a, b]$ or $x \in]a, b]$	 $\{x : 3 < x \leq 8\}$ or $x \in (3, 8]$ $x \in]3, 8]$
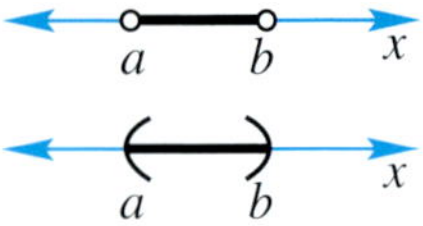	$\{x : a < x < b\}$	$x \in (a, b)$ or $x \in]a, b[$	 $\{x : 3 < x < 8\}$ or $x \in (3, 8)$ $x \in]3, 8[$
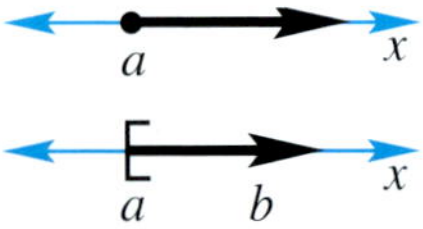	$\{x : x \geq a\}$ or $\{x : a \leq x < \infty\}$	$x \in [a, \infty)$ or $x \in [a, \infty[$	 $\{x : x \geq 3\}$ or $x \in [3, \infty)$ $x \in [3, \infty[$
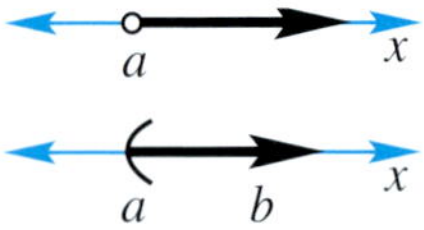	$\{x : x > a\}$ or $\{x : a < x < \infty\}$	$x \in (a, \infty)$ or $x \in]a, \infty[$	 $\{x : x > 3\}$ or $x \in (3, \infty)$ $x \in]3, \infty[$
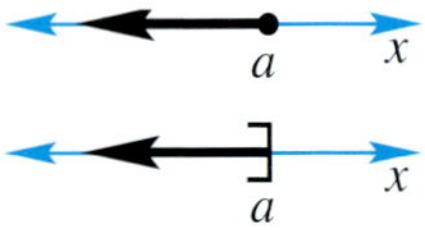	$\{x : x \leq a\}$ or $\{x : -\infty < x \leq a\}$	$x \in (-\infty, a]$ or $x \in]-\infty, a]$	 $\{x : x \leq 8\}$ or $x \in (-\infty, 8]$ $x \in]-\infty, 8]$
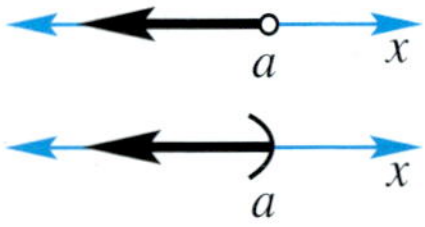	$\{x : x < a\}$ or $\{x : -\infty < x < a\}$	$x \in (-\infty, a)$ or $x \in]-\infty, a[$	 $\{x : x < 8\}$ or $x \in (-\infty, 8)$ $x \in]-\infty, 8[$

We also make the following point in relation to set notation. Rather than using the colon ':' in expressions such as $\{x : x < 3\}$ we can also use the separator '|'. That is, $\{x \mid x < 3\}$. Either notation can be used.

2.1.4 NUMBER SYSTEMS

The set of real numbers can be broken down into two subsets, namely, the set of **rational numbers** and the set of **irrational numbers**. The set of rational numbers can itself be broken down into two sets, the set of **integers** and the set of fractions. The set of integers can then be broken down into the set of positive integers, the set of negative integers and the set that includes the number zero. Each of these sets can be represented by a different symbol.

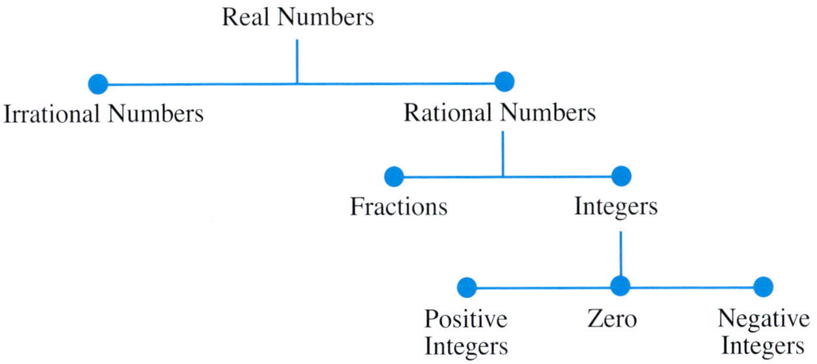

In this book we will use the following notation and definitions:

Set of positive integers and zero $= \mathbb{N} = \{0, 1, 2, 3, \ldots\}$.
The set of integers $= \mathbb{Z} = \{0, \pm1, \pm2, \pm3, \ldots\}$
The set of positive integers $= \mathbb{Z}^+ = \{1, 2, 3, \ldots\}$ [Also known as Natural numbers]
The set of rational numbers $= \mathbb{Q}$

Definition: $\mathbb{Q} = \left\{x \mid x = \frac{a}{b}, b \neq 0 \text{ and } a \text{ and } b \text{ are integers}\right\}$

The set of positive rational numbers $= \mathbb{Q}^+$
The set of positive real numbers $= \mathbb{R}^+ = \{x \mid x \in \mathbb{R}, x > 0\}$
The empty set $= \varnothing =$ The set with *no* members.

EXAMPLE 2.1

(a) Write each of the following using interval notation.
(b) Represent the sets on the real number line.

i. $\{x \mid -1 < x \leq 4\}$
ii. $\{x \mid x \geq 3\} \cap \{x \mid x < 6\}$
iii. $\{x : x < 8\} \backslash \{5\}$

Solution

i. (a) $\{x | -1 < x \leq 4\} =]-1, 4]$ or using round bracket: $(-1, 4]$.

(b) 

ii. (a) $\{x | x \geq 3\} \cap \{x | x < 6\} = [3, \infty[\cap]\infty, 6[= [3, 6[$ or $[3, 6)$ – using round bracket.

(b) 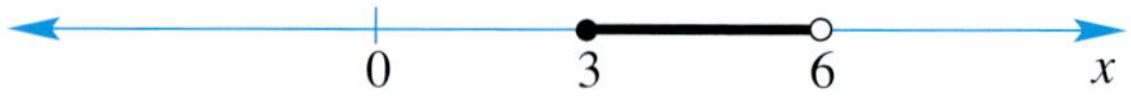

iii. (a) $\{x : x < 8\} \backslash \{5\} =]-\infty, 8[\backslash \{5\}$ or $(-\infty, 8) \backslash \{5\}$ – using round brackets.

(b) 

2.1.5 IRRATIONAL NUMBERS

We provided a definition of rational numbers earlier. The question then remains, what is an irrational number? The obvious answer is 'Whatever is not a rational number'! So, what do these numbers look like? The best way to answer this is to say that it is a number that cannot be written in the form $\frac{a}{b}$, $b \neq 0$ where a and b are integers. Examples of irrational numbers are $\sqrt{2}$, $\sqrt{3}$, π and so on.

How do we know that $\sqrt{2}$ is an irrational number?

We can show this as follows – a process known as *reductio ad absurdum* – meaning to prove by contradiction:

Assume that $\sqrt{2}$ is a rational number. Then by the definition of rational numbers there exist integers a and b (where a and b have no common factors) such that

$$\sqrt{2} = \frac{a}{b}, b \neq 0$$

Upon squaring we have:

$$2 = \frac{a^2}{b^2}$$

$$\Leftrightarrow a^2 = 2b^2$$

Then, a^2 must be even [because $2b^2$ is even – i.e., any number multiplied by 2 is even] and so a must also be even.

This means we can express a as $2k$. So, setting $a = 2k$ we have:

$$a^2 = 4k^2$$

But, $a^2 = 2b^2$, so that

$$2b^2 = 4k^2$$

$$\Leftrightarrow b^2 = 2k^2$$

And so, b^2 must also be even, meaning that b is even.

Then, since both a and b are even it follows that a and b have a common factor (i.e., 2).

This is contrary to our original hypothesis. Therefore $\sqrt{2}$ is not a rational number and must therefore be an irrational number.

One subset of irrational numbers is known as the set of **surds**. Surds can be expressed in the form $\sqrt[n]{a}$ where $n \in \mathbb{Z}^+$ and $a \in \mathbb{Q}^+$. A commonly encountered surd is $\sqrt{a}$ (i.e., the square root).

The laws of operations apply to surds in the same way that they apply to real numbers. We summarise some results involving surds:

$$\sqrt{a} \times \sqrt{b} = \sqrt{ab}$$
$$\frac{\sqrt{a}}{\sqrt{b}} = \sqrt{\frac{a}{b}}$$
$$\sqrt{a^2 b} = a\sqrt{b},\ a > 0$$
$$a\sqrt{b} \times c\sqrt{d} = ac\sqrt{bd}$$
$$(\sqrt{a} + \sqrt{b})(\sqrt{a} - \sqrt{b}) = a - b$$

Notice that the last result shows that we obtain a rational number! The surds $\sqrt{a} - \sqrt{b}$ and $\sqrt{a} + \sqrt{b}$ are know as **conjugate pairs**. Whenever conjugate pairs are multiplied together they produce a rational number.

EXAMPLE 2.2

Expand the following

(a) $(\sqrt{2} + \sqrt{3})(\sqrt{6} - 3)$ (b) $(\sqrt{5} - 2\sqrt{3})(\sqrt{5} - \sqrt{3})$

Solution

(a)
$$\begin{aligned}(\sqrt{2} + \sqrt{3})(\sqrt{6} - 3) &= \sqrt{2} \times \sqrt{6} - 3 \times \sqrt{2} + \sqrt{3} \times \sqrt{6} - 3 \times \sqrt{3} \\ &= \sqrt{2} \times (\sqrt{2} \times \sqrt{3}) - 3\sqrt{2} + \sqrt{3} \times (\sqrt{3} \times \sqrt{2}) - 3\sqrt{3} \\ &= 2\sqrt{3} - 3\sqrt{2} + 3\sqrt{2} - 3\sqrt{3} \\ &= -\sqrt{3}\end{aligned}$$

(b)
$$\begin{aligned}(\sqrt{5} - 2\sqrt{3})(\sqrt{5} - \sqrt{3}) &= \sqrt{5} \times \sqrt{5} - \sqrt{5} \times \sqrt{3} - 2\sqrt{3} \times \sqrt{5} + 2\sqrt{3} \times \sqrt{3} \\ &= 5 - \sqrt{15} - 2\sqrt{15} + 2 \times 3 \\ &= 11 - 3\sqrt{15}\end{aligned}$$

EXAMPLE 2.3

Rationalise the denominator of (a) $\frac{1}{2 + \sqrt{2}}$ (b) $\frac{1 + \sqrt{2}}{2\sqrt{3} - 1}$

Solution

(a)
$$\begin{aligned}\frac{1}{2 + \sqrt{2}} &= \frac{1}{2 + \sqrt{2}} \times \frac{2 - \sqrt{2}}{2 - \sqrt{2}} \quad \text{[multiplying numerator and denominator by conjugate]} \\ &= \frac{2 - \sqrt{2}}{4 - 2} \\ &= 1 - \frac{1}{2}\sqrt{2}\end{aligned}$$

(b)
$$\begin{aligned}\frac{1 + \sqrt{2}}{2\sqrt{3} - 1} = \frac{1 + \sqrt{2}}{2\sqrt{3} - 1} \times \frac{2\sqrt{3} + 1}{2\sqrt{3} + 1} &= \frac{2\sqrt{3} + 1 + 2\sqrt{6} + \sqrt{2}}{4 \times 3 - 1} \\ &= \frac{2\sqrt{3} + 1 + 2\sqrt{6} + \sqrt{2}}{11} \quad \text{[cannot be simplified further]}\end{aligned}$$

2.1.6 THE ABSOLUTE VALUE

The **absolute value** or **modulus** of a number x, denoted by $|x|$, is defined as follows:

If $x \geq 0 \Rightarrow |x| = x$ and if $x < 0 \Rightarrow |x| = -x$.

This means that the absolute value of any number will always be positive.
E.g., $|4| = 4$:– seeing as $4 > 0$, we use the value of 4. Whereas, $|-2| = -(-2) = 2$:– by taking the negative of a negative number we obtain a positive number.

EXAMPLE 2.4

(a) Find $\{x : |x| = 7\}$.

(b) Use the number line to represent the sets

i. $\{x : |x| \leq 3\}$ ii. $\{x : |x| > 1\}$

Express these sets using interval notation.

Solution

(a) We are looking for value(s) of x such that when we take the absolute value of x it is 7. From the definition of the absolute value, we must have that $x = 7$ or $x = -7$.
That is, $|x| = 7 \Leftrightarrow x = \pm 7$. Therefore, the solution set is $\{7, -7\}$

(b) i This time we want all values of x such that their absolute value is less than or equal to 3. For example, if $x = -2.5$ then $|-2.5| = 2.5$ which is less than 3. However, if $x = -4$ then $|-4| = 4$ which is greater than 3. So we cannot have $x = -4$.
Working along these lines we must have:

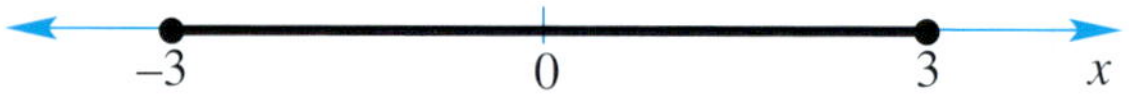

Using interval notation we have $\{x : |x| \leq 3\} = \{x : -3 \leq x \leq 3\} = [-3, 3]$

ii. This time we want numbers for which their absolute value is greater than 1.
For example, $|1.2| = 1.2 > 1$ and $|-3.2| = 3.2 > 1$. We then have:

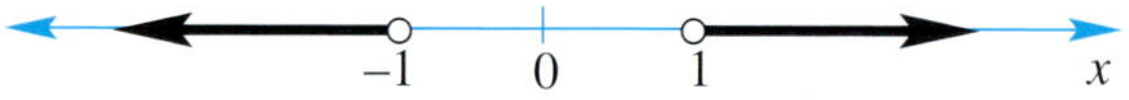

Using interval notation we have $\{x : |x| > 1\} =]-\infty, -1[\cup]1, \infty[$

EXERCISES 2.1

1. Show the following sets on the real number line.

(a) $\{x | 2 \leq x \leq 8\}$ (b) $\{x | x > 7\}$ (c) $\{x | -2 < x \leq 6\} \backslash \{4\}$

(d) $]2, 7] \cap]4, 8[$ (e) $(-\infty, 4) \cap [-2, 5)$ (f) $\{x : x < -6\}$

2. Write the following using interval notation.

(a) $\{x | -2 \leq x \leq 7\}$ (b) $\{x | x > 9\}$

(c) $\{x | 0 < x \leq 5\}$ (d) $\{x : x \leq 0\}$

(e) $\{x : x < 8\} \cap \{x : x > -4\}$ (f) $\{x : x < -1\} \cup \{x : x > 2\}$

3. Simplify the following.

(a) $3\sqrt{5}+\sqrt{20}$ (b) $2\sqrt{3}-\sqrt{27}$ (c) $\sqrt{2}+\sqrt{3}+\sqrt{8}-\sqrt{18}$

4. Simplify the following.

(a) $(\sqrt{5}+1)(\sqrt{5}-1)$ (b) $(2\sqrt{3}-\sqrt{2})(\sqrt{2}+\sqrt{3})$

(c) $(3\sqrt{2}-\sqrt{6})(\sqrt{3}+3)$ (d) $(2+3\sqrt{3})^2$

5. Rationalise the denominator in each of the following.

(a) $\dfrac{1}{2+\sqrt{3}}$ (b) $\dfrac{3}{\sqrt{7}-2}$ (c) $\dfrac{\sqrt{3}}{\sqrt{5}-2}$

(d) $\dfrac{2\sqrt{5}+1}{\sqrt{3}-2}$ (e) $\dfrac{\sqrt{2}+\sqrt{3}}{\sqrt{3}-\sqrt{5}}$ (f) $\dfrac{2\sqrt{3}}{2\sqrt{5}-3\sqrt{2}}$

6. (a) If $x = \sqrt{5}+\sqrt{3}$, find the value of i. $x+\dfrac{1}{x}$ ii. $x^2+\dfrac{1}{x^2}$

(b) If $x = 4+\sqrt{3}$, find the value of i. $x-\dfrac{1}{x}$ ii. $x^2+\dfrac{1}{x^2}$

7. Find the value of x if

(a) $\{x \mid |x| = 3\}$ (b) $\{x \mid |x| = 10\}$ (c) $\{x \mid |x| = -2\}$

(d) $\{x \mid |x+1| = 3\}$ (e) $\{x \mid |x+2| = 10\}$ (f) $\{x \mid |x-2| = 2\}$

8. Represent each of the following on the real number line.

(a) $\{x : |x| \le 5\}$ (b) $\{x \mid |x| > 2\}$

(c) $\{x : 2 \le |x| < 5\}$ (d) $\{x : 2|x| \ge 8\}$

9. Write the following using interval notation.

(a) $\{x \mid x-1>0\}$ (b) $\left\{x \mid \frac{1}{2}x > 2\right\}$ (c) $\{x : x>4\} \cap \{x : 2x<12\}$

2.2 LINEAR ALGEBRA

2.2.1 REVIEW OF LINEAR EQUATIONS

A **linear equation** in the **variable** x (say) takes on the form $ax+b = c$ where a, b and c are real constants. The equation is linear because x is raised to the power of one. To solve such equations we use the rules of transposition:

$$ax+b = c \Leftrightarrow ax = c-b$$
$$\Leftrightarrow x = \frac{c-b}{a}$$

Solving $ax+b = c$ produces a solution that can be represented on the real number line.

EXAMPLE 2.5

Solve the following linear equations

(a) $4x+5 = 21$ (b) $9-2x = 7$ (c) $3(5x-2) = 12$

Solution

(a) $4x+5 = 21 \Leftrightarrow 4x = 16 \Leftrightarrow x = 4$

(b) $9-2x = 7 \Leftrightarrow -2x = -2 \Leftrightarrow x = 1$

(c) $3(5x-2) = 12 \Leftrightarrow 15x-6 = 12$ [don't forget to multiply the 3 and –2]

$$\Leftrightarrow 15x = 18$$

$$\Leftrightarrow x = \frac{6}{5}$$

Sometimes equations might not appear to be linear, but with some algebra, they form into linear equations. The following examples shows how this works.

EXAMPLE 2.6

(a) Solve for x, $\frac{x-3}{2} - 1 = x$ (b) Find $\left\{x \mid \frac{x}{3} - \frac{2-x}{2} = 1\right\}$.

Solution

(a) $\frac{x-3}{2} - 1 = x \Leftrightarrow \frac{x-3}{2} = x+1 \Leftrightarrow x-3 = 2(x+1)$

$$\Leftrightarrow x-3 = 2x+2$$

$$\Leftrightarrow -x = 5$$

$$\therefore x = -5$$

(b) $\frac{x}{3} - \frac{2-x}{2} = 1 \Leftrightarrow \frac{2x}{6} - \frac{3(2-x)}{6} = 1 \Leftrightarrow 2x-3(2-x) = 6$

$$\Leftrightarrow 2x-6+3x = 6$$

$$\Leftrightarrow 5x = 12$$

$$\Leftrightarrow x = \frac{12}{5}$$

EXAMPLE 2.7

Solve the following literal equations for x, where a and b are real constants.

(a) $bx-b^2 = ab$ (b) $bx = a(b-x)$

Solution

(a) $bx-b^2 = ab \Leftrightarrow bx = ab+b^2 \Leftrightarrow bx = b(a+b)$ [taking b out as a common factor]

$\Leftrightarrow x = a+b$ [dividing both sides by b]

(b) $bx = a(b-x) \Leftrightarrow bx = ab-ax \Leftrightarrow bx+ax = ab$

$$\Leftrightarrow (b+a)x = ab$$

$$\Leftrightarrow x = \frac{ab}{b+a}$$

Linear Equations involving Absolute values

Recall that if $|x| = a$ (where $a \geq 0$) then, $x = a$ or $-a$. Using this result we can solve similar linear equations.

That is, $|ax+b| = c \Leftrightarrow ax+b = c$ or $ax+b = -c$

$$\Leftrightarrow ax = c-b \text{ or } ax = -c-b$$

$$\Leftrightarrow x = \frac{c-b}{a} \text{ or } x = -\frac{(b+c)}{a}$$

Notice that this time we have two solutions!

EXAMPLE 2.8

Solve the following.

(a) $|2x| = 6$ (b) $|x-1| = 5$ (c) $\left|3-\frac{1}{2}x\right| = 2$

Solution

(a) $|2x| = 6 \Leftrightarrow 2x = 6$ or $2x = -6$

$\Leftrightarrow x = 3$ or $x = -3$

(b) $|x-1| = 5 \Leftrightarrow x-1 = 5$ or $x-1 = -5$

$\Leftrightarrow x = 6$ or $x = -4$

(c) $\left|3-\frac{1}{2}x\right| = 2 \Leftrightarrow 3-\frac{1}{2}x = 2$ or $3-\frac{1}{2}x = -2$

$\Leftrightarrow -\frac{1}{2}x = -1$ or $-\frac{1}{2}x = -5$

$\Leftrightarrow x = 2$ or $x = 10$

The next example illustrates a demanding algebraic solution.

EXAMPLE 2.9

Solve for x where $|x| = 2x+1$

Solution

(a) By definition, $|x| = x$ if $x \geq 0$ and $-x$ if $x < 0$. Therefore we have two separate equations to solve, one for $x \geq 0$ and one for $x < 0$.

Case 1 ($x \geq 0$): $x = 2x+1, x \geq 0$

$\Leftrightarrow -x = 1, x \geq 0$

$\Leftrightarrow x = -1, x \geq 0$

Now, our solution is that $x = -1$, however, we must also satisfy the condition that $x \geq 0$. As both statements do not agree with each other, we conclude that for $x \geq 0$ there is no solution.

Case 2 ($x < 0$):

$$-x = 2x + 1,\ x < 0$$
$$\Leftrightarrow -3x = 1,\ x < 0$$
$$\Leftrightarrow x = -\frac{1}{3},\ x < 0$$

This time our solution is that $x = -\frac{1}{3}$, and we must also satisfy the condition that $x < 0$.

As both statements agree with each other, we conclude that for $x < 0$ there is a solution.

Namely, $x = -\frac{1}{3}$. Therefore, $|x| = 2x + 1$ has only one solution, $x = -\frac{1}{3}$.

Solving Equations with the TI-83

Equations such as the we have just looked at can also be solved using the **solve(** option on the TI–83. We do this by calling up the **Catalogue** and then

1. locating the **solve(** option
2. enter the relevant equation
   [The equation must be entered in the form **Equation = 0**. So, to solve the equation $2x + 6 = 15$, we must rewrite it as $2x + 6 - 15 = 0$ so that the equation that is entered into the TI–83 is $2x + 6 - 15 = 0$]
3. indicate the variable we are solving for
4. provide a reasonable guess (for the answer)

To obtain the **solve(** option we use the following sequence:

[2nd] [0] [LN] then use the arrow key to reach **solve(** and then press [ENTER]:

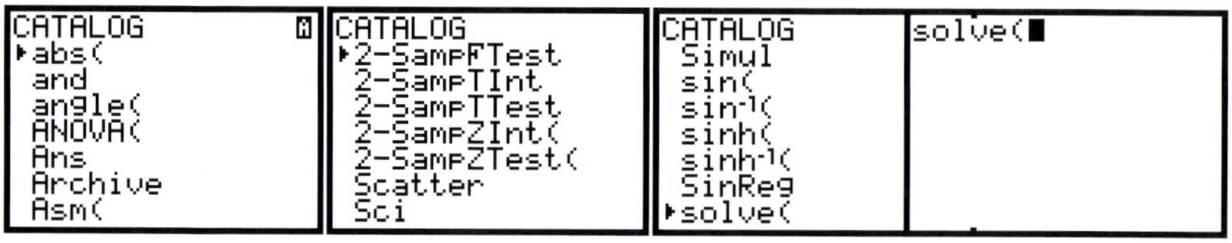

We look at some of the problems we have aleady solved:

EXAMPLE 2.10

Solve the following linear equations

(a) $4x + 5 = 21$ (b) $3(5x - 2) = 12$ (c) $\frac{x-3}{2} - 1 = x$ (d) $|x| = 2x + 1$

Solution

(a)

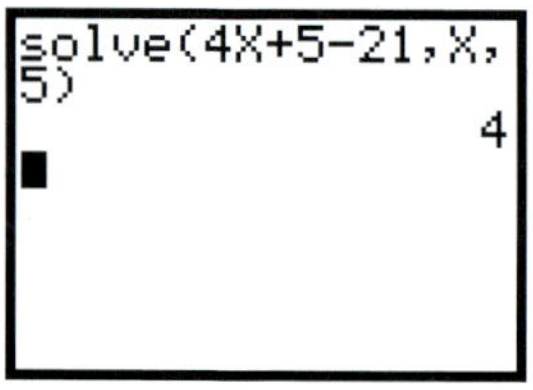

(b)

(c)

(d)

```
solve(abs(X)-2X-
1,X,5)
      -.3333333333
Ans▸Frac
              -1/3
█
```

Notice that in each case we have used a guess of 5.

Another method is to use the **Equation solver** facility. The expression must still be entered in the form **Equation = 0**. To call up the Equation solver screen

1. press [MATH] [0]
2. Enter the equation in the form **Equation = 0**
3. Move the cursor over the variable for which you want to solve and then press [ALPHA] [ENTER].

It is important that you become familiar with both modes of solving equations, although eventually you will prefer one method over the other.

EXERCISES 2.2.1

1. Solve the following linear equations.

(a) $2x = 8$ (b) $5x - 3 = 12$ (c) $2 - \frac{1}{3}x = 4$

(d) $\frac{3-2x}{7} = 2$ (e) $\frac{5x}{3} + \frac{1}{2} = \frac{2}{3}$ (f) $2x + \frac{1}{4} = 1$

2. Solve the following equations.

(a) $5(x-1) = 12$ (b) $3\left(2 - \frac{1}{2}x\right) = 4$ (c) $-2(2x+1) = 1$

(d) $2x - 1 = 3 - x$ (e) $3\left(5 - \frac{1}{3}x\right) = x - 20$ (f) $5(2x-3) = 8 - \frac{1}{4}x$

3. Solve the following equations.

(a) $\frac{2-u}{6} + 8 = 1 - u$ (b) $\frac{u-1}{4} - 3 = \frac{u}{3}$ (c) $\frac{x-2}{3} + 1 = \frac{1-x}{4}$

(d) $\frac{5}{x+1} + 2 = \frac{1}{x+1}$ (e) $\frac{1}{y-1} + 1 = \frac{2}{y-1}$ (f) $\frac{3(u+1)}{5} - 2 = \frac{(u+1)}{5}$

4. Solve the following equations for x.

(a) $x - b = b - 2$ (b) $a(x-b) = b + a$ (c) $ax = b(a-x)$

(d) $\frac{x}{a} - a = b$ (e) $\frac{x}{b} - a = \frac{x}{a} - b$ (f) $\frac{1}{a} + \frac{1}{x} = \frac{1}{b}$

(g) $\frac{b-x}{a+x} = \frac{b+x}{a-x}$ (h) $\frac{1-ax}{b} + \frac{1-bx}{a} = 0$ (i) $\frac{a}{b-x} = \frac{b}{a-x}$

5. Solve for x.

(a) $|2x| = 8$ (b) $|5x-3| = 12$ (c) $\left|2-\frac{1}{3}x\right| = 4$

(d) $\left|\frac{3-2x}{7}\right| = 2$ (e) $\left|\frac{5x}{3}+\frac{1}{2}\right| = \frac{2}{3}$ (f) $\left|2x+\frac{1}{4}\right| = 1$

(g) $|5(x-1)| = 12$ (h) $\left|3\left(2-\frac{1}{2}x\right)\right| = 4$ (i) $\left|\left(\frac{2}{3}x+1\right)\right| = 1$

(j) $a-2|x| = b$ (k) $a-\left|\frac{1}{b}x\right| = b$ (l) $|2ax-b| = 3b$

2.2.2 LINEAR INEQUATIONS

Inequalities are solved in the same way as equalities, with the exception that when both sides are mulitplied or divided by a negative number, the direction of the inequality sign reverses.

EXAMPLE 2.11

Find (a) $\{x : x+1<4\}$ (b) $\{x|2x-5<1\}$

Solution

(a) $x+1<4 \Leftrightarrow x<3$.

Therefore, the solution set (s.s) is $\{x : x<3\}$.

(b) $2x-5<1 \Leftrightarrow 2x<6$

$\Leftrightarrow x<3$

Therefore, the solution set (s.s) is $\{x : x<3\}$.

EXAMPLE 2.12

Find (a) $\{x : x+2>3-2x\}$ (b) $\left\{x : \frac{3-2x}{7} \le \frac{4x-3}{2}\right\}$

Solution

(a) $x+2>3-2x \Leftrightarrow 3x+2>3$

$\Leftrightarrow 3x>1$

$\Leftrightarrow x>\frac{1}{3}$. Therefore, s.s. $= \left\{x : x>\frac{1}{3}\right\}$

(b) $\frac{3-2x}{7} \le \frac{4x-3}{2} \Leftrightarrow 14\left(\frac{3-2x}{7}\right) \le 14\left(\frac{4x-3}{2}\right)$ [multiply both sides by 14]

$\Leftrightarrow 2(3-2x) \le 7(4x-3)$

$\Leftrightarrow 6-4x \le 28x-21$

$\Leftrightarrow -32x \le -27$

$\Leftrightarrow x \ge \frac{27}{32}$ [notice the reversal of the inequality – as we divided by a negative number]

Therefore, s.s. is $\left\{x : x \geq \frac{27}{32}\right\}$.

When dealing with inequalities that involve absolute values, we need to keep in mind the following:

1. $|x| < a \Leftrightarrow -a < x < a$

2. $|x| > a \Leftrightarrow x < -a$ or $x > a$

EXAMPLE 2.13

Find (a) $\{x : |x + 1| < 4\}$ (b) $\{x \,|\, |2x - 5| \leq 1\}$

Solution

(a) $|x + 1| < 4 \Leftrightarrow -4 < x + 1 < 4$

$\Leftrightarrow -5 < x < 3$ [subtracting 1 from both inequalities]

Therefore, s.s. is $\{x : -5 < x < 3\}$.

(b) $|2x - 5| \leq 1 \Leftrightarrow -1 \leq 2x - 5 \leq 1$

$\Leftrightarrow 4 \leq 2x \leq 6$ [adding 5 to both sides of inequality]

$\Leftrightarrow 2 \leq x \leq 3$ [dividing both sides by 2]

EXAMPLE 2.14

Find (a) $\left\{x : \left|1 - \frac{1}{2}x\right| > 3\right\}$ (b) $\{x \,|\, |3x - 2| - 1 \geq 5\}$

Solution

(a) $\left|1 - \frac{1}{2}x\right| > 3 \Leftrightarrow 1 - \frac{1}{2}x > 3$ or $1 - \frac{1}{2}x < -3$

$\Leftrightarrow -\frac{1}{2}x > 2$ or $-\frac{1}{2}x < -4$

$\Leftrightarrow x < -4$ or $x > 8$ [Note the reversal of inequality sign, i.e., × by –2]

Therefore, s.s. is $\{x : x < -4\} \cup \{x : x > 8\}$.

(b) $|3x - 2| - 1 \geq 5 \Leftrightarrow |3x - 2| \geq 6$

$\Leftrightarrow 3x - 2 \geq 6$ or $3x - 2 \leq -6$

$\Leftrightarrow 3x \geq 8$ or $3x \leq -4$

$\Leftrightarrow x \geq \frac{8}{3}$ or $x \leq -\frac{4}{3}$

Therefore, s.s. is $\left\{x : x \leq -\frac{4}{3}\right\} \cup \left\{x : x \geq \frac{8}{3}\right\}$.

EXERCISES 2.2.2

1. Solve the following inequalities.

(a) $2x + 1 < x - 3$ (b) $\frac{x-4}{3} \geq 2x - 1$ (c) $x + 1 > \frac{x+3}{2}$

(d) $x \geq 3(x + 4)$ (e) $\frac{x-4}{5} > \frac{2-x}{2}$ (f) $1 - 3x < 5x - 2$

2. Solve the following inequalities.

(a) $\frac{2x+1}{5} - \frac{2-x}{3} > 3$ (b) $\frac{1+x}{2} + \frac{1-x}{4} \leq 1$ (c) $\frac{x}{5} + \frac{2-3x}{3} \geq -2$

3. Solve the following inequalities.

(a) $a(x + 1) > 2a, a < 0$ (b) $\frac{a-x}{2} + 1 > a, a > 0$

(c) $\frac{x}{a} + \frac{b}{a^2} < \frac{4x}{a} - \frac{b}{a^2}, b > a > 0$ (d) $x + \frac{x-1}{a+1} \geq \frac{x+1}{a+1} - ax, a > 0$

4. Solve the following inequalities.

(a) $|4x + 2| \leq 6$ (b) $|2x - 1| \leq 5$ (c) $|4x - 2| \leq 8$

(d) $|4x + 2| \leq 0$ (e) $|x - 1| \leq 8$ (f) $|3x + 3| \leq 12$

(g) $\left|3 - \frac{x}{2}\right| \leq 5$ (h) $\left|2 - \frac{x}{4}\right| \leq 9$ (i) $\left|3x + \frac{1}{2}\right| \leq \frac{3}{4}$

5. Solve the following inequalities.

(a) $|2x - 1| > 4$ (b) $|5 - 2x| > 2$ (c) $\left|1 - \frac{x}{2}\right| \geq 7$

(d) $\left|3 + \frac{1}{3}x\right| \geq 5$ (e) $3|6 - 4x| + 1 > 10$ (f) $12 - |4 - x| > 2$

(g) $\left|2 - \frac{x}{4}\right| > 9$ (h) $\left|3x + \frac{1}{2}\right| > \frac{3}{4}$ (i) $\left|3 - \frac{x}{2}\right| \geq 5$

6. For what value(s) of p does $\left|\frac{3x}{2} - 7\right| \leq p - 3$ have no solutions?

2.3 LINEAR FUNCTIONS

2.3.1 GRAPH OF THE LINEAR FUNCTION

The study of functions and relations is dealt with in detail in Chapter 5, however, we give a basic definition of the term function at this point.

A **function** $y = f(x)$ is an algebraic expression that will generate only one value of y for any one value of x.

For example, if $x = 5$, then $f(5)$ will only generate one value of y. Consider the function $f(x) = x + 3$ then, $f(5) = 5 + 3 = 8$. i.e., only one y–value has been generated.

A **linear function** has the form

$$y = f(x) = mx + c$$

Its graph is a straight line such that
1. m is the slope of the line.
2. c is the y–intercept (i.e., where the line cuts the y–axis).

There are three possible outcomes when a linear function is graphed:

Case 1: $m > 0$

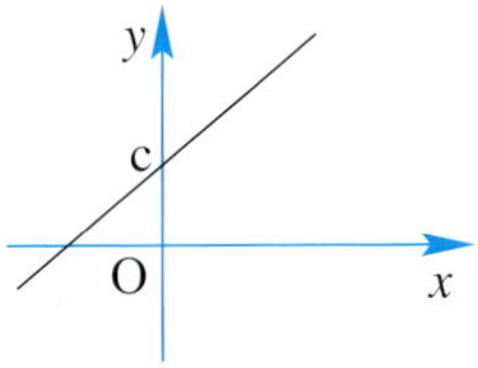

Case 2: $m < 0$

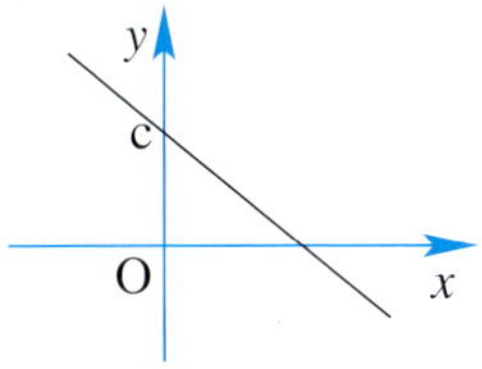

Case 3: $m = 0$

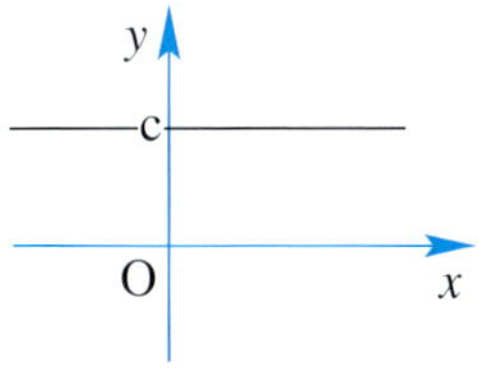

However, sometimes the linear function is expressed in different forms. The reason is that it is sometimes more convenient to express it in a form other than the standard form. For example, the following linear functions are all the same:

$$f(x) = 5x - 2, \quad y = 5x - 2, \quad y + 2 = 5x, \quad -5x + y + 2 = 0$$

They have simply been rewritten into different formats.

EXAMPLE 2.15

Sketch the graphs of the following linear functions.

(a) $f(x) = 2x + 1$ (b) $y = 3 - \frac{1}{2}x$ (c) $4x - 2y = 5$

Solution

(a) The function $f(x) = 2x + 1$ represents a straight line with a gradient of 2 and a y–intercept of 1.
i.e., the graph will cut the y–axis at the point (0, 1).
The x–intercept is obtained by solving $f(x) = 0$:
$2x + 1 = 0 \Leftrightarrow 2x = -1 \Leftrightarrow x = -0.5$.

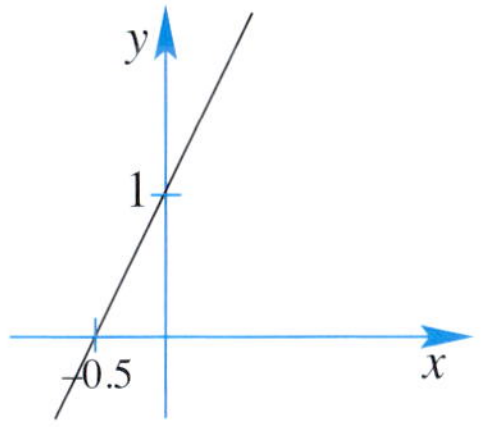

(b) The function $y = 3 - \frac{1}{2}x$ represents a straight line

with a gradient of $-\frac{1}{2}$ and a y–intercept of 3.

i.e., the graph will cut the y–axis at the point (0, 3).
The x–intercept is obtained by solving $y = 0$:

$3 - \frac{1}{2}x = 0 \Leftrightarrow -\frac{1}{2}x = -3 \Leftrightarrow x = 6$.

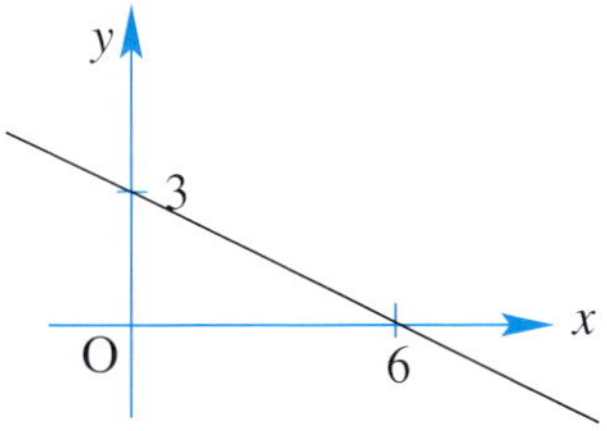

(c) We first rewrite the equation $4x - 2y = 5$:

$4x - 2y = 5 \Leftrightarrow 2y = 4x - 5 \Leftrightarrow y = 2x - \frac{5}{2}$.

So, we have a straight line with gradient 2 and y–intercept at –2.5.

The x–intercept is obtained by solving $y = 0$.

Using the original equation we have $4x - 2(0) = 5 \Leftrightarrow x = \frac{5}{4} = 1.25$.

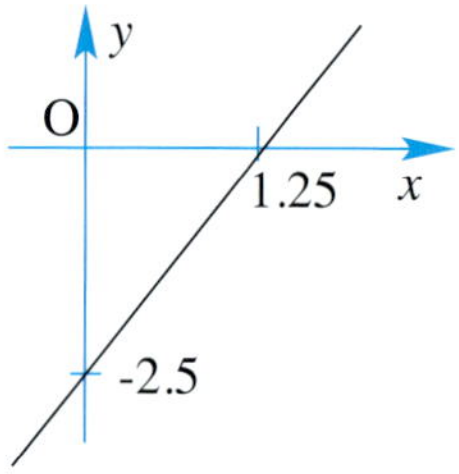

Geometrical interpretation of solving equations

It is interesting to note that although we have been solving equations like $2x + 1 = 7$, we had not provided (apart from using the real number line) a geometrical representation corresponding to the equation. We can now provide such an interpretation.

Solving $2x + 1 = 7$ for x, is the same as considering the function $f(x) = 2x + 1$ and then asking the question:

"When will the straight line $y = 2x + 1$ have a y–value of 7".

Or,

"When will the straight line $y = 2x + 1$ meet the straight line $y = 7$".

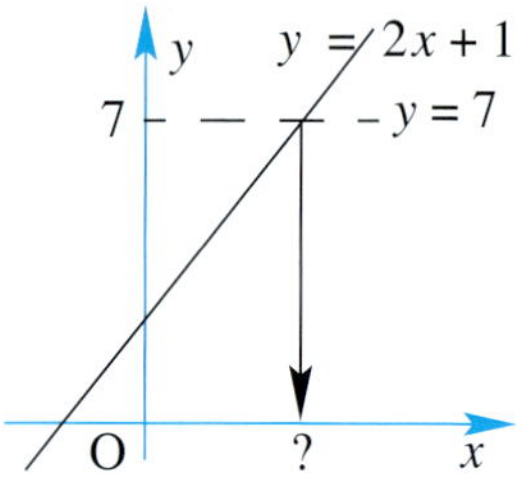

In general we have the following geometrical interpretation for any equation.

Solving $f(x) = k$ for x is equivalent to finding the x–value where the graphs of $y = f(x)$ and $y = k$ intersect.

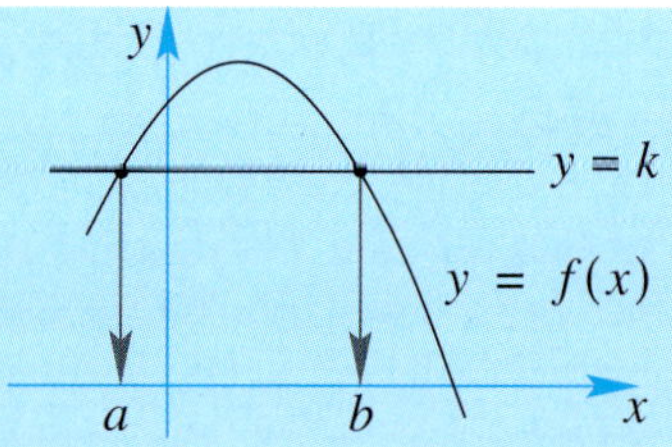

Note that although with linear equations there will only be one solution, for non-linear functions there may very well be more than one solution.

We complete this section by providing a summary of other properties of straight lines:

Properties of straight lines

1. Gradient of a line

The gradient, m, of the line through two points (x_1, y_1) and (x_2, y_2) is given by $m = \frac{y_2 - y_1}{x_2 - x_1}$ or $m = \frac{\text{Rise}}{\text{Run}}$

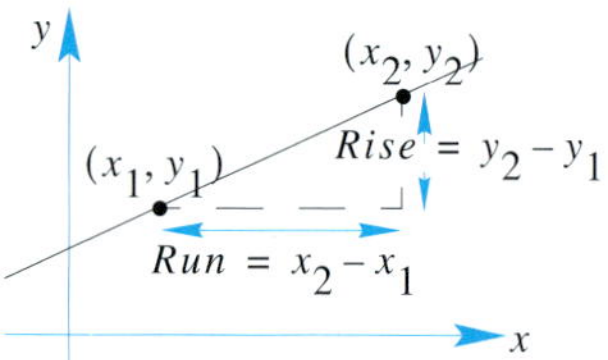

From this we can obtain the point–gradient form of a line.

That is, if (x, y) is any point on a straight line having a gradient m, and (x_1, y_1) is another fixed point on that line then the equation of that line is given by

$$y - y_1 = m(x - x_1)$$

2. Parallel lines

The straight line l_1 with gradient m_1 is parallel to the straight line l_2 with gradient m_2 if and only if $m_1 = m_2$.

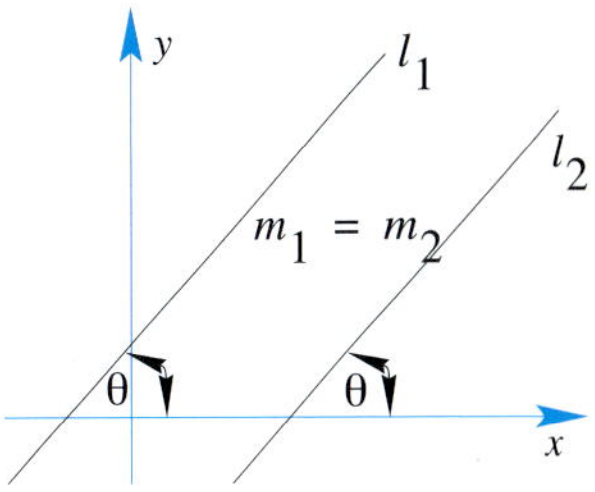

That is, $l_1 \parallel l_2$ iff $m_1 = m_2$

Notice that if the two lines are parallel, they also make equal angles with the x–axis.

3. Perpendicular lines

The straight line l_1 with gradient m_1 is perpendicular to the straight line l_2 with gradient m_2 if and only if $m_1 \times m_2 = -1$.

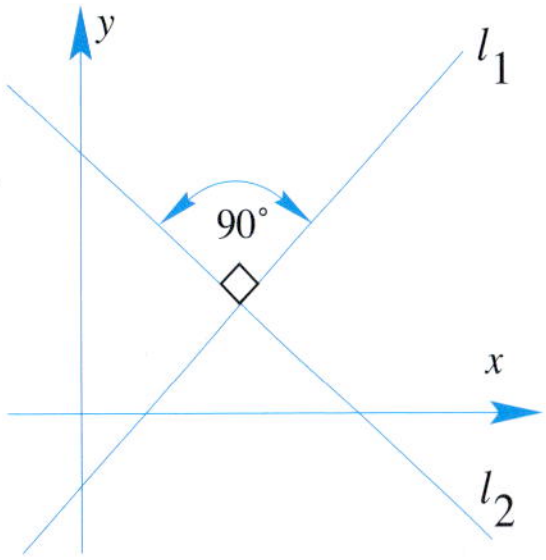

That is, $l_1 \perp l_2$ iff $m_1 \times m_2 = -1$ or $m_1 = -\frac{1}{m_2}$.

EXAMPLE 2.16

Find the equation of the line that passes through the point (–1,3) and is parallel to the line with equation $2x - y + 7 = 0$.

Solution

The gradient of the line $2x - y + 7 = 0$ is found by rearranging to the form $y = mx + c$ to get: $y = 2x + 7$. The gradient is 2 and so all the lines parallel to this will also have gradient 2. The equation of the required line is $y = 2x + c$. The value of the constant c can be found by using the fact that the line passes through the point (–1,3). That is, $y = 2x + c \therefore 3 = 2 \times -1 + c \Leftrightarrow c = 5$. Therefore the equation of the straight line is $y = 2x + 5$.

EXAMPLE 2.17

Find the equation of the line which passes through the point (–1,4) and which is perpendicular to the line with equation $2x + 5y + 2 = 0$.

Solution

The gradient form of $2x + 5y + 2 = 0$ is $5y = -2x - 2 \Rightarrow y = -\frac{2}{5}x - \frac{2}{5}$.

So the gradient is $-\frac{2}{5}$. The gradient of all lines perpendicular to this line is found using the fact that the product of the gradients of perpendicular lines is –1: $\left(-\frac{2}{5}\right)m = -1 \Rightarrow m = \frac{5}{2} = 2.5$.

Then, the equation of the line is $y = \frac{5}{2}x + c$. The constant c is found in the same way as the previous example: Using the point (–1, 4) we have $4 = \frac{5}{2} \times -1 + c \Leftrightarrow c = 6.5$.

Therefore the equation of the straight line is $y = \frac{5}{2}x + 6\frac{1}{2}$.

EXERCISES 2.3.1

1. Sketch the graph of the following straight lines.

(i) $y = x+1$ (ii) $f(x) = x-2$ (iii) $y = 2x-3$

(iv) $f(x) = 2-3x$ (v) $y = \frac{x+1}{2}$ (vi) $f(x) = 3+4x$

(vii) $x+f(x) = 3$ (viii) $x+2y = 4$ (ix) $x-3y = 6$

(x) $\frac{x}{2}+\frac{y}{5} = 1$ (xi) $x-\frac{y}{3} = 1$ (xii) $\frac{2x}{5}-3y = 2$

(xiii) $x+\frac{5y}{4} = -1$ (xiv) $\frac{4+t}{2} = q$ (xv) $x+4y = 2-x$

2. Find the gradient of the line joining the points

(a) (3, 2) and (5, 6) (b) (4, 5) and (6, 11) (c) (–1, 3) and (2, 8)

3. Use the gradient–point method to find the equation of the straight line if

(a) it passes through the point (1, 1) and has a gradient of 2.

(b) it passes through the point (–2, 3) and has a gradient of 3.

(c) it passes through the point (3, –4) and has a gradient of –1.

4. Find the gradient of the straight line that is perpendicular to the straight line with gradient equal to

(a) 2. (b) – 3. (c) $-\frac{2}{3}$. (d) $\frac{5}{4}$

5. Find the equation of the straight line that passes through the origin and the point (2,4).

6. Find the equation of the straight line that passes through the points (–1,2) and (0,1).

7. A straight line passes through the point (4,3) and is perpendicular to the line joining the points (–1,3) and (1,–1). Find the equation of this line.

8. The lines $px+4y-2 = 0$ and $2x-y+p = 0$ are perpendicular. Find the value of p.

9. Find equations for each of the following lines.

(a)

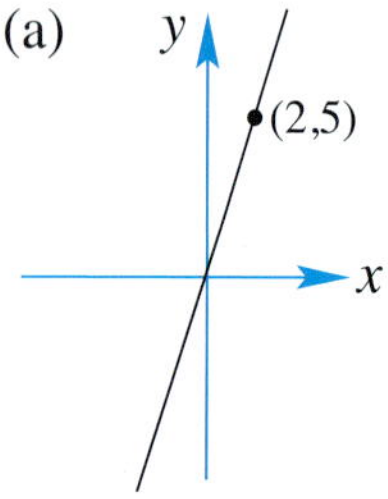

(b)

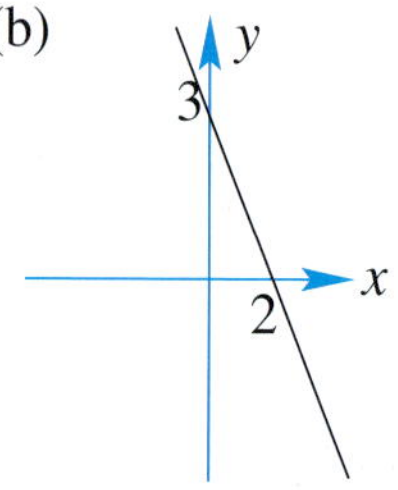

(c)

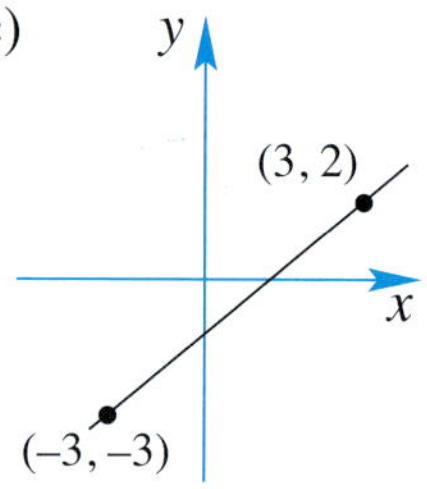

(d)

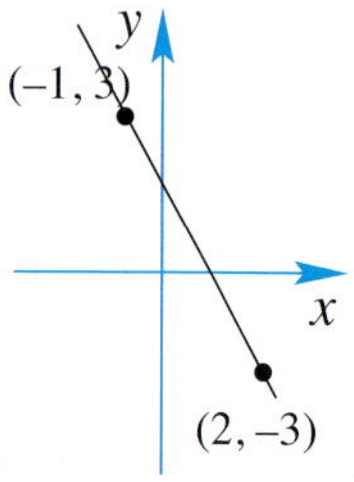

10. Sketch the graph of the following functions.

(a) $f(x) = ax-b, a<0, b>0$ (b) $f(x) = a^2x+b, b<0, a\neq 0$

(c) $f(x) = \frac{a}{a+1}x-a, a>0$ (d) $f(x) = 2a+\frac{1}{a}x, a>0$

2.3.2 SIMULTANEOUS LINEAR EQUATIONS IN TWO UNKNOWNS

Pairs of simultaneous equations in two unknowns may be solved in two ways, either algebraically or graphically. To solve means to find where the two straight lines intersect once they have been sketched. So, we are looking for the point of intersection.

Method 1: Graphical

EXAMPLE 2.18

Solve the system of linear equations $y = -x + 7$ and $y = 2x + 1$.

Solution

We sketch both lines on the same set of axes:

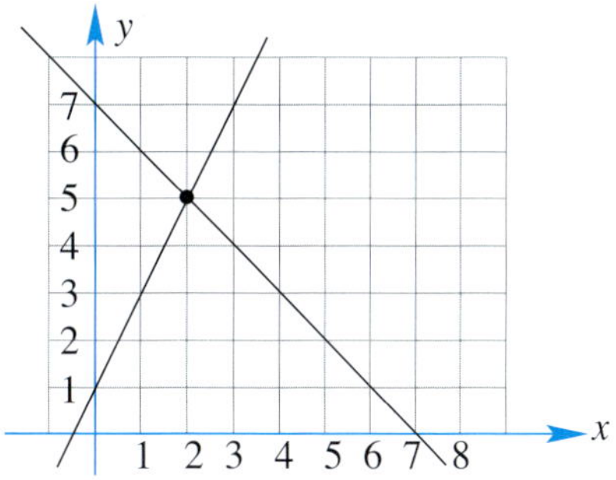

Reading off the grid we can see that the straight lines meet at the point with coordinates (2, 5).

So, the solution to the given system of equations is $x = 2$ and $y = 5$.

The graphical approach has a disadvantage. Sometimes it can only provide an approximate answer. This depends on the accuracy of the sketch or simply on the equations, for example, sketching the pair of straight lines with equations $y = \sqrt{2}x - 1$ and $y = -x + \sqrt{3}$ can only result in an approximate solution.

Then there is the graphics calculator. There are a number of ways that the graphics calculator can be used. Using the TI–83 we can make use of the **TRACE** function or the **intersection** option under the **CALC** menu. We solve the previous problem using the intersection option from the CALC menu.

Step 1: Enter both equations in the required form, i.e., $y = \ldots$

Step 2: Choose an appropriate window setting, in this case we have [–2,8] by [–1,8].

Step 3: Sketch the straight lines using the **GRAPH** key.

Step 4: Call up the **CALC** menu (i.e., press **2nd TRACE**) and choose option **5: intersect**.

Step 5:
- 5.1 Move the cursor to where the lines intersect and press **ENTER** – this confirms that you have selected your first equation.
- 5.2 Press **ENTER** again, this confirms that you are using the second equation.
- 5.3 Because you have already placed your cursor near the point of intersection, when prompted to Guess? simple press **ENTER**.

Step 1:

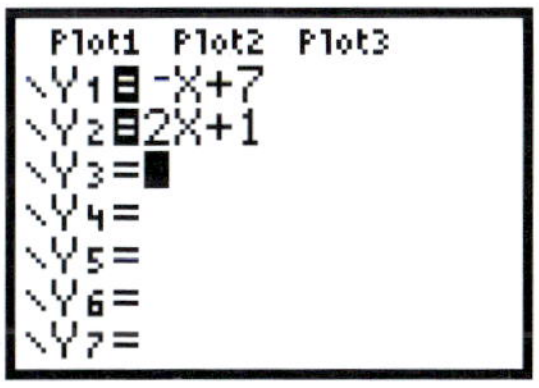

Step 2:

Step 3:

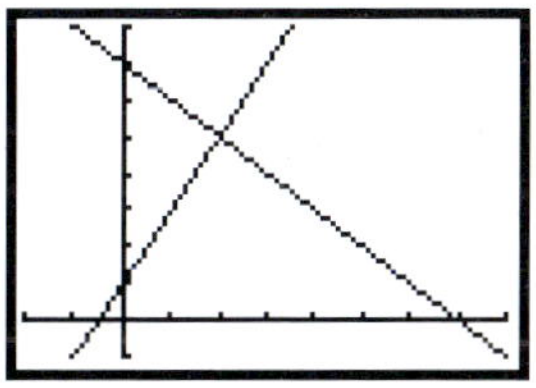

Step 4:

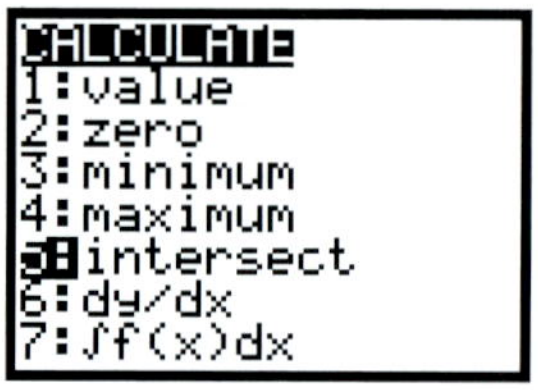

Step 5.1:

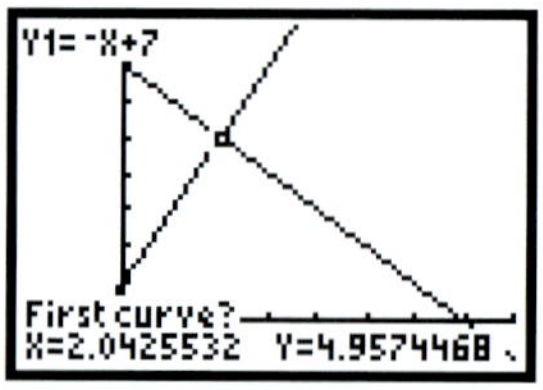

Step 5.2:

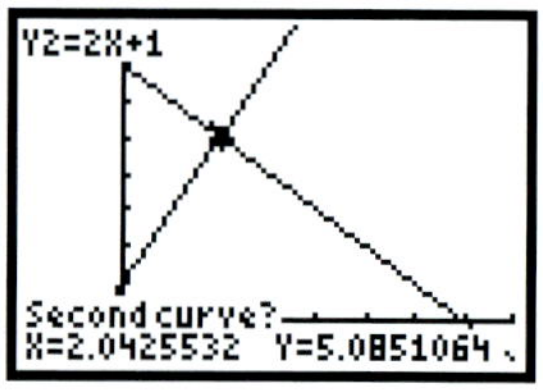

Step 5.3:

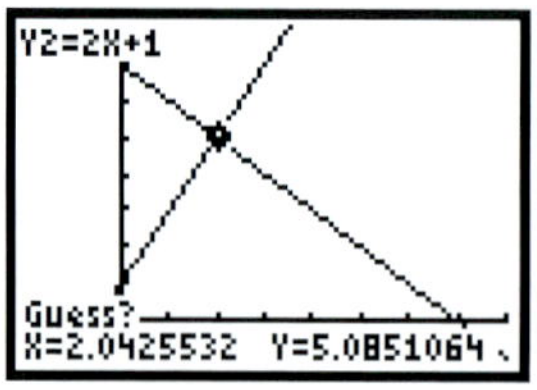

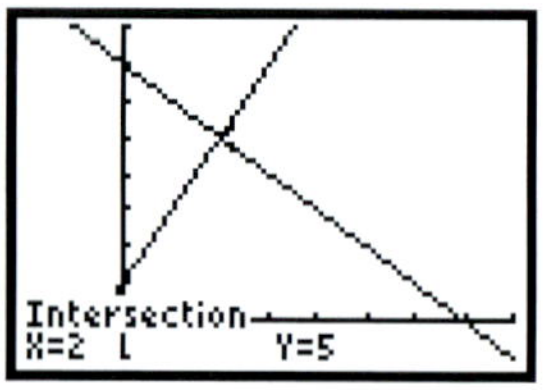

The final screen shows the solution as $x = 2$ and $y = 5$.

This particular example worked out rather neatly, however, the solution of the pairs of equations $y = \sqrt{2}x - 1$ and $y = -x + \sqrt{3}$ would produce the following result:

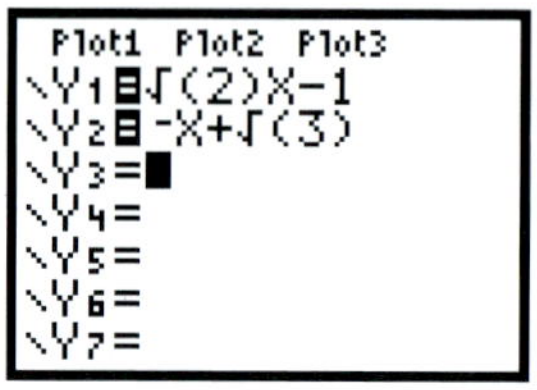

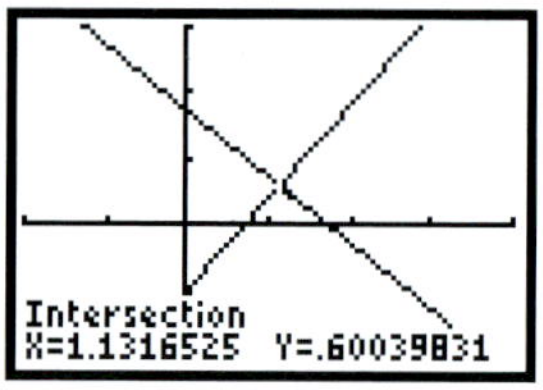

i.e., we only obtain an approximate solution!

Of course, depending on the application, an approximate solution might suffice. However, at this stage we are interested in the mathematical process. Because we cannot always obtain an exact answer using a graphical means we need to consider an algebraic approach.

Method 2: Algebraic

There are two possible approaches when dealing with simultaneous equations algebraically. They are the process of **a. Elimination** **b. Substitution**

Elimination Method

The **key step** in using the elimination method is to obtain, for one of the variables (in both equations), coefficients that are the same (or only differ in sign). Then:

if the coefficients are the same, you subtract one equation from the other – this will **eliminate** one of the variables – leaving you with only one unknown.

However,

if the coefficients only differ in sign, you add the two equations – this will **eliminate** one of the variables – leaving you with only one unknown.

EXAMPLE 2.19

Use the elimination method to solve $\begin{aligned} x - 2y &= -7 \\ 2x + 3y &= 0 \end{aligned}$

Solution

As it is easier to add than subtract, we try to eliminate the variable which differs in sign. In this case the variable 'y' is appropriate. However, the coefficients still need to be manipulated. We label the equations as follows:

$$x - 2y = -7 \quad -(1)$$
$$2x + 3y = 0 \quad -(2)$$

3×(1): $3x - 6y = -21 \quad -(3)$

2×(2): $4x + 6y = 0 \quad -(4)$

Adding (3) + (4): $7x + 0 = -21$

$$\Leftrightarrow x = -3$$

Substituting into (1) we can now obtain the y–value: $-3 - 2y = -7 \Leftrightarrow -2y = -4 \Leftrightarrow y = 2$.
Therefore, the solution is $x = -3$, $y = 2$.

Once you have found the solution, always check with one of the original equations.
Using equation (2) we have: L.H.S $= 2 \times -3 + 3 \times 2 = 0 =$ R.H.S.

Note that we could also have multiplied equation (1) by 2 and then subtracted the result from equation (2). Either way, we have the same answer.

Substitution Method

The substitution method relies on making one of the variables the subject of one of the equations. Then we substitute this equation for its counterpart in the other equation. This will then produce a new equation that involves only one unknown. We can solve for this unknown and then substitute its value back into the first equation. This will then provide a solution pair.

EXAMPLE 2.20

Use the substitution method to solve $\begin{aligned} 5x - y &= 4 \\ x + 3y &= 4 \end{aligned}$

Solution

Label the equations as follows: $5x - y = 4 \quad -(1)$

$x + 3y = 4 \quad -(2)$

From equation (1) we have that $y = 5x - 4 \quad -(3)$

Substituting (3) into (2) we have: $x + 3(5x - 4) = 4$

$$\Leftrightarrow 16x - 12 = 4$$
$$\Leftrightarrow 16x = 16$$
$$\Leftrightarrow x = 1$$

Substituting $x = 1$ into equation (3) we have: $y = 5 \times 1 - 4 = 1$.
Therefore, the solution is given by $x = 1$ and $y = 1$.
Check: Using equation (2) we have: L.H.S $= 1 + 3 \times 1 = 4 =$ R.H.S

Not all simultaneous equations have unique solutions. Some pairs of equations have no solutions while others have infinite solution sets. You will need to be able to recognise the 'problem' in the processes of both algebraic and graphical solutions when dealing with such equations.

The following examples illustrate these possibilities:

EXAMPLE 2.21

Solve: (a) $\begin{aligned} 2x + 6y &= 8 \\ 3x + 9y &= 12 \end{aligned}$ (b) $\begin{aligned} 2x + 6y &= 8 \\ 3x + 9y &= 15 \end{aligned}$

Solution

(a) Algebraic solution:

Label the equations as follows:

$2x + 6y = 8$ – (1)

$3x + 9y = 12$ – (2)

3×(1): $6x + 18y = 24$ – (3)

2×(2): $6x + 18y = 24$ – (4)

Graphical Solution:

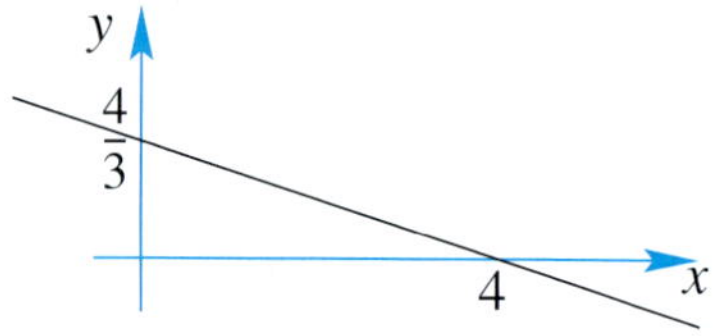

In this case, we have the same equation. That is, the straight lines are **coincident**.

If we were to 'blindly' continue with the solution process, we would have:

3×(1) – 2×(2): $0 = 0$!

The algebraic method produces an equation that is always true, i.e., zero will always equal zero. This means that any pair of numbers that satisfy either equation will satisfy both and are, therefore, solutions to the problem. Examples of solutions are: $x = 4, y = 0, x = 1, y = 1$ & $x = 7, y = -1$. In this case we say that there is an **infinite** number of solutions.

Graphically, the two equations produce the same line. The coordinates of any point on this line will be solutions to both equations.

(a) Algebraic solution:

Label the equations as follows:

$2x + 6y = 8$ – (1)

$3x + 9y = 15$ – (2)

3×(1): $6x + 18y = 24$ – (3)

2×(2): $6x + 18y = 30$ – (4)

(4) – (3): $0 = 6$

Graphical Solution:

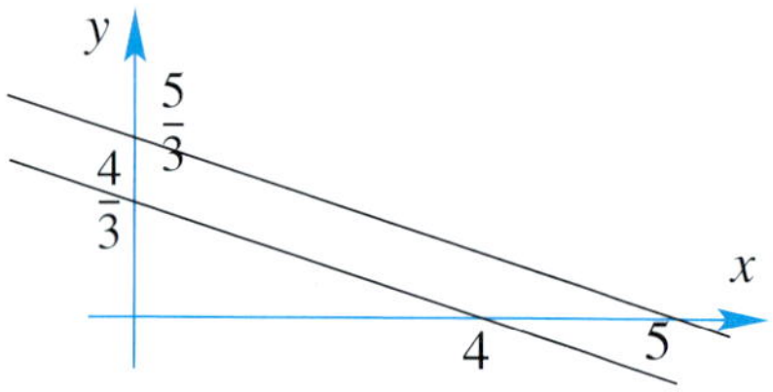

The algebraic method produces an equation that is never true. This means that there are no solutions to the equations.

Graphically, the two lines are parallel and produce no points of intersection.

There is a **matrix** method for solving systems of simultaneous equations that will be discussed in Chapter 23. This method is particularly useful when using graphics calculators which can perform the matrix arithmetic necessary to solve simultaneous equations.

EXERCISES 2.3.2

1. Solve these simultaneous equations, giving exact answers.

(i) $3x - 2y = -1$, $5x + 2y = 9$ (ii) $3x + 5y = 34$, $3x + 7y = 44$ (iii) $2x + 4y = 6$, $4x - 3y = -10$

(iv) $3x + 2y = 2$, $2x - 6y = -6$ (v) $5x + 4y = -22$, $3x - y = -3$ (vi) $5x - 9y = -34$, $2x + 3y = -7$

2. Solve these simultaneous equations, giving fractional answers where appropriate.

(i) $3x - y = 2$, $5x + 2y = 9$ (ii) $4x + 2y = 3$, $x - 3y = 0$ (iii) $-3x + y = 0$, $2x - 4y = 0$

(iv) $\frac{x}{2} - 3y = 4$, $4x + \frac{3y}{2} = -1$ (v) $5x + \frac{2y}{3} = -4$, $4x + y = 2$ (vi) $\frac{3x}{5} - 4y = \frac{1}{2}$, $x - 2y = \frac{1}{3}$

3. Find the values of m such that these equations have no solutions.

(i) $3x - my = 4$, $x + y = 12$ (ii) $5x + y = 12$, $mx - y = -2$ (iii) $4x - 2y = 12$, $3x + my = 2$

4. Find the values of m and a such that these equations have infinite solution sets.

(i) $4x + my = a$, $2x + y = 4$ (ii) $5x + 2y = 12$, $mx + 4y = a$ (iii) $3x + my = a$, $2x - 4y = 6$

5. Find the solution sets of the following simultaneous equations, solving for x and y.

(a) $bx + y = a$, $ax - y = b$ (b) $bx + y = a$, $ax + y = b$ (c) $ax + by = 1$, $ax - by = 1$

(d) $ax + y = ab$, $bx - y = b^2$ (e) $ax + by = a - b$, $bx + ay = a - b$ (f) $ax + y = b$, $bx + ay = 2ab - a^3$

2.3.3 SIMULTANEOUS LINEAR EQUATIONS IN THREE UNKNOWNS

So far we have looked at linear equation in two unknowns. However, this can be extended to linear equations in three unknowns. Equations such as these, involving the variables x, y and z take on the general form $ax + by + cz = k$ where a, b, c and k are real constants.

Just as for the case with two unknowns, where we required two equations to (hopefully) obtain a unique solution to the system of simultaneous equations, when dealing with three unknowns we will require a minimum of three equations to (hopefully) obtain a unique solution.

The solution process for a system of linear equations in three unknowns will require, primarily, the use of the elimination method. The method usually involves the reduction of a system of three equations in three unknowns to one of two equations in two unknowns. This will then enable the use of the methods already discussed to solve the 'reduced' system. Once two of the unknowns have been determined from this 'reduced' system, we substitute back into one of the original three equations to solve for the third unknown.

EXAMPLE 2.22

Solve the simultaneous equations
$$\begin{aligned} x + 3y - z &= 13 \\ 3x + y - z &= 11 \\ x + y - 3z &= 11 \end{aligned}$$

Solution

We label the equations as follows;
$$\begin{aligned} x + 3y - z &= 13 \quad -(1) \\ 3x + y - z &= 11 \quad -(2) \\ x + y - 3z &= 11 \quad -(3) \end{aligned}$$

Step 1: Reduce the system to one involving two equations and two unknowns.
We first eliminate the variable z:

$(2) - (1)$: $\quad 2x - 2y = -2 \quad -(4)$

$3\times(2) - (3)$: $\quad 8x + 2y = 22 \quad -(5)$

Step 2: Solve the reduced system of equations.

$(4) + (5)$: $\quad 10x = 20$

$\Leftrightarrow x = 2$

Substitute into (5): $\quad 2 \times 2 - 2y = -2 \Leftrightarrow -2y = -6 \Leftrightarrow y = 3$.

Step 3: Solve for the third unknown.
Subtituting $x = 2$ and $y = 3$ into (1):

$$2 + 3 \times 3 - z = 13 \Leftrightarrow z = -2$$

Therefore the solution is given by $x = 2$, $y = 3$ and $z = -2$.
Check: Using equation (2): L.H.S. = 2 + 3 –3×–2 = 11 = R.H.S

EXERCISES 2.3.3

1. Solve the simultaneous equations

(a) $$\begin{aligned} 6x + 4y - z &= 3 \\ x + 2y + 4z &= -2 \\ 5x + 4y &= 0 \end{aligned}$$

(b) $$\begin{aligned} x + y + z &= 2 \\ 4x + y &= 4 \\ -x + 3y + 2z &= 8 \end{aligned}$$

(c) $$\begin{aligned} 4x + 9y + 13z &= 3 \\ -x + 3y + 24z &= 17 \\ 2x + 6y + 14z &= 6 \end{aligned}$$

(d) $$\begin{aligned} x - 2y - 3z &= 3 \\ x + y - 2z &= 7 \\ 2x - 3y - 2z &= 0 \end{aligned}$$

(e) $$\begin{aligned} x - y - z &= 2 \\ 3x + 3y - 7z &= 7 \\ x + 2y - 3z &= 3 \end{aligned}$$

(f) $$\begin{aligned} x - 2y &= -1 \\ -x - y + 3z &= 1 \\ y - z &= 0 \end{aligned}$$

2.4 QUADRATICS

2.4.1 QUADRATIC EQUATION

A **quadratic equation** in the **variable** x (say) takes on the form $ax^2 + bx + c = 0$ where a, b and c are real constants. The equation is a quadratic because x is raised to the power of two. The solution(s) to such equations can be obtained in one of two ways.

Method 1: Factorise the quadratic and use the Null Factor Law.

Method 2: Use the quadratic formula.

We look at each of these methods.

Method 1 Factorisation and the Null Factor Law

First of all we must have one side of the equation as 0, otherwise the Null Factor Law cannot be used. Next, when factorising the quadratic, you will need to rely on your ability to recognise the form of the quadratic and hence which approach to use. A summary of the factorisation process for quadratics is shown below:

Case 1: $a = 1$

Quadratic expression $x^2 + bx + c$ →	Example
Trial and error: $(x + \alpha)(x + \beta)$	$x^2 + 12x + 32 = (x + 4)(x + 8)$ $x^2 + 12x - 28 = (x + 14)(x - 2)$
Perfect square: $(x + \alpha)^2$ or $(x - \alpha)^2$	$x^2 + 6x + 9 = (x + 3)^2$ $x^2 - 4x + 4 = (x - 2)^2$
Difference of two squares: $(x + \alpha)(x - \alpha)$	$x^2 - 16 = (x + 4)(x - 4)$ $x^2 - 3 = (x - \sqrt{3})(x + \sqrt{3})$

Note that sometimes you might need to use a perfect square approach to part of the quadratic and then complete the factorisation process by using the difference of two squares.

EXAMPLE 2.23 Solve the quadratic $x^2 + 6x + 7 = 0$

Solution

In this instance it is not obvious what the factors are and so trial and error is not appropriate. However, we notice that $x^2 + 6x + 7$ can be 'broken up' into $x^2 + 6x + 9 - 2$.
That is, part of the quadratic has been expressed as a perfect square, so that

$$x^2 + 6x + 9 - 2 = (x + 3)^2 - 2$$

Then, we are left with a difference of perfect squares: $(x + 3)^2 - 2 = (x + 3 + \sqrt{2})(x + 3 - \sqrt{2})$.
Therefore, $x^2 + 6x + 7 = 0 \Leftrightarrow (x + 3 + \sqrt{2})(x + 3 - \sqrt{2}) = 0 \Leftrightarrow x = -3 \pm \sqrt{2}$.

Case 2: $a \neq 1$

Quadratic expression $ax^2 + bx + c$	Method	Example
→	Trial and error:	$2x^2 + 3x + 1 = (2x+1)(x+1)$
→	Perfect square:	$4x^2 + 12x + 9 = (2x+3)^2$
→	Difference of two squares:	$9x^2 - 25 = (3x+5)(3x-5)$

That is, the methods for the case $a \neq 1$ are the same as for when $a = 1$, they require a little more mental arithmetic to 'juggle' the correct numbers. Sometimes they can get messy, as the next example shows.

EXAMPLE 2.24

Solve the quadratic $3x^2 + 2x - 2 = 0$.

Solution

None of the above methods provides a quick solution, so, as for Example 2.23, we will combine the methods of perfect squares with the difference of two squares:

$$3x^2 + 2x - 2 = 3\left(x^2 + \frac{2}{3}x\right) - 2 = 3\left(x^2 + \frac{2}{3}x + \frac{1}{9}\right) - 3 \times \frac{1}{9} - 2 \text{ [add **and** subtract } \frac{3}{9}\text{]}$$

$$= 3\left(x + \frac{1}{3}\right)^2 - \frac{7}{3}$$

Part of the quadratic is now a perfect square. Next, we use the difference of perfect squares:

$$= 3\left(x + \frac{1}{3}\right)^2 - \frac{7}{3}$$

$$= 3\left[\left(x + \frac{1}{3}\right)^2 - \frac{7}{9}\right]$$

$$= 3\left(x + \frac{1}{3} - \frac{\sqrt{7}}{3}\right)\left(x + \frac{1}{3} + \frac{\sqrt{7}}{3}\right)$$

Therefore, $3x^2 + 2x - 2 = 0 \Leftrightarrow 3\left(x + \frac{1}{3} - \frac{\sqrt{7}}{3}\right)\left(x + \frac{1}{3} + \frac{\sqrt{7}}{3}\right) = 0 \Leftrightarrow x = -\frac{1}{3} \pm \frac{\sqrt{7}}{3}$.

Now, that was quite a bit of work!

Of course, coming up with the 'magic' number, $\frac{1}{9}$ did make life a little easier. The rest was simply being careful with the arithmetic. So, how did we pull $\frac{1}{9}$ out of the hat? Well, once we have made the coefficient of the x^2 one, i.e., by factorising the '3' out, we look at the coefficient of the x term. Then we halve it, square the result and add it. i.e., $\frac{1}{2} \times \frac{2}{3} = \frac{1}{3} \to \left(\frac{1}{3}\right)^2 = \frac{1}{9}$.

Then, so that the equation is unaltered, we subtract this result – but be careful, do not forget to multiply it by the factor '3' at the front of the brackets. The rest then follows.

The only way to be proficient with these methods is practice, practice and more practice. However, there is a short cut to solving quadratic equations. We look at this next.

Method 2 Quadratic Formula and the Discriminant

A formula that allows us to solve any quadratic equation $ax^2 + bx + c = 0$ (if real solutions exist), is given by $x = \dfrac{-b \pm \sqrt{b^2 - 4ac}}{2a}$. So that obtaining solutions requires that we make the appropriate substitution for a, b and c.

To derive this expression we proceed in exactly the same way as we did in Example 2.24. We show some of the key steps in obtaining this result and leave the proof for you to complete.

$$ax^2 + bx + c = a\left(x^2 + \frac{b}{a}x\right) + c = a\left[\left(x^2 + \frac{b}{a}x + \frac{b^2}{4a^2}\right)\right] - a \times \frac{b^2}{4a^2} + c$$

$$= a\left(x + \frac{b}{2a}\right)^2 - a\left(\frac{b^2}{4a^2} - \frac{c}{a}\right)$$

$$= a\left[\left(x + \frac{b}{2a}\right)^2 - \left(\frac{b^2 - 4ac}{4a^2}\right)\right]$$

$$= a\left(x + \frac{b + \sqrt{b^2 - 4ac}}{2a}\right)\left(x + \frac{b - \sqrt{b^2 - 4ac}}{2a}\right)$$

Therefore, $ax^2 + bx + c = 0 \Leftrightarrow x = \dfrac{-b \pm \sqrt{b^2 - 4ac}}{2a}$.

EXAMPLE 2.25

Use the formula to solve the quadratic equations:

(a) $x^2 - x - 4 = 0$ (b) $2x^2 = 4 - x$

Solution

(a) For the equation $x^2 - x - 4 = 0$, we have $a = 1$, $b = -1$ and $c = -4$.

Using $x = \dfrac{-b \pm \sqrt{b^2 - 4ac}}{2a}$ we have $x = \dfrac{1 \pm \sqrt{(-1)^2 - 4 \times 1 \times -4}}{2 \times 1} = \dfrac{1 \pm \sqrt{17}}{2}$

(b) Similarly, if $2x^2 = 4 - x$, then, $2x^2 + x - 4 = 0$ so that $a = 2$, $b = 1$ and $c = -4$, so that

using $x = \dfrac{-b \pm \sqrt{b^2 - 4ac}}{2a}$ we have $x = \dfrac{-1 \pm \sqrt{(-1)^2 - 4 \times 2 \times -4}}{2 \times 2} = \dfrac{-1 \pm \sqrt{33}}{4}$.

The Discriminant

Closer inspection of this formula indicates that much can be deduced from the term under the square root sign, i.e., $b^2 - 4ac$. The expression $b^2 - 4ac$ is known as the **discriminant** and is often represented by the delta symbol $\Delta = b^2 - 4ac$.

In particular, there are three cases to address:

Case 1. $b^2 - 4ac > 0$

Case 2. $b^2 - 4ac = 0$

Case 3. $b^2 - 4ac < 0$

Case 1. $b^2 - 4ac > 0$

In this case, the expression $x = \frac{-b \pm \sqrt{b^2 - 4ac}}{2a}$ produces **two real solutions**.

This is because taking the square root of a positive number will produce another positive real number. This in turn implies that there will be one solution corresponding to the '+' term and one solution corresponding to the '–' term.

That is, say that $\sqrt{b^2 - 4ac} = K$, where K is a real number. We then have that

$x = \frac{-b \pm K}{2a}$, i.e., $x_1 = \frac{-b + K}{2a}$, $x_2 = \frac{-b - K}{2a}$, giving **two distinct real solutions**.

Case 2. $b^2 - 4ac = 0$

In this case, the expression $x = \frac{-b \pm \sqrt{b^2 - 4ac}}{2a}$ produces only **one real solution**.

This is because taking the square root of zero gives zero. This in turn implies that there will be only one solution because adding and subtracting '0' to the '$-b$' term in the numerator will not alter the answer.

That is, if $\sqrt{b^2 - 4ac} = 0$, we then have that $x = \frac{-b \pm 0}{2a} = -\frac{b}{2a}$ meaning that we have only **one real solution** (or two repeated solutions).

Case 3. $b^2 - 4ac < 0$

In this case, the expression $x = \frac{-b \pm \sqrt{b^2 - 4ac}}{2a}$ produces **no real solution**.

This is because the square root of a negative number will not produce a **real** number. This in turn implies that the formula cannot be utilised (if we are dealing with quadratic equations under the real numbers).

Summary

Discriminant $\Delta = b^2 - 4ac$	Number of solutions for $ax^2 + bx + c = 0$
$\Delta > 0$	Can be factorised to obtain **2 real** and unique solutions.
$\Delta = 0$	Can be factorised to obtain **1 real** (repeated) solution.
$\Delta < 0$	Cannot be factorised and so **no real** solutions exist.

EXAMPLE 2.26

(a) Find the value(s) of m for which the equation $2x^2 + mx + 1 = 0$ has one real solution.

(b) Find the value(s) of k for which the equation $x^2 + 4x + k = 0$ has two real solutions.

Solution

(a) For one real solution to exist, we must have that $\Delta = b^2 - 4ac = 0$.
For this quadratic we have that $a = 2, b = m$ and $c = 1$.
Therefore, we need that $m^2 - 4 \times 2 \times 1 = 0$

$$\Leftrightarrow m^2 - 8 = 0$$

$$\Leftrightarrow (m - \sqrt{8})(m + \sqrt{8}) = 0$$

$$\therefore m = 2\sqrt{2} \text{ or } m = -2\sqrt{2}$$

(b) For two real solutions, we must have that $\Delta = b^2 - 4ac > 0$.
For this quadratic we have that $a = 1, b = 4$ and $c = k$.
Therefore, we need that $4^2 - 4 \times 1 \times k > 0$

$$\Leftrightarrow 16 - 4k > 0$$

$$\Leftrightarrow 16 > 4k$$

$$\Leftrightarrow 4 > k.$$

i.e., the quadratic $x^2 + 4x + k = 0$ will have two real solutions as long as $k < 4$.

EXERCISES 2.4.1

1. By using a factorisation process, solve for the given variable.

(a) $x^2 + 10x + 25 = 0$ (b) $x^2 - 10x + 24 = 0$

(c) $3x^2 + 9x = 0$ (d) $x^2 - 4x + 3 = 0$

(e) $(3 - u)(u + 6) = 0$ (f) $3x^2 + x - 10 = 0$

(g) $3v^2 - 12v + 12 = 0$ (h) $y(y - 3) = 18$

(i) $(x + 3)(x + 2) = 12$ (j) $(2a - 1)(a - 1) = 1$

2. Without using the quadratic formula, solve for the given variable.

(a) $u + \frac{1}{u} = -2$ (b) $x + 2 = \frac{35}{x}$ (c) $5x - 13 = \frac{6}{x}$

(d) $\frac{x}{2} - \frac{1}{x + 1} = 0$ (e) $y + 1 = \frac{4}{y + 1}$ (f) $v + \frac{20}{v} = 9$

3. By completing the square, solve for the given variable.

(a) $x^2 + 2x = 5$ (b) $x^2 + 4 = 6x$ (c) $x^2 - 2x = 4$

(d) $4x^2 + x = 2$ (e) $2y^2 = 9y - 1$ (f) $3a^2 - a = 7$

4. Use the quadratic formula to solve these equations.

(a) $x^2 - 3x - 7 = 0$ (b) $x^2 - 5x = 2$ (c) $x^2 - 3x - 6 = 0$

(d) $x^2 = 7x + 2$ (e) $x(x + 7) = 4$ (f) $x^2 + 2x - 8 = 0$

(g) $x^2 + 2x - 7 = 0$ (h) $x^2 + 5x - 7 = 0$ (i) $x^2 - 3x - 7 = 0$

(j) $x^2 - 3x + 9 = 0$ (k) $x^2 + 9 = 8x$ (l) $4x^2 - 8x + 9 = 0$

(m) $4x^2 = 8x + 9$ (n) $5x^2 - 6x - 7 = 0$ (o) $5x^2 - 12x + 1 = 0$

5. For what value(s) of p does the equation $x^2 + px + 1 = 0$ have
(a) no real solutions (b) one real solution (c) two real solutions.

6. Find the values of m for which the quadratic $x^2 + 2x + m = 0$ has
(a) one real solution (b) two real solutions (c) no real solutions.

7. Find the values of m for which the quadratic $x^2 + mx + 2 = 0$ has
(a) one real solution (b) two real solutions (c) no real solutions.

8. Find the values of k for which the quadratic $2x^2 + kx + 9 = 0$ has
(a) one real solution (b) two real solutions (c) no real solutions.

9. Consider the equation $x^2 + 2x = 7$. Prove that this equation has two real roots.

2.4.2 QUADRATIC FUNCTION

A **quadratic function** has the general form $f(x) = ax^2 + bxc, a \neq 0$ and $a, b, c \in \mathbb{R}$.
All quadratic functions have **parabolic graphs** and have a **vertical axis of symmetry**.

If **a > 0**, the parabola is concave up: 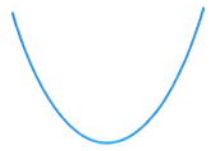

If **a < 0** the parabola is concave down: 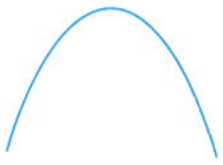

General Properties of the graph of $f(x) = ax^2 + bx + c, a \neq 0$

1. *y*–intercept

This occurs when $x = 0$, so that $y = f(0) = a(0)^2 + b(0) + c = c$.
That is, the curve passes through the point $(0, c)$

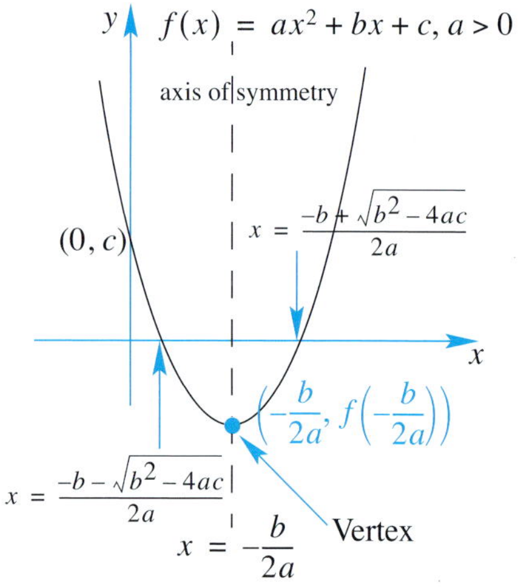

2. *x*–intercept(s)

This occurs where $f(x) = 0$.

Therefore we need to solve $ax^2 + bx + c = 0$.
To solve we either factorise and solve or use the quadratic formula, which would provide the solution(s) $x = \frac{-b \pm \sqrt{b^2 - 4ac}}{2a}$.

3. Axis of symmetry

This occurs at $x = -\frac{b}{2a}$.

4. Vertex (turning point)

The vertex occurs when $x = -\frac{b}{2a}$. Then, to find the y–value, find $f\left(-\frac{b}{2a}\right)$.

Geometrical Interpretation and The Discriminant

Just as it was the case for linear functions, solving the quadratic $ax^2 + bx + c = k$ is geometrically equivalent to finding where the parabola with function $f(x) = ax^2 + bx + c$ meets the graph (horizontal straight line) $y = k$. Then, when $k = 0$, we are finding where the parabola meets the line $y = 0$, i.e., we are finding the x–intercept(s). Based on our results of the discriminant about the number of solutions to the equation $ax^2 + bx + c = 0$, we can extend these results to the following:

The number of x–intercepts for the function $f(x) = ax^2 + bx + c$	=	The number of solutions to the equation $ax^2 + bx + c = 0$

If $\Delta = b^2 - 4ac > 0$, then there are **two x–intercepts**.	If $\Delta = b^2 - 4ac = 0$, then there is **one x–intercept**.	If $\Delta = b^2 - 4ac < 0$, then there are **no x–intercepts**.
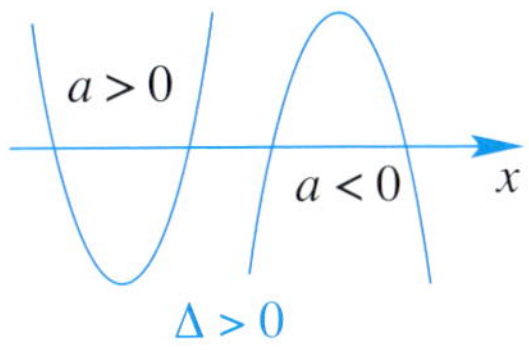	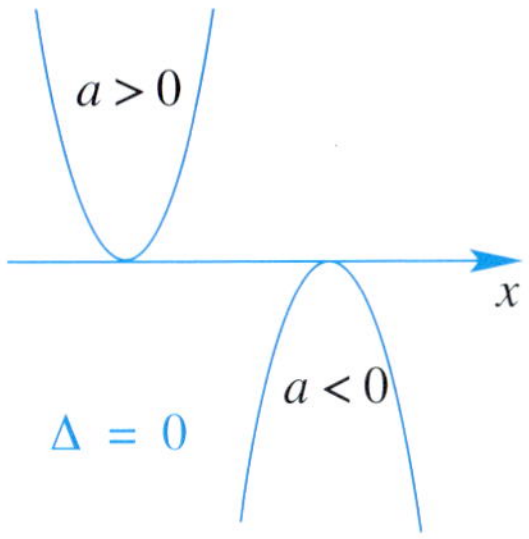	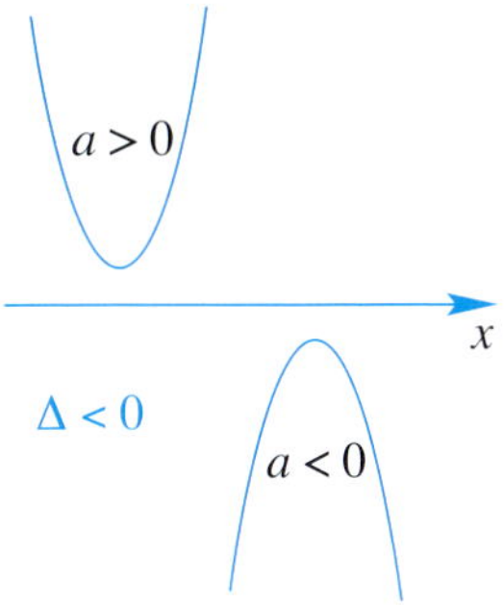

Sketching the graph of a quadratic function

Two methods for sketching the graph of $f(x) = ax^2 + bx + c, a \neq 0$ are:

Method 1: The intercept method
i.e., expressing $f(x) = ax^2 + bx + c, a \neq 0$ in the form $f(x) = a(x - p)(x - q)$

This involves

Step 1 Finding the x–intercepts [by solving $ax^2 + bx + c = 0$]
Step 2 Finding the y–intercept [finding $f(0)$]
Step 3 Sketch parabola passing through the three points

Method 2: The turning-point form
i.e., expressing $f(x) = ax^2 + bx + c, a \neq 0$ in the form $f(x) = a(x - h)^2 + k$

This involves

Step 1 Expressing $f(x) = ax^2 + bx + c, a \neq 0$ in the form $a(x - h)^2 + k$ [by completing the square]
Step 2 Use the turning point (h, k)
Step 3 Finding the y–intercept [finding $f(0)$]
Step 4 Sketch the parabola passing through the two points

EXAMPLE 2.27

Sketch the graph of

(a) $f(x) = x^2 - 6x + 8$ (b) $f(x) = 3x^2 + 12x + 4$ using

i. the intercept method ii. the turning point method

Solution

(a) i. For the intercept method we start by factorising the quadratic:

$f(x) = x^2 - 6x + 8 = (x-4)(x-2)$.

Then, for the x–intercepts we have:

$$f(x) = 0 \Leftrightarrow (x-4)(x-2) = 0 \Leftrightarrow x = 2 \text{ or } x = 4$$

Next, the y–intercept:

$$x = 0 \Rightarrow f(0) = 0^2 - 6(0) + 8 = 8.$$

Summary: Parabola passes through $(2, 0)$, $(4, 0)$ and $(0, 8)$.

Note, the axis of symmetry is the midpoint of the x–intercepts,

i.e., $x = \frac{2+4}{2} = \frac{6}{2} = 3$. Or use $x = -\frac{b}{2a} = -\frac{-6}{2} = 3$

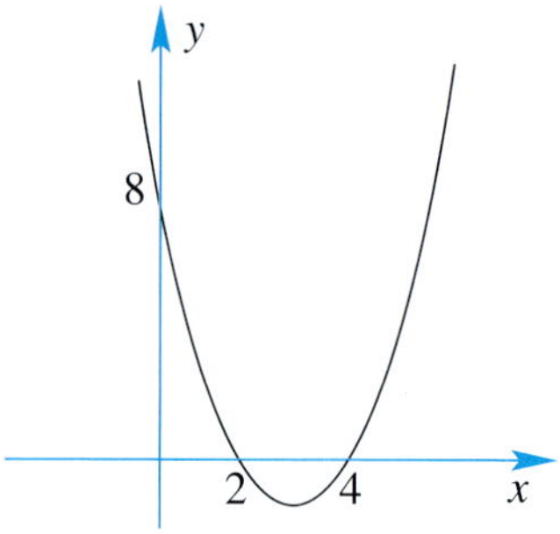

ii. We first complete the square:

$f(x) = x^2 - 6x + 8 = (x^2 - 6x + 9) - 1 = (x-3)^2 - 1$

Therefore, the parabola has a turning point at $(3, -1)$

Then find the y–intercept:

$$x = 0 \Rightarrow f(0) = 0^2 - 6(0) + 8 = 8$$

Summary: Parabola passes through $(3, -1)$ and $(0, 8)$.

Notice that both graphs are the same but display different information. Each display has its strengths and weaknesses – it depends on what information is important for the question. Of course, including all the information will rid us of any problems.

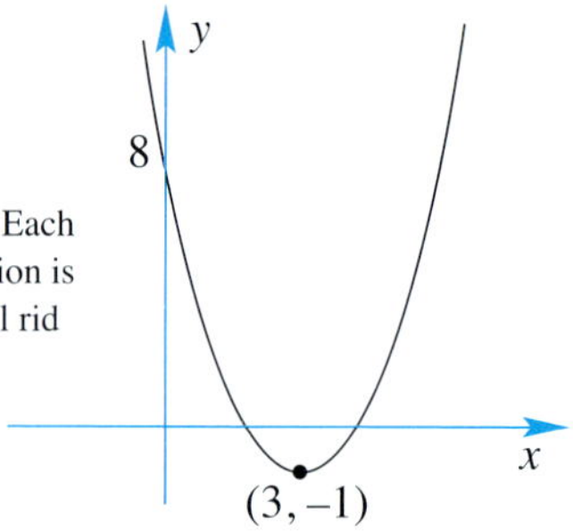

(b) i. For the intercept method we start by factorising the quadratic:

$f(x) = 3x^2 + 12x + 4 = 3(x^2 + 4x + 4) - 12 + 4$ [completing the square]

$= 3(x+2)^2 - 8$ [next use difference of two squares]

$= [\sqrt{3}(x+2) - 2\sqrt{2}][\sqrt{3}(x+2) + 2\sqrt{2}]$.

Then, for the x–intercepts we have:

$$f(x) = 0 \Rightarrow [\sqrt{3}(x+2) - 2\sqrt{2}][\sqrt{3}(x+2) + 2\sqrt{2}] = 0$$

$$\therefore x = -2 + 2\sqrt{\frac{2}{3}} \text{ or } x = -2 - 2\sqrt{\frac{2}{3}}$$

Note: It would have been quicker to find the intercepts if we had used the quadratic formula!

Next, the y–intercept:

$$x = 0 \Rightarrow f(0) = 3(0)^2 + 12(0) = 4 = 4.$$

Summary: Parabola passes through $\left(-2+2\sqrt{\frac{2}{3}}, 0\right)$, $\left(-2-2\sqrt{\frac{2}{3}}, 0\right)$ and $(0, 4)$.

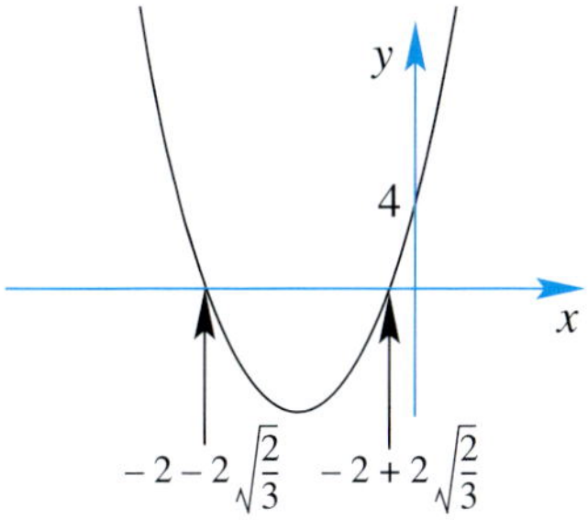

ii. We first complete the square:

$$f(x) = 3x^2 + 12x + 4 = 3(x^2 + 4x + 4) - 12 + 4 \text{ [completing the square]}$$
$$= 3(x+2)^2 - 8$$

Therefore, the parabola has a turning point at $(-2, -8)$

Then find the y–intercept:

$$x = 0 \Rightarrow f(0) = 3(0)^2 + 12(0) = 4 = 4.$$

Summary: Parabola passes through $(-2, -8)$ and $(0, 4)$.

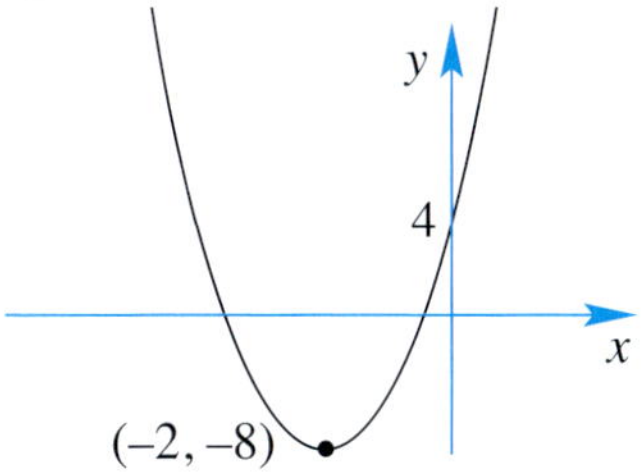

EXAMPLE 2.28

For what value(s) of k will the function $f(x) = x^2 + 6x + k$

(a) cut the x–axis twice (b) touch the x–axis (c) have no x–intercepts

Solution

We will need to use the relationship between the number of x–intercepts of the function $f(x) = x^2 + 6x + k$ and the number of solutions to the equation $x^2 + 6x + k = 0$. To do this we first find the discriminant: $\Delta = b^2 - 4ac = (6)^2 - 4 \times 1 \times k = 36 - 4k$.

(a) If the function cuts twice, then $\Delta > 0 \Rightarrow 36 - 4k > 0 \Leftrightarrow k < 9$.

(b) If the function touches the x–axis, then we have repeated solutions (or roots), in this case we have only one solution, so $\Delta = 0 \Rightarrow 36 - 4k = 0 \Leftrightarrow k = 9$.

(c) If the function has no x–intercepts, then $\Delta < 0 \Rightarrow 36 - 4k < 0 \Leftrightarrow k > 9$.

So far we have sketched a graph from a given function, but what about finding the equation of a given graph?

Finding the equation from a graph

If sufficient information is provided on a graph, then it is possible to obtain the equation that corresponds to that graph. When dealing with quadratics there are some standard approaches that can be used (depending on the information provided).

Information provided	Process
Graph cuts the x–axis at two points: 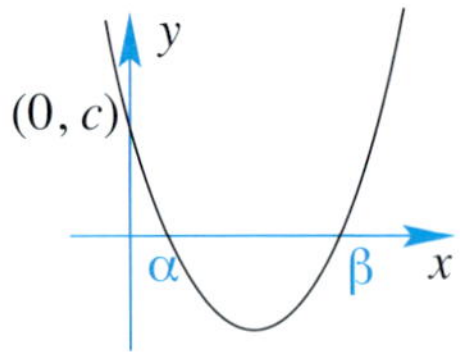	Use the function $f(x) = k(x-\alpha)(x-\beta)$ and then use the point $(0, c)$ to solve for k.
Graph touches the x–axis at $x = \alpha$: 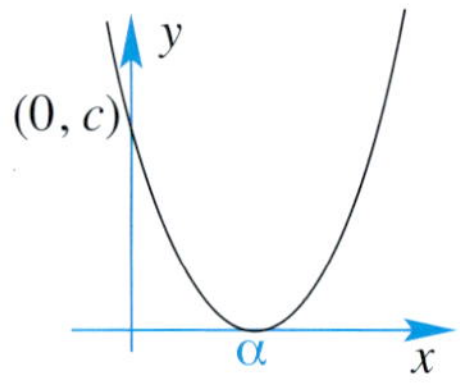	Use the function $f(x) = k(x-\alpha)^2$ and then use the point $(0, c)$ to solve for k.
Graph does not meet the x–axis: 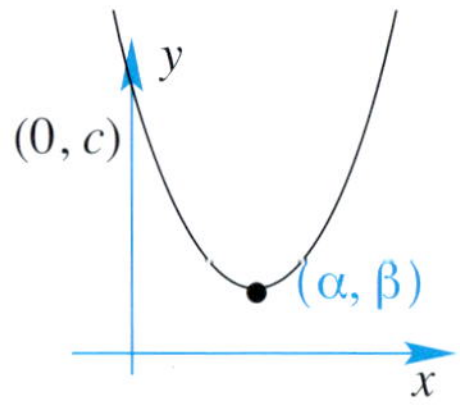	Use the function $f(x) = k(x-\alpha)^2 + \beta$ and then use the point $(0, c)$ to solve for k.
Three arbitrary points are given: 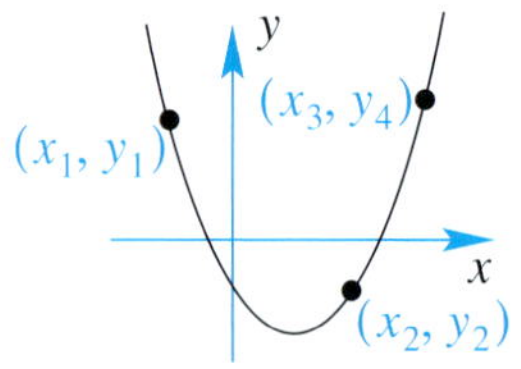	Use the function $f(x) = ax^2 + bx + c$ and then set up and solve the system of simultaneous equations by substituting each coordinate into the function: $ax_1^2 + bx_1 + c = y_1$ $ax_2^2 + bx_2 + c = y_2$ $ax_3^2 + bx_3 + c = y_3$

Note: the process is identical for a downward concave parabola.

EXAMPLE 2.29

Find the equation defining each of the following graphs.

(a)

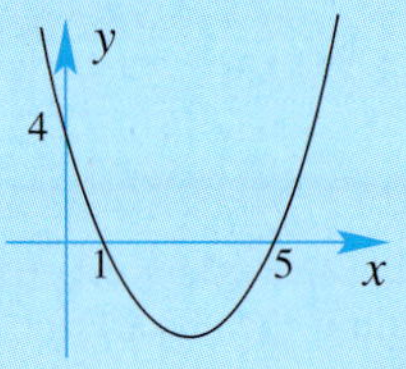

(b)

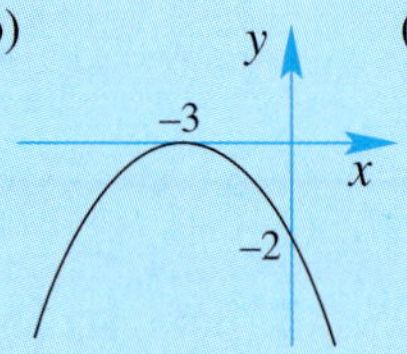

(c)

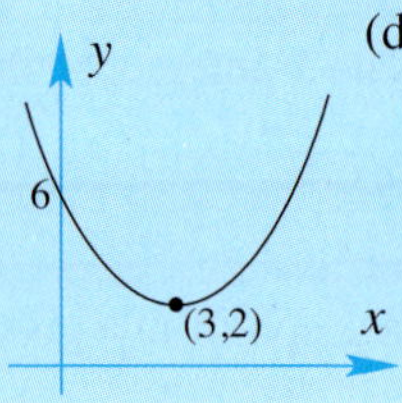

(d)

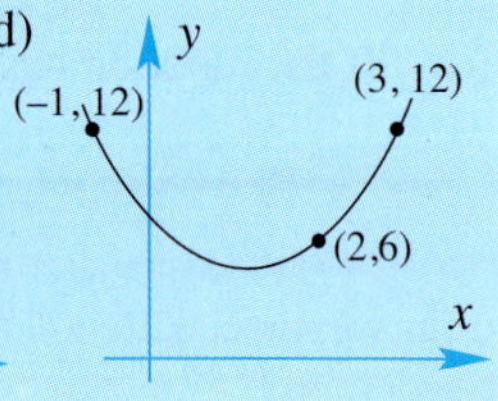

Solution

(a) Using the form $f(x) = k(x-\alpha)(x-\beta)$ we have $f(x) = k(x-1)(x-5)$.

Next, when $x = 0$, $y = 4$, therefore, $4 = k(0-1)(0-5) \Leftrightarrow 4 = 5k \Leftrightarrow k = \frac{4}{5}$.

Therefore, $f(x) = \frac{4}{5}(x-1)(x-5)$.

(b) Graph touches x–axis at $x = -3$, therefore use $f(x) = k(x+3)^2$.

As graph passes through $(0, -2)$, we have $-2 = k(0+3)^2 \Leftrightarrow -2 = 9k \Leftrightarrow k = -\frac{2}{9}$.

Therefore, $f(x) = -\frac{2}{9}(x+3)^2$.

(c) Graph shows turning point and another point, so use the form $f(x) = k(x-\alpha)^2 + \beta$.

So we have, $f(x) = k(x-3)^2 + 2$.

Then, as graph passes through $(0, 6)$, we have $6 = k(0-3)^2 + 2 \Leftrightarrow 4 = 9k \Leftrightarrow k = \frac{4}{9}$.

Therefore, $f(x) = \frac{4}{9}(x-3)^2 + 2$.

(d) As we are given three arbitrary points, we use the general equation $f(x) = ax^2 + bx + c$.

From $(-1, 12)$ we have $12 = a(-1)^2 + b(-1) + c$ i.e., $12 = a - b + c$ –(1)

From $(2, 6)$ we have $6 = a(2)^2 + b(2) + c$ i.e., $6 = 4a + 2b + c$ –(2)

From $(3, 12)$ we have $12 = a(3)^2 + b(3) + c$ i.e., $12 = 9a + 3b + c$ –(3)

Solving for a, b and c we have: (2) – (1): $-6 = 3a + 3b$

i.e., $-2 = a + b$ –(3)

(3) – (2): $6 = 5a + b$ –(4)

(4) – (3): $8 = 4a \Leftrightarrow a = 2$

Substitute $a = 2$ into (3): $-2 = 2 + b \Leftrightarrow b = -4$.

Substitute results into (1): $12 = 2 - (-4) + c \Leftrightarrow c = 6$.

Therefore function is $f(x) = 2x^2 - 4x + 6$

EXERCISES 2.4.2

1. Express the following functions in turning point form and hence sketch their graphs.

(a) $y = x^2 - 2x + 1$ (b) $y = x^2 + 4x + 2$ (c) $y = x^2 - 4x + 2$

(d) $y = x^2 + x - 1$ (e) $y = x^2 - x - 2$ (f) $y = x^2 + 3x + 1$

(g) $y = -x^2 + 2x + 1$ (h) $y = -x^2 - 2x + 2$ (i) $y = 2x^2 - 2x - 1$

(j) $y = -\frac{1}{2}x^2 + 3x - 2$ (k) $y = -\frac{x^2}{3} + x - 2$ (l) $y = 3x^2 - 2x + 1$

2. Find the axial intercepts of these quadratic functions (correct to 2 decimal places) and hence sketch their graphs.

(a)	$y = x^2 + 3x + 2$	(b)	$y = x^2 - x - 6$	(c)	$y = 2x^2 - 5x - 3$
(d)	$y = x^2 - 4$	(e)	$y = x^2 + x - 5$	(f)	$y = -x^2 + x + 6$
(g)	$y = -x^2 + x + 1$	(h)	$y = -2x^2 - 3x + 5$	(i)	$y = 2x^2 + 5x - 3$
(j)	$y = \dfrac{x^2}{3} - 2x + 3$	(k)	$y = -\dfrac{x^2}{2} + x + 4$	(l)	$y = 3x^2 - 2x - 4$

3. For the quadratic function $f(x) = 2x^2 - 4x + 1$, find
(a) the equation of the axis of symmetry.
(b) the coordinates of the vertex.
(c) the i. x–intercept(s) i i. y–intercept.
Hence, sketch the graph of the function.

4. For the quadratic function $f(x) = 7 + 4x - 2x^2$, find
(a) the equation of the axis of symmetry.
(b) the coordinates of the vertex.
(c) the i. x–intercept(s) ii. y–intercept.
Hence, sketch the graph of the function.

5. For what value(s) of k will the graph of $y = x^2 - 3x + k$
(a) touch the x–axis (b) cut the x–axis (c) never meet the x–axis?

6. For what value(s) of k will the graph of $y = kx^2 + 5x + 2$
(a) touch the x–axis (b) cut the x–axis (c) never meet the x–axis?

7. For what value(s) of k will the graph of $y = kx^2 - 2x + k$
(a) touch the x–axis (b) cut the x–axis (c) never meet the x–axis?

8. Find the equation of the quadratic function with graph

(a)

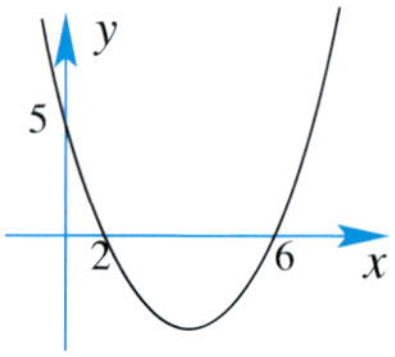

(b)

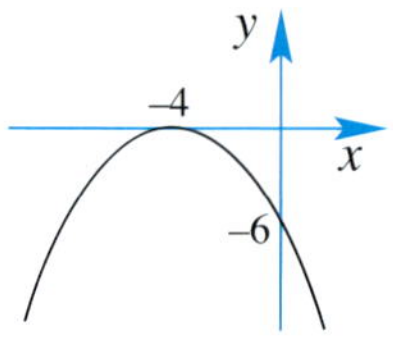

(c)

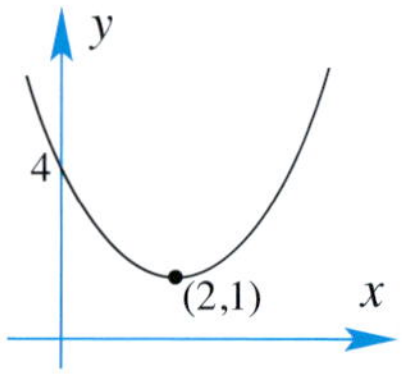

(d)

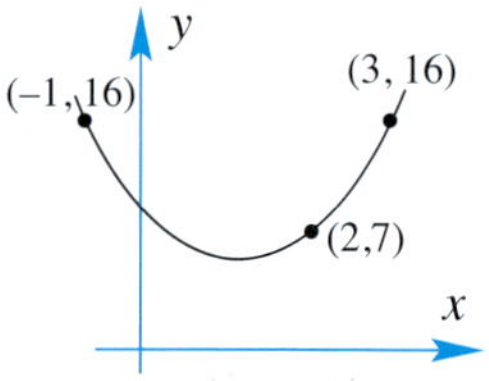

9. Find the equation of the quadratic function with graph

(a)

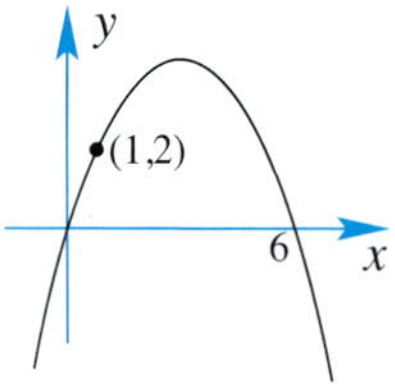

(b)

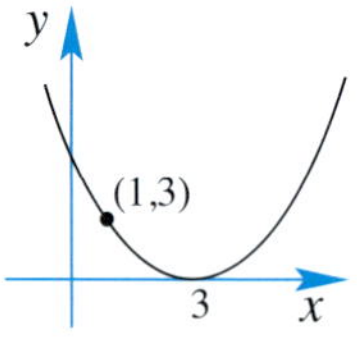

(c)

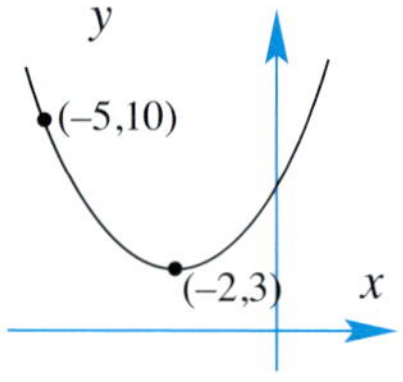

(d)

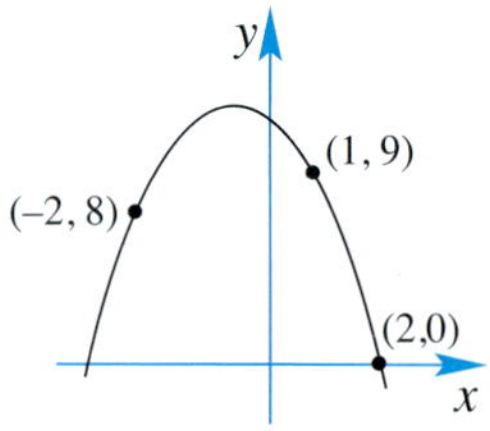

2.4.3 QUADRATIC INEQUALITIES

Quadratic inequations arise from replacing the '=' sign in a quadratic by an inequality sign. Solving inequations can be carried out in two ways, either algebraically or graphically.

Method 1 Algebraic Method
This method relies on factorising the quadratic and then using the fact that when two terms, a and b are multiplied, the following rules apply:

1. $ab > 0 \Leftrightarrow a > 0$ and $b > 0$ or $a < 0$ and $b < 0$

2. $ab < 0 \Leftrightarrow a > 0$ and $b < 0$ or $a < 0$ and $b > 0$

The same rules apply if we replace ' > ' with ' ≥ ' and ' < ' with ' ≤ '

EXAMPLE 2.30 Find

(a) $\{x|\ x^2 - 6x + 8 > 0\}$ (b) $\{x|\ 2x^2 + 5x - 3 \le 0\}$ (c) $\{x|\ x^2 - 3 < 2x\}$

Solution

(a) We start by factorising the quadratic: $x^2 - 6x + 8 = (x-2)(x-4)$

Then, $x^2 - 6x + 8 > 0 \Leftrightarrow (x-2)(x-4) > 0$

Which means either $x - 2 > 0$ **and** $x - 4 > 0$ i.e., $x > 2$ **and** $x > 4 \Rightarrow x > 4$ – (1)

or $x - 2 < 0$ **and** $x - 4 < 0$ i.e., $x < 2$ **and** $x < 4 \Rightarrow x < 2$ – (2)

Then, combining (1) and (2) we have $\{x|\ x^2 - 6x + 8 > 0\} = \{x|\ x < 2\} \cup \{x|\ x > 4\}$.

(b) Now, $2x^2 + 5x - 3 \le 0 \Leftrightarrow (2x-1)(x+3) \le 0$.

Meaning that either $2x - 1 \le 0$ **and** $x + 3 \ge 0$ i.e., $x \le \frac{1}{2}$ **and** $x \ge -3$ – (1)

or $2x - 1 \ge 0$ **and** $x + 3 \le 0$ i.e., $x \ge \frac{1}{2}$ **and** $x \le -3$ – (2)

From result (1) we have that $-3 \le x \le \frac{1}{2}$.

However, the inequalities in result (2) are inconsistent, i.e., we cannot have that x is both greater than or equal to $\frac{1}{2}$ **and** less than or equal to –3 simultaneously. Therefore we discard this inequality. Therefore, $\{x|\ 2x^2 + 5x - 3 \le 0\} = \left\{x|\ -3 \le x \le \frac{1}{2}\right\}$

(c) This time we need some rearranging:

$$x^2 - 3 < 2x \Leftrightarrow x^2 - 2x - 3 < 0 \Leftrightarrow (x+1)(x-3) < 0$$

Then, we must have that $x + 1 < 0$ **and** $x + 3 > 0$ i.e., $x < -1$ **and** $x > 3$ – (1)

$x + 1 > 0$ **and** $x + 3 < 0$ i.e., $x > -1$ **and** $x < 3$ – (2)

This time (1) is inconsistent, so we discard it and from (2) we have $-1 < x < 3$.

Therefore, $\{x|\ x^2 - 3 < 2x\} = \{x|\ -1 < x < 3\}$.

Method 2 Graphical Method

This method relies on examining the graph of the corresponding quadratic function and then

1. quoting the x–values that produce y–values that lie above (or on) the x–axis (i.e., $y > 0$ or $y \geq 0$)

or 2. quoting x–values that produce y–values that lie below (or on) the x–axis (i.e., $y < 0$ or $y \leq 0$)

We consider inequations from Example 2.30

EXAMPLE 2.31

Use a graphical method to find

(a) $\{x|\ x^2 - 6x + 8 > 0\}$ (b) $\{x|\ 2x^2 + 5x - 3 \leq 0\}$ (c) $\{x|\ x^2 - 3 < 2x\}$

Solution

(a) The corresponding function in this case is $f(x) = x^2 - 6x + 8$. That part of the graph corresponding to $f(x) > 0$ is highlighted in red. The values of x that correspond to these parts are

$$x < 2 \text{ as well as } x > 4$$

Therefore, s.s. = $\{x|\ x < 2\} \cup \{x|\ x > 4\}$

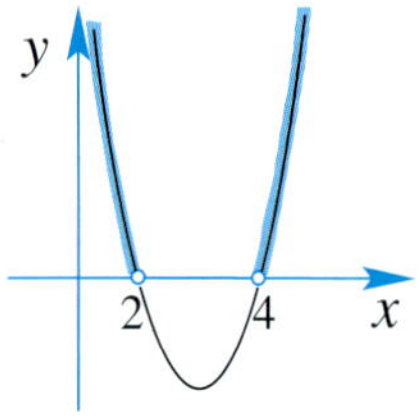

(b) The corresponding function in this case is $f(x) = 2x^2 + 5x - 3$. That part of the graph corresponding to $f(x) \leq 0$ is highlighted in red. The values of x that correspond to these parts are

$$-3 \leq x \leq \frac{1}{2}$$

Therefore, s.s. = $\left\{x|\ -3 \leq x \leq \frac{1}{2}\right\}$

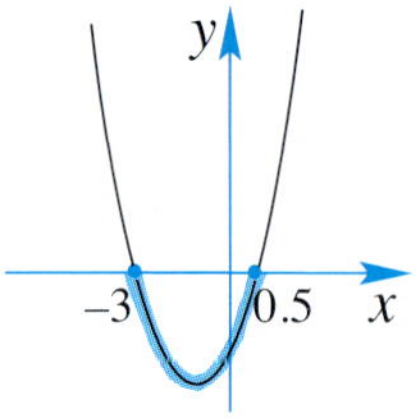

(c) This time we have two functions, that of $f(x) = x^2 - 3$ and $g(x) = 2x$, and we want to find those values of x where $f(x) < g(x)$.

We do this by sketching both graphs on the same set of axes and then finding those values of x for which $f(x) < g(x)$ i.e., where the graph of $y = g(x)$ lies above that of $y = f(x)$.

Once we have found the point of intersection, i.e., once we have solved $f(x) = g(x)$, we refer to the graph.

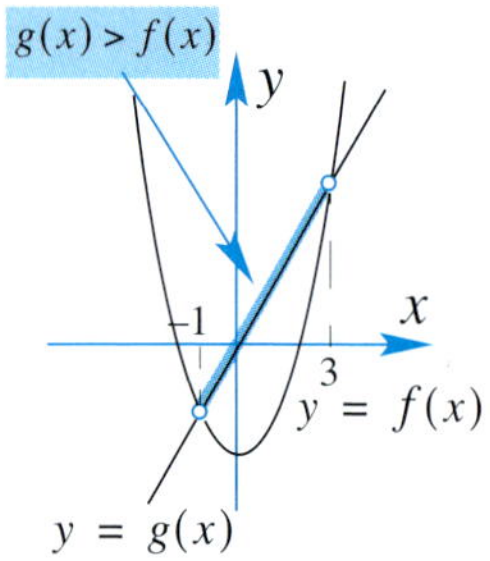

Now, $f(x) = g(x) \Leftrightarrow x^2 - 3 = 2x \Leftrightarrow x^2 - 2x - 3 = 0$

$$\Leftrightarrow (x-3)(x+1) = 0$$

$$\Leftrightarrow x = 3 \text{ or } x = -1$$

Then, $f(x) < g(x)$ for $-1 < x < 3$.

i.e., $\{x|\ x^2 - 3 < 2x\} = \{x|\ -1 < x < 3\}$.

Part (c) in Example 2.31 leads us to a process that deals with any expression of the form $f(x) < g(x)$, $f(x) > g(x)$, $f(x) \leq g(x)$ or $f(x) \geq g(x)$. Basically, we have the following:

To solve an inequality between two functions, $f(x)$ and $g(x)$,
i.e., to solve for $f(x) < g(x)$, $f(x) > g(x)$, $f(x) \le g(x)$ or $f(x) \ge g(x)$
We proceed as follows:

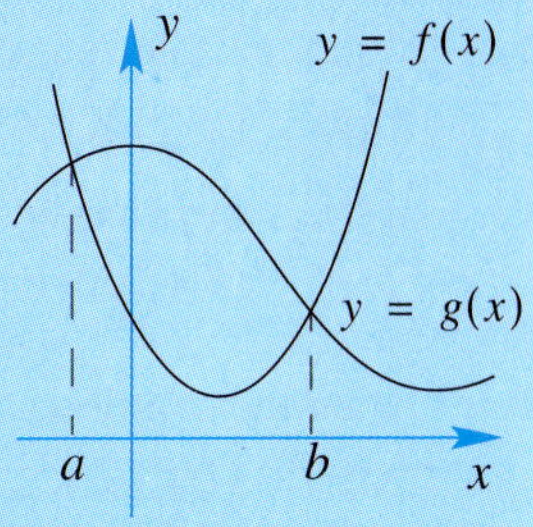

1. Sketch the corresponding graphs of both $f(x)$ and $g(x)$ on the same set of axes.

2. Find where the two graphs intersect, i.e., solve $f(x) = g(x)$ so that we find $x = a$ and $x = b$.

3. Identify where one function lies above or below the other. Then, depending on the inequality, quote the values of x that correspond to that region (or those regions).
   e.g., $f(x) > g(x) \Leftrightarrow x \in \left]-\infty, a\right] \cup \left]b, \infty\right[$ and $f(x) \le g(x) \Leftrightarrow x \in [a, b]$

EXAMPLE 2.32

Use a graphical method to find $\{x \mid |x| < 2 - x^2 + 2x\}$

Solution

We make use of the TI–83 to solve this problem.

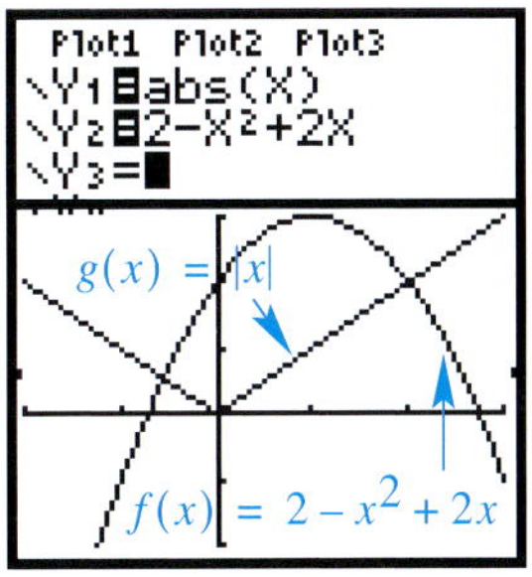

Let $g(x) = |x|$ and $f(x) = 2 - x^2 + 2x$, we sketch these graphs on the same set of axes.

Next we need to find where the two graphs intersect. That is we need to solve the equation $g(x) = f(x)$. So, we have:

$$|x| = 2 - x^2 + 2x$$

Using the **intersect** option from the **CALC** menu we have:

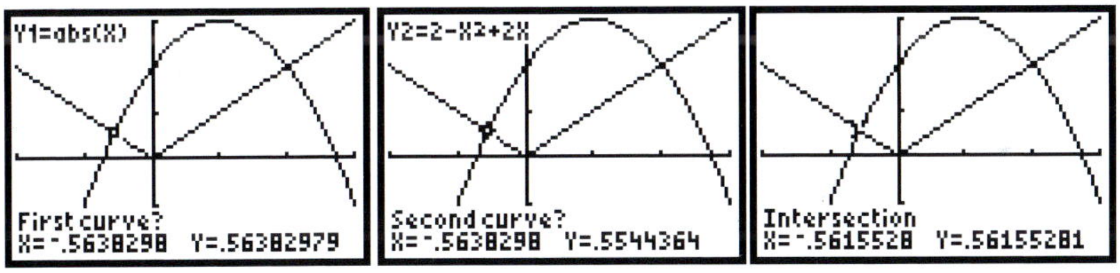

Having found one point of intersection, we now find the other:

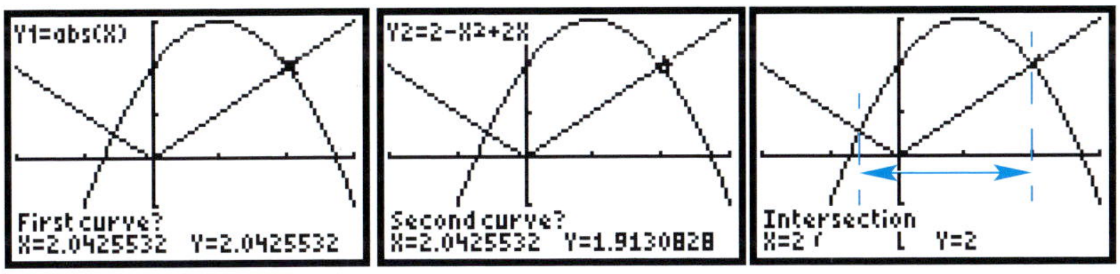

Therefore, from our results we have that $\{x \mid |x| < 2 - x^2 + 2x\} = \{x \mid -0.5616 < x < 2\}$.
Notice once again, the the graphics calculator could only provide an approximate answer for one of the points of intersection.

EXERCISES 2.4.3

1. Find the solution set for each of the following inequalities.

(a) $(x-1)(x+2)>0$ (b) $(x+3)(x-2)\le 0$

(c) $x(4-x)\le 0$ (d) $(1-3x)(x-3)>0$

(e) $(3+2x)(x+1)\ge 0$ (f) $(5-2x)(3-4x)<0$

2. Find the solution set for each of the following inequalities.

(a) $x^2+3x+2>0$ (b) $x^2-x-6<0$ (c) $2x^2-5x-3\ge 0$

(d) $x^2-4\le 0$ (e) $x^2+x-5<0$ (f) $-x^2+x+6\le 0$

(g) $-x^2+x+1\ge 0$ (h) $-2x^2-3x+5\ge 0$ (i) $2x^2+5x-3>0$

(j) $x^2-4x+3<0$ (k) $2x^2+x-1<0$ (l) $x^2+3<0$

(m) $-x^2-2>0$ (n) $2x^2-7x\le 15$ (o) $3x^2+5x>2$

3. (a) For what value(s) of k is the inequation $x^2+2kx-k>0$ true for all values of x?

(b) For what value(s) of k is the inequation $x^2-kx+2\ge 0$ true for all values of x?

(c) For what value(s) of n is the inequation $x^2+2x\ge 2n$ true for all values of x?

4. By sketching on the same set of axes, the graphs of the functions $f(x)$ and $g(x)$, solve the inequalities

i. $f(x)<g(x)$ ii. $f(x)\ge g(x)$

(a) $f(x)=x+2$, $g(x)=x^2$ (b) $f(x)=x-1$, $g(x)=x^2-4x+5$

(c) $f(x)=x^2+2$, $g(x)=4x-1$ (d) $f(x)=3x^2-1$, $g(x)=x+1$

(e) $f(x)=5-x^2$, $g(x)=x^2-3$ (f) $f(x)=x^2-3x-3$, $g(x)=x-4$

5. On the same set of axes sketch the graphs of $f(x)=|x-1|$ and $g(x)=1-x^2$. Hence find $\{x : |x-1|<1-x^2\}$.

6. Given that $f(x)=x^2+3x+2$ and $g(x)=4-x^2$, find $\{x|\ f(x)\le g(x)\}$.

8. Find (a) $\left\{x : \dfrac{x-2}{x+3}>0\right\}$ (b) $\left\{x : \dfrac{4-x}{x+1}>0\right\}$

2.4.4 SIMULTANEOUS EQUATIONS INVOLVING LINEAR – QUADRATIC EQUATIONS

In part (c) of Examples 2.30 and 2.31 we have already found the need to solve simultaneous equations where one equation was a quadratic, i.e., $f(x)=x^2-3$ and the other was linear, i.e., $g(x)=2x$. In this instance, we equated the two functions, $f(x)=g(x)$ so that $x^2-3=2x$, then transposed to get a new quadratic, i.e., $x^2-2x-3=0$ which could readily be solved for x. In this section we formalise the process of solving simultaneous equations involving a quadratic and linear expression or two quadratic expressions.

To solve a **linear–quadratic** system of equations it is often the case that the method of substitution is most appropriate. The process is as follows:

Step 1: Arrange the equations so that they are both in the form $y = \ldots$ i.e., y is expressed **explicitly** in terms of x.

Step 2: Label the two equations
$$y = ax^2 + bx + c \quad -(1)$$
$$y = mx + k \quad -(2)$$

Step 3: Equate (1) and (2): $mx + k = ax^2 + bx + c$

Step 4: Transpose to obtain a new quadratic:
$$ax^2 + (b-m)x + (c-k) = 0$$

Step 5: Solve for x and then find y by substituting into (1) or (2).

EXAMPLE 2.33

Solve the simultaneous sytem of equations

(a) $y = x^2 + 3x - 6$, $y = 2x - 4$ (b) $y = -2x^2 + 4x + 9$, $x + 2y = 7$

Solution

(a) We label the equations as follows;
$$y = x^2 + 3x - 6 \quad -(1)$$
$$y = 2x - 4 \quad -(2)$$

Equating (1) to (2) gives: $2x - 4 = x^2 + 3x - 6$

Solving we have:
$$0 = x^2 + x - 2$$
$$\Leftrightarrow \quad = (x+2)(x-1)$$
$$\Leftrightarrow x = -2, 1$$

Substituting $x = -2$ into (2): $y = 2(-2) - 4 = -8$

$x = 1$ into (2): $y = 2(1) - 4 = -2$

The solution can be expressed as two coordinate pairs: (–2,–8), (1,–2).

(b) The first step in this case is to make y the subject of the second equation and then to substitute this into the first equation.

i.e., $x + 2y = 7 \Leftrightarrow 2y = 7 - x \Leftrightarrow y = \frac{1}{2}(7 - x)$.

Therefore we have,
$$y = -2x^2 + 4x + 9 \quad -(1)$$
$$y = \frac{1}{2}(7 - x) \quad -(2)$$

Substituting [or equating] (2) into (1) gives:
$$\frac{7-x}{2} = -2x^2 + 4x + 9$$
$$\Leftrightarrow 7 - x = -4x^2 + 8x + 18$$
$$\Leftrightarrow 4x^2 - 9x - 11 = 0$$

$$\therefore x = \frac{9 \pm \sqrt{(-9)^2 - 4 \times 4 \times (-11)}}{2 \times 4}$$

$$= \frac{9 \pm \sqrt{257}}{8}$$

$$\approx -0.88, 3.13$$

Substituting into (2): $y = \frac{1}{2}\left(7 - \frac{9 \pm \sqrt{257}}{8}\right) = \frac{47}{16} \pm \frac{\sqrt{257}}{16}$

$$\approx 3.94, 1.94$$

The approximate coordinate pairs for the solution to this problem are (–0.88,3.94), (3.13,1.94).

Again, we see that the discriminant can be used to determine the geometrical relationship between the parabola and the straight line.

When solving the simultaneous system of equations

$$y = px^2 + qx + r \quad \text{– (1)}$$

$$y = mx + k \quad \text{– (2)}$$

which results in solving the quadratic $ax^2 + bx + c = 0$ (after equating (1) to (2)) we have three possible outcomes:

Case 1	Case 2	Case 3
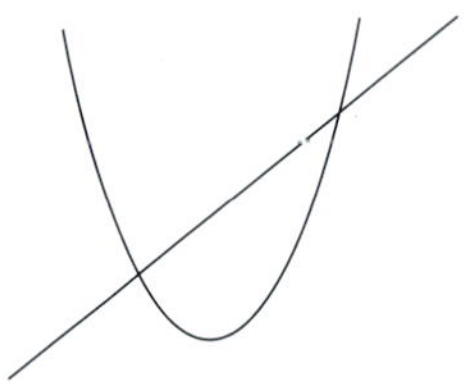	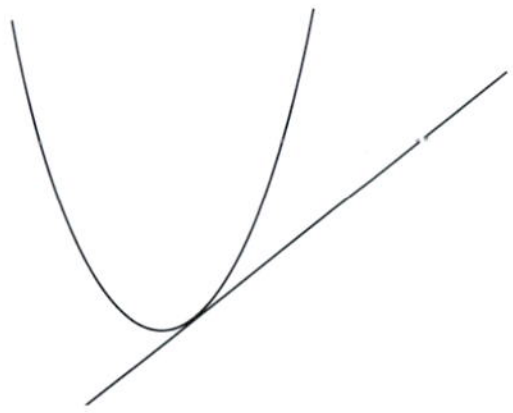	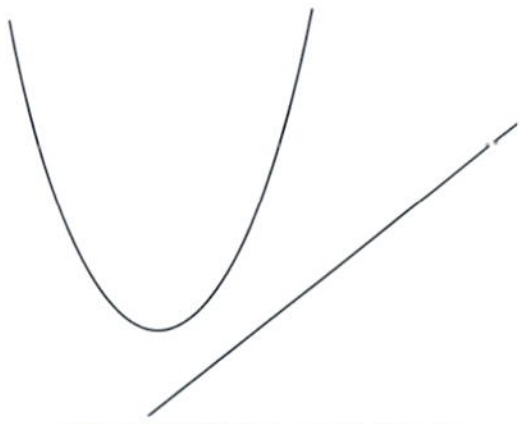
i.e., $\Delta = b^2 - 4ac > 0$ The straight line cuts the parabola twice.	i.e., $\Delta = b^2 - 4ac = 0$ The straight line touches the parabola at one point. We say the straight line is a **tangent** to the parabola.	i.e., $\Delta = b^2 - 4ac < 0$ The straight line neither touches nor cuts the parabola anywhere.

EXAMPLE 2.34

Find the value(s) of m for which the straight line with equation $y = mx - 2$ is a tangent to the parabola with equation $y = x^2 - 3x + 7$.

Solution

We start by solving the system of equations as we have done previously:

$$y = x^2 - 3x + 7 \quad \text{– (1)}$$

$$y = mx - 2 \quad \text{– (2)}$$

equating (1) to (2): $x^2 - 3x + 7 = mx - 2$

$$\Leftrightarrow x^2 - (m+3)x + 9 = 0$$

Then, for the straight line to be a tangent, it means that the line and the parabola touch. This in turn implies that the discriminant is zero.

That is, $\Delta = [-(m+3)]^2 - 4 \times 1 \times 9 = (m+3)^2 - 36$

$= (m+3+6)(m+3-6)$ [using the diff. of two squares]

$= (m+9)(m-3)$

Then, setting $\Delta = 0$ we have

$(m+9)(m-3) = 0 \Leftrightarrow m = -9$ or $m = 3$

Geometrically we have:

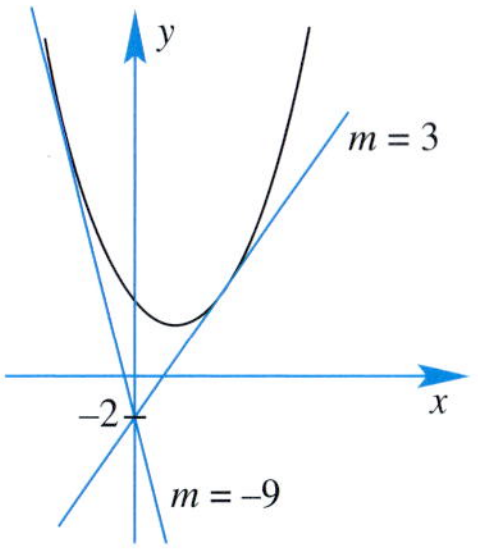

To solve a **quadratic–quadratic** system of equations we use the same method of substitution that was used for the linear–quadratic set up.

EXAMPLE 2.35

Solve simultaneously the system of equations $\begin{aligned} y &= 2x^2 + 3x + 1 \\ y &= 2x - x^2 + 3 \end{aligned}$

Solution

Set up the equations as follows:

$y = 2x^2 + 3x + 1$ – (1)

$y = 2x - x^2 + 3$ – (2)

Equating (1) to (2):

$$2x^2 + 3x + 1 = 2x - x^2 + 3$$
$$\Leftrightarrow 3x^2 + x - 2 = 0$$
$$\Leftrightarrow (3x-2)(x+1) = 0$$
$$\therefore x = \frac{2}{3} \text{ or } x = -1$$

Using (2): When $x = \frac{2}{3}$, $y = 2\left(\frac{2}{3}\right) - \left(\frac{4}{9}\right) + 3 = \frac{35}{9}$ and when $x = -1$, $y = -2 - 1 + 3 = 0$

Therefore, the pairs that satisfy this system of equations are $\left(\frac{2}{3}, \frac{35}{9}\right)$ and $(-1, 0)$.

EXERCISES 2.4.4

1. Solve the following pairs of simultaneous equations.

(a) $y = 2x + 1$, $y = x^2 + 2x - 3$

(b) $y = x + 1$, $y = x^2 + 2x - 1$

(c) $y = 3x - 1$, $y = 3x^2 - 2x - 3$

(d) $3x - 2y = 3$, $y = x^2 + 2x - 3$

(e) $x - 2y = 5$, $y = x^2 + 4x - 7$

(f) $x + 2y = 0$, $x^2 + 3y = 4$

(g) $y = 2x$, $y = x^2 + x - 3$

(h) $x + 2y = 3$, $y = -x^2 + 2x - 3$

(i) $y = 3x + 1$, $y = x^2 + 2x - 3$

(j) $x - y = 1$, $y = x^2 + x - 5$

(k) $3x - 2y = 3$, $y = -x^2 + 2x - 3$

2. Solve the following simultaneous equations.

(a) $y = x^2 - 4x + 7$, $y = 2x^2 + 2x$ (b) $y = 4x^2 - 16x + 8$, $y = 9x - 8x^2 - 4$ (c) $y = 3x^2 - 2x + 2$, $y = 2x^2 + x + 2$

(d) $y = 4x^2 + 3ax - 2a^2$, $y = 2x^2 + 2ax - a^2$ where a is a real number (e) $y = x^2 - 2x + 4$, $y = 2 - 2x - x^2$

(f) $y = x^2 + 3x - 2$, $y = 2x^2 - x + 2$ (g) $y = x^2 + 4x + 3$, $y = 4x - x^2$ (h) $y = 5x^2 + x + 4$, $y = x^2 + 5x + 3$

3. For what value(s) of m will the straight line with equation $y = mx - 6$

(a) touch (b) intersect (c) never meet

the parabola with equation $y = x^2$.

4. Find the distance between the points of intersection of the line with equation $y = 2x + 1$ and the parabola with equation $y = x^2 - 4x + 6$.

5. Find the value of a such that the line $y = 2x + a$ has exactly one intersection point with the parabola with equation $y = x^2 + 3x + 2$.

6. If the graph of the function $f(x) = 4x^2 - 20x - 4$ intersects the graph of the function $g(x) = 1 + 4x - x^2$ at the point (m, k), show that the other point of intersection occurs where $x = -m^{-1}$.

7. For what value(s) of k is the line $2x = 3y + k$ a tangent to the parabola $y = x^2 - 3x + 4$?

11. Find the value for c, for some fixed value of m, so that the straight line $y = mx + c$ is a tangent to the parabola $y^2 = 4ax$, $a \neq 0$.

12. The parabola with equation $y = ax^2 + x + ab$ meets the straight line with equation $y = a^2bx + 2ab$ where a and b are real constants at the points where $x = x_1$ and $x = x_2$. Show that $x_1x_2 + b = 0$.

13. (a) Solve the simultaneous equations

i. $y = 3x^2$, $y = 14 - 11x$ ii. $y = 3x^2$, $y = x + 14$

(b) The functions $f(x) = x + |2x|$ and $g(x) = -x^2 - \frac{2}{3}x + \frac{14}{3}$ intersect at the points A(a, b) and B(c, d) shown in the diagram below.

i. Show that $3a^2 + 11a - 14 = 0$.

ii. Show that $3c^2 - c - 14 = 0$.

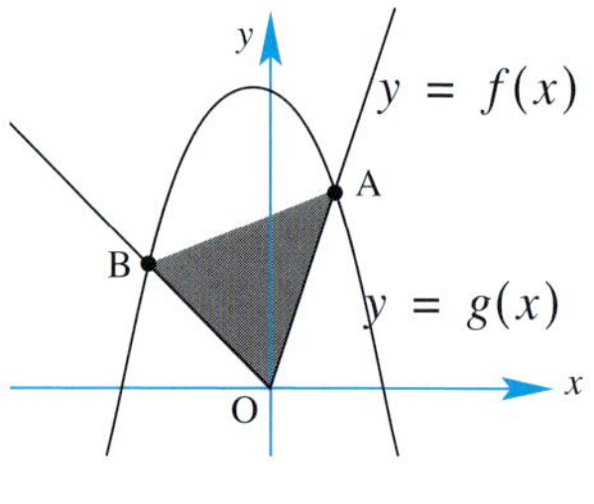

(c) i. Find the equation of the straight line passing through the points A and B

ii. Find the area of the region enclosed by the triangle OAB.

3.1 MODELLING – AN INTRODUCTION

This Chapter is dedicated to modelling. In particular, modelling that requires the use of work covered in Chapter 2. We will concentrate on modelling using linear functions and quadratic functions. In the last section we will look at a broader approach to modelling

However, before we start on the topic of modelling, we will consider a precursor to modelling and look at worded problems – which in themselves incorporate elements of modelling. We will initially deal with worded problems based solely on linear and quadratic equations. We begin this chapter with worded problems that involve quadratic equations

We will take a three phase approach towards modelling.

Phase 1:	Constructing equations based on verbal descriptions
Phase 2:	Constructing functions/equations based on applications and verbal descriptions
Phase 3:	Constructing models based on applications and data.

3.1.1 MODELLING – PHASE 1

In this phase of our process we will encounter problems that are presented via some form of description of the situation at hand. The problems or situations can be described in one sentence or could entail an elaborate account of the problem. Whatever the details of the problem presented, the task at hand is to translate english into mathematics.

Much of what is required for solving these problems, you already possess. The rest is based on your experiences and general knowledge. For example, if the problem contains elements that refer to areas, then you will need to recall facts that allow you to work out areas of different shapes. On the other hand, if the problem involves a situation that is referring to time and distances then you will need to recall facts about the relationship between speed, distance and time. Once you have recalled all necessary knowledge, you can go about constructing an equation that can then be solved.

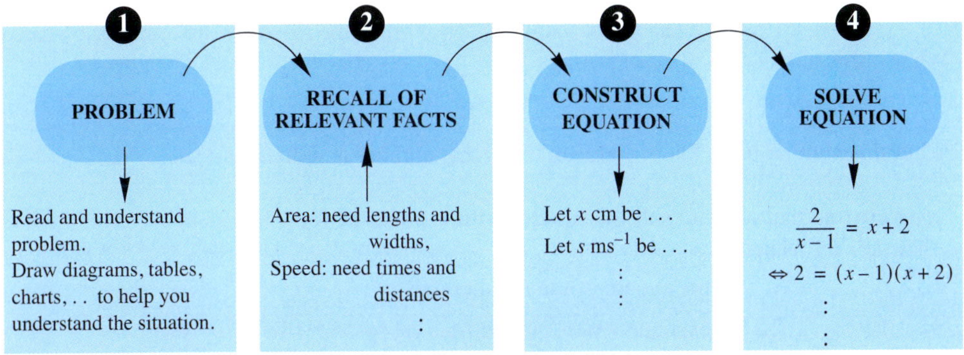

Let's consider some examples to highlight this process.

EXAMPLE 3.1

Find two positive numbers that differ by 2 and have a product of 63.

Solution

Having read the problem, it seems that we are in search of two numbers that possess particular features:

1. their difference is 2
2. their product is 63

Let these numbers be x and y, so that from 1. $y - x = 2$

2. $xy = 63$

That is we have formulated two equations with two unknowns. As such, it would then seem appropriate to consider this a system of equations that requires to be solved using one of the methods encountered in earlier chapters.
We label them as follows:

$$y - x = 2 \quad - (1)$$
$$xy = 63 \quad - (2)$$

From (1) we have $y = 2 + x$ – (3)
Substituting (3) into (2) gives:

$$x(2+x) = 63 \Rightarrow 2x + x^2 = 63$$

We have obtained a quadratic equation, $2x + x^2 = 63$ which we can now solve.

$$2x + x^2 = 63 \Leftrightarrow x^2 + 2x - 63 = 0$$
$$\Leftrightarrow (x+9)(x-7) = 0$$
$$\Leftrightarrow x = -9 \text{ or } x = 7$$

If $x = -9$, from (3) we have $y = -7$
If $x = 7$, from (3) we have $y = 9$.
Therefore we have two pairs of numbers satisfying the conditions.
First pair: $x = -9$, $y = -7$ and the second pair is $x = 7$, $y = 9$.

However, the problem stipulates that the numbers are positive numbers and so we must discard the pair $x = -9$, $y = -7$. Leaving the solution as $x = 7$, $y = 9$.

It is also a good idea to check you answer: $9 - 7 = 2$ √ and $9 \times 7 = 63$ √.

EXAMPLE 3.2

Eighteen times a certain number exceeds the square of the number by forty five. What is this number?

Solution

Let this number be a. Then, eighteen times this number is written as $18a$.

We are told that $18a$ exceeds the square of the number by forty five.
The key terms here are:

1. the square of the number $\Rightarrow a^2$.
2. exceeds $\Rightarrow$ is greater than a^2 by 45 $\therefore 18a = a^2 + 45$.

We have formulated the equation $18a = a^2 + 45$ – (1)
All that remains is to solve for a:

$$18a = a^2 + 45 \Leftrightarrow a^2 - 18a + 45 = 0$$
$$(a - 15)(a - 3) = 0$$

Therefore, $a = 15$ or $a = 3$.

We still need to check our answers.

Case 1. $a = 15$, L.H.S $= 18a = 18 \times 15 = 270$ [substituting into (1)]

R.H.S $= a^2 + 45 = 15^2 + 45 = 270$

Therefore, as L.H.S = R.H.S, $a = 15$ is a solution.

Case 2. $a = 3$, L.H.S $= 18a = 18 \times 3 = 54$ [substituting into (1)]

R.H.S $= a^2 + 45 = 3^2 + 45 = 54$

Therefore, as L.H.S = R.H.S, $a = 3$ is a solution.

Therefore, there are two such numbers, 3 and 15.

EXAMPLE 3.3 Florence purchased a number of hand-held electrical games for \$180. She decided to keep one for herself and sell the rest. She sold the remaining games for \$1.00 more than she bought them for. After keeping one game for herself, Florence made a profit of \$10.00. How many games did she purchase?

Solution

Let n be the number of games Florence bought, meaning that each game cost her $\$\frac{180}{n}$.

She then sold each game for \$1.00 more than she paid for.

i.e., she sold them for $\$\left(\frac{180}{n} + 1\right)$.

As she kept one for herself, she had $(n - 1)$ to sell.

Therefore, the revenue from selling the games is $\$\left(\frac{180}{n} + 1\right)(n - 1)$.

We also know that Profit = revenue – cost

Therefore, $$10 = \left(\frac{180}{n} + 1\right)(n - 1) - 180$$

Or, $$\left(\frac{180}{n} + 1\right)(n - 1) = 190. \quad \text{– (1)}$$

Now that we have our equation we proceed to solve it:

$$\left(\frac{180}{n} + 1\right)(n - 1) = 190 \Leftrightarrow \left(\frac{180 + n}{n}\right)(n - 1) = 190$$

$$\Leftrightarrow (180 + n)(n - 1) = 190n$$

$$\Leftrightarrow n^2 + 179n - 180 = 190n$$

$$\Leftrightarrow n^2 - 11n - 180 = 0$$

$$\Leftrightarrow (n - 20)(n + 9) = 0$$

$$\therefore n = 20 \text{ or } n = -9$$

As $n > 0$, $n = 20$ is the only plausible solution.
Checking with equation (1) shows that $n = 20$ is a solution.
Therefore, Florence bought 20 games.

EXAMPLE 3.4

Francesco walks 24 km everyday. He always maintains a constant speed. If he had walked 2 kmh^{-1} faster than he usually does, he would have completed his walk one hour earlier. At what speed does Francesco usually walk?

Solution

The problems is looking for a speed and so, we need to have a variable that defines this speed. Let x kmh^{-1} be Francesco's usual speed.
We are given information about time and speed and distance. These quantities are related by the following equation; $\text{speed} = \frac{\text{distance}}{\text{time}}$.
As we are comparing to what ususally happens it is worthwhile finding out the time that it usually takes Francesco to walk 24 km:

Let t_1 hr be the time it usually takes to walk 24 km. $\therefore t_1 = \frac{24}{x}$. – (1)

Let t_2 hr be the time it takes if he walks 2 kmh^{-1} faster. $\therefore t_2 = \frac{24}{x+2}$. – (2)

However, it takes him 1 hour less if he walks 2 kmh^{-1} faster, i.e., $t_2 = t_1 - 1$

Using equations (1) and (2) we have

$$\frac{24}{x+2} = \frac{24}{x} - 1 \quad -(3)$$
$$\Leftrightarrow 24x = 24(x+2) - x(x+2)$$
$$\Leftrightarrow 24x = 24x + 48 - x^2 - 2x$$
$$\Leftrightarrow x^2 + 2x - 48 = 0$$
$$\Leftrightarrow (x+8)(x-6) = 0$$
$$\therefore x = -8 \text{ or } x = 6$$

As $x > 0$, discard $x = -8$.

Check $x = 6$: Using (3), we have, L.H.S $= \frac{24}{6+2} = \frac{24}{8} = 3$. R.H.S $= \frac{24}{6} - 1 = 4 - 1 = 3$.

Therefore $x = 6$ is the solution.
That is, Francesco's usual speed is 6 kmh^{-1}.

EXAMPLE 3.5

A man $18x$ years old has a son who is $2x^2$ years old. When the man was $3x^2$ years old, the son was only $(x+4)$ years old. How old would the man be today?

Solution

In this scenario we have a mixture of english and mathematics within the question. The variable has already been defined for us. So, we begin by setting up a table to display the information:

	Age today	Age then	Difference in age
Father	$18x$	$3x^2$	$18x - 3x^2$
Son	$2x^2$	$(x+4)$	$2x^2 - (x+4)$

The number of years that has passed from 'then' to 'now' is the same for both the father and the son. Therefore, the difference in their respective ages must be the same.

Therefore, we have that $18x - 3x^2 = 2x^2 - (x+4)$ – (1)

$$\Leftrightarrow 5x^2 - 19x - 4 = 0$$
$$\Leftrightarrow (5x+1)(x-4) = 0$$
$$\text{i.e., } x = -\frac{1}{5} \text{ or } x = 4$$

Now, as $x > 0$, the only possible solution is $x = 4$.
Therefore the father must be $18 \times 4 = 72$ years old.
Checking with (1) will show this to be the case.

EXERCISES 3.1.1

1. What positive number, when subtracted from its own square gives an answer of 56?

2. Eight times the sum of a number and its reciprocal results in 34. Find this number.

3. The square of the difference between a number and 12 is twice the number. Find it.

4. The product of two numbers is 88. What are the numbers if one number is 3 more than the other?

5. The sum of four times a number and its square is 60. What is the number?

6. Twice the square of Graham's height is 6 metres more than his height. How tall is he?

7. Rachelle has taken $(4x - 6)$ hours to travel 102 km at a speed of $(25x + 1)$ kmh^{-1}. How fast was she travelling?

8. The product of two consecutive odd numbers is 143. If x is the smaller of the two numbers, show that $x^2 + 2x = 143$. Find the numbers

9. It will take an apprentice 15 days longer than a qualified painter to complete a job. However, if 3 apprentices and two qualified painters work together the job can be completed in $3\frac{1}{8}$ days. How long would it take one apprentice to do the job?

10. At a local screening of a movie, 450 people are seated in rows such that every row contains an equal number of people. However, if there were three more people in each row the number of rows needed would be reduced by 5. How many rows are there?

11. Derek sold walking sticks. Had he charged \$10 extra for each stick he would have made \$1,800. However, had he sold 10 more sticks at the original price he would have made \$2,000.
(a) How many sticks did he sell?
(b) How much was each stick?

12. Bella set off on a 96 km journey, all the while she maintained a constant speed. On her return she again kept a constant speed but this time increased it by 2 kmh^{-1} , reducing her travel time by 4 hours. What was her original speed?

13. The hiring cost for transportation for a class of students was \$144. If two students from a different class were to attend, the cost per student would have decreased by \$1.00. How many students are there in the class?

14. A person walked a distance of 4 km at a constant speed of v kmh^{-1} . If this person increased his speed by 2 kmh^{-1} , the journey would have been reduced by 10 minutes. Find v.

15. A train travels at a constant speed for 120 km. Had the train been travelling 5 kmh^{-1} faster, it would have completed the journey 20 minutes earlier. How long would it take the train to travel 120 km?

16. A group of students are to pay equal amounts to a total of £72. However, Leanne, Josh and Andrew have successfully argued that they should not pay, in which case the others would each have to pay an extra £4. How many people were there in the group?

[Questions 17 – 20 are more challenging. These are often referred to as productivity problems.]

17. Two taps having different rates of flow are used to fill a large water tank. If tap A is used on its own it will take 5 hours longer to fill the tank than it would tap B to fill it on its own. Together, the taps would fill the tap in 6 hours. Assuming that the taps are running at full capacity, find

(a) how long will it take for tap A to fill the tank.

(b) how long will it take for tap B to fill the tank.

18. Two chair manufacturers have been in competition for many years. To protect the innocent in this story we will refer to these companies as Chair-one and Chair-two. Chair-one had to fulfill an order for 810 chairs within a set time, while Chair-two had to fulfill an order for 900 chairs in the same amount of time. Competition led Chair-one and Chair-two to complete the task 3 days and 6 days in advance respectively. We know that Chair-one makes 4 chairs less, per day, than Chair-two. How many chairs does each manufacturer produce per day, assuming that they are working at maximum capacity.

19. The local council has sent out two teams to complete work on a stretch of road. Team A works on the left lane while team B works on the right lane. Although Team B started one day after Team A, each team did manage to repair a 10 km stretch of the road. The teams daily joint quota was 4.5 km per day. How many kilometres per day did

(a) Team A repair?

(b) Team B repair?

20. Two workers need to dig a trench. One worker (working alone) can do the job three hours quicker than the second worker. The total time taken by the workers if they were to work separately on this job is $\frac{144}{35}$ times longer than the time it would take them to do the job if they had worked together.

How long would it take each worker to do the job if they were to do it on their own?

3.1.2 MODELLING – PHASE 2

There are no real changes in moving from Phase 1 to Phase 2. The changes, where they exist, are rather subtle. They might involve the use of diagrams, an aspect that did not feature in Phase 1. They might involve the use of trial and error or the construction of a function as opposed to an equation.

For example, if the problem is referring to the area of a square, then after defining the side length of the square as having a length x cm, you might decide that it is appropriate to intoduce the function $A(x)$ cm^2 to define the area of this square.

One difference, however, can occur in the way information is provided and therefore, how that information and/or data is used. Here is an example:

A common task that is peformed by farmers is setting up enclosures on their farm. A particular farmer has 30 metres of fencing with which to enclose an area of 100 m^2. The farmer uses a wall to act as one side of the area to be enclosed. What are the possible dimensions of this enclosure?

We begin by using a diagram that includes the information provided:

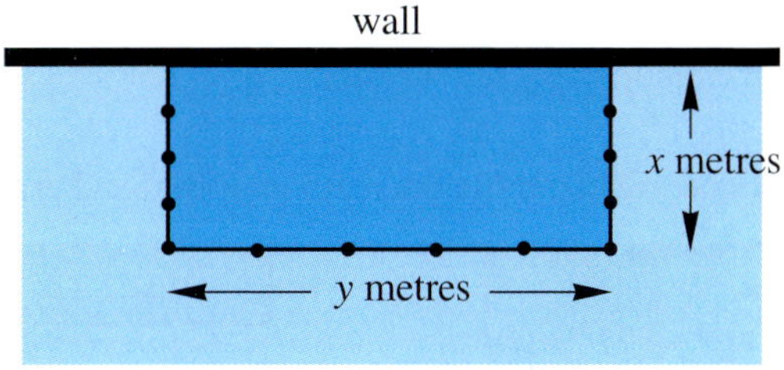

Next we introduce variables to define the length and width, so,

let the length be y m and the width x m.

Then, as we are given 30 metres of fencing, we must have that $x + y + x = 30$

i.e., $2x + y = 30$ – (1)

Also, we are told what the area is to be, 100 m^2.
As the area is a feature of this problem we also define it by introducing a function to represent the area:
Let the area enclosed be defined by A m^2, so that $A = x \times y$.

Now, at this point in time we observe that there are three variables, two independent ones, x and y and the dependent variable A.

Somehow we need to reduce this to one independent variable and one dependent variable. We keep A and now look to keep one of x or y. The choice as to which one is kept is not important (in this example), so we choose to keep x, meaning that we need an expression for y in terms of x. From (1) we have that $2x + y = 30 \Leftrightarrow y = 30 - 2x$.

This means that we can now express A in terms of one variable, namely

$$A = x \times y = x \times (30 - 2x)$$

However, we know that the area must be 100 m^2, therefore,

$$100 = x(30 - 2x)$$

$$100 = 30x - 2x^2$$

That is, we now have a quadratic equation to solve.

$$\begin{aligned} 2x^2 - 30x + 100 &= 0 \\ \Leftrightarrow 2(x-5)(x-10) &= 0 \\ \therefore x &= 5 \text{ or } x = 10 \end{aligned}$$

When $x = 10$, the length is $30 - 2(10) = 10$, giving an area of $10 \times 10 = 100 \text{ m}^2$. √

When $x = 5$, the length is $30 - 2(5) = 20$, giving an area of $5 \times 20 = 100 \text{ m}^2$. √

It is always a good idea to check your answer.

So that the possible dimension satisfying the restictions are:10 m by 10 m and 5 m by 20 m

Comparing this solution with those in the previous section, we see that there is a build-up to the solution, in the sense that we had three variables to start with and then somehow had to express one in terms of the other. Although we did finish by solving an equation, we first created a function. Now, these are subtle differences, but differences nonetheless. This does not mean that all problems in this section will have three variables to start with. Each question will need to be dealt with on its own merit.

Before moving onto some examples, lets consider a slight alteration to the question above. This time, we want:

Of all possible rectangular enclosures, which one will have the maximum area?

Of course we could use the method of trial and error. Some possible scenarioes are show below.

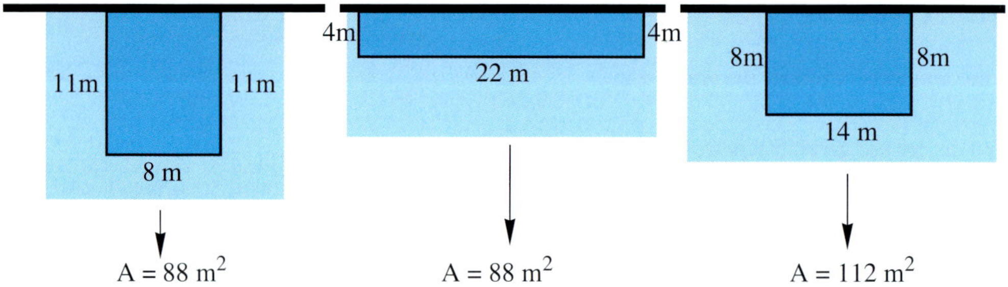

As we can see, different dimensions produce different areas (in some cases, the same area!).

To determine the dimensions that will provide the largest area we make use of the area function, $A = x(30-2x)$. In fact, because A is a function of x only, we can write it as

$$A(x) = x(30-2x)$$

All that remains is to sketch the graph and use it to determine its maximum value.

Notice that if $x = 0$, $A = 0$ and if $x = 15$, $A = 0$.

So, we need to restrict the values of x to $0 < x < 15$.

Our definition of A now becomes, $A(x) = x(30-2x)$, $0 < x < 15$.

Using the **CALC** menu and option **4:maximum** we were able to obtain the maximum area as 112.5 m^2.

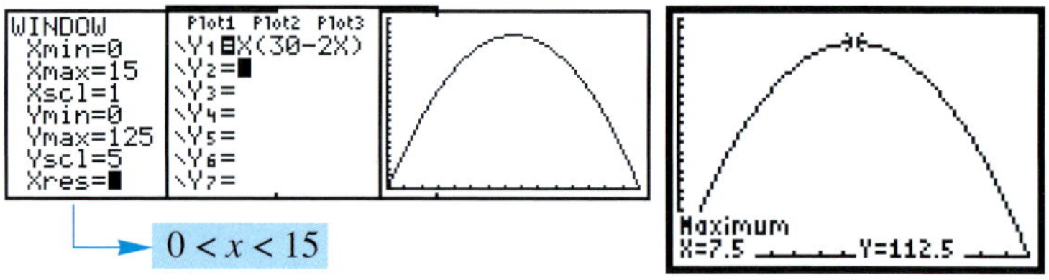

EXAMPLE 3.6

The height, h m, above ground level, reached by a ball, t seconds from when it is thrown, is given by the equation $h = 20t - 5t^2 + 1$.

(a) How long did it take for the ball to first reach a height of 16 metres?
(b) What is the maximum height reached by the ball?
(c) For how long is the ball in flight?

Solution

Unlike previous problems, we are supplied with all the variables and the equation! In fact, there is no need to define or introduce any new variables. However, here is a problems where our task is to interpret the information that has been given.

We start by visualising and interpreting the situation:

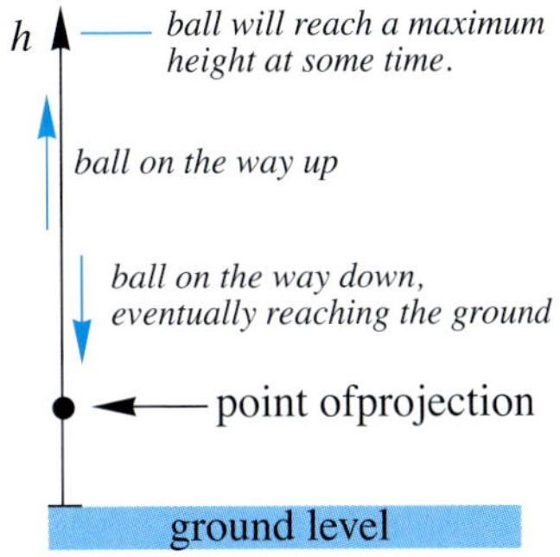

As t represents time, then we know that $t \geq 0$.
Also, when $t = 0$, we have that

$$h = 20 \times 0 - 5 \times 0^2 + 1 = 1$$

This tells us that initially, the ball is 1m above ground level (the point of projection).

Having obtained some 'feel' for what is happening, we now start answering the questions.

(a) We wish to determine t when $h = 16$, so, we solve the equation

$$16 = 20t - 5t^2 + 1 \Leftrightarrow 5t^2 - 20t + 15 = 0$$
$$\Leftrightarrow 5(t^2 - 4t + 3) = 0$$
$$\Leftrightarrow 5(t-3)(t-1) = 0$$
$$\therefore t = 3 \text{ or } t = 1$$

As we are asked for the first time, then $t = 1$. That is, it took 1 sec to reach 16 m.

But what of $t = 3$, what does it represent?
As the ball must come down, $t = 1$ represents the time when the ball is 16 m above ground level on its way up. While $t = 3$ represents the time taken for the ball to be 16 m from the ground, but this time it is on its way back down. This also means that it took 2 seconds for the ball come back to a height of 16 m from the time it first reached that height on its way up.

(b) We recognise the equation as a quadratic, so we can sketch its graph and use it to determine the maximum height by locating its turning point.

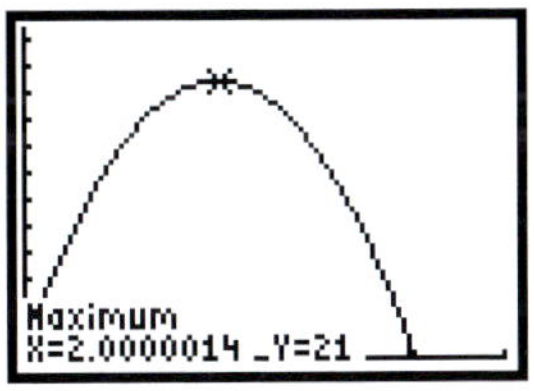

Using the **CALC** menu and option **4:maximum** we have that the maximum height reached is 21 metres, 2 seconds after it was projected.

(c) When $h = 0$, the ball will have reached ground level and so, we need to find that value of t for which $h = 0$.

i.e., $0 = 20t - 5t^2 + 1$

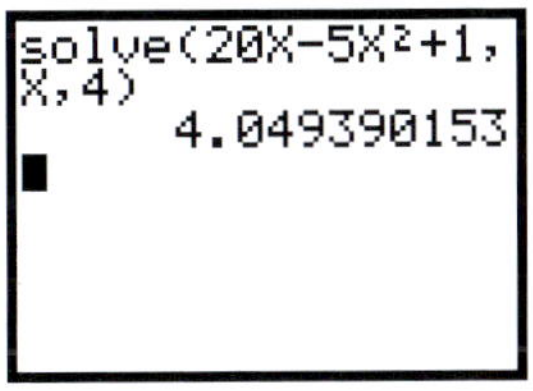

There are a number of options that can be used.
e.g., use **CALC** menu and select option **2:zero** or the **TRACE** option or as we have done, use solve from the catalogue menu.
Giving $t = 4.05$ (to 2 d.pl.).

Had we used the quadratic formula in part (c), we would have obtained the following solutions,

$$0 = 20t - 5t^2 + 1 \therefore t = \frac{-20 \pm \sqrt{20^2 - 4 \times -5 \times 1}}{2 \times -5}$$

$$= 4.0494, -0.0494$$

However, as $t \geq 0$ we would have rejected $t = -0.0494$.

We find that linear models are used in many facets of life as was seen in Chapter 6 and will be seen in Chapter 24. One area where this is very common is that of economics and business. Examples can be found in equations that model company profits and revenue as well as the costs inccured during the running of a business. There are a number of terms that are often referred to when dealing with economic and business situations. Some of these terms are; profit, revenue, cost, depreciation, market demand, market supply, market equilibrium and break-even point. Analysing economic behaviour through mathematical interpretation is a skill which is sought by many of the leading firms around the world. Modelling such economic behaviour displays the power inherent in mathematics.

EXAMPLE 3.7

The total cost, $\$C$, to a manufacturer of calculators is given by the equation

$$C = 12n + 40000$$

where n is the total number of calculators produced. The manufacturer sells these calculators to suppliers at a fixed price of \$32 each.

(a) Find the manufacturer's cost when 1000 calculators are produced.
(b) Will the manufacturer make a profit when 1000 calculators are produced?
(c) How many calculators must this manufacturer produce in order to break-even?

Solution

(a) When $n = 1000$, $C = 12(1000) + 40000$
$= 52000$
That is, it will cost the manufacturer \$52,000.

(b) Let R, dollars, denote the revenue. If the manufacturer has sold 1000 calculators, the revenue is given by $R = 32 \times 1000 = 32000$.

Therefore the manufacturer makes a profit of \$32,000 – \$52,000 = – \$20,000.
A negative profit means that the manufacturer has made a loss.
Therefore the manufacturer has lost \$20,000!

(c) The break-even point occurs when no gain or loss is made, that is, when $R = C$.

Because the **profit** is defined as the **revenue – cost**, i.e., $\boldsymbol{P = R - C}$, the **break-even** point occurs when $\boldsymbol{P = 0}$.

Therefore, if n calculators are produced, we have $32n = 12n + 40000$

$$\Leftrightarrow 20n = 40000$$

$$\Leftrightarrow n = 2000$$

That is, to break-even, the manufacturer needs to produce (and sell) 2000 calculators.
Although we have answered the questions, it will be of great use if we can visualise this situation

by sketching the relationships involved:

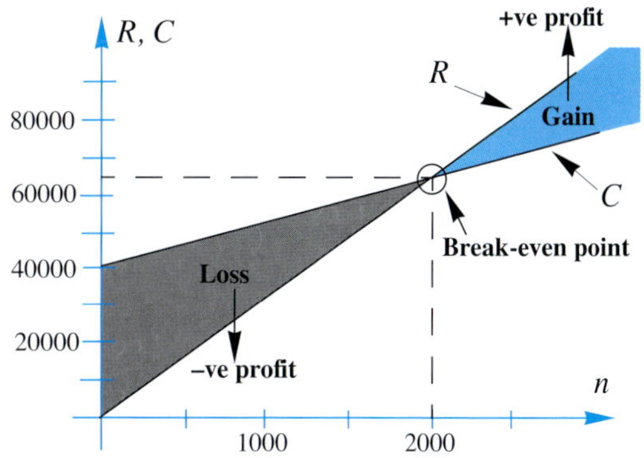

From the graphs, we see that the break-even point corresponds to the point of intersection of the two straight lines representing

A. the cost function, $C = 12n + 40000$

and

B. the revenue function, $R = 32n$

Notice how we have only sketched the graphs for values of n that are greater than (or equal to) zero.

The above example illustrates that the amount of items a manufacturer sells and the price of the item determine the total amount of money earned by selling the items. If the item is more expensive the number sold commonly decreases. The cost of producing items often depends on a fixed (set-up) cost and a price per item (resource cost). Three other terms that are of importance in economic modelling are:

Market demand The quantity of a product (goods or services) that consumers are willing to buy at various prices.

Market supply The quantity of a product (goods or services) that producers are prepared to offer for sale at various prices.

Market equilibrium This occurs when the quantity demanded equasl the quantity supplied.

EXAMPLE 3.8

Suppose that we have found that x units can be sold daily for a price of p dollars per unit where $x = 800 - p$ and the cost of making these x units is given by the function

$$C(x) = \begin{cases} 500 + 20x & \text{if } x > 0 \\ 0 & \text{if } x = 0 \end{cases}$$

Find (a) the daily revenue function, $R(x)$.

(b) the daily profit function $P(x)$.

Assuming the production capacity is at most 400 units per day, find

(c) how many units should be made and sold to maximize profit.

(d) the maximum profit per day.

(e) the unit price to be charged to maximize the profit.

Solution

(a) The revenue is $R(x) = x \times p = x(800 - x)$.

(b) Profit is $P(x) = R(x) - C(x)$

$$= x(800 - x) - (500 + 20x)$$

$$= 780x - 500 - x^2$$

(c) The profit function is given by a quadratic function, meaning that it will form an arc of a parabola (or at least part of an arc). The reason we state '*part of an arc*' is that we need to consider the restrictions on x.
First of all $x \geq 0$. Then, we are told that the production capacity is 400 units, so we have that $x \leq 400$. That is, the profit function has a domain given by $0 \leq x \leq 400$.

We can then use the turning-point form of the parabola:

$$\begin{aligned} P(x) &= 780x - 500 - x^2 \\ &= -(x^2 - 780x + 500) \\ &= -[(x-390)^2 - 152100 + 500] \\ &= -(x-390)^2 + 151600 \end{aligned}$$

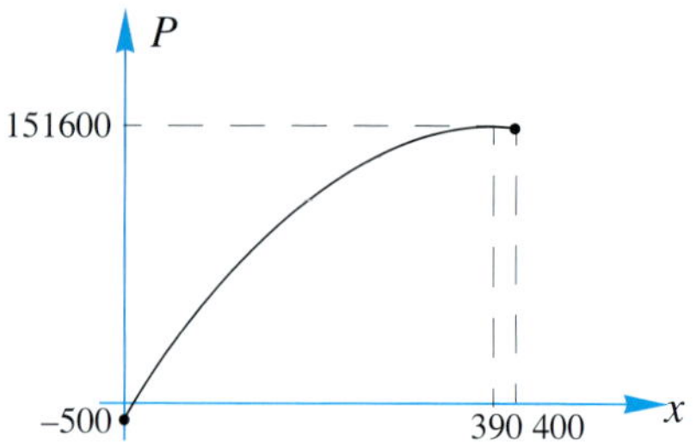

From this equation we can see immediately that the maximum occurs at $x = 390$.

(d) The maximum profit is then \$151600.

(e) The unit price is $800 - 390 = 410$ dollars.

Note: In this case we have not used the full capacity. If our production limit were 350 units per day then the optimum would have occurred at $x = 350$ which would not be at the vertex of the arc but at an endpoint.

EXAMPLE 3.9

The graph shows the cost function, $C(x)$, and the revenue function, $R(x)$, for a transistor manufacturer, where x is the number of units sold.

(a) Determine the number of transistors that need to be sold for the manufacturer to break-even.

(b) For what values of x will the manufacturer be
   i. in the black?
   ii. in the red?

(c) Determine the maximum profit that the manufacturer can make.

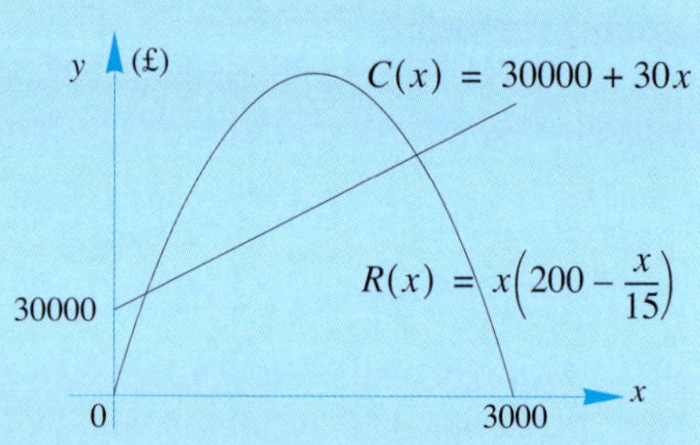

Solution

(a) The manufacturer will break-even when $R(x) = C(x)$.

That is, $x\left(200 - \frac{x}{15}\right) = 30000 + 30x \Leftrightarrow 200x - \frac{x^2}{15} = 30000 + 30x$

$$\Leftrightarrow \frac{x^2}{15} - 170x + 30000 = 0$$

$\therefore x \approx 190.74$ or $x \approx 2359.26$ [using the quadratic formula]

Therefore, as x is an integer, we can say that at $x = 191$ and $x = 2359$, the manufacturer will break-even.

(b) i. To be in the black means to be making a positive profit –i.e., a gain.
This occurs when $R(x) > C(x)$ or simply, $P(x) > 0$.
From part (a) and the graph, this would occur when $191 \le x \le 2359$.

ii. To be in the red means to be making a negative profit – i.e., a loss.
This occurs when $R(x) < C(x)$ or simply, $P(x) < 0$.
From part (a) and the graph, this would occur when $0 \le x \le 190$ or $2360 \le x \le 3000$.

These results can be shown on the graph below:

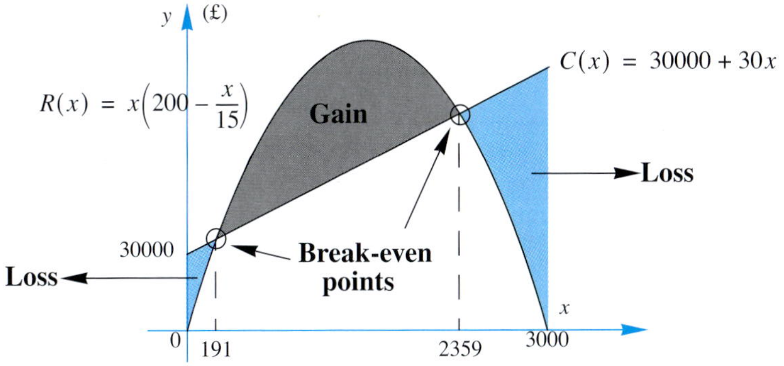

(c) The profit equation is given by, $P(x) = R(x) - C(x)$.

$$\therefore P(x) = x\left(200 - \frac{x}{15}\right) - (30000 + 30x)$$

$$= -\frac{x^2}{15} + 170x - 30000,\ 0 \le x \le 3000$$

We determine the maximum value of this quadratic function using the graphics calculator:

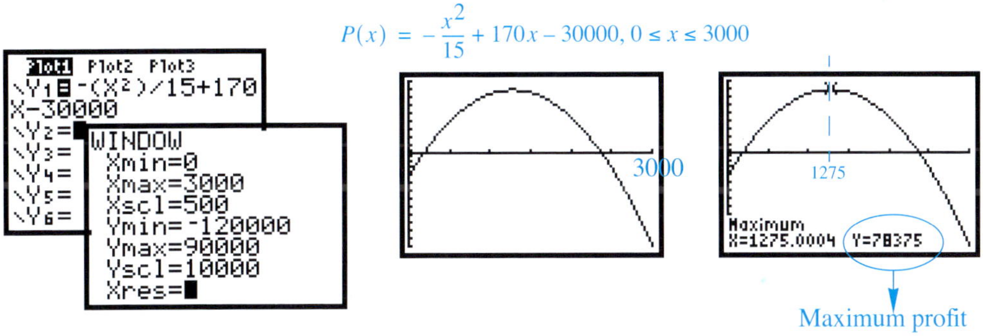

We can then say that the maximum profit is \$78,375 and occurs when 1275 items are sold.

Note the importance of setting the appropriate window in order to obtain a good visual display of the behaviour of the profit function.

As we have seen in this section, there are times when information, including equations is given. This information has been obtained by some means and is presented as a summary – which, as we have seen can come in different forms. However, it is then up to you to use this information and from it be able to analyse and provide findings that are consistent with the information presented.

We conclude this section with a quick note on aspects of linear functions that have not been dealt with specifically in previous chapters but have a number of applications in many standard practices that involve the usage of utilities such as water, electricity and so forth as well as payments for taxi fares, postage and the like.

PIECEWISE LINEAR RELATIONS

Some functions are made up of several different rules for different subsets of their domains. Such functions are often referred to as hybrid functions. A special case of hybrid functions is when each rule is a linear function, in this case we have what is known as a **piecewise linear function**. An example of such a function is

$$f(x) = \begin{cases} x-2 & \text{if } x > 4 \\ 2x-6 & \text{if } x \le 4 \end{cases}$$

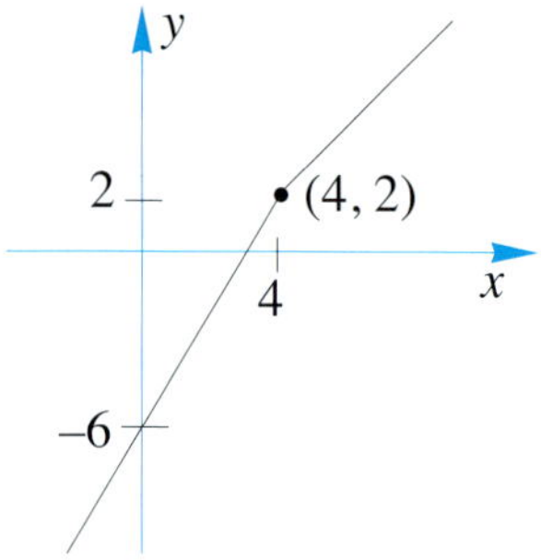

To sketch this graph we sketch the graph of $y = x - 2$ for $x > 2$ and then sketch the graph of $y = 2x - 6$ for $x \le 4$:

Notice then, that because the two lines meet at the point (4, 2), the resulting function is called a **piecewise continuous linear function**.

Hint: The easiest way to sketch these graphs (other than using a graphics calculator) is to sketch each straight line as if there were no restrictions and then rub out those sections that are not required.

On the TI–82 or TI–83 graphics calculator, we can sketch the above function as follows: After entering the equation editor, i.e., **Y =** screen, type the following:

Y = (X – 2)(X > 4) + (2x – 6)(X ≤ 4)

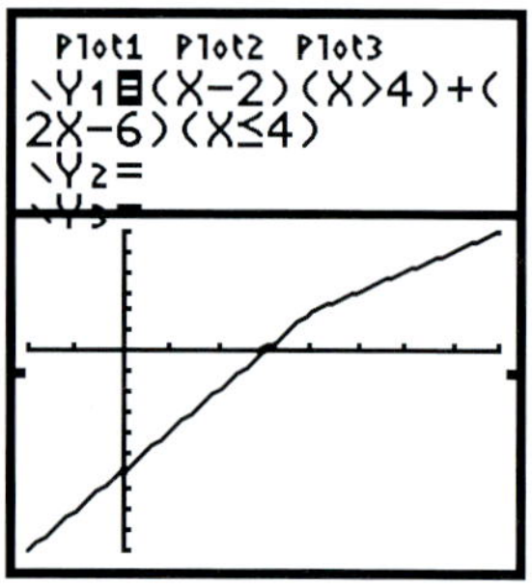

The expressions (**X > 4**) and (**X ≤ 4**) are found under the **TEST/ LOGIC** window. To access this window press **2nd MATH**.
i.e., type (**X 2nd MATH** [then use the arrow key to select **>**, press ENTER] **4**) + (**2 X – 6**)(**X** 2nd MATH [then use the arrow key to select **≤**, press ENTER] **4**) and then press **GRAPH**.

EXAMPLE 3.10

A Water Supply company decides to change its rates to increase the user-pays component of water supply and it phases in the procedure. The new method included a fixed charge (possibly different for each property depending on its assessed value) and a rate of 45 cents per kilolitre for the first part used and 65 cents per kilolitre for the second part.

In a particular bill, the charge for a quarter is \$150 plus a charge for up to 50kl at the lower rate and the rest at the higher rate. If the water used is 63kl what will be the total bill?

Solution

Let the volume of water used be denoted by v kL, then, based on the information given, we have that

1. for $v \le 50$, the total cost is given by $150 + 0.45v$

2. for $v > 50$, the total cost is given by $172.5 + 0.65(v - 50)$

As there are two linear graphs that make up the function, we have a piecewise linear function. The graph then has the form shown.

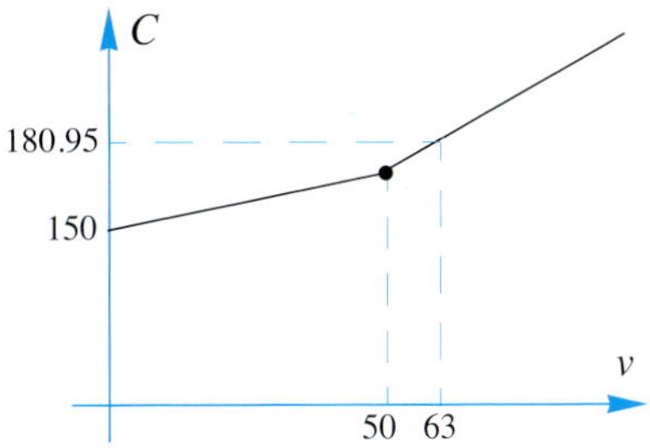

Here for $v = 63$ we have the second rule applying and therefore, $C = 180.95$

GREATEST INTEGER FUNCTION

Another type of piecewise linear function is the **greatest integer function** (y = [x]) or the **step function**.

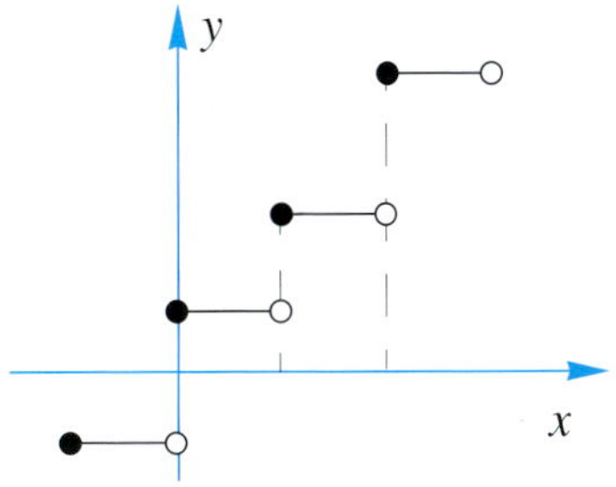

These graphs are made up of discontinuous horizontal lines (segments) that resemble a stair case.

Such graphs are useful when modelling the cost of mailing parcels of different weights.

For example, if it costs \$3.00 to send parcels weighing between at least 1kg but less than 3 kg, \$ 7.00 for parcels weighing at least 3 kg but less that 5 kg, \$ 13.00 for parcels weighing at least 5 kg but less that 8 kg and so on, then, we can summarise that information using the graph shown:

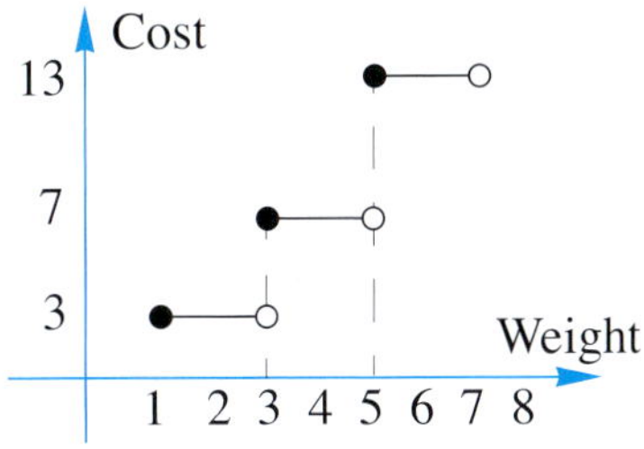

EXERCISES 3.1.2

1. Bobby has decided to start a vegetable garden so that he may sell his produce at the local Sunday market. He buys 100 metres of wirenetting to fence off a rectangular area using an existing fence for one side of the rectangle.

Bobby wishes to fence off the largest possible area. Let the length of one side of the rectangle be x m as shown in the diagram.

(a) i. Find the length of the other dimension in terms of x.
ii. What values can x take?

(b) Let A m^2 denote the area of the enclosure.
i. Find an expression for A in terms of x.
ii. What are the dimensions of the enclosure if it has an area of 800 m^2 ?
iii. What is the largest possible area that the enclosure can have?
iv. What are the dimensions of the enclosure that produce the largest area?

2. A small rectangular enclosure is to be placed in a paddock. There is 24 m of fencing available for the enclosure.

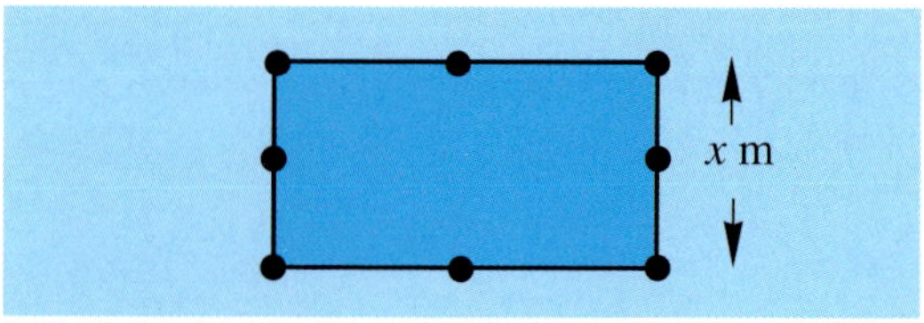

(a) i. If x m is the length of one of the sides, show that the area, A m^2, is given by $A = x(12-x)$.
ii. What restrictions need to be placed on x?

(b) Find the area of this enclosure when
i. $x = 2$ ii. $x = 4$ iii. $x = 8$

(c) Sketch the graph of the function $A(x)$ for its specified domain.

(d) Find the dimensions of the enclosure that will have the largest possible area.

3. The cost, $\$C$, of new small cars in a particular state have been increasing according to the relation $C(t) = 12,000 + 25t^2$, where t is the time in years since 2000.

(a) Sketch a graph of this relation from 2000 to 2008.

(b) How much would you expect to pay for a new small car from this state in 2006?

4. The demand equation for a certain product is given by the equation $p = 40 - 0.0004x$, where p is the price per unit and x is the number of units sold.

(a) Find the equation for the total revenue, $\$R$, when x units are sold.

(b) i. What is the revenue when 40000 are sold?
ii. How many units must be sold to produce a revenue of \$600,000?
iii. What is the maximum revenue that the product will return?

5. Kow Boy has decided to use the 200 m of left over fencing to create a rectangular enclosure that consists of two adjacent sections as shown below.

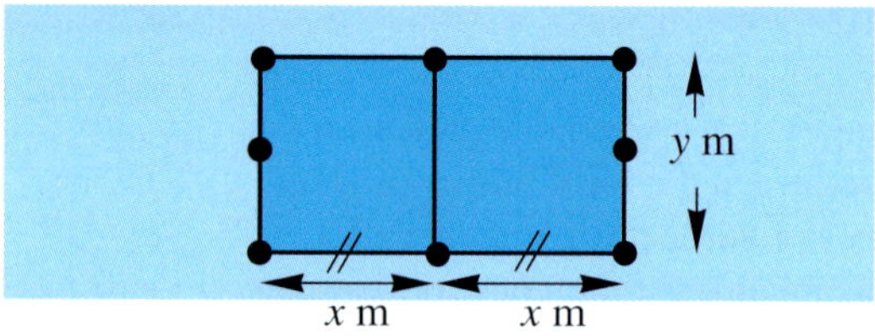

(a) Find an expression for y in terms of x.

(b) Let the area of the combined enclosures be A m^2.
i. Show that $A = \frac{8}{3}x(50-x)$.
ii. What restrictions need to be placed on x?

(c) i. What is the maximum area that the combined enclosures have?
ii. What are the dimensions of the enclosure with the maximum area?

6. The stopping distance, d metres, of a heavily weighted vehicle travelling at v kmh^{-1} can be approximated by the formula $d = v + 0.05v^2$. At what speed was the vehicle travelling if it took the vehicle 100 m to stop? (Assume $v > 0$)

7. The percentage of normal level oxygen, $P\%$, in a lake, t weeks after organic waste is dumped into it is approximated by the model $\frac{1}{100}P = \frac{t^2 - 2t + 3}{t^2 + 3}, t \geq 0$.

(a) What is the level of oxygen in the lake prior to the waste being dumped?
(b) How long will it take for the level of oxygen in the lake to reduce by 15%?
(c) Using your graphics calculator, sketch the graph of $\frac{1}{100}P(t), t \geq 0$.
(d) i. What is the lowest level of oxygen that the lake will reach?
ii. How long after the waste has been dumped will this level occur?
(e) Describe the behaviour of the level of oxygen in the lake over time. Assume that waste is dumped only once.

8. A Gas company charges a fixed connection fee of \$20 and 0.7 cents per MegaJoule (MJ) up to 5000 MJ usage and thereafter charges 0.9 cents per MJ.
(a) Sketch a graph of the total bill as a function of usage (in MJ).
(b) Determine the bill for a household using 9473 MJ in a particular billing period.

9. Following the construction of two new roads, traffic lights need to be put in place where the roads meet.
One of the roads, along O, B, C, is modelled by the equation $y = \frac{3}{5}x$, while the other, along the path A, B, C, is modelled by the equation $y = -(x-5)^2 + 5$, where x and y are measured in kilometres.

The section of the road has been set up on a set of axes as shown in the diagram. Where must the traffic lights, relative to O, be placed?

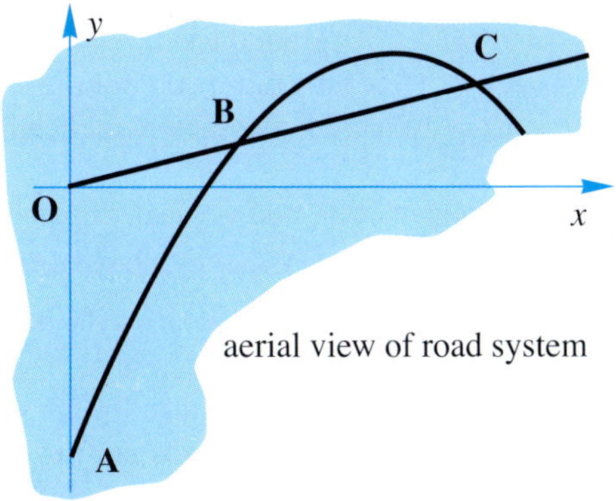

aerial view of road system

10. Tele-Kom use the following scale of charges for a 5 minute S.T.D call, between the hours of 7:30 am and 6:30 pm

Distance	Cost (\$)
Less than 50 km	0.25
As of 50 km but less than 80 km	0.65
As of 80 km but less than 160 km	1.30
As of 160 km but less than 320 km	1.95
As of 320 km but less than 500 km	2.50
500 km or more	3.40

(a) Construct a function that will give the cost, $\$C$, in terms of the distance, x km, up to a distance of 1000 km for each 5 minute call.
(b) Sketch the graph of your function.
(c) Peter makes three 5 minute S.T.D calls in succession. The first to Town A, 250 km away, the next to Town B, 45 km away and the third to Town C, 450 km away. How much did these three calls cost all together.
(d) If the cost is calculated on a sliding scale relative to the time of the call. How much would Peter have paid for the three calls if the calls only lasted 2 minutes each?

11. ABCD is a trapezium where [AB] and [CD] are parallel and 4 m apart. Angle DAB is a right angle. The points X and Y lie on [AD] and [BC] respectively with [XY] being parallel to [AB]. The lines [XY] and [AB] are x m apart.

(a) What values can x take?

(b) Given that AB = 3 and that CD = 5, find an expression for the area, A m^2, of the quadrilateral ABYX, in terms of x.

(c) i. Sketch the graph of $A(x)$.

ii. Find the area of the quadrilateral ABYX when $x = 1$.

iii. Find the value of x so that the quadrilateral ABYX has an area of 3 m^2.

12. The height, $h(t)$ metres, of a stone above ground level, t seconds, after it has been projected vertically upwards is given by the equation

$$h(t) = 120t - 10t^2, t \geq 0.$$

(a) Calculate its height

i. after 2 seconds ii. after 4 seconds

(b) i. At what times will the stone reach a height of 40 metres?

ii. How long was it between the times when the stone was 40 metres from ground level?

(c) How long did it take for the stone to come back to ground level?

(d) What was the maximum height reached by the stone?

13. A stone is projected vertically upwards from the top of a building 250 m high. The height of the stone, $h(t)$ metres, above ground level, t seconds after it is thrown is given by the equation $h(t) = 250 + 100t - 10t^2, t \geq 0$.

(a) How long did it take for the stone to reach a height of 50 m above the building?

(b) How long did it take for the stone to be back at the point of projection?

(c) What is the maximum height that the stone reached (above ground level).

(d) The stone is allowed to continue down to ground level. How long did it take for the stone to reach the ground?

(e) What distance did the stone travel by the time it reached the ground?

14. The total cost, \$$C$, to manufacturer A, of electrical parts is given by the equation

$$C = 15n + 50000$$

where n is the total number of electrical parts produced. Manufacturer A sells these parts to manufacturer B at a fixed price of \$35 each.

(a) Find manufacturer A's cost when 1500 electrical parts are produced.

(b) Will manufacturer A make a profit when 1500 parts are produced?

(c) How many parts must manufacturer A produce to break-even?

15. A company manufactures and sells x cheap radios per month.
The cost, \$$C$, involved in producing x radios per month is given by the equation

$$C = 60x + 70000, 0 \leq x \leq 6000.$$

The revenue equation, \$$R$, based on the sales of x radios per month, is given by

$$R = -\frac{1}{30}x^2 + 200x, 0 \leq x \leq 6000.$$

(a) Accurately draw the graphs of the cost and revenue functions on the same set of axes.

(b) What is i. the minimum cost involved?
ii. the maximum revenue?

(c) Why is there a cost involved when no radios are produced?

(d) On your graph, identify the break-even points.

(e) What profit does the company make when 2000 radios are produced and sold?

(f) i. Find an expression in terms of x for the profit, $\$P$, this company makes on the sales of their radios.
ii. What is the maximum profit the company can hope to make?
iii. How many radios would they need to sell to achieve this maximum profit.

(g) For what values of x will the company be
i. in the red?
ii. in the black.

(h) Clearly identify each of the regions in part (g) on your graph from part (a).

16. The market research department of a company recommends that the company manufacture and market a new transistor radio. After extensive surveys, the research department submits the following demand and cost equations respectively:

$$x = 6000 - 30p \qquad (x \text{ is the demand at } \$p \text{ per radio})$$
$$C(x) = 72000 + 6x\,.$$

The revenue equation, $\$R$, is given by the product of the number of units sold and the price per unit.

(a) Show that the revenue equation is given by $R(x) = x\left(200 - \frac{x}{30}\right)$

(b) i. Find the profit equation, $\$P$, where $P(x) = R(x) - C(x)$.
ii. Define the domain for the function $P(x)$.

(c) On the same set of axes, sketch the graph of the revenue, $\$R$ and cost $\$C$ functions.

(d) Provide a brief analysis based on the graph in part c., making sure to discuss issues such as i. break-even points.
ii. regions of gain and loss.

(e) How many units need to be sold in order to maximize their profit?

3.1.3 MODELLING – PHASE 3

And so we arrive at the last phase of our process – sometimes known as 'Are we there yet?' The question then is, how does this phase differ from the previous two? Looking at the first two stages we progressed from constructing equations based on some verbal account, that could be solved, to constructing functions (or being presented with functions) that had to be interpreted and/or analysed. In the next stage we are still required to construct a model or a function, but this time we do so based on information presented in the form of data.

On many occasions, the form of a function may be derived from data, perhaps from an experiment. The character of a polynomial function can be determined from data which is sufficiently ordered. To see this clearly for a linear function let's look at the following examples.

EXAMPLE 3.11

The results of an experiment (a rather precise experiment) are shown in the table below. What type of model best describes these results?

l (cm)	5.0	5.1	5.2	5.3	5.4
Q (cm)	13.6	14.9	16.2	17.5	18.8

Solution

We start by plotting the data and trying to establish the type of trend displayed.
After turning the **STAT PLOT** feature on, we create two lists, **L1** (for l) and **L2** (for Q):
To create the lists:

❶ select **STAT**, press **ENTER** and select the scatter-diagram icon, then **QUIT.**

❷ choose **4:ClrList** (to identify which lists you wish to clear), then press **ENTER.**

❸ select **STAT**, choose **1:Edit...**, and then enter the data, first **L1** (after every input press **ENTER**) and then use the right arrow key to move to the **L2** column and repeat the process as for **L1**.

❶
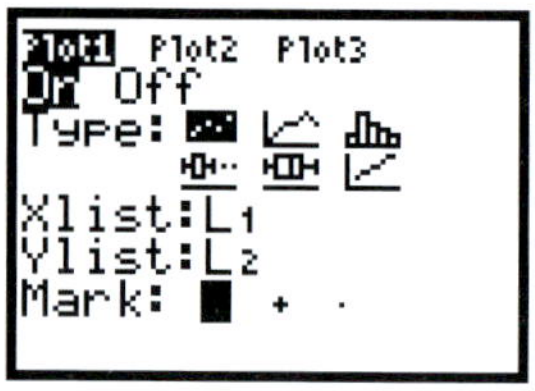

❷

❸
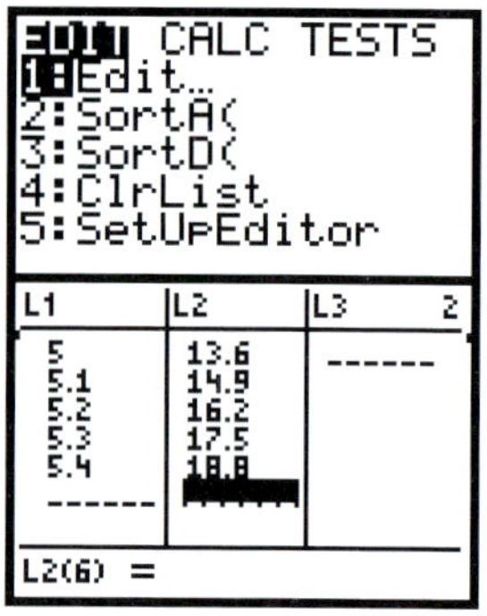

We can now plot the data and look for a trend. Using two different windows sometimes helps.

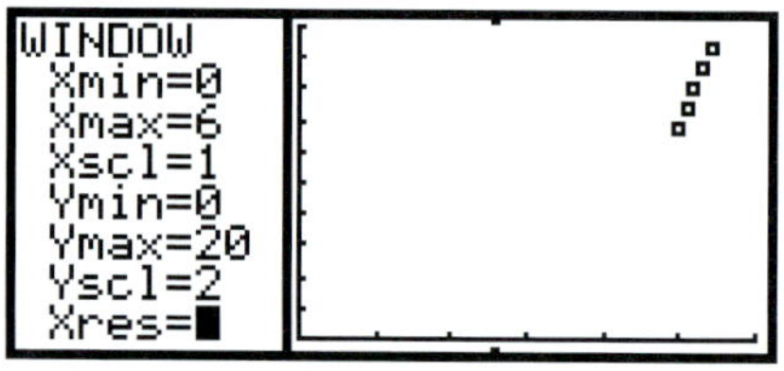

Although the first plot indicated a linear trend, after altering the window, as in the second plot, it becomes clearer that there is a linear trend. We can now suggest that the variables l and Q can be **modelled** by the equation $Q = al + b$, where a and b are constants to be determined.

To solve for a and b,we use two points from our data. We choose (5.0, 13.6) and (5.1, 14.9). We can then set up a system of equations:

From (5.0, 13.6) we have $13.6 = 5a + b$ – (1)

From (5.1, 14.9) we have $14.9 = 5.1a + b$ – (2)

(2) – (1): $1.3 = 0.1a$

Therefore, $a = 13$

Sub. into (1): $13.6 = 5 \times 13 + b \Rightarrow b = -51.4$

Therefore we have the data modelled by the equation $Q = 13l - 51.4$.

There is another method that can be used to find the equation of a straight line for which a linear trend is suspected. This method is known as the **least squares regression line**. This is actually outside the domain of this course and there are some cautions that need to be taken into account, however, in this case, as the data seems to fit a straight line perfectly, we use it for demonstrative purposes.

From the **STAT** menu use the right arrow key and select **CALC** then use the down arrow key and select option **4:Linreg(ax+b)**: then enter **L1, L2** in that order, followed by , **Y1**:

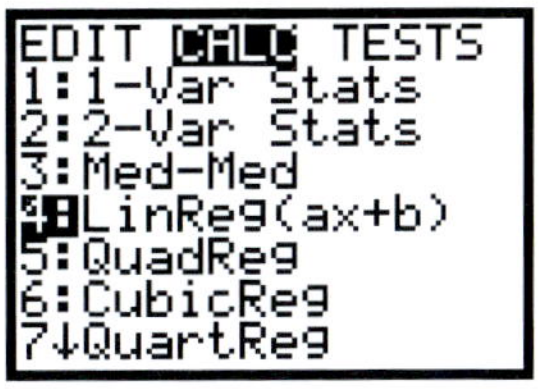

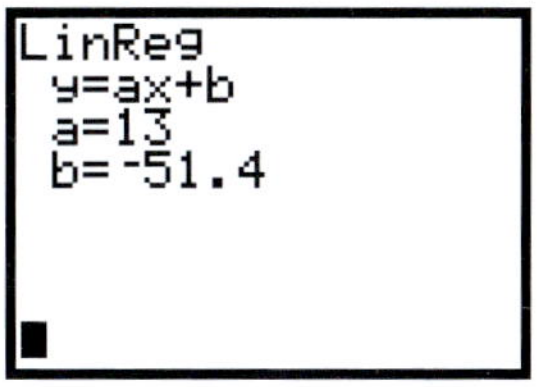

As we can see, the last screen provides the same results as before.

Now, the reason why caution needs to be taken when using the least squares regression equation is that this method uses a process whereby a great many points can be used to obtain the **line of best fit**, even when the points from the data set do not lie perfectly on a straight line.

We can also sketch the line on the same screen as the scatter diagram. The second screen above tells us that the linear regression line has been stored as **Y1**. If we then press **GRAPH** we have:

A glance at the table also reveals a constant first differences:

l (cm)	5.0	5.1	5.2	5.3	5.4
Q (cm)	13.6	14.9	16.2	17.5	18.8
First difference	14.9 – 13.6 = 1.3	16.2-14.9 = 1.3	17.5 – 16.2 = 1.3	18.8 – 17.5 = 1.3	

As the first difference is constant we then know that the data fits a linear model. Then, using substitution of data solves, the details of the model can be determined.

This practice can be extended to higher polynomial functions and in particular to quadratic functions. The quadratic functions exhibit constant second differences - that is, the differences between successive first differences are constant. To see this, consider the function defined by

$f(x) = ax^2 + bx + c$. We will tabulate the function for the first few non-negative integer values.

x	$f(x)$	1st difference	2nd difference
0	c		
	$\mapsto$	$a+b$	
1	$a+b+c$	$\mapsto$	$2a$
	$\mapsto$	$3a+b$	
2	$4a+2b+c$	$\mapsto$	$2a$
	$\mapsto$	$5a+b$	
3	$9a+3b+c$	$\mapsto$	$2a$
	$\mapsto$	$7a+b$	
4	$16a+4b+c$		

EXAMPLE 3.12

The following results where obtained from a medical laboratory. What appropriate model would fit the data?

r	3	3.2	3.4	3.6	3.8
T	3	3.84	4.76	5.76	6.84

Solution

Our first step is to visualise the data. So, as in Example 9.11 we plot the data points. Using a window of [0,4] by [0,7] with a scale of 1 we have:

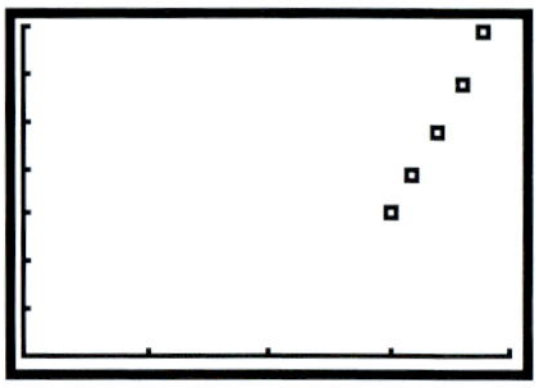

It appears that the data displays a linear relationship.
However, closer inspection reveals that the first differences of the *T*-values are not constant.

First difference: $3.84 - 3.00 = 0.84$
$4.76 - 3.84 = 0.92$
$5.76 - 4.76 = 1$
$6.84 - 5.76 = 1.08$

Therefore a linear model is not supported by the data.
On the otherhand, we see that the second differences are constant.

Second difference: $0.92 - 0.84 = 0.08$
$1.00 - 0.92 = 0.08$
$1.08 - 1.00 = 0.08$

And so, the data supports a quadratic model!

We can therefore write the model as $T = ar^2 + br + c$ and use the raw data to determine a, b and c by substitution. As we have three constants to evaluate, we need three data pairs to solve the problem which will then be a set of three linear equations in the unknowns, a, b, c. The first three data pairs give:

$$3 = 9a + 3b + c \quad -(1)$$
$$3.84 = 10.24a + 3.2b + c \quad -(2)$$
$$4.76 = 11.56a + 3.4b + c \quad -(3)$$

We can solve this system of linear equations in the same way that we solved systems of linear equations in Chapter 2. However, we first eliminate c using (1).

$$c = 3 - 9a - 3b \quad -(4)$$

Then, we substitute this into (2) and (3):

$$3.84 = 10.24a + 3.2b + (3 - 9a - 3b) \Leftrightarrow 0.84 = 1.24a + 0.2b \quad -(5)$$

$$4.76 = 11.56a + 3.4b + (3 - 9a - 3b) \Leftrightarrow 1.76 = 2.56a + 0.4b \quad -(6)$$

Then, (6) – 2×(5) gives: $0.08 = 0.08a$

$$\Leftrightarrow a = 1$$

By substitution in (5), $b = -2$. Finally substitution of these into (4) gives $c = 0$.
Therefore, our quadratic model is given by, $T = r^2 - 2r$.

We can fit a least squares quadratic equation by selecting **5:QuadReg**, saving it as **Y1** and then sketching it. From the last screen, it appears to be a pretty good fit!

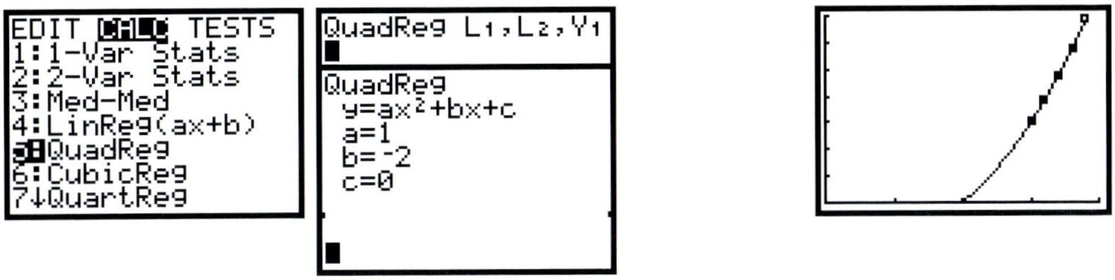

Again, caution needs to be taken when using the least squares method. In this case, because the data was a perfect fit, the equation we obtained using simultaneous equations and that obtained using the method of least squares are the same. On the otherhand, if our data looked like the one shown below, the least squares approach would still find the equation of best fit, whereas, depending on which three points we decide to use to obtain 3 equations and then solve, our result and that of the TI–83 will most probably not be the same (but hopefully not too different).

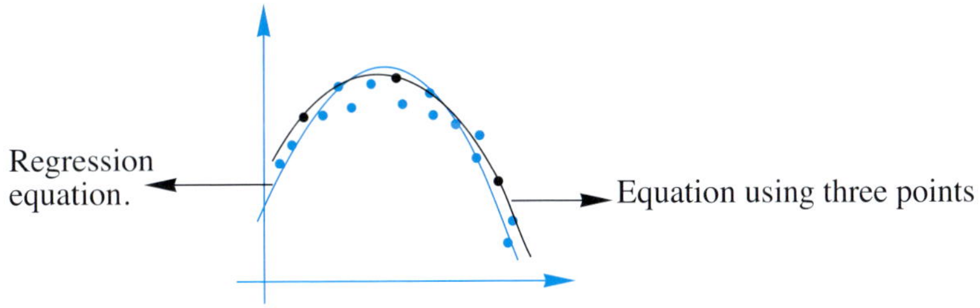

Information and data also come in forms other than tabular – although once we extract the data from the information presented, we can then organise it in tabular form. Consider the following example.

EXAMPLE 3.13 A piece of machinery was bought for $12,000 new and was then depreciated uniformly to an estimated scrap value of $1200 after 6 years of use.

(a) Use the information to sketch a graph of the machine's values against time.
(b) How much does the machine depreciate by annually?
(c) Construct a model for the value of the machine during its operating life.

Solution

In this problem there are probably some terms that are unfamiliar. When this occurs, make sure that you research the meaning of the terms that are not familiar to you.
We are given that at the begining, the machine is worth \$12,000 and that in 6 years time it will be worth \$1200. The key term is that it depreciates (i.e., losses its value) uniformally, that is, each year its value will decrease by the same amount. A linear model displays this attribute, and so we will use that as the base for building our model.

(a) We start by setting up a table of values.
Let t years denote the numbers of years since the machine was bought and $\$V$ the value of the machine at time t.

Time (t years)	0	6
Value ($\$V$)	12,000	1200

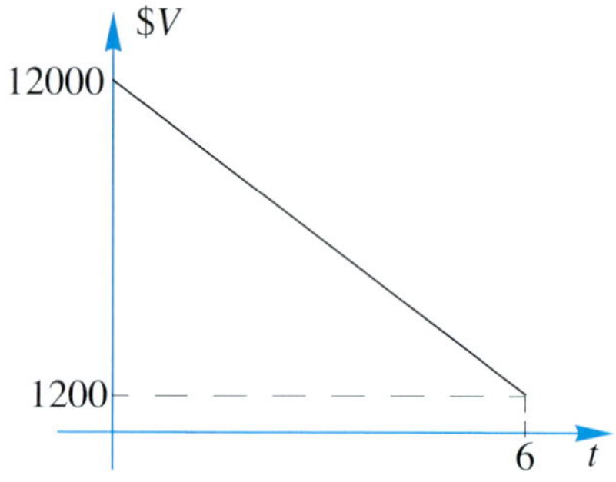

We can use the above information to sketch the graph.

(b) Over a period of 6 years the machine has depreciated by \$10,800.

Therefore, the rate of **depreciation** is $\frac{10800}{6} = 1800$. (Why don't we write '–1800'?)

Meaning that every year the machine depreciates by \$1800

(c) We can now use our work in Chapter 5 to determine the equation of the straight line.
We have a gradient of –1800 and a V-intercept of 12,000 therefore the equation becomes

$$V = -1800t + 12000$$

However, to fully specify the model we need to take into account any restrictions. In this case we have a restriction on the time, i.e., $0 \le t \le 6$.

Therefore, our model becomes: $V = -1800t + 12000, 0 \le t \le 6$

Of course, data can be presented in different formats, but we still need to extract all (and only) relevant information so that we can proceed with constructing a model. Consider the following example.

EXAMPLE 3.14

A suspension bridge, one very much like the Golden Gate Bridge or the Severn Suspension Bridge, has its roadway supported by parabolic cables hanging from support towers.The cost involved in constructing such a bridge is a complicated affair, however, our task is to provide a cost for the suspender cables, running vertically from the road level to the parabolic cable, assuming that they are 2 metres apart. All measurements are shown on the plan below. A marker at A is known to be 55 m above the road level and 40 m from the right support tower.

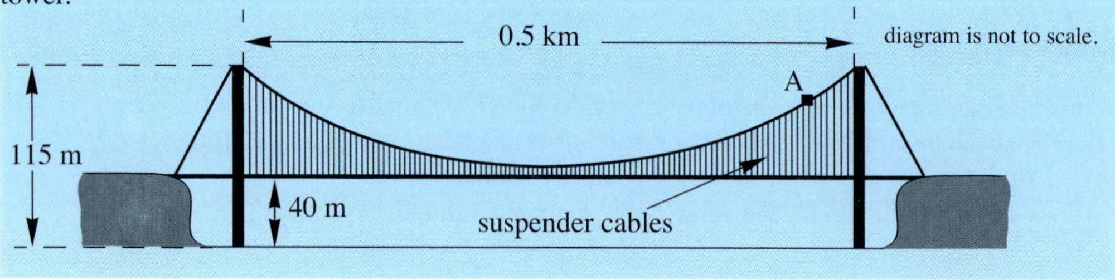

Solution

Part of the modelling process is to make certain assumptions. One assumption we will make is that the cost of the suspender cable is only dependent on the material used and the length of cable used.

So, we let the cost of the suspender cable per unit length (for a particular material) be $\$k$. Then, once we know the total length of suspender cable required, L m say, the total cost for the cable would be $\$kL$.

Notice that we have made no mention of waste. Extra cable will be needed so that it can be 'attached' to the parabolic cable as well as to the
road level. However, this is among others all part of the simplifying assumptions we are making.

We now need to find a value for L. What we need is the length of each suspender cable and the sum their lengths –this will provide us with the value of L.
We start by 'redrawing' the bridge (without concerning ourselves with accurate scales):

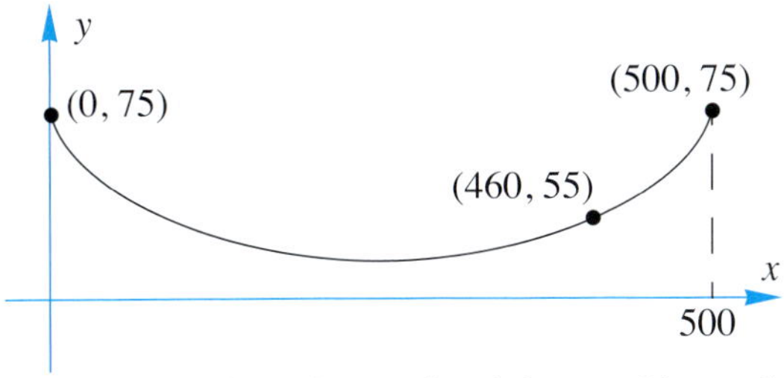

We can make use of the function

$$y = ax^2 + bx + c$$

to model the parabolic cable and from it find the lengths of the suspending cables.

The data consists of the three points lying on this parabola. As we know that we are dealing with a parabola and we have three points on the parabola, we can obtain the equation by using the least squares method (as was done in Example 9.12). Otherwise we can make use of simultaneous equations. So, using the TI–83, we enter our data and obtain the equation:

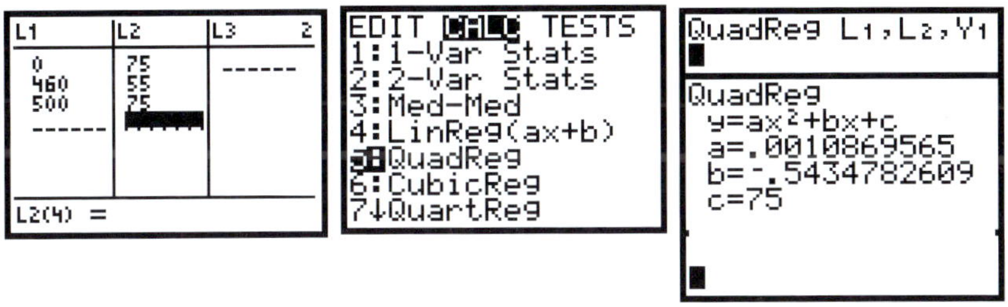

This gives the equation $y = 0.0010869x^2 - 0.5434782x + 50$.
So that when $x = 2$ (the first suspender cable)
it will have a length of 73.917 m

when $x = 4$ (the second suspender cable)
it will have a length of 72.843 m
Obviously, a table of values will make life much easier.

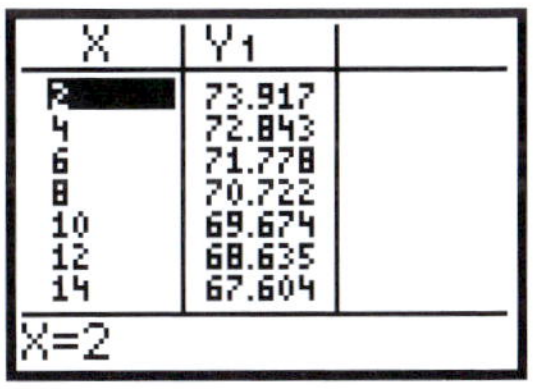

All that remains is to add these lengths. This is a rather tedious task and unless you have access to a spreadsheet package, we need to come up with a way of doing this. We do this using standard arithmetic procedures on our lists:

❶ Set up a list of even consecutive integers from 2 to 498. (Store it as L3).
First press **MODE** and select the **Seq** setting. Then press **QUIT**.
Next go to the **LIST** menu, select **OPS** and then option **5:seq(** and store sequence in L3.

❷ Evaluate the y-value (using the quadratic regression equation) for each number in L3. Store these values in L4.

❸ Go back to **LIST** select **MATH** and then option **5:sum(**
This will add the values in L4 and hence give the sum of the lengths of the cables)

❶ 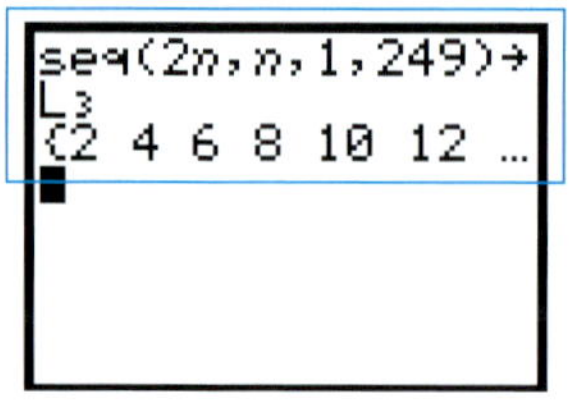

❷ 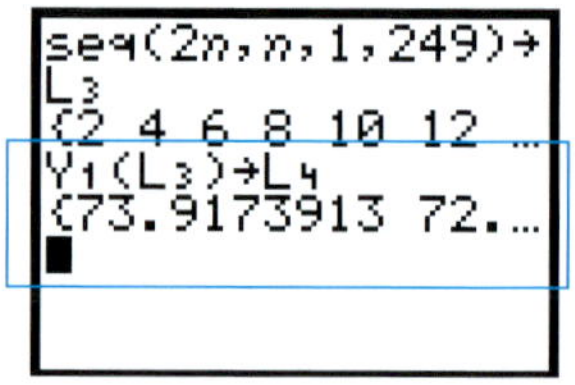

❸ 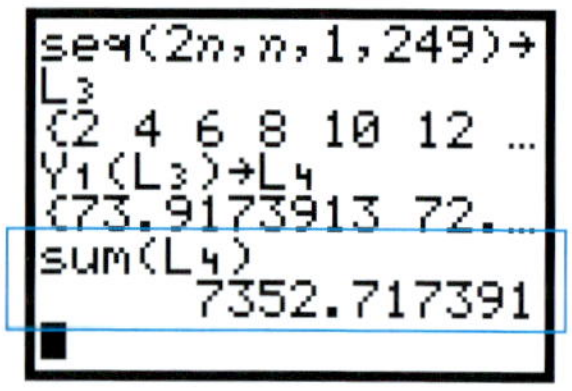

That is, we will need 7352.72 m of suspender cable for one side of the suspension bridge which will then need to be doubled (as we need an equal amount for the other side).

Therefore, the cost for the suspender cable is \$14705.43k. (\$14706k)

Of course, we should always question our findings and then validate them. It is possible to obtain a model that acurrately represents as set of data and yet, the model might not be realistic.

A somewhat simplified approach to phase 3 of our modelling process can be summarised using the following diagram:

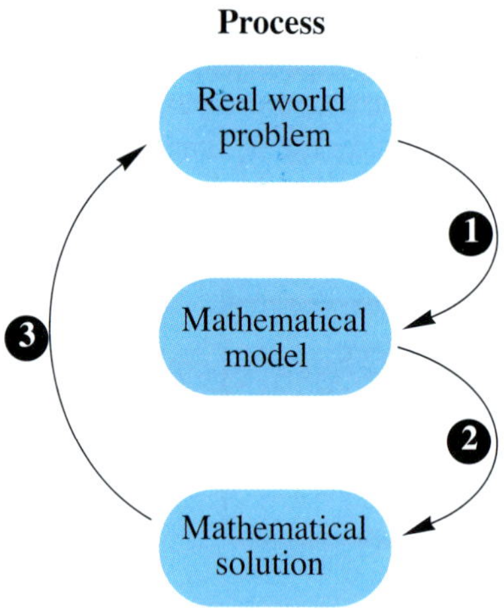

Stage ❶ Collect information and analyse data to construct a mathematical statement about the problem. Based on this statement, build a mathematical model.

Stage ❷ Solve the mathematical model.

Stage ❸ Interpret your solution in light of the original problem. Validate your solution. Then, if necessary refine your mathematical model and repeat the process.

Our work on modelling so far has concentrated on linear and quadratic modelling. In the next section we will consider a broader approach to modelling. In all, the work covered in this chapter will be very useful while undertaking your portfolio task. Chapter 23 – Portfolio Work will revisit modelling.

EXERCISES 3.1.3

1. (a) Show that the following data supports a linear relationship between the variables x and y.

(b) Find the linear relationship in each case.

i.

x	2	3	4	5	6
y	8	8.4	8.8	9.2	9.6

ii.

x	0	2.4	4.8	7.2	8.6
y	6	1.2	–3.6	–8.4	–13.2

iii.

x	–6	–4	–2	0	2
y	–0.2	1.2	2.2	3.2	4.2

2. Show that the following data supports a quadratic relationship between x and y.

x	2	2.4	2.8	3.2	3.6
y	6	9.12	12.88	17.28	22.32

3. (a) Show that the following data supports a quadratic relationship between x and y.

x	0	1	2	3	4
y	2	7	14	23	34

(b) Determine the rule for this quadratic.

4. A parabolic curve passes through the points (3, 16), (–1, 12) and (2, 9).

(a) Plot these points on a set of axes.

(b) Find the equation of this curve.

(c) Draw the graph of the curve found in (b) on the set of axes in (a).

5. Find the equation of the parabola passing through the points A(1, 4), B(2, 9) and C(–1, 6).

6. The tables below show the demand and supply data for a particular product.

Demand:

q	5	10	15	20
p	3	2	1	0

Supply:

q	5	10	15	20
p	1.11	1.46	1.81	2.16

Where q represents the quantity in thousands and p the price per item.

(a) On the same set of axes, plot the

i. data representing the demand.

ii. data representing the supply.

(b) Show that both demand and supply can be modelled by a linear relationship.
(c) Find the equation for
i. demand. ii. supply.
(d) Describe the optimum scenario.

7. The graph shows the temperature readings taken by a student on a winter's day.

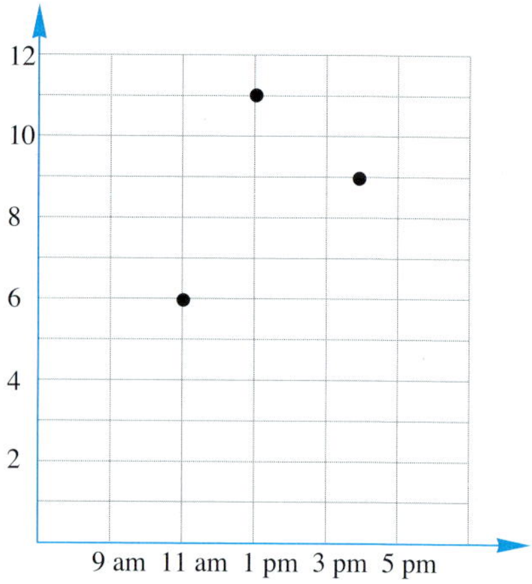

The readings were taken at hourly intervals after 9:00 am. Unfortunately a number of the readings were smudged and therefore could not be transferred onto the graph.

(a) What type of graph would most probably fit the set of data shown?

(b) What would you expect the temperature to be at 11:00 pm?

8. During a baseball game the ball is hit 1 metre above ground level with a bat. After having travelled a horizontal distance of 30 m, it is 20 m above ground level. An outfielder runs to meet the ball and catches it as he dives to the ground. The outfielder is 80 metres from home base when he catches the ball. Asuming that the path of the ball is parabolic, what is the greatest height that the ball reaches?

9. A spring is attached to a bar. The extension, x m, of this spring when different masses, M kg, are attached at the free end, is recorded and tabulated:

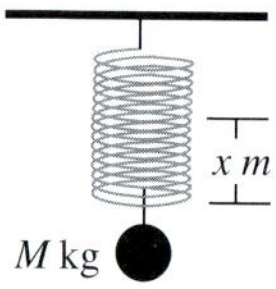

M (kg)	1	2	4	6	10	12
x (cm)	1.5	2.0	3.0	4.0	6.0	7.0

(a) Plot the graph of x versus M.
(b) What model would appear to suit the data?
(c) i. Determine the equation that best fits the data.
ii. What is the spring's natural length?

10. An object is dropped from a building, 125 m high. During its descent, the distance, x m, above ground level is recorded as tabulated:

t (sec)	0	1	2	3	4	5
x (m)	125	80	45	20	5	0

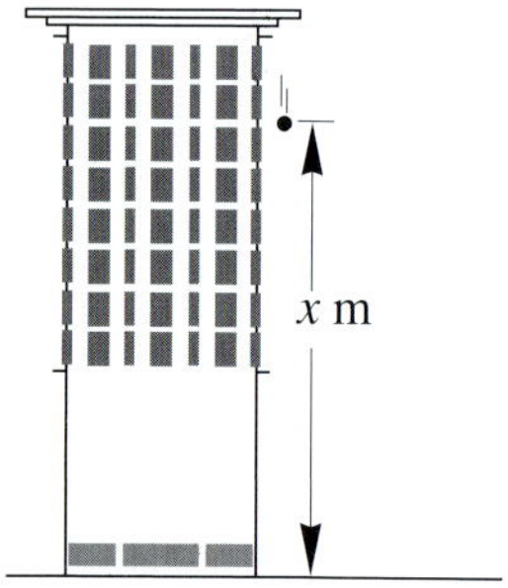

(a) Plot the data on a set of axes.
(b) Fit, by eye, a curve that fits the data.
(c) i. What type of curve would fit this data?
ii. Use a difference table to verify your statement.
(d) Find the equation of the curve that best models the data.

11. (a) A parabolic curve passes through the points (1, 2), (–1, 0) and (2, –3). Determine the equation of this curve.

(b) Find the *family* of quadratics that pass through the points (1, 2) and (–1, 0).

12. Over the years 2000 to 2004, the sales of a company have increased at a constant rate. The table below shows the variation over this period.

Year	2000	2004
Sales	\$500 000	\$680 000

Assuming that a linear model of the form $S(t) = at + b, k_1 \le t \le k_2$ exists, since 2000

(a) i. find the values of a and b.

ii. determine the values of k_1 and k_2 for which this model is assumed to hold.

(b) Sketch the graph showing the relationship between the value of sales \$$S$ and the time, t years, since 2000.

13. In 1988, Natalya Lisovskaya (of the then USSR) won the gold medal for the shot putt at the Olympic Games.

The height of the shot putt from ground level, relative to the horizontal distance it travelled from the point where it left Natalya's hand was recorded and tabulated.

Horizontal distance, x m	12.19	18.29	21.34
Height, h m	6.58	3.74	1.01

(a) Plot these points on a set of axes.

(b) i. What type of curve would best model this situation?

ii. Determine the equation of such a curve.

(c) i. What is the maximum height reached by the shot putt?

ii. How far did Natalya throw the shot putt?

14. An archway is constructed as shown in the diagram below.

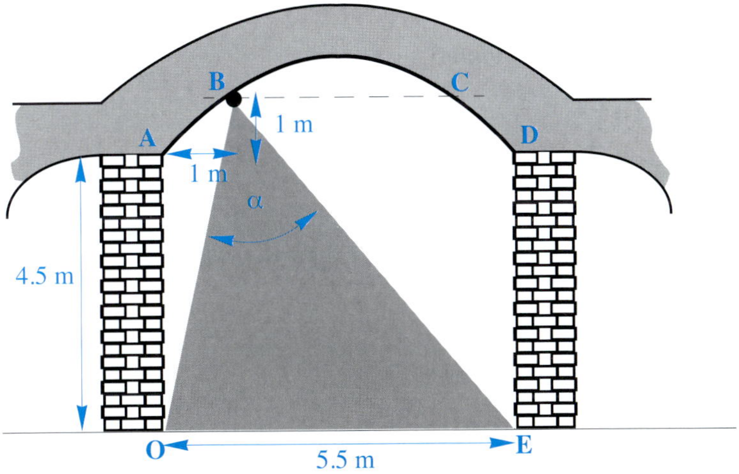

A light source is positioned at B as shown. The archway can be modelled by a parabolic curve.

(a) Using O as the origin, determine the quadratic equation that produces a parabolic curve to pass through the points A, B and D.

(b) Find the equations of the straight lines that form the 'edges' of the light beams
  i. BO.
  ii. BE.

(b) Find the tilting angle, α, so that light can spread from O to E, where the pillars meet the floor.

15. The total cost, $C(x)$ in thousands of dollars, for manufacturing specialised cars was recorded so that it could then be analysed.

Number of cars produced (x)	Cost ('000) $C(x)$
0	580
10	805
20	980
30	1105
40	1180
50	1205

(a) On a set of axes, plot the data points.
(b) Verify that a quadratic model is appropriate in this situation.
(c) Determine the equation of the quadratic that satisfies this data.
(d) What is the average rate of increase in production costs for the first 10 cars?
(e) i. How much did it cost to produce 11 cars?
  ii. How much did it cost to produce the eleventh car?

16. Stella has just been employed as Marketing Manager at Nattel Toys Group. During the first few weeks of her employment she has undertaken to research the sales of one of their products, know as 'Talking Tutu". Stella provided the following data:

Price, $\$p$, per Talking Tutu:

x	0	10	100	1000
p	10	9.99	9.9	9

Cost, $\$C$, for producing x Talking Tutus:

x	0	10	100	1000
C	7000	7020	7200	9000

Revenue, $\$R$, for selling x Talking Tutus:

x	0	10	100	1000
R	0	99.9	990	9000

(a) Show that the relation between
  i. p and x is linear
  ii. C and x is linear
  iii. R and x is quadratic.
(b) i. On the same set of axes, plot the graphs of p, C and R versus x.
  ii. Determine the equation that best models each of p, C and R in terms of x.
(c) Determine the maximum profit Stella can expect based on assumed future sales.

3.2 MATHEMATICAL MODELS

3.2.1 MATHEMATICAL INVESTIGATION AND MODELLING

"Assume that you have a perfectly smooth elephant whose weight may be neglected ...."

This is a classic line, designed to make fun of mathematicians approach to solving problems in mechanics. In this sentence we see the essence of applied mathematical modelling reduced to a caricature. We all know elephants are neither smooth nor of negligible weight. But in some cases we can still use mathematics which calls on these assumptions and draw meaningful conclusions. Suppose we are investigating the braking on a large train. In studying how objects move we learn that we can model deceleration using forces. The mass of the train will enter these calculations. If we add an elephant to the train the total weight will change very little and so the effect on the braking will be neglible. If we added the elephant to a truck however to examine the effect on braking its weight could hardly be neglected then!!!!

In much of the mathematical work done in mechanics, the assumption is that we are using particles - 'point masses' - which simplifies analysis and removes certain issues like lines of action of forces, rotation etc. Of course we sometimes need to consider these features and so we add more to our model. The complexity of the model depends on what details we need.

Mathematics provides us with a convenient way to investigate and represent activities and processes in the physical world and the world of finance. Although many parts of mathematics arose from a 'pure' base where an investigation of patterns led a mathematician to abstract ideas, it is common for most of mathematics to end up having some application in real phenomena. The trail of investigations can wind back and forth from 'pure' to 'applied' mathematics with each prompting the other. For example, the solution of equations by algebraic means led to the invention of groups which seemed to be very abstract mathematical objects of little use, yet, they have proved very helpful in building models in particle physics. Applications may not be obvious at the time the basic work is done. Number theory seems an extremely abstract field, yet work on factorization with large primes has led to valuable applications in code encryption which has been of major interest to governments and financial institutions in recent times.

A mathematical model aims to represent at least part of a process. The model built will be universal in some cases or dependent on assumptions supporting some problem in other cases. The model will frequently be reduced to an equation or set of equations describing aspects of the process. We find that some problems can be modelled by 'deterministic' means – the state of a system is determined completely and predictably by knowing some features of it. An example of this is Hooke's law which models the restoring force in a spring or elastic string. This law tells us that the force is proportional to the extension and opposes the extension. Once we know an input variable, the extension, we can predict the output, which is the force if we know the spring constant for that spring – something experiment can tell us.

Other problems require 'probabilistic' models where elements of chance play a role and we cannot predict the system with certainty. A model which aims to forecast the price of electricity on a daily basis will always be subject to randomness for example.

The character of a mathematical model varies substantially depending on how it is to be used. Sometimes we need only a crude model to gain insight and few numbers are involved. Consider this whimsical example originally attributed to Galileo.

EXAMPLE 3.15

Why are there are no real giants?

Solution

The mass and weight of objects in scale depend on volume when made of the same material. Volume increases as the *cube* of a linear dimension. Hence we know that when a sphere has its radius doubled its volume increases by a factor of 8. Thus the weight of a giant, who is 3 times the height of a man, but made of the same flesh and bone, would be 27 times as great. However bone strength depends on its cross sectional area, which increases only as the *square* of a linear dimension. Such a giant would have bone strength only 9 times as great. He would find his increased weight a crushing burden unless his bones were increased in size out of proportion …or made of different material!

Arguments from scale are very popular in modelling. The use of dimensions helps to give quick 'order of magnitude' rough estimates.

3.2.2 OPEN AND CLOSED PROBLEM SOLVING

A mathematical investigation can be very focussed or quite broad depending on whether the problem set is 'open' or 'closed'. Consider a closed problem with a fixed answer – few assumptions need to be drawn in building the model. A typical closed problem might be to determine the water capacity held in a tank, which has a circular cross-section of three metres and a depth of 4 metres. We can immediately use our knowledge of cylindrical volumes to determine the capacity of the tank. (The author was actually set this problem by a farmer once!)

Other times detailed results are called for and much data is needed as an input for the model. In fact we often build models from data where mathematical relationships can be found without a clear knowledge of the underlying cause. This is especially true with probabilistic models and those which draw on statistics.

For an open problem we may need to list a number of conditions which apply so as to create a suitable model with real application. When testing the model these assumptions may need to be refined.

A typical example of an open problem related to the same mathematics as our tank problem is the following.

EXAMPLE 3.16

Many things we buy are stored and sold in steel cans – what is the 'best' height and radius of such cans?

Solution

We can see that the volume of a can will depend on the height and the radius.
We also observe that the amount of metal used to make the can will depend on the surface area of the can.
We can make a first approximation to this problem by making some assumptions.

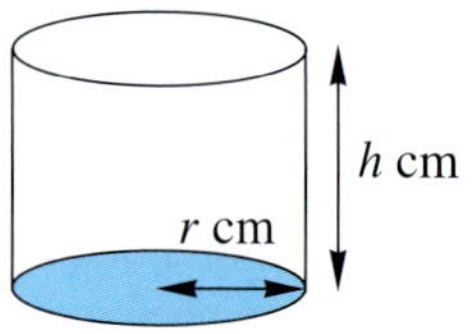

1. The cost of making a steel can is the cost of the material used in making it.

2. The can is the shape of a cylinder and metal used is only that given by that geometry.

These two assumptions can be questioned but they give us a starting point. We begin with the basic geometry implied by assumption 2.

The volume, V cubic centimetres of the can is given by $V = \pi r^2 h$ where the radius is r and the height is h in centimetres.

If we unfold a can we see that it is made up of two circles (the top and bottom) and a rectangle making the curved surface. The surface area, A cm^2 of the can is seen to be $A = 2\pi r^2 + 2\pi rh$

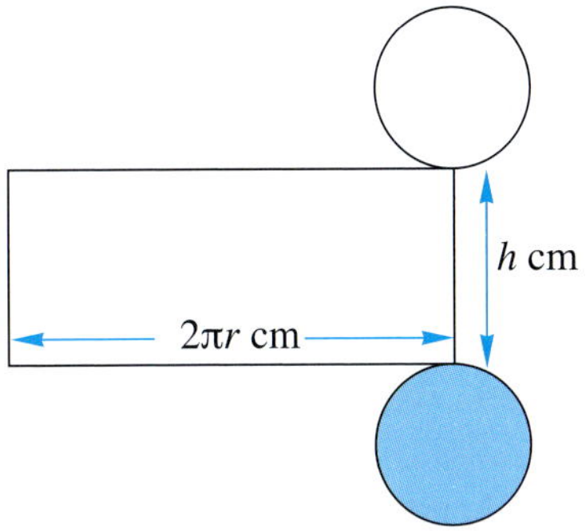

Generally we decide in advance how much is to be sold – that is – the volume is fixed. We can now see that if we do fix the volume of the can then the radius and height of the can are directly related and that as the radius increases the height decreases and vice versa. We also know that V varies for different things. A soft drink can holds typically between 325 and 375 ml (each millilitre is a cubic centimetre) whereas paint may be sold in containers of between 1 litre and 4 litres. But when we know the value of V we can determine h if we know r for example.

Since h and r are both measured in the same units it will be useful to look at the ratio h/r, which will be dimensionless.

Let us simplify the problem by looking at a 1 litre container and set $V = 1$. (This actually has the effect of measuring h and r in decimetres but we will not use actual lengths, just their ratio.)

Now, from $V = \pi r^2 h$ we have $1 = \pi r^2 h \Rightarrow h = \dfrac{1}{\pi r^2}$

and so, after substituting $h = \dfrac{1}{\pi r^2}$ into the equation for the surface area $A = 2\pi r^2 + 2\pi rh$, we

have
$$A = 2\pi r^2 + 2\pi r \times \frac{1}{\pi r^2}$$
$$= 2\pi r^2 + \frac{2}{\pi r}$$

Giving us a simple equation, showing how the area varies with the can radius for a fixed volume.

We can easily graph this using our calculators and show that it has a unique minimum. The calculator allows us to do this numerically and then we can find the value of h and the ratio.

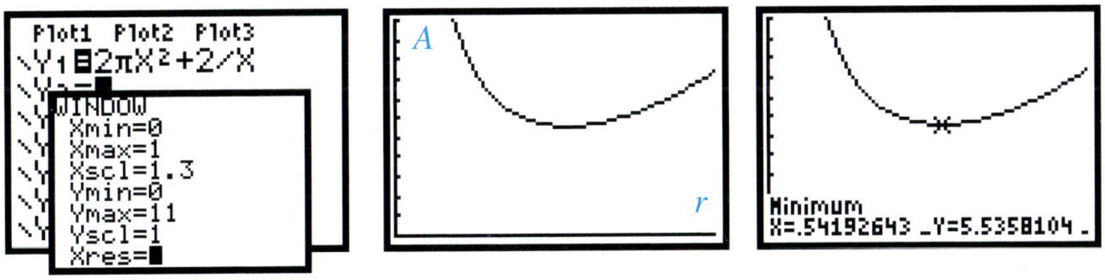

In fact it is possible (using advanced mathematics called calculus) to show that this occurs when

$r = \sqrt[3]{\frac{1}{2\pi}}$ and that this gives $\frac{h}{r} = 2$. We can actually show also that this ratio holds regardless of the volume of the steel can. If we use a general V instead of $V = 1$ we get the same ratio.

So far we have converted our open problem into a very closed problem using our assumptions. We could immediately look at some real steel cans and see what ratios we actually find. In practice we find $\frac{h}{r} > 2$ is far more common. In fact, it may often exceed 3! Does this mean the manufacturers like to waste money?

We now open up the problem by looking at the assumptions. We assume that the metal used is all in the can geometry. But the circular ends at least are stamped from sheets, which leave a lot of waste.

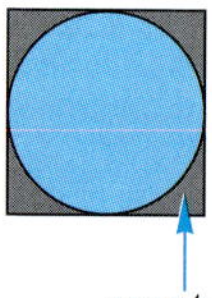

Each circle may be stamped from a square of side $2r$ and leaves waste $4r^2 - \pi r^2$.

Then this means that the actual metal used really has an area of

$$A = 8r^2 + 2\pi rh = 8r^2 + \frac{2}{r}$$

From this we can review our analysis and show that $r = \frac{1}{2}$ (we can check this by using our graphing calculators) and so we find that $\frac{h}{r} = 2.55$ (approximately). In fact, theory shows that this gives $\frac{h}{r} = \frac{8}{\pi}$. This is larger than our previous value and shows that the waste metal generated is an issue.

It may be inflating the ratio too much however, despite the fact that real observed ratios are generally larger. We can look at this stamping process more closely. We imagined that the circles were stamped out of metal sheets in squares as shown but they are more efficiently stamped if the process is offset to stamp them out of hexagons which can also cover the metal sheet (with some edge effects we can neglect overall).

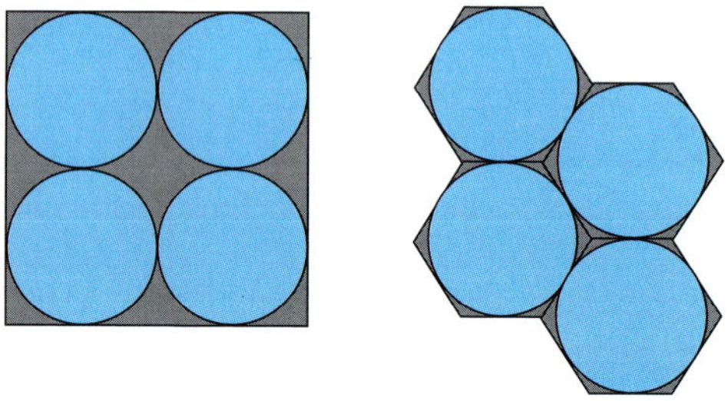

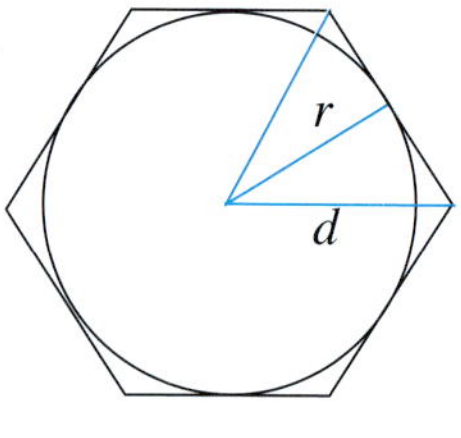

This can reduce the ratio again by cutting waste. We look at the simple geometry involved. Each regular hexagon can be seen to comprise 6 equilateral triangles as shown in the diagram. These triangles have side length d where, by Pythagoras's theorem, $d^2 = r^2 + \left(\frac{1}{2}d\right)^2 \therefore d = \frac{2r}{\sqrt{3}}$

Each triangle has area $\frac{1}{2}dr$ and there are 6 triangles used to cover the hexagon used to stamp out the circle. The area stamped is then $2\sqrt{3}r^2$. We use two of these in each can and so the rule for the steel can surface area is now $A = 4\sqrt{3}r^2 + 2\pi rh = 4\sqrt{3}r^2 + \frac{2}{r}$

We again graph this using our calculator and estimate the values of r and hence h. We see this ratio $\frac{h}{r}$ is approximately 2.21. It is possible to show that this is exactly $\frac{4\sqrt{3}}{\pi}$ but we need not prove this here.

This is less than before so we have saved on waste metal. What else have we neglected?

We could look at the problem in more detail. In practice the steel can must have circles stamped of larger radius than the can radius to enable a 'lip' to be produced to fold the ends over and connect them to the can side. This increases the metal usage and so the cost. The actual production process will also include this sealing up of the can lips and there are also likely to be fixed costs associated with steel can production runs anyway. It is possible to see that these require a modification of our assumptions and that they suffice to increase the h/r ratio.

So far we note that this problem has only used simple geometry and graphing to get numerical estimates of the ratios. Many real problems will call on higher mathematics like calculus to get results as we are commonly dealing with rates of change or averages.

If we look closely at the real situation for the steel can dimensions problem we would find that in practice the optimal ratio will depend on volume V - for larger cans we find that the ratio is generally lower and nearer our present theory. The reason for this depends on sealing costs, but it is not important to see why this is so at present ... only to note that we can make increasingly more complex models by changing assumptions in an open problem in order to get more realistic results. It is clear that even our simplest model gave some idea of the real situation however.

We see from this example that a mathematical investigation may lead to open problem solving where answers depend on assumptions and we often need to compare theory with observation to improve model representation. From this example we can revisit our general rules about modelling and mathematical investigation.

1. **Formulate the problem.**
   We first need to decide what we want to know.

2. **Outline a model.**
   Identify input and output variables and their relationships. Set up equations.

3. **Examine model for usefulness.**
   Is the model as built able to be used. This means can we get the input variables and other data needed and can we solve the equations at all or possibly within a fixed time frame which we may have (some models need to be run on tight timelines – a typical example may be a scheduling or rostering system)

4. **Test the Model.**
   Does the output predicted by the model match the results observed.

5. **Refine the Model.**

If the answer to step 4 is NO we may need to refine the model and move back to steps 3 and 4.

There are many 'closed' problems in Chapter 23 that, like the example we have just completed, can be transformed into an open ended problem. Many of which are ideal for your portfolio work.

4.1 THE BINOMIAL THEOREM

4.1.1 THE BINOMIAL THEOREM

Bracketed expressions such as $(2x-3)^7$ are called 'binomial' because there are two terms in the bracket (the prefix bi- means two). Such expressions can be expanded using the distributive law. In a simple case such as $(a+b)^2$ the distributive law gives:

$$\begin{aligned}(a+b)^2 &= (a+b)(a+b)\\ &= a^2+ab+ba+b^2\\ &= a^2+2ab+b^2\end{aligned}$$

The distributive law states that each term in the first bracket must be multiplied by each term in the second bracket.

The next most complicated binomial can be evaluated using the previous result:

$$\begin{aligned}(a+b)^3 &= (a+b)(a+b)^2\\ &= (a+b)(a^2+2ab+b^2)\\ &= a^3+2a^2b+ab^2+a^2b+2ab^2+b^3\\ &= a^3+3a^2b+3ab^2+b^3\end{aligned}$$

Similarly, the fouth power of this simple binomial expression can be expanded as:

$$\begin{aligned}(a+b)^4 &= (a+b)(a+b)^3\\ &= (a+b)(a^3+3a^2b+3ab^2+b^3)\\ &= a^4+3a^3b+3a^2b^2+ab^3+a^3b+3a^2b^2+3ab^3+b^4\\ &= a^4+4a^3b+6a^2b^2+4ab^3+b^4\end{aligned}$$

The calculations are already fairly complex and it is worth looking at these results for the underlying pattern. There are three main features to the pattern. Looking at the fourth power example above, these patterns are:

1. **The powers of *a*.**

These start at 4 and decrease: a^4, a^3, a^2, a^1, a^0. Remember that $a^0 = 1$

2. **The powers of *b*.**

These start at 0 and increase: b^0, b^1, b^2, b^3, b^4. Putting these two patterns together gives the final pattern of terms in which the sum of the indices is always 4:

$$\underset{4}{\ldots a^{4}} + \underset{3+1=4}{\ldots a^{3}b^{1}} + \underset{2+2=4}{\ldots a^{2}b^{2}} + \underset{1+3=4}{\ldots a^{1}b^{3}} + \underset{4}{\ldots b^{4}}$$

3. **The coefficients complete the pattern.**

Thcsc coefficients arise because there is more than one way of producing most of the terms. Following the pattern begun above, produces a triangular pattern of coefficients known as Pascal's Triangle. Blaise Pascal (1623-1662) developed early probability theory but is lucky to have this triangle named after him as it had been studied by Chinese mathematicians long before he was born.

So, if we continue our expansions (up to and including the sixth power) we have the following:

$$(x+a)^0 = 1$$
$$(x+a)^1 = 1x + 1a$$
$$(x+a)^2 = 1x^2 + 2ax + 1a^2$$
$$(x+a)^3 = 1x^3 + 3x^2a + 3xa^2 + 1a^3$$
$$(x+a)^4 = 1x^4 + 4x^3a + 6x^2a^2 + 4xa^3 + 1a^4$$
$$(x+a)^5 = 1x^5 + 5x^4a + 10x^3a^2 + 10x^2a^3 + 5xa^4 + 1a^5$$
$$(x+a)^6 = 1x^6 + 6x^5a + 15x^4a^2 + 20x^3a^3 + 15x^2a^4 + 6xa^5 + 1a^6$$

: : :

etc. etc. etc.

Now consider only the coefficients for the above expansions. Writing down these coefficients we reproduce Pascal's triangle:

1
1 1
1 2 1
1 3 3 1
1 4 6 4 1
1 5 10 10 5 1
1 6 15 20 15 6 1
etc.

The numbers in the body of the triangle are found by adding the two numbers immediately above and to either side.

An alternative to using Pascal's Triangle to find the coefficients is to use combinatorial numbers. If expanding $(a+b)^5$ the set of coefficients are:

$$\binom{5}{0} = 1, \binom{5}{1} = 5, \binom{5}{2} = 10, \binom{5}{3} = 10, \binom{5}{4} = 5, \binom{5}{5} = 1$$

which are the same as those given by Pascal's Triangle.

Most calculators can do this.

The TI-83 uses the **MATH** main menu command followed by the **PRB** (probability) sub-menu.

To calculate a combinatorial number such as $\binom{5}{2} = {}^5C_2$, return to the ready mode and enter 5. Select **nCr** as previously described, press 2 and press **ENTER**.

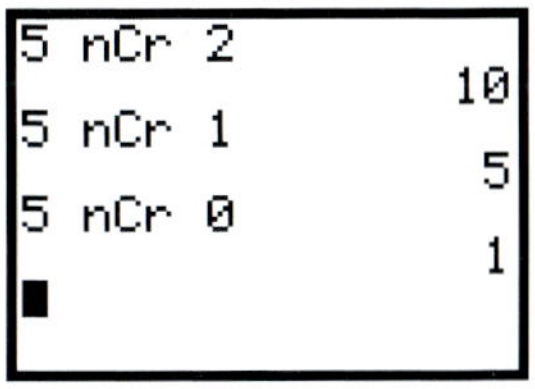

In a case such as this in which it may be necessary to calculate a sequence of related combinatorial numbers, it is much quicker to use the replay and edit feature of the calculator (2nd **ENTRY**).

Or, we can represent **nCr** as a function. For example, we can set **Y = 5nCrX**, meaning that we now have a function of x, $y = \binom{5}{x}$, $x = 0, 1, ..., 5$.

EXAMPLE 4.1

Expand $(x+y)^6$

Solution

(a) **Step 1:** Making use of Pascal's triangle we first determine the required coefficients:

1
1 2 1
1 3 3 1
1 4 6 4 1
1 5 10 10 5 1
1 6 15 20 15 6 1

Step 2: Write down each term:

$$x^6y^0 \quad x^5y^1 \quad x^4y^2 \quad x^3y^3 \quad x^2y^4 \quad x^1y^5 \quad x^0y^6$$

Finally, combine the two steps:

$$x^6y^0 + 6x^5y^1 + 15x^4y^2 + 20x^3y^3 + 15x^2y^4 + 6x^1y^5 + x^0y^6$$

Obviously, we can also combine the two steps into one:

$$(x+y)^6 = {}^6C_0x^6y^0 + {}^6C_1x^5y^1 + {}^6C_2x^4y^2 + {}^6C_3x^3y^3 + {}^6C_4x^2y^4 + {}^6C_5x^1y^5 + {}^6C_6x^0y^6$$
$$= x^6y^0 + 6x^5y^1 + 15x^4y^2 + 20x^3y^3 + 15x^2y^4 + 6x^1y^5 + x^0y^6$$

And, with practice, you should be able to expand such expressions as we have just done.

What hapens if we have the difference of two terms rather than the sum? For example, what about expanding $(x-a)^5$? Well the process is the same except that this time we rewrite $(x-a)^5$ as follows:

$$(x-a)^5 = (x+(-a))^5$$

So, how do we proceed? Essentially in exactly the same way.

Step 1: Making use of Pascal's triangle we first determine the required coefficients:

1
1 2 1
1 3 3 1
1 4 6 4 1
1 5 10 10 5 1
1 6 15 20 15 6 1

Notice that Pascal's triangle does not change due to the change in sign and so the coefficients remain the same whether it is a sum or difference.

Step 2: Write down each term:

$(x)^5(a)^0 \quad (x)^4(-a) \quad (x)^3(-a)^2 \quad (x)^2(-a)^3 \quad (x)^1(-a)^4 \quad (x)^0(-a)^5$

i.e., $x^5 \quad -x^4a \quad x^3a^2 \quad -x^2a^3 \quad xa^4 \quad -a^5$

Combining the coefficients and the terms we have:

$$(x-a)^5 = x^5 - 5x^4a + 10x^3a^2 - 10x^2a^3 + 5xa^4 - a^5$$

So basically the only difference is in the alternating negative sign.

EXAMPLE 4.2

Expand (a) $(4x-3)^3$ (b) $\left(2x - \frac{2}{x}\right)^3$

Solution

(a) **Step 1:** Making use of Pascal's triangle we first determine the required coefficients:

1
1 2 1
1 3 3 1
1 4 6 4 1
1 5 10 10 5 1
1 6 15 20 15 6 1

Step 2: Write down each term:

$(4x)^3 \quad (4x)^2(-3) \quad (4x)(-3)^2 \quad (-3)^3$

$64x^3 \quad -48x^2 \quad 36x \quad -27$

Combining steps one and two we have:

$$\begin{aligned}(4x-3)^3 &= 1 \times 64x^3 + 3 \times -48x^2 + 3 \times 36x + 1 \times -27\\ &= 64x^3 - 144x^2 + 108x - 27\end{aligned}$$

(b) As with (a), the term pattern must be built on $(2x)$ and $-\frac{2}{x}$:

$$\begin{aligned}\left(2x-\frac{2}{x}\right)^3 &= 1 \times (2x)^3\left(-\frac{2}{x}\right)^0 + 3 \times (2x)^2\left(-\frac{2}{x}\right)^1 + 3 \times (2x)^1\left(-\frac{2}{x}\right)^2 + 1 \times (2x)^0\left(-\frac{2}{x}\right)^3\\ &= 8x^3 - 24x + \frac{24}{x} - \frac{8}{x^3}\end{aligned}$$

What about an expansion of the form $(x + 2y + 1)^4$? How do we deal with this situation? Again, we use our two steps, however, as we did when we introduced a difference rather than a sum, we need to rearrange our expression slightly. We do this by writing $(x + 2y + 1)^4$ as $(x + (2y + 1))^4$

and treating the sum as one made up of '(x)' and '$(2y+1)$'. So, we have:

Step 1: Making use of Pascal's triangle we first determine the required coefficients:

1
1 2 1
1 3 3 1
1 4 6 4 1
1 5 10 10 5 1
1 6 15 20 15 6 1

Step 2: Write down each term:

$$(x)^4 \quad (x)^3(2y+1) \quad (x)^2(2y+1)^2 \quad (x)(2y+1)^3 \quad (2y+1)^4$$

Now, the first two terms are easy enough

$$x^4 \qquad 2x^3y + x^3$$

In fact, the third term is not all that bad either:

$$x^2(4y^2+4y+1) = 4x^2y^2+4x^2y+x^2$$

The fourth term, well we need to use Pascal's triangle again, which will lead to:

$$x(8y^3+12y^2+6y+1) = 8xy^3+12xy^2+6xy+x$$

Similarly for the fifth term:

$$(2y+1)^4 = 16y^4+32y^3+24y^2+8y+1$$

Then, combining the coefficients with step 2 we have:

$$1\times x^4 + 4\times[2x^3y+x^3] + 6\times[4x^2y^2+4x^2y+x^2] + 4\times[8xy^3+12xy^2+6xy+x]$$
$$+\ 1\times[16y^4+32y^3+24y^2+8y+1]$$

Which still needs to be expanded and then (hopefully) simplified. However, we stop here and leave the final simplification to you – if you feel up to it. The purpose was more to do with how to deal with an expansion that involves the sum of three terms instead of two. Of course, we can then extend this to an expansion involving the sum of four terms and so on.

EXERCISES 4.1.1

1. Expand the following binomial expressions:

(a)	$(b+c)^2$	(b)	$(a+g)^3$	(c)	$(1+y)^3$
(d)	$(2+x)^4$	(e)	$(2+2x)^3$	(f)	$(2x-4)^3$
(g)	$\left(2+\frac{x}{7}\right)^4$	(h)	$(2x-5)^3$	(i)	$(3x-4)^3$
(j)	$(3x-9)^3$	(k)	$(2x+6)^3$	(l)	$(b+3d)^3$
(m)	$(3x+2y)^4$	(n)	$(x+3y)^5$	(o)	$\left(2p+\frac{5}{p}\right)^3$
(p)	$\left(x^2-\frac{2}{x}\right)^4$	(q)	$\left(q+\frac{2}{p^3}\right)^5$	(r)	$\left(x+\frac{1}{x}\right)^3$

4.1.2 THE GENERAL TERM

We have already seen the relationship between Pascal's triangle and its combinatorial equivalent. From this relationship we were able to produce the general expansion for $(x+a)^n$. That is,

$$(x+a)^n = \binom{n}{0}x^n + \binom{n}{1}x^{n-1}a + \binom{n}{2}x^{n-2}a^2 + \ldots + \binom{n}{r}x^{n-r}a^r + \ldots + a^n$$

Where the **first** term, $t_1 = \binom{n}{0}x^n$

the **second** term, $t_2 = \binom{n}{1}x^{n-1}a$

the **third** term, $t_3 = \binom{n}{2}x^{n-2}a^2$

⋮

the ***r*th** term, $t_r = \binom{n}{r-1}x^{n-(r-1)}a^{r-1}$

and the **(*r* + 1)th** term, $t_{r+1} = \binom{n}{r}x^{n-r}a^r$

The $(r+1)$th term is also know as the **general term**. That is

$$t_{r+1} = \binom{n}{r}x^{n-r}a^r$$

It is common in examinations for questions to only ask for a part of an expansion. This is because the previous examples are time consuming to complete.

EXAMPLE 4.3

Find the 5th term in the expansion $\left(x+\frac{2}{x}\right)^{10}$, when expanded in descending powers of x.

Solution

The fifth term is given by t_5. Using the general term, this means that $r+1 = 5 \Leftrightarrow r = 4$. For this expansion we have that $n = 10$, therefore,

$$t_5 = \binom{10}{4}(x)^{10-4}\left(\frac{2}{x}\right)^4 = 210 \times x^6 \times \frac{16}{x^4}$$

$$= 3360x^2$$

Therefore the fifth term is $t_5 = 3360x^2$.

EXAMPLE 4.4

Find the coefficient of x^{12} in the expansion $(x^2-2)^8$.

Solution

The general term for this expansion is given by $t_{r+1} = \binom{8}{r}(x^2)^{8-r}(-2)^r$

$$= \binom{8}{r}x^{16-2r}(-2)^r$$

Seeing as we want the term involving x^{12} we equate the power of x in the general term to 12:
So, $16-2r = 12 \Leftrightarrow 2r = 4 \Leftrightarrow r = 2$.

Then, $t_3 = \binom{8}{2}x^{16-4}(-2)^2 = 2(8)\times x^{12}\times 4 = 112x^{12}$. So, the coefficient of x^{12} is in fact 112.

EXAMPLE 4.5

Find the term independent of x in the expansion $\left(2x-\frac{1}{x^2}\right)^6$.

Solution

In this case we want the term independent of x, that is, the term that involves x^0.

Again, we first find an expression for the general term, $t_{r+1} = \binom{6}{r}(2x)^{6-r}\left(-\frac{1}{x^2}\right)^r$

$$= \binom{6}{r}(2)^{6-r}(x)^{6-r}(-1)^r(x^{-2})^r$$

$$= \binom{6}{r}2^{6-r}(-1)^r x^{6-r-2r}$$

$$= \binom{6}{r}2^{6-r}(-1)^r x^{6-3r}$$

Notice how we had to separate the constants and the x term.
Next, we equate the power of x in the expansion to 0: $6-3r = 0 \Leftrightarrow r = 2$.

We therefore want $t_3 = \binom{6}{2}2^{6-2}(-1)^2x^{6-6} = 15\times 16\times 1\times x^0 = 240$.

So, the term independent of x is 240.

EXAMPLE 4.6

Write the expansion of $(1+x)^6$

i. Use the first three terms of the series to approximate 1.01^6

ii. Find the absolute, relative and percentage errors in making this approximation.

Solution

$$(1+x)^6 = 1+6x+15x^2+20x^3+15x^4+6x^5+x^6$$

i. 1.01^6 can be approximated using the expansion of $(1+x)^6$ and an x value of 0.01
$1.01^6 \approx 1+6\times 0.01+15\times 0.01^2 = 1.0615$ using the first three terms of the series.

ii. The correct answer can be found using the power key of a calculator and is

1.061520150601. It is probable that most calculators will not be able to display the 12 decimal places of the full answer. An answer such as 1.0615201506 is sufficient for the remaining part of this problem

The absolute error $= |1.0615 - 1.0615201| \approx 2.015 \times 10^{-5}$

The relative error $\approx \dfrac{2.015 \times 10^{-5}}{1.01^6} \approx 1.898 \times 10^{-5}$

The percentage error $\approx 1.898 \times 10^{-5} \approx 1.898 \times 10^{-3}\,\%$

EXERCISES 4.1.2

1. Find the terms indicated in the expansions of the following expressions:

	Expression	Term
(a)	$(x+4)^5$	x^3
(b)	$(x+y)^7$	x^5y^2
(c)	$(2x-1)^8$	x^3
(d)	$(3x-2)^5$	x^4
(e)	$(2-3p^2)^4$	p^4
(f)	$(2p-3q)^7$	p^2q^5
(g)	$\left(3p-\frac{2}{p}\right)^7$	p

2. Find the coefficients of the terms indicated in the expansions of the following expressions:

	Expression	Term
(a)	$(2x-5)^8$	x^3
(b)	$(5x-2y)^6$	x^2y^4
(c)	$(x+3)^6$	x^3
(d)	$(2p-3q)^5$	p^4q
(e)	$\left(2x-\frac{3}{p}\right)^8$	$\frac{x^2}{p^6}$
(f)	$\left(q+\frac{2}{p^3}\right)^5$	$\frac{q^3}{p^6}$

3. Use the first three terms in the expansion of $(1+x)^4$ to find an approximate value for 1.01^4. Find the percentage error in using this approximation.

4. (i) Write the expansion of $(5+2x)^6$.

(ii) Use the first three terms of the expansion to approximate 5.2^6.

(iii) Find the absolute error in this approximation.
(iv) Find the percentage error in this approximation.

5. Find the coefficient of x^{-3} in the expansion of $(x-1)^3\left(\frac{1}{x}+x\right)^6$.

6. Find the constant term in the expansion of $\left(x-\frac{1}{2x}\right)^{10}$.

7. Find the constant term in the expansion of $\left(3x-\frac{1}{6x}\right)^{12}$.

8. Find the term independent of x in the expansion of $(2-x)^3\left(\frac{1}{3x}-x\right)^6$.

9. Find the term independent of x in the expansion of $\left(2x-\frac{1}{x}\right)^6\left(\frac{1}{2x}+x\right)^6$.

10. In the expansion of $\left(x-\frac{a}{x}\right)^5\left(x+\frac{a}{x}\right)^5$, where a is a non-zero constant, the coefficient of the term in x^{-2} is '–9' times the coefficient in x^2. Find the value of the constant a.

11. If the coefficient of the x^2 in the expansion of $(1-3x)^n$ is 90, find n.

12. Three consecutive coefficients in the expansion of $(1+x)^n$ are in the ratio 6 : 14 : 21. Find the value of n.

13. Find the independent term in the following expansions

(a) $\left(y+\frac{1}{y}\right)^3\left(y-\frac{1}{y}\right)^5$ (b) $\left(2x+1-\frac{1}{2x^2}\right)^6$

14. In the expansion of $(1+ax)^n$ the first term is 1, the second term is $24x$ and the third term is $252x^2$. Find the values of a and n.

15. In the expansion of $(x+a)^3(x-b)^6$ the coefficient of x^7 is –9 and there is no x^8 term. Find a and b.

16. By considering the expansion of $(1+x)^{m+n}$, prove that:

(a) ${}^{m+n}C_1 = {}^{m}C_1 + {}^{n}C_1$

(b) ${}^{m+n}C_2 = {}^{m}C_2 + {}^{m}C_1{}^{n}C_1 + {}^{n}C_2$

4.2 PROOF

Note: This proof by induction is outside the scope of the syllabus.

A formal statement of the binomial expansion is:

$$(a+b)^k = {}^nC_0a^kb^0 + {}^nC_1a^{k-1}b^1 + \ldots + {}^nC_ra^{k-r}b^r + \ldots + {}^nC_ka^0b^k = \sum_{r=0}^{n} {}^nC_ra^{n-r}b^r$$

The binomial theorem for positive integral index may be proved using mathematical induction.

A preliminary result from combinatorics is required, namely ${}^nC_r + {}^nC_{r-1} = {}^{n+1}C_r$. We leave its proof as an exercise. We can now move on to the main induction proof:

(i) Check the case $k = 1$: $(a+b)^1 = {}^1C_0a^1 + {}^1C_0b^1 = a+b$, which is true.

(ii) Assume the theorem is true for $n = k$:

$$(a+b)^k = {}^kC_0a^kb^0 + {}^kC_1a^kb^1 + \ldots + {}^kC_ra^{k-r}b^r + \ldots + {}^kC_ka^0b^k$$

(iii) Look at the $n = k+1$. This involves multiplying each term in the expansion from (ii) first by a and then by b. To see what happens to the general term, it is a good idea to look at two consecutive terms in the middle of the expansion from (ii):

$$(a+b)^k = {}^kC_0a^kb^0 + \ldots + {}^kC_{r-1}a^{k-r+1}b^{r-1} + {}^kC_ra^{k-r}b^r + \ldots + {}^kC_ka^0b^k$$

When this expansion has been multiplied by a, the result is:

$${}^kC_0a^{k+1}b^0 + \ldots + {}^kC_{r-1}a^{k-r+2}b^{r-1} + {}^kC_ra^{k-r+1}b^r + \ldots + {}^kC_ka^1b^k$$

and when it is multiplied by b, the result is:

$${}^kC_0a^kb^1 + \ldots + {}^kC_{r-1}a^{k-r+1}b^r + {}^kC_ra^{k-r}b^{r+1} + \ldots + {}^kC_ka^0b^{k+1}$$

The expansion of $(a+b)^{k+1}$ begins with ${}^kC_0a^{k+1}b^0 = a^{k+1}b^0 = a^{k+1}$, which is correct. The expansion ends with ${}^kC_ka^0b^{k+1} = a^0b^{k+1} = b^{k+1}$, which is also correct.

It now remains to prove that the general term in the middle of the expansion is also correct. Lining up like terms from the two parts of the expansion gives:

$${}^kC_{r-1}a^{k-r+2}b^{r-1} + {}^kC_ra^{k-r+1}b^r$$

$${}^kC_{r-1}a^{k-r+1}b^{r-1} + {}^kC_ra^{k-r}b^{r+1}$$

The general term is: ${}^kC_ra^{k-r+1}b^r + {}^kC_{r-1}a^{k-r+1}b^r = [{}^kC_r + {}^kC_{r-1}](a^{k-r+1}b^r)$

$$= {}^{k+1}C_ra^{k-r+1}b^r$$

[Using the combinatorial result given at the start of this section].

(iv) We can conclude that the binomial theroem gives the correct expansion for $n = 1$ from part (i). Part (iii) indicates that the theorem gives the correct expansion for an index of 2, 3 etc. Hence the theroem holds for all positive integral indices.

5.1 RELATIONS

5.1.1 RELATIONS

Consider the relationship between the weight of five students and their ages as shown below.

We can represent this information as a **set of ordered pairs**. An age of 10 years would correspond to a weight of 31kg. An age of 16 years would correspond to a weight of 53kg and so on.

Age (years)	Weight (kg)
10	31
12	36
14	48
16	53
18	65

This type of information represents a **relation** between two sets of data. This information could then be represented as a set of ordered pairs,

$$\{(10, 31), (12, 36), (14, 48), (16, 53), (18, 65)\}$$

The **set of all first elements** of the ordered pair is called the **domain** of the relation and is referred to as the **independent variable**. The **set of all second elements** is called the **range** and is referred to as the **dependent variable**.

For the above example, the domain $= \{10, 12, 14, 16, 18\}$
and the range $= \{31, 36, 48, 53, 65\}$.

Notice that $(10, 31)$ and $(31,10)$ are not the same! This is because the ordered pair $(10, 31)$ provides the correct relation between age and weight, i.e., at age 10 years the weight of the student is 31kg. On the other hand, the ordered pair $(31,10)$ would be informing us that at age 31 years the weight of the student is 10kg!

Summary:

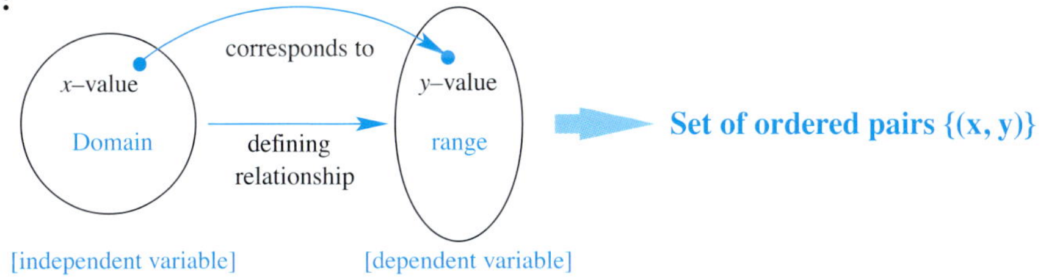

EXAMPLE 5.1 Determine the domain and range for each of the following relations:

(a) $\{(0,0), (1,1), (2,4), (3,9), (4,16), (5,25)\}$.

(b) $\{(-3,4), (-1,0), (2,-2), (-2,2)\}$.

Solution

(a) The domain is the set of all first elements, i.e., $\{0, 1, 2, 3, 4, 5\}$.
The range is the set of all second elements, i.e., $\{0, 1, 4, 9, 16, 25\}$.

(b) The domain is the set of all first elements, i.e., $\{-3, -1, 2, -2\}$.
The range is the set of all second elements, i.e., $\{4, 0, -2, 2\}$

The letter 'X' is often used to denote the domain and the letter 'Y' to denote the range. For part (a) this means that we could write X = {0, 1, 2, 3, 4, 5} and Y = {0, 1, 4, 9, 16, 25} and for (b) we could write X = {–3, –1, 2, –2} and Y = {4, 0, –2, 2}.
This is a convention, nothing more.

Rather than giving a verbal description of how the independent variable and the dependent variable are related, it is much clearer to provide a **mathematical rule** that shows how the elements in the range relate to the elements in the domain.

EXAMPLE 5.2

A relation is defined by the rule $y = x + 2$, where $x \in \{0, 1, 2, 3, 4\}$.

(a) Determine the range of this relation.
(b) Express this relation as a set of ordered pairs.

Solution

(a) The domain of this relation is given by the x–values, i.e., {0, 1, 2, 3, 4}. We can therefore substitute these values into the equation $y = x + 2$ and determine their corresponding y–values. This will provide the range of the relation.

Substituting we have, $x = 0 \Rightarrow y = 0 + 2 = 2$

$x = 1 \Rightarrow y = 1 + 2 = 3$

$x = 2 \Rightarrow y = 2 + 2 = 4$, and so on.

This produces a set of y–values {2, 3, 4, 5, 6} that defines the range.

(b) The set of ordered pairs would be {(0, 2), (1, 3), (2, 4), (3, 5), (4, 6)}.

Notice that we can describe the set of ordered pairs more formally as:

$$\{(x,y): y = x + 2, x \in \{0, 1, 2, 3, 4\}\}.$$

which is read as;

"The set of ordered pairs x and y, such that $y = x + 2$, where x is an element of the set of values {0, 1, 2, 3, 4}."

The information in Example 5.2 can be displayed in different ways. Both those shown below are visual displays – they show the mappings in different ways:

Mapping diagram

The mapping diagram below displays which y–value corresponds to a given x–value.

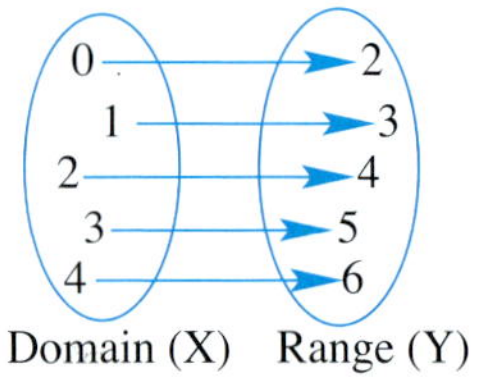

However it is often not easy to see the 'pattern' between the variables with this style of diagram.

Cartesian Plane

The Cartesian plane is made up of a horizontal axis (independent variable, X) and a vertical axis (dependent variable,Y).

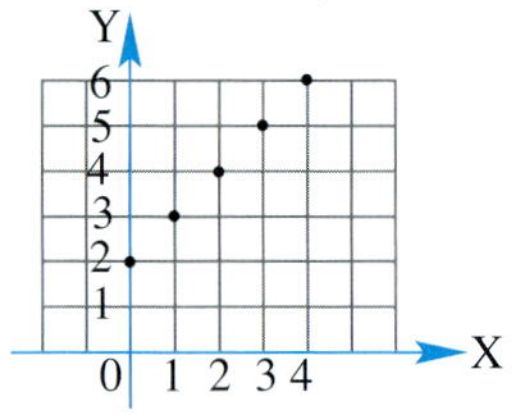

We plot the points on the grid, so that (3, 5) is 3 units to the right and 5 units up.

Notice that in the mapping diagram that uses the Cartesian plane, we have not joined the points together in a straight line. This is because the domain specifies that the only values of x that can be used must be from the set $\{0, 1, 2, 3, 4\}$, and so a value such as $x = 2.4$ cannot be used.

Both these visual representations are useful in displaying which values in the domain generate a given value in the range. However, the Cartesian plane more readily gives a quick overview of what the underlying relationship between the two variables is. It is very easy (and quick) to see that as the x–values increase, so too do the y–values. We can do this by simply looking at the points on the graph and observing the 'trend' without really concerning ourselves with what the actual values are.

We now provide a formal definition of the Cartesian plane and a relation.

5.1.2 THE CARTESIAN PLANE

The **Cartesian plane** is formed by constructing two real lines that intersect at a right-angle where the point of intersection of these two lines becomes the **origin**. The horizontal real line is usually referred to as the x–axis and the vertical real line is usually called the y-axis. This also implies that the plane has been divided into four **quadrants**. Each point on this plane is represented by an ordered pair (x, y) where x and y are real numbers and are the **coordinates** of the point.

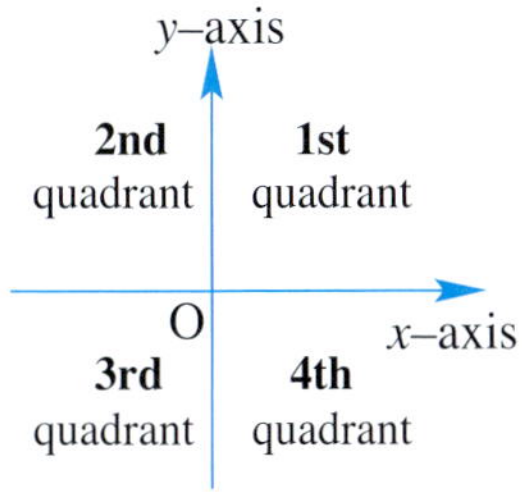

The set of all ordered pairs (x, y), where $x \in X$ and $y \in Y$ can also be defined by making use of the Cartesian product, $X \times Y = \{(x, y): x \in X, y \in Y\}$.

Relation

A **relation** is any subset of the Cartesian plane and can be represented by a set of ordered pairs $\{(x, y)\} \subseteq \mathbb{R}\times\mathbb{R}$, where the Cartesian product $\mathbb{R}\times\mathbb{R}$ ($= \mathbb{R}^2$) represents the region covered by the whole of the Cartesian plane.

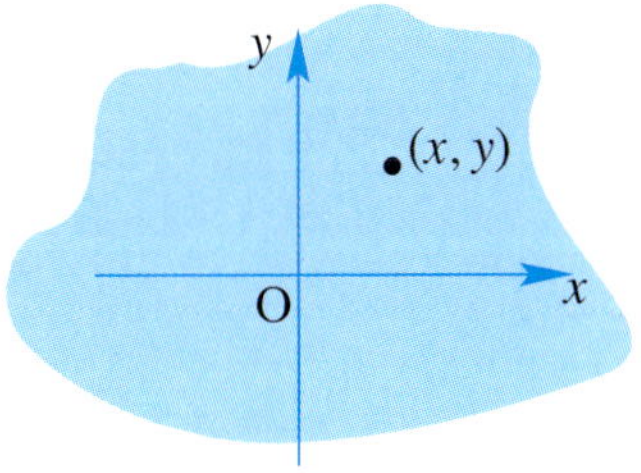

We now consider some further examples.

EXAMPLE 5.3 Determine the domain and range of the following relations

(a) $y = 2 + x, x \geq 0$ (b) $y = 4 - x, -1 \leq x \leq 2$ (c) $y = 1 - x^2, x \geq 0$

Solution

Note: When finding the range of a relation, it is always a good idea to sketch its graph.

(a) The equation $y = 2 + x, x \geq 0$ represents a straight line with the restriction that $x \geq 0$. So in this case, the domain is $[0, \infty)$ (or $[0, \infty[$).

From the graph, the range is given by $[2, \infty)$ (or $[2, \infty[$).

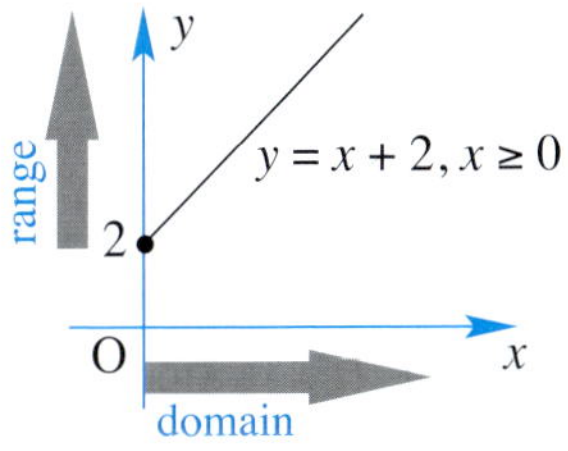

(b) The equation $y = 4 - x, -1 \le x \le 2$ represents a straight line with the restriction that $-1 \le x \le 2$. So in this case, the domain is $[-1, 2]$.
when $x = -1, y = 5$ and when $x = 2, y = 2$
From the graph, the range is given by $[2,5]$.

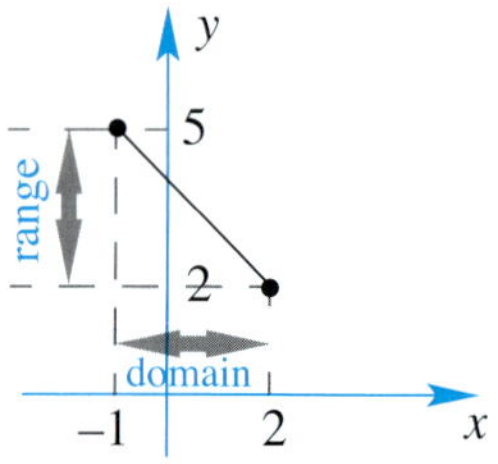

(c) For this relation the domain is specified as $\{x : x \ge 0\}$ or simply $[0, \infty)$. So we can only sketch the graph of $y = 1 - x^2$, for these values of x. Using the graph we can see that the range is $\{y : y \le 1\}$ or simply $(-\infty, 1]$.

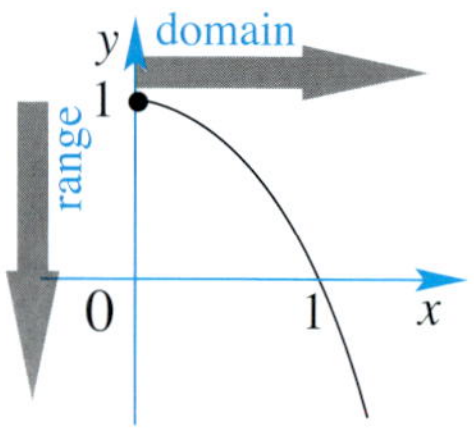

5.1.3 IMPLIED DOMAIN

So far we have looked at examples for which a domain has been specified. For example, if asked to find the range of the relation $y = 1 + x^2, x \ge 3$. Then, after sketching its graph, we would determine its range to be $[10, \infty)$. However, what if we only wanted to know the range of the relation $y = 1 + x^2$? In this case, because we have not been provided with any restriction on the x–values, we will need to assume that we can **use the largest possible set of x–values for which the relation is defined** – this domain is known as the **implied domain** (or **maximal domain**) – in this case that would be the real number set, $\mathbb{R}$. Then, after sketching the graph of $y = 1 + x^2$ for all real values of x we would have a range defined by $[1, \infty)$.

EXAMPLE 5.4

Determine the domain and range of the following relations

(a) $y = \sqrt{x-3}$ (b) $y = \dfrac{2}{\sqrt{x-3}}$ (c) $y = \dfrac{3}{2-x}$

Solution

(a) Using the TI–83 to sketch the graph of $y = \sqrt{x-3}$ (i.e., the square root relation) we observe that its domain is $[3, \infty)$.

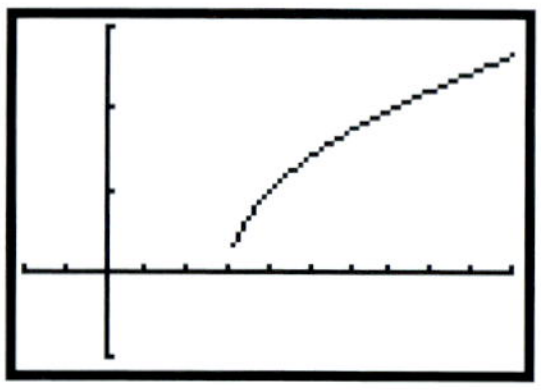

Now, let's take a closer look at why that is the case.
Because we are dealing with an expression that involves a square root, then, the term 'inside' the square root must be greater than or equal to zero (as we cannot take the square root of a negative number).

So, we must have that $x - 3 \ge 0 \Leftrightarrow x \ge 3$. Therefore, the implied domain is $\{x : x \ge 3\}$.
From the graph, the range can be seen to be $[0, \infty)$.

It should be noted that the TI–83 uses the implied domain when graphing. Also realise that from the sketch, we could be misled into thinking that there is a 'gap' at the point $(3, 0)$. Be careful with this – use the graphics calculator as an aid, then, double check to make sure.

(b) The equation $y = \dfrac{2}{\sqrt{x-3}}$ represents the reciprocal of a square root relation. As in (a), we must have that $x - 3 \geq 0 \Leftrightarrow x \geq 3$.

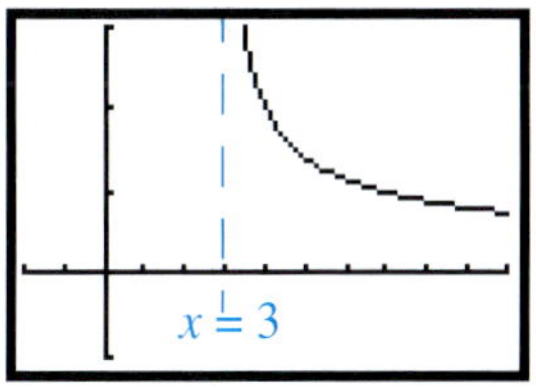

However, this time we have another restriction – we cannot divide by zero and so we cannot include $x = 3$ in our domain. So , at $x = 3$, we draw an **asymptote**.

We then have $x - 3 > 0 \Leftrightarrow x > 3$. This leads to a range of $(0, \infty)$ (or $]0, \infty[$).

(c) The only restriction that can be readily seen for the relation $y = \dfrac{3}{2-x}$ is that we cannot divide by zero and so, we must have that $2 - x \neq 0$. That is, $x \neq 2$.

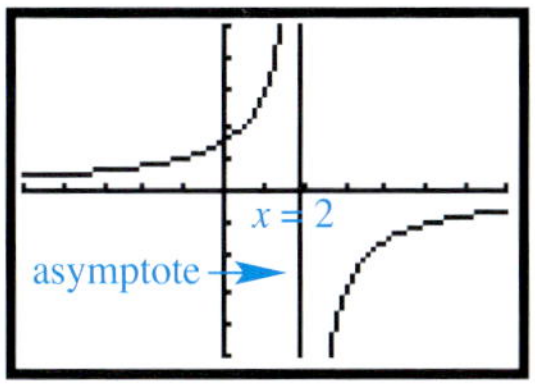

As it is a reciprocal relation, we have an asymptote at $x = 2$. So, the domain is given by $]-\infty, 2[\ \cup\ ]2, \infty[$ or simply, $\mathbb{R}\backslash\{2\}$.
The range can then be seen to be $\mathbb{R}\backslash\{0\}$

5.1.4 TYPES OF RELATIONS

Relations fall into one of four categories. These are:

1. One to one relations [one x to one y]

For any one value of x, there will be only one corresponding value of y.

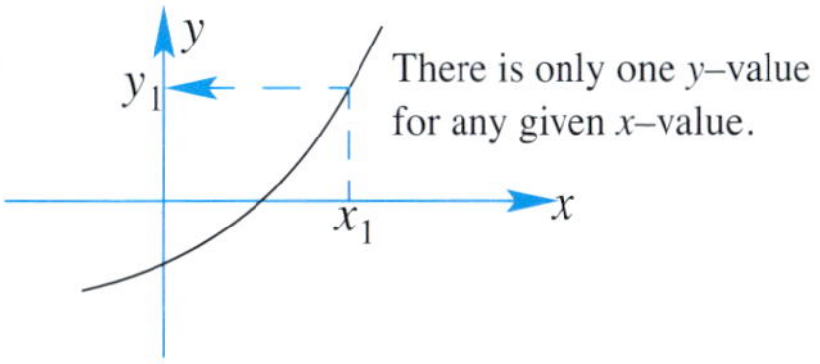

2. One to many relation [one x to many y]

There is at least one value of x for which there exists more than one corresponding value of y.

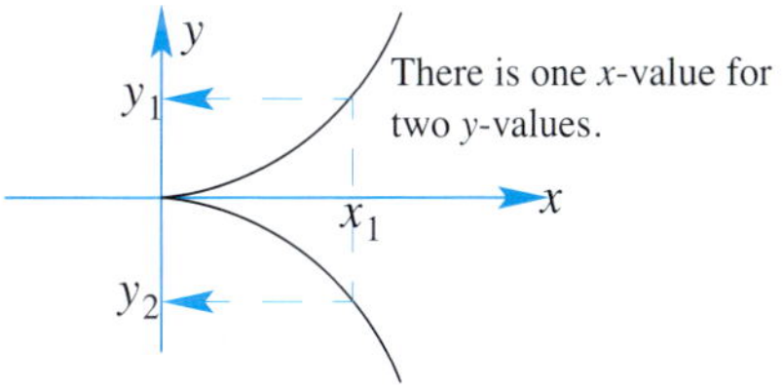

3. Many to one relation [many x to one y]

There are at least two different values of x that will correspond to only one value of y.

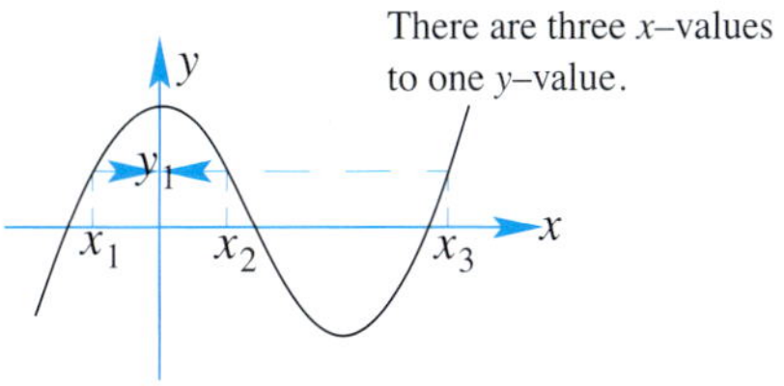

4. Many to many [many x to many y]

There are at least two different values of x that will correspond to at least two different values of y.

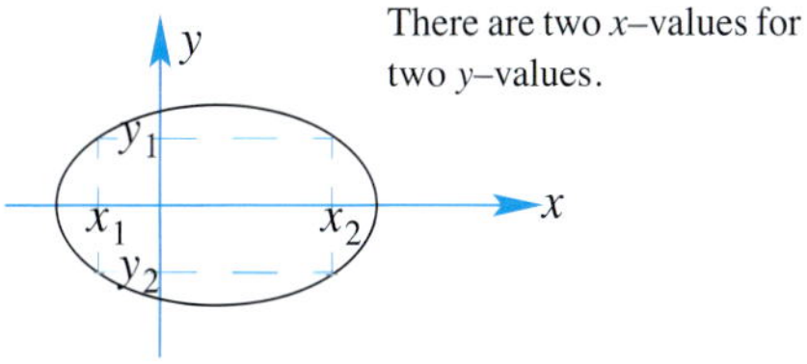

5.1.5 SKETCHING WITH THE GRAPHICS CALCULATOR

Why do graphs sometimes appear distorted on the graphics calculator display screen?

On most graphics calculators, the display screen is two thirds as high as it is wide. Because of this, when we graph a relation that describes a circle we obtain a diagram that is not a true geometric representation. To obtain a true geometric representation we need to use a window that produces a square setting. This is done by using the fact that $\frac{Y_{max} - Y_{min}}{X_{max} - X_{min}} = \frac{2}{3}$.

For example, the window alongside shows such a setting:

i.e., $\frac{Y_{max} - Y_{min}}{X_{max} - X_{min}} = \frac{4-(-4)}{6-(-6)} = \frac{8}{12} = \frac{2}{3}$

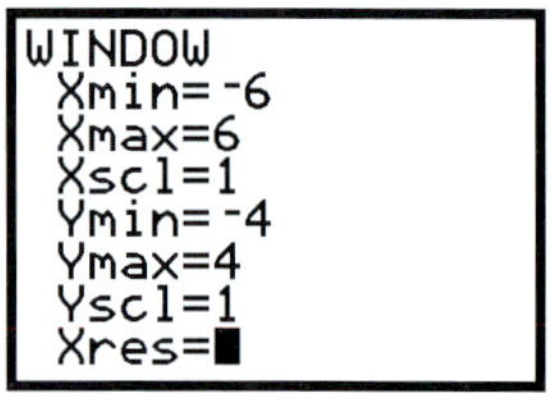

These settings enable us to obtain a true geometric representation of the circle with equation $x^2 + y^2 = 16$.

Re–arranging the equation to make y the subject, we have, $x^2 + y^2 = 16 \Rightarrow y = \pm\sqrt{16 - x^2}$.

We can now graph the equations $y = \sqrt{16 - x^2}$ and $y = -\sqrt{16 - x^2}$.

We enter both the positive and the negative equations [we do this using the **VARS** option]. Then we use the above window settings. Finally press the **GRAPH** command.

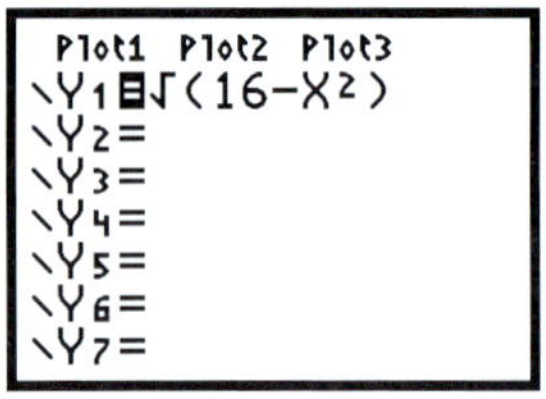

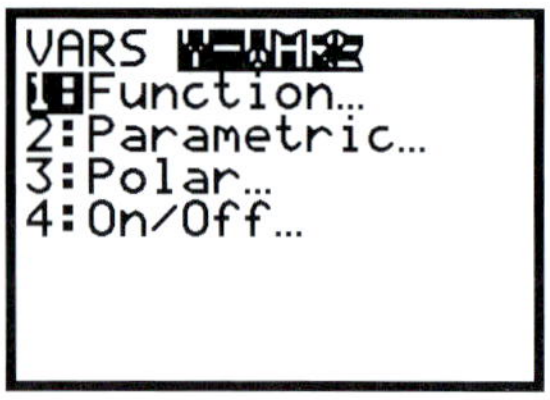

The final output is:

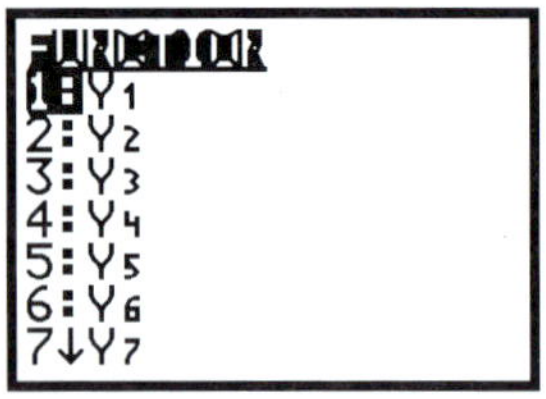

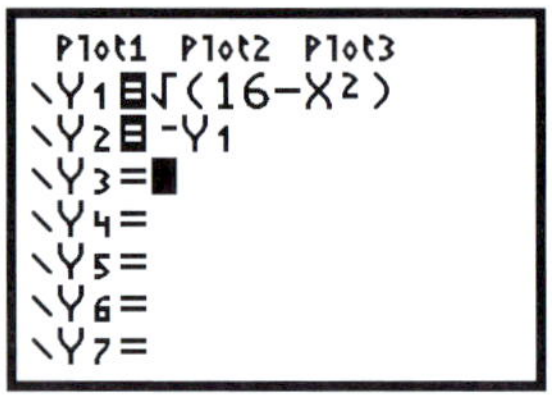

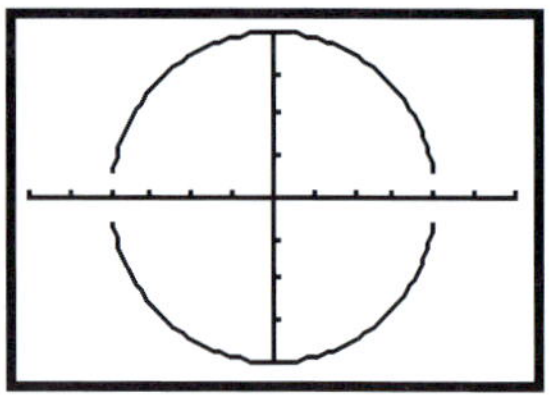

Using the window setting shown below, we obtain a distorted graph, as shown below:

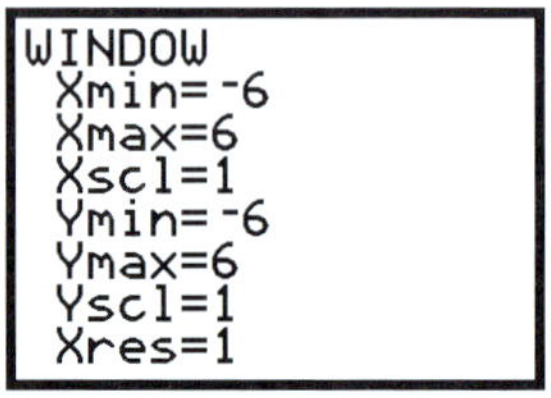

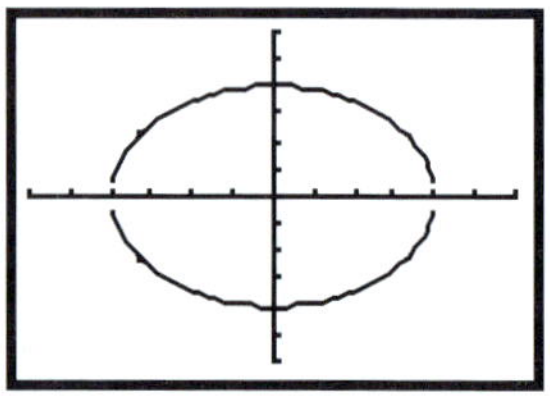

NB: You can also use the **ZOOM** options **4:ZDecimal** or **5:ZSquare** to generate correct graphs.

EXAMPLE 5.5

Determine the domain and range of the following relations.

(a) $x^2 + y^2 = 4$.

(b) $y > 2 - x, x < 0$.

(c) $y = \frac{2}{x^2 + 1}, -1 \le x < 2$

Solution

(a) The relation $x^2 + y^2 = 4$ represents a circle of radius 2 units with its centre at the origin.

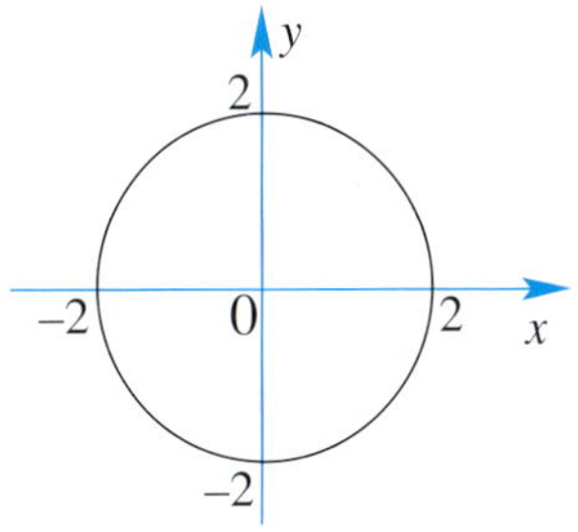

Note that we can only use values of x between –2 and 2.
For example, if we have $x = 3$, then we must have
$9 + y^2 = 4 \Rightarrow y^2 = -5$ for which there are no real solutions.
Therefore, the domain is $-2 \le x \le 2$ (or $[-2, 2]$).
The range is $-2 \le y \le 2$ (or $[-2, 2]$).

(b) This time we have a subset of the cartesian plane. That is, we are defining a region in the plane as opposed to a set of points lying on a curve.

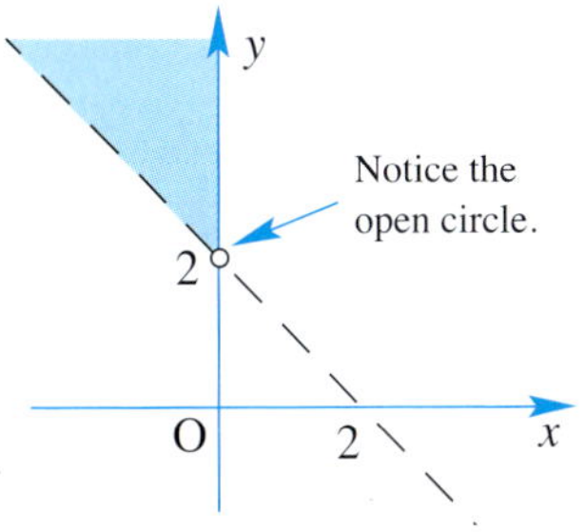

The domain has already been set as $\{x : x < 0\}$ (or $]-\infty, 0[$).
From our sketch, we see that the range is $\{y : y > 2\}$ (or $]2, \infty[$).

(c) This time we make use of the TI–83 to sketch the graph:

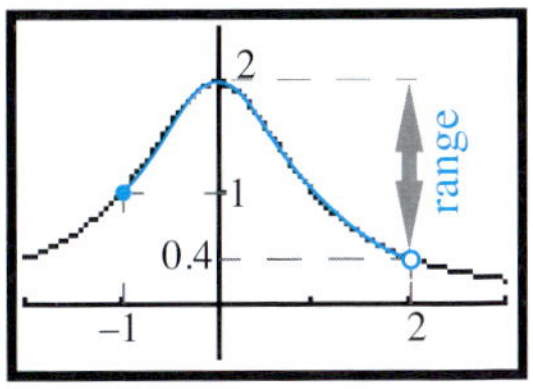

The domain has already been specified as $[-1, 2[$, meaning that we include the value $x = -1$ (which generates a y–value of 1) but exclude the value $x = 2$ (which would have produced a y–value of $\frac{2}{5} = 0.4$).

From the sketch, the range is therefore $]0.4, 2]$.

NB: Had we been given the relation $y = \frac{2}{x^2 + 1}$ without a specified domain, then its implied domain would be $]-\infty, \infty[$. This time, unlike Example 5.4 (b), as $x^2 + 1$ will always be greater than or equal to one, there is no danger of ever dividing by zero (for any value of x). This would have then produced a range $]0, 2]$.

One of the key steps in determining properties of a relation is the ability to sketch its graph. We have already dealt with polynomials, however, there are many other relations that you will need to be able to sketch. The fact that you have access to a graphics calculator will make it easier for you to sketch graphs, however, it is still important that you have a 'feel' for the shape of some standard relations. We will deal with a number of relations in detail throughout this book, so, at this stage we will provide a 'bank' of the graphs for some standard relations.

Relation	Properties
$y = mx + c$ 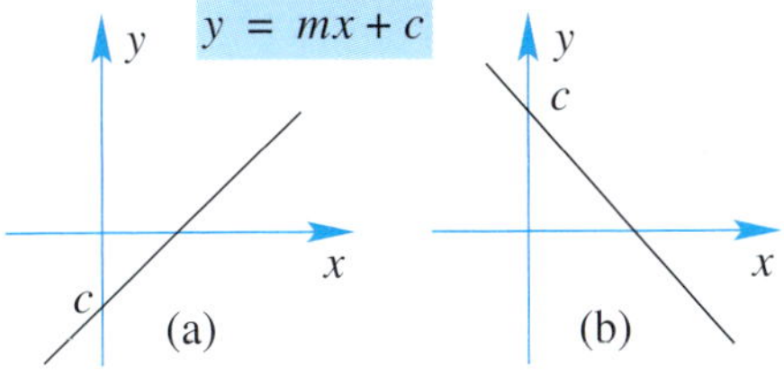	1. Linear relation [straight line] 2. Case (a): $m > 0$, y–intercept at (0, c) increasing Case (b): $m < 0$ y–intercept at (0, c) decreasing

Relation	Properties
$y = ax^2 + bx + c$ 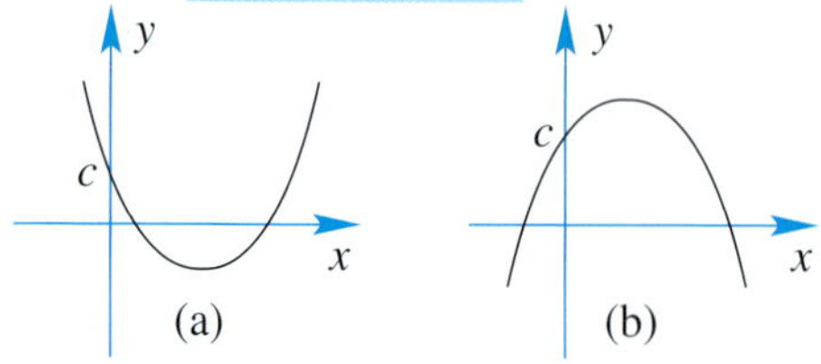 	**1.** Quadratic relation [parabola] **2. Case (a):** $a > 0$, y–intercept at $(0, c)$ axis of symmetry at $x = -\frac{b}{2a}$. **Case (b):** $a < 0$ y–intercept at $(0, c)$ axis of symmetry at $x = -\frac{b}{2a}$.
$y = ax^3 + bx^2 + cx + d$ 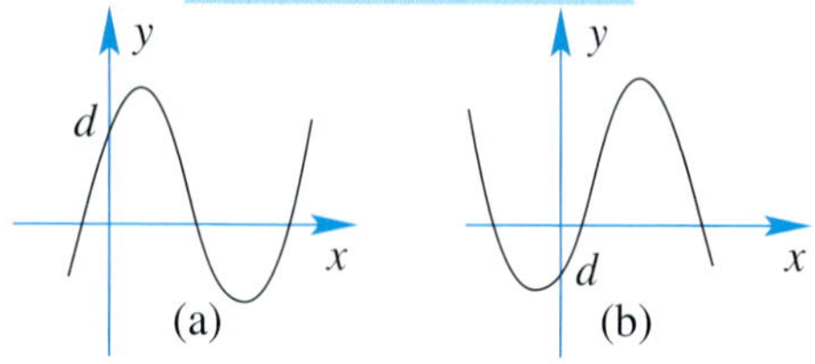 	**1.** Polynomial relation (of order 3) [cubic] **2. Case (a):** $a > 0$, y–intercept at $(0, d)$ **Case (b):** $a < 0$ y–intercept at $(0, d)$
$y = a \times b^{kx} + c$ 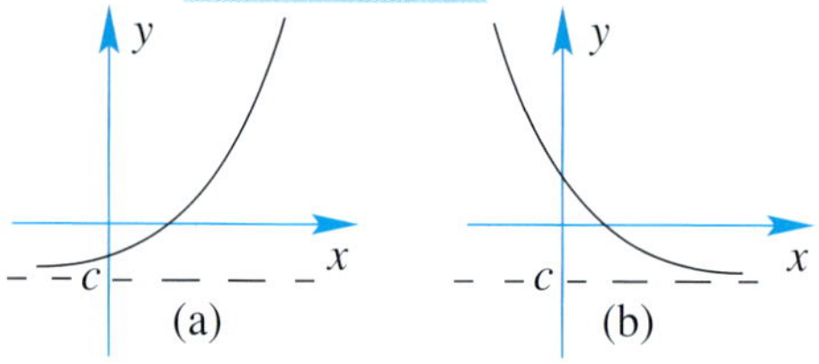 	**1.** Exponential relation [exponential] **2. Case (a):** $a > 0, k > 0$ *asymptote* at $y = c$ increasing **Case (b):** $a < 0$ *asymptote* at $y = c$ decreasing NB: If $a < 0$, then graphs of $y = a \times b^{kx}$ is first reflected about the x – axis.
$y = \frac{a}{x-b} + c$ 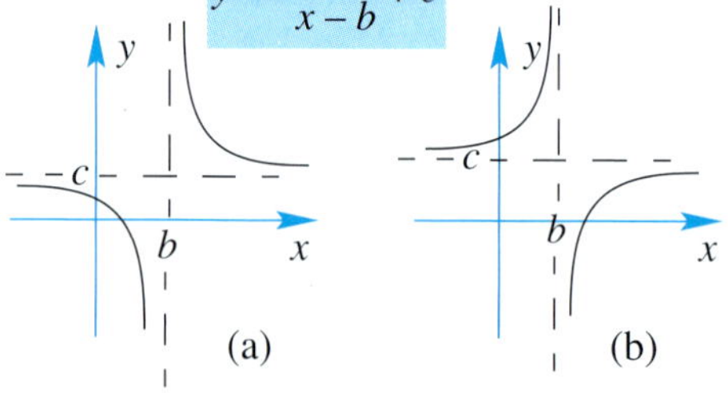 	**1.** Reciprocal relation [rectangular hyperbola] **2. Case (a):** $a > 0$, *asymptote* at $y = c$ and $x = b$ **Case (b):** $a > 0, k < 0$ *asymptote* at $y = c$ and $x = b$
$y = a\sqrt{x} + b$ 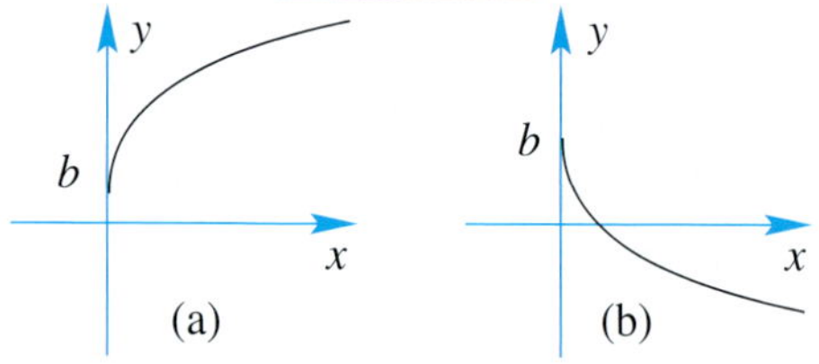 	**1.** Square root relation [sideways parabola] **2. Case (a):** $a > 0$, y–intercept at $(0, b)$ increasing **Case (b):** $a < 0$ y–intercept at $(0, b)$ decreasing NB: $y^2 = x$ is a sideways parabola, $y = \sqrt{x}$ is only 'half' of a sideways parabola.

Relation	Properties
$y = a\log_b(x-c)$ 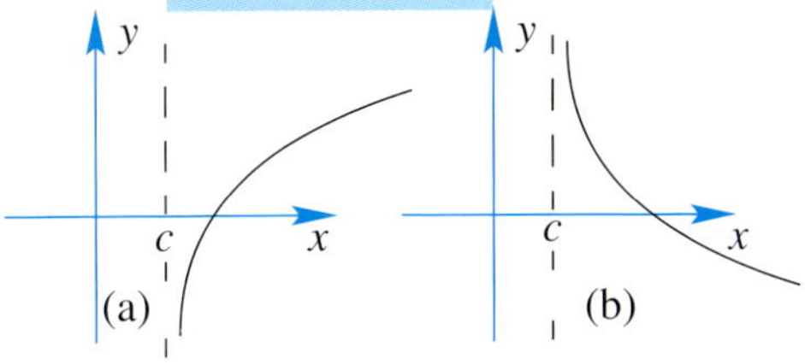	**1.** Logarithmic relation [log curve] **2. Case (a):** $a > 0, b > 1$ asymptote at $x = c$ increasing **Case (b):** $a < 0, b > 1$ asymptote at $x = c$ decreasing
$y = \dfrac{a}{(x-b)^2} + c$ 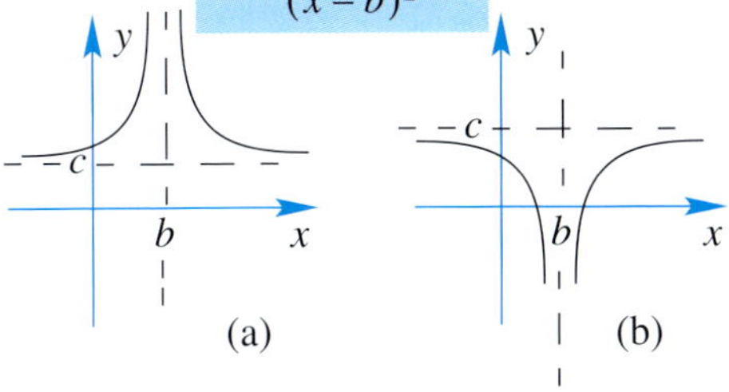	**1.** Reciprocal squared relation [truncus] **2. Case (a):** $a > 0$, *asymptote* at $y = c$ and $x = b$ **Case (b):** $a < 0$ *asymptote* at $y = c$ and $x = b$
$(x-a)^2 + (y-b)^2 = r^2$ 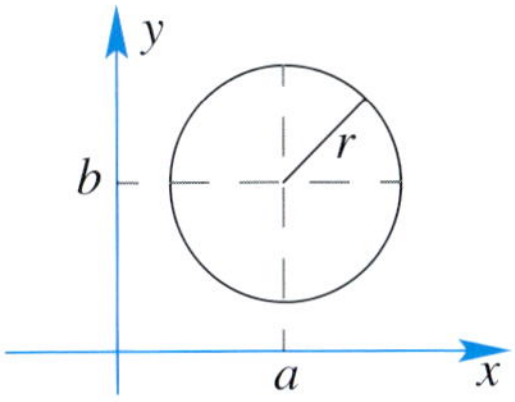	**1.** Circular relation [circle] **2.** Radius r **3.** Centre at (a, b) NB: The equation $(x-a)^2 + (y-b)^2 = r^2$ can be rewritten as $y = \pm\sqrt{r^2 - (x-a)^2} + b$ with the positive root representing the top half and the negative root the bottom half.

Remember, this 'bank' of relations is a mere sample of possible relations and are not necessarily given in their most general form.

EXERCISES 5.1

1. State the domain and range of the following relations.

(a) $\{(2,4), (3,-9), (-2,4), (3,9)\}$

(b) $\{(1,2), (2,3), (3,4), (5,6), (7,8), (9,10)\}$

(c) $\{(0,1), (0, 2), (1,1), (1,2)\}$

2. Find the range for each of the following.

(a) $\{(x,y): y = x + 1, x \in \mathbb{R}^+\}$ (b) $\{(x,y): y \geq x, x \geq 0\}$

(c) $y = x^2 + 2x + 1, x > 2$ (d) $y = 2x - x^2, x \in \mathbb{R}$

(e) $x^2 + y^2 = 9, -3 \leq x \leq 3$ (f) $x^2 - y^2 = 9, x \geq 3$

(g) $y = x - 1, 0 < x \leq 1$ (h) $y = 4 - x^2, -2 \leq x < 1$

(i) $y = \sqrt{x}, x \geq 0$ (j) $y = \sqrt{x}, 1 \leq x \leq 25$

(k) $y = \dfrac{4}{x+1}, x > 0$ (l) $\{(x,y): y^2 = x, x \geq 1\}$

3. State the range and domain for each of the following relations.

(a)

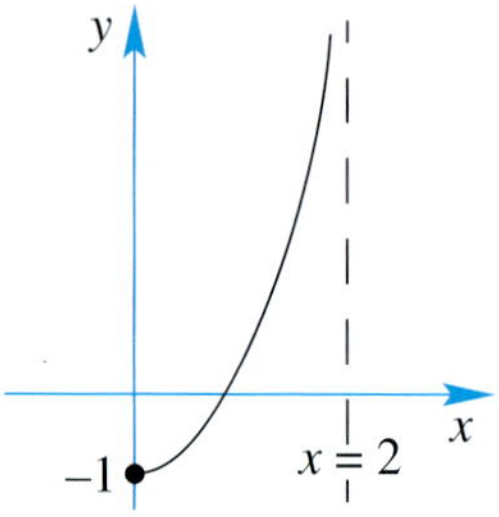

(b)

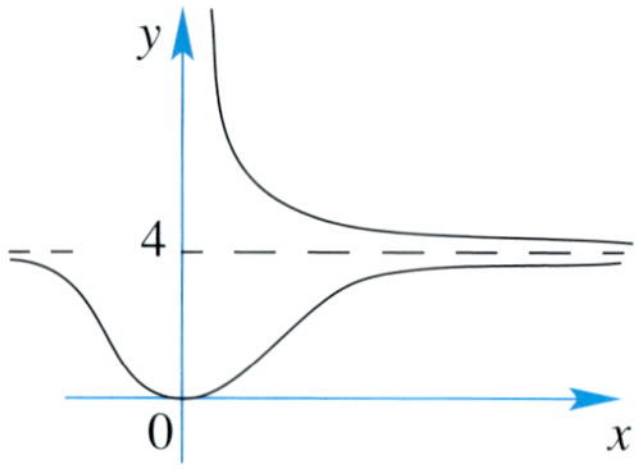

(c)

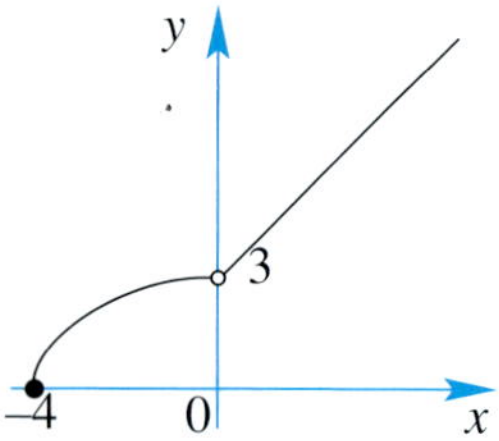

(d)

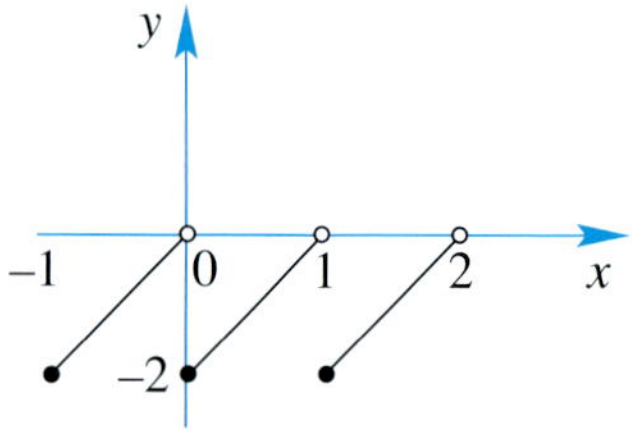

(e)

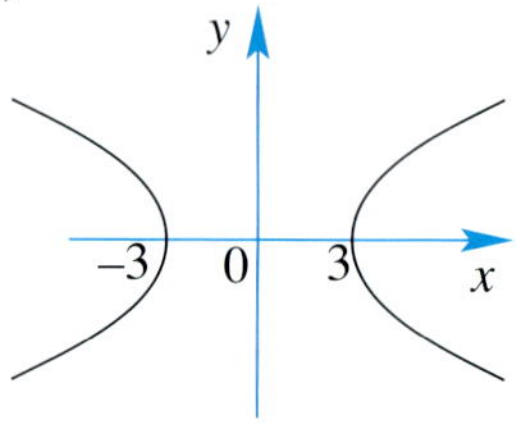

(f)

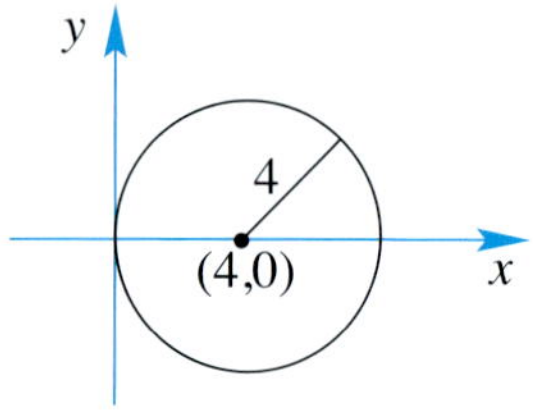

4. What types of relation are the following?

(a)

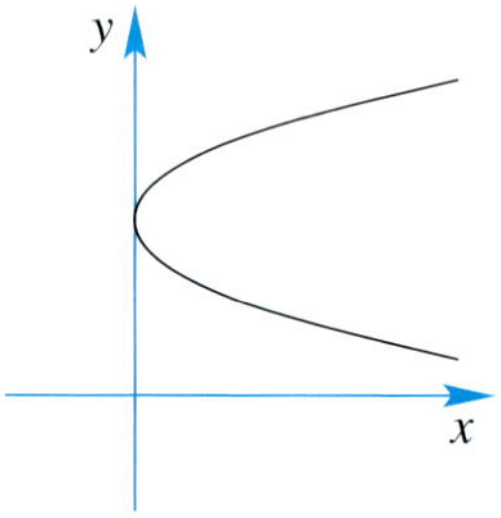

(b)

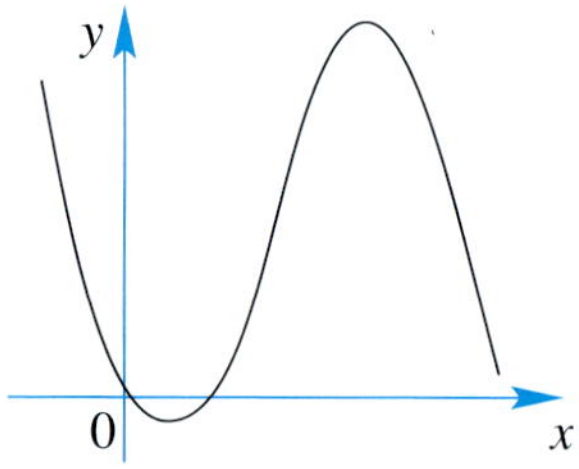

(c)

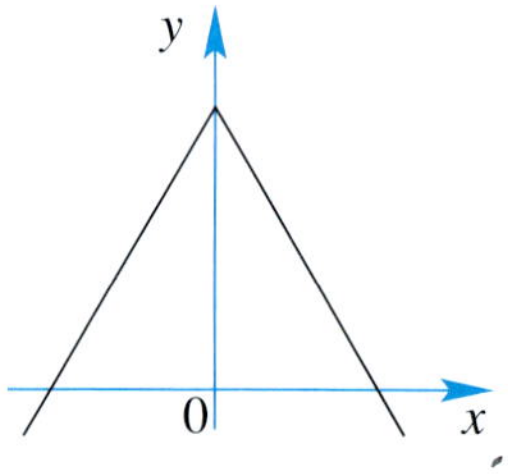

(d)

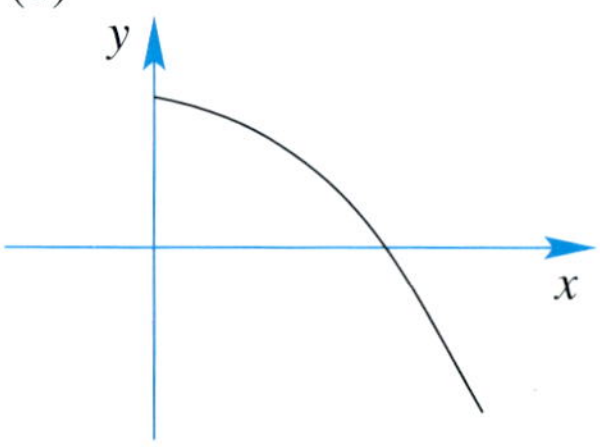

(e)

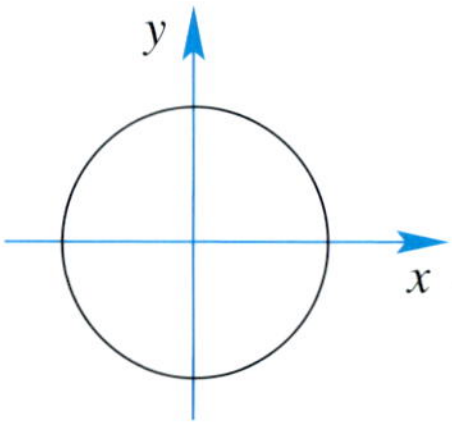

(f)

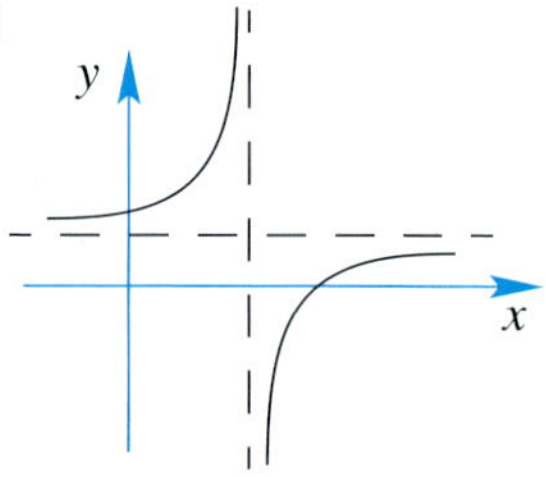

5. Determine the implied domain for each of the following relations.

(a) $y = \dfrac{2x}{x+2}$ (b) $y = \dfrac{3}{\sqrt{9-x}}$ (c) $y = \sqrt{16-x^2}$

(d) $y = \sqrt{x^2-4}$ (e) $xy - x = 3$ (f) $y = \dfrac{2}{x^2+1}$

(g) $y = \dfrac{2}{x^3+1}$ (h) $y = \sqrt{x+a}, a > 0$ (i) $y = \dfrac{a}{\sqrt{x}-a}, a > 0$

(j) $x^2 - y^2 = a^2$ (k) $y^2 - x^2 = a^2$ (l) $axy + y - x = a, a > 0$

6. Find the range of the following relations.

(a) $y = x - a, x < 0, a > 0$

(b) $y = \dfrac{ab}{x+1}, x \geq 0, ab > 0$

(c) $y = a^2x - ax^2, x \geq \dfrac{1}{2}a, a > 0$

(d) $y = a^2x - ax^2, x \geq \dfrac{1}{2}a, a < 0$

(e) $y = \dfrac{a}{x} + a, a > 0$

(f) $y = a - \dfrac{a}{x^2}, a > 0$

(g) $y = 2\sqrt{x-a} - a, a > 0$

(h) $y = \dfrac{2a}{\sqrt{a^2 - x}}, a < 0$

5.2 FUNCTIONS

5.2.1 DEFINITIONS

There is a special group of relations which are known as **functions**. This means that every set of ordered pairs is a relation, but **every relation is not a function**. Functions then make up a subset of all relations.

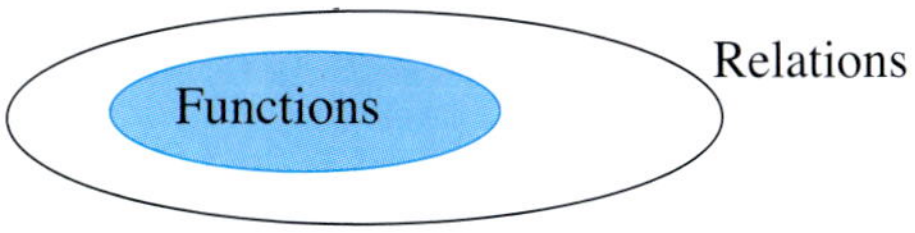

A function is defined as a relation that is either one to one or many to one. That is a function is a relation for which no ordered pairs have the same first element.

There are two ways to determine if a relation is a function.

Method 1: Algebraic approach

For Method One we use the given equation and determine the number of y–values that can be generated from one x–value.

EXAMPLE 5.6

Determine which (if any) of the following are functions.

(a) $y^3 - x = 2$ (c) $y^2 + x = 2$

Solution

(a) From $y^3 - x = 2$ we have that $y = \sqrt[3]{2+x}$, then for any given value of x, say $x = a$, we have that $y = \sqrt[3]{2+a}$ which will only ever produce one unique y–value. Therefore, the relation $y^3 - x = 2$ is a function. In fact, it is a one to one function.

(b) From $y^2 + x = 2$ we have that $y^2 = 2 - x \Leftrightarrow y = \pm\sqrt{2-x}$. Then, for any given value of x, say $x = a$ (where $a \leq 2$), we have that $y = \pm\sqrt{2-a}$, meaning that we have two different y–values; $y_1 = \sqrt{2-a}$ and $y_2 = -\sqrt{2-a}$, for the same x–value. Therefore, this relation is not a function. In fact, it is a one–to–many relation.

Method 2: Vertical line Test

The method is quite simple:

Step 1 Sketch the graph of the relation.

Step 2 Make a visual check of the number of times a vertical line would cut the graph.

Step 3 If the vertical line only ever cuts at one place for every value in the domain the relation is a function.

Function

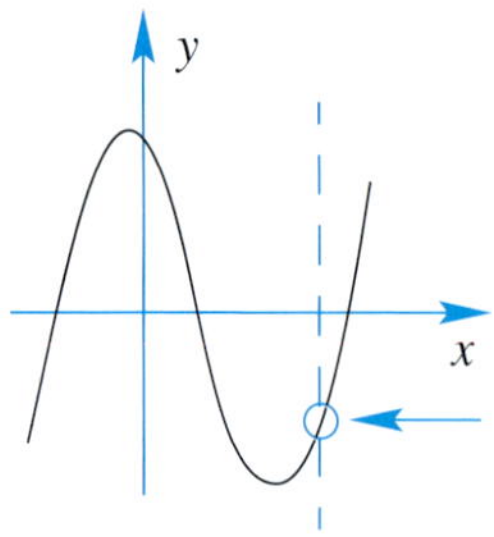

Not a function

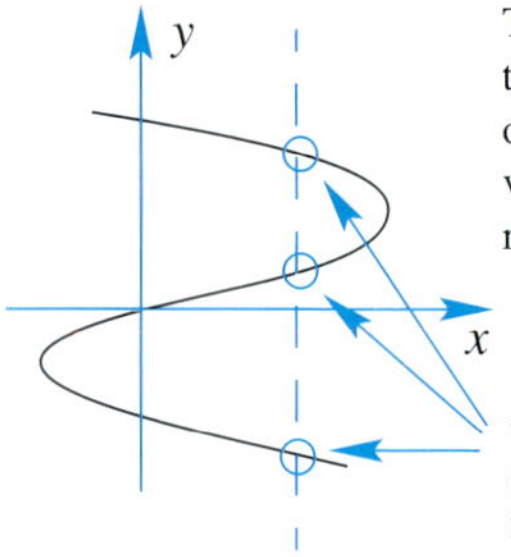

EXAMPLE 5.7

Which of the following defines a function?

(a) $\{(0,2), (1,2), (2,1)\}$ (b) $\{(x, y): y = x^3 + 1, x \in \mathbb{R}\}$

(c) $y^2 = x, x \geq 0$ (d) $\{(x,y): x^2 + y^2 = 16\}$

Solution

(a) Clearly, we have every first element of the ordered pairs different. This means that this relation is also a function.

(b) Using the TI–83 to provide a visual check:
From the graph shown, a vertical line drawn anywhere on the domain for which the relation is defined, will cut the graph at only one place.
This relation is therefore a function.

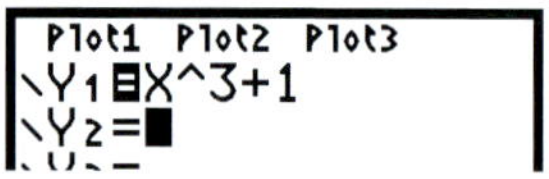

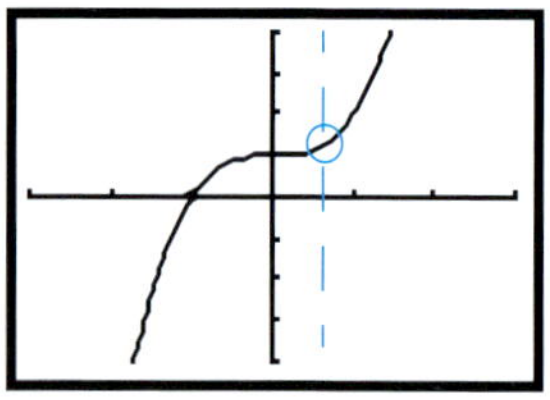

(c) Again we make use of a visual approach to determine if the relation is a function.
First we write the relation in a form that will enable us to enter it into the TI–83: $y^2 = x \Rightarrow y = \pm\sqrt{x}$
We can therefore define the relation $Y_1 = \sqrt{X}$ and $Y_2 = -\sqrt{X}$ and sketch both on the same set of axes.
Placing a vertical line over sections of the domain shows that the line cuts the graph in two places (except at the origin). Therefore this relation is not a function.

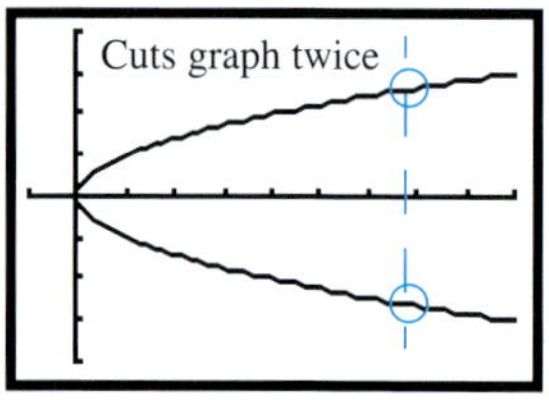

Algebraic proof

We can also determine if a relation is a function by using algebraic means.

Begin by choosing a value of x that lies in the domain. For example $x = 4$.

This gives the following equation: $y^2 = 4 \Rightarrow y = \pm\sqrt{4}$

From which we can say that when $x = 4$, $y = 2$ **and** $y = -2$, so that there are two ordered pairs, (4,2) and (4, –2). As we have two different y-values for one x-value this relation is not a function.

(d) This relation describes the equation of a circle with radius 4 units and centre at the origin. The graph of this relation is shown alongside. The graph fails the vertical line test, and so is not a function.

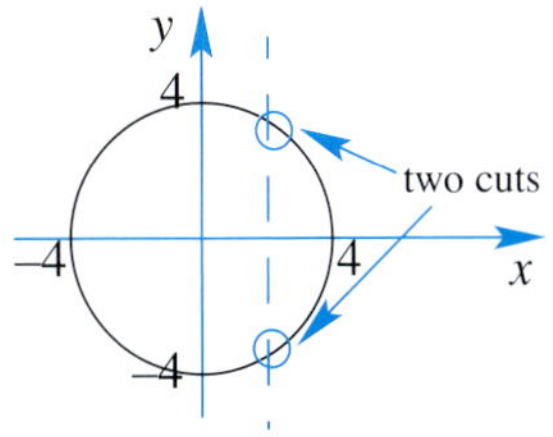

We now provide a formal definition together with commonly used notation for a funcion.

> A function f, (or a mapping f), from a set X to a set Y is a relation that assigns to each element x in the set X a unique element y in the set Y.

The set X is called the **domain** of f and the set Y the **co-domain**. The element y is called the **image** of x under f and we denote this image by $f(x)$, the value of the function f at x (read as; f of x).

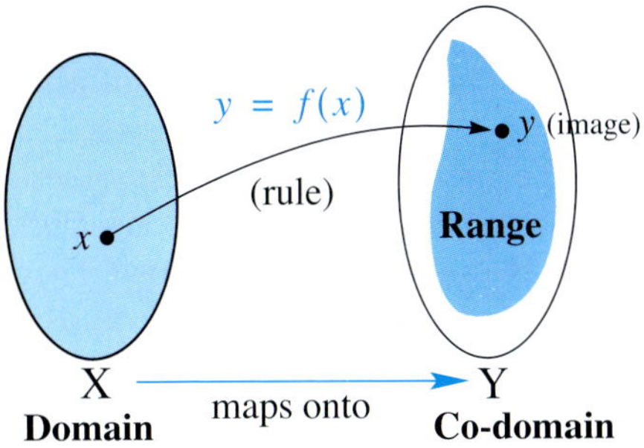

We write this mapping as: $f : x \mapsto f(x)$

We can also write this mapping as follows

1. $f : X \mapsto Y$, where $y = f(x)$
2. $f : X \mapsto Y$, $y = f(x)$
3. $y = f(x)$, $x \in X$

It is important to realise that the range of f is not necessarily the set Y. The **range** of f is actually a subset of Y (sometimes it could also be equal toY). Set Y, i.e., the co-domain, describes the **types** of numbers that will be produced when f is applied to different x-values — not necessarily which numbers will result! The range of f is given by the values of $f(x)$.

Translating the mathematical notation into English we have the following:

> $f : X \mapsto Y$, where $f(x)$ = rule in terms of x
>
> "f is such that the set X maps onto the set Y where f of x is equal to . . ."

Notice that f describes not only the rule, $f(x)$, but also the domain, X.

EXAMPLE 5.8

For the function $f(x) = x^3 + 1, x \in \mathbb{R}$ find

(a) $f(-1)$, $f(2)$. (b) the element of the domain that has an image of 28.

Solution

(a) $f(x) = x^3 + 1, x \in \mathbb{R} \Rightarrow f(-1) = (-1)^3 + 1 = -1 + 1 = 0$.

Similarly, $f(2) = 2^3 + 1 = 9$.

(b) If the image is 28, then we want the value of x for which $f(x) = 28$.

We then have, $f(x) = 28 \Leftrightarrow x^3 + 1 = 28$

$\Leftrightarrow x^3 = 27$ [Taking the cube root of both sides]

$\Leftrightarrow x = 3$

Therefore, the element of the domain that has an image of 28 is 3.

EXAMPLE 5.9

Determine the range of the function $f:\{x:x \geq 0\} \mapsto \mathbb{R}$, $f(x) = x + 1$.

Solution

First note that the co-domain is given by $\mathbb{R}$, the set of real numbers – meaning that all image values will be real. To determine the actual range of this function we sketch its graph.

From the graph, the only possible values of y are those for which $y \geq 1$. In this case, because $x = 0$ is included in the domain, we also include the value y = 1 in the range. Therefore, we have a closed circle at the end point.

The range of f is then given by $\{y:y \geq 1\}$ (or $[1, \infty[$).

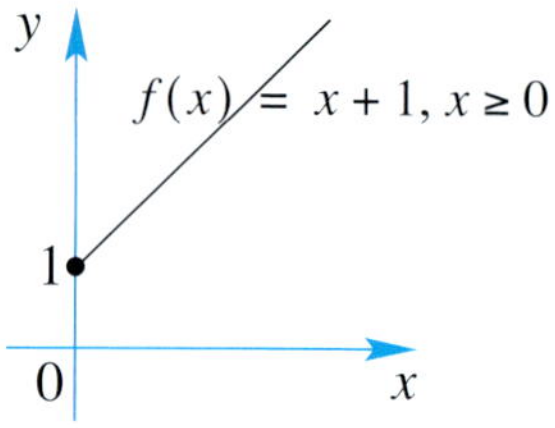

EXAMPLE 5.10

For the function $f(x) = x^2 - 3, x \in \mathbb{R}$ find

(a) $f(6)$ (b) $f(x+1)$ (c) $f(x+h) - f(x)$

Solution

(a) To determine the value of $f(6)$, we 'replace' the x–term in the rule of $f(x)$ with the number '6' i.e., $f(6) = (6)^2 - 3 = 36 - 3 = 33$.

(b) This time we 'replace' the x–term in the rule of $f(x)$ with '$x + 1$':

So, $f(x+1) = (x+1)^2 - 3 = x^2 + 2x + 1 - 3 = x^2 + 2x - 2$.

(c) $f(x+h) - f(x) = (x+h)^2 - 3 - (x^2 - 3)$

$= (\cancel{x^2} + 2xh + h^2) - \cancel{3} - \cancel{x^2} + \cancel{3}$

$= 2xh + h^2$

EXAMPLE 5.11

Consider the function $g:\{x: -1 \leq x \leq 2\} \mapsto \mathbb{R}$, where $g(x) = x^2 + 2$,

(a) Find $g(-1)$, $g(0)$, and $g(2)$. (b) Determine the range of g.

Solution

(a) $g(-1) = (-1)^2 + 2 = 3$

$g(0) = (0)^2 + 2 = 2$

$g(2) = (2)^2 + 2 = 6$

(b) To determine the range of the function g, we first need to sketch its graph.

We begin by sketching the graph of $y = x^2 + 2$ for all real values of x. Then, as the domain is restricted to $x \in [-1, 2]$. This means that we 'remove' the parts of the graph that lie outside this domain. This leaves the required part of the graph.

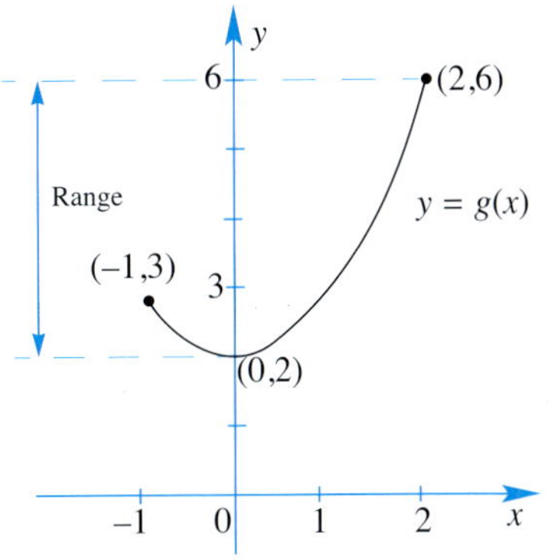

From our graph of g, the range of this function is given by $\{y: 2 \leq y \leq 6\}$ (or $[2, 6]$).

EXAMPLE 5.12

Consider the function $f:\mathbb{R} \mapsto \mathbb{R}$, where $f(x) = 2 - x^2$.

(a) Find i. $f(5)$ ii. $\{x : f(x) = -7\}$ iii. $\{x : f(1-x) = -7\}$

(b) What type of function is f?

Solution

(a) i. $f(5) = 2 - 5^2 = 2 - 25 = -23$

ii. $f(x) = -7 \Leftrightarrow 2 - x^2 = -7 \Leftrightarrow x^2 = 9$

$\therefore x = \pm 3$

So, the solution set is $\{-3, 3\}$.

iii. $f(1-x) = -7 \Leftrightarrow 2 - (1-x)^2 = -7$

$\Leftrightarrow (1-x)^2 = 9$

$\Leftrightarrow 1 - x = \pm 3$

Therefore, the solution set is $\{-2, 4\}$.

(b) From (a) we obtained two values of x for one value of y and so $f(x)$ is a many–to –one function.

EXERCISES 5.2

1. A function is defined as follows, $f: x \mapsto 2x + 3, x \geq 0$.

(a) Find the value of $f(0)$, $f(1)$.

(b) Evaluate the expressions i. $f(x + a)$ ii. $f(x + a) - f(x)$

(c) Find $\{x : f(x) = 9\}$.

2. If $f(x) = \dfrac{x}{x+1}, x \in [0, 10]$, find

(a) $f(0), f(10)$.

(b) $\{x : f(x) = 5\}$.

(c) the range of $f(x) = \dfrac{x}{x+1}, x \in [0, 10]$.

3. For the mapping $x \mapsto 2 - \frac{1}{2}x^2, x \in \mathbb{R}$, find

(a) $f(x + 1)$, $f(x - 1)$.

(b) a, given that $f(a) = 1$.

(c) b, given that $f(b) = 10$.

4. A function is defined as follows, $y = x^3 - x^2, x \in [-2, 2]$.

(a) Find the value(s) of x such that $y = 0$.

(b) Sketch the graph of $y = x^3 - x^2, x \in [-2, 2]$ and determine its range.

5. The function f is defined as $f:]-\infty, \infty[\mapsto \mathbb{R}$, where $f(x) = x^2 - 4$.

(a) Sketch the graph of i. f

ii. $y = x + 2, x \in]-\infty, \infty[$

(b) Find i. $\{x : f(x) = 4\}$ ii. $\{x : f(x) = x + 2\}$

6. Which of the following relations are also functions?

(a)

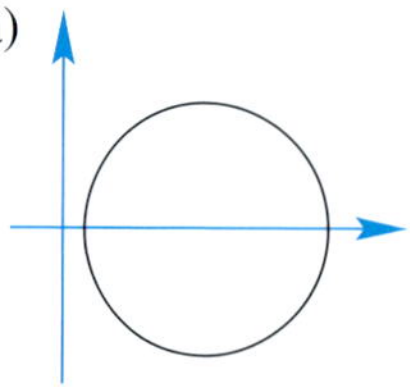

(b)

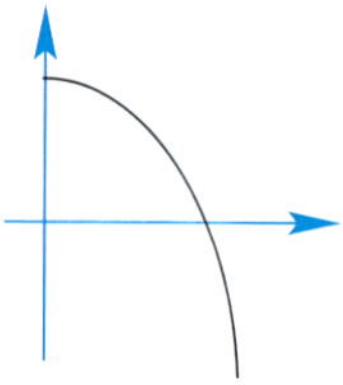

(c)

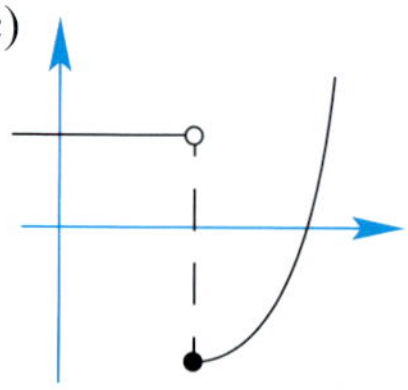

(d)

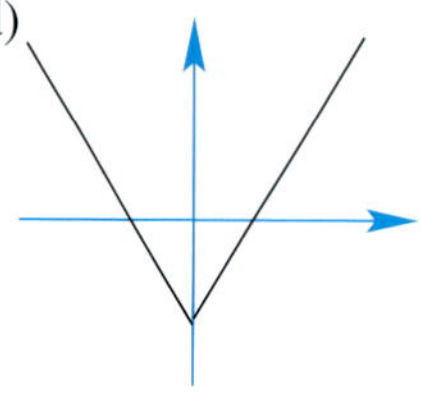

(e)

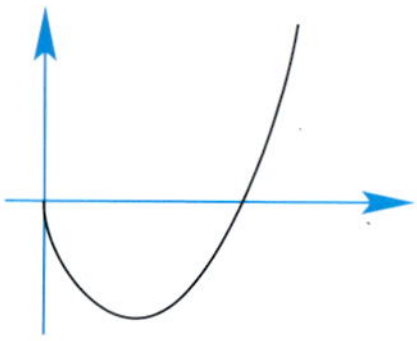

(f)

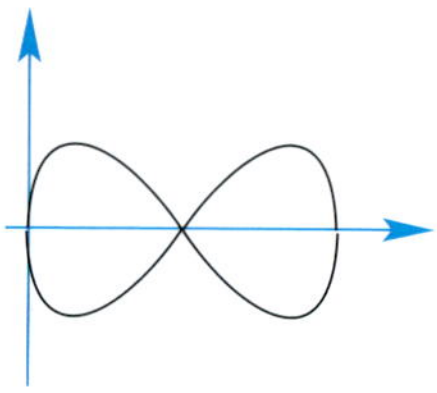

7. Use both *visual tests* and *algebraic tests* to show that the following relations are also functions.

(a) $x \mapsto x^3 + 2, x \in]0,5[$

(b) $x \mapsto \sqrt{x} + 1, x \in [0, 9)$

(c) $\{(x,y) : y^3 = x + 1, x \in \mathbb{R}\}$

(d) $\{(x,y) : y = x^2 + 1, x \in \mathbb{R}\}$

8. Use an algebraic method to decide which of the following relations are also functions.

(a) $f:x \mapsto \frac{1}{x}, x \in \mathbb{R}\backslash\{0\}$ (b) $\{(x,y):y^2 - x = 9,\ x \geq -9\}$

(c) $\{(x,y):y^2 - x^2 = 9,\ x \geq -9\}$ (d) $f(x) = \frac{1}{x^2} + 1, x \neq 0$

(e) $f(x) = 4 - 2x^2, x \in \mathbb{R}$ (f) $f:x \mapsto \frac{4}{x+1}, x \in \mathbb{R}\backslash\{-1\}$

9. Sketch the graph of $f: \mapsto \frac{x^2}{x^2+2}, x \in \mathbb{R}$ and use it to

(a) show that f is a function.
(b) determine its range.

10. A function is defined by $f:x \mapsto \frac{x+10}{x-8}, x \neq 8$ and $x \geq 0$.

(a) Determine the range of f.
(b) Find the value of a such that $f(a) = a$.

11. Consider the functions $h(x) = \frac{1}{2}(2^x + 2^{-x})$ and $k(x) = \frac{1}{2}(2^x - 2^{-x})$.

(a) Show that $2[h(x)]^2 = h(2x) + 1$.
(b) If $[h(x)]^2 - [k(x)]^2 = a$, find the constant a.

12. Which of the following functions are identical? Explain

(a) $f(x) = \frac{x}{x^2}$ and $h(x) = \frac{1}{x}$. (b) $f(x) = \frac{x^2}{x}$ and $h(x) = x$.

(c) $f(x) = x$ and $h(x) = \sqrt{x^2}$. (d) $f(x) = x$ and $h(x) = (\sqrt{x})^2$.

13. Find the the largest possible subset X of $\mathbb{R}$, so that the following relations are one to one increasing functions

(a) $f : X \to \mathbb{R}$, where $f(x) = x^2 + 6x + 10$
(b) $f : X \to \mathbb{R}$, where $f(x) = \sqrt{9 - x^2}$
(c) $f : X \to \mathbb{R}$, where $f(x) = \sqrt{x^2 - 9}$
(d) $f : X \to \mathbb{R}$, where $f(x) = \frac{1}{3x - x^2}$, $x \neq 0, 3$

14. An isosceles triangle ABC has two side lengths measuring 4 cm and a variable altitude. Let the altitude be denoted by x cm.

(a) Find, in terms of x, a relation for
  i. its perimeter, $p(x)$ cm and specify its implied domain.
  ii. its area, $A(x)$ cm^2 and specify its implied domain.
(b) Sketch the graph of
  i. $p(x)$ and determine its range.
  ii. $A(x)$ and determine its range.

5.3 SOME STANDARD FUNCTIONS

5.3.1 HYBRID FUNCTIONS AND CONTINUITY

A hybrid function is a relation that consists of more than one function, where each function is defined over mutually exclusive domains. Generally, these functions take on the following form:

$$f(x) = \begin{cases} g_1(x) & x \in X_1 \\ g_2(x) & x \in X_2 \\ \vdots & \vdots \end{cases} \quad \text{where } X_i \cap X_j = \varnothing, i \neq j$$

EXAMPLE 5.13

Sketch the graph of $f(x) = \begin{cases} -2 & x \le 1 \\ x+1 & x > 1 \end{cases}$, stating its doman and range.

Solution

We first look at the domain of the function. In this instance we have that $x \ge 1$ and $x < 1$, so in fact we have that $x \in]-\infty, \infty[$ (or simply, $x \in \mathbb{R}$).

To determine the range we will need to sketch the graph of *f*.
Hybrid functions can be sketched using the TI–83 (use the **TEST** menu)

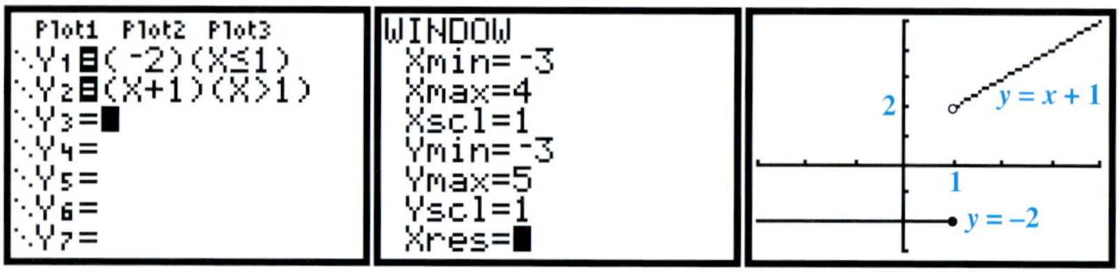

From the graph the range is given by the set $\{y : y > 2\} \cup \{-2\}$.
A few important points to note:

1. Even though $x = 1$ is not in the domain of the $y = x + 1$ part of the function, we have still used $x = 1$ to find the value of $y = 2$.
2. At $x = 1$, we have what is known as a **discontinuity**.

EXAMPLE 5.14

Sketch the graph of $f(x) = \begin{cases} x^2 - 4 & x < -2 \\ \sqrt{4 - x^2} & -2 \le x \le 2 \\ 4 - x^2 & x > 2 \end{cases}$

(a) State the domain of *f*.
(b) Find the range of *f*.

Solution

This time we have a hybrid function with three branches; $y_1 = x^2 - 4, x < -2$ (a parabola), $y_2 = \sqrt{4 - x^2}, -2 \le x \le 2$ (a semi-circle) and $y_3 = 4 - x^2, x > 2$ (a parabola). Individually they are straight foward to sketch. In fact, it is not that much more difficult to sketch the hybrid function either. The 'trick' is to sketch each branch in its entirety and then remove those sections that do not belong.

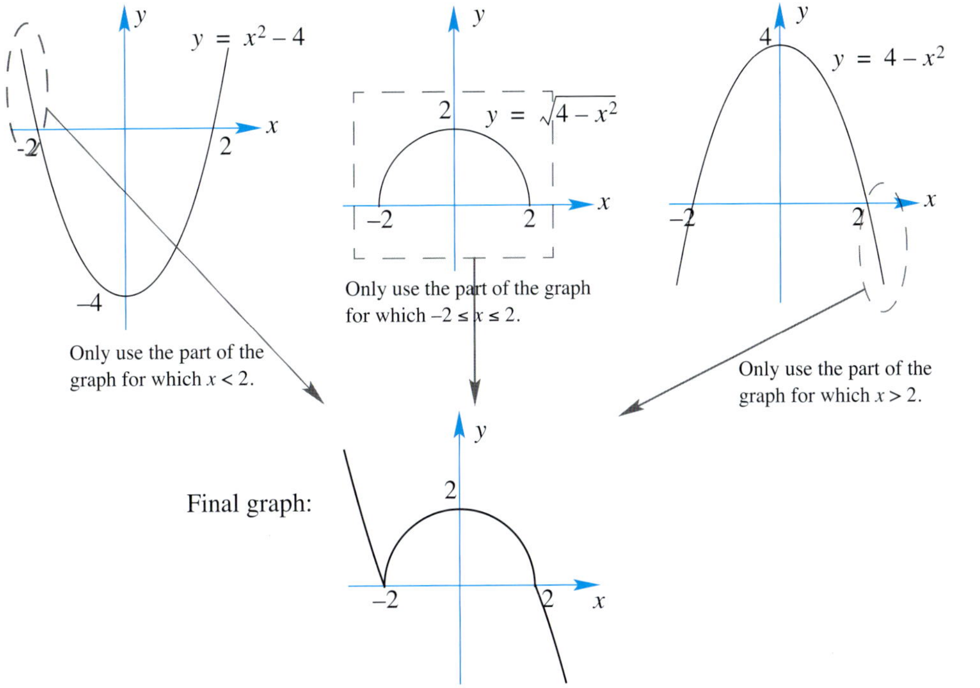

(a) The domain is given by $]-\infty, 2\,[\ \cup [-2, 2] \cup\ ]2, \infty[\ = \ ]-\infty, \infty[$ (or simply, $\mathbb{R}$).

(b) From the graph, the range can be seen to be $\mathbb{R}$.

Nb:
1. At $x = 2, y = 0$ and at $x = -2, y = 0$.
2. Unlike the previous example, the graph is **continous** at $x = \pm 2$.
3. We could have still made use of the TI–83 for this problem:

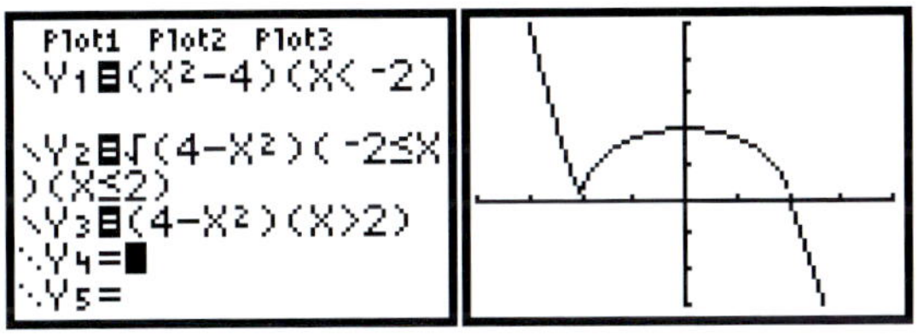

Notice the distortion of the semi-circle (see §5.1.5).

Continuity

> A graph is said to be continuous over a domain if it can be drawn without lifting the pen from the paper in that domain. In other words, there are no breaks.

Eventhough this is a rather weak 'definition' it does provide the basics behind the concept of continuity. If a relation is **not continuous** at some point (i.e., there is a break in the graph), we then say that the graph is **discontinuous** at that point.

The three types of discontinuities that we will encounter can be summarised as follows:

'Jump' discontinuity

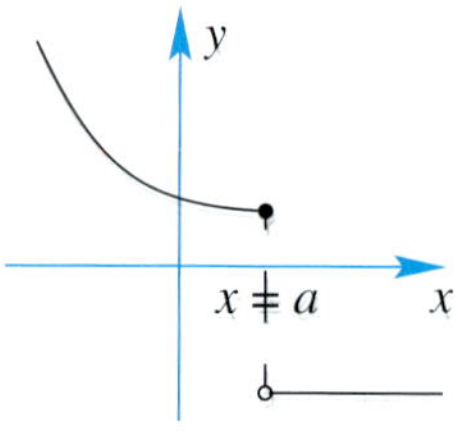

This function is continuous for all real values of x except at $x = a$. At $x = a$, there exists a 'jump' discontinuity.

'Asymptotic' discontinuity

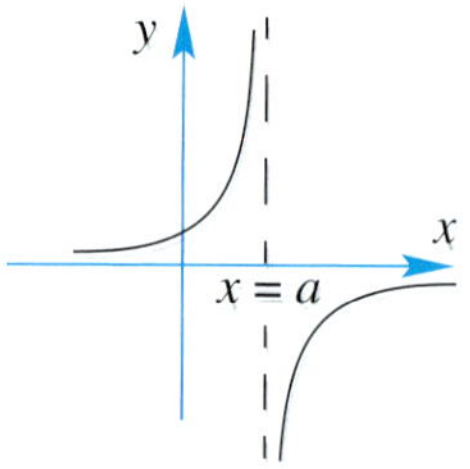

This function is continuous for all real values of x except at $x = a$. At $x = a$, there exists an 'asymptotic' discontinuity.

'Missing point' discontinuity

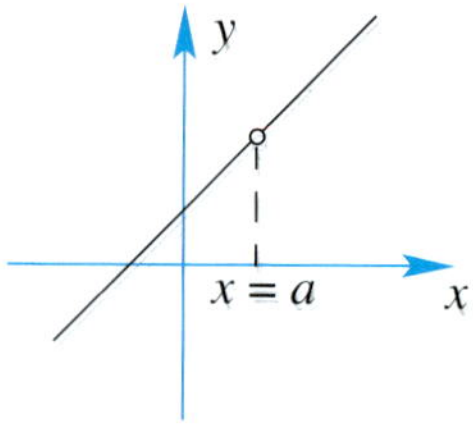

This function is continuous for all real values of x except at $x = a$. At $x = a$, there exists an 'missing point' discontinuity.

EXAMPLE 5.15

For what value of a will the function $y = \begin{cases} x^2 + a, & x > 1 \\ -x + 4, & x \le 1 \end{cases}$ be continuous?

Solution

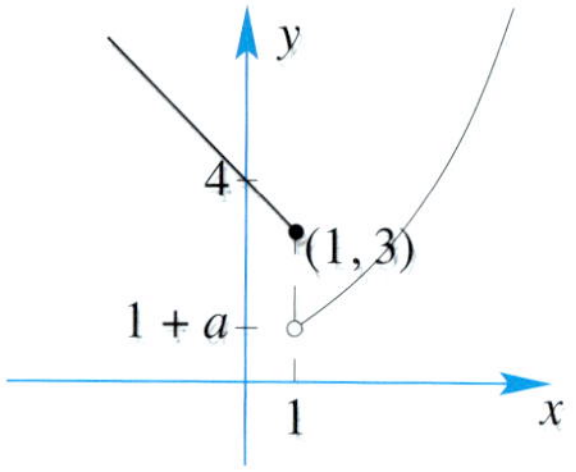

We begin by treating this function in the same way that we have treated other hybrid functions. We first sketch its graph (for an unknown value of a). In this instance we have drawn the graph with $y = 1 + a$ lying below $y = 4$ (it would make no difference if we had placed $y = 1 + a$ above $y = 4$).

Now, for the function to be continuous there must be no jump at $x = 1$, that is, we must have the two branches meeting at $x = 1$.

Therefore we must have that $1 + a = 3 \Leftrightarrow a = 2$.

EXAMPLE 5.16

For what value of b will $f(x) = \begin{cases} -x + 4, & x > 3 \\ 2 - \sqrt{x - b}, & x \le 3 \end{cases}$ be continuous?

Solution

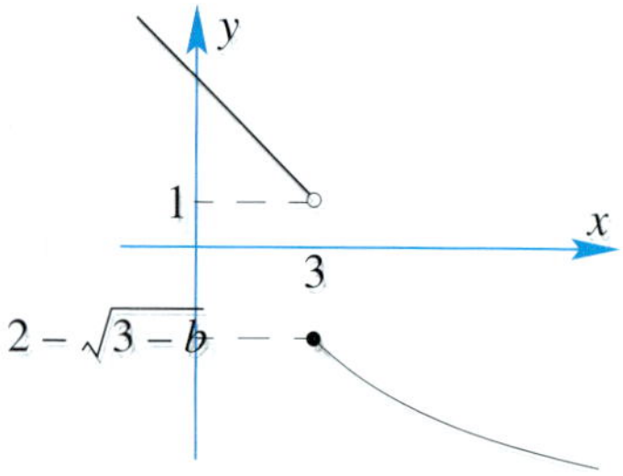

At $x = 3$, we have $f(3) = 2 - \sqrt{3 - b}$.

If $f(x)$ is to be continuous then the two branches of the graph must meet at $x = 3$.

That is, we must have that $2 - \sqrt{3 - b} = 1 \Leftrightarrow \sqrt{3 - b} = 1$

$$\therefore 3 - b = 1$$

$$\Leftrightarrow b = 2$$

EXERCISES 5.3.1

1. Sketch the graphs of the following functions

(a) $f(x) = \begin{cases} -x+2, & x > 3 \\ 1, & x \le 3 \end{cases}$

(b) $f(x) = \begin{cases} x^2+2, & x > -1 \\ 3, & x \le -1 \end{cases}$

(c) $f(x) = \begin{cases} \sqrt{x}, & x \ge 4 \\ 6-x, & x < 4 \end{cases}$

(d) $f(x) = \begin{cases} \frac{1}{x}, & x > 1 \\ 2-x^2, & x \le 1 \end{cases}$

2. Sketch the graphs of the following functions

(a) $f(x) = \begin{cases} -2, & x < 0 \\ x-2, & 0 \le x \le 4 \\ 2 & x > 4 \end{cases}$

(b) $h(x) = \begin{cases} \sqrt{-x}, & x \le -1 \\ 1-x^2, & -1 < x < 1 \\ -\sqrt{x}, & x \ge 1 \end{cases}$

3. Sketch the graphs of the following functions

(a) $h(x) = \begin{cases} x^3+1, & x > 0 \\ -1, & x \le 0 \end{cases}$

(b) $g(x) = \begin{cases} x+2, & x > 1 \\ x^2-1, & x \le 1 \end{cases}$

(c) $f(x) = \begin{cases} \frac{x}{x+1}, & x \ge 0 \\ 1, & x < 0 \end{cases}$

(d) $f(x) = \begin{cases} 2-\sqrt{x}, & x > 0 \\ x+3, & x \le 0 \end{cases}$

4. Sketch the graphs of the following functions

(a) $f(x) = \begin{cases} -4, & x < -2 \\ x^2-4, & -2 \le x \le 2 \\ 4 & x > 2 \end{cases}$

(b) $h(x) = \begin{cases} \sqrt{2-x}, & x \le -2 \\ -2x, & -2 < x < 2 \\ -\sqrt{x+2}, & x \ge 2 \end{cases}$

5. Sketch the graphs of the following functions

(a) $f(x) = \begin{cases} \frac{1}{x+1}, & x \ge 0 \\ a, & x < 0 \end{cases}, a > 1$

(b) $f(x) = \begin{cases} a+x^2, & x > 0 \\ x+3, & x \le 0 \end{cases}, a < -2$

6. For what value(s) of a will the following functions be continuous? Sketch their graphs.

(a) $f(x) = \begin{cases} ax+1, & x > 1 \\ 5, & x \le 1 \end{cases}$

(b) $h(x) = \begin{cases} 2x-4, & x \ge 2 \\ a-2x, & x < 2 \end{cases}$

(c) $f(x) = \begin{cases} ax^3-1, & x > 2 \\ 3+x^2, & x \le 2 \end{cases}$

(d) $h(x) = \begin{cases} \frac{1}{a}x^2+1, & x \ge 2 \\ ax+1, & x < 2 \end{cases}$

7. Given that $-\frac{1}{2} \le \frac{x}{x^2+1} \le \frac{1}{2}$, sketch the graph of $f(x) = \frac{2ax}{x^2+1}$ for $x \in]-\infty,\infty[$, where $a > 0$. For what value(s) of a will the function $h(x) = \begin{cases} \frac{2ax}{x^2+1}, & x \ge 1 \\ 4 & x < 1 \end{cases}$ be continuous?

Sketch the graph of h.

5.3.2 THE ABSOLUTE VALUE FUNCTION

* Parts of this section lies outside the syllabus. However, simple absolute value functions such as Example 5.17 and Example 5.18 (a) and (c) should form part of your repertoire.

The absolute value function is defined as

$$y = |x| = \sqrt{x^2} = \begin{cases} x & \text{if } x \ge 0 \\ -x & \text{if } x < 0 \end{cases}$$

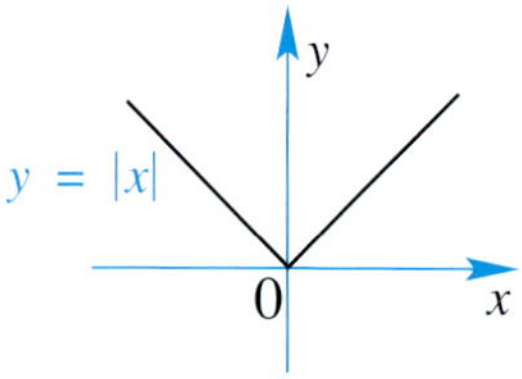

That is, sketch the graph of $y = x$ for $x \ge 0$, and then sketch the graph of $y = -x$ for $x < 0$.

Similarly, the function $x \mapsto |ax+b|$, represents the absolute value of the linear function $y = ax + b$.

EXAMPLE 5.17

Sketch the graph of (a) $y = |x-2|$

(b) $y = |x| + 1$

(c) $y = |2x+1|$

Solution

Parts (a) and (b) are best done by considering the functions as translations of the basic absolute value function. That is, the graph of $y = |x-2|$ is in fact the graph of $y = |x|$ translated two units to the left. While the graph of $y = |x| + 1$ is in fact the graph of $y = |x|$ translated one unit vertically up. So, we have:

(a)

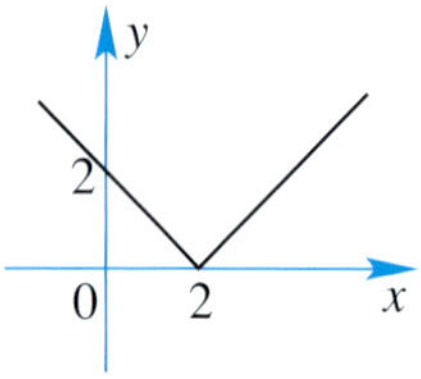

(b)

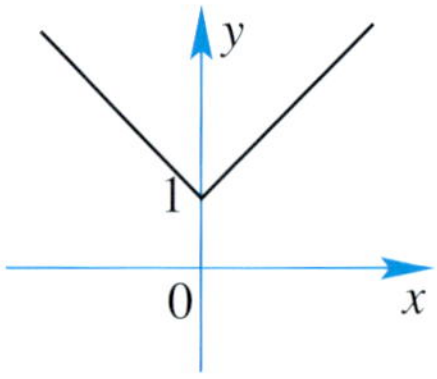

We could have used the definition to sketch the graphs above. This is done as follows:

(a) By definition, $y = |x-2| = \begin{cases} x-2 & \text{if } x-2 \ge 0 \\ -(x-2) & \text{if } x-2 < 0 \end{cases} = \begin{cases} x-2 & \text{if } x \ge 2 \\ 2-x & \text{if } x < 2 \end{cases}$.

Then, sketch the straight line $y = x - 2$ for $x \ge 2$ and $y = 2 - x$ for $x < 2$.

(b) By definition, $y = |x| + 1 = \begin{cases} x & \text{if } x \ge 0 \\ -x & \text{if } x < 0 \end{cases} + 1 = \begin{cases} x+1 & \text{if } x \ge 0 \\ -x+1 & \text{if } x < 0 \end{cases}$.

Notice that because there are no restrictions on '+ 1', we add it to both branches. Then, sketch the straight line $y = x + 1$ for $x \ge 0$ and $y = 1 - x$ for $x < 0$.

(c) By definition $y = |2x+1| = \begin{cases} 2x+1 & \text{if } 2x+1 \geq 0 \\ -(2x+1) & \text{if } 2x+1 < 0 \end{cases} = \begin{cases} 2x+1 & \text{if } x \geq -\frac{1}{2} \\ -2x-1 & \text{if } x < -\frac{1}{2} \end{cases}$.

Then, sketch the straight lines $y = 2x+1$ for $x \geq -\frac{1}{2}$

and $y = -2x-1$ for $x < -\frac{1}{2}$

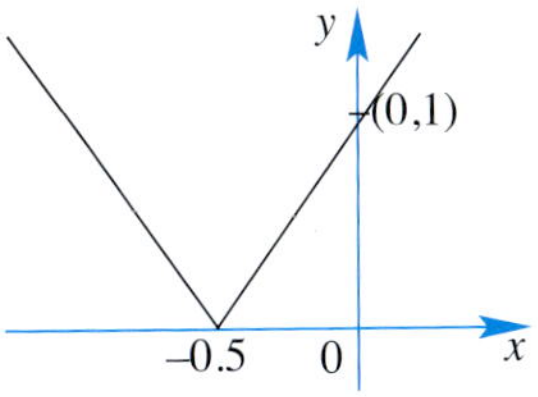

Notice that in fact, all we have done in part (c) is to sketch the graph of $y = 2x + 1$, and then reflect (about the x-axis) any part of the graph that was drawn below the x-axis.

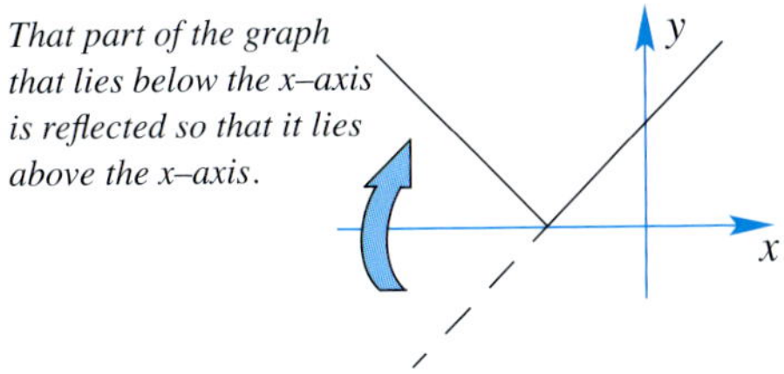

We can also make use of the TI–83 to sketch graphs of absolute value functions:

EXAMPLE 5.18 Find the range of the following functions

(a) $x \mapsto |3-2x|, x \in \mathbb{R}$

(b) $y = |x+1| + |x-1|, x \in \mathbb{R}$

(c) $y = |x-4| - 2, x \in \mathbb{R}$

Solution

(a) Use the **MATH** menu and then selecting the **NUM** option, we can choose the **abs(** option. After 'pasting' the **abs(** command, enter the equation as shown on the screen.

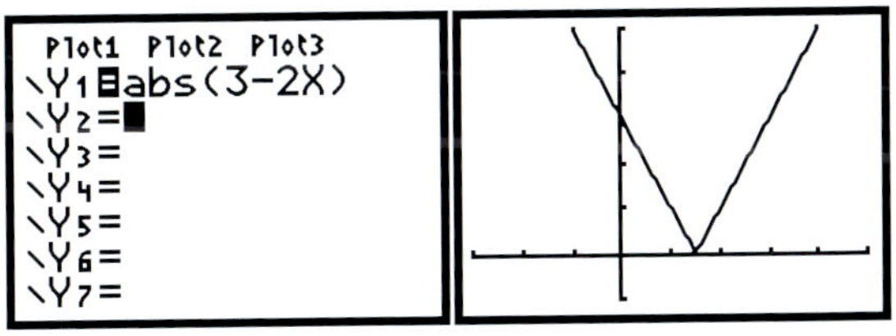

From the given graph, the range is defined as $\{y : y \geq 0\}$. As before, we enter the required options and obtain the following:

(b)

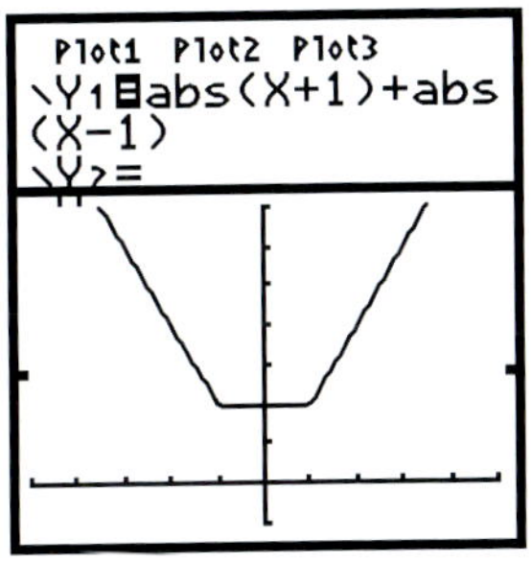

range = [2, ∞ [

(c)

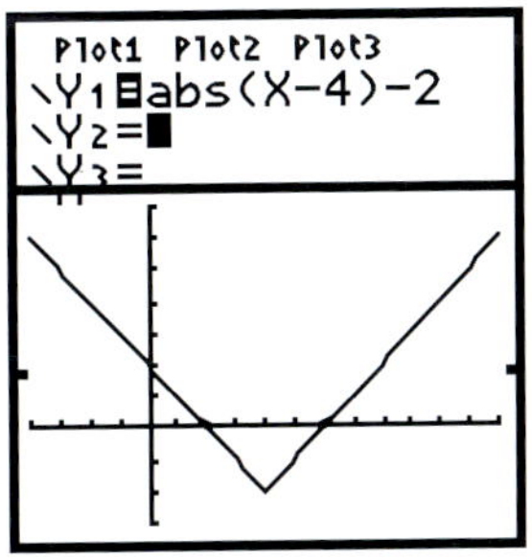

range = [–2, ∞ [

Notice that the TI–83 has made sketching the graph of $y = |x+1| + |x-1|, x \in \mathbb{R}$ a relatively easy task. Without using the TI–83, we would need to look at the overlapping domains and then sketch those parts of each equation that correspond to the overlapping domains.

In this instance we have three overlapping regions:

$$y = |x+1| + |x-1| = \begin{cases} x+1 & \text{if } x \geq -1 \\ -x-1 & \text{if } x < -1 \end{cases} + \begin{cases} x-1 & \text{if } x \geq 1 \\ 1-x & \text{if } x < 1 \end{cases}$$

$$= \begin{cases} (x+1)+(x-1) & \text{if } x \geq 1 \\ (x+1)+(1-x) & \text{if } -1 \leq x < 1 \\ (-x-1)+(1-x) & \text{if } x < -1 \end{cases} = \begin{cases} 2x & \text{if } x \geq 1 \\ 2 & \text{if } -1 \leq x < 1 \\ -2x & \text{if } x < -1 \end{cases}$$

We now look at the more general form of the absolute value function, i.e., $y = |f(x)|$. By definition we have that

$$y = |f(x)| = \begin{cases} f(x) & \text{if } f(x) \geq 0 \\ -f(x) & \text{if } f(x) < 0 \end{cases}$$

However, depending on the expression of $f(x)$, solving the inequalities $f(x) \geq 0$ and $f(x) < 0$ can be time consuming. So, to sketch a graph of $y = |f(x)|$ it is often convenient to first sketch the graph of $y = f(x)$ and then we reflect (about the x–axis) whatever parts of this graph lie below the x–axis. That is:

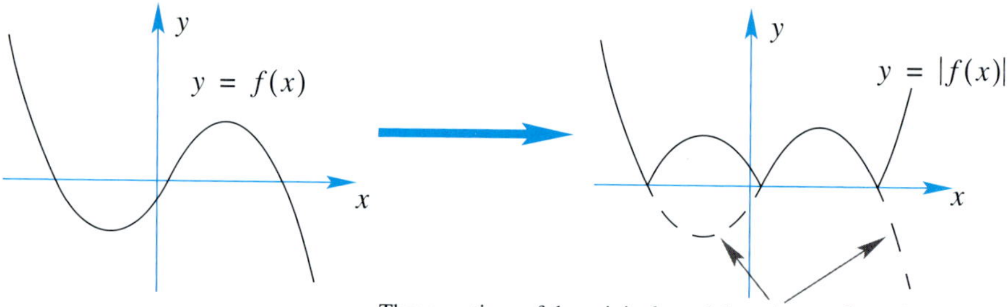

These sections of the original graph have been reflected about the x–axis.

EXAMPLE 5.19

Sketch the graph of $x \mapsto |x^2 - 4x|, x \in \mathbb{R}$

Solution

We start by sketching the graph of $f(x) = x^2 - 4x$ and then follow the above process.

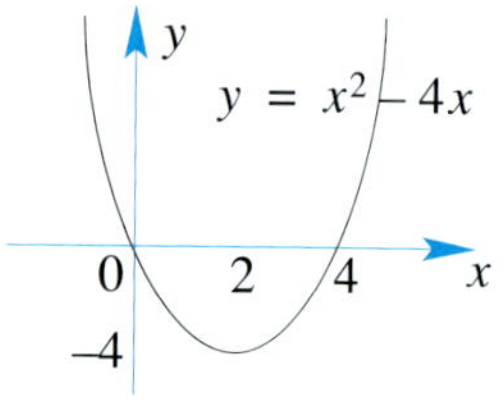

That is, once we have sketched the graph of $f(x) = x^2 - 4x$ we then reflect (about the x–axis) the part of the graph that lies below the x–axis:

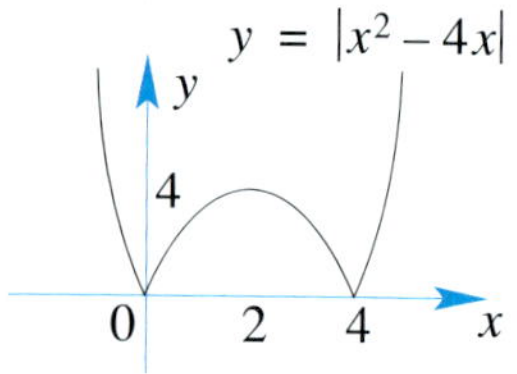

Sometimes, rather than considering the absolute value of the whole expression, we have equations for which we consider the absolute value of the independent variable only. That is, rather than sketching the graph of $y = |f(x)|$ we want to sketch the graph of $y = f(|x|)$. So, what does the graph of $y = f(|x|)$ look like. Obviously it will depend on the original function $f(x)$, however, we can deduce the general properties or transformation in getting from $f(x)$ to $f(|x|)$.

We begin by considering a number of examples using the TI–83:

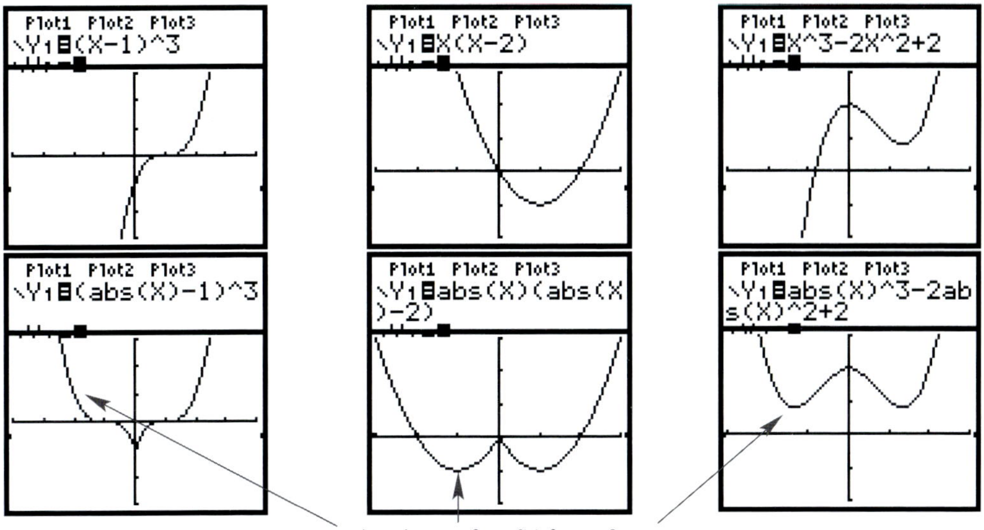

mirror image of $y = f(x)$ for $x \geq 0$.

From these graphs it is clear that the graph of $y = f(|x|)$ is obtained from the graph of $y = f(x)$ by **replacing** the part of $y = f(x)$ that lies in the domain $x \leq 0$ with a mirror image (about the y–axis) of the part of $y = f(x)$ that lies in the domain $x \geq 0$.

To sketch the graph of $y = f(|x|)$ we discard that part of $y = f(x)$ that is sketched for $x < 0$ and replace it with the reflection (about the y–axis) of $y = f(x)$ that is sketched for $x \geq 0$.

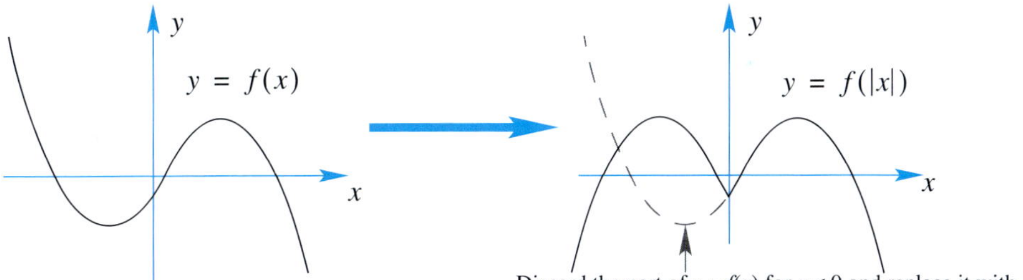

Discard the part of $y = f(x)$ for $x < 0$ and replace it with the reflection about the y–axis of $y = f(x)$ for $x \geq 0$.

EXAMPLE 5.20

For each of the functions $y = f(x)$ sketch the graph of $y = f(|x|)$

(a) $f(x) = x^2 - 4x$

(b) $f(x) = 1 - \sqrt{x+2}$

Solution

(a) We start by sketching the graph of $f(x) = x^2 - 4x$ and then follow the process just described:

Step 1: Discard the part of $y = f(x)$ for $x < 0$.
Step 2: Replace it with the mirror image (about the y–axis) of $y = f(x)$ for $x \geq 0$.

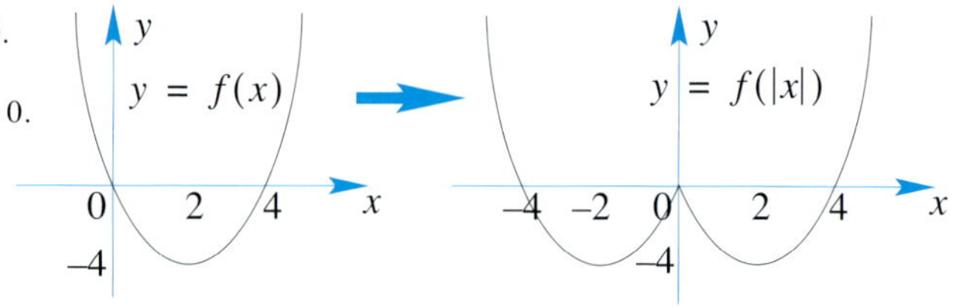

(b) Again we start by sketching the graph of $f(x) = 1 - \sqrt{x+2}$ and then follow the process just described:

Step 1: Discard the part of $y = f(x)$ for $x < 0$.
Step 2: Replace it with the mirror image (about the y–axis) of $y = f(x)$ for $x \geq 0$.

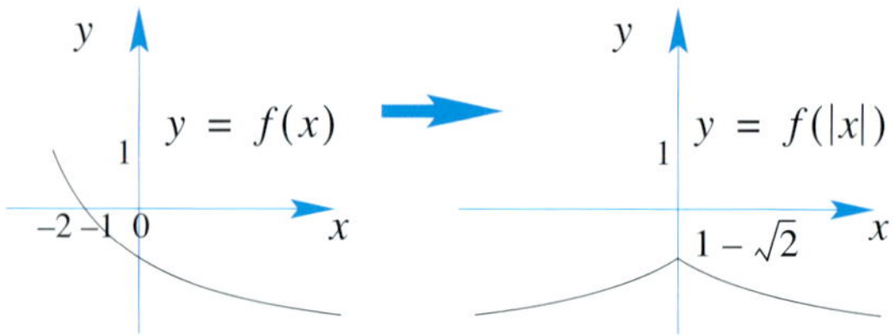

EXERCISES 5.3.2

1. On separate sets of axes, sketch the graphs of the following functions for $x \in \mathbb{R}$.

(a) $f(x) = |2x|$ (b) $f(x) = |x| + 2$ (c) $f(x) = |x + 1|$
(d) $f(x) = |4x - 2|$ (e) $f(x) = |2x| - 4$ (f) $f(x) = -|x| + 1$
(g) $g(x) = |3 - x|$ (h) $f(x) = \left|5 - \frac{1}{2}x\right|$ (i) $g(x) = \left|\frac{4-x}{2}\right|$

2. On separate sets of axes, sketch the graphs of the following functions for $x \in \mathbb{R}$.

(a) $g(x) = |x^2 - 9|$ (b) $h(x) = |2x - x^2|$ (c) $h(x) = |(x-1)(2+x)|$
(d) $f(x) = |x(3 - x)|$ (e) $g(x) = |x||x + 1|$ (f) $f(x) = |x^3 - 8x^2|$
(g) $f(x) = |8 - x^3|$ (h) $f(x) = 8 - |x^3|$ (i) $f(x) = |x^3 - 1| - 1$

3. On separate sets of axes, sketch the graphs of the following functions for $x \in \mathbb{R}$ and determine the range of each function.

(a) $f(x) = |x + 1| + |x - 1|$ (b) $f(x) = |x + 2| + |x - 2|$
(c) $f(x) = x + |x|$ (d) $f(x) = x - |x|$
(e) $f(x) = |x + 2| - |x - 2|$ (f) $f(x) = |x + 1| - |x - 1|$

4. On separate sets of axes, sketch the graphs of the following functions

(a) $f(x) = x|x|$ (b) $f(x) = \frac{|x|}{x}, x \neq 0$
(c) $f(x) = \left|\frac{1}{x} + 1\right|$ (d) $f(x) = \left|\frac{1}{x} - 1\right|$

5. For each of the functions $y = f(x)$ sketch the graph of $y = f(|x|)$.

(a) $f(x) = 2x^2 - x$ (b) $f(x) = 3 + (x-2)^2$ (c) $f(x) = x^3 - 8$

(d) $f(x) = x - 4$ (e) $f(x) = \frac{1}{x} - 1$ (f) $f(x) = \sqrt{x+4}$

(g) $f(x) = x(4-x)$ (h) $f(x) = \frac{1}{x-1}, x \neq 1$ (i) $f(x) = |x-2|$

6. For each of the functions $y = f(x)$ sketch the graph of $y = f(|x|)$.

(a) $f(x) = \begin{cases} x-4, & x > 3 \\ 8 - x^2, & x \leq 3 \end{cases}$ (b) $f(x) = \begin{cases} x^3 - 1, & x > 1 \\ 1 - x^2, & x \leq 1 \end{cases}$

(c) $f(x) = \begin{cases} \frac{1}{x-1}, & x < 1 \\ \sqrt{x} + 2, & x \geq 1 \end{cases}$ (d) $f(x) = \begin{cases} |x^2 - 2x|, & x \geq 1 \\ |x| - 1, & x < 1 \end{cases}$

7. (a) Sketch the graphs of $f(x) = |x-2|$ and $g(x) = |x+2|$ on the same set of axes. Hence, sketch the graph of $y = f(x) + g(x)$.

(b) Find $\{x : |x-2| + |x+2| = k\}$ where

i. $k = 2$ ii. $k = 4$ iii. $k = 8$

8. On the same set of axes, sketch the graphs of $f(x) = 2|x|$ and $g(x) = |3 - x|$. Hence, find $\{x : f(x) \geq g(x)\}$.

5.3.3 THE EXPONENTIAL FUNCTION

The **exponential function** takes the form $f(x) = a^x, x \in \mathbb{R}, a > 0, a \neq 1$, where the independent variable is the exponent.

Graphs with $a > 1$

An example of an exponential function is $f(x) = 2^x, x \in \mathbb{R}$. So, how does the graph of $f(x) = 2^x$ compare to that of $f(x) = x^2$?

We know that the graph of $f(x) = x^2$ represents a parabola with its vertex at the origin, and is symmetrical about the y–axis. To determine the properties of the exponential function we set up a table of values and use these values to sketch a graph of $f(x) = 2^x$.

x	–3	–2	–1	0	1	2	3	4	5
$y = x^2$	9	4	1	0	1	4	9	16	25
$y = 2^x$	2^{-3}	2^{-2}	2^{-1}	2^0	2^1	2^2	2^3	2^4	2^5
	$= \frac{1}{8}$	$= \frac{1}{4}$	$= \frac{1}{2}$	= 1	= 2	= 4	= 8	= 16	= 32

We can now plot both graphs on the same set of axes and compare their properties:

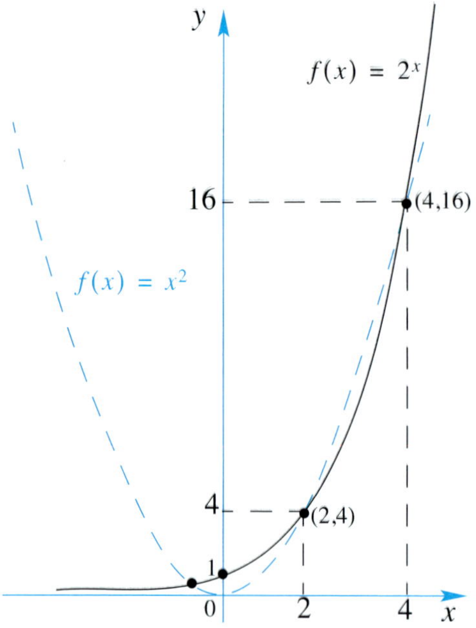

Properties of $f(x) = 2^x$

1. The function increases for all values of x (i.e., as x increases so too do the values of y).
2. The function is always positive (i.e., it lies above the x-axis).
3. As $x \to \infty$ then $y \to \infty$
   $x \to -\infty$ then $y \to 0$.
   i.e., the x-axis is an asymptote.
4. When
   i. $x > 0$ then $y > 1$,
   ii. $x = 0$ then $y = 1$
   iii. $x < 0$ then $0 < y < 1$.

Notice how different the graphs of the two functions are, even though their rules appear similar. The difference being that for the quadratic function, the variable x is the base, whereas for the exponential, the variable x is the power.

We can now investigate the exponential function for different bases. Consider the exponential functions $f(x) = 3^x$ and $g(x) = 4^x$:

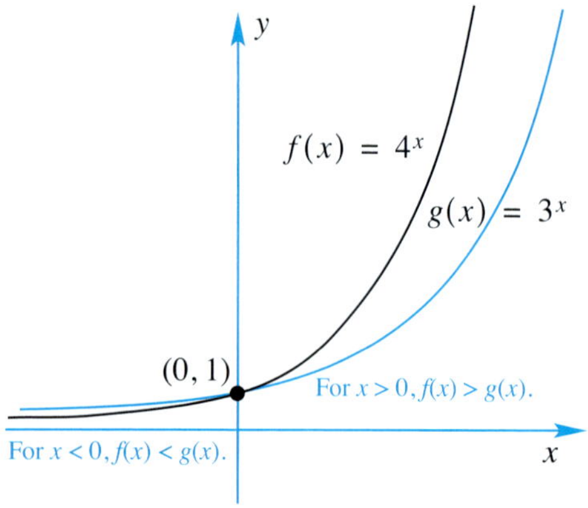

From the graphs we can see that $f(x) = 4^x$ increases much faster than $g(x) = 3^x$ for $x > 0$. For example, at $x = 1$, $f(1) = 4$, $g(1) = 3$ and then, at $x = 2$, $f(2) = 16$, $g(2) = 9$.

However, for $x < 0$ we have the opposite, $f(x) = 4^x$ decreases faster than $g(x) = 3^x$.

Notice then that at $x = 0$, both graphs pass through the point $(0, 1)$.

From the graphs we can see that for values of x less than zero, the graph of $f(x) = 4^x$ lies below that of $f(x) = 3^x$. Whereas for values of x greater than zero, then the graph of $f(x) = 4^x$ lies above that of $f(x) = 3^x$.

Exponential functions that display these properties are referred to as **exponential growth** functions.

What happens when $0 < a < 1$?

We make use of the TI–83 to investigate such cases. Consider the case where $a = \frac{1}{2}$.

Rather than using a table of values we provide a sketch of the curve. The graph shows that the function is decreasing – such exponential functions are referred to as **exponential decay**. In fact, from the second screen we can see that the graph of

$y = \left(\frac{1}{2}\right)^x$ is a reflection of $y = 2^x$ about the y–axis.

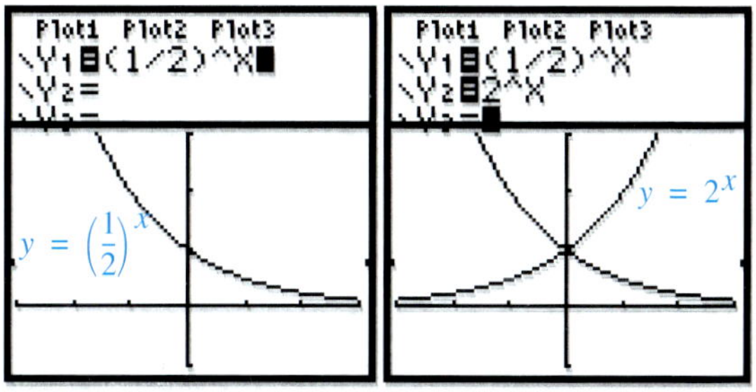

We note that the function $y = \left(\frac{1}{2}\right)^x$ can also be written as $y = (2^{-1})^x = 2^{-x}$. Meaning that there are two ways to represent an **exponential decay** function, either as $f(x) = a^x, 0 < a < 1$ or $f(a) = a^{-x}, a > 1$. For example, the functions $f(x) = \left(\frac{1}{4}\right)^x$ and $g(x) = 4^{-x}$ are identical.

We can summarise the exponential function as follows:

$x \mapsto a^x, a > 1, x \in \mathbb{R}$

(e.g., $f(x) = 2^x$, $f(x) = 3^x$)

Properties

Domain : $\mathbb{R} = (-\infty, \infty)$
Range : $\mathbb{R}^+ = (0, \infty)$
Asymptote : $y = 0$ (or x–axis)
Intercepts : Cuts y–axis at (0,1)
Other : Increases [growth]
Continuous

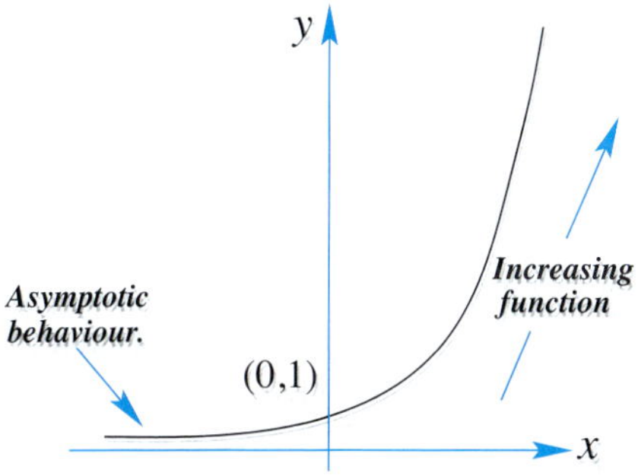

$x \mapsto a^x, 0 < a < 1, x \in \mathbb{R}$

(e.g., $f(x) = \left(\frac{1}{2}\right)^x$, $f(x) = \left(\frac{1}{3}\right)^x$)

Properties

Domain : $\mathbb{R} = (-\infty, \infty)$
Range : $\mathbb{R}^+ = (0, \infty)$
Asymptote : $y = 0$ (or x–axis)
Intercepts : Cuts y–axis at (0,1)
Other : Decreases [decay]
Continuous

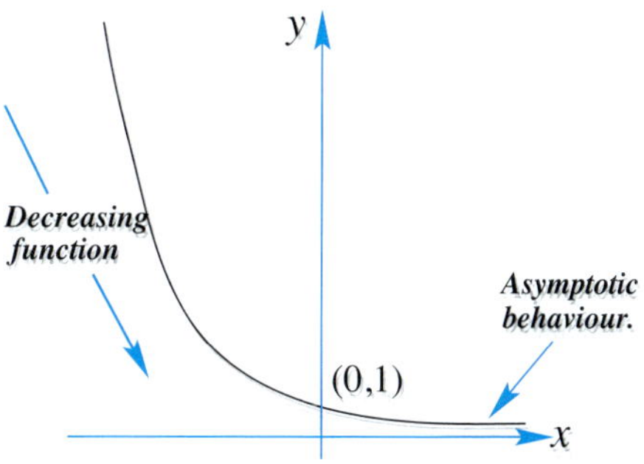

There also exists an important exponential function known as the **natural exponential function**. This function is such that the base has the special number '**e**'. The number '*e*', which we will consider in more detail in Chapter 7 has a value that is given by the expression

$$e = \lim_{n \to \infty}\left(1 + \frac{1}{n}\right)^n \approx 2.71828\ldots$$

However, at this stage it suffices to realise that the number 'e' is greater than one. This means that a function of the form $f(x) = e^x$ will have the same properties as that of $f(x) = a^x$ for $a > 1$. That is, it will depict an exponential growth. Whereas the function $f(x) = e^{-x}$ will depict an exponential decay.

In Chapter 6 we will look at transformations of functions and consider terms such as 'stretching', 'translations', 'dilations' and so on – terms that are applicable to the examples we are about to examine. However, at this stage, we will consider sketching exponential curves from first principles only (and make general observations).

EXAMPLE 5.21

On the same set of axes sketch the following.

(a) $f(x) = 2^x, g(x) = 2^{x-1}$ (b) $f(x) = 3^x, g(x) = 3^{x+2}$

Solution

(a) Making use of the TI–83 we have:

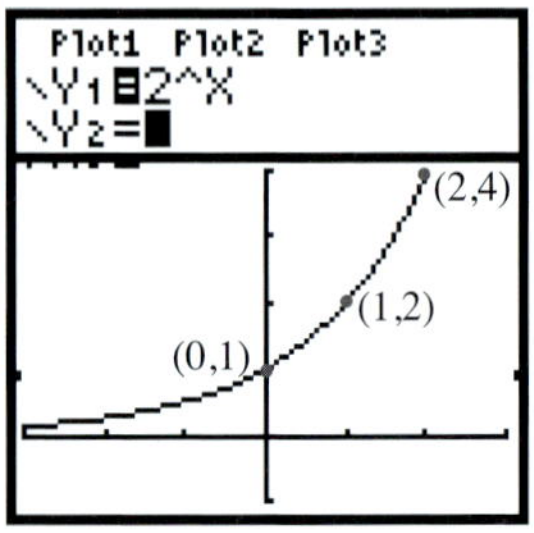

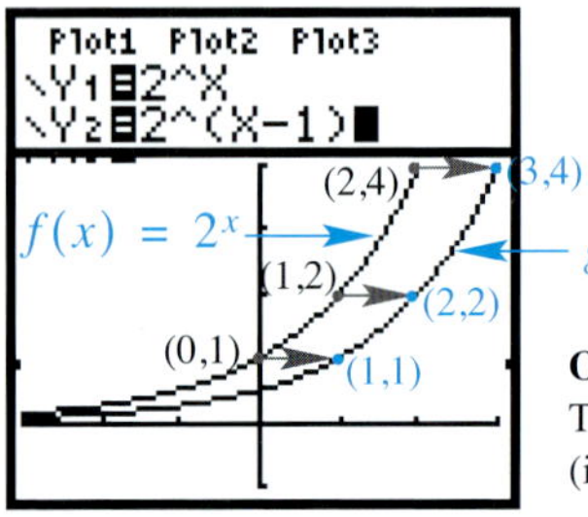

Observations:
The graph of $g(x)$ is the graph of $f(x)$ moved (i.e., translated) one unit to the right.

(b)

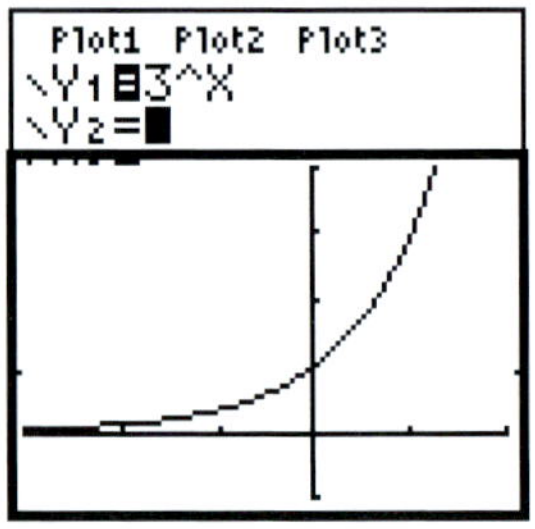

Notice that we had to change the window settings to see both graphs on the same screen.

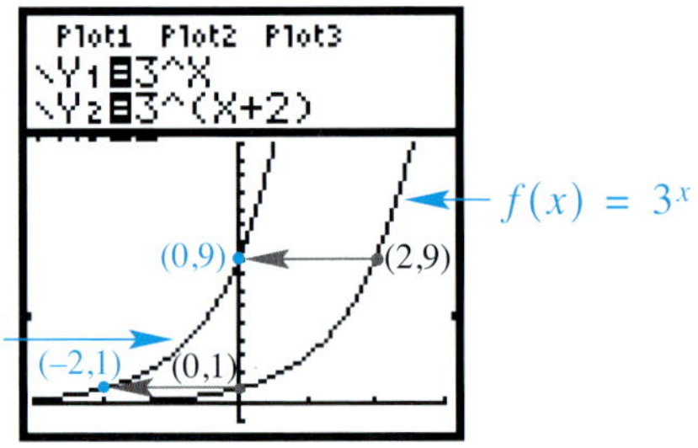

Observations:
The graph of $g(x)$ is the graph of $f(x)$ moved (i.e., translated) two units to the left.

From our observations we can make the following generalisation about the graph of $y = a^{x \pm k}$.

1. The graph of $y = a^{x-k}, k > 0$ is identical to $y = a^x$ but moved 'k' units to the right.
2. The graph of $y = a^{x+k}, k > 0$ is identical to $y = a^x$ but moved 'k' units to the left.

Because of the extremities between the large and small values encountered with exponential functions, sketching these graphs to scale is often difficult. When sketching exponential functions, it is important to include the main features of the graph – for example, make sure that the intercepts and asymptotes are clearly labelled and then provide the general shape of the curve.

EXAMPLE 5.22

On the same set of axes sketch the following.

(a) $f(x) = 2^x, g(x) = 2^x - 1$ (b) $f(x) = 3^x, g(x) = 3^x + 2$

Solution

(a)

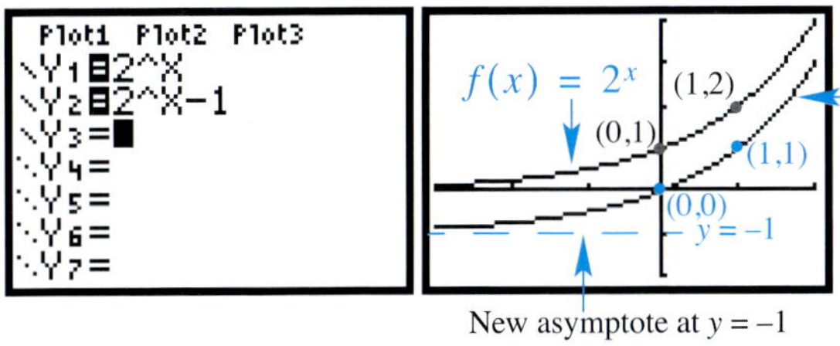

Observations:
The graph of $g(x)$ is the graph of $f(x)$ moved (i.e., translated) one unit down. Notice too that the asymptote has also moved down one unit, from $y = 0$ to $y = -1$.

(b)

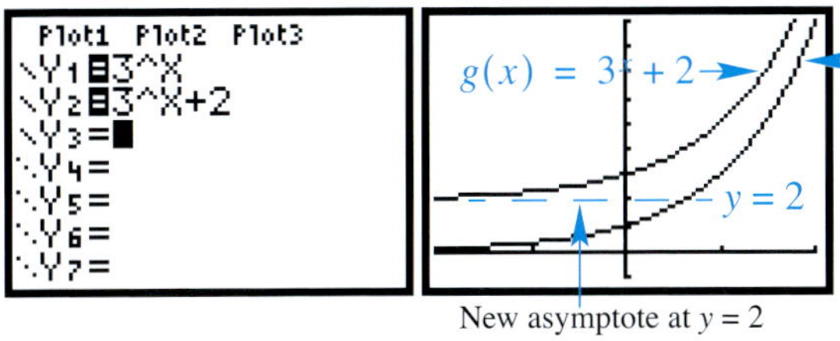

Observations:
The graph of $g(x)$ is the graph of $f(x)$ moved (i.e., translated) two units up. Notice too that the asymptote has also moved up two units, from $y = 0$ to $y = 2$.

From our observations we can make the following generalisation about the graph of $y = a^x \pm k$.

1. The graph of $y = a^x - k, k > 0$ is identical to $y = a^x$ but moved 'k' units down.
2. The graph of $y = a^x + k, k > 0$ is identical to $y = a^x$ but moved 'k' units up.

EXAMPLE 5.23

On the same set of axes sketch the following.

(a) $f(x) = 2^x, g(x) = 3 \times 2^x$ (b) $f(x) = 3^x, g(x) = -\left(\frac{1}{2}\right) \times 3^x$

Solution

(a)

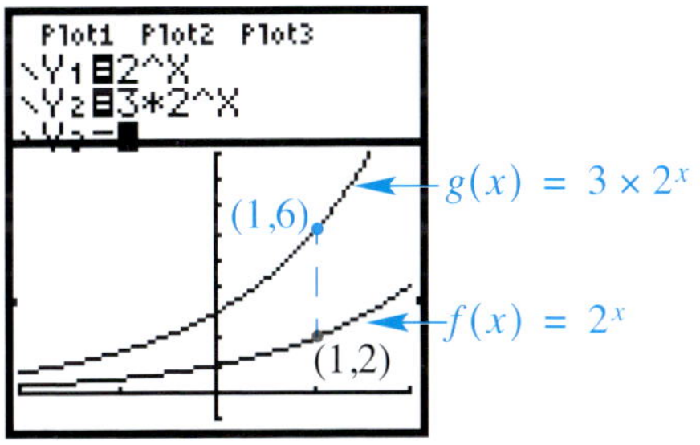

Observations:
The graph of $g(x)$ is the graph of $f(x)$ stretched by a factor of 3 along the y–axis. Notice too that the asymptote has not changed.

(b)

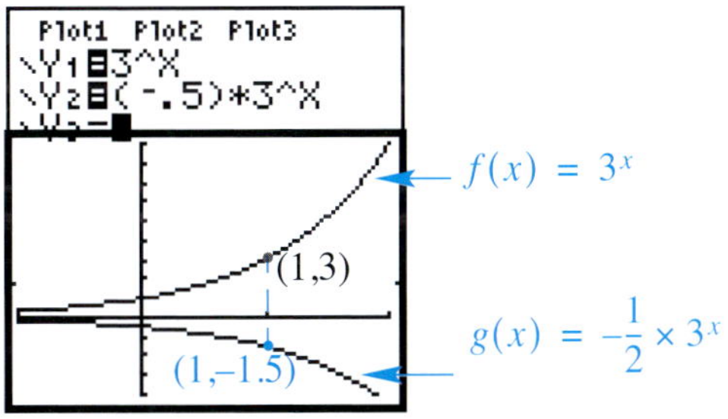

Observations:
The graph of $g(x)$ is the graph of $f(x)$ shrunk by a factor of 2 along the y–axis and reflected about the x–axis. Notice too that the asymptote has not changed.

From our observations we can make the following generalisation about the graph of $y = k \times a^x$.

1. The graph of $y = k \times a^x, k > 0$ is identical to $y = a^x$ but
- i. Stretched along the y–axis if $k > 1$.
- ii. Shrunk along the y–axis if $0 < k < 1$.

2. The graph of $y = k \times a^x, k < 0$ is identical to $y = a^x$ but
- i. Reflected about the x–axis and stretched along the y–axis if $k < -1$.
- ii. Reflected about the x–axis and shrunk along the y–axis if $-1 < k < 0$.

Obviously we can use a combination of these 'effects' on the basic exponential function:

EXAMPLE 5.24

Sketch the following.

(a) $f(x) = 2 \times e^x - 2$ (b) $f(x) = \frac{1}{2}e^{2x-1}$

Solution

Notice that in both cases the general shape of the exponential growth remains unaltered. Only the main features of the graph are of interest when sketching is involved.

(a)

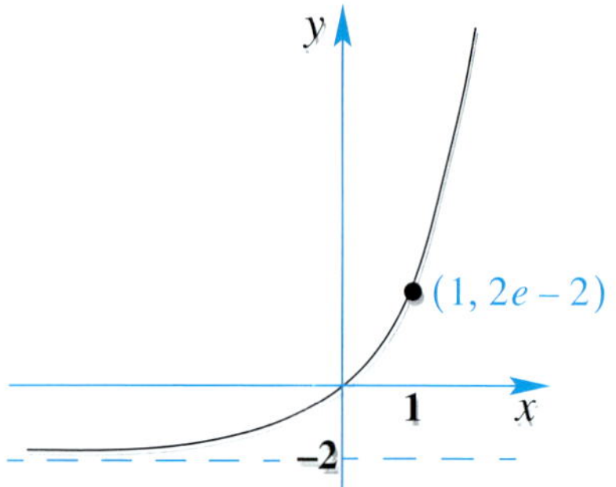

When $x = 0$, $f(0) = 2 \times e^0 - 2 = 0$.

When $x = 1$, $f(1) = 2 \times e^1 - 2$.

(b)

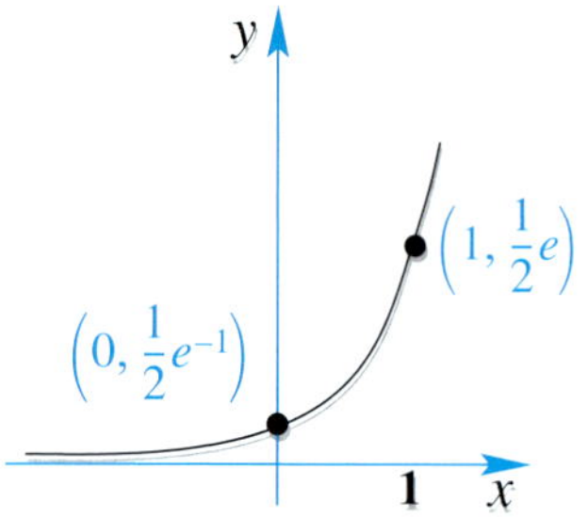

When $x = 0$, $f(0) = \frac{1}{2}e^{0-1} = \frac{1}{2}e^{-1}$.

When $x = 1$, $f(1) = \frac{1}{2}e^{2-1} = \frac{1}{2}e^1$.

EXERCISES 5.3.3

1. On separate sets of axes sketch the graphs of the following functions and determine the range of each function.

(a) $f(x) = 4^x$	(b) $f(x) = 3^x$	(c) $f(x) = 5^x$
(d) $f(x) = (2.5)^x$	(e) $f(x) = (3.2)^x$	(f) $f(x) = (1.8)^x$
(g) $f(x) = \left(\frac{1}{2}\right)^x$	(h) $f(x) = \left(\frac{1}{3}\right)^x$	(i) $f(x) = \left(\frac{1}{5}\right)^x$
(j) $f(x) = \left(\frac{3}{4}\right)^x$	(k) $f(x) = \left(\frac{5}{8}\right)^x$	(l) $f(x) = (0.7)^x$

2. Sketch the following on the same set of axes, clearly labelling the y–intercept.

(a) $f(x) = 3^x + c$ where i. $c = 1$ ii. $c = -2$

(b) $f(x) = 2^{-x} + c$ where i. $c = 0.5$ ii. $c = -0.5$

3. Sketch the following on the same set of axes, clearly labelling the y–intercept.

(a) $f(x) = b \times 3^x$ where i. $b = 2$ ii. $b = -2$

(b) $f(x) = b \times \left(\frac{1}{2}\right)^x$ where i. $b = 3$ ii. $b = -2$

4. On the same set of axes, sketch the following graphs

(a) $f(x) = 3^x$ and $f(x) = 3^{-x}$ (b) $f(x) = 5^x$ and $f(x) = 5^{-x}$

(c) $f(x) = 10^x$ and $f(x) = 10^{-x}$ (d) $f(x) = \left(\frac{1}{3}\right)^x$ and $f(x) = \left(\frac{1}{3}\right)^{-x}$

5. Find the range of the following functions:

(a) $f:[0,4] \mapsto \mathbb{R}, y = 2^x$ (b) $f:[1,3] \mapsto \mathbb{R}, y = 3^x$

(c) $f:[-1,2] \mapsto \mathbb{R}, y = 4^x$ (d) $f:[-1,2] \mapsto \mathbb{R}, y = 2^x$

(e) $f:[2,3] \mapsto \mathbb{R}, y = 2^{-x}$ (f) $f:[-1,1] \mapsto \mathbb{R}, y = 10^{-x}$

6. Sketch the graphs of the following functions, stating their range.

(a) $f : \mathbb{R} \mapsto \mathbb{R}$, where $f(x) = 2e^x + 1$ (b) $f : \mathbb{R} \mapsto \mathbb{R}$, where $f(x) = 3 - e^{x-1}$

(c) $f : \mathbb{R} \mapsto \mathbb{R}$, where $f(x) = e - e^{-x}$ (d) $f : \mathbb{R} \mapsto \mathbb{R}$, where $f(x) = 2 + \frac{1}{2}e^{-x}$

7. (a) Solve for x, if $x^2 - 4x - 5 = 0$.

(b) On the same set of axes, sketch the graphs of $f(x) = 5 \times 5^{-x}$ and $g(x) = 5^x - 4$.

(c) Find i. $\{(x, y) : f(x) = g(x)\}$.

ii. $\{x : f(x) > g(x)\}$.

8. Find the range of the following functions

(a) $f :]0, \infty[\mapsto \mathbb{R}$, where $f(x) = e^{-(x+1)} + 2$.

(b) $g(x) = -2 \times e^x + 1, x \in]-\infty, 0]$.

(c) $x \mapsto xe^{-x} + 1, x \in [-1, 1]$.

9. (a) Sketch the graph of $f(x) = |2^x - 1|$, clearly labelling all intercepts with the axes and the equation of the asymptote.

(b) Solve for x, where $|2^x - 1| = 3$.

10. Sketch the graphs of the following functions

(a) $f(x) = |1 - 2^x|$ (b) $g(x) = |4^x - 2|$ (c) $h(x) = 1 - |2^x|$

11. Sketch the graphs of the following functions

(a) $f(x) = 1 - 2^{-|x|}$ (b) $g(x) = -4 + 2^{|x|}$ (c) $h(x) = |3^{-|x|} - 3|$

12. Sketch the graphs of the following functions and find their range.

(a) $f(x) = \begin{cases} 2^x, & x < 1 \\ 3, & x \geq 1 \end{cases}$ (b) $f(x) = \begin{cases} 3 - e^x, & x > 0 \\ x + 3, & x \leq 0 \end{cases}$

(c) $f(x) = \begin{cases} \frac{2}{x+1}, & x \geq 1 \\ 3 - 2^{2-x}, & x < 1 \end{cases}$ (d) $g(x) = \begin{cases} 4 - 3^{-|x|}, & -1 < x < 1 \\ 4 - \frac{1}{3}|x|, & 1 \leq |x| \leq 12 \end{cases}$

13. Sketch the graphs of the following, and hence state the range in each case.

(a) $f:\mathbb{R} \mapsto \mathbb{R}, y = 2^x + \left(\frac{1}{2}\right)^x$ (b) $f:\mathbb{R} \mapsto \mathbb{R}, y = 3^x + \left(\frac{1}{3}\right)^x$

(c) $f:\mathbb{R} \mapsto \mathbb{R}, y = 2^x - \left(\frac{1}{2}\right)^x$ (d) $f:\mathbb{R} \mapsto \mathbb{R}, y = \left|2^x - \left(\frac{1}{2}\right)^x\right|$

14. Sketch the graph of the functions

(a) $g(x) = 2^{(x-a)}, a > 0$ (b) $h(x) = 2^x - a, 0 < a < 1$

(c) $f(x) = 2 \times a^x - 2a, a > 1$ (d) $f(x) = 2 \times a^x - 2a, 0 < a < 1$

(e) $g(x) = a - a^x, a > 1$ (f) $h(x) = -a + a^{-x}, a > 1$

15. (a) On the same set of axes, sketch $f(x) = 2 \times a^x$ and $g(x) = 4 \times a^{-x}$ where $a > 1$. Hence, sketch the graph of the function $h(x) = a^x + 2a^{-x}$, where $a > 1$.

(b) On the same set of axes, sketch $f(x) = x - a$ and $g(x) = a^{x+1}$, where $a > 1$. Hence, deduce the graph of $h(x) = (x - a) \times a^{x+1}$, where $a > 1$.

16. Sketch the graph of the following functions and determine their range.

(a) $f(x) = a^{-x^2}, a > 1$ (b) $f(x) = a^{-x^2}, 0 < a < 1$

(c) $g(x) = (a-1)^{-x}, a > 1$ (d) $h(x) = 2 - a^{-|x|}, a > 1$

(e) $f(x) = \frac{2}{a^x - 1}, a > 1$ (f) $g(x) = |a^{x^2} - a|, a > 1$

5.3.4 THE LOGARITHMIC FUNCTION

The logarithmic function, with base 'a' is represented by the expression $f(x) = \log_a x, x > 0$. To determine the shape of its graph we start by constructing a table of values for the function $f(x) = \log_2 x$ and comparing it with the table of values for $g(x) = 2^x$:

x	$\frac{1}{8}$	$\frac{1}{4}$	$\frac{1}{2}$	0	1	2	4	8	16
$y = \log_2 x$	–3	–2	–1	–	0	1	2	3	4

x	–3	–2	–1		0	1	2	3	4
$y = 2^x$	$\frac{1}{8}$	$\frac{1}{4}$	$\frac{1}{2}$		1	2	4	8	16

From the table of values we observe that the x and y values have interchanged! Plotting these results on the same set of axes, we observe that the graph of the logarithmic function is a reflection of the exponential function about the line $y = x$. So, whereas for the exponential function, the asymptote is the x–axis (i.e., $y = 0$), for the logarithmic function, the asymptote is the y–axis (i.e., $x = 0$).

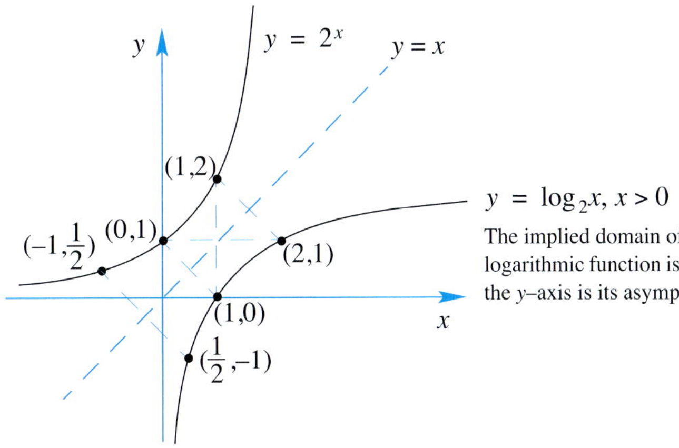

The implied domain of the basic logarithmic function is $]0, \infty[$ and the y–axis is its asymptote.

So, how do the graphs of $y = \log_e x$, $y = \log_{10} x$, $y = \log_3 x$ compare to $y = \log_2 x$. The best way to see this is to sketch the graphs on the same set of axes:

The implied domain of the basic logarithmic function with any positive base is $]0, \infty[$ and has the y–axis as its asymptote.

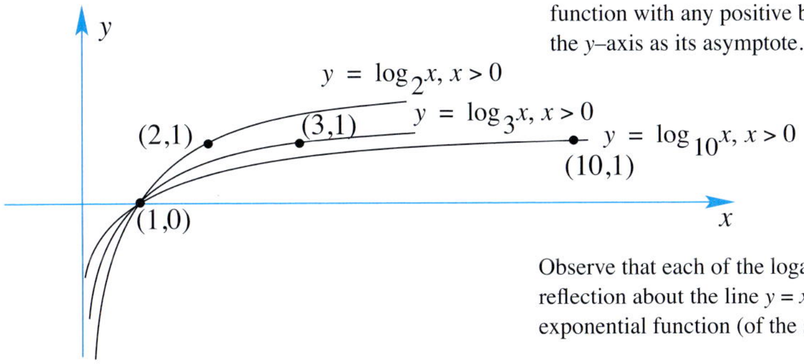

Observe that each of the logarithmic functions is a reflection about the line $y = x$ of its corresponding exponential function (of the same base).

Notice that for all $a > 0$ $\log_a 1 = 0$ and $\log_a a = 1$.

As is the case for the exponential functions, the base 'e' also plays an important role when dealing with logarithmic functions. When using the number 'e' as the base for the logarithmic function, we refer to it as the **natural logarithmic function** and can write it in one of two ways:

$$f(x) = \log_e x, x > 0 \text{ or } f(x) = \ln x, x > 0$$

EXAMPLE 5.25

Sketch the following, specifying the implied domain in each case.

(a) $f(x) = \log_3(x-2)$ (b) $g(x) = \log_2(2x+3)$

Solution

(a) We begin by looking at the domain: As we cannot have the logarithm of a negative number we must have that $x-2>0 \Leftrightarrow x>2$.
Therefore, domain of $f =]2, \infty[$.
Note also then, that by default we have obtained the equation of the vertical asymptote, in this case it is $x = 2$.
When $x = 3$, $f(3) = \log_3 1 = 0$ and
when $x = 5$, $f(5) = \log_3 3 = 1$.

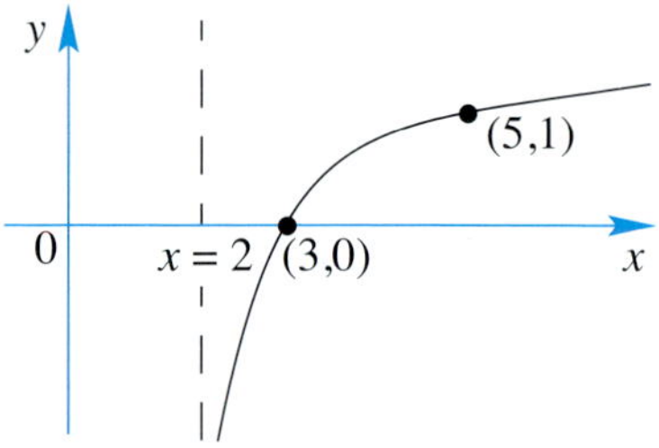

(b) This time we need to have $2x+3>0 \Leftrightarrow x>-\frac{3}{2}$.

Therefore, the implied domain of $g = \left(-\frac{3}{2}, \infty\right)$.

The vertical asymptote has the equation $x = -\frac{3}{2}$.

When $x = -1$, $g(-1) = \log_2(-2+3) = \log_2 1 = 0$.
When $x = 0$, $g(0) = \log_2(0+3) = \log_2 3$.

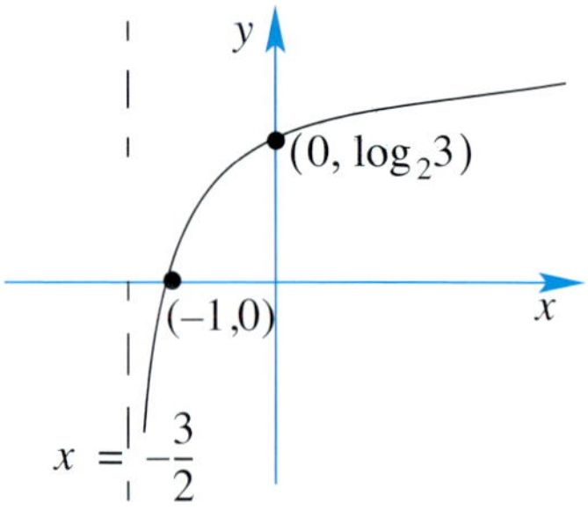

We make the following general observations:

1. The graph of $y = \log_a(x-k), k>0$ is identical to $y = \log_a x$ but moved 'k' units to the right and has a vertical asymptote at $x = k$.
2. The graph of $y = \log_a(x+k), k>0$ is identical to $y = \log_a x$ but moved 'k' units to the left and has a vertical asymptote at $x = -k$.

Notice that in part (b) of example 5.25, we have $g(x) = \log_2(2x+3) = \log_2 2\left(x+\frac{3}{2}\right)$, where we can still 'see' the equation of the asymptote as $x = -\frac{3}{2}$. The extra factor of '2' can be viewed as either a dilation or a translation – we will leave further discussion of this to Chapters 6 and 7.

EXAMPLE 5.26

Sketch the following, specifying the implied domain in each case.

(a) $f(x) = -2\log_3 x$ (b) $f(x) = \frac{1}{3}\log_e x$

Solution

(a) The implied domain in this case is $x > 0$.
So, the vertical asymptote has the equation $x = 0$.
We note that the negative sign in front of the $\log_3 x$ will have the effect of reversing the sign of the $\log_3 x$ values.
That is, the graph of $f(x) = -\log_3 x$ is a reflection about the x–axis of the graph of $y = \log_3 x$.

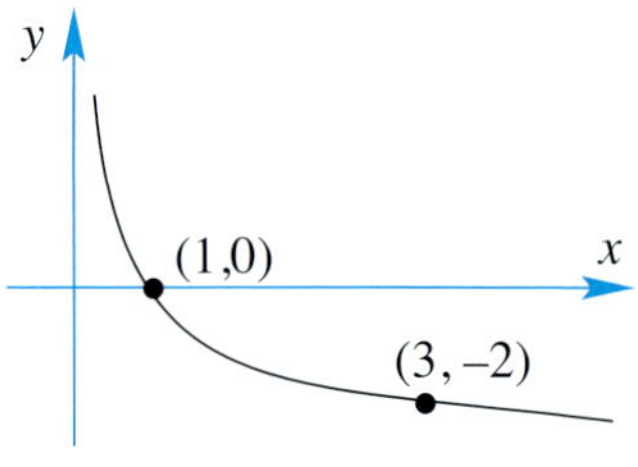

The factor of 2 will have the effect of 'stretching' the graph of $y = \log_3 x$ by a factor of 2 along the y–axis.
Also, we have that $f(1) = -2\log_3 1 = 0$ and $f(3) = -2\log_3 3 = -2$.

(b) This time the implied domain is $]0, \infty[$.
Therefore, the equation of the asymptote is $x = 0$. The one third factor in front of $\log_e x$ will have the effect of 'shrinking' the graph of $y = \log_e x$ by a factor of 3.

Then, $f(1) = \frac{1}{3}\log_e 1 = 0$ and $f(e) = \frac{1}{3}\log_e e = \frac{1}{3}$.

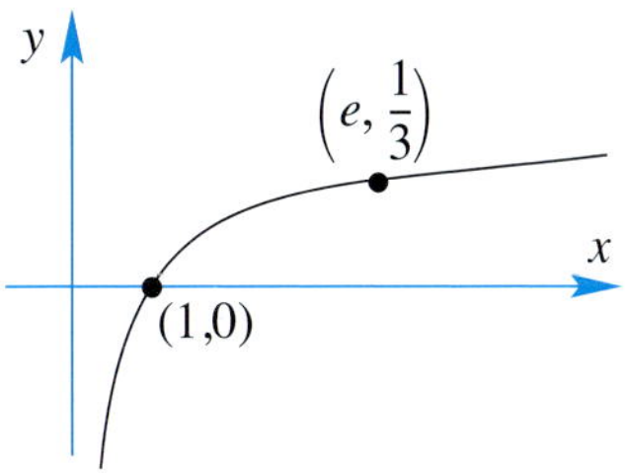

Again, we have the following observations:

1. The graph of $y = k \times \log_a x, k > 0$ is identical to $y = \log_a x$ but
 i. Stretched along the y–axis if $k > 1$.
 ii. Shrunk along the y–axis if $0 < k < 1$.

2. The graph of $y = k \times \log_a x, k < 0$ is identical to $y = \log_a x$ but
 i. Reflected about the x–axis and stretched along the y–axis if $k < -1$.
 ii. Reflected about the x–axis and shrunk along the y–axis if $-1 < k < 0$.

EXAMPLE 5.27

Sketch the following, specifying the implied domain in each case.

(a) $g(x) = \log_2 x + 3$ (b) $g(x) = \log_2 x - 2$ (c) $g(x) = \log_2(3 - x)$

Solution

(a) The effect of adding 3 to the graph of $y = \log_2 x$ will result in $g(x) = \log_2 x + 3$ being moved up 3 units.
Its implied domain is $]0, \infty[$ and its asymptote has equation $x = 0$.

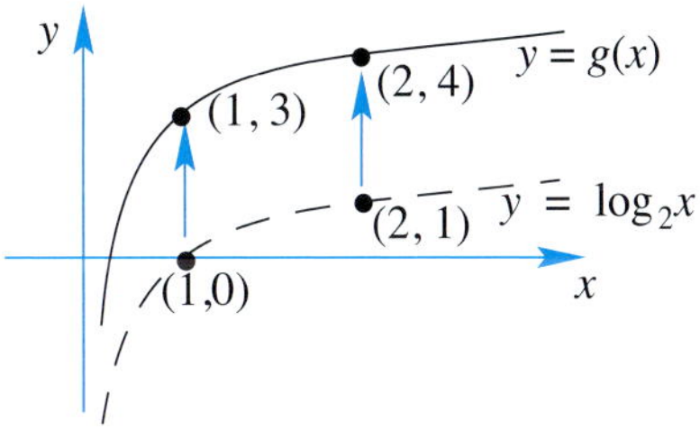

(b) The effect of subtracting 2 from the graph of $y = \log_2 x$ will result in $g(x) = \log_2 x - 2$ being moved down 2 units.
Its implied domain is $]0, \infty[$ and its asymptote has equation $x = 0$.

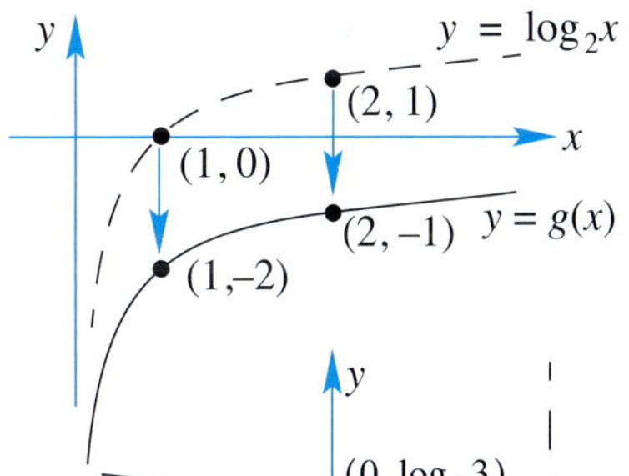

(c) As we cannot have the logarithm of a negative number we must have that $3 - x > 0 \Leftrightarrow x < 3$.
This means that the vertical asymptote is given by $x = 3$ and the graph must be drawn to the left of the asymptote.

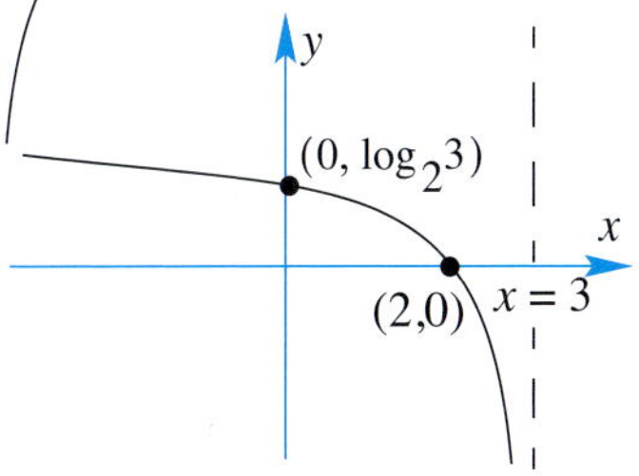

We make the following general observations:

1. The graph of $y = \log_a x - k, k > 0$ is identical to $y = \log_a x$ but moved 'k' units vertically down and has a vertical asymptote at $x = 0$.
2. The graph of $y = \log_a x + k, k > 0$ is identical to $y = \log_a x$ but moved 'k' units vertically up and has a vertical asymptote at $x = 0$.

1. The graph of $y = \log_a(-x)$ is identical to $y = \log_a x$ but reflected about the y–axis and has a vertical asymptote at $x = 0$.
2. The graph of $y = \log_a(k - x), k > 0$ is identical to $y = \log_a x$ but reflected about the y–axis, moved 'k' units to the right and has a vertical asymptote at $x = k$.

EXERCISES 5.3.4

1. Sketch the graph of the following functions, clearly stating domains and labelling asymptotes.

(a) $f(x) = \log_4(x-2)$ (b) $f(x) = \log_2(x+3)$

(c) $h(x) = \log_{10}x + 2$ (d) $g(x) = -3 + \log_3 x$

(e) $f(x) = \log_5(2x-1)$ (f) $h(x) = \log_2(2-x)$

(g) $g(x) = 2\log_{10}x$ (h) $f(x) = -\log_{10}x + 1$

2. Sketch the graph of the following functions, clearly stating domains and labelling asymptotes.

(a) $f(x) = 2\log_2 x + 3$ (b) $f(x) = 10 - 2\log_{10}x$

(c) $h(x) = 2\log_2 2(x-1)$ (d) $g(x) = -\frac{1}{2}\log_{10}(1-x)$

(e) $f(x) = \log_2(3x+2) - 1$ (f) $h(x) = 3\log_2\left(\frac{1}{2}x - 1\right) + 1$

3. Sketch the graph of the following functions, clearly stating domains and labelling asymptotes.

(a) $f(x) = 2\ln x$ (b) $g(x) = -5\ln x$ (c) $f(x) = \ln(x-e)$

(d) $f(x) = \ln(1-ex)$ (e) $f(x) = 5 - \ln x$ (f) $h(x) = \ln x - e$

4. Sketch the graph of the following functions, clearly stating domains and labelling asymptotes.

(a) $f(x) = \log_2\sqrt{x}$ (b) $f(x) = \log_{10}x^2$ (c) $h(x) = \ln|x|$

(d) $g(x) = \ln\left(\frac{1}{x}\right)$ (e) $h(x) = \ln(1-x^2)$ (f) $f(x) = \log_2(x^2-4)$

5. Sketch the graph of the following functions, clearly stating domains and labelling asymptotes.

(a) $f(x) = |\log_{10}x|$ (b) $g(x) = |\log_2(x-1)|$ (c) $h(x) = |\ln x - 1|$

(d) $h(x) = 2 - |\ln x|$ (e) $f(x) = \log_2|x+2|$ (f) $f(x) = \log_5\left|\frac{1}{x}\right|$

6. Given the function $y = f(x)$, sketch the graph of

i. $y = |f(x)|$ ii. $y = f(|x|)$

(a) $f(x) = \log_{10}(-x)$ (b) $f(x) = \ln\left(\frac{1}{x} - e\right)$

(c) $f(x) = 2 - \ln(ex - 1)$ (d) $f(x) = \log_2(x^2 - 2x)$

7. (a) On the same set of axes, sketch the graphs of $f(x) = \ln x - 1$ and $g(x) = \ln(x - e)$.

(b) Find $\{x : \ln x > \ln(x-e) + 1\}$.

8. Sketch the graph of the following functions and find their range.

(a) $f(x) = \begin{cases} \log_{10}x, & x \geq 1 \\ 1 - x, & x < 1 \end{cases}$ (b) $f(x) = \begin{cases} \log_2(x^2 - 1), & |x| \geq 1 \\ 1 - x^2, & |x| < 1 \end{cases}$

(c) $f(x) = \begin{cases} 2 - \ln x, & x \geq e \\ \dfrac{x^3}{e^3}, & x < e \end{cases}$ (d) $g(x) = \begin{cases} 1 + \sqrt{x-1}, & x > 1 \\ |\log_2 x| + 1, & 0 < x \leq 1 \\ 1 & x \leq 0 \end{cases}$

9. Sketch the graph of the following functions.

(a) $f(x) = \log_{\frac{1}{2}}x$ (b) $f(x) = \log_{\frac{1}{2}}(x-2)$ (c) $f(x) = \log_{\frac{1}{3}}x + 1$

10. Sketch the graph of the following functions, clearly stating domains and labelling asymptotes.

(a) $f(x) = 2\log_a(x-a), a > 1$ (b) $f(x) = -\ln(ax - e), a > e$

(c) $g(x) = |\log_{10}(10 - ax)|, 1 < a < 10$ (d) $g(x) = \ln|x - ae|, a > 1$

(e) $g(x) = |\ln|x - ae||, a > 1$ (f) $h(x) = \log_a\left(1 - \frac{x}{a}\right), 0 < a < 1$

11. Sketch the graph of $f(x) = \frac{1}{a}\log_a(ax - 1), 0 < a < 1$ clearly labelling its asymptote, and intercept(s) with the axes. Hence, find $\left\{x : f(x) > \frac{1}{a}\right\}$.

12. Sketch the graph of (a) $f(x) = \frac{\ln x}{x}, x > 0$ (b) $g(x) = \frac{x}{\ln x}, x > 0$

Given that $f(x) \leq e^{-1}$ for all real $x > 0$, state the range of $g(x)$.

5.3.5 EQUATIONS OF THE FORM $y = x^n, n = -1, -2$

Case 1: $n = -1$

With $n = -1$ we have the equation $y = x^{-1} = \frac{1}{x}, x \neq 0$ – this graph has a shape known as a **rectangular hyperbola**.

Making use of the TI–83 we make the following observations:

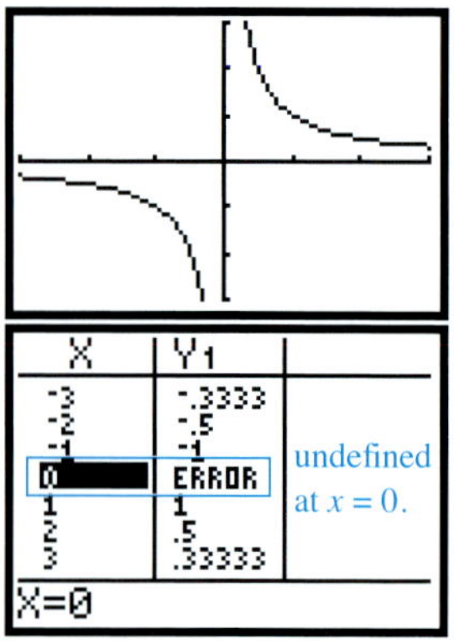

1. The sign of y is the same as the sign of x.
2. As x increases, values of y decrease.
3. The function is **undefined** at $x = 0$.
   At $x = 0$ there exists a **vertical asymptote**.
4. The graph never crosses (or touches) the x–axis.
   The x–axis then becomes a **horizontal asymptote**, having the equation $y = 0$.

We now formalise some of these observations and introduce some new mathematical terms.

We first consider the interval $x > 0$:

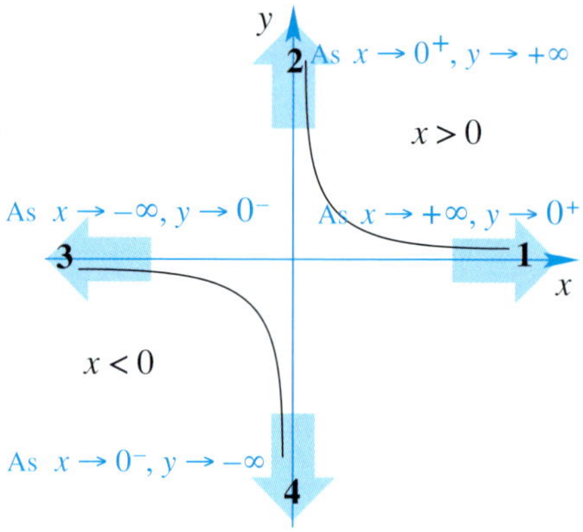

1. As x **tends** to (i.e., approaches) **positive infinity**, the values of y **tend** to (i.e., approach) **zero from above**. i.e., As $x \to +\infty, y \to 0^+$.

2. As x **tends** to (i.e., approaches) **zero from above**, the values of y **tend** to (i.e., approach) **positive infinity**. i.e., As $x \to 0^+, y \to +\infty$.

Next we consider the interval $x < 0$:

3. As x **tends** to (i.e., approaches) **negative infinity**, the values of y **tend** to (i.e., approach) **zero from below**. i.e., As $x \to -\infty, y \to 0^-$.

4. As x **tends** to (i.e., approaches) **zero from below**, the values of y **tend** to (i.e., approach) **negative infinity**. i.e., As $x \to 0^-, y \to -\infty$.

The expression '*approaching zero from above*' means that a variable is approaching the value zero, but the values are always greater than zero. e.g., the sequence 0.1, 0.001, 0.0001, . . . approaches zero from above (as the values are always greater than zero).

The expression '*approaching zero from below*' means that a variable is approaching the value zero, but the values are always less than zero. e.g., the sequence –0.1, –0.001, –0.0001, . . . approaches zero from below (as the values are always less than zero).

As was the case for the exponential and logarithmic functions, the horizontal and vertical asymptotes of the basic function $f(x) = \frac{1}{x}, x \neq 0$ can also be 'relocated'. We summarise these results now:

1. The graph of $y = \frac{1}{x-k}, k > 0$ is identical to $y = \frac{1}{x}$ but moved 'k' units to the right and so has a vertical asymptote at $x = k$.

2. The graph of $y = \frac{1}{x+k}, k > 0$ is identical to $y = \frac{1}{x}$ but moved 'k' units to the left and so has a vertical asymptote at $x = -k$.

1. The graph of $y = \frac{1}{x} + k, k > 0$ is identical to $y = \frac{1}{x}$ but moved 'k' units up and so has a horizontal asymptote at $y = k$.

2. The graph of $y = \frac{1}{x} - k, k > 0$ is identical to $y = \frac{1}{x}$ but moved 'k' units down and so has a horizontal asymptote at $y = -k$.

Case 2: $n = -2$

With $n = -2$ we have the equation $y = x^{-2} = \frac{1}{x^2}, x \neq 0$ – this graph has a shape known as a **truncus**.

The reason for its name becomes obvious once we sketch its graph – it looks like the trunk of a tree. As before, we can make use of a table of values and plot its graph, however, this time we list the properties of this function and its graph:

1. Function is **undefined** at $x = 0$.
2. Asymptotes are: vertical, $x = 0$
   horizontal, $y = 0$
3. The graph is symmetrical about the y–axis.

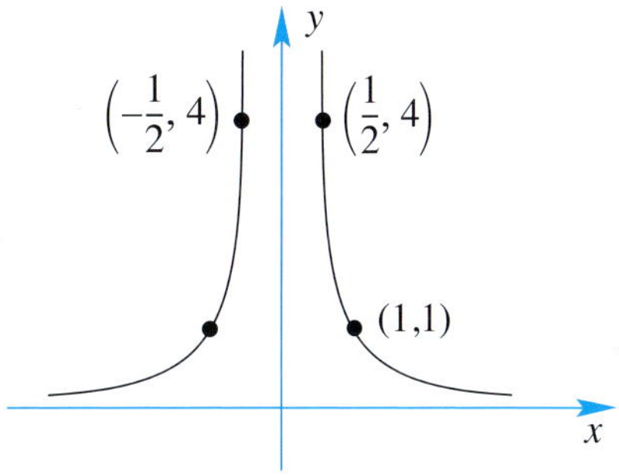

We can also make the following observations:

1. The graph of $y = \frac{1}{(x-k)^2}, k > 0$ is identical to $y = \frac{1}{x^2}$ but moved 'k' units to the right and so has a vertical asymptote at $x = k$.

2. The graph of $y = \frac{1}{(x+k)^2}, k > 0$ is identical to $y = \frac{1}{x^2}$ but moved 'k' units to the left and so has a vertical asymptote at $x = -k$.

1. The graph of $y = \frac{1}{x^2} + k, k > 0$ is identical to $y = \frac{1}{x^2}$ but moved 'k' units up and so has a horizontal asymptote at $y = k$.
2. The graph of $y = \frac{1}{x^2} - k, k > 0$ is identical to $y = \frac{1}{x^2}$ but moved 'k' units down and so has a horizontal asymptote at $y = -k$.

EXAMPLE 5.28

Sketch the following, specifying the implied domain in each case.

(a) $f(x) = \frac{1}{x} + 1$ (b) $g(x) = \frac{1}{(x+1)^2}$ (c) $h(x) = \frac{1}{x-1} - 2$

Solution

(a) $d_f =]-\infty, \infty[\backslash\{0\}$ (b) $d_g =]-\infty, \infty[\backslash\{-1\}$ (c) $d_h =]-\infty, \infty[\backslash\{1\}$

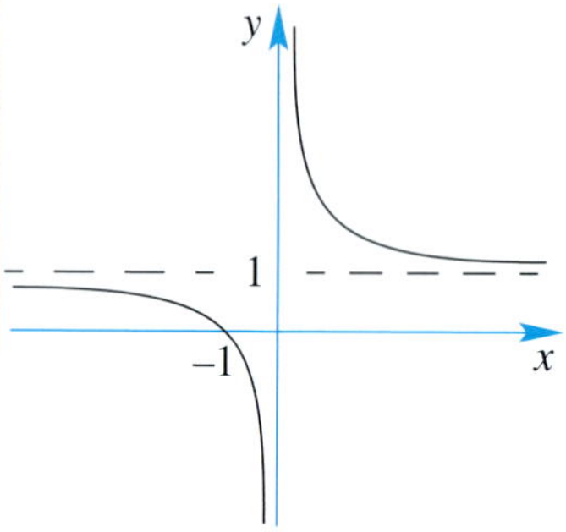

Asymptotes: vertical, $x = 0$
horizontal, $y = 1$

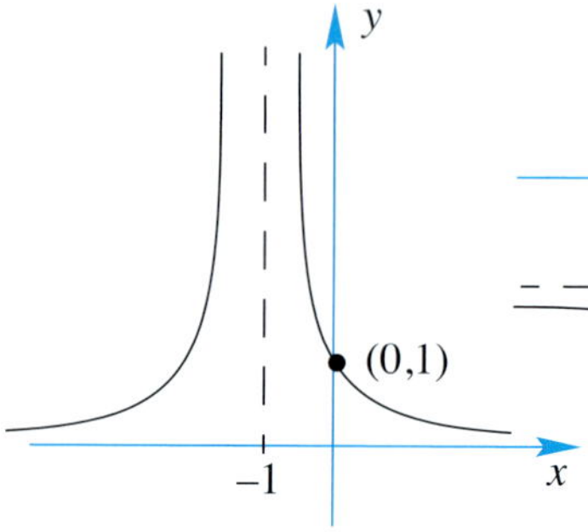

Asymptotes: vertical, $x = -1$
horizontal, $y = 0$

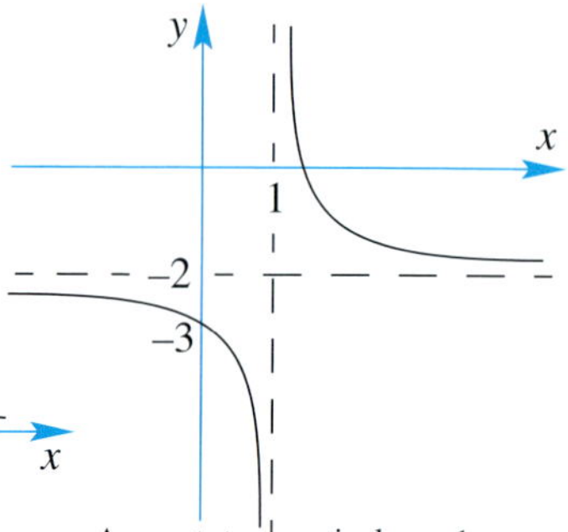

Asymptotes: vertical, $x = 1$
horizontal, $y = -2$

EXAMPLE 5.29

Sketch the following, specifying the implied domain in each case.

(a) $f(x) = 2 - \frac{2}{x}$ (b) $g(x) = -\frac{1}{2x^2} + 2$

Solution

(a) The efffect of the '2' in the $\frac{2}{x}$ term is to stretch the graph of $y = \frac{1}{x}$ along the y–axis by a factor of 2. The '–ve' in front of the $\frac{2}{x}$ term will reflect the the graph of $y = \frac{2}{x}$ about the x–axis. Adding '2' to the graph of $y = -\frac{2}{x}$ will move the graph up '2' units.

The domain of this function is given by $]-\infty, \infty[\backslash \{0\}$ and it has two asymptotes. The vertical asymptote is at $x = 0$ and the horizontal asymptote is at $y = 2$.

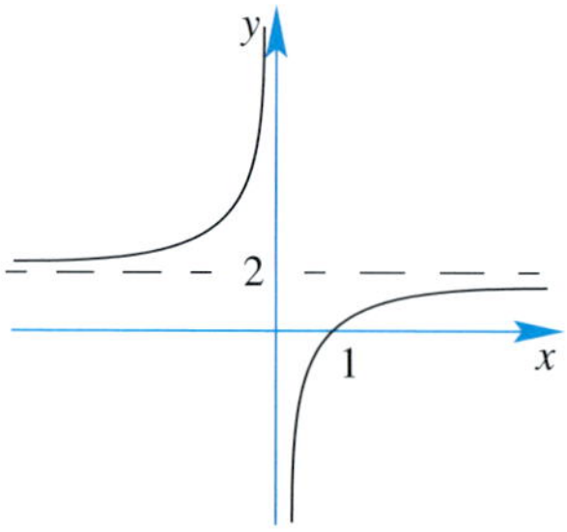

To find the x–intercept, set $y = 0$:

$$y = 0 \Leftrightarrow 2 - \frac{2}{x} = 0 \Leftrightarrow 2 = \frac{2}{x} \Leftrightarrow x = 1$$

(b) Using the same argument as in part (a), except this time the original graph of $y = \frac{1}{x^2}$ is 'shrunk' by a factor of '2' (or stretched by a factor of 0.5) along the y–axis, reflected about the x–axis, and moved up '2' units, we have:

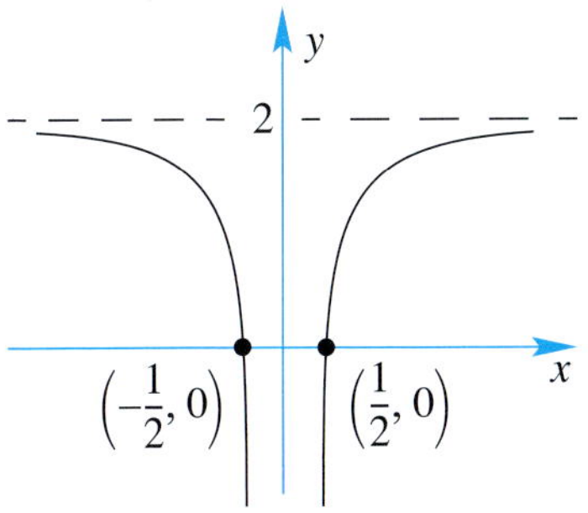

To find the x–intercept, set $y = 0$:

$$y = 0 \Leftrightarrow -\frac{1}{2x^2} + 2 = 0 \Leftrightarrow \frac{1}{2x^2} = 2$$

$$\Leftrightarrow x^2 = \frac{1}{4}$$

$$\Leftrightarrow x = \pm\frac{1}{2}$$

EXERCISES 5.3.5

1. Sketch the graphs of the following functions, clearly identifying the equations of all asymptotes and intercepts with the axes.

(a) $f(x) = \frac{1}{x+2}, x \neq -2$ (b) $g(x) = \frac{2}{x-2}, x \neq 2$

(c) $f(x) = 3 + \frac{1}{x}, x \neq 0$ (d) $h(x) = 2 - \frac{3}{x}, x \neq 0$

(e) $f(x) = \frac{4}{3-x}, x \neq 3$ (f) $f(x) = \frac{3}{2(x-1)}, x \neq 1$

(g) $g(x) = \frac{1}{2} - \frac{2}{x+1}, x \neq -1$ (h) $f(x) = \frac{6}{2+x} - 3, x \neq -2$

2. Sketch the graphs of the following functions, clearly identifying the equations of all asymptotes and intercepts with the axes.

(a) $g(x) = \frac{2}{x^2} + 1, x \neq 0$ (b) $f(x) = -\frac{4}{(x+1)^2}, x \neq -1$

(c) $f(x) = 4 - \frac{16}{x^2}, x \neq 0$ (d) $g(x) = -2 + \frac{4}{(x-1)^2}, x \neq 1$

3. Given that $\frac{2x+3}{x+1} = a + \frac{b}{x+1}$, find a and b. Hence sketch the graph of $f: x \mapsto \left|\frac{2x+3}{x+1}\right|$.

4. For the given function, f, sketch the graph of i. $y = |f(x)|$ ii. $y = f(|x|)$

(a) $f(x) = \frac{1}{x} - 1$ (b) $f(x) = \frac{2}{2-x} + 1$ (c) $f(x) = 2 - \frac{1}{x+1}$

(d) $f(x) = \frac{4}{(2-x)^2}$ (e) $f(x) = 2 - \frac{1}{2x^2}$ (f) $f(x) = \frac{2}{(1+x)^2}$

5.4 ALGEBRA OF FUNCTIONS

5.4.1 BASIC OPERATIONS AND COMPOSITE FUNCTIONS

For any two real functions $f: d_f \mapsto \mathbb{R}, y = f(x)$ and $g: d_g \mapsto \mathbb{R}, y = g(x)$, defined over domains d_f and d_g respectively, the following rules of algebra apply:

Operation	Rule and Domain	Example
Addition	**rule:** $(f+g)(x) = f(x)+g(x)$ **domain:** $d_{f+g} = d_f \cap d_g$	For $f:[0, \infty) \mapsto \mathbb{R}, f(x) = \sqrt{x}$ and $g:\mathbb{R} \mapsto \mathbb{R}, g(x) = x$, we have: $d_{f+g} = d_f \cap d_g = [0, \infty) \cap \mathbb{R}$ $= [0, \infty)$ $\therefore f+g:[0, \infty) \mapsto \mathbb{R}, (f+g)(x) = \sqrt{x}+x$.
Subtraction	**rule:** $(f-g)(x) = f(x)-g(x)$ **domain:** $d_{f-g} = d_f \cap d_g$	The process is the same as that for $(f+g)(x)$. Using the same functions as those above, we have: $\therefore f-g:[0, \infty) \mapsto \mathbb{R}, (f-g)(x) = \sqrt{x}-x$.
Multiplication	**rule:** $(f \times g)(x) = f(x) \times g(x)$ or $(fg)(x) = f(x)g(x)$ **domain:** $d_{f \times g} = d_f \cap d_g$ or $d_{fg} = d_f \cap d_g$	If $f(x) = \frac{2}{x-1}$, where $x \in \mathbb{R}\backslash\{1\}$ and $g(x) = 1-x^2$, where $x \in [0, \infty)$, we have: $d_{f \times g} = d_f \cap d_g$ $= \mathbb{R}\backslash\{1\} \cap [0, \infty)$ $= [0, \infty)\backslash\{1\}$ $\therefore (f \times g)(x) = \frac{2}{x-1} \times (1-x^2) = -2(1+x)$ That is, $(fg)(x) = -2(1+x)$, where $x \in [0, \infty)\backslash\{1\}$
Division	**rule:** $\left(\frac{f}{g}\right)(x) = \frac{f(x)}{g(x)}, g(x) \neq 0$ **domain:** $d_{\frac{f}{g}} = d_f \cap d_g \backslash\{x:g(x) = 0\}$	If $f:\mathbb{R} \mapsto \mathbb{R}, f(x) = x^2$ and $g:\mathbb{R} \mapsto \mathbb{R}, g(x) = x+1$, we have: domain $d_{\frac{f}{g}} = d_f \cap d_g \backslash\{x:g(x) = 0\}$ $= \mathbb{R} \cap \mathbb{R}\backslash\{x:x+1 = 0\}$ $= \mathbb{R}\backslash\{-1\}$ $\therefore \left(\frac{f}{g}\right)(x) = \frac{x^2}{x+1}, x \in \mathbb{R}\backslash\{-1\}$

COMPOSITE FUNCTIONS

We now investigate another way in which we can combine functions, namely **composition**. Consider the two functions $f(x) = 3x$ and $g(x) = x^2 + 1$. Observe what happens to the value $x = 2$ as we first apply the function $f(x)$ and then the function $g(x)$ to the image of the first mapping, i.e.,

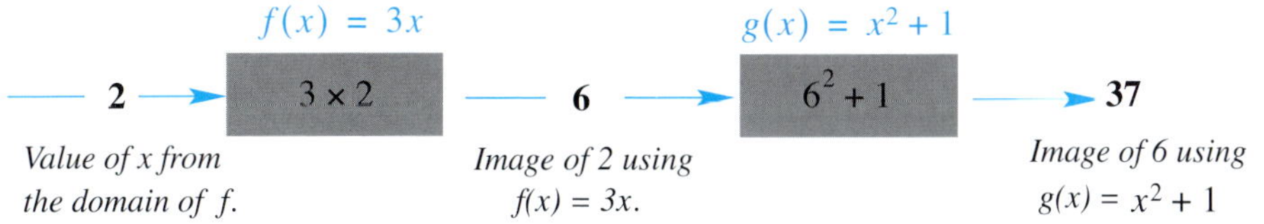

Such a combination of functions leads to the question

"Is there a third function that will enable us to produce the same result in one step?"

We consider any value x that belongs to the domain of f and follow 'its path':

1. This value of x, has as its image the value $f(x) = 3x$.
2. The resulting number, $3x$, now represents an element of the domain of g.
3. The image of $3x$ under the mapping g is given by $g(3x) = (3x)^2 + 1 = 9x^2 + 1$.

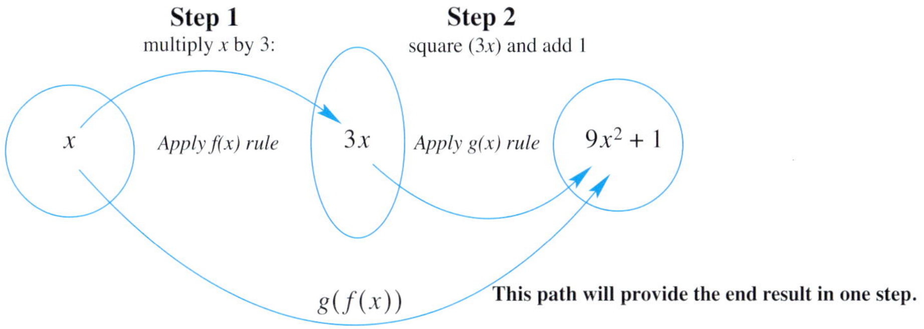

We can now test our result by using the value of $x = 2$ with the mapping $x \mapsto 9x^2 + 1$.

For $x = 2$, we have $9(2)^2 + 1 = 9 \times 4 + 1 = 37$, which agrees with our previous result.

The two **critical steps** in this process are:

1. That the image under the first mapping must belong to the domain of the second mapping.
2. The expression $g(f(x))$ exists.

Notation

The expression $g(f(x))$ is called the composite function of f and g and is denoted by $g \circ f$.

(Notice that although f is applied first, it is placed second in the expression $g \circ f$.)

EXAMPLE 5.30

Given the functions $f(x) = x^2 + 1$ and $g(x) = \ln(x - 1)$, find the composite function $(g \circ f)(x)$.

Solution

The composite function $(g \circ f)(x) = g(f(x))$

$$= \ln(f(x) - 1)$$
$$= \ln(x^2 + 1 - 1)$$
$$= \ln x^2$$

In Example 5.30 we have $(g \circ f)(-1) = \ln(-1)^2 = \ln 1 = 0$ and $(g \circ f)(2) = \ln 2^2 = \ln 4$.

EXAMPLE 5.31

Given the functions $f(x) = 2 - x$ and $g(x) = \sqrt{x-1}$, find the composite function $(g \circ f)(x)$.

Solution

The composite function $(g \circ f)(x) = g(f(x))$

$$= \sqrt{f(x) - 1}$$
$$= \sqrt{(2 - x) - 1}$$
$$= \sqrt{1 - x}$$

In Example 5.31 we have $(g \circ f)(-1) = \sqrt{1-(-1)} = \sqrt{2}$ but $(g \circ f)(2) = \sqrt{1-2} = \sqrt{-1}$ is undefined! So what went wrong? To answer this question let's take another look at the composition process.

The process is made up of two stages:

Stage 1: An element from the domain of the first function, $f(x)$ is used to produce an image. That is, using $x = a$ we produce the image $f(a)$.

Stage 2: Using the second function, $g(x)$, the image, $f(a)$, is used to produce a second image $g(f(a))$.

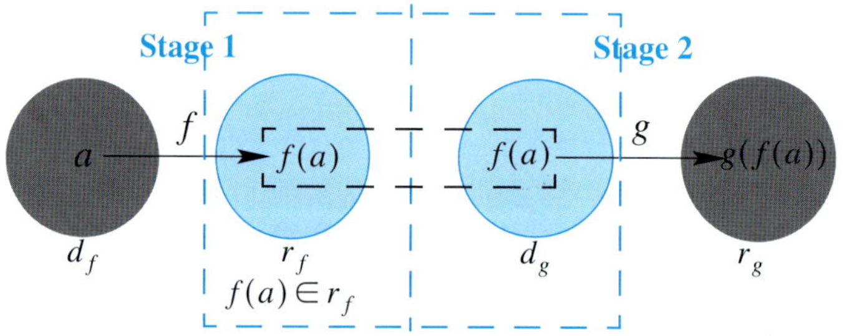

From the diagram, the result of stage 1 is $f(a)$ (which belongs to the range of f) we also observe that at stage 2, when using the value $f(a)$ (produced from stage 1) we have assumed that $f(a)$ belongs to the domain of $g(x)$. This is where problems can arise – as seen in Example 5.31.

To overcome this difficulty we need to strengthen our definition of composition of functions as well as ensuring the existence of composite functions.

What we need to prove is that all values produced from stage 1, i.e., $f(a)$, are values in the domain of the function in stage 2, i.e., $f(a) \in d_g$. Making use of a mapping diagram, we show the inter-relation between the range of *f*, r_f, and the domain of *g*, d_g.

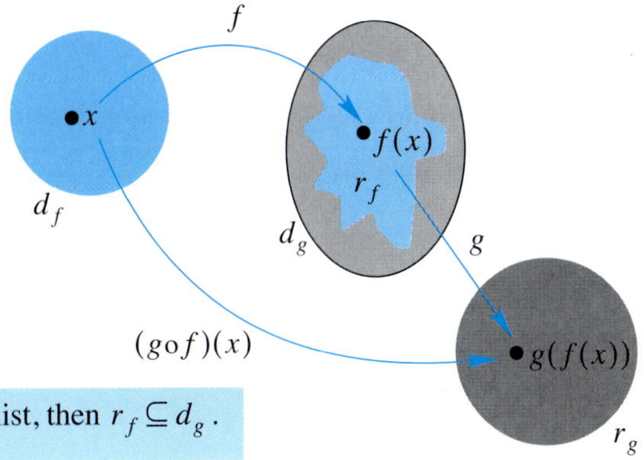

For $(g \circ f)(x) = g(f(x))$ to exist, then $r_f \subseteq d_g$.

What is the domain of $g \circ f$?

If we refer to the diagram alongside, we see that

$$\text{if } r_f \subseteq d_g \text{, then } d_{g \circ f} = d_f.$$

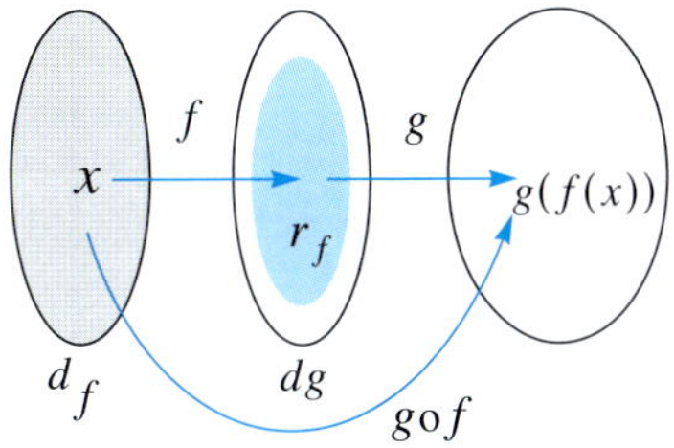

This means that we can substitute values of *x* that belong to the domain of *f* directly into the expression $g(f(x))$ (once we have established that it exists).

EXAMPLE 5.32

If $f(x) = \sqrt{x+1}, x \in (0,\infty)$ and $g(x) = x^3, x \in \mathbb{R}$, determine if $g \circ f$ exists, and find an expression for $g \circ f$ if it does.

Solution

For $g \circ f$ to exist we must have that $r_f \subseteq d_g$.

Using the TI–83 we obtain the range of f from its sketch, in this case, $r_f = (1,\infty)$.

The domain of g is $(-\infty,\infty)$ (i.e., the real field).

Then, given that $(1,\infty) \subseteq (-\infty,\infty)$, $g \circ f$ does exist.

We are now able to determine $g \circ f$.

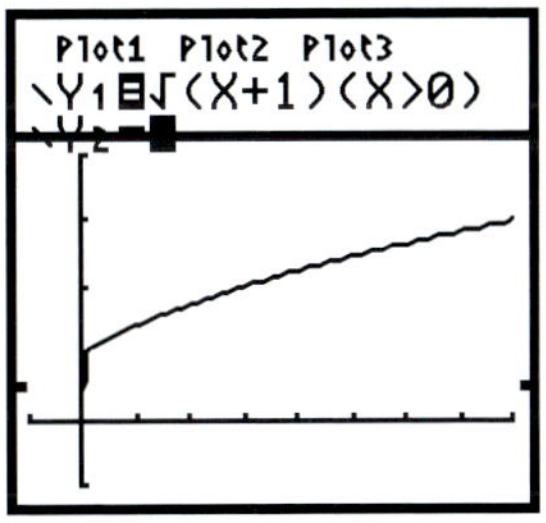

First we determine the equation $g(f(x))$: $g(f(x)) = g(\sqrt{x+1})$

$$= (\sqrt{x+1})^3$$

$$= (x+1)^{3/2}.$$

Next we need the domain of $g \circ f$. As we have seen, $d_{g \circ f} = d_f$, $\therefore d_{g \circ f} = (0,\infty)$.

Therefore, $g \circ f : (0,\infty) \mapsto \mathbb{R}$, $(g \circ f)(x) = (x+1)^{3/2}$.

Hint on setting out

When solving problems that involve the use of composition, it is useful to set up a **domain–range table** in order to help us determine the existence of the composition. Such a table includes information about the domain and range of both the functions under consideration:

	domain	range
f	d_f	r_f
g	d_g	r_g

The existence of $g\circ f$ can then be established by looking at r_f and d_g. Similarly, the existence of $f\circ g$ can be established by comparing r_g and d_f.

EXAMPLE 5.33

Find $g\circ f$ and its range, given that $g(x) = \frac{1}{x+1}, x \in \mathbb{R}\backslash\{-1\}$ and $f(x) = 2^x, x \in \mathbb{R}$.

Solution

We first sketch the graphs of both functions to help us complete the domain–range table:

$g(x) = \frac{1}{x+1}, x \in \mathbb{R}\backslash\{-1\}$, $f(x) = 2^x, x \in \mathbb{R}$. We now complete the table:

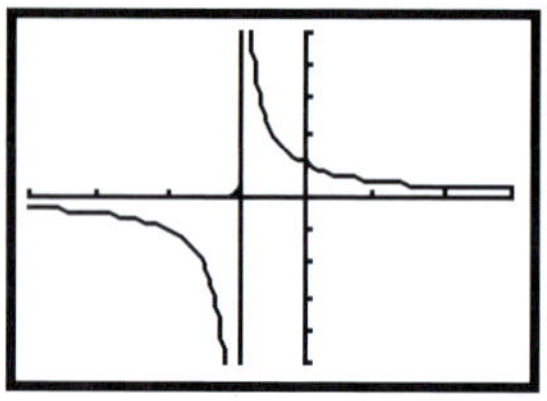

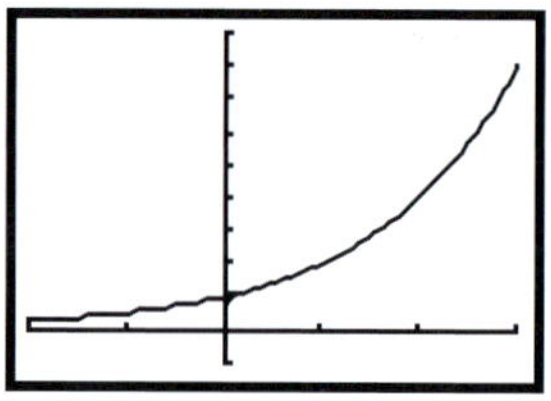

	domain	range
f	$\mathbb{R}$	$]0, \infty[$
g	$\mathbb{R}\backslash\{-1\}$	$\mathbb{R}\backslash\{0\}$

Using the table we see that $r_f \subseteq d_g \Rightarrow g\circ f$ exists.

We can now determine $g\circ f$: $g(f(x)) = \frac{1}{f(x)+1} = \frac{1}{2^x+1}$.

We also have that $d_{g\circ f} = d_f$, $\therefore d_{g\circ f} = \mathbb{R}$.

Therefore, $g\circ f : \mathbb{R} \mapsto \mathbb{R}$, where $(g\circ f)(x) = \frac{1}{2^x+1}$.

Making use of the TI–83 we see that the range of $g\circ f$ is $]0, 1[$.

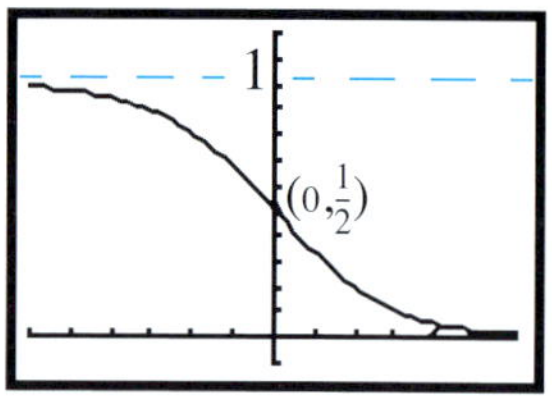

EXAMPLE 5.34

Given that $f(x) = \sqrt{x-1}$ and $g(x) = \ln x$. Does $f\circ g$ exist? If so, define fully the function $f\circ g$. If not, find a suitable restriction on the domain of g so that $f\circ g$ exists.

Solution

For $f\circ g$ to exist it is necessary that $r_g \subseteq d_f$. To determine the range of g we need to know the domain of g. Using the implied domain we have that $d_g =]0, \infty[$ and so, $r_g = \mathbb{R}$. However, the implied domain of f is $[1, \infty[$. Then, as $r_g \not\subseteq d_f$, $f\circ g$ does not exist.

In order that $f\circ g$ exists we need to have $r_g \subseteq [1, \infty[$, i.e., we must have that $g(x) \geq 1$. What remains then, is to find those values of x such that $g(x) \geq 1$.

Now, $g(x) = \ln x$ therefore, $g(x) \geq 1 \Leftrightarrow \ln x \geq 1 \Leftrightarrow x \geq e$.

So, if the domain of g is restricted to $[e, \infty[$ or any subset of $[e, \infty[$, then $f \circ g$ will exist.

Does $g \circ f = f \circ g$?

In general the answer is **no**! However, there exist situations when $(f \circ g)(x) = (g \circ f)(x)$ – we will look at such cases in the next section.

Consider Example 5.33, where $g(x) = \frac{1}{x+1}, x \in \mathbb{R}\backslash\{-1\}$ and $f(x) = 2^x, x \in \mathbb{R}$.

From our previous working, we have that $(g \circ f)(x) = \frac{1}{2^x + 1}$.

To determine if $(f \circ g)(x)$ exists, we will need to determine if $r_g \subseteq d_f$. Using the domain–range table we have that $r_g = \mathbb{R}\backslash\{0\}$ and $d_f = \mathbb{R}$. Therefore as $r_g \subseteq d_f \Rightarrow f \circ g$ does exist.

We then have, $$(f \circ g)(x) = f(g(x)) = 2^{g(x)} = 2^{\frac{1}{x+1}}.$$

To determine the domain of $f \circ g$, we use the fact that, $d_{f \circ g} = d_g$ so that $d_g = \backslash\{\mathbb{R}1\}$.

Then, $f \circ g : \mathbb{R}\backslash\{-1\} \mapsto \mathbb{R}$, where $(f \circ g)(x) = 2^{\frac{1}{x+1}}$.

Clearly then, in this case, $(f \circ g)(x) \neq (g \circ f)(x)$.

EXAMPLE 5.35

Given $f(x) = x^2 + 1$, where $x \geq 0$ and $g(x) = x - 1$, where $x \geq 1$, determine the functions $(f \circ g)(x)$ and $(g \circ f)(x)$ (if they exist). For the composite functions that exist, find the image of $x = 3$.

Solution

We first set up the domain–range table:

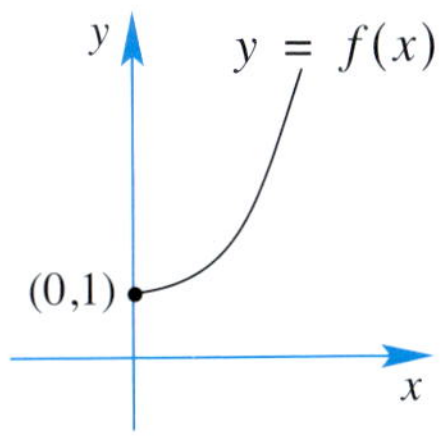

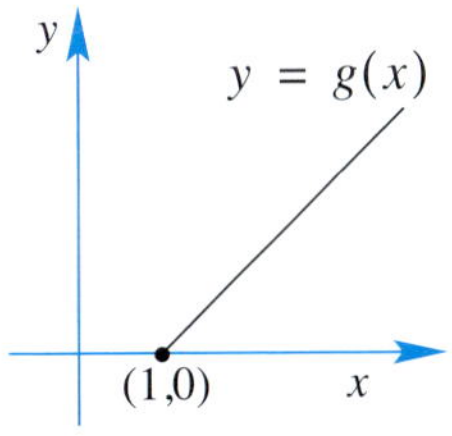

	domain	range
f	$[0, \infty[$	$[1, \infty[$
g	$[1, \infty[$	$[0, \infty[$

From the table we see that $r_g \subseteq d_f$ and $r_f \subseteq d_g$, and so, both $(f \circ g)(x)$ and $(g \circ f)(x)$ exist. We can now determine both composite functions.

We start with $(f \circ g)(x)$:

$$(f \circ g)(x) = f(g(x)) = (g(x))^2 + 1 = (x-1)^2 + 1$$

As $d_{fog} = d_g = [1, \infty[$, we have that $(fog)(x) = (x-1)^2 + 1$, where $x \geq 1$

Next, we find $(gof)(x)$:

$$(gof)(x) = g(f(x)) = f(x) - 1 = (x^2 + 1) - 1$$
$$= x^2$$

Similarly, $d_{gof} = d_f = [0, \infty[$, and so, $(gof)(x) = x^2$, where $x \geq 0$.

To find the image of 3, we substitute $x = 3$ into the final equations.
$(fog)(3) = (3-1)^2 + 1 = 2^2 + 1 = 5$, whereas, $(gof)(3) = (3)^2 = 9$.

EXERCISES 5.4.1

1. Fully define the functions (a) $f + g$ (b) fg

given that i. $f(x) = x^2$ and $g(x) = \sqrt{x}$

ii. $f(x) = \ln x$ and $g(x) = \frac{1}{x}$

iii. $f(x) = \sqrt{9 - x^2}$ and $g(x) = \sqrt{x^2 - 4}$

Find the range for case (a).

2. Fully define the functions (a) $f - g$ (b) f/g

given that i. $f(x) = e^x$ and $g(x) = 1 - e^x$

ii. $f(x) = x + 1$ and $g(x) = \sqrt{x+1}$

iii. $f(x) = |x - 2|$ and $g(x) = |x + 2|$

Find the range for case (a).

3. All of the following functions are mappings of $\mathbb{R} \mapsto \mathbb{R}$ unless otherwise stated.

(a) Determine the composite functions $(fog)(x)$ and $(gof)(x)$, if they exist.

(b) For the composite functions in (a) that do exist, find their range.

i. $f(x) = x + 1, g(x) = x^3$

ii. $f(x) = x^2 + 1, g(x) = \sqrt{x}, x \geq 0$

iii. $f(x) = (x+2)^2, g(x) = x - 2$

iv. $f(x) = \frac{1}{x}, x \neq 0, g(x) = \frac{1}{x}, x \neq 0$,

v. $f(x) = x^2, g(x) = \sqrt{x}, x \geq 0$

vi. $f(x) = x^2 - 1, g(x) = \frac{1}{x}, x \neq 0$

vii. $f(x) = \frac{1}{x}, x \neq 0, g(x) = \frac{1}{x^2}, x \neq 0$

viii. $f(x) = x - 4, g(x) = |x|$

ix. $f(x) = x^3 - 2, g(x) = |x + 2|$

x. $f(x) = \sqrt{4 - x}, x \leq 4, g(x) = x^2$

xi. $f(x) = \frac{x}{x+1}, x \neq -1, g(x) = x^2$

xii. $f(x) = x^2 + x + 1, g(x) = |x|$

xiii. $f(x) = 2^x, g(x) = x^2$ xiv. $f(x) = \frac{1}{x+1}, x \neq -1, g(x) = x - 1$

xv. $f(x) = \frac{2}{\sqrt{x-1}}, x > 1, g(x) = x^2 + 1$ xvi. $f(x) = 4^x, g(x) = \sqrt{x}$

4. Given the functions $f:x \mapsto 2x + 1, x \in]-\infty,\infty[$ and $g:x \mapsto x + 1, x \in]-\infty,\infty[$
Find the functions (a) $(f \circ g)$ (b) $(g \circ f)$ (c) $(f \circ f)$

5. Given that $f:x \mapsto x + 1, x \in \mathbb{R}$ and $g \circ f:x \mapsto x^2 + 2x + 2, x \in \mathbb{R}$, determine the function g.

6. The functions f and g are defined by $f:x \mapsto x + 1, x \in \mathbb{R}$ and $g:x \mapsto x + \frac{1}{x}, x \in \mathbb{R} \backslash \{0\}$.
Find the composite functions (where they exist) of
(a) $f \circ g$ (b) $g \circ f$ (c) $g \circ g$ stating the range in each case.

7. If $g:x \mapsto x^3 + 1, x \in \mathbb{R}$ and $f:x \mapsto \sqrt{x}, x \in [0, \infty[$, evaluate (a) $(g \circ f)(4)$
(b) $(f \circ g)(2)$

8. Given that $f:x \mapsto x + 5, x \in \mathbb{R}$ and $h:x \mapsto x - 7, x \in \mathbb{R}$, show that $(f \circ h)(x)$ is equal to $(h \circ f)(x)$ for all $x \in \mathbb{R}$.

9. Solve the equation $(f \circ g)(x) = 0$, where
(a) $f:x \mapsto x + 5, x \in \mathbb{R}$ and $g:x \mapsto x^2 - 6, x \in \mathbb{R}$.
(b) $f:x \mapsto x^2 - 4, x \in \mathbb{R}$ and $g:x \mapsto x + 1, x \in \mathbb{R}$.

10. Given that $f:x \mapsto 2x + 1, x \in \mathbb{R}$ determine the two functions g, given that
(a) $(g \circ f)(x) = \frac{1}{2x+1}$ (b) $(f \circ g)(x) = \frac{1}{2x+1}$

11. Find $(h \circ f)(x)$, given that $h(x) = \begin{cases} x^2 + 4, & x \geq 1 \\ 4 - x, & x < 1 \end{cases}$ and $f:x \mapsto x - 1, x \in \mathbb{R}$.

Sketch the graph of $(h \circ f)(x)$ and use it to find its range.

12. (a) Given three functions, f, g and h, when would $h \circ g \circ f$ exist?
(b) If $f:x \mapsto x + 1, x \in \mathbb{R}$, $g:x \mapsto x^2, x \in \mathbb{R}$ and $h:x \mapsto 4x, x \in \mathbb{R}$, find $(h \circ g \circ f)(x)$.

13. Given the functions $f(x) = e^{2x-1}$ and $g(x) = \frac{1}{2}(\ln x + 1)$ find, where they exist
(a) $(f \circ g)$ (b) $(g \circ f)$ (c) $(f \circ f)$
In each case find the range of the composite function.

14. Given that $h(x) = \log_{10}(4x - 1), x > \frac{1}{4}$ and $k(x) = 4x - 1, x \in]-\infty, \infty[$, find, where they exist, (a) $(h \circ k)$ (b) $(k \circ h)$.

15. Given the functions $f(x) = \sqrt{x^2 - 9}, x \in \mathbf{S}$ and $g(x) = |x| - 3, x \in \mathbf{T}$, find the largest positive subsets of $\mathbb{R}$ so that (a) $g \circ f$ exists (b) $f \circ g$ exists.

16. For each of the following functions
(a) determine if $f \circ g$ exists and sketch the graph of $f \circ g$ when it exists.
(b) determine if $g \circ f$ exists and sketch the graph of $g \circ f$ when it exists.

i.
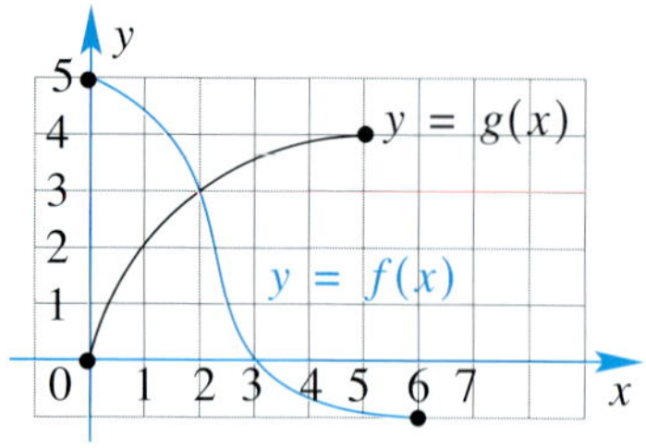

ii.
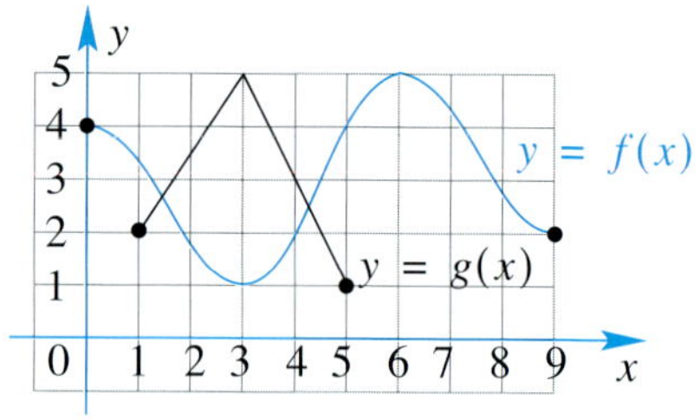

17. Given the functions $f : S \to \mathbb{R}$ where $f(x) = e^{x+1}$ and $g : S \to \mathbb{R}$ where $g(x) = \ln 2x$ where $S =]0, \infty[$.
(a) State the domain and range of both f and g.
(b) Giving reasons, show that $g \circ f$ exists but $f \circ g$ does not exist.
(c) Fully define $g \circ f$, sketch its graph and state its range.

18. The functions f and g are given by $f(x) = \begin{cases} \sqrt{x-1} & \text{if } x \geq 1 \\ x - 1 & \text{if } 0 < x < 1 \end{cases}$ and $g(x) = x^2 + 1$.

(a) Show that $f \circ g$ is defined. (b) Find $(f \circ g)(x)$ and determine its range.

19. Let $f : \mathbb{R}^+ \mapsto \mathbb{R}^+$ where $f(x) = \begin{cases} \dfrac{1}{x^2}, & 0 < x \leq 1 \\ \dfrac{1}{\sqrt{x}}, & x > 1 \end{cases}$.

(a) Sketch the graph of f.
(b) Define the composition $f \circ f$, justifying its existence.
(c) Sketch the graph of $f \circ f$, giving its range.

20. Consider the functions $f :]1, \infty[\to \mathbb{R}$ where $f(x) = \sqrt{x}$ and
$g : \mathbb{R} \backslash \{0\} \to \mathbb{R}$ where $g(x) = x^2$.
(a) Sketch the graphs of f and g on the same set of axes.
(b) Prove that $g \circ f$ exists and find its rule.
(c) Prove that $f \circ g$ cannot exist.
(d) If a new function $g^* : S \to \mathbb{R}$ where $g^*(x) = g(x)$ is now defined, find the largest positive subset of $\mathbb{R}$ so that $f \circ g^*$ does exist. Find $f \circ g^*$, sketch its graph and determine its range.

21. Given that $f(x) = \dfrac{ax - b}{cx - a}$, show that $f \circ f$ exists and find its rule.

22. (a) Sketch the graphs of $f(x) = \frac{1}{a}x^2$ and $g(x) = \sqrt{2a^2 - x^2}$, where $a > 0$.

(b) Show that fog exists, find its rule and state its domain.

(c) Let S be the largest subset of $\mathbb{R}$ so that gof exists.

i. Find S.

ii. Fully define gof, sketch its graph and find its range.

5.4.2 IDENTITY AND INVERSE FUNCTIONS

Before we start our discusion of inverse functions, it is worthwhile looking back at a fundamental area of algebra – algebraic operations. The relevant algebraic properties for real numbers $a \in \mathbb{R}$, $b \in \mathbb{R}$ and $c \in \mathbb{R}$ are:

	Under addition	Under multiplication
Closure	$a + b \in \mathbb{R}$	$a \times b \in \mathbb{R}$
Commutativity	$a + b = b + a$	$a \times b = b \times a$
Associativity	$(a + b) + c = a + (b + c)$	$(a \times b) \times c = a \times (b \times c)$
Existence of the identity	$0 : a + 0 = 0 + a = a$	$1 : a \times 1 = 1 \times a = a$
Inverse element	$-a : a + (-a) = 0 = (-a) + a$	$\frac{1}{a} : \left(\frac{1}{a}\right) \times a = 1 = a \times \left(\frac{1}{a}\right)$

Just as there exists an **identity element** for addition, i.e., 0 and for multiplication, i.e., 1 under the real number system, it seems reasonable to assume that an identity element exists when dealing with functions.

It should be noted that without an identity element, equations such as $x + 2 = 7$ and $2x = 10$ could not be solved under the real number system. Because we take the process for solving these equations for granted, sometimes we lose sight of the underpinning algebraic process that led to their solution.

For example, to solve $x + 2 = 7$ we would write $x = 5$ as the next step. However, if we break the process down we have the following:

$$x + 2 = 7 \Leftrightarrow x + 2 + (-2) = 7 + (-2) \quad \text{[Inverse element]}$$
$$\Leftrightarrow x + 0 = 5 \quad \text{[Identity]}$$
$$\Leftrightarrow x = 5$$

So that without the identity element, we would not be able to make the last statement!

Consider the two functions $f : x \mapsto \mathbb{R}, \sqrt[3]{x}, x \in \mathbb{R}$ and $g : x \mapsto \mathbb{R}, x^3, x \in \mathbb{R}$. The composite functions fog and gof exist. The composite functions are then given by:

$$(fog)(x) = f(g(x)) = \sqrt[3]{g(x)} = \sqrt[3]{x^3} = x.$$

and

$$(gof)(x) = g(f(x)) = (f(x))^3 = (\sqrt[3]{x})^3 = x.$$

For this particular example we have the result that $f(g(x)) = x = g(f(x))$.

Using an analogy to the algebraic properties for the real number system, we introduce the identity function:

Identity function

We define the identity function, I, as $I(x) = x$

If an identity function, I, exists then it must have the property that for any given function f

$$foI = Iof = f \quad \text{or} \quad (foI)(x) = (Iof)(x) = f(x)$$

As we have already seen, it is found that a function with the rule $I(x) = x$ has the required properties. However, unlike its counterpart in the real number system, where the identities are unique, the domain of the identity function is chosen to match that of the function f.

For example, if $f(x) = \sqrt{x}, x \geq 0$ then $I(x) = x$ where $x \geq 0$.
Whereas if $f(x) = 2x, x \in]-\infty, \infty[$ then $I(x) = x$ where $x \in]-\infty, \infty[$.

The existence of the identity function leads us to investigate the existence of an **inverse function**.

Inverse function

Using an analogy to the real number system, the concept of an inverse requires that given some function f, there exists an inverse function, f^{-1}, such that

$$fof^{-1} = I = f^{-1}of \quad \text{or} \quad (fof^{-1})(x) = x = (fof^{-1})(x)$$

The '–1' used in f^{-1} should not be mistaken for an exponent, i.e., $f^{-1}(x) \neq \dfrac{1}{f(x)}$!

The reciprocal function, $\dfrac{1}{f(x)}$, can be written in exponent form as $(f(x))^{-1}$. Pay close attention to where the '–1' is positioned. We highlight this difference geometrically:

Consider the function $f(x) = \sqrt{x+1}$. Using the **DRAW** menu on the TI–83 to sketch the graphs of f^{-1} and the reciprocal of f, we immediately see how different the two graphs are.

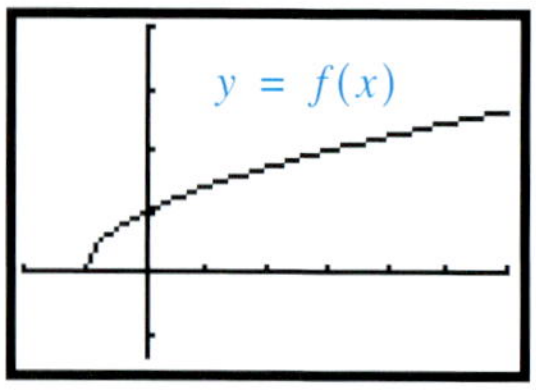

Clearly, the inverse function and the reciprocal function are not the same.

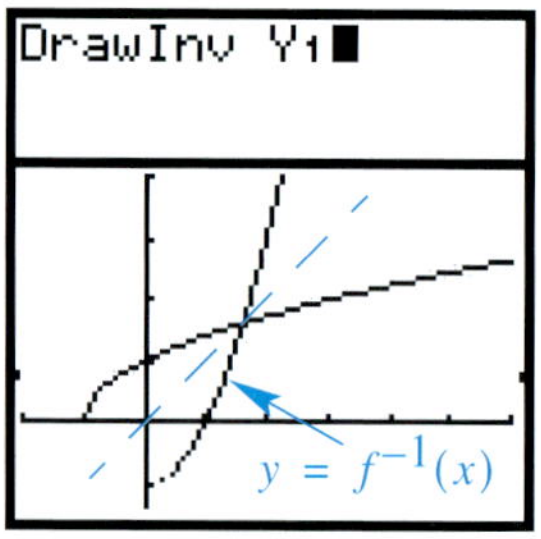

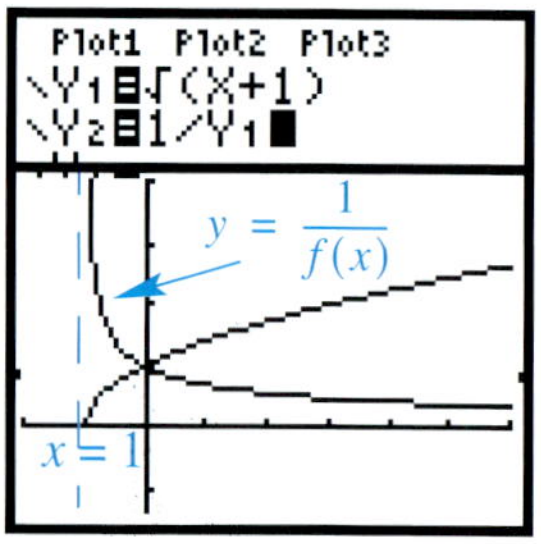

Looking back at the two functions $f : x \mapsto \mathbb{R}, \sqrt[3]{x}, x \in \mathbb{R}$ and $g : x \mapsto \mathbb{R}, x^3, x \in \mathbb{R}$ we notice that the function g is the reverse operation of function f, and function f is the reverse operation of function g. Making use of a mapping diagram we can 'visualise' the process:

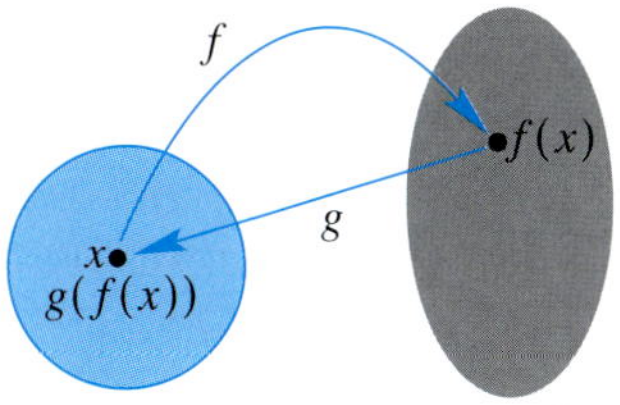

1. Use an element (x) from the domain of the function f and obtain its image $f(x)$.
2. Using this image, which must be an element of the domain of g, we then apply g to $f(x)$ and obtain its image, $g(f(x))$.
Resulting in the value x that we started with.

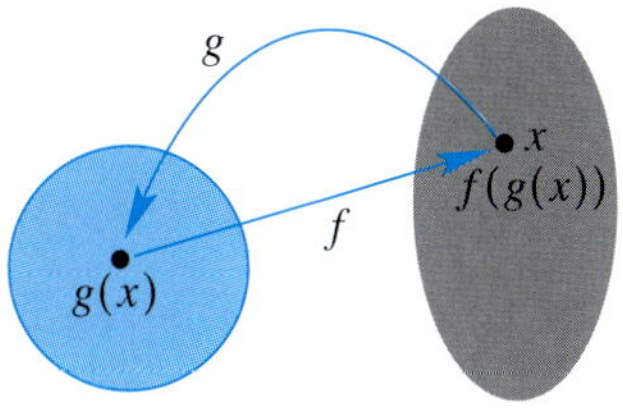

1. Use an element (x) from the domain of the function g and obtain its image $g(x)$.
2. Using this image, which must be an element of the domain of f, we then apply f to $g(x)$ and obtain its image, $f(g(x))$.
Resulting in the value x that we started with.

We can now make some observations:

1. The domain of f, d_f, must equal the range of g, r_g and the domain of g, d_g must equal the range of f, r_f.

2. For the uniqueness of 'x' to be guarranteed both **f and g must be one-to-one functions**.

We therefore have the result:

> If f and g are **one–to–one functions**, such that $f(g(x)) = x = g(f(x))$, then g is known as the inverse of f and f as the inverse of g.

In our case we write, $g(x) = f^{-1}(x)$ and $f(x) = g^{-1}(x)$. This then brings us back to the notation we first introduced for the inverse function. We summarise our findings:

> For the inverse function of f, f^{-1} (read as f inverse) to exists, then
>
> **1.** f must be a one to one function
>
> **2.** i. the domain of f is equal to the range of f^{-1}, i.e., $d_f = r_{f^{-1}}$.
>
> ii. the range of f is equal to the domain of f^{-1}, i.e., $d_{f^{-1}} = r_f$.
>
> **3.** $f(f^{-1}(x)) = x = f^{-1}(f(x))$.

How do we find the inverse function?

A guideline for determining the inverse function can be summarised as follows:

Step 1. Check that the function under investigation is a one–to–one function. This is best done by using a sketch of the function.

Step 2. Use the expression $f(f^{-1}(x)) = x$ to solve for $f^{-1}(x)$.

Step 3. Use the fact that $d_{f^{-1}} = r_f$, and $d_f = r_{f^{-1}}$ to complete the problem.

EXAMPLE 5.36

Find the inverse of the function $f : x \mapsto 5x + 2$, where $x \in \mathbb{R}$.

Solution

We start by checking if the function is a one–to–one function.
From the graph, it is clearly the case that $f(x)$ is a one–to–one function.

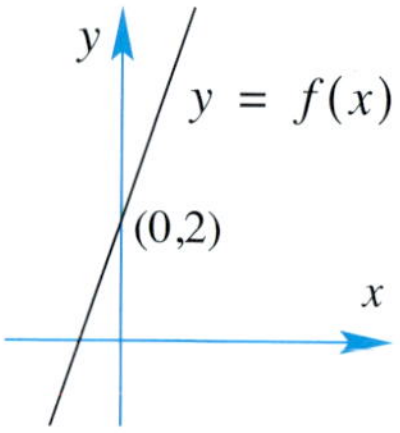

Making use of the result that $f(f^{-1}(x)) = x$ to solve for $f^{-1}(x)$:

$$f(f^{-1}(x)) = 5f^{-1}(x) + 2$$

Then, $f(f^{-1}(x)) = x \Rightarrow 5f^{-1}(x) + 2 = x$

$$\Leftrightarrow \quad 5f^{-1}(x) = x - 2$$

$$\Leftrightarrow \quad f^{-1}(x) = \frac{1}{5}(x-2)$$

To complete the question we need the domain of f^{-1}. We already know that $d_{f^{-1}} = r_f$, therefore all we now need is the range of f, so $d_{f^{-1}} = \mathbb{R}$ (using the graph of $f(x)$).

That is, $f^{-1} : \mathbb{R} \mapsto \mathbb{R}$ where $f^{-1}(x) = \frac{1}{5}(x-2)$.

EXAMPLE 5.37

Find the inverse function of $f(x) = \sqrt{x+2}, x \geq -2$ and sketch its graph.

Solution

A quick sketch of the graph of f verifies that it is a one–to–one function.
We can now determine the inverse function, $f^{-1}(x)$:

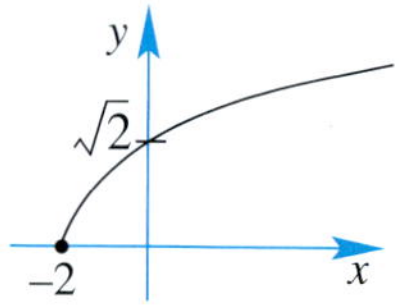

Now, $f(f^{-1}(x)) = \sqrt{f^{-1}(x) + 2}$, so, using the fact that $f(f^{-1}(x)) = x$ we have: $\sqrt{f^{-1}(x) + 2} = x \Leftrightarrow f^{-1}(x) + 2 = x^2$ [After squaring both sides]

$$\Leftrightarrow \quad f^{-1}(x) = x^2 - 2.$$

To fully define f^{-1} we need to determine the domain of f^{-1}.

We do this by using the result that $d_{f^{-1}} = r_f$. Using the graph shown above, we have that $r_f = [0, \infty[\; \therefore d_{f^{-1}} = [0, \infty[$. We are now in a position to fully define the inverse function.

We have, $f^{-1} : [0, \infty) \mapsto \mathbb{R}, f^{-1}(x) = x^2 - 2$.

We can now sketch the graph of the inverse function:

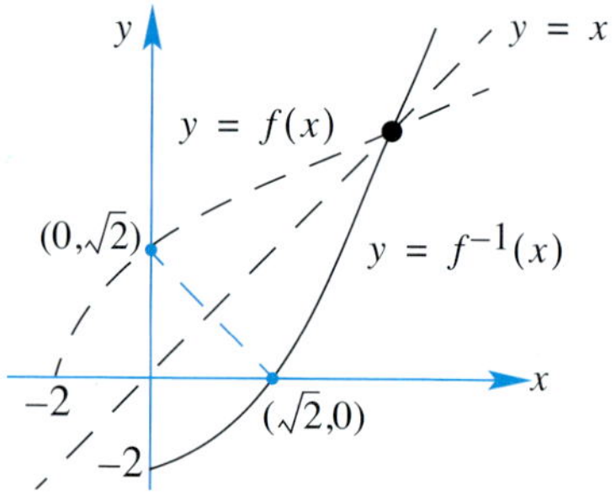

We make two important observations:

1. The graph of $y = f^{-1}(x)$ is the graph of $y = f(x)$ reflected about the line $y = x$.

2. Points of intersection of the graphs $y = f^{-1}(x)$ and $y = f(x)$ will always occur where both curves meet the line $y = x$.

The relationship between the graph of a function f and the graph of its inverse function f^{-1} can be explained rather neatly, because all that has actually happened is that we have interchanged the x– and y– values, i.e., $(a, b) \leftrightarrow (b, a)$. In doing so, we find that $d_{f^{-1}} = r_f$, and $d_f = r_{f^{-1}}$. This then has the result that the graph of the inverse function $f^{-1}(x)$ is a reflection of the graph of the original function $f(x)$ about the line with equation $y = x$.

Graphing the inverse function

The graph of $f^{-1}(x)$ is a reflection of the graph of $f(x)$ about the line $y = x$.

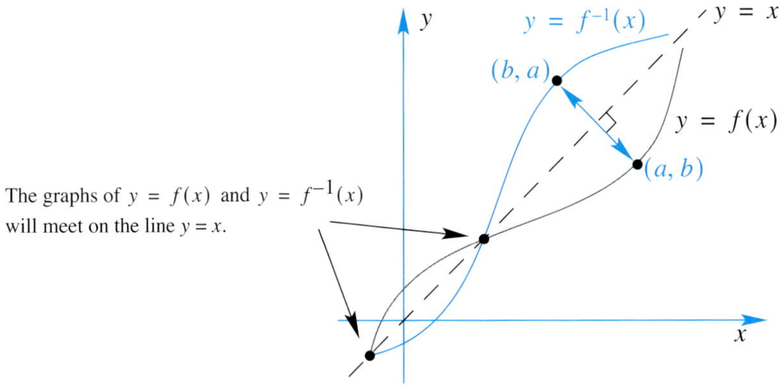

Because of the nature of the functions $f^{-1}(x)$ and $f(x)$, when finding the points of intersection of the two graphs, rather than solving $f^{-1}(x) = f(x)$, it might be easier to solve the equations $f^{-1}(x) = x$ or $f(x) = x$. The only caution when solving the latter two, is to always keep in mind the domain of the original function.

One interesting function is $f(x) = \frac{1}{x}, x \neq 0$. It can be established that its inverse function is given by $f^{-1}(x) = \frac{1}{x}, x \neq 0$ i.e., $f(f^{-1}(x)) = x \Rightarrow \frac{1}{f^{-1}(x)} = x \Leftrightarrow f^{-1}(x) = \frac{1}{x}$. Then, as the two functions are identical, it is its **self-inverse**. Sketching the graph of $f(x) = \frac{1}{x}, x \neq 0$ and reflecting it about the line $y = x$ will show that the two graphs overlap each other. Note then, that self-inverse functions also intersect at points other than just those on the line $y = x$ – basically because they are the same functions!

There is a second method that we can use to find the inverse of a function.The steps required are:

Step 1. Let y denote the expression $f(x)$.
Step 2. Interchange the variables x and y.
Step 3. Solve for y.

Step 4. The expression in step 3 gives the inverse function, $f^{-1}(x)$.

We work through an example using this method.

EXAMPLE 5.38

Find the inverse function, f^{-1} of $f :]-1, \infty[\mapsto \mathbb{R}$ where $f(x) = \dfrac{2}{x-1} + 1$ and sketch its graph.

Solution

Let $y = f(x)$, giving $y = \dfrac{2}{x-1} + 1$.

Next, we interchange the variables x and y: $\quad x = \dfrac{2}{y-1} + 1$

We now solve for y : $x = \dfrac{2}{y-1} + 1 \Leftrightarrow x - 1 = \dfrac{2}{y-1}$

$$\Leftrightarrow \frac{1}{x-1} = \frac{y-1}{2} \quad \text{[inverting both sides]}$$

$$\Leftrightarrow y - 1 = \frac{2}{x-1}$$

Therefore, we have that $\quad y = \dfrac{2}{x-1} + 1$

That is, $f^{-1}(x) = \dfrac{2}{x-1} + 1, x > 1$.

We notice that the original function $y = f(x)$ and its inverse function $y = f^{-1}(x)$ are the same. This becomes obvious when we sketch both graphs on the same set of axes.

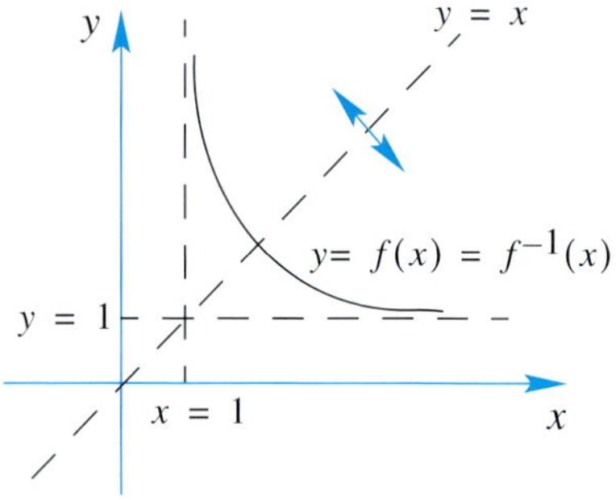

When reflected about the line $y = x$, they are identical!

Again, we have an example of a self-inverse function

In our work with exponential and logarithmic functions we have already observed the 'inverse relationship' that exists between these two functions. That is, we observed that the graph of the curve $y = \log_a x$ was simply the reflection of the graph of $y = a^x$ about the line $y = x$.

When dealing with exponential and logarithmic functions we use the defining relationship between the exponential and logarithmic representation to help us find their inverses.
That is, we use the relationship $N = b^x \Leftrightarrow x = \log_b N$ – when in the form $N = b^x$, x is called the index, whereas when in the form $x = \log_b N$, x is called the logarithm.

We then have:

if $f: x \mapsto \mathbb{R}$ where $f(x) = a^x$

then, $f^{-1}: x \mapsto \mathbb{R}^+, f^{-1}(x) = \log_a x$

Similarly,

if $f: x \mapsto \mathbb{R}^+, f(x) = \log_a x$

then $f^{-1}: x \mapsto \mathbb{R}$ where $f^{-1}(x) = a^x$

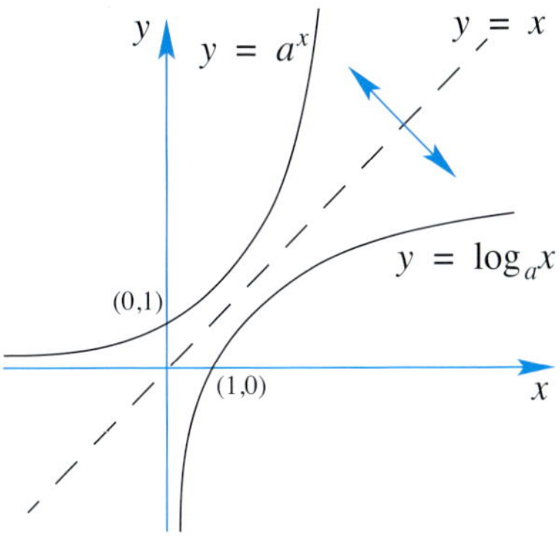

EXAMPLE 5.39

Find the inverse of the function $g: x \mapsto 4^x - 2, x \in \mathbb{R}$.

Solution

The function $g: x \mapsto 4^x - 2, x \in \mathbb{R}$ is a one–to–one function and so its inverse function exists.

Using the result that $g(g^{-1}(x)) = x$ we have:

$$4^{g^{-1}(x)} - 2 = x$$

$$\Leftrightarrow 4^{g^{-1}(x)} = x + 2$$

$$\therefore g^{-1}(x) = \log_4(x+2) \text{ [using } N = b^x \Leftrightarrow x = \log_b N \text{]}$$

Now, $d_{g^{-1}} = r_g =]-2,\infty[$ (we have obtained the range by using a sketch of $g(x)$)

Therefore the inverse, g^{-1}, is given by $g^{-1}: x \mapsto \log_4(x+2), x > -2$

EXAMPLE 5.40

Find the inverse of $g : \left(\frac{1}{2}, \infty\right) \mapsto \mathbb{R}$, $\log_{10}(2x-1) + 2$.

Solution

This time we make use of the second method of finding the inverse function.

Let $y = \log_{10}(2x-1) + 2$, interchanging x and y, we have:

$$x = \log_{10}(2y-1) + 2$$

$$\Leftrightarrow \quad x - 2 = \log_{10}(2y-1)$$

$$\Leftrightarrow \quad 2y - 1 = 10^{x-2} \text{ [using } N = b^x \Leftrightarrow x = \log_b N \text{]}$$

$$\Leftrightarrow \quad y = \frac{1}{2}(10^{x-2} + 1).$$

Therefore, we have that $f^{-1}(x) = \frac{1}{2}(10^{x-2} + 1)$.

Next, $d_{f^{-1}} = r_f = (-\infty, \infty)$, so the inverse function is $f^{-1}: x \mapsto \frac{1}{2}(10^{x-2} + 1), x \in (-\infty, \infty)$.

EXERCISES 5.4.2

1. Find the inverse function for each of the following.

(a) $f(x) = 2x + 1, x \in \mathbb{R}$ (b) $f(x) = x^3, x \in \mathbb{R}$

(c) $g(x) = \frac{1}{3}x - 3, x \in \mathbb{R}$ (d) $g(x) = \frac{2}{5}x + 2, x \in \mathbb{R}$

(e) $h(x) = \sqrt{x+1}, x > -1$ (f) $f(x) = \sqrt{x} + 1, x \geq 0$

(g) $f(x) = \frac{1}{x+1}, x > -1$ (h) $h(x) = \frac{1}{\sqrt{x}} - 1, x > 0$

2. Using the graph of the original function, sketch the graph of the corresponding inverse function for each part in Question 1.

3. Find and sketch the inverse function of

(a) $f(x) = x^2 - 3, x \geq 0$. (b) $f(x) = x^2 - 3, x \leq 0$.

4. Show that $f(x) = \frac{x}{\sqrt{x^2+1}}, x \in \mathbb{R}$ is a one–to–one function, hence find its inverse, f^{-1}.

5. Sketch the inverse of the followingfunctions:

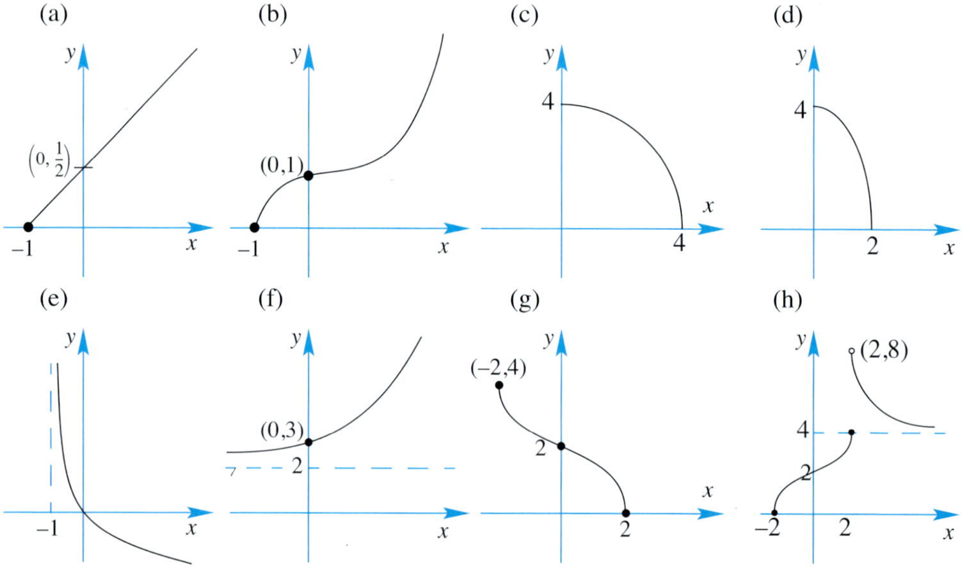

6. Find the inverse function (if it exists) of the following

(a) $f(x) = 3^x + 1, x \in]-\infty,\infty[$ (b) $f(x) = 2^x - 5, x \in]-\infty, \infty[$

(c) $f(x) = 3^{2x+1}, x \in]-\infty, \infty[$ (d) $g(x) = 3 - 10^{x-1}, x \in]-\infty,\infty[$

(e) $h(x) = \frac{2}{3^x - 1}, x \neq 0$ (f) $g(x) = \frac{1}{2^x} - 1, x \in]-\infty,\infty[$

7. Using the graph of the original functions in Question 6, sketch the graph of their inverses.

8. Find the inverse function (if it exists) of the following functions

(a) $f(x) = \log_2(x+1), x > -1$ (b) $f(x) = \log_{10}(2x), x > 0$

(c) $h(x) = 1 - \log_2 x, x > 0$ (d) $g(x) = \log_3(x-1) - 1, x > 1$

(e) $h(x) = 2\log_5(x-5), x > 5$ (f) $f(x) = 2 - \frac{1}{3}\log_{10}(1-x), x < 1$

9. Find the inverse function of $f(x) = x^2 + 2x, x \geq -1$, stating both its domain and range. Sketch the graph of f^{-1} .

10. Find the inverse function of (a) $f(x) = -x + a, x \in]-\infty,\infty[$, where a is real.

(b) $h(x) = \dfrac{2}{x-a} + a, x > a$, where a is real.

(c) $f(x) = \sqrt{a^2 - x^2}, 0 \leq x \leq a$, where a is real.

11. Find the inverse of $h(x) = -x^3 + 2, x \in \mathbb{R}$. Sketch both $h(x)$ and $h^{-1}(x)$ on the same set of axes.

12. Find the largest possible set of positive real numbers **S**, that will enable the inverse function h^{-1} to exist, given that $h(x) = (x-2)^2, x \in \mathbf{S}$.

13. Determine the largest possible positive valued domain, X, so that the inverse function, $f^{-1}(x)$, exists, given that $f(x) = \dfrac{3x+2}{2x-3}, x \in \text{X}$.

14. (a) Sketch the graph of $f(x) = x - \frac{1}{x}, x > 0$. Does the inverse function, f^{-1} exist? Give a reason for your answer.

(b) Consider the function $g : \text{S} \to \mathbb{R}$ where, $g(x) = x - \frac{1}{x}$. Find the two largest sets S so the the inverse function, g^{-1} , exists. Find both inverses and on separate axes, sketch their graphs.

15. Find f^{-1} given that $f(x) = \sqrt{\frac{x}{a} - 1}$ where $a > 1$. On the same set of axes sketch both the graphs of $y = f(x)$ and $y = f^{-1}(x)$. Find $\{x : f(x) = f^{-1}(x)\}$.

16. Find and sketch the inverse, f^{-1} , of the functions

(a) $f(x) = \begin{cases} -\frac{1}{2}(x+1), & x > 1 \\ -x^3, & x \leq 1 \end{cases}$ (b) $f(x) = \begin{cases} e^{x+1}, & x \leq 0 \\ x + e, & x > 0 \end{cases}$

(c) $f(x) = \begin{cases} -\ln(x-1), & x > 2 \\ 2 - x, & x \leq 2 \end{cases}$ (d) $f(x) = \begin{cases} \sqrt{x} + 4, & x > 0 \\ x + 4, & -4 < x < 0 \end{cases}$

17. (a) On the same set of axes sketch the graph of $f(x) = \frac{1}{a}\ln(x-a), x > a > 0$ and its reciprocal.

(b) Find and sketch the graph of f^{-1}.

18. Consider the functions f and g:

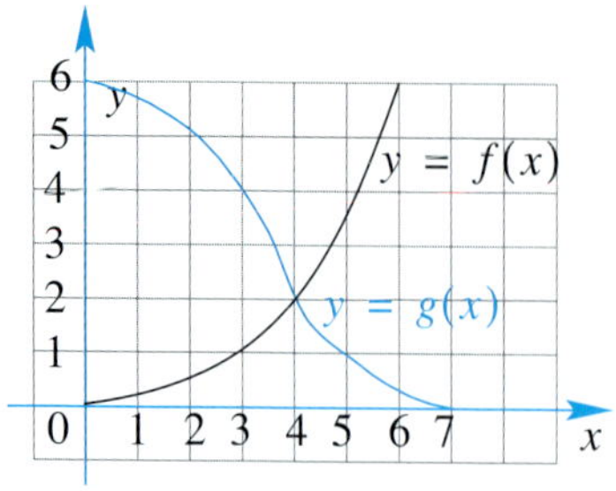

(a) Does gof exist? Justify your answer.

(b) Does $(gof)^{-1}$ exist? Justify your answer.
If it does exist, sketch the graph of $(gof)^{-1}$.

19. (a) On the same set of axes, sketch the graph of $f(x) = -x^3$ and its inverse, $f^{-1}(x)$.

(b) The function g is given $g(x) = \begin{cases} 2x+1, & x < -1 \\ -x^3, & -1 \le x \le 1 \\ 2x-1, & x > 1 \end{cases}$.

i. Sketch the graph of g.

ii. Fully define its inverse, g^{-1}, stating why it exists.

iii. Sketch the graph of g^{-1}.

iv. Find $\{x : g(x) = g^{-1}(x)\}$.

20. Consider the functions $t(x) = e^x$ and $m(x) = \sqrt{x}$.

(a) Find, where they exist, the composite functions i. $(tom)(x)$ ii. $(mot)(x)$

(b) With justification, find and sketch the graphs of i. $(tom)^{-1}(x)$ ii. $(mot)^{-1}(x)$

(c) Find i. $t^{-1}om^{-1}(x)$ ii. $m^{-1}ot^{-1}(x)$

(d) What conclusion(s) can you make from your results of parts (b) and (c)?

(e) Will your results of part (d) work for any two functions f and g? Explain.

21. (a) Find $\{x : x^3 + x - 2 = 0\}$.

(b) If $f(x) = \frac{1}{\sqrt{x}} - 2$, sketch the graph of $y = f(x)$ and find $\{x : f(x) = f^{-1}(x)\}$.

22. Consider the functions $f(x) = |x|, x \in \mathbf{A}$ and $g(x) = e^x - 2, x \in \mathbf{B}$.

(a) Sketch the graphs of i. f if $\mathbf{A} = \mathbb{R}$ ii. g if $\mathbf{B} = \mathbb{R}$.

(b) With **A** and **B** as given in (a), give reasons why $(fog)^{-1}$ will not exist.

(c) i. Find the largest set **B** which includes positive values, so that $(fog)^{-1}$ exists.

ii. Fully define $(fog)^{-1}$.

iii. On the same set of axes, sketch the graphs of $(fog)(x)$ and $(fog)^{-1}(x)$.

6.1 TRANSLATIONS

6.1.1 HORIZONTAL TRANSLATION

A horizontal translation takes on the form $f(x) \rightarrow f(x-a)$.

We start by looking at the transformation of the basic parabolic graph with equation $y = x^2$. The horizontal translation of $y = x^2$ is given by $y = (x-a)^2$. This transformation represents a **translation along the *x*–axis**.

For example, the graph of $y = (x-4)^2$ represents the parabola $y = x^2$ translated 4 units to the right ($a = 4$):

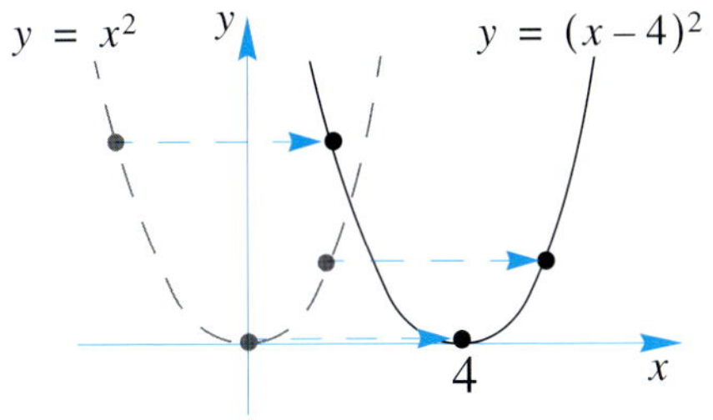

Similarly, the graph of $y = (x+2)^2$ represents the parabola $y = x^2$ translated 2 units to the left ($a = -2$):

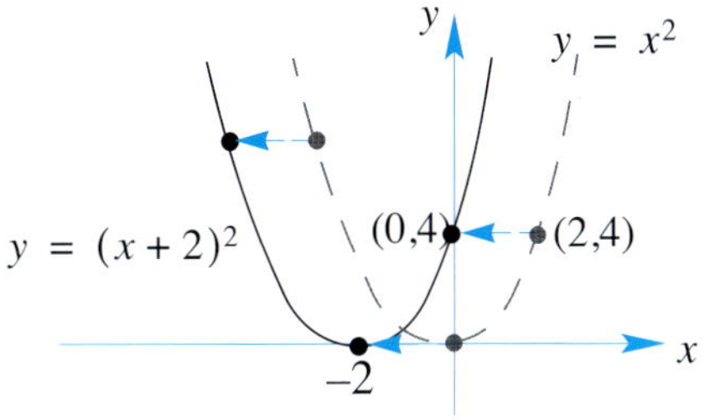

In both situations, it appears as if the graphs have been translated in the 'opposite direction' to the sign of 'a'. That is, $y = (x+2)^2$ has been translated 2 units back (i.e., in the negative direction), while $y = (x-2)^2$ has been translated 2 units forward (i.e., in the positive direction).

The reason for this is that **the transformation is applied to the *x*–values, not the graph!** That is, given $y = f(x)$, the graph of $y = f(x+2)$ is telling us to 'Add two units to all the *x*–values'. In turn, this means, that the combined *x*/*y*–axes should be moved in the positive direction by two units (whilst the graph of $y = f(x)$ remains exactly where it is):

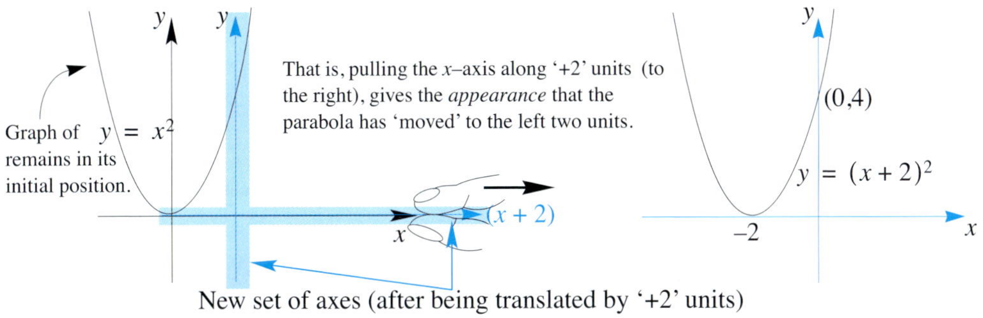

Similarly, given $y = f(x)$, the graph of $y = f(x-1)$ is telling us to 'Subtract one unit from all the *x*–values'. In turn, this means, that the combined *x*/*y*–axes should be moved in the negative direction by one unit (while maintaining the graph of $y = f(x)$ fixed at its original position):

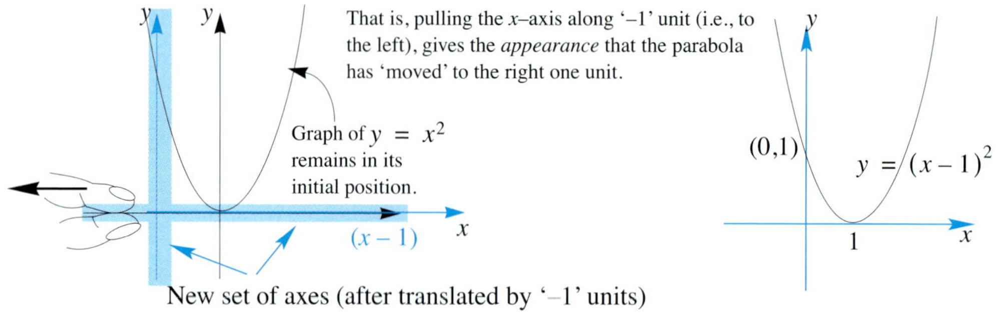

However, whenever we are asked to sketch a graph, rather than drawing the axes after the graph has been sketched, the first thing we do is draw the set of axes and then sketch the graph. This is why there appears to be a sense in which we seem to do the 'opposite' when sketching graphs that involve transformations.

EXAMPLE 6.1

Given the graph of $y = f(x)$ sketch the graph of

(a) $y = f(x+1)$ (b) $y = f\left(x - \frac{1}{2}\right)$

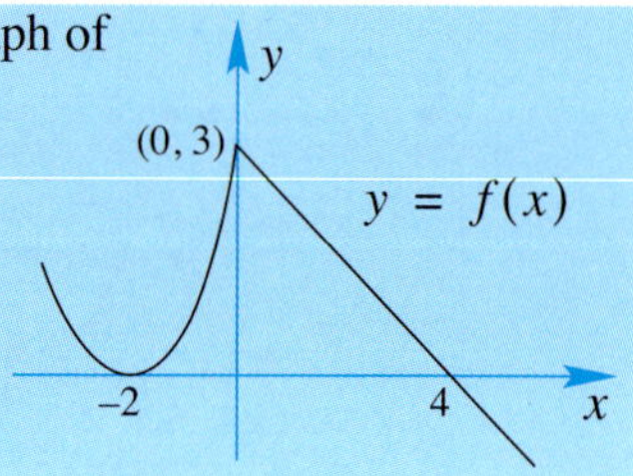

Solution

(a) The graph of $y = f(x+1)$ represents a translation along the x–axis of the graph of $y = f(x)$ by 1 unit to the left.

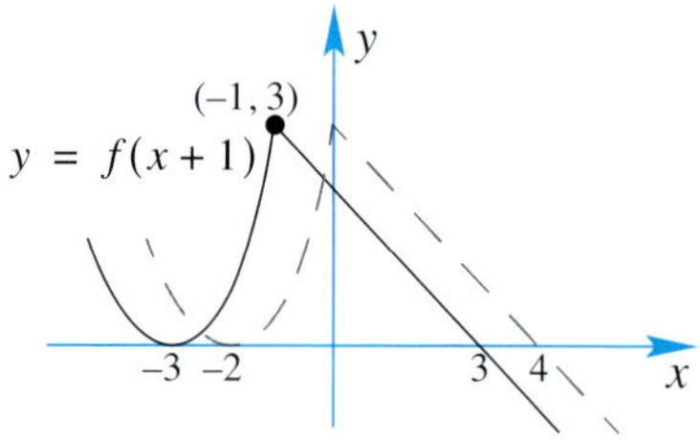

(b) The graph of $y = f\left(x - \frac{1}{2}\right)$ represents a translation along the x–axis of the graph of $y = f(x)$ by $\frac{1}{2}$ unit to the right.

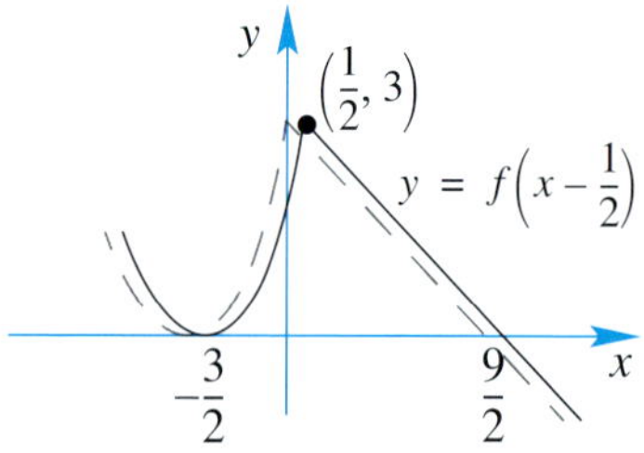

The mapping from the original coordinates (x, y) to the new coordinates (x', y') can also be presented in vector form. That is, if the point (x, y) is translated 'a' units along the x–axis, the new coordinates would be given by $(x + a, y)$

That is, $x' = x + a$ and $y' = y$.

The vector notation for such a translation is given by:

$$\begin{pmatrix} x' \\ y' \end{pmatrix} = \begin{pmatrix} x \\ y \end{pmatrix} + \begin{pmatrix} a \\ 0 \end{pmatrix}$$

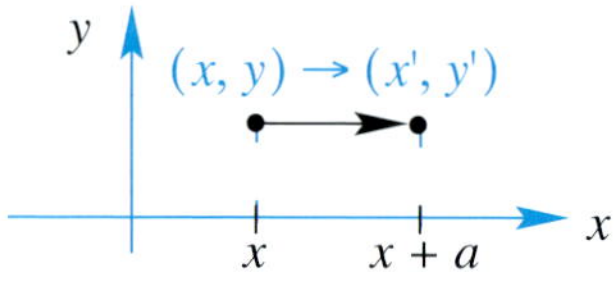

From $\begin{pmatrix} x' \\ y' \end{pmatrix} = \begin{pmatrix} x \\ y \end{pmatrix} + \begin{pmatrix} a \\ 0 \end{pmatrix}$ we than have that

$$\begin{pmatrix} x \\ y \end{pmatrix} = \begin{pmatrix} x' \\ y' \end{pmatrix} - \begin{pmatrix} a \\ 0 \end{pmatrix} \Leftrightarrow \begin{pmatrix} x \\ y \end{pmatrix} = \begin{pmatrix} x'-a \\ y' \end{pmatrix} \therefore x = x'-a \text{ and } y = y'$$

Substituting these results into the equation $y = f(x)$ we obtain the transformed equation

$$y' = f(x'-a)$$

As x' and y' are only dummy variables, we can rewrite this last eqation as $y = f(x-a)$.

> Under the vector translation $\begin{pmatrix} a \\ 0 \end{pmatrix}$, we have the mapping $f(x) \to f(x-a)$

EXAMPLE 6.2

Find the equation of the relation under the translation vector indicated.

(a) $x^2+y^2 = 9 ; \begin{pmatrix} 3 \\ 0 \end{pmatrix}$ (b) $xy = 4 ; \begin{pmatrix} -2 \\ 0 \end{pmatrix}$.

Sketch both the original and transformed graph on the same set of axes.

Solution

(a) Under the vector transformation we have $\begin{pmatrix} x' \\ y' \end{pmatrix} = \begin{pmatrix} x \\ y \end{pmatrix} + \begin{pmatrix} 3 \\ 0 \end{pmatrix}$.

Meaning that $x' = x+3 \Rightarrow x = x'-3$ – (1) and $y = y'$ – (2).

Substituting (1) and (2) into the equation $x^2+y^2 = 9$ gives $(x'-3)^2+(y')^2 = 9$.

Dropping off the dashes we then have the transformed equation $(x-3)^2+y^2 = 9$.

(b) Under the vector transformation we have $\begin{pmatrix} x' \\ y' \end{pmatrix} = \begin{pmatrix} x \\ y \end{pmatrix} + \begin{pmatrix} -2 \\ 0 \end{pmatrix}$.

Meaning that $x' = x-2 \Rightarrow x = x'+2$ – (1) and $y = y'$ – (2).

Substituting (1) and (2) into the equation $xy = 4$ gives $(x'+2)y' = 4$.

Dropping off the dashes we then have the transformed equation $(x+2)y = 4$ or $y = \dfrac{4}{x+2}$.

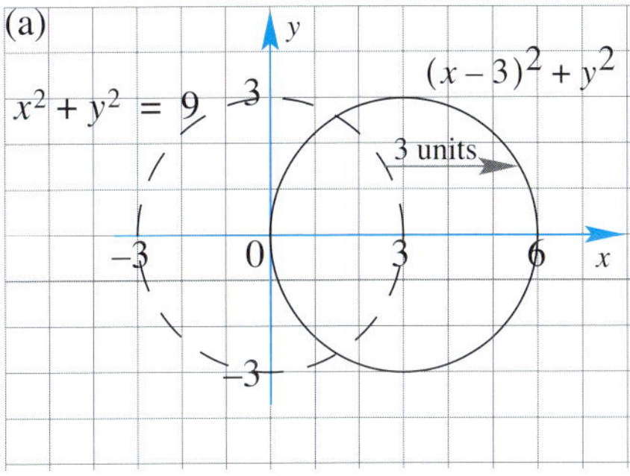

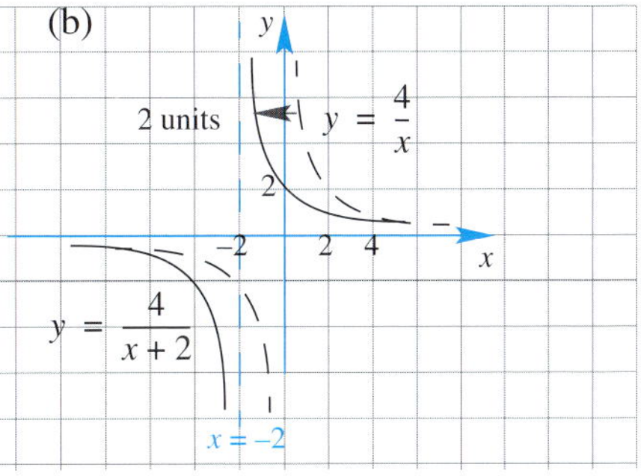

6.1.2 VERTICAL TRANSLATION

A vertical translation takes on the form $f(x) \to f(x) + b$.

Again we consider the transformation of the basic parabolic graph with equation $y = x^2$. The vertical translation of $y = x^2$ is given by $y = x^2 + b$. This transformation represents a **translation along the *y*–axis**.

For example, the graph of $y = x^2 - 2$ represents the parabola $y = x^2$ translated 2 units down:

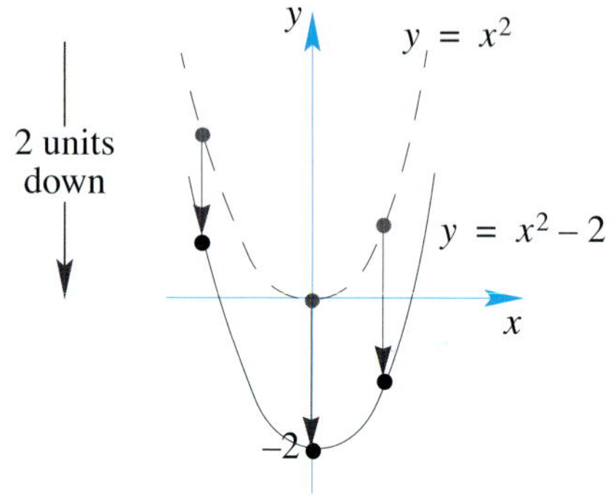

Similarly, the graph of $y = x^2 + 1$ represents the parabola $y = x^2$ translated 1 unit up:

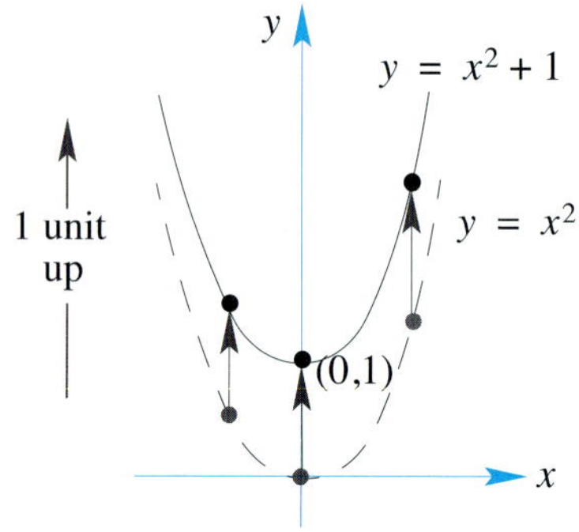

Note that this time, when applying a vertical translation, we are consistent with the sign of *b*!

The reason being that although we are sketching the graph of $y = f(x) + b$, we are in fact sketching the graph of $y - b = f(x)$. So that this time, **the transformation is applied to the *y*–values, and not the graph!**

So, if we consider the two previous examples, we have $y = x^2 - 2 \Leftrightarrow y + 2 = x^2$, and so, in this case we would be pulling the *y*–axis **UP** 2 units, which gives the appearance that the parabola has been moved down 2 units.

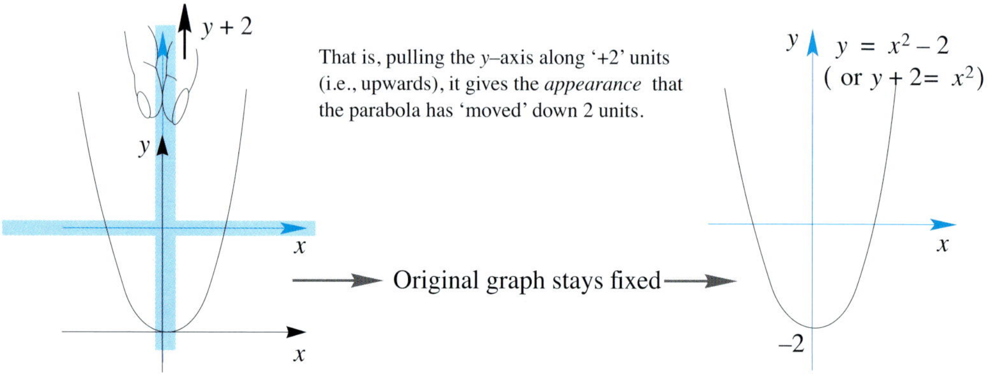

Similarly, $y = x^2 + 1 \Leftrightarrow y - 1 = x^2$, so that in this case we would be pulling the *y*–axis **DOWN** 1 unit, which gives the appearance that the parabola has been moved up 1 unit.

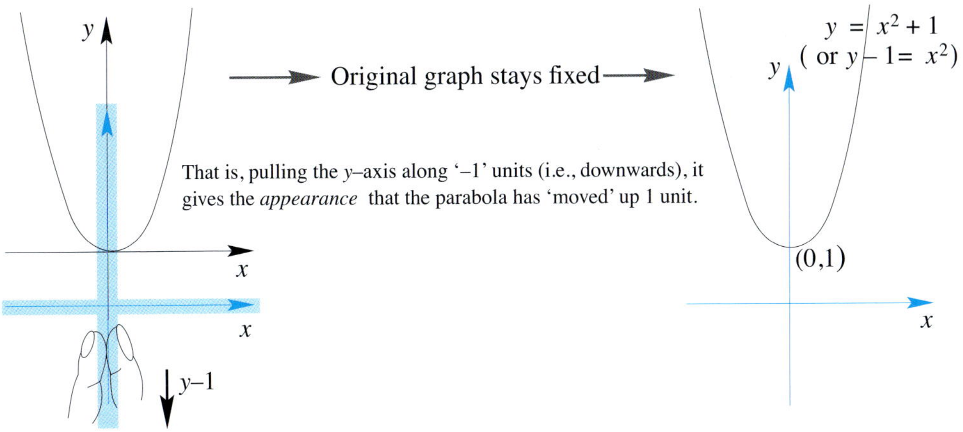

EXAMPLE 6.3

Given the graph of $y = f(x)$ sketch the graph of

(a) $y = f(x) - 3$ (b) $y = f(x) + 2$

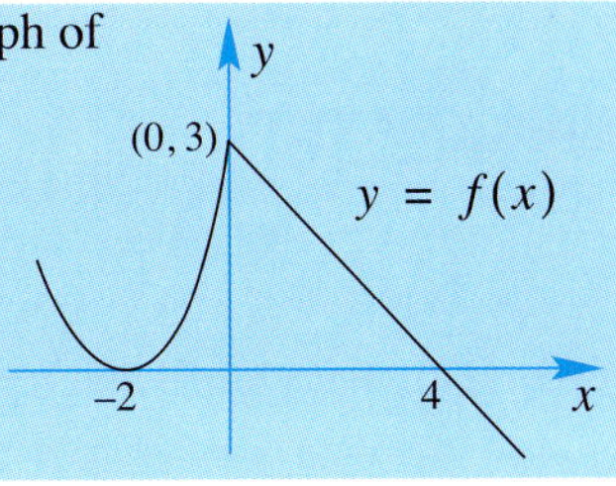

Solution

(a) The graph of $y = f(x) - 3$ represents a translation along the y–axis of the graph of $y = f(x)$ by 3 units in the downward direction.

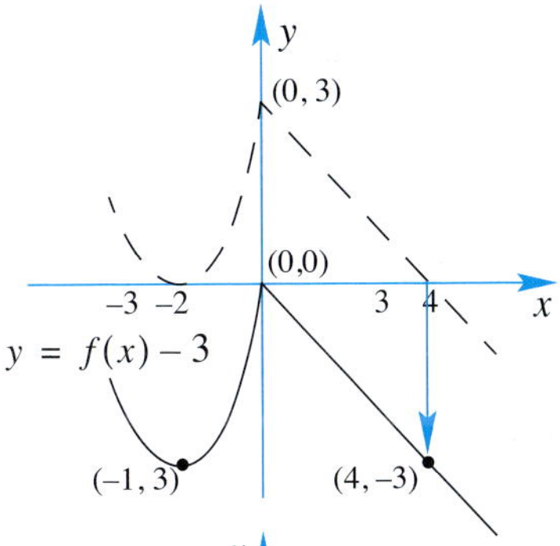

(b) The graph of $y = f(x) + 2$ represents a translation along the y–axis of the graph of $y = f(x)$ by 2 units in the upward direction.

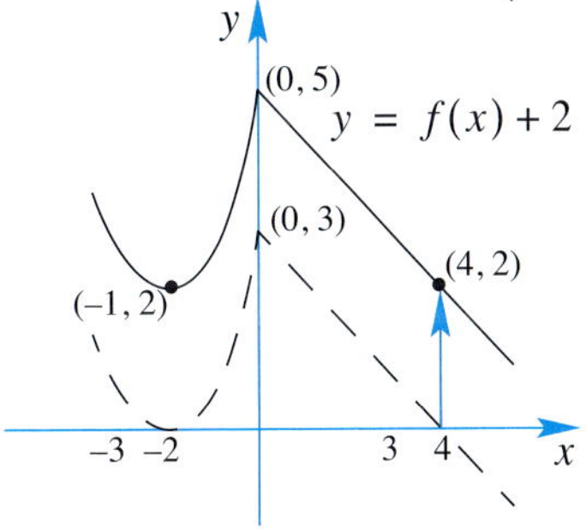

The mapping from the original coordinates (x, y) to the new coordinates (x', y') can also be presented in vector form. That is, if the point (x, y) is translated 'b' units along the y–axis, the new coordinates would be given by $(x, y + b)$

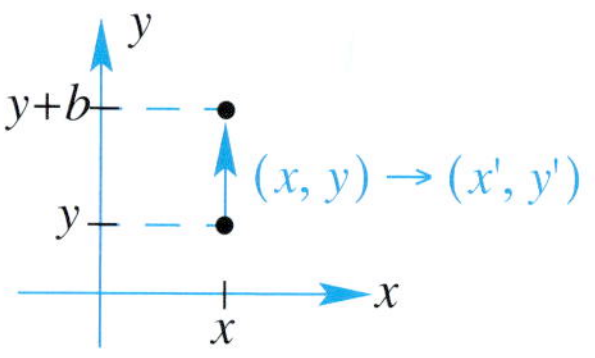

That is, $x' = x$ and $y' = y + b$.
The vector notation for such a translation is given by:

$$\begin{pmatrix} x' \\ y' \end{pmatrix} = \begin{pmatrix} x \\ y \end{pmatrix} + \begin{pmatrix} 0 \\ b \end{pmatrix}$$

From $\begin{pmatrix} x' \\ y' \end{pmatrix} = \begin{pmatrix} x \\ y \end{pmatrix} + \begin{pmatrix} 0 \\ b \end{pmatrix}$ we then have that

$$\begin{pmatrix} x \\ y \end{pmatrix} = \begin{pmatrix} x' \\ y' \end{pmatrix} - \begin{pmatrix} 0 \\ b \end{pmatrix} \Leftrightarrow \begin{pmatrix} x \\ y \end{pmatrix} = \begin{pmatrix} x' \\ y' - b \end{pmatrix} \therefore x = x' \text{ and } y = y' - b$$

Then, substituting these results into the equation $y = f(x)$ we obtain the transformed equation

$$y' - b = f(x') \Rightarrow y' = f(x') + b$$

As x' and y' are only dummy variables, we can rewrite this last eqation as $y = f(x) + b$.

> Under the vector translation $\begin{pmatrix} 0 \\ b \end{pmatrix}$, we have the mapping $f(x) \to f(x) + b$

EXAMPLE 6.4

Find the equation of the relation under the translation vector indicated.

(a) $x^2 + y^2 = 9$; $\begin{pmatrix} 0 \\ 3 \end{pmatrix}$ (b) $xy = 4$; $\begin{pmatrix} 0 \\ -2 \end{pmatrix}$.

Sketch both the original and transformed graph on the same set of axes.

Solution

(a) Under the vector transformation we have $\begin{pmatrix} x' \\ y' \end{pmatrix} = \begin{pmatrix} x \\ y \end{pmatrix} + \begin{pmatrix} 0 \\ 3 \end{pmatrix}$.

Meaning that $y' = y + 3 \Rightarrow y = y' - 3$ – (1) and $x = x'$ – (2).

Substituting (1) and (2) into the equation $x^2 + y^2 = 9$ gives $(x')^2 + (y' - 3)^2 = 9$.

Dropping off the dashes we then have the transformed equation $x^2 + (y - 3)^2 = 9$.

(b) Under the vector transformation we have $\begin{pmatrix} x' \\ y' \end{pmatrix} = \begin{pmatrix} x \\ y \end{pmatrix} + \begin{pmatrix} 0 \\ -2 \end{pmatrix}$.

Meaning that $y' = y - 2 \Rightarrow y = y' + 2$ – (1) and $x = x'$ – (2).

Substituting (1) and (2) into the equation $xy = 4$ gives $x'(y' + 2) = 4$.

Dropping off the dashes we then have the transformed equation $x(y + 2) = 4$ or $y = \frac{4}{x} - 2$.

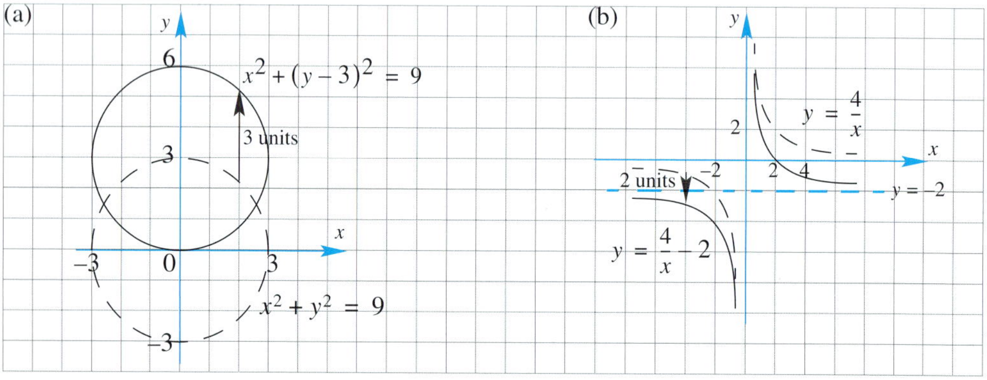

SUMMARY

$y = f(x-a)$, $a > 0$: **translation of $f(x)$ along the *x*–axis of *a* units to the right.**

$y = f(x+a)$, $a > 0$: **translation of $f(x)$ along the *x*–axis of *a* units to the left.**

$y = f(x)+b$, $b > 0$: **translation of $f(x)$ along the *y*–axis of *b* units up.**

$y = f(x)-b$, $b > 0$: **translation of $f(x)$ along the *y*–axis of *b* units down.**

Of course it is also possible to apply both a vertical and horizontal translation to the one graph at the same time. That is, the graph of $y = (x-1)^2+2$ would represent the graph of $y = x^2$ after it had been translated one unit to the right and two units up.

Such a combination takes on the form $f(x) \to f(x-a)+b$, representing a horizontal translation of 'a' along the x–axis and a vertical translation of 'b' along the y–axis.

EXAMPLE 6.5

Sketch the graph of $y = f(x-3)-1$ for the graph shown.

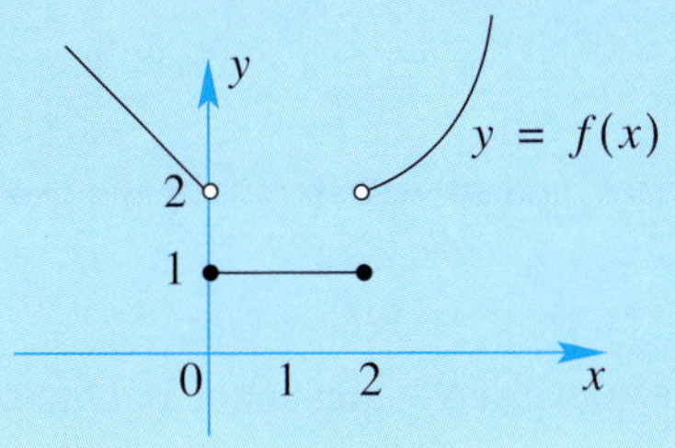

Solution

The graph of $y = f(x-3)-1$ represents the graph of $y = f(x)$ after a translation along the x–axis of 3 units to the right followed by a translation along the y–axis 1 unit down.

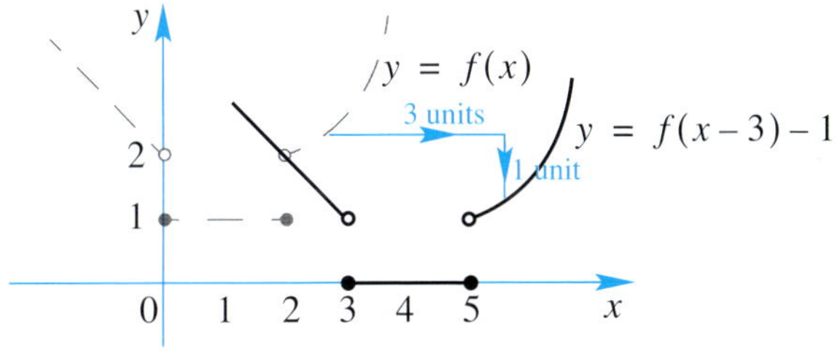

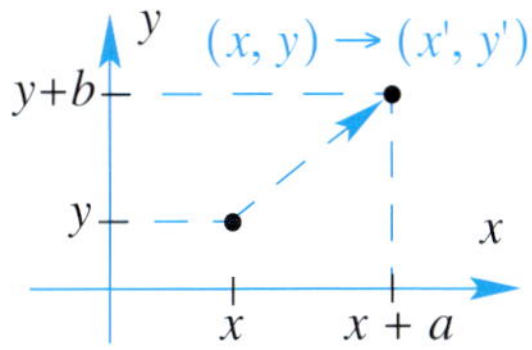

The mapping from the original coordinates (x, y) to the new coordinates (x', y') can also be presented in vector form. That is, if the point (x, y) is translated 'a' units along the x–axis and 'b' units along the y–axis, the new coordinates would be given by $(x+a, y+b)$ That is, $x' = x+a$ and $y' = y+b$.

The vector notation for such a translation is given by: $\begin{pmatrix} x' \\ y' \end{pmatrix} = \begin{pmatrix} x \\ y \end{pmatrix} + \begin{pmatrix} a \\ b \end{pmatrix}$

From $\begin{pmatrix} x' \\ y' \end{pmatrix} = \begin{pmatrix} x \\ y \end{pmatrix} + \begin{pmatrix} a \\ b \end{pmatrix}$ we than have that

$$\begin{pmatrix} x \\ y \end{pmatrix} = \begin{pmatrix} x' \\ y' \end{pmatrix} - \begin{pmatrix} a \\ b \end{pmatrix} \Leftrightarrow \begin{pmatrix} x \\ y \end{pmatrix} = \begin{pmatrix} x' - a \\ y' - b \end{pmatrix} \therefore x = x' - a \text{ and } y = y' - b$$

Substituting these results into the equation $y = f(x)$ we obtain the transformed equation

$$y' - b = f(x' - a) \Rightarrow y' = f(x' - a) + b$$

As x' and y' are only dummy variables, we can rewrite this last eqation as $y = f(x - a) + b$.

> Under the vector translation $\begin{pmatrix} a \\ b \end{pmatrix}$, we have the mapping $f(x) \to f(x - a) + b$

EXAMPLE 6.6

Find the equation of $y = x^3 - 1$ under the translation $\begin{pmatrix} 3 \\ 5 \end{pmatrix}$.

Solution

Let the new set of axis be u and v so that $\begin{pmatrix} u \\ v \end{pmatrix} = \begin{pmatrix} x \\ y \end{pmatrix} + \begin{pmatrix} 3 \\ 5 \end{pmatrix} \Rightarrow \left.\begin{matrix} u = x + 3 \\ v = y + 5 \end{matrix}\right\}$ or $\left.\begin{matrix} x = u - 3 \\ y = v - 5 \end{matrix}\right\}$.

Substituting these into $y = x^3 - 1$, we have $v - 5 = (u - 3)^3 - 1 \Rightarrow v = (u - 3)^3 + 4$.

However, u and v are dummy variables, and so we can rewrite the last equation in terms of x and y, i.e., $y = (x - 3)^3 + 4$.

This represents the graph of $y = x^3$ translated 3 units to the right (along the x–axis) and 4 units up (along the y–axis).

Notice that the original relation is in fact $y = x^3 - 1$, so that relative to this graph, the graph of $y = (x - 3)^3 + 4$ has been translated 3 units to the right (along the x–axis) and 5 units up (along the y–axis).

EXERCISES 6.1

1. Find the equation of the given relation under the translation indicated.

(a) $y = x^2$; $\begin{pmatrix} 4 \\ 0 \end{pmatrix}$	(b) $y = x^2$; $\begin{pmatrix} -2 \\ 0 \end{pmatrix}$	(c) $y = x^2$; $\begin{pmatrix} 0 \\ 5 \end{pmatrix}$
(d) $x^2 + y = 2$; $\begin{pmatrix} 2 \\ 0 \end{pmatrix}$	(e) $x^2 + y = 2$; $\begin{pmatrix} 0 \\ 2 \end{pmatrix}$	(f) $x^2 + y = 2$; $\begin{pmatrix} 0 \\ -2 \end{pmatrix}$
(g) $xy = 8$; $\begin{pmatrix} 4 \\ 0 \end{pmatrix}$	(h) $xy = 8$; $\begin{pmatrix} 0 \\ -1 \end{pmatrix}$	(i) $x^2 + y^2 = 4$; $\begin{pmatrix} -1 \\ 0 \end{pmatrix}$
(j) $xy^2 = 9$; $\begin{pmatrix} 3 \\ 0 \end{pmatrix}$	(k) $xy^2 = 9$; $\begin{pmatrix} 0 \\ -3 \end{pmatrix}$	(l) $x + y^2 = 4$; $\begin{pmatrix} 4 \\ 0 \end{pmatrix}$

2. Consider the graphs shown below.

(a)

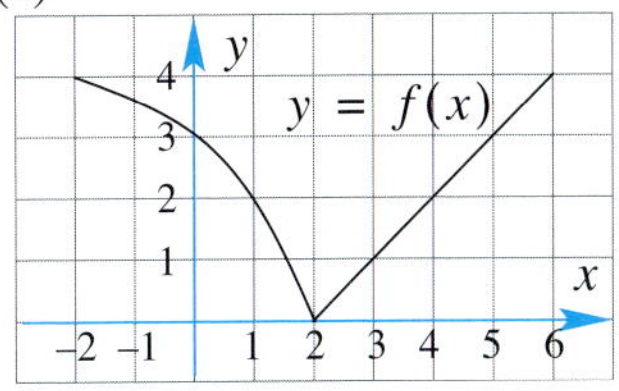

(b)

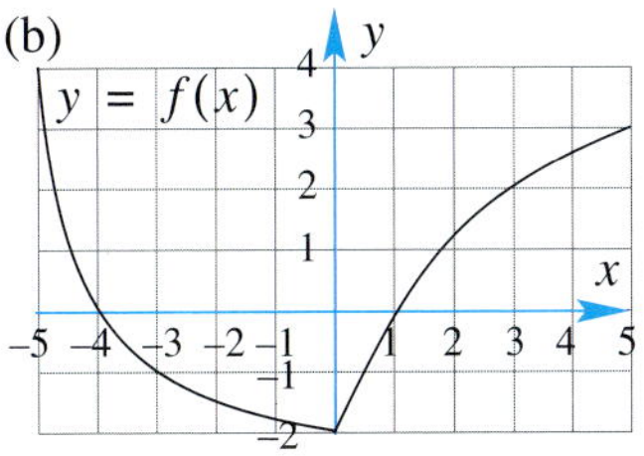

In each case, sketch the graph of

i. $y = f(x+2)$ ii. $y = f(x-1)$

iii. $y = f(x)-3$ iv. $y = f(x)+1$

3. Using translations on the graph of $f(x) = \sqrt{x}$, sketch the graph of the following

(a) $y = f(x-4)$ (b) $y = f(x)-2$ (c) $y = f(x-2)+3$

4. Using translations on the graph of $f(x) = \frac{1}{x}$, sketch the graph of the following

(a) $y = f(x+1)$ (b) $y = f(x)-4$ (c) $y = f(x+2)-3$

5. Using translations on the graph of $f(x) = \frac{1}{x^2}$, sketch the graph of the following

(a) $y = f(x)-1$ (b) $y = f(x-1)$ (c) $y = 2+f\left(x-\frac{3}{2}\right)$

6. On the same set of axes sketch the graphs of

(a) $y = x^2-4$ and $y = (x-4)^2$ (b) $y = x^2+5$ and $y = (x+5)^2$

(c) $y = x^2+2$ and $y = x^2-2$ (d) $y = \left(x+\frac{3}{2}\right)^2$ and $y = \left(x-\frac{3}{2}\right)^2$

(e) $y = x^3-8$ and $y = (x-8)^3$ (f) $y = (x+1)^3$ and $y = x^3+1$

(g) $y = \frac{1}{(x-2)}$ and $y = \frac{1}{x}-2$ (h) $y = \frac{1}{(x+3)}$ and $y = \frac{1}{x}+3$

(i) $y = \sqrt{x-2}$ and $y = \sqrt{x}-2$ (j) $y = \sqrt{x+4}$ and $y = \sqrt{x}+4$

7. Sketch the graph of the following functions, making sure to include all axial intercepts and labelling the equations of asymptotes (where they exist).

(a) $y = (x-2)^2+3$ (b) $y = \frac{1}{x+1}+2$ (c) $y = (x-1)^3-1$

(d) $y = \frac{1}{x-2}+\frac{1}{2}$ (e) $y = (x+2)^3-8$ (f) $y = (x+3)^2-9$

(g) $y = \sqrt{x-2}+2$ (h) $y = \sqrt{4+x}+2$ (i) $y = (1+x)^2-1$

(j) $y = 1+\frac{1}{3+x}$ (k) $y = \frac{1}{(x-1)^2}-1$ (l) $y = -2+\frac{1}{(2-x)^2}$

(m) $y = 2-\frac{1}{3-x}$ (n) $y = 8-(2-x)^3$ (o) $y = -2+\sqrt{x-4}$

8. Find the vector translation necessary for the following mappings

(a)	$x^2 \mapsto x^2 + 4$	(b)	$x^3 \mapsto x^3 - 2$	(c)	$\sqrt{x} \mapsto \sqrt{x+1}$
(d)	$\frac{1}{x} \mapsto \frac{1}{x-2}$	(e)	$x^4 \mapsto (x+2)^4$	(f)	$\frac{1}{x^2} \mapsto \frac{1}{x^2} - 4$
(g)	$x^3 \mapsto (x-2)^3 - 2$	(h)	$\frac{1}{x} \mapsto 3 + \frac{1}{x+2}$	(i)	$x^2 \mapsto 2 + (x-4)^2$
(j)	$\sqrt{x} \mapsto 3 + \sqrt{x-2}$	(k)	$\frac{1}{x^3} \mapsto \frac{1}{(x-3)^3} - 1$	(l)	$f(x) \mapsto h + f(x+k)$
(m)	$x^2 - 4 \mapsto (x+2)^2$	(n)	$x^3 + 1 \mapsto (x-1)^3$	(o)	$\frac{1}{x} - 2 \mapsto \frac{1}{x+1}$

9. Express, in terms of $f(x)$, the transformation(s) required to map $f(x)$ to $g(x)$.

(a) $f(x) = x^2$, $g(x) = x^2 - 2x + 2$

(b) $f(x) = x^2$, $g(x) = x^2 + 4x$

(c) $f(x) = x^3$, $g(x) = x^3 - 6x^2 + 12x - 8$

(d) $f(x) = \frac{1}{x^2}$, $g(x) = \frac{1}{x^2 - 2x + 1} + 1$

(e) $f(x) = x^3$, $g(x) = x^3 - 3x^2 + 3x + 2$

10. Consider the relations shown below.

(a)
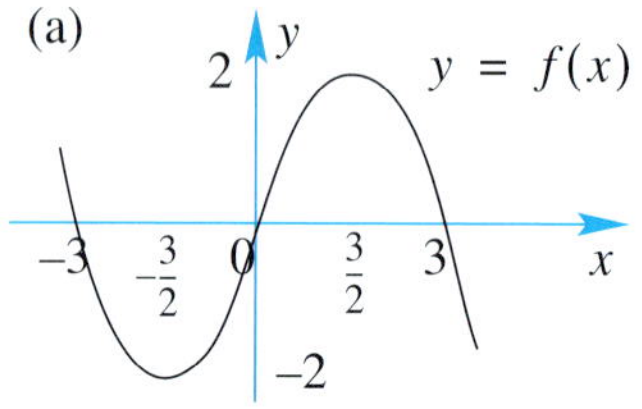

(b)
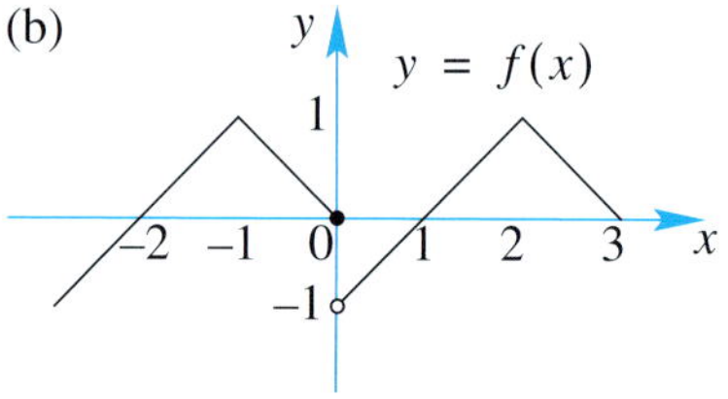

(c)
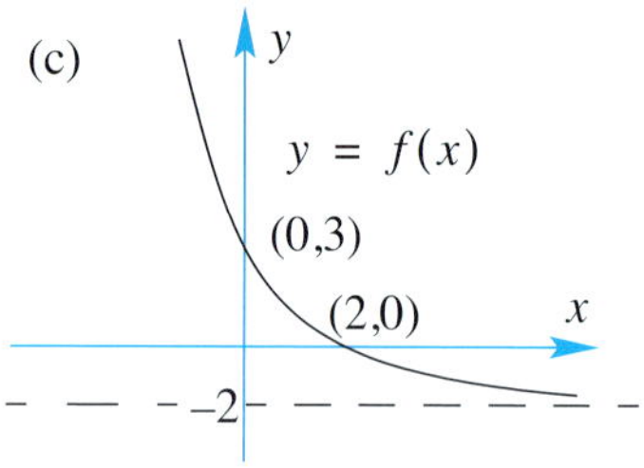

(d)
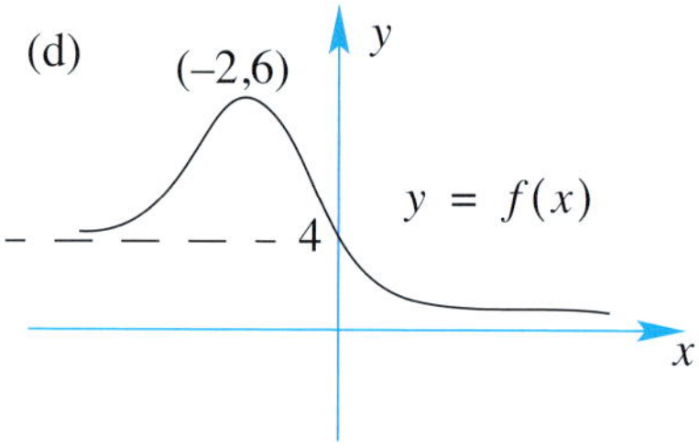

Sketch the following

i. $y = f(x+2) - 2$ ii. $y = f(x-2) - 4$

iii. $y = 3 + f(x-3)$ iv. $y = 1 + f(x+1)$

11. Express, in terms of $f(x)$, the equation of the graph represented in Figure 2, given that the graph in Figure 1 has the equation $y = f(x)$, $-1 \le x \le 1$.

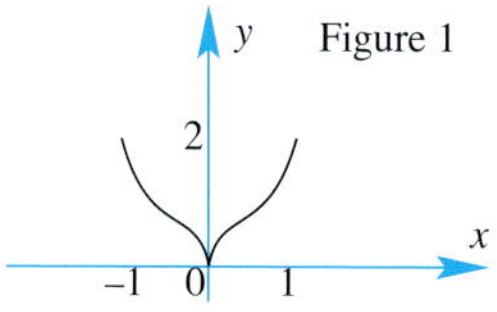

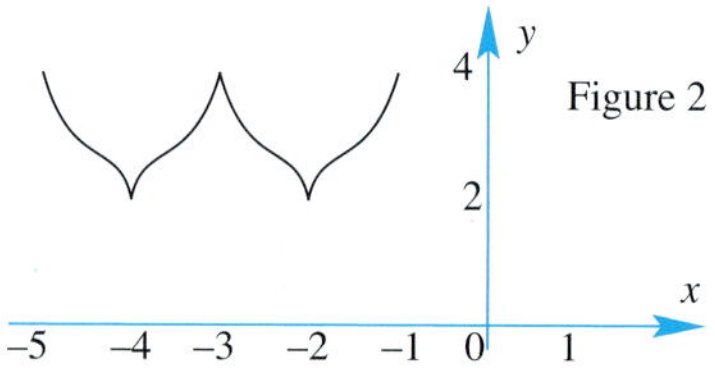

6.2 DILATIONS

6.2.1 DILATION FROM THE X–AXIS

Before we start our discussion it should be pointed out that other commonly used expressions for *dilations from the x–axis* are; *dilation along the y–axis* and *dilation parallel to the y–axis* – either of these three expressions can be used when describing this dilation.

The equation $y = pf(x)$ can be written as $\frac{y}{p} = f(x)$. We have rearranged the expression so that we can more clearly see the effects that p has on the y–values. That is, the term $\frac{y}{p}$ represents a transformation on the y–axis as opposed to a transformation on the graph of $f(x)$. The effect of 'p' in the term $\frac{y}{p}$ is that of a **dilation from the x–axis**.

If $|p| > 1$, we *shrink* the y–axis (seeing as we are dividing the y–values by a number larger than one). Whereas if $0 < |p| < 1$ we *stretch* the y–axis.

However, we still need to describe the effect this transformation has on the final appearance of the graph of $f(x)$.
We summarise these results, stating the **effects on the graph of $f(x)$**:

For the curve with equation $y = pf(x)$,

$|p| > 1$, represents '**stretching**' $f(x)$ by a factor p from the x–axis.

$0 < |p| < 1$, represents '**shrinking**' $f(x)$ by a factor $\frac{1}{p}$ from the x–axis

Realise that quoting a 'stretch' of factor $\frac{1}{3}$ is the same as quoting a 'shrink' of factor 3. However, it is more common that when referring to a dilation from the x–axis we are referring to a stretch. So a dilation from the x–axis of factor 3 would imply a stretching effect whereas a dilation from the x–axis of factor $\frac{1}{3}$ would imply a shrinking effect.

The relationship between the the original coordinates (x, y) and the new coordinates (x', y') can be seen in the diagram below.

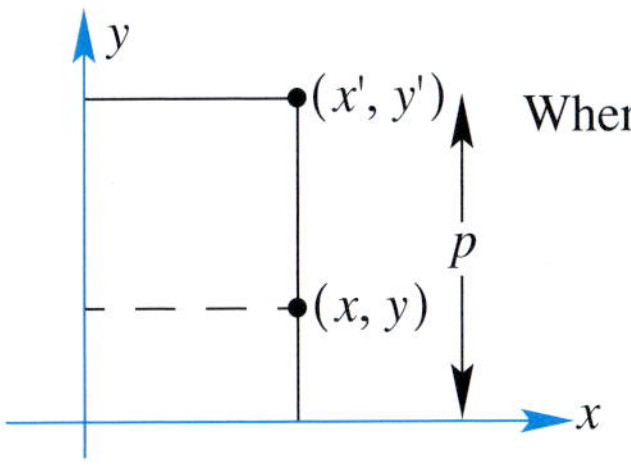

Where: $x' = x$
$y' = py$

Unfortunately, unlike the translation vector of §6.1, we have no dilation vector to describe this transformation [although there does exist a *dilation matrix*].

EXAMPLE 6.7

Describe the effects on the graph of $y = f(x)$ when the following graphs are sketched (a) $y = 2f(x)$ (b) $y = \frac{1}{4}f(x)$ (c) $2y = f(x)$

Solution

(a) $y = 2f(x)$ represents a dilation of factor 2 from the x–axis. This means that the graph of $y = f(x)$ would be stretched by a factor of two along the y–axis.

(b) $y = \frac{1}{4}f(x)$ represents a dilation of factor $\frac{1}{4}$ from the x–axis. This means that the graph of $y = f(x)$ would shrink by a factor of four along the y–axis.

(c) $2y = f(x)$ needs to first be written as $y = \frac{1}{2}f(x)$, which represents a dilation of factor $\frac{1}{2}$ from the x–axis. This means that the graph of $y = f(x)$ would shrink by a factor of two along the y–axis.

EXAMPLE 6.8

Given the graph of $y = f(x)$ sketch the graph of

(a) $y = 2f(x)$ (b) $3y = f(x)$

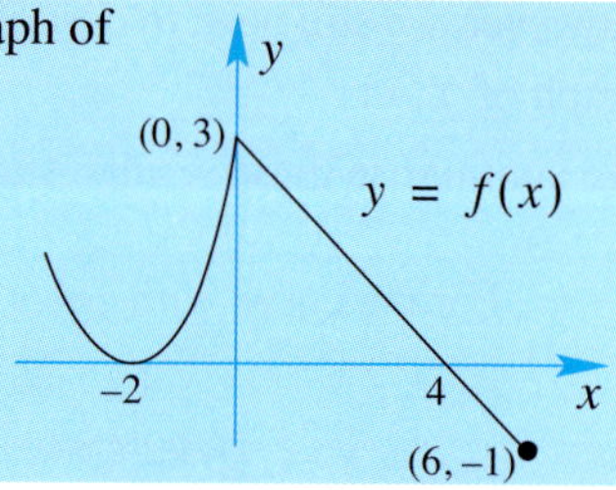

Solution

(a) The graph of $y = 2f(x)$ represents a dilation of factor 2 along the y–axis – i.e., the graph of $y = f(x)$ will be stretched by a factor of 2.
Notice how the x–intercepts are invariant – i.e., they have not altered after the transformation.
This is because the y–value at these points is zero, and multiplying zero by any number will still be zero.

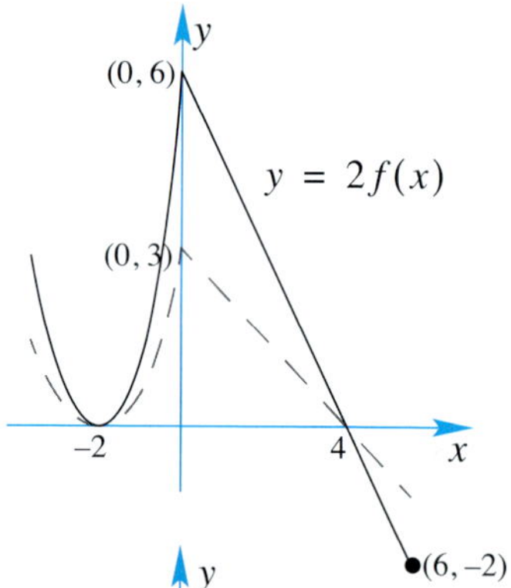

(b) $3y = f(x) \Leftrightarrow y = \frac{1}{3}f(x)$ which represents a dilation of factor $\frac{1}{3}$.
– i.e., the graph of $y = f(x)$ will shrink by a fator of 3.

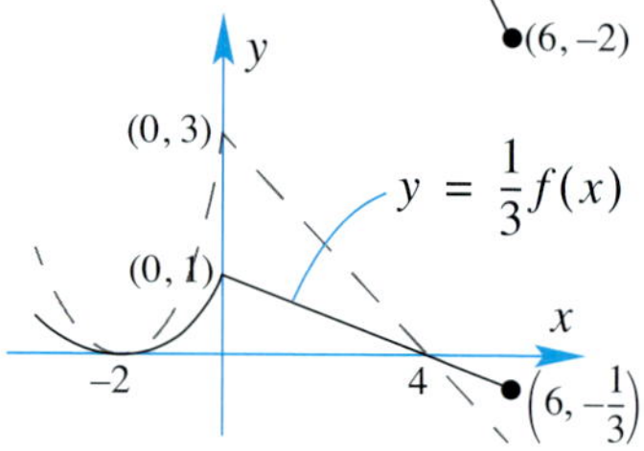

6.2.2 DILATION FROM THE Y-AXIS

The equation $y = f\left(\frac{x}{q}\right)$ represents a transformation applied to the x–values. We now need to consider how this factor 'q' affects the graph of $f(x)$. The term $\frac{x}{q}$ represents a transformation on the x–values as opposed to a transformation on the graph of $f(x)$. The effect of 'q' in the term $\frac{x}{q}$ is that of a **dilation from the *y*–axis**.

If $|q| > 1$, we *shrink* the x–axis (seeing as we are dividing the x–values by a number larger than one). Whereas if $0 < |q| < 1$ we *stretch* the x–axis (because we are dividing by a number less than one but greater than zero).

However, we still need to describe the effect this transformation has on the final appearance of the graph of $f(x)$.
We summarise these results, stating the **effects on the graph of $f(x)$**:

> For the curve with equation $y = f(x/q)$,
>
> $|q| > 1$, represents '**stretching**' $f(x)$ by a factor q from the y–axis.
>
> $0 < |q| < 1$, represents '**shrinking**' $f(x)$ by a factor $\frac{1}{q}$ from the y–axis

So a dilation from the y–axis of factor 2 (e.g., $y = f(x/2)$) would imply a stretching effect whereas a dilation from the y–axis of factor $\frac{1}{2}$ (e.g., $y = f(2x)$) would imply a shrinking effect. We show this in the diagram below:

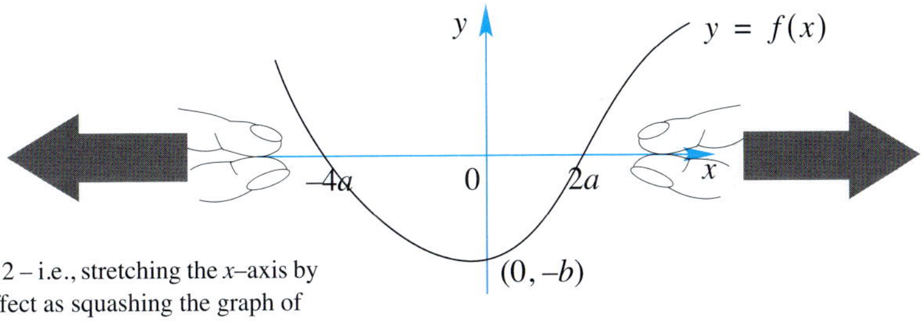

Multiplying the x–values by 2 – i.e., stretching the x–axis by a factor of 2 has the same effect as squashing the graph of $y = f(x)$ by a factor of 2.

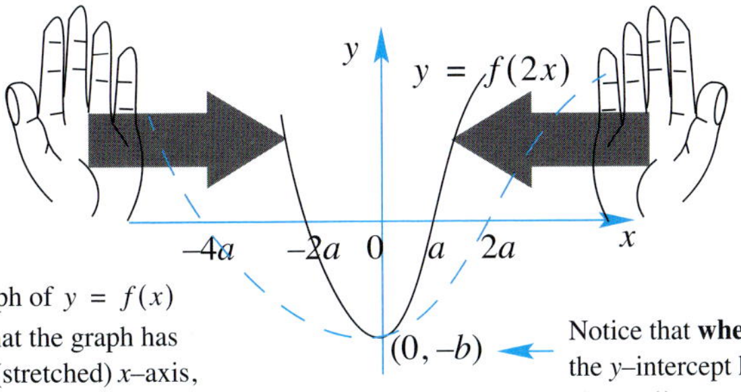

Again, it must be remembered that the graph of $y = f(x)$ has not changed, it is simply the illusion that the graph has been squashed. That is, relative to the new (stretched) x–axis, the graph of $y = f(x)$ appears to have been squashed.

Notice that **where $x = 0$**, the y–intercept has not changed!

The relationship between the the original coordinates (x, y) and the new coordinates (x', y') can be seen in the diagram below.

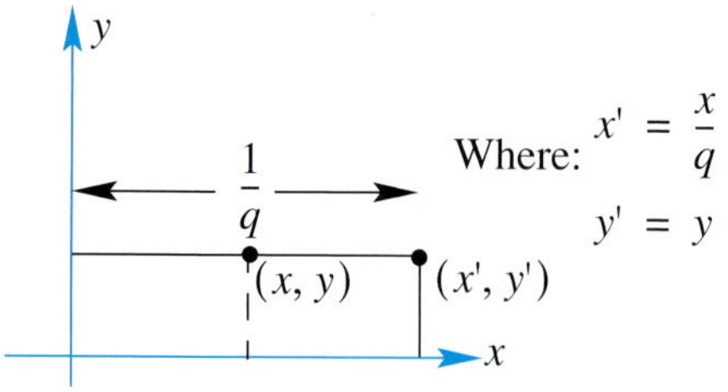

Again, we have no dilation vector to describe this transformation [although there does exist a *dilation matrix*].

EXAMPLE 6.9

Given the graph of $y = f(x)$ sketch the graph of

(a) $y = f(2x)$ (b) $y = f\left(\frac{x}{3}\right)$

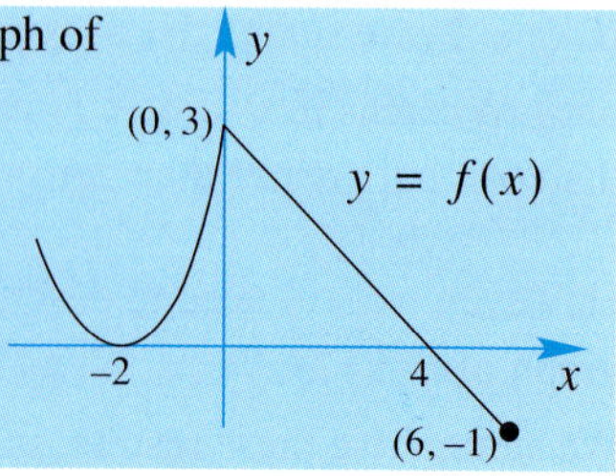

Solution

(a) We first express the dilation in the form $y = f(x/q)$: $y = f(2x) \Rightarrow y = f\left(x/\left(\frac{1}{2}\right)\right)$.

This means that we have a dilation **from** the y–axis of factor $\frac{1}{2}$. That is, the graph of $y = f(x)$ will 'shrink'[or rather be squashed] **along** the x-axis by a factor of 2.
i.e., because the x–values are doubled (from the $2x$ term in the expression $y = f(2x)$) it seems reasonable to deduce that on the new set of axes the graph will be squashed by a factor of 2.

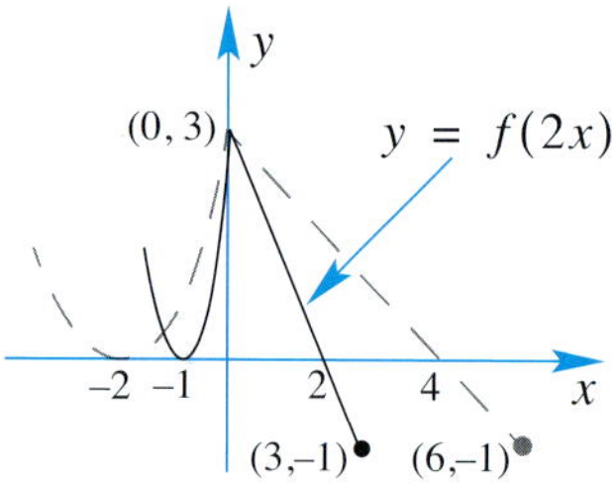

(b) The term $\frac{x}{3}$ in the expression $y = f\left(\frac{x}{3}\right)$ implies that the new x–values will be a third of the original x–values, this means that the new x–axis will be compressed by a factor of 3. This in turn will have a stretching effect on $y = f(x)$ of factor 3 (along the x–axis).

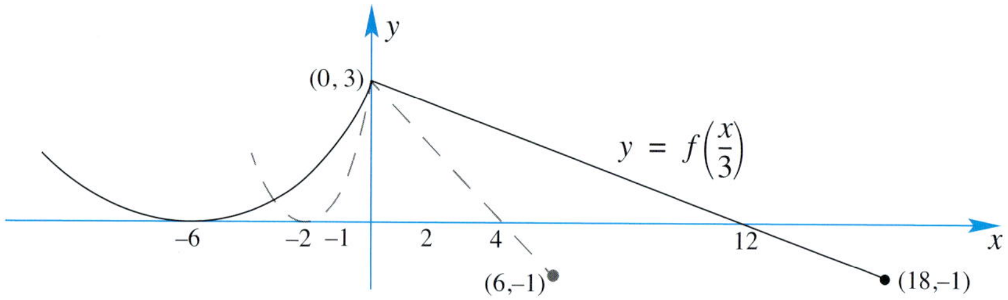

EXERCISES 6.2

1. On the same set of axes sketch the graphs of

(a) $f(x) = x^2,\ y = f(2x)$

(b) $f(x) = \sqrt{x},\ y = f(4x)$

(c) $f(x) = \frac{1}{x},\ y = f\left(\frac{x}{3}\right)$

(d) $f(x) = x^3,\ y = f\left(\frac{x}{2}\right)$

2. On the same set of axes sketch the graphs of

(a) $f(x) = x^2,\ y = 2f(x)$

(b) $f(x) = \sqrt{x},\ y = 4f(x)$

(c) $f(x) = \frac{1}{x},\ y = \frac{1}{3}f(x)$

(d) $f(x) = x^3,\ y = \frac{1}{2}f(x)$

3. Consider the graphs shown below.

(a)

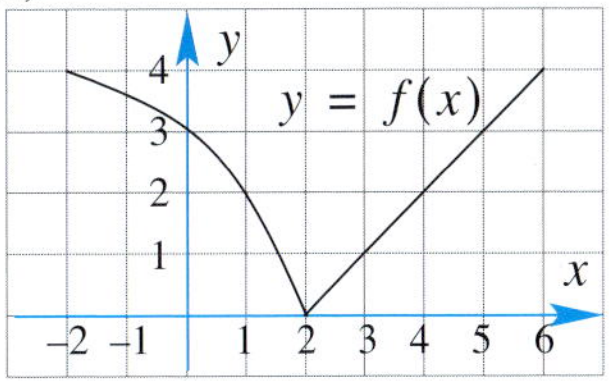

(b)

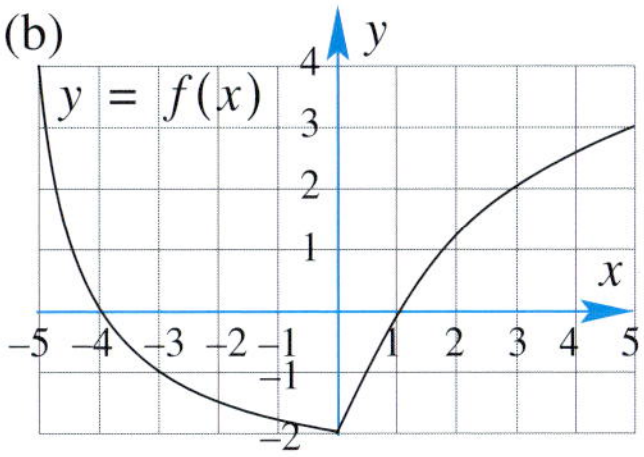

In each case, sketch the graph of

i. $y = f(0.5x)$

ii. $y = f(2x)$

iii. $y = 0.5f(x)$

iv. $y = 2f(x)$

4. Consider the relations shown below.

(a)

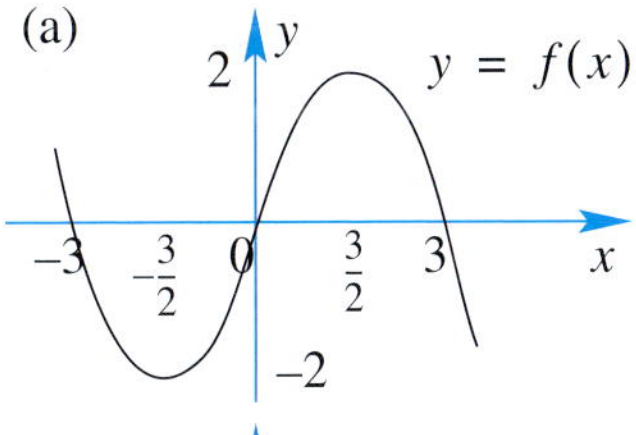

(b)

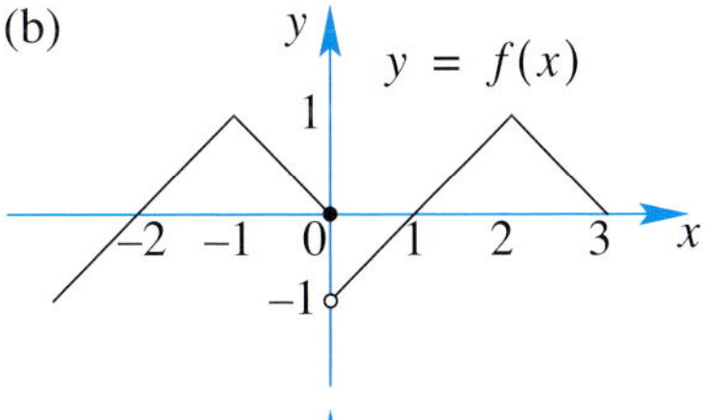

(c)

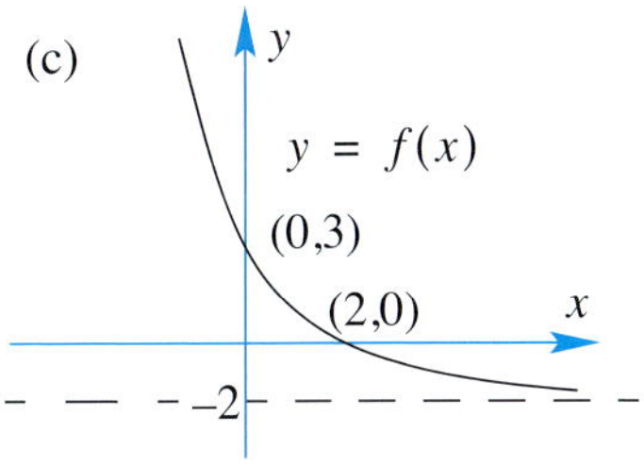

(d)

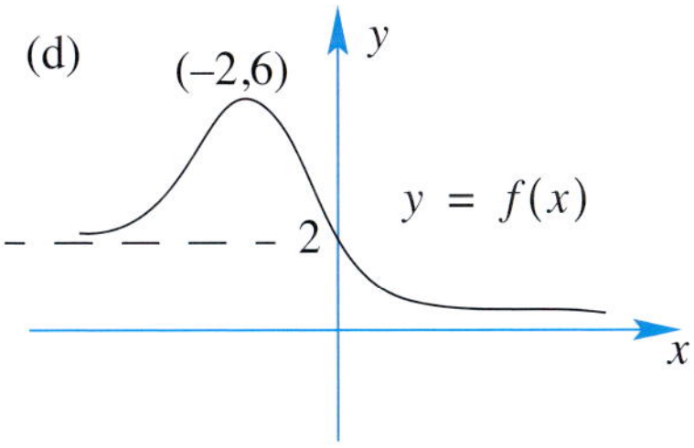

Sketch the following

i. $y = f\left(\frac{2}{3}x\right)$

ii. $y = 4f(x)$

5. Describe the transformation(s) under the following mappings

(a) $|x| \mapsto |2x| + 1$ (b) $x^2 \mapsto \frac{1}{2}(x-2)^2 - 3$ (c) $\frac{1}{x} \mapsto \frac{1}{2x-1}$

(d) $x^3 \mapsto (3x-2)^3$ (e) $x^4 \mapsto \frac{1}{2}(4x-2)^4 - 2$ (f) $\sqrt{x} \mapsto \frac{1}{2}\sqrt{8x} + 2$

6. For the graph shown below, sketch each of the following on a separate set of axes.

(a) $y = 2f(x) - 1$ (b) $y = 2f(x-1)$ (c) $y = f(2x-1)$

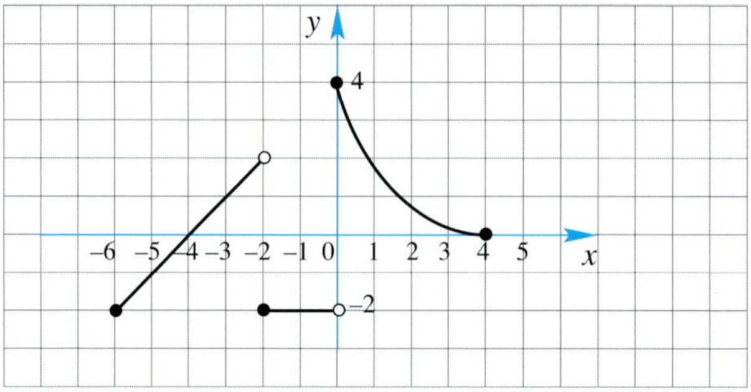

7. Consider the function $g(x) = \begin{cases} x^2 & \text{if } x \geq 2 \\ 6 - x & \text{if } x < 2 \end{cases}$.

(a) Find an expression for

i. $f(x) = g(x+2)$ ii. $h(x) = g(x) - 3$

iii. $h(x) = 2g(x)$ iv. $k(x) = g(2x)$

v. $k(x) = g(2x-1)$ vi. $f(x) = \frac{1}{2}g(4x+2)$

(b) On separate sets of axes, sketch the graphs of each of the functions in part (a).

8. Given the relation $f(x) = \begin{cases} -\sqrt{4-(x-2)^2} & \text{if } 1 < x \leq 4 \\ \sqrt{3}x & \text{if } x \leq 1 \\ -\sqrt{3}x & \text{if } x \leq 1 \\ \sqrt{4-(x-2)^2} & \text{if } 1 < x \leq 4 \end{cases}$, sketch the graphs of

(a) $y = \frac{1}{2}f(x)$ (b) $y = f\left(\frac{1}{2}x\right)$

9. Given the function $f(x) = \sqrt{x}$, sketch the graph of

(a) $y = af(x), a > 0$ (b) $y = f(ax), a > 0$

(c) $y = bf(x+b), b > 0$ (d) $y = \frac{1}{a}f(a^2x), a \neq 0$

10. Given the function $f(x) = \frac{1}{x^2}$, sketch the graph of

(a) $y = bf(\sqrt{a}x) - a, a, b > 0$ (b) $y = bf(\sqrt{a}x) - \frac{a}{b}, a, b > 0$

6.3 REFLECTIONS

We first consider reflections about the x–axis and about the y–axis. The effects of reflecting a curve about these axes can be seen in the diagrams below:

Reflection about the x–axis

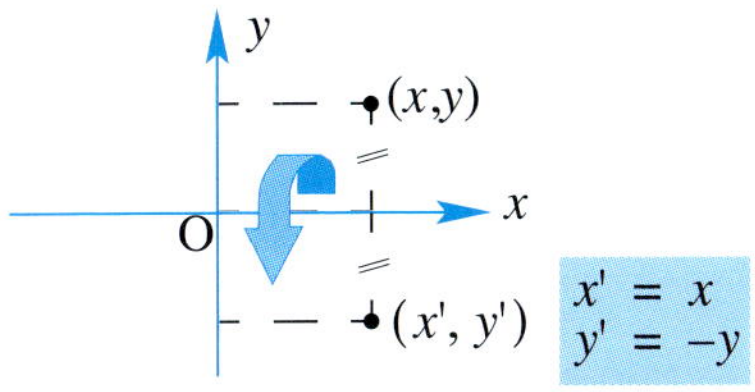

When reflecting about the x–axis we observe that the coordinates (x,y) are mapped to the coordinates $(x, -y)$. Meaning that the x–values remain the same but the y–values change sign.

Reflection about the y–axis

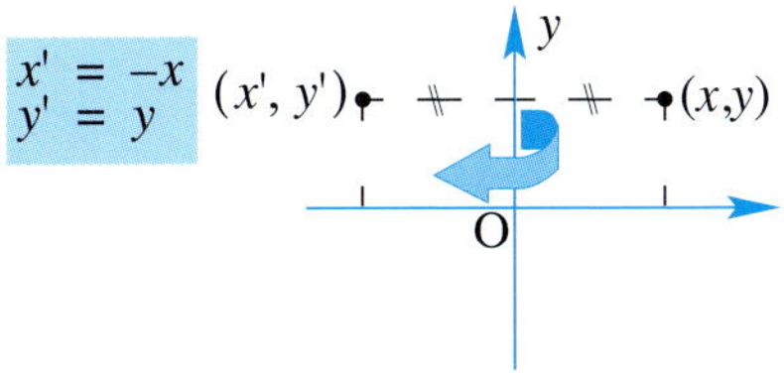

When reflecting about the y–axis we observe that the coordinates (x,y) are mapped to the coordinates $(-x, y)$. Meaning that the y–values remain the same but the x–values change sign.

We summarise the effects of these two transformations of the graph of $y = f(x)$

$y = -f(x)$, represents a **reflection of** $f(x)$ **about the x–axis**.
$y = f(-x)$, represents a **reflection of** $f(x)$ **about the y–axis**.

Another type of reflection is the reflection about the line $y = x$, when sketching the inverse of a function. Inverse functions are dealt with in detail in Chapter 5, so here we give a summary of those results.

Reflection about the line $y = x$

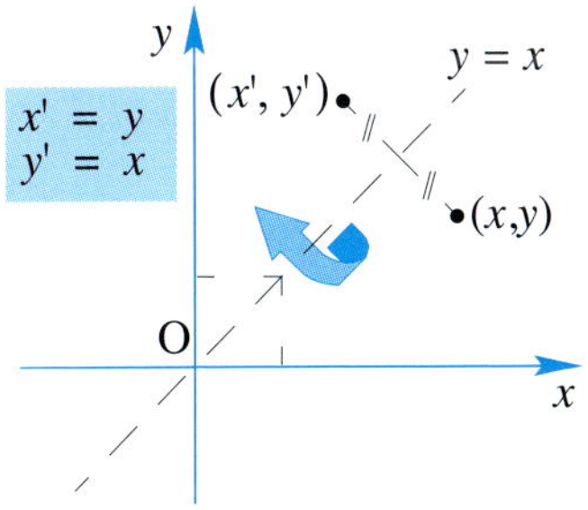

When reflecting about the line y = x we observe that the coordinates (x,y) are mapped to the coordinates (y, x). Meaning that the x–values and the y–values are interchanged.

If a **one-one function** $y = f(x)$ undergoes such a reflection, we call its transformed graph the **inverse function** and denote it by $y = f^{-1}(x)$.

EXAMPLE 6.10

Given the graph of $y = f(x)$ sketch the graph of

(a) $y = f(-x)$ (b) $y = -f(x)$

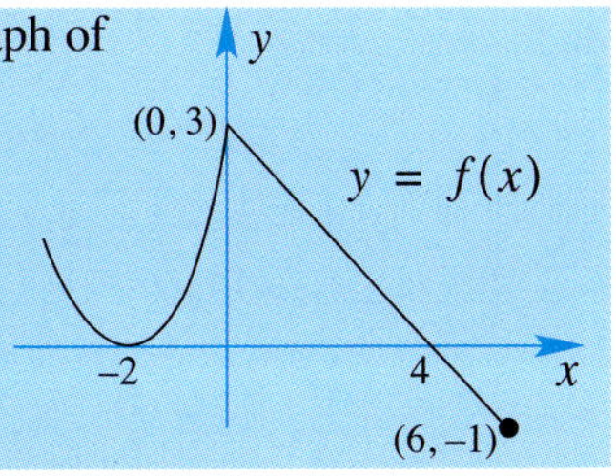

Solution

(a) The graph of $y = f(-x)$ represents a reflection of the graph of $y = f(x)$ about the y–axis:

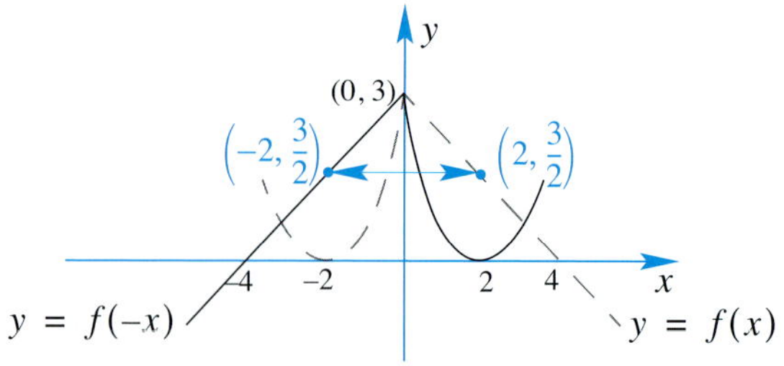

(b) The graph of $y = -f(x)$ represents a reflection of the graph of $y = f(x)$ about the x–axis:

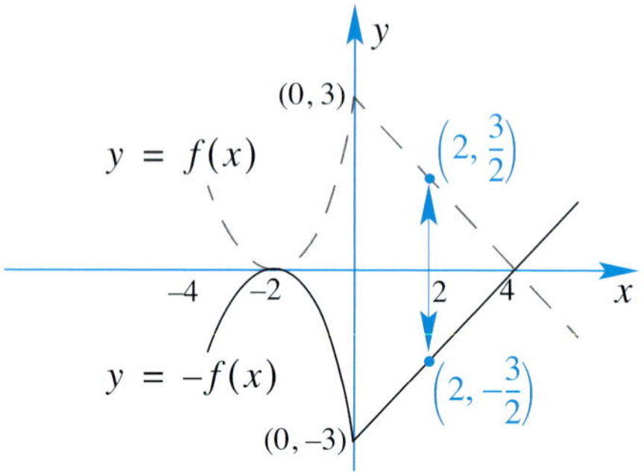

We now consider a combination of the transformations we have looked at so far.

EXAMPLE 6.11

Sketch the graph of $y = 3 - 2\sqrt{4-x}$.

Solution

We start by considering the function $f(x) = \sqrt{x}$, then, the expression $y = 3 - 2\sqrt{4-x}$ can be written in terms of $f(x)$ as follows: $y = 3 - 2f(4-x)$ or $y = -2f(4-x) + 3$.

This represents

1. reflection about the y–axis [due to the '$-x$' term]
2. translation of 4 units to the right [due to the '$4 - x$' term]
   Note: $4 - x = -(x - 4)$
3. dilation of factor 2 along the y–axis [due to the '$2f(x)$' term]
4. reflection about the x–axis [due to the '–' in front of the '$2f(x)$' term]
5. translation of 3 units up [due to the '+3' term]

We produce the final graph in stages:

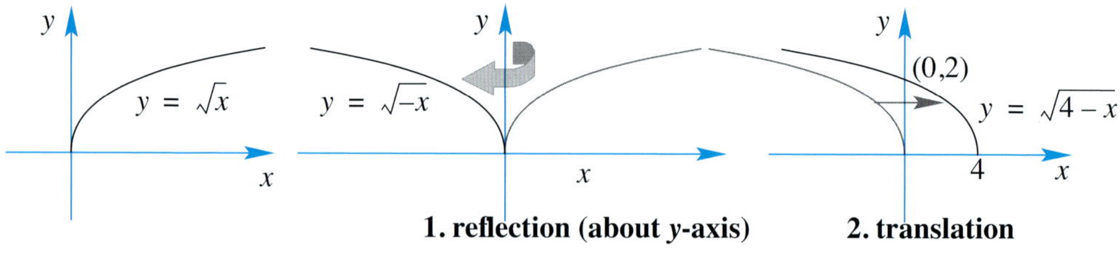

3. dilation

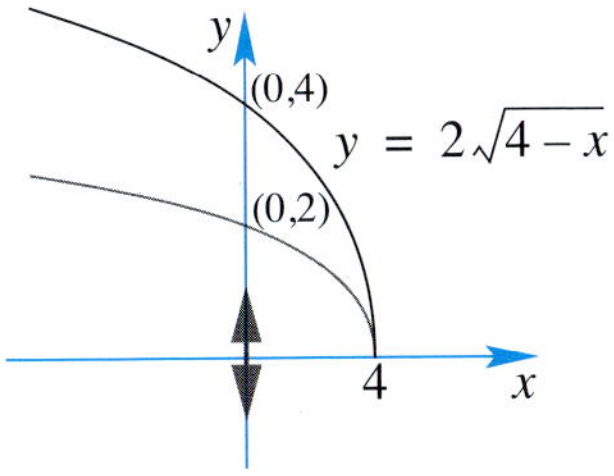

4. reflection (about x–axis)

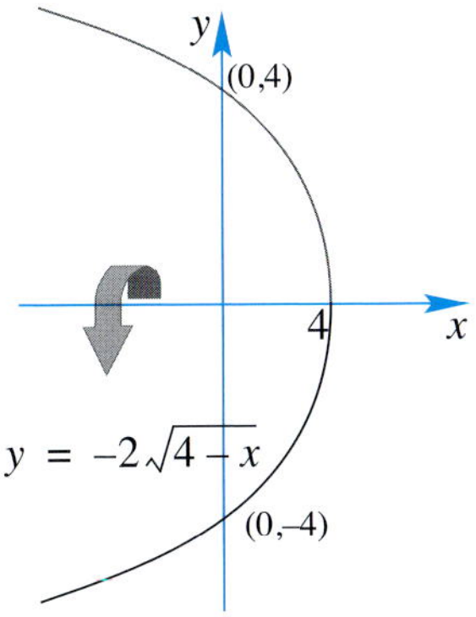

5. translation

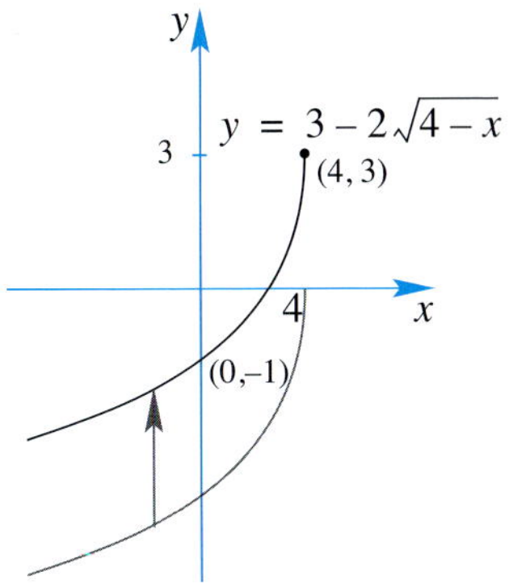

The above example gave a step by step account of how to produce the final graph, however, there is no need to draw that many graphs to produce the final outcome. We can reduce the amount of work involved by including all the transformation on one set of axes and then produce the final graph on a new set of axes.

EXAMPLE 6.12

Sketch the graph of $y = 2 - \frac{1}{2}(x+1)^2$.

Solution

If we consider the function $f(x) = x^2$, the graph of $y = 2 - \frac{1}{2}(x+1)^2$ can be written as

$y = 2 - \frac{1}{2}f(x+1)$. This represents a 'shrinking' effect of factor 2 from the x–axis followed by a reflection about the x–axis then a translation of 1 to the left and finally a translation of 2 units up.

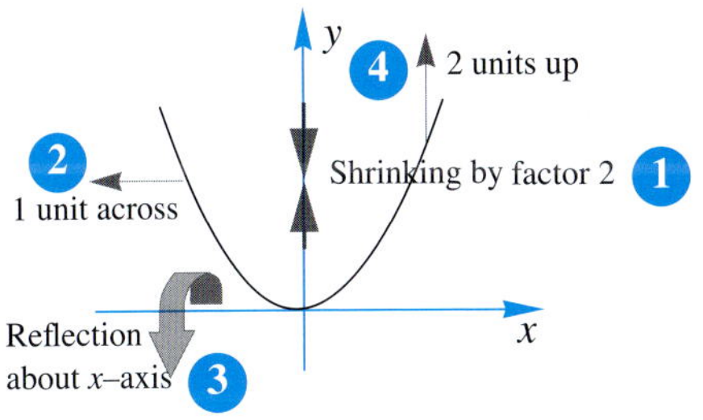

Final graph:

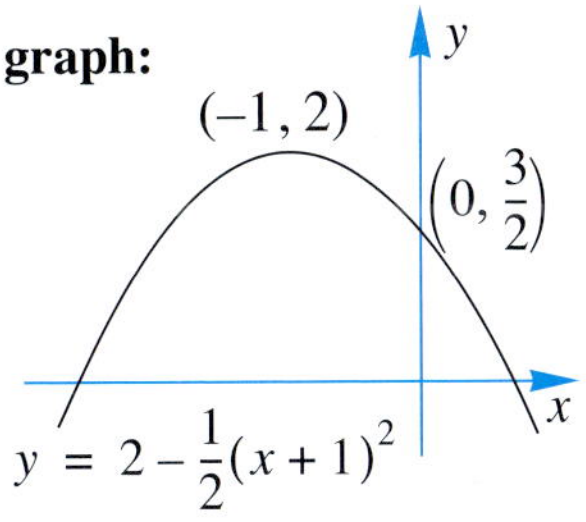

At this stage we have not looked at the x– or y–intercepts, although these should always be determined.

Important note!

Note the order in which we have carried out the transformations – although there is some freedom in which order the transformations are carried out, there are some transformations that must be carried out before others.

You should try and alter the order in which the transformations in Example 6.12 have been carried out. For example, does it matter if we apply Step 2 before Step 1. Can Step 2 be carried out after Step 4?

EXERCISES 6.3

1. Sketch the graphs of for each of the following i. $y = f(-x)$ ii. $y = -f(x)$

(a)

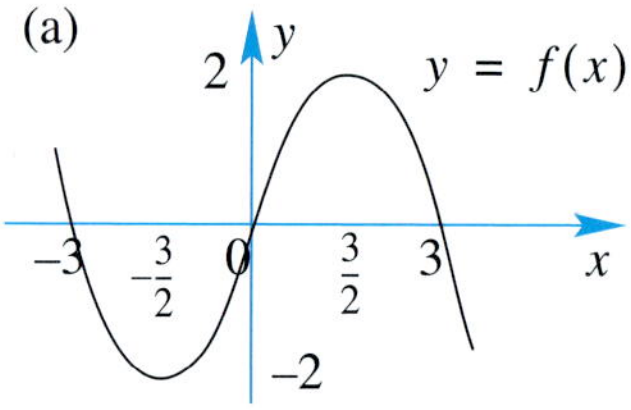

(b)

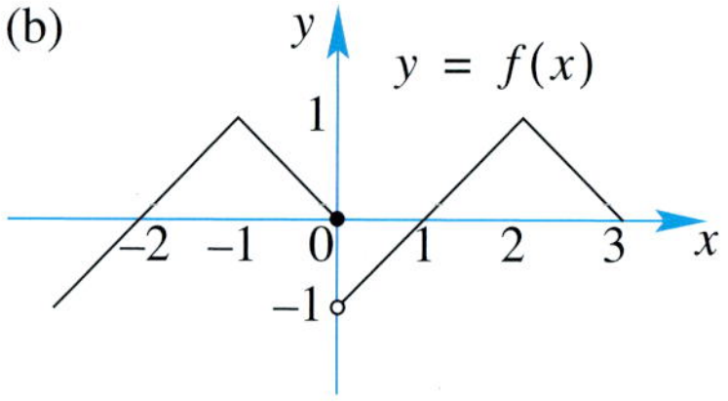

(c)

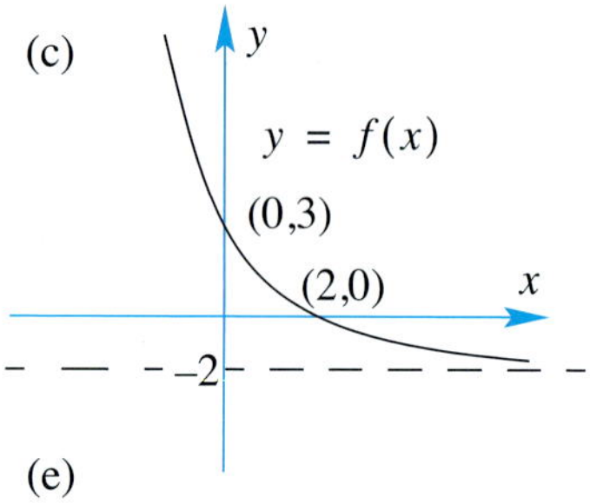

(d)

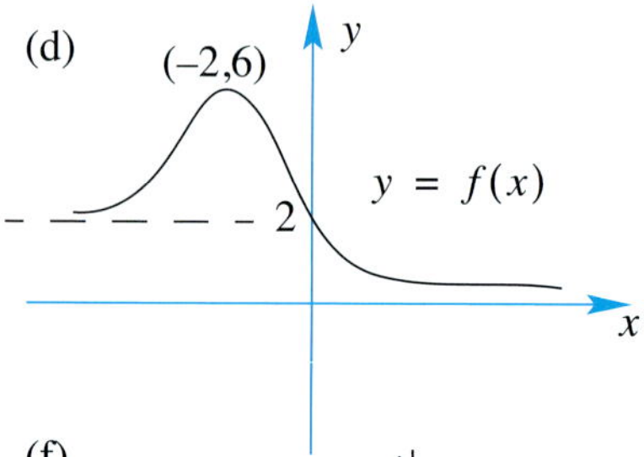

(e)

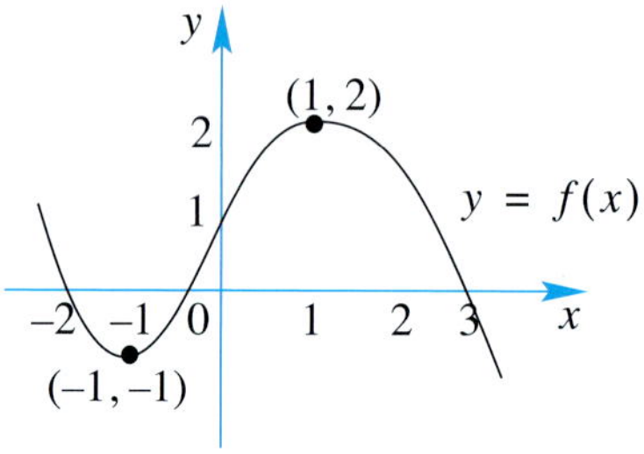

(f)

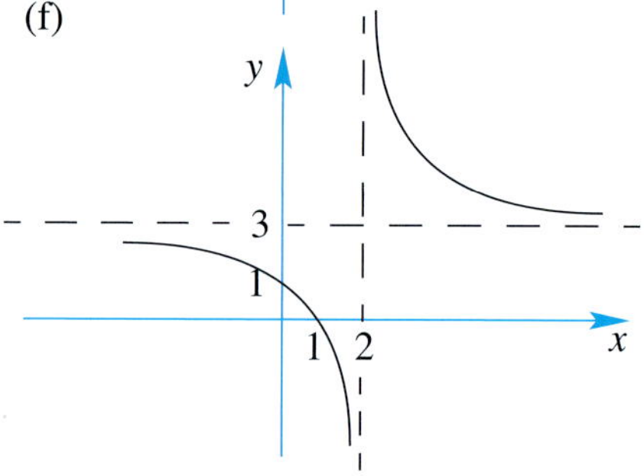

2. The diagram shows the graph of the function $y = f(x)$.

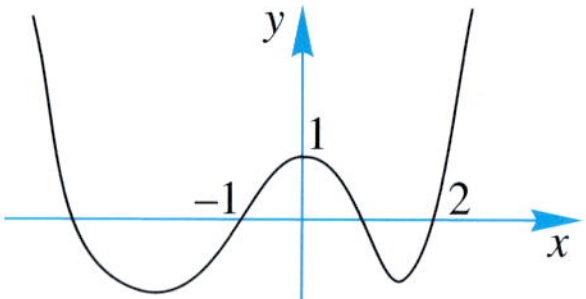

Find the equation in terms of $f(x)$ for each of the following graphs.

(a)

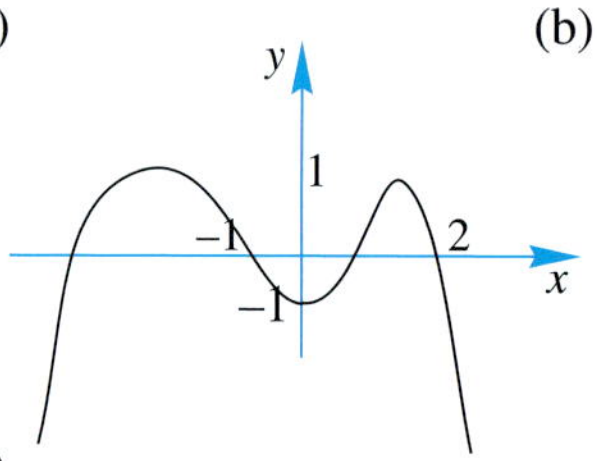

(b)

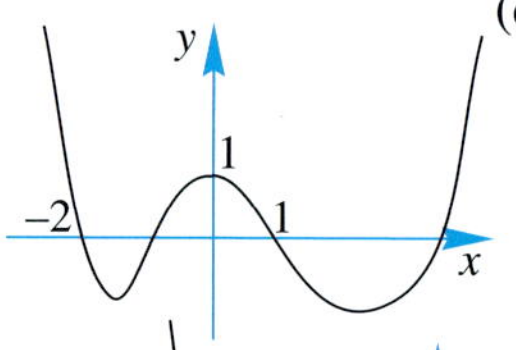

(c)

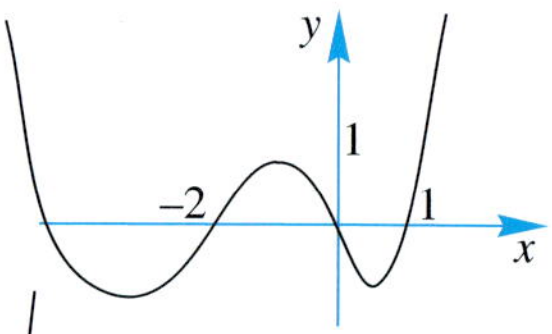

(d)

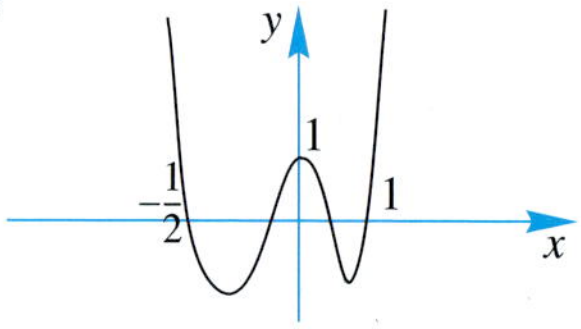

(e)

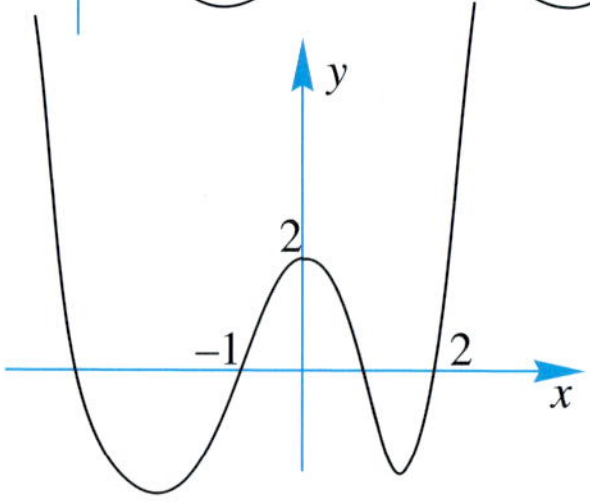

3. Sketch the graphs of the following

(a) $f(x) = -3x^2 + 9$ (b) $f(x) = 4 - \frac{1}{2}x^2$ (c) $f(x) = 1 - \frac{1}{8}x^3$

(d) $f(x) = -(x+2)^2 + 3$ (e) $f(x) = \frac{2}{1-x}$ (f) $f(x) = 1 - \frac{1}{x+2}$

(g) $f(x) = -(x-2)^3 - 2$ (h) $f(x) = \frac{1}{2(2-x)}$ (i) $f(x) = 4 - \frac{2}{x^2}$

(j) $f(x) = 3 - \frac{2}{1-x}$ (k) $f(x) = -(1-x)^2$ (l) $f(x) = 4\sqrt{9-x}$

(m) $f(x) = -\frac{2}{(2x-1)^2}$ (n) $f(x) = \frac{1}{2}\sqrt{2 - \frac{1}{4}x}$ (o) $f(x) = |2-x|$

(p) $f(x) = 2(1 - \sqrt{x-2})$ (q) $f(x) = 4 - |8 - x^3|$ (r) $f(x) = \frac{2}{2 - \sqrt{x}}$

4. The graph of $y = f(x)$ is shown.
Use it to sketch the graphs of

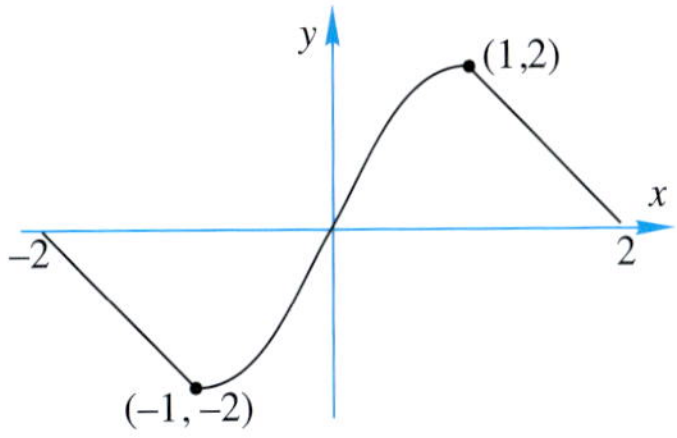

(a) $y = f(x-1)$ (b) $y = f(x) - 1$

(c) $y = f(x+1)$ (d) $y = 1 - f(x)$

(e) $y = 1 + f(-x)$ (f) $y = 2 - f(-x)$

(g) $y = -\frac{1}{2}f(x)$ (h) $y = f(-2x)$

(i) $y = 2 - f\left(\frac{1}{2}x\right)$ (j) $y = -2 + f(1-x)$

5. Sketch the graphs of the following functions relative to the graph of

(a) $f(x) = \frac{1}{x}$
i. $y = f(ax) + b, a > 1, b < 0$
ii. $y = bf(a-x), b > 0, a > 0$

(b) $f(x) = \frac{1}{x^2}$.
i. $y = bf(\sqrt{a}x) - a, a > 1, b < 0$
ii. $y = f\left(\frac{x}{b}\right) - a, a > 0, b < 0$

(c) $f(x) = \sqrt{x}$.
i. $y = af(a^2 - ax), a > 0$
ii. $y = af(a^2 - ax), a < 0$

6. Consider the following transformations

A: Reflection about x-axis. B: Reflection about y-axis C: $\begin{pmatrix} 1 \\ 2 \end{pmatrix}$

D: Squash by factor 2 along x-axis E: Stretch by factor 3 along y-axis

Sketch the graph of $f(x)$ under the transformations in their given order

(a) $f(x) = |x-2|$, A; C. (b) $f(x) = |x-2|$, C; A.

(c) $f(x) = |x^2 - 2x|$, D; B; C. (d) $f(x) = |x^2 - 2x|$, C; D; B.

(e) $f(x) = x^3$, A; E; C. (f) $f(x) = x^3$, E; C; A.

6.4 RECIPROCAL OF A FUNCTION

In this section we sketch the graph of $y = \frac{1}{f(x)}$ based on the graph of $y = f(x)$. We make observations of the behaviour of $y = f(x)$ and its reciprocal – we use the TI–83 to begin.

Example 1: Graph of $y = f(x)$ and $y = \frac{1}{f(x)}$ where $f(x) = x$:

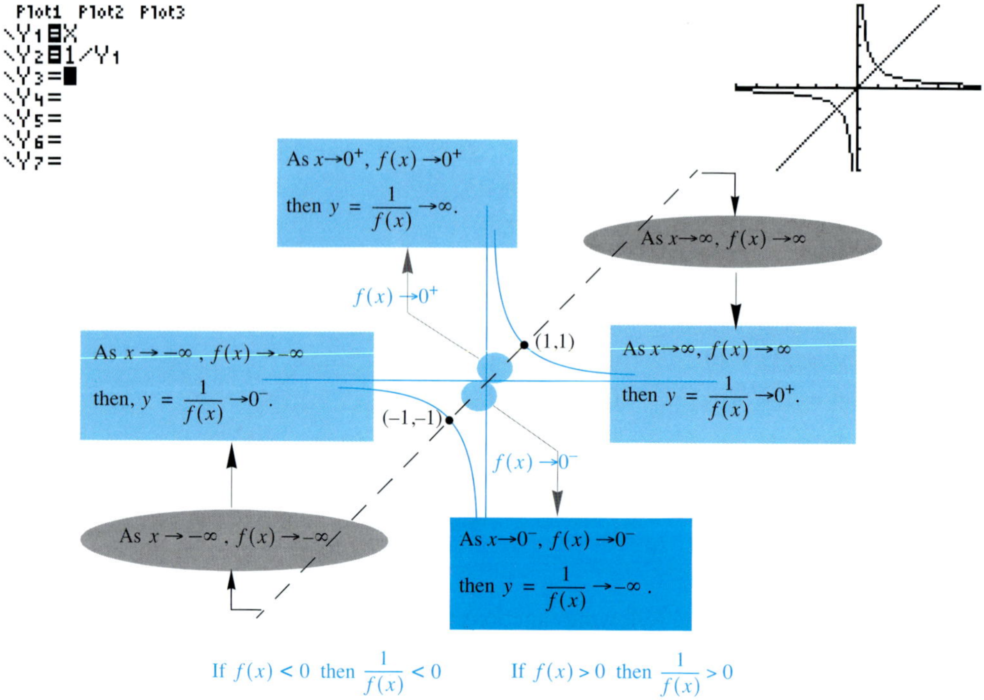

Example 2: Graph of $y = f(x)$ and $y = \frac{1}{f(x)}$ where $f(x) = x^2$:

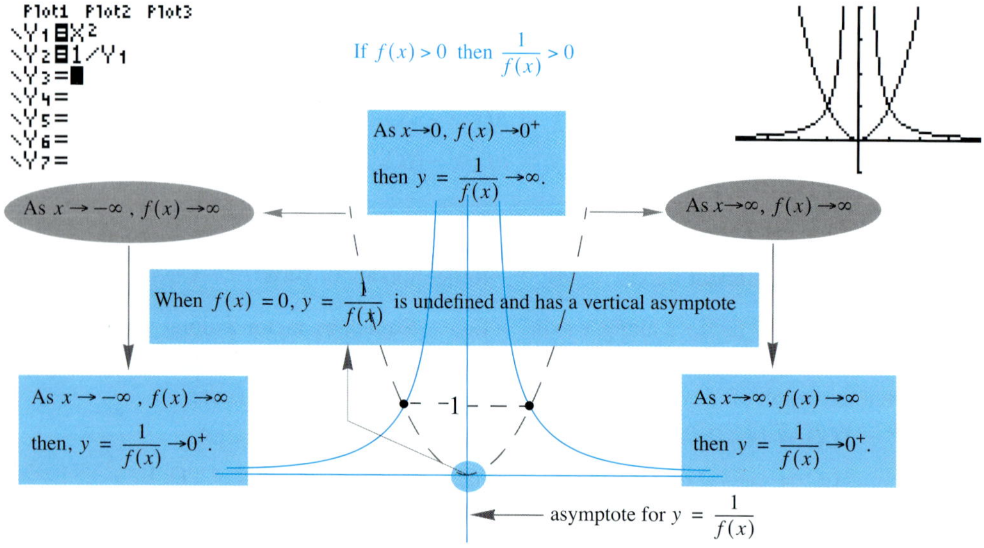

Examples 1 and 2 display some similar properties. Such properties will also exist in the examples that follow. Rather than listing all of the observations we now list selected observations.

Example 3: Graph of $y = f(x)$ and $y = \dfrac{1}{f(x)}$ where $f(x) = (x-2)(x+1)$:

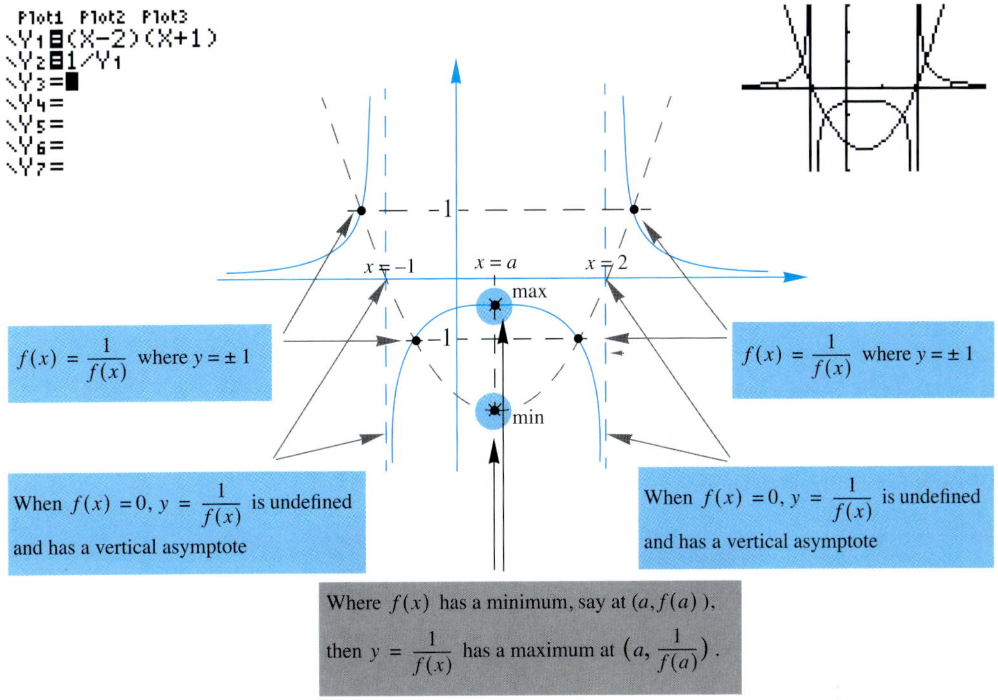

Example 4: Graph of $y = f(x)$ and $y = \dfrac{1}{f(x)}$ where $f(x) = \dfrac{1}{x} + 2$:

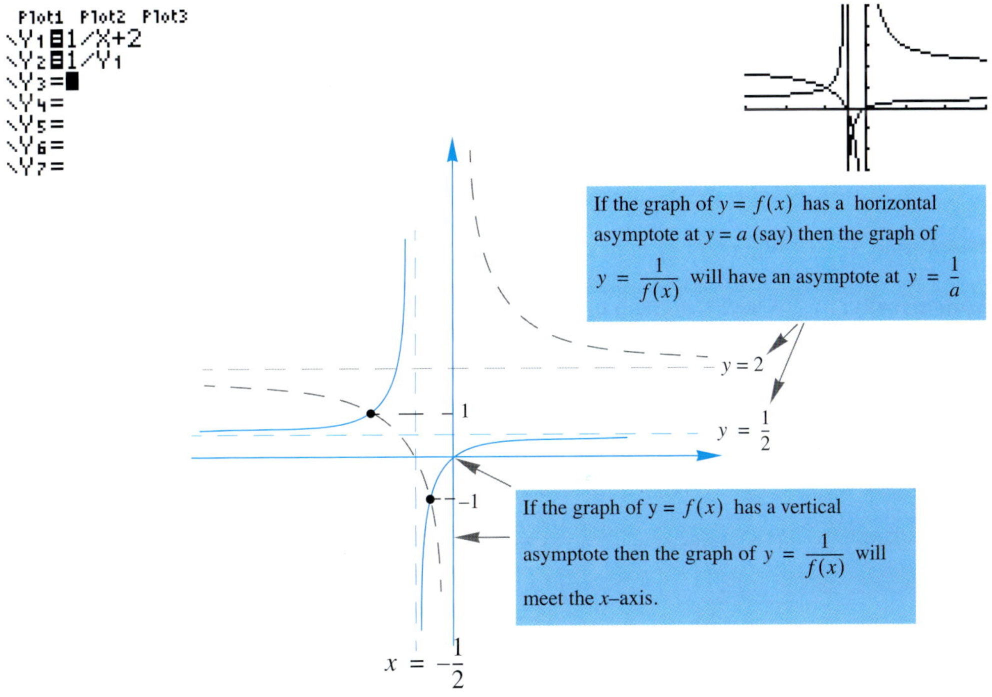

Example 5: Graph of $y = f(x)$ and $y = \dfrac{1}{f(x)}$ where $f(x) = (x-1)^3 + 2$:

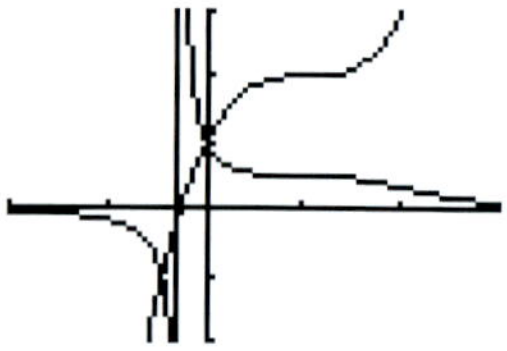

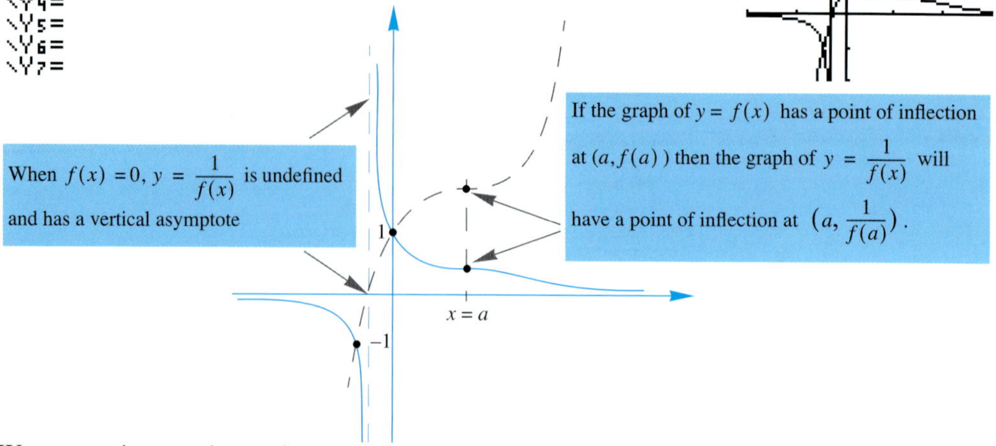

We summarise our observations as follows:

Observation 1: If $f(x) > 0$ then $\dfrac{1}{f(x)} > 0$. If $f(x) < 0$ then $\dfrac{1}{f(x)} < 0$.

Geometrically: Where the graph of $y = f(x)$ lies above the x–axis, so too does the graph of $y = \dfrac{1}{f(x)}$. Where the graph of $y = f(x)$ lies below the x–axis, so too does the graph of $y = \dfrac{1}{f(x)}$.

Observation 2: Where $f(x) = 0$ then $\dfrac{1}{f(x)}$ is undefined.

Geometrically: Where the graph of $y = f(x)$ cuts (or touches) the x–axis, the graph of $y = \dfrac{1}{f(x)}$ has a vertical asymptote.

Observation 3: Where $f(x)$ is of the form $\dfrac{a}{0}$, $a \neq 0$ then $\dfrac{1}{f(x)} = 0$

Geometrically: Where the graph of $y = f(x)$ has a vertical asymptote, the graph of $y = \dfrac{1}{f(x)}$ cuts (or touches) the x–axis.

Observation 4: $f(x) = \dfrac{1}{f(x)} \Leftrightarrow [f(x)]^2 = 1 \Leftrightarrow f(x) = \pm 1$

Geometrically: The graph of $y = f(x)$ and the graph of $y = \dfrac{1}{f(x)}$ intersect along the line $y = \pm 1$.

Observation 5: If $f(x) \to a$ as $x \to \infty$ (or $x \to -\infty$) then $\frac{1}{f(x)} \to \frac{1}{a}$.

Geometrically: If the graph of $y = f(x)$ has a horizontal asymptote at $y = a$, then the graph of $y = \frac{1}{f(x)}$ has a horizontal asymptote at $y = \frac{1}{a}$.

Observation 6: As $|f(x)| \to \infty$, $\left|\frac{1}{f(x)}\right| \to 0$ and as $|f(x)| \to 0$, $\left|\frac{1}{f(x)}\right| \to \infty$.

Geometrically: As the graph of $y = f(x)$ increases, the graph of $y = \frac{1}{f(x)}$ decreases.

As the graph of $y = f(x)$ decreases, the graph of $y = \frac{1}{f(x)}$ increases.

Observation 7: If $y = f(x)$ has a stationary point at $x = a$ and $f(a) \neq 0$, then $y = \frac{1}{f(x)}$ also has a stationary point at $x = a$.
[A full discussion of stationary points is found in Chapter 20].

Geometrically:

1. If the graph of $y = f(x)$ has a maximum at $(a, f(a))$, then the graph of $y = \frac{1}{f(x)}$ has a minimum at $\left(a, \frac{1}{f(a)}\right)$.

2. If the graph of $y = f(x)$ has a minimum at $(a, f(a))$, then the graph of $y = \frac{1}{f(x)}$ has a maximum at $\left(a, \frac{1}{f(a)}\right)$.

3. If the graph of $y = f(x)$ has a point of inflection at $(a, f(a))$, then the graph of $y = \frac{1}{f(x)}$ has a point of inflection at $\left(a, \frac{1}{f(a)}\right)$.
The point of inflection can be either stationary or non-stationary.

Notice that we have not worked out the axial intercepts in the examples that were used to derive our observations, however, you should always find the coordinates of all axial intercepts.

EXAMPLE 6.13

Sketch the graph of $y = \frac{1}{f(x)}$ for each of the functions shown below.

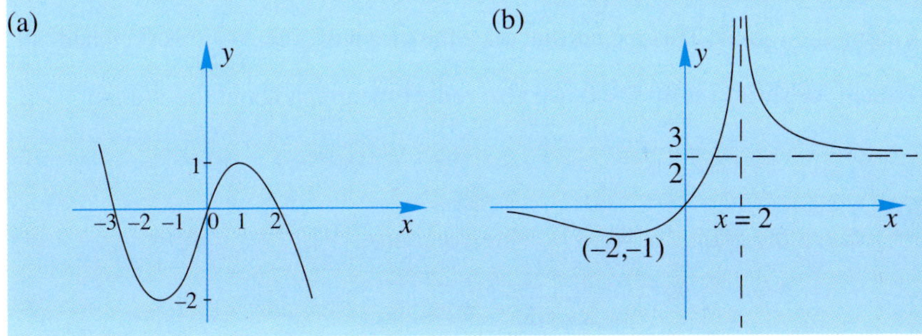

Solution

i(a) Using the observations made, we have:

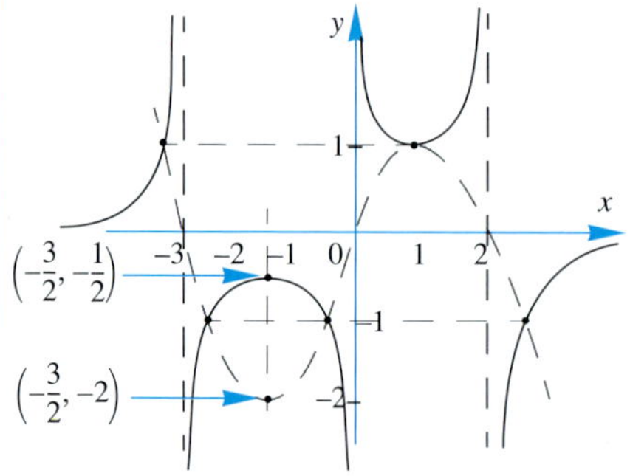

Based on observations 1–7, the graph of $y = \frac{1}{f(x)}$ has the following properties:

1. vertical asymptotes at $x = -3$, $x = 0$ and $x = 2$

2. (local) maximum at $\left(-\frac{3}{2},-\frac{1}{2}\right)$ and (local) minimum at (1,1).

3. horizontal asymptote at $y = 0$ (i.e., x–axis).

To determine the points of intersection between the two graphs we recall that the graphs will meet along the line $y = \pm 1$. This means we would need to solve two equations, $f(x) = 1$ and $f(x) = -1$. As we have no equations, we leave the graph above as is, noting that one point of intersection occurs at (1, 1) while the other two occur in the domains $-1 < x < 0$ and $-3 < x < -1$.

(b)

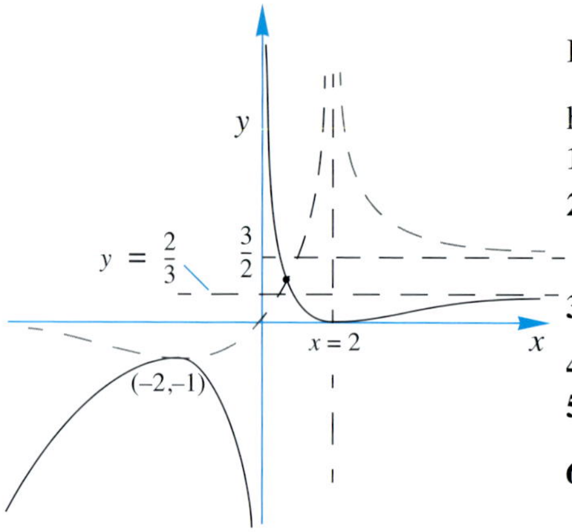

Based on observations 1–7, the graph of $y = \frac{1}{f(x)}$ has the following properties:

1. vertical asymptotes at $x = 0$.

2. (local) maximum at $(-2, -1)$ and (local) minimum at (2,0).

3. horizontal asymptote at $y = \frac{2}{3}$.

4. meets the x–axis at $(2, 0)$.

5. passes through the points $(1, 1)$ and $(-2, -1)$.

6. as $x \to -\infty$, $\frac{1}{f(x)} \to -\infty$.

So far we have considered sketching the graph of $y = \frac{1}{f(x)}$ based on the graph of $y = f(x)$, however, we could just as easily sketch the graph of $y = \frac{k}{f(x)}$ based on the graph of $y = f(x)$. The equation $y = \frac{k}{f(x)}$ can be rewritten as $y = k \times \frac{1}{f(x)}$ so that the factor 'k' represents a dilation of factor k along the y–axis. This means that once the graph of $y = \frac{1}{f(x)}$ is sketched, all that remains is to stretch (or shrink) it. If $k < 0$, we also reflect the graph about the x–axis.

EXAMPLE 6.14

Sketch the graphs of the following functions

(a) $y = \frac{4}{x^2 - 2x}$ (b) $y = \frac{4}{x^2 - 2x + 5}$

Solution

In this example we sketch the results and then briefly comment on some observations.

(a)

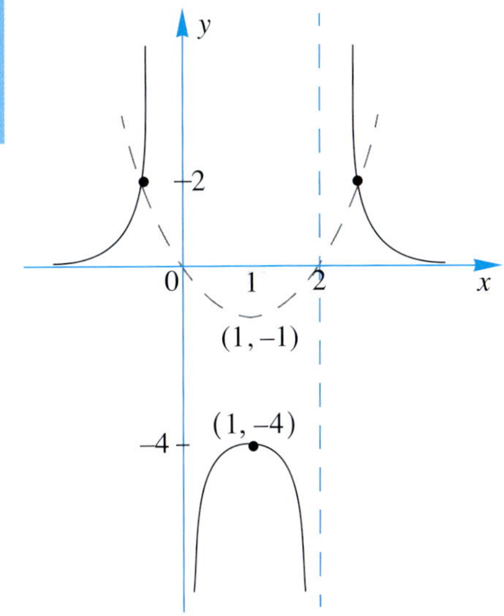

We first sketched the graph of $y = x^2 - 2x$, located where it cuts the x–axis ($x = 0$ and $x = 2$) and found its turning point $(1, -1)$. From this graph we could deduce the following properties for

$y = \frac{1}{x^2 - 2x}$:

1. place asymptotes at $x = 0$ and $x = 2$
2. have a turning point at $(1, -1)$. However, because of the factor of '4', the turning point would now be located at $(1, -4)$.

Note: To find where the graphs of $y = f(x)$ and $y = \frac{4}{f(x)}$ meet we need to solve $f(x) = \frac{4}{f(x)} \Leftrightarrow (f(x))^2 = 4$

$$\Leftrightarrow f(x) = \pm 2$$

3. as $f(x) \to \infty$, $\frac{4}{f(x)} \to 0^+$.
4. as $f(x) \to 0^+$, $\frac{4}{f(x)} \to \infty$ and as $f(x) \to 0^-$, $\frac{4}{f(x)} \to -\infty$.

(b)

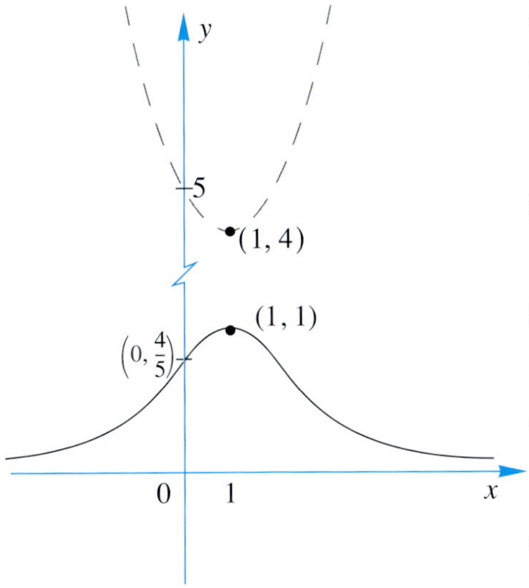

We note that

1. the graph of $f(x) = x^2 - 2x + 5$ lies above the x–axis, meaning that the graph of $y = \frac{4}{f(x)}$ will also lie above the x–axis.
2. as the graph of $y = f(x)$ has no x–intercepts, the graph of $y = \frac{4}{f(x)}$ has no vertical asymptotes.
3. the minimum of $y = f(x)$ is 4, therefore the maximum of $y = \frac{4}{f(x)} = \frac{4}{4} = 1$.
4. the graph of $y = f(x)$ cuts the y–axis at $y = 5$, so the graph of $y = \frac{4}{f(x)}$ will cut the y–axis at $y = \frac{4}{5}$.
5. as $f(x) \to \infty$, $\frac{4}{f(x)} \to 0^+$.

Two other 'transformations', involving the use of the absolute value function, namely, sketching the graph of $y = |f(x)|$ and $y = f(|x|)$ from the graph of $y = f(x)$ have already been dealt with in detail in Chapter 5.

Of course there are other types of graphs that can be sketched based on the graph of $y = f(x)$. For example, $y = \sqrt{f(x)}$, where, for example, we could only consider that part of $y = f(x)$ where $f(x) \geq 0$ or $y = [f(x)]^2$. We leave graphs such as these for the exercises.

EXERCISES 6.4

1. Sketch the graphs of $y = \dfrac{1}{f(x)}$ for each of the following

(a)

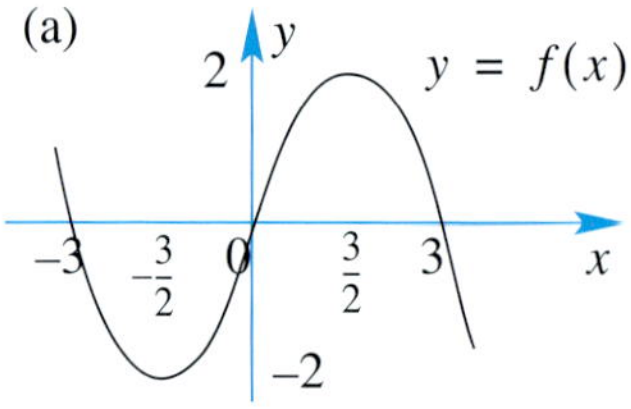

(b)

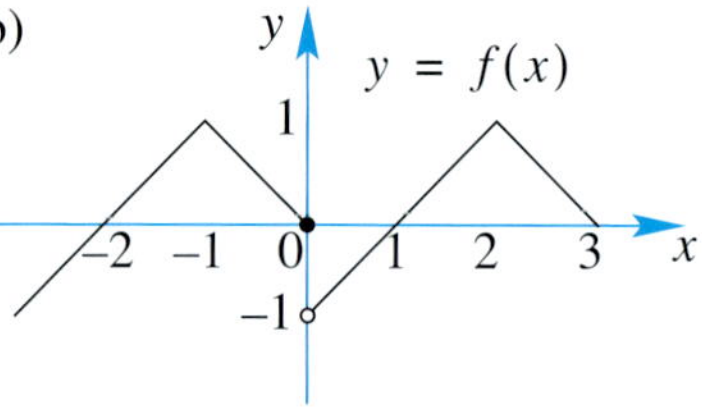

(c)

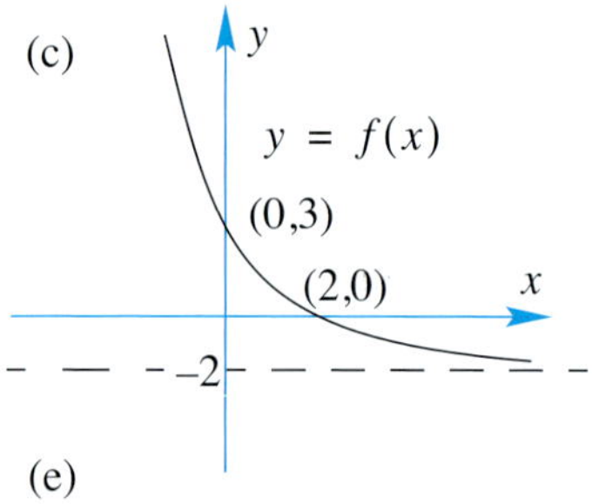

(d)

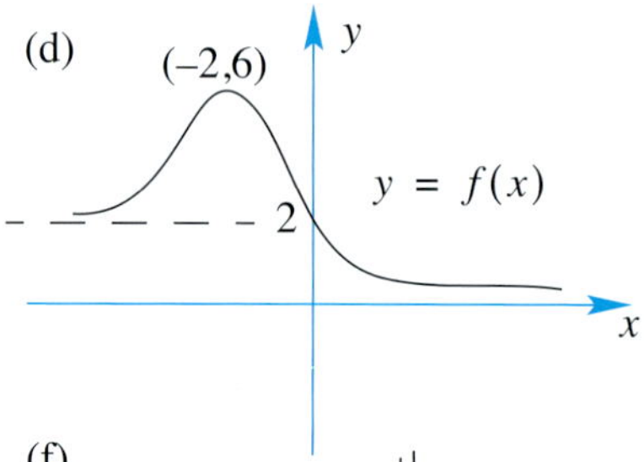

(e)

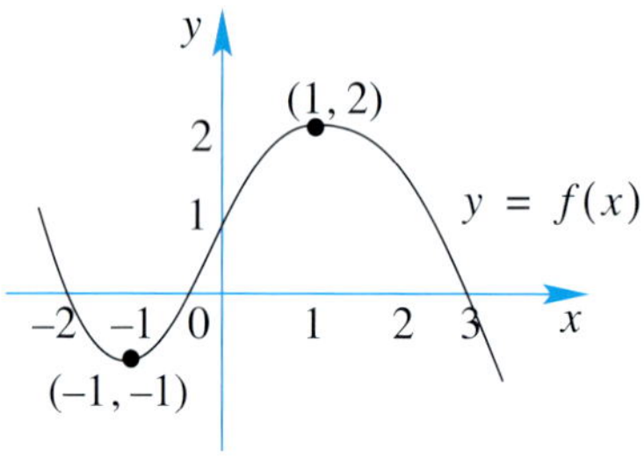

(f)

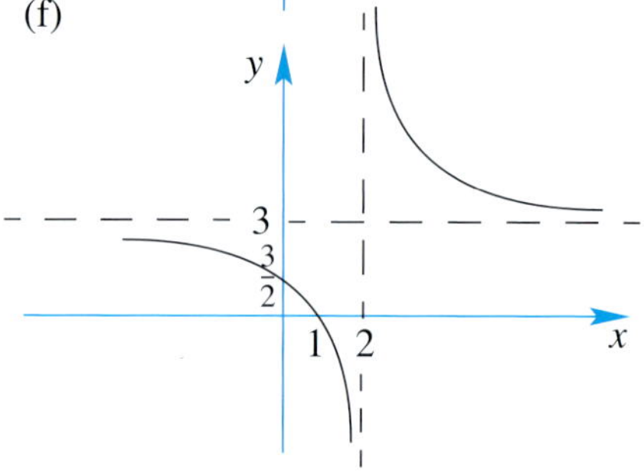

2. On the same set of axes sketch the graphs of $y = f(x)$ and $y = \dfrac{1}{f(x)}$ where

(a) $f(x) = \sqrt{x} - 2$ (b) $f(x) = x^2 + 4x$ (c) $f(x) = x^3 + 2$

(d) $f(x) = e^x - 1$ (e) $f(x) = \ln x$ (f) $f(x) = |x| - 1$

(g) $f(x) = (2 - x)^3$ (h) $f(x) = 2 - 2^x$ (i) $f(x) = \dfrac{x + 1}{x}$

(j) $f(x) = \left|\dfrac{1}{x} - 2\right|$ (k) $f(x) = \dfrac{1}{|x^2 - 4|}$ (l) $f(x) = \log_{10}(3 - x)$

3. (a) Sketch the graph of $f(x) = x^2 - 4x + 3$.

(b) Using your graph in part (a), sketch the graph of $y = \dfrac{1}{x-3} - \dfrac{1}{x-1}$

4. Sketch the graph of $f(x) = 2\left(\dfrac{1}{x-5} - \dfrac{1}{x-3}\right)$.

EXERCISES 6.5 – MISCELLANEOUS QUESTIONS

1. For the functions shown below sketch the graphs of

(a) $y = \dfrac{1}{f(x-1)}$

(b) $y = \dfrac{2}{f(x)}$

(c) $y = \dfrac{2}{f(2x)}$

i.

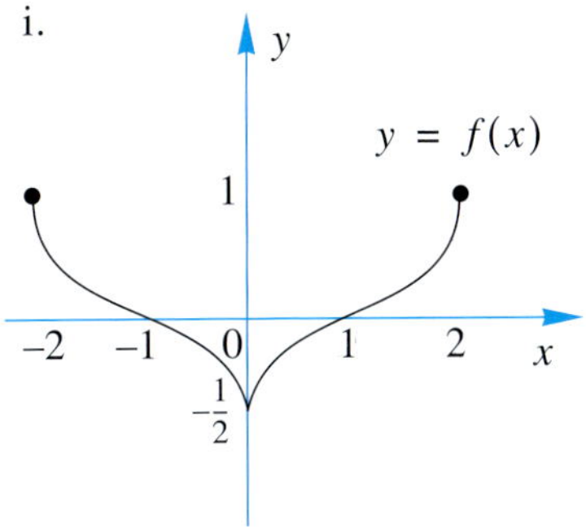

ii.

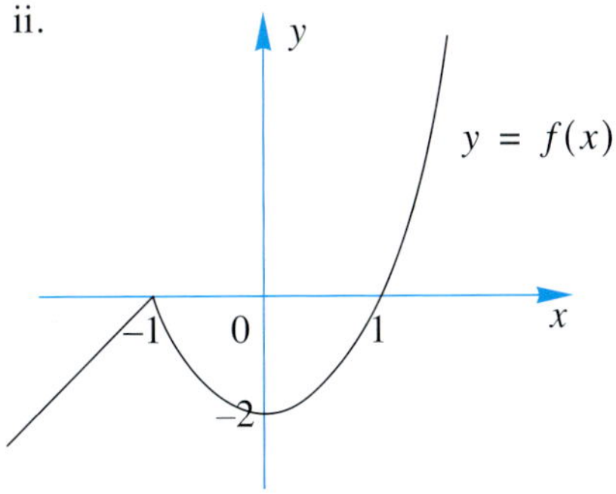

2. On the same set of axes sketch the graphs of

(a) $f(x) = x^2 - 6x + 8$ and $y = \dfrac{1}{f(x)}$.

(b) $f(x) = x^2 - 6x + 9$ and $y = \dfrac{1}{f(x)}$.

(c) $f(x) = x^2 - 6x + 10$ and $y = \dfrac{1}{f(x)}$.

(d) $f(x) = x^2 - 6x + 11$ and $y = \dfrac{2}{f(x)}$.

3. Consider the functions $f(x) = a^x, a > 1$ and $g(x) = a^{-x}, a > 1$.

(a) On the same set of axes sketch the graphs of $y = f(x)$ and $y = g(x)$.

(b) i. Sketch the graph of $y = f(x) + g(x)$.

ii. Sketch the graph of $y = f(x) - g(x)$.

Let $u(x) = a^x + a^{-x}, a > 1$ and $v(x) = a^x - a^{-x}, a > 1$.

(c) Sketch the graph of

i. $y = \dfrac{2}{u(x)}$ ii. $y = \dfrac{1}{v(x)}$

iii. $y = \dfrac{2}{u(|x|)}$ iv. $y = \dfrac{1}{v(|x|)}$

v. $y = \left|\dfrac{2}{u(x)}\right|$ vi. $y = \left|\dfrac{1}{v(x)}\right|$

4. Consider the graph shown below

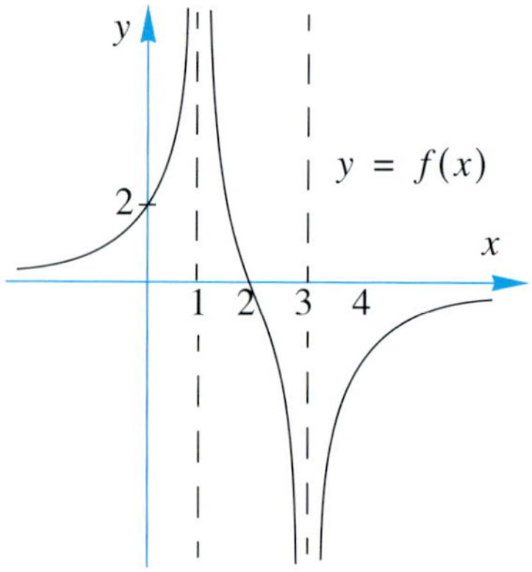

Sketch the graph of

i. $y = \frac{4}{f(x)}$ ii. $y = \frac{1}{f(|x|)}$ iii. $y = \frac{2}{|f(x)|}$

5. Consider the funcion $u(x) = \begin{cases} x^2 - ax \text{ if } x \geq 0 \\ a - ae^x \text{ if } x < 0 \end{cases}$, where $a > 0$.

(a) Sketch the graph of $y = u(x)$.

(b) Sketch the graph of $y = \frac{a}{u(x)}$ for

(c) If $v(x) = |x|$ sketch the graph of

i. $y = (u \circ v)(x)$ ii. $y = -(v \circ u)(x)$.

6. (a) Given that the curve $y = x^3 - 12x$ has turning points at $(2, -16)$ and $(-2, 16)$ sketch its graph.

(b) Hence, sketch the graph of

i. $y = \frac{12}{x^3 - 12x}$.

ii. $y^2 = x^3 - 12x$.

7. On the same set of axes sketch the graphs of $f(x) = \frac{1}{a}\log_a(x - a), a > 1$ and $y = \frac{1}{f(x)}$.

8. (a) On the same set of axes sketch the graphs of $f(x) = \frac{1}{a}x$ and $g(x) = \frac{a}{x}, a > 0$.

Hence sketch the graph of $y = \frac{x^2 + a^2}{ax}, a > 0, x \neq 0$.

(b) Using your results from (a), sketch the graph of $h(x) = \frac{1}{f(x) + g(x)}$.

(c) Hence, show that $-\frac{1}{2} \leq \frac{ax}{x^2 + a^2} \leq \frac{1}{2}$.

7.1 EXPONENTS

In §5.3.3 and §5.3.4 we looked at the exponential function, $f(x) = a^x, a > 0$ and the logarithmic function, $f(x) = \log_a x, a > 0$ and considered their general behaviour. In this chapter we will look in more detail at how to solve exponential and logarithmic equations as well as applications of both the exponential and logarithmic functions.

7.1.1 BASIC RULES OF INDICES

We start by looking at the notation involved when dealing with indices (or exponents).
The expression

$$\underbrace{a \times a \times a \times \ldots \times a}_{n \text{ times}}$$

can be written in index form, a^n, where **n is the index** (or power or **exponent**) and a is the **base**. This expression is read as "*a to the power of n.*" or more briefly as "*a to the n*".

For example, we have that $3^5 = 3 \times 3 \times 3 \times 3 \times 3$ so that 3 is the base and 5 the exponent (or index).

The laws for positive integral indices are summarised below.

If a and b are real numbers and m and n are positive integers, we have that

Law	Rule	Example
1. Multiplication [same base]	: $a^m \times a^n = a^{m+n}$	$3^4 \times 3^6 = 3^{4+6} = 3^{10}$
2. Division [same base]	: $a^m \div a^n = \frac{a^m}{a^n} = a^{m-n}$	$7^9 \div 7^5 = 7^{9-5} = 7^4$
3. Power of a power [same base]	: $(a^m)^n = a^{m \times n}$	$(2^3)^5 = 2^{3 \times 5} = 2^{15}$
4. Power of a power [same power]	: $a^m \times b^m = (ab)^m$	$3^4 \times 7^4 = (3 \times 7)^4 = 21^4$
5. Division [same power]	: $a^m \div b^m = \frac{a^m}{b^m} = \left(\frac{a}{b}\right)^m$	$5^3 \div 7^3 = \left(\frac{5}{7}\right)^3$
6. Negative one to a power	: $(-1)^n = \begin{cases} -1 & \text{if } n \text{ is odd} \\ 1 & \text{if } n \text{ is even} \end{cases}$	$(-1)^3 = (-1)^5 = \ldots = -1$ $(-1)^2 = (-1)^4 = \ldots = 1$

There are more laws of indices that are based on rational indices, negative indices and the zero index. A summary of these laws is provided next.

Law	Rule	Example
1. Fractional index Type 1 [nth root]	: $a^{1/n} = \sqrt[n]{a}, n \in \mathbb{N}$. Note: if n is even, then $a \geq 0$. if n is odd, then $a \in \mathbb{R}$.	$8^{1/3} = \sqrt[3]{8} = 2$ $(-27)^{1/3} = \sqrt[3]{-27} = -3$
2. Fractional index Type 2	: $a^{m/n} = \sqrt[n]{a^m}$ Note: if n is even, then $a^m \geq 0$ if n is odd, then $a \in \mathbb{R}$.	$16^{3/4} = \sqrt[4]{16^3} = 8$
3. Negative index	: $a^{-1} = \frac{1}{a}, a \neq 0$ $a^{-n} = \frac{1}{a^n}, a \neq 0, n \in \mathbb{N}$	$2^{-1} = \frac{1}{2} = 0.5$ $3^{-2} = \frac{1}{3^2} = \frac{1}{9}$
4. Zero index	: $a^0 = 1, a \neq 0$ Note: $0^n = 0, n \neq 0$	$12^0 = 1$

We make the following note about fractional indices:

As $\frac{m}{n} = m \times \frac{1}{n} = \frac{1}{n} \times m$, we have that for $b \geq 0$ i. $b^{\frac{m}{n}} = b^{m \times \frac{1}{n}} = (b^m)^{\frac{1}{n}} = \sqrt[n]{b^m}$

ii. $b^{\frac{m}{n}} = b^{\frac{1}{n} \times m} = \left(b^{\frac{1}{n}}\right)^m = (\sqrt[n]{b})^m$

Then, If $b \geq 0$, then $b^{\frac{m}{n}} = \sqrt[n]{b^m} = (\sqrt[n]{b})^m, m \in \mathbb{Z}, n \in \mathbb{N}$

If $b < 0$, then $b^{\frac{m}{n}} = \sqrt[n]{b^m} = (\sqrt[n]{b})^m, m \in \mathbb{Z}, n \in \{1, 3, 5, \ldots\}$

EXAMPLE 7.1

Simplify the following

(a) $\left(\frac{4x^2}{5y^4}\right)^2 \times (2x^3y)^3$ (b) $\frac{3^{n+1} + 3^2}{3}$

Solution

(a) $$\left(\frac{4x^2}{5y^4}\right)^2 \times (2x^3y)^3 = \frac{4^2x^{2\times 2}}{5^2y^{4\times 2}} \times 2^3x^{3\times 3}y^{1\times 3}$$
$$= \frac{16x^4}{25y^8} \times 8x^9y^3$$
$$= \frac{128}{25}x^{4+9}y^{3-8}$$
$$= \frac{128}{25}x^{13}y^{-5}$$
$$= \frac{128x^{13}}{25y^5}$$

(b) $$\frac{3^{n+1} + 3^2}{3} = \frac{3(3^n + 3)}{3}$$
$$= 3^n + 3$$

EXAMPLE 7.2

Simplify the following

(a) $\dfrac{4x^2(-y^{-1})^{-2}}{(-2x^2)^3(y^{-2})^2}$ (b) $\dfrac{x^{-1}+y^{-1}}{x^{-1}y^{-1}}$

Solution

(a) $$\frac{4x^2(-y^{-1})^{-2}}{(-2x^2)^3(y^{-2})^2} = \frac{4x^2 \times y^{-1\times -2}}{-8x^{2\times 3} \times y^{-2\times 2}} = -\frac{x^2y^2}{2x^6y^{-4}}$$
$$= -\frac{y^{2-(-4)}}{2x^{6-(2)}}$$
$$= -\frac{y^6}{2x^4}$$

(b) $$\frac{x^{-1}+y^{-1}}{x^{-1}y^{-1}} = \frac{\frac{1}{x}+\frac{1}{y}}{\frac{1}{xy}} = \left(\frac{1}{x}+\frac{1}{y}\right) \times \frac{xy}{1} = \frac{xy}{x}+\frac{xy}{y}$$
$$= y + x$$

EXAMPLE 7.3

Simplify the following

(a) $\dfrac{2^{n-3}\times 8^{n+1}}{2^{2n-1}\times 4^{2-n}}$ (b) $\dfrac{(a^{1/3}\times b^{1/2})^{-6}}{\sqrt[4]{a^8b^9}}$

Solution

(a) $$\frac{2^{n-3}\times 8^{n+1}}{2^{2n-1}\times 4^{2-n}} = \frac{2^{n-3}\times (2^3)^{n+1}}{2^{2n-1}\times (2^2)^{2-n}} = \frac{2^{n-3}\times 2^{3n+3}}{2^{2n-1}\times 2^{4-2n}}$$
$$= \frac{2^{n-3+(3n+3)}}{2^{2n-1+(4-2n)}}$$
$$= \frac{2^{4n}}{2^3}$$
$$= 2^{4n-3}$$

(b) $$\frac{(a^{1/3}\times b^{1/2})^{-6}}{\sqrt[4]{a^8b^9}} = \frac{a^{\frac{1}{3}\times -6}\times b^{\frac{1}{2}\times -6}}{(a^8b^9)^{\frac{1}{4}}} = \frac{a^{-2}\times b^{-3}}{a^2b^{\frac{9}{4}}}$$
$$= a^{-2-2}\times b^{-3-\frac{9}{4}}$$
$$= a^{-4}b^{-\frac{21}{4}}$$
$$= \frac{1}{a^4b^{\frac{21}{4}}}$$

EXERCISES 7.1.1

1. Simplify the following

(a) $\left(\frac{3y^2}{4x^3}\right)^3 \times (2x^2y^3)^3$ (b) $\left(\frac{2}{3a^2}\right)^3 + \frac{1}{8a^6}$ (c) $\frac{2^{n+1}+2^2}{2}$

(d) $\left(\frac{2x^3}{3y^2}\right)^3 \times (xy^2)^2$ (e) $\left(\frac{2x^3}{4y^2}\right)^2 \times \frac{12y^6}{8x^4}$ (f) $\frac{3^{n+2}+9}{3}$

(g) $\frac{4^{n+2}-16}{4}$ (h) $\frac{4^{n+2}-16}{2}$ (i) $\left(\frac{1}{2b}\right)^4 - \frac{b^2}{16}$

2. Simplify the following

(a) $\frac{20^6}{10^6}$ (b) $\frac{12^{2x}}{(6^3)^x}$ (c) $\frac{16^{2y+1}}{8^{2y+1}}$

(d) $\frac{(ab)^{2x}}{a^{2x}b^{4x}}$ (e) $\frac{(xy)^6}{64x^6}$ (f) $\frac{27^{n+2}}{6^{n+2}}$

3. Simplify the following

(a) $\left(\frac{x}{y}\right)^3 \times \left(\frac{y}{z}\right)^2 \times \left(\frac{z}{x}\right)^4$ (b) $3^{2n} \times 27 \times 243^{n-1}$ (c) $\frac{25^{2n} \times 5^{1-n}}{(5^2)^n}$

(d) $\frac{9^n \times 3^{n+2}}{27^n}$ (e) $\frac{2^n \times 4^{2n+1}}{2^{1-n}}$ (f) $\frac{2^{2n+1} \times 4^{-n}}{(2^n)^3}$

(g) $\frac{x^{4n+1}}{(x^{n+1})^{(n-1)}}$ (h) $\frac{x^{4n^2+n}}{(x^{n+1})^{(n-1)}}$ (i) $\frac{(3^x)(3^{x+1})(3^2)}{(3^x)^2}$

4. Simplify $\frac{(x^m)^n(y^2)^m}{(x^m)^{(n+1)}y^2}$.

5. Simplify the following, leaving your answer in positive power form

(a) $\frac{(-3^4) \times 3^{-2}}{(-3)^{-2}}$ (b) $\frac{9y^2(-x^{-1})^{-2}}{(-2y^2)^3(x^{-2})^3}$ (c) $\frac{x^{-1}-y^{-1}}{x^{-1}y^{-1}}$

(d) $\frac{x^{-2}+2x^{-1}}{x^{-1}+x^{-2}}$ (e) $\frac{(-2)^3 \times 2^{-3}}{(x^{-1})^2 \times x^2}$ (f) $\frac{(-a)^3 \times a^{-3}}{(b^{-1})^{-2}b^{-3}}$

6. Simplify the following

(a) $\frac{(x^{-1})^2+(y^2)^{-1}}{x^2+y^2}$ (b) $\frac{(x^2)^{-2}+2y}{1+2yx^4}$ (c) $\frac{(x+h)^{-1}-x^{-1}}{h}$

(d) $(x^2-1)^{-1} \times (x+1)$ (e) $\frac{(x-1)^{-3}}{(x+1)^{-1}(x^2-1)^2}$ (f) $\frac{y(x^{-1})^2+x^{-1}}{x+y}$

7. Simplify the following

(a) $5^{n+1}-5^{n-1}-2 \times 5^{n-2}$ (b) $a^{x-y} \times a^{y-z} \times a^{z-x}$ (c) $\left(\frac{a^{-\frac{1}{2}}b^3}{ab^{-1}}\right)^2 \times \frac{1}{ab}$

(d) $\left(\frac{a^{m+n}}{a^n}\right)^m \times \left(\frac{a^{n-m}}{a^n}\right)^{m-n}$ (e) $\frac{p^{-2}-q^{-2}}{p^{-1}-q^{-1}}$ (f) $\frac{1}{1+a^{\frac{1}{2}}} - \frac{1}{1-a^{\frac{1}{2}}}$

(g) $\frac{2^{n+4}-2(2^n)}{2(2^{n+3})}$ (h) $\sqrt{a\sqrt{a\sqrt{a}}}$

8. Simplify the following

(a) $\frac{\sqrt{x} \times \sqrt[3]{x^2}}{\sqrt[4]{x}}$ (b) $\frac{b^{n+1} \times 8a^{2n-1}}{(2b)^2(ab)^{-n+1}}$ (c) $\frac{2^n - 6^n}{1-3^n}$

(d) $\frac{7^{m+1}-7^m}{7^n - 7^{n+2}}$ (e) $\frac{5^{2n+1}+25^n}{5^{2n}+5^{1+n}}$ (f) $\left(x - 2x^{\frac{1}{2}} + 1\right)^{\frac{1}{2}} \times \frac{x+1}{\sqrt{x}-1}$

7.1.2 INDICIAL EQUATIONS

Solving equations of the form $x^{\frac{1}{2}} = 3$, where the **variable is the base**, requires that we square both sides of the equation so that $\left(x^{\frac{1}{2}}\right)^2 = 3^2 \Rightarrow x = 9$. However, when the **variable is the power** and not the base we need to take a different approach.

> Indicial (exponential) equations take on the general form $b^x = a$, where the unknown (variable), x, is the power

Consider the case where we wish to solve for x given that $2^x = 8$.

In this case we need to think of a value of x so that when 2 is raised to the power of x the answer is 8. Using trial and error, it is not too difficult to arrive at $x = 3$ $(2^3 = 2 \times 2 \times 2 = 8)$.

Next consider the equation $3^{x+1} = 27$.
Again, we need to find a number such that when 3 is raised to that number, the answer is 27. Here we have that $27 = 3^3$. Therefore we can rewrite the equation as $3^{x+1} = 3^3$.
As the base on both sides of the equality is the same we can then equate the powers, that is,

$$3^{x+1} = 27 \Leftrightarrow 3^{x+1} = 3^3$$
$$\Leftrightarrow x + 1 = 3$$
$$\Leftrightarrow x = 2$$

Such an approach can be used for a variety of equations. We summarise this process for simple exponential equations:

	Solve for x : $b^x = N$	Example: Solve $5^x = 625$
Step 1:	Express the number N in the form b^{number}	$625 = 5^4$
Step 2:	Write the equation $b^x = b^{\text{number}}$	$\therefore 5^x = 5^4$
Step 3:	Equate exponents, x = number	$\Leftrightarrow x = 4$

EXAMPLE 7.4

Solve the following

(a) $3^x = 81$ (b) $2 \times 5^u = 250$ (c) $2^x = \frac{1}{32}$

Solution

(a) $3^x = 81 \Leftrightarrow 3^x = 3^4$

$\Leftrightarrow x = 4$

(b) $2 \times 5^u = 250 \Leftrightarrow 5^u = 125$

$\Leftrightarrow 5^u = 5^3$

$\Leftrightarrow u = 3$

(c) $2^x = \frac{1}{32} \Leftrightarrow 2^x = \frac{1}{2^5}$

$\Leftrightarrow 2^x = 2^{-5}$

$\Leftrightarrow x = -5$

EXAMPLE 7.5

Find

(a) $\left\{x \mid \left(\frac{1}{2}\right)^x = 16\right\}$ (b) $\{x \mid 3^{x+1} = 3\sqrt{3}\}$ (c) $\{x \mid 4^{x-1} = 64\}$

Solution

(a) $\left(\frac{1}{2}\right)^x = 16 \Leftrightarrow (2^{-1})^x = 16$

$\Leftrightarrow 2^{-x} = 2^4$

$\Leftrightarrow -x = 4$

$\Leftrightarrow x = -4$

i.e., solution set is $\{-4\}$.

(b) $3^{x+1} = 3\sqrt{3} \Leftrightarrow 3^{x+1} = 3 \times 3^{1/2}$

$\Leftrightarrow 3^{x+1} = 3^{3/2}$

$\Leftrightarrow x + 1 = \frac{3}{2}$

$\Leftrightarrow x = \frac{1}{2}$

i.e., solution set is $\{0.5\}$.

(c) $4^{x-1} = 64 \Leftrightarrow (2^2)^{x-1} = 2^6$

$\Leftrightarrow 2^{2x-2} = 2^6$

$\Leftrightarrow 2x - 2 = 6$

$\Leftrightarrow 2x = 8$

$\Leftrightarrow x = 4$

i.e., solution set is $\{4\}$.

EXERCISES 7.1.2

1. Solve the following equations.

(a) $\{x \mid 4^x = 16\}$ (b) $\left\{x \mid 7^x = \frac{1}{49}\right\}$ (c) $\{x \mid 8^x = 4\}$

(d) $\{x \mid 3^x = 243\}$ (e) $\{x \mid 3^{x-2} = 81\}$ (f) $\left\{x \mid 4^x = \frac{1}{32}\right\}$

(g) $\{x \mid 3^{2x-4} = 1\}$ (h) $\{x \mid 4^{2x+1} = 128\}$ (i) $\{x \mid 27^x = 3\}$

2. Solve the following equations.

(a) $\{x \mid 7^{x+6} = 1\}$ (b) $\left\{x \mid 8^x = \frac{1}{4}\right\}$ (c) $\{x \mid 10^x = 0.001\}$

(d) $\{x \mid 9^x = 27\}$ (e) $\{x \mid 2^{4x-1} = 1\}$ (f) $\{x \mid 25^x = \sqrt{5}\}$

(g) $\left\{x \mid 16^x = \frac{1}{\sqrt{2}}\right\}$ (h) $\{x \mid 4^{-x} = 32\sqrt{2}\}$ (i) $\{x \mid 9^{-2x} = 243\}$

7.1.3 EQUATIONS OF THE FORM $b^{f(x)} = b^{g(x)}$.

This is an extension of the previous section, in that now we will consider exponential equations of the form $b^{f(x)} = N$ where N can be expressed as a number having base b so that $N = b^{g(x)}$.

Consider the equation $2^{x^2-1} = 8$. Our first step is to express 8 as 2^3 so that we can then write

$$2^{x^2-1} = 8 \Leftrightarrow 2^{x^2-1} = 2^3$$

Then, equating powers we have: $\Leftrightarrow x^2 - 1 = 3$

So that, $\Leftrightarrow x^2 - 4 = 0$

$$\Leftrightarrow (x-2)(x+2) = 0$$

$$\therefore x = 2 \text{ or } x = -2$$

Checking these values by substituting back into the origial equation shows them to be correct.

i.e., when $x = 2$, L.H.S $= 2^{2^2-1} = 2^{4-1} = 2^3 = 8 =$ R.H.S

when $x = -2$, L.H.S $= 2^{(-2)^2-1} = 2^{4-1} = 2^3 = 8 =$ R.H.S

However, had the equation been, $2^{x^2-1} = 2^{5-x}$, then the solution would have been

$$2^{x^2-1} = 2^{5-x} \Leftrightarrow x^2 - 1 = 5 - x \quad \text{[equating powers]}$$

$$\Leftrightarrow x^2 + x - 6 = 0$$

$$\Leftrightarrow (x-2)(x+3) = 0$$

$$\therefore x = 2 \text{ or } x = -3$$

Again, we can check that these solutions satisfy the original equation.

The thing to note here is that the solution process has not altered. Rather than having one of the powers represented by a constant, we now have both powers containing the variable.

EXAMPLE 7.6

Find $\{x|\ 3^{x^2-5x+2} = 9^{x+1}\}$.

Solution

We need to first express the equation in the form $b^{f(x)} = b^{g(x)}$ where, in this case, $b = 3$:

$$3^{x^2-5x+2} = 9^{x+1} \Leftrightarrow 3^{x^2-5x+2} = (3^2)^{x+1}$$
$$\Leftrightarrow 3^{x^2-5x+2} = 3^{2x+2}$$
$$\Leftrightarrow x^2-5x+2 = 2x+2$$
$$\Leftrightarrow x^2-7x = 0$$
$$\Leftrightarrow x(x-7) = 0$$
$$\Leftrightarrow x = 0 \text{ or } x = 7$$

Again, checking our solutions we have, $x = 0$: L.H.S $= 3^{0-0+2} = 9 = 9^{0+1}$ = R.H.S

$x = 7$: L.H.S $= 3^{7^2-5\times7+2} = 3^{16} = 9^{7+1}$ = R.H.S

Therefore, the solution set is $\{0, 7\}$

We now have a more general statement for solving exponential equations:

$$b^{f(x)} = b^{g(x)} \Leftrightarrow f(x) = g(x), \text{ where } b > 0 \text{ and } b \neq 1.$$

It is important to realise that this will only be true if **the base is the same on both sides** of the equality sign.

EXERCISES 7.1.3

1. Solve the following for the unknown

(a) $9^{2x-1} = 3^{2x+5}$ (b) $4^{x+1} = 8^{2x-4}$ (c) $25^{2x+3} = 125^{x+1}$

(d) $2^{4x+1} = 4^{x+2}$ (e) $16^{2x-1} = 8^{2x+1}$ (f) $\sqrt{3}\times 27^{x+1} = 9^{2x+1}$

(g) $(\sqrt{3})^{x-1} = 9^{-x+2}$ (h) $8^x = \dfrac{1}{16^{x+1}}$ (i) $4^{x+2}\times 8^{x-1} = 2$

2. Solve for the unknown

(a) $8^{x+1} = \dfrac{1}{2^x}$ (b) $8^{x+1} = 2^{x^2-1}$ (c) $3^{x-1} = 3^{x^2-1}$

(d) $4^{x^2-7x+12} = 1$ (e) $6^{\sqrt{n^2-3n}} = 36$ (f) $(5^x)^2 = 5^{x^2}$

3. Solve the following

(a) $(x^2-x-1)^{x^2} = x^2-x-1$

(b) $(x-2)^{x^2-x-12} = 1$

(c) $(3x-4)^{2x^2} = (3x-4)^{5x-2}$

(d) $|x|^{x^2-2x} = 1$

(e) $(x^2+x-57)^{3x^2+3} = (x^2+x-57)^{10x}$

7.1.4 WHAT IF THE BASE IS NOT THE SAME?

Consider the equation $2^x = 10$. It is not possible (at this stage) to express the number 10, in exponent form with a base of 2. This means that our previous methods will not work.

However, we could try a numerical or even graphical approach to this problem. Clearly the value of x must be somewhere in the range [3,4] as $2^3 = 8$ and $2^4 = 16$.We explore this problem using the graphics calculator.

We begin by defining the two relevant equations,

$$y = 2^x \text{ and } y = 10.$$

Then we enter these functions using the equation editor screen:

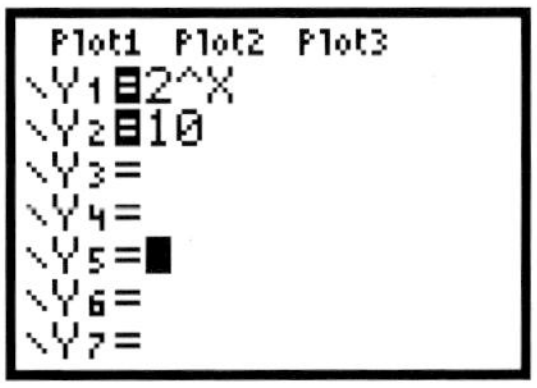

Next, we set our domian and range.
As we have already decided that $x \in [3, 4]$, we can set the domain to be $0 \le x \le 4$.

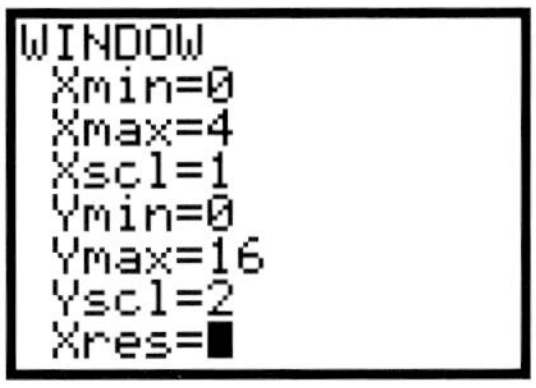

We can now obtain a graphical display of the equation $2^x = 10$.

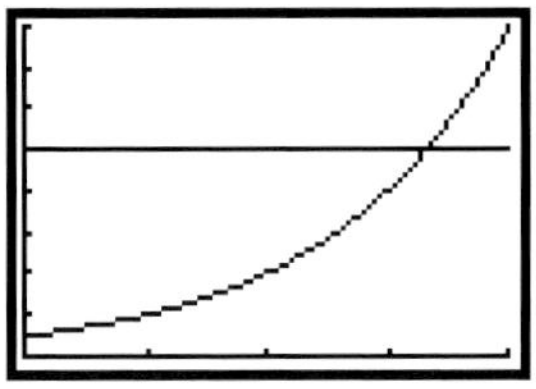

We can now find the point of intersection.
To do this we use the **CALC** menu and choose option **5: int** (this will determine the intersection of the two curves).

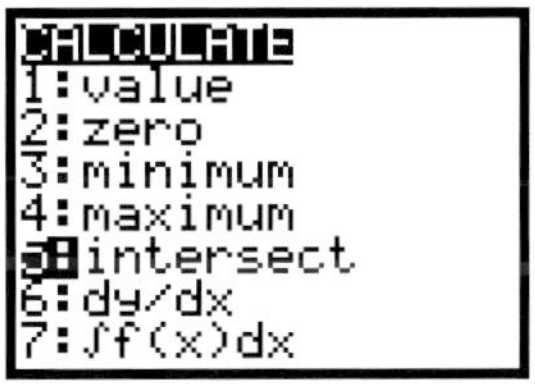

When asked for **First curve?** press **ENTER**. Similarly, for **Second curve?** When asked to **Guess?** move the cursor as close as possible to the point where both graphs meet and press **ENTER**.

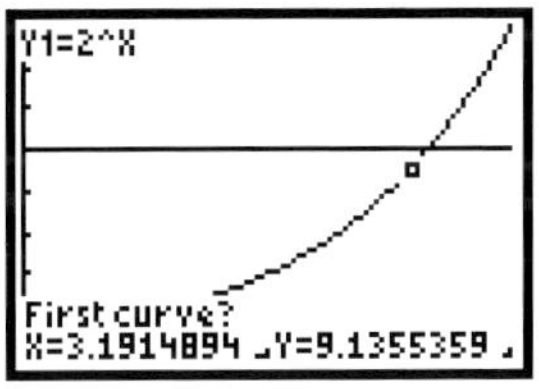

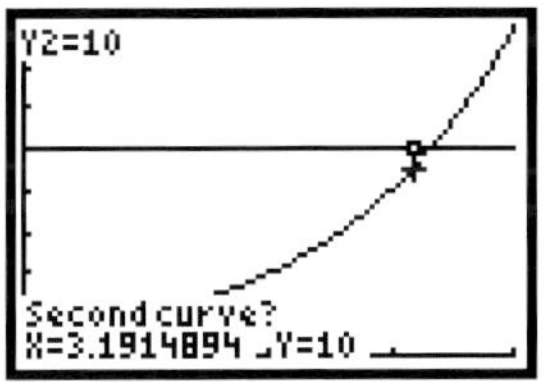

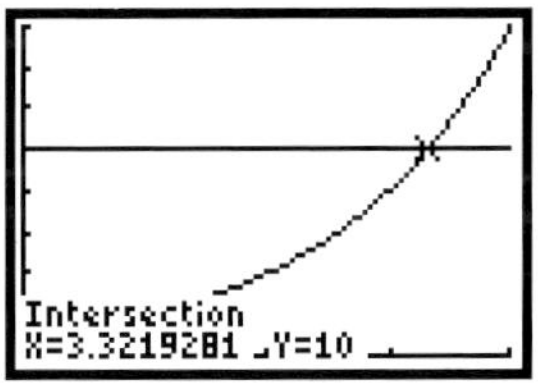

We have a solution (to four decimal places) for x, i.e., $x = 3.3219$. We could also have used the **ZOOM** facility to obtain the same result. We do this in the next example.

At this stage, the key to being able to solve equations of the form $a^x = b$ (where b cannot be easily expressed as a number having base 'a'), is to accurately sketch the graphs of $y = a^x$ and $y = b$, and then to determine where the two graphs meet.

EXAMPLE 7.7

Solve for x, $2^x = 12$

Solution

Let $y_1 = 2^x$ and $y_2 = 12$.

We enter the functions: **Y_1 = 2^X**

Y_2 = 12

Using the TI–83 to sketch the given graphs, we have:

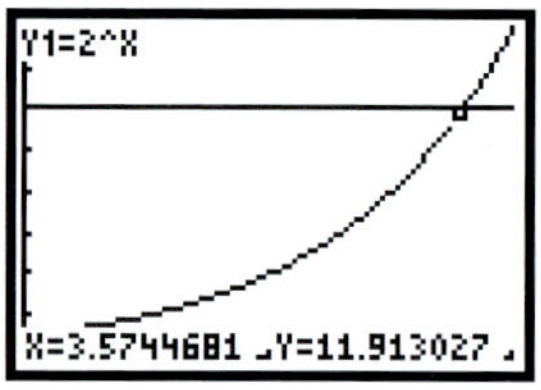

Using the **TRACE** key, we can move the cursor along the graph so that the square lies at the point where the two graphs meet. At this stage, our 'best solution' is $x = 3.57$

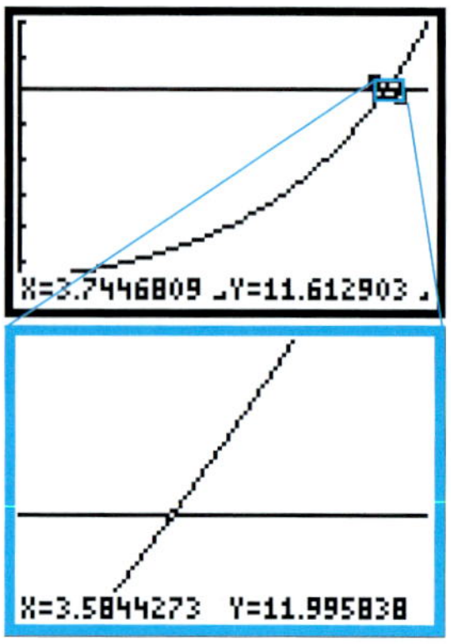

We can obtain a more accurate answer by using the **ZOOM** facility and then selecting **1:ZBOX** option (to 'close–in' on the point of intersection). Repeated use of the **Zoom** facility will continue to provide a more accurate solution.

After using the zoom facility once we have $x = 3.5844$ (The actual answer is $x = 3.5849...$).

Note: We can use the **0:solve** function on the TI–83,
i.e., **solve** (2^x–12,x,3) = 3.5849. . . .

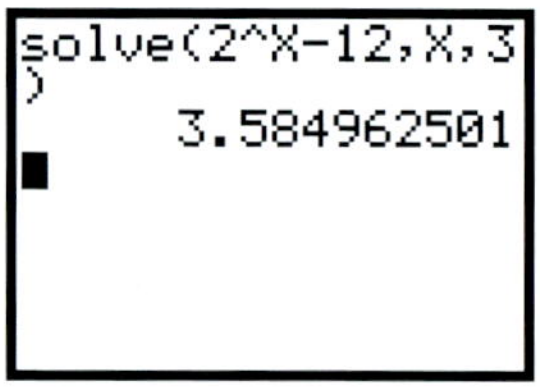

However, it is the graphical approach that we wish to emphaise at this stage, so that the relationship between the solution, the roots and the graphical representation can be clearly observed.

EXERCISES 7.1.4

1. Use a graphical approach to solve the following (give your answer correct to 2 d.p.).

(a) i. $2^x = 40$ ii. $2^x = 1020$ iii. $2^x = 6$

(b) i. $3^x = 12$ ii. $3^x = 80$ iii. $3^x = 500$

(c) i. $\left(\frac{1}{2}\right)^x = \frac{1}{10}$ ii. $\left(\frac{1}{2}\right)^x = 20$ iii. $\left(\frac{1}{2}\right)^x = 80$

(d) i. $5^x = 0.1$ ii. $5^x = 15$ iii. $5^x = 0.01$

2. Use a graphical approach to solve the following (give your answer correct to 2 d.p.).

(a) $\{x \mid 2^x = 1 - x\}$ (b) $\{x \mid 2^x = -x + 2\}$

(c) $\{x \mid 2^{-x} = 4x + 2\}$ (d) $\{x \mid 3^x = 1 - x^2\}$

(e) $\{x \mid 3^{-x} = x^2 - 1\}$ (f) $\{x \mid 5^x = 2 - (x - 1)^2\}$

7.1.5 A SPECIAL BASE (e)

Of all the expressions a^x , that for which $a = e$ is known as **the exponential function**. The exponential function is also known as the natural exponential function, in recognition of the important role that the value 'e' has. The importance of 'e' is that it occurs in many applications that arise as a result of natural phenomena. The question then remains; **What is 'e'**?

We consider how an investment can earn continuously compounded interest:

If a principal amount $\$P$ is invested at an annual percentage rate r, compounded once a year, the amount in the balance, $\$A$, after one year is given by $A = P + P \times r = P(1 + r)$. We can then have more frequent (quarterly, monthly, daily) compounding interest.

For example, if we have quarterly compounding interest then each quarter will have an effective rate of $\frac{r}{4}$, which will be compounded 4 times. This means that by the end of the year, the balance will be given by $A = \left(1 + \frac{r}{4}\right)^4$.

If we next consider the situation where there are n compoundings per year, so that the rate per compounding becomes $\frac{r}{n}$, we then have that the amount in the balance after a year (i.e., after n compoundings) is given by $A = \left(1 + \frac{r}{n}\right)^n$.

If we allow the number of compoundings n, to increase without bound, we obtain what is known as continuous compounding. We can set up a table of values for the case when $r = 1$.

n	$\left(1 + \frac{1}{n}\right)^n$
1	$\left(1 + \frac{1}{1}\right)^1 = 2$
10	$\left(1 + \frac{1}{10}\right)^{10} = 2.593742\ldots$
100	$\left(1 + \frac{1}{100}\right)^{100} = 2.704813\ldots$
1000	$\left(1 + \frac{1}{1000}\right)^{1000} = 2.716923\ldots$
10 000	$\left(1 + \frac{1}{10000}\right)^{10000} = 2.718145\ldots$

From the table of values, we have that as the value of n increases, the value of $\left(1 + \frac{1}{n}\right)^n$ approaches a fixed number. This number is given by 2.718145. . ., which happens to be an approximate value for the number 'e'.

That is, $$\lim_{n \to \infty}\left(1 + \frac{1}{n}\right)^n = e = 2.71828\ldots$$

This limiting expression is also known as Euler's number.

This means that the natural base, e, is an irrational number just as the number π is. Notice then, the number e can be used in the same way that π is used in calculations.

EXAMPLE 7.8

Evaluate the following to 4 decimal places

(a) e^2 (b) $\sqrt{e}$ (c) $e^3 - 2$

Solution

Remember, treat 'e' as you would any other number.

(a) $e^2 = 7.3891$

(b) $\sqrt{e} = 1.6487$

(c) $e^3 - 2 = 18.0855$

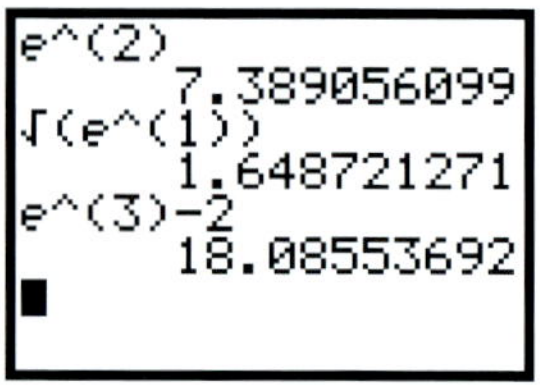

EXAMPLE 7.9

Solve the following

(a) $e^{2x} = e$ (b) $e^{2x-3} = \frac{1}{e}$ (c) $e^{x^2-1} = e^{3x-3}$

Solution

(a) $e^{2x} = e \Leftrightarrow e^{2x} = e^1 \Leftrightarrow 2x = 1$

$\Leftrightarrow x = \frac{1}{2}$

(b) $e^{2x-3} = \frac{1}{e} \Leftrightarrow e^{2x-3} = e^{-1} \Leftrightarrow 2x - 3 = -1$

$\Leftrightarrow 2x = 2$

$\Leftrightarrow x = 1$

(c) $e^{x^2-1} = e^{3x-3} \Leftrightarrow x^2 - 1 = 3x - 3$

$\Leftrightarrow x^2 - 3x + 2 = 0$

$\Leftrightarrow (x-1)(x-2) = 0$

$\Leftrightarrow x = 1$ or $x = 2$

EXERCISES 7.1.5

1. Solve for x

(a) $e^x = e^2$ (b) $e^x = \frac{1}{e}$ (c) $e^x = \sqrt{e}$ (d) $e^{-2x} = \frac{1}{e}$

2. Solve for x

(a) $e^{2x} = e^{3-x}$ (b) $e^{2+x} = e^{5-4x}$ (c) $\frac{1}{e^{3x+1}} = e^{2x-1}$

3. Solve for x

(a) $(e^x)^2 = e^x$ (b) $(e^x)^2 = \sqrt{e^{x+2}}$

4. Solve for x

(a) $e^{x^2-x} = e^2$ (b) $e^{x^2-x} = e^6$ (c) $e^{x^2+3x-1} = e^{x-2}$

(d) $e^{x^2+4x} = e^{6-x}$ (e) $\left(\frac{1}{e}\right)^x = e^{-x^2}$ (f) $e^{-x+2} = e^{\frac{1}{x}}$

5. Solve for x, giving your answer correct to 4 decimal places

(a) $e^x = 4$ (b) $e^x = 9$ (c) $e^x = 25$ (d) $e^x = -4$

6. Solve for x, giving your answer correct to 4 decimal places

(a) $3e^{2x} = 7$ (b) $4e^{2x} = 9$ (c) $2e^{3x} = 5$ (d) $7e^{3x} = 2$

7. Solve for x, giving your answer correct to 4 decimal places

(a) $e^x = -x+1$ (b) $2e^x = -3x-1$ (c) $e^x = x+1$

8. (a) Show that if $e^{2x}+6 = 5e^x$ then $e^x = 3$ or $e^x = 2$.

(b) Show that if $e^{2x}-5e^x = 6$ then $e^x = 6$.

9. The graph of the function $f(x) = a \times 2^x + b$ is shown alongside.
Find the value of $f(3)$.

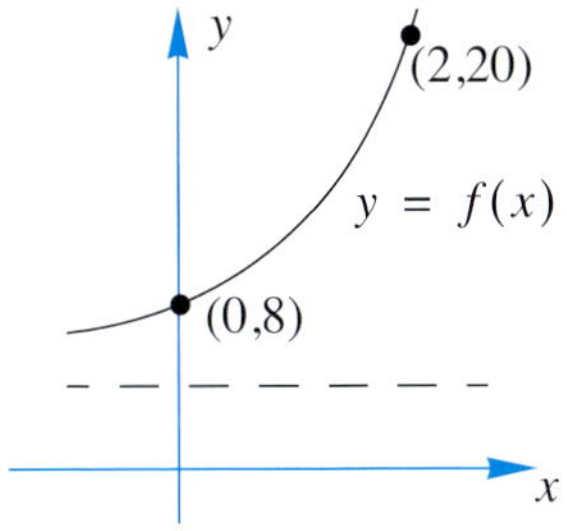

10. Find the values of a and k if the graph with equation $f(x) = ae^{-kx}$ passes through the points $(1, e)$ and $(-1, 2e)$.

7.2 EXPONENTIAL MODELLING

There are many situations and examples where an exponentional function is an appropriate function to model a particular **growth** or **decay** process. For example:

1. When looking at the bacteria count of an experiment, the growth in the number of bacteria present in the colony is accurately represented by an exponential growth model.

If there are initially 100 bacteria in a colony and the size doubles every day, we model this situation by making use of the exponential function,

$$f: [0, a[\mapsto \mathbb{R}, \text{ where } f(t) = 100 \times 2^t, a \in \mathbb{R}.$$

The graph of such a model is given below.

Exponential growth

$f: [0, a[\mapsto \mathbb{R}$, where $f(t) = 100 \times 2^t, a \in \mathbb{R}$

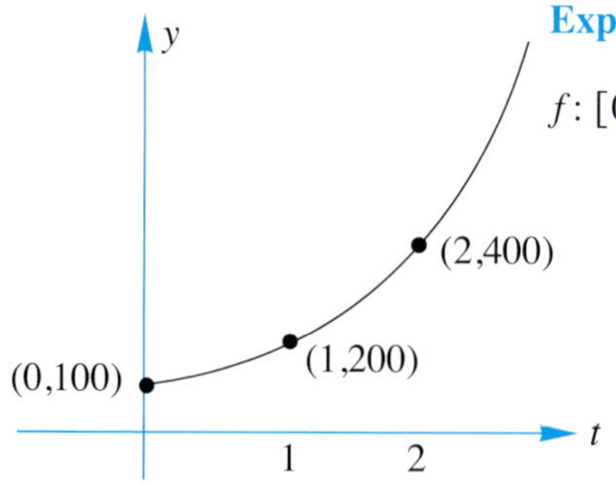

2. Certain physical quantities decrease exponentially, for example, the decay of a radioactive substance, or isotope. Associated with this is the half–life, that is, the time that it takes for the substance to decay to one half of its original amount.

A radioactive bismuth isotope has a half–life of 5 days. If there is 100 milligrams initially, then we can model this situation by making use of the exponential function,

$$f : [0, \infty) \mapsto \mathbb{R}, \text{ where } f(t) = 100 \times \left(\frac{1}{2}\right)^{t/5}$$

Exponential decay

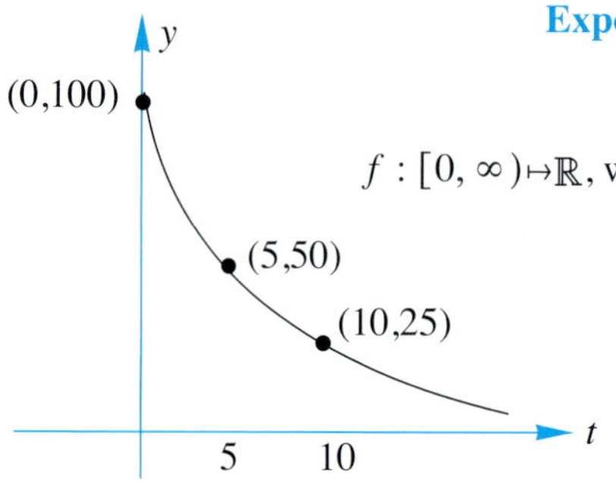

$$f : [0, \infty) \mapsto \mathbb{R}, \text{ where } f(t) = 100 \times \left(\frac{1}{2}\right)^{t/5}$$

Other areas where the use of exponential modelling appears include, medicine (drug dosage), economics (compound interest), oceanography (light penetration in an ocean), environment (endangered species) and many more. We shall look at a few examples of exponential modelling in detail.

Notice that whenever making use of an exponential function to model a real life situation, the domain of consideration is always restricted to $[0, \infty)$. Corresponding to time, $t = 0$ (or $x = 0$), there exists an initial amount. This initial amount is usually denoted by a capital letter with a subscript of '0'. For example, if we are referring to the population size of bacteria, N or the number of radioactive particles P, then their initial amounts would be represented by N_0 and P_0 respectively, so that when $t = 0$, $N = N_0$ and $P = P_0$

Such equations would then be given by

1.	$N = N_0 \times a^t, t \geq 0, a > 1$ [growth]
2.	$P = P_0 \times a^{-t}, t \geq 0, a > 1$ [decay]

EXAMPLE 7.10

During the chemical processing of a particular type of mineral, the amount M kg of the mineral present at time t hours since the process started, is given by

$$M(t) = M_0(2)^{kt}, t \geq 0, k < 0$$

where M_0 is the original amount of mineral present. If 128 kilograms of the mineral are reduced to 32 kilograms in the first six hours of the process find,

(a) i. the value of k.

ii. the quantity of the mineral that remains after 10 hours of processing.

(b) Sketch a graph of the amount of mineral present at time t hours after the process started.

Solution

(a) i. We have that when $t = 0, M = 128 \Rightarrow M_0 = 128$ (the initial amount of mineral).

The equation then becomes

$$M(t) = 128 \times (2)^{kt}, t \geq 0, k < 0 .$$

Next, when $t = 6$, $M = 32$, so that when we substitute this information into the equation, we have, $32 = 128 \times (2)^{6k} \Leftrightarrow 2^{6k} = \frac{1}{4}$

$$\Leftrightarrow 2^{6k} = 2^{-2}$$

$$\Leftrightarrow 6k = -2$$

$$\Leftrightarrow k = -\frac{1}{3}$$

Therefore, the equation is given by, $M(t) = 128 \times (2)^{-\frac{1}{3}t}, t \geq 0$

ii. After 10 hours, we have, $M(10) = 128 \times (2)^{-\frac{1}{3} \times 10}$

$$= 12.699$$

That is, there is approximately 12.70 kg of mineral left after 10 hours of processing.

(b) We notice that the equation is of the form $f : [0, \infty) \mapsto \mathbb{R}$,where $f(t) = a^{-x}, a > 1$ i.e., an exponential decay.

Hence, we have a decreasing function:

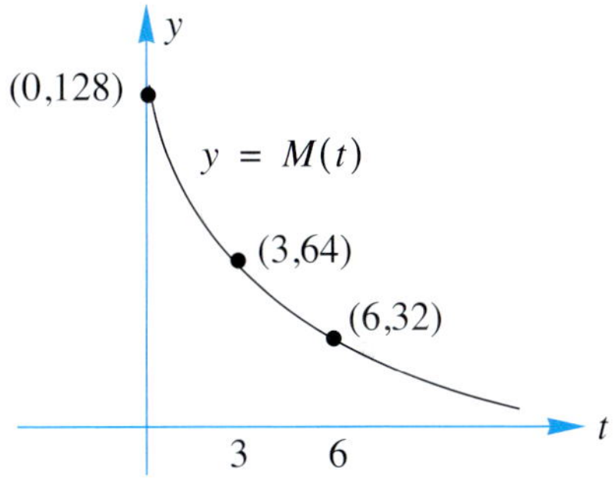

EXAMPLE 7.11

The scrap value , $\$V$, of some machinery after t years is given by

$$V(t) = 50000(0.58)^t, t \geq 0 .$$

(a) What was the initial cost of the machine?

(b) What is the scrap value of the machine after 4 years?

(c) How long would it be before the scrap value reaches \$20000?

(d) The machine needs to be sold at some time when the scrap value of the machine lies somewhere between 10000 and 15000. What time–frame does the owner have?

Solution

(a) When $t = 0$, we have $V(0) = 50000(0.58)^0 = 50000$.
That is, the machine initially cost \$50000.

(b) After 4 years, we have $V(4) = 50000(0.58)^4 = 5658.25$
That is, after 4 years, the scrap value of the machine would be \$5658.25.

(c) We need to determine the value of t when $V = 20000$:

$$20000 = 50000(0.58)^t \Leftrightarrow 0.4 = 0.58^t$$

Then, using the TI–83 we have (using the solve facility):
That is, $t = 1.68$ (to 2 d.p)

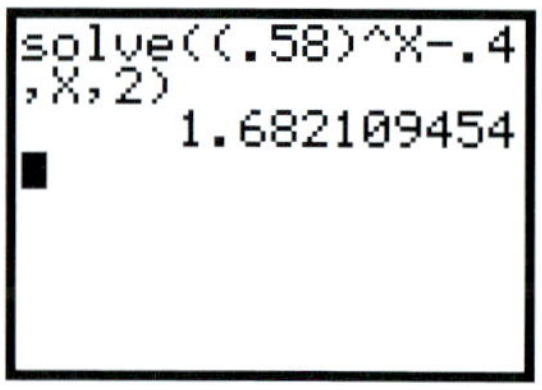

(d) This time we want to solve for t where $10000 \le V(t) \le 15000$.

Now, $10000 \le V(t) \le 15000 \Leftrightarrow 10000 \le 50000(0.58)^t \le 15000$

$$\Leftrightarrow 0.2 \le (0.58)^t \le 0.3$$

Solving the corresponding equalities, we have:
Giving $2.21 \le t \le 2.95$.

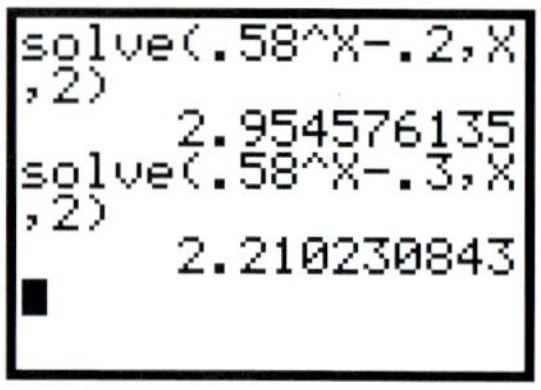

Notice that the graph helped in guessing the values of t that were used in determining the solutions.

Using the TI–83, we can easily sketch the graph of $y = (0.58)^t,\ t \ge 0$:

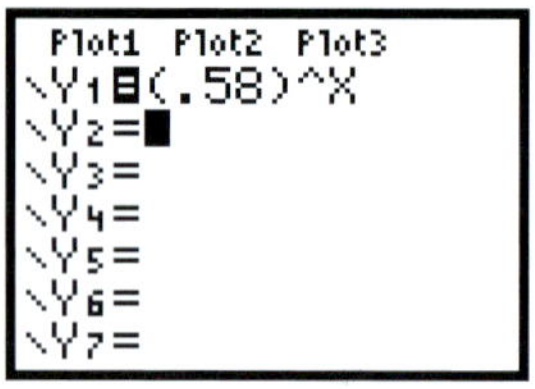

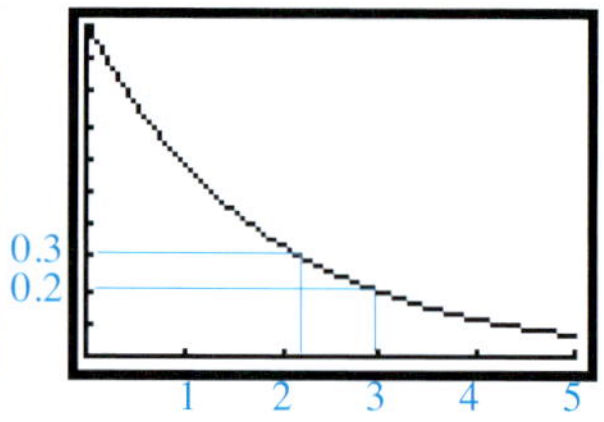

EXERCISES 7.2

1. The number of bacteria in a culture, N, is modelled by the exponential function

$$N = 1000 \times 2^{0.2t},\ t \ge 0$$

where t is measured in days.

(a) Find the initial number of bacteria in this culture.

(b) Find the number of bacteria after i 3 days.
ii. 5 days.

(c) How long does it takes for the number of bacteria to grow to 4000?

2. The ‘growth’ of crystals, measured in kilograms, in a chemical solution, has been approximately modelled by the exponential function $W = 2 \times 10^{kt},\ t \ge 0$, where W is measured in kilograms and t in years. After 1 year in a chemical solution, the amount of crystal in the chemical increased by 6 grams.

(a) Find the value of k.

(b) Find the amount of crystal in the chemical solution after 10 years.

(c) How long does it takes for this crystal to double in ‘size’?

(d) Sketch the graph showing the amount of crystal in the chemical solution at time t.

3. It is found that the intensity of light decreases as it passes through water. The intensity I units at a depth x metres from the surface is given by

$$I = I_0(10)^{-kx}, x \geq 0$$

where I_0 units is the intensity at the surface.

Based on recordings taken by a diving team, it was found that $I = 0.2I_0$ at a depth of 50 metres.

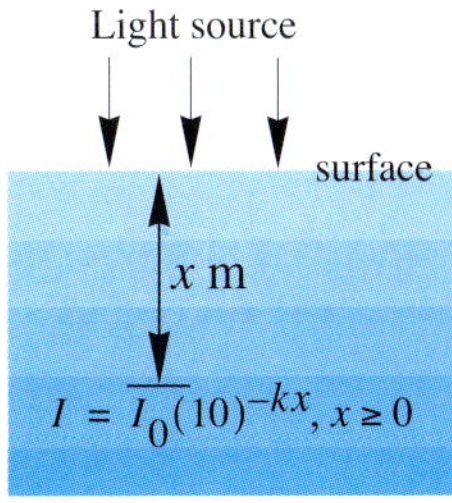

(a) Find the value of k (to 5 d.p.).
(b) Find the percentage of light remaining at a depth of 20 metres.
(c) How much further would the divers need to descend, to reach a level at which the intensity of light would be given by $I = 0.1I_0$?
(d) Find the depth at which the intensity would be a half of that at the surface.
(e) Sketch the graph representing the intensity of light at a depth of x metres.

4. An endangered species of animal is placed into a game reserve. 150 such animals have been introduced into this reserve. The number of animals, $N(t)$, alive t years after being placed in this reserve is predicted by the exponential growth model $N(t) = 150 \times 1.05^t$.
(a) Find the number of animals that are alive after
i. 1 year ii. 2 years iii. 5 years
(b) How long will it take for the population to double?
(c) How long is it before there are 400 of this species in the reserve?
(d) Sketch a graph depicting the population size of the herd over time.
Is this a realistic model?

5. The processing of a type of mineral in a chemical solution has been found to reduce the amount of that mineral left in the solution. Using this chemical process, the amount W kg of the mineral left in the solution at time t hours is modelled by the exponential decay function $W = W_0 \times 10^{-kt}$, where W_0 kg is the original amount of mineral.
It is found that 50 kilograms of mineral are reduced to 30 kilograms in 10 hours.
(a) Write down the value of W_0.
(b) Find the value of k (to 4 decimal places).
(c) How much of the mineral will be in the solution after 20 hours?
(d) Sketch the graph representing the amount of mineral *left* in the solution.
(e) Sketch the graph representing the amount by which the mineral is *reduced*.

6. The temperatures of distant dying stars have been modelled by exponential decay functions. A distant star known to have an initial surface temperature of 15000˚C, is losing heat according to the function $T = T_0 \times 10^{-0.1t}$, where T_0 ˚C is its present temperature, and T ˚C the temperature at time t (in millions of years).
(a) Determine the value of T_0.
(b) Find the temperature of this star in i. one million years,
ii. 10 million years.
(c) How long will it be before the star reaches a temperature that is half its original surface temperature?
(d) Sketch a graph representing this situation.

7. The amount of radioactive material, Q grams, decays according to the model given by the equation $Q = 200 \times 10^{-kt}, t \geq 0$, where t is measured in years. It is known that after 40 years, the amount of radioactive material present is 50 grams.
(a) Find the value of k (to 4 d.p.).
(b) Find the amount of radioactive material present after 80 years.
(c) What is the half life for this radioactive substance ? *The half–life is the time taken for the radioactive material to reach half its original amount.*
(d) Sketch the graph representing the amount of radioactive material present as a function of time, t years.

8. The resale value, V dollars, of a structure, decreases according to the function

$$V = 2000000(10)^{-0.01t}, t \geq 0$$

where t is the number of years since the structure was built.
(a) How much would the structure have sold for upon completion?
(b) How much would the structure have sold for 10 years after completion?
(c) How long will it take for the structure to lose half its value? (Answer to 1 d.p)
(d) Sketch the graph of the structure's value since completion.

9. The population number N in a small town in northern India is approximately modelled by the equation $N = N_0 \times 10^{kt}, t \geq 0$, where N_0 is the initial popluation and t is the time in years since 1980.
The population was found to increase from 100,000 in 1980 to 150,000 in 1990.
(a) Show that $N_0 = 100000$ and that $1.5 = 10^{10k}$.
(b) Hence find the value of k (to 5 d.p.).
(c) Find the population in this town in 1997.
(d) How long (since 1980) will it be before the population reaches 250,000?

10. The healing process of certain types of wounds is measured by the decrease in the surface area that the wound occupies on the skin. A certain skin wound has its surface area modelled by the equation $S = 20 \times 2^{-0.01t}, t \geq 0$ where S square centimetres is the unhealed area t days after the skin received the wound.
(a) What area did the wound originally cover?
(b) What area will the wound occupy after 2 days?
(c) How long will it be before the wound area is reduced by 50%?
(d) How long will it be before the wound area is reduced by 90%?

11. In a certain city the number of inhabitants, N, at time t years since the 1st of January 1970, is modelled by the equation $N = 120000(1.04)^{kt}, t \geq 0, k > 0$.
On the 1st of January 1980, the inhabitants numbered 177629.
(a) Determine the value of k.
(b) How many people will be living in this city by
i. 1st January 2007? ii. 1st April 2007?
(c) How long will it take for the population to reach 1000000?

12. Suppose you deposited $700 into an account that pays 5.80% interest per annum.
(a) How much money will you have in the account at the end of 5 years if
i. the interest is compounded quarterly?
ii. the interest is compounded continuously?
(b) With continuous compounding, how long will it take to double your money?
(c) Sketch the graph showing the amount of money in the account for (b).

13. On the 1st of January 1988, a number of antelopes were introduced into a wildlife reservation, free of predators. Over the years, the number of antelopes in the reservation was recorded:

Date (day/month/year)	1/1/88	1/1/90	1/6/94	1/1/98	1/6/02	1/6/04
Number of antelopes	–	120	190	260	400	485

Although the exact number of antelopes that were placed in the reserve was not available, it is thought that an exponential function would provide a good model for the number of antelopes present in the reserve.

Assume an exponential growth model of the form $N = N_0 \times 2^{kt}, t \geq 0, k > 0$, where N represents the number of antelopes present at time t years since 1/1/80, an N_0 is the initial population size of the herd, and k is a positive real constant.

(a) Determine the number of antelopes introduced into the reserve.
(b) Determine the equation that best models this situation.
(c) Based on this model, predict the number of antelopes that will be present in the reserve by the year 2008.

14. Betty, the mathematician, has a young baby who was recently ill with fever. Betty noticed that the baby's temperature, T, was increasing linearly, until an hour after being given a dose of penicillin. It peaked, then decreased very quickly, possibly exponentially.

Betty approximated the baby's temperature, above 37°C by the function

$$T(t) = t \times 0.82^t, t \geq 0$$

where t refers to the time in hours after 7.00pm.

(a) Sketch the graph of $T(t)$.
(b) Determine the maximum temperature and the time when this occured (giving your answer correct to to 2 d.p)

15. An equation of the form $N(t) = \dfrac{a}{1 + be^{-ct}}, t \geq 0$, where a, b and c are positive constants represents a logistic curve. Logistic curves have been found useful when describing a population N that initially grows rapidly, but whose growth rate decreases after t reaches a certain value.

A study of the growth of protozoa was found to display these characteristics. It was found that the population was well described if $c = 1.12$, $a = 100$, and t measured time in days.

(a) If the initial population was 5 protozoa, find the value of b.
(b) It was found that the growth rate was a maximum when the population size reached 50. How long did it take for this to occur ?
(c) Determine the optimum population size for the protozoa.

16. The height of some particular types of trees can be approximately modelled by the logistic function $h = \dfrac{36}{1 + 200e^{-0.2t}}, t \geq 0$ where h is the height of the tree measured in metres and t the age of the tree (in years) since it was planted.

(a) Determine the height of the tree when planted.
(b) By how much will the tree have grown in the first year ?
(c) How tall will the tree be after 10 years ?
(d) How tall will it be after 100 years ?

(e) How long will it take for the tree to grow to a height of
i. 10 metres ?
ii. 20 metres ?
iii. 30 metres ?
(f) What is the maximum height that a tree, whose height is modelled by this equation, will reach? Explain your answer.
(g) Sketch a graph representing the height of trees against time for trees whose height can be modelled by the above function.

17. Certain prescription drugs, e.g. tablets that are taken orally, which enter the bloodstream at a rate R, are approximately modelled by the equation $R = a \times b^t, t \geq 0$ where t is measured in minutes and a and b are appropriate constants.

When an adult is administered a 100-milligram tablet, the rate is modelled by the function $R = 5 \times 0.95^t, t \geq 0$ mg/min.
The amount A mg of the drug in the bloodstream at time t minutes can then be approximated by a second function, $A = 98(1 - 0.95^t)$ mg.
(a) What is the initial rate at which the drug enters the bloodstream?
(b) How long will it take before the rate at which the drug enters the bloodstream is halved?
(c) How long does it takes for
i. 10 milligrams of the drug to enter the bloodstream.
ii. 50 milligrams of the drug to enter the bloodstream.
iii. 95 milligrams of the drug to enter the bloodstream.
(d) How much of the drug is in the bloodstream when the drug is entering at a rate of 4 mg/min.
(e) Sketch the graph of R and A, on the same set of axes.
(f) Will the patient ever feel the full effects of the 100-milligram drug?

18. As consumers, we know from experience that the demand for a product tends to decrease as the price increases. This type of information can be represented by a demand function. The demand function for a particular product is given by $p = 500 - 0.6 \times e^{0.0004x}$, where p is the price per unit and x is the total demand in number of units.

(a) Find the price p to the nearest dollar for a demand of
i. 1000 units.
ii. 5000 units.
iii. 10 000 units.
(b) Sketch the graph of this demand function.
(c) What level of demand will produce a price per unit of $200?

The total revenue, R, obtained by selling x units of this product is given by $R = xp$.

(d) Find the revenue by selling
i. 1000 units
ii. 5000 units
iii. 10 000 units
(e) Sketch the graph of the revenue equation.
(f) Find the number of units that must be sold in order to maximize the total revenue.
(g) Determine the maximum revenue. Giving your answer to 2 d.p.

7.3 LOGARITHMS

7.3.1 WHAT ARE LOGARITHMS?

Consider the following sequence of numbers:

Sequence N	1	2	3	4	5	6	7	8	9	10	. . .
Sequence y	2	4	8	16	32	64	128	256	512	1024	

The relationship between the values of N and y is given by $y = 2^N$.

Using the above table, evaluate the product 16×64. Using the above table? What for? Clearly there is no use for such a table. Surely this can be done using mental arithmetic (or even using a calculator!). There really is no need to use the above table. However, let's explore this question further, in the hope that we might find something more than the answer.

We start by setting up a table of values that correspond to the numbers in question:

N	4	6	Sum(4 + 6) = 10
y	16	64	Product = 1024

From the first table of sequences, we notice that the sum of the 'N–sequence' (i.e., 10), corresponds to the value of the 'y–sequence' (i.e., 1024).

We next consider the product 8×32, again. Setting up a table of values for the numbers in the sequences that are under investigation we have:

N	3	5	Sum(3 + 5) = 8
y	8	32	Product = 256

What about 4×64 ? As before, we set up the required table of values:

N	2	6	Sum(2 + 6) = 8
y	4	64	Product = 256

In each case the **product** of two terms of the sequence y corresponds to the **sum** obtained by adding corresponding terms of the sequence N.

Notice then that **dividing** two numbers from the sequence y corresponds to the result when **subtracting** the two corresponding numbers from the sequence N,

e.g., for the sequence y: $512 \div 32 = 16$.

for the sequence N: $9 - 5 = 4$

This remarkable property was observed as early as 1594 by **John Napier**. John Napier was born in 1550 (when his father was all of sixteen years of age!) He lived most of his life at the family estate of Merchiston Castle, near Edinburgh, Scotland. Although his life was not without controversy, in matters both religious and political, Napier (when relaxing from his political and religious polemics) would endulge in the study of mathematics and science. His amusement with the study of mathematics led him to the invention of logarithms. In 1614 Napier published his discussion of logarithms in a brochure entitled Mirifici logarithmorum canonis descriptio (A description of the Wonderful Law of Logarithms). Napier died in1617.

It is only fair to mention that the Swiss instrument maker **Jobst Bürgi** (1552–1632) conceived and constructed a table of logarithms independently of Napier, publishing his results in 1620, six years after Napier had announced his discovery.

One of the anomalies in the history of mathematics is the fact that logarithms were discovered before exponents were in use.

Although in this day and age of technology, the use of electronic calculators and computers, render the evaluation of products and quotients to a task that involves the simple push of a few buttons, logarithms are an efficient means of converting a product to a sum and a quotient to a difference. So, what are logarithms?

Nowadays, a logarithm is universally regarded as an exponent.

Thus, if $y = b^N$ we say that N is the logarithm of y to the base b.

From the sequence table, we have that $2^7 = 128$, so that 7 is the **logarithm** of 128 to the **base** 2. Similarly, $3^4 = 81$, and so 4 is the **logarithm** of 81 to the **base** 3.

We use the following notation when using logarithms: $y = b^N \Leftrightarrow N = \log_N y$

That is, ***N* is the logarithm of y to the base b**, which corresponds to the power that the base b must be raised so that the result is y.

EXAMPLE 7.12

Find the following logarithms

(a) $\log_2 32$ (b) $\log_{10} 1000$ (c) $\log_3 729$ (d) $\log_2 \frac{1}{16}$

Solution

(a) To determine the number $\log_2 32$, we ask ourselves the following question:
"To what power must we raise the number 2, so that our result is 32?"
Letting $x = \log_2 32$, we must find the number x such that $2^x = 32$.
Clearly then, $x = 5$, and so we have that $\log_2 32 = 5$.
One convention in setting out such questions is: $\log_2 32 = x \Leftrightarrow 2^x = 32$

$$\Leftrightarrow x = 5$$

(b) As in part (a), we ask the question
"To what power must we raise the number 10, so that our result is 1000 ?"
That is, $x = \log_{10} 1000 \Leftrightarrow 10^x = 1000$

$$\Leftrightarrow x = 3$$

So that $\log_{10} 1000 = 3$.

(c) Now, $x = \log_3 729 \Leftrightarrow 3^x = 729$

$\Leftrightarrow x = 6$ [This was obtained by trial and error.]

Therefore, $\log_3 729 = 6$.

(d) Although we have a fraction, this does not alter the process:

$$x = \log_2 \frac{1}{16} \Leftrightarrow 2^x = \frac{1}{16}$$

$$\Leftrightarrow 2^x = \frac{1}{2^4} (= 2^{-4})$$

$$\Leftrightarrow x = -4$$

7.3.2 CAN WE FIND THE LOGARITHM OF A NEGATIVE NUMBER?

To evaluate $\log_a(-4)$ for some base $a > 0$, we need to solve the equivalent statement:

$$x = \log_a(-4) \Leftrightarrow a^x = -4 .$$

However, the value of a^x where $a > 0$, will always be positive, therefore there is no value of x for which $a^x = -4$. This means that

we can not evaluate the logarithm of a negative number.

We can now make our definition a little stronger: $N = \log_b y \Leftrightarrow y = b^N, y > 0$

Note: we also require that $b \neq 1$, otherwise we will have that $y = 1^N = 1$ for any value of N.

As we saw earlier, there exists a natural exponential whose base is 'e'. In the same way we also have the natural logarithm, whose base is also 'e'.

In this instance, we refer to this logarithm base 'e' as the **natural logarithm**.

We denote the natural logarithm by $\log_e x$ or $\ln x$ (read as '*el n*'). i.e., $y = e^N \Leftrightarrow N = \log_e y$.

For example, $\log_e e^2 = 2$. That is, $N = \log_e e^2 \Leftrightarrow e^N = e^2$

$$\Leftrightarrow N = 2$$

EXAMPLE 7.13

Find the value of x given that

(a) $\log_e x = 3$ (b) $\log_e(x-2) = 0.5$ (c) $\log_e 5 = x$

Solution

(a) $\log_e x = 3 \Leftrightarrow x = e^3 \approx 20.09$

(b) $\log_e(x-2) = 0.5 \Leftrightarrow x - 2 = e^{0.5}$

$$\Leftrightarrow x = 2 + \sqrt{e}$$

$$\therefore x \approx 3.65$$

```
e^(3)
          20.08553692
2+e^(.5)
          3.648721271
ln(5)
          1.609437912
```

(c) $x = \log_e 5 \approx 1.61$

Notice that your calculator has two logarithmic functions:
The **log** button stands for $\log_{10} x$.

The **ln** button stands for $\log_e x$.

EXERCISES 7.3

1. Use the definition of a logarithm to determine

(a) $\log_6 36$ (b) $\log_7 49$ (c) $\log_3 243$ (d) $\log_4 64$

(e) $\log_2\left(\frac{1}{8}\right)$ (f) $\log_3\left(\frac{1}{9}\right)$ (g) $\log_4 1$ (h) $\log_{10} 1$

(i) $\log_{\frac{1}{2}} 2$ (j) $\log_{\frac{1}{3}} 9$ (k) $\log_3 \sqrt{3}$ (l) $\log_{10} 0.01$

2. Change the following exponential expressions into their equivalent logarithmic form.

(a) $10^4 = 10000$ (b) $10^{-3} = 0.001$ (c) $10^y = x + 1$

(d) $10^7 = p$ (e) $2^y = x - 1$ (f) $2^{4x} = y - 2$

3. Change the following logarithmic expressions into their equivalent exponential form.

(a) $\log_2 x = 9$ (b) $\log_b y = x$ (c) $\log_b t = ax$

(d) $\log_{10} z = x^2$ (e) $\log_{10} y = 1 - x$ (f) $\log_2(ax - b) = y$

4. Solve for x in each of the following.

(a) $\log_2 x = 4$ (b) $\log_3 9 = x$ (c) $\log_4 x = \frac{1}{2}$

(d) $\log_x 3 = \frac{1}{2}$ (e) $\log_x 2 = 4$ (f) $\log_5 x = 3$

(g) $\log_x 16 = 2$ (h) $\log_x 81 = 2$ (i) $\log_x\left(\frac{1}{3}\right) = 3$

(j) $\log_2(x - 5) = 4$ (k) $\log_3 81 = x + 1$ (l) $\log_3(x - 4) = 2$

5. Solve for x in each of the following, giving your answer to 4 d.p.

(a) $\log_e x = 4$ (b) $\log_e 4 = x$ (c) $\log_e x = \frac{1}{2}$

(d) $\log_x e = \frac{1}{2}$ (e) $\log_x e = 2$ (f) $\log_x e = -1$

(g) $\log_e(x + 2) = 4$ (h) $\log_e(x - 2) = 1$ (i) $\log_x e = -2$

7.4 THE ALGEBRA OF LOGARITHMS

The following logarithmic laws are a direct consequence of the definition of a logarithm and the index laws already established.

First law: **The logarithm of a product** $\log_a(x \times y) = \log_a x + \log_a y, x > 0, y > 0$

Proof: Let $M = \log_a x$ and $N = \log_a y$ so that $x = a^M$ and $y = a^N$.

Then, $x \times y = a^M \times a^N$

$\Leftrightarrow x \times y = a^{M+N}$

$\Leftrightarrow \log_a(x \times y) = M + N$

$\Leftrightarrow \log_a(x \times y) = \log_a x + \log_a y$

EXAMPLE 7.14

Simplify the expression

(a) $\log_3 x + \log_3(4x)$ (b) $\log_2 x + \log_2(4x)$

Solution

(a) $\log_3 x + \log_3(4x) = \log_3(x \times 4x)$

$= \log_3 4x^2$

(b) This time we note that because the base is '2' and there is a '4' in one of the logarithmic expressions, we could first try to 'remove the '4'.

$\log_2 x + \log_2(4x) = \log_2 x + (\log_2 4 + \log_2 x)$

$= \log_2 x + 2 + \log_2 x$

$= 2\log_2 x + 2$

EXAMPLE 7.15

Given that $\log_a p = 0.70$ and $\log_a q = 2$, evaluate the following

(a) $\log_a p^2$ (b) $\log_a(p^2 q)$ (c) $\log_a(apq)$

Solution

(a) $\log_a p^2 = \log_a(p \times p) = \log_a p + \log_a p$

$= 2\log_a p$

$= 2 \times 0.70$

$= 1.40$

(b) $\log_a(p^2 q) = \log_a p^2 + \log_a q$

$= 2\log_a p + \log_a q$

$= 1.40 + 2$

$= 3.40$

(c) $\log_a(apq) = \log_a a + \log_a p + \log_a q$

$= 1 + 0.70 + 2$

$= 3.70$

EXAMPLE 7.16

Find $\{x \mid \log_2 x + \log_2(x+2) = 3\}$.

Solution

$$\log_2 x + \log_2(x+2) = 3 \Leftrightarrow \log_2[x \times (x+2)] = 3$$
$$\Leftrightarrow x(x+2) = 2^3$$
$$\Leftrightarrow x^2 + 2x = 8$$
$$\Leftrightarrow x^2 + 2x - 8 = 0$$
$$\Leftrightarrow (x+4)(x-2) = 0$$
$$\Leftrightarrow x = -4 \text{ or } x = 2$$

Next, we must check our solutions.

When $x = -4$, substituting into the **original equation**, we have:

L.H.S = $\log_2(-4) + \log_2(-4+2)$ – which cannot be evaluated (as the logarithm of a negative number does not exist).

Therefore, $x = -4$, is not a possible solution.

When $x = 2$, substituting into the **original equation**, we have:

L.H.S $= \log_2(2) + \log_2(2+2)$

$= \log_2 8$

$= 3$

$=$ R.H.S

Therefore, $\{x \mid \log_2 x + \log_2(x+2) = 3\} = \{2\}$.

Second law: The logarithm of a quotient

$$\log_a\left(\frac{x}{y}\right) = \log_a x - \log_a y,\ x > 0,\ y > 0$$

Proof: Let $M = \log_a x$ and $N = \log_a y$ so that $x = a^M$ and $y = a^N$.

Then, $\frac{x}{y} = \frac{a^M}{a^N}$

$$\Leftrightarrow \frac{x}{y} = a^{M-N}$$

$$\Leftrightarrow \log_a\left(\frac{x}{y}\right) = M - N$$

$$\Leftrightarrow \log_a\left(\frac{x}{y}\right) = \log_a x - \log_a y$$

EXAMPLE 7.17

Simplify

(a) $\log_{10}100x - \log_{10}xy$ (b) $\log_2 8x^3 - \log_2 x^2 + \log_2\left(\frac{y}{x}\right)$

Solution

(a) $$\log_{10}100x - \log_{10}xy = \log_{10}\left(\frac{100x}{xy}\right)$$

$$= \log_{10}\left(\frac{100}{y}\right)$$

Note: We could then express $\log_{10}\left(\frac{100}{y}\right)$ as $\log_{10}100 - \log_{10}y = 2 - \log_{10}y$.

(b) $$\log_2 8x^3 - \log_2 x^2 + \log_2\left(\frac{y}{x}\right) = \log_2\left(\frac{8x^3}{x^2}\right) + \log_2\left(\frac{y}{x}\right)$$

$$= \log_2 8x + \log_2\left(\frac{y}{x}\right)$$

$$= \log_2\left(8x \times \frac{y}{x}\right)$$

$$= \log_2 8y$$

Note: We could then express $\log_2 8y$ as $\log_2 8 + \log_2 y = 3 + \log_2 y$.

EXAMPLE 7.18

Find $\{x \mid \log_{10}(x+2) - \log_{10}(x-1) = 1\}$

Solution

$$\log_{10}(x+2) - \log_{10}(x-1) = 1 \Leftrightarrow \log_{10}\left(\frac{x+2}{x-1}\right) = 1$$

$$\Leftrightarrow \left(\frac{x+2}{x-1}\right) = 10^1$$

$$\Leftrightarrow x + 2 = 10x - 10$$

$$\Leftrightarrow 12 = 9x$$

$$\Leftrightarrow x = \frac{4}{3}$$

Next, we check our answer. Substituting into the original equation, we have:

$$\text{L.H.S} = \log_{10}\left(\frac{4}{3}+2\right) - \log_{10}\left(\frac{4}{3}-1\right) = \log_{10}\frac{10}{3} - \log_{10}\frac{1}{3} = \log_{10}\left(\frac{10}{3} \div \frac{1}{3}\right)$$

$$= \log_{10}10$$

$$= 1 = \text{R.H.S}$$

Therefore, $\{x \mid \log_{10}(x+2) - \log_{10}(x-1) = 1\} = \left\{\frac{4}{3}\right\}$

Third law: The logarithm of a power $\log_a x^n = n\log_a x,\ x > 0$

Proof: This follows from repeated use of the First Law or it can be shown as follows:

Let $M = \log_a x \Leftrightarrow a^M = x$

$$\Leftrightarrow (a^M)^n = x^n \quad \text{[Raising both sides to the power of } n\text{]}$$

$$\Leftrightarrow a^{nM} = x^n \quad \text{[Using the index laws]}$$

$$\Leftrightarrow nM = \log_a x^n \quad \text{[Converting from exponential to log form]}$$

$$\Leftrightarrow n\log_a x = \log_a x^n$$

EXAMPLE 7.19

Given that $\log_a x = 0.2$ and $\log_a y = 0.5$, evaluate

(a) $\log_a x^3 y^2$ (b) $\log_a \sqrt{\frac{x}{y^4}}$

Solution

(a)
$$\log_a x^3 y^2 = \log_a x^3 + \log_a y^2$$
$$= 3\log_a x + 2\log_a y$$
$$= 3 \times 0.2 + 2 \times 0.5$$
$$= 1.6$$

(b)
$$\log_a \sqrt{\frac{x}{y^4}} = \log_a \left(\frac{x}{y^4}\right)^{1/2} = \frac{1}{2}\log_a \left(\frac{x}{y^4}\right)$$
$$= \frac{1}{2}[\log_a(x) - \log_a y^4]$$
$$= \frac{1}{2}[\log_a x - 4\log_a y]$$
$$= \frac{1}{2}[0.2 - 4 \times 0.5]$$
$$= -0.9$$

Fourth law: Change of base $\log_a b = \frac{\log_k b}{\log_k a},\ a, k \in \mathbb{R}^+ \backslash \{1\}$

Proof: Let $\log_a b = N$ so that $a^N = b$

Taking the logarithms to base k of both sides of the equation we have:

$$\log_k(a^N) = \log_k b \Leftrightarrow N\log_k a = \log_k b$$

$$\Leftrightarrow N = \frac{\log_k b}{\log_k a}$$

However, we have that $\log_a b = N$, therefore, $\log_a b = \frac{\log_k b}{\log_k a}$.

Other observations include:

1. $\log_a a = 1$
2. $\log_a 1 = 0$
3. $\log_a x^{-1} = -\log_a x,\ x > 0$
4. $\log_{\frac{1}{a}} x = -\log_a x$
5. $a^{\log_a x} = x,\ x > 0$

MISCELLANEOUS EXAMPLES

EXAMPLE 7.20

Express y in terms of x if (a) $2 + \log_{10} x = 4\log_{10} y$

(b) $\log x = \log(a - by) - \log a$

Solution

(a) Given that $2 + \log_{10} x = 4\log_{10} y$ then $2 = 4\log_{10} y - \log_{10} x$

$$\Leftrightarrow 2 = \log_{10} y^4 - \log_{10} x$$

$$\Leftrightarrow 2 = \log_{10}\left(\frac{y^4}{x}\right)$$

$$\Leftrightarrow 10^2 = \frac{y^4}{x}$$

$$\Leftrightarrow y^4 = 100x$$

$$\Leftrightarrow y = \sqrt[4]{100x} \quad (\text{ as } y > 0)$$

(b) Given that $\log x = \log(a - by) - \log a$ then $\log x = \log\dfrac{a - by}{a}$

$$\Leftrightarrow x = \frac{a - by}{a}$$

$$\Leftrightarrow ax = a - by$$

$$\Leftrightarrow by = a - ax$$

$$\Leftrightarrow y = \frac{a}{b}(1 - x)$$

EXAMPLE 7.21

Find x if (a) $\log_x 64 = 3$ (b) $\log_{10} x - \log_{10}(x - 2) = 1$

Solution

(a) $\log_x 64 = 3 \Leftrightarrow x^3 = 64$

$$\Leftrightarrow x^3 = 4^3$$

$$\Leftrightarrow x = 4$$

(b) $\log_{10} x - \log_{10}(x - 2) = 1 \Leftrightarrow \log_{10}\left(\dfrac{x}{x - 2}\right) = 1$

$$\Leftrightarrow \frac{x}{x-2} = 10^1$$

$$\Leftrightarrow x = 10x - 20$$

$$\Leftrightarrow -9x = -20$$

$$\Leftrightarrow x = \frac{20}{9}$$

We still need to check our answer: substituting $x = \frac{20}{9}$ into the original equation we have:

$$\text{L.H.S} = \log_{10}\frac{20}{9} - \log_{10}\left(\frac{20}{9} - 2\right) = \log_{10}\frac{20}{9} - \log_{10}\left(\frac{2}{9}\right) = \log_{10}\left(\frac{20}{9} \times \frac{9}{2}\right)$$

$$= \log_{10}10$$

$$= 1$$

$$= \text{R.H.S}$$

Therefore. solution is $x = \frac{20}{9}$.

EXAMPLE 7.22

Find $\{x|\ 5^x = 2^{x+1}\}$. Give both an exact answer and one to 2 d.p.

Solution

Taking the logarithm of base 10 of both sides $5^x = 2^{x+1}$ gives:

$$5^x = 2^{x+1} \Leftrightarrow \log_{10}5^x = \log_{10}2^{x+1}$$

$$\Leftrightarrow x\log_{10}5 = (x+1)\log_{10}2$$

$$\Leftrightarrow x\log_{10}5 - x\log_{10}2 = \log_{10}2$$

$$\Leftrightarrow x(\log_{10}5 - \log_{10}2) = \log_{10}2$$

$$\Leftrightarrow x = \frac{\log_{10}2}{\log_{10}5 - \log_{10}2}$$

And so, $x = 0.75647\ldots = 0.76$ (to 2 d.p).

Exact answer $= \left\{\frac{\log_{10}2}{\log_{10}5 - \log_{10}2}\right\}$, answer to 2 d.p $= \{0.76\}$

EXAMPLE 7.23

Find x, where $6e^{2x} - 17 \times e^x + 12 = 0$

Solution

We first note that $6e^{2x} - 17 \times e^x + 12$ can be written as $6 \times e^{2x} - 17 \times e^x + 12$.

Which in turn can be expressed as $6 \times (e^x)^2 - 17 \times e^x + 12$.

Therefore, making the substitution $y = e^x$, we have that

$6 \times (e^x)^2 - 17 \times e^x + 12 = 6y^2 - 17y + 12$ (i.e., we have a 'hidden' quadratic)

Solving for y, we have:

$$6y^2 - 17y + 12 = 0 \Leftrightarrow (2y-3)(3y-4) = 0$$

So that $$y = \frac{3}{2} \text{ or } y = \frac{4}{3}$$

However, we wish to solve for x, and so, we need to substitute back:

$$e^x = \frac{3}{2} \text{ or } e^x = \frac{4}{3}$$

$$\Leftrightarrow x = \ln\frac{3}{2} \text{ or } x = \ln\frac{4}{3}$$

EXAMPLE 7.24

Solve for x, where $8^{2x+1} = 4^{5-x}$

Solution

Taking logs of both sides of the equation $8^{2x+1} = 4^{5-x}$, we have

$$\log 8^{2x+1} = \log 4^{5-x} \Leftrightarrow (2x+1)\log 8 = (5-x)\log 4$$

$$\Leftrightarrow (2x+1)\log 2^3 = (5-x)\log 2^2$$

$$\Leftrightarrow 3(2x+1)\cancel{\log 2} = 2(5-x)\cancel{\log 2}$$

Therefore, we have that $6x + 3 = 10 - 2x \Leftrightarrow 8x = 7$

$$\therefore x = \frac{7}{8}$$

EXERCISES 7.4

1. Without using a calculator, evaluate the following.

(a) $\log_2 8 + \log_2 4$ (b) $\log_6 18 + \log_6 2$ (c) $\log_5 2 + \log_5 12.5$

(d) $\log_3 18 - \log_3 6$ (e) $\log_2 20 - \log_2 5$ (f) $\log_2 10 - \log_2 5$

2. Write down an expresion for $\log a$ in terms of $\log b$ and $\log c$ for the following.

(a) $a = bc$ (b) $a = b^2 c$ (c) $a = \frac{1}{c^2}$

(d) $a = b\sqrt{c}$ (e) $a = b^3 c^4$ (f) $a = \frac{b^2}{\sqrt{c}}$

3. Given that $\log_a x = 0.09$, find

(a) $\log_a x^2$ (b) $\log_a \sqrt{x}$ (c) $\log_a\left(\frac{1}{x}\right)$

4. Express each of the following as an equation that does not involve a logarithm.

(a) $\log_2 x = \log_2 y + \log_2 z$
(b) $\log_{10} y = 2\log_{10} x$
(c) $\log_2(x+1) = \log_2 y + \log_2 x$
(d) $\log_2 x = y + 1$
(e) $\log_2 y = \frac{1}{2}\log_2 x$
(f) $3\log_2(x+1) = 2\log_2 y$

5. Solve the following equations

(a) $\log_2(x+1) - \log_2 x = \log_2 3$
(b) $\log_{10}(x+1) - \log_{10} x = \log_{10} 3$
(c) $\log_2(x+1) - \log_2(x-1) = 4$
(d) $\log_{10}(x+3) - \log_{10} x = \log_{10} x + \log_{10} 2$
(e) $\log_{10}(x^2+1) - 2\log_{10} x = 1$
(f) $\log_2(3x^2+28) - \log_2(3x-2) = 1$
(g) $\log_{10}(x^2+1) = 1 + \log_{10}(x-2)$
(h) $\log_2(x+3) = 1 - \log_2(x-2)$
(i) $\log_6(x+5) + \log_6 x = 2$
(j) $\log_3(x-2) + \log_3(x-4) = 2$
(k) $\log_2 x - \log_2(x-1) = 3\log_2 4$
(l) $\log_{10}(x+2) - \log_{10} x = 2\log_{10} 4$

6. Simplify the following

(a) $\log_3(2x) + \log_3 w$
(b) $\log_4 x - \log_4(7y)$
(c) $2\log_a x + 3\log_a(x+1)$
(d) $5\log_a x - \frac{1}{2}\log_a(2x-3) + 3\log_a(x+1)$
(e) $\log_{10} x^3 + \frac{1}{3}\log x^3 y^6 - 5\log_{10} x$
(f) $2\log_2 x - 4\log_2\left(\frac{1}{y}\right) - 3\log_2 xy$

7. Solve the following

(a) $\log_2(x+7) + \log_2 x = 3$
(b) $\log_3(x+3) + \log_3(x+5) = 1$
(c) $\log_{10}(x+7) + \log_{10}(x-2) = 1$
(d) $\log_3 x + \log_3(x-8) = 2$
(e) $\log_2 x + \log_2 x^3 = 4$
(f) $\log_3\sqrt{x} + 3\log_3 x = 7$

8. Solve for x.

(a) $\log_2 x^2 = (\log_2 x)^2$
(b) $\log_3 x^3 = (\log_3 x)^3$
(c) $\log_4 x^4 = (\log_4 x)^4$
(d) $\log_5 x^5 = (\log_5 x)^5$

Investigate the solution to $\log_n x^n = (\log_n x)^n$

9. Solve the following, giving an exact answer and an answer to 2 d.p.

(a) $2^x = 14$ (b) $10^x = 8$ (c) $3^x = 125$

(d) $\frac{1}{1-2^x} = 12$ (e) $3^{4x+1} = 10$ (f) $0.8^{x-1} = 0.4$

(g) $10^{-2x} = 2$ (h) $2.7^{0.3x} = 9$ (i) $0.2^{-2x} = 20$

(j) $\frac{2}{1+0.4^x} = 5$ (k) $\frac{2^x}{1-2^x} = 3$ (l) $\frac{3^x}{3^x+3} = \frac{1}{3}$

10. Solve for x

(a) $(\log_2 x)^2 - \log_2 x - 2 = 0$ (b) $\log_2(2^{x+1} - 8) = x$

(c) $\log_{10}(x^2 - 3x + 6) = 1$ (d) $(\log_{10} x)^2 - 11\log_{10} x + 10 = 0$

(e) $\log_x(3x^2 + 10x) = 3$ (f) $\log_{x+2}(3x^2 + 4x - 14) = 2$

11. Solve the following simultaneous equations

(a) $x^y = 5x - 9$, $\log_x 11 = y$ (b) $\log_{10} x - \log_{10} y = 1$, $x + y^2 = 200$ (c) $xy = 2$, $2\log_2 x - \log_2 y = 2$

12. Express each of the following as an equation that does not involve a logarithm.

(a) $\log_e x = \log_e y - \log_e z$ (b) $3\log_e x = \log_e y$ (c) $\ln x = y - 1$

13. Solve the following for x

(a) $\ln(x+1) - \ln x = 4$ (b) $\ln(x+1) - \ln x = \ln 4$

(c) $\log_e(x+1) + \log_e x = 0$ (d) $\log_e(x+1) - \log_e x = 0$

14. Solve the following for x

(a) $e^x = 21$ (b) $e^x - 2 = 8$ (c) $-5 + e^{-x} = 2$

(d) $200e^{-2x} = 50$ (e) $\frac{2}{1-e^{-x}} = 3$ (f) $70e^{-\frac{1}{2}x} + 15 = 60$

(g) $\ln x = 3$ (h) $2\ln(3x) = 4$ (i) $\ln(x^2) = 9$

(j) $\ln x - \ln(x+2) = 3$ (k) $\ln\sqrt{x+4} = 1$ (l) $\ln(x^3) = 9$

15. Solve the following for x

(a) $e^{2x} - 3e^x + 2 = 0$ (b) $e^{2x} - 4e^x - 5 = 0$

(c) $e^{2x} - 5e^x + 6 = 0$ (d) $e^{2x} - 2e^x + 1 = 0$

(e) $e^{2x} - 6e^x + 5 = 0$ (f) $e^{2x} - 9e^x - 10 = 0$

16. Solve each of the following

(a) $4^{x-1} = 132$ (b) $5^{5x-1} = 3^{1-2x}$

(c) $3^{2x+1} - 7 \times 3^{x} + 4 = 0$ (d) $2^{2x+3} - 7 \times 2^{x+1} + 5 = 0$

(e) $3 \times 4^{2x+1} - 2 \times 4^{x+2} + 5 = 0$ (f) $3^{2x} - 3^{x+2} + 8 = 0$

(g) $2\log x + \log 4 = \log(9x-2)$ (h) $2\log 2x - \log 4 = \log(2x-1)$

(i) $\log_3 2x + \log_3 81 = 9$ (j) $\log_2 x + \log_x 2 = 2$

7.5 LOGARITHMIC MODELLING

Some examples of where the logarithmic functions are used can be found in:

i. the measurement of the magnitude of an earthquake (better known as the Richter scale), where the magnitude R of an earthquake of intensity I is given by $R = \log_{10}\left(\frac{I}{I_0}\right)$, where I_0 is a certain minimum intensity.

ii. the measurement of children's weight (better known as The Ehrenberg relation) is given by $\log_{10}W = \log_{10}2.4 + 0.8h$, where W kg is the average weight for children aged 5 through to 13 years and h is the height measured in metres.

iii. the brightness of stars, given by the function $m = 6 - 2.5\log_{10}\left(\frac{L}{L_0}\right)$, where L_0 is the light flux of the faintest star visible to the naked eye (having magnitude 6), and m is the magnitude of brighter stars having a light flux L.

EXAMPLE 7.25

After working through an area of study, students in year 7 sat for a test based on this unit. Over the following two years, the same students were retested on several occasions. The average score was found to be modelled by the function

$S = 90 - 20\log_{10}(t+1), 0 \le t \le 24$ where t is measured in months.

(a) What was the average score on the first test?

(b) What was the score after i. 6 months? ii. 2 years?

(c) How long should it be before the test is re–issued, if the average score is to be 80?

Solution

(a) The first test occurs at time $t = 0$, so that

$$S = 90 - 20\log_{10}(0+1) = 90 - 20\log 1 = 90$$

That is, the average score on the first test was 90%.

(b) i. After six months we have that $t = 6$. Therefore, $S = 90 - 20\log_{10}(6+1) \approx 73$.
That is, the average score on the test after six months was 73%.

ii. After 2 years we have that $t = 24$. Therefore, $S = 90 - 20\log_{10}(24+1) \approx 62$.
That is, the average score on the test after two years was 62%.

(c) We need to find t when $S = 80$. Using the given equation we have:

$$80 = 90 - 20\log_{10}(t+1) \Leftrightarrow 20\log_{10}(t+1) = 10$$

$$\Leftrightarrow \log_{10}(t+1) = \frac{1}{2}$$

$$\Leftrightarrow t+1 = \sqrt{10}$$

$$\Leftrightarrow t = \sqrt{10} - 1$$

That is, $t \approx 2.16$

Therefore, the students should be retested in approximately 2 months time.

EXERCISES 7.5

1. The loudness of a sound, as experienced by the human ear, is based on its intensity level. This intensity level is modelled by the logarithmic function $d = 10\log_{10}\left(\frac{I}{I_0}\right)$ where d is measured in decibels and corresponds to a sound intensity I and I_0 (known as the threshold intensity) is the value of I that corresponds to be the weakest sound that can be detected by the ear under certain conditions.

(a) Find the value of d when I is 10 times as great as I_0 (i.e., $I = 10I_0$).

(b) Find the value of d when I is 1000 times as great as I_0.

(c) Find the value of d when I is 10000 times as great as I_0.

2. A model, for the relationship between the average weight W kilograms and the height h metres for children aged 5 through to 13 years has been closely approximated by the function $\log_{10}W = \log_{10}2.4 + 0.80h$

(a) Based on this model, determine the average weight of a 10-year-old child who is 1.4 metres tall.

(b) How tall would an 8 year old child weighing 50 kg be?

(c) Find an expression for the weight, W, as a function of h.

(d) Sketch the graph of W kg versus h m.

(e) Hence, or otherwise, sketch the graph of h m versus W kg .

3. A measure of the 'energy' of a star can be related to its brightness. To determine this 'energy' stars are classified into categories of brightness called magnitudes. Those considered to be the least 'energetic' are labelled as the faintest stars. Such stars have a light flux given by L_0, and are assigned a magnitude 6. Other brighter stars having a light flux L are assigned a magnitude m by means of the formula $m = 6 - 2.5\log_{10}\left(\frac{L}{L_0}\right)$.

(a) Find the magnitude m of a star, if relative to the faintest star, its light flux L is such that $L = 10^{0.5}L_0$.

(b) Find an equation for L in terms of m and L_0.

(c) Sketch the general shape of the function for L (as a function of m).

(d) Hence, or otherwise, sketch the graph of the function $m = 6 - 2.5\log_{10}\left(\frac{L}{L_0}\right)$.

4. For some manufacturers, it is important to consider the failure time of their computer chips. For Multi–Chips Pty Ltd, the time taken before a fraction x of their computer chips fail has been approximated by the logarithmic function $t = -\frac{1}{c}\log_{10}(1-x)$, where c is some positive constant and time t is measured in years.

(a) Define the domain for this function.

(b) Determine how long will it be before 40% of the chips fail, when

i. $c = 0.1$ ii. $c = 0.2$ iii. $c = 0.3$

(c) How does the value of c affect the reliability of a chip?

(d) Find an expression for the fraction x of chips that will fail after t years.

(e) For the case where c = 0.10, sketch the graph of x versus t. Hence, sketch the graph of $t = -\frac{1}{c}\log_{10}(1-x)$ where $c = 0.10$.

5. Logarithms have been found useful in modelling economic situations in some countries. Pareto's law for capitalist countries states that the relationship between annual income, $ I and the number, n, of individuals whose income exceeds $ I is approximately modelled by the function $\log_{10}I = \log_{10}a - k\log_{10}n$ where a and k are real positive constants.

(a) Find and expression for I that does not involve logarithms.

(b) By varying the values of a and b, describe their effects on

i. the income $ I.

ii. the number of people whose income exceeds $ I.

6. After prolonged observations of our environment, it became obvious that the thickness of the ozone layer had being affected by the production of waste that had taken place over many years. The thickness of the ozone layer has been estimated by making use of the function $\log_{10}\lambda_0 - \log_{10}\lambda = kx$, where λ_0 is the intensity of a particular wavelength of light from the sun before it reaches the atmosphere, λ is the intensity of the same wavelength after passing through a layer of ozone x centimetres thick, and k is the absorption constant of ozone for that wavelength.

The following table has some results based on available data for one region of the Earth's atmosphere:

λ_0	k	$\frac{\lambda_0}{\lambda}$
3200×10^{-8}	$k \approx 0.40$	1.10

(a) Based on the above table, find the approximate thickness of the ozone layer in this region of the atmosphere, giving your answer to the nearest hundredth of a centimetre.

(b) Obtain an expression for the intensity λ, in terms of k, λ_0 and x.

(c) What would the percentage decrease in the intensity of light with a wavelength of 3200×10^{-8} cm be, if the ozone layer is 0.20 centimetre thick?

(d) For a fixed value of λ_0, how does k relate to the intensity λ?

REVISION SET A - PAPER 1 & PAPER 2 STYLE QUESTIONS

1. Find the coefficient of x^2 in the expansion of $(2x-1)^7$.

2. A function f is defined by $f:\mathbb{R}\mapsto\mathbb{R}$, where $f(x) = e^x - 1$.

(a) Sketch the graph of f.

(b) Find i. the domain of the inverse function, f^{-1}.

ii. the rule of the inverse function, f^{-1}.

(c) Sketch the graph of f^{-1} on the set of axes used in (a).

3. Find the coefficient of x^6 in the expansion of $\left(2x - \frac{1}{2}\right)^{10}$.

4. Let $f(x) = \sqrt{x}$ and $g(x) = 4 - x^2$.

(a) Find i. $f(g(2))$ ii. $g(f(2))$

(b) State all values of x for which $f(g(x))$ is defined.

(c) State all values of x for which $g(f(x))$ is defined.

5. Let $f(x) = 1 - \frac{3}{x}, x \neq 0$ and $g(x) = x^2 - 3x$

(a) Find the coordinates of the points of intersection of the graphs f and g.

(b) Sketch on the same set of axes the graphs of f and g.

6. (a) Let $g(x) = e^x$ and $h(x) = \sqrt{x}$. Find $h(g(\log_e 4))$.

(b) Consider the function f, where $f(x) = g(h(x))$ and the domain **S** of f is the largest set of real numbers for which $f(x)$ is defined.

Specify **S**, and the range of f.

(c) Define completely the inverse function f^{-1}, given that it exists.

7. For the graph shown, sketch, on different sets of axes, the graphs of

(a) $y = 2 + f(x)$

(b) $y = f(2x)$

(c) $y = f(x-1)$

(e) $y = f^{-1}(x)$

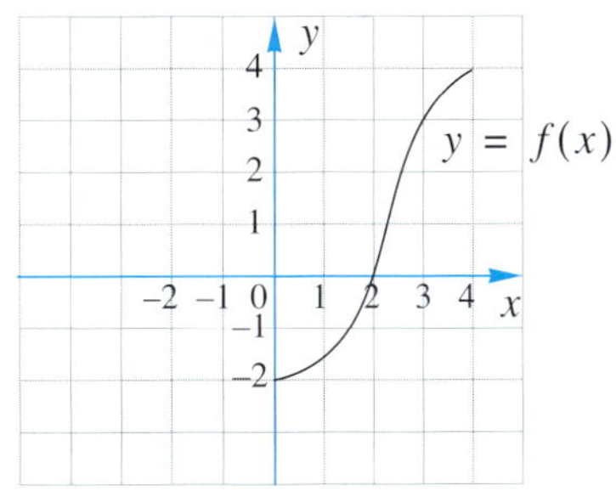

8. Given that $f(x) = x^3$

(a) i. Find $f(8)$

ii. $\{x: f(x) = 8\}$

(b) i. Find $f(x+h) - f(x)$. ii. Simplify $\frac{f(x+h) - f(x)}{h}, h \neq 0$.

9. (a) Solve for x, where i. $(x-1)(x-4) = 10$

ii. $\log_e(x+3) - \log_e x = 1$

(b) Consider the function f defined by, $f : \mathbb{R}\backslash\{2\} \mapsto \mathbb{R}$, where $f(x) = 3 + \frac{1}{x-2}$

Find i. the range of f.

ii. $\frac{1}{f(2.5)}$ iii. $f^{-1}(2.5)$

10. (a) Solve for x, where i. $(x-3)(x-4) = x$

ii. $\log_e(4+3x) = 2$

(b) Consider the function $f(x) = \log_e x - \log_e(1-x)$

Find i. the values of x for which f is defined.

ii. the range of f.

iii. $(f(0.8))^{-1}$ (to 2 d.p.)

iv. $f^{-1}(0.8)$ (to 2 d.p.)

11. (a) Given that $f(x) = \frac{1-x}{1+x}$, find $g(f(x))$ where $g(x) = 1 - \frac{1}{x}$, and specify all values of x for which $g(f(x))$ is defined.

(b) The curves with equations $y = \sqrt{8x}$ and $y = x^2$ meet at O and P, where O is the origin. Find the coordinates of P.

12. (a) Solve each of the following equations for x, giving exact values in terms of the natural logarithm, '*ln*' or in terms of '*e*'.

i. $3^x = 6$

ii. $\log_e(3x+1) - \log_e(4-x) = \log_e 4$

(b) If $f(x) = x + \frac{1}{x+1}$, $x > 0$, find $f^{-1}(3)$.

(c) i. If $f(x) = \sqrt{x-1}$ and $g(x) = \frac{1}{x^2}$, write down expressions for $f(g(x))$ and $g(f(x))$.

ii. For what values of x is $f(g(x))$ defined?

13. Find the value of c for which the coefficient of x^4 in the expansion of $(2x+c)^7$ is 70.

14. The graph of $y = f(x)$ is shown in the diagram. Using different sets of axes for each graph, sketch the graphs of each of the following, showing clearly any intercepts with the axes and any asymptotes:

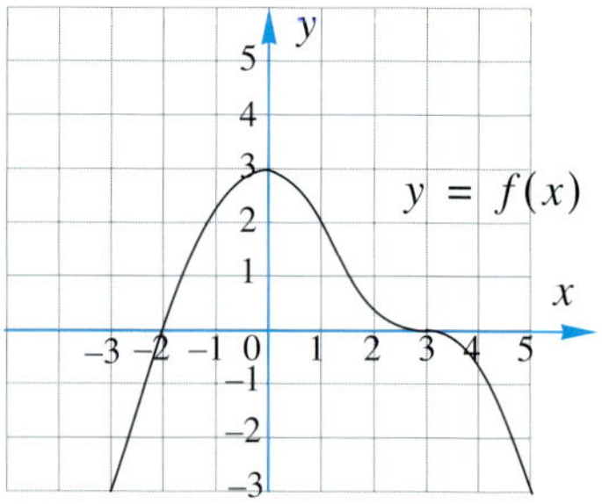

(a) $y = \frac{2}{3}f(x)$

(b) $y = f(-x)$

(c) $y = f(x) + 1$

15. (a) Find k such that the equation $2x^2 + kx + 2k = 0$ has exactly one solution.

(b) Solve $|2x - 3| < 3$

16. (a) For what value of x is $\frac{2}{(x-1)} > \frac{1}{2}$?

(b) Find the term independent of a in the expansion $\left(a - \frac{1}{a}\right)^8$.

17. Find the value of a if the equations $2x + 3y = 6$ and $6x + ay = 9$ are

i. parallel. ii. perpendicular.

18. The coefficient of x in the expansion of $\left(x + \frac{1}{ax^2}\right)^7$ is $\frac{7}{3}$. Find the value(s) of a.

19. (a) Find the equation of the line which passes through both the intersection of $x + y = 2$ and $2x + 3y = 8$ and the point $(0, 0)$.

(b) Simplify $\frac{x^3 - y^3}{x^3 + y^3} \times \frac{(x-y)^2 + xy}{x^2 + xy + y^2}$

20. Consider the system of equations $5^x \cdot 25^{2y} = 1$ and $3^{5x} \cdot 9^y = \frac{1}{9}$.

(a) Show that this system of equations **imply** that $x + 4y = 0$ and $5x + 2y + 2 = 0$.

(b) Hence solve the system of equations.

21. (a) Show that $N \le 11$, where $N = 10 + 2x - x^2$.

(b) i. If $p^2 = 1 + p$ prove that $p^3 = 1 + 2p$.

ii. Hence express p^5 and p^4 in the form $a + bp$.

22. (a) Solve the equation $8^x = 0.25^{3x-1}$.

(b) Find the term independent of $\left(2x - \frac{1}{x}\right)^{12}$.

23. Given that n is a positive integer and that a and b are real constants, find a and b if $(1 + ax)^n = 1 - 6x + \frac{81}{5}x^2 + bx^3 + \ldots$.

24. (a) i. Show that if $a + \frac{1}{a} = \frac{10}{3}$ then $3a^2 - 10a + 3 = 0$.

ii. Hence find $\left\{x \mid 3^x + 3^{-x} = \frac{10}{3}\right\}$.

(b) i. Find an expression for y in terms of x, if $\log 6 + \log(x-3) = 2\log y$.

ii. Solve the system of equations $\begin{aligned} \log 6 + \log(x-3) &= 2\log y \\ 2y - x &= 3 \end{aligned}$.

25. Find the term of x^5 in the expansion $(1 + 2x)^8$.

26. Solve for x: (a) $|2x - 1| = 4$
(b) $(2x - 1)^2 = 4$
(c) $\log_2(2x - 1) = 4$

27. Write down the largest possible domain and range for the following functions
(a) $f(x) = |4 - x|$ (b) $g(x) = \sqrt{4 - x}$ (c) $h(x) = \log_a(4 - x)$

28. ABCD is a square of side length one unit. Points E and F are taken on [AB] and [AD] respectively such that AE = AF = x.
(a) Show that the area, y sq units, of the quadrilateral CDFE is gven by
$y = \frac{1}{2}(1 + x - x^2)$.
(b) Find the maximum area of the quadrilateral.

29. The diagram shows a sketch of the part of the graph of $y = f(x)$ for $0 \leq x \leq 2c$.
The line $x = 2c$ is a line of symmetry of the graph.
Sketch on separate axes the graphs of
i. $y = f(x)$, for $0 \leq x \leq 4c$.
ii. $y = f(x - c)$, for $2c \leq x \leq 4c$.

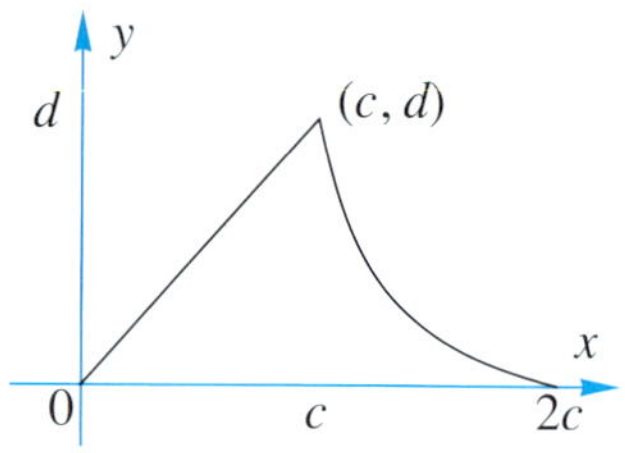

30. (a) On the same set of axes, sketch the graphs of each of the following
i. $y = x^2 - 2x + 2$ ii. $y = \frac{1}{x^2 - 2x + 2}$
(b) Rumours of an immanent take-over by a large electronics company has forced the value of shares in Smith Electronics to rise. Unfortunately, only one week after the rumour started, the large electronics company declared that the take-over would **not** take place. The value of shares in Smith electronics t weeks after the rumour started can be represented by
$$V(t) = \frac{400}{t^2 - 2t + 2}, t \geq 0.$$
where $V(t)$ is measured in cents.
i. Sketch the graph of the function V.
ii. What was the value of shares in Smith Electronics before the rumour started?
iii. What is the maximum value that shares in Smith Electronics reaches?
iv. Mr Brown bought shares in Smith Electronics before the rumour of a take-over. If he is prepared to sell them at 50% profit, when should he sell his shares?

31. (a) On the same set of axes, sketch the graphs of $f : [0, a] \mapsto \mathbb{R}$, where $f(x) = \frac{1}{3}x^2$ and $g : [0, a] \mapsto \mathbb{R}$, where $g(x) = 4x - x^2$, where $a > 0$ and $g(a) = f(a)$.
(b) Find a.

32. Find and sketch the inverse of $g(x) = \frac{1}{2}\log_e(x-e),\ x > e$.

33. The minimum value of the function, $f(x) = ax\log_e(bx)$ is $-\frac{a}{be}$, $b > 0, a > 0$.

(a) State the i. maximal domain of *f*.
ii. range of *f*.

(b) Sketch the graph of *f*, giving the coordinates of the *x*-intercept.

(c) Find the coordinates of the point where the graph of *f* meets the straight line with equation $y = (a\log_e b)x$.

(d) Find the relationship between *b* and *x* at the point where the graph of *f* meets the straight line with equation $y = a\log_e b$.

34. A biologist is observing the growth of two bacterial cultures during an experiment on a new drug. After a number of experiments the biologist observes that the growth of one culture and the decrease in the other culture can be approximated by mathematical functions.

$A(t)$ is the number of cells in the first culture after *t* hours. The number of cells started with in this culture is 900 and the biologist notes that all the cells have died after 5 hours. $B(t)$ is the number of cells in the second culture after *t* hours and grows according to the function, $B(t) = \frac{1000}{1 + 49e^{-2t}}$.

(a) If $A(t)$ is a function of the form $at^2 + b$, find the values of *a* and *b*.

(b) Copy and complete the table for *t* and $B(t)$.

t	1	2	3.5	5
$B(t)$				

(c) What is the initial number of cells for the second culture?

(d) As time increases, what is the *limiting* value of the number of cells for $B(t)$?

(e) After what time is the number of cells in the second culture greater than 500?

(f) Using the same diagram, draw the graphs of $A(t)$ and $B(t)$.

35. The water tank shown is used to top up the level of water in a bird pond. The tank is initially full. When the tap is opened, water flows from a hose connected to the bottom of the tank into the bird pond. The height of the water in the tank, *h* cm, at a time *t* hours after the tap has been opened is modelled by the function *h*, with the rule

$$h(t) = (0.13t - 12.25)^2$$

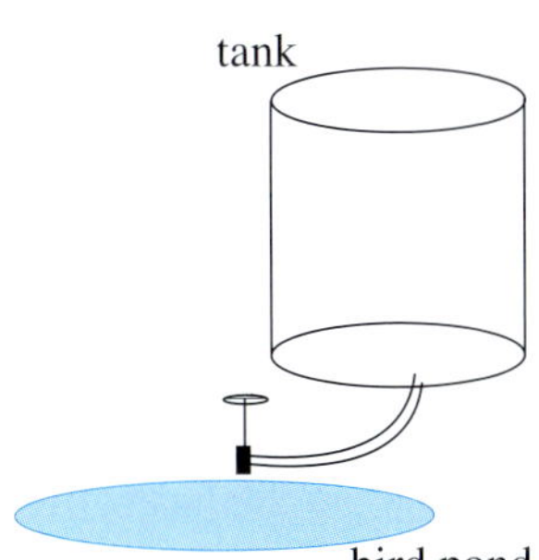

(a) What is the initial height of water in the tank?

(b) What is the value of *h* after 4 hours?

(c) When, to the nearest hour, will the tank be empty if water is allowed to continue to flow out?

(d) Give the maximal domain and range for the function *h*.

(e) Explain briefly why an inverse function, h^{-1} exists. Find its rule and domain.

(f) On the same set of axes, draw the graphs of h and h^{-1}.

(g) When (to the nearest 0.1 of an hour) will the tank be two-thirds full?

36. Let $f(x) = \begin{cases} 1 & x < 0 \\ 2 - x & 0 \le x \le 1 \\ x^{-1} & x > 1 \end{cases}$

(a) Sketch the graph of f.
(b) State the range of f.
(c) Is f a continuous function?

37. Find the coefficient of x^4 in the expansion of $(1 + x + x^2)(1 - x)^9$.

38. Let functions f and g be defined by $f : [-1, 1] \mapsto \mathbb{R}$, where $f(x) = \sqrt{1 - x^2}$ and $g : [a, b] \mapsto \mathbb{R}$, where $g(x) = 1 - x^2$.

(a) If a is a negative real number, what is the greatest real number b such that g has an inverse function?

(b) What is the least value of a for which $f \circ g$ exist?

(c) If a and b have the values found in (a) and (b), state, with reason, whether or not $g \circ f$ exists.

39. Find the middle term in the expansion of $\left(x^2 - \frac{1}{2x}\right)^{10}$.

40. (a) Find the inverse, g^{-1} of $g(x) = x^2 + 2x + 3, x \ge -1$.

(b) On the same set of axes, sketch the graphs of $g(x)$ and $g^{-1}(x)$.

(c) Will there exists a value of x for which $g(x) = g^{-1}(x)$?
If so, find it.

41. Let $f : \mathbb{R} \mapsto \mathbb{R}$ where $f(x) = 4 - x^2$ and $g : [0, \infty[\mapsto \mathbb{R}$, where $g(x) = \sqrt{x}$

i. Show that one of the $f \circ g$ and $g \circ f$ exists but the other does not.

ii. If h is the one which does exist, define h and state its range.

42. A function f is defined by $f : \mathbb{R} \mapsto \mathbb{R}$, where $f(x) = 1 - e^{-x}$

i. Sketch the graph of f and show that an inverse function f^{-1} exists.

ii. Find the domain of f^{-1} and find $f^{-1}(x)$.

iii. Sketch the graph of f^{-1} on the same set of axes as the graph of f.

43. Find the coefficient of x^3 in the expansion of $(1 - x)^5(1 + x)^7$.

44. (a) Let $f(x) = x^2$ and $g:]-\infty, 3] \mapsto \mathbb{R}$, where $g(x) = \sqrt{3 - x}$. State, with reasons, whether or not $f \circ g$ and $g \circ f$ exist.

(b) Find all real values of x for which $\log_e(\log_e(x^2 - 3))$ is real.

(c) Let $f : S \mapsto \mathbb{R}$, where $f(x) = \log_e\left(\frac{2-x}{3+x}\right)$. If S is the set of all real values of x for which $f(x)$ is defined, find S.

45. (a) Let $f : [0, \infty) \mapsto \mathbb{R}$, where $f(x) = 2 - \sqrt{x}$. Define the inverse function f^{-1}.

(b) Let $g : [0, \infty) \mapsto \mathbb{R}$, where $g(x) = \sqrt{x}$. Explain why one of $f^{-1} o g$ and $g o f^{-1}$ is defined and one is not.

(c) If F is the one which is defined, find $F(x)$ and sketch the graph of F.

46. When a colony of wasps was studied, its population was found to be approximated by the model $P(t) = 50e^{0.1t}, t \geq 0$, where P is the population of wasps, and t days is the time from the start of the study.

(a) i. What was the population of the wasp colony when the study began?

ii. What was the population of the wasp colony 10 days after the study began?

(b) Sketch, on a set of axes, a graph of the population P against the time t.

(c) In a sentence, describe how the population of the wasp colony was changing.

Over the same period of time a second wasp colony was also studied. Its population, Q, was found to be approximated by the model $Q(t) = 500 - 450e^{-0.1t}, t \geq 0$.

(d) i. What was the population of this second colony when the study began?

ii. What was its population 10 days after the study began?

(e) On the same set of axes as (b), sketch the graph of the population Q against time.

(f) In a sentence, describe how the population of the wasp colony was changing.

(g) Using you graph, estimate the population when the populations of the two wasps colonies are the same.

(h) i. By solving an appropriate equation, show that when the two wasp colonies have equal numbers, then $k^2 - 10k + 9 = 0$ where $k = e^{0.1t}$.

ii. Hence, find the exact time when the two wasp colonies have equal numbers.

8.1 ARITHMETIC SEQUENCES & SERIES

8.1.1 ARITHMETIC SEQUENCES

A **sequence** is a set of quantities arranged in a definite order.

$1, 2, 3, 4, 5, 6, \ldots\ldots\ldots\ldots$ $-1, 2, -4, 8, -16\ldots.1,$ $1, 2, 3, 5, 8, 13\ldots.$

are all examples of sequences. When the terms of a sequence are added, we obtain a series. Sequences and series are used to solve a variety of practical problems in, for example, business.

There are two major types of sequence, **arithmetic** and **geometric**. This section will consider arithmetic sequences (also known as arithmetic progressions, or simply A.P). The characteristic of such a sequence is that there is a common difference between successive terms. For example:

$1, 3, 5, 7, 9, 11, \ldots$ (the odd numbers) has a first term of 1 and a common difference of 2.
$18, 15, 12, 9, 6, \ldots$ has a first term of 18 and a common difference of –3 (sequence is decreasing).

The terms of a sequence are generally labelled $u_1, u_2, u_3, u_4, \ldots u_n$. The '$n$th term' of a sequence is labelled u_n. In the case of an arithmetic sequence which starts with a and has a **common difference** of d, the **nth term** can be found using the formula:

$$u_n = a + (n-1)d \quad \text{where } d = u_2 - u_1 = u_3 - u_2 = \ldots$$

CHAPTER 8

EXAMPLE 8.1

For the sequence 7, 11, 15, 19, . . . , find the 20th term.

Solution

In this case $a = 7$ and $d = 4$ because the sequence starts with a 7 and each term is 4 bigger than the one before it, i.e., $d = 11 - 7 = 4$. Therefore the nth term is given by

$$u_n = 7 + (n-1) \times 4$$

That is, $u_n = 4n + 3$

$\therefore u_{20} = 4 \times 20 + 3 = 83$ [$n = 20$ corresponds to the 20th term]

EXAMPLE 8.2

An arithmetic sequence has a first term of 120 and a 10th term of 57. Find the 15th term.

Solution

The data is: $a = 120$ and when $n = 10$, $u_{10} = 57$ [i.e., 10th term is 57].

This gives, $u_{10} = 120 + (10-1)d = 57 \Leftrightarrow 120 + 9d = 57$

$\therefore d = -7$

Using $u_n = a + (n-1)d$, we then have $u_n = 120 + (n-1) \times (-7) = 127 - 7n$.

Therefore, when $n = 15$, $u_{15} = 127 - 7 \times 15 = 22$.

EXAMPLE 8.3

An arithmetic sequence has a 7th term of 16.5 and a 12th term of 24. Find the 24th term.

Solution

In this instance we know neither the first term nor the common difference and so we will need to set up equations to be solved simultaneously.

The data is:

$$u_7 = a + 6d = 16.5 \quad -(1)$$

$$u_{12} = a + 11d = 24 \quad -(2)$$

We first solve for 'd': (2) – (1) : $5d = 7.5 \Leftrightarrow d = 1.5$

Substituting into (1): $a + 6 \times 1.5 = 16.5 \Leftrightarrow a = 7.5$

To find the 24th term we use the general term $u_n = a + (n-1)d$ with $n = 24$:

$$u_{24} = 7.5 + (24 - 1) \times 1.5 = 42$$

EXAMPLE 8.4

A car whose original value was \$25600 decreases in value by \$90 per month. How long will it take before the car's value falls below \$15000?

Solution

The values can be seen as a sequence: \$25600, \$25510, \$25420 etc.

In this case $a = 25600$ and $d = 25510 - 25600 = -90$ so that:

$$\begin{aligned} u_n &= 25600 + (n-1) \times (-90) \\ &= 25690 - 90n \\ \therefore 15000 &= 25690 - 90n \\ \Leftrightarrow 90n &= 25690 - 15000 \\ \Leftrightarrow n &= 118.777 \end{aligned}$$

The car will be worth less than \$15000 after 119 months

Using a Graphics Calculator

Most graphic calculators have an automatic memory facility (usually called **Ans**) that stores the result of the last calculation as well as an ability to remember the actual calculation. This can be very useful in listing a sequence.

EXAMPLE 8.5

List the arithmetic sequence 5, 12, 19, 26,......

Solution

The sequence has a first term of 5. Enter this and press **ENTER** or **EXE**.
The common difference of the sequence is 7 so enter + 7.
The display will show Ans + 7 which means 'add 7 to the previous answer.
From here, every time you press **ENTER** (or **EXE**), you will repeat the calculation, generating successive terms of the sequence.

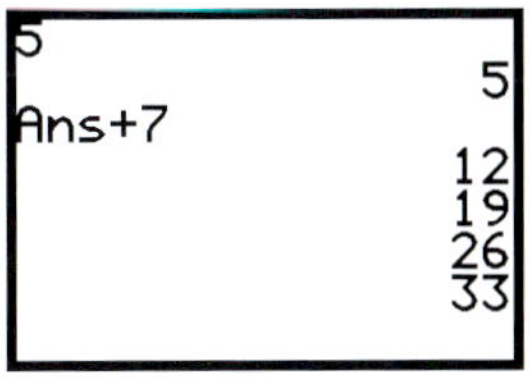

However, the TI–83 is much more sophisticated than this. It is possible to set up a sequence rule on the TI–83. To do this we use the **MODE** key to switch to **Seq** mode and this changes the Equation editor screen from **Y=** to a sequence version (instead of the usual function form).

There are three sequence forms; $u(n)$, $v(n)$ and $w(n)$, which can be accessed on the home screen using the 2nd function key with 7, 8 and 9 respectively. Once these equations are defined we can plot their sequence graph.

We now consider Example 8.2, where we obtained the sequence $u_n = 127 - 7n$ and wished to determine the 15th term.

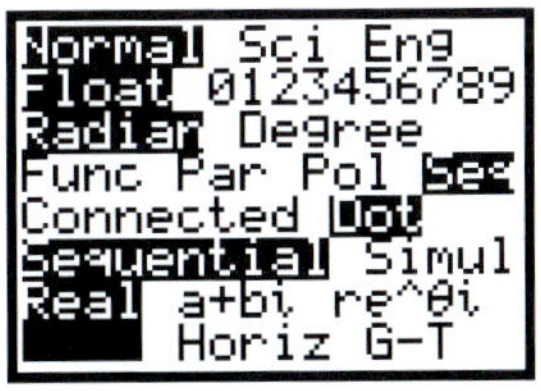

Setting into sequence mode

Define sequence equation

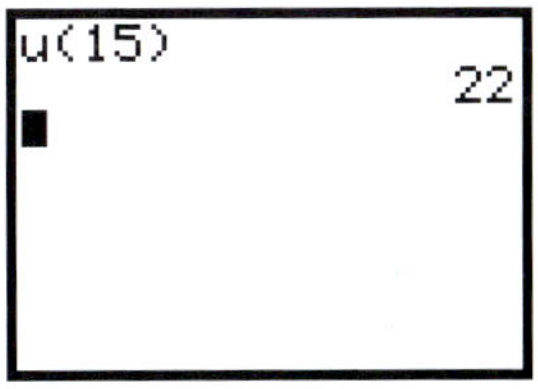

Use 2nd key '7' to call up *u*.

We can also use other features of the TI–83. For example, set up the sequence in a table format:

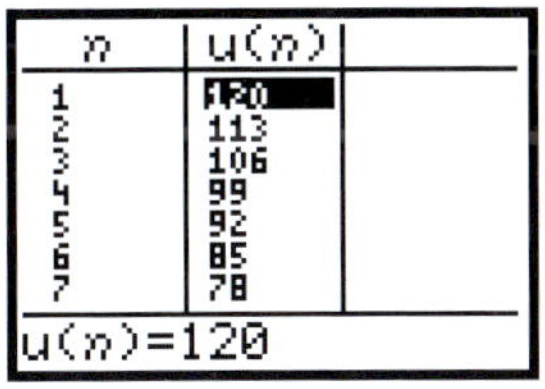

We can plot the sequence:

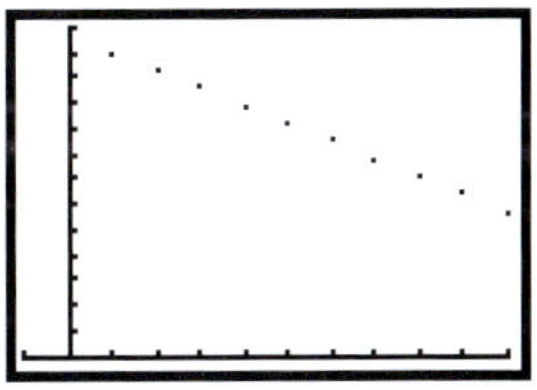

The TI–83 has many features that can be used with sequences. Become familiar with all of them.

EXERCISES 8.1.1

1. (a) Show that the following sequences are arithmetic.
(b) Find the common difference.
(c) Define the rule that gives the nth term of the sequence.

i. $\{2, 6, 10, 14, \ldots\}$
ii. $\{20, 17, 14, 11, \ldots\}$
iii. $\{1, -4, -9, \ldots\}$
iv. $\{0.5, 1.0, 1.5, 2.0, \ldots\}$
v. $\{y + 1, y + 3, y + 5, \ldots\}$
vi. $\{x + 2, x, x - 2, \ldots\}$

2. Find the 10th term of the sequence whose first four terms are 8, 4, 0, –4.

3. Find the value of x and y in the arithmetic sequence $\{5, x, 13, y, \ldots\}$.

4. An arithmetic sequence has 12 as its first term and a common difference of –5. Find its 12th term.

5. An arithmetic sequence has –20 as its first term and a common difference of 3. Find its 10th term.

6. The 14th term of an arithmetic sequence is 100. If the first term is 9, find the common difference.

7. The 10th term of an arithmetic sequence is –40. If the first term is 5, find the common difference.

8. If $n + 5$, $2n + 1$ and $4n - 3$ are three consecutive terms of an arithmetic sequences, find n.

9. The first three terms of an arithmetic sequence are 1, 6, 11.
(a) Find the 9th term.
(b) Which term will equal 151?

10. Find x and y given that $4 - \sqrt{3}, x, y$ and $2\sqrt{3} - 2$ are the first four terms of an arithmetic sequence.

11. For each of the following sequences
(a) $u_n = -5 + 2n, n \geq 1$. (b) $u_n = 3 + 4(n + 1), n \geq 1$.
determine
i. its common difference
ii. its first term

12. The third and fifth terms of an A.P are $x + y$ and $x - y$ respectively. Find the 12th term.

13. The sum of the fifth term and twice the third of an arithmetic sequence equals the twelth term. If the seventh term is 25 find an expression for the general term, u_n.

14. For a given arithmetic sequence, $u_n = m$ and $u_m = n$. Find
(a) the common difference.
(b) u_{n+m}.

8.1.2 ARITHMETIC SERIES

If the **terms of a sequence are added**, the result is known as a **series**.

The	sequence:	1, 2, 3, 4, 5, 6, . . .
gives the	series:	1 + 2 + 3 + 4 + 5 + 6 + . . .
and the	sequence:	–1, –2, –4, –8, –16 . . .
gives the	series:	(–1) + (–2) + (–4) + (–8) + (–16) + . . .
		[or – 1– 2 – 4 – 8 – 16 – . . .

The sum of the terms of a series is referred to as S_n, the **sum of n terms of a series**.

For an arithmetic series, we have

$$S_n = u_1 + u_2 + u_3 + \ldots + u_n$$
$$= a + (a+d) + (a+2d) + \ldots\ldots\ldots\ldots + a + (n-1)d$$

For example, if we have a sequence defined by $u_n = 6 + 4n,\ n \geq 1$, then the sum of the first 8 terms is given by

$$S_8 = u_1 + u_2 + u_3 + \ldots + u_8$$
$$= 10 + 14 + 18 + \ldots + 38$$
$$= 192$$

```
seq(6+4n,n,1,8)
{10 14 18 22 26…
Ans→L1
{10 14 18 22 26…
sum(L1)
                192
```

Again, the screen display of the TI–83 shows how readily we can obtain the sum. Once the sequence has been stored as a **List**, use the **sum(** operation to obtain the answer.

There will be many cases in which we can add the terms of a series in this way. If, however, there are a large number of terms to add, a formula is more appropriate.

There is a story that, when the mathematician Gauss was a child, his teacher was having problems with him because he always finished all his work long before the other students. In an attempt to keep Gauss occupied for a period, the teacher asked him to add all the whole numbers from 1 to 100.

'5050' Gauss replied immediately.

It is probable that Gauss used a method similar to this:

1	2	3	4	5	6	......,	96	97	98	99	100
100	99	98	97	96	95	......,	5	4	3	2	1
101	101	101	101	101	101	......,	101	101	101	101	101

Adding each of the pairings gives 100 totals of 101 each. This gives a total of 10100. This is the sum of two sets of the numbers 1 + 2 + 3 + ... + 98 + 99 +100 and so dividing the full answer by 2 gives the answer 5050, as the young Gauss said, 5050.

It is then possible to apply the same approach to such a sequence, bearing in mind that the sequence of numbers must be arithmetic.

Applying this process to the general arithmetic series we have:

a	$a+d$	$a+2d$	...	$a+(n-3)d$	$a+(n-2)d$	$a+(n-1)d$
$a+(n-1)d$	$a+(n-2)d$	$a+(n-3)d$	...	$a+2d$	$a+d$	a

Each of the pairings comes to the same total.

Here are some examples:

1st pairing: $a+[a+(n-1)d] = 2a+(n-1)d$

2nd pairing: $(a+d)+[a+(n-2)d] = 2a+(n-1)d$

3rd pairing: $(a+2d)+[a+(n-3)d] = 2a+(n-1)d$

$\vdots$

There are n such pairings so $S_n + S_n = n\times[2a+(n-1)d]$

That is, $2S_n = n[2a+(n-1)d]$

Giving the formula, for the sum of n terms of a sequence

$$S_n = \frac{n}{2}[2a+(n-1)d]$$

This formula can now be used to sum large arithmetic series:

EXAMPLE 8.6

Find the sum of 20 terms of the series –2 + 1 + 4 + 7 + 10 +. . .

Solution

We have the following information: $a = u_1 = -2$ and $d = u_2 - u_1 = 1-(-2) = 3$.

Then, the sum to n terms is given by $S_n = \frac{n}{2}[2a+(n-1)d]$

So that the sum to 20 terms is given by

$$S_{20} = \frac{20}{2}[2\times(-2)+(20-1)\times 3]$$
$$= 10[-4+19\times 3]$$
$$= 530$$

EXAMPLE 8.7

Find the sum of 35 terms of the series $-\frac{3}{8}-\frac{1}{8}+\frac{1}{8}+\frac{3}{8}+\frac{5}{8}+\ldots\ldots$

Solution

We have the following information: $a = u_1 = -\frac{3}{8}$ and $d = u_2 - u_1 = -\frac{1}{8}-\left(-\frac{3}{8}\right) = \frac{1}{4}$.

Then, with $n = 35$ we have $S_{35} = \frac{35}{2}\left[2\times-\frac{3}{8}+(35-1)\frac{1}{4}\right] = 17.5\left[-\frac{3}{4}+34\times\frac{1}{4}\right]$

$$= 135\frac{5}{8}$$

EXAMPLE 8.8

An arithmetic series has a third term of 0. The sum of the first 15 terms is –300. What is the first term and the sum of the first ten terms?

Solution

From the given information we have: $u_3 = a + 2d = 0$ – (1)

&: $S_{15} = \frac{15}{2}[2a + 14d] = -300$

i.e., $15a + 105d = -300$

$\therefore a + 7d = -20$ – (2)

The pair of equations can now be solved simultaneously:

(2) – (1): $5d = -20 \Leftrightarrow d = -4$

Substituting result into (1) we have : $a + 2 \times -4 = 0 \Leftrightarrow a = 8$

This establishes that the series is $8 + 4 + 0 + (-4) + (-8) + \ldots$

So the first term is 8 and the sum of the first ten terms is $S_{10} = \frac{10}{2}[16 + 9 \times -4] = -100$.

Using the TI–83 we have, using the general term $u_n = 8 + (n-1) \times -4 = 12 - 4n$

```
seq(12-4n,n,1,10
)
{8 4 0 -4 -8 -1…
Ans→L1
{8 4 0 -4 -8 -1…
sum(L1)
                -100
■
```

EXAMPLE 8.9

A new business is selling home computers. They predict that they will sell 20 computers in their first month, 23 in the second month, 26 in the third and so on, in arithmetic sequence. How many months will pass before the company expects to sell their thousandth computer.

Solution

The series is: 20 + 23 + 26 + ...

The question implies that the company is looking at the **total** number of computers sold, so we are looking at a series, not a sequence.

The question asks how many terms (months) will be needed before the total sales reach more than 1000. From the given information we have: $a = 20$, $d = 23 - 20 = 3$.

Therefore, we have the sum to n terms given by $S_n = \frac{n}{2}[2 \times 20 + (n-1) \times 3]$

$$= \frac{n}{2}[3n + 37]$$

Next, we determine when $S_n = 1000$: $\frac{n}{2}[3n + 37] = 1000 \Leftrightarrow 3n^2 + 37n = 2000$

$$\Leftrightarrow 3n^2 + 37n - 2000 = 0$$

We solve for n use either of the following methods:

Method 1: Quadratic formula

$$n = \frac{-37 \pm \sqrt{37^2 - 4 \times 3 \times -2000}}{2 \times 3}$$

$$= 20.37 \text{ or } (-32.7)$$

Method 2: Graphics Calculator **Solve** function

```
solve(3X²+37X-20
00,X,20)
       20.37941486
■
```

Method 3: Table of values

Notice that we have entered the expression for S_n as the sequence rule for $u(n)$. In fact, the series itself is made up of terms in a sequence of so-called **partial sums**, often called a **sum sequence**. That is, we have that $\{S_1, S_2, S_3, \ldots\} = \{15, 33, 54, \ldots\}$ forms a sequence.

The answer then, is that the company will sell its thousandth computer during the 20th month.

EXERCISES 8.1.2

1. Find the sum of the first ten terms in the arithmetic sequences
(a) $\{1, 4, 7, 10, \ldots\}$ (b) $\{3, 9, 15, 21, \ldots\}$ (c) $\{10, 4, -2, \ldots\}$.

2. For the given arithmetic sequences, find the sum, S_n, to the requested number of terms.
(a) $\{4, 3, 2, \ldots\}$ for $n = 12$
(b) $\{4, 10, 16, \ldots\}$ for $n = 15$
(c) $\{2.9, 3.6, 4.3, \ldots\}$ for $n = 11$

3. Find the sum of the following sequences:
(a) $\{5, 4, 3, \ldots, -15\}$
(b) $\{3, 9, 15, \ldots, 75\}$
(c) $\{3, 5, 7, \ldots, 29\}$

4. The weekly sales of washing machines from a retail store that has just opened in a new housing complex increases by 2 machines per week. In the first week of January 1995, 24 machines were sold.
(a) How many are sold in the last week of December 1995?
(b) How many machines did the retailer sell in 1995?
(c) When was the 500th machine sold?

5. The fourth term of an arithmetic sequence is 5 while the sum of the first 6 terms is 10. Find the sum of the first nineteen terms.

6. Find the sum of the first 10 terms for the sequences defined by
(a) $u_n = -2 + 8n$ (b) $u_n = 1 - 4n$

7. The sum of the first eight terms of the sequence $\{\ln x, \ln x^2y, \ln x^3y^2, \ldots\}$ is given by $4(a\ln x + b\ln y)$. Find a and b.

8.1.3 SIGMA NOTATION

There is a second notation to denote the sum of terms. This other notation makes use the Greek letter $\sum\ldots$ as the symbol to inform us that we are carrying out a summation.
In short, $\sum\ldots$ stands for '**The sum of . . .**'.

This means that the expression $\sum_{i=1}^{n} u_i = u_1 + u_2 + u_3 + \ldots + u_{n-1} + u_n$.

For example, if $u_i = 2 + 5(i-1)$, i.e., an A.P with first term $a = 2$ and common difference $d = 5$, the expression $S_n = \sum_{i=1}^{n} [2 + 5(i-1)]$ would respesent the sum of the first n terms of the sequence. So, the sum of the first 3 terms would be given by

$$S_3 = \sum_{i=1}^{3} [2 + 5(i-1)] = \underbrace{[2 + 5(1-1)]}_{i=1} + \underbrace{[2 + 5(2-1)]}_{i=2} + \underbrace{[2 + 5(3-1)]}_{i=3}$$
$$= 2 + 7 + 12$$
$$= 21$$

Properties of $\sum$

1. $\sum$ is distributive. That is, $\sum_{i=1}^{n} [u_i + v_i] = \sum_{i=1}^{n} u_i + \sum_{i=1}^{n} v_i$.

2. $\sum_{i=1}^{n} ku_i = k\sum_{i=1}^{n} u_i$, for some constant value k.

3. $\sum_{i=1}^{n} k = kn$, i.e., adding a constant term, k, n times is the same as multiplying k by n.

EXAMPLE 8.10

Given that $u_i = 5 + 2i$ and that $v_i = 2 - 5i$ find

(a) $\sum_{i=1}^{5} u_i$ (b) $\sum_{i=1}^{5} [2u_i - v_i]$ (c) $\sum_{i=1}^{1000} [5u_i + 2v_i]$

Solution

(a) $\sum_{i=1}^{5} u_i = u_1 + u_2 + u_3 + u_4 + u_5 = [5 + 2] + [5 + 4] + [5 + 6] + [5 + 8] + [5 + 10]$
$= 7 + 9 + 11 + 13 + 15$
$= 55$

(b) $\sum_{i=1}^{5} [2u_i - v_i] = \sum_{i=1}^{5} (2u_i) + \sum_{i=1}^{5} (-v_i) = 2\sum_{i=1}^{5} u_i - \sum_{i=1}^{5} v_i$

Now, $2\sum_{i=1}^{5} u_i = 2 \times 55 = 110$

and $\sum_{i=1}^{5} v_i = \sum_{i=1}^{5}(2-5i) = \sum_{i=1}^{5}(2) - 5\sum_{i=1}^{5} i = 2\times 5 - 5[1+2+3+4+5]$ [Using properties]

$= -65$

Therefore, $\sum_{i=1}^{5}[2u_i - v_i] = 110 - (-65) = 175$

(c) $\sum_{i=1}^{1000}[5u_i + 2v_i] = \sum_{i=1}^{1000}[5(5+2i) + 2(2-5i)]$

$= \sum_{i=1}^{1000}[25 + 10i + 4 - 10i]$

$= \sum_{i=1}^{1000} 29$

$= 29\ 000$ [i.e., 29×1000]

In this example we have tried to show that there are a number of ways to obtain a sum. It is not always necessary to enumerate every term and then add them. Often, an expression can first be simplified.

EXERCISES 8.1.3 – MISCELLANEOUS QUESTIONS

1. Find the twentieth term in the sequence 9, 15, 21, 27, 33, . . .

2. Fill the gaps in this arithmetic sequence: –3, _, _, _, _, _, 12.

3. An arithmetic sequence has a tenth term of 17 and a fourteenth term of 30. Find the common difference.

4. If $u_{59} = \frac{1}{10}$ and $u_{100} = -1\frac{19}{20}$ for an arithmetic sequence, find the first term and the common difference.

5. Find the sum of the first one hundred odd numbers.

6. An arithmetic series has twenty terms. The first term is –50 and the last term is 83, find the sum of the series.

7. Thirty numbers are in arithmetic sequence. The sum of the numbers is 270 and the last number is 38. What is the first number?

8. How many terms of the arithmetic sequence: 2, 2.3, 2.6, 2.9, . . . must be taken before the terms exceed 100?

9. Brian and Melissa save \$50 in the first week of a savings program, \$55 in the second week, \$60 in the third and so on, in arithmetic progression. How much will they save in ten weeks? How long will they have to continue saving if their target is to save \$5000?

10. A printing firm offers to print business cards on the following terms:
\$45 for design and typesetting and then \$0.02 per card.
(i) What is the cost of 500 cards from this printer?
(ii) How many cards can a customer with \$100 afford to order?

11. A children's game consists of the players standing in a line with a gap of 2 metres between each. The child at the left hand end of the line has a ball which s/he throws to the next child in the line, a distance of 2 metres. The ball is then thrown back to the first child who then throws the ball to the third child in the line, a distance of 4 metres. The ball is then returned to the first child, and so on until all the children have touched the ball at least once.

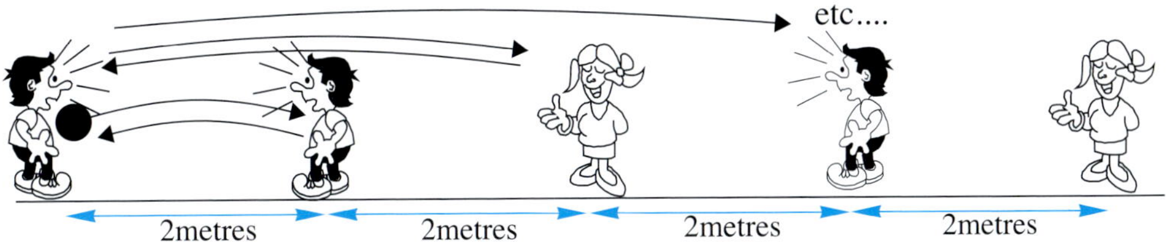

(a) If a total of five children play and they make the least number of throws so that only the leftmost child touches the ball more than once:
i. What is the largest single throw?
ii. What is the total distance travelled by the ball?
(b) If seven children play, what is the total distance travelled by the ball?
(c) If n children play, derive a formula for the total distance travelled by the ball.
(d) Find the least number of children who need to play the game before the total distance travelled by the ball exceeds 100 metres.
(e) The children can all throw the ball 50 metres at most
i. What is the largest number of children that can play the game?
ii. What is the total distance travelled by the ball?

12. Find each sum,

(a) $\sum_{k=1}^{100} k$ (b) $\sum_{k=1}^{100} (2k+1)$ (c) $\sum_{k=1}^{51} (3k+5)$

13. If $u_i = -3+4i$ and $v_i = 12-3i$ find

(a) $\sum_{i=1}^{10} (u_i+v_i)$ (b) $\sum_{i=1}^{10} (3u_i+4v_i)$ (c) $\sum_{i=1}^{10} u_i v_i$

14. (a) Show that for an arithmetic sequence, $u_n = S_n - S_{n-1}$, where u_n is the nth term and S_n is the sum of the first n terms.

(b) Find the general term, u_n, of the A.P given that $\sum_{i=1}^{n} u_i = \frac{n}{2}(3n-1)$.

8.2 GEOMETRIC SEQUENCES & SERIES

8.2.1 GEOMETRIC SEQUENCES

Sequences such as 2, 6, 18, 54, 162, . . . and 200, 20, 2, 0.2, . . . in which each term is obtained by multiplying the previous one by a fixed quantity are known as **geometric sequences**.

The sequence: 2, 6, 18, 54, 162, . . . is formed by starting with 2 and then multiplying by 3 to get the second term, by 3 again to get the third term, and so on.

For the sequence 200, 20, 2, 0.2, . . . , begin with 20 and multiplied by 0.1 to get the second term, by 0.1 again to get the third term and so on.

The constant multiplier of such a sequence is known as the **common ratio**.

The common ratio of 2, 6, 18, 54, 162,.... is 3 and of 200, 20, 2, 0.2,...... it is 0.1.

The **nth term of a geometric sequence** is obtained from the first term by multiplying by $n-1$ common ratios. This leads to the formula for the

nth term of a geometric sequence: $u_n = a \times r^{n-1}$ where $r = \frac{u_2}{u_1} = \ldots = \frac{u_n}{u_{n-1}}$

and n is the term number, a the first term and r is the common ratio.

EXAMPLE 8.11

Find the tenth term in the sequence 2, 6, 18, 54, 162, . . .

Solution

The first term is $a = 2$. The common ratio $r = 3 = \frac{6}{2} = \frac{18}{6}$ and n, the required term, is 10.

Use the formula to solve the problem:

$$\begin{aligned} u_n &= a \times r^{n-1} \\ u_{10} &= 2 \times 3^{(10-1)} \\ &= 2 \times 3^9 \\ &= 39366 \end{aligned}$$

EXAMPLE 8.12

Find the fifteenth term in the sequence 200, 20, 2, 0.2, . . .

Solution

In this case, $a = 20$, $r = \frac{20}{200} = \frac{1}{10} = 0.1$ and $n = 15$.

Using the general term $u_n = a \times r^{n-1}$, the 15th term is given by

$$\begin{aligned} u_{15} &= 200 \times 0.1^{(15-1)} \\ &= 200 \times 0.1^{14} \\ &= 2 \times 10^{-12} \end{aligned}$$

EXAMPLE 8.13

Find the eleventh term in the sequence $1, -\frac{1}{2}, \frac{1}{4}, -\frac{1}{8}, \frac{1}{16}, \ldots$

Solution

The sequence $1, -\frac{1}{2}, \frac{1}{4}, -\frac{1}{8}, \frac{1}{16}, \ldots$ has a common ratio of $r = \frac{-1/2}{1} = -\frac{1}{2}$.

Using the general term $u_n = a \times r^{n-1}$, we have $u_{11} = 1 \times \left(-\frac{1}{2}\right)^{(11-1)}$

$$= \left(-\frac{1}{2}\right)^{10}$$

$$\approx 0.000977$$

Many questions will be more demanding in terms of the way in which you use this formula. You should also recognise that the formula can be applied to a range of practical problems.

Many of the practical problems involve growth and decay and can be seen as similar to problems studied in Chapter 7.

EXAMPLE 8.14

A geometric sequence has a fifth term of 3 and a seventh term of 0.75. Find the first term, the common ratio and the tenth term.

Solution

From the given information we can set up the following equations:

$$u_5 = a \times r^4 = 3 \text{ – (1)}$$

$$\& \quad u_7 = a \times r^6 = 0.75 \text{ – (2)}$$

As with similar problems involving arithmetic sequences, the result is a pair of simultaneous equations. In this case these can best be solved by dividing (2) by (1) to get:

$$\frac{a \times r^6}{a \times r^4} = \frac{0.75}{3} \Leftrightarrow r^2 = 0.25 \Leftrightarrow r = \pm 0.5$$

Substituting results into (1) we have: $a\left(\pm\frac{1}{2}\right)^4 = 3 \Leftrightarrow a = 48$

Therefore, the 10th term is given by $u_{10} = 48 \times (\pm 0.5)^9 = \pm\frac{3}{32}$

There are two solutions: 48, 24, 12, 6, . . . (for the case $r = 0.5$) & 48, –24, 12, –6, . . . ($r = -0.5$).

EXAMPLE 8.15

Find the number of terms in the geometric sequence:
0.25, 0.75, 2.25, . . . , 44286.75.

Solution

The sequence 0.25, 0.75, 2.25, . . . , 44286.75 has a first term $a = 0.25$ and a common ratio $r = \frac{0.75}{0.25} = 3$. In this problem it is n that is unknown. Substitution of the data into the formula gives: $u_n = 0.25 \times 3^{(n-1)} = 44286.75$

The equation that results can be solved using logarithms (see Chapter 7).

$$\begin{aligned} 0.25 \times 3^{(n-1)} &= 44286.75 \\ \therefore 3^{(n-1)} &= 177147 \\ \Leftrightarrow \log_{10}3^{(n-1)} &= \log_{10}177147 \\ \Leftrightarrow (n-1)\log_{10}3 &= \log_{10}177147 \\ \Leftrightarrow n-1 &= \frac{\log_{10}177147}{\log_{10}3} \\ \therefore n-1 &= 11 \\ \Leftrightarrow n &= 12 \end{aligned}$$

Or, by making use of the TI–83

```
solve(.25*3^(X-1
)-44286.75,X,5)
              12
```

EXAMPLE 8.16

A car originally worth \$34000 loses 15% of its value each year.

(a) Write a geometric sequence that gives the year by year value of the car.

(b) Find the value of the car after 6 years.

(c) After how many years will the value of the car fall below \$10000?

Solution

(a) If the car loses 15% of its value each year, its value will fall to 85% (100% – 15%) of its value in the previous year. This means that the common ratio is 0.85 (the fractional equivalent of 85%). Using the formula, the sequence is: $u_n = 34000 \times 0.85^{(n-1)}$
i.e. \$34000, \$28900, \$24565, \$20880.25, . . .

(b) The value after six years have passed is the **seventh** term of the sequence. This is because the first term of the sequence is the value after **no** years have passed.
$u_7 = 34000 \times 0.85^6 \approx 12823$ or \$12823.

(c) This requires solution of the equation $10000 = 34000 \times 0.85^n$:

$$\begin{aligned} 10000 &= 34000 \times 0.85^n \\ 0.85^n &= 0.2941 \\ \log_{10}(0.85^n) &= \log_{10}0.2941 \\ n\log_{10}0.85 &= \log_{10}0.2941 \\ n-1 &= \frac{\log_{10}0.2941}{\log_{10}0.85} \\ n &\approx 7.53 \end{aligned}$$

Or, by making use of the TI–83

```
solve(34000*(.85
)^X-10000,X,5)
     7.530048564
```

This means that the car's value will fall to \$10000 after about 7 years 6 months.

EXAMPLE 8.17 The number of people in a small country town increases by 2% per year. If the population at the start of 1970 was 12500, what is the predicted population at the start of the year 2010?

Solution

A quantity can be increased by 2% by multiplying by 1.02. Note that this is different from finding 2% of a quantity which is done by multiplying by 0.02.

The sequence is: 12500, 12500 × 1.02, 12500 × 1.02^2 etc. with $a = 12500$, $r = 1.02$.

It is also necessary to be careful about which term is required. In this case, the population at the start of 1970 is the first term, the population at the start of 1971 the second term, and so on. The population at the start of 1980 is the **eleventh** term and at the start of 2010 we need the forty-first term:

$$u_{41} = 12500 \times 1.02^{40}$$
$$\approx 27600$$

In all such cases, you should round your answer to the level given in the question or, if no such direction is given, round the answer to a reasonable level of accuracy. In this question, the original population appears to have been given to the nearest 100 and so it is hardly reasonable to give a higher level of accuracy in the answer.

Using a Graphics Calculator

As with arithmetic sequences, geometric sequences such as 50, 25, 12.5, . . . can be listed using a graphics calculator. For this sequence we have $a = 50$ and $r = 0.5$, so, $u_n = 50(0.5)^{n-1}$

We first set the **MODE** to **Seq** and then enter the sequence rule:

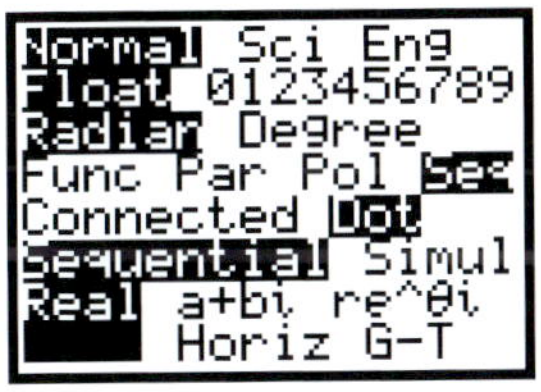

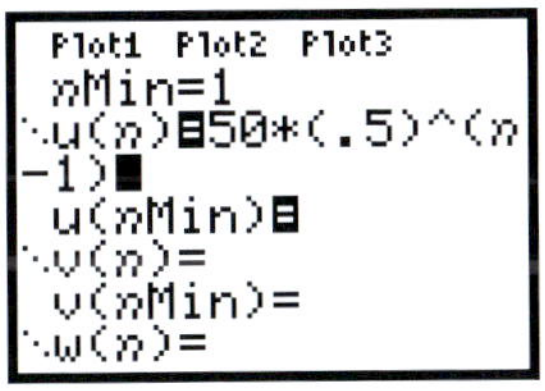

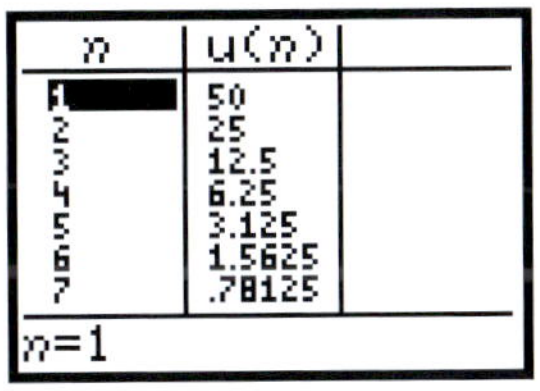

Plot of sequence:

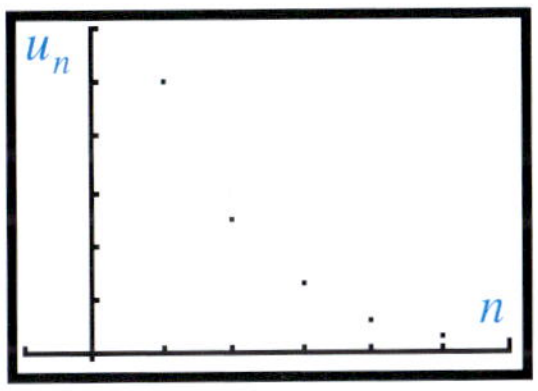

EXERCISES 8.2.1

1. Find the common ratio, the 5th term and the general term of the following sequences

(a) 3, 6, 12, 24, . . . (b) $3, 1, \frac{1}{3}, \frac{1}{9}, \ldots$ (c) $2, \frac{2}{5}, \frac{2}{25}, \frac{2}{125}, \ldots$

(d) –1, 4, –16, 64, ... (e) $ab, a, \frac{a}{b}, \frac{a}{b^2}, \ldots$ (f) $a^2, ab, b^2, \ldots$

2. Find the value(s) of x if each of the following are in geometric sequence

(a) $3, x, 48$

(b) $\frac{5}{2}, x, \frac{1}{2}$

3. The third and seventh terms of a geometric sequence are $\frac{3}{4}$ and 12 respectively.

(a) Find the 10th term.

(b) What term is equal to 3072?

4. A rubber ball is dropped from a height of 10 m and bounces to reach $\frac{5}{6}$ of its previous height after each rebound. Let u_n is the ball's maximum height **before** its nth rebound.

(a) Find an expression for u_n .

(b) How high will the ball bounce **after** its 5th rebound.

(c) How many times has the ball bounced by the time it reaches a maximum height of $\frac{6250}{1296}$ m.

5. The terms $k + 4, 5k + 4, k + 20$ are in a geometric sequence. Find the value(s) of k.

6. A computer depreciates each year to 80% of its value from the previous year. When bought the computer was worth $8000.

(a) Find its value after

i. 3 years

ii. 6 years

(b) How long does it take for the computer to depreciate to a quarter of its purchase price.

7. The sum of the first and third terms of a geometric sequence is 40 while the sum of its second and fourth terms is 96. Find the sixth term of the sequence.

8. The sum of three successive terms of a geometric sequence is $\frac{35}{2}$ while their product is 125. Find the three terms.

9. The population in a town of 40,000 increases at 3% per annum. Estimate the town's population after 10 years.

10. Following new government funding it is expected that the unemployed workforce will decrease by 1.2% per month. Initially there are 120,000 people unemployed. How large an unemployed workforce can the government expect to report in 8 months time.

11. The cost of erecting the ground floor of a building is $44,000, for erecting the first floor it costs $46,200, to erect the second floor costs $48,510 and so on.
Using this cost structure

(a) How much will it cost to erect the 5th floor?

(b) What will be to total cost of erecting a building with six floors?

8.2.2 GEOMETRIC SERIES

When the terms of a geometric sequence are added, the result is a **geometric series**.
For example:
The sequence 3, 6, 12, 24, 48, . . . gives rise to the series: 3 + 6 + 12 + 24 + 48 + . . .

and, the sequence $24, -16, 10\frac{2}{3}, -7\frac{1}{9}, \ldots$ leads to the series $24 - 16 + 10\frac{2}{3} - 7\frac{1}{9} + \ldots$

Geometric series can be summed using the formula that is derived by first multiplying the series by r:

$$S_n = a + ar + ar^2 + ar^3 + \ldots\ldots\ldots + ar^{n-3} + ar^{n-2} + ar^{n-1}$$

$$r \times S_n = ar + ar^2 + ar^3 + \ldots\ldots\ldots + ar^{n-3} + ar^{n-2} + ar^{n-1} + ar^n$$

$$S_n - r \times S_n = a - ar^n \quad \text{(subtracting the second equation from the first)}$$

$$S_n(1-r) = a(1-r^n)$$

$$S_n = \frac{a(1-r^n)}{1-r}$$

This formula can also be written as: $S_n = \frac{a(r^n-1)}{r-1}, r \neq 1$. It is usual to use the version of the formula that gives a positive value for the denominator. And so, we have:

The sum of the first n terms of a geometric series, S_n, where $r \neq 1$ is given by

$$S_n = \frac{a(1-r^n)}{1-r}, |r| < 1 \text{ or } S_n = \frac{a(r^n-1)}{r-1}, |r| > 1$$

EXAMPLE 8.18

Sum the following series to the number of terms indicated.

(a)	$2 + 4 + 8 + 16 + \ldots$	9 terms.
(b)	$5 - 15 + 45 - 135 + \ldots$	7 terms.
(c)	$24 + 18 + \frac{27}{2} + \frac{81}{8} + \ldots$	12 terms.
(d)	$20 - 30 + 45 - 67.5 + \ldots$	10 terms.

Solution

(a) In this case $a = 2$, $r = 2$ and $n = 9$.
Because $r = 2$ it is more convenient to use:

$$S_n = \frac{a(r^n-1)}{r-1}$$

$$S_9 = \frac{2(2^9-1)}{2-1}$$

$$= 1022$$

Using this version of the formula gives positive values for the numerator and denominator. The other version is correct but gives negative numerator and denominator and hence the same answer.

(b) $a = 5, r = -3$ and $n = 7$.

$$S_n = \frac{a(1-r^n)}{1-r} \qquad \text{or} \qquad S_n = \frac{a(r^n-1)}{r-1}$$

$$S_7 = \frac{5(1-(-3)^7)}{1-(-3)} \qquad \qquad S_7 = \frac{5((-3)^7-1)}{(-3)-1}$$

$$= 2735 \qquad \qquad = 2735$$

(c) $a = 24, r = 0.75$ and $n = 12$.

$$S_n = \frac{a(1-r^n)}{1-r}$$

$$S_{12} = \frac{24\left(1-\left(\frac{3}{4}\right)^{12}\right)}{1-\left(\frac{3}{4}\right)}$$ This version gives the positive values.

$$= 92.95907$$

(d) $a = 20, r = -1.5$ and $n = 10$.

$$S_n = \frac{a(1-r^n)}{1-r}$$

$$S_{10} = \frac{20(1-(-1.5)^{10})}{1-(-1.5)}$$

$$= -453.32031$$

When using a calculator to evaluate such expressions, it is advisable to use brackets to ensure that correct answers are obtained. For both the graphics and scientific calculator, the negative common ratio must be entered using the +/– or (–) key.

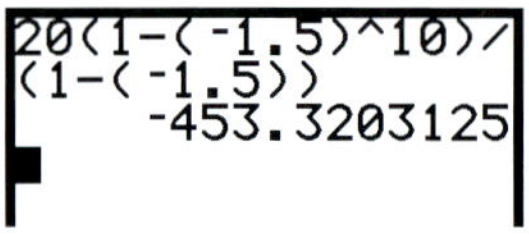

Other questions that may be asked in examinations could involve using both formulas. A second possibility is that you may be asked to apply sequence and series theory to some simple problems.

EXAMPLE 8.19

The 2nd term of a geometric series is –30 and the sum of the first two terms is –15. Find the first term and the common ratio.

Solution

From the given information we have: $u_2 = -30 \therefore ar = -30$ – (1)

$$S_2 = -15 \therefore \frac{a(r^2-1)}{r-1} = -15 \quad \text{– (2)}$$

The result is a pair of simultaneous equations in the two unknowns. The best method of solution is substitution:

From (1): $a = \frac{-30}{r}$. Substituting into (2): $\frac{\frac{-30}{r}(r^2-1)}{r-1} = -15 \Leftrightarrow \frac{(-30)(r^2-1)}{r(r-1)} = -15$

$$\therefore \frac{-30(r+1)(r-1)}{r(r-1)} = -15$$

$$\Leftrightarrow -30(r+1) = -15r$$

$$\Leftrightarrow -30r - 30 = -15r$$

$$\Leftrightarrow r = -2$$

$$\therefore a = \frac{-30}{r} = \frac{-30}{-2} = 15$$

The series is 15 – 30 + 60 – 120 + 240 – . . . which meets the conditions set out in the question.

EXAMPLE 8.20

A family decide to save some money in an account that pays 9% annual compound interest calculated at the end of each year. They put $2500 into the account at the beginning of each year. All interest is added to the account and no withdrawals are made. How much money will they have in the account on the day after they have made their tenth payment?

Solution

The problem is best looked at from the last payment of $2500 which has just been made and which has not earned any interest.

The previous payment has earned one lot of 9% interest and so is now worth 2500×1.09.

The previous payment has earned two years' worth of compound interest and is worth 2500×1.09^2.

This process can be continued for all the other payments and the various amounts of interest that each has earned. They form a geometric series:

Last payment   First payment

$$2500 + 2500 \times 1.09 + 2500 \times 1.09^2 + \ldots\ldots + 2500 \times 1.09^9$$

The total amount saved can be calculated using the series formula:

$$S_n = \frac{a(r^n - 1)}{r - 1}$$

$$S_{10} = \frac{2500(1.09^{10} - 1)}{1.09 - 1}$$

$$= 37982.32$$

The family will save about $37,982.

EXERCISES 8.2.2

1. Find the common ratios of these geometric sequences:

(a) $7, 21, 63, 189, \ldots$ (b) $12, 4, \frac{4}{3}, \frac{4}{9}, \ldots$ (c) $1, -1, 1, -1, 1, \ldots$

(d) $9, -3, 1, -\frac{1}{3}, \frac{1}{9}, \ldots$ (e) $64, 80, 100, 125, \ldots$ (f) $27, -18, 12, -8, \ldots$

2. Find the term indicated for each of these geometric sequences:

(a) $11, 33, 99, 297, \ldots$ 10th term.
(b) $1, 0.2, 0.04, 0.008, \ldots$ 15th term.
(c) $9, -6, 4, -\frac{8}{3}, \ldots$ 9th term.
(d) $21, 9, \frac{27}{7}, \frac{81}{49}, \ldots$ 6th term.
(e) $-\frac{1}{3}, -\frac{1}{4}, -\frac{3}{16}, -\frac{9}{64}, \ldots$ 6th term.

3. Find the number of terms in each of these geometric sequences and the sum of the numbers in each sequence:

(a) $4, 12, 36, \ldots, 236196$
(b) $11, -22, 44, \ldots, 704$
(c) $100, -10, 1, \ldots, -10^{-5}$
(d) $48, 36, 27, \ldots, \frac{6561}{1024}$
(e) $\frac{1}{8}, -\frac{9}{32}, \frac{81}{128}, \ldots, \frac{6561}{2048}$
(f) $100, 10, 1, \ldots, 10^{-10}$

4. Write the following in expanded form and evaluate.

(a) $\sum_{k=1}^{7} \left(\frac{1}{2}\right)^k$
(b) $\sum_{i=1}^{6} 2^{i-4}$
(c) $\sum_{j=1}^{4} \left(\frac{2}{3}\right)^j$
(d) $\sum_{s=1}^{4} (-3)^s$
(e) $\sum_{t=1}^{6} 2^{-t}$

5. The third term of a geometric sequence is 36 and the tenth term is 78,732. Find the first term in the sequence and the sum of these terms.

6. A bank account offers 9% interest compounded annually. If \$750 is invested in this account, find the amount in the account at the end of the twelfth year.

7. When a ball is dropped onto a flat floor, it bounces to 65% of the height from which it was dropped. If the ball is dropped from 80 cm, find the height of the fifth bounce.

8. A computer loses 30% of its value each year.
(a) Write a formula for the value of the computer after n years.
(b) How many years will it be before the value of the computer falls below 10% of its original value?

9. A geometric sequence has a first term of 7 and a common ratio of 1.1. How many terms must be taken before the value of the term exceeds 1000?

10. A colony of algae increases in size by 15% per week. If 10 grams of the algae are placed in a lake, find the weight of algae that will be present in the lake after 12 weeks. The lake will be considered 'seriously polluted' when there is in excess of 10000 grams of algae in the lake. How long will it be before the lake becomes seriously polluted?

11. A geometric series has nine terms, a common ratio of 2 and a sum of 3577. Find the first term.

12. A geometric series has a third term of 12, a common ratio of $-\frac{1}{2}$ and a sum of $32\frac{1}{16}$. Find the number of terms in the series.

13. A geometric series has a first term of 1000, seven terms and a sum of $671\frac{7}{8}$. Find the common ratio.

14. A geometric series has a third term of 300, and a sixth term of 37500. Find the common ratio and the sum of the first fourteen terms (in scientific form correct to two significant figures).

15. A \$10000 loan is offered on the following terms: 12% annual interest on the outstanding debt calculated monthly. The required monthly repayment is \$270. How much will still be owing after nine months.

16. As a prize for inventing the game of chess, its originator is said to have asked for one grain of wheat to be placed on the first square of the board, 2 on the second, 4 on the third, 8 on the fourth and so on until each of the 64 squares had been covered. How much wheat would have been the prize?

8.2.3 COMBINED A.Ps AND G.Ps

There will be occasions on which questions will be asked that relate to both arithmetic and geometric sequences and series.

EXAMPLE 8.21

A geometric sequence has the same first term as an arithmetic sequence. The third term of the geometric sequence is the same as the tenth term of the arithmetic sequence with both being 48. The tenth term of the arithmetic sequence is four times the second term of the geometric sequence. Find the common difference of the arithmetic sequence and the common ratio of the geometric sequence.

Solution

When solving these sorts of question, write the data as equations, noting that a is the same for both sequences. Let u_n denote the general term of the A.P and v_n the general term of the G.P. We then have:

$$u_{10} = a + 9d = 48,\ v_3 = ar^2 = 48,$$

i.e.,

$$a + 9d = ar^2 = 48 \quad -(1)$$

$$u_{10} = 4v_2 \Rightarrow a + 9d = 4ar \quad -(2)$$

(1) represents the information 'The third term of the geometric sequence is the same as the tenth term of the arithmetic sequence with both being 48'.
(2) represents 'The tenth term of the arithmetic sequence is four times the second term of the geometric sequence'.

There are three equations here and more than one way of solving them. One of the simplest is:

From (1) $a + 9d = 48$ and so substituting into (2): $48 = 4ar \Leftrightarrow ar = 12$ – (3)

Also from (1) we have: $ar^2 = 48 \Leftrightarrow (ar)r = 48$ – (4)

Substituting (3) into (4): $12r = 48 \Leftrightarrow r = 4$

Substituting result into (1): $a \times 16 = 48 \Leftrightarrow a = 3$

Substituting result into (1): $3 + 9d = 48 \Leftrightarrow d = 5$

The common ratio is 4 and the common difference is 5.

It is worth checking that the sequences are as specified:

Geometric sequence: 3, 12, 48
Arithmetic sequence: 3, 8, 13, 18, 23, 28, 33, 38, 43, 48

EXERCISES 8.2.3

1. Consider the following sequences:
Arithmetic: 100, 110, 120, 130, . . .
Geometric: 1, 2, 4, 8, 16, . . .
Prove that:
The terms of the geometric sequence will exceed the terms of the arithmetic sequence after the 8th term.
The sum of the terms of the geometric sequence will exceed the sum of the terms of the arithmetic after the 10th term.

2. An arithmetic series has a first term of 2 and a fifth term of 30. A geometric series has a common ratio of –0.5. The sum of the first two terms of the geometric series is the same as the second term of the arithmetic series. What is the first term of the geometric series?

3. An arithmetic series has a first term of –4 and a common difference of 1. A geometric series has a first term of 8 and a common ratio of 0.5. After how many terms does the sum of the arithmetic series exceed the sum of the geometric series?

4. The first and second terms of an arithmetic and a geometric series are the same and are equal to 12. The sum of the first two terms of the arithmetic series is four times the first term of the geometric series. Find the first term of each series.

5. Bo-Youn and Ken are to begin a savings program. Bo-Youn saves \$1 in the first week \$2 in the second week, \$4 in the third and so on, in geometric progression. Ken saves \$10 in the first week, \$15 in the second week, \$20 in the third and so on, in arithmetic progression. After how many weeks will Bo-Youn have saved more than Ken?

6. Ari and Chai begin a training program. In the first week Chai will run 10km, in the second he will run 11km and in the third 12km, and so on, in arithmetic progression. Ari will run 5km in the first week and will increase his distance by 20% in each succeeding week.
(a) When does Ari's weekly distance first exceed Chai's?
(b) When does Ari's total distance first exceed Chai's?

7. The Fibonacci sequence: 1, 1, 2, 3, 5, 8, 13, 21, . . . in which each term is the sum of the previous two terms is neither arithmetic nor geometric. However, after the eighth term (21) the sequence becomes approximately geometric. If we assume that the sequence is geometric:

(a) What is the common ratio of the sequence (to four significant figures)?

(b) Assuming that the Fibonacci sequence can be approximated by the geometric sequence after the eighth term, what is the approximate sum of the first 24 terms of the Fibonacci sequence?

8.2.4 CONVERGENT SERIES

If a geometric series has a common ratio between –1 and 1, the terms get smaller and smaller as n increases.

The sum of these terms is still given by the formula: $S_n = \frac{a(1-r^n)}{1-r}, r \neq 1$

For $-1 < r < 1, r^n \to 0$ as $n \to \infty \Rightarrow S_n = \frac{a}{1-r}$.

If $|r| < 1$, the infinite sequence has a sum given by $S_\infty = \frac{a}{1-r}$.

This means that if the common ratio of a geometric series is between –1 and 1, the sum of the series will approach a value of $\frac{a}{1-r}$ as the number of terms of the series becomes large i.e., the **series is convergent**.

EXAMPLE 8.22

Find the sum to infinity of the series

(a) $16 + 8 + 4 + 2 + 1 + \ldots$

(b) $9 - 6 + 4 - \frac{8}{3} + \frac{16}{9} - \ldots$

Solution

(a) $16 + 8 + 4 + 2 + 1 + \ldots$

In this case $a = 16, r = \frac{1}{2} \Rightarrow S_\infty = \frac{a}{1-r} = \frac{16}{1-\frac{1}{2}} = 32$

(b) $9 - 6 + 4 - \frac{8}{3} + \frac{16}{9} -$

$a = 9, r = -\frac{2}{3} \Rightarrow S_\infty = \frac{a}{1-r} = \frac{9}{1-\left(-\frac{2}{3}\right)} = 5.4$

There are a variety of applications for convergent geometric series. The following examples illustrate two of these.

EXAMPLE 8.23

Use an infinite series to express the recurring decimal $0.\dot{4}6\dot{2}$ as a rational number.

Solution

$0.\dot{4}6\dot{2}$ can be expressed as the series: $0.462 + 0.000462 + 0.000000462 + \ldots$

or $\frac{462}{1000} + \frac{462}{1000000} + \frac{462}{1000000000} + \ldots$

This is a geometric series with $a = \frac{462}{1000}, r = \frac{1}{1000}$

It follows that $S_\infty = \frac{a}{1-r} = \frac{\frac{462}{1000}}{1 - \frac{1}{1000}} = \frac{\frac{462}{1000}}{\frac{999}{1000}} = \frac{462}{999}$

EXAMPLE 8.24

A ball is dropped from a height of 10 metres. On each bounce the ball bounces to three quartes of the height of the previous bounce. Find the distance travelled by the ball before it comes to rest (if it does not move sideways).

Solution

The ball bounces in a vertical line and does not move sideways. On each bounce after the drop, the ball moves both up and down and so travels twice the distance of the height of the bounce.

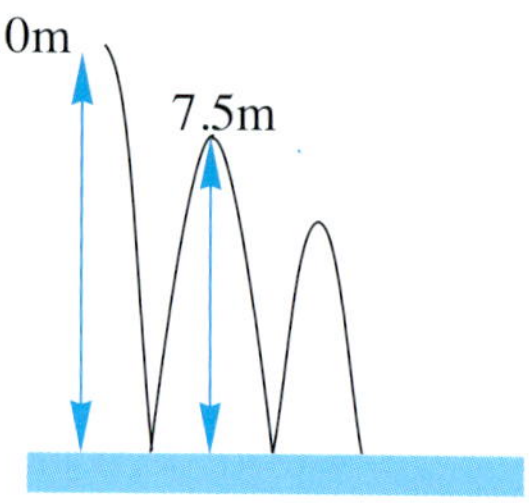

$$\text{Distance} = 10 + 15 + 15 \times \frac{3}{4} + 15 \times \left(\frac{3}{4}\right)^2 + \ldots$$

All but the first term of this series are geometric $a = 15, r = \frac{3}{4}$

$$\text{Distance} = 10 + S_\infty = 10 + \frac{15}{1 - \frac{3}{4}} = 70\text{m}$$

EXERCISES 8.2.4

1. Evaluate:

(i) $27 + 9 + 3 + \frac{1}{3} + \ldots$

(ii) $1 - \frac{3}{10} + \frac{9}{100} - \frac{27}{1000} + \ldots$

(iii) $500 + 450 + 405 + 364.5 + \ldots$

(iv) $3 - 0.3 + 0.03 - 0.003 + 0.0003 - \ldots$

2. Use geometric series to express the recurring decimal 23.232323... as a mixed number.

3. Biologists estimate that there are 1000 trout in a lake. If none are caught, the population will increase at 10% per year. If more than 10% are caught, the population will fall. As an approximation, assume that if 25% of the fish are caught per year, the population will fall by 15% per year. Estimate the total catch before the lake is 'fished out'. If the catch rate is reduced to 15%, what is the total catch in this case? Comment on these results.

4. Find the sum to infinity of the sequence 45, –30, 20, . . .

5. The second term of a geometric sequence is 12 while the sum to infinity is 64. Find the first three terms of this sequence.

6. Express the following as rational numbers

(a) $0.3\dot{6}$ (b) $0.\dot{3}\dot{7}$ (c) $2.1\dot{2}$

7. A swinging pendulum covers 32 centimetres in its first swing, 24 cm on its second swing, 18 cm on its third swing and so on. What is the total distance this pendulum swings before coming to rest?

8. The sum to infinity of a geometric sequence is $\frac{27}{2}$ while the sum of the first three terms is 13. Find the sum of the first 5 terms.

9. Find the sum to infinity of the sequence $1+\sqrt{3}, 1, \frac{1}{\sqrt{3}+1}, \ldots$

10. (a) Find i. $\sum_{i=0}^{n}(-t)^i, |t| < 1$. ii. $\sum_{i=0}^{\infty}(-t)^i, |t| < 1$.

(b) i. Hence, show that $\ln(1+x) = x - \frac{1}{2}x^2 + \frac{1}{3}x^3 - \frac{1}{4}x^4 + \ldots, |x| < 1$

ii. Using the above result, show that $\ln 2 = 1 - \frac{1}{2} + \frac{1}{3} - \frac{1}{4} + \ldots$.

11. (a) Find i. $\sum_{i=0}^{n}(-t^2)^i, |t| < 1$. ii. $\sum_{i=0}^{\infty}(-t^2)^i, |t| < 1$.

(b) i. Hence, show that $\arctan x = x - \frac{1}{3}x^3 + \frac{1}{5}x^5 - \frac{1}{7}x^7 + \ldots, |x| < 1$

ii. Using the above result, show that $\frac{\pi}{4} = 1 - \frac{1}{3} + \frac{1}{5} - \frac{1}{7} + \ldots$.

EXERCISES 8.2.5 – MISCELLANEOUS QUESTIONS

1. $2k+2$, $5k+1$ and $10k+2$ are three successive terms of a geometric sequence. Find the value(s) of k.

2. Evaluate $\dfrac{1+2+3+\ldots+10}{1+\frac{1}{2}+\frac{1}{4}+\ldots+\frac{1}{512}}$

3. Find a number which, when added to each of 2, 6 and 13 gives three numbers in geometric sequence.

4. Find the fractional equivalent of

(a) $2.3\dot{8}$ (b) $4.\dot{6}\dot{2}$ (c) $0.41717\ldots$

5. Find the sum of all intergers between 200 and 400 that are divisible by 6.

6. Find the sum of the first 50 terms of an arithmetic progression given that the 15th term is 34 and the sum of the first 8 terms is 20.

7. Find the value of p so that $p + 5, 4p + 3$ and $8p - 2$ will form successive terms of an arithmetic progression.

8. For the series defined by $S_n = 3n^2 - 11n$, find t_n and hence show that the sequence is arithmetic.

9. How many terms of the series $6 + 3 + \frac{3}{2} + \ldots$ must be taken to give a sum of $11\frac{13}{16}$?

10. If $1 + 2x + 4x^2 + \ldots = \frac{3}{4}$, find the value of x.

11. Logs of wood are stacked in a pile so that there are 15 logs on the top row, 16 on the next row, 17 on the next and so on. If there are 246 logs in total,

(a) how many rows are there?

(b) how many logs are there in the bottom row?

12. The lengths of the sides of a right angled triangle form the terms of an arithmetic sequence. If the hypotenuse is 15 cm in length, what is the length of the other two sides?

13. The sum of the first 8 terms of a geometric series is 17 times the sum of its first four terms. Find the common ratio.

14. Three numbers a, b and c whose sum is 15 are successive terms of a G.P, and b, a, c are successive terms of an A.P. Find a, b and c.

15. The sum of the first n terms of an arithmetic series is given by $S_n = \frac{n(3n+1)}{2}$.

(a) Calculate S_1 and S_2.

(b) Find the first three terms of this series.

(c) Find an expression for the nth term.

16. An ant walks along a straight path. After travelling 1 metre it stops, turns through an angle of 90° in an anticlockwise direction and sets off in a straight line covering a distance of half a metre. Again, the ant turns through an angle of 90° in an anticlockwise direction and sets off in a straight line covering a quarter of a metre. The ant continues in this manner indefinitely.

(a) How many turns will the ant have made after covering a distance of $\frac{63}{32}$ metres?

(b) How far will the ant 'eventually' travel?

8.3 COMPOUND INTEREST AND SUPERANNUATION

8.3.1 COMPOUND INTEREST

We have already come across some practical examples of the use of G.Ps in the area of finance. In this section we further develop these ideas and look at the area of compound interest and superannuation.

EXAMPLE 8.25 Find what \$600 amounts to in 20 years if it is invested at 8% p.a. compounding annually.

Solution

End of year 1 value $= \$600 + 8\% \times \$600$

$= \$600(1.08)$

End of year 2 value $= \$600(1.08) + 8\% \times \$600(1.08)$

$= \$600(1.08) + 0.08 \times \$600(1.08)$

$= \$600(1.08)[1 + 0.08]$

$= \$600(1.08)^2$

End of year 3 value $= \$600(1.08)^2 + 8\% \times \$600(1.08)^2$

$= \$600(1.08)^2 + 0.08 \times \$600(1.08)^2$

$= \$600(1.08)^2[1 + 0.08]$

$= \$600(1.08)^3$

↓

End of year 20 value $= \$\,600(1.08)^{20}$

Thus, after 20 years the \$600 amounts to \$2796.57.

Looking closely at the terms of the sequence, they form a G.P:

$$600(1.08), 600(1.08)^2, 600(1.08)^3, \ldots, 600(1.08)^{20}$$

where $a = 600$ and $r = 1.08$.

Developing a formula for compound interest

In general, if \$$P$ is invested at r% p.a. compound interest, it grows according to the sequence

$$P\left(1 + \frac{r}{100}\right), P\left(1 + \frac{r}{100}\right)^2, P\left(1 + \frac{r}{100}\right)^3, \ldots, P\left(1 + \frac{r}{100}\right)^n$$

where $a = P\left(1 + \frac{r}{100}\right)$ and $r = \left(1 + \frac{r}{100}\right)$ so that

$$A_n = P\left(1 + \frac{r}{100}\right)^n$$ where A_n is the amount after n time periods.

8.3.2 SUPERANNUATION*

EXAMPLE 8.26

A woman invests \$1000 at the beginning of each year in a superannuation scheme. If the interest is paid at the rate of 12% p.a. on the investment (compounding annually), how much will her investment be worth after 20 years?

Solution

t_1 = the 1st \$1000 will be invested for 20 years at 12% p.a.

t_2 = the 2nd \$1000 will be invested for 19 years at 12% p.a.

t_3 = the 3rd \$1000 will be invested for 18 years at 12% p.a.

⋮

t_{20} = the 20th \$1000 will be invested for 1 year at 12% p.a.

Finding the amount compounded annually using $A = P\left(1 + \frac{r}{100}\right)^n$, we have:

$$t_1 = 1000\left(1 + \frac{12}{1000}\right)^{20} = 1000(1.12)^{20}$$

$$t_2 = 1000\left(1 + \frac{12}{1000}\right)^{19} = 1000(1.12)^{19}$$

$$t_3 = 1000\left(1 + \frac{12}{1000}\right)^{18} = 1000(1.12)^{18}$$

⋮

$$t_{20} = 1000\left(1 + \frac{12}{1000}\right)^{1} = 1000(1.12)^{1}$$

To find the total of her investment after 20 years, we need to add the separate amounts:

$$\begin{aligned} \text{Total} &= 1000(1.12)^{20} + 1000(1.12)^{19} + 1000(1.12)^{18} + \ldots + 1000(1.12)^{1} \\ &= 1000[(1.12)^{20} + (1.12)^{19} + (1.12)^{18} + \ldots + (1.12)^{1}] \\ &= 1000\left[\frac{1.12(1-(1.12)^{20})}{1-1.12}\right] \qquad \text{Using } S_n = \frac{a(1-r^n)}{1-r} \\ &= \$80,698.74 \end{aligned}$$

Thus her total investment amounts to \$80,698.74

* This section lies outside the syllabus. It has been included as extension material.

EXAMPLE 8.27 Linda borrows \$2000 at 1% per month reducible interest. If she repays the loan in equal monthly instalments over 4 years, how much is each instalment?

Solution

Amount borrowed = \$ 2000, $r = 1\%$ per month = 0.01 and $n = 4 \times 12 = 48$ months

Let the monthly instalment be = \$ M and the amount owing after n months = \$ A_n

Our aim is to find \$$M$ i.e., the amount of each instalment.

After 1 month (after paying the 1st instalment), we have:

$$A_1 = 2000 + \text{ interest } - M = 2000 + 2000 \times 0.01 - M$$
$$= 2000(1.01) - M$$

After 2 months

$$A_2 = A_1 \times 1.01 - M = [2000(1.01) - M] \times 1.01 - M$$
$$= 2000(1.01)^2 - 1.01 \times M - M$$
$$= 2000(1.01)^2 - M(1.01 + 1)$$

After 3 months

$$A_3 = A_2 \times 1.01 - M = [2000(1.01)^2 - M(1.01 + 1)] \times 1.01 - M$$
$$= 2000(1.01)^3 - M(1.01 + 1) \times 1.01 - M$$
$$= 2000(1.01)^3 - M[1.01^2 + 1.01 + 1]$$

After 4 months

$$A_4 = A_3 \times 1.01 - M = [2000(1.01)^3 - M(1.01^2 + 1.01 + 1)] \times 1.01 - M$$
$$= 2000(1.01)^4 - M(1.01^3 + 1.01^2 + 1.01) - M$$
$$= 2000(1.01)^4 - M[1.01^3 + 1.01^2 + 1.01 + 1]$$

$$\vdots \qquad\qquad \vdots$$

After n months, we then have

$$A_n = 2000(1.01)^n - M[1 + 1.01 + 1.01^2 + 1.01^3 + \ldots + 1.01^{n-1}]$$

thus, $A_{48} = 2000(1.01)^{48} - M[1 + 1.01 + 1.01^2 + 1.01^3 + \ldots + 1.01^{47}]$.

Now, the loan is repaid after 48 months, meaning that $A_{48} = 0$, therefore, solving for M we have

$$0 = 2000(1.01)^{48} - M[1 + 1.01 + 1.01^2 + 1.01^3 + \ldots + 1.01^{47}]$$

$$\Leftrightarrow 2000(1.01)^{48} = M[1 + 1.01 + 1.01^2 + 1.01^3 + \ldots + 1.01^{47}]$$

$$\Leftrightarrow M = \frac{2000(1.01)^{48}}{[1 + 1.01 + 1.01^2 + 1.01^3 + \ldots + 1.01^{47}]}$$

The denominator is a G.P with $a = 1$, $r = 1.01$ and $n = 48$, so that

$$[1 + 1.01 + 1.01^2 + 1.01^3 + \ldots + 1.01^{47}] = S_{48} = \frac{1(1 - 1.01^{48})}{1 - 1.01} = 61.22261$$

Therefore, $M = \dfrac{2000(1.01)^{48}}{61.22261} = 52.67$

That is, each instalment must be \$52.67.

The total paid = 52.67 × 48 = 2528.16 so that the interest paid = 2528.16 – 2000 = 528.16
That is, she ends up paying $ 528.16 in interest.

Although it is important that you understand the process used in the examples shown, it is also important that you can make use of technology. We now look at how the TI–83 can help us ease the pain of long calculations.

The TI–83 has a number of financial functions which enable computational ease. In particular, it has a **TVM Solver**. The **TVM** Solver displays the time-value-of-money (TVM) variables. In short, given four variable values, the **TVM Solver** solves for the fifth variable. To access the finance screen simply press

2nd [FINANCE] ENTER

Then, it is a matter of entering the 4 known pieces of information and then letting the **TVM Solver** do the rest.

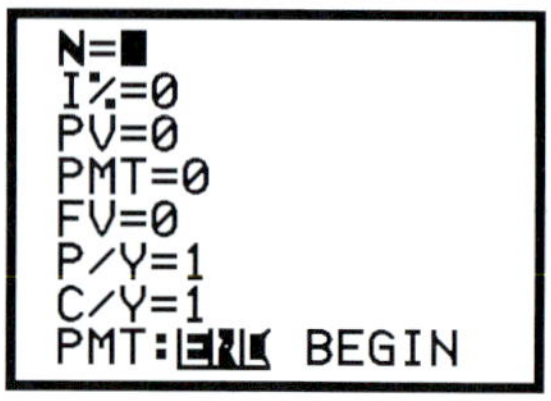

Note: When using the TI–83 financial functions, you must enter cash inflows (cash received) as positive numbers and cash outflows (cash paid) as negative numbers.

Once you have entered your data, there are two ways in which you can then obtain the value of the unknown variable.

Method 1: Place the cursor (using the arrow keys) on the TVM variable for which you want to solve. Press ALPHA [SOLVE]. The answer is computed, displayed in the **TVM Solver** screen and stored to the appropriate TVM variable.

Method 2: You need to leave this window by pressing 2nd QUIT , and return again to the finance menu, by pressing 2nd [FINANCE]. Select the variable you wish to solve for and then press ENTER .

We now illustrate this process using the previous example. In this example we have the known quantities: Linda borrowed $2000, $\therefore PV = -2000$

I% = 1 and N (= 4 × 12) = 48

$FV = 0$ (i.e., loan is fully repaid)

PMT = ? (the monthly repayment required)

That is, once we have entered the information, we then make use of method 1 while at the TVM Solver screen:

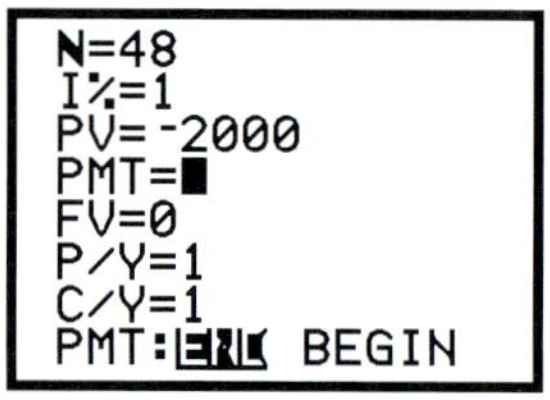

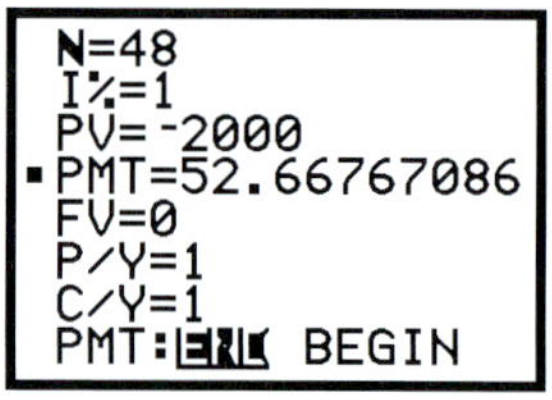

Notice that in the second screen there is a square next to the PMT variable. This is to indicate which variable has just been calculated.

EXERCISES 8.3

1. To how much will \$1000 grow to if it is invested at 12% p.a. for 9 years, compounding annually?

2. A bank advertises an annual interest rate of 13.5% p.a. but adds interest to the account monthly, giving a monthly interest rate of 1.125%. Scott deposits \$3500 with the bank. How much will he withdraw in 20 months time?

3. To what amount will \$900 grow to if it is invested at 10% p.a. for 7 years, compounding every 6 months?

4. A man borrows \$5000 at 18% p.a. over a period of 5 years, with the interest compounding every month. Find to the nearest dollar the amount owing after 5 years.

5. Find the total amount required to pay off a loan of \$20,000 plus interest at the end of 5 years if the interest is compounded half yearly and the rate is 12%.

6. A man invests \$500 at the beginning of each year in a superannuation fund. If the interest is paid at the rate of 12% p.a. on the investment (compounding annually), how much will his investment be worth after 20 years?

7. A woman invests \$2000 at the beginning of each year into a superannuation fund for a period of 15 years at a rate of 9% p.a. (compounding annually). Find how much her investment is worth at the end of the 15 years.

8. A man deposits \$3 000 annually to accumulate at 9% p.a. compound interest. How much will he have to his credit at the end of 25 years? Compare this to depositing \$750 every three months for the same length of time and at the same rate. Which of these two options gives the better return?

9. A woman invests \$200 at the beginning of each month into a superannuation scheme for a period of 15 years. Interest is paid at the rate of 7% p.a. and is compounded monthly. How much will her investment be worth at the end of the 15 year period?

10. Peter borrows \$5000 at 1.5% per month reducible interest. If he repays the loan in equal monthly instalments over 8 years, how much is each instalment, and what is the total interest charged on the loan? Compare this to taking the same loan, but at a rate of 15% p.a. flat rate.

11. Kevin borrows \$7500 to be paid back at 12.5% p.a. monthly reducible over a period of 7 years. What is the amount of each monthly instalment and what is the total interest charged on the loan. Find the equivalent flat rate of interest.

9.1 TRIGONOMETRIC RATIOS

9.1.1 REVIEW OF TRIGONOMETRIC FUNCTIONS FOR RIGHT-ANGLED TRIANGLES

The trigonometric functions are defined as **ratio functions** in a right-angled triangle. As such they are often referred to as the **trigonometric ratios**.

The trigonometric ratios are based on the right-angled triangle shown alongside. Such right-angled triangles are defined in reference to a nominated angle. In the right-angled triangle ABC the longest side [AB] (opposite the right-angle) is the **hypotenuse**. Relative to the angle $\angle BAC$ of size $\theta°$, the side BC is called the **opposite** side while the side AC is called the **adjacent** side.

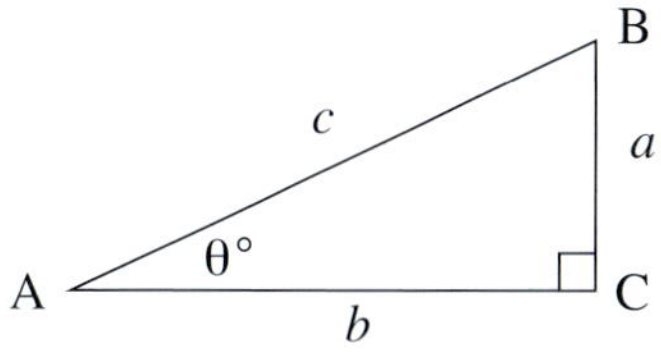

The trigonometric ratios are defined as

$$\sin\theta° = \frac{\text{opposite}}{\text{hypotenuse}} = \frac{a}{c}, \quad \cos\theta° = \frac{\text{adjacent}}{\text{hypotenuse}} = \frac{b}{c}, \quad \tan\theta° = \frac{\text{opposite}}{\text{adjacent}} = \frac{a}{b}$$

Note then, that $\tan\theta° = \frac{\sin\theta°}{\cos\theta°}, \cos\theta \neq 0$.

There also exists another important relation between the side lengths of a right-angled triangle.

This relationship, using **Pythagoras's Theorem** is $a^2 + b^2 = c^2$

Do not forget to adjust the mode of your calculator to degree mode when necessary. On the TI–83, this is done by pressing **MODE** and then selecting the **Degree** mode. As angles can be quoted in degrees '°', minutes '′' and seconds '″' we make use of the **DMS** option under the **ANGLE** menu (accessed by pressing **2nd APPS**) to convert an angle quoted as a decimal into one quoted in degrees, minutes and seconds.

9.1.2 EXACT VALUES

There are a number of special right-angled triangles for which exact values of the trigonometric ratios exist. Two such triangles are shown below:

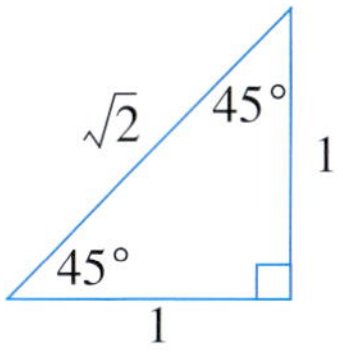

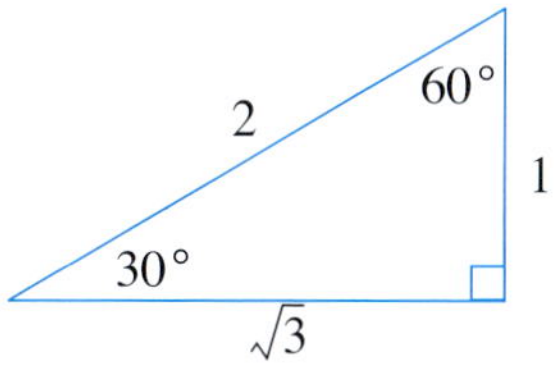

From these triangles we can tabulate the trigonometric ratios as follows:

θ	$\sin\theta°$	$\cos\theta°$	$\tan\theta°$
30	$\frac{1}{2}$	$\frac{\sqrt{3}}{2}$	$\frac{1}{\sqrt{3}}$
45	$\frac{1}{\sqrt{2}}$	$\frac{1}{\sqrt{2}}$	1
60	$\frac{\sqrt{3}}{2}$	$\frac{1}{2}$	$\sqrt{3}$

EXAMPLE 9.1

Find x in each of the following triangles (correct to 4 d.p)

(a)

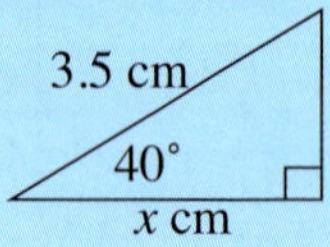

(b)

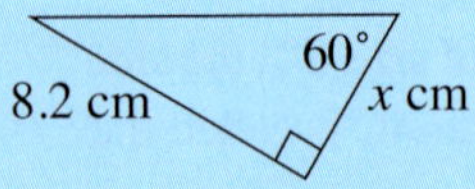

(c)

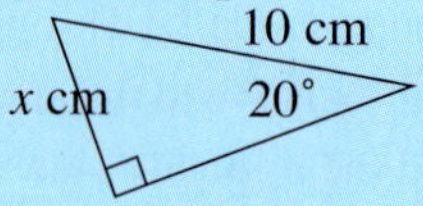

Solution

(a) We label the sides relative to the given angle:
As the sides involved are the adjacent (adj) and the hypotenuse (hyp). The appropriate ratio is the cosine ratio, i.e., $\cos\theta = \frac{\text{adj}}{\text{hyp}}$. Then, substituting the information into the expression we can solve for x:

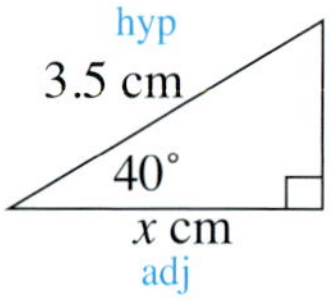

$$\cos 40° = \frac{x}{3.5} \Leftrightarrow 3.5 \times \cos 40° = x$$

$$\therefore x = 2.6812 \text{ (to 4 d.p)}$$

A quick word about using the TI–83. Below we show that depending on the mode setting we obtain different values. In particular, note that in Case B, even though the mode setting was in radians, we were able to over ride this by using the degree measure, '°', under the **ANGLE** menu.

Case A:

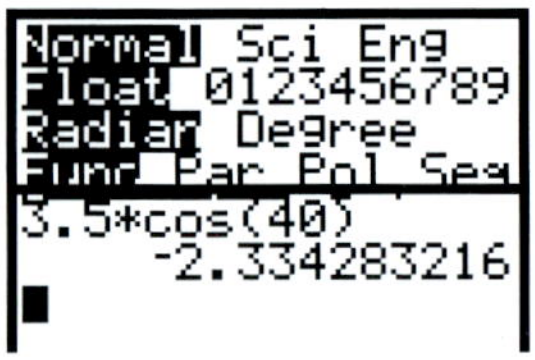

Case B:

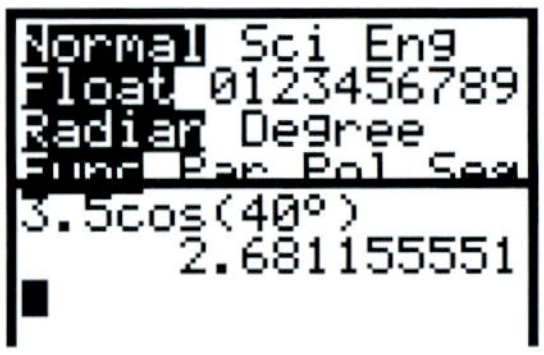

Case C:

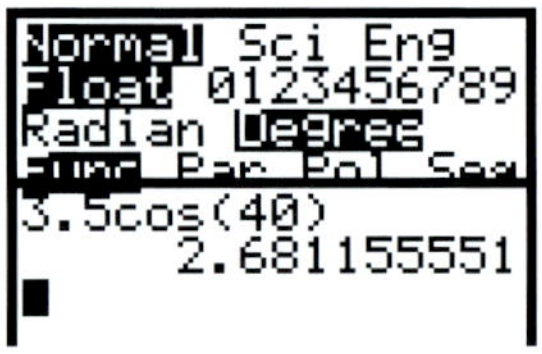

(b) We label the sides relative to the given angle.
The sides involved are the adjacent (adj) and the opposite (opp) the appropriate ratio is the tangent ratio, i.e., $\tan\theta = \frac{\text{opp}}{\text{adj}}$. Then, substituting the information into the expression we can solve for x:

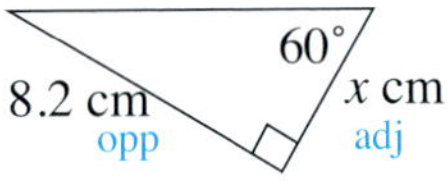

$$\tan 60° = \frac{8.2}{x} \Leftrightarrow x\tan 60° = 8.2 \Leftrightarrow x = \frac{8.2}{\tan 60°}$$

$$\therefore x = 4.7343 \text{ (to 4 d.p)}$$

(c) We label the sides relative to the given angle.
The sides involved are the opposite (opp) and the hypotenuse (hyp).
The appropriate ratio is the sine ratio, i.e., $\sin\theta = \frac{\text{opp}}{\text{hyp}}$. Then, substituting the information into the expression we can solve for x:

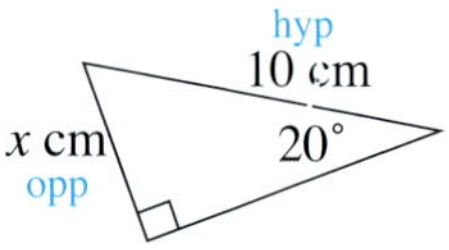

$$\sin 20° = \frac{x}{10} \Leftrightarrow 10 \times \sin 20° = x$$

$$\therefore x = 3.4202 \text{ (to 4 d.p)}$$

EXAMPLE 9.2

Find x and θ in the following triangles

(a)

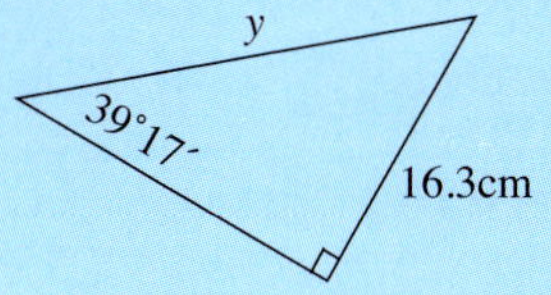

(b)

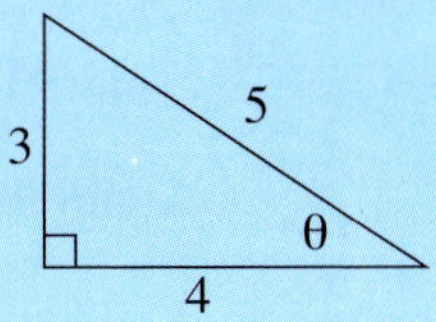

Solution

(a) The important sides are the opposite and hypotenuse. So,

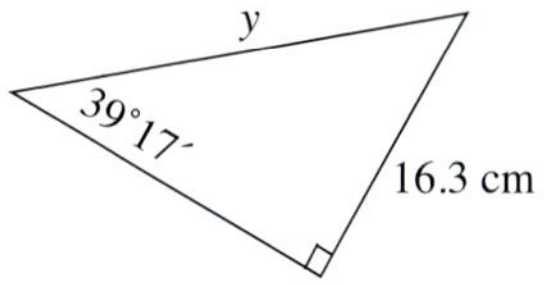

$$\sin 39°17' = \frac{16.3}{y} \Leftrightarrow y \times \sin 39°17' = 16.3$$

$$\Leftrightarrow y = \frac{16.3}{\sin 39°17'}$$

$$\therefore y = 25.7 \text{ cm}$$

The TI–82/83 calculators accept angle inputs using the **2nd ANGLE** menu. **Option 1** allows entry of angles in degrees irrespective of the **MODE** setting of the calculator. **Option 2** allows the entry of degrees, minutes, seconds.

The problem would be solved using the keying sequence

16.3÷sin 39 **2nd ANGLE 1** 17 **2nd ANGLE 2** ENTER.

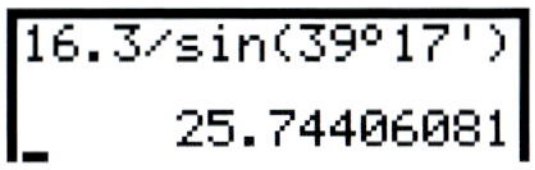

(b) When using a calculator to find an angle, **option 4** of the **2nd ANGLE** menu will allow you to display an answer in degree, minute, second format.

Any of the three trigonometric ratios will do, but when finding angles, it is generally best to use the cosine ratio. The reason for this should become apparent as this chapter progresses.

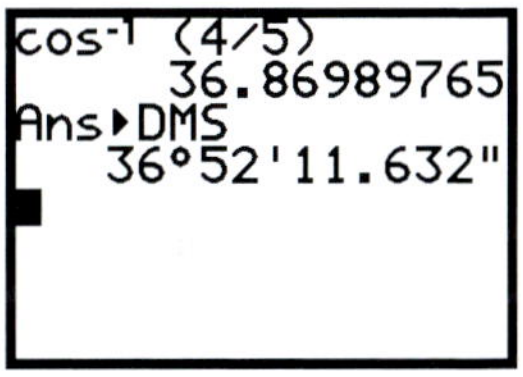

$$\cos\theta = \frac{4}{5} \Rightarrow \theta = \cos^{-1}\left(\frac{4}{5}\right) \therefore \theta \approx 36°52'12''$$

rounded to the nearest second.

EXAMPLE 9.3

Using the triangle shown, find

(a) i. AB

ii. $\cos\alpha$

iii. $\tan\alpha$

(b) If $\cos\alpha = 0.2$ find $\sin(90° - \alpha)$.

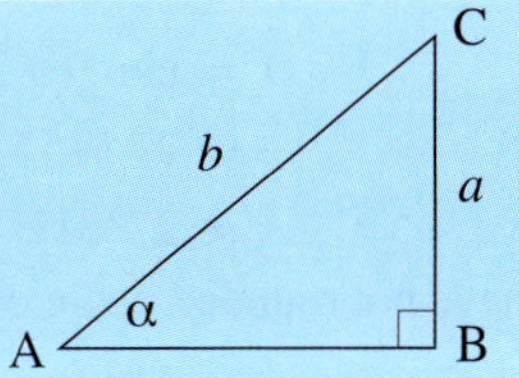

Solution

(a) Using Pythagoras's Theorem we have $AC^2 = AB^2 + BC^2 \therefore b^2 = AB^2 + a^2$

$$\Leftrightarrow AB^2 = b^2 - a^2$$

$$\Rightarrow AB = \sqrt{b^2 - a^2}$$

(b) $\cos\alpha = \dfrac{AB}{AC} = \dfrac{\sqrt{b^2 - a^2}}{b}$

(c) $\tan\alpha = \dfrac{CB}{AB} = \dfrac{a}{\sqrt{b^2 - a^2}}$

(d) $\sin(90° - \alpha) = \dfrac{AB}{AC}$, but $\cos\alpha = \dfrac{AB}{AC} \therefore \sin(90° - \alpha) = \cos\alpha$.

$$\text{i.e., } \sin(90° - \alpha) = 0.2$$

Often we are dealing with non right-angled triangles. However, these can be 'broken up' into at least two right-angled triangles, which then involves solving simultaneous equations. This is illustrated in the next example.

EXAMPLE 9.4

Find the angle θ in the diagram shown.

Note that $\angle ACB \neq 90°$.

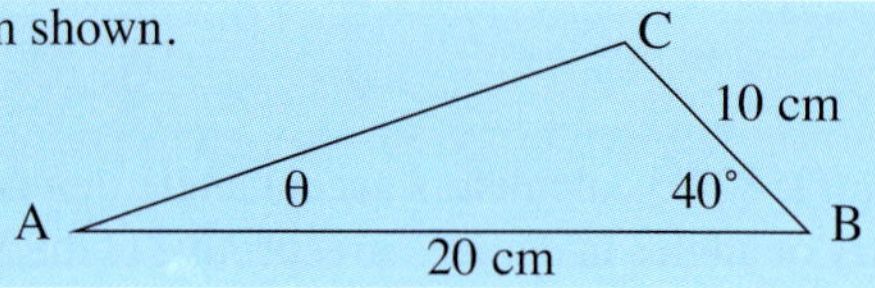

Solution

We start by 'breaking-up' the triangle into two right-angled triangles as follows:

Using ΔACP:

$$\tan\theta = \frac{PC}{AP} = \frac{y}{20 - x} \quad -(1)$$

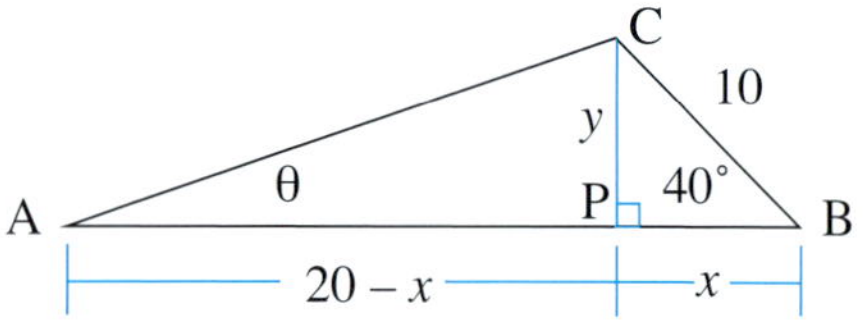

We now need to determine x and y.

Using ΔBPC:

$$\sin 40° = \frac{PC}{BC} = \frac{y}{10}$$

$$\Leftrightarrow y = 10\sin 40° \quad -(2)$$

and $\quad \cos 40° = \dfrac{BP}{BC} = \dfrac{x}{10}$

$$\Leftrightarrow x = 10\cos 40° \quad -(3)$$

Therefore, substituting (3) and (2) into (1) we have:

$$\tan\theta = \frac{10\sin 40°}{20 - 10\cos 40°}$$

$$= 0.5209$$

$$\therefore \theta = \tan^{-1}(0.5209)$$

$$= 27.5157$$

$$= 27°31'$$

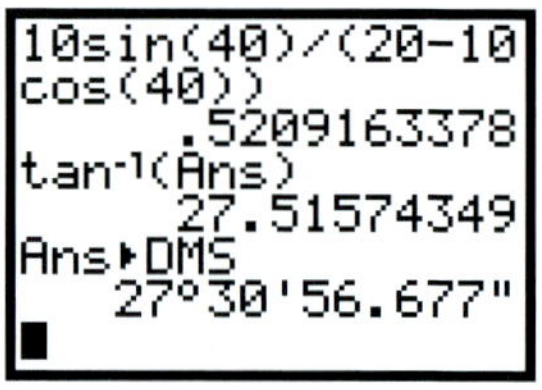

Note that we have not rounded down our answer until the very last step.

EXERCISES 9.1

1. The parts of this question refer to the triangle shown. Complete the blank spaces in this table, giving lengths correct to three significant figures and angles correct to the nearest degree.

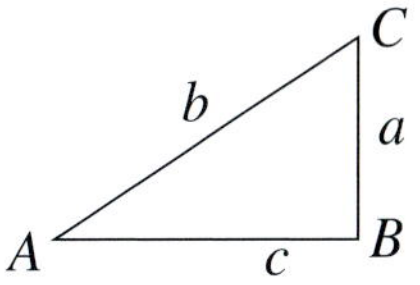

	a cm	*b* cm	*c* cm	*A*	*B*	*C*
1.			1.6		90°	23°
2.		98.3			90°	34°
3.			33.9		90°	46°
4.		30.7			90°	87°
5.	2.3				90°	33°
6.		77			90°	51°
7.	44.4		68.4		90°	57°
8.			12.7	13°	90°	
9.		94.4		52°	90°	
10.	71.8		64.6	48°	90°	
11.		34.1		43°	90°	
12.			2.3	87°	90°	
13.	71.5			63°	90°	
14.	33.5		6.5		90°	
15.	6.1	7.2			90°	
16.		30	7.3		90°	
17.	29.0		2.0		90°	
18.	34.5	88.2			90°	
19.	24.0	29.7			90°	
20.		46.2			90°	27°
21.	59.6		41.8		90°	35°
22.		6.8			90°	37°
23.			14.9	41°	90°	49°
24.			16.1	41°	90°	49°
25.			33.3	68°	90°	22°

2. Find the exact value of x in each of the following

(a)

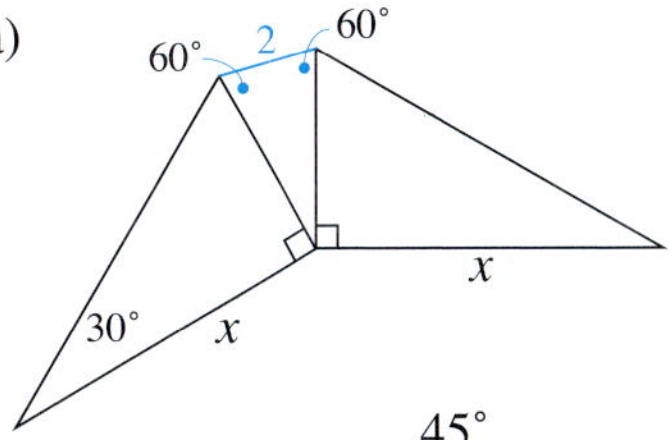

(b)

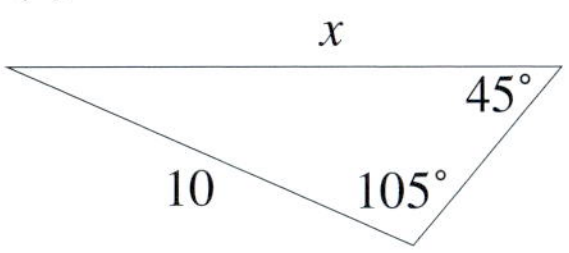

(c)

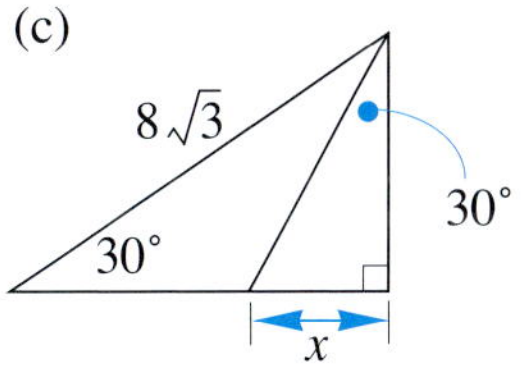

(d)

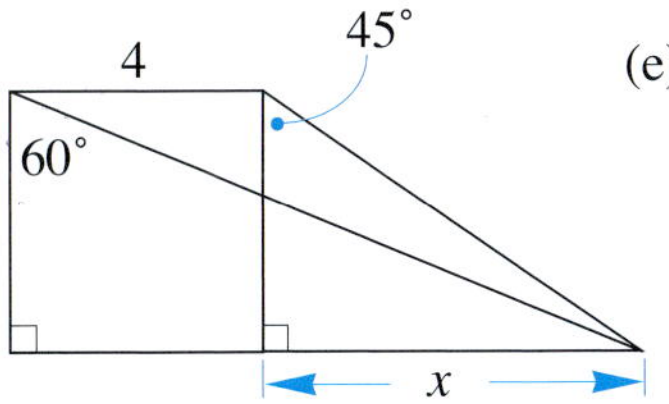

(e)

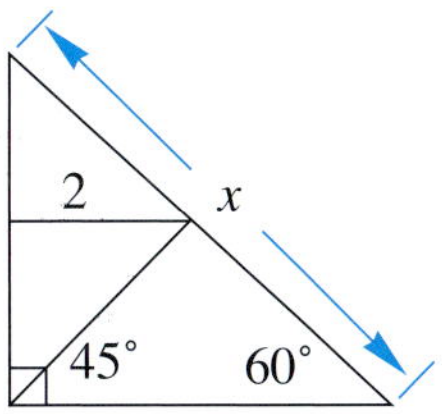

(f)

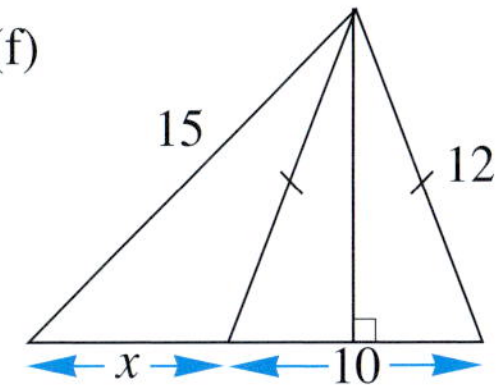

3. Using the triangle on the right, show that

(a) $\sin(90° - \theta) = \cos\theta$

(b) $\cos(90° - \theta) = \sin\theta$

(c) $\tan(90° - \theta) = \dfrac{1}{\tan\theta}$

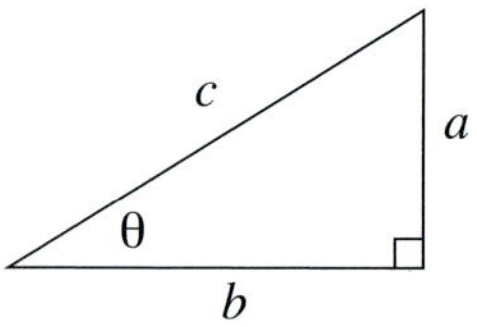

4. Find the exact value of x in each of the following.

(a)

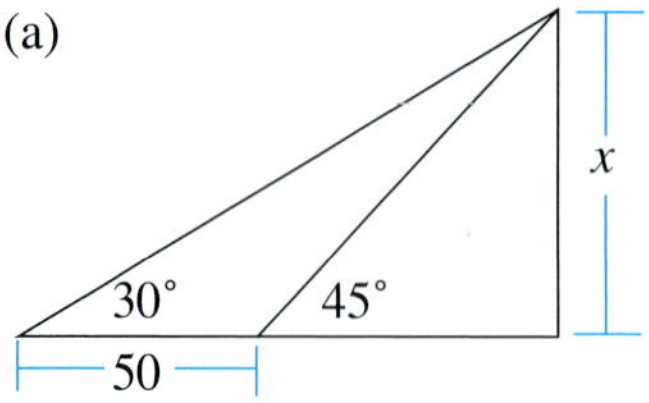

(b)

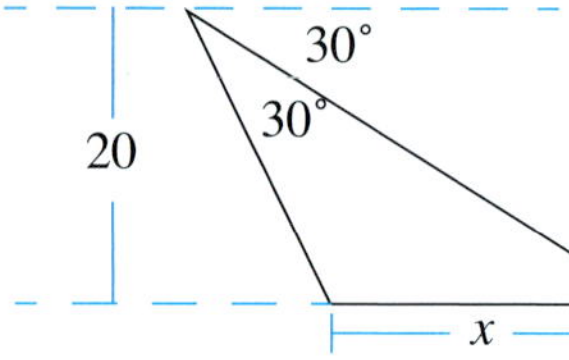

5. Show that $OZ = \dfrac{x\tan\theta(\sin\alpha + \cos\alpha\tan\beta)}{\tan\theta - \tan\beta}$

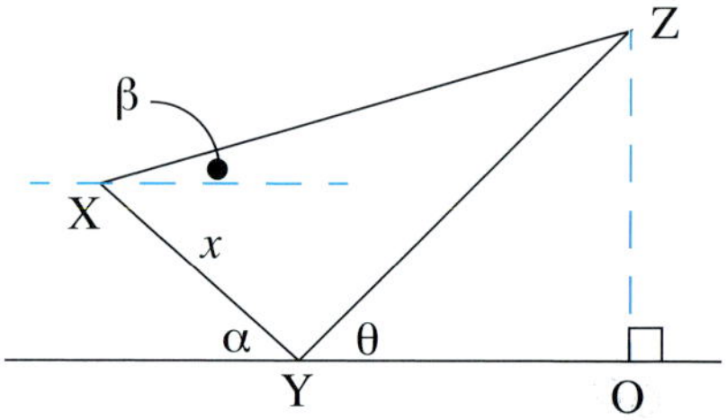

9.2 APPLICATIONS

Applications that require the use of trigonometric ratios and right-angled triangles are many and varied. In this section we consider a number of standard applications to highlight this.

9.2.1 ANGLE OF ELEVATION AND DEPRESSION

The **angle of elevation** is the angle of the line of sight **above** the horizontal of an object seen above the horizontal

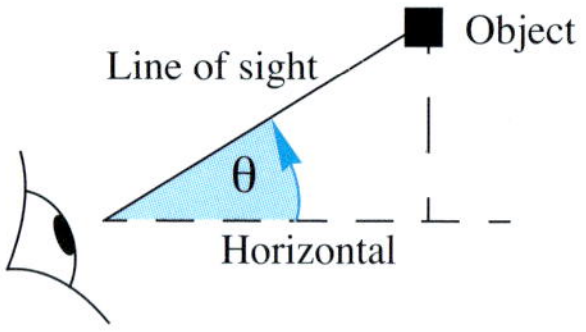

The **angle of depression** is the angle of the line of sight **below** the horizontal of an object seen below the horizontal

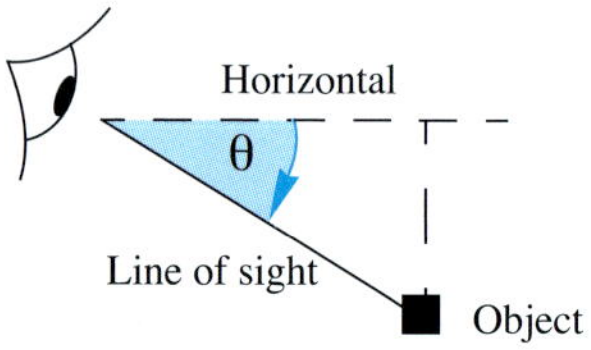

Note that the angle of depression and elevation for the same line of sight are **alternate angles**.

EXAMPLE 9.5

An observer standing on the edge of a cliff 82 m above sea level sees a ship at an angle of depression of 26°. How far from the base of the cliff is the ship situated?

Solution

We first draw a diagram to represent this situation:
Let the ship be at point S, x metres from the base of the cliff, B, and let O be where the observer is standing.

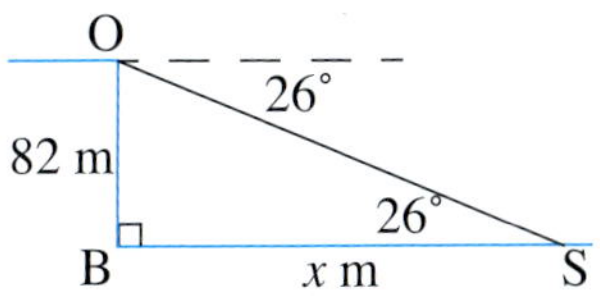

Using the right-angled triangle OBS we have:

$$\tan 26° = \frac{82}{x} \Leftrightarrow x\tan 26° = 82$$

$$\Leftrightarrow x = \frac{82}{\tan 26°}$$

$$= 168.1249\ldots$$

Therefore, the ship is 168 m from the base of the cliff.

EXAMPLE 9.6

The angle of depression from the roof of building A to the foot of a second building, B, across the same street and 40 metres away is 65°. The angle of elevation of the roof of building B to the roof of building A is 35°. How tall is building B?

Solution

Let the height of building B be x m and that of building A be y m.

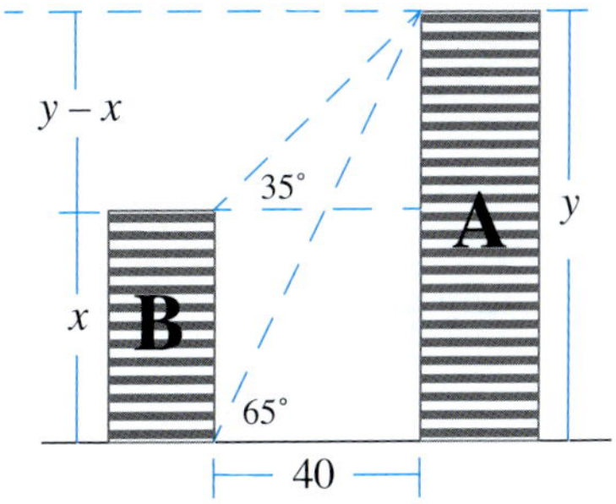

Note that we are using the fact that for the same line of sight, the angle of depression equals the angle of elevation.

The height difference between the two buildings must then be $(y - x)$ m.

We now have two right-angled triangles to work with:

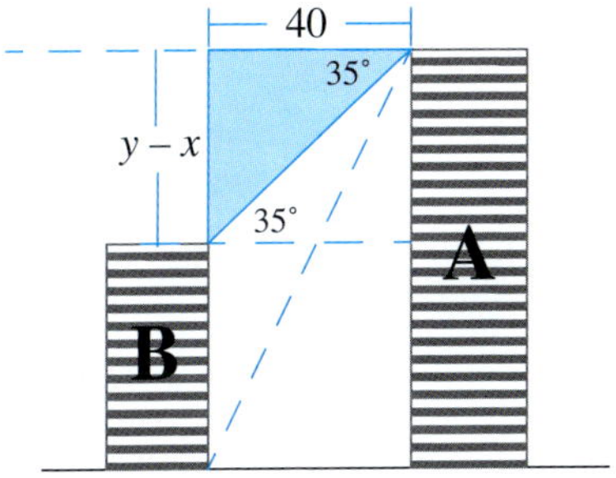

$$\tan 35° = \frac{y-x}{40}$$

$$\Leftrightarrow y - x = 40\tan 35° \quad -(1)$$

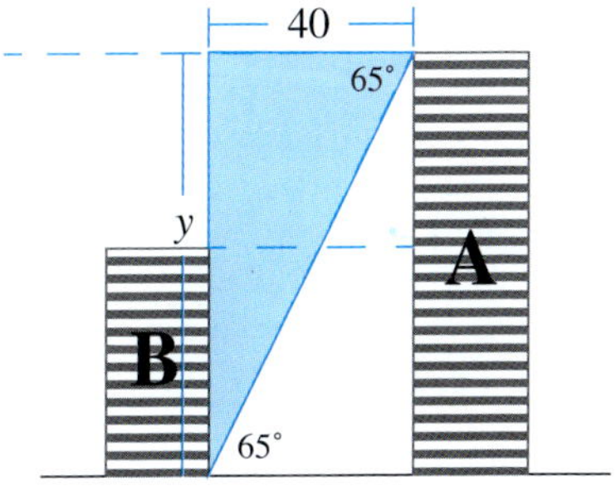

$$\tan 65° = \frac{y}{40}$$

$$\Leftrightarrow y = 40\tan 65° \quad -(2)$$

Substituting (2) into (1) we have:

$$40\tan 65° - x = 40\tan 35°$$

$$\Leftrightarrow x = 40\tan 65° - 40\tan 35°$$

$$\therefore x = 57.7719\ldots$$

That is, building B is 57.77 m

9.2.2 BEARINGS

In the sport of orienteering, participants need to be skilled in handling bearings and reading a compass. Bearings can be quoted by making reference to the North, South, East and West directions or using true bearings. We look at each of these.

Compass bearings

These are quoted in terms of an angle measured East, West, North or South, or somewhere in between. For example, North 30° East, expressed as N30°E, informs us that from the North direction we rotate 30° towards the East and then follow that line of direction. The following diagrams display this for a number of bearings.

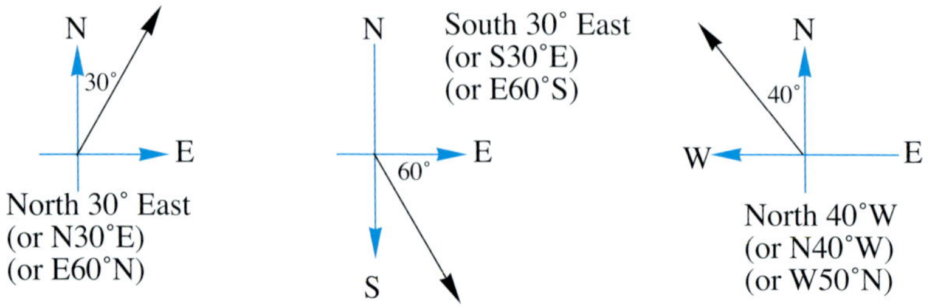

True bearings

These are quoted in terms of an angle measured in a clockwise direction from North (and sometimes a capital T is attached to the angle to highlight this fact). So, for example, a bearing of 030°T would represent a bearing of 30° in a clockwise direction from the North – this corresponds to a compass bearing of N30°E. Using the above compass bearings we quote the equivalent true bearings:

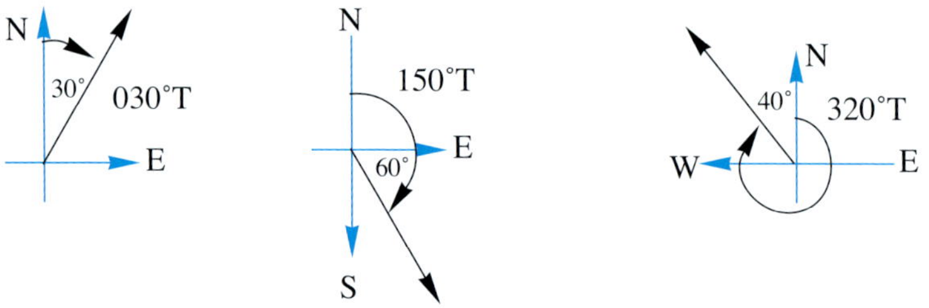

EXAMPLE 9.7

Janette walks for 8 km North-East and then 11 km South-East. Find the distance and bearing from her starting point.

Solution

First step is to draw a diagram.

As $\angle OAB = 90°$ we can make use of Pythagoras's Theorem:

$$x^2 = 8^2 + 11^2$$

$$\therefore x = 13.60 \text{ [taking +ve square root]}$$

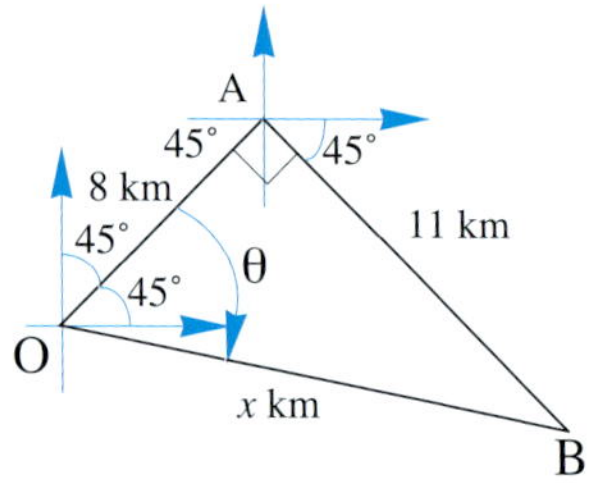

Let $\theta = \angle AOB$ so that $\tan\theta = \frac{11}{8}$. $\therefore \theta = \tan^{-1}\left(\frac{11}{8}\right)$

$$= 53.97°$$

Therefore, bearing is $45° + \theta = 45° + 53.97° = 98.97°$

That is, B has a bearing of 98.97°T from O and is (approx) 13.6 km away.

EXAMPLE 9.8

The lookout, on a ship sailing due East, observes a light on a bearing of 056°T. After the ship has travelled 4.5 km, the lookout now observes the light to be on a bearing of 022°T. How far is the light source from the ship at its second sighting?

Solution

As always, we start with a diagram.

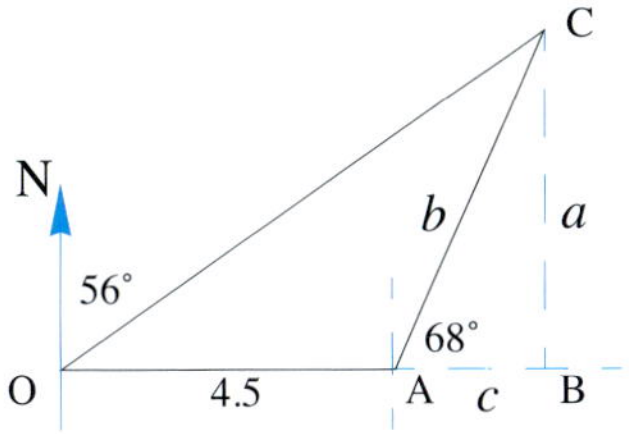

Using ΔOBC we have, $\tan 34° = \dfrac{BC}{OB} = \dfrac{a}{4.5+c}$

$$\therefore a = (4.5+c)\tan 34° \quad -(1)$$

Using ΔABC we have, $\tan 68° = \dfrac{BC}{AB} = \dfrac{a}{c}$

$$\therefore a = c\tan 68° \quad -(2)$$

Equating (1) and (2) we have, $c\tan 68° = (4.5+c)\tan 34°$

$$c\tan 68° = 4.5\tan 34° + c\tan 34°$$

$$\Leftrightarrow c(\tan 68° - \tan 34°) = 4.5\tan 34°$$

$$\Leftrightarrow c = \frac{4.5\tan 34°}{(\tan 68° - \tan 34°)}$$

$$\therefore c = 1.6857$$

Substituting this result into (2) we have,

$$a = \frac{4.5\tan 34°}{(\tan 68° - \tan 34°)} \times \tan 68°$$

$$\therefore a = 4.1723$$

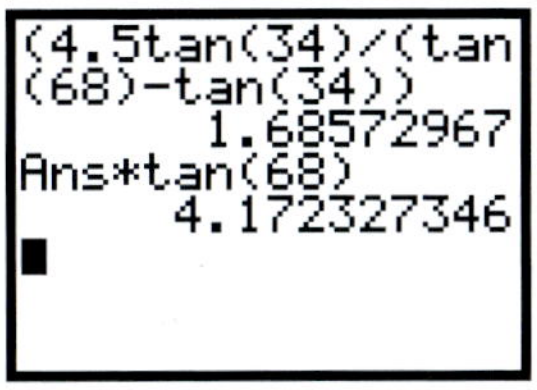

Then, using ΔABC and Pythagoras's Theorem, we have

$$b^2 = a^2 + c^2$$

$$= 4.1723^2 + 1.6857^2$$

$$\therefore b = \sqrt{20.2496}$$

$$= 4.4999$$

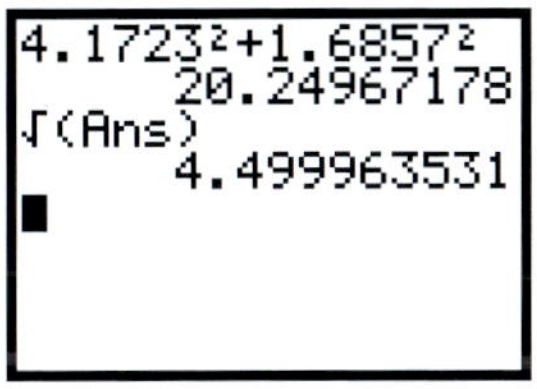

That is, the light is 4.5 km from the ship (at the second sighting).
Can you see a much quicker solution? Hint – think isosceles triangle.

EXERCISES 9.2

1. (a) Change the following compass bearings into true bearings

i. N30°E ii. N30°W iii. S15°W iv. W70°S

(b) Change the following true bearings into compass bearings

i. 025°T ii. 180°T iii. 220°T iv. 350°T

2. The angle of depression from the top of a building 60 m high to a swing in the local playground is 58°. How far is the swing from the foot of the building?

3. From a point A on the ground, the angle of elevation to the top of a tree is 52°. If the tree is 14.8 m away from point A, find the height of the tree.

4. Find the angle of elevation from a bench to the top of an 80 m high building if the bench is 105 m from the foot of the building.

5. Patrick runs in a direction N60°E and after 45 minutes finds himself 3900 m North of his starting point. What is Patrick's average speed in ms^{-1}.

6. A ship leaves Oldport and heads NW. After covering a distance of 16 km it heads in a direction of N68°22'W travelling a distance of 22 km where it drops anchor. Find the ship's distance and bearing from Oldport after dropping anchor.

7. From two positions 400 m apart on a straight road, running in a northerly direction, the bearings of a tree are N36°40'E and E33°22'S. What is the shortest distance from the tree to the road?

8. A lamp post leaning away from the sun and at 6° from the vertical, casts a shadow 12 m long when the sun's angle of elevation is 44°. Assuming that the level of the ground where the pole is situated is horizontal, find its length.

9. From a window, 29.6 m above the ground, the angle of elevation of the top of a building is 42°, while the angle of depression to the foot of the building is 32°. Find the height of the building.

10. Two towns P and Q are 50 km apart, with P due west of Q. The bearing of a station from town P is 040°T while the bearing of the station from town Q is 300°T. How far is the station from town P?

11. When the sun is 74° above the horizon, a vertical flagpole casts a shadow 8.5 m onto a horizontal ground. Find the shadow cast by the sun when it falls to 62° above the horizontal.

12. A hiker walks for 5km on a bearing of 053° true (North 53° East). She then turns and walks for another 3km on a bearing of 107° true (East 17° South).

(a) Find the distance that the hiker travels North/South and the distance that she travels East/West on the first part of her hike.

(b) Find the distance that the hiker travels North/South and the distance that she travels East/West on the second part of her hike.

(c) Hence find the total distance that the hiker travels North/South and the distance that she travels East/West on her hike.

(d) If the hiker intends to return directly to the point at which she started her hike, on what bearing should she walk and how far will she have to walk?

13. A surveying team are trying to find the height of a hill. They take a 'sight' on the top of the hill and find that the angle of elevation is 23°27´. They move a distance of 250 metres on level ground directly away from the hill and take a second 'sight'. From this point, the angle of elevation is 19°46´. Find the height of the hill, correct to the nearest metre.

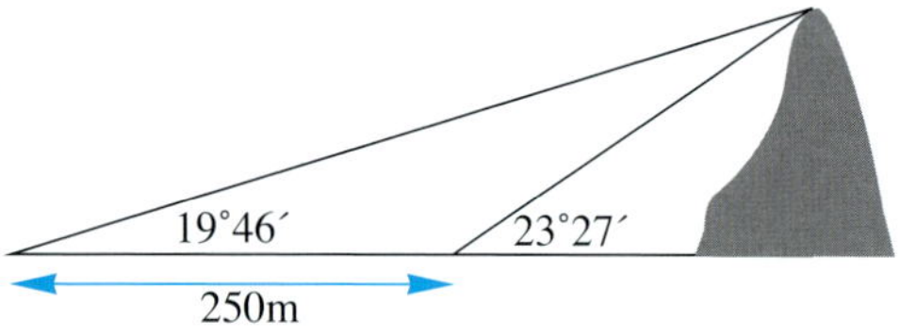

9.3 RIGHT ANGLES IN 3–DIMENSIONS

When dealing with problems in three dimensions, we draw the figures in perspective, so that a model can be more accurately visualised. This does not mean that you must be an artist, simply that you take a little time (and a lot of practice) when drawing such diagrams. The key to many 3–D problems is locating the relevant right–angled triangles within the diagram. Once this is done, all of the trigonometric work that has been covered in the previous two sections can be applied. As such, we will not be learning new theory, but rather developing new drawing and modelling skills. Some typical examples of solids that may be encountered are:

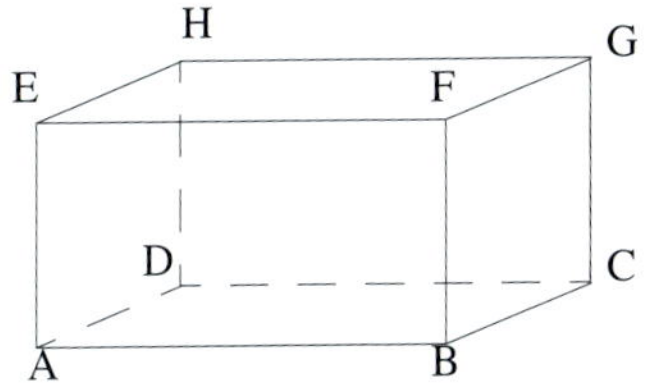

Cuboid ABCD, EFGH

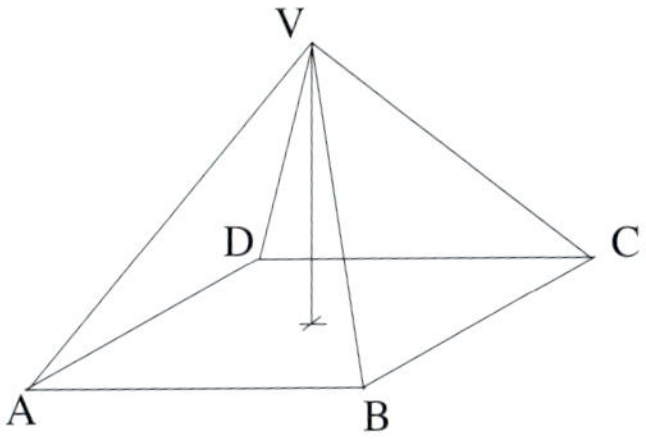

Right Pyramid V, ABCD

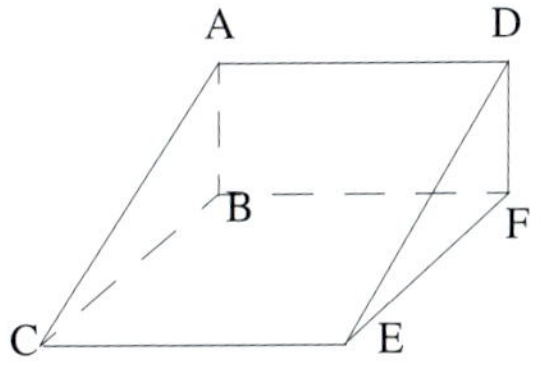

Wedge ABC, DEF

We look at two basic concepts and drawing techniques to help us.

1. A line and a plane:

A line will always cut a plane at some point (unless the line is parallel to the plane). To find the angle between a line and a plane construct a perpendicular from the line to the plane and complete a right–angled triangle. In our diagram, we have that the segment $[AB]$ is projected onto the plane. A perpendicular, $[BC]$ is drawn, so that a right–angled triangle, ABC is completed. The angle that the line then makes with the plane is given by θ (which can be found by using the trig–ratios).

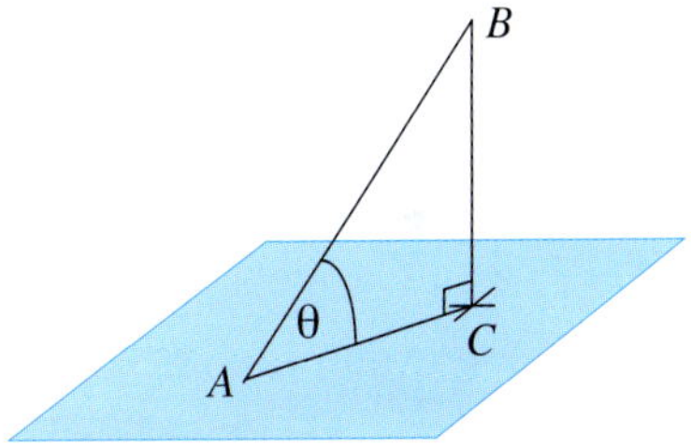

2. A plane and a plane:

To find the angle between two planes ABCD and ABEF (assuming that they intersect), take any point P on the intersecting segment $[AB]$ and draw $[PQ]$ and $[PR]$ on each plane in such a way that they are perpendicular to $[AB]$. Then, the angle between $[PQ]$ and $[PR]$ (θ) is the angle between the two planes.

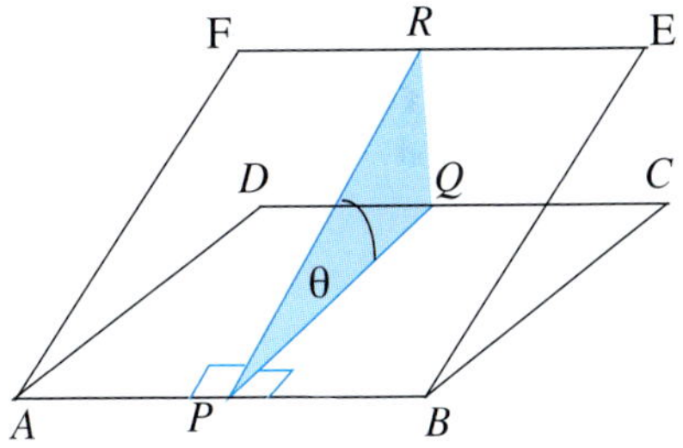

EXAMPLE 9.9

A cube ABCD, EFGH has a side length measuring 6 cm.

(a) Find the length of the segment $[AC]$.

(b) The length of the diagonal $[AG]$.

(c) The angle that the diagonal $[AG]$ makes with the base.

Solution

First we need to draw a cube:

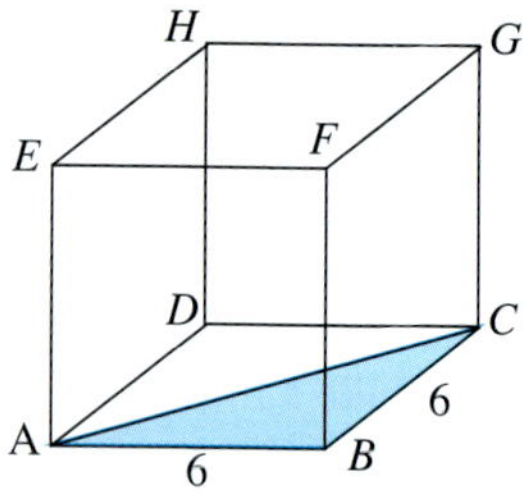

(a) Now the base of the cube is a square, so that $\angle ABC = 90°$, i.e., we have a right–angled triangle, so we can use Pythagoras's Theorem:

$$AC^2 = AB^2 + BC^2$$
$$= 6^2 + 6^2$$
$$= 72$$
$$\therefore AC = \sqrt{72} \approx 8.49$$

(b) This time we have that $\angle ACG = 90°$, therefore,

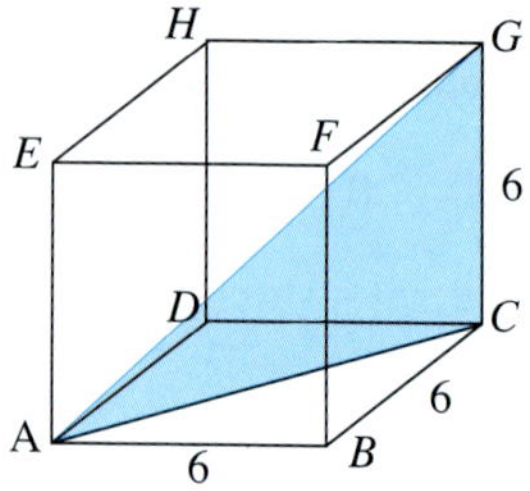

$$AG^2 = AC^2 + CG^2$$
$$= (\sqrt{72})^2 + 6^2$$
$$= 108$$
$$\therefore AC = \sqrt{108} \approx 10.39$$

(c) Using triangle ACG, $\tan\theta = \frac{CG}{AC} \therefore \tan\theta = \frac{6}{\sqrt{72}}$

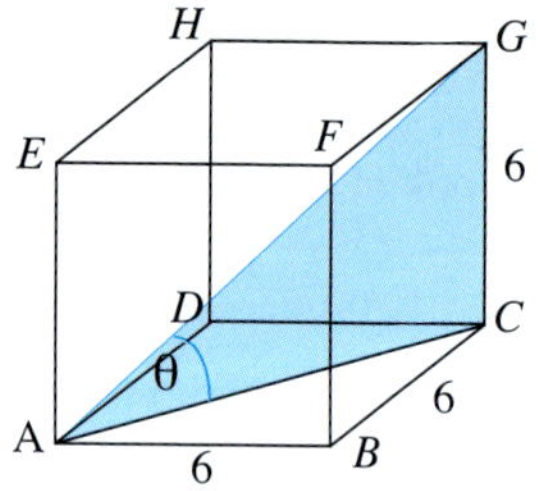

$$\theta = \tan^{-1}\left(\frac{6}{\sqrt{72}}\right)$$
$$= 35.26°$$
$$= 35°16'$$

EXAMPLE 9.10

From a point X, 200 m due South of a cliff, the angle of elevation of the top of the cliff is 30°. From a point Y, due East of the cliff, the angle of elevation of the top of the cliff is 20°. How far apart are the points X and Y?

Solution

We start by illustrating this information on a 3–D diagram (Note that North–South and West–East are drawn on a plane. It is necessary to do this otherwise the diagram will not make sense).

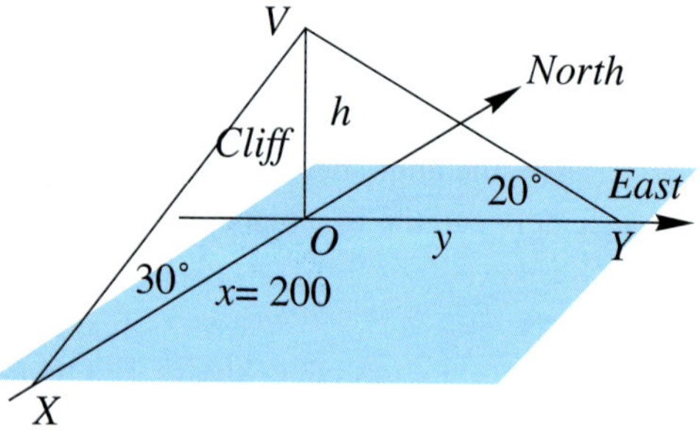

Let the cliff be h metres high. The distance from X to the base of the cliff be x metres and the distance from Y to the base of the cliff be y metres.

As $\angle XOY = 90°$, then $XY^2 = x^2 + y^2$

$$= 200^2 + y^2$$

But, $\tan 20° = \frac{h}{y}$, of which we know neither h or y.

However, using triangle XOV, we have that $\tan 30° = \frac{h}{200} \Rightarrow h = 200 \times \tan 30°$.

Therefore, we have that $\tan 20° = \dfrac{200 \times \tan 30°}{y} \Leftrightarrow y = \dfrac{200 \times \tan 30°}{\tan 20°}$

That is, $y = 317.25$

Therefore, $XY^2 = x^2 + y^2 = 200^2 + \left(\dfrac{200 \times \tan 30°}{\tan 20°}\right)^2$

$= 140648.4289$

$XY = 375.0312$.

That is, X and Y are approximately 375 m apart.

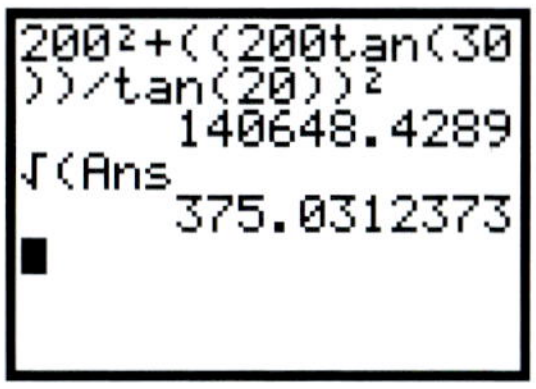

EXERCISES 9.3

1. For the diagram shown, determine the angle of inclination between the plane

(a) ABCD and the base, EABH (Figure 1).

(b) ABC and the base EBFA (Figure 2).

Figure 1.

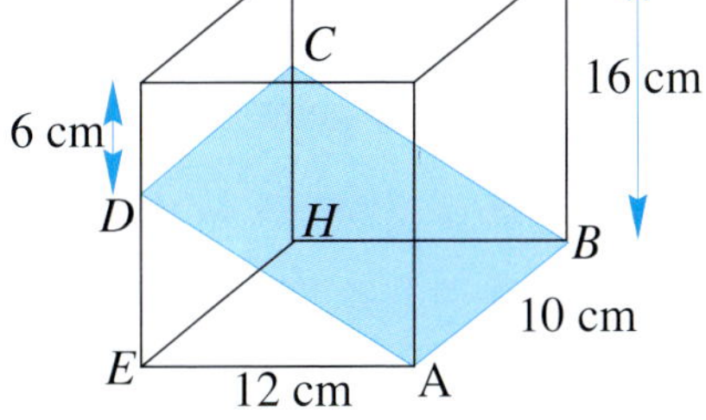

Figure 2.

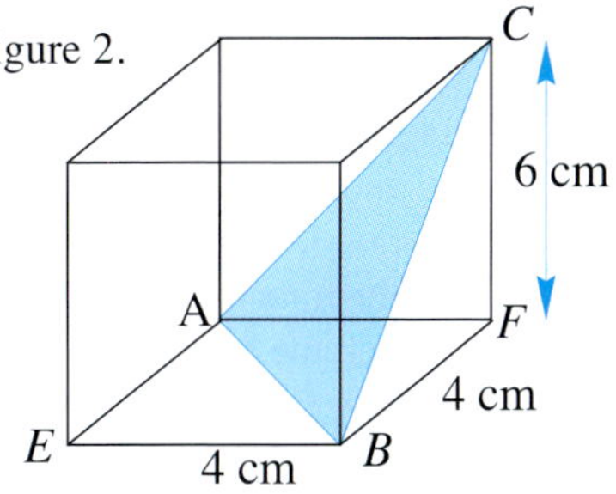

2. A right pyramid with a rectangular base and a vertical height of 60 cm is shown in the diagram alongside. The points X and Y are the midpoints of the sides [AB] *and* [BC] *respectively*

Find

(a) the length, AP.

(b) the length of the edge [AV].

(c) the angle that the edge AV makes with the base ABCD.

(d) the length, YV.

(e) The angle that the plane BCV makes with the base.

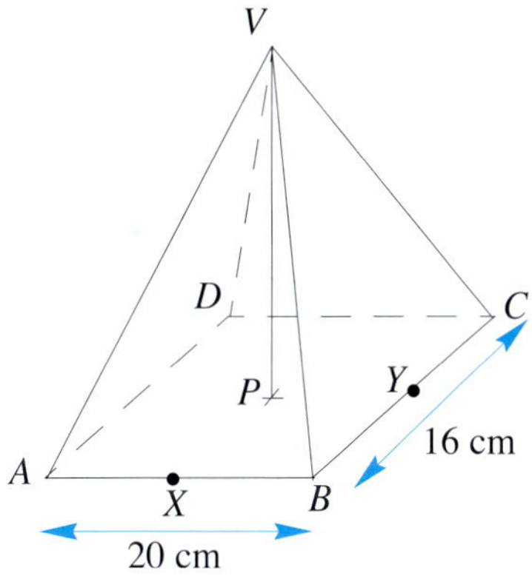

3. The diagram alongside shows a rectangular box with side lengths AB = 8 cm, BC = 6 cm and CG = 4 cm. Find the angle between

(a) the line [BH] and the plane $ABCD$.

(b) the lines [BH and [BA].

(c) the planes $ADGF$ and $ABCD$.

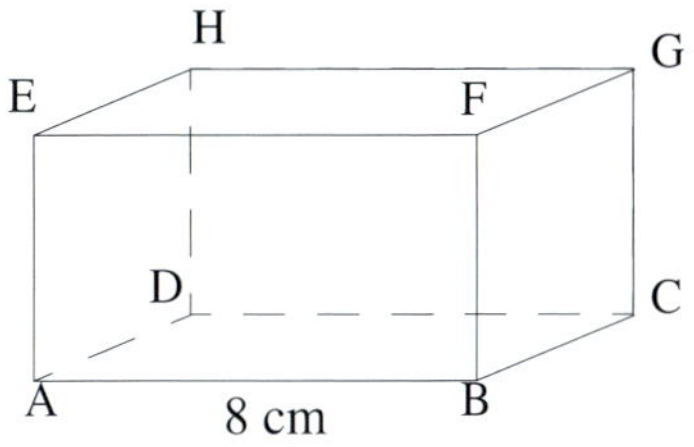

4. For the wedge shown alongside, given that the angle between the lines EA and ED is 50° find

(a) the length of [AE].

(b) the $\angle AEB$.

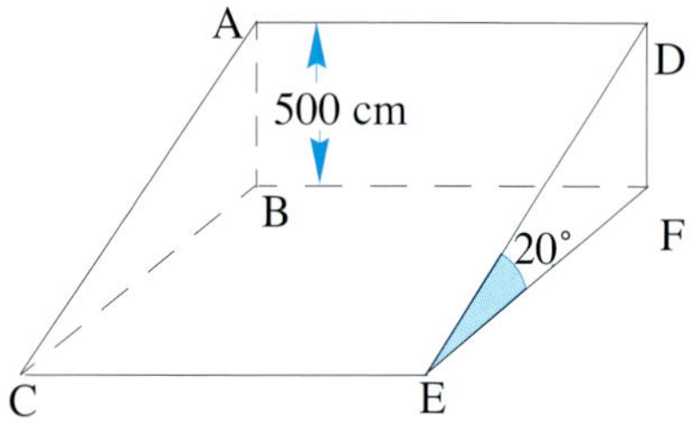

5. From a point A, 100 m due South of a tower, the angle of elevation of the top of the tower is 40°. From a point B, due East of the tower, the angle of elevation of the top of the tower is 20°. How far apart are the points A and B?

6. For the triangular prism shown alongside find

(a) the value of h

(b) the value of α

(c) the angle that the plane ABV makes with the base ABC.

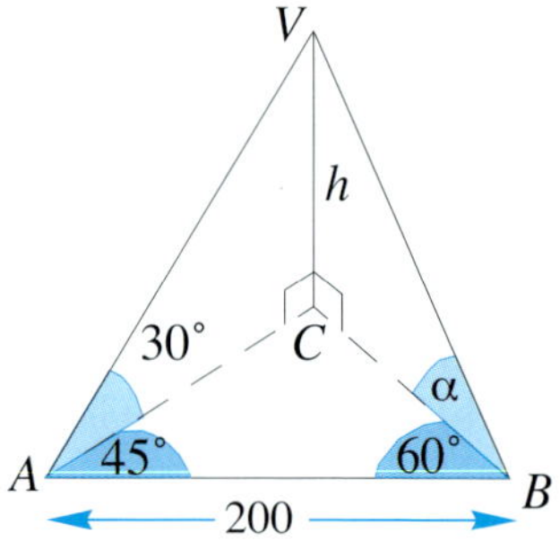

7. The angle of depression from the top of a tower to a point X South of the tower, on the ground and 120 m from the foot of the tower is 24°. From point Y due West of X the angle of elevation to the top of the tower is 19°.

(a) Illustrate this information on a diagram.

(b) Find the height of the tower.

(c) How far is Y from the foot of the tower?

(d) How far apart are the points X and Y?

8. A mast is held in a vertical position by four ropes of length 60 metres. All four ropes are attached at the same point at the top of the mast so that their other ends form the vertices of a square when pegged into the (level) ground. Each piece of rope makes an angle of 54° with the ground.

(a) Illustrate this information on a diagram.

(b) How tall is the mast?

9. A symmetrical sloping roof has dimensions as shown in the diagram. Find

(a) the length of [FM].

(b) the angle between the plane BCEF and the ground.

(c) the angle between the plane ABF and the ground

(d) the total surface area of the roof.

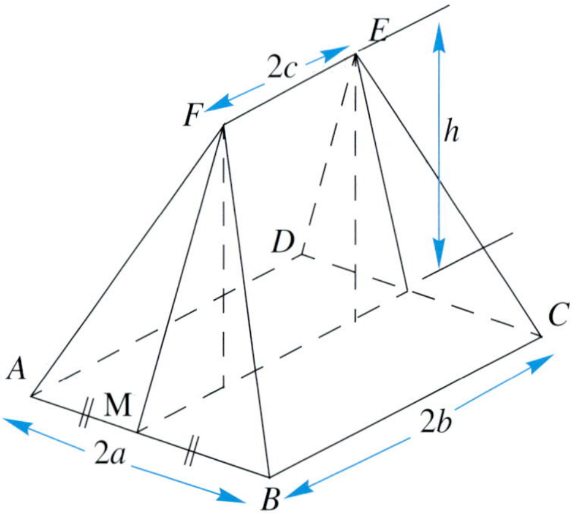

10. The angle of elevation of the top of a tower from a point A due South of it is $68°$. From a point B, due East of A, the angle of elevation of the top is $54°$. If A is 50 m from B, find the height of the tower.

11. A tower has been constructed on the bank of a long straight river. From a bench on the opposite bank and 50 m downstream from the tower, the angle of elevation of the top of the tower is $30°$. From a second bench on the same side of the tower and 100 m upstream from the tower, the angle of elevation of the top of the tower is $20°$. Find
(a) the height of the tower.
(b) the width of the river.

12. A right pyramid of height 10 m stands on a square base of side lengths 5 m. Find
(a) the length of the slant edge.
(b) the angle betwen a sloping face and the base.
(c) the angle between two sloping faces.

13. A camera sits on a tripod with legs 1.5 m long. The feet rest on a horizontal flat surface and form an equilateral triangle of side lengths 0.75 m. Find
(a) the height of the camera above the ground.
(b) the angles made by the legs with the ground.
(c) the angle between the sloping faces formed by the tripod legs.

14. From a point A due South of a mountain, the angle of elevation of the mountaintop is α. When viewed from a point B, x m due East of A, the angle of elevation of the mountaintop is β. Show that the height, h m, of the mountain is given by $h = \dfrac{x\sin\alpha\sin\beta}{\sqrt{\sin^2\alpha - \sin^2\beta}}$.

9.4 AREA OF A TRIANGLE

Given **any** triangle with sides a and b, and the included angle θ, the area, A, is given by

$$A = \frac{1}{2}bh$$

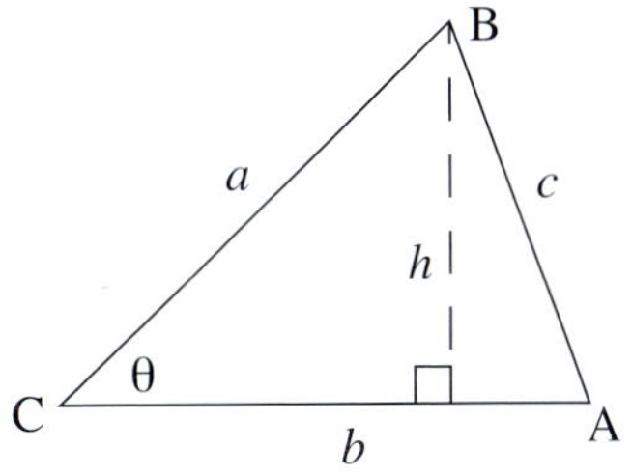

However, $\sin\theta = \frac{h}{a} \Leftrightarrow h = a \times \sin\theta$ and so, we have that

$$A = \frac{1}{2}b \times a \times \sin\theta$$

where θ is the angle between sides a and b.

Note that the triangle need not be a right-angled triangle.

Because of the standard labelling system for triangles, the term $\sin\theta$ is often replaced by $\sin C$, given the expresion Area $= \frac{1}{2}ab\sin C$.

A similar argument can be used to generate the formulas: Area $= \frac{1}{2}bc\sin A = \frac{1}{2}ac\sin B$

EXAMPLE 9.11

Find the area of the triangle PQR given that PQ = 9 cm, QR = 10 cm and $\angle PQR = 40°$.

Solution

Based on the given information we can contruct the following triangle:

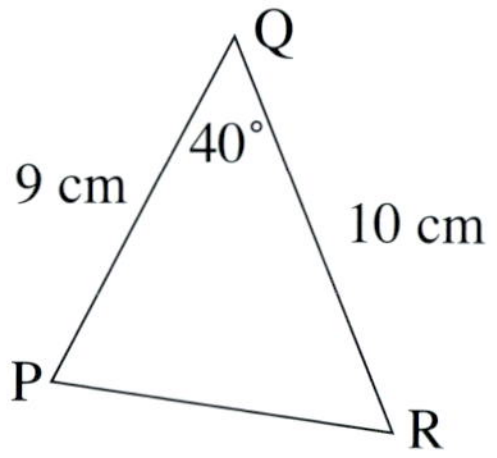

Meaning that, the required area, A, is given by

$$A = \frac{1}{2}ab\sin\theta = \frac{1}{2} \times 9 \times 10 \times \sin 40°$$
$$= 28.9$$

That is, the area is 28.9 cm^2.

EXAMPLE 9.12

The diagram shows a triangular children's playground. Find the area of the playground

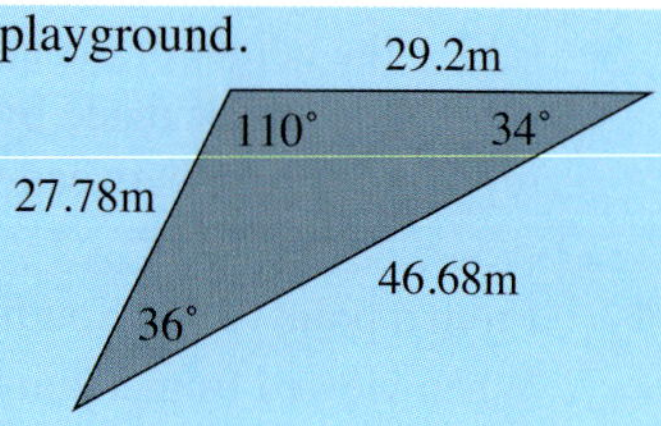

Solution

Since all the measurements of the triangle are known, any one of the three formulas could be used. Many people remember the formula as 'Area equals half the product of the lengths of two sides times the sine of the angle between them'.

$$\text{Area} = \frac{1}{2} \times 27.78 \times 46.68 \times \sin 36° = 381\text{m}^2$$

$$\text{Area} = \frac{1}{2} \times 27.78 \times 29.2 \times \sin 110° = 381\text{m}^2$$

$$\text{Area} = \frac{1}{2} \times 29.2 \times 46.68 \times \sin 34° = 381\text{m}^2$$

EXERCISES 9.4

1. Find the areas of these triangles that are labelled using standard notation.

	a cm	*b* cm	*c* cm	*A*	*B*	*C*
(a)	35.94	128.46	149.70	12°	48°	120°
(b)	35.21	54.55	81.12	20°	32°	128°
(c)	46.35	170.71	186.68	14°	63°	103°
(d)	33.91	159.53	163.10	12°	78°	90°
(e)	42.98	25.07	48.61	62°	31°	87°
(f)	39.88	24.69	34.01	84°	38°	58°
(g)	43.30	30.26	64.94	34°	23°	123°
(h)	12.44	2.33	13.12	68°	10°	102°

	a cm	b cm	c cm	A	B	C
(i)	43.17	46.44	24.15	67°	82°	31°
(j)	23.16	32.71	24.34	45°	87°	48°
(k)	50.00	52.91	38.64	64°	72°	44°
(l)	44.31	17.52	48.77	65°	21°	94°
(m)	12.68	23.49	22.34	32°	79°	69°
(n)	42.37	42.37	68.56	36°	36°	108°
(o)	40.70	15.65	41.26	77°	22°	81°

2. A car park is in the shape of a parallelogram. The lengths of the sides of the car park are given in metres.

What is the area of the car park?

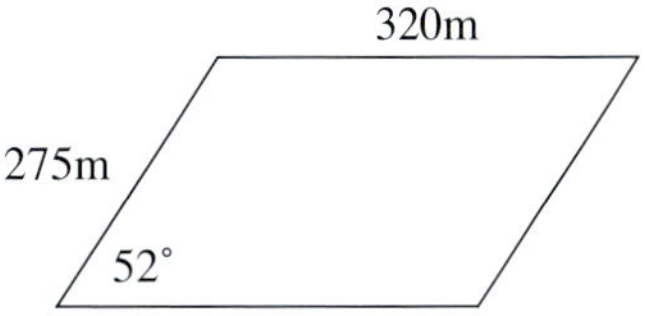

3. The diagram shows a circle of radius 10 cm. AB is a diameter of the circle. AC = 6 cm.

Find the area of the shaded region, giving an exact answer.

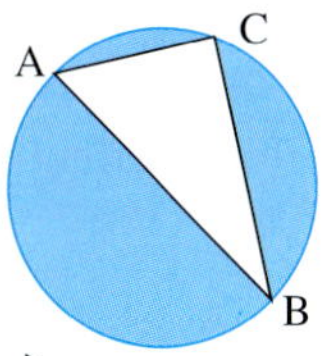

4. The triangle shown has an area of $110\ \text{cm}^2$. Find x.

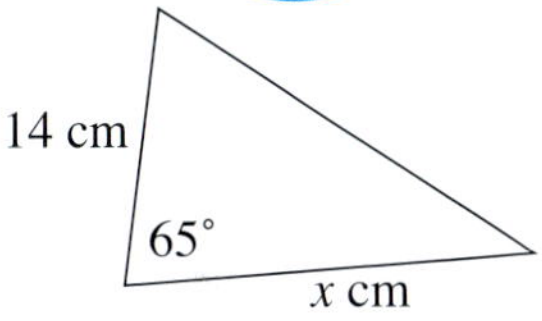

5. Find the area of the following

(a)

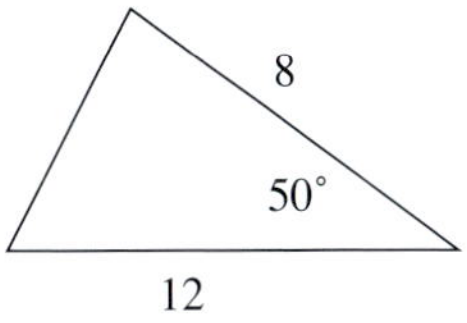

(b)

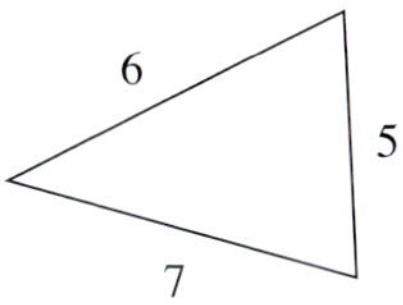

(c)

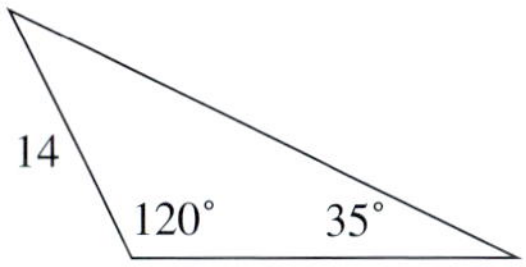

6. A napkin is in the shape of a quadrilateral with diagonals 9 cm and 12 cm long. The angle between the diagonals is 75°. What area does the napkin cover when laid out flat?

7. A triangle of area $50\ \text{cm}^2$ has side lengths 10 cm and 22 cm. What is the magnitude of the included angle?

8. A variable triangle OAB is formed by a straight line passing through the point $P(a, b)$ on the Cartesian plane and cutting the x–axis and y–axis at A and B respectively. If $\angle OAB = \theta$, find the area of ΔOAB in terms of a, b and θ.

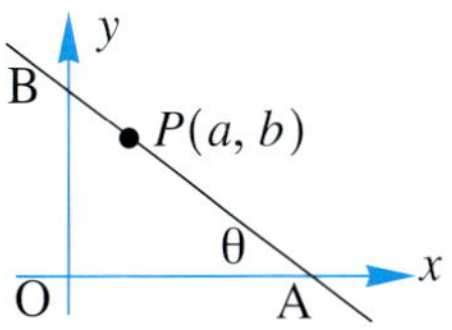

9. Find the area of ΔABC for the given diagram.

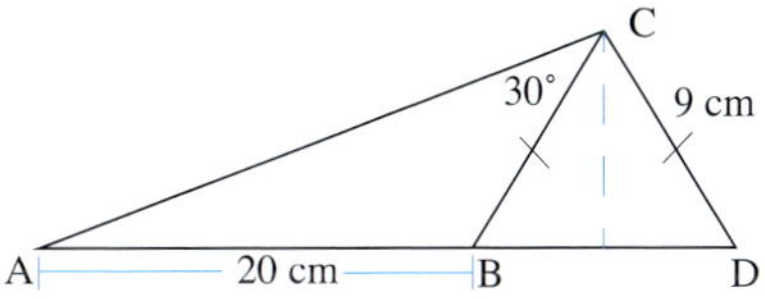

9.5 NON RIGHT-ANGLED TRIANGLES

9.5.1 THE SINE RULE

Previous sections dealt with the trigonometry of right-angled triangles. The trigonometric ratios can be used to solve non right-angled triangles. There are two main methods for solving non right-angled triangles, the **sine rule** and the **cosine rule** (which we look at later in this chapter). Both are usually stated using a standard labelling of the triangle. This uses capital letters to label the vertices and the corresponding small letters to label the sides opposite these vertices.

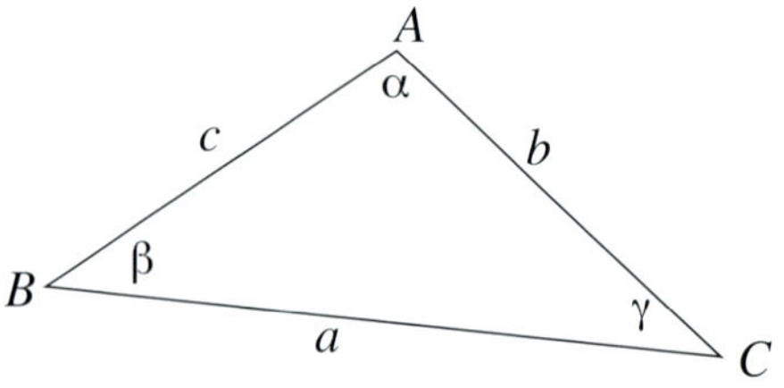

$$\sin A = \sin\alpha$$
$$\sin B = \sin\beta$$
$$\sin C = \sin\gamma$$

Using this labelling of a triangle, the sine rule can be stated as:

$$\frac{a}{\sin A} = \frac{b}{\sin B} = \frac{c}{\sin C} \quad \text{or} \quad \frac{\sin A}{a} = \frac{\sin B}{b} = \frac{\sin C}{c}$$

Note: the sine rule can only be used in a triangle in which an angle and the side **opposite** that angle are known.

So, why does this work?

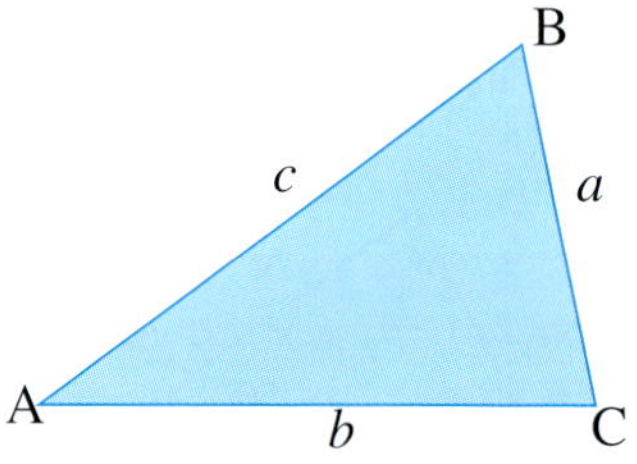

Using the results of the last section and labelling a triangle in the standard manner we have:

$$Area = \frac{1}{2}bc\sin A,$$

$$Area = \frac{1}{2}ac\sin B$$

and

$$Area = \frac{1}{2}ab\sin C$$

However, each of these are equal, meaning that

$$\frac{1}{2}ac\sin B = \frac{1}{2}bc\sin A \Leftrightarrow a\sin B = b\sin A \Leftrightarrow \frac{a}{\sin A} = \frac{b}{\sin B}$$

Similarly,

$$\frac{1}{2}ac\sin B = \frac{1}{2}ab\sin C \Leftrightarrow c\sin B = b\sin C \Leftrightarrow \frac{c}{\sin C} = \frac{b}{\sin B}$$

Combining these results we have that

$$\frac{a}{\sin A} = \frac{b}{\sin B} = \frac{c}{\sin C}$$

So, when should/can we make use of the sine rule?

Although the sine rule can be used for right-angled triangles, it is more often used for situations when we do not have a right-angled triangle, and when the given triangle has either of the following pieces of information:

(a) Two angles and one side

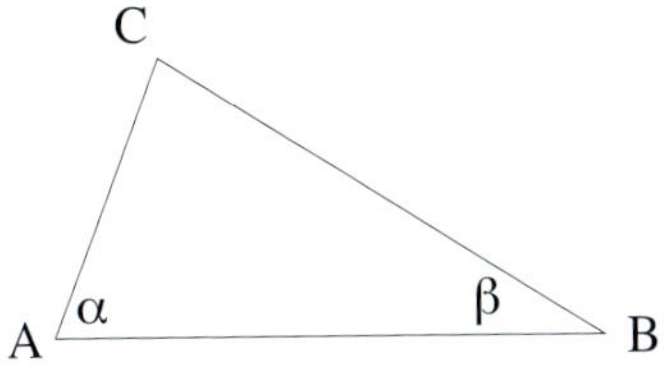

Either the length CB, AB or AC can be given and the triangle can be 'solved'. i.e., we can find the length of every side and every angle.

In this case, if we are give the length AB, we need $\angle ACB$, which can be found using $\angle ACB = 180° - \alpha - \beta$

(b) Two sides and a non-included angle

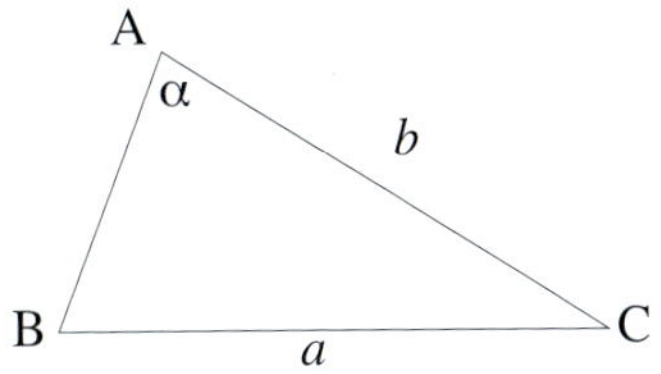

In this case, we first need to determine angle B using $\frac{\sin\alpha}{a} = \frac{\sin B}{b}$.

Once we have angle B, we can the find angle C and then the length AB.

EXAMPLE 9.13

Solve the following triangles giving the lengths of the sides in centimetres, correct to one decimal place and angles correct to the nearest degree.

(a)

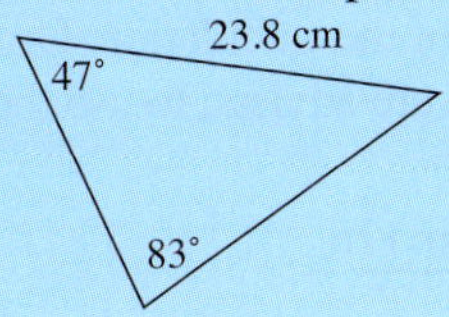

(b)

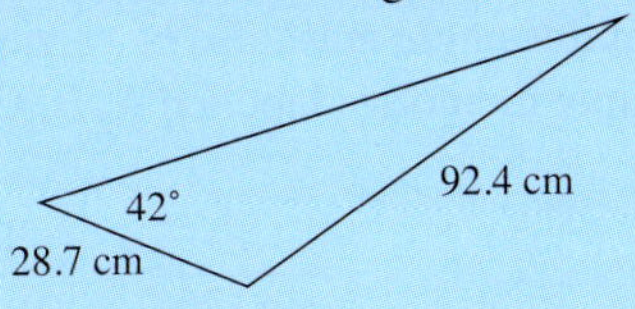

Solution

(a) Firstly, label the triangle using the standard method of lettering. 'Solve the triangle' means find all the angles and the lengths of all the sides. Since two of the angles are known, the third is; $C = 180° - 47° - 83° = 50°$.
The lengths of the remaining sides can be found using the known pairing of side and angle, *b* and *B*.

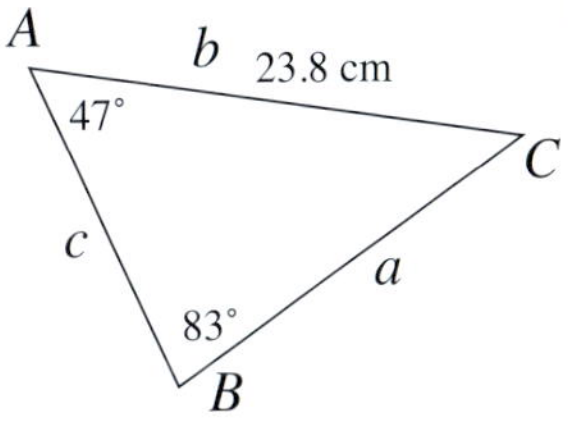

$$\frac{a}{\sin A} = \frac{b}{\sin B} \Leftrightarrow \frac{a}{\sin 47°} = \frac{23.8}{\sin 83°}$$

$$a = \frac{23.8 \times \sin 47°}{\sin 83°}$$

$$= 17.5369\ldots$$

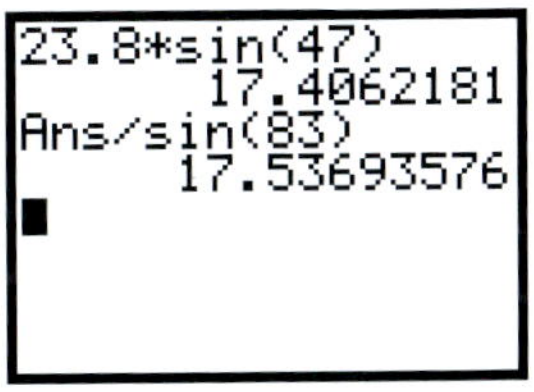

That is, BC is 17.5 cm (correct to one d.p).

Similarly, the remaining side can be calculated: $\frac{c}{\sin C} = \frac{b}{\sin B} \Leftrightarrow \frac{c}{\sin 50°} = \frac{23.8}{\sin 83°}$

$$\therefore c = \frac{23.8 \times \sin 50°}{\sin 83°}$$

$$= 18.3687\ldots$$

That is, AB is 18.4 cm (correct to one d.p).

(b) This triangle is different from the previous example in that only one angle is known. It remains the case that a pair of angles and an opposite side are known and that the sine rule can be used. The angle A must be found first.

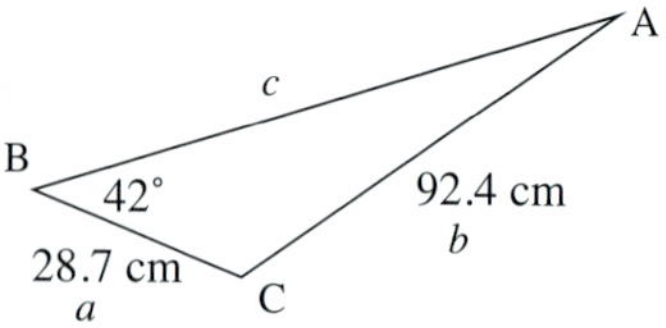

$$\frac{\sin A}{a} = \frac{\sin B}{b} \Leftrightarrow \frac{\sin A}{28.7} = \frac{\sin 42°}{92.4}$$

$$\Leftrightarrow \sin A = \frac{28.7 \times \sin 42°}{92.4}$$

$$= 0.207836$$

$$\therefore A = \sin^{-1} 0.207836$$

$$= 11.9956°$$

$$= 11°59'44''$$

The answer to the first part of the question is 12° correct to the nearest degree. It is important, however, to carry a much more accurate version of this angle through to subsequent parts of the calculation. This is best done using the calculator memory.

The third angle can be found because the sum of the three angles is 180°.
So, $C = 180° - 12° - 42° = 126°$

An accurate version of this angle must also be carried to the next part of the calculation. Graphics calculators have multiple memories labelled A, B, C etc. and students are advised to use these in such calculations.

The remaining side is: $\frac{c}{\sin 126°} = \frac{28.7}{\sin 12°} \Leftrightarrow c = \frac{28.7 \sin 126°}{\sin 12°}$

$$\therefore c = 111.6762...$$

That is, AB is 111.7 cm (correct to one d.p)

EXERCISES 9.5.1

Use the sine rule to complete the following table, which refers to the standard labelling of a triangle.

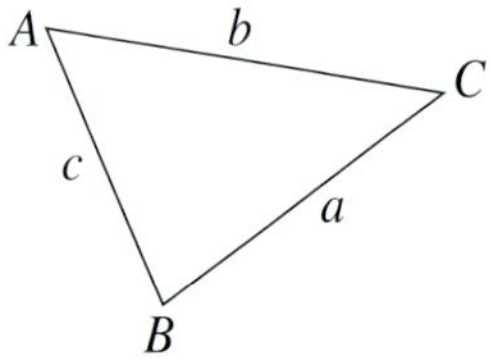

	a cm	*b* cm	*c* cm	*A*	*B*	*C*
1.			48.2		29°	141°
2.		1.2		74°	25°	
3.			11.3	60°		117°
4.			51.7	38°		93°
5.	18.5	11.4		68°		
6.	14.6	15.0			84°	
7.		7.3			16°	85°
8.			28.5	39°		124°
9.	0.8		0.8	82°		
10.			33.3	36°		135°
11.	16.4			52°	84°	

	a cm	*b* cm	*c* cm	*A*	*B*	*C*
12.			64.3		24°	145°
13.	30.9	27.7		75°		
14.			59.1	29°		102°
15.		9.8	7.9		67°	
16.			54.2	16°		136°
17.	14.8		27.2			67°
18.			10.9		3°	125°
19.			17.0		15°	140°
20.			40.1	30°		129°

EXAMPLE 9.14

For the triangle shown, find the angle ABC.

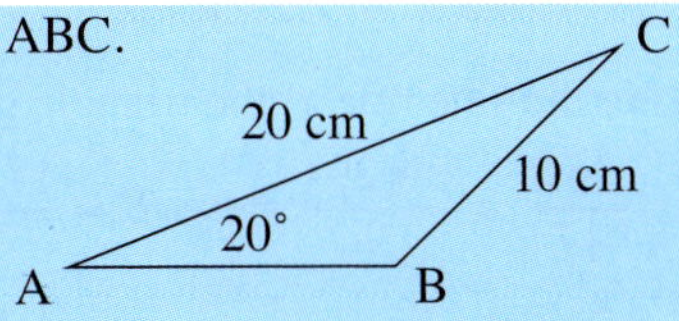

Solution

Making use of the sine rule we have:

$$\frac{\sin A}{a} = \frac{\sin B}{b} \Leftrightarrow \frac{\sin 20°}{10} = \frac{\sin B}{20}$$

$$\Leftrightarrow \sin B = \frac{20 \sin 20°}{10}$$

$$\therefore B = \sin^{-1}(2 \sin 20°)$$

$$= 43.1601\ldots$$

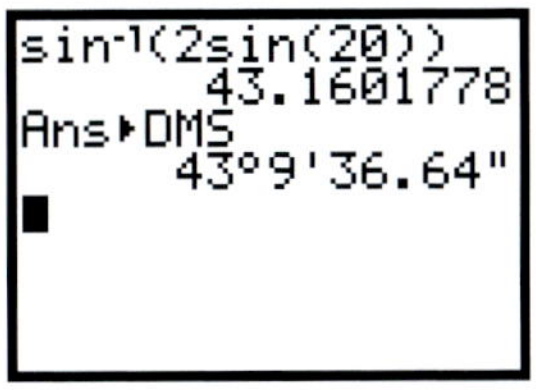

That is, $B = 43°10'$

However, from our diagram, the angle ABC should have been greater than 90°! That is, we should have obtained an **obtuse angle** (90° < B < 180°) rather than an **acute angle** (0° < B < 90°).

So, what went wrong?

This example is a classic case of what is known as the **ambiguous case**, in that, from the given information it is possible to draw two different diagrams, both having the same data. we show both these triangles:

Case 1

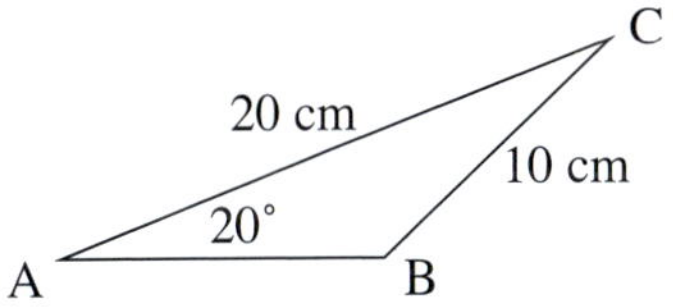

Case 2

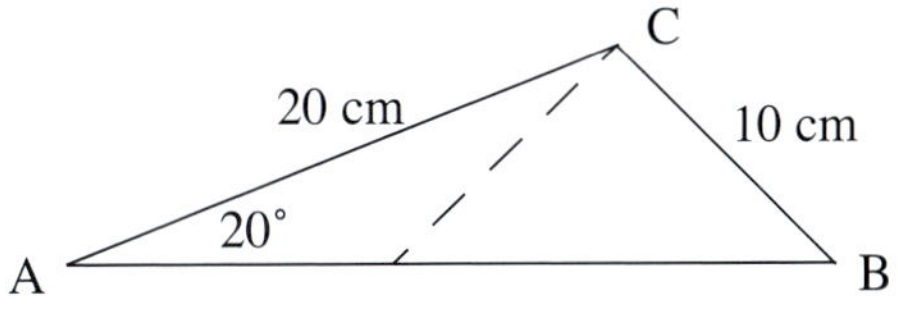

Notice that the side BC can be pivoted about the point C and therefore two different triangles can be formed with BC = 10. This is why there are two possible triangles based on the same information.

In the solution above, $B = 43°10'$ – representing Case 2. However, our diagram is represented by Case 1! Therefore, the correct answer is $180 - 43°10' = 136°50'$.

9.5.2 THE AMBIGUOUS CASE

From Example 9.14, it can be seen that an ambiguous case can arise when using the sine rule. In the given situation we see that the side CB can be pivoted about its vertex, forming two posible triangles.

We consider another such triangle.

EXAMPLE 9.15 Draw diagrams showing the triangles in which AC = 17 cm, BC = 9 cm and A = 29° and solve these triangles.

Solution

Applying the sine rule to the triangle gives:

$$\frac{\sin B}{17} = \frac{\sin 29°}{9} \Leftrightarrow \sin B = \frac{17 \times \sin 29°}{9}$$

$$= 0.91575$$

$$\therefore B = 66°$$

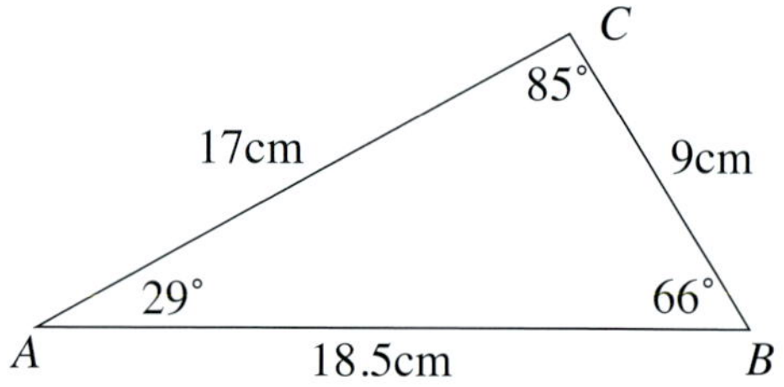

Next, we have,

$$C = 180° - 29° - 66° = 85°$$

$$\frac{c}{\sin 85°} = \frac{9}{\sin 29°} \Leftrightarrow c = 18.5$$

There is, however, a second solution that results from drawing an isosceles triangle BCE. This creates the triangle AEC which also fits the data. The third angle of this triangle is 37° and the third side is:
$$\frac{\text{AE}}{\sin 37°} = \frac{9}{\sin 29°} \Leftrightarrow \text{AE} = 11.2$$

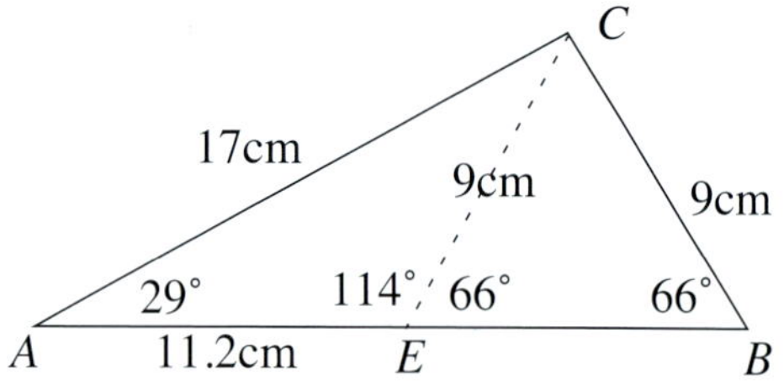

The original data is ambiguous in the sense that there are two triangles that are consistent with it.

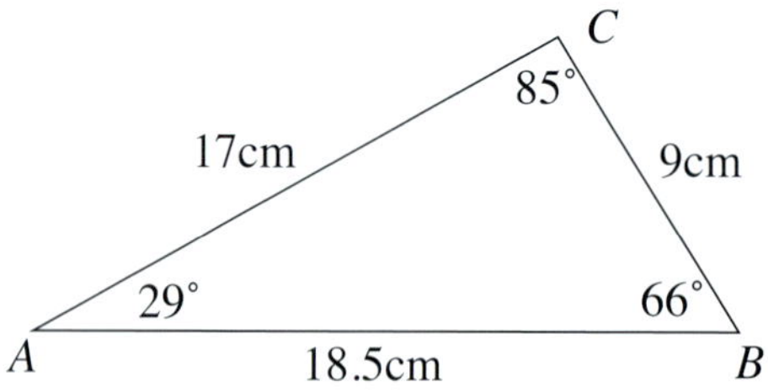

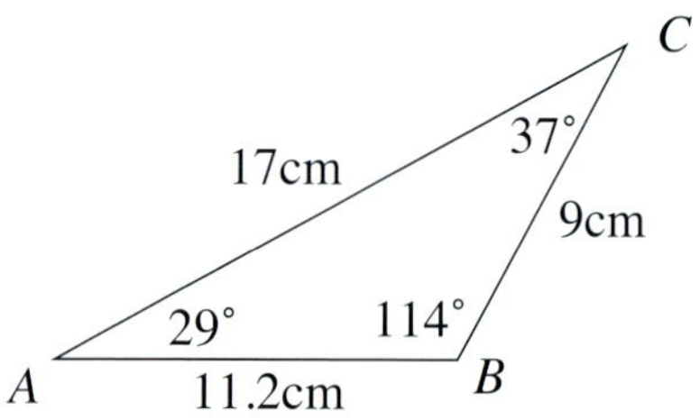

You should also notice that the two angles in the solution are 66° and 114° and that sin66° = sin114°. (That is, sin 66° = sin(180° – 66°) = sin 114°. This will be developed further in Chapter 10).

In fact, we can go one step further and make the following statement:

If we are given two sides of a triangle and the magnitude of an angle opposite one of the sides, there may exist one, two or no possible solutions for the given information.

We summarise our findings:

Type 1

Given the **acute angle** α and the side lengths a and b, there are four possible outcomes:

Notice that the number of solutions **depends on** the length a relative to the perpendicular height, h, of the triangle as well as the length b. Where the height h is based on the right-angled triangle formed in each case, i.e., $\sin\alpha = \frac{h}{b} \Leftrightarrow h = b\sin\alpha$.

So that

- if $a < h$, then the triangle cannot be completed.
- if $a = h$, then we have a right-angled triangle.
- if $a > b$, then we have a triangle that is consistent with the given information.
- if $h < a < b$, then the side BC can be pivoted about the vertex C, forming two triangles.

Number of Δ s	Necessary condition	Type of triangle that can be formed	
None	$a < h$	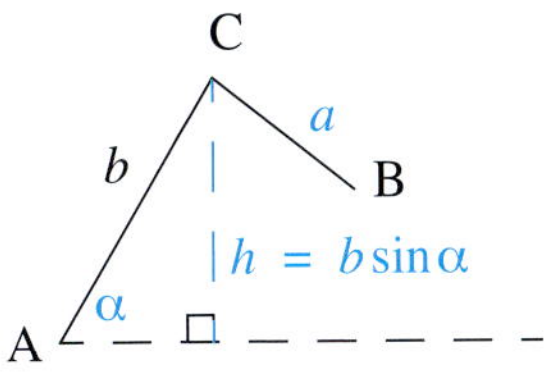	In this case, the triangle cannot be constructed.
One	$a = h$	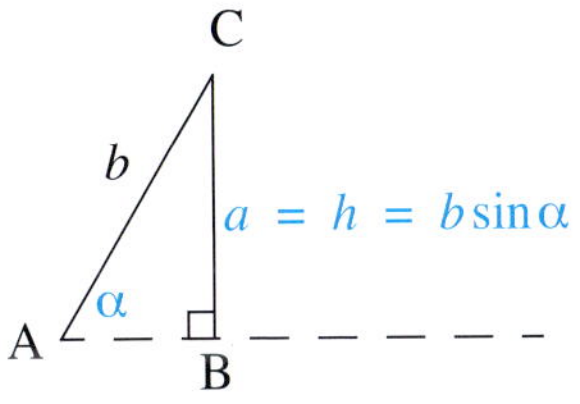	In this case we have formed a right-angled triangle.
One	$a > b$	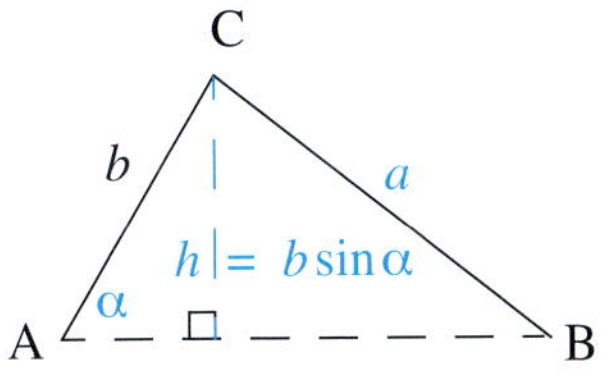	In this case there can be only one triangle that is consistent with the given information.
Two	$h < a < b$	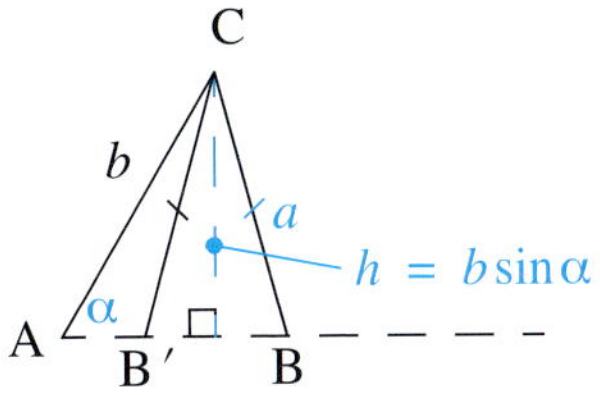	In this case there are two possible triangles, ΔABC and $\Delta AB'C$. This is because BC can be pivoted about C and still be consistent with the given information.

The table above reflects the case where α is acute. What if α is obtuse?

Type 2

Given the **obtuse angle** α and the side lengths a and b, there are two possible outcomes:

Number of Δ s	Necessary condition	Type of triangle that can be formed	
None	$a \le b$	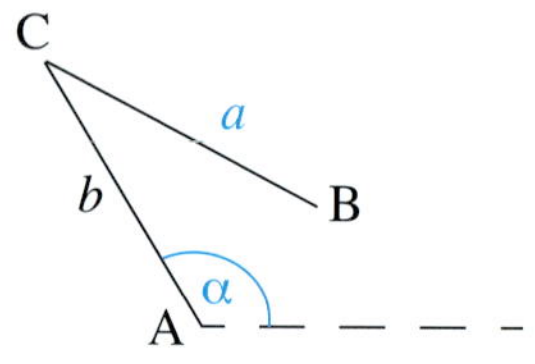	In this case, the triangle cannot be constructed.
One	$a > b$	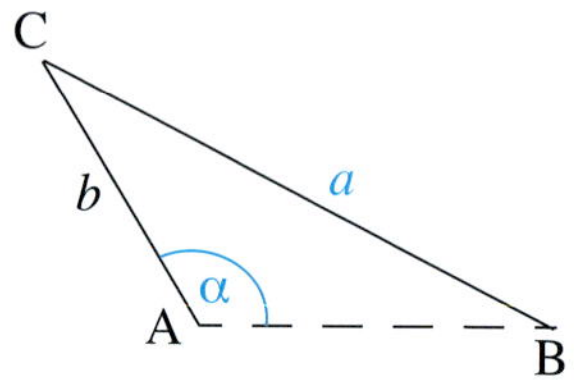	In this case there can be only one triangle that is consistent with the given information.

EXAMPLE 9.16

Find $\angle ABC$ for the triangle ABC given that $a = 50$, $b = 80$ and $A = 35°$.

Solution

We first determine the value of $b\sin\alpha$ and compare it to the value a :

Now, $b\sin\alpha = 80\sin 35° = 45.89$

Therefore we have that $b\sin\alpha\ (= 45.89) < a\ (= 50) < b\ (= 80)$

meaning that we have an ambiguous case.

Case 1:

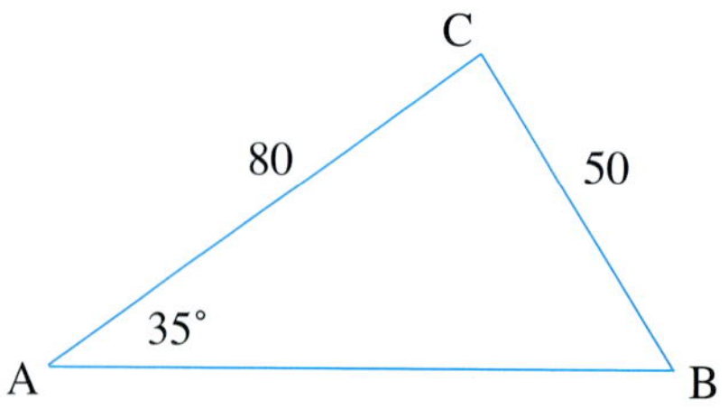

Using the sine rule, $\frac{\sin A}{a} = \frac{\sin B}{b}$, we have

$$\frac{\sin 35°}{50} = \frac{\sin B}{80} \Leftrightarrow \sin B = \frac{80\sin 35°}{50}$$

$$\therefore B = 66°35'$$

Case 2:

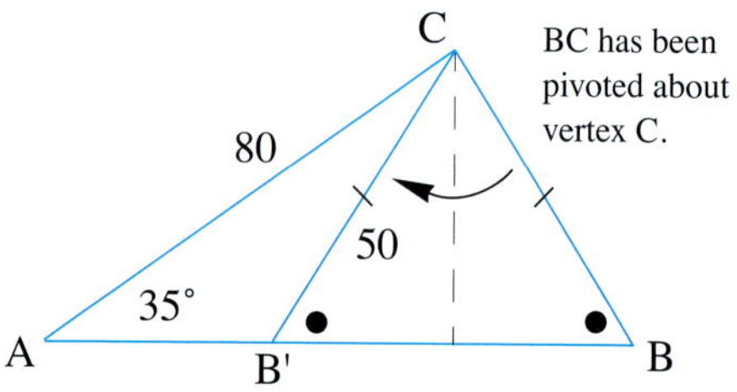

From case 1, the obtuse angle B' is given by $180° - 66°35' = 113°25'$.

This is because $\Delta B'CB$ is an **isosceles** triangle, so that $\angle AB'C = 180° - \angle CB'B$

EXAMPLE 9.17

Find $\angle ACB$ for the triangle ABC given that $a = 70$, $c = 90$ and $A = 75°$.

Solution

We start by drawing the triangle with the given information:

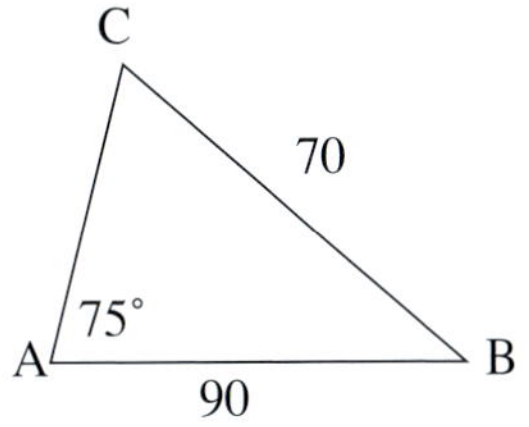

Using the sine rule we have

$$\frac{\sin C}{90} = \frac{\sin 75}{70} \Leftrightarrow \sin C = \frac{90\sin 75}{70}$$

$$\therefore \sin C = 1.241\ldots$$

Which is imposible to solve for as the sine of an angle can never be greater than one.

Therefore no such triangle exists.

EXERCISES 9.5.2

Find the two solutions to these triangles which are defined using the standard labelling:

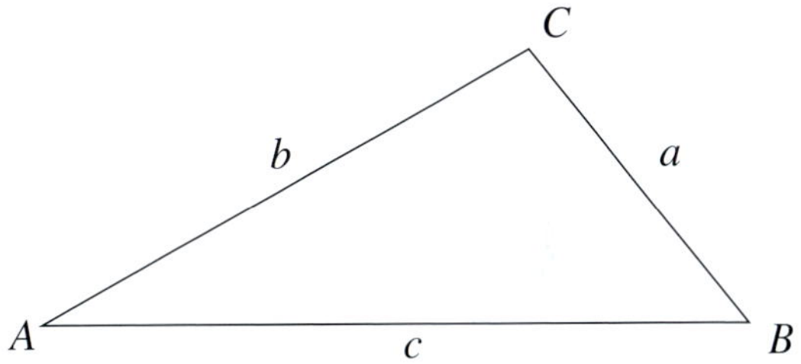

	a cm	*b* cm	*A*
1.	7.4	18.1	20°
2.	13.3	19.5	14°
3.	13.5	17	28°
4.	10.2	17	15°
5.	7.4	15.2	20°
6.	10.7	14.1	26°
7.	11.5	12.6	17°
8.	8.3	13.7	24°
9.	13.7	17.8	14°
10.	13.4	17.8	28°
11.	12.1	16.8	23°
12.	12	14.5	21°
13.	12.1	19.2	16°
14.	7.2	13.1	15°
15.	12.2	17.7	30°
16.	9.2	20.9	14°
17.	10.5	13.3	20°
18.	9.2	19.2	15°
19.	7.2	13.3	19°
20.	13.5	20.4	31°

21. Solve the following triangles

(a) $\alpha = 75°, a = 35, c = 45$

(b) $\alpha = 35°, a = 30, b = 80$

(c) $\beta = 40°, a = 22, b = 8$

(d) $\gamma = 50°, a = 112, c = 80$

9.5.3 APPLICATIONS OF THE SINE RULE

Just as in the case of right-angled triangles, the sine rule becomes very useful. In particular, it means that previous problems that required the partitioning of a non right-angled triangle into two (or more) right-angled triangles can be solved using the sine rule.

We start by considering Q.13., Exercise 9.2:

EXAMPLE 9.18

A surveying team are trying to find the height of a hill. They take a 'sight' on the top of the hill and find that the angle of elevation is 23°27′. They move a distance of 250 metres on level ground directly away from the hill and take a second 'sight'. From this point, the angle of elevation is 19°46′. Find the height of the hill, correct to the nearest metre.

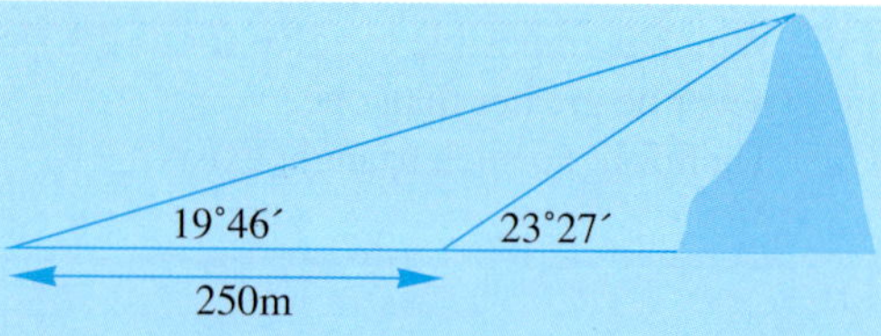

Solution

Labelling the given diagram using the standard notation we have:

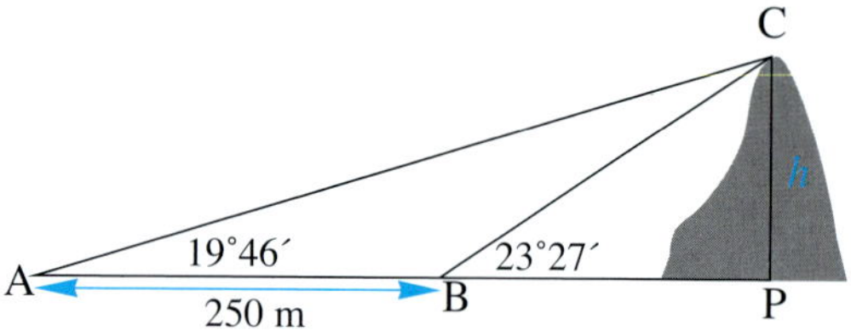

With $\beta = 180 - 23°27' = 156°33'$

and $\gamma = 180 - 19°46' - 156°33' = 3°41'$

Then, using the sine rule,

$$\frac{b}{\sin 156°33'} = \frac{250}{\sin 3°41'}$$

$$\Leftrightarrow b = \frac{250 \sin 156°33'}{\sin 3°41'}$$

$$= 1548.63\ldots$$

Then, using ΔACP we have,

$$\sin 19°46' = \frac{h}{b} \Leftrightarrow h = b \sin 19°46'$$

$$= 523.73$$

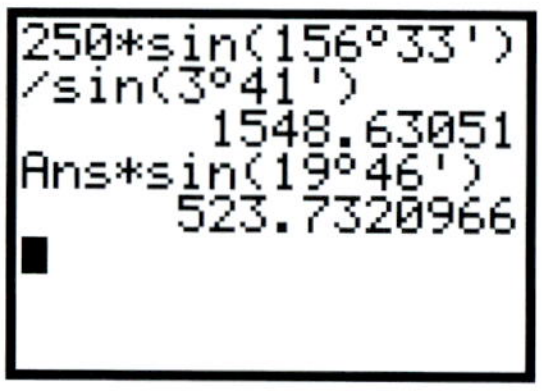

So, the hill is 524 m high (to nearest metre).
A much neater solution (as opposed to solving simultaneous equations – as was required previously).

EXERCISES 9.5.3

1. A short course biathlon meet requires the competitors to run in the direction S60°W to their bikes and then ride S40°E to the finish line, situated 20 km due South of the starting point. What is the distance of this course?

2. A pole is slanting towards the sun and is making an angle of 10° to the vertical. It casts a shadow 7 metres long along the horizontal ground. The angle of elevation of the top of the pole to the tip of its shadow is 30°. Find the length of the pole, giving your answer to 2 d.p.

3. A statue A, is observed from two other statues B and C which are 330 m apart. The angle between the lines of sight AB and BC is 63° and the angle between the lines of sight AC and CB is 75°. How far is statue A from statue B?

4. Town A is 12 km from town B and its bearing is 132°T from B. Town C is 17 km from A and its bearing is 063°T from B. Find the bearing of A from C.

5. The angle of elevation of the top of a building from a park bench on level ground is 18°. The angle of elevation from a second park bench, 300 m closer to the base of the building is 30°. Assuming that the two benches and the building all lie on the same vertical plane, find the height of the building.

6. (a) A man standing 6 metres away from a lamp post casts a shadow10 metres long on a horizontal ground. The angle of elevation from the tip of the shadow to the lamp light is 12°. How high is the lamp light?

(b) If the shadow is cast onto a road sloping at 30° upwards, how long would the shadow be if the man is standing at the foot of the sloping road and 6 metres from the lamp post?

7. At noon the angle of elevation of the sun is 72° and is such that a three metre wall AC, facing the sun, is just in the shadow due to the overhang AB.

The angle that the overhang makes with the vertical wall is 50°.

(a) Copy and illustrate this information on the diagram shown.
(b) Find the length of the overhang.

At 4 p.m. the angle of elevation of the sun is 40° and the shadow due to the overhang just reaches the base of the window.

(c) How far from the ground is the window?

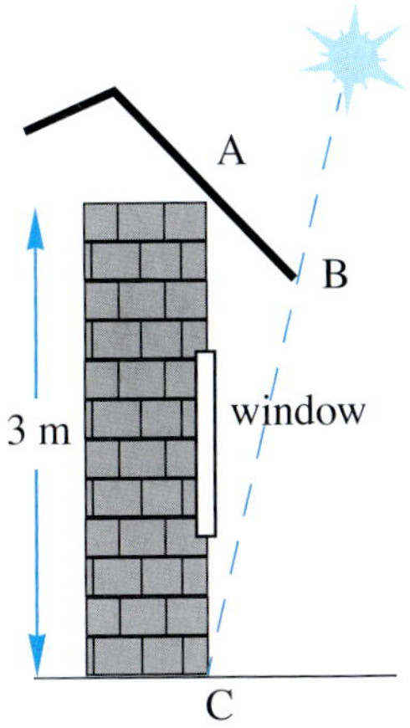

8. The lookout on a ship sailing due East at 25 km/h observes a reef N62°E at a distance of 30 km.
(a) How long will it be before the ship is 15 km from the reef, assuming that it continues on its easterly course.
(b) How long is it before it is again 15 km from the reef?
(c) What is the closest that the ship will get to the reef?

9. The framework for an experimental design for a kite is shown. Material for the kite costs $12 per square cm. How much will it cost for the material if it is to cover the framework of the kite.

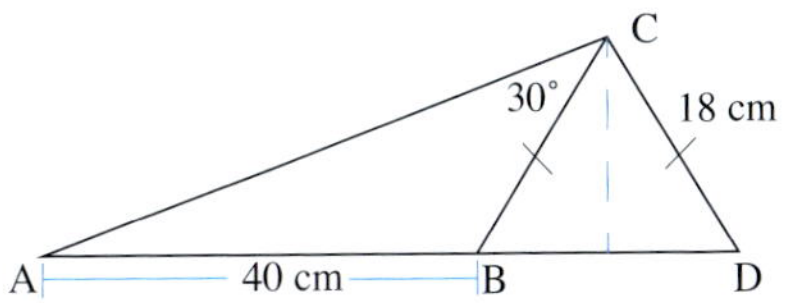

10. A boy walking along a straight road notices the top of a tower at a bearing of 284°T. After walking a further 1.5 km he notices that the top of the tower is at a bearing of 293°T. How far from the road is the tower?

9.5.4 THE COSINE RULE

Sometimes the sine rule is not enough to help us solve for a non right-angled triangle. For example, in the triangle shown, we do not have enough information to use the sine rule. That is, the sine rule only provided the following:

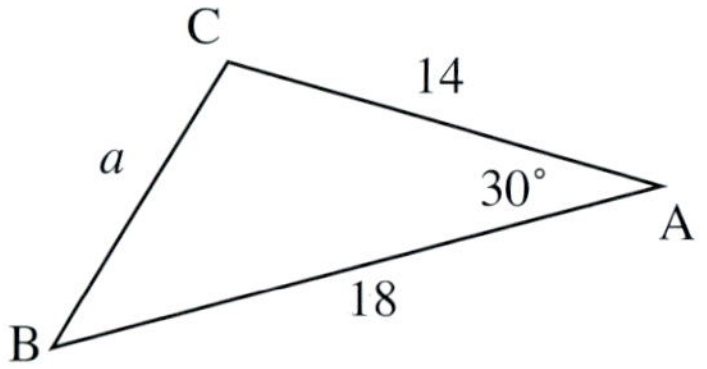

$$\frac{a}{\sin 30°} = \frac{14}{\sin B} = \frac{18}{\sin C}$$

where there are too many unknowns.

For this reason we derive another useful result, known as the cosine rule. The cosine rule may be used when

1. **two sides and an included angle are given**:
This means that the third side can be determined and then we can make use of the sine rule (or the cosine rule again).

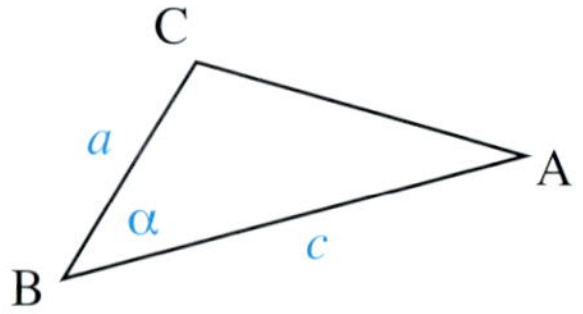

2. **three sides are given**:
This means we could then determine any of the angles.

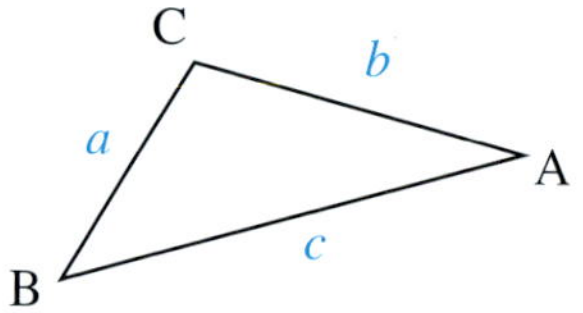

The cosine rule, with the standard labelling of the triangle has three versions:

$$a^2 = b^2 + c^2 - 2bc\cos A$$
$$b^2 = a^2 + c^2 - 2ac\cos B$$
$$c^2 = a^2 + b^2 - 2ab\cos C$$

The cosine rule can be remembered as a version of Pythagoras's Theorem with a correction factor. We now show why this works.

Consider the case where there is an acute angle at A. Draw a perpendicular from C to N as shown in the diagram.

In ΔANC we have $b^2 = h^2 + x^2$

$$\Leftrightarrow h^2 = b^2 - x^2 \quad -(1)$$

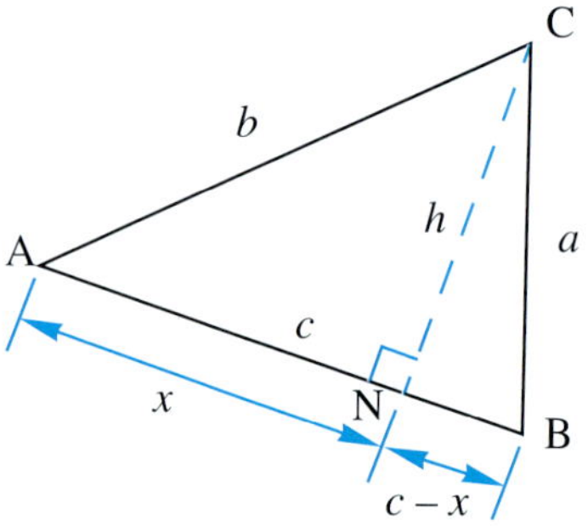

In ΔBNC we have $a^2 = h^2 + (c-x)^2$

$$\Leftrightarrow h^2 = a^2 - (c-x)^2 \quad -(2)$$

Equating (1) and (2) we have,

$$a^2 - (c-x)^2 = b^2 - x^2$$
$$\Leftrightarrow a^2 - c^2 + 2cx - x^2 = b^2 - x^2$$
$$\Leftrightarrow a^2 = b^2 + c^2 - 2cx$$

However, from ΔANC we have that $\cos A = \frac{x}{b} \Leftrightarrow x = b\cos A$

Substituting this result for x, we have

$$a^2 = b^2 + c^2 - 2cb\cos A$$

Although we have shown the result for an acute angle at A, the same rule applies if A is obtuse.

EXAMPLE 9.19

Solve the following triangles giving the lengths of the sides in centimetres, correct to one decimal place and angles correct to the nearest degree.

(a)

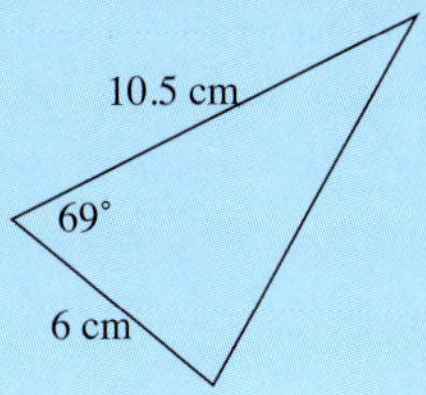

(b)

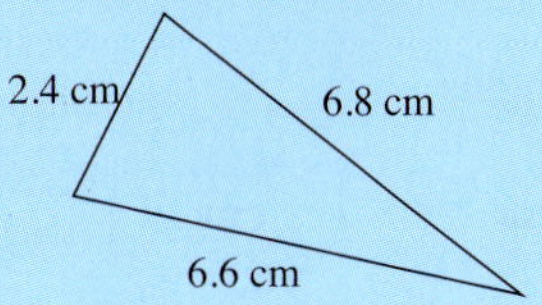

Solution

(a) The data does not include an angle and the opposite side so the sine rule cannot be used. The first step, as with the sine rule, is to label the sides of the triangle. Once the triangle has been labelled, the correct version of the cosine rule must be chosen. In this case, the solution is:

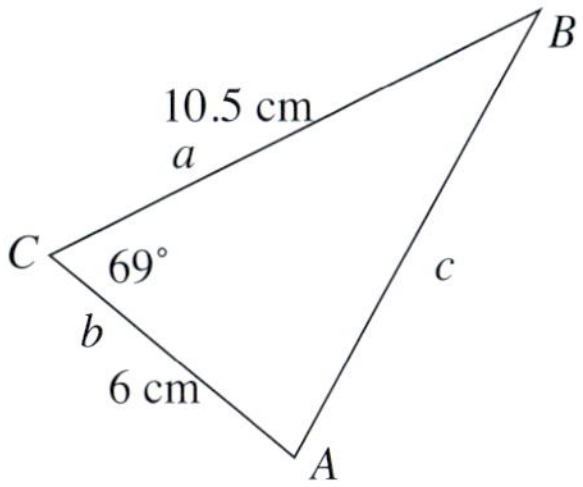

$$c^2 = a^2 + b^2 - 2ab\cos C$$
$$c^2 = 10.5^2 + 6^2 - 2 \times 10.5 \times 6 \times \cos 69°$$
$$= 101.0956$$
$$a = 10.1$$

The remaining angles can be calculated using the sine rule. Again, it is important to carry a high accuracy for the value of c to the remaining problem:

$$\frac{\sin B}{b} = \frac{\sin C}{c} \Leftrightarrow \sin B = \frac{6 \times \sin 69°}{10.0546} \quad \therefore B = 34°$$

Finally, $A = 180° - 34° - 69° = 77°$

(b) In this case, there are no angles given. The cosine rule can be used to solve this problem as follows:

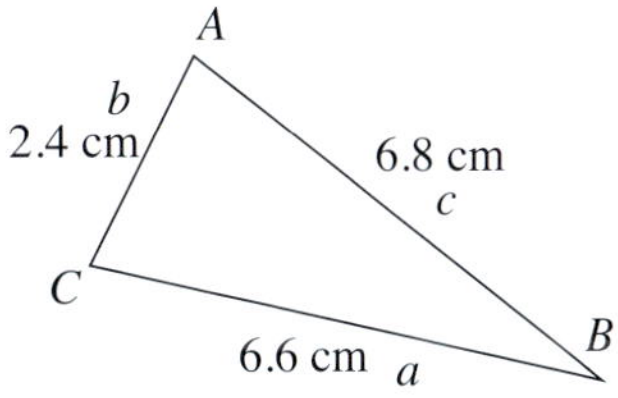

$$a^2 = b^2 + c^2 - 2bc\cos A$$
$$6.6^2 = 2.4^2 + 6.8^2 - 2 \times 2.4 \times 6.8 \times \cos A$$
$$2 \times 2.4 \times 6.8 \times \cos A = 2.4^2 + 6.8^2 - 6.6^2$$
$$\cos A = \frac{2.4^2 + 6.8^2 - 6.6^2}{2 \times 2.4 \times 6.8}$$
$$= 0.25858$$
$$A = 75.014°$$
$$= 75°1'$$

Next, use the sine rule: $\dfrac{\sin B}{b} = \dfrac{\sin A}{a} \Leftrightarrow \sin B = \dfrac{2.4 \times \sin 75}{6.6} \quad \therefore B = 20°34'$

So that $C = 180° - 75° - 21° = 84°$

The three angles, correct to the nearest degree are $A = 75°$, $B = 21°$ & $C = 84°$.

EXERCISES 9.5.4

Solve the following triangles

	a cm	b cm	c cm	A	B	C
1.	13.5		16.7		36°	
2.	8.9	10.8				101°
3.	22.8		12.8		87°	
4.	21.1	4.4				83°
5.		10.6	15.1	74°		
6.		13.6	20.3	20°		
7.	9.2		13.2		46°	
8.	23.4	62.5				69°
9.		9.6	15.7	41°		
10.	21.7	36.0	36.2			
11.	7.6	3.4	9.4			
12.	7.2	15.2	14.3			
13.	9.1		15.8		52°	
14.	14.9	11.2	16.2	63°	42°	75°
15.	2.0	0.7	2.5			
16.	7.6	3.7	9.0			
17.	18.5	9.8	24.1			
18.	20.7	16.3	13.6			
19.		22.4	29.9	28°		
20.	7.0		9.9		42°	
21.	21.8	20.8	23.8			
22.	1.1		1.3		89°	
23.		1.2	0.4	85°		
24.	23.7	27.2				71°
25.	3.4	4.6	5.2			

9.5.5 APPLICATIONS OF THE COSINE RULE

EXAMPLE 9.20

A cyclist rode her bike for 22 km on a straight road heading in a westerly direction towards a junction. Upon reaching the junction, she headed down another straight road bearing 200°T for a distance of 15 km. How far is the cyclist from her starting position?

Solution

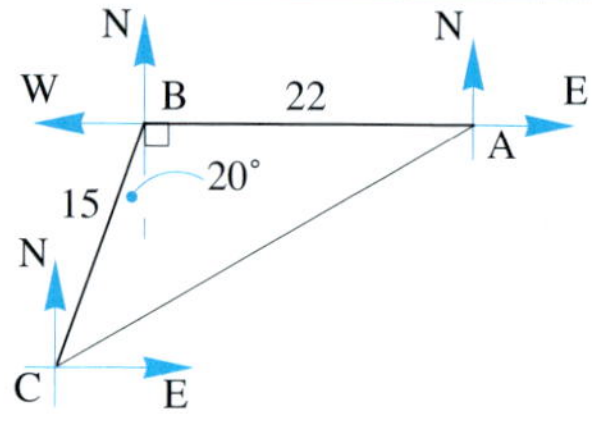

We start with a diagram:

Note that $\angle ABC = 90° + 20° = 110°$

Using the cosine rule we have,

$$AC^2 = 15^2 + 22^2 - 2 \times 15 \times 22\cos 110°$$

$$\Rightarrow AC = \sqrt{225 + 484 - 660 \times (-0.3420\ldots)}$$

$$\therefore AC = 30.5734\ldots$$

That is, she is (approximately) 30.57 km from her starting point.

EXAMPLE 9.21 A yacht starts from a harbour and sails for a distance of 11 km in a straight line. The yacht then makes a turn to port (left) of 38° and sails for 7 km in a straight line in this new direction until it arrives at a small island. Draw a diagram that shows the path taken by the yacht and calculate the distance from the harbour to the island.

Solution

The question does not give the bearing of the first leg of the trip so the diagram can show this in any direction. H is the harbour, I the island and T the point where the yacht makes its turn.

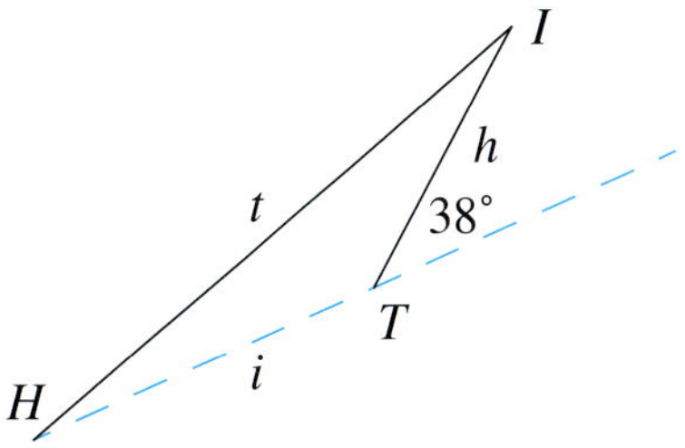

The angle in the triangle at T is 180° – 38° = 142°. The problem does not contain an angle and the opposite side and so must be solved using the cosine rule.

$$\begin{aligned} t^2 &= h^2 + i^2 - 2hi\cos T \\ &= 7^2 + 11^2 - 2 \times 7 \times 11 \times \cos 142° \\ &= 291.354 \\ \therefore t &= 17.1 \end{aligned}$$

That is, distance from the harbour to the port is 17.1 km (to one d.p)

EXAMPLE 9.22 A triangular sandpit having side lengths 5 m, 4 m and 8 m is to be constructed to a depth of 20 cm. Find the volume of sand required to fill this sandpit.

Solution

We will need to find an angle. In this case we determine the largest angle, which will be the angle opposite the longest side.
From our diagram we have

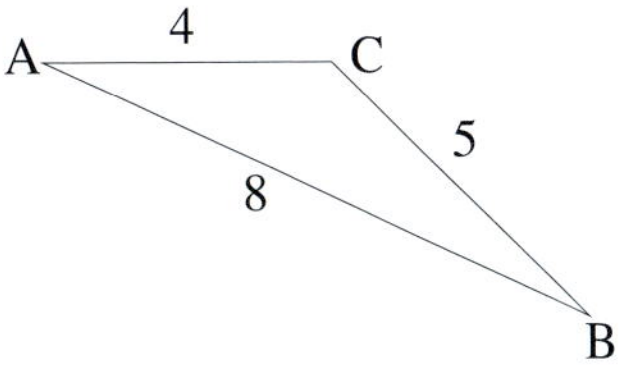

$$\begin{aligned} 8^2 &= 4^2 + 5^2 - 2 \times 4 \times 5\cos C \\ \therefore 64 &= 16 + 25 - 40\cos C \\ \Leftrightarrow \cos C &= \frac{16 + 25 - 64}{40} \\ &= -\frac{23}{40} \\ \therefore C &= 125°6' \end{aligned}$$

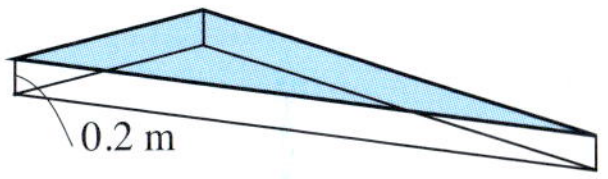

To find the volume of sand we first need to find the surface area of the sandpit.

$$\text{Area} = \frac{1}{2}ab\sin C = \frac{1}{2} \times 4 \times 5 \times \sin(125°6') = 8.1815 \text{ m}^2.$$

Therefore, volume of sand required is $0.2 \times 8.1815 = 1.64 \text{ m}^3$.

EXERCISES 9.5.5

1. Thomas has just walked 5 km in a direction N70˚E when he realises that he needs to walk a further 8 km in a direction E60˚S.
(a) How far from the starting point will Thomas have travelled?
(b) What is his final bearing from his starting point?

2. Two poles, 8 m apart are facing a rugby player who is 45 m from the left pole and 50 from the right one. Find the angle that the player makes with the goal mouth.

3. The lengths of the adjacent sides of a parallelogram are 4.80 cm and 6.40 cm. If these sides have an inclusive angle of $40°$, find the length of the shorter diagonal.

4. During an orienteering venture, Patricia notices two rabbit holes and estimates them to be 50 m and 70 m away from her. She measures the angle between the line of sight of the two holes as 54˚. How far apart are the two rabbit holes?

5. To measure the length of a lake, a surveyor chooses three points. Starting at one end of the lake she walks in a straight line for 223.25 m to some point X, away from the lake. She then heads towards the other end of the lake in a straight line and measures the distance covered to be 254.35 m. If the angle between the paths she takes is 82˚25', find the width of the lake.

6. A light airplane flying N87˚W for a distance of 155 km, suddenly needs to alter its course and heads S 34˚E for 82 km to land on an empty field.
(a) How far from its starting point did the plane land.
(b) What was the plane's final bearing from its starting point?

EXERCISES 9.5.6 – MISCELLANEOUS EXERCISES

1. The diagram shows a triangular building plot. The distances are given in metres. Find the length of the two remaining sides of the plot giving your answers correct to the nearest hundredth of a metre.

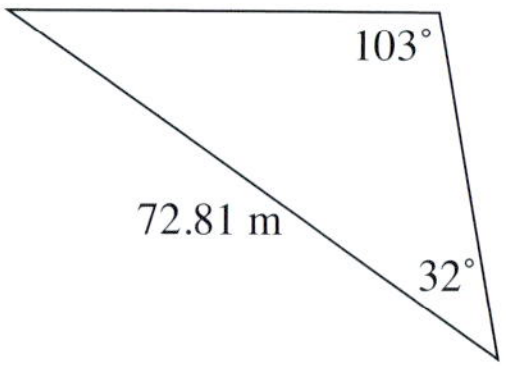

2. Xiang is standing on level ground. Directly in front of him and 32 metres away is a flagpole. If Xiang turns $61°$ to his right, he sees a post box 26.8 metres in front of him. Find the distance between the flagpole and the post box.

3. A triangular metal brace is part of the structure of a bridge. The lengths of the three parts are shown in metres. Find the angles of the brace.

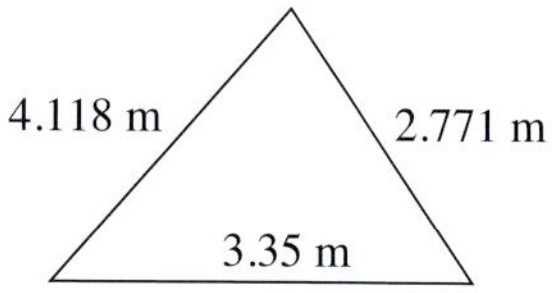

4. Find the smallest angle in the triangle whose sides have length 35.6 cm, 58.43 cm and 52.23 cm.

5. Ayton is directly north of Byford. A third town, Canfield, is 9.93km from Ayton on a bearing of 128° true. The distance from Byford to Canfield is 16.49km. Find the bearing of Canfield from Byford.

6. A parallelogram has sides of length 21.90 cm and 95.18 cm. The angle between these sides is 121°. Find the length of the long diagonal of the parallelogram.

7. A town clock has 'hands' that are of length 62cm and 85cm.
(a) Find the angle between the hands at half past ten.
(b) Find the distance between the tips of the hands at half past ten.

8. A shop sign is to be made in the shape of a triangle. The lengths of the edges are shown.
Find the angles at the vertices of the sign.

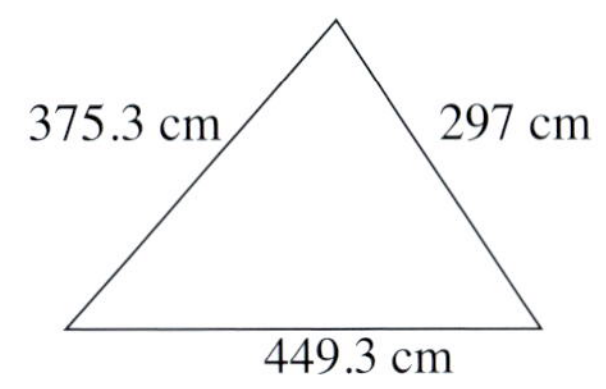

9. An aircraft takes off from an airstrip and then flies for 16.2 km on a bearing of 066° true. The pilot then makes a left turn of 88° and flies for a further 39.51 km on this course before deciding to return to the airstrip.
(a) Through what angle must the pilot turn to return to the airstrip?
(b) How far will the pilot have to fly to return to the airstrip?

10. A golfer hits two shots from the tee to the green.
How far is the tee from the green?

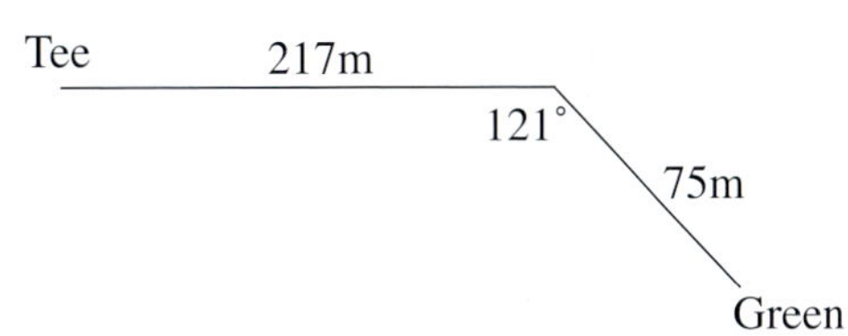

11. The diagram shows a parallelogram. Find the length of the longer of the two diagonals.

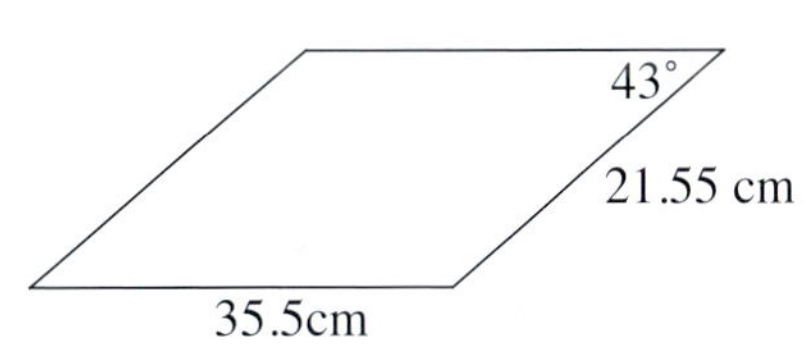

12. A triangle has angles 64°, 15° and 101°. The shortest side is 49 metres long. What is the length of the longest side?

13. The diagram shows a part of the support structure for a tower. The main parts are two identical triangles, ABC and ADE.
AC = DE = 27.4cm and BC = AE = 23.91cm
The angles ACB and AED are 58°.
Find the distance BD.

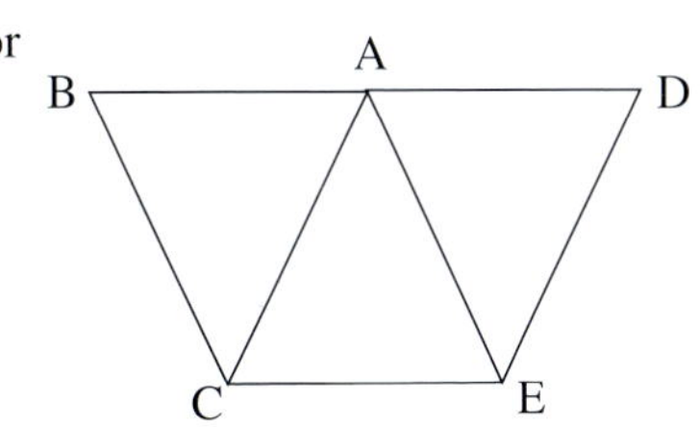

14. The diagram shows a design for the frame of a piece of jewellery. The frame is made of wire.

Find the length of wire needed to make the frame.

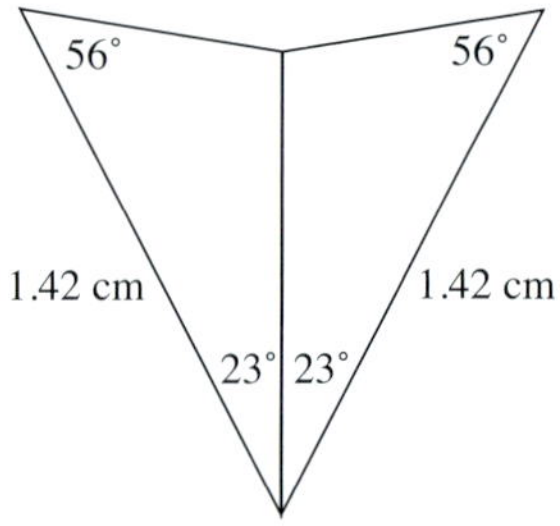

15. A triangular cross-country running track begins with the runners running North for 2050 metres. The runners then turn right and run for 5341 metres on a bearing of 083° true. Finally, the runners make a turn to the right and run directly back to the starting point.

(a) Find the length of the final leg of the run.

(b) Find the total distance of the run.

(c) What is the angle through which the runners must turn to start the final leg of the race?

(d) Find the bearing that the runners must take on the final leg of the race.

16. Show that for any standard triangle ABC, $\frac{\cos A}{a} + \frac{\cos B}{b} + \frac{\cos C}{c} = \frac{a^2 + b^2 + c^2}{2abc}$.

17. A sandpit in the shape of a pentagon ABCDE is to be built in such a way that each of its sides are of equal length, but its angles are not all equal.
The pentagon is symmetrical about DX, where X is the midpoint of AB.
The angle AXE and BXC are both 45° and each side is 2 m long.

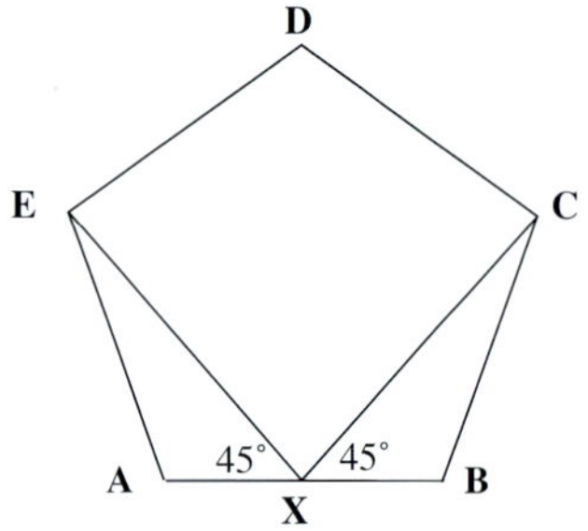

(a) Find $\angle XEA$.

(b) Find the length of EX.

(c) How much sand is required if the sandpit is 30 cm deep? Give your answer to three decimal place.

18. A triangular region has been set aside for a housing development which is to be divided into two sections. Two adjacent street frontages, AB and AC measuring 100 m and 120 m respectively. With the 100 m frontage running in an easterly direction, while the 120 m frontage runs in a north-east direction. A plan for this development is shown alongside. Give all answers to the nearest metre.

(a) Find the area covered by the housing development.

During the development stages, an environmental group specified that existing trees were not to be removed from the third frontage. This made it difficult for the surveyors to measure the length of the third frontage.

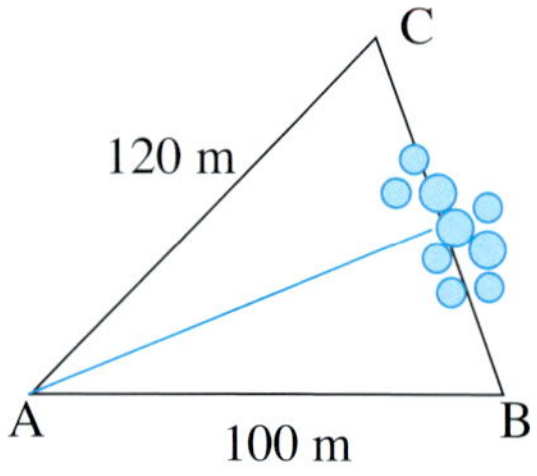

(b) Calculate the length of the third frontage, BC.

The estate is to be divided into 2 regions, by bisecting the angle at A with a stepping wall running from A to the frontage BC.

(c) How long is this stepping wall?

9.6 MORE APPLICATIONS IN 3–D*

EXAMPLE 9.23

A vertical tower PA is due West of a point B. From C, bearing 210° and 500 m from B, the bearing of the foot of the tower A is 290°, and the angle of elevation of the top of the tower P is 1.5°. The points A, B and C are on level ground. Given that h is the height of the tower, show that $h = \frac{250\sqrt{3}\tan 1.5°}{\sin 20°}$ and find the height to the nearest metre.

Solution

We start by drawing a diagram to depict the situation, producing both a perspective and aerial diagram:

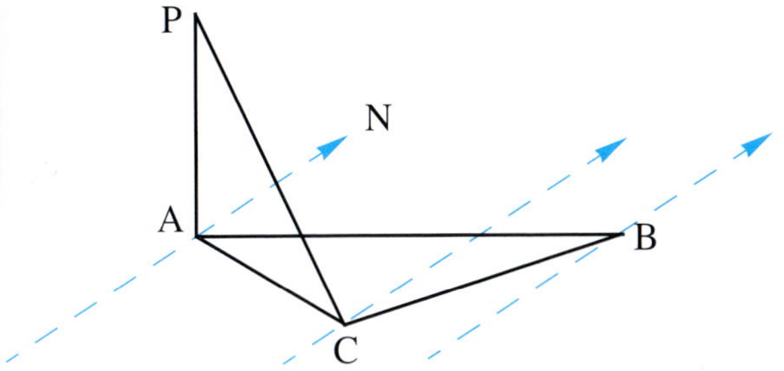

Figure 1

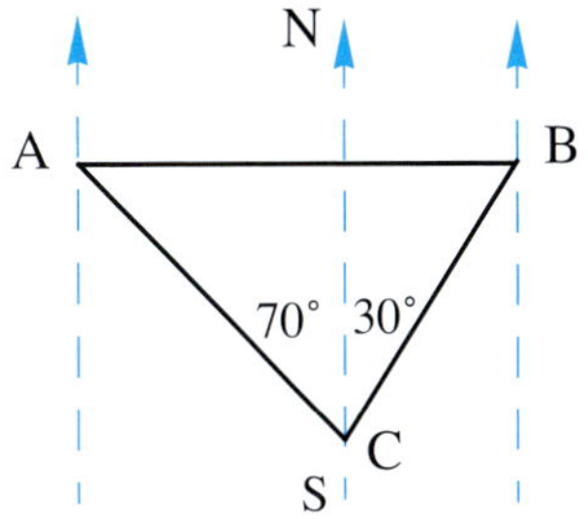

Figure 2

In figure 1, $PA = h$, $\angle PCA = 1.5°$ and thus, $h = AC \times \tan 1.5°$.
In figure 2, $\angle NCA = 70°$ (= 360° – bearing of A from C), $\angle NCB = 30°$ (alternate to $\angle CBS$) and $\angle ABC = 60°$ (complementary to $\angle CBS$) and thus, $\angle CAB = 20°$ (using the angle sum of a triangle).

In ΔABC we have (using the sine rule) that $\frac{AC}{\sin 60°} = \frac{500}{\sin 20°} \Leftrightarrow AC = \frac{500}{\sin 20°} \times \sin 60°$

But, $AC = \frac{h}{\tan 1.5°}$ (from above), $\therefore \frac{h}{\tan 1.5°} = \frac{500}{\sin 20°} \times \sin 60°$

$$\Leftrightarrow h = \tan 1.5° \times \frac{500}{\sin 20°} \times \sin 60°$$

$$= \tan 1.5° \times \frac{500}{\sin 20°} \times \frac{\sqrt{3}}{2}$$

$$= \frac{250\sqrt{3}\tan 1.5°}{\sin 20°}$$

= 33 (to the nearest metre)

That is, the height of the tower (to the nearest metre) is 33 m.

* The problems in this section are more demanding than those of section 9.3 and can be left out.

EXERCISES 9.6

1. A rectagular box is constructed as shown, with measurements HG = 10 cm, $\angle FHE = 30°$, $\angle CEG = 15°$. Find the height of the box.

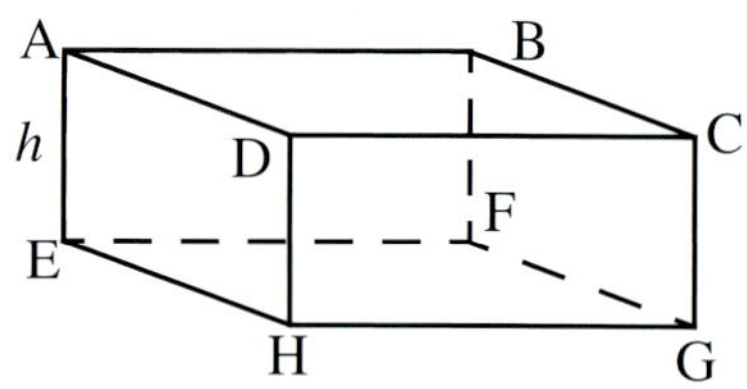

2. From a point A due South of a vertical tower, the angle of elevation of the top of the tower is 15°, and from a point B due West of the tower it is 32°. If the distance from A to B is 50 metres, find the height of the tower.

3. From a point P, the angle of elevation of the top of a radio mast due North of P is 17°. From Q, due West of the radio mast, the angle of elevation is 13°. Given that P and Q are 130 m apart, show that h, the height of the mast, can be given by

$$h = \frac{130}{\sqrt{\tan^2 73 + \tan^2 77}}$$

and find h to the nearest m.

4. From a point due South of a radio tower, an observer measures the angle of elevation to the top of the tower to be 41 °. A second observer is standing on a bearing of 130° from the base of the tower, and on a bearing of 50° from the first observer. If the height of the tower is 45m, find the distance between the two observers, and the angle of elevation of the top of the tower as measured by the second observer.

5. A small plane is flying due east at a constant altitude of 3 km and a constant speed of 120 km/h. It is approaching a small control tower that lies to the South of the plane's path. At time t_0 the plane is on a bearing of 300° from the tower, and elevated at 4.5°. How long does it take for the plane to be due North of the tower, and what is its angle of elevation from the tower at this time?

6. A plane is flying at a constant altitude h with a constant speed of 250km/h. At 10:30 AM it passes directly over a town T heading towards a second town R. A fisherman located next to a river 50km due South of T observes the angle of elevation to the plane to be 4.5°. Town R lies on a bearing of 300° from where the fisherman is standing, and when the plane flies directly over R, the angle of depression to the fisherman is 2.5°. At what time does the plane pass directly over town R?

7. Frank and Stella are walking along a straight road heading North, when they spot the top of a tower in the direction N$\theta°$ E, behind low lying trees. The angle of elevation to the top of the tower is $\alpha°$. After walking d m along the road they notice that the tower is now N$\phi°$ E of the road and that the angle of elevation of the top of the tower is now $\beta°$. Let the height of the tower be h m.

(a) Find the distance of the tower from the
  i. first sighting.
  ii. second sighting.

(b) Find an expression of the height, h m, of the tower.

(c) How much further must Stella and Frank walk before the tower is located in an easterly direction?

9.7 ARCS, SECTORS AND SEGMENTS

9.7.1 RADIAN MEASURE OF AN ANGLE

So far we have been dealing with angles that have been measured in degrees. However, while this has been very useful, such measurements are not suitable for many topics in mathematics. Instead, we introduce a new measure, called the **radian measure**.

The degree measure of angle is based on dividing the complete circle into 360 equal parts known as degrees. Each degree is divided into sixty smaller parts known as minutes, and each minute is divided into sixty seconds.

Decimal parts of a degree can be converted into degrees, minutes and seconds using the **2nd ANGLE** menu and selecting option 4.

This can be useful as calculators generally produce answers in the decimal format. It should also be noted that the degree, minute, second angle system is the same as the hours minutes seconds system that we use to measure time. The above screen could be interpreted as 2.456° and is equal to 2 degrees 27 minutes and 21.6 seconds or as 2.456 hours which is the same as 2 hours 27 minutes and 21.6 seconds.

The degree system is arbitrary in the sense that the decision was made (in the past and due to astronomical measurements) to divide the complete circle into 360 parts. The radian system is an example of a natural measurement system.

One radian is defined as the size of angle needed to cut off an arc of the same length as the radius.

Two radians is the angle that gives an arc length of twice the radius, etc., giving a natural linear conversion between the measure of a radian, the arc length and the radius of a circle.

A complete circle has an arc length of $2\pi r$. It follows that a complete circle corresponds to $\frac{2\pi r}{r} = 2\pi$ radians. This leads to the conversion factor between these two systems:

$$360^\circ = 2\pi \text{ radians or } 180^\circ = \pi \text{ radians (often written as } \pi^c\text{)}$$

So, exactly how large is a radian?

Using the conversion above, if $360^\circ = 2\pi^c$, then $1^c = \frac{360}{2\pi} \approx 57.2957^\circ$

That is, the angle which subtends an **arc of length 1 unit** in a **circle of radius 1** unit, is **1 radian**.

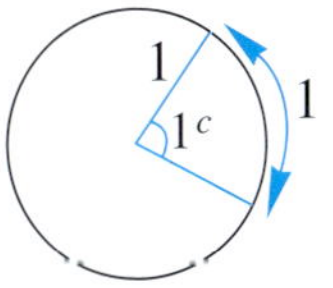

More generally we have:

To convert from **degrees to radians**, multiply angle by $\frac{\pi^c}{180}$.

To convert from **radians to degrees**, multiply angle by $\frac{180}{\pi^c}$.

All conversions between the two systems follow this ratio. It is not generally necessary to convert between the systems as problems are usually worked either entirely in the degree system (as in the previous sections) or in radians (as in the functions and calculus chapters). In the case of arc length and sector areas, it is generally better to work in the radian system.

EXAMPLE 9.24

Convert

(a) 70° into radians (b) 2.34^c into degrees (c) $\frac{\pi^c}{6}$ into degrees

Solution

Using the above conversion factors we have:

(a) $70^\circ = 70 \times \frac{\pi^c}{180} = \frac{7\pi^c}{18}$ or 1.2217^c

(b) $2.34^c = 2.34 \times \frac{180^\circ}{\pi} = 134.0721^\circ = 134^\circ 4' 20''$

(c) $\frac{\pi^c}{6} = \frac{\pi}{6} \times \frac{180^\circ}{\pi} = 30^\circ$

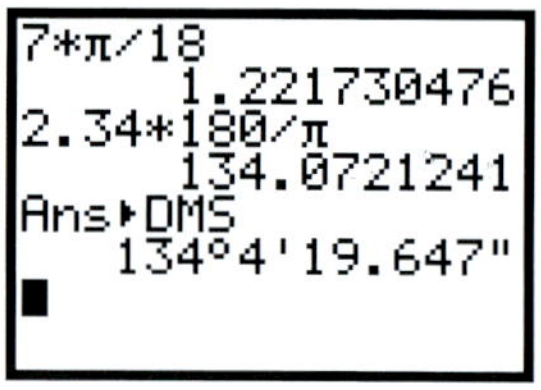

9.7.2 ARC LENGTH

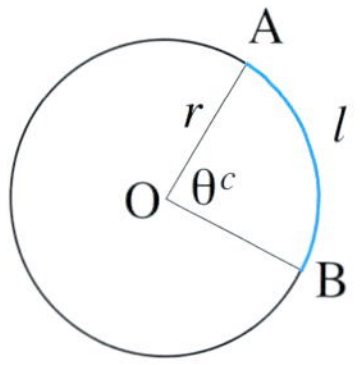

As the arc length AB of a circle is directly proportional to the angle which AB subtends at its centre, then, the arc length AB is a fraction of the circumference of the circle of radius r.

So, if the angle is θ^c, then the arc length is $\frac{\theta}{2\pi}$ of the circumference.

Then, the **(minor) arc length**, AB, denoted by l, is given by $l = \frac{\theta}{2\pi} \times 2\pi r = r\theta$.

i.e.,

$$l = r\theta^c$$

The longer arc AB, called the **major arc**, has a length of $2\pi r - l$.

EXAMPLE 9.25

Using the circle shown, find the arc length AB.

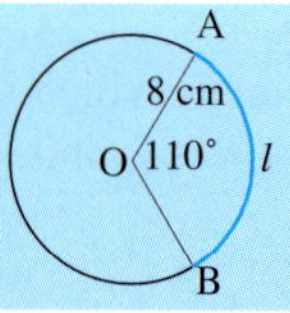

Solution

First we need to convert 110° into radian measure.

$$110^\circ = 110 \times \frac{\pi^c}{180} = \frac{11\pi^c}{18}$$

Then, the arc length, l, is given by, $l = r\theta = 8 \times \frac{11\pi}{18} = \frac{44\pi}{9}$

$$= 15.3588\ldots$$

Therefore, the arc length is 15.36 cm.

9.7.3 AREA OF SECTOR

The formula for the area of a sector is derived as follows:

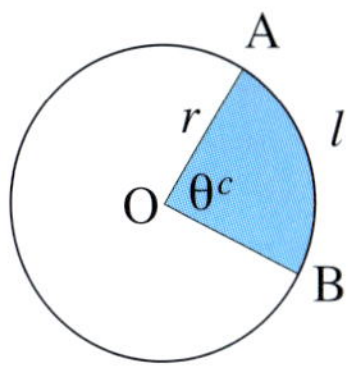

If a sector is cut from a circle of radius r using an angle at the centre of θ radians, the area of the complete circle is πr^2. The fraction of the circle that forms the sector is $\frac{\theta}{2\pi}$ of the complete circle, so the area of the sector is $\pi r^2 \times \frac{\theta}{2\pi} = \frac{1}{2}r^2\theta$.

i.e.,

$$A = \frac{1}{2}r^2\theta^c$$

EXAMPLE 9.26

Find the area and perimeter of the sector shown

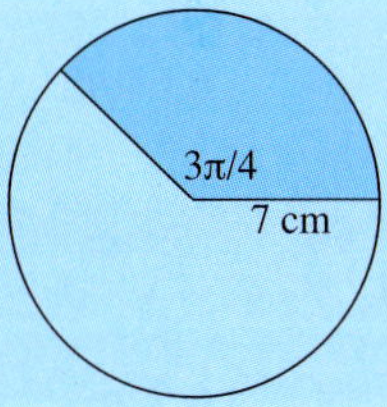

Solution

Area of sector $= \frac{1}{2} \times 7^2 \times \frac{3\pi}{4} = \frac{147\pi}{8}$ cm^2

The perimeter is made up from two radii (14 cm) and the arc $l = r\theta = 7 \times \frac{3\pi}{4} = \frac{21\pi}{4}$

The perimeter is $14 + \frac{21\pi}{4}$ cm.

EXAMPLE 9.27

Find the area and perimeter of the shaded part of the diagram. The radius of the inner circle is 4 cm and the radius of the outer circle is 9 cm.

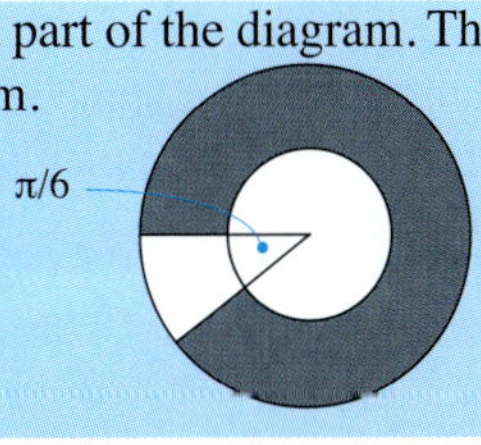

Solution

The angle of the shaded segment = $2\pi - \frac{\pi}{6} = \frac{11\pi^c}{6}$

The shaded area can be found by subtracting the area of the sector in the smaller circle from that in the larger circle.

Shaded area = $\frac{1}{2} \times 9^2 \times \frac{11\pi}{6} - \frac{1}{2} \times 4^2 \times \frac{11\pi}{6} = \frac{11\pi}{12}(9^2 - 4^2) = 59\frac{7}{12}\pi \text{ cm}^2$

The perimeter is made up from two straight lines (each 9 – 4 = 5 cm long) and two arcs.

Perimeter = $10 + 4 \times \frac{11\pi}{6} + 9 \times \frac{11\pi}{6} = 10 + \frac{143\pi}{6}$ cm.

EXERCISES 9.7

1. Find the areas and perimeters of the following sectors:

	Radius	Angle		Radius	Angle
i.	2.6cm	$\frac{\pi}{3}$	ii.	11.5cm	$\frac{\pi}{4}$
iii.	44cm	$\frac{\pi}{4}$	iv.	6.8m	$\frac{2\pi}{3}$
v.	0.64cm	$\frac{3\pi}{4}$	vi.	7.6cm	$\frac{5\pi}{6}$
vii.	324m	$\frac{\pi}{10}$	viii.	8.6cm	$\frac{7\pi}{6}$
ix.	6.2cm	$\frac{4\pi}{3}$	x.	76m	$\frac{11\pi}{6}$
xi.	12cm	30°	xii.	14m	60°
xiii.	2.8cm	120°	xiv.	24.8cm	270°
xv.	1.2cm	15°			

2. A cake has a circumference of 30cm and a uniform height of 7cm. A slice is to be cut from the cake with two straight cuts meeting at the centre. If the slice is to contain 50cm^3 of cake, find the angle between the two cuts, giving the answer in radians to 2 significant figures and in degrees correct to the nearest degree.

3. The diagram shows a part of a Norman arch. The dimensions are shown in metres.

Find the volume of stone in the arch, giving your answer in cubic metres, correct to three significant figures.

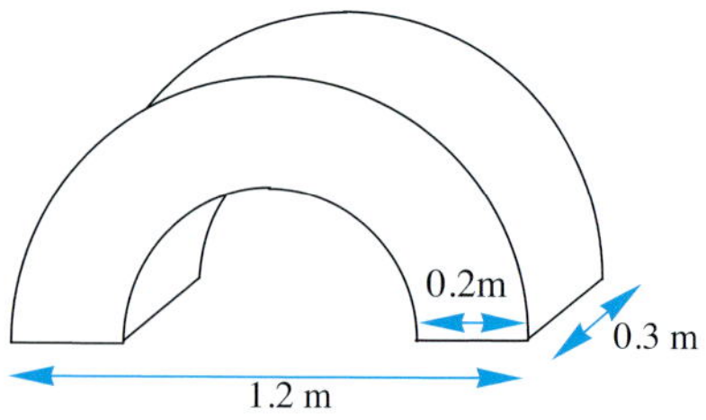

4. In the diagram, find the value of the angle A in radians, correct to three significant figures, if the perimeter is equal to 40cm.

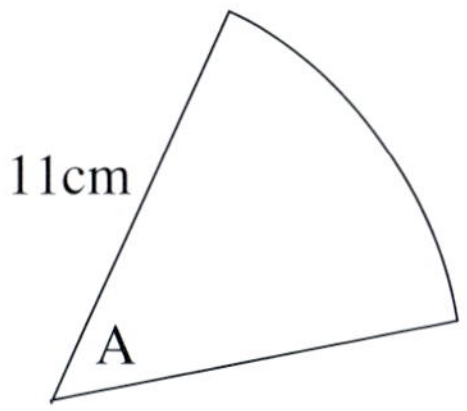

5. The diagram shows a design for a shop sign. The arcs are each one quarter of a complete circle. The radius of the smaller circle is 7 cm and the radius of the larger circle is 9cm.

Find the perimeter of the shape, correct to the nearest centimetre.

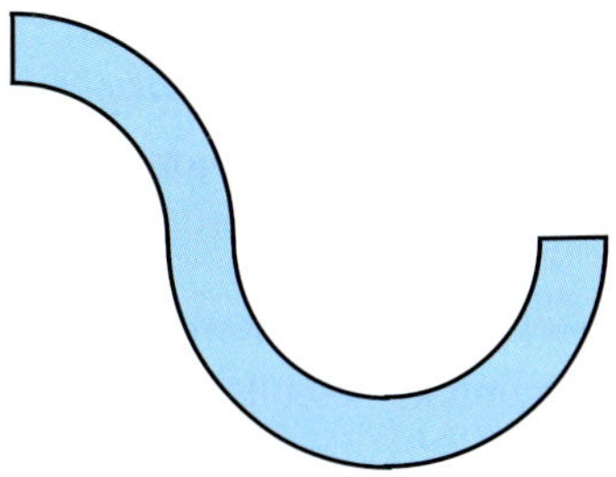

6. Find the shaded area in the diagram. The dimensions are given in centimetres. O is the centre of the circle and AT is a tangent.

Give your answer correct to three significant figures.

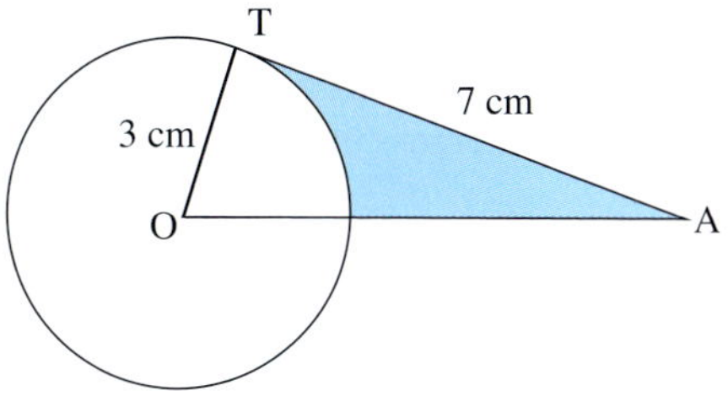

7. The diagram shows a running track. The perimeter of the inside line is 400 metres and the length of each straight section is 100 metres.

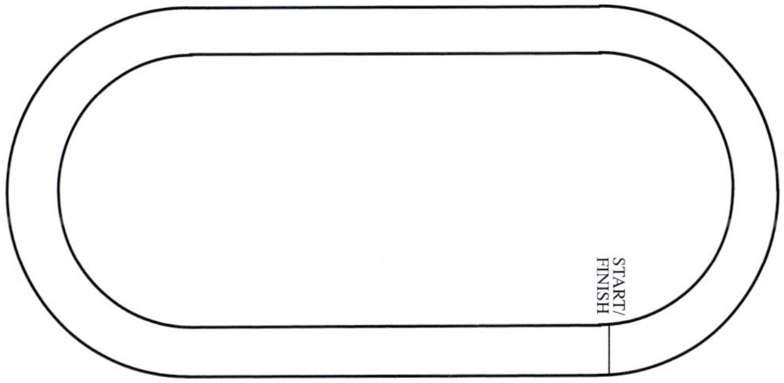

(a) Find the radius of each of the semi-circular parts of the inner track.

(b) If the width of the lane shown is 1 metre, find the perimeter of the outer boundary of the lane.

A second lane is added on the outside of the track. The starting positions of runners who have to run (anti-clockwise) in the two lanes are shown.

(c) Find the value of angleA° (to the nearest degree) if both runners are to run 400 metres.

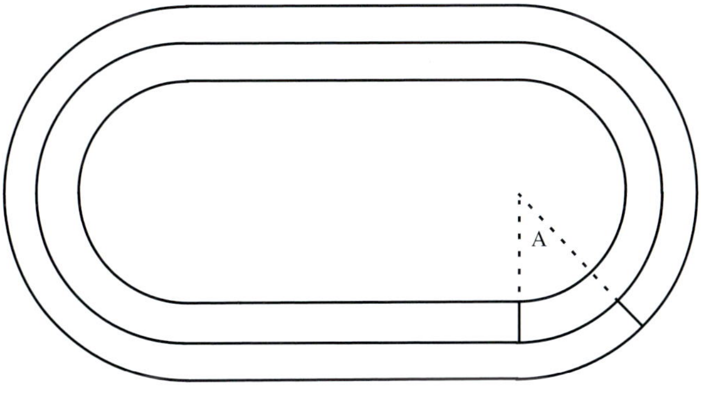

8. Find the angle subtended at the centre of a circle of radius length 12 cm by an arc which forms a sector of area 80 sq. cm

9. Find the angle subtended at the circumference of a circle of radius length 10 cm by an arc which forms a sector of area 75 sq. cm

10. A chord of length 32 cm is drawn in a circle of radius 20 cm.

(a) Find the angle it subtends at the centre.

(b) Find :

i. the minor arc length.

ii. the major arc length.

(c) Find the area of the minor sector.

11. Two circles of radii 6 cm and 8 cm have their centres 10 cm apart. Find the area common to both circles.

12. Two pulleys of radii 16 cm and 20 cm have their centres 40 cm apart. Find the length of the piece of string that will be required to pass tightly round the circles if the string does not cross over.

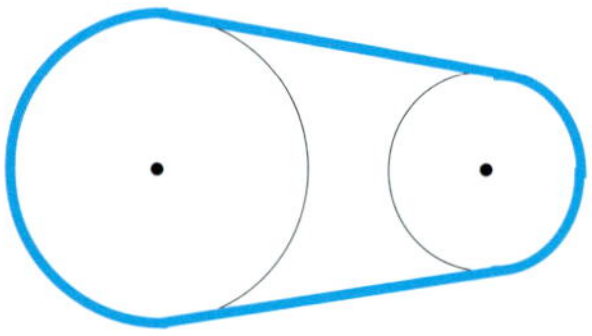

13. Two pulleys of radii 7 cm and 11 cm have their centres 24 cm apart. Find the length of the piece of string that will be required to pass tightly round the circles if

(a) the string cannot cross over.

(b) the string crosses over itself.

14. A sector of a circle has a radius of 15 cm and an angle of $216°$. The sector is folded in such a way that it forms a cone, so that the two straight edges of the sector do not overlap.

(a) Find the base radius of the cone.

(b) Find the vertical height of the cone.

(c) Find the semi–vertical angle of the cone.

15. A taut belt passes over two discs of radii 12 cm and 4 cm as shown in the diagram.

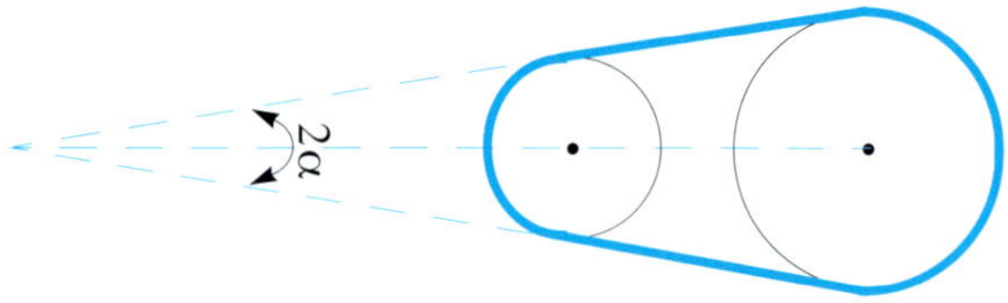

(a) If the total length of the belt is 88 cm, show that $1 = (5.5 - \pi - \alpha)\tan\alpha$

(b) On the same set of axes, sketch the graphs of

i. $y = \dfrac{1}{\tan\alpha}$ ii. $y = 5.5 - \pi - \alpha$.

(c) Hence find $\{\alpha : 1 = (5.5 - \pi - \alpha)\tan\alpha\}$, giving your answer to two d.p.

16. The diagram shows a disc of radius 40 cm with parts of it painted. The smaller circle (having the same centre as the disc) has a radius of 10 cm. What area of the disc has not been painted in red?

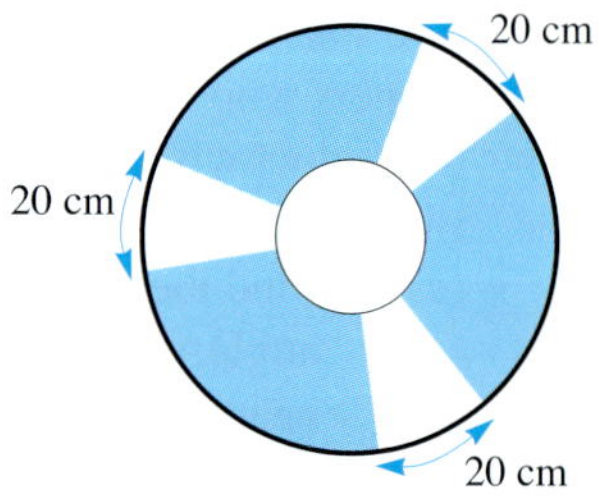

10.1 TRIGONOMETRIC RATIOS

10.1.1 THE UNIT CIRCLE

We saw in Chapter 9 that we were able to find the sine, cosine and tangent of acute angles contained within a right-angled triangle. We extended this to enable us to find the sine and cosine ratio of obtuse angles. To see why this worked, or indeed why it would work for an angle of any magnitude, we need to reconsider how angles are measured. To do this we start by making use of the unit circle and introduce some definitions.

From this point on we define the angle θ as a real number that is measured in either degrees or radians. So that, an expression such as $\sin(180° - \theta)$ will imply that θ is measured in degrees as opposed to the expression $\sin(\pi^c - \theta)$ which would imply that θ is measured in radians. In both cases, it should be clear from the context of the question which one it is.

From the work in §9.7 we have the following conversions between degrees and radians and the exact value of their trigonometric ratios:

θ	$\sin\theta$	$\cos\theta$	$\tan\theta$
$30° = \frac{\pi^c}{6}$	$\frac{1}{2}$	$\frac{\sqrt{3}}{2}$	$\frac{1}{\sqrt{3}}$
$45° = \frac{\pi^c}{4}$	$\frac{1}{\sqrt{2}}$	$\frac{1}{\sqrt{2}}$	1
$60° = \frac{\pi^c}{3}$	$\frac{\sqrt{3}}{2}$	$\frac{1}{2}$	$\sqrt{3}$
$90° = \frac{\pi^c}{2}$	1	0	–

Note that $\tan 90°$ is undefined. We will shortly see why this is the case.

By convention, an angle θ is measured in terms of the rotation of a ray OP from the positive direction of the x–axis, so that a rotation in the **anticlockwise** direction is described as a **positive** angle, whereas a rotation in the **clockwise** direction is described as a **negative** angle.

Let the point $P(x, y)$ be a point on the circumference of the unit circle, $x^2 + y^2 = 1$, with centre at the origin and radius 1 unit.

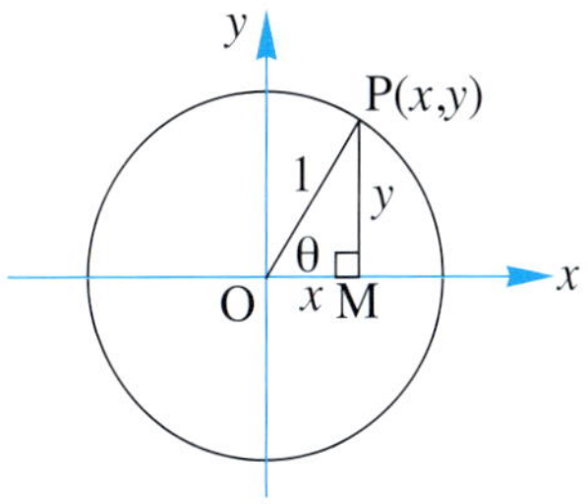

With OP making an angle of θ with the positive direction of the x–axis, we draw the perpendicular from P to meet the x–axis at M. This then provides the following definitions:

$$\sin\theta = \frac{MP}{OP} = \frac{y\text{-coordinate of P}}{OP} = \frac{y}{1} = y$$

$$\cos\theta = \frac{MP}{OP} = \frac{x\text{-coordinate of P}}{OP} = \frac{x}{1} = x$$

$$\tan\theta = \frac{MP}{OM} = \frac{y\text{-coordinate of P}}{x\text{-coordinate of P}} = \frac{y}{x}$$

Note that this means that the y-coordinate corresponds to the sine of the angle θ. That the x-coordinate corresponds to the cosine of the angle θ and that the tangent,. . . , well, for the tangent, let's revisit the unit circle, but this time we will make an addition to the diagram.

Using the existing unit circle, we draw a tangent at the point where the circle cuts the positive x-axis, Q.

Next, we extend the ray OP to meet the tangent at R.

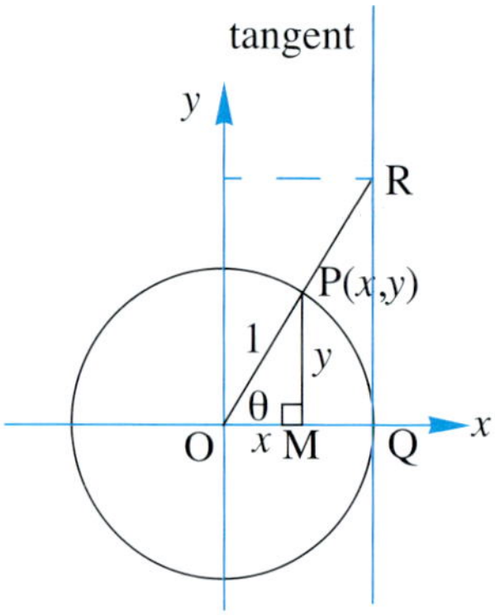

Using similar triangles, we have that $\frac{PM}{OM} = \frac{RQ}{OQ} = \frac{RQ}{1}$.

That is, $\tan\theta = RQ$ – which means that the value of the tangent of the angle θ corresponds to the y-coordinates of point R cut off on the tangent at Q by the extended ray OP.

Also, it is worth noting that $\tan\theta = \frac{PM}{OM} = \frac{y}{x} = \frac{\sin\theta}{\cos\theta}$ (as long as $\cos\theta \neq 0$).

That is, $\tan\theta = \frac{\sin\theta}{\cos\theta}, \cos\theta \neq 0$

From our table of exact values, we note that $\tan 90°$ was undefined. This can be observed from the above diagram. If $\theta = 90°$, P lies on the y-axis, meaning that OP would be parallel to QR, and so, P would never cut the tangent, meaning that no y-value corresponding to R could ever be obtained.

Using a table of values for $\tan\theta$ on the TI-83, we see how the tangent ratio increases as θ increases to 90° and in particular how it is undefined for $\theta = 90°$.

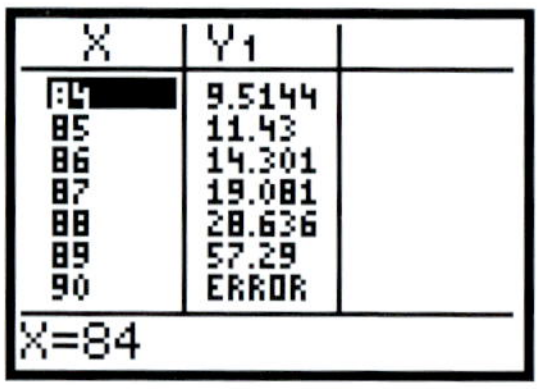

10.1.2 ANGLE OF ANY MAGNITUDE

From the unit circle we have seen how the trigonometric ratios of an acute angle can be obtained – i.e., for the sine ratio we read off the y–axis, for the cosine ratio, we read off the x-axis and for the tangent ratio we read off the tangent. As the point P is located in the first quadrant, then $x \geq 0$, $y \geq 0$ and $\frac{y}{x} \geq 0, x \neq 0$. Meaning that we obtain positive trigonometric ratios.

So, what if P lies in the second quadrant?

We start by drawing a diagram for such a situation:

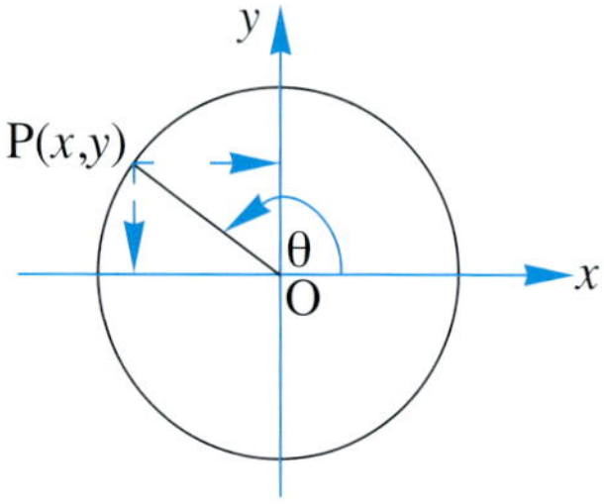

From our diagram we see that if P lies in the second quadrant, the y-value is still positive, the x-value is negative and therefore the ratio, $\frac{y}{x}$ is negative.

This means that, $\sin\theta > 0$, $\cos\theta < 0$ and $\tan\theta < 0$.

In a similar way, we can conclude that if $180° < \theta < 270°$, i.e., the point P is in the **third quadrant**, then,

y-value is negative $\Rightarrow \sin\theta < 0$,

x-value is negative $\Rightarrow \cos\theta < 0$

and therefore the ratio $\frac{y}{x}$-value is positive $\Rightarrow \tan\theta > 0$

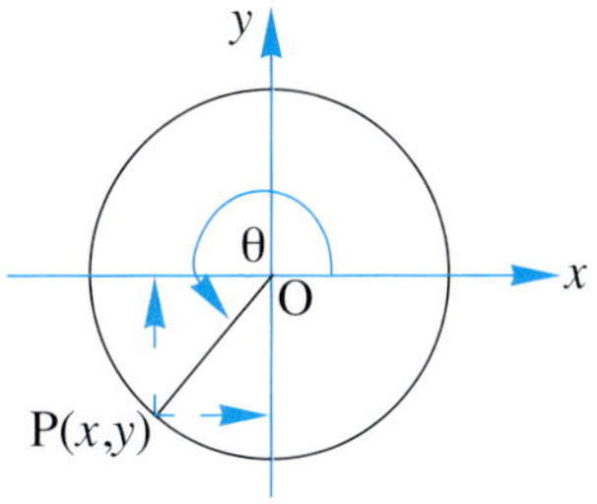

For the **fourth quadrant** we have, $270 < \theta < 360$, so that

y-value is negative $\Rightarrow \sin\theta < 0$,

x-value is positive $\Rightarrow \cos\theta > 0$

and therefore the ratio $\frac{y}{x}$-value is negative $\Rightarrow \tan\theta < 0$

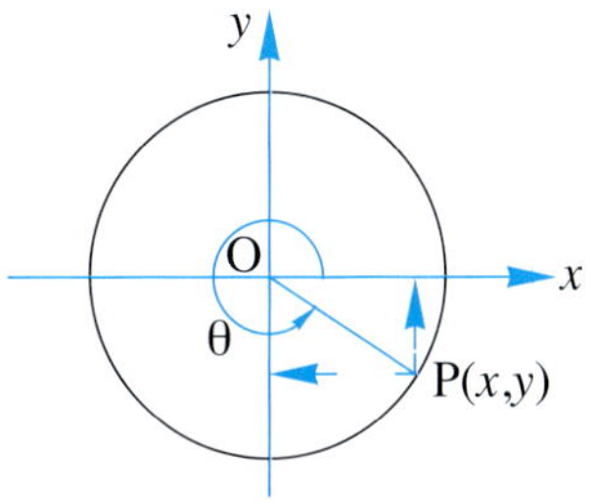

So far, so good. We now know that depending on which quadrant an angle lies in, the sign of the trigonometric ratio will be either positive or negative. In fact, we can summarise this as follows:

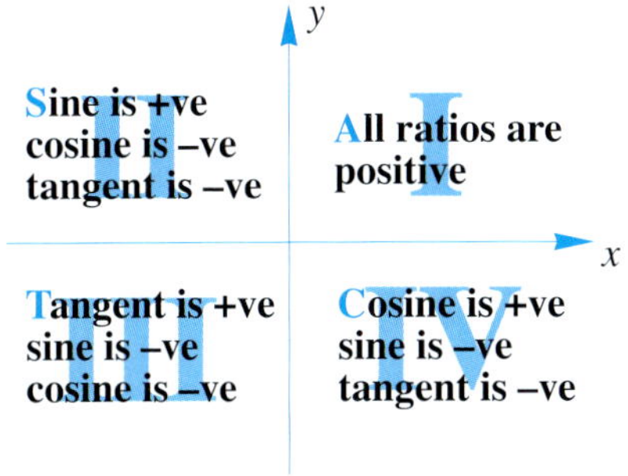

This may be usefully recalled from the following diagram, or by use of a mnemonic sentence

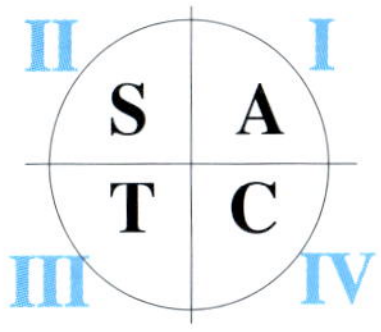

All Stations To City

However, knowing the sign of a trigonometric ratio reflects only half the information. We still need to determine the numerical value. We start by considering a few examples:

Consider the value of $\sin 150°$. Using the unit circle we have:
By symmetry we see that the y-coordinate of Q and the y-coordinate of P are the same and so, $\sin 150° = \sin 30°$.

Therefore, $\sin 150° = \frac{1}{2}$

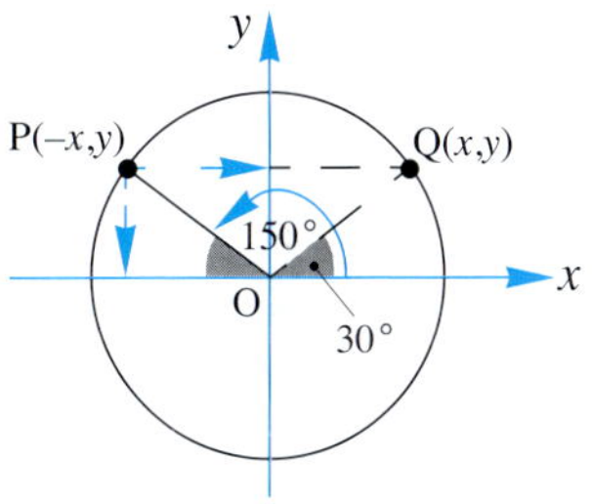

Note that $150° = 150 \times \frac{\pi}{180} = \frac{5\pi}{6}$ and $30° = \frac{\pi}{6}$, so that in radian form we have, $\sin\frac{5\pi}{6} = \sin\frac{\pi}{6} = \frac{1}{2}$.

In other words, we were able to express the sine of an angle in the second quadrant in terms of the sine of an angle in the first quadrant. In particular, we have that

$$\text{If } 0° < \theta < 90°, \sin(180 - \theta) = \sin\theta$$

Or,

$$\text{If } 0^c < \theta < \frac{\pi^c}{2}, \sin(\pi^c - \theta) = \sin\theta$$

Next, consider the value of $\cos 225°$. Using the unit circle we have:

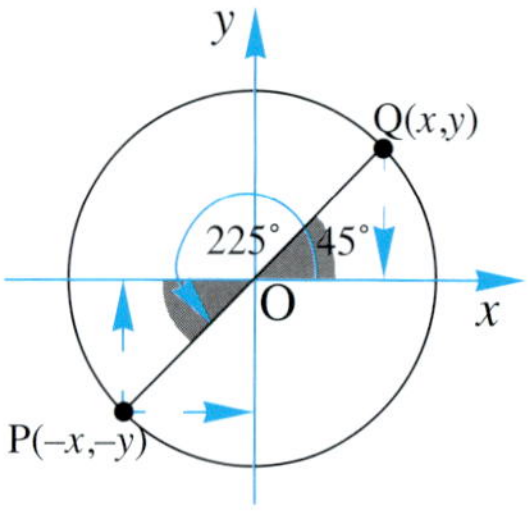

By symmetry we see that the x-coordinate of P has the same magnitude as the x-coordinate of Q but is of the opposite sign.

So, we have that $\cos 225° = -\cos 45°$.

Therefore, $\cos 225° = -\frac{1}{\sqrt{2}}$.

Similarly, as $225° = \frac{3\pi^c}{4}$ and $45° = \frac{\pi^c}{4}$, $\cos\frac{3\pi^c}{4} = -\cos\frac{\pi^c}{4} = -\frac{1}{\sqrt{2}}$.

In other words, we were able to express the cosine of an angle in the third quadrant in terms of the cosine of an angle in the first quadrant. In particular, we have that

If $0° < \theta < 90°$, $\cos(180 + \theta) = -\cos\theta$

Or, If $0^c < \theta < \frac{\pi^c}{2}$, $\cos(\pi^c + \theta) = -\cos\theta$

As a last example we consider the value of $\tan 300°$. This time we need to add a tangent to the unit circle cutting the positive x–axis:

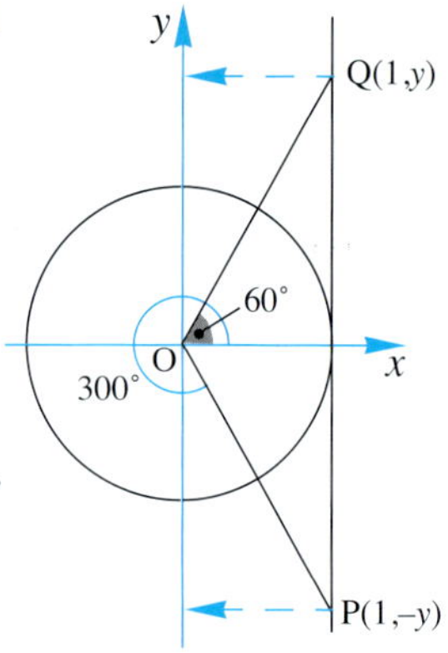

By symmetry we see that the y-coordinate of P has the same magnitude as the y-coordinate of Q but is of the opposite sign.

So, we have that $\tan 300° = -\tan 60°$.

Therefore, $\tan 300° = -\sqrt{3}$.

Similarly, as $300° = \frac{5\pi^c}{3}$ and $60° = \frac{\pi^c}{3}$, $\tan\frac{5\pi^c}{3} = -\tan\frac{\pi^c}{3} = -\sqrt{3}$.

In other words, we were able to express the tangent of an angle in the fourth quadrant in terms of the tangent of an angle in the first quadrant. In particular, we have that

If $0° < \theta < 90°$, $\tan(360 - \theta) = -\tan\theta$

Or, If $0^c < \theta < \frac{\pi^c}{2}$, $\tan(2\pi^c - \theta) = -\tan\theta$

In summary we have:

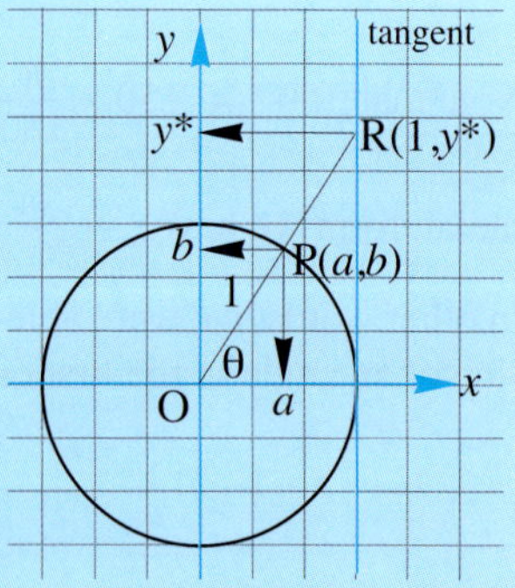

To find the sine of θ , i.e., $\sin\theta$ we read off the y–value of P.

So that, $\sin\theta = b$

To find the cosine of θ , i.e., $\cos\theta$ we read off the x–value of P.

So that, $\cos\theta = a$

To find the tangent of θ , i.e., $\tan\theta$ we read off the y^*–value of R.

So that, $\tan\theta = \frac{b}{a} = y^*$

EXAMPLE 10.1

Find the exact values of

(a) $\cos 120°$ (b) $\sin 210°$ (c) $\cos\frac{7\pi}{4}$ (d) $\tan\frac{5\pi}{4}$

Solution

(a) **Step 1:** Start by drawing the unit circle:
Step 2: Trace out an angle of 120°
Step 3: Trace out the **reference angle** in the first quadrant. In this case it is 60°.
Step 4: Use the symmetry between the reference angle and the given angle.
Step 5: State relationship and give answer;

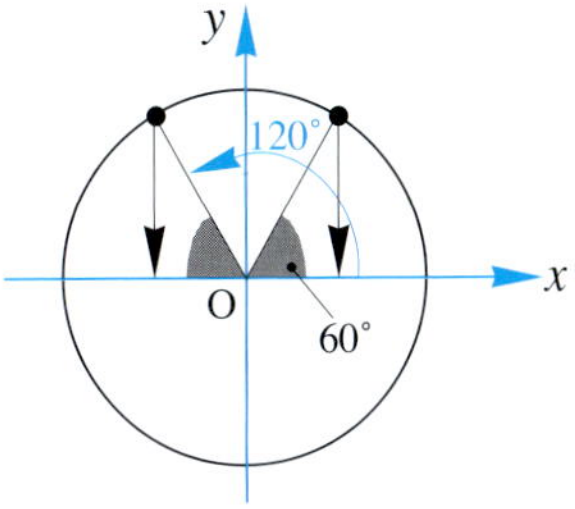

$$\cos 120° = -\cos 60° = -\frac{1}{2}$$

(b) **Step 1:** Start by drawing the unit circle:
Step 2: Trace out an angle of 210°
Step 3: Trace out the **reference angle** in the first quadrant. In this case it is 30°.
Step 4: Use the symmetry between the reference angle and the given angle.
Step 5: State relationship and give answer;

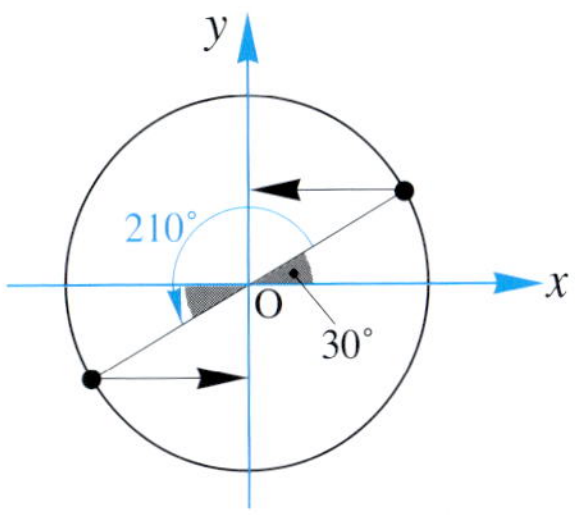

$$\sin 210° = -\sin 30° = -\frac{1}{2}$$

(c) **Step 1:** Start by drawing the unit circle:
Step 2: Trace out an angle of $\frac{7\pi}{4}$ (= 315°)
Step 3: Trace out the **reference angle** in the first quadrant. In this case it is $\frac{\pi}{4}$.
Step 4: Use the symmetry between the reference angle and the given angle.
Step 5: State relationship and give answer;

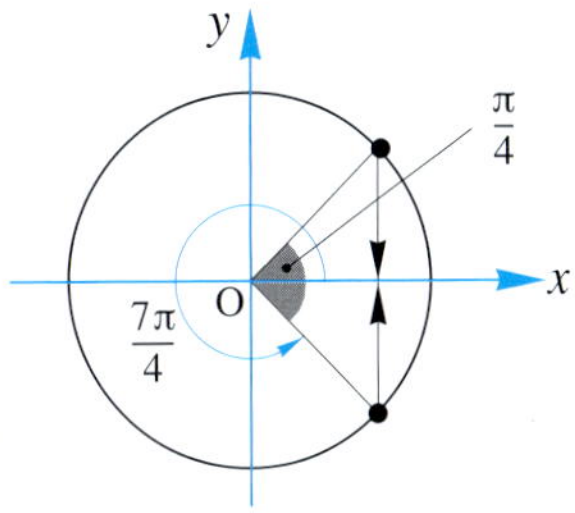

$$\cos\frac{7\pi}{4} = \cos\frac{\pi}{4} = \frac{1}{\sqrt{2}}$$

(c) **Step 1:** Start by drawing the unit circle:
Step 2: Trace out an angle of $\frac{5\pi}{4}$ (= 225°)
Step 3: Trace out the **reference angle** in the first quadrant. In this case it is $\frac{\pi}{4}$.
Step 4: Use the symmetry between the reference angle and the given angle.
Step 5: State relationship and give answer;

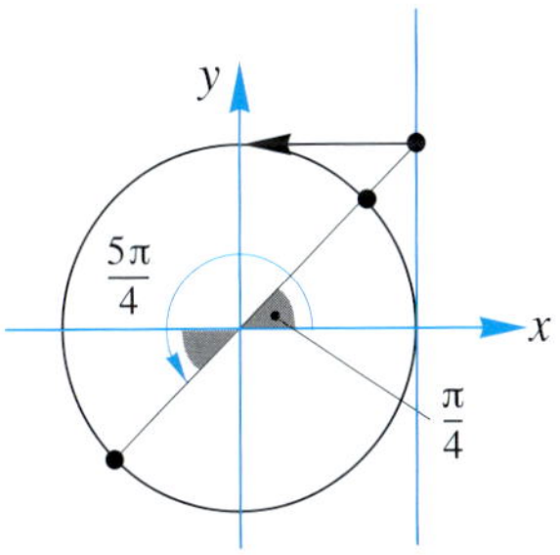

$$\tan\frac{5\pi}{4} = \tan\frac{\pi}{4} = 1$$

The results we have obtained, that is, expressing trigonometric ratios of any angle in terms of trigonometric ratios of acute angles in the first quadrant (i.e., reference angles) are known as **trigonometric reduction formulae**. There are too many formulae to commit to memory, and so it is advisible to draw a unit circle and then use symmetry properties as was done in Examples 10.1. We list a number of these formulae in the table below, where $0 < \theta < \frac{\pi}{2}$ (= 90°).

Note: From this point on angles without the degree symbol or radian symbol will mean an angle measured in radian mode.

Quadrant	θ in degrees	θ in radians
2	$\sin(180° - \theta) = \sin\theta$ $\cos(180° - \theta) = -\cos\theta$ $\tan(180° - \theta) = -\tan\theta$	$\sin(\pi - \theta) = \sin\theta$ $\cos(\pi - \theta) = -\cos\theta$ $\tan(\pi - \theta) = -\tan\theta$
3	$\sin(180° + \theta) = -\sin\theta$ $\cos(180° + \theta) = -\cos\theta$ $\tan(180° + \theta) = \tan\theta$	$\sin(\pi + \theta) = -\sin\theta$ $\cos(\pi + \theta) = -\cos\theta$ $\tan(\pi + \theta) = \tan\theta$
4	$\sin(360° - \theta) = -\sin\theta$ $\cos(360° - \theta) = \cos\theta$ $\tan(360° - \theta) = -\tan\theta$	$\sin(2\pi - \theta) = -\sin\theta$ $\cos(2\pi - \theta) = \cos\theta$ $\tan(2\pi - \theta) = -\tan\theta$

There is another set of results that is suggested by symmetry through the fourth quadrant:

Quadrant	θ in degrees or θ in radians
4	$\sin(-\theta) = -\sin\theta$ $\cos(-\theta) = \cos\theta$ $\tan(-\theta) = -\tan\theta$

There are other trigonometric reduction formulae, where $0 < \theta < \frac{\pi}{2}$ or $0 < \theta < 90°$. These formulae however, are expressed in terms of their variation from the vertical axis.
That is:

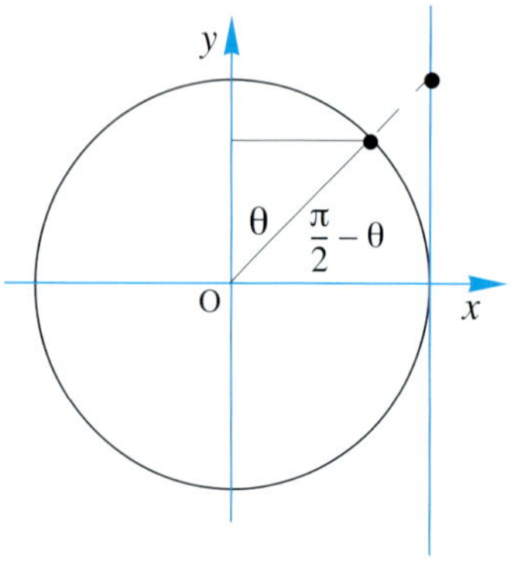

Quadrant	θ in radians or θ in degrees
1	$\sin\left(\frac{\pi}{2} - \theta\right) = \cos\theta$, $\sin(90° - \theta) = \cos\theta$ $\cos\left(\frac{\pi}{2} - \theta\right) = \sin\theta$, $\cos(90° - \theta) = \sin\theta$ $\tan\left(\frac{\pi}{2} - \theta\right) = \cot\theta$, $\tan(90° - \theta) = \cot\theta$

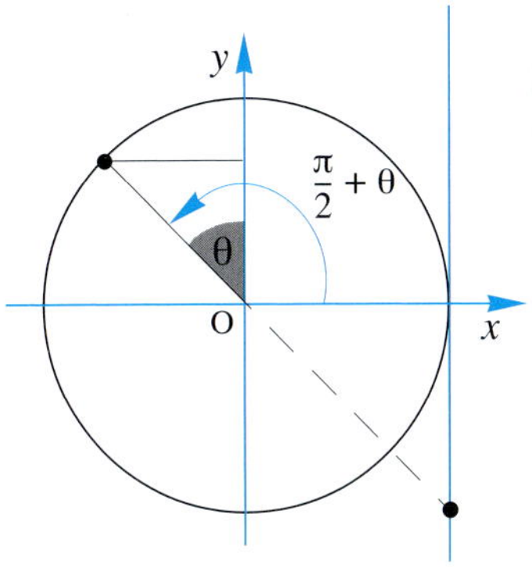

Quadrant	θ in radians or θ in degrees
	$\sin\left(\frac{\pi}{2}+\theta\right) = \cos\theta$, $\sin(90°+\theta) = \cos\theta$
2	$\cos\left(\frac{\pi}{2}+\theta\right) = -\sin\theta$, $\cos(90°+\theta) = -\sin\theta$
	$\tan\left(\frac{\pi}{2}+\theta\right) = -\cot\theta$, $\tan(90°+\theta) = -\cot\theta$

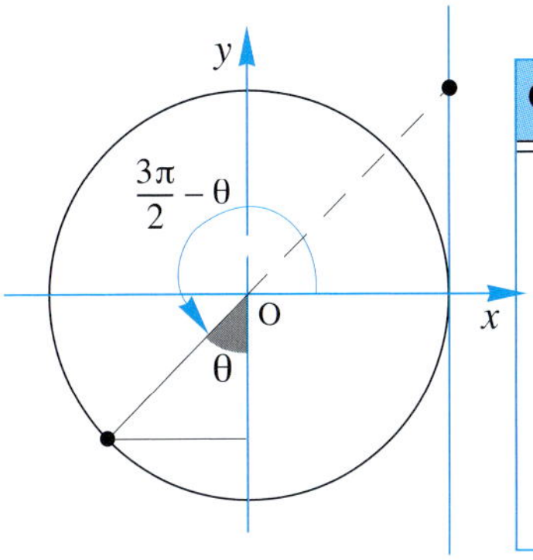

Quadrant	θ in radians or θ in degrees
	$\sin\left(\frac{3\pi}{2}-\theta\right) = -\cos\theta$, $\sin(270°-\theta) = -\cos\theta$
3	$\cos\left(\frac{3\pi}{2}-\theta\right) = -\sin\theta$, $\cos(270°-\theta) = -\sin\theta$
	$\tan\left(\frac{3\pi}{2}-\theta\right) = \cot\theta$, $\tan(270°-\theta) = \cot\theta$

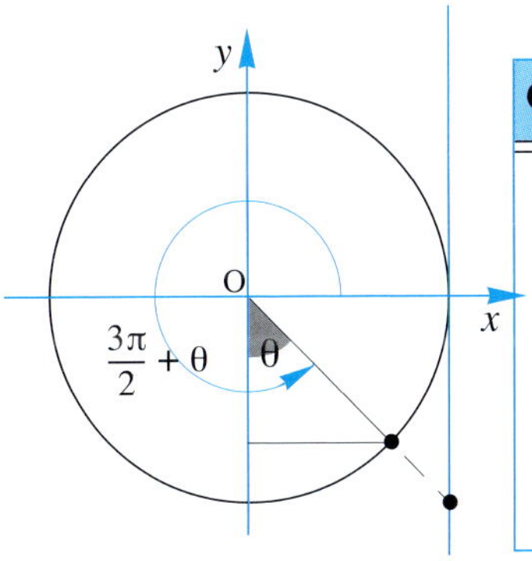

Quadrant	θ in radians or θ in degrees
	$\sin\left(\frac{3\pi}{2}+\theta\right) = -\cos\theta$, $\sin(270°+\theta) = -\cos\theta$
4	$\cos\left(\frac{3\pi}{2}+\theta\right) = \sin\theta$, $\cos(270°+\theta) = \sin\theta$
	$\tan\left(\frac{3\pi}{2}+\theta\right) = -\cot\theta$, $\tan(270°+\theta) = -\cot\theta$

Note the introduction of a new trigonometric ratio, $\cot\theta$. This is one of a set of three other trigonometric ratios known as the **reciprocal trigonometric ratios**, namely **cosecant**, **secant** and **cotangent** ratios. (**Nb**:**These are not required for examination purposes**). They are defined as

cosecant ratio : $\operatorname{cosec}\theta = \frac{1}{\sin\theta}$, $\sin\theta \neq 0$

secant ratio : $\sec\theta = \frac{1}{\cos\theta}$, $\cos\theta \neq 0$

cotangent ratio : $\cot\theta = \frac{1}{\tan\theta}$, $\tan\theta \neq 0$

Note then, that $\cot\theta = \frac{1}{\tan\theta} = \frac{\cos\theta}{\sin\theta}$, $\sin\theta \neq 0$

EXAMPLE 10.2

Given that $\sin\theta = 0.3$, where $0 < \theta < \frac{\pi}{2}$ find

(a) $\sin(\pi + \theta)$ (b) $\sin(2\pi - \theta)$ (c) $\cos\left(\frac{\pi}{2} - \theta\right)$

Solution

(a) From the reduction formulae, we have that $\sin(\pi + \theta) = -\sin\theta$.
Therefore, $\sin(\pi + \theta) = -0.3$.

(b) From the reduction formulae, we have that $\sin(2\pi - \theta) = -\sin\theta$.
Therefore, $\sin(\pi + \theta) = -0.3$.

(c) From the reduction formulae, we have that $\cos\left(\frac{\pi}{2} - \theta\right) = \sin\theta$.

Therefore, $\cos\left(\frac{\pi}{2} - \theta\right) = 0.3$.

EXAMPLE 10.3

Given that $\cos\theta = k$ and $0 < \theta < \frac{\pi}{2}$ find

(a) $\cos(\pi + \theta)$ (b) $\cos(2\pi - \theta)$ (c) $\cos\left(\frac{\pi}{2} + \theta\right)$

Solution

(a) $\cos(\pi + \theta) = -\cos\theta \therefore \cos(\pi + \theta) = -k$.

(b) $\cos(2\pi - \theta) = \cos\theta \therefore \cos(2\pi - \theta) = k$.

(c) $\cos\left(\frac{\pi}{2} + \theta\right) = -\sin\theta$. However, we only have a value for $\cos\theta$.

To determine the value of $\sin\theta$ that corresponds to $\cos\theta = k$ we make use of a right-angled triangle where $\cos\theta = k$.

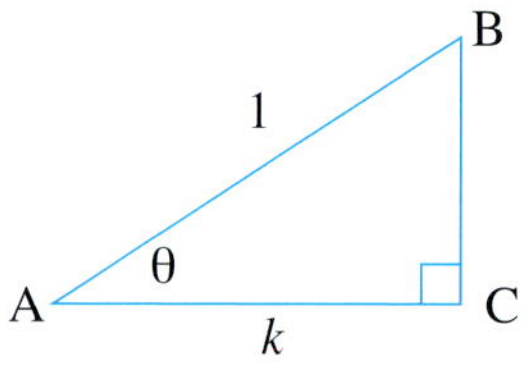

Construct a right-angled triangle ABC, where $\angle BAC = \theta$ so that $AC = k$ and $AB = 1$ $\left(\text{i.e., } \cos\theta = \frac{AC}{AB} = \frac{k}{1} = k\right)$.

Then, from Pythagoras's theorem, we have

$$1^2 = k^2 + BC^2 \Leftrightarrow BC = \pm\sqrt{1 - k^2}$$

Therefore, as $\sin\theta = \frac{BC}{AB} \Rightarrow \sin\theta = \frac{\pm\sqrt{1-k^2}}{1} = \pm\sqrt{1-k^2}$.

However, as $0 < \theta < \frac{\pi}{2}$, then θ is in the first quadrant and so, $\sin\theta > 0 \therefore \sin\theta = \sqrt{1-k^2}$.

Now that we have the value of $\sin\theta$ we can complete the question:

$\sin\left(\frac{\pi}{2} + \theta\right) = -\sin\theta \therefore \sin\left(\frac{\pi}{2} + \theta\right) = -\sqrt{1-k^2}$

Part (c) in Example 10.3 shows a useful approach, i.e., contstructing a right-angled triangle to help in determining the trigonometric ratio of one of the six trig ratios based on any one of the remaining five trig ratios.

EXAMPLE 10.4

Given that $\sin\theta = k$ and $0 < \theta < \frac{\pi}{2}$ find $\tan\theta$

Solution

As we are looking for trigonometric ratios based solely on that of the sine ratio, we start by constructing a right-angled triangle satisfying the relationship, $\sin\theta = k$

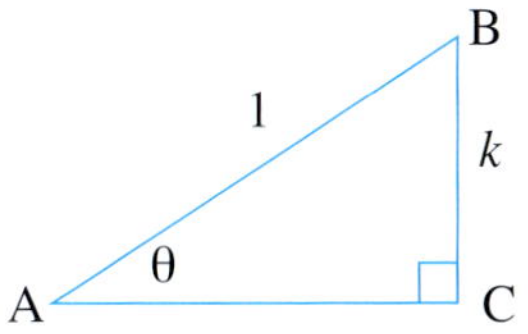

In this case, as $\sin\theta = \frac{\text{opp}}{\text{hyp}} = k \Rightarrow \frac{\text{opp}}{\text{hyp}} = \frac{BC}{AB} = \frac{k}{1}$

(using the simplest ratio).
Using Pythagoras's theorem, we have

$$1^2 = k^2 + AC^2 \Leftrightarrow AC = \pm\sqrt{1-k^2}$$

$$\tan\theta = \frac{\text{opp}}{\text{adj}} = \frac{k}{\pm\sqrt{1-k^2}}.$$

However, as $0 < \theta < \frac{\pi}{2}$, $\tan\theta > 0$ $\therefore \tan\theta = \frac{k}{\sqrt{1-k^2}}$.

EXAMPLE 10.5

Find (a) $\sin\theta = \frac{1}{2}, 0° < \theta < 360°$

(b) $\tan\theta = -\sqrt{3}, 0 < \theta < 2\pi$

(c) $\cos\theta = 1, 0 < \theta < 2\pi$

Solution

(a) This time we are searching for those values of θ for which $\sin\theta = \frac{1}{2}$.

To do this we make use of the unit circle:
From the unit circle, we draw a horizontal chord passing the y–axis at $y = 0.5$.
Then, in the first quadrant, from our table of exact values we have that $\theta = 30°$.

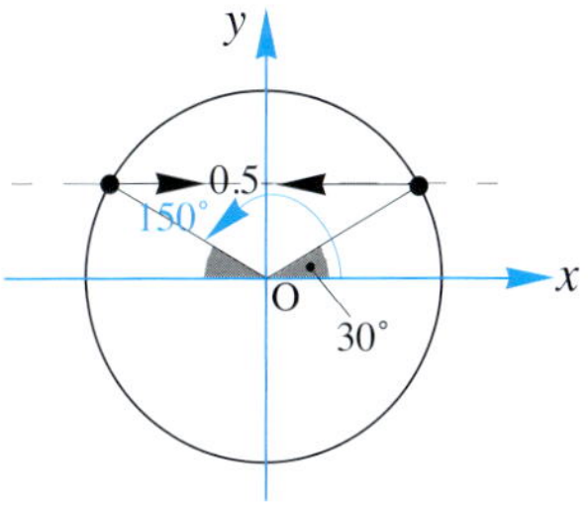

However, by symmetry, we also have that $\sin 150° = \frac{1}{2}$.

Therefore, $\sin\theta = \frac{1}{2}, 0° < \theta < 360°$ if $\theta = 30°$ or $150°$.

(b) This time we are searching for those values of θ for which $\tan\theta = -\sqrt{3}$. From the unit circle, we extend the ray OP so that it cuts the tangent line at R.

Using the exact values, we have $\tan\left(\pi - \frac{\pi}{3}\right) = \tan\frac{2\pi}{3} = -\sqrt{3}$ (as our first value).

And, by symmetry, we also have that $\tan\left(2\pi - \frac{\pi}{3}\right) = -\sqrt{3}$.

Therefore, $\theta = \frac{2\pi}{3}$ or $\theta = \frac{5\pi}{3}$.

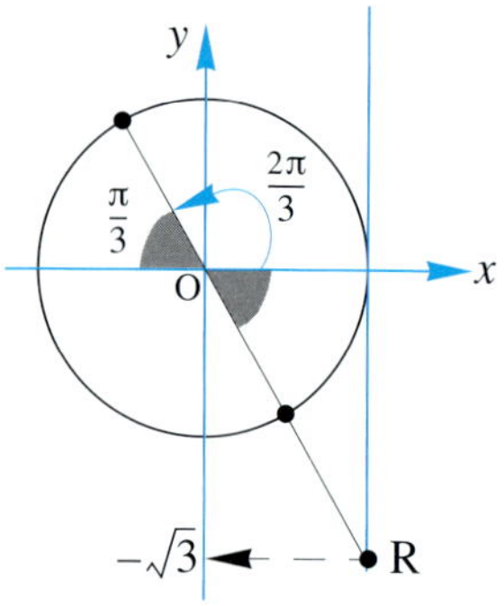

(c) $\cos\theta = 1$, therefore, $\theta = 0$ or $\theta = 2\pi$

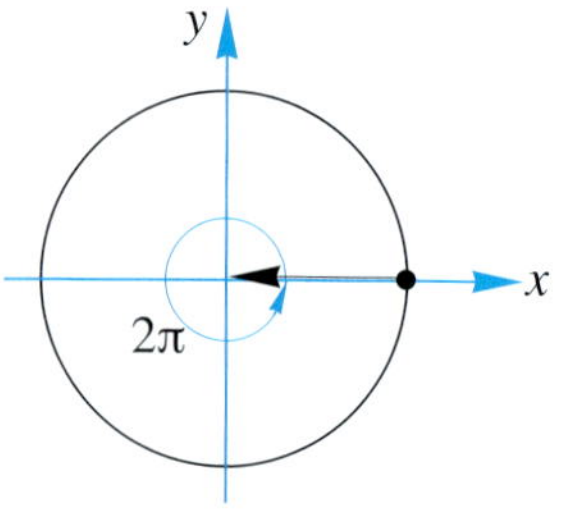

EXAMPLE 10.6

Simplify

(a) $\dfrac{\sin(\pi+\theta)}{\cos(2\pi-\theta)}$ (b) $\dfrac{\sin\left(\frac{\pi}{2}+\theta\right)\cos\left(\frac{\pi}{2}-\theta\right)}{\cos(\pi+\theta)}$, where $0 < \theta < \frac{\pi}{2}$.

Solution

(a) $\dfrac{\sin(\pi+\theta)}{\cos(2\pi-\theta)} = \dfrac{-\sin\theta}{\cos\theta}$

$= -\tan\theta$

(b) $\dfrac{\sin\left(\frac{\pi}{2}+\theta\right)\cos\left(\frac{\pi}{2}-\theta\right)}{\cos(\pi+\theta)} = \dfrac{\cos\theta\sin\theta}{-\cos\theta}$

$= -\sin\theta$

EXERCISES 10.1

1. Convert the following angles to degrees.

(a) $\frac{2\pi}{3}$ (b) $\frac{3\pi}{5}$ (c) $\frac{12\pi}{10}$ (d) $\frac{5\pi}{18}$

2. Convert the following angles to radians.

(a) 180° (b) 270° (c) 140° (d) 320°

3. Find the exact value of

(a)	$\sin 120°$	(b)	$\cos 120°$	(c)	$\tan 120°$	(d)	$\sin 210°$
(e)	$\cos 210°$	(f)	$\tan 210°$	(g)	$\sin 225°$	(h)	$\cos 225°$
(i)	$\tan 225°$	(j)	$\sin 315°$	(k)	$\cos 315°$	(l)	$\tan 315°$
(m)	$\sin 360°$	(n)	$\cos 360°$	(o)	$\tan 360°$		

4. Find the exact value of

(a)	$\sin\pi$	(b)	$\cos\pi$	(c)	$\tan\pi$	(d)	$\sin\frac{3\pi}{4}$
(e)	$\cos\frac{3\pi}{4}$	(f)	$\tan\frac{3\pi}{4}$	(g)	$\sin\frac{7\pi}{6}$	(h)	$\cos\frac{7\pi}{6}$
(i)	$\tan\frac{7\pi}{6}$	(j)	$\sin\frac{5\pi}{3}$	(k)	$\cos\frac{5\pi}{3}$	(l)	$\tan\frac{5\pi}{3}$
(m)	$\sin\frac{7\pi}{4}$	(n)	$\cos\frac{7\pi}{4}$	(o)	$\tan\frac{7\pi}{4}$		

5. Find the exact value of

(a)	$\sin(-210°)$	(b)	$\cos(-30°)$	(c)	$\tan(-135°)$	(d)	$\cos(-420°)$

6. Find the exact value of

(a)	$\sin\left(-\frac{\pi}{6}\right)$	(b)	$\cos\left(-\frac{3\pi}{4}\right)$	(c)	$\tan\left(-\frac{2\pi}{3}\right)$	(d)	$\sin\left(-\frac{7\pi}{6}\right)$
(e)	$\cos\left(-\frac{7\pi}{6}\right)$	(f)	$\tan\left(-\frac{11\pi}{6}\right)$	(g)	$\sin\left(-\frac{7\pi}{3}\right)$		

7. Find the coordinates of the point P on the following unit circles.

(a) (b) (c) (d)

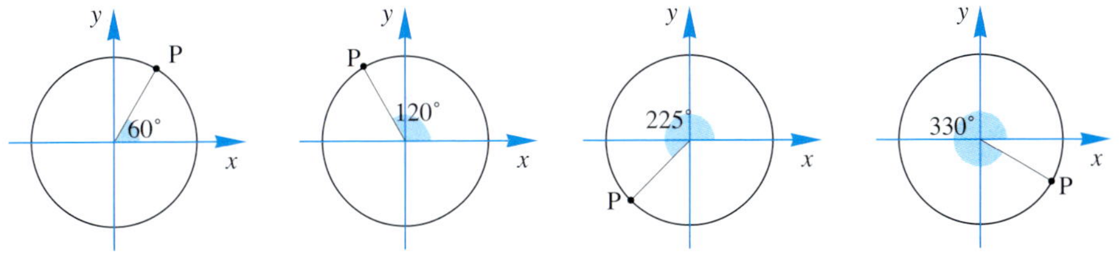

8. Find the exact value of

(a) $\sin\frac{11\pi}{6}\cos\frac{5\pi}{6} - \sin\frac{5\pi}{6}\cos\frac{11\pi}{6}$

(b) $2\sin\frac{\pi}{6}\cos\frac{\pi}{6}$

(c) $\dfrac{\tan\frac{\pi}{3} - \tan\frac{\pi}{6}}{1 + \tan\frac{\pi}{3}\tan\frac{\pi}{6}}$

(d) $\cos\frac{\pi}{4}\cos\frac{\pi}{3} + \sin\frac{\pi}{4}\sin\frac{\pi}{3}$

9. Show that the following relationships are true

(a) $\sin 2\theta = 2\sin\theta\cos\theta$, where $\theta = \frac{\pi}{3}$

(b) $\cos 2\theta = 2\cos^2\theta - 1$, where $\theta = \frac{\pi}{6}$.

(c) $\tan 2\theta = \frac{2\tan\theta}{1-\tan^2\theta}$, where $\theta = \frac{2\pi}{3}$.

(d) $\sin(\theta - \phi) = \sin\theta\cos\phi - \sin\phi\cos\theta$, where $\theta = \frac{2\pi}{3}$ and $\phi = -\frac{\pi}{3}$.

10. Given that $\sin\theta = \frac{2}{3}$ and $0 < \theta < \frac{\pi}{2}$ find

(a) $\sin(\pi + \theta)$ (b) $\sin(2\pi - \theta)$ (c) $\cos\left(\frac{\pi}{2} + \theta\right)$

11. Given that $\cos\theta = \frac{2}{5}$ and $0 < \theta < \frac{\pi}{2}$ find

(a) $\cos(\pi - \theta)$ (b) $\sin\left(\frac{\pi}{2} - \theta\right)$

12. Given that $\tan\theta = k$ and $0 < \theta < \frac{\pi}{2}$ find

(a) $\tan(\pi + \theta)$ (b) $\tan\left(\frac{\pi}{2} + \theta\right)$ (c) $\tan(-\theta)$

13. Given that $\sin\theta = \frac{2}{3}$ and $0 < \theta < \frac{\pi}{2}$ find

(a) $\cos\theta$ (b) $\cos(\pi + \theta)$

14. Given that $\cos\theta = -\frac{4}{5}$ and $\pi < \theta < \frac{3\pi}{2}$ find

(a) $\sin\theta$ (b) $\tan\theta$ (c) $\cos(\pi + \theta)$

15. Given that $\tan\theta = -\frac{4}{3}$ and $\frac{\pi}{2} < \theta < \pi$ find

(a) $\sin\theta$ (b) $\tan\left(\frac{\pi}{2} + \theta\right)$

16. Given that $\cos\theta = k$ and $\frac{3\pi}{2} < \theta < 2\pi$ find

(a) $\cos(\pi - \theta)$ (b) $\sin\theta$

17. Given that $\sin\theta = -k$ and $\pi < \theta < \frac{3\pi}{2}$ find

(a) $\cos\theta$ (b) $\tan\theta$

18. Simplify the following

(a) $\dfrac{\sin(\pi-\theta)\cos\left(\frac{\pi}{2}+\theta\right)}{\sin(\pi+\theta)}$ (b) $\dfrac{\sin\left(\frac{\pi}{2}+\theta\right)\cos\left(\frac{\pi}{2}-\theta\right)}{\sin^2\theta}$ (c) $\dfrac{\sin\left(\frac{\pi}{2}-\theta\right)}{\cos\theta}$

19. If $0 \le \theta \le 2\pi$, find all values of x such that

(a) $\sin x = \dfrac{\sqrt{3}}{2}$ (b) $\cos x = \dfrac{1}{2}$ (c) $\tan x = \sqrt{3}$

(d) $\cos x = -\dfrac{\sqrt{3}}{2}$ (e) $\tan x = -\dfrac{1}{\sqrt{3}}$ (f) $\sin x = -\dfrac{1}{2}$

10.2 TRIGONOMETRIC IDENTITIES

10.2.1 THE FUNDAMENTAL IDENTITY

We have seen a number of important relationships between trigonometric ratios. Relationships that are true for all values of θ are known as **identities**. To signal an identity (as opposed to an equation) the **equivalence** symbol is used, i.e., ' $\equiv$ '.

For example, we can write $(x+1)^2 \equiv x^2+2x+1$, as this statement is true for all values of x. However, we would have to write $(x+1)^2 = x+1$, as this relationship is only true for some values of x (which need to be determined).

One trigonometric identity is based on the unit circle

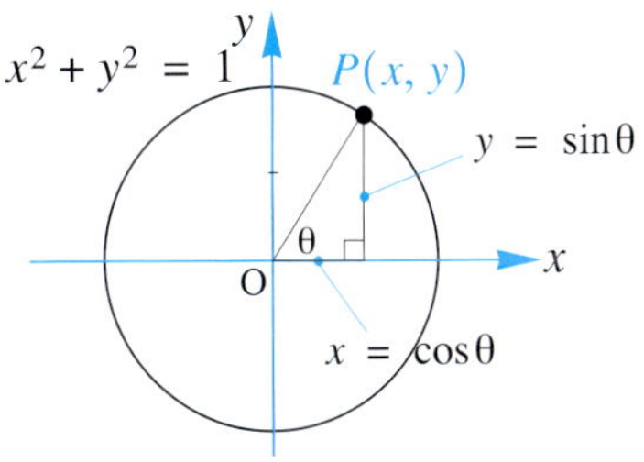

Consider the point $P(x, y)$ on the unit circle,

$$x^2 + y^2 = 1 \quad -(1)$$

From the previous section, we know that

$$x = \cos\theta \quad -(2)$$

$$y = \sin\theta \quad -(3)$$

Substituting (2) and (3) into (1) we have: $(\cos\theta)^2 + (\sin\theta)^2 = 1$

or

$$\sin^2\theta + \cos^2\theta = 1 \quad -(4)$$

This is known as the fundamental trigonometric identity. Note that we have not used the identity symbol, i.e., we have not written $\sin^2\theta + \cos^2\theta \equiv 1$. This is because more often than not, it will be 'obvious' from the setting as to whether a relationship is an identity or an equation. And so, there is a tendency to forgo the formal use of the identity statement and restrict ourselves to the equality statement.

By rearranging the identity we have that $\sin^2\theta = 1-\cos^2\theta$ and $\cos^2\theta = 1-\sin^2\theta$. Similarly we obtain the following two new identities:

Divide both sides of (4) by $\cos^2\theta$: $\dfrac{\sin^2\theta + \cos^2\theta}{\cos^2\theta} = \dfrac{1}{\cos^2\theta} \Leftrightarrow \dfrac{\sin^2\theta}{\cos^2\theta} + \dfrac{\cos^2\theta}{\cos^2\theta} = \dfrac{1}{\cos^2\theta}$

$\Leftrightarrow \tan^2\theta + 1 = \sec^2\theta$ – (5)

Divide both sides of (4) by $\sin^2\theta$: $\dfrac{\sin^2\theta + \cos^2\theta}{\sin^2\theta} = \dfrac{1}{\sin^2\theta} \Leftrightarrow \dfrac{\sin^2\theta}{\sin^2\theta} + \dfrac{\cos^2\theta}{\sin^2\theta} = \dfrac{1}{\sin^2\theta}$

$\Leftrightarrow 1 + \cot^2\theta = \operatorname{cosec}^2\theta$ – (6)

In summary we have:

$$\sin^2\theta + \cos^2\theta = 1$$
$$\tan^2\theta + 1 = \sec^2\theta$$
$$1 + \cot^2\theta = \operatorname{cosec}^2\theta$$

EXAMPLE 10.7

If $\cos\theta = -\dfrac{3}{5}$, where $\pi \le \theta \le \dfrac{3\pi}{2}$, find (a) $\sin\theta$ (b) $\tan\theta$

Solution

(a) Although we solved problems like this in §10.1 by making use of a right-angled triangle, we now solve this question by making use of the trigonometric identities we have just developed.

From $\sin^2\theta + \cos^2\theta = 1$ we have $\sin^2\theta + \left(-\dfrac{3}{5}\right)^2 = 1 \Leftrightarrow \sin^2\theta + \dfrac{9}{25} = 1$

$\Leftrightarrow \sin^2\theta = \dfrac{16}{25}$

$\therefore \sin\theta = \pm\dfrac{4}{5}$

Now, as $\pi \le \theta \le \dfrac{3\pi}{2}$, this means the angle is in the third quadrant, where the sine value is negative. Therefore, we have that $\sin\theta = -\dfrac{4}{5}$

(b) Using the identity $\tan\theta = \dfrac{\sin\theta}{\cos\theta}$, we have $\tan\theta = \dfrac{(-4/5)}{(-3/5)} = \dfrac{4}{3}$.

EXAMPLE 10.8

If $\tan\theta = \dfrac{5}{12}$, where $\pi \le \theta \le \dfrac{3\pi}{2}$, find $\cos\theta$

Solution

(a) From the identity $\tan^2\theta + 1 = \sec^2\theta$ we have $\left(\dfrac{5}{12}\right)^2 + 1 = \sec^2\theta \Leftrightarrow \sec^2\theta = \dfrac{25}{144} + 1$

$\therefore \sec^2\theta = \dfrac{169}{144}$

$\therefore \sec\theta = \pm\dfrac{13}{12}$

Therefore, as $\cos\theta = \dfrac{1}{\sec\theta} \Rightarrow \cos\theta = \pm\dfrac{12}{13}$. However, $\pi \le \theta \le \dfrac{3\pi}{2}$, meaning that θ is in the third quadrant. And so, the cosine is negative. That is, $\cos\theta = -\frac{12}{13}$.

EXAMPLE 10.9

Simplify the following expressions

(a) $\cos\theta + \tan\theta\sin\theta$ (b) $\frac{\cos\theta}{1+\sin\theta} - \frac{1-\sin\theta}{\cos\theta}$

Solution

(a)
$$\begin{aligned}\cos\theta + \tan\theta\sin\theta &= \cos\theta + \frac{\sin\theta}{\cos\theta}\sin\theta \\ &= \cos\theta + \frac{\sin^2\theta}{\cos\theta} \\ &= \frac{\cos^2\theta + \sin^2\theta}{\cos\theta} \\ &= \frac{1}{\cos\theta}\end{aligned}$$

Which can also be written as $\sec\theta$

(b)
$$\begin{aligned}\frac{\cos\theta}{1+\sin\theta} - \frac{1-\sin\theta}{\cos\theta} &= \frac{\cos^2\theta}{(1+\sin\theta)\cos\theta} - \frac{(1-\sin\theta)(1+\sin\theta)}{(1+\sin\theta)\cos\theta} \\ &= \frac{\cos^2\theta}{(1+\sin\theta)\cos\theta} - \frac{1-\sin^2\theta}{(1+\sin\theta)\cos\theta} \\ &= \frac{\cos^2\theta - 1 + \sin^2\theta}{(1+\sin\theta)\cos\theta} \\ &= \frac{(\cos^2\theta + \sin^2\theta) - 1}{(1+\sin\theta)\cos\theta} \\ &= \frac{1-1}{(1+\sin\theta)\cos\theta} \\ &= 0\end{aligned}$$

EXAMPLE 10.10

Show that $\frac{1-2\cos^2\theta}{\sin\theta\cos\theta} = \tan\theta - \cot\theta$.

Solution

$$\begin{aligned}\text{R.H.S} = \tan\theta - \cot\theta &= \frac{\sin\theta}{\cos\theta} - \frac{\cos\theta}{\sin\theta} \\ &= \frac{\sin^2\theta - \cos^2\theta}{\sin\theta\cos\theta} \\ &= \frac{(1-\cos^2\theta) - \cos^2\theta}{\sin\theta\cos\theta} \\ &= \frac{1-2\cos^2\theta}{\sin\theta\cos\theta}\end{aligned}$$

EXERCISES 10.2.1

1. Prove the identity

(a) $\sin\theta + \cot\theta\cos\theta = \operatorname{cosec}\theta$

(b) $\frac{\sin\theta}{1+\cos\theta} + \frac{1+\cos\theta}{\sin\theta} = \frac{2}{\sin\theta}$

(c) $\frac{\sin^2\theta}{1-\cos\theta} = 1+\cos\theta$

(d) $3\cos^2 x - 2 = 1 - 3\sin^2 x$

(e) $\tan^2 x\cos^2 x + \cot^2 x\sin^2 x = 1$

(f) $\sec\theta - \sec\theta\sin^2\theta = \cos\theta$

(g) $\sin^2\theta(1+\cot^2\theta) - 1 = 0$

(h) $\frac{1}{1-\sin\phi} + \frac{1}{1+\sin\phi} = 2\sec^2\phi$

(i) $\frac{\cos\theta}{1+\sin\theta} + \tan\theta = \sec\theta$

(j) $\frac{1-\sin\theta}{\cos\theta} = \frac{\cos\theta}{1+\sin\theta}$

(k) $\frac{1}{\sec x + \tan x} = \sec x - \tan x$

(l) $\sin x + \frac{\cos^2 x}{1+\sin x} = 1$

(m) $\frac{\sec\phi + \operatorname{cosec}\phi}{\tan\phi + \cot\phi} = \sin\phi + \cos\phi$

(n) $\frac{\sin x + 1}{\cos x} = \frac{\sin x - \cos x + 1}{\sin x + \cos x - 1}$

(o) $\tan x + \sec x = \frac{\tan x + \sec x - 1}{\tan x - \sec x + 1}$

2. Prove the following

(a) $(\sin x + \cos x)^2 + (\sin x - \cos x)^2 = 2$

(b) $\sec^2\theta\operatorname{cosec}^2\theta = \sec^2\theta + \operatorname{cosec}^2\theta$

(c) $\sin^4 x - \cos^4 x = (\sin x + \cos x)(\sin x - \cos x)$

(d) $\sec^4 x - \sec^2 x = \tan^4 x + \tan^2 x$

(e) $\frac{\sin^3 x + \cos^3 x}{\sin x + \cos x} = 1 - \sin x\cos x$

(f) $(\cot x - \operatorname{cosec} x)^2 = \frac{\sec x - 1}{\sec x + 1}$

(g) $(2b\sin x\cos x)^2 + b^2(\cos^2 x - \sin^2 x)^2 = b^2$

3. Eliminate θ from each of the following pairs

(a) $x = k\sin\theta,\ y = k\cos\theta$

(b) $x = b\sin\theta,\ y = a\cos\theta$

(c) $x = 1+\sin\theta,\ y = 2-\cos\theta$

(d) $x = 1-b\sin\theta,\ y = 2+a\cos\theta$

(e) $x = \sin\theta + 2\cos\theta,\ y = \sin\theta - 2\cos\theta$

4. (a) If $\tan\theta = \frac{3}{4}, \pi \le \theta \le \frac{3\pi}{2}$, find i. $\cos\theta$ ii. $\operatorname{cosec}\theta$

(b) If $\sin\theta = -\frac{3}{4}, \frac{3\pi}{2} \le \theta \le 2\pi$, find i. $\sec\theta$ ii. $\cot\theta$

5. Solve the following, where $0 \le \theta \le 2\pi$

(a) $4\sin\theta = 3\operatorname{cosec}\theta$

(b) $2\cos^2\theta + \sin\theta - 1 = 0$

(c) $2 - \sin\theta = 2\cos^2\theta$

(d) $2\sin^2\theta = 2 + 3\cos\theta$

6. Prove $\sin^2 x(1 + n\cot^2 x) + \cos^2 x(1 + n\tan^2 x) = \sin^2 x(n + \cot^2 x) + \cos^2 x(n + \tan^2 x)$.

7. If $k\sec\phi = m\tan\phi$, prove that $\sec\phi\tan\phi = \dfrac{mk}{m^2 - k^2}$.

8. If $x = k\sec^2\phi + m\tan^2\phi$ and $y = l\sec^2\phi + n\tan^2\phi$, prove that $\dfrac{x-k}{k+m} = \dfrac{y-l}{l+n}$.

9. Given that $\tan\theta = \dfrac{2a}{a^2-1}, 0 < \theta < \dfrac{\pi}{2}$, find (a) $\sin\theta$ (b) $\cos\theta$

10. (a) If $\sin x + \cos x = 1$, find the values of

i. $\sin^3 x + \cos^3 x$ ii. $\sin^4 x + \cos^4 x$

(b) Hence, deduce the value of $\sin^k x + \cos^k x$, where k is a positive integer.

11. If $\tan\phi = -\dfrac{1}{\sqrt{x^2-1}}, \dfrac{\pi}{2} < \phi < \pi$, find, in terms of x,

(a) $\sin\phi + \cos\phi$ (b) $\sin\phi - \cos\phi$ (c) $\sin^4\phi - \cos^4\phi$

12. Find (a) the maximum value of (b) the minimum value of

i. $\cos^2\theta + 5$ ii. $\dfrac{5}{3\sin^2\theta + 2}$ iii. $2\cos^2\theta + \sin\theta - 1$

13. (a) Given that $b\sin\phi = 1$ and $b\cos\phi = \sqrt{3}$, find b.

(b) Hence, find all values of ϕ that satisfy the relationship described in (a).

14. Find (a) the maximum value of (b) the minimum value of

i. $5^{3\sin\theta - 1}$ ii. $3^{1 - 2\cos\theta}$

15. Given that $\sin\theta\cos\theta = k$, find (a) $(\sin\theta + \cos\theta)^2$, $\sin\theta + \cos\theta > 0$.

(b) $\sin^3\theta + \cos^3\theta$, $\sin\theta + \cos\theta > 0$.

16. (a) Given that $\sin\phi = \dfrac{1-a}{1+a}, 0 < \phi < \dfrac{\pi}{2}$, find $\tan\phi$.

(b) Given that $\sin\phi = 1 - a, \dfrac{\pi}{2} < \phi < \pi$, find i. $2 - \cos\phi$

ii. $\cot\phi$

17. Find

(a) the value(s) of $\cos x$, where $\cot x = 4(\operatorname{cosec} x - \tan x)$, $0 < x < \pi$.

(b) the values of $\sin x$, where $3\cos x = 2 + \dfrac{1}{\cos x}, 0 \le x \le 2\pi$.

18. Given that $\sin 2x = 2\sin x\cos x$, find all values of x, such that $2\sin 2x = \tan x, 0 \le x \le \pi$.

10.2.2 DOUBLE ANGLE IDENTITIES

As we have seen in the previous section, there are numerous trigonometric identities. However, they were all derived from the fundamental identities. In fact there are many more identities that can be derived, however, in this section we shall concentrate on a special set of identities known as the **double angle identities**.

$$\sin 2\theta = 2\sin\theta\cos\theta$$
$$\cos 2\theta = \cos^2\theta - \sin^2\theta$$

Both identities can be confirmed by considering their graphical representation – which is dealt with in the next section. However, using a table of values to compare the values of $\sin 2\theta$ and $2\sin\theta\cos\theta$ for the first identity and then doing the same to compare the values of $\cos 2\theta$ to those of $\cos^2\theta - \sin^2\theta$ will suffice at this stage.

Tables of values for $\sin 2\theta$ and $2\sin\theta\cos\theta$:

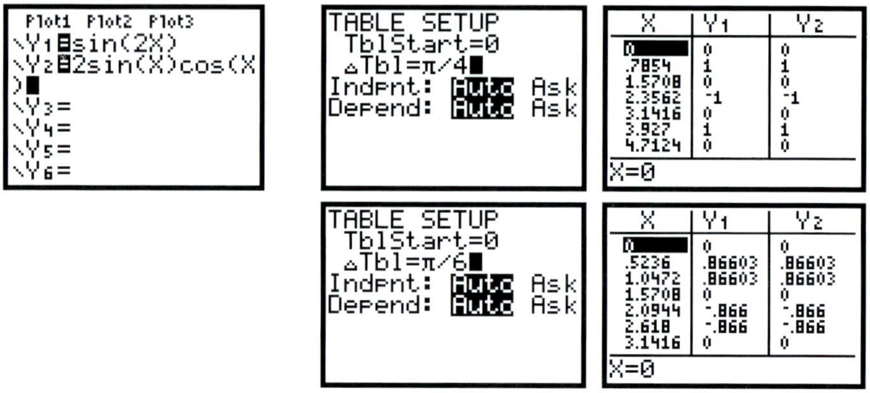

Tables of values for $\cos 2\theta$ and $\cos^2\theta - \sin^2\theta$:

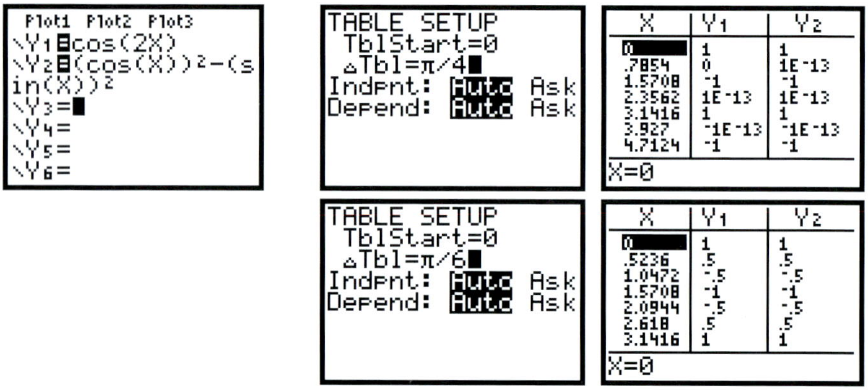

The second of these can be further developed to give:

$$\cos 2\theta = \cos^2\theta - \sin^2\theta = \cos^2\theta - (1-\cos^2\theta) = 2\cos^2\theta - 1$$

and $$\cos 2\theta = \cos^2\theta - \sin^2\theta = (1-\sin^2\theta) - \sin^2\theta = 1 - 2\sin^2\theta$$

Summary of double angle identities:

$$\begin{aligned} \sin 2\theta &= 2\sin\theta\cos\theta \\ \cos 2\theta &= \cos^2\theta - \sin^2\theta \\ &= 2\cos^2\theta - 1 \\ &= 1 - 2\sin^2\theta \end{aligned}$$

EXAMPLE 10.11

If $\sin\theta = \frac{2}{7}$, where $\frac{\pi}{2} \le \theta \le \pi$ find

(a) $\sin 2\theta$ (b) $\cos 2\theta$ (c) $\tan 2\theta$

Solution

We start by drawing the relevant right-angled triangle:

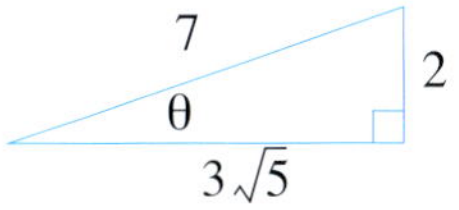

(a) $\sin 2\theta = 2\sin\theta\cos\theta = 2 \times \frac{2}{7} \times -\frac{3\sqrt{5}}{7}$

$= -\frac{12\sqrt{5}}{49}$

(b) $\cos 2\theta = 1 - 2\sin^2\theta = 1 - 2 \times \left(\frac{2}{7}\right)^2 = \frac{41}{49}$

(c) $\tan 2\theta = \frac{\sin 2\theta}{\cos 2\theta} = \frac{-\frac{12\sqrt{5}}{49}}{\frac{41}{49}} = -\frac{12\sqrt{5}}{41}$

EXAMPLE 10.12

Find the exact value of $\cos 15°$

Solution

We start by noticing that $30 = 2 \times 15$ so that using the identiy $\cos 2\theta = 2\cos^2\theta - 1$ we have

$$\cos 30° = \cos(2 \times 15°) = 2\cos^2 15° - 1 .$$

Or, after rearranging: $\cos^2 15° = \frac{1}{2}(\cos 30° + 1)$

$= \frac{1}{2}\left(\frac{\sqrt{3}}{2} + 1\right)$

$= \frac{1}{4}(\sqrt{3} + 2)$

$\therefore \cos 15° = \frac{1}{2}\sqrt{(\sqrt{3} + 2)}$ (taking the positive root)

Of course, having the square root of an expression involving another square root isn't very nice.

We leave it as an exercise for you to show that $\frac{1}{2}\sqrt{(\sqrt{3}+2)} = \frac{1}{4}(\sqrt{3}+1)\sqrt{2}$.

This then gives the result that $\cos 15° = \frac{1}{4}(\sqrt{3}+1)\sqrt{2}$

EXAMPLE 10.13

Prove that $\frac{\sin 2\phi + \sin\phi}{\cos 2\phi + \cos\phi + 1} = \tan\phi$

Solution

(a) $$\text{L.H.S} = \frac{\sin 2\phi + \sin\phi}{\cos 2\phi + \cos\phi + 1} = \frac{2\sin\phi\cos\phi + \sin\phi}{2\cos^2\phi - 1 + \cos\phi + 1}$$
$$= \frac{\sin\phi(2\cos\phi + 1)}{\cos\phi(2\cos\phi + 1)}$$
$$= \frac{\sin\phi}{\cos\phi}$$
$$= \tan\phi$$
$$= \text{R.H.S}$$

EXAMPLE 10.14

Prove that $\sin 2\alpha\tan\alpha + \cos 2\alpha = 1$

Solution

(a) $$\text{L.H.S} = \sin 2\alpha\tan\alpha + \cos 2\alpha = 2\sin\alpha\cos\alpha \times \frac{\sin\alpha}{\cos\alpha} + (1 - 2\sin^2\alpha)$$
$$= 2\sin^2\alpha + 1 - 2\sin^2\alpha$$
$$= 1$$
$$= \text{R.H.S}$$

(b) $$\text{R.H.S} = \cot\beta - \tan\beta = \frac{\cos\beta}{\sin\beta} - \frac{\sin\beta}{\cos\beta}$$
$$= \frac{\cos^2\beta - \sin^2\beta}{\sin\beta\cos\beta}$$
$$= \frac{\cos 2\beta}{\frac{1}{2}\sin 2\beta}$$
$$= 2\frac{\cos 2\beta}{\sin 2\beta}$$
$$= 2\cot 2\beta$$
$$= \text{L.H.S}$$

Notice that when proving identities, when all else fails, then express everything in terms of sine and cosine. This will always lead to the desired result – even though sometimes the working seems like it will only grow and grow – eventually, it does simplify. Be persistent.

To prove a given identity, any one of the following approaches can be used:

1. Start with the L.H.S and then show that L.H.S = R.H.S
2. Start with the R.H.S and then show that R.H.S = L.H.S
3. Show that L.H.S = p, show that R.H.S = p $\Rightarrow$ L.H.S = R.H.S
4. Start with L.H.S = R.H.S $\Rightarrow$ L.H.S – R.H.S = 0 .

When using approaches 1., and 2., choose whichever side has more to work with.

EXAMPLE 10.15

Find all values of x, such that $\sin 2x = \cos x$, where $0 \le x \le 2\pi$.

Solution

$$\sin 2x = \cos x \Leftrightarrow 2\sin x\cos x = \cos x$$
$$\Leftrightarrow 2\sin x\cos x - \cos x = 0$$
$$\Leftrightarrow \cos x(2\sin x - 1) = 0$$
$$\Leftrightarrow \cos x = 0 \text{ or } \sin x = \frac{1}{2}$$

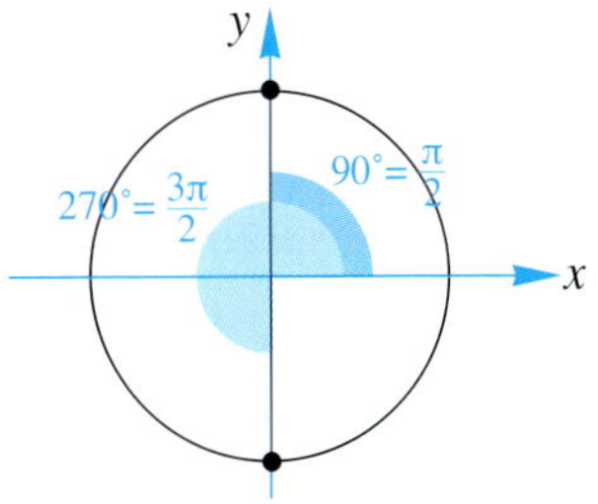

Now, $\cos x = 0, 0 \le x \le 2\pi \Leftrightarrow x = \frac{\pi}{2}, \frac{3\pi}{2}$

and $\sin x = \frac{1}{2}, 0 \le x \le 2\pi \Leftrightarrow x = \frac{\pi}{6}, \frac{5\pi}{6}$

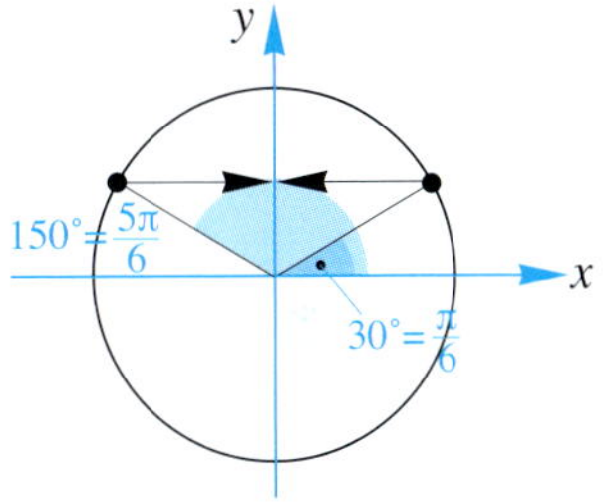

EXERCISES 10.2.2

1. Given that $\sin\theta = -\frac{5}{6}, \frac{3\pi}{2} \le \theta \le 2\pi$, evaluate

   (a) $\sin 2\theta$ (b) $\cos 2\theta$ (c) $\tan 2\theta$ (d) $\sin 4\theta$

2. Given that $\tan x = -3, \frac{\pi}{2} \le x \le \pi$, evaluate

   (a) $\sin 2x$ (b) $\cos 2x$ (c) $\tan 2x$ (d) $\tan 4x$

3. Find the exact value of $\sin\frac{\pi}{12}$

4. Given that $\tan x = \frac{a}{b}, \pi \le x \le \frac{3\pi}{2}$, evaluate

   (a) $\sin 2x$ (b) $\operatorname{cosec} 2x$ (c) $\cos 4x$ (d) $\tan 2x$

5. Prove the following identities:

(a) $\tan(\theta + \phi) + \tan(\theta - \phi) = \dfrac{2\sin 2\theta}{\cos 2\theta + \cos 2\phi}$

(b) $\dfrac{1 + \cos 2y}{\sin 2y} = \dfrac{\sin 2y}{1 - \cos 2y}$

(c) $\cos^4\alpha - \sin^4\alpha = 1 - 2\sin^2\alpha$

(d) $\dfrac{1}{\sin y \cos y} - \dfrac{\cos y}{\sin y} = \tan y$

(e) $\dfrac{1 + \sin 2\theta}{\cos 2\theta} = \dfrac{\cos\theta + \sin\theta}{\cos\theta - \sin\theta}$

(f) $\cos 2x = \dfrac{1 - \tan^2 x}{1 + \tan^2 x}$

(g) $\cos\beta + \sin\beta = \dfrac{\cos 2\beta}{\cos\beta - \sin\beta}$

(h) $\sin^2\dfrac{\theta}{2} = \dfrac{1 - \cos\theta}{2}$

(i) $\dfrac{\sin^3 x + \cos^3 x}{\sin x + \cos x} = 1 - \dfrac{1}{2}\sin 2x$

(j) $\cos 4x = 8\cos^4 x - 8\cos^2 x + 1$

6. For the right-angled triangle shown, prove that

(a) $\sin 2\alpha = \dfrac{2ab}{c^2}$

(b) $\cos 2\alpha = \dfrac{b^2 - a^2}{c^2}$

(c) $\sin\dfrac{1}{2}\alpha = \sqrt{\dfrac{c - b}{2c}}$

(d) $\cos\dfrac{1}{2}\alpha = \sqrt{\dfrac{c + b}{2c}}$

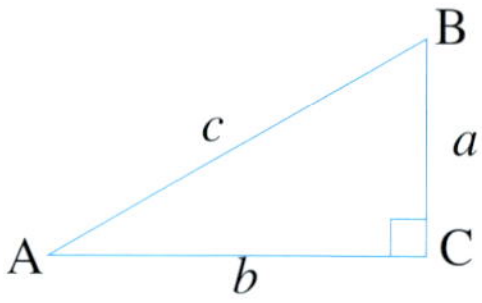

7. Solve the following for $0 \le x \le 2\pi$

(a) $\sin x = \sin 2x$

(b) $\sin x = \cos 2x$

(c) $\tan 2x = 4\tan x$

10.3 TRIGONOMETRIC FUNCTIONS

10.3.1 THE SINE, COSINE AND TANGENT FUNCTIONS

As we saw at the begining of this chapter, there is an infinite set of angles all of which give values (when they exist) for the main trigonometric ratios. We also noticed that the trigonometric ratios behave in such a way that values are repeated over and over. This is known as **periodic** behaviour. Many real world phenomena are periodic in the sense that the same patterns repeat over time. The **trigonometric functions** are often used to model such phenomena which include sound waves, light waves, alternating current electricity and other more approximately periodic events such as tides and even our moods (i.e., biorythms).

Notice how we have introduced the term trigonometric function, replacing the term trigonometric ratio. By doing this we can extend the use of the trigonometric ratios to a new field of problems described at the end of the previous paragraph.

When the trigonometric functions are used for these purposes, the angles are almost always measured in **radians** (which is a different way of measuring angles). However there is no reason why we cannot use degrees. It will always be obvious from the equation as to which mode of angle we are using, an expression such as $\sin x$ will imply (by default) that the angle is measured in radians, otherwise it will be written as $\sin x°$, implying that the angle is measured in degrees.

What do trigonometric functions look like?

The sine and cosine values have displayed a periodic nature. Meaning that if we were to plot a graph of the sine values versus their angles or the cosine values versus their angles, we could expect their graphs to demonstrate periodic behaviour. We start by plotting points.

The sine function:

θ	0	30	45	60	90	120	135	150	180	. . .	330	360
$\sin\theta°$	0.0	0.5	0.71	0.87	1.0	0.86	0.71	0.5	0.0	. . .	–0.5	0.0

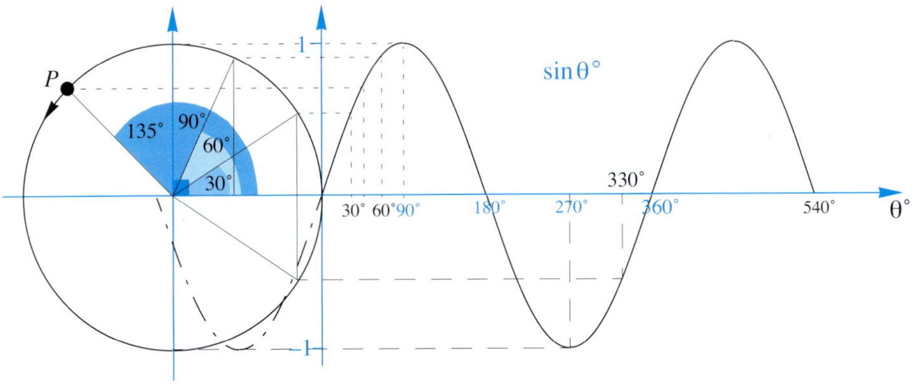

Notice that as the sine of angle θ corresponds to the y–value of a point P on the unit circle, as P moves around the circle in an anti–clockwise direction, we can measure the y–value and plot it on a graph as a function of θ (as shown above).

Feature of sine graph:
1. Maximum value = 1, Minimum value = –1
2. Period = 360° (i.e., graph repeats itself every 360°)
3. If P moves in a clockwise direction, y–values continue in their periodic nature (see dashed part of graph).

The cosine function:

θ	0	30	45	60	90	120	135	150	180	...	330	360
cos θ°	1.0	0.87	0.71	0.5	0.0	–0.5	–0.71	–0.87	–1.0	...	0.87	1.0

Plotting these points on a $\cos\theta°$ versus θ–axis, we have:

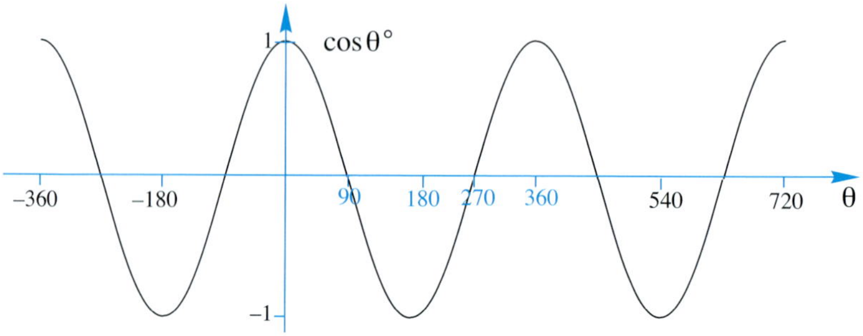

Feature of cosine graph:
1. Maximum value = 1, Minimum value = –1
2. Period = 360° (i.e., graph repeats itself every 360°)
3. If P moves in a clockwise direction, x–values continue in their periodic nature.

There is a note to be made about using the second method (the one used to obtain the sine graph) when dealing with the cosine graph. As the cosine values correspond to the x–values on the unit circle, the actual cosine graph should have been plotted as shown below. However, for the sake of consistency, we convert the 'vertical graph' to the more standard 'horizontal graph':

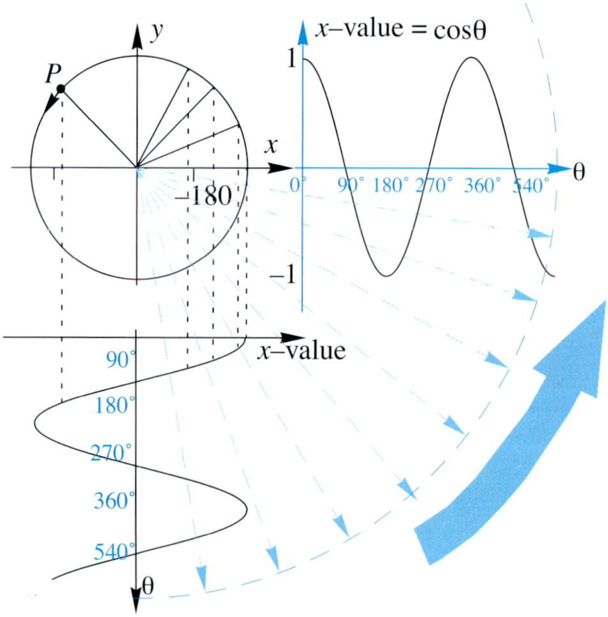

There are some common observations to be made from these two graphs;

1. We have that the **period** of each of these functions is 360°.
This is the length that it takes for the curve to start repeating itself.

2. The **amplitude** of the function is the distance between the centre line (in this case the θ -axis) and one of the maximum points. In this case, the amplitude is 1.

The sine and cosine functions are useful for modelling wave phenomena such as sound, light, water waves etc.

The third trigonometric function (tangent) is defined as $\tan\theta = \frac{\sin\theta}{\cos\theta}$ and so is defined for all angles for which the cosine function is non-zero.

The angles for which the tangent function are not defined correspond to the x-axis intercepts of the cosine function which are ±90°, ±270°, ±450°, … . At these points the graph of the tangent function has **vertical asymptotes**.

The period of the tangent function is 180°, which is half that of the sine and cosine functions. Since the tangent function has a vertical asymptote, it cannot be said to have an amplitude. It is also generally true that the tangent function is less useful than the sine and cosine functions for modelling applications. The graph of the basic tangent function is:

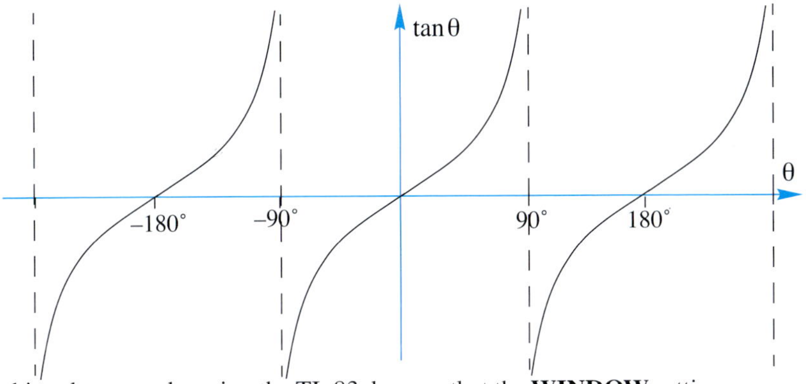

When sketching these graphs using the TI–83, be sure that the **WINDOW** settings are appropriate for the **MODE** setting. In the case of degrees we have:

Step 1. Make sure that the calculator is in degree mode. Failure to do this could be disastrous!
Step 2. Select an appropriate range
Step 3. Enter the function rule.

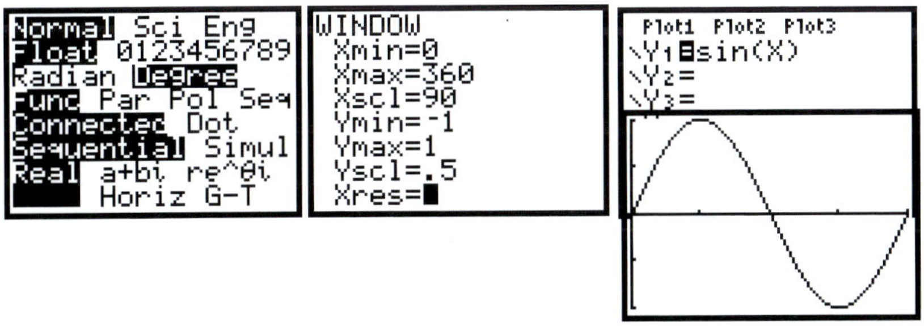

As we mentioned at the start of this section, angles measured in radians are much more useful when modelling situations that are cyclic or repetitive. And, although we have sketched the graphs of the sine, cosine and tangent functions using angles measured in degrees, it is reassuring to know that the shape of the graph does not alter when radians are used. The only difference between the graphs then is that 90° would be replaced by $\frac{\pi}{2}$, 180° by π and so on. We provide these graphs with the corresponding observations.

$f:\mathbb{R}\mapsto\mathbb{R}, f(x) = \sin x$

period 2π
amplitude 1

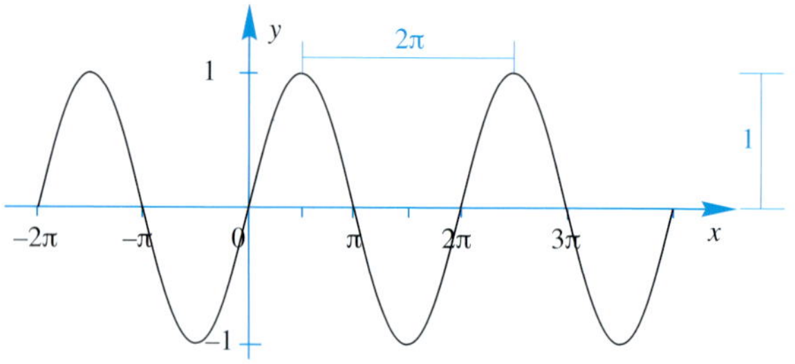

$f:\mathbb{R}\mapsto\mathbb{R}, f(x) = \cos x$

period 2π
amplitude 1

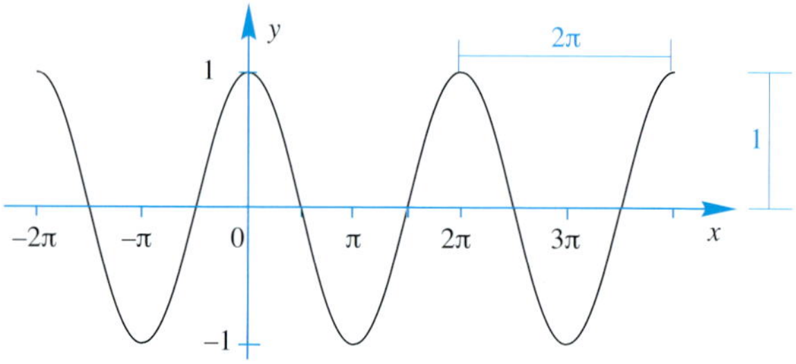

$f:\mathbb{R}^*\mapsto\mathbb{R}, f(x) = \tan x$

period π
no amplitude

$\mathbb{R}^* = \mathbb{R}\backslash\left\{x : x = \pm\frac{\pi}{2}, \pm\frac{3\pi}{2}, \pm\frac{5\pi}{2}, \ldots\right\}$

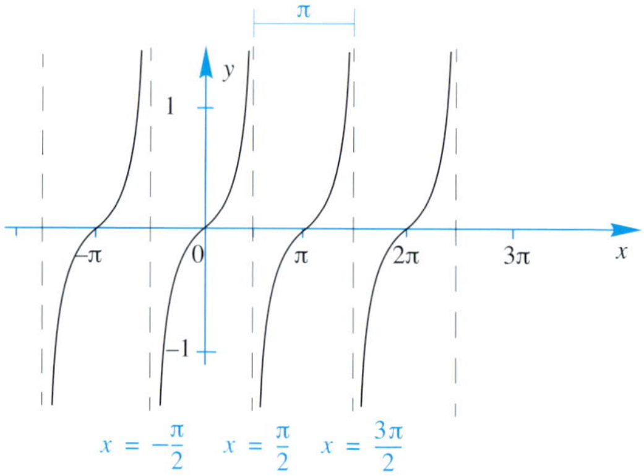

10.3.2 TRANSFORMATIONS OF TRIGONOMETRIC FUNCTIONS

We now consider some of the possible transformations that can be applied to the standard sine and cosine function and look at how these transformations affect the basic properties of both these graphs

1. Vertical translations

Functions of the type $f(x) = \sin(x) + c$, $f(x) = \cos(x) + c$ and $f(x) = \tan(x) + c$ represent vertical translations of the curves of $\sin(x)$, $\cos(x)$ and $\tan(x)$ respectively. If $c > 0$ the graph is moved vertically up and if $c < 0$ the graph is moved vertically down.
That is, adding or subtracting a fixed amount to a trigonometric function translates the graph parallel to the y-axis.

EXAMPLE 10.16 Sketch the graphs of the functions for x values in the range -2π to 4π.

(a) $y = \sin(x) + 1$ (b) $y = \cos(x) - 2$

Solution

(a)

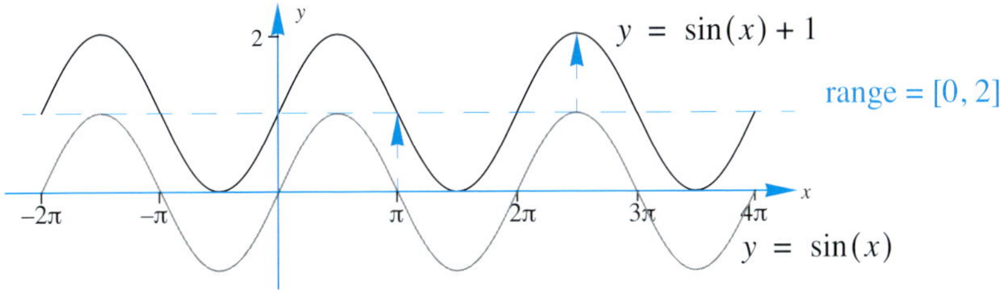

(b)

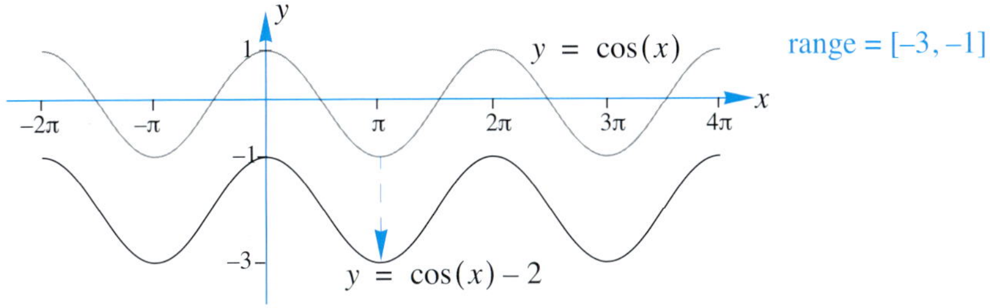

A graph sketch should show all the important features of a graph. In this case, the axes scales are important and should show the correct period (2π) and range $[-3,-1]$.

2. Horizontal translations

Functions of the type $f(x) = \sin(x \pm \alpha)$, $f(x) = \cos(x \pm \alpha)$ and $f(x) = \tan(x \pm \alpha)$ where $\alpha > 0$ are horizontal translations of the curves $\sin x$, $\cos x$ and $\tan x$ respectively.

So that $f(x) = \sin(x - \alpha)$, $f(x) = \cos(x - \alpha)$ and $f(x) = \tan(x - \alpha)$ are translations to the **right**

while that of $f(x) = \sin(x + \alpha)$, $f(x) = \cos(x + \alpha)$ and $f(x) = \tan(x + \alpha)$ are translations to the **left**.

EXAMPLE 10.17 For $-2\pi \le x \le 4\pi$, sketch the graphs of the curves with equations

(a) $y = \cos\left(x - \frac{\pi}{4}\right)$ (b) $y = \cos\left(x + \frac{\pi}{3}\right)$ (c) $y = \tan\left(x - \frac{\pi}{6}\right)$

Solution

(a)

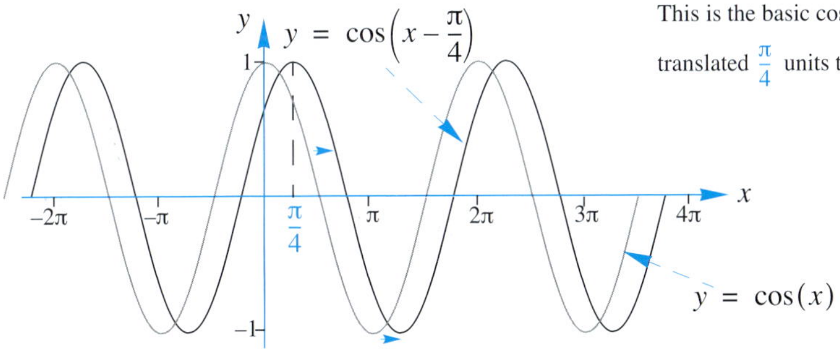

This is the basic cosine graph translated $\frac{\pi}{4}$ units to the **right**.

(b)

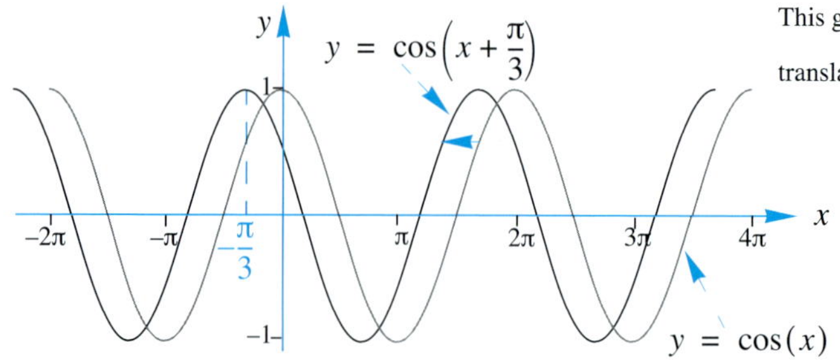

This graph is the basic cosine graph translated $\frac{\pi}{3}$ units to the **left**.

(c)

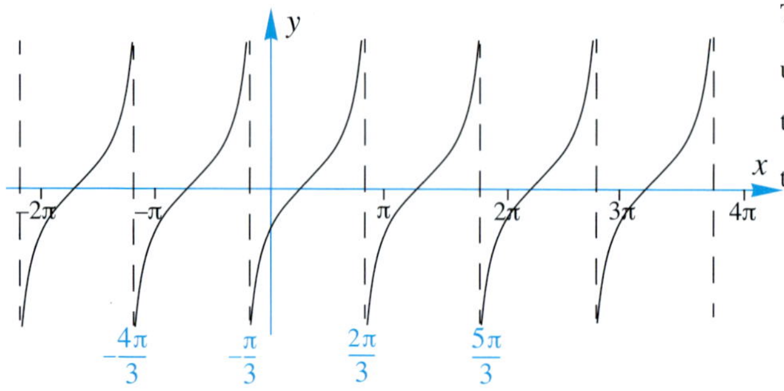

This is the tangent graph translated $\frac{\pi}{6}$ units to the **right**. This also translates the **asymptotes** $\frac{\pi}{6}$ units to the **right**.

Of course, it is also possible to combine vertical and horizontal translations, as the next example shows.

EXAMPLE 10.18

Sketch the graphs of the function $f(x) = \sin\left(x - \frac{\pi}{4}\right) + 2$, $-2\pi \le x \le 4\pi$.

Solution

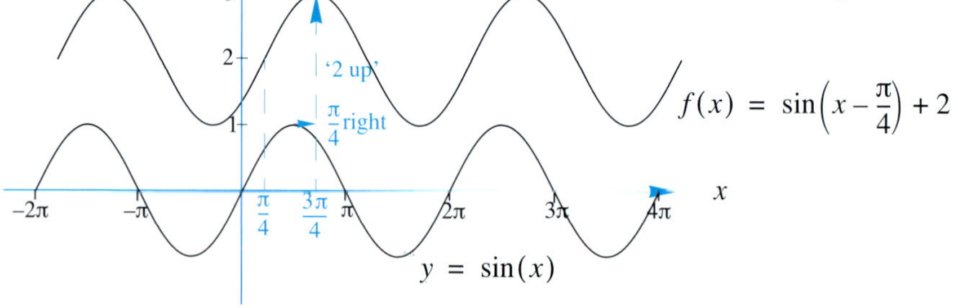

Basic sine graph translated $\frac{\pi}{4}$ units right and 2 up.

---

3. Dilations

Functions of the form $f(x) = a\sin x$, $f(x) = a\cos(x)$ and $f(x) = a\tan(x)$ are **dilations** of the curves $\sin(x)$, $\cos(x)$ and $\tan(x)$ respectively, **parallel to the y-axis**.

In the case of the sine and cosine functions, the **amplitude** becomes $|a|$ and not 1. This dilation does not affect the shape of the graph. Also, as the tangent function extends indefinitely, the term amplitude has no relevance.

EXAMPLE 10.19

Sketch the graphs of the following functions for $-2\pi \leq x \leq 4\pi$.

(a) $f(x) = 2\cos x$ (b) $f(x) = \frac{1}{3}\sin x$

Solution

(a)

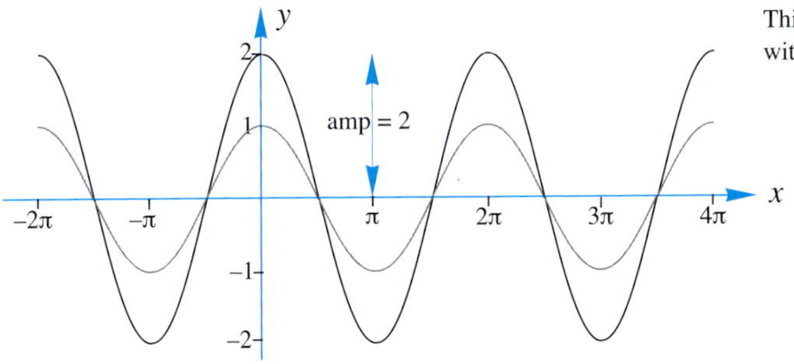

This is the cosine graph (broken line) with an amplitude of 2.

(b)

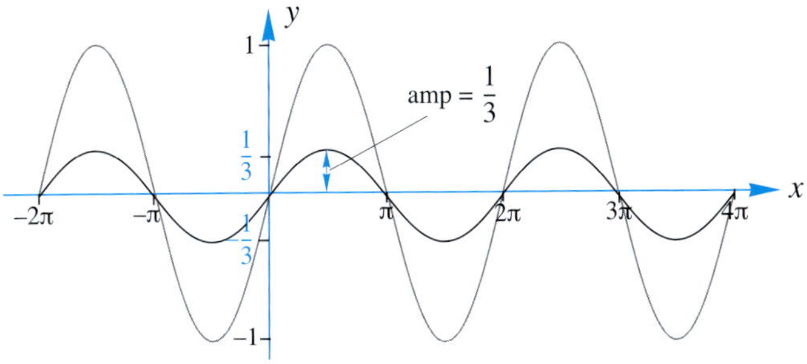

This is the sine graph (broken line) with an amplitude of $\frac{1}{3}$.

Functions of the form $f(x) = \cos(bx)$, $f(x) = \sin(bx)$ and $f(x) = \tan(bx)$ are **dilations** of the curves $\sin(x)$, $\cos(x)$ and $\tan(x)$ respectively, **parallel to the x-axis**.

This means that the period of the graph is altered. It can be valuable to remember and use the formula that relates the value of b to the period τ of the dilated function:

1. The graph of $f(x) = \cos(bx)$ will show b cycles in 2π radians, meaning that its period will be given by $\tau = \frac{2\pi}{b}$.

2. The graph of $f(x) = \sin(bx)$ will show b cycles in 2π radians, meaning that its period will be given by $\tau = \frac{2\pi}{b}$.

3. The graph of $f(x) = \tan(bx)$ will show b cycles in π radians, meaning that its period will be given by $\tau = \frac{\pi}{b}$.

Note: In the case of the tangent function, whose original period is π, the period is $\tau = \frac{\pi}{n}$.

EXAMPLE 10.20 Sketch graphs of the following functions for x values in the range -2π to 4π.

(a) $f(x) = \sin(2x)$ (b) $f(x) = \cos\left(\frac{x}{2}\right)$ (c) $f(x) = \tan(2x)$

Solution

(a) $f(x) = \sin(2x)$. The value of n is 2 so the period is $\tau = \frac{2\pi}{n} = \frac{2\pi}{2} = \pi$. Note that this means that the period is **half** that of the basic sine function.

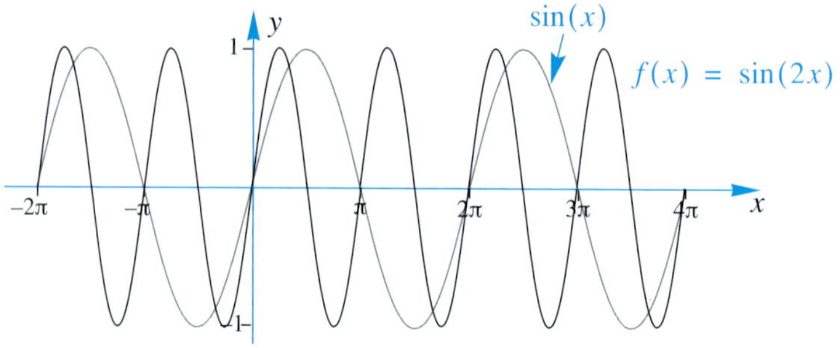

(b) $f(x) = \cos\left(\frac{x}{2}\right)$. In this case the value of $n = \frac{1}{2}$ and the period $\tau = \frac{2\pi}{n} = \frac{2\pi}{1/2} = 4\pi$. The graph is effectively stretched to twice its original period.

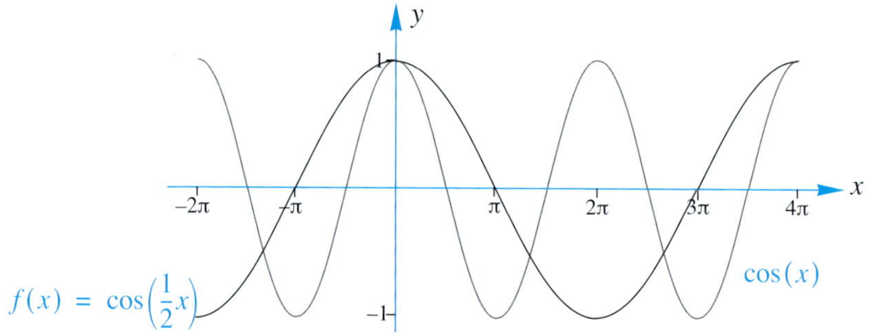

(c) In this case, with $f(x) = \tan\left(\frac{x}{4}\right)$, the value of $n = \frac{1}{4}$ and the period $\tau = \frac{\pi}{n} = 4\pi$.

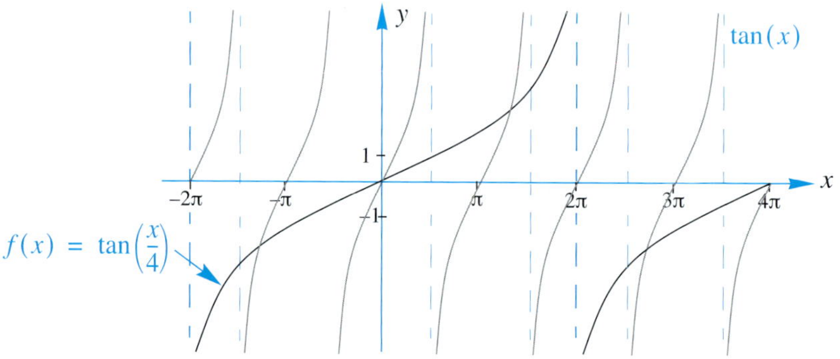

4. Reflections

Recall that the graph of $y = -f(x)$ is the graph of $y = f(x)$ reflected about the x–axis, while that of $y = f(-x)$ is the graph of $y = f(x)$ reflected about the y–axis.

$f(x) = -\cos x$ is the basic cosine graph (broken line) **reflected in the x-axis**.

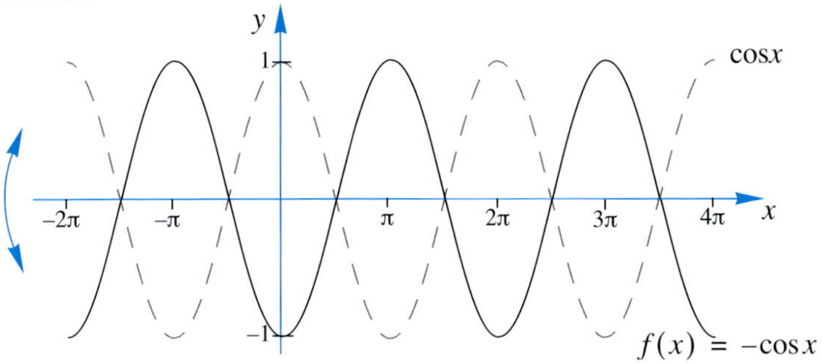

$f(x) = \sin(-x)$ is the basic sine graph **reflected in the y-axis**.

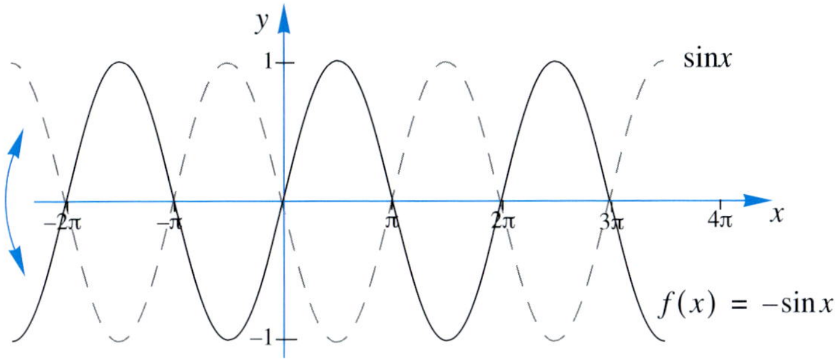

5. Combined transformations

You may be required to combine some or all of these transformations in a single function. The functions of the type

$$f(x) = a\sin[b(x+c)]+d \quad \text{and} \quad f(x) = a\cos[b(x+c)]+d$$

have:

1. an amplitude of $|a|$ (i.e., the absolute value of a).

2. a period of $\frac{2\pi}{b}$

3. a horizontal translation of c units, $c > 0 \Rightarrow$ to the left, $c < 0 \Rightarrow$ to the right

4. a vertical translation of d units, $d > 0 \Rightarrow$ up, $d < 0 \Rightarrow$ down

Care must be taken with the horizontal translation.

For example, the function $f(x) = 2\cos\left(3x+\frac{\pi}{2}\right)-1$ has a horizontal translation of $\frac{\pi}{6}$ to the left, not $\frac{\pi}{2}$! This is because $f(x) = 2\cos\left(3x+\frac{\pi}{2}\right)-1 = 2\cos\left[3\left(x+\frac{\pi}{6}\right)\right]-1$. i.e., if the coefficient of x is not one, we must first express the function in the form $a\cos[b(x+c)]+d$.

Similarly we have

$$f(x) = a\tan[b(x+c)]+d$$

has:

1. no amplitude (as it is not appropriate for the tan function).
2. a period of $\frac{\pi}{b}$
3. a horizontal translation of c units, $c>0 \Rightarrow$ to the left, $c<0 \Rightarrow$ to the right
4. a vertical translation of d units, $d>0 \Rightarrow$ up, $d<0 \Rightarrow$ down

EXAMPLE 10.21

Sketch graphs of the following functions for x values in the range -2π to 4π.

(a) $f(x) = 2\sin\left[2\left(x-\frac{\pi}{4}\right)\right]+1$ (b) $f(x) = -\cos\frac{1}{2}\left(x-\frac{\pi}{3}\right)+2$

(c) $f(x) = -\frac{1}{2}\tan\frac{1}{2}\left(x-\frac{\pi}{2}\right)$

Solution

(a) $f(x) = 2\sin\left[2\left(x-\frac{\pi}{4}\right)\right]+1$. This graph has an amplitude of 2, a period of π, a horizontal translation of $\frac{\pi}{4}$ units to the right and a vertical translation of 1 unit up.

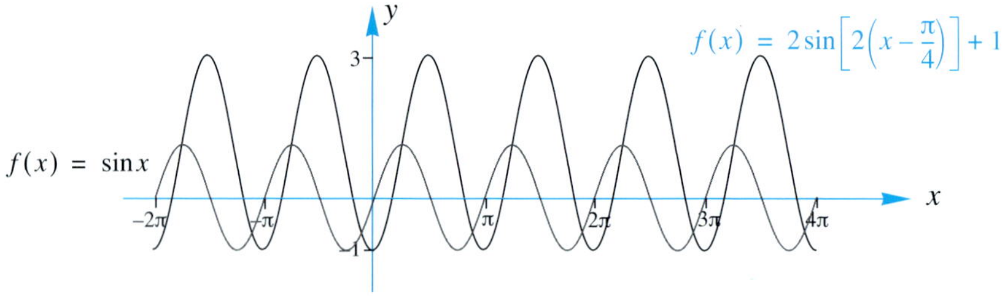

(b) $f(x) = -\cos\frac{1}{2}\left(x-\frac{\pi}{3}\right)+2$. The transformations are a reflection in the x-axis, a dilation of factor 2 parallel to the x-axis and a translation of $\frac{\pi}{3}$ right and 2 up.

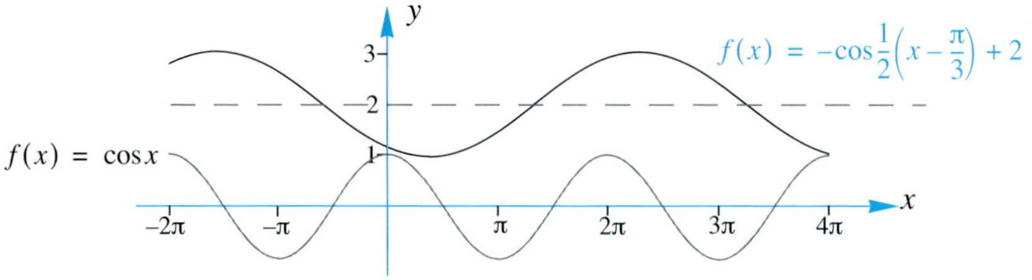

(b) $f(x) = -\frac{1}{2}\tan\frac{1}{2}\left(x - \frac{\pi}{2}\right)$. The transformations are a reflection in the x-axis, a vertical dilation with factor $\frac{1}{2}$, a horizontal dilation with factor 2 and a translation of $\frac{\pi}{2}$ to the right.

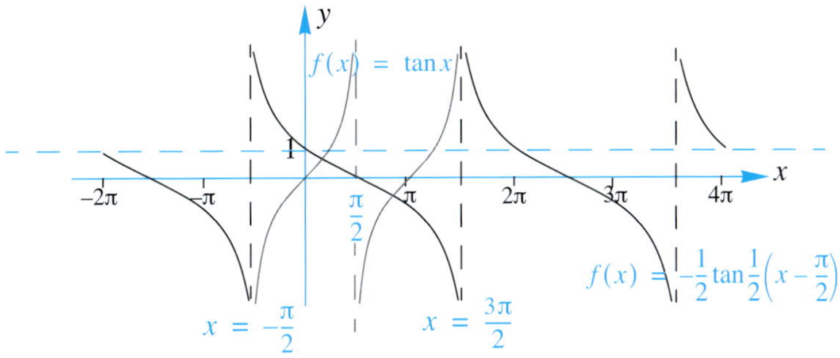

Again we see that a graphics calculator is very useful in such situations – in particular it allows for a checking process.

Example 10.21 (a) could be sketched as follows:

Step 1. Make sure that the calculator is in radian mode. Failure to do this could be disastrous!

Step 2. Enter the function rule. Remember to use the π key where necessary.
Do not use approximations such as 3.14

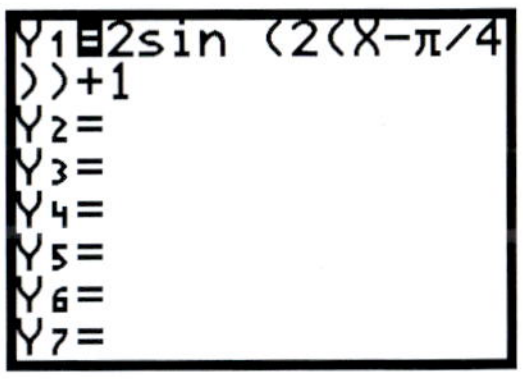

Step 3. Select an appropriate range. This may not be the **ZTrig** option (7) from the **ZOOM** menu.
However, this is probably a good place to start.

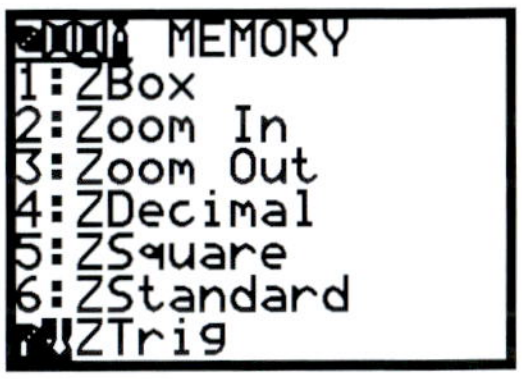

In this case, the viewing window is suitable, though it should be noted that it is not the specified range of x values $[-2\pi, 4\pi]$.

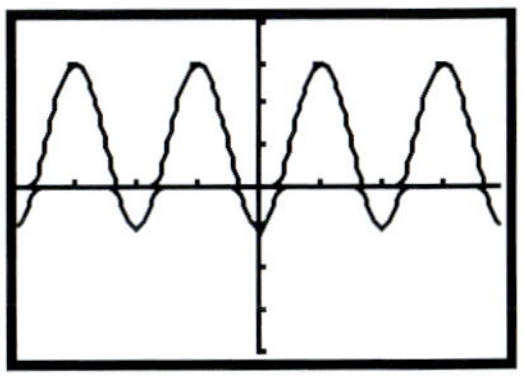

Step 4. Adjust the viewing window using the **WINDOW** menu. In this case it would be wise to select the correct set of x values.

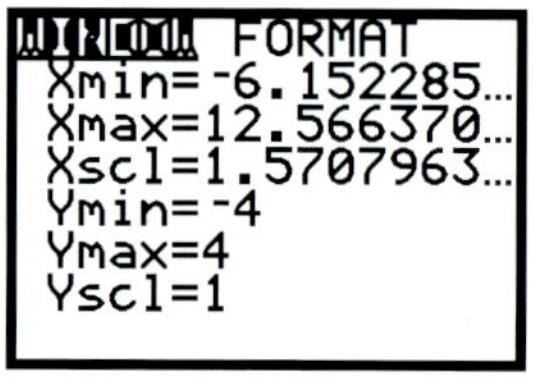

Finally use **GRAPH** to display the graph.

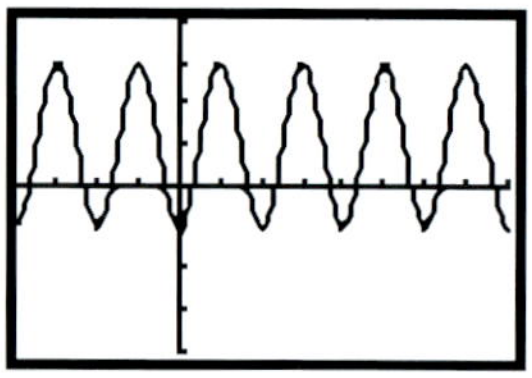

EXERCISES 10.3

1. State the period of the following functions

(a) $f(x) = \sin\frac{1}{2}x$

(b) $f(x) = \cos 3x$

(c) $f(x) = \tan\frac{x}{3}$

(d) $g(x) = \cos\left(\frac{x}{2} - \pi\right)$

(e) $g(x) = 4\sin(\pi x + 2)$

(f) $g(x) = 3\tan\left(\frac{\pi}{2} - 2x\right)$

2. State the amplitude of the following functions

(a) $f(x) = 5\sin 2x$

(b) $g(x) = -3\cos\frac{x}{2}$

(c) $g(x) = 4 - 5\cos(2x)$

(d) $f(x) = \frac{1}{2}\sin(3x)$

3. Find the period and, where appropriate, the amplitude of the following functions.

(a) $y = 2\sin x$

(b) $y = 3\cos\frac{x}{3}$

(c) $y = 3\tan x$

(d) $2\tan(x - 2\pi)$

(e) $y = -4\sin\left[2\left(x + \frac{\pi}{6}\right)\right] + 1$

(f) $y = 2 - 3\cos(2x - \pi)$

(g) $y = -2\tan\frac{x}{6}$

(h) $y = \frac{1}{4}\cos\left[3\left(x - \frac{3\pi}{4}\right)\right] + 5$

(i) $y = 4\tan\left(\frac{x-4}{3}\right) - 3$

(j) $y = -\frac{2}{3}\cos\left(\frac{3}{4}\left(x + \frac{3\pi}{5}\right)\right) + 5$

4. Sketch the graph of the curve with equation given by

(a) $y = 3\cos x, 0 \le x \le 2\pi$

(b) $y = \sin\frac{x}{2}, -\pi \le x \le \pi$

(c) $y = 2\cos\left(\frac{x}{3}\right), 0 \le x \le 3\pi$

(d) $y = -\frac{1}{2}\sin 3x, 0 \le x \le \pi$

(e) $y = 4\tan\left(\frac{x}{2}\right), 0 \le x \le 2\pi$

(f) $y = \tan(-2x), -\pi \le x \le \frac{\pi}{4}$

(g) $y = \frac{1}{3}\cos(-3x), -\frac{\pi}{3} \le x \le \frac{\pi}{3}$

(h) $y = 3\sin(-2x), -\pi \le 0 \le \pi$

5. Sketch the graph of the curve with equation given by

(a) $y = 3\cos x + 3, 0 \le x \le 2\pi$

(b) $y = \sin\frac{x}{2} - 1, -\pi \le x \le \pi$

(c) $y = 2\cos\left(\frac{x}{3}\right) - 2, 0 \le x \le 3\pi$

(d) $y = -\frac{1}{2}\sin 3x + 2, 0 \le x \le \pi$

(e) $y = 4\tan\left(\frac{x}{2}\right) - 1, 0 \le x \le 2\pi$

(f) $y = \tan(-2x) + 2, -\pi \le x \le \frac{\pi}{4}$

(g) $y = \frac{1}{3}\cos(-3x) + \frac{1}{3}, -\frac{\pi}{3} \le x \le \frac{\pi}{3}$

(h) $y = 3\sin(-2x) - 2, -\pi \le 0 \le \pi$

6. Sketch the graph of the curve with equation given by

(a) $y = 3\cos\left(x + \frac{\pi}{2}\right), 0 \le x \le 2\pi$

(b) $y = \sin\left(\frac{x}{2} - \pi\right), -\pi \le x \le \pi$

(c) $y = 2\cos\left(\frac{x}{3} + \frac{\pi}{6}\right), 0 \le x \le 3\pi$

(d) $y = -\frac{1}{2}\sin(3x + 3\pi), 0 \le x \le \pi$

(e) $y = 4\tan\left(\frac{x}{2} - \frac{\pi}{4}\right), 0 \le x \le 2\pi$

(f) $y = \tan(-2x + \pi), -\pi \le x \le \frac{\pi}{4}$

(g) $y = \frac{1}{3}\cos(-3x - \pi), -\frac{\pi}{3} \le x \le \frac{\pi}{3}$

(h) $y = 3\sin\left(-2x - \frac{\pi}{2}\right), -\pi \le 0 \le \pi$

7. Sketch graphs of the following functions for x values in the interval $[-2\pi, 2\pi]$.

(a) $y = \sin(2x)$

(b) $y = -\cos\left(\frac{x}{2}\right)$

(c) $y = 3\tan\left(x - \frac{\pi}{4}\right)$

(d) $y = 2\sin\left(x - \frac{\pi}{2}\right)$

(e) $y = 1 - 2\sin(2x)$

(f) $y = -2\cos\left(\frac{x - \pi}{2}\right)$

(g) $y = 3\tan\left[\frac{1}{2}\left(x + \frac{\pi}{4}\right)\right] - 3$

(h) $y = 3\cos\left(x + \frac{\pi}{4}\right)$

(i) $y = 2\sin\left[\frac{1}{3}\left(x + \frac{2\pi}{3}\right)\right] - 1$ (j) $y = 3\tan(2x + \pi)$

(k) $y = 4\sin\left(\frac{x + \frac{\pi}{2}}{2}\right)$ (l) $y = 2 - \sin\left(\frac{2(x - \pi)}{3}\right)$

(m) $y = 2\cos(\pi x)$ (n) $y = 2\sin[\pi(x + 1)]$

8. (a) i. Sketch one cycle of the graph of the function $f(x) = \sin x$.

ii. For what values of x is the function $y = \frac{1}{f(x)}$ not defined?

iii. Hence, sketch one cycle of the graph of the function $g(x) = \operatorname{cosec} x$.

(b) i. Sketch one cycle of the graph of the function $f(x) = \cos x$.

ii. For what values of x is the function $y = \frac{1}{f(x)}$ not defined?

iii. Hence, sketch one cycle of the graph of the function $g(x) = \sec x$.

(c) i. Sketch one cycle of the graph of the function $f(x) = \tan x$.

ii. For what values of x is the function $y = \frac{1}{f(x)}$ not defined?

iii. Hence, sketch one cycle of graph of the function $g(x) = \cot x$.

10.4 TRIGONOMETRIC EQUATIONS

We have already encountered solutions to trigonometric equations as part of a general observation in this chapter. There are two basic methods that can be used when solving trigonometric equations;

Method 1. Use the unit circle as a visual aid.
Method 2. Use the graph of the function as a visual aid.

The method you choose depends entirely on what you feel comfortable with. However, it is recommended that you become familiar with both methods.

10.4.1 SOLUTION TO $\sin x = a$, $\cos x = a$ & $\tan x = a$

The equation $\sin x = \frac{1}{2}$ produces an infinite number of solutions. This can be seen from the graph of the sine function.

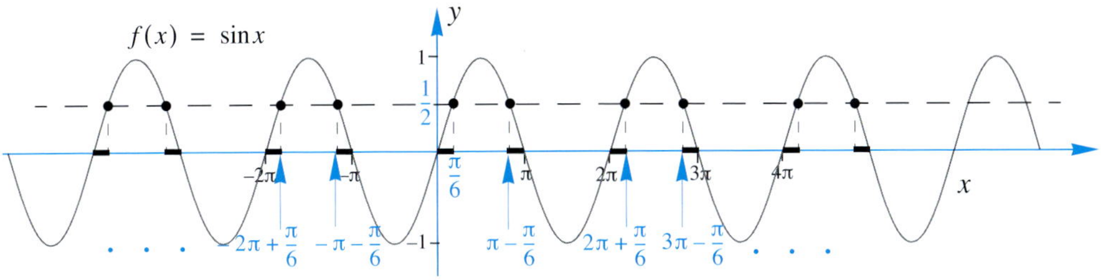

Using the **principal angle** $\left(\frac{\pi}{6}\right)$ and **symmetry**, the solutions generated are

For $x \geq 0$: $x = \frac{\pi}{6}, \pi - \frac{\pi}{6}, 2\pi + \frac{\pi}{6}, 3\pi - \frac{\pi}{6}, \ldots$

$= \frac{\pi}{6}, \frac{5\pi}{6}, \frac{13\pi}{6}, \frac{17\pi}{6}, \ldots$

For $x < 0$: $x = -\pi - \frac{\pi}{6}, -2\pi + \frac{\pi}{6}, -3\pi - \frac{\pi}{6}, \ldots$

$= -\frac{7\pi}{6}, -\frac{11\pi}{6}, -\frac{19\pi}{6}, \ldots$

That is, $\sin x = \frac{1}{2} \Leftrightarrow x = \ldots, -\frac{7\pi}{6}, -\frac{11\pi}{6}, -\frac{19\pi}{6}, \frac{\pi}{6}, \frac{5\pi}{6}, \frac{13\pi}{6}, \frac{17\pi}{6}, \ldots$

or, $x = n\pi + (-1)^n \times \frac{\pi}{6}$, where n is an integer (including zero).

The solution $x = n\pi + (-1)^n \times \frac{\pi}{6}$ is known as the **general solution**. However, in this course, there will always be some restriction on the domain. For example, solve $\sin x = \frac{1}{2}, -2\pi < x < \pi$, which would then give $x = -\frac{7\pi}{6}, -\frac{11\pi}{6}, \frac{\pi}{6}, \frac{5\pi}{6}$ as its solutions.

The same problem could have been solved using Method 1. We start by drawing a unit circle and we continue to move around the circle until we have all the required solutions within the domain restriction. Again, the use of symmetry plays an important role in solving these equations.
For $x \geq 0$:

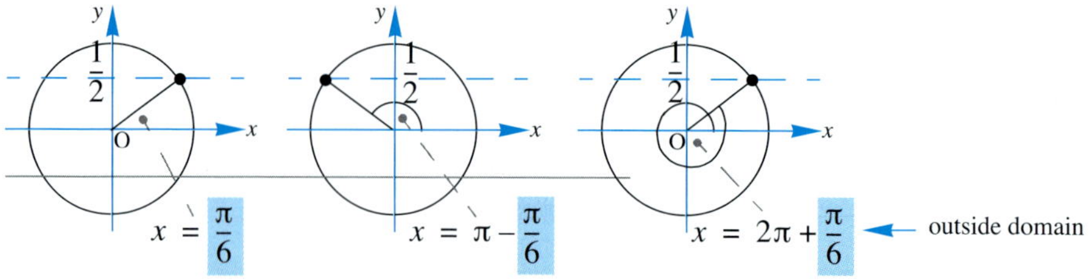

For $x < 0$:

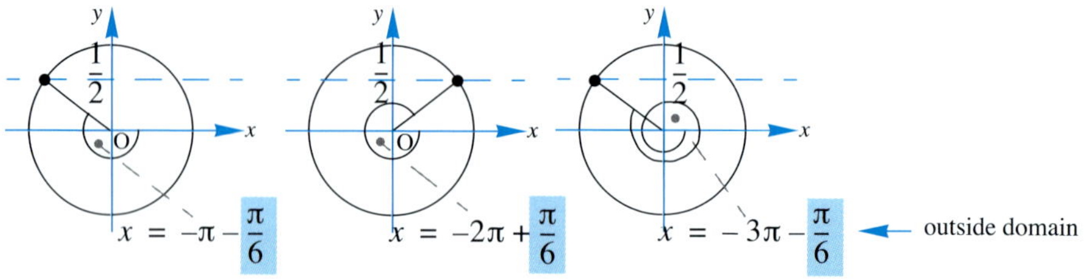

Again, for the restricted domain $-2\pi < x < \pi$, we have $\sin x = \frac{1}{2}$ if $x = -\frac{7\pi}{6}, -\frac{11\pi}{6}, \frac{\pi}{6}, \frac{5\pi}{6}$.

The process is identical for the cosine and tangent function.

EXAMPLE 10.22

Solve the following, for $0 \leq x \leq 4\pi$, giving answers to 4 decimal places if no exact answers are available.

(a) $\cos x = 0.4$ (b) $\tan x = -1$ (c) $5\cos x - 2 = 0$

Solution

(a) **Step 1:** Find the **reference angle**: $x = \text{Cos}^{-1}(0.4) = 1.1593$.

Step 2: Sketch the cosine graph (or use the unit circle):

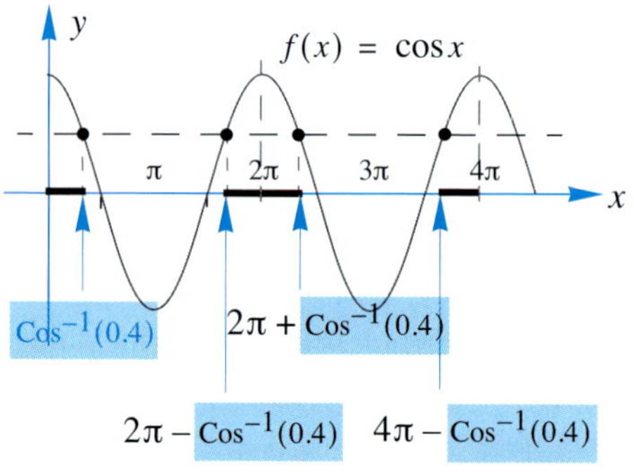

Step 3: Use the reference angle and symmetry to obtain solutions.

Therefore, solutions are, $x = \text{Cos}^{-1}(0.4), 2\pi - \text{Cos}^{-1}(0.4), 2\pi + \text{Cos}^{-1}(0.4), 4\pi - \text{Cos}^{-1}(0.4)$

$= 1.1593, 5.1239, 7.4425, 11.4071$

Step 4: Check that i. all solutions are within the domain
ii. you have obtained all the solutions in the domain. (Use the graphics calculator to check).

Using the TI–83 we count the number of intersections for $y = 0.4$ and $y = \cos x$:

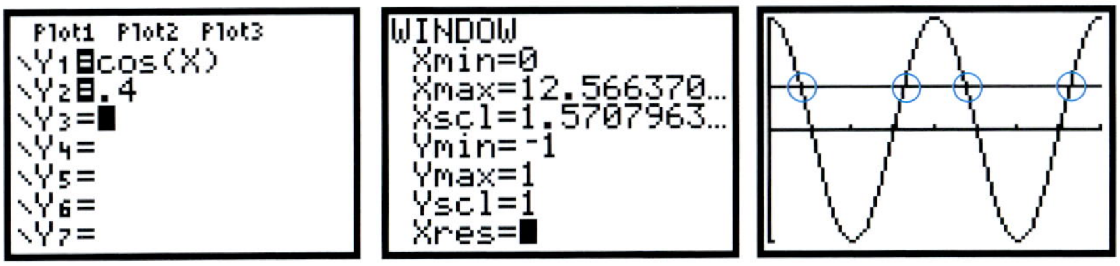

4 solutions are required.

(b) **Step 1:** Find reference angle (in first quadrant): $\text{Tan}^{-1}(1) = \frac{\pi}{4}$.

Step 2: Sketch the tangent graph (or use the unit circle):

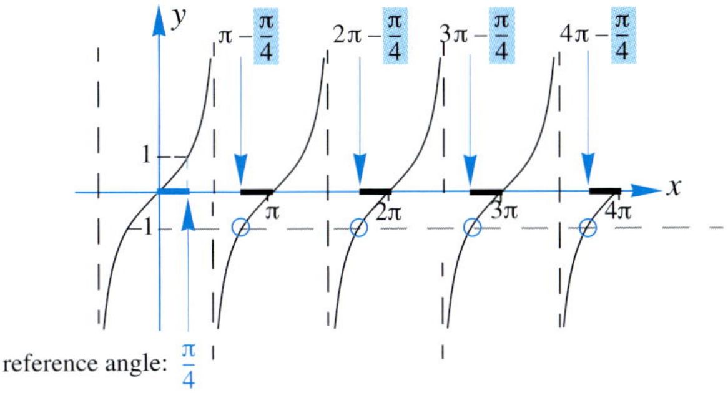

Step 3: Use the reference angle and symmetry to obtain solutions.

Reference angle is $\frac{\pi}{4}$.

Therefore solutions are $x = \pi - \frac{\pi}{4}, 2\pi - \frac{\pi}{4}, 3\pi - \frac{\pi}{4}, 4\pi - \frac{\pi}{4}$

$= \frac{3\pi}{4}, \frac{7\pi}{4}, \frac{11\pi}{4}, \frac{15\pi}{4}$

From the TI–83 display, we see that there are four solutions. Then, as the four solutions obtained all lie in the interval $[0, 4\pi]$ Step 4 is satisfied.

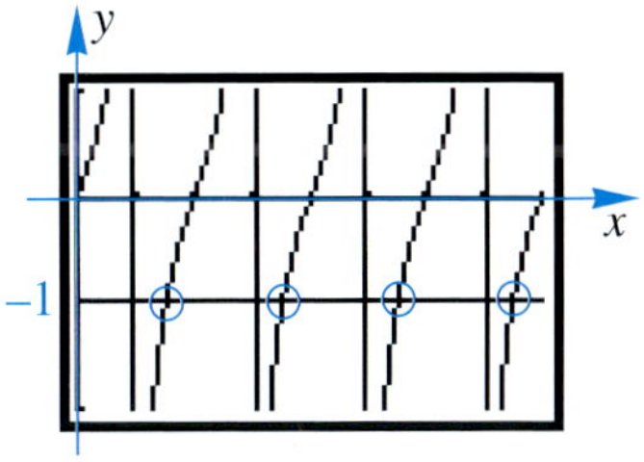

(c) $5\cos x - 2 = 0 \Leftrightarrow 5\cos x = 2 \Leftrightarrow \cos x = \frac{2}{5}$

i.e., $\cos x = 0.4$

Which is in fact, identical to the equation in part (a) and so, we have that

$x = 1.1593, 5.1239, 7.4425, 11.4071$

Part (c) in Example 10.36 highlights the fact that it is possible to transpose a trigonometric equation into a simpler form, which can readily be solved. Rather than remembering (or tryingt to commit to memory) the different possible forms of trigonometric equations and their specific solution processes, the four steps used (with possibly some algrebraic manupulation) will always transform a (seemingly) difficult equation into one having a simpler form, as in Example 10.36.

Some forms of trigonometric equations are:
$\sin(kx) = a$, $\cos(x + c) = a$, $\tan(kx + c) = a$, $b\cos(kx + c) = a$, $b\sin(kx + c) + d = a$

And of course, then there are equations involving the secant, cosecant and cotangent functions. But even the most involved of these equations, e.g., $b\sin(kx + c) + d = a$, can be reduced to a simpler form:

1. Transpose:

$$b\sin(kx + c) + d = a \Leftrightarrow b\sin(kx + c) = a - d$$
$$\Leftrightarrow \sin(kx + c) = \frac{a - d}{b}$$

2. Substitute:

Then, setting $kx + c = \theta$ and $\frac{a - d}{b} = m$, we have $\sin\theta = m$ which can be readily solved.

3. Solve for new variable:

So that the solutions to $\sin\theta = m$ are $\theta = \theta_1, \theta_2, \theta_3, \ldots$.

4. Solve for original variable:

We substitute back for θ and solve for x:

$$kx + c = \theta_1, \theta_2, \theta_3, \ldots$$
$$\Leftrightarrow kx = \theta_1 - c, \theta_2 - c, \theta_3 - c, \ldots$$
$$\Leftrightarrow x = \frac{\theta_1 - c}{k}, \frac{\theta_2 - c}{k}, \frac{\theta_3 - c}{k}, \ldots$$

All that remains is to check that all the solutions have been obtained and that they all lie in the restricted domain.

The best way to see how this works is through a number of examples.

EXAMPLE 10.23

Solve the following, for $0 \le x \le 2\pi$

(a) $\cos(2x) = 0.4$ (b) $\tan\left(\frac{1}{2}x\right) = -1$

Solution

(a) Let $\theta = 2x$, so that we now solve the equation $\cos\theta = 0.4$.
From Example 10.36 (a) we already have the solutions, namely;

$$\theta = \text{Cos}^{-1}(0.4), 2\pi - \text{Cos}^{-1}(0.4), 2\pi + \text{Cos}^{-1}(0.4), 4\pi - \text{Cos}^{-1}(0.4)$$
$$= 1.1593, 5.1239, 7.4425, 11.4071$$

However, we want to solve for x not θ. So, we substitute back for x:

i.e., $2x = 1.1593, 5.1239, 7.4425, 11.4071$

$\therefore x = 0.5796, 2.5620, 3.7212, 5.7045$

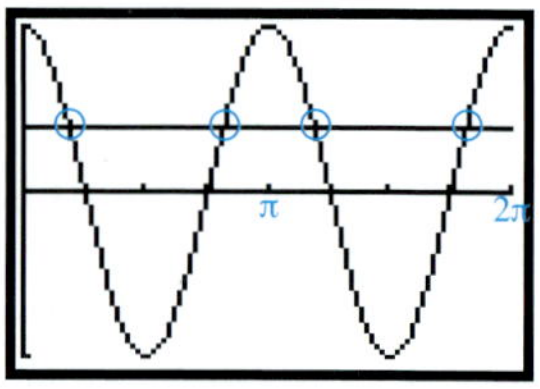

To check that we have all the solutions, we sketch the graphs of $y = \cos(2x)$ and $y = 0.4$ over the domain $0 \le x \le 2\pi$.
The diagram shows that there should be four solutions.

(b) This time, to solve $\tan\left(\frac{1}{2}x\right) = -1$ we first let $\theta = \frac{1}{2}x$ so that we now need to solve the simpler equation $\tan\theta = -1$.

From Example 10.35 (b) we have that

$$\theta = \pi - \frac{\pi}{4}, 2\pi - \frac{\pi}{4}, 3\pi - \frac{\pi}{4}, 4\pi - \frac{\pi}{4}$$
$$= \frac{3\pi}{4}, \frac{7\pi}{4}, \frac{11\pi}{4}, \frac{15\pi}{4}$$

However, we want to solve for x, not θ. So, we substitute for x:

$$\frac{1}{2}x = \frac{3\pi}{4}, \frac{7\pi}{4}, \frac{11\pi}{4}, \frac{15\pi}{4}$$
$$\therefore x = \frac{3\pi}{2}, \frac{7\pi}{2}, \frac{11\pi}{2}, \frac{15\pi}{2}$$

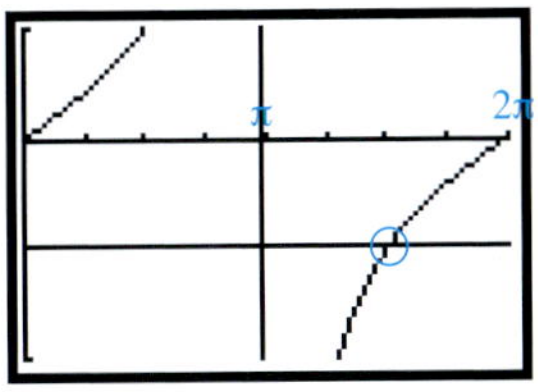

To check that we have all the solutions, we sketch the graphs of $y = \tan\left(\frac{1}{2}x\right)$ and $y = -1$ over the domain $0 \le x \le 2\pi$.

The diagram shows that there should be only one solution.

Therefore, the only solution is $x = \frac{3\pi}{2}$.

There is of course another step that could be used to help us predetermine which solutions are valid. This requires that we make a substitution not only into the equation, but also into the restricted domain statement.

In Example 10.36 (b), after setting $\theta = \frac{1}{2}x$ to give $\tan\theta = -1$, we next adjust the restricted domain as follows: $\theta = \frac{1}{2}x \Leftrightarrow x = 2\theta$.

So, from $0 \le x \le 2\pi$ we now have $0 \le 2\theta \le 2\pi \Leftrightarrow 0 \le \theta \le \pi$.

That is, we have the equivalent equation:

$$\tan\left(\frac{1}{2}x\right) = -1, 0 \le x \le 2\pi:$$

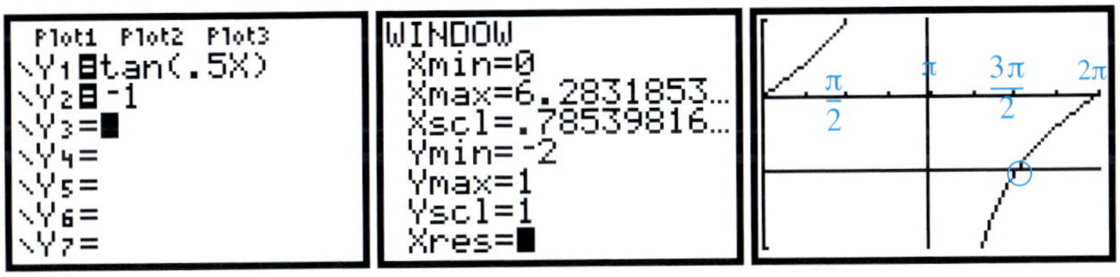

$\therefore x = \frac{3\pi}{2}$

$$\tan\theta = -1, 0 \le \theta \le \pi:$$

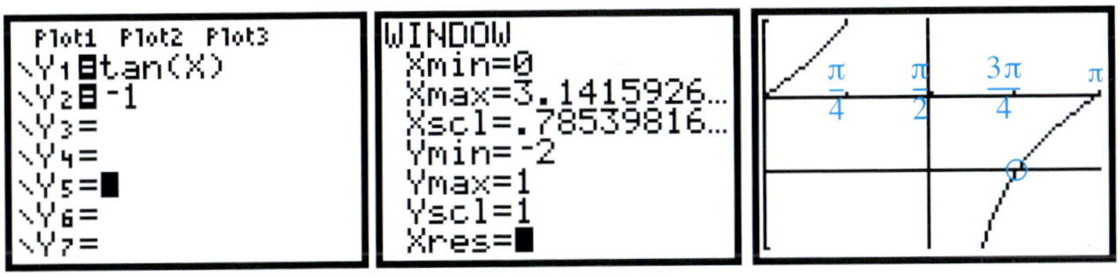

$\theta = \frac{3\pi}{4} \therefore \frac{1}{2}x = \frac{3\pi}{4}$

$\therefore x = \frac{3\pi}{2}$

EXAMPLE 10.24

Solve $4\sin(3x) = 2$, $0° \le x \le 360°$

Solution

(a) $4\sin(3x) = 2 \Leftrightarrow \sin(3x) = 0.5, 0° \le x \le 360°$.

Let $\theta = 3x \Rightarrow \sin\theta = 0.5$.

New domain: $0° \le x \le 360° \Leftrightarrow 0° \le \frac{\theta}{3} \le 360° \Leftrightarrow 0° \le \theta \le 1080°$

Therefore, we have, $\sin\theta = 0.5, 0° \le \theta \le 1080°$

The reference angle is $30°$, then, by symmetry we have

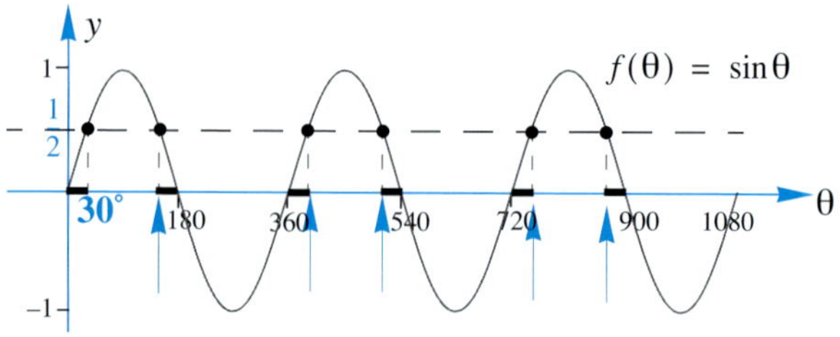

$$\therefore \theta = 30°, 180° - 30°, 360° + 30°, 540° - 30°, 720° + 30°, 900° - 30°$$
$$\therefore 3x = 30°, 150°, 390°, 510°, 750°, 870°$$
$$\therefore x = 10°, 50°, 130°, 170°, 250°, 290°$$

All solutions lie within the **original** specified domain, $0° \le x \le 360°$.

EXAMPLE 10.25

Solve $2\cos\left(\frac{x}{2} + \frac{\pi}{2}\right) - \sqrt{3} = 0$, for $-\pi \le x \le 4\pi$.

Solution

$2\cos\left(\frac{x}{2} + \frac{\pi}{2}\right) - \sqrt{3} = 0 \Leftrightarrow \cos\left(\frac{x}{2} + \frac{\pi}{2}\right) = \frac{\sqrt{3}}{2}, -\pi \le x \le 4\pi$.

Let $\frac{x}{2} + \frac{\pi}{2} = \theta \Rightarrow \cos\theta = \frac{\sqrt{3}}{2}$

Next, $\frac{x}{2} + \frac{\pi}{2} = \theta \Leftrightarrow \frac{x}{2} = \theta - \frac{\pi}{2} \Leftrightarrow x = 2\theta - \pi$. [Obtain x in terms of θ]

New domain: $-\pi \le x \le 4\pi \Leftrightarrow -\pi \le 2\theta - \pi \le 4\pi \Leftrightarrow 0 \le 2\theta \le 5\pi$

$$\Leftrightarrow 0 \le \theta \le \frac{5\pi}{2}$$

Therefore, our equivalent statement is .

$\cos\theta = \frac{\sqrt{3}}{2}, 0 \le \theta \le \frac{5\pi}{2}$

$$\therefore \theta = \frac{\pi}{6}, 2\pi - \frac{\pi}{6}, 2\pi + \frac{\pi}{6}$$
$$= \frac{\pi}{6}, \frac{11\pi}{6}, \frac{13\pi}{6}$$

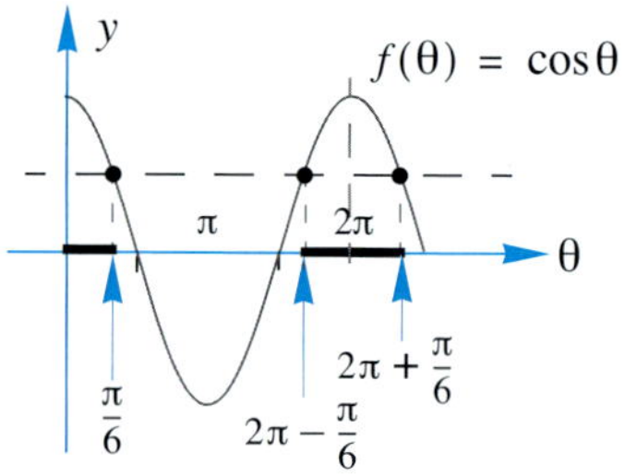

But we still need to find the x-values:

Therefore, substituting $\theta = \frac{x}{2}+\frac{\pi}{2}$ back into the solution set, we have:

$$\frac{x}{2}+\frac{\pi}{2} = \frac{\pi}{6}, \frac{11\pi}{6}, \frac{13\pi}{6}$$

$$\Leftrightarrow x+\pi = \frac{\pi}{3}, \frac{11\pi}{3}, \frac{13\pi}{3}$$

$$\Leftrightarrow x = -\frac{2\pi}{3}, \frac{8\pi}{3}, \frac{10\pi}{3}$$

And, we notice that all solutions lie within the **original** specified domain, $-\pi \le x \le 4\pi$.

EXAMPLE 10.26

Solve the following, for $0 \le x \le 2\pi$.

(a) $2\sin x = 3\cos x$ (b) $2\sin 2x = 3\cos x$ (c) $\sin x = \cos 2x$

Solution

(a) $2\sin x = 3\cos x \Leftrightarrow \frac{\sin x}{\cos x} = \frac{3}{2}, 0 \le x \le 2\pi$

$\Leftrightarrow \tan x = 1.5, 0 \le x \le 2\pi$

The reference angle is $\text{Tan}^{-1}(1.5) \approx 0.9828$

Therefore, $x = \text{Tan}^{-1}(1.5), \pi + \text{Tan}^{-1}(1.5)$

$\approx 0.9828, 4.1244$

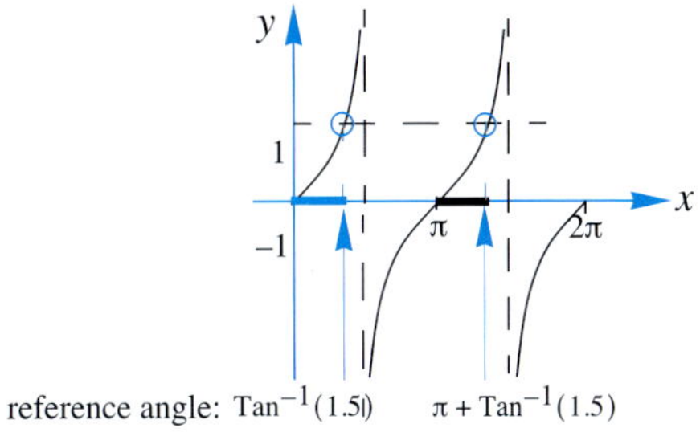

(b) In this case we make use of the double angle identity, $\sin 2x = 2\sin x\cos x$.

$\therefore 2\sin 2x = 3\cos x \Leftrightarrow 2(2\sin x\cos x) = 3\cos x$

$\Leftrightarrow 4\sin x\cos x - 3\cos x = 0$

$\Leftrightarrow \cos x(4\sin x - 3) = 0$

$\cos x = 0, \sin x = \frac{3}{4}$

Solving for $\cos x = 0 : x = \frac{\pi}{2}, \frac{3\pi}{2}$. Solving for $\sin x = \frac{3}{4} : x = \text{Sin}^{-1}\left(\frac{3}{4}\right), \pi - \text{Sin}^{-1}\left(\frac{3}{4}\right)$

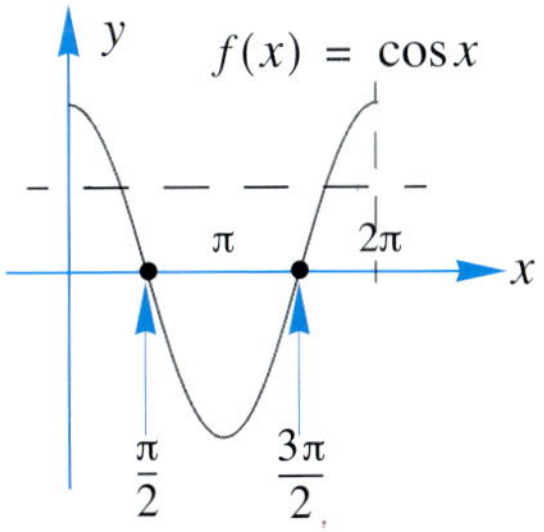

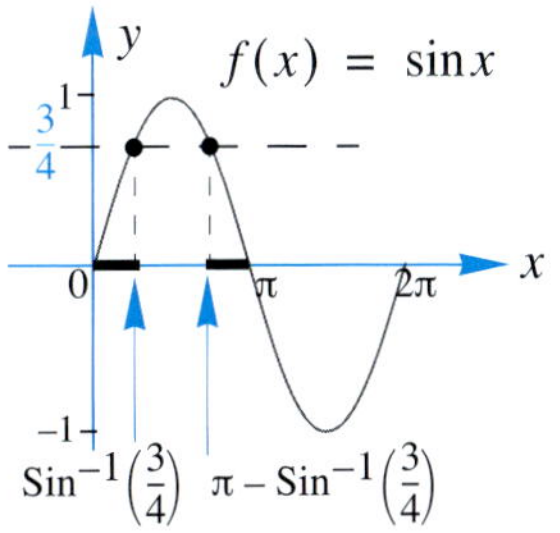

We solved two separate equations, giving the solution, $x = 0, \text{Sin}^{-1}\left(\frac{3}{4}\right), \frac{\pi}{2}, \pi - \text{Sin}^{-1}\left(\frac{3}{4}\right), \frac{3\pi}{2}$.

(c) This time we make use of the cosine double angle formula, $\cos 2x = 1 - 2\sin^2 x$.

$$\begin{aligned} \therefore \quad \sin x &= \cos 2x \\ \Leftrightarrow \sin x &= 1 - 2\sin^2 x \\ \Leftrightarrow 2\sin^2 x + \sin x - 1 &= 0 \\ \Leftrightarrow (2\sin x - 1)(\sin x + 1) &= 0 \\ \Leftrightarrow \sin x &= \frac{1}{2} \text{ or } \sin x = -1 \end{aligned}$$

Again, we have two equations to solve.

Solving for $\sin x = \frac{1}{2}$:

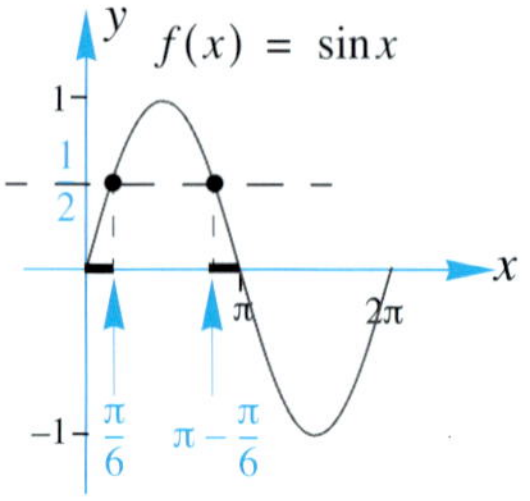

Solving for $\sin x = -1$:

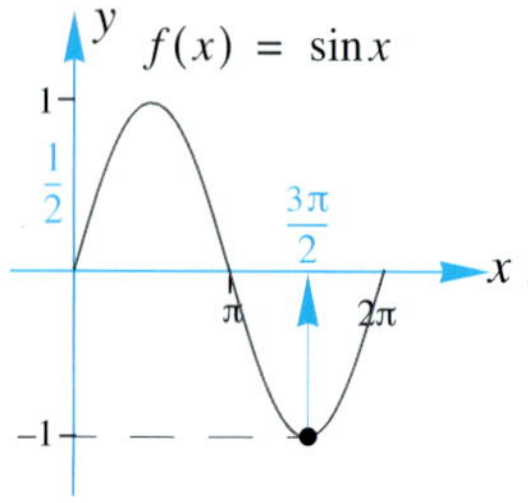

Therefore, solution set is $x = \frac{\pi}{6}, \frac{5\pi}{6}, \frac{3\pi}{2}$.

As the next example shows, the amount of working required to solve trigonometric equations can be significantly reduced, especially if you know the exact values (for the basic trigonometric angles) as well as the symmetry properties (without making use of a graph). However, we encourage you to use a visual aid when solving such equations.

EXAMPLE 10.27

Solve the equation $4\sin x\cos x = \sqrt{3}, -2\pi \le x \le 2\pi$

Solution

$$\begin{aligned} 4\sin x\cos x &= \sqrt{3}, -2\pi \le x \le 2\pi \\ 2\sin 2x &= \sqrt{3} \quad \text{using } \sin 2\theta \equiv 2\sin\theta\cos\theta \\ \sin 2x &= \frac{\sqrt{3}}{2} \\ 2x &= \ldots \frac{-11\pi}{3}, \frac{-10\pi}{3}, \frac{-5\pi}{3}, \frac{-4\pi}{3}, \frac{\pi}{3}, \frac{2\pi}{3}, \frac{7\pi}{3}, \frac{8\pi}{3}, \ldots \\ x &= \frac{-11\pi}{6}, \frac{-5\pi}{3}, \frac{-5\pi}{6}, \frac{-2\pi}{3}, \frac{\pi}{6}, \frac{\pi}{3}, \frac{7\pi}{6}, \frac{4\pi}{3} \end{aligned}$$

EXERCISES 10.4

1. If $0 \le x \le 2\pi$, find:

(a) $\sin x = \frac{1}{\sqrt{2}}$ (b) $\sin x = -\frac{1}{2}$ (c) $\sin x = \frac{\sqrt{3}}{2}$

(d) $\sin 3x = \frac{1}{2}$ (e) $\sin\left(\frac{x}{2}\right) = \frac{1}{2}$ (f) $\sin(\pi x) = -\frac{\sqrt{2}}{2}$

2. If $0 \le x \le 2\pi$, find:

(a) $\cos x = \frac{1}{\sqrt{2}}$ (b) $\cos x = -\frac{1}{2}$ (c) $\cos x = \frac{\sqrt{3}}{2}$

(d) $\cos\left(\frac{x}{3}\right) = \frac{1}{2}$ (e) $\cos(2x) = \frac{1}{2}$ (f) $\cos\left(\frac{\pi}{2}x\right) = -\frac{\sqrt{2}}{2}$

3. If $0 \le x \le 2\pi$, find:

(a) $\tan x = \frac{1}{\sqrt{3}}$ (b) $\tan x = -1$ (c) $\tan x = \sqrt{3}$

(d) $\tan\left(\frac{x}{4}\right) = 2$ (e) $\tan(2x) = -\sqrt{3}$ (f) $\tan\left(\frac{\pi}{4}x\right) = -1$

4. If $0 \le x \le 2\pi$ or $0 \le x \le 360$, find:

(a) $\sin(x° + 60°) = \frac{1}{2}$ (b) $\cos(x° - 30°) = -\frac{\sqrt{3}}{2}$

(c) $\tan(x° + 45°) = -1$ (d) $\sin(x° - 20°) = \frac{1}{\sqrt{2}}$

(e) $\cos\left(2x - \frac{\pi}{2}\right) = \frac{1}{2}$ (f) $\tan\left(\frac{\pi}{4} - x\right) = 1$

5. If $0 \le x \le 2\pi$ or $0 \le x \le 360$, find:

(a) $\cos x° = \frac{1}{2}$ (b) $2\sin x + \sqrt{3} = 0$ (c) $\sqrt{3}\tan x = 1$

(d) $5\sin x° = 2$ (e) $4\sin^2 x - 3 = 0$ (f) $\frac{1}{\sqrt{3}}\tan x + 1 = 0$

(g) $2\sin\left(x + \frac{\pi}{3}\right) = -1$ (h) $5\cos(x + 2) - 3 = 0$ (i) $\tan\left(x - \frac{\pi}{6}\right) = \frac{1}{\sqrt{3}}$

(j) $2\cos 2x + 1 = 0$ (k) $\tan 2x - \sqrt{3} = 0$ (l) $2\sin x° = 5\cos x°$

6. Solve the following equations for the intervals indicated, giving exact answers:

(a) $\sin\theta\cos\theta = \frac{1}{2}, -\pi \le \theta \le \pi$ (b) $\cos^2\theta - \sin^2\theta = -\frac{1}{2}, -\frac{\pi}{2} \le \theta \le \frac{\pi}{2}$

(c) $\tan A = \frac{1 - \tan^2 A}{2}, -\pi \le A \le \pi$ (d) $\frac{\sin\theta}{1 + \cos\theta} = -1, -\pi \le \theta \le \pi$

(e) $\cos^2 x = 2\cos x, -\pi \le x \le \pi$ (f) $\sec 2x = \sqrt{2}, 0 \le x \le 2\pi$

(g) $2\sin^2 x - 3\cos x = 2, 0 \le x \le 2\pi$ (h) $\sin 2x = 3\cos x, 0 \le x \le 2\pi$

7. Find (a) $3\tan^2 x + \tan x = 2, 0 \le x \le 2\pi$.

(b) $\tan^3 x + \tan^2 x = 3\tan x + 3, 0 \le x \le 2\pi$.

8. If $0 \le x \le 2\pi$ find:

(a) $\sin^2 2x - \frac{1}{4} = 0$ (b) $\tan^2\left(\frac{x}{2}\right) - 3 = 0$ (c) $\cos^2(\pi x) = 1$

9. (a) Sketch the graph of $f(x) = \sin x, 0 \le x \le 4\pi$.

(b) Hence, find i. $\left\{x | \sin x > \frac{1}{2}\right\} \cap \{x | 0 < x < 4\pi\}$.

ii. $\{x | \sqrt{3}\sin x < -1\} \cap \{x | 0 < x < 4\pi\}$.

10. (a) i. On the same set of axes sketch the graphs of $f(x) = \sin x$ and $g(x) = \cos x$ for $0 \le x \le 2\pi$.

ii. Find $\{x | \sin x < \cos x, 0 \le x \le 2\pi\}$.

(b) i. On the same set of axes sketch the graphs of $f(x) = \sin 2x$ and $g(x) = \cos x$ for $0 \le x \le 2\pi$.

ii. Find $\{x | \sin 2x < \cos x, 0 \le x \le 2\pi\}$.

11. (a) Solve $\left\{x° : 3\sin x° - \frac{1}{\sin x°} = 2, 0 \le x \le 360\right\}$

(b) Hence, find $\left\{x° : 3\sin x° < \frac{1}{\sin x°} + 2, 0 \le x \le 360\right\}$.

10.5 APPLICATIONS

Functions of the type considered in the previous section are useful for modelling periodic phenomena. These sorts of applications usually start with data that has been measured in an experiment. The next task is to find a function that 'models' the data in the sense that it produces function values that are similar to the experimental data. Once this has been done, the function can be used to predict values that are missing from the measured data (interpolation) or values that lie outside the experimental data set (extrapolation).

EXAMPLE 10.28 The table shows the depth of water at the end of a pier at various times (measured, in hours after midnight on the first day of the month.)

t (hr)	0	3	6	9	12	15	18	21	24	27	30	33
d (m)	16.20	17.49	16.51	14.98	15.60	17.27	17.06	15.34	15.13	16.80	17.42	15.89

Plot the data as a graph. Use your results to find a rule that models the depth data. Use your model to predict the time of the next high tide.

Solution

We start by entering the data as lists and then plotting them using the TI–83:

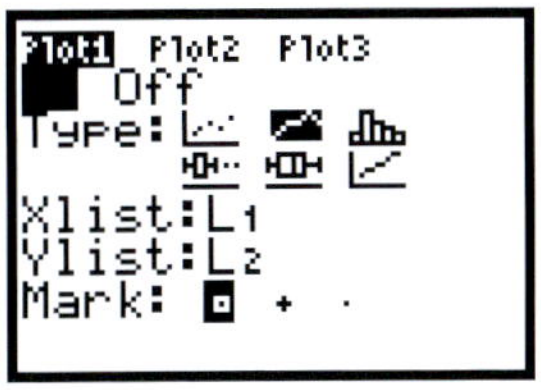

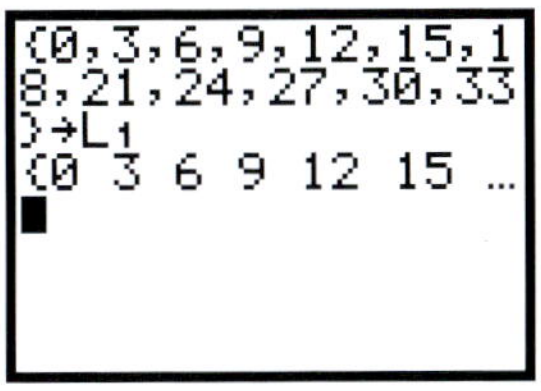

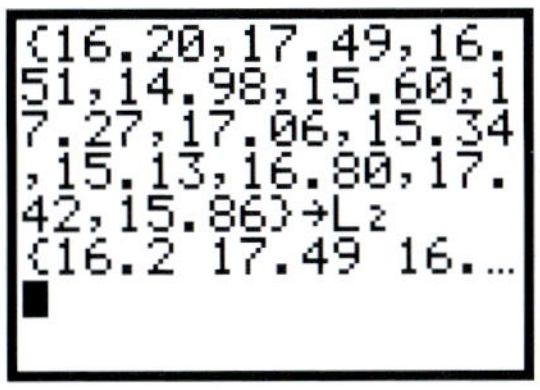

Using two different scales can provide graphs that do not really resemble each other.

Although both display a periodic nature, the second seems to reflect the need for a circular trigonometric function to model the behaviour.

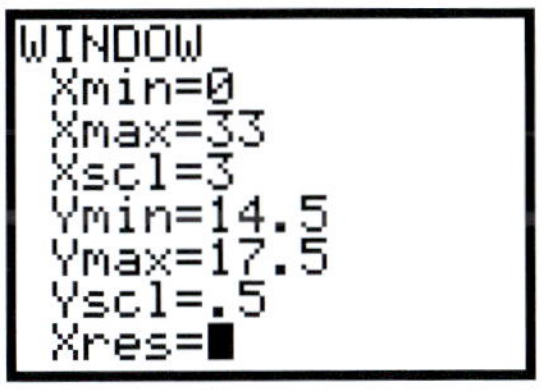

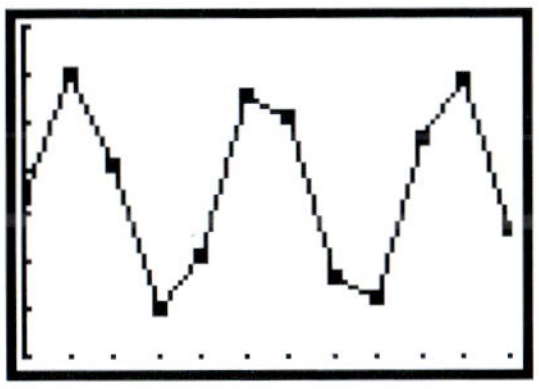

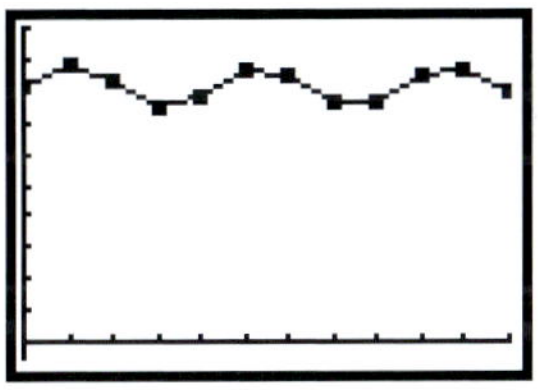

This does suggest that the depth is varying periodically. It appears that the period is approximately 13 hours. This is found by looking at the time between successive high tides. This is not as easy as it sounds as the measurements do not appear to have been made exactly at the high tides. This means that an estimate will need to be made based upon the observation that successive high tides appear to have happened after 3, 16 and 32 hours. Next, we look at the amplitude and vertical translation. Again, because we do not have exact readings at high and low tides, these will need to be estimated. The lowest tide recorded is 14.98 and the highest is 17.49.

A first estimate of the vertical translation is $\frac{17.49 + 14.98}{2} = 16.235$ and the amplitude is

$17.7 - 16.235 = 1.465$. Since the graph starts near the mean depth and moves up it seems likely that the first model to try might be: $y = 1.465 \times \sin\left(\frac{2\pi t}{13}\right) + 16.235$

Notice that the dilation factor (along the x–axis) is found by using the result that if

$$\tau = 13 \Rightarrow \frac{2\pi}{n} = 13 \therefore n = \frac{2\pi}{13}.$$

The model should now be 'evaluated' which means testing how well it fits the data. This can be done by making tables of values of the data and the values predicted by the model and working to make the differences between these as small as possible. This can be done using a scientific or graphics calculator.

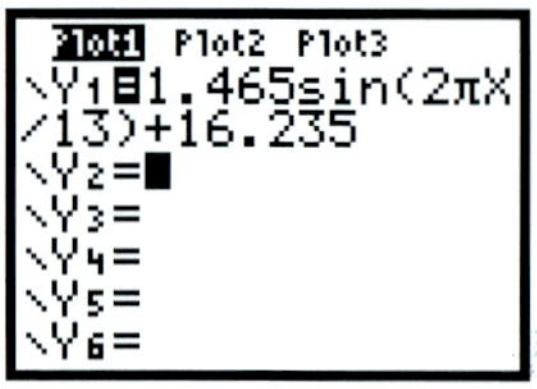

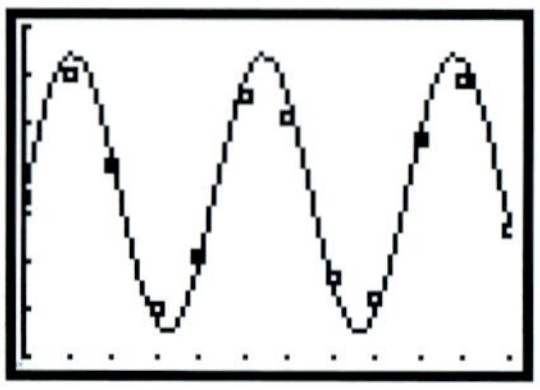

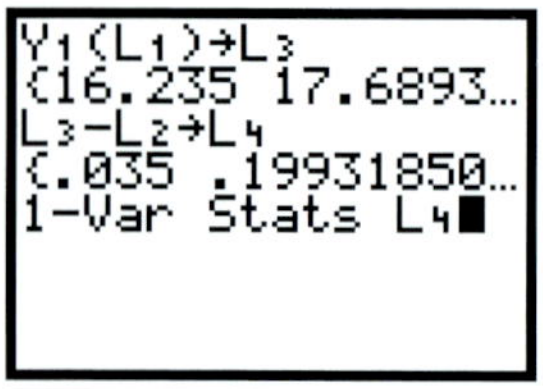

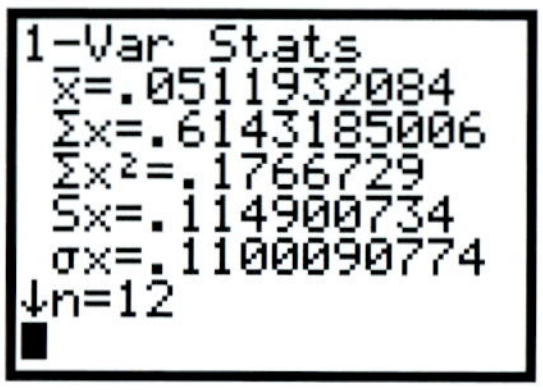

The model shown is quite good in that the errors are small with some being positive and some being negative. The function used is $d = 1.465 \times \sin\left(\frac{2\pi t}{13}\right) + 16.235$ and this can now be used to predict the depth for times that measurements were not made. Also, the graph of the modelling function can be added to the graph of the data (as shown).

The modelling function can also be used to predict depths into the future (extrapolation). The next high tide, for example can be expected to be 13 hours after the previous high tide at about 29.3 hours. This is after 42.3 hours.

EXAMPLE 10.29

During the summer months, a reservoir supplies water to an outer suburb based on the water demand, $D(t) = 120 + 60\sin\left(\frac{\pi}{90}t\right), 0 \le t \le 90$, where t measures the number of days from the start of Summer (which lasts for 90 days).

(a) Sketch the graph of $D(t)$.

(b) What are the maximum and minimum demands made by the community over this period?

Solution

(a) The features of this function are:

Period $= \frac{2\pi}{\left(\frac{\pi}{90}\right)} = 180$ days

Amplitude $= 60$

Translation $= 120$ units up.

We 'pencil in' the graph of $y = 60\sin\left(\frac{\pi}{90}t\right)$ and then move it up 120 units:

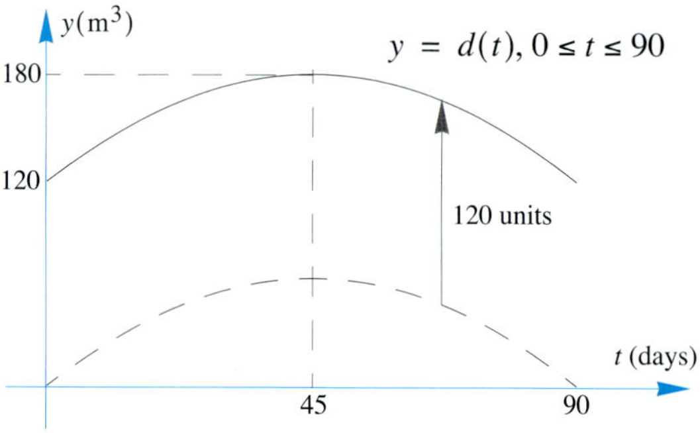

(b) The minimum is 120 m^3 and the maximum is 180 m^3.

EXAMPLE 10.30 When a person is at rest, the blood pressure, P millimetres of mercury at any time t seconds can be approximately modelled by the equation

$$P(t) = -20\cos t\left(\frac{5\pi}{3}t\right) + 100,\ t \geq 0$$

(a) Determine the amplitude and period of P.
(b) What is the maximum blood pressure reading that can be recorded for this person?
(c) Sketch the graph of $P(t)$, showing one full cycle.
(d) Find the first two times when the pressure reaches a reading of 110 mmHg.

Solution

(a) The amplitude is 20 mmHg and the period is given by $\dfrac{2\pi}{\left(\frac{5\pi}{3}\right)} = \dfrac{6}{5} = 1.2$ seconds.

(b) The maximum is given by (100 + amplitude) = 100 + 20 =120.

(c) One full cycle is 1.2 seconds long:

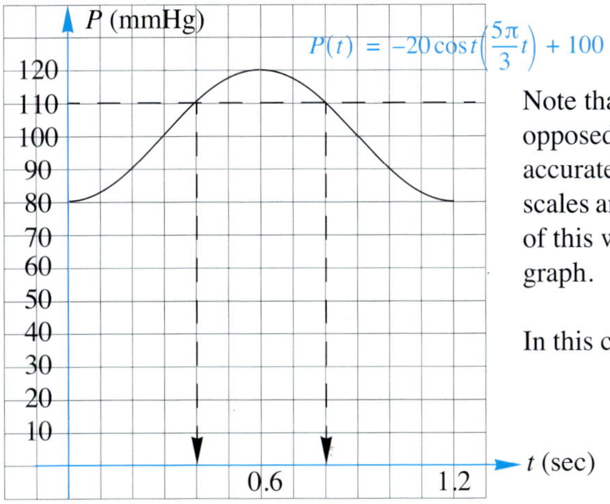

Note that the graph has been drawn as opposed to sketched. That is, it has been accurately sketched, meaning that the scales and the curve are accurate. Because of this we can read directly from the graph.

In this case, $P = 110$ when $t = 0.4$ and 0.8.

(d) Even though we have drawn the graph, we will now solve the relevant equation:

$$P(t) = 110 \Leftrightarrow 110 = -20\cos\left(\frac{5\pi}{3}t\right) + 100$$

$$\Leftrightarrow 10 = 20\cos\left(\frac{5\pi}{3}t\right)$$

$$\Leftrightarrow \cos\left(\frac{5\pi}{3}t\right) = -\frac{1}{2}$$

$$\therefore \frac{5\pi}{3}t = \pi - \text{Cos}^{-1}\left(\frac{1}{2}\right), \pi + \text{Cos}^{-1}\left(\frac{1}{2}\right) \quad \left[\text{Reference angle is } \text{Cos}^{-1}\left(\frac{1}{2}\right) = \frac{\pi}{3}\right]$$

$$\Leftrightarrow \frac{5\pi}{3}t = \frac{2\pi}{3}, \frac{4\pi}{3}$$

$$\Leftrightarrow t = \frac{2}{5}, \frac{4}{5}$$

EXERCISES 10.5

1. The table shows the temperature in an office block over a 36 hour period.

t (hr)	0	3	6	9	12	15	18	21	24	27	30	33	36
T °C	18.3	15.0	14.1	16.0	19.7	23.0	23.9	22.0	18.3	15.0	14.1	16.0	19.7

(a) Estimate the amplitude, period, horizontal and vertical translations.
(b) Find a rule that models the data.
(c) Use your rule to predict the temperature after 40 hours.

2. The table shows the light level (L) during an experiment on dye fading.

t (hr)	0	1	2	3	4	5	6	7	8	9	10
L	6.6	4.0	7.0	10.0	7.5	4.1	6.1	9.8	8.3	4.4	5.3

(a) Estimate the amplitude, period, horizontal and vertical translations.
(b) Find a rule that models the data.

3. The table shows the value in $s of an industrial share over a 20 month period.

Month	0	2	4	6	8	10	12	14	16	18	20
Value	7.0	11.5	10.8	5.6	2.1	4.3	9.7	11.9	8.4	3.2	2.5

(a) Estimate the amplitude, period, horizontal and vertical translations.
(b) Find a rule that models the data.

4. The table shows the population (in thousands) of a species of fish in a lake over a 22 year period.

Year	0	2	4	6	8	10	12	14	16	18	20	22
Pop	11.2	12.1	13.0	12.7	11.6	11.0	11.6	12.7	13.0	12.1	11.2	11.2

(a) Estimate the amplitude, period, horizontal and vertical translations.
(b) Find a rule that models the data.

5. The table shows the average weekly sales (in thousands of $s) of a small company over a 15 year period.

Time	0	1.5	3	4.5	6	7.5	9	10.5	12	13.5	15
Sales	3.5	4.4	7.7	8.4	5.3	3.3	5.5	8.5	7.6	4.3	3.6

(a) Estimate the amplitude, period, horizontal and vertical translations.
(b) Find a rule that models the data.

6. The table shows the average annual rice production, P, (in thousands of tonnes) of a province over a 10 year period.

t (yr)	0	1	2	3	4	5	6	7	8	9	10
P	11.0	11.6	10.7	10.5	11.5	11.3	10.4	11.0	11.6	10.7	10.5

(a) Estimate the amplitude, period, horizontal and vertical translations.
(b) Find a rule that models the data.

7. The table shows the depth of water (D metres) over a 5 second period as waves pass the end of a pier.

t (sec)	0	0.5	1	1.5	2	2.5	3	3.5	4	4.5	5
D	11.3	10.8	10.3	10.2	10.4	10.9	11.4	11.7	11.8	11.5	11.0

(a) Estimate the amplitude, period, horizontal and vertical translations.
(b) Find a rule that models the data.

8. The population (in thousands) of a species of butterfly in a nature sanctuary is modelled by the function:

$$P = 3 + 2\sin\left(\frac{3\pi t}{8}\right), 0 \le t \le 12$$

where t is the time in weeks after scientists first started making population estimates.

(a) What is the initial population?
(b) What are the largest and smallest populations?
(c) When does the population first reach 4 thousand butterflies?

9. A water wave passes a fixed point. As the wave passes, the depth of the water (D metres) at time t seconds is modelled by the function:

$$D = 7 + \frac{1}{2}\cos\left(\frac{2\pi t}{5}\right), t > 0$$

(a) What are the greatest and smallest depths?
(b) Find the first two times at which the depth is 6.8 metres.

10. The weekly sales (S) (in hundreds of cans) of a soft drink outlet is modelled by the function:

$$S = 13 + 5.5\cos\left(\frac{\pi t}{6} - 3\right), t > 0$$

t is the time in months with $t = 0$ corresponding to January 1st 1990,

(a) Find the minimum and maximum sales during 1990.
(b) Find the value of t for which the sales first exceed 1500 ($S = 15$).
(c) During which months do the weekly sales exceed 1500 cans?

11. The rabbit population, $R(t)$ thousands, in a northern region of South Australia is modelled by the equation $R(t) = 12 + 3\cos\left(\frac{\pi}{6}t\right), 0 \le t \le 24$, where t is measured in months after the first of January.

(a) What is the largest rabbit population predicted by this model?
(b) How long is it between the times when the population reaches consecutive peaks?
(c) Sketch the graph of $R(t)$ for $0 \le t \le 24$.
(d) Find the longest time span for which $R(t) \ge 13.5$.
(c) Give a possible explanation for the behaviour of this model.

12. A hill has its cross-section modelled by the function,

$$h : [0, 2] \mapsto \mathbb{R},\ h(x) = a + b\cos(kx),$$

where $h(x)$ measures the height of the hill relative to the horizontal distance x m from O.

(a) Determine the values of
  i. k.
  ii. b.
  iii. a.

(b) How far, horizontally from O, would an ant climbing this hill from O be, when it first reaches a height of 1 metre?

(c) How much further, horizontally, will the ant have travelled when it reaches the same height of 1 metre once over the hill and on its way down?

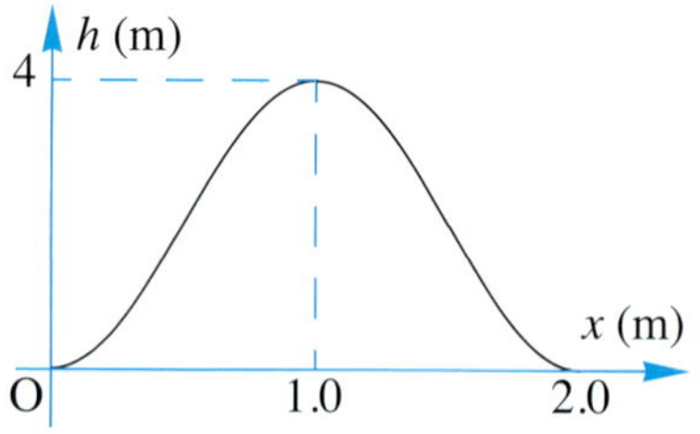

13. A nursery has been infested by two insect pests: the Fruitfly and the Greatfly. These insects appear at about the same time that a particular plant starts to flower. The number of Fruitfly (in thousands), t weeks after flowering has started is modelled by the function

$$F(t) = 6 + 2\sin(\pi t),\ 0 \le t \le 4.$$

Whereas the number of Greatfly (in thousands), t weeks after flowering has started is modelled by the function

$$G(t) = 0.25t^2 + 4,\ 0 \le t \le 4$$

(a) Copy and complete the following table of values, giving your answers correct to the nearest hundred.

t	0	0.5	1	1.5	2	2.5	3	3.5	4
$F(t)$									
$G(t)$									

(b) On the same set of axes **draw** the graphs of
  i. $F(t) = 6 + 2\sin(\pi t),\ 0 \le t \le 4$.
  ii. $G(t) = 0.25t^2 + 4,\ 0 \le t \le 4$.

(c) On how many occassions will there be equal numbers of each insect?

(d) For what percentage of the time will there be more Greatflies than Fruitflies?

14. The depth, $d(t)$ metres, of water at the entrance to a harbour at t hours after midnight on a particular day is given

$$d(t) = 12 + 3\sin\left(\frac{\pi}{6}t\right),\ 0 \le t \le 24$$

(a) Sketch the graph of $d(t)$ for $0 \le t \le 24$.

(b) For what values of t will
  i. $d(t) = 10.5,\ 0 \le t \le 24$.
  ii. $d(t) \ge 10.5,\ 0 \le t \le 24$.

Boats requiring a minimum depth of b metres are only permitted to enter the harbour when the depth of water at the entrance of the harbour is at least b metres for a continuous period of one hour.

(c) Find the largest value of b, correct to two decimal place, which satisfies this condition.

REVISION SET B - PAPER 1 & PAPER 2 STYLE QUESTIONS

* some of these questions go beyond the level expected in examinations.

1. (a) In the sequence $2, 2\frac{1}{2}, 3, 3\frac{1}{2}, \ldots$, what term will the number 96 be?

(b) In a city there was on average a crime rate every 90 minutes. This represented an increase of 10% over the previous year. In the previous year there was on average a crime every x minutes. Find x.

(c) Find x given that $3 - 6 + 12 - 24 + \ldots + x = -63$.

(d) A series is defined as $\log_2 3 + \log_2 3^2 + \log_2 3^3 + \log_2 3^4 + \ldots$. What is the smallest value of n such that $S_n > 1000$?

2. (a) Given that $\frac{1}{b+c}, \frac{1}{a+c}$ and $\frac{1}{a+b}$ are three consecutive terms of an arithmetic sequence. Show that a^2, b^2 and c^2 are also three consecutive terms of an arithmetic sequence.

(b) The 3rd term of an A.P is 6 and the 5th term is –4. Find the sum of the first 10 terms of the A.P.

3. Three towns are positioned so that Blacktown is 15 km north of Acton and Capetown is 32 km north-east of Acton. (Giving all answers to 1 d.p.)

(a) Draw a triangle representing the locations of the three towns Acton, Blacktown and Capetown labelling the towns A, B and C respectively. Making sure to label all distances and angles.

(b) Calculate the straight-line distance between Blacktown and Capetown.

(c) Determine the angle BCA.

An aeroplane is 3.0 km directly above Blacktown at point P. The points A, P and C form a triangle.

(d) Determine how far the plane is from the point A which represents the town Acton.

(e) Calculate the size of angle APC.

4. Three companies A, B and C are competitors and are keen to compare sales figures for the period July 2003 to June 2004 inclusive.
Company A had sales of \$35,400 in July 2003 and increased sales by an average 3% per month over the period.
Company B had sales of \$32,000 in July 2003 and increased sales by an average of \$1859 per month over the period.
Company C had sales of \$48,000 in July 2003 and unfortunately found that their sales decreased by an average of 8% per month over the period.

(a) Calculate to the nearest hundred dollars;

i. the sales figures for each of the companies in June 2004.

ii. the total sales figures for each of the companies for the twelve month period July 2003 to June 2004 inclusive.

(b) Calculate, correct to one decimal place, the average percentage increase in sales per month over the period for company B.

(c) Assuming that the given sales trends continue beyond June 2004:

i. calculate how many months from July 2003 Company A will take to reach a total sales target of \$600,000.

ii. will Company C ever reach a total sales target of \$600,000? Give reasons

5. (a) In an infinite G.P, the sum of the first three terms is equal to seven eighths of the sum to infinity. Find the common ratio, r.

(b) A plant is 50 cm high when it is first observed. One week later it has grown 10 cm, and each week thereafter it grows 80% of the previous week's growth. Given that this pattern continues, what will be the plants ultimate height?

6. (a) Show that $\frac{1-\cos 2A}{1+\cos 2A} = \tan^2 A$

(b) Solve the equation $2\sec^2\theta + \tan\theta - 3 = 0, 0 \le \theta \le 180°$.

7. (a) Prove that $\frac{\sin 2x}{1+\cos 2x} = \tan x$.

(b) Solve $2\cos^2\theta + \sin\theta = 1, 0 \le \theta \le 2\pi$.

8. For a certain A.P, $u_3 = 7$ and $u_6 = 16$. Find the value of u_{10}.

9. (a) Solve for θ if $4\sin^2\theta = 3, 0 \le \theta \le 2\pi$.

(b) Find $\{\theta : \sin^2\theta = \sin\theta, 0 \le \theta \le 2\pi\}$

10. (a) If $g(x) = \frac{17}{5-3\sin^2 x}, x \in \mathbb{R}$, state the greatest and least value of $g(x)$ and the positive values of x for which these occur.

(b) Solve for θ in the range $]0, 2\pi[$ where $2\sin^2\theta = 3\cos\theta$.

(c) Prove that $\frac{\sin 2\theta}{1-\cos 2\theta} = \frac{1}{\tan\theta}$.

11. The first three terms of a sequence are 68, 62, 56.

(a) Write a formula for the nth term, u_n

(b) If the last term of the sequence is p, how many terms are there in the sequence?

(c) Find a simplified expression for the sum of n terms of the sequence in terms of p, and hence find the value of p that gives the maximum possible sum.

12. In the triangle ABC, AB = 5, BC = 3, CA = 4. D is a point on [AB] such that $\angle BCD = \frac{\pi}{6}$.

Find the exact value of CD.

13. Two vertical masts BD and CE, each of height h, have their base B and C on level ground with C to the east of B. Point A lies on the same level as B and C and is due south of B. The angle of elevation of D from A is θ and angle DAE = ϕ. If the angle of elevation of E from A is α, show that $\sin\alpha = \sin\theta\cos\phi$.

14. (a) Solve for θ where $2\tan\theta - 4\cot\theta = \operatorname{cosec}\theta, 0° \le \theta \le 360°$.

(b) i. Simplify $\frac{\sin\theta}{1-\cos\theta} + \frac{\sin\theta}{1+\cos\theta}$.

ii. Hence solve $\frac{\sin\theta}{1-\cos\theta} + \frac{\sin\theta}{1+\cos\theta} = \frac{4}{\sqrt{3}}, 0 \le \theta \le 2\pi$.

15. (a) A geometric sequence has a first term 2 and a common ratio 1.5. Find the sum of the first 11 terms.

(b) Find the number of terms of the sequence 3, 7, 11, . . . required to sum to 820.

(c) For what values of x will the sum $1 + (1-x) + (1-x)^2 + (1-x)^3 + \ldots$ exist?

(d) Find the first four terms of the sequence defined by $u_{n+1} = 2u_n + n, u_1 = 1$.

(e) Find the nth term of the sequence defined by the difference equation $u_{n+1} = u_n - 3$, with $u_1 = 20$.

(f) Robyn borrows $10000 from a bank. Interest at 18% per annum is calculated monthly on the amount of money still owing. Robyn repays $600 each month. How much does Robyn owe the bank after the first 12 repayments?

16. Let $f(x) = 2x + \frac{\pi}{6}$ and $g(x) = 1 - 3\sin x$.

(a) Find $g(f(0))$.

(b) Write down the maximum value of $g(f(x))$.

17. The sides of a triangle have lengths 7cm, 8cm and 13cm respectively. Find

(a) the size of its largest angle, (b) the area of the triangle.

18. A photographer's tripod stands on a horizontal floor, with its feet A, B and C at the vertices of an equilateral triangle of side length 0.6 metres. Each leg is 1.2 metres long.

The line [AM] meets the side [BC] at right angles where M is the mid-point of [BC]. The point G is the centroid of triangle ABC, that is,

$$AG = \frac{2}{3}AM.$$

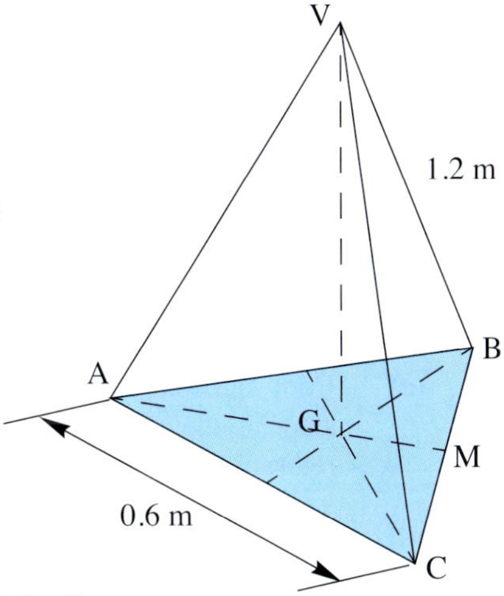

(a) Calculate i. the length of [AM]

ii. the length of [AG]

(b) Calculate the height above the floor of the top, V, of the tripod.

(c) Calculate the angle which a leg makes with the floor.

19. (a) Find the values of x between 0 and 2π for which $\sin x = \sqrt{3}\cos x$.

(b) Sketch, on the same set of axes, the graphs of $y = \sin x$ and $y = \sqrt{3}\cos x$ for values of x from 0 to 2π. Hence find $\{x \mid \sin x > \sqrt{3}\cos x, 0 \le x \le 2\pi\}$.

20. A rectangular pyramid VABCD has a base measuring 24 cm by 18 cm Each slant edge measures 17 cm.

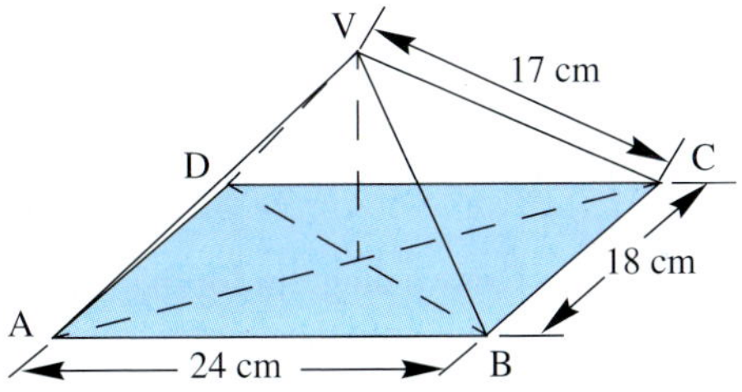

Find

(a) the height of the pyramid.

(b) the angle which a slant edge makes with the base.

21. If $\sin\theta = -\frac{3}{5}$ and $\cos\theta < 0$, find $\cos\theta$ and $\tan\theta$ and hence, evaluate $5 - \frac{2}{\sin^2\theta} + \frac{2}{\tan^2\theta}$.

22. (a) Find the values of x which satisfies the equation: $\cos 2\left(x - \frac{\pi}{3}\right) = 0, 0 \le x \le 2\pi$.

(b) Make use of these values in sketching the graph of $y = \cos 2\left(x - \frac{\pi}{3}\right)$ for $0 \le x \le 2\pi$ labelling the intercepts with the axes. Also find one turning point.

23. (a) Mr F. Nurke has been working at the firm of Snyde & Shyster for 6 years. He started work on an annual salary of \$10,000, and each year this has increased by 10%. Find the total amount he has earned with the firm.

(b) The first term in a geometric sequence is a negative number, the second is 3 and the fourth is 9. Find the first and third terms.

24. (a) The graph on the right represents the function definded by $f(x) = a\cos(bx)$. Find the values of a and b.

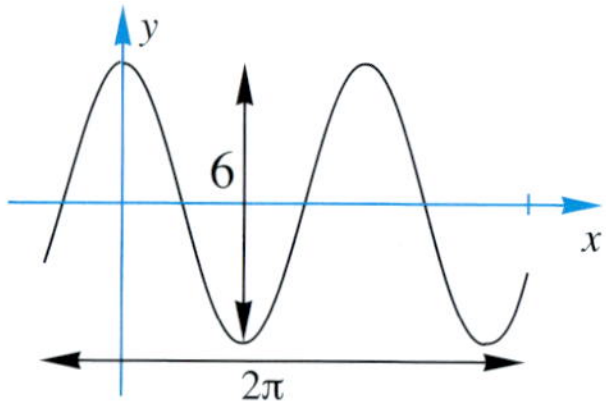

(b) Find $\left\{x : 2\cos(2x) = \sqrt{3}, \pi < x < \frac{3\pi}{2}\right\}$.

(c) Find $n(\mathbf{S})$, if $\mathbf{S} = \{x : \sin(3x) = \cos(3x), 0 \le x \le \pi\}$

25. A, P and Q are three points in that order in a straight line. The bearing of P and Q from A is 310°. From another point B, 1 kilometre due west of A, the bearing of P is 053°; and Q is due north of B.

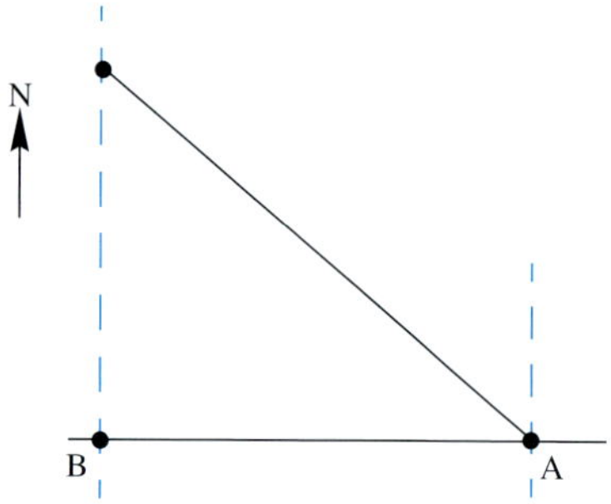

(a) Copy and complete the diagram, clearly showing the position of P and Q and all relevant angles and distances.

(b) Calculate, to the nearest metre,
  i. the distance from point P from point B?
  ii. distance from P to Q,

26. A sector OAB is to be cut from a circle of cardboard with radius 25 cm, and then folded so that radii OA and OB are joined to form a cone, with a slant height 25 cm.
If the vertical height of the cone is to be 20 cm, what must the angle $\alpha°$ be (in degrees)?

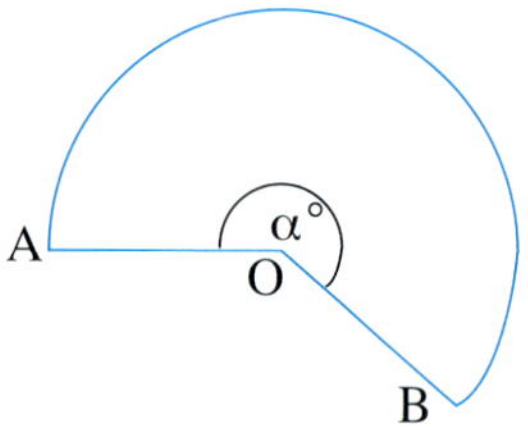

27. A girl walking due east along a straight horizontal road observes a church spire on a true bearing of 076°. After walking 1500 metres further she observes the spire on a true bearing of 067°.

(a) Draw a diagram for this situation.

(b) How far is the church from the road (to the nearest metre)?

28. A monument consists of a truncated square pyramid ABCD, EFGH, of height 2.2 metres, surmounted by a square pyramid V ABCD, of height 0.2 metres, as shown in the diagram. The square ABCD has edge 0.5 metres and the square EFGH has edge 0.8 metres.

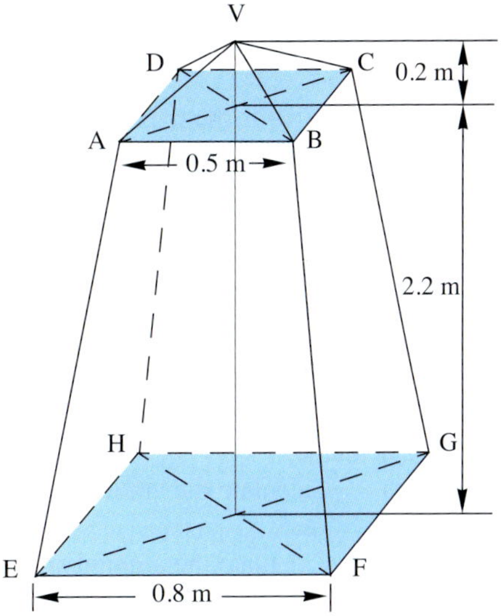

Find:

(a) the inclination of a sloping triangular face to the base.

(b) the surface area of one of the sloping triangular faces.

(c) the total surface area of all the sloping faces of the monument.

(d) the monument needs to be rendered with two coats of a cement mix. The cost per square metre for this cement mix is set at \$32.00. How much will it cost to render the monument if it labour costs will total \$300.00?

29. (a) Given that $\tan 2\alpha = 2, \frac{\pi}{2} < \alpha < \pi$, find the exact value of $\tan\alpha$.

(b) Find the range of the function $f(x) = 3 - \frac{1}{2}\sin 2x, \frac{\pi}{2} < x < \pi$.

(c) The equation of the graph shown is given by

$$g(x) = a(1 - \cos kx).$$

Find the values of i. a.

ii. k.

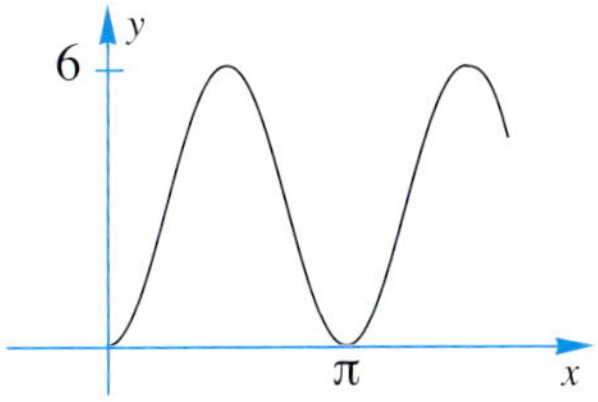

30. As part of a long term research project, economists from the Department of Agriculture have been recording the price of a bale of wool and a bushel of wheat on a weekly basis. The economists started collecting the information on the 1st January 2004.

After some time they realized that the the price for each item could be reasonably approximated by mathematical functions.

The price, in dollars per bale of wool is given by $W(t) = 15 + 5\cos\left(\frac{\pi}{25}t\right)$.

The price, in dollars per bushel of wheat is given by $P(t) = 3te^{-0.05t} + 5$.

In each case, t is the number of weeks from the 1st January 2004 (so that 1st January corresponds to $t = 0$, 8th January corresponds to $t = 1$ and so on).

(a) Determine the price of wool and wheat on the following dates:

i. 29 January 2004 (after 4 weeks).

ii. 20th May 2004 (after 20 weeks).

iii. 2nd August 2004 (after 35 weeks).

(b) Give the amplitude and period of the function W.

(c) Use the following information to help you sketch the graphs of W and P.

t	5	15	25	50
$W(t)$	19.05	13.45	10	20

t	10	15	25	50
$P(t)$	23.2	26.26	26.48	17.31

(d) What is the maximum price that the wheat reaches?

(e) Estimate, **from the graph**, during which weeks, after the 1st January 2004, the price of wool and the price of wheat are equal.

31. The owner of 'Sandra's Health Foods' decides that even though there is a recession, now is the time to expand by opening another shop. To complete the expansion she intends to borrow \$50000. After discussions with her bank manager, they come to an agreement where she can borrow \$50000 at 10% compounded annually. Sandra agrees to repay the loan at a fixed rate of \$6000 per year at the end of the year (i.e., after the interest has been calculated).

(a) Calculate the amount she owes the bank at the end of the first, second and third year.

(b) Calculate the amount she owes at the end of the tenth year.

(c) How many years will it take her to repay the debt?

Being conscious of her financial affairs, Sandra decides that on her retirement she wishes to establish an annuity. She wants an annuity which will pay her a fixed amount of \$2000 per month at the end of each month for thirty years. She can invest with a broker at 9% p.a. compounded monthly.

(d) How much should she invest in order to establish this annuity?

32. Towers of cards can be built as follows:
The first stage has 2 cards in one layer.
The second stage has 4 cards in the bottom layer, a card resting on top which supports another 2 cards. The third stage has 6 cards in the bottom layer, 2 cards resting on top which supports a second stage tower.

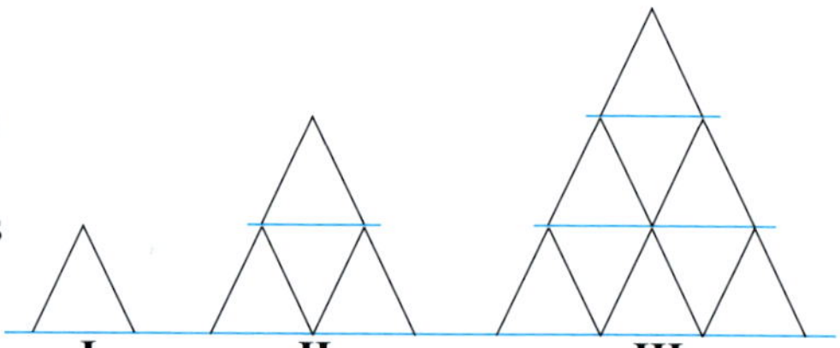

(a) i. Draw a four stage tower.
ii. How many cards are used in building this tower?

(b) The following sequence represents the number of cards used to build each tower: 2, 7, 15, ... Find the next four terms of the sequence.

(c) If t_n = the number of cards used to build an n-stage tower then there is a difference equation of the form $t_n = t_{n-1} + an + b$. Find a and b.

(d) How many cards would be needed to make a ten-stage tower?

(e) Find a formula for t_n in terms of n.

(f) Prove by mathematical induction, your proposition in (e).

33. The depth ($h(t)$ metres) of the water at a certain point on the coast, at time t hours after noon on a particular day, is given by

$$h(t) = 2.5 + \frac{1}{2}\cos\frac{4\pi}{25}(t+2),\ t \ge 0$$

(a) What is the depth of the water at noon?

(b) What is the depth of the water at i. high tide?

ii. low tide?

(c) At what time on that afternoon will low tide occur?

(d) Sketch the graph of $h(t)$ for $0 \le t \le 12.5$.

(e) On that afternoon the local people are building a bicentennial bonfire on a rock shelf at the point. They estimate that they can go to and fro to work on the rock shelf only when the depth of water at the point is less than 2.25 metres. Between what times can they plan to work on the rock shelf at building the bonfire?

34. A block of land has measurements as shown in the diagram. Calculate its area in hectares, correct to 3 decimal places. (1 hectare = 10000 square metres)

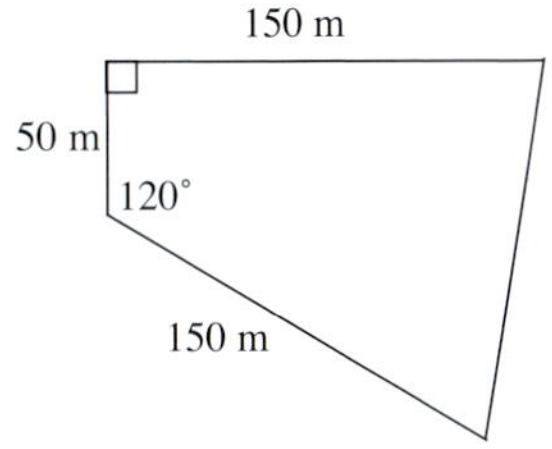

35. Find $\left\{x \middle| 1 + \sqrt{3}\tan 2\left(x - \frac{\pi}{3}\right) = 0, 0 < x < \pi\right\}$

36. (a) Find the values of x between $-\pi$ and π for which $2\cos\frac{x}{2} - 1 = 0$.

(b) Sketch the graph of $y = 2\cos\frac{x}{2} - 1, -\pi \le x \le \pi$.

(c) Hence find the value of x between $-\pi$ and π for which $2\cos\frac{x}{2} - 1 > 0$.

37. From an observation point A, the true bearing of a landmark C is 051° and the bearing of another landmark D is 284°. From another observation point B, 1000 metres due north of A, the bearing of C is 115° and the bearing of D is 224°.
Find the distance between C and D, correct to the nearest metre.

38. The temperature A°C inside a house at time t hours after 4:00 am is given by

$$A = 21 - 3\cos\left(\frac{\pi t}{12}\right), 0 \le t \le 24$$

and the temperature B°C outside the house at the same time is given by

$$B = 22 - 5\cos\left(\frac{\pi t}{12}\right), 0 \le t \le 24 .$$

(a) Find the temperature inside the house at 8:00 am.

(b) Write down an expresion for $D = A - B$, the difference between the inside and outside temperatures.

(c) Sketch the graph of D for $0 \le t \le 24$.

(d) Determine when the inside temperature is less than the outside temperature.

39. A surveyor measures the angle of elevation of the top of a mountain from a point at sea level as 20°. She then travels 1000 metres along a road which slopes uniformily uphill towards the mountain. From this point, which is 100 metres above sea level, she measures the angle of elevation as 23°. Find the height of the mountain above sea level, correct to he nearest metre.

40. The number of cancer cells in a solution is believed to increase in such a way that at the end of every hour, there are α % more cells than at the end of the previous hour. Dr. Bac

Teria, who is in charge of this experiment, has been recording the cell counts. Unfortunately, because of his preoccupation with another experiment he has been rather neglectful. The only available readings are shown in the table below:

t	1	2	3	4	5	6
$N(t)$			2420			3221

(a) The number of cells, $N(t)$, in the solution at the end of every hour is thought to be modelled by a geometric progression. That is, $N(t) = N_0 \times r^{t-1}, t = 1, 2, \ldots$

i. Show that $r = 1 + \frac{\alpha}{100}$.

ii. Find the values of N_0 and α.

(b) Copy and complete the table above.

(c) When is the first time that the number of cells in the solution exceeds 258259?

(d) At the end of every hour a new identical solution is set up in the same way that the first one was. How many cells will there be altogether at the end of a 24 hours run of introducing a new solution at the end of every hour.

41. (a) In Figure 1., A square having side length 2 cm has a quarter of a circle drawn within it as shown in the diagram below.
Find the area of the shaded region in Figure 1.

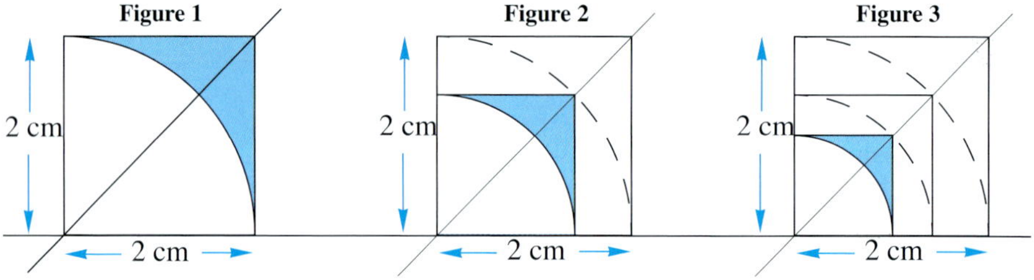

(b) Figure 2 consists of the same diagram as Figure 1 but this time a second square is drawn so that its diagonal line coincides with that of the original square and its corner meets the quarter circle. Once again, a quarter of a circle is drawn within the smaller square as shown.
Find the area of the shaded region in Figure 2.

The process is then continued, as in Figure 3, so that the shaded areas in each of the figures, form a sequence, $A_1, A_2, A_3, \ldots$, where A_1 is the measure of the shaded area in Figure 1, A_2 is the measure of the shaded area in Figure 2, and so on.

(c) i. Show that the sequence $A_1, A_2, A_3, \ldots$ is in fact geometric.

ii. Find the common ratio.

iii. Write down an expression for the nth term, A_n.

iv. What would the area of 5th shaded region be if we continued drawing figures using the same method?

(d) i. Find the sum of the first 5 shaded regions.

ii. If the process was to continue indefinitely, what would the areas of the shaded regions add to?

(e) What type of a sequence would the arc lengths of the quarter circle form? Explain.

11.1 INTRODUCTION TO MATRICES

11.1.1 DEFINITIONS

1. Matrix

Matrices are in many senses similar to vectors except that whilst vectors are represented by columns of numbers, matrices are generally blocks of numbers made up of a number of **rows** and **columns**. In their simplest form, matrices are often used to store information.

Consider the following family shopping list:

	Mon	Tues	Wed	Thurs	Fri	Sat	Sun
Bread (loaf)	1	1	0	0	0	2	0
Milk (litre)	3	1	3	2	0	5	1
Eggs	6	6	0	0	6	6	0
Butter (100g)	2	0	1	0	0	3	1

If the family always makes the list in this way, that is beginning with Monday as far as the columns are concerned, and always recording the produce items in the same rows, then the headings become unnecessary and the information can be stored as a **matrix** (plural - matrices):

$$\begin{bmatrix} 1 & 1 & 0 & 0 & 0 & 2 & 0 \\ 3 & 1 & 3 & 2 & 0 & 5 & 1 \\ 6 & 6 & 0 & 0 & 6 & 6 & 0 \\ 2 & 0 & 1 & 0 & 0 & 3 & 1 \end{bmatrix}$$

This matrix has four **rows** and seven **columns** and is said to have an **order** of 4 by 7.

To enter this matrix on the TI-82/3, press the **MATRIX** key, select **EDIT** and matrix **[A]**.

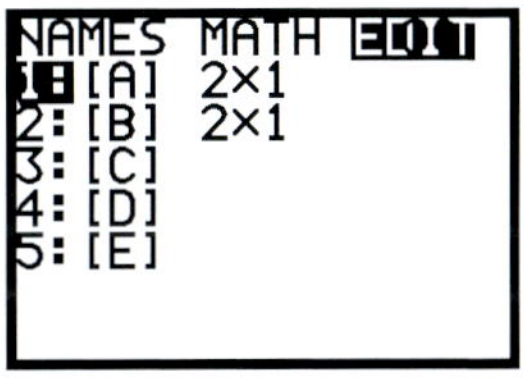

Next, enter the order of the matrix 4×7, and then enter the values given in the table. You can move from position to position by using the arrow keys or **ENTER** key. The screen is not large enough to display all the entries of this matrix. The other entries will 'scroll' into view when you use the right arrow.

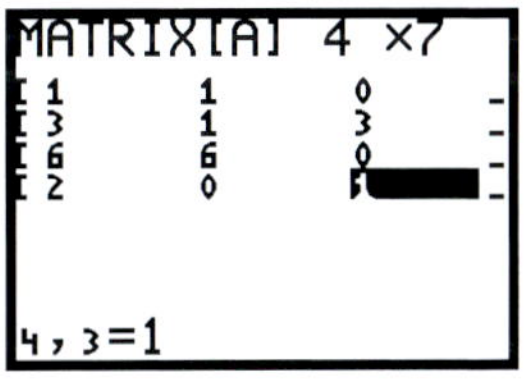

Once the information has been stored in the matrix memory it can only be accessed using the **MATRIX**, **NAMES** submenu.

It is also useful to remember that spreadsheets are ideally suited to handle information in matrix form. Spreadsheets can generally handle larger matrices and do more with them, but you need a computer and these are not allowed in the exams (yet!).

2. Equal Matrices

Two matrices are said to be equal iff
1. they have the same dimensions.
2. corresponding elements are equal.

e.g., $\begin{bmatrix} 1 & 4 \\ 2 & 5 \\ 6 & x \end{bmatrix} = \begin{bmatrix} 1 & 4 \\ y & 5 \\ 6 & 7 \end{bmatrix}$ iff $x = 7$ and $y = 2$;

$\begin{bmatrix} 1 & 5 \\ 2 & 9 \end{bmatrix} \neq \begin{bmatrix} 1 & 5 \\ 2 & 9 \\ 0 & 0 \end{bmatrix}$ as they have different orders. $\begin{bmatrix} 1 & 5 \\ 2 & 9 \end{bmatrix}$ has order 2×2, $\begin{bmatrix} 1 & 5 \\ 2 & 9 \\ 0 & 0 \end{bmatrix}$ has order 3×2

3. Addition of Matrices

Two matrices A and B can only be added if they have of the same order. Then, they can be added by adding corresponding elements.

e.g., $\begin{bmatrix} 1 & 2 & 3 \\ 4 & 5 & 6 \end{bmatrix} + \begin{bmatrix} 3 & -4 & 3 \\ 2 & 8 & -1 \end{bmatrix} = \begin{bmatrix} 1+3 & 2-4 & 3+3 \\ 4+2 & 5+8 & 6-1 \end{bmatrix} = \begin{bmatrix} 4 & -2 & 6 \\ 6 & 13 & 5 \end{bmatrix}$;

$\begin{bmatrix} 2 & 5 \\ 7 & 1 \end{bmatrix} + \begin{bmatrix} -4 & 3 & 8 \\ 2 & 9 & 4 \end{bmatrix}$ cannot be added as they are of different order.

4. Scalar multiplication

If A is any matrix, then kA, where k is a real number, is a new matrix for which each element of A has been multiplied by k.

e.g., $3\begin{bmatrix} 2 & 4 \\ -1 & 5 \end{bmatrix} = \begin{bmatrix} 3\times 2 & 3\times 4 \\ 3\times -1 & 3\times 5 \end{bmatrix} = \begin{bmatrix} 6 & 12 \\ -3 & 15 \end{bmatrix}$;

$-\frac{1}{2}\begin{bmatrix} 4 & 8 & 10 \\ 6 & -4 & -2 \end{bmatrix} = \begin{bmatrix} -2 & -4 & -5 \\ -3 & 2 & 1 \end{bmatrix}$

5. Subtraction of Matrices

The difference, $A - B$ of two matrices A and B (of the same order) may be defined as $A + (-B)$. So, the difference, $A - B$, is a new matrix whose elements are the differences between corresponding elements of A and B.

e.g., $\begin{bmatrix} 4 & 6 \\ 2 & -8 \\ 1 & 0 \end{bmatrix} - \begin{bmatrix} 2 & 1 \\ 5 & 3 \\ -2 & 4 \end{bmatrix} = \begin{bmatrix} 4-2 & 6-1 \\ 2-5 & -8-3 \\ 1-(-2) & 0-4 \end{bmatrix} = \begin{bmatrix} 2 & 5 \\ -3 & -11 \\ 3 & -4 \end{bmatrix}$

6. Square Matrix

A square matrix is a matrix with the same number of rows as columns, so that an $n \times n$ matrix is a square matrix of order n.

e.g., $\begin{bmatrix} 2 & -3 \\ 4 & 6 \end{bmatrix}$ is a square matrix of order 2; $\begin{bmatrix} 1 & 4 & 0 \\ -3 & 2 & 1 \\ 0 & -2 & -1 \end{bmatrix}$ is a square matrix of order 3.

7. Zero Matrix

A matrix with every element equal to zero is called a zero matrix. A zero matrix can have any dimension and is usually written as O.

e.g., each of the following are zero matrices; $\begin{bmatrix} 0 & 0 \end{bmatrix}$; $\begin{bmatrix} 0 & 0 \\ 0 & 0 \end{bmatrix}$; $\begin{bmatrix} 0 & 0 \\ 0 & 0 \\ 0 & 0 \end{bmatrix}$.

8. Additive Identity Matrix

For any matrix A, the zero matrix O (having the same order as A) is the additive identity of A.

i.e., $A + O = A = O + A$

9. Additive Inverse

The additive inverse of a matrix A is the matrix $-A$.
This can be seen from the result that $A + (-A) = O = (-A) + A$.

EXAMPLE 11.1

Three matrices are defined as follows:

$$A = \begin{bmatrix} 2 & 3 & -4 \\ 1 & 0 & -2 \end{bmatrix} \quad B = \begin{bmatrix} 2 & 1 \\ 5 & 3 \\ 4 & 5 \end{bmatrix} \quad C = \begin{bmatrix} -1 & 3 & 0 \\ -3 & -1 & 6 \end{bmatrix}$$

Evaluate: (a) $2B$ (b) $-3C$ (c) $A + C$
(d) $A + B$ (e) $2A + 3C$ (f) $3A - 4C$

Solution

(a) $2B = 2 \times \begin{bmatrix} 2 & 1 \\ 5 & 3 \\ 4 & 5 \end{bmatrix} = \begin{bmatrix} 4 & 2 \\ 10 & 6 \\ 8 & 10 \end{bmatrix}$

(b) $-3C = -3 \times \begin{bmatrix} -1 & 3 & 0 \\ -3 & -1 & 6 \end{bmatrix} = \begin{bmatrix} 3 & -9 & 0 \\ 9 & 3 & -18 \end{bmatrix}$

(c) $\begin{bmatrix} 2 & 3 & -4 \\ 1 & 0 & -2 \end{bmatrix} + \begin{bmatrix} -1 & 3 & 0 \\ -3 & -1 & 6 \end{bmatrix} = \begin{bmatrix} 2-1 & 3+3 & -4+0 \\ 1-3 & 0-1 & -2+6 \end{bmatrix} = \begin{bmatrix} 1 & 6 & -4 \\ -2 & -1 & 4 \end{bmatrix}$

(d) $A + B$ cannot be calculated as the matrices are of different order.

(e) $$2A+3C = 2\times\begin{bmatrix}2&3&-4\\1&0&-2\end{bmatrix}+3\times\begin{bmatrix}-1&3&0\\-3&-1&6\end{bmatrix}$$

$$= \begin{bmatrix}4&6&-8\\2&0&-4\end{bmatrix}+\begin{bmatrix}-3&9&0\\-9&-3&18\end{bmatrix}$$

$$= \begin{bmatrix}1&15&-8\\-7&-3&14\end{bmatrix}$$

(f) $$3A-4C = 3\times\begin{bmatrix}2&3&-4\\1&0&-2\end{bmatrix}-4\times\begin{bmatrix}-1&3&0\\-3&-1&6\end{bmatrix}$$

$$= \begin{bmatrix}6&9&-12\\3&0&-6\end{bmatrix}-\begin{bmatrix}-4&12&0\\-12&-4&24\end{bmatrix}$$

$$= \begin{bmatrix}10&-3&-12\\15&4&-30\end{bmatrix}$$

As a second example we look at the matrix of the family shopping list that was introduced at the start of this chapter.

If a second family makes a shopping list using the same row and column headings, they will also store their information in a matrix of order 4 by 7. Suppose that this second family want to buy the items indicated in this second matrix.

$$\begin{bmatrix}1&1&2&0&0&2&0\\3&0&1&2&0&5&1\\6&2&0&1&2&6&0\\2&1&1&0&1&3&1\end{bmatrix}$$

If the two families wanted to place a single order, they would add the two matrices together. In this application, it would only make sense to add corresponding entries (top left to top left etc.

$$\begin{bmatrix}1&1&0&0&0&2&0\\3&1&3&2&0&5&1\\6&6&0&0&6&6&0\\2&0&1&0&0&3&1\end{bmatrix}+\begin{bmatrix}1&1&2&0&0&2&0\\3&0&1&2&0&5&1\\6&2&0&1&2&6&0\\2&1&1&0&1&3&1\end{bmatrix}=\begin{bmatrix}1+1&1+1&0+2&0+0&0+0&2+2&0+0\\3+3&1+0&3+1&2+2&0+0&5+5&1+1\\6+6&6+2&0+0&0+1&6+2&6+6&0+0\\2+2&0+1&1+1&0+0&0+1&3+3&1+1\end{bmatrix}$$

$$=\begin{bmatrix}2&2&2&0&0&4&0\\6&1&4&4&0&10&2\\12&8&0&1&8&12&0\\4&1&2&0&1&6&2\end{bmatrix}$$

EXAMPLE 11.2

Find x, y and z if $A = B$, where $A = \begin{bmatrix}2x&y+1\\5&9\end{bmatrix}$, $B = \begin{bmatrix}x^2+1&3\\3z-4&z^2\end{bmatrix}$

Solution

If $A = B$ then $\begin{bmatrix} 2x & y+1 \\ 5 & 9 \end{bmatrix} = \begin{bmatrix} x^2+1 & 3 \\ 3z-4 & z^2 \end{bmatrix} \Leftrightarrow 2x = x^2+1, y+1 = 3, 5 = 3z-4, z^2 = 9$

Now, $2x = x^2+1 \Leftrightarrow x^2-2x+1 = 0 \Leftrightarrow (x-1)^2 = 0 \therefore x = 1$ – (1)

$y+1 = 3 \Leftrightarrow y = 2$ – (2)

$5 = 3z-4 \Leftrightarrow 9 = 3z \Leftrightarrow z = 3$ – (3)

$z^2 = 9 \Leftrightarrow z = \pm 3$ – (4)

From (3) and (4) we conclude that $z = 3$. Therefore, we have that $x = 1$, $y = 2$ and $z = 3$.

EXERCISES 11.1.1

1. Two matrices are defined as: $P = \begin{bmatrix} 1 & 4 \\ -1 & 8 \end{bmatrix}$, $Q = \begin{bmatrix} 2 & -3 \\ -1 & 5 \end{bmatrix}$, evaluate:

(i) $2P$ (ii) $-Q$ (iii) $3Q$
(iv) $2P+Q$ (v) $3P+2Q$ (vi) $Q-4P$

2. If $A = \begin{bmatrix} 1 & 1 & 2 \\ 0 & -6 & 9 \end{bmatrix}$, $B = \begin{bmatrix} 1 & 0 \\ 2 & 1 \\ 0 & -1 \end{bmatrix}$, $C = \begin{bmatrix} -3 & -5 \\ 0 & 1 \\ 3 & 2 \end{bmatrix}$, evaluate, where possible:

(i) $-A$ (ii) $-2A$ (iii) $B+C$
(iv) $2B$ (v) $2B-C$ (vi) $3(B+C)$

3. If $X = \begin{bmatrix} 1 & 1 & 2 \\ 0 & 1 & 4 \\ 0 & 2 & 1 \end{bmatrix}$, $Y = \begin{bmatrix} -1 & -2 & 3 \\ 2 & 0 & 1 \\ 2 & 0 & -4 \end{bmatrix}$, evaluate:

(i) $3X$ (ii) $2Y$ (iii) $X+Y$
(iv) $X-Y$ (v) $3X+2Y$ (vi) $2Y-X$

4. The diagram shows a spreadsheet used to store the sales of a small shop.

	A	B	C	D	E	F	G	H	I	J	K	L	M
1	**Week 1**							**Week 2**					
2		**Mon**	**Tues**	**Wed**	**Thurs**	**Fri**			**Mon**	**Tues**	**Wed**	**Thurs**	**Fri**
3	Nuts	4	27	20	18	32		Nuts	45	4	44	4	53
4	Bolts	34	33	39	50	5		Bolts	11	10	10	48	8
5	Pegs	37	23	51	23	29		Pegs	23	10	39	51	9
6	Pins	0	48	45	10	23		Pins	36	22	44	5	38
7	Clips	49	21	55	17	25		Clips	53	17	56	28	11
8	Taps	53	10	49	9	49		Taps	8	16	33	47	26
9	Plugs	12	12	45	16	50		Plugs	46	33	0	25	9

(i) What was the number of sales of pins on Friday of Week 1?
(ii) Which item sold least on Thursday of Week 2?
(iii) Which item sold most on Monday of Week 1?
(iv) On what day were there no sales of Plugs?

A place on the spreadsheet at which the total sales in the two weeks will be calculated by formula is needed. What formula should be used to calculate the Monday sales of nuts? If this is copied to the cell immediately below, what will the formula read?

5. A stock control system stores the stocks of seven different products as the rows of a matrix. The company works a five day week and records the closing stocks as the columns of a matrix.
What is the order of the matrix?

6. If $A = \begin{bmatrix} a & 1 \\ -1 & -a \end{bmatrix}$, $B = \begin{bmatrix} 2a & 2 \\ 1 & a \end{bmatrix}$, express $A + B$ as a matrix.

7. The zero matrix has entries that are all zero. Find the value(s) of a such that $P + Q$ is the zero matrix. $P = \begin{bmatrix} a & 2 \\ 5 & -3 \end{bmatrix}$, $Q = \begin{bmatrix} -3 & a-5 \\ -5 & a \end{bmatrix}$.

8. A diagonal matrix has all zero elements except for those on the top left to bottom right diagonal. Find the values of a such that the matrix $A + B$ is diagonal where:

$$A = \begin{bmatrix} 5 & 1 & a \\ 3 & 7 & -4 \\ -2 & a+2 & 2 \end{bmatrix} \qquad B = \begin{bmatrix} 3 & -1 & -2 \\ -3 & -2 & 2a \\ 2 & -4 & 3 \end{bmatrix}$$

9. If $A = \begin{bmatrix} \sin^2\theta & 0 \\ 0 & \cos^2\theta \end{bmatrix}$ and $B = \begin{bmatrix} \cos^2\theta & 0 \\ 0 & \sin^2\theta \end{bmatrix}$ simplify

(a) $A + B$ (b) $A - B$ (c) $2A - \begin{bmatrix} 1 & 0 \\ 0 & 1 \end{bmatrix}$ (d) $A + B - \begin{bmatrix} 1 & 0 \\ 0 & 1 \end{bmatrix}$

10. If $A = \begin{bmatrix} \sin^2\theta & \sin\theta \\ -\cos\theta & \cos^2\theta \end{bmatrix}$ and $B = \begin{bmatrix} \cos^2\theta & -\sin\theta \\ \cos\theta & \sin^2\theta \end{bmatrix}$ simplify

(a) $A + B$ (b) $A - B$

11. Find a and b if aA + B = A where $A = \begin{bmatrix} 2 & 5 \\ -3 & 4 \end{bmatrix}$ and $B = \begin{bmatrix} -b & -10 \\ 6 & -2b \end{bmatrix}$.

12. Find x if $x^2\begin{bmatrix} 1 & -1 \\ 2 & 1 \end{bmatrix} + x\begin{bmatrix} 2 & 4 \\ 1 & -6 \end{bmatrix} + \begin{bmatrix} 3 & 1 \\ 4 & 5 \end{bmatrix} = \begin{bmatrix} 6 & 4 \\ 7 & 0 \end{bmatrix}$.

11.1.2 MATRIX MULTIPLICATION

Matrix multiplication has a special definition that was developed because it appears to be the most practical. This definition can be useful in business applications, solving simultaneous equations and in transformation geometry. This last application has helped in the development of one of the newest braches of mathematics, fractal geometry.

The definition involves taking the rows of the left hand matrix and pairing these with the columns of the second matrix. If the problem is to find the product:

$$\begin{bmatrix} 2 & 5 & 0 \\ 1 & 0 & -2 \\ 1 & -3 & 1 \end{bmatrix} \times \begin{bmatrix} 1 & 3 & -1 \\ 0 & 2 & 0 \\ 1 & -3 & 1 \end{bmatrix}$$

The first step is to take the first row of the left hand matrix [2 5 0], write it as a column and pair it up with the first column of the 2nd matrix. The pairs are then multiplied and the products totalled to give a single number, in this case, '2'.

$$\begin{aligned} 2 \times 1 &= 2 \\ 5 \times 0 &= 0 \\ 0 \times 1 &= 0 \\ \hline &\boxed{2} \end{aligned}$$

This number is the result of combining the first row and the first column. It becomes the element in the first row and first column of the answer. The calculation must be completed for all the combinations of rows and columns. The nine calculations are:

$$\begin{bmatrix} 2 & 5 & 0 \\ 1 & 0 & -2 \\ 1 & -3 & 1 \end{bmatrix} \times \begin{bmatrix} 1 & 3 & -1 \\ 0 & 2 & 0 \\ 1 & -3 & 1 \end{bmatrix} : \quad 2 \times 1 = 2,\ 5 \times 0 = 0,\ 0 \times 1 = 0,\ \boxed{2}$$

$$\begin{bmatrix} 2 & 5 & 0 \\ 1 & 0 & -2 \\ 1 & -3 & 1 \end{bmatrix} \times \begin{bmatrix} 1 & 3 & -1 \\ 0 & 2 & 0 \\ 1 & -3 & 1 \end{bmatrix} : \quad 2 \times 3 = 6,\ 5 \times 2 = 10,\ 0 \times -3 = 0,\ \boxed{16}$$

$$\begin{bmatrix} 2 & 5 & 0 \\ 1 & 0 & -2 \\ 1 & -3 & 1 \end{bmatrix} \times \begin{bmatrix} 1 & 3 & -1 \\ 0 & 2 & 0 \\ 1 & -3 & 1 \end{bmatrix} : \quad 2 \times -1 = -2,\ 5 \times 0 = 0,\ 0 \times 1 = 0,\ \boxed{-2}$$

$$\begin{bmatrix} 2 & 5 & 0 \\ 1 & 0 & -2 \\ 1 & -3 & 1 \end{bmatrix} \times \begin{bmatrix} 1 & 3 & -1 \\ 0 & 2 & 0 \\ 1 & -3 & 1 \end{bmatrix} : \quad 1 \times 1 = 1,\ 0 \times 0 = 0,\ -2 \times 1 = -2,\ \boxed{-1}$$

$$\begin{bmatrix} 2 & 5 & 0 \\ 1 & 0 & -2 \\ 1 & -3 & 1 \end{bmatrix} \times \begin{bmatrix} 1 & 3 & -1 \\ 0 & 2 & 0 \\ 1 & -3 & 1 \end{bmatrix} : \quad 1 \times 3 = 3,\ 0 \times 2 = 0,\ -2 \times -3 = 6,\ \boxed{9}$$

$$\begin{bmatrix} 2 & 5 & 0 \\ 1 & 0 & -2 \\ 1 & -3 & 1 \end{bmatrix} \times \begin{bmatrix} 1 & 3 & -1 \\ 0 & 2 & 0 \\ 1 & -3 & 1 \end{bmatrix} : \quad 1 \times -1 = -1,\ 0 \times 0 = 0,\ -2 \times 1 = -2,\ \boxed{-3}$$

$$\begin{bmatrix} 2 & 5 & 0 \\ 1 & 0 & -2 \\ 1 & -3 & 1 \end{bmatrix} \times \begin{bmatrix} 1 & 3 & -1 \\ 0 & 2 & 0 \\ 1 & -3 & 1 \end{bmatrix} : \quad 1 \times 1 = 1,\ -3 \times 0 = 0,\ 1 \times 1 = 1,\ \boxed{2}$$

$$\begin{bmatrix} 2 & 5 & 0 \\ 1 & 0 & -2 \\ 1 & -3 & 1 \end{bmatrix} \times \begin{bmatrix} 1 & 3 & -1 \\ 0 & 2 & 0 \\ 1 & -3 & 1 \end{bmatrix} : \quad 1 \times 3 = 3,\ -3 \times 2 = -6,\ 1 \times -3 = -3,\ \boxed{-6}$$

$$\begin{bmatrix} 2 & 5 & 0 \\ 1 & 0 & -2 \\ 1 & -3 & 1 \end{bmatrix} \times \begin{bmatrix} 1 & 3 & -1 \\ 0 & 2 & 0 \\ 1 & -3 & 1 \end{bmatrix} : \quad 1 \times -1 = -1,\ -3 \times 0 = 0,\ 1 \times 1 = 1,\ \boxed{0}$$

In matrix form, this answer is: $\begin{bmatrix} 2 & 16 & -2 \\ -1 & 9 & -3 \\ 2 & -6 & 0 \end{bmatrix}$

It is not only square matrices that can be multiplied. All that is necessary is for the columns of the first matrix to match the number of rows in the second matrix. That is,

> If A is a matrix of order $l \times m$ and B is a matrix of order $m \times n$, then their product is the matrix C = AB of order $l \times n$. So that
>
> | | | | | | |
> |---|---|---|---|---|---|
> | matrix: | A | $\times$ | B | = | C |
> | order: | $(l \times m)$ | $\times$ | $(m \times n)$ | = | $(l \times n)$ |
>
> (the inner dimensions m are the same; the outer dimensions l and n give the order of C)
>
> same

e.g., the following products are possible:

$$\begin{bmatrix} 2 & 4 & 7 \\ 1 & 5 & -2 \end{bmatrix}\begin{bmatrix} 1 & -2 \\ -4 & 5 \\ 1 & 6 \end{bmatrix} = \text{matrix of order } (2\times2)$$

$$(2 \times 3) \times (3 \times 2)$$

$$\begin{bmatrix} 4 & -5 & 10 & 6 \end{bmatrix}\begin{bmatrix} 3 & 1 \\ -5 & 2 \\ -11 & 3 \\ 9 & 4 \end{bmatrix} = \text{matrix of order } (1\times2)$$

$$(1 \times 4) \times (4 \times 2)$$

the following product is **not** possible:

$$\begin{bmatrix} -1 & 5 & 8 \\ 9 & 3 & 12 \end{bmatrix}\begin{bmatrix} 3 & 6 \\ 2 & 10 \end{bmatrix}$$

$$(2 \times 3) \times (2 \times 2)$$

Note: reversing the order in which the matrices in our example are multiplied does not yield the same result. That is,

$$\begin{bmatrix} 1 & 3 & -1 \\ 0 & 2 & 0 \\ 1 & -3 & 1 \end{bmatrix}\begin{bmatrix} 2 & 5 & 0 \\ 1 & 0 & -2 \\ 1 & -3 & 1 \end{bmatrix} = \begin{bmatrix} 4 & 8 & -7 \\ 2 & 0 & -4 \\ 0 & 2 & 7 \end{bmatrix} \neq \begin{bmatrix} 2 & 16 & -2 \\ -1 & 9 & -3 \\ 2 & -6 & 0 \end{bmatrix}$$

This shows us that matrix multiplication is not commutative. This means that the order in which the matrices are multiplied does matter.

EXAMPLE 11.3

For the matrices: $A = \begin{bmatrix} 1 & 0 \\ 2 & 6 \end{bmatrix}$, $B = \begin{bmatrix} 2 & -1 & -1 \\ 2 & 0 & -3 \end{bmatrix}$, $C = \begin{bmatrix} 2 & -1 \\ -3 & 2 \\ 1 & 4 \end{bmatrix}$, calculate, where possible:

(a) AB (b) BC (c) CA
(d) BA (e) A^2 (f) ABC

Solution

(a) $AB = \begin{bmatrix} 1 & 0 \\ 2 & 6 \end{bmatrix}\begin{bmatrix} 2 & -1 & -1 \\ 2 & 0 & -3 \end{bmatrix} = \begin{bmatrix} 2 & -1 & -1 \\ 16 & -2 & -20 \end{bmatrix}$

Note, in this case, multiplying a 2 by 2 matrix into a 2 by 3 matrix produces a 2 by 3 matrix.

(b $BC = \begin{bmatrix} 2 & -1 & -1 \\ 2 & 0 & -3 \end{bmatrix}\begin{bmatrix} 2 & -1 \\ -3 & 2 \\ 1 & 4 \end{bmatrix} = \begin{bmatrix} 6 & -8 \\ 1 & -14 \end{bmatrix}$

(c) $CA = \begin{bmatrix} 2 & -1 \\ -3 & 2 \\ 1 & 4 \end{bmatrix}\begin{bmatrix} 1 & 0 \\ 2 & 6 \end{bmatrix} = \begin{bmatrix} 0 & -6 \\ 1 & 12 \\ 9 & 24 \end{bmatrix}$

(d) $BA = \begin{bmatrix} 2 & -1 & -1 \\ 2 & 0 & -3 \end{bmatrix}\begin{bmatrix} 1 & 0 \\ 2 & 6 \end{bmatrix}$ cannot be calculated as the orders are not compatible.

(e) $A^2 = \begin{bmatrix} 1 & 0 \\ 2 & 6 \end{bmatrix}\begin{bmatrix} 1 & 0 \\ 2 & 6 \end{bmatrix} = \begin{bmatrix} 1 & 0 \\ 14 & 36 \end{bmatrix}$

(f) $ABC = \begin{bmatrix} 1 & 0 \\ 2 & 6 \end{bmatrix}\begin{bmatrix} 2 & -1 & -1 \\ 2 & 0 & -3 \end{bmatrix}\begin{bmatrix} 2 & -1 \\ -3 & 2 \\ 1 & 4 \end{bmatrix} = \begin{bmatrix} 1 & 0 \\ 2 & 6 \end{bmatrix}\begin{bmatrix} 6 & -8 \\ 1 & -14 \end{bmatrix} = \begin{bmatrix} 6 & -8 \\ 18 & -100 \end{bmatrix}$

There are two ways of calculating this triple product. The second is:

$$ABC = \begin{bmatrix} 1 & 0 \\ 2 & 6 \end{bmatrix}\begin{bmatrix} 2 & -1 & -1 \\ 2 & 0 & -3 \end{bmatrix}\begin{bmatrix} 2 & -1 \\ -3 & 2 \\ 1 & 4 \end{bmatrix} = \begin{bmatrix} 2 & -1 & -1 \\ 16 & -2 & -20 \end{bmatrix}\begin{bmatrix} 2 & -1 \\ -3 & 2 \\ 1 & 4 \end{bmatrix} = \begin{bmatrix} 6 & -8 \\ 18 & -100 \end{bmatrix}$$

This second method involves multiplying the first two matrices at the start, getting an answer and then multiplying this into the third. The order of the matrices has not been altered, just the way in which they have been grouped. This is called the **associative property** of matrix multiplication

Laws of Matrix Multiplication

If the matrices A B, C are conformable under multiplication and belong to the set of matrices, M (say), then

1.	as $AB \in M$,	i.e., closure holds in matrix multiplication.
2.	$AB \neq BA$,	i.e., multiplication is not commutative
3.	$(AB)C = A(BC)$	i.e., multiplication is associative.
4.	$A(B + C) = AB + AC$	i.e., multiplication is distributive.
5.	$AI = A = IA$	i.e., there exists a multiplicative identity.
6.	$AA^{-1} = I = A^{-1}A$	i.e., there exists a multiplicative inverse.

Note: Under multiplication, the identity matrix, I, is a square matrix with all of its elements zero except for those on its leading diagonal, where the elements are all one. Also, if a matrix has an inverse under multiplication, it must be a square matrix.

Examples of the identity matrix under matrix multiplication are

$$I = \begin{bmatrix} 1 & 0 \\ 0 & 1 \end{bmatrix} \text{ (also written as } I_2 \text{)}; \qquad I = \begin{bmatrix} 1 & 0 & 0 \\ 0 & 1 & 0 \\ 0 & 0 & 1 \end{bmatrix} \text{ (also written as } I_3 \text{)}.$$

Similarly, if the inverse of a matrix A exists, then A must first of all be a square matrix.

Using a Graphics Calculator

Matrices can be stored as described earlier in this chapter. The names can be recovered using the **MATRIX NAME** command. To calculate AB in the example discussed above, enter the correct matrices and then draw out the two names to produce the result.

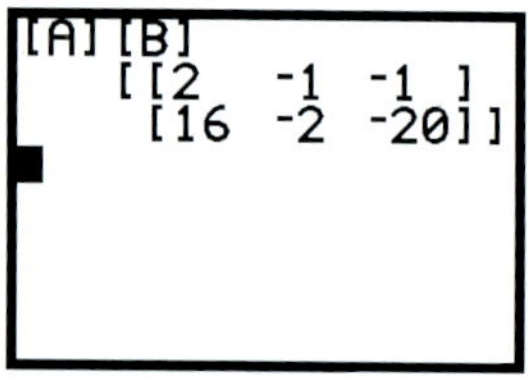

EXAMPLE 11.4

For the matrices: $A = \begin{bmatrix} 1 & 0 \\ 2 & 6 \end{bmatrix}$, $B = \begin{bmatrix} 2 & -1 & -1 \\ 2 & 0 & -3 \end{bmatrix}$, calculate, where possible:

(a) AB (b) A^3 (c) B^2

Solution

We start by entering the data into the TI–83:

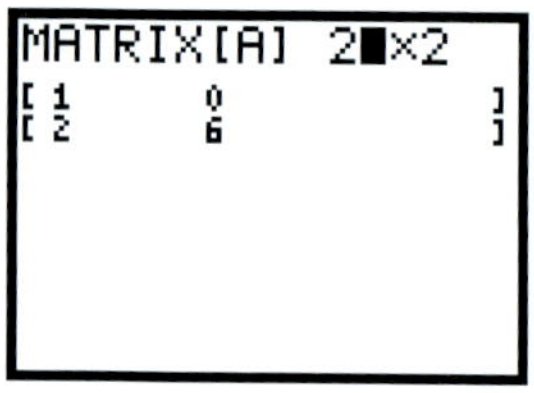

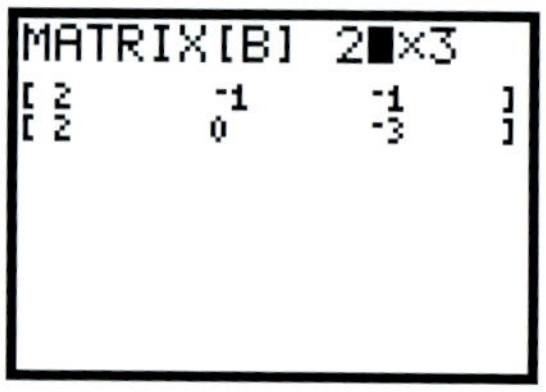

(a)

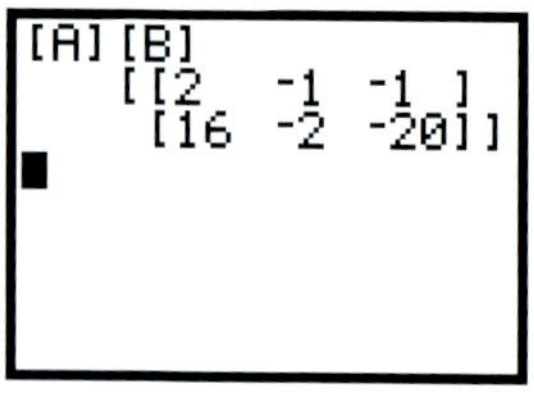

(b)

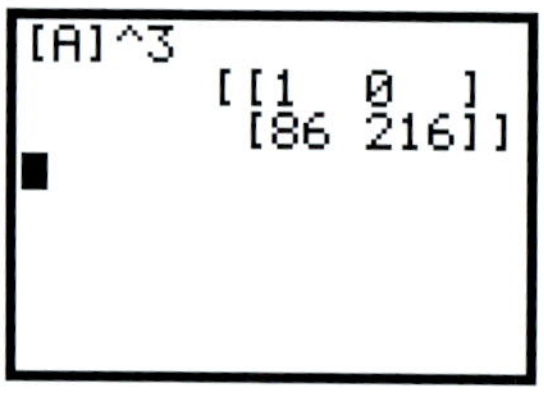

(c)

As B is not a square matrix, then B^2 cannot be evaluated.

EXERCISES 11.1.2

1. Evaluate the following matrix products:

(i) $\begin{bmatrix} 2 & 3 \\ -1 & 0 \end{bmatrix}\begin{bmatrix} -2 & 1 \\ 3 & -2 \end{bmatrix}$ (ii) $\begin{bmatrix} 2 & -2 \\ 1 & 5 \end{bmatrix}\begin{bmatrix} 0 & 1 \\ -3 & 5 \end{bmatrix}$ (iii) $\begin{bmatrix} -6 & 5 \\ 0 & 3 \end{bmatrix}\begin{bmatrix} 2 & -2 \\ 1 & 1 \end{bmatrix}$

(iv) $\begin{bmatrix} -2 & 3 \\ -5 & 2 \end{bmatrix}\begin{bmatrix} 0 & 8 \\ -5 & 2 \end{bmatrix}$ (v) $\begin{bmatrix} \frac{1}{3} & 7 \\ \frac{1}{2} & \frac{3}{4} \end{bmatrix}\begin{bmatrix} -1 & 5 \\ \frac{1}{2} & 4 \end{bmatrix}$ (vi) $\begin{bmatrix} 1.5 & 2 \\ 3 & -1.5 \end{bmatrix}\begin{bmatrix} -1.6 & 1 \\ 2.5 & 2 \end{bmatrix}$

(vii) $\begin{bmatrix} 4 & -2 \\ 0 & -3 \end{bmatrix}\begin{bmatrix} 2 & -1 & 1 \\ -2 & 0 & 1 \end{bmatrix}$ (viii) $\begin{bmatrix} 5 & -1 \\ -2 & 0 \end{bmatrix}\begin{bmatrix} -2 & 1 & 1 \\ -1 & 2 & 3 \end{bmatrix}$ (ix) $\begin{bmatrix} 1 & 2 \\ 4 & 7 \end{bmatrix}\begin{bmatrix} \frac{1}{2} & -\frac{1}{2} & 3 \\ \frac{1}{3} & 1 & -\frac{1}{3} \end{bmatrix}$

(x) $\begin{bmatrix} 2 & 1 \\ -3 & 0 \end{bmatrix}\begin{bmatrix} 1.3 & 1.4 & 1.9 \\ -1 & -3.5 & 0 \end{bmatrix}$ (xi) $\begin{bmatrix} 1 & -2 & 0 \\ -2 & -3 & 1 \\ 3 & 0 & 1 \end{bmatrix}\begin{bmatrix} 1 & -2 \\ -5 & 0 \\ -2 & -1 \end{bmatrix}$ (xii) $\begin{bmatrix} 1 & 6 & -1 \\ 2 & -3 & -9 \\ 1 & 2 & 0 \end{bmatrix}\begin{bmatrix} 6 & 2 \\ 0 & 1 \\ -1 & 3 \end{bmatrix}$

(xiii) $\begin{bmatrix} 1 & 0 & 2 \\ 2 & 1 & 0 \\ 3 & 0 & 1 \end{bmatrix}\begin{bmatrix} 6 & 0 & 2 \\ 0 & 1 & 0 \\ 3 & 3 & 0 \end{bmatrix}$ (xiv) $\begin{bmatrix} -1 & 3 & -2 \\ 0 & -9 & 2 \\ -3 & -2 & 1 \end{bmatrix}\begin{bmatrix} 1 & -2 & -1 \\ 0 & 0 & 2 \\ 2 & -2 & 3 \end{bmatrix}$

(xv) $\begin{bmatrix} 1 & x & x \\ 2 & x & 2x \\ -x & 1 & 2 \end{bmatrix}\begin{bmatrix} x & 1 \\ x & -x \\ 2x & 0 \end{bmatrix}$ (xvi) $\begin{bmatrix} a & 2x & -a \\ a & -2a & 0 \\ 0 & a & -x \end{bmatrix}\begin{bmatrix} 1 & 2 & 3 \\ x & 2 & -x \\ a & 0 & 1 \end{bmatrix}$

2. A matrix is defined by $M = \begin{bmatrix} -1 & 0 \\ 0 & -1 \end{bmatrix}$

(i) Find M^2 (ii) Find M^3, M^4, M^5 (iii) Find a rule for M^n

3. The following information gives the numbers of journeys made by the delivery department of a small company:

From:→ To↓	A	B	C	D
A	0	3	4	2
B	1	0	0	6
C	1	2	0	2
D	4	3	0	0

The company has four different depots and the above table indicates that there was one journey made from depot A to depot B and three journeys made from B to A etc.

This second table gives the loading costs per load for goods leaving each depot.

Depot	Cost
A	\$120
B	\$105
C	\$110
D	\$100

It costs \$120 to load for one trip at depot A, \$105 to load for one trip at depot B, etc.

(i) Write these two sets of information as matrices.

(ii) Find the product of the two matrices.

(iii) Explain the meaning of the entries in the product matrix.

4. Two matrices are defined as follows: $A = \begin{bmatrix} \frac{1}{2} & -\frac{\sqrt{3}}{2} \\ \frac{\sqrt{3}}{2} & \frac{1}{2} \end{bmatrix}$, $B = \begin{bmatrix} \frac{\sqrt{3}}{2} & -\frac{1}{2} \\ \frac{1}{2} & \frac{\sqrt{3}}{2} \end{bmatrix}$

Find the product AB and hence find a value of n such that $(AB)^n = \begin{bmatrix} 1 & 0 \\ 0 & 1 \end{bmatrix}$.

5. Find, if possible, the order of matrix A so that the following equations can be solved

(a) $\begin{bmatrix} a & b \\ c & d \end{bmatrix} A = \begin{bmatrix} x \\ y \end{bmatrix}$ (b) $A \begin{bmatrix} a & b \\ c & d \end{bmatrix} = \begin{bmatrix} x \\ y \end{bmatrix}$ (c) $\begin{bmatrix} a & b & c \end{bmatrix} A = \begin{bmatrix} x & y \end{bmatrix}$

(d) $\begin{bmatrix} a & b & c \end{bmatrix} A = \begin{bmatrix} m & n \\ x & y \\ u & v \end{bmatrix}$ (e) $A \begin{bmatrix} a & b & c \end{bmatrix} = \begin{bmatrix} m & n \\ x & y \\ u & v \end{bmatrix}$ (f) $\begin{bmatrix} a \\ b \\ c \end{bmatrix} A = \begin{bmatrix} x & y & z \\ k & l & m \\ n & o & p \end{bmatrix}$

6. (a) If $A = \begin{bmatrix} a & b \\ 1 & 1 \end{bmatrix}$ and $B = \begin{bmatrix} a & 1 \\ -b & 1 \end{bmatrix}$ does $(A-B)(A+B) = A^2 - B^2$?

(b) If $A = \begin{bmatrix} 1 & 0 \\ 0 & 1 \end{bmatrix}$ and $B = \begin{bmatrix} a & 1 \\ 1 & b \end{bmatrix}$ does $(A-B)(A+B) = A^2 - B^2$?

(c) What conditions must be placed on the matrices A and B so that $(A-B)(A+B) = A^2 - B^2$?

7. Find all possible matrices X which satisfy the equation $\begin{bmatrix} 5 & 4 \\ 10 & 8 \end{bmatrix} X = \begin{bmatrix} 12 \\ 24 \end{bmatrix}$

8. Find the real numbers a and b if $aX^2 + bX = 2I$ and $X = \begin{bmatrix} 2 & 1 \\ 1 & 3 \end{bmatrix}$.

9. Three divisions, A, B and C, of an animation studio are required to produce a certain amount of conceptual paintings, P, footage of film, F, and storyboards, S. The amounts produced by each division can be described using the following matrix:

	P	F	S
A	13	120	30
B	5	95	15
C	11	115	30

The cost of producing a conceptual painting is \$50. The profit made per footage of film is \$15 and the cost per storyboard is \$20.

Using a matrix method, which division is the most profitable?

10. Let S be the matrix $\begin{bmatrix} -\sin\theta & \cos\theta \\ \cos\theta & \sin\theta \end{bmatrix}$ and T be the matrix $\begin{bmatrix} -\sin 2\theta & \cos 2\theta \\ \cos 2\theta & \sin 2\theta \end{bmatrix}$.

(a) Find i. S^2 ii. S^3.

(b) Deduce i. S^{2n} ii. S^{2n+1}, where n is a natural number ≥ 1.

(c) For what value(s) of θ will $\begin{bmatrix} -\sin\theta & \cos\theta \\ \cos\theta & \sin\theta \end{bmatrix} - \begin{bmatrix} -\sin 2\theta & \cos 2\theta \\ \cos 2\theta & \sin 2\theta \end{bmatrix} = O$?

11. If $A = \begin{bmatrix} a & b \\ c & d \end{bmatrix}$, $b \neq 0$ and $B = \begin{bmatrix} w & x \\ y & z \end{bmatrix}$, find the conditions that the elements of B must satisfy so that the matrices A and B are commutative under matrix multiplication.

12. If $AX = kX, k \in \mathbb{R}$ where $A = \begin{bmatrix} 4 & -2 \\ -2 & 4 \end{bmatrix}$, $X = \begin{bmatrix} x \\ y \end{bmatrix}$ and $x \neq 0 \neq y$, find k.

13. Find the matrix A and express a, b and c in terms of p, if the system of equations

$$\begin{aligned} x + z &= 2p \\ x - y + p^2 z &= p - 1 \\ px + y - (2p+1)z &= p \end{aligned}$$

is expressed in the matrix form $A\begin{bmatrix} x \\ y \\ z \end{bmatrix} = \begin{bmatrix} a \\ b \\ c \end{bmatrix}$

14. If $A = \begin{bmatrix} -2i & 1 \\ 1 & 2i \end{bmatrix}$, where $i^2 = -1$. Show that $A^2 + 4I = \begin{bmatrix} 1 & 0 \\ 0 & 0 \end{bmatrix}$.

15. Given that $A = \begin{bmatrix} -3 \\ 2 \end{bmatrix}$, $X = \begin{bmatrix} x & y \end{bmatrix}$, $B = \begin{bmatrix} a & 6 \\ 6 & b \end{bmatrix}$ and $AX = B$, find x, y, a, b.

16. For the matrices X and Y, show that if $XY = X$ and $YX = Y$ then $X^2 = X$.

17. The Fibonacci numbers 0, 1, 1, 2, 3, 5, . . . are generated by the the recurrence relationship $F_{n+2} = F_{n+1} + F_n$, where $n \geq 0$ and F_n is the nth number such that $F_0 = 0, F_1 = 1$.

Given the matrix $A = \begin{bmatrix} 1 & 1 \\ 1 & 0 \end{bmatrix}$

(a) find i. A^2 ii. A^3 iii. A^4

(b) deduce an expression for A^n where $n \in \mathbb{Z}$.

11.2 INVERSES AND DETERMINANTS

11.2.1 INVERSE AND DETERMINANT OF 2 BY 2 MATRICES

Matrices with two rows and two columns can be manipulated and many useful calculations can be completed with them. We begin by providing some definitions.

1. The identity matrix

Recall, the **identity matrix**, I, is defined as the matrix such that $AI = A = IA$ for all square matrices A. For 2 by 2 matrices, $I = \begin{bmatrix} 1 & 0 \\ 0 & 1 \end{bmatrix}$.

2. Determinant

The determinant of a 2 by 2 matrix is defined by $\det\begin{bmatrix} a & b \\ c & d \end{bmatrix} = ad - bc$ or $\begin{vmatrix} a & b \\ c & d \end{vmatrix} = ad - bc$.

The determinant is a scalar. If a matrix has a non-zero determinant, it is said to be non-singular, and if the **determinant is zero**, the matrix is said to be **singular**. We shall see that whether or not a matrix is singular is significant for some applications.

3. Inverse

The inverse of a square matrix A is written A^{-1}.

The product of a matrix and its inverse is the identity matrix, I, $AA^{-1} = A^{-1}A = I$. For 2 by 2 matrices, the inverse can be found by interchanging the top left and bottom right entries and changing the signs of the top right and bottom left entries. Finally, the inverse is found by dividing by the determinant. This can be done by dividing the matrix as a whole or by dividing each element separately. That is,

$$\text{If } A = \begin{bmatrix} a & b \\ c & d \end{bmatrix}\text{, then, } A^{-1} = \begin{bmatrix} \frac{d}{ad-bc} & \frac{-b}{ad-bc} \\ \frac{-c}{ad-bc} & \frac{a}{ad-bc} \end{bmatrix} = \frac{1}{ad-bc}\begin{bmatrix} d & -b \\ -c & a \end{bmatrix}$$

Given that we need to divide by the determinant in order to find the inverse, if the determinant is zero we cannot divide and thus we cannot find the inverse. This implies that **a singular matrix has no inverse**.

PROOF

Let $A = \begin{bmatrix} a & b \\ c & d \end{bmatrix}$ and its inverse, $A^{-1} = \begin{bmatrix} u & v \\ x & y \end{bmatrix}$, then, if A^{-1} exists, we have $AA^{-1} = I = A^{-1}A$.

Using the L.H.S, we have, $\begin{bmatrix} a & b \\ c & d \end{bmatrix}\begin{bmatrix} u & v \\ x & y \end{bmatrix} = \begin{bmatrix} 1 & 0 \\ 0 & 1 \end{bmatrix} \Leftrightarrow \begin{bmatrix} au+bx & av+by \\ cu+dx & cv+dy \end{bmatrix} = \begin{bmatrix} 1 & 0 \\ 0 & 1 \end{bmatrix}$

That is, we have the system of equations: $\begin{matrix} au+bx = 1 \\ cu+dx = 0 \end{matrix}$ and $\begin{matrix} av+by = 0 \\ cv+dy = 1 \end{matrix}$.

By solving these sets of equations simultaneously, we obtain;

$$u = \frac{d}{ad-bc}, v = -\frac{b}{ad-bc}, x = -\frac{c}{ad-bc}, y = \frac{a}{ad-bc}$$

or $$u = \frac{d}{|A|}, v = -\frac{b}{|A|}, x = -\frac{c}{|A|}, y = \frac{a}{|A|}$$

That is, If $A = \begin{bmatrix} a & b \\ c & d \end{bmatrix}$ then $A^{-1} = \begin{bmatrix} \frac{d}{ad-bc} & -\frac{b}{ad-bc} \\ -\frac{c}{ad-bc} & \frac{a}{ad-bc} \end{bmatrix} = \frac{1}{ad-bc}\begin{bmatrix} d & -b \\ -c & a \end{bmatrix} = \frac{1}{|A|}\begin{bmatrix} d & -b \\ -c & a \end{bmatrix}$

Of course, this only works if A^{-1} exists, i.e., if $|A| = ad-bc \neq 0$.

EXAMPLE 11.5

Find the determinant and, where possible, the inverses of these matrices:

(a) $\begin{bmatrix} 2 & 1 \\ 5 & 3 \end{bmatrix}$ (b) $\begin{bmatrix} -3 & 2 \\ 0 & -1 \end{bmatrix}$ (c) $\begin{bmatrix} 2 & 6 \\ 1 & 3 \end{bmatrix}$

Solution

(a) $\det\begin{bmatrix} 2 & 1 \\ 5 & 3 \end{bmatrix} = 2\times3-1\times5 = 1$ so the matrix is non-singular.

To find the inverse matrix, we interchange the 2 and 3 and reverse the signs of the other two elements. Since the determinant is 1, this gives us the inverse matrix directly:

$$\begin{bmatrix} 3 & -1 \\ -5 & 2 \end{bmatrix}.$$

To check that this is correct, we calculate the products:

$$\begin{bmatrix} 2 & 1 \\ 5 & 3 \end{bmatrix}\begin{bmatrix} 3 & -1 \\ -5 & 2 \end{bmatrix} = \begin{bmatrix} 1 & 0 \\ 0 & 1 \end{bmatrix}, \begin{bmatrix} 3 & -1 \\ -5 & 2 \end{bmatrix}\begin{bmatrix} 2 & 1 \\ 5 & 3 \end{bmatrix} = \begin{bmatrix} 1 & 0 \\ 0 & 1 \end{bmatrix}.$$

In both cases, the product is the identity matrix confirming that the inverse is correct.

(b) Let $B = \begin{bmatrix} -3 & 2 \\ 0 & -1 \end{bmatrix}$. Then, $\det(B) = \begin{vmatrix} -3 & 2 \\ 0 & -1 \end{vmatrix} = -3\times-1-2\times0 = 3$ so the matrix is non-singular.

The necessary interchanges give $\begin{bmatrix} -1 & -2 \\ 0 & -3 \end{bmatrix}$, and, after dividing each element by the determinant the inverse, $B^{-1} = \frac{1}{3}\begin{bmatrix} -1 & -2 \\ 0 & -3 \end{bmatrix} = \begin{bmatrix} -\frac{1}{3} & -\frac{2}{3} \\ 0 & -1 \end{bmatrix}$.

The check gives: $\begin{bmatrix} -3 & 2 \\ 0 & -1 \end{bmatrix} \begin{bmatrix} -\frac{1}{3} & -\frac{2}{3} \\ 0 & -1 \end{bmatrix} = \begin{bmatrix} 1 & 0 \\ 0 & 1 \end{bmatrix}$, $\begin{bmatrix} -\frac{1}{3} & -\frac{2}{3} \\ 0 & -1 \end{bmatrix} \begin{bmatrix} -3 & 2 \\ 0 & -1 \end{bmatrix} = \begin{bmatrix} 1 & 0 \\ 0 & 1 \end{bmatrix}$ (which is correct).

(c) Now, $\det \begin{bmatrix} 2 & 6 \\ 1 & 3 \end{bmatrix} = 2 \times 3 - 6 \times 1 = 0$ so the matrix is singular.

Because the calculation of an inverse involves division by the determinant and division by zero is a forbidden process, singular matrices do not have inverses.

The answers in Example 25.5 can be checked using the TI–83:
First delete any existing matrices from the memory of the calculator – i.e., Press **2nd** '**+**' '**2**' '**5**' and delete the matrices you no longer wish to use. Then define the above matrices as

$[A] = \begin{bmatrix} 2 & 1 \\ 5 & 3 \end{bmatrix}$, $[B] = \begin{bmatrix} -3 & 2 \\ 0 & -1 \end{bmatrix}$ and $[C] = \begin{bmatrix} 2 & 6 \\ 1 & 3 \end{bmatrix}$ and use the **MATH** menu after calling up the **MATRIX** option:

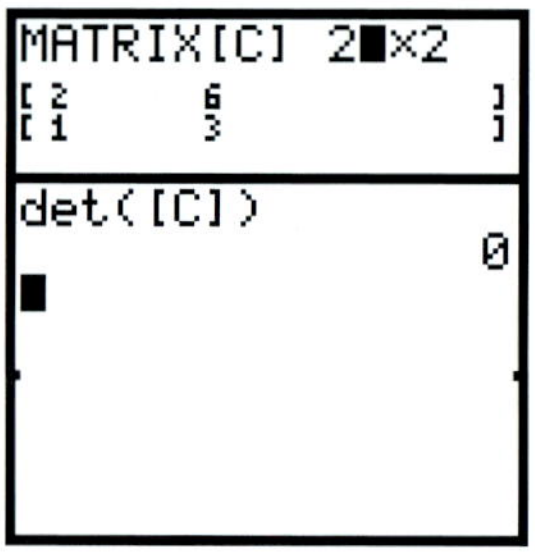

EXAMPLE 11.6

For what value(s) of k will the matrix $A = \begin{bmatrix} k & 2 \\ 3 & k+1 \end{bmatrix}$ be singular?

Solution

If the matrix A is singular, then $|A| = \det(A) = 0$.

Now, $|A| = \begin{vmatrix} k & 2 \\ 3 & k+1 \end{vmatrix} = k(k+1) - 2 \times 3$

$= k^2 + k - 6$

Then, we must have $k^2 + k - 6 = 0 \Leftrightarrow (k+3)(k-2) = 0$

$\Leftrightarrow k = -3$ or $k = 2$.

That is, matrix A will be singular iff $k = -3$ or 2.

EXAMPLE 11.7

Express the determinant, $\begin{vmatrix} k^2 & 2 \\ 3k & 5 \end{vmatrix}$ in terms of the matrix $A = \begin{bmatrix} 5k & 2 \\ 3 & 1 \end{bmatrix}$.

Solution

$$\begin{vmatrix} k^2 & 2 \\ 3k & 5 \end{vmatrix} = 5k^2 - 6k = k(5k-6)$$

$$= k\left|\begin{bmatrix} 5k & 2 \\ 3 & 1 \end{bmatrix}\right|$$

$$= k|A|$$

11.2.2 INVERSE AND DETERMINANT OF 3 BY 3 MATRICES

In the same way that we determined the determinant of a 2 by 2 matrix, we now determine the determinant of a 3 by 3 matrix.

The determinant of a 3 by 3 matrix can be defined in terms of the determinant of 2 by 2 submatrices.

$$\left|\begin{bmatrix} a & b & c \\ d & e & f \\ g & h & i \end{bmatrix}\right| = a \times \begin{vmatrix} e & f \\ h & i \end{vmatrix} - b \times \begin{vmatrix} d & f \\ g & i \end{vmatrix} + c \times \begin{vmatrix} d & e \\ g & h \end{vmatrix}$$

The three terms are made from the elements of the first row paired with the 2 by 2 matrix in the other columns:

$$\left|\begin{bmatrix} a & b & c \\ d & e & f \\ g & h & i \end{bmatrix}\right| = a \times \begin{vmatrix} e & f \\ h & i \end{vmatrix} - b \times \begin{vmatrix} d & f \\ g & i \end{vmatrix} + c \times \begin{vmatrix} d & e \\ g & h \end{vmatrix}$$

$$\left|\begin{bmatrix} a & b & c \\ d & e & f \\ g & h & i \end{bmatrix}\right| = a \times \begin{vmatrix} e & f \\ h & i \end{vmatrix} - b \times \begin{vmatrix} d & f \\ g & i \end{vmatrix} + c \times \begin{vmatrix} d & e \\ g & h \end{vmatrix}$$

$$\left|\begin{bmatrix} a & b & c \\ d & e & f \\ g & h & i \end{bmatrix}\right| = a \times \begin{vmatrix} e & f \\ h & i \end{vmatrix} - b \times \begin{vmatrix} d & f \\ g & i \end{vmatrix} + c \times \begin{vmatrix} d & e \\ g & h \end{vmatrix}$$

The terms are **signed** according to the sum of the row and column in which the single element lies. In the first term, a lies in the first row and first column. This is even and so the term is positive. The second part of the expression uses b. This is in the second row and first column. This is an odd total and so the term is negative. This calculation can also be done working along any row or column that contains ones and/or zeros (if they exist). In short, the sign attached to each term is given by the **alternating sign rule**:

$$\begin{vmatrix} + & - & + \\ - & + & - \\ + & - & + \end{vmatrix}$$

EXAMPLE 11.8

Find the determinant of $\begin{bmatrix} 2 & -1 & 4 \\ 1 & 2 & 0 \\ 2 & 1 & 2 \end{bmatrix}$

Solution

Expanding along the first row gives:

$$\left|\begin{bmatrix} 2 & -1 & 4 \\ 1 & 2 & 0 \\ 2 & 1 & 2 \end{bmatrix}\right| = 2\left|\begin{bmatrix} 2 & 0 \\ 1 & 2 \end{bmatrix}\right| - (-1)\left|\begin{bmatrix} 1 & 0 \\ 2 & 2 \end{bmatrix}\right| + 4\left|\begin{bmatrix} 1 & 2 \\ 2 & 1 \end{bmatrix}\right|$$

$$= 2(2 \times 2 - 0 \times 1) + 1(1 \times 2 - 2 \times 0) + 4(1 \times 1 - 2 \times 2)$$

$$= 2 \times 4 + 1 \times 2 + 4 \times -3$$

$$= 8 + 2 - 12$$

$$= -2$$

Similarly, expanding down the first column gives:

$$\left|\begin{bmatrix} 2 & -1 & 4 \\ 1 & 2 & 0 \\ 2 & 1 & 2 \end{bmatrix}\right| = 2\left|\begin{bmatrix} 2 & 0 \\ 1 & 2 \end{bmatrix}\right| - 1\left|\begin{bmatrix} -1 & 4 \\ 1 & 2 \end{bmatrix}\right| + 2\left|\begin{bmatrix} -1 & 4 \\ 2 & 0 \end{bmatrix}\right|$$

$$= 2(2 \times 2 - 0 \times 1) - 1(-1 \times 2 - 4 \times 1) + 2(-1 \times 0 - 4 \times 2)$$

$$= 2 \times 4 - 1 \times -6 + 2 \times -8$$

$$= 8 + 6 - 16$$

$$= -2$$

That is, it does not matter which column or row is used to 'expand' the determinant.
As the determinant is not zero, the matrix is, therefore, non-singular.
A quicker calculation would be that obtained from expanding the determinant along the second row, as this contains a 1 and a 0. (This is left for you to verify).

An important result of determinant, is the following:

For any two square matrices, A and B, of the same order, $\det(AB) = \det(A)\det(B)$.

EXAMPLE 11.9

For what value(s) of k will the matrix $\begin{bmatrix} (1-k) & 1 & 1 \\ 1 & (1-k) & 1 \\ 1 & 1 & k \end{bmatrix}$ be singular?

Solution

For the matrix to be singular, then $\begin{vmatrix} (1-k) & 1 & 1 \\ 1 & (1-k) & 1 \\ 1 & 1 & k \end{vmatrix} = 0$

Expanding along the first row we have:

$$(1-k)\begin{vmatrix} (1-k) & 1 \\ 1 & k \end{vmatrix} - 1\begin{vmatrix} 1 & 1 \\ 1 & k \end{vmatrix} + 1\begin{vmatrix} 1 & (1-k) \\ 1 & 1 \end{vmatrix} = 0$$

$$\therefore (1-k)[k(1-k)-1]-[k-1]+[1-(1-k)] = 0$$

$$\Leftrightarrow (1-k)(k-k^2-1)-k+1+k = 0$$

$$\Leftrightarrow k^3-2k^2-2k = 0$$

$$\Leftrightarrow k(k^2-2k-2) = 0$$

$$\Leftrightarrow k(k-1+\sqrt{3})(k-1-\sqrt{3}) = 0$$

$$\therefore k = 0 \text{ or } 1-\sqrt{3} \text{ or } 1+\sqrt{3}$$

Checking these solutions by substituting back into the original matrix, we find that for $k = 0$, the determinant is zero, for $k = 1-\sqrt{3}$ the determinant is $4(1-\sqrt{3})$ while for $k = 1+\sqrt{3}$ the determinant is $4(1+\sqrt{3})$. Therefore, only solution is $k = 0$.

Properties of determinants

There are a number of useful properties of determinants that can reduce the amount of work involved when evaluating them. We list some of these now.

1. If all of the elements of a row (or column) of a matrix A are zero, then $\det(A) = 0$.
2. If two rows (or columns) of a matrix A are identical or in proportion, then $\det(A) = 0$.
3. If any two rows (or columns) of a matrix A are interchanged (call this new matrix B), then, $\det(B) = -\det(A)$.
4. If we add any non-zero multiple of one row (or column) of a matrix A to a different row (or column), (call this new matrix B), then $\det(B) = \det(A)$.
5. If each of the elements in a row (or column) of a matrix A are multiplied by some scalar, k (call this new matrix B), then, $\det(B) = k\det(A)$.

EXAMPLE 11.10

Show that $\begin{vmatrix} 1 & 1 & 1 \\ x & y & z \\ x^2 & y^2 & z^2 \end{vmatrix} = (x-y)(y-z)(z-x)$.

Solution

We subtract column 2 from column 1 (i.e., $C_1 - C_2$) and column 3 from column 2 (i.e., $C_2 - C_3$).

$$\text{i.e., } \begin{vmatrix} 1 & 1 & 1 \\ x & y & z \\ x^2 & y^2 & z^2 \end{vmatrix} = \begin{vmatrix} 0 & 0 & 1 \\ x-y & y-z & z \\ x^2-y^2 & y^2-z^2 & z^2 \end{vmatrix}$$

Next, we expand along the first row (as two of the elements are zero):

$$\begin{vmatrix} 0 & 0 & 1 \\ x-y & y-z & z \\ x^2-y^2 & y^2-z^2 & z^2 \end{vmatrix} = 0 - 0 + 1\begin{vmatrix} x-y & y-z \\ x^2-y^2 & y^2-z^2 \end{vmatrix}$$

$$= (x-y)(y^2-z^2) - (y-z)(x^2-y^2)$$
$$= (x-y)(y-z)(y+z) - (y-z)(x-y)(x+y)$$
$$= (x-y)(y-z)[(y+z)-(x+y)]$$
$$= (x-y)(y-z)(z-x)$$

Searching for the inverse of a 3 by 3 matrix*

* For examination purposes, the inverse of a 3×3 matrix should be found using a Graphics calculator. We begin this section by considering the following example.

EXAMPLE 11.11

Find the product AB where $A = \begin{bmatrix} 1 & 1 & 2 \\ 3 & 4 & 1 \\ 1 & 3 & -2 \end{bmatrix}$ and $B = \begin{bmatrix} -11 & 8 & -7 \\ 7 & -4 & 5 \\ 5 & -2 & 1 \end{bmatrix}$

Hence find A^{-1}.

Solution

Using the TI–83 we have:

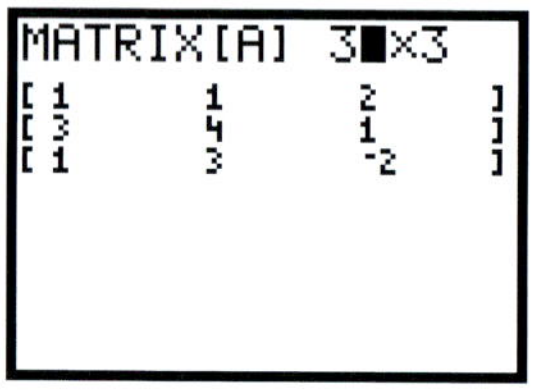

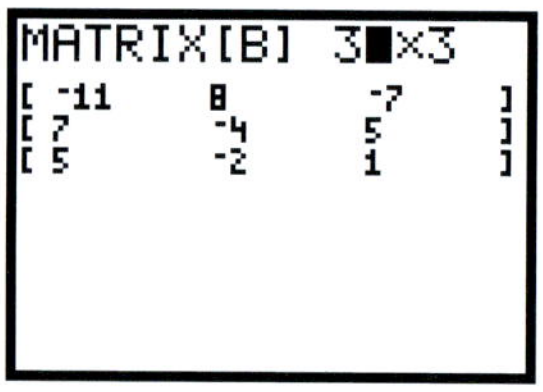

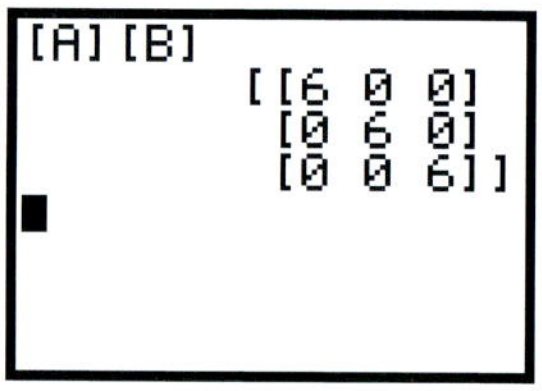

So, $AB = \begin{bmatrix} 1 & 1 & 2 \\ 3 & 4 & 1 \\ 1 & 3 & -2 \end{bmatrix}\begin{bmatrix} -11 & 8 & -7 \\ 7 & -4 & 5 \\ 5 & -2 & 1 \end{bmatrix} = \begin{bmatrix} 6 & 0 & 0 \\ 0 & 6 & 0 \\ 0 & 0 & 6 \end{bmatrix}$. That is, $AB = 6I$.

We can rewrite $AB = 6I$ as $A \times \left(\frac{1}{6}B\right) = I$.

However, we also know that $A \times A^{-1} = I$ and so we have that $A^{-1} = \frac{1}{6}B$.

That is, $A^{-1} = \frac{1}{6}\begin{bmatrix} -11 & 8 & -7 \\ 7 & -4 & 5 \\ 5 & -2 & 1 \end{bmatrix}$.

In Example 11.11 we were able to find the inverse of a matrix based on the result that if the product of two matrices produces the identity matrix (or a multiple of the identity matrix) then the matrices must be inverses of each other. The problem here is that we need to be given the two matrices such that their product will result in the identity matrix (or a multiple of the identity matrix). So, what happens when we do not have two such matrices?

It should also be noted at this stage that the TI–83 can also produce the inverse of a 3 by 3 matrix. Using Example 25.11 we have:

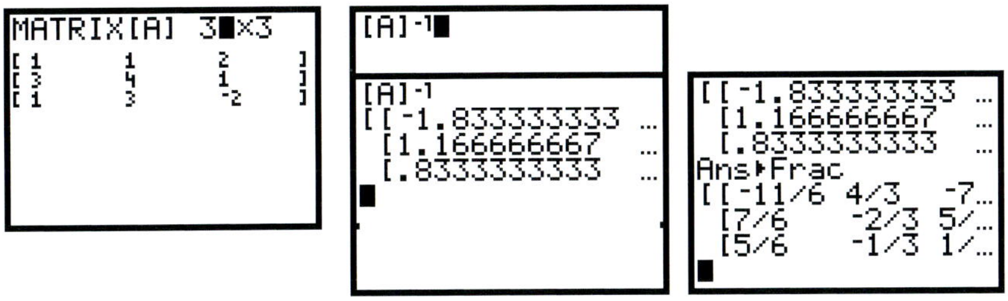

In §25.2.1 we were able to obtain the inverse of non-singular 2 by 2 square matrices. A similar result can also be obtained for a 3 by 3 matrix, however, the process for obtaining the general form for the inverse of a 3 by 3 matrix is beyond the scope of this course. Nonetheless, we can still obtain the inverse of a 3 by 3 (or other squares matrices of higher order) by using a method known as the **Gaussian method**, (also known as the Gauss–Jordan Elimination Method) named after the great mathematician, Carl Friedrich Gauss (1777–1855) and the German geodesist Wilhelm Jordan (1842–1899) who later generalised Gauss's method to solve problems in large-scale surveying. **Note: This method is not required for examination purposes.**

The method requires the pairing of the matrix A and the identity matrix, in a form known as the **augmented matrix form**. Then, by making use of basic arithmetic applied to the rows of the augmented matrix we convert the matrix A to the identity matrix – and in the process the identity matrix is converted to A^{-1}. That is,

Step 1: Set up augmented matrix form: $[\ A \mid I\ ]$

Step 2: Use a sequence of **elementary row operations** (i.e., $R_i \to \alpha R_i + \beta R_j$, replaces R_i with $\alpha R_i + \beta R_j$)

Step 3: Continue Step 2 until the augmented matrix $[\ I \mid B\ ]$ is achieved.

Step 4: $B = A^{-1}$.

EXAMPLE 11.12

Find the inverse of $A = \begin{bmatrix} 2 & -1 & 4 \\ 1 & 2 & 0 \\ 2 & 1 & 2 \end{bmatrix}$.

Solution

We set up the augmented matrix form, $[\ A \mid I\ ]$:

$$\left[\begin{array}{ccc|ccc} 2 & -1 & 4 & 1 & 0 & 0 \\ 1 & 2 & 0 & 0 & 1 & 0 \\ 2 & 1 & 2 & 0 & 0 & 1 \end{array}\right]$$

The aim is now to use linear combinations of the rows to reduce the left hand half of the tableau to the identity matrix. The first step is to halve the first row to make the top left entry 1.

$$R_1 \to \frac{1}{2} \times R_1 \left[\begin{array}{ccc|ccc} 1 & -\frac{1}{2} & 2 & \frac{1}{2} & 0 & 0 \\ 1 & 2 & 0 & 0 & 1 & 0 \\ 2 & 1 & 2 & 0 & 0 & 1 \end{array}\right]$$

The next step is to replace row 2 with row 2 – row 1 and row 3 with row 3 – twice row 1. This will make the first column correct.

$$\begin{array}{l} \\ R_2 \to R_2 - R_1 \\ R_3 \to R_3 - 2R_1 \end{array} \left[\begin{array}{ccc|ccc} 1 & -\frac{1}{2} & 2 & \frac{1}{2} & 0 & 0 \\ 0 & 2\frac{1}{2} & -2 & -\frac{1}{2} & 1 & 0 \\ 0 & 2 & -2 & -1 & 0 & 1 \end{array}\right]$$

The next thing to do is reduce the term in the second row and second column to 1 by dividing each element in the row by $2\frac{1}{2}$.

$$\begin{array}{l} \\ R_2 \to R_2 \div 2\frac{1}{2} \\ \\ \end{array} \left[\begin{array}{ccc|ccc} 1 & -\frac{1}{2} & 2 & \frac{1}{2} & 0 & 0 \\ 0 & 1 & -\frac{4}{5} & -\frac{1}{5} & \frac{2}{5} & 0 \\ 0 & 2 & -2 & -1 & 0 & 1 \end{array}\right]$$

Next, we use the term in the second row and second column as the pivot (the basis for reducing the next column). Row 1 is replaced by row 1 + half row 2. Row 3 is replaced by row 3 – twice row 2.

$$\begin{array}{l} R_1 \to R_1 + \frac{1}{2}R_2 \\ \\ R_3 \to R_3 - 2R_2 \end{array} \left[\begin{array}{ccc|ccc} 1 & 0 & 1\frac{3}{5} & \frac{2}{5} & \frac{1}{5} & 0 \\ 0 & 1 & -\frac{4}{5} & -\frac{1}{5} & \frac{2}{5} & 0 \\ 0 & 0 & -\frac{2}{5} & -\frac{3}{5} & -\frac{4}{5} & 1 \end{array}\right]$$

Next, reduce the bottom right element to 1 by dividing the bottom row by $-\frac{2}{5}$:

$$\begin{array}{l} \\ \\ R_3 \to R_3 \div \left(-\frac{2}{5}\right) \end{array} \left[\begin{array}{ccc|ccc} 1 & 0 & 1\frac{3}{5} & \frac{2}{5} & \frac{1}{5} & 0 \\ 0 & 1 & -\frac{4}{5} & -\frac{1}{5} & \frac{2}{5} & 0 \\ 0 & 0 & 1 & \frac{3}{2} & 2 & -2\frac{1}{2} \end{array}\right]$$

Finally, use the 1 to reduce the elements above it to zero. Replace row 1 by row 1 minus $1\frac{3}{5}$ times row 3. Row 2 is replaced by row 2 + $\frac{4}{5}$ ×row 3.

$$R_1 \to R_1 - 1\frac{3}{5} \times R_3 \qquad R_2 \to R_2 + \frac{4}{5} \times R_3 \qquad \left[\begin{array}{ccc|ccc} 1 & 0 & 0 & -2 & -3 & 4 \\ 0 & 1 & 0 & 1 & 2 & -2 \\ 0 & 0 & 1 & \frac{3}{2} & 2 & -2\frac{1}{2} \end{array}\right]$$

From this we can conclude that the inverse of $\begin{bmatrix} 2 & -1 & 4 \\ 1 & 2 & 0 \\ 2 & 1 & 2 \end{bmatrix}$ is $\begin{bmatrix} -2 & -3 & 4 \\ 1 & 2 & -2 \\ \frac{3}{2} & 2 & -2\frac{1}{2} \end{bmatrix}$.

EXERCISES 11.2

1. Find the determinants of the following matrices:

(i) $\begin{bmatrix} 2 & 6 \\ 2 & 4 \end{bmatrix}$ (ii) $\begin{bmatrix} 3 & 0 \\ 2 & 3 \end{bmatrix}$ (iii) $\begin{bmatrix} -3 & 2 \\ 0 & -4 \end{bmatrix}$ (iv) $\begin{bmatrix} 4 & -2 \\ 1 & 0 \end{bmatrix}$

(v) $\begin{bmatrix} 3 & -3 \\ 4 & 1 \end{bmatrix}$ (vi) $\begin{bmatrix} \frac{1}{3} & -\frac{2}{3} \\ \frac{1}{3} & \frac{2}{3} \end{bmatrix}$ (vii) $\begin{bmatrix} 0.4 & -0.1 \\ -0.2 & 1.3 \end{bmatrix}$ (viii) $\begin{bmatrix} \frac{1}{\sqrt{2}} & -\frac{3}{\sqrt{2}} \\ \frac{1}{\sqrt{2}} & \frac{5}{\sqrt{2}} \end{bmatrix}$

(ix) $\begin{bmatrix} \cos\theta & -\sin\theta \\ \sin\theta & \cos\theta \end{bmatrix}$ (x) $\begin{bmatrix} x & 2 \\ 1 & x \end{bmatrix}$ (xi) $\begin{bmatrix} \sqrt{a} & a \\ 1 & \sqrt{a} \end{bmatrix}$ (xii) $\begin{bmatrix} \frac{\sqrt{3}}{2} & 1 \\ \frac{1}{2} & \sqrt{3} \end{bmatrix}$

2. Find the inverses of these matrices, where they exist.

(i) $\begin{bmatrix} -2 & 4 \\ 1 & 0 \end{bmatrix}$ (ii) $\begin{bmatrix} 3 & 2 \\ -4 & -2 \end{bmatrix}$ (iii) $\begin{bmatrix} -2 & -2 \\ 4 & -3 \end{bmatrix}$ (iv) $\begin{bmatrix} 1 & 0 \\ -2 & -4 \end{bmatrix}$

(v) $\begin{bmatrix} -4 & -1 \\ 1 & 0 \end{bmatrix}$ (vi) $\begin{bmatrix} 3 & 2 \\ 6 & 4 \end{bmatrix}$ (vii) $\begin{bmatrix} -1 & -3 \\ 4 & -3 \end{bmatrix}$ (viii) $\begin{bmatrix} 3 & -2 \\ 4 & -4 \end{bmatrix}$

(ix) $\begin{bmatrix} -0.2 & 0.2 \\ -0.2 & -0.8 \end{bmatrix}$ (x) $\begin{bmatrix} 0 & 1 \\ \frac{1}{3} & -\frac{1}{3} \end{bmatrix}$ (xi) $\begin{bmatrix} \frac{1}{4} & \frac{1}{4} \\ \frac{1}{6} & -\frac{1}{6} \end{bmatrix}$ (xii) $\begin{bmatrix} \frac{2}{9} & \frac{1}{3} \\ \frac{1}{9} & -\frac{1}{3} \end{bmatrix}$

(xiii) $\begin{bmatrix} 3 & 4 \\ 2 & 3 \end{bmatrix}$ (xiv) $\begin{bmatrix} x & 1 \\ 2x & 3 \end{bmatrix}$ (xv) $\begin{bmatrix} 2 & x \\ 3 & x \end{bmatrix}$ (xvi) $\begin{bmatrix} x & x-1 \\ x+1 & \frac{1}{x-1} \end{bmatrix}$

3. Find the value(s) of x such that $\begin{bmatrix} x & 2 \\ 2 & x \end{bmatrix}$ is singular.

4. Find the value(s) of x such that $\frac{1}{x}\begin{bmatrix} -3 & 4 \\ 3 & 4 \end{bmatrix}\begin{bmatrix} 4 & -4 \\ -3 & -3 \end{bmatrix} = \begin{bmatrix} 1 & 0 \\ 0 & 1 \end{bmatrix}$

5. If $\begin{bmatrix} 2 & 4 \\ 1 & 1 \end{bmatrix}\begin{bmatrix} a & 2 \\ -a & -1 \end{bmatrix} = \begin{bmatrix} 1 & 0 \\ 0 & 1 \end{bmatrix}$, find the value of a.

6. Find the determinant of the following matrices

(a) $\begin{bmatrix} 2 & -1 & 3 \\ 1 & 3 & 4 \\ 2 & 2 & 1 \end{bmatrix}$ (b) $\begin{bmatrix} 4 & -2 & 3 \\ 3 & -1 & -2 \\ 2 & 4 & -1 \end{bmatrix}$ (c) $\begin{bmatrix} 3 & -2 & 1 \\ 2 & 1 & -4 \\ 1 & 1 & -3 \end{bmatrix}$ (d) $\begin{bmatrix} 0 & 3 & -1 \\ 3 & -2 & 1 \\ -2 & 0 & 1 \end{bmatrix}$

7. Using the Gaussian method find the inverse (where it exists) of each of the matrices in Q.6.

8. Find the determinant and using the Gaussian method, find the inverse of $\begin{bmatrix} 1 & 0 & 2 \\ 0 & 2 & 1 \\ 1 & 1 & 0 \end{bmatrix}$.

9. For the matrix: $\begin{bmatrix} -1 & 3 & p \\ 3 & -2 & 1 \\ 4 & -1 & -2 \end{bmatrix}$, find the value of p such that the determinant of the matrix is 35. Find the inverse of this matrix.

10. Prove that for two square matrices A & B, $(AB)^{-1} = B^{-1}A^{-1}$, where $|A| \neq 0 \neq |B|$,

11. Find the determinant and, using the Gaussian method, find the inverse of $\begin{bmatrix} -1 & 0 & -2 \\ 2 & 1 & -1 \\ \frac{1}{2} & \frac{3}{4} & 3 \end{bmatrix}$.

12. Find the value of x such that the matrix $\begin{bmatrix} x & -3 & x \\ -1 & 0 & 2 \\ 2 & 1 & -5 \end{bmatrix}$ is singular.

13. Find the determinant and inverse of the matrix $\begin{bmatrix} 2 & 5 & -2 \\ -1 & 5 & 2 \\ 3 & 1 & -4 \end{bmatrix}$.

14. If A is a 2×2 matrix, and k is a real non-zero number, prove that

(a) $(kA)^{-1} = k^{-1}A^{-1}$ (b) $(A^m)^{-1} = (A^{-1})^m$

15. (a) If $A = \begin{bmatrix} 1 & -2 & 4 \\ 2 & 1 & -2 \\ -1 & 2 & 1 \end{bmatrix}$ and $B = \begin{bmatrix} 1 & 2 & 0 \\ 0 & 1 & 2 \\ 1 & 0 & 1 \end{bmatrix}$, find the product AB. Hence find A^{-1}.

(b) If $A = \begin{bmatrix} 5 & -8 & -1 \\ 2 & -3 & -1 \\ -3 & 5 & 1 \end{bmatrix}$ and $B = \begin{bmatrix} 2 & 3 & 5 \\ 1 & 2 & 3 \\ 1 & -1 & 1 \end{bmatrix}$, find the product AB. Hence find A^{-1}.

(c) If $A = \begin{bmatrix} 8 & -1 & -3 \\ 2 & -1 & 0 \\ -7 & 2 & 3 \end{bmatrix}$ and $B = \begin{bmatrix} 1 & 1 & 1 \\ 2 & -1 & 2 \\ 1 & 3 & 2 \end{bmatrix}$, find the product AB. Hence find A^{-1}.

16. Find x for each of the following

(a) $\begin{vmatrix} x-1 & 1 & 2 \\ 1 & 1-x & 1 \\ -1 & 1 & 2-x \end{vmatrix} = 0$ (b) $\begin{vmatrix} 1 & x+1 & x+1 \\ x+1 & 0 & 2 \\ x+1 & 2 & 0 \end{vmatrix} = 0$

17. If $\alpha X^2 + \beta X + \gamma I = O$ and $A = Y^{-1}XY$ prove that $\alpha A^2 + \beta A + \gamma I = O$.

18. (a) If $A = A^{-1}$ prove that $|A|^2 - 1 = 0$.

(b) Show that for any square matrix A i. $|A^n| = |A|^n$.

ii. $|kA| = k|A|$.

(c) Prove that $\begin{vmatrix} a+k & b \\ c+k & d \end{vmatrix} = \begin{vmatrix} a & b \\ c & d \end{vmatrix} + \begin{vmatrix} k & b \\ k & d \end{vmatrix}$

(d) If A is a 2×2 matrix, show that $A^2 - (a+d)A + |A|I = O$.

Hence find the real numbers α and β such that $A^2 = \alpha A + \beta I$ and $A = \begin{bmatrix} 5 & 6 \\ 8 & 9 \end{bmatrix}$.

(e) If $A^2 = A$ show that i. show that $(I + A)^2 = I + 3A$.

ii. show that $(I + A)^n = I + (2^n - 1)A$.

iii. find $|A|$.

19. If $\begin{vmatrix} 1 & ab + cx & c^2 \\ 1 & bc + ax & a^2 \\ 1 & ca + bx & b^2 \end{vmatrix} = 0$ and a, b and c are unequal real numbers, show that $a + b + c = x$.

20. Prove that if $A^2 = I$ then $A = I$ if $|A + I| \neq 0$ and $A = -I$ if $|A - I| \neq 0$

21. Find the value(s) of k for each of the following

(a) $\begin{vmatrix} x & y & x+y \\ y & x+y & x \\ x+y & x & y \end{vmatrix} = k(x^3 + y^3)$ (b) $\begin{vmatrix} x & y & y \\ y & x & y \\ y & y & x \end{vmatrix} = (x - 2ky)(x + ky)^2$

11.3 SIMULTANEOUS EQUATIONS

The special definition of matrix multiplication makes it possible to apply matrix methods to the problem of solving simultaneous equations. A system of two simultaneous equations in two unknowns can be represented as a single matrix equation as follows.

EXAMPLE 11.13

Represent the following pairs of equations in matrix form.

(a) $\begin{aligned} 4x+2y &= -2 \\ 4x-y &= -11 \end{aligned}$ (b) $\begin{aligned} x+y &= 1 \\ y &= \frac{3x-31}{4} \end{aligned}$

Solution

(a) The matrix version of the pair of equations $\begin{aligned} 4x+2y &= -2 \\ 4x-y &= -11 \end{aligned}$ is $\begin{bmatrix} 4 & 2 \\ 4 & -1 \end{bmatrix}\begin{bmatrix} x \\ y \end{bmatrix} = \begin{bmatrix} -2 \\ -11 \end{bmatrix}$.

The left hand side is represented by a 2 by 2 matrix containing the coefficients of the equations multiplied by a column matrix containing the variables. The right hand side is represented by a column matrix containing the right hand sides of the original equations.

The reason that the two representations are equivalent can be seen when the matrix version is expanded: $\begin{bmatrix} 4 & 2 \\ 4 & -1 \end{bmatrix}\begin{bmatrix} x \\ y \end{bmatrix} = \begin{bmatrix} -2 \\ -11 \end{bmatrix} \Leftrightarrow \begin{bmatrix} 4x+2y \\ 4x-y \end{bmatrix} = \begin{bmatrix} -2 \\ -11 \end{bmatrix}$

The two matrices given as equal above are only equal if corresponding elements are identical. So, the top elements must be equal giving $4x+2y = -2$ and the bottom elements must be equal giving $4x-y = -11$, which are the two equations we were given.

(ii) The second of the equations $\begin{aligned} x+y &= 1 \\ y &= \frac{3x-31}{4} \end{aligned}$ must be rearranged to $3x-4y = 31$

before writing in matrix form to get: $\begin{bmatrix} 1 & 1 \\ 3 & -4 \end{bmatrix}\begin{bmatrix} x \\ y \end{bmatrix} = \begin{bmatrix} 1 \\ 31 \end{bmatrix}$.

Now that we can convert a pair of simultaneous equations to matrix form, we are in a position to solve the system. The process for using matrices to solve simultaneous equations can be summarised as follows:

> If a system of equations is represented by the matrix equation, $AX = B$, where the matrix A is non-singular (i.e., A^{-1} exists), then,
>
> $$AX = B \Leftrightarrow A^{-1}AX = A^{-1}B \Leftrightarrow X = A^{-1}B$$

This is equivalent to the method used to solve simple equations such as:

$$2x = 12 \Leftrightarrow \frac{1}{2} \times 2x = \frac{1}{2} \times 12 \Leftrightarrow x = 6$$

Central to the method is to multiply both sides of the equation by the multiplicative inverse of the coefficient of x. This coefficient is 2 and its inverse is $\frac{1}{2}$. The result of this process is that the left hand side is reduced to x and the solution is produced.

The process described for solving a matrix equation is very similar to the process used to solve the equation $2x = 12$. The first step is to find the inverse of the matrix of coefficients.

EXAMPLE 11.14

Solve the following pairs of equations using the matrix method.

(a) $\begin{aligned} 4x + 2y &= -2 \\ 4x - y &= -11 \end{aligned}$ (b) $\begin{aligned} x + y &= 1 \\ y &= \frac{3x - 31}{4} \end{aligned}$

Solution

(a) The matrix version of this pair of equations is: $\begin{bmatrix} 4 & 2 \\ 4 & -1 \end{bmatrix}\begin{bmatrix} x \\ y \end{bmatrix} = \begin{bmatrix} -2 \\ -11 \end{bmatrix}$ (i.e., $AX = B$)

This can be solved by pre-multiplying both sides of this equation by the inverse of the 2 by 2 coefficients matrix. In this case the coefficient matrix is $\begin{bmatrix} 4 & 2 \\ 4 & -1 \end{bmatrix}$ and the inverse is

$\begin{bmatrix} \frac{1}{12} & \frac{1}{6} \\ \frac{1}{3} & -\frac{1}{3} \end{bmatrix}$. So, we have $\begin{bmatrix} \frac{1}{12} & \frac{1}{6} \\ \frac{1}{3} & -\frac{1}{3} \end{bmatrix}\begin{bmatrix} 4 & 2 \\ 4 & -1 \end{bmatrix}\begin{bmatrix} x \\ y \end{bmatrix} = \begin{bmatrix} \frac{1}{12} & \frac{1}{6} \\ \frac{1}{3} & -\frac{1}{3} \end{bmatrix}\begin{bmatrix} -2 \\ -11 \end{bmatrix}$ (i.e., $AX = B \Leftrightarrow A^{-1}AX = A^{-1}B$).

Remember that matrix multiplication is non-commutative and that we must be careful to **pre-multiply by the inverse** on both sides of the equation. This version of the problem looks much more complicated than other methods such as substitution. This, however, is not really the case as the left hand side (much as it does in the case of simple algebraic equations) should simplify to the matrix of the unknowns.

$$\begin{bmatrix} 1 & 0 \\ 0 & 1 \end{bmatrix}\begin{bmatrix} x \\ y \end{bmatrix} = \begin{bmatrix} \frac{1}{12} & \frac{1}{6} \\ \frac{1}{3} & -\frac{1}{3} \end{bmatrix}\begin{bmatrix} -2 \\ -11 \end{bmatrix} \Leftrightarrow \begin{bmatrix} x \\ y \end{bmatrix} = \begin{bmatrix} \frac{1}{12} & \frac{1}{6} \\ \frac{1}{3} & -\frac{1}{3} \end{bmatrix}\begin{bmatrix} -2 \\ -11 \end{bmatrix} \text{ (i.e., } X = A^{-1}B\text{)}$$

This should always be the case, provided the inverse has been found correctly, and so does not require any work. All that is necessary is to perform the matrix multiplication on the right hand side.

$$\begin{bmatrix} x \\ y \end{bmatrix} = \begin{bmatrix} -2 \\ 3 \end{bmatrix}$$

The solution is $x = -2$, $y = 3$ or, as a number pair $(-2,3)$.

(b) In matrix form, this is: $\begin{bmatrix} 1 & 1 \\ 3 & -4 \end{bmatrix}\begin{bmatrix} x \\ y \end{bmatrix} = \begin{bmatrix} 1 \\ 31 \end{bmatrix}$. The inverse of $\begin{bmatrix} 1 & 1 \\ 3 & -4 \end{bmatrix}$ is $\frac{1}{7}\begin{bmatrix} 4 & 1 \\ 3 & -1 \end{bmatrix}$.

Premultiplication by this inverse gives: $\frac{1}{7}\begin{bmatrix} 4 & 1 \\ 3 & -1 \end{bmatrix}\begin{bmatrix} 1 & 1 \\ 3 & -4 \end{bmatrix}\begin{bmatrix} x \\ y \end{bmatrix} = \frac{1}{7}\begin{bmatrix} 4 & 1 \\ 3 & -1 \end{bmatrix}\begin{bmatrix} 1 \\ 31 \end{bmatrix} \Leftrightarrow \begin{bmatrix} x \\ y \end{bmatrix} = \begin{bmatrix} 5 \\ -4 \end{bmatrix}$

The solution is $x = 5, y = -4$ or, as a number pair (5,–4).

Such a system of equations can also be solved by adapting the Guassian method encountered in the previous section. This adaptation requires that we write the system of equations, $AX = B$, in the augmented matrix form $[\ A \ |\ B\ ]$.

That is,

$$\begin{matrix} a_1x + b_1y = c_1 \\ a_2x + b_2y = c_2 \end{matrix} \Leftrightarrow \begin{bmatrix} a_1 & b_1 \\ a_2 & b_2 \end{bmatrix}\begin{bmatrix} x \\ y \end{bmatrix} = \begin{bmatrix} c_1 \\ c_2 \end{bmatrix} \Leftrightarrow \left[\begin{array}{cc|c} a_1 & b_1 & c_1 \\ a_2 & b_2 & c_2 \end{array}\right]$$

Then, once we have the augmented form we use elementary row operations to achieve the

solution form: $\left[\begin{array}{cc|c} 1 & 0 & d_1 \\ 0 & 1 & d_2 \end{array}\right]$, i.e., $x = d_1$ and $y = d_2$.

EXAMPLE 11.15

Use the Guassian method to solve the system of equations $\begin{matrix} 4x + 2y = -2 \\ 4x - y = -11 \end{matrix}$.

Solution

The system of equations has the augmented matrix form: $\left[\begin{array}{cc|c} 4 & 2 & -2 \\ 4 & -1 & -11 \end{array}\right]$

Using elementary row operations we have:

$$\begin{matrix} R_1 \to \frac{1}{4}R_1 \\ R_2 \to R_2 - R_1 \end{matrix} \left[\begin{array}{cc|c} 1 & \frac{1}{2} & -\frac{1}{2} \\ 0 & -3 & -9 \end{array}\right]$$

$$R_2 \to -\frac{1}{3}R_2 \left[\begin{array}{cc|c} 1 & \frac{1}{2} & -\frac{1}{2} \\ 0 & 1 & 3 \end{array}\right]$$

$$R_1 \to R_1 - \frac{1}{2}R_2 \left[\begin{array}{cc|c} 1 & 0 & -2 \\ 0 & 1 & 3 \end{array}\right]$$

From which we then have the solution $x = -2$ and $y = 3$.

Using a Graphics Calculator

Both methods that were used in Examples 25.13 and 25.14 can also be solved using the TI–83.

Matrix Form

The matrix method is particularly suited to implementation on a graphics calculator. Returning to the first of these examples, $\begin{bmatrix} 4 & 2 \\ 4 & -1 \end{bmatrix}\begin{bmatrix} x \\ y \end{bmatrix} = \begin{bmatrix} -2 \\ -11 \end{bmatrix}$, the solution would proceed as follows:

The matrices $A = \begin{bmatrix} 4 & 2 \\ 4 & -1 \end{bmatrix}$, $B = \begin{bmatrix} -2 \\ -11 \end{bmatrix}$ must be entered under the **MATRIX EDIT** menu. Remember to use the key **(-)** for negative quantities, not the subtraction key.

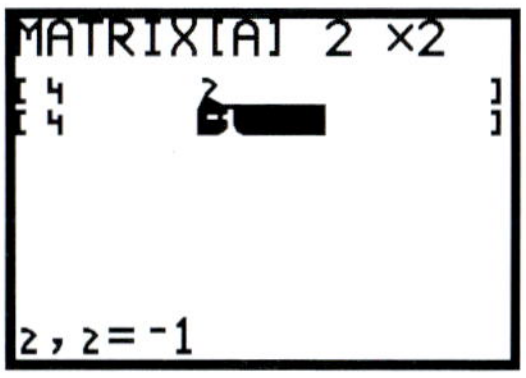

Now pre-multiply B by the inverse of A. Using the TI-82/3, the names of the matrices are accessed using the **MATRIX NAMES** menu. The inverse of A is calculated using the x^{-1} key. The sequence of operations is to first return to the home screen by using **2nd QUIT**. Then use **MATRIX NAMES** and select matrix A. Press x^{-1} and finally use **MATRIX NAMES** and select matrix B.

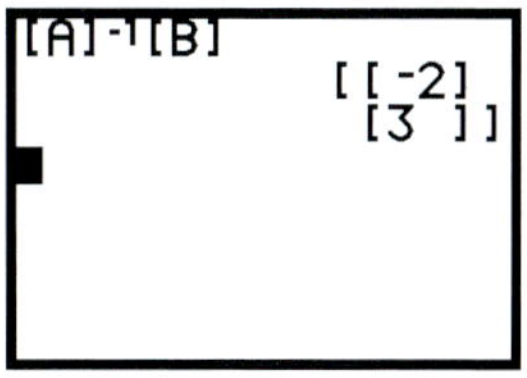

Augmented Matrix form

The TI–83 can also be used to set up our matrix equation, $AX = B$, into its augmented matrix form.

Step 1: Define the matrix A and B.

1.

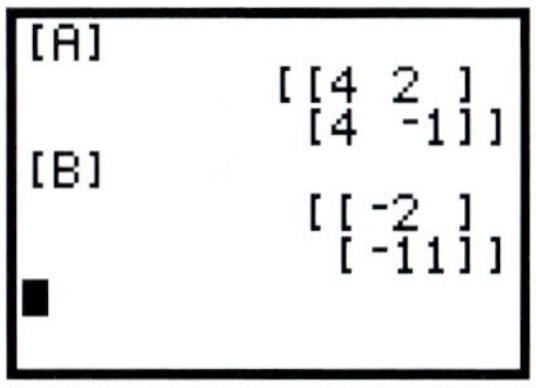

Step 2: Call up the **MATRIX** window and then select the **MATH** menu followed by option **7:augment(** Enter the matrices in the correct order, A first followed by ',' and then B followed by ')'.

2.

Step 3: Store the augmented matrix by pressing **STO▶** followed by calling up the **MATRIX** window and selecting a matrix [C] (say) and then press **ENTER**.

3.

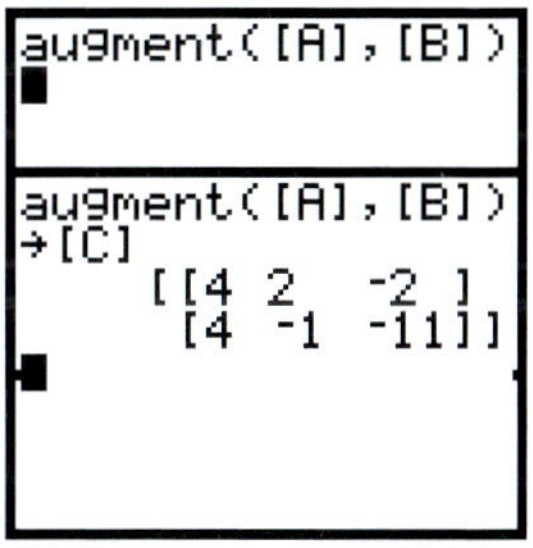

Step 4: Call up the **MATRIX** window and select the **MATH** menu from which option **B:rref(** is selected (make use of the down arrow key). Then press **ENTER**.

Step 5: Select the **MATRIX** Window and choose matrix [C] press ')' and then ENTER. This will provide the **reduced row-echelon** form which will enable us to quote the solutions directly from the screen.

4. & 5.

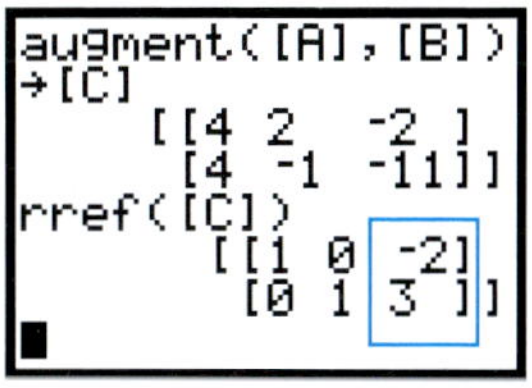

From the last screen, we have the solution $x = -2$ and $y = 3$.

There are occasions on which simultaneous equations either have no solutions or infinite solution sets. This problem was covered in Chapter 2, §2.3. If we use the matrix method to solve one of these types of equations, we will find that the matrix of coefficients is singular and has no inverse. In this case, the simple matrix method fails. If you encounter a pair of equations of this type, it is best to handle them using an algebraic method as the matrix method, i.e., $AX = B$, will not tell you whether the equations have no solutions or an infinite solution set.

However, using the augmented matrix form $[\ A \mid B\ ]$, then depending on the end form of the reduced row-echelon matrix we have the following results:

Form 1
Unique solution

$$\left[\begin{array}{cc|c} 1 & 0 & d_1 \\ 0 & 1 & d_2 \end{array}\right]$$

where $d_2 \neq 0$.

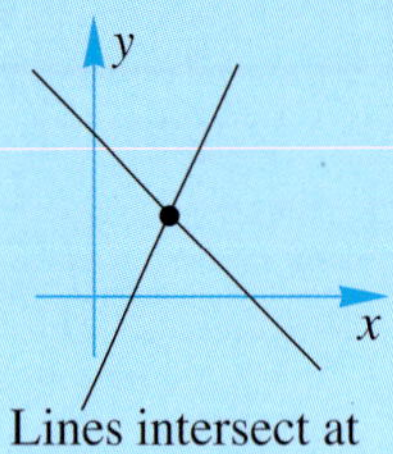

Lines intersect at one unique point.

Form 2
Infinitely many solutions

$$\left[\begin{array}{cc|c} 1 & k & d_1 \\ 0 & 0 & 0 \end{array}\right]$$

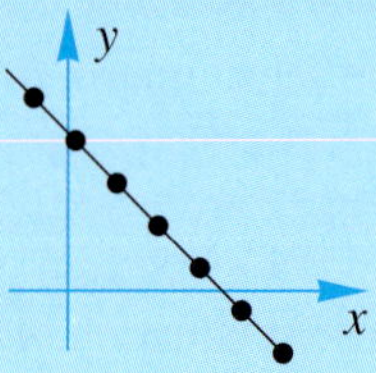

Lines are coincident. i.e. they are identical.

Form 3
No solution

$$\left[\begin{array}{cc|c} 1 & k & d_1 \\ 0 & 0 & d_2 \end{array}\right]$$

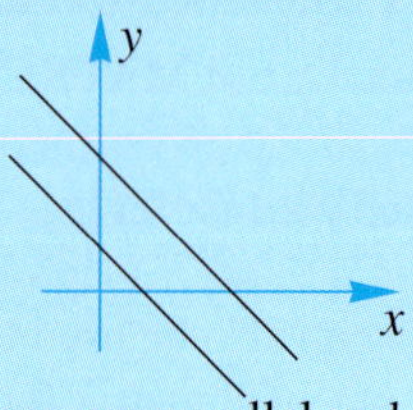

Lines are parallel and so never meet.

The methods described for 2 by 2 matrices hold equally for 3 by 3 matrices.

EXAMPLE 11.16

Find the product AB where $A = \begin{bmatrix} 1 & 1 & 2 \\ 3 & 4 & 1 \\ 1 & 3 & -2 \end{bmatrix}$ and $B = \begin{bmatrix} -11 & 8 & -7 \\ 7 & -4 & 5 \\ 5 & -2 & 1 \end{bmatrix}$

Hence solve the system of equations $\begin{aligned} x + y + 2z &= 9 \\ 3x + 4y + z &= -2 \\ x + 3y - 2z &= 4 \end{aligned}$

Solution

From Example 25.11 we have that $AB = \begin{bmatrix} 1 & 1 & 2 \\ 3 & 4 & 1 \\ 1 & 3 & -2 \end{bmatrix}\begin{bmatrix} -11 & 8 & -7 \\ 7 & -4 & 5 \\ 5 & -2 & 1 \end{bmatrix} = \begin{bmatrix} 6 & 0 & 0 \\ 0 & 6 & 0 \\ 0 & 0 & 6 \end{bmatrix} = 6I$

Then, as $AB = 6I \Rightarrow A^{-1} = \frac{1}{6}B$ and $B^{-1} = \frac{1}{6}A$.

In the matrix form we have the system of equations as $\begin{bmatrix} 1 & 1 & 2 \\ 3 & 4 & 1 \\ 1 & 3 & -2 \end{bmatrix}\begin{bmatrix} x \\ y \\ z \end{bmatrix} = \begin{bmatrix} 9 \\ -2 \\ 4 \end{bmatrix}$ or $A\begin{bmatrix} x \\ y \\ z \end{bmatrix} = \begin{bmatrix} 9 \\ -2 \\ 4 \end{bmatrix}$.

So, pre-muliplying both sides by A^{-1} we have $A^{-1}A\begin{bmatrix} x \\ y \\ z \end{bmatrix} = A^{-1}\begin{bmatrix} 9 \\ -2 \\ 4 \end{bmatrix}$ or $I\begin{bmatrix} x \\ y \\ z \end{bmatrix} = A^{-1}\begin{bmatrix} 9 \\ -2 \\ 4 \end{bmatrix}$.

But, $I\begin{bmatrix} x \\ y \\ z \end{bmatrix} = \begin{bmatrix} x \\ y \\ z \end{bmatrix}$ and $A^{-1}\begin{bmatrix} 9 \\ -2 \\ 4 \end{bmatrix} = \frac{1}{6}B\begin{bmatrix} 9 \\ -2 \\ 4 \end{bmatrix} = \frac{1}{6}\begin{bmatrix} -11 & 8 & -7 \\ 7 & -4 & 5 \\ 5 & -2 & 1 \end{bmatrix}\begin{bmatrix} 9 \\ -2 \\ 4 \end{bmatrix} = \frac{1}{6}\begin{bmatrix} -143 \\ 91 \\ 53 \end{bmatrix}$

And so we have $\begin{bmatrix} x \\ y \\ z \end{bmatrix} = \frac{1}{6}\begin{bmatrix} -143 \\ 91 \\ 53 \end{bmatrix} \Leftrightarrow x = -\frac{143}{6}, y = \frac{91}{6}, z = \frac{53}{6}$.

(Of course, we used a graphics calculator to evaluate the required products.)

EXAMPLE 11.17

Solve the system of equations
$$\begin{aligned} x + 2y + z &= 6 \\ x - y - z &= -1 \\ 3x - y - 2z &= 2 \end{aligned}.$$

Solution

In this case we will make use of the Gaussian method. Setting up the augmented matrix form we have:

$$\left[\begin{array}{ccc|c} 1 & 2 & 1 & 6 \\ 1 & -1 & -1 & -1 \\ 3 & -1 & -2 & 2 \end{array}\right]$$

Applying elementary row operations we have:

$$\begin{matrix} \\ R_2 \to R_2 - R_1 \\ R_3 \to R_3 - 3R_1 \end{matrix} \left[\begin{array}{ccc|c} 1 & 2 & 1 & 6 \\ 0 & -3 & -2 & -7 \\ 0 & -7 & -5 & -16 \end{array}\right]$$

$$\begin{matrix} \\ \\ R_3 \to 3R_3 - 7R_2 \end{matrix} \left[\begin{array}{ccc|c} 1 & 2 & 1 & 6 \\ 0 & -3 & -2 & -7 \\ 0 & 0 & -1 & 1 \end{array}\right]$$

$$\begin{matrix} R_1 \to R_1 + R_3 \\ R_2 \to R_2 - 2R_3 \\ R_3 \to -R_3 \end{matrix} \left[\begin{array}{ccc|c} 1 & 2 & 0 & 7 \\ 0 & -3 & 0 & -9 \\ 0 & 0 & 1 & -1 \end{array}\right]$$

$$\begin{matrix} R_1 \to R_1 + \frac{2}{3}R_2 \\ R_2 \to \frac{1}{3}R_2 \\ \\ \end{matrix} \left[\begin{array}{ccc|c} 1 & 0 & 0 & 1 \\ 0 & 1 & 0 & 3 \\ 0 & 0 & 1 & -1 \end{array}\right]$$

And so the solution set is $x - 1, y - 3$ and $z = -1$.

However, making use of the TI–83 is so much quicker!

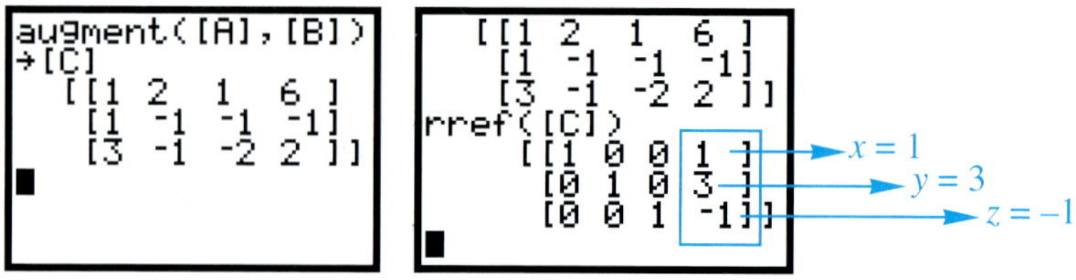

A similar summary to that provided for a 2 by 2 system can be presented – so, we conclude our work on solutions to systems of equations in three unknowns with a summary for the 3 by 3 (or really 3 by 4) augmented matrix form. Suffice to say, however, that if the inverse matrix exists, then a unique solution will exist.

Form 1 Unique solution	**Form 2** Infinitely many solutions	**Form 3** No solution
$\left[\begin{array}{ccc\|c} 1 & 0 & 0 & d_1 \\ 0 & 1 & 0 & d_2 \\ 0 & 0 & 1 & d_3 \end{array}\right]$	$\left[\begin{array}{ccc\|c} 1 & m & n & d_1 \\ 0 & 1 & k & d_2 \\ 0 & 0 & 0 & 0 \end{array}\right]$	$\left[\begin{array}{ccc\|c} 1 & m & n & d_1 \\ 0 & 1 & k & d_2 \\ 0 & 0 & 0 & d_3 \end{array}\right]$ Where $d_3 \neq 0$
Geometrically: Three planes meet at one point.	**Geometrically:** Three planes meet along one line.	**Geometrically:** At least two planes are parallel.

NOTE: There are other geometrical outcomes:

No solutions:

1. Two of the planes meet along a line and this line is parallel to (and does not meet) the third plane.

Infinitely many solutions:

1. The three planes are coincident.

EXERCISES 11.3

1. Solve the following pairs of simultaneous equations using the matrix method. Give your answers as ordered pairs, (x,y).

(i) $\begin{aligned} 3x - 2y &= 5 \\ x + y &= 0 \end{aligned}$ (ii) $\begin{aligned} 2x + 2y &= 2 \\ 4x + 3y &= 7 \end{aligned}$ (iii) $\begin{aligned} -4x + y &= 9 \\ 2x &= -6 \end{aligned}$

(iv) $\begin{aligned} x + 3y &= -5 \\ x - y &= -1 \end{aligned}$ (v) $\begin{aligned} 5x - 3y &= -5 \\ -3x + 3y &= 9 \end{aligned}$ (vi) $\begin{aligned} y &= \frac{5 - 5x}{2} \\ 6x - 4y &= 6 \end{aligned}$

(vii) $\begin{aligned} -3x + 4y &= 12 \\ 4x - 2y &= -6 \end{aligned}$ (viii) $\begin{aligned} x + 3y &= 10 \\ 4x - 2y &= -16 \end{aligned}$ (ix) $\begin{aligned} 2x + 3y &= 7 \\ -2x + 4y &= 7 \end{aligned}$

(x) $\begin{aligned} 0.2x + 0.4y &= 0.64 \\ -0.4x - 3y &= -2.6 \end{aligned}$ (xi) $\begin{aligned} y &= \frac{x-11}{4} \\ y - x &= -2 \end{aligned}$ (xii) $\begin{aligned} x + 2y &= 2 \\ 2x + 3y &= 2 \end{aligned}$

(xiii) $\begin{aligned} \frac{2}{3}x + \frac{1}{2}y &= 2\frac{1}{6} \\ \frac{1}{3}x + 5y &= 2\frac{2}{3} \end{aligned}$ (xiv) $\begin{aligned} -x - 2.5y &= 0 \\ 1.5x + 2y &= 3.5 \end{aligned}$ (xv) $\begin{aligned} 1.6x - 0.8y &= 4.8 \\ 5x - 0.2y &= 24.2 \end{aligned}$

2. Find the values of a and b such that the equations $\begin{aligned} 3x + ay &= 2 \\ -6x + 4y &= b \end{aligned}$ have:

(i) an infinite set of solutions (ii) no solutions.

3. If the equations $\begin{aligned} ax + y &= 7 \\ 4x + ay &= 19 \end{aligned}$ have no solutions, what are the possible value(s) of a.

4. Use a graphics calculator to solve the simultaneous equations:

(a) $\begin{aligned} 3x + 7y + 4z &= -10 \\ 4x - 2y + 3z &= 16 \\ -x + 3y - 2z &= -14 \end{aligned}$ (b) $\begin{aligned} x + 2z &= 3 \\ 2x + 3y - z &= 2 \\ x + y - 2z &= 4 \end{aligned}$ (c) $\begin{aligned} x - y + 4z &= -2 \\ 3x - y + z &= 1 \\ 2x - 2y + 3z &= 3 \end{aligned}$

5. Find AB if $A = \begin{bmatrix} 5 & -8 & -1 \\ 2 & -3 & -1 \\ -3 & 5 & 1 \end{bmatrix}$ and $B = \begin{bmatrix} 2 & 3 & 5 \\ 1 & 2 & 3 \\ 1 & -1 & 1 \end{bmatrix}$. Hence solve $\begin{aligned} 2a + 3b + 5c &= 8 \\ a + 2b + 3c &= -5 \\ a - b + c &= 2 \end{aligned}$.

6. Find AB if $A = \begin{bmatrix} 3 & 4 & 1 \\ 13 & -1 & -3 \\ 5 & 3 & -2 \end{bmatrix}$ and $B = \begin{bmatrix} 1 & 1 & -1 \\ 1 & -1 & 2 \\ 4 & 1 & -5 \end{bmatrix}$. Hence solve $\begin{aligned} 3x + 4y + z &= 5 \\ 13x - y - 3z &= 2 \\ 5x + 3y - 2z &= 3 \end{aligned}$.

7. For what value(s) of m will the system of equations $\begin{aligned} x + my &= 2 \\ (m-1)x + 2y &= m \end{aligned}$ have

(a) a unique solution?
(b) no solution?
(c) an infinite number of solutions?

Give a geometrical interpretation for each of these cases.

8. Find the determinant of $A = \begin{bmatrix} p & 3 \\ 2 & \frac{1}{p-1} \end{bmatrix}$. For what values of p and k will the system of equations $\begin{aligned} px + 3y &= k \\ 2(p-1)x + y &= 2 \end{aligned}$ have

(a) a unique solution?
(b) no solution?
(c) an infinite number of solutions?

9. Use the Gaussian method to solve the following systems of equations.

(a) $\begin{aligned} 3x + 4y &= 12 \\ 2x - 3y &= 8 \end{aligned}$ (b) $\begin{aligned} -x + 2y &= 6 \\ 2x - 4y &= 9 \end{aligned}$ (c) $\begin{aligned} 4x + y &= 5 \\ -x - 3y &= 7 \end{aligned}$

(d) $\begin{aligned} x + 3y - z &= 13 \\ 3x + y - z &= 11 \\ x + y - 3z &= 11 \end{aligned}$ (e) $\begin{aligned} x + 2y &= 10 \\ 3x + 2y - 4z &= 18 \\ y + z &= 3 \end{aligned}$ (f) $\begin{aligned} x - 2y - 3z &= 3 \\ x + y - 2z &= 7 \\ 2x - 3y - 2z &= 0 \end{aligned}$

(g) $\begin{aligned} x - y - z &= 2 \\ 3x + 3y - 7z &= 7 \\ x + 2y - 3z &= 3 \end{aligned}$ (h) $\begin{aligned} x - 2y &= -1 \\ -x - y + 3z &= 1 \\ y - z &= 0 \end{aligned}$ (i) $\begin{aligned} -2x + y - 2z &= 5 \\ x + 4z &= 1 \\ x + y + 10z &= 10 \end{aligned}$

10. A parabola passes through the points (–1, –19), (2, 2) and (3, 13). Set up a system of equations that satisfy these conditions. Use the Gaussian method to find its equation.

11. A parabola passes through the points (–1, –19), (2, 2). Set up a system of equations that satisfy these conditions. Use the Gaussian method to find its equation.

12. (a) If $A = \begin{bmatrix} 3 & 1 & 2 \\ -2 & 3 & -1 \end{bmatrix}$ and $X = \begin{bmatrix} x \\ y \\ z \end{bmatrix}$, solve the system of equations $AX = B$ where

i. $B = \begin{bmatrix} 0 \\ 0 \end{bmatrix}$ ii. $B = \begin{bmatrix} 2 \\ -1 \end{bmatrix}$

(b) Solve the system of equations $\begin{aligned} 3x + y + 2z &= 0 \\ -2x + 3y - z &= 0 \\ x + y + z &= 0 \end{aligned}$.

13. (a) Solve for $\begin{vmatrix} t & 1 & 1 \\ 1 & t & 1 \\ 1 & 1 & t \end{vmatrix} = 0$.

(b) For what value(s) of λ will the system of equations $\begin{aligned} \lambda x + y + z &= 3 \\ x + \lambda y + z &= 2 \\ x + y + \lambda z &= 1 \end{aligned}$ have

i. no solutions?
ii. a unique solution?
iii. an infinite number of solutions?

12.1 INTRODUCTION TO VECTORS

12.1.1 SCALAR AND VECTOR QUANTITIES

Numerical measurement scales are in widespread use. It is important to be able to distinguish between two distinct types of measurement scales, **scalars** and **vectors**.

Scalar Quantities

A scalar is a quantity that has **magnitude** (size) but no **direction**. For example, we measure the mass of objects using a variety of scales such as 'kilograms' and 'pounds'. These measures have **magnitude** in that more massive objects (such as the sun) have a larger numerical mass than small objects (such as this book). Giving the mass of this book does not, however, imply that this mass has a direction. This does **not** mean that scalar quantities must be positive. **Signed** scalar quantities, such as temperature as measured by the Celsius or Fahrenheit scales (which are commonly used) also exist.

Vector Quantities

Some measurements have both **magnitude** and **direction**. When we pull on a door handle, we exert what is known as a **force**. The force that we exert has both magnitude (we either pull hard or we pull gently) and direction (we open or close the door). Both the size of the pull and its direction are important in determining its effect. Such quantities are said to be **vectors**. Other examples of vectors are velocity, acceleration and displacement. The mathematics that will be developed in this section can be applied to problems involving any type of vector quantity.

EXERCISES 12.1

The following situations need to be described using an appropriate measure. Classify the measure as a scalar (s) or a vector (v).

1. A classroom chair is moved from the front of the room to the back.

2. The balance in a bank account.

3. The electric current passing through an electric light tube.

4. A dog, out for a walk, is being restrained by a lead.

5. An aircraft starts its takeoff run.

6. The wind conditions before a yacht race.

7. The amount of liquid in a jug.

8. The length of a car.

12.2 REPRESENTING VECTORS

12.2.1 DIRECTED LINE SEGMENT

There are a number of commonly used notations for vectors:

Notation 1:

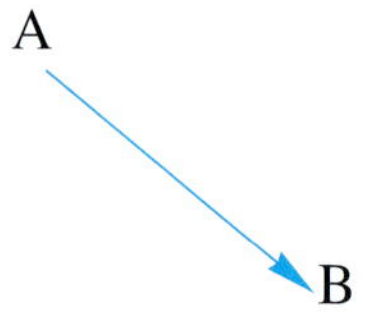

This vector runs from A to B and is depicted as $\overrightarrow{AB}$ or **AB** with the arrow giving the direction of the vector. Point A is known as the tail of the vector $\overrightarrow{AB}$ and point B is known as the head of vector $\overrightarrow{AB}$.

We also say the $\overrightarrow{AB}$ is the position vector of B relative to (from) A.

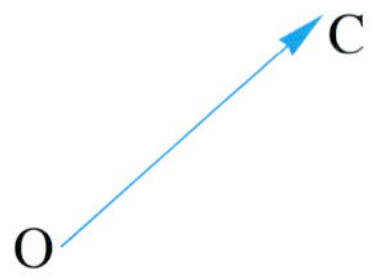

In the case where a vector starts at the origin (O), the vector running from O to another point C is simply called the position vector of C, $\overrightarrow{OC}$ or **OC**.

Notation 2:

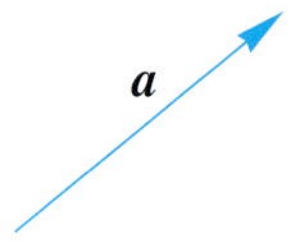

Rather than using two reference points, A and B, as in notation 1, we can also refer to a vector by making reference to a single letter attached to an arrow. In essence we are ‘naming’ the vector.

The vector $\boldsymbol{a}$ can be expressed in several ways. In text books they are often displayed in bold type, however, in written work, the following notations are generally used:

$$\boldsymbol{a} = \underset{\sim}{a} = \vec{a}$$

We will consider another vector notation later in this chapter.

12.2.2 MAGNITUDE OF A VECTOR

The magnitude or modulus of a vector is its length, which is the distance between its tail and head. We denote the magnitude of **AB** by $|\mathbf{AB}|$ (or more simply by AB).

Similarly, if we are using vector notation 2, we may denote the magnitude of $\boldsymbol{a}$ by $|\boldsymbol{a}| = a$.

Note then that $|\boldsymbol{a}| \geq 0$.

12.2.3 EQUAL VECTORS

Two vectors $\boldsymbol{a}$ and $\boldsymbol{b}$ are said to be equal if they have the same direction and the same magnitude.
i.e., $\boldsymbol{a} = \boldsymbol{b}$ if they point in the same direction and $|\boldsymbol{a}| = |\boldsymbol{b}|$.

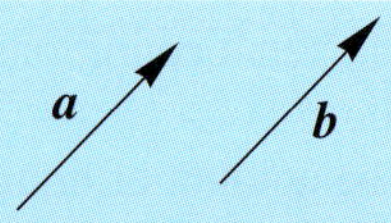

Notice that if $\boldsymbol{a} = \boldsymbol{b}$, then vector $\boldsymbol{b}$ is simply a translation of vector $\boldsymbol{a}$. Using this notation, where there is no reference to a fixed point in space, we often use the word *free* vectors. That is, **free vectors** are vectors that have no specific position associated with them. In the diagram below, although the four vectors occupy a different space, they are all equal.

Note that we can also have that the vectors **AB** = **CD**, so that although they do not have the same starting point (or ending point) they are still equal because their magnitudes are equal and they have the same direction.

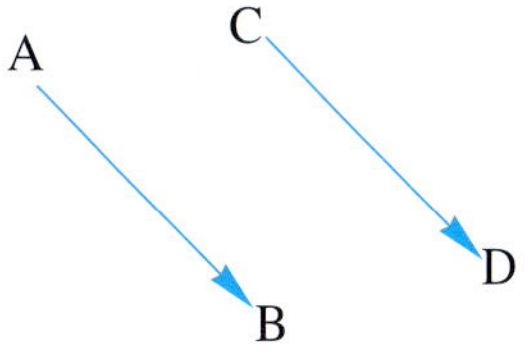

12.2.4 NEGATIVE VECTORS

The **negative of a vector** $\boldsymbol{a}$, denoted by '$-\boldsymbol{a}$' is the vector $\boldsymbol{a}$ but pointing in the opposite direction to $\boldsymbol{a}$.

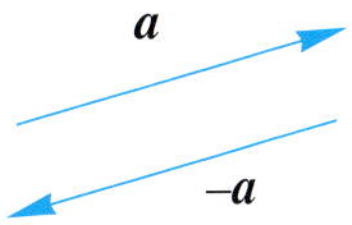

Similarly, the negative of **AB** is –**AB** or **BA**, because rather than starting at A and ending at B the negative of **AB** starts at B and ends at A.

Note that $|\boldsymbol{a}| = |-\boldsymbol{a}|$ and $|\mathbf{AB}| = |-\mathbf{AB}| = |\mathbf{BA}|$.

12.2.5 ZERO VECTOR

The zero vector has zero magnitude, $|\mathbf{0}| = 0$ and has no definite direction. It is represented geometrically by joining a point onto itself. Note then that for any non-zero vector $\boldsymbol{a}$, $|\boldsymbol{a}| > 0$

12.2.6 ORIENTATION AND VECTORS

Vectors are very useful when representing positions relative to some starting point. Consider

(a) the position of a man who has walked 2.8 km across a field in a direction East 30° S or

(b) a car moving at 20 km/h in a direction West 40° N for 2 hours.

Each of these descriptions can be represented by a vector.

We start by setting up a set of axes and then we represent the above vectors showing the appropriate direction and magnitude. Representing the magnitude can be done using a scale drawing or labelling the length of the vector.

(a)

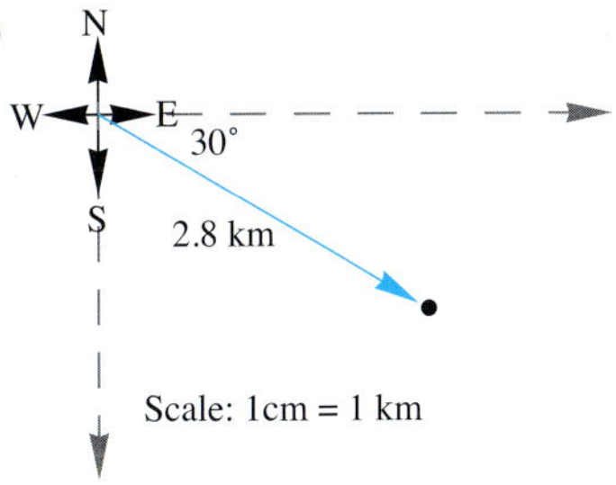

(b)

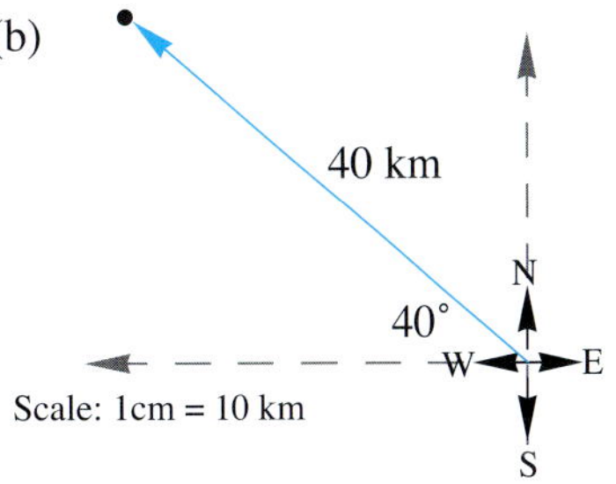

EXAMPLE 12.1

Find the position of a bushwalker if on the first part of her journey she walks 2.8 km across a field in a direction East 30° South and then continues for a further 4 km in a Northerly direction.

Solution

We start by representing her journey using a vector diagram:

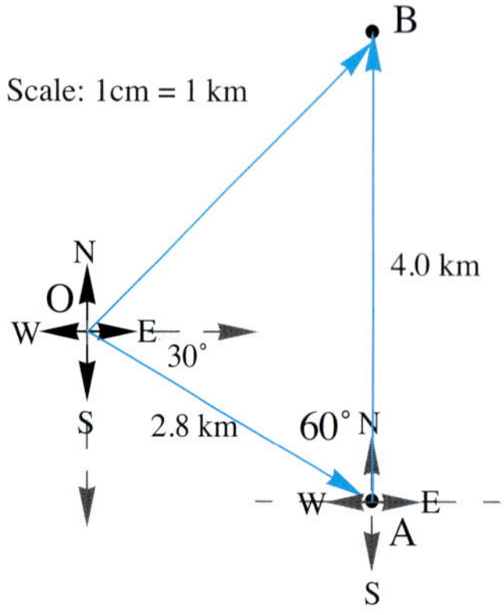

The first part of her journey is represented by vector **OA** and the second part by **AB**. Note then that because her final position is at point B, her final position, relative to O, is given by the vector **OB**.

All that remains is to find the direction of **OB** and its magnitude. To do this we make use of trigonometry.

Finding $|\mathbf{OB}|$:

Using the cosine rule we have

$$OB^2 = OA^2 + AB^2 - 2(AB)(OA)\cos(60°)$$
$$OB^2 = 2.8^2 + 4.0^2 - 2 \times 2.8 \times 4.0 \times 0.5$$
$$= 12.64$$
$$\therefore OB = 3.56$$

Next, we find the angle BOA:

$$AB^2 = OA^2 + OB^2 - 2(OA)(OB)\cos(\angle BOA)$$
$$4.0^2 = 2.8^2 + 12.64 - 2(2.8)(\sqrt{12.64})\cos(\angle BOA)$$
$$\therefore \cos(\angle BOA) = \frac{2.8^2 + 12.64 - 4.0^2}{2(2.8)(\sqrt{12.64})}$$
$$\angle BOA = \cos^{-1}(0.2250)$$
$$= 76°59'45''$$
$$\approx 77°$$

That is, the bushwalker is 3.56 km E 47° N from her starting point.

Although we will investigate the algebra of vectors in the next section, with Example 26.1 we have already looked at adding two vectors informally. That is, the final vector **OB** was found by *joining* the vectors **OA** and **AB** . Writing this in vector form we have, **OB** = **OA** + **AB** .

To add two vectors, ***a*** and ***b***, geometrically we

1. first draw ***a***,
2. draw vector ***b*** so that its tail meets the arrow end of vector ***a***,
3. draw a line segment from the tail of vector ***a*** to the arrow end of vector ***b***.

This vector then represents the result ***a*** + ***b***.

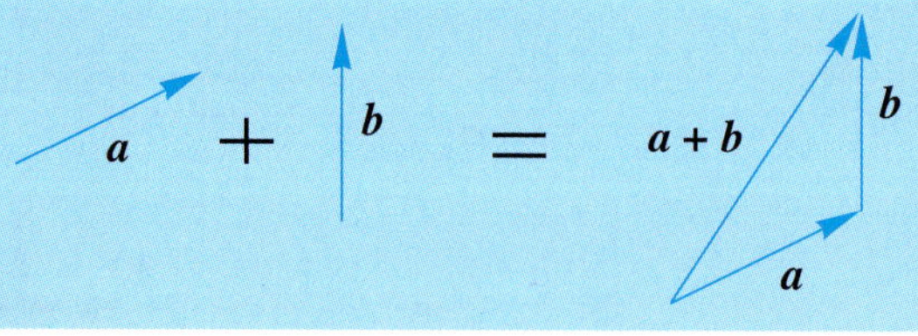

EXAMPLE 12.2

For the equilateral triangle shown, express in terms of ***a*** and ***b***

(a) **CA** (c) **BC** (d) |**AB** + **BC**|

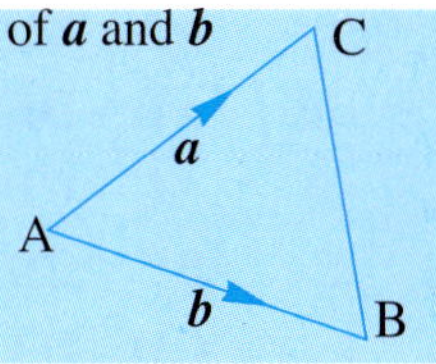

Solution

(a) **CA** = –**AC** = –***a***.

(b) To get from B to C we first get from B to A and then from A to C. That is, we 'join' the vectors **BA** and **AC**. In vector notation we have:

$$\mathbf{BC} = \mathbf{BA} + \mathbf{AC}$$

However, $\mathbf{AB} = \boldsymbol{b} \Rightarrow \mathbf{BA} = -\mathbf{AB} = -\boldsymbol{b}$

$$\therefore \mathbf{BC} = -\boldsymbol{b} + \boldsymbol{a}$$

(c) $\mathbf{AB} + \mathbf{BC} = \mathbf{AC} = \boldsymbol{a} \therefore |\mathbf{AB} + \mathbf{BC}| = |\boldsymbol{a}|$.

EXERCISES 12.2

1. Using a scale of 1 cm representing 10 units sketch the vectors that represent

(a) 30 km in a Westerly direction.

(b) 20 newtons applied in a NS direction.

(c) 15 m/s N 60° E.

(d) 45 km/h W 30° S.

2. The vector  represents a velocity of 20 m/s due West. Represent the following vectors:

(a) 20 m/s due East

(b) 40 m/s due West

(c) 60 m/s due East

(d) 40 m/s due NE

3. State which of the vectors shown

(a) have the same magnitude.

(b) are in the same direction.

(c) are in the opposite direction.

(d) are equal.

(e) are parallel.

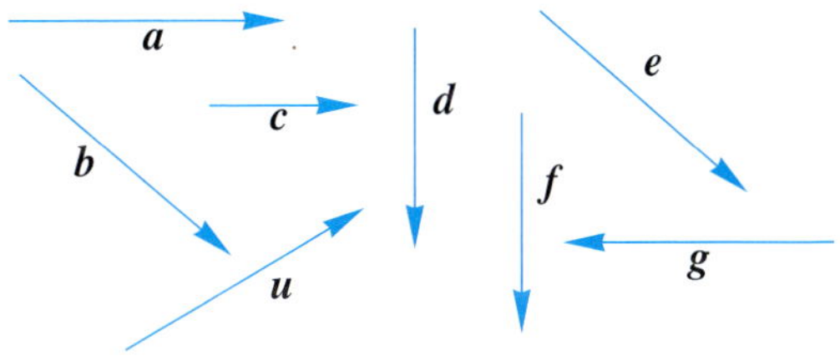

4. For each of the following pairs of vectors, find ***a*** + ***b***.

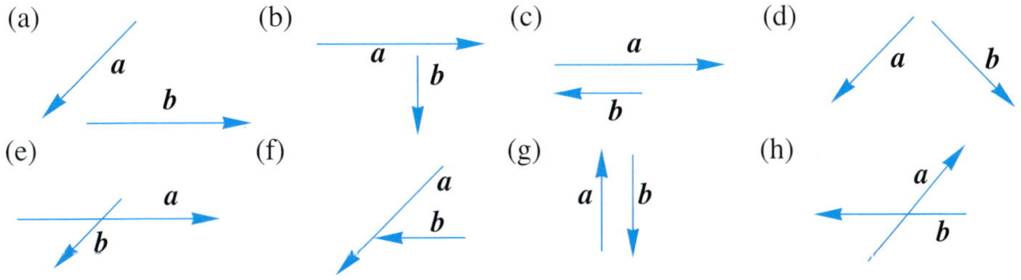

5. For the shape shown, find a single vector which is equal to

(a) **AB** + **BC**

(b) **AD** + **DB**

(c) **AC** + **CD**

(d) **BC** + **CD** + **DA**

(e) **CD** + **DA** + **AB** + **BC**

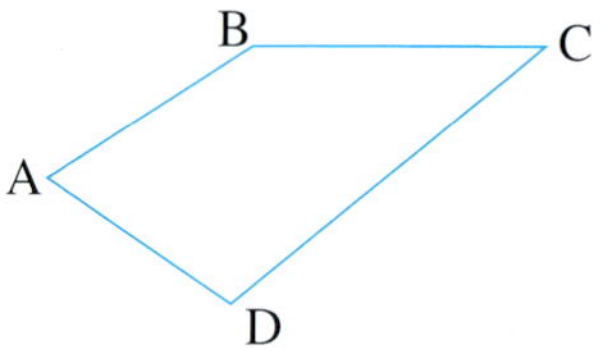

6. Consider the parralelogram shown alongside.
Which of the following statements are true?

(a) **AB** = **DC** (b) $|\boldsymbol{a}| = |\boldsymbol{b}|$

(c) **BC** = $\boldsymbol{b}$ (d) $|\mathbf{AC} + \mathbf{CD}| = |\boldsymbol{b}|$

(e) **AD** = **CB**

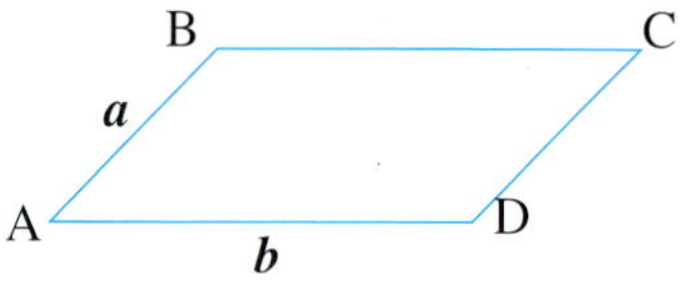

7. For each of the following

i. complete the diagram by drawing the vector **AB** + **BC**.

ii. find $|\mathbf{AB} + \mathbf{BC}|$.

(a)

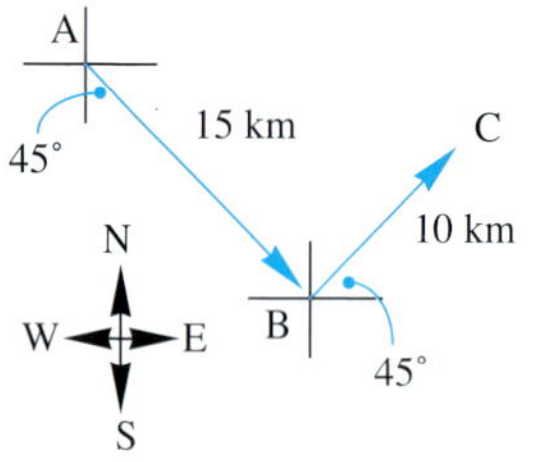

(b)

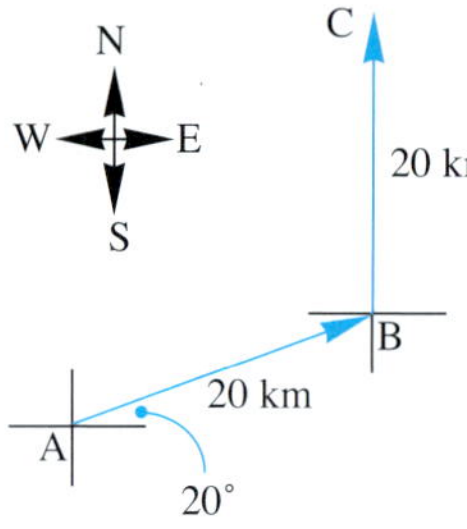

(c)

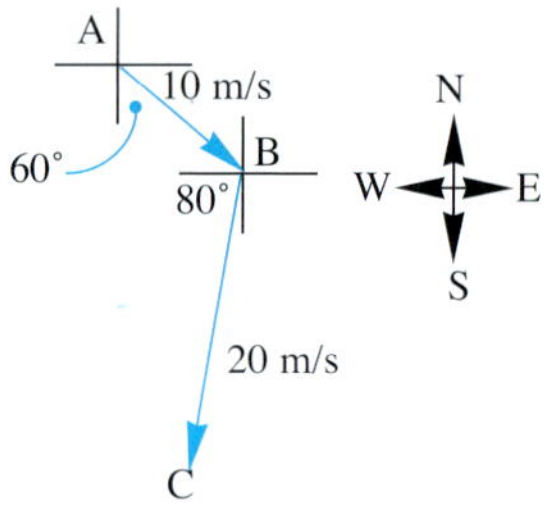

8. Two forces, one of 40 Newtons acting in a Northerly direction and one of 60 Newtons acting in an Easterly direction, are applied at a point A. Draw a vector diagram representing the forces. What is the resulting force at A?

9. Two trucks, on opposite sides of a river, are used to pull a barge along a straight river. They are connected to the barge at one point by ropes of equal length. The angle between the two ropes is 50°. Each truck is pulling with a force of 1500 Newtons.

(a) Draw a vector diagram representing this situation.

(b) Find the magnitude and direction of the force acting on the barge.

10. An aircraft is flying at 240 km/h in a Northerly direction when it encounters a 40 km/h wind from

i. the North.

ii. the North-East.

(a) Draw a vector diagram representing these situations.

(b) In each case, find the actual speed and direction of the aircraft.

11. Patrick walks for 200 m to point P due East of his cabin at point O, then 300 m due North where he reaches a vertical cliff, point Q. Patrick then climbs the 80 m cliff to point R.

(a) Draw a vector diagram showing the vectors **OP**, **PQ** and **QR**.

(b) Find i. $|\mathbf{OQ}|$

ii. $|\mathbf{OR}|$

12.3 ALGEBRA AND GEOMETRY OF VECTORS

12.3.1 ADDITION OF VECTORS

We have already had an 'informal' look at the addition of vectors in §26.2.5. We now take this a step further.

The vector sum of two vectors $\boldsymbol{a}$ and $\boldsymbol{b}$ is given by the unique vector $\boldsymbol{c}$ (also known as the resultant) by using the **Triangle Law of Addition** or the **Parallelogram Law of Addition**.

Triangle Law of Addition

Method:

1. Vector $\boldsymbol{b}$ is translated so that its tail coincides with the head of vector $\boldsymbol{a}$.
2. Then, the vectors sum of $\boldsymbol{a}$ and $\boldsymbol{b}$ is the vector $\boldsymbol{c}$ which closes the triangle.

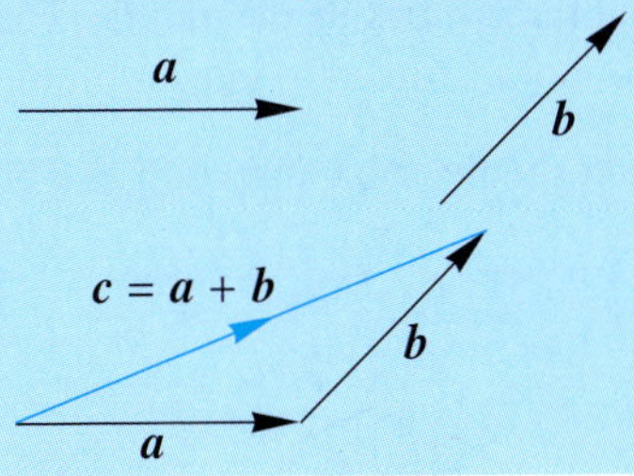

Parallelogram Law of Addition

Method:

1. Vector $\boldsymbol{b}$ is translated so that its tail coincides with the tail of vector $\boldsymbol{a}$.
2. The vectors $\boldsymbol{a}$ and $\boldsymbol{b}$ then form adjacent sides of a parallelogram, so that their sum, vector $\boldsymbol{c}$, is represented by the diagonal as shown.

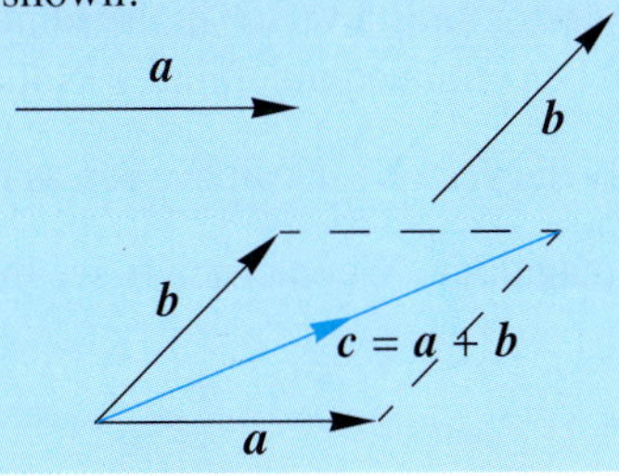

It should be noted:

1. $|\boldsymbol{a}+\boldsymbol{b}|$ does not necessarily equal $|\boldsymbol{a}|+|\boldsymbol{b}|$. In fact, $|\boldsymbol{a}+\boldsymbol{b}| \leq |\boldsymbol{a}|+|\boldsymbol{b}|$ – which is known as the Triangle Inequality.
   Under what circumstances will $|\boldsymbol{a}+\boldsymbol{b}| = |\boldsymbol{a}|+|\boldsymbol{b}|$?

2. **Properties of vector addition**

Closure	Since $\boldsymbol{a}+\boldsymbol{b}$ is a vector, the operation of vector addition **is closed**.
Commutative	$\boldsymbol{a}+\boldsymbol{b} = \boldsymbol{b}+\boldsymbol{a}$
Associative	$(\boldsymbol{a}+\boldsymbol{b})+\boldsymbol{c} = \boldsymbol{a}+(\boldsymbol{b}+\boldsymbol{c})$
Additive identity	There exists a unique zero vector, $\mathbf{0}$ (identity vector under addition), such that $\boldsymbol{a}+\mathbf{0} = \mathbf{0}+\boldsymbol{a} = \boldsymbol{a}$.
Additive Inverse	The inverse vector under addition is that vector which, when added to any vector, gives the zero vector. For any vector $\boldsymbol{a}$ then, the vector $-\boldsymbol{a}$ (the negative of $\boldsymbol{a}$) is called the additive inverse of $\boldsymbol{a}$, so that $\boldsymbol{a}+(-\boldsymbol{a}) = (-\boldsymbol{a})+\boldsymbol{a} = \mathbf{0}$

12.3.2 SUBTRACTION OF VECTORS

Using the properties of vector addition we are now in a position to define the difference $\boldsymbol{a} - \boldsymbol{b}$ of two vectors $\boldsymbol{a}$ and $\boldsymbol{b}$. Rather than subtracting vector $\boldsymbol{b}$ from vector $\boldsymbol{a}$, we add vector $-\boldsymbol{b}$ to vector $\boldsymbol{a}$.

That is, $$\boldsymbol{a} - \boldsymbol{b} = \boldsymbol{a} + (-\boldsymbol{b})$$

Therefore, to subtract vector $\boldsymbol{b}$ from vector $\boldsymbol{a}$ we reverse the direction of vector $\boldsymbol{b}$ and then add it to vector $\boldsymbol{a}$ according to the triangle law (or parallelogram law):

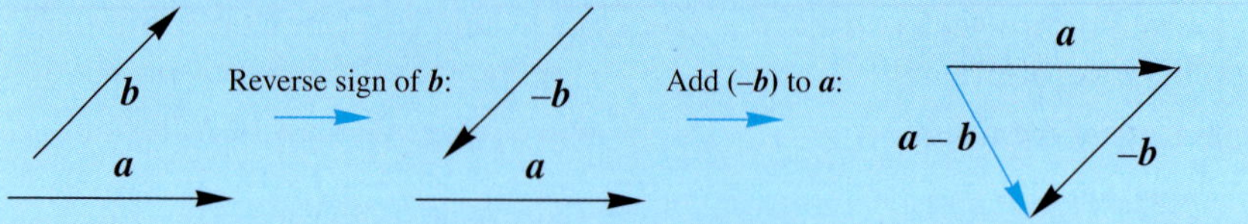

12.3.3 MULTIPLICATION OF VECTORS BY SCALARS

Multiplication of a vector by a scalar is best seen as the repeated use of the addition of the same vector. That is, the vector $\boldsymbol{a} + \boldsymbol{a}$, is by definition a vector having the same direction as $\boldsymbol{a}$ but twice its length and so, we can write $\boldsymbol{a} + \boldsymbol{a}$ as $2\boldsymbol{a}$. We can then extend this to $\boldsymbol{a} + \boldsymbol{a} + \boldsymbol{a} = 2\boldsymbol{a} + \boldsymbol{a} = 3\boldsymbol{a}$ so that the vector $3\boldsymbol{a}$ is a vector in the same direction as $\boldsymbol{a}$ but three times its magnitude. The process can then be continued indefinitely. Notice then that this is not restricted to integer multiples.

Similarly the vector $\frac{1}{2}\boldsymbol{a}$ can be seen as representing a vector in the same direction as $\boldsymbol{a}$ but having half its magnitude. Geometrically we then have:

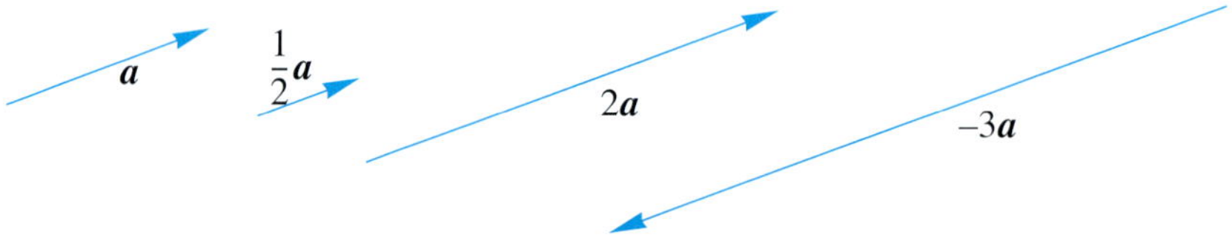

We can formalise scalar multiplication as follows:

For any scalar k, the product $k\boldsymbol{a}$ is a vector parallel to $\boldsymbol{a}$ whose magnitude is $|k|$ times that of $\boldsymbol{a}$. i.e., $|k\boldsymbol{a}| = |k||\boldsymbol{a}|$.
In particular,

1. if $k > 0$ then $k\boldsymbol{a}$ and $\boldsymbol{a}$ have the same direction.
2. if $k < 0$ then $k\boldsymbol{a}$ and $\boldsymbol{a}$ are in opposite direction.
3. if $k = 0$ then $k\boldsymbol{a} = \boldsymbol{0}$, the zero vector.

It should be noted:

1. From the above definition, we can restate the condition for two vectors to be parallel:
   i.e., $\boldsymbol{a} \parallel \boldsymbol{b}$ if $\boldsymbol{a} = k\boldsymbol{b}$
   Where, $\boldsymbol{a}$ and $\boldsymbol{b}$ are in the same direction if $k > 0$ (i.e., parallel) and in opposite directions if $k < 0$ (i.e., anti-parallel).

2. **Properties of multiplication by a scalar**

   Closure Since $k\boldsymbol{a}$ is a vector, multiplication by a scalar **is closed**.

Commutative $a \times k = k \times a$

Associative $(k \times m)\boldsymbol{a} = k \times (m\boldsymbol{a})$

Distributive
1. $(k+m)\boldsymbol{a} = k\boldsymbol{a} + m\boldsymbol{a}$
2. $k(\boldsymbol{a}+\boldsymbol{b}) = k\boldsymbol{a} + k\boldsymbol{b}$

Geometrically, the distributive properties can be seen as follows:

1. $(k+m)\boldsymbol{a} = k\boldsymbol{a} + m\boldsymbol{a}$

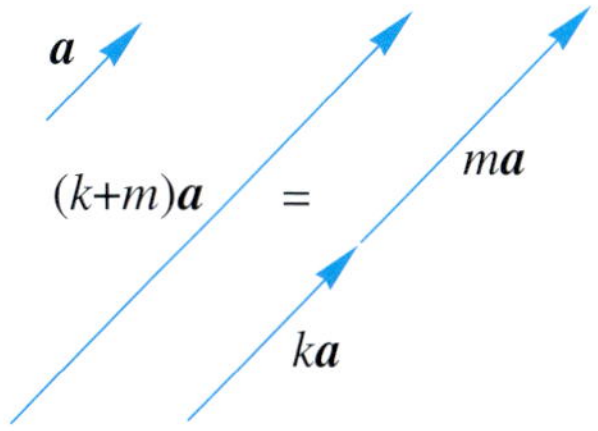

2. $k(\boldsymbol{a}+\boldsymbol{b}) = k\boldsymbol{a} + k\boldsymbol{b}$

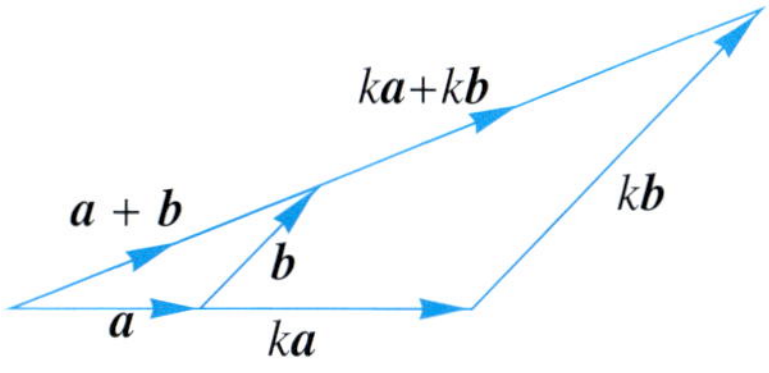

12.3.4 ANGLE BETWEEN VECTORS

For the vectors $\boldsymbol{a} = \mathbf{OA}$ and $\boldsymbol{b} = \mathbf{OB}$, the **angle between** the vectors $\boldsymbol{a}$ and $\boldsymbol{b}$, θ, is given by the angle AOB, and is taken to be positive when measured in an anticlockwise direction.

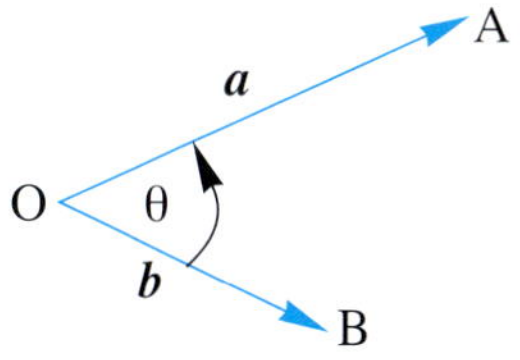

Note then that to find such an angle, the **vectors must first be joined tail to tail**.

12.3.5 APPLICATIONS TO GEOMETRY

As already seen in §26.2 vectors have clear applications to problems that involve geometry and trigonometry. In this section we concentrate on how the operations on vectors defined so far can be used to prove geometrical properties. Unlike geometric proofs encountered in the earlier years of schooling, vectors may be used very neatly to prove many of these theorems. We shall illustrate this by means of several examples.

EXAMPLE 12.3

Using the vectors shown, draw the vector diagrams showing

(a) $\boldsymbol{a} + 2\boldsymbol{b}$ (b) $-\boldsymbol{a} + \frac{1}{2}\boldsymbol{b}$ (c) $3\boldsymbol{a} - 2\boldsymbol{b}$

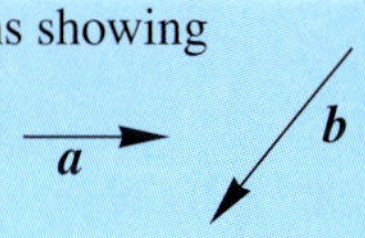

Solution

(a) We first need the vector $2\boldsymbol{b}$ and then we add it to the vector $\boldsymbol{a}$.

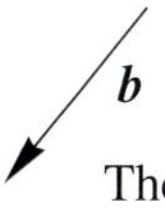

Then, $2\boldsymbol{b}$ is

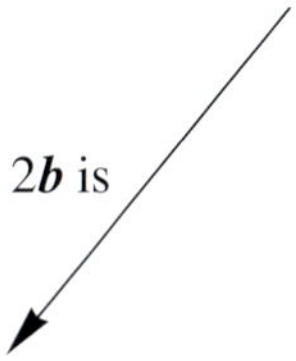

Method 1:

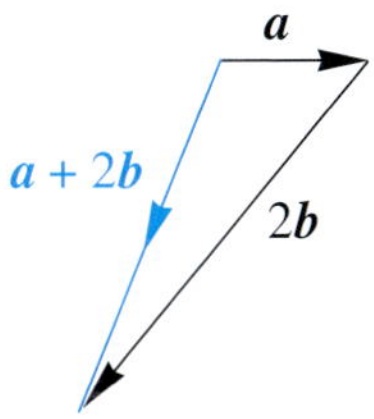

Method 2:

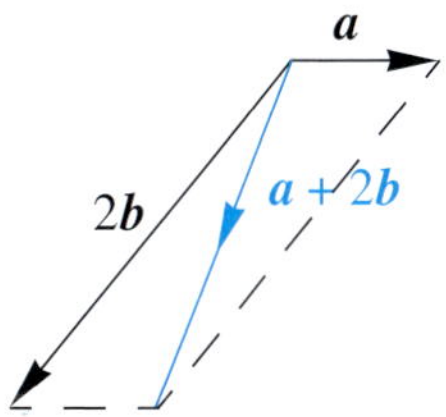

(b) This time we need the vectors $-\boldsymbol{a}$ and $\frac{1}{2}\boldsymbol{b}$

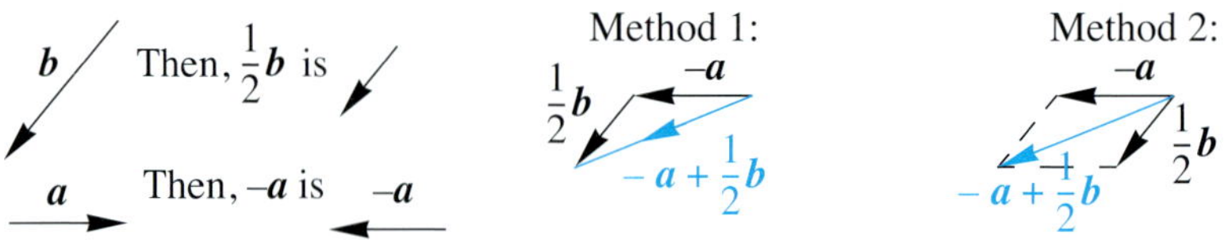

(c) For this problem we need the vectors $3\boldsymbol{a}$ and $-2\boldsymbol{b}$.

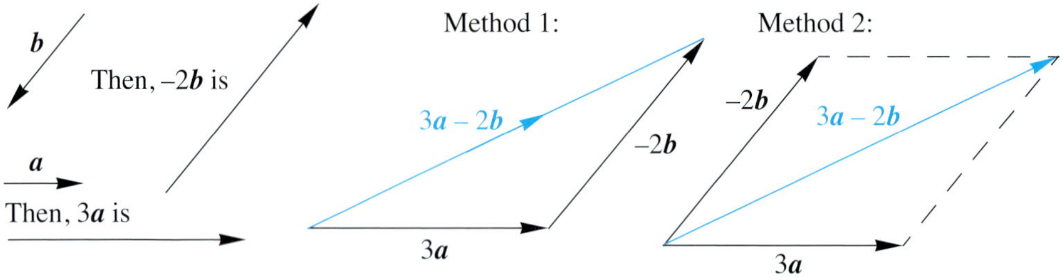

EXAMPLE 12.4

The rectangle OABC is such that OA = 8, AB = 6 and **OA** = $\boldsymbol{a}$ and **OC** = $\boldsymbol{b}$.

(a) Express the two diagonals in terms of $\boldsymbol{a}$ and $\boldsymbol{b}$.

(b) Find i. $|\mathbf{OB}|$ ii. $|\mathbf{OB}+\mathbf{CA}|$.

(c) Find $\mathbf{OA}+\mathbf{AC}+\mathbf{CO}$.

Solution

(a) We start by drawing the figure under discussion:

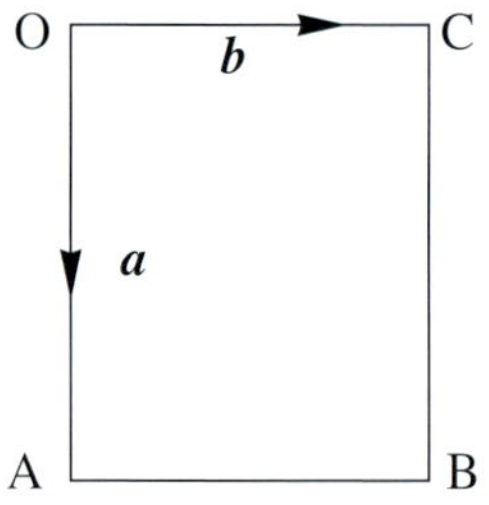

The diagonals are:

$$\begin{aligned}\mathbf{OB} &= \mathbf{OA}+\mathbf{AB}\\ &= \mathbf{OA}+\mathbf{OC}\\ &= \boldsymbol{a}+\boldsymbol{b}\end{aligned}$$

and

$$\begin{aligned}\mathbf{AC} &= \mathbf{AB}+\mathbf{BC}\\ &= \mathbf{OC}+(-\mathbf{CB})\\ &= \mathbf{OC}+(-\mathbf{OA})\\ &= \boldsymbol{b}-\boldsymbol{a}\end{aligned}$$

(b) i. $|\mathbf{OB}| = \sqrt{|\mathbf{OC}|^2+|\mathbf{CB}|^2} = \sqrt{8^2+6^2} = 10$ [using Pythagoras's Theorem]

ii.

$$\begin{aligned}\mathbf{OB}+\mathbf{CA} &= (\mathbf{OC}+\mathbf{CB})+(\mathbf{CO}+\mathbf{OA})\\ &= (\mathbf{OC}+\mathbf{CO})+\mathbf{CB}+\mathbf{OA}\\ &= (\mathbf{OC}+(-\mathbf{OC}))+\boldsymbol{a}+\boldsymbol{a}\\ &= (\mathbf{0})+2\boldsymbol{a}\\ &= 2\boldsymbol{a}\end{aligned}$$

$\therefore |\mathbf{OB}+\mathbf{CA}| = |2\boldsymbol{a}| = 2|\boldsymbol{a}| = 2\times 8 = 16$

(c)

$$\begin{aligned}\mathbf{OA}+\mathbf{AC}+\mathbf{CO} &= \mathbf{OA}+(\mathbf{AO}+\mathbf{OC})+\mathbf{CO}\\ &= (\mathbf{OA}+\mathbf{AO})+(\mathbf{OC}+\mathbf{CO})\\ &= \mathbf{0}+\mathbf{0}\\ &= \mathbf{0}\end{aligned}$$

Notice that in Example 26.4 we consistently kept re-expressing the diagonals in terms of the sides of the rectangle. In fact you will find that for many problems that involve geometric shapes, it helps to express vectors in terms of the sides of the shape under discussion.

EXAMPLE 12.5

Simplify the following expressions

(a) **AB** + **BC** + **DE** + **CD**

(b) 2**AB** + 4**AC** + 4**BA** + 6**CD** + 2**DA** + 2**DC**

Solution

(a) In this case we have no given shape to make reference to, however, we can still simplify the sum by referring to the head and tail of each vector and 'grouping' a head to a corresponding tail:

AB + **BC** + **DE** + **CD** = **AB** + **BC** + **CD** + **DE** [rearranging]
= **AC** + **CE** [grouping]
= **AE**

That is, when adding **AB** and **BC** we observe that one vector ends at B while the second starts from B, this becomes a common point and so we 'group' those two vectors, then we complete the triangle ABC, i.e., **AB** + **BC** = **AC**.

(b) 2**AB** + 4**AC** + 4**BA** + 6**CD** + 2**DA** + 2**DC**
= 2**AB** + 4**BA** + 4**AC** + 6**CD** + 2**DA** + 2**DC**
= 2**AB** + 2**BA** + 2**BA** + 4**AC** + 4**CD** + 2**CD** + 2**DA** + 2**DC**
= 2**AB** – 2 **AB** + 2**BA** + 4(**AC** + **CD**) + 2(**CD** + **DC**) + 2**DA**
= **0** + 2**BA** + 4**AD** + 2(**0**) + 2**DA**
= 2**BA** + 2**AD** + 2**AD** + 2**DA**
= 2(**BA** + **AD**) +2(**DA** + **AD**)
= 2**BD** + 2(**0**)
= 2**BD**

EXAMPLE 12.6

Prove that the line segment joining the mid-points of two sides of a triangle is parallel to the third side and half its length.

Solution

Consider the triangle ABC where D and E are the mid-points of $\overline{AC}$ and $\overline{BC}$ respectively.

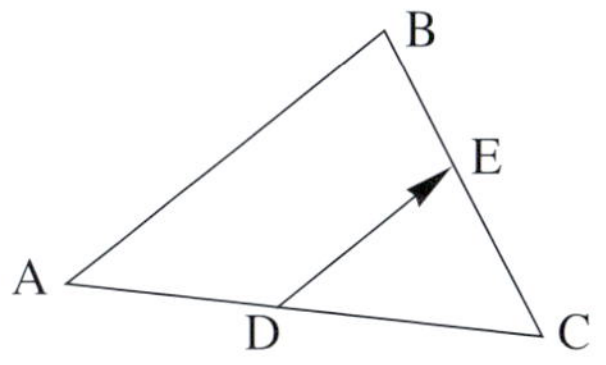

We then need to express **DE** in terms of **AB**.

First we note that $\mathbf{DC} = \frac{1}{2}\mathbf{AC}$ and $\mathbf{CE} = \frac{1}{2}\mathbf{CB}$.

Then, we have
$$\mathbf{DE} = \mathbf{DC} + \mathbf{CE}$$
$$= \frac{1}{2}\mathbf{AC} + \frac{1}{2}\mathbf{CB}$$
$$= \frac{1}{2}(\mathbf{AC} + \mathbf{CB})$$
$$= \frac{1}{2}\mathbf{AB}$$

Now, as $\mathbf{DE} = \frac{1}{2}\mathbf{AB}$ we can make two observations

1. $\mathbf{DE} \parallel \mathbf{AB}$
2. $|\mathbf{DE}| = \frac{1}{2}|\mathbf{AB}|$, i.e., **DE** is half the length of **AB**.

Hence we have shown that the line segment joining the mid-points of two sides of a triangle is parallel to the third side and half its length.

EXAMPLE 12.7

Prove that if one pair of opposite sides of a quadrilateral is equal and parallel, then the quadrilateral is a parallelogram.

Solution

Start by drawing a random quadrilateral, i.e., one without any pre-conceived properties. The reason we do this is because had we drawn two opposite sides as being equal and parallel then we would have a parallelogram and then, we might make assumptions during our proof (asumptions that we are trying to prove!)

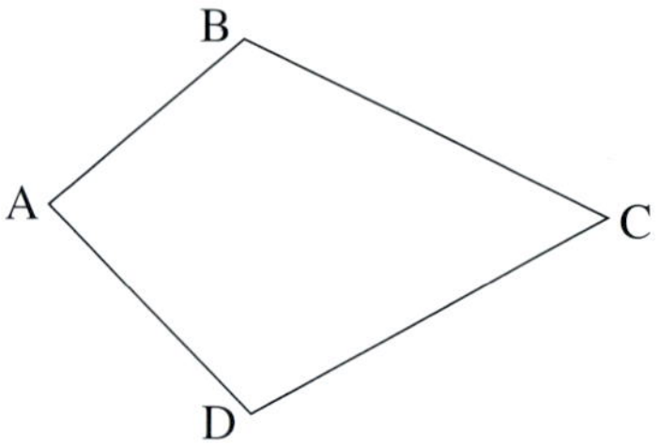

Consider the quadrilateral ABCD as shown:

From the information given, we have that $\mathbf{AB} = \mathbf{DC}$.

Note that we could have chosen any two sides to be parallel and equal (even though the diagram does not reflect this).

Now, $\mathbf{AD} = \mathbf{AB} + \mathbf{BD}$
$= \mathbf{DC} + \mathbf{BD}$ [since $\mathbf{AB} = \mathbf{DC}$]
$= \mathbf{BD} + \mathbf{DC}$ [rearranging terms]
$= \mathbf{BC}$

That is, we have shown that $\mathbf{AD} = \mathbf{BC}$, meaning that the other two sides of the quadrilateral are equal in length and parallel. Then, by definition, the quadrilateral is a parallelogram.

EXAMPLE 12.8

Find the position vector of the point P which divides the line segment AB in the ratio $m:n$.

Solution

As the problem makes reference to the position vector of point P we need to introduce an origin. Let O be this origin. Then, let $\mathbf{OA} = \boldsymbol{a}$, $\mathbf{OB} = \boldsymbol{b}$ and $\mathbf{OP} = \boldsymbol{p}$ where AP:PB = $m:n$.

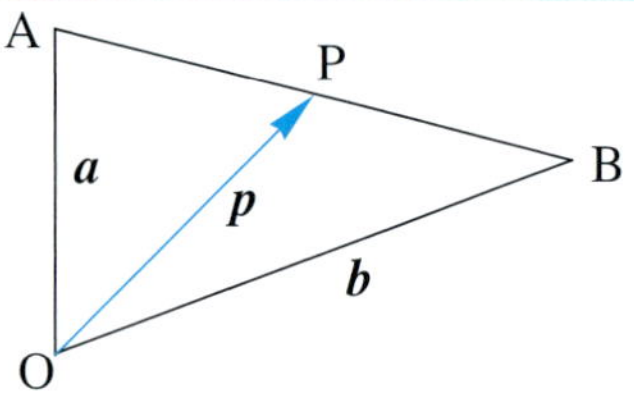

Then, $\mathbf{OP} = \mathbf{OA} + \mathbf{AP} = \mathbf{OA} + \frac{m}{m+n}\mathbf{AB}$

$$= \boldsymbol{a} + \frac{m}{m+n}(\mathbf{AO} + \mathbf{OB})$$

$$= \boldsymbol{a} + \frac{m}{m+n}(-\boldsymbol{a} + \boldsymbol{b})$$

That is, $\mathbf{OP} = \left(\frac{m+n}{m+n}\right)\boldsymbol{a} + \frac{m}{m+n}(-\boldsymbol{a}+\boldsymbol{b})$

$= \frac{1}{m+n}(m\boldsymbol{a} + n\boldsymbol{a} - m\boldsymbol{a} + m\boldsymbol{b})$

$= \frac{1}{m+n}(n\boldsymbol{a} + m\boldsymbol{b})$

EXAMPLE 12.9

Prove that the diagonals of a parallelogram bisect each other.

Solution

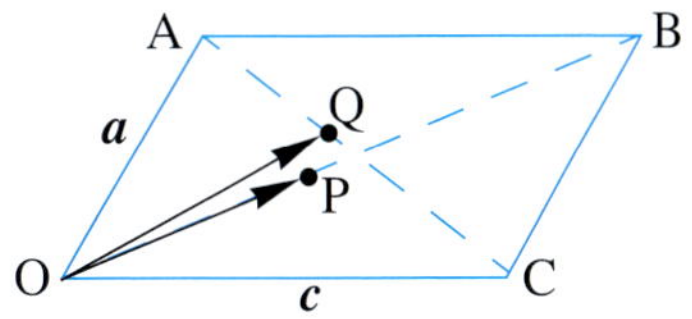

Consider the parallelogram OABC where $\mathbf{OA} = \boldsymbol{a}$ and $\mathbf{OC} = \boldsymbol{c}$. Let P be the mid-point of $\overline{\text{OB}}$ and Q be the mid-point of $\overline{\text{AC}}$.

$\mathbf{OQ} = \mathbf{OA} + \mathbf{AQ}$ [Triangular Law of addition]

$= \mathbf{OA} + \frac{1}{2}\mathbf{AC}$ [Q is the mid-point of $\overline{\text{AC}}$]

$= \boldsymbol{a} + \frac{1}{2}(-\boldsymbol{a} + \boldsymbol{c})$ [$\mathbf{AC} = \mathbf{AO} + \mathbf{OC} = -\mathbf{OA} + \mathbf{OC}$]

$\therefore \mathbf{OQ} = \frac{1}{2}(\boldsymbol{a} + \boldsymbol{c})$ – (1)

Next, $\mathbf{OP} = \frac{1}{2}\mathbf{OB}$ [P is the mid-point of $\overline{\text{OB}}$]

$= \frac{1}{2}(\mathbf{OC} + \mathbf{CB})$ [Triangular Law of addition]

$\therefore \mathbf{OP} = \frac{1}{2}(\boldsymbol{c} + \boldsymbol{a})$ – (2)

From (1) and (2) $\mathbf{OP} = \mathbf{OQ}$ meaning that P and Q are the same point and so, as they are also the midpoints of the diagonals, this means that the diagonals of a parallelogram bisect each other.

EXERCISES 12.3

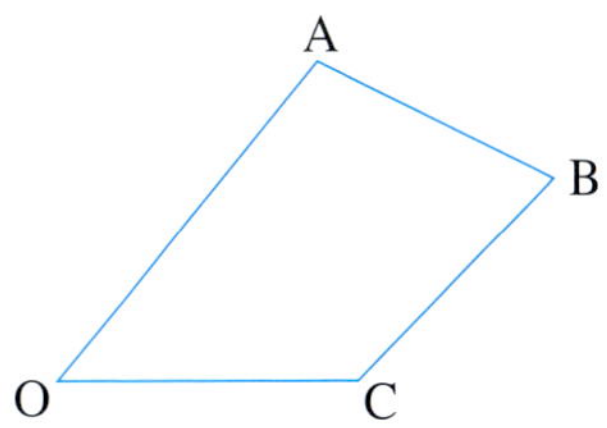

1. For the quadrilateral OABC, where $\mathbf{OA} = \boldsymbol{a}$, $\mathbf{OB} = \boldsymbol{b}$ and $\mathbf{OC} = \boldsymbol{c}$, find in terms of $\boldsymbol{a}$, $\boldsymbol{b}$ and $\boldsymbol{c}$ an expression for

(a) **AC**

(b) **BC**

(c) the mid-point of $\overline{\text{AB}}$ relative to O

2. Using the vectors shown below, draw the vectors

(a) $\boldsymbol{a} - \boldsymbol{b}$

(b) $\boldsymbol{b} - 2\boldsymbol{a}$

(c) $2\boldsymbol{b} - 3\boldsymbol{a}$

(d) $\frac{1}{2}(\boldsymbol{b} + 2\boldsymbol{a})$

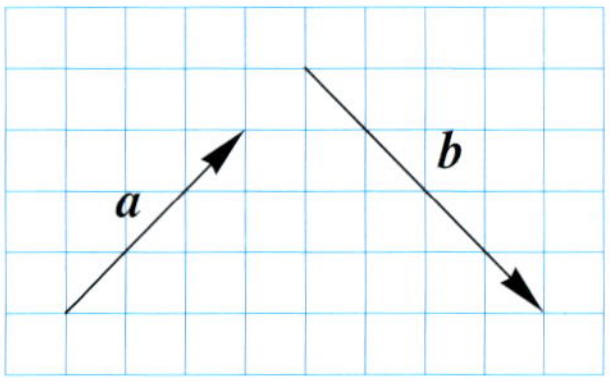

3. Simplify

(a) **WX** + **XY** + **YW**

(b) **PQ** – **SR** + **QR**

(c) **AX** – **BX** + **BZ** + **YD** – **YZ** – **YD**

(d) 3**OA** + 6**BC** + 2**AO** + **AB** + 5**OB**

4. Consider the triangle ABC whose vertices have position vectors $\boldsymbol{a}$, $\boldsymbol{b}$ and $\boldsymbol{c}$ respectively. Find the position vector of

(a) P, the mid-point of $\overline{AB}$.

(b) Q, the point of trisection of $\overline{AB}$, with Q closer to B.

(c) R, the mid-point of the median CP.

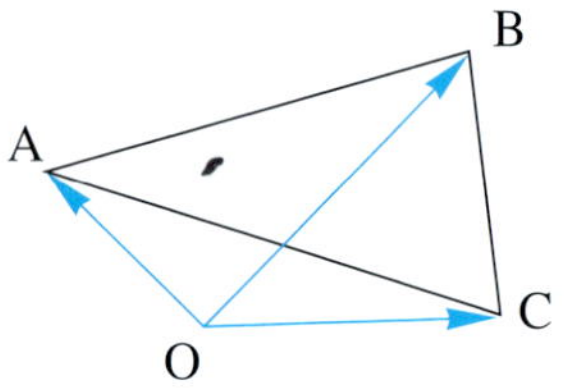

5. Prove that

(a) $\overrightarrow{AB} + \overrightarrow{BO} + \overrightarrow{OA} = \mathbf{0}$

(b) $\overrightarrow{XY} + \overrightarrow{YO} + \overrightarrow{OZ} + \overrightarrow{ZX} = \mathbf{0}$

(c) if $\overrightarrow{AO} + \overrightarrow{OB} = \overrightarrow{BO} + \overrightarrow{OC}$ then A, B, C are collinear

6. If M is the mid-point of $\overline{AB}$ and N is the mid-point of $\overline{CD}$, show that 2**MN** = **AC** + **BD**.

7. Consider the cuboid ABCD, EFGH with vectors as shown on the diagram. Express, in terms of $\boldsymbol{a}$, $\boldsymbol{b}$ and $\boldsymbol{c}$ the following:

(a) **BC**

(b) **AG**

(c) **BH**

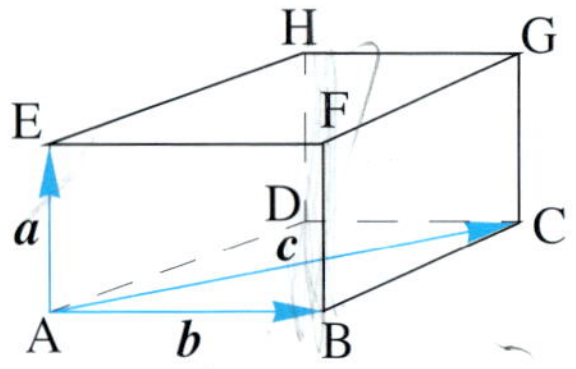

8. In question 7., $|\boldsymbol{a}| = 2$, $|\boldsymbol{b}| = 4$ and $|\boldsymbol{c}| = 10$. Find (a) BC (b) BH

9. Prove that the sum of the vectors from the centre to the vertices of a regular hexagon is **0**.

10. In the diagram alongside, P and Q are the mid-points of WY and XZ respectively.

Show that 4**PQ** = **WX** + **WZ** + **YX** + **YZ**

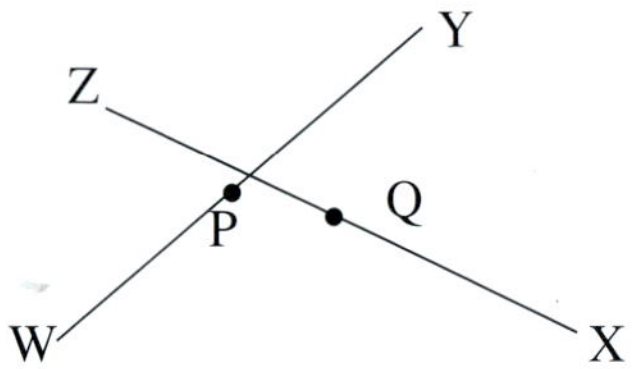

11. Show that the line segments joining the mid-points of the sides of any quadrilateral form a parallelogram.

12. Show that a quadrilateral whose diagonals bisect each other is a parallelogram.

13. A, B C are the mid-points of the sides PQ, QR and RP respectively of the triangle PQR. Show that **OP** + **OQ** + **OR** = **OA** + **OB** + **OC** where O is some origin.

14. Prove that
(a) A and B are coincident points if **MB** + **NM** = **NA** + **BA**.
(b) A, B, C are collinear if **OC** = 3**OB** – 2**OA**.
(c) if ABCD is a quadrilateral then **AC** – **BD** = **AB** – **CD**.

15. If the vectors $\boldsymbol{a} = 2\boldsymbol{u} - 3\boldsymbol{v}$ and $\boldsymbol{b} = 5\boldsymbol{u} + 4\boldsymbol{v}$, find the scalars m and n such that $\boldsymbol{c} = m\boldsymbol{a} + n\boldsymbol{b}$ where $\boldsymbol{c} = 12\boldsymbol{u} + 7\boldsymbol{v}$ and $\boldsymbol{u}$ and $\boldsymbol{v}$ are non-parallel vectors.

16. Consider the parallelogram ABCD where the point P is such that AP:PD = 1:2 and **BD** intersects **CP** at Q where DQ:QB = 1:3.
Find the scalar m if **CP** = m**CQ**.

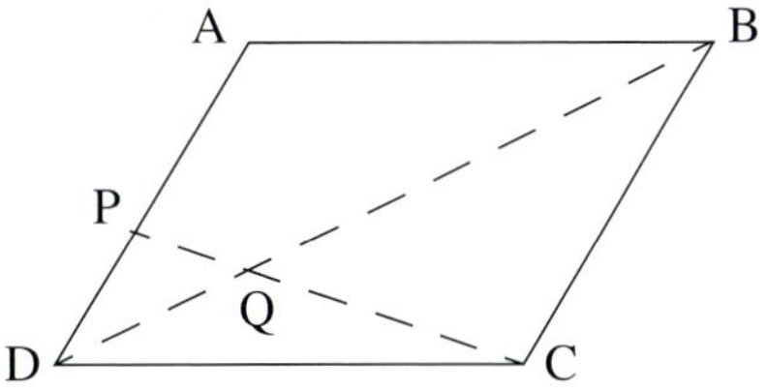

12.4 CARTESIAN REPRESENTATION OF VECTORS IN 2-D AND 3-D

12.4.1 REPRESENTATION IN 2-D

When describing vectors in two-dimensional space it is often helpful to make use of a rectangular Cartesian coordinate system.

As such, the position vector of the point P, **OP**, has the coordinates (x, y).

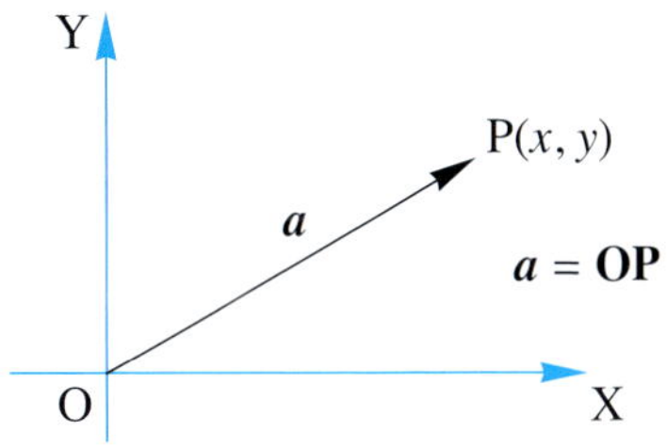

The vector $\boldsymbol{a}$ can be expressed as a **column vector** $\begin{pmatrix} x \\ y \end{pmatrix}$. That is,

> $\boldsymbol{a} = \begin{pmatrix} x \\ y \end{pmatrix}$ is the position vector **OP** where P has the coordinates (x,y).

12.4.2 UNIT VECTOR AND BASE VECTOR NOTATION

We define the unit vector $\boldsymbol{i} = \begin{pmatrix} 1 \\ 0 \end{pmatrix}$ as the position vector of the point having coordinates $(1, 0)$,

and the unit vector $\boldsymbol{j} = \begin{pmatrix} 0 \\ 1 \end{pmatrix}$ as the position vector of the point having coordinates $(0, 1)$.

The term **unit vector** refers to the fact that the vector **has a magnitude of one**.

$$a = \begin{pmatrix} x \\ y \end{pmatrix} = \begin{pmatrix} x \\ 0 \end{pmatrix} + \begin{pmatrix} 0 \\ y \end{pmatrix} = x\begin{pmatrix} 1 \\ 0 \end{pmatrix} + y\begin{pmatrix} 0 \\ 1 \end{pmatrix} = xi + yj$$

i.e., the position vector of any point can be expressed as the sum of two vectors, one parallel to the x–axis and one parallel to the y–axis.

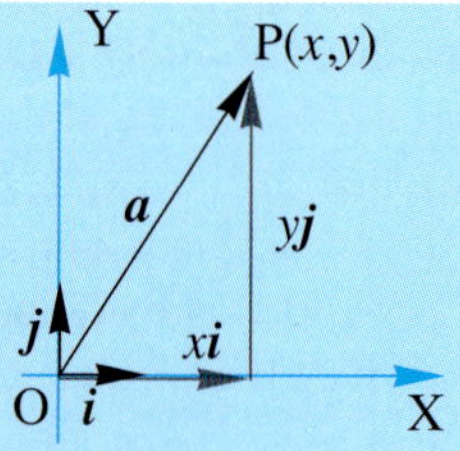

The unit vectors $\boldsymbol{i}$ and $\boldsymbol{j}$ are also known as the **base vectors**. If we confine ourselves to vectors that exist in the plane of this page, the most commonly used basis is:

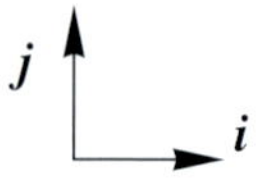

where $|\boldsymbol{i}| = |\boldsymbol{j}| = 1$

Notice the definite direction of the base vectors, i.e., $\boldsymbol{i}$ points in the positive x–axis direction while $\boldsymbol{j}$ points in the positive y–axis direction.

Vectors can now be expressed in terms of these base vectors.

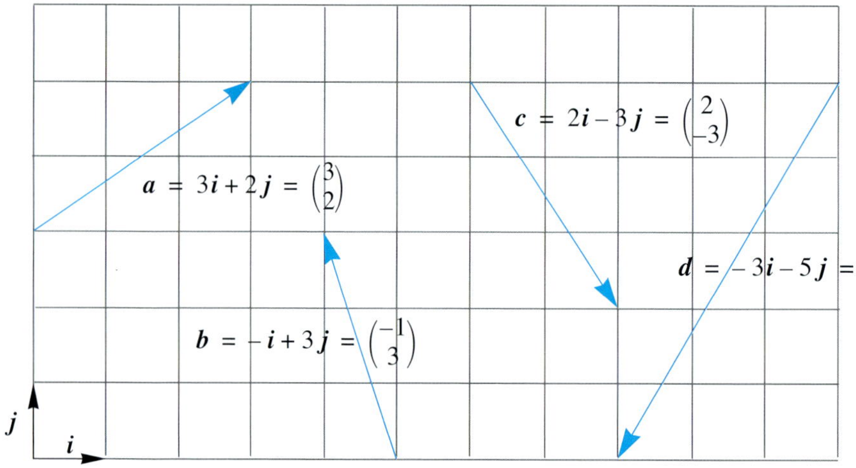

The vector $\boldsymbol{a}$ is 'three steps to the right and two steps up' and can be written in terms of the standard basis as $\boldsymbol{a} = 3\boldsymbol{i} + 2\boldsymbol{j}$.

The vector $\boldsymbol{b}$ is 'one step to the left and three steps up'. 'One step to the left' is in the opposite direction of the basis element $\boldsymbol{i}$ and is written $-\boldsymbol{i}$, giving the definition of the vector $\boldsymbol{b} = -\boldsymbol{i} + 3\boldsymbol{j}$. The vectors $-\boldsymbol{i}$ and $3\boldsymbol{j}$ are known as **components** of the vector $\boldsymbol{b}$.

The other definitions follow in a similar way.

12.4.3 REPRESENTATION IN 3-D

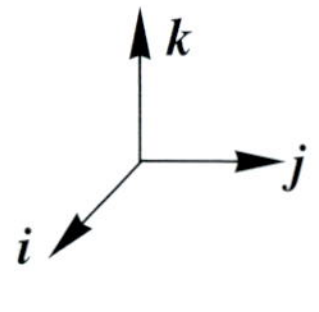

When vectors are represented in three dimensional space, a third vector must be added to the basis, in this case it is a unit vector $\boldsymbol{k}$ and is such that the three unit vectors are mutually perpendicular as shown.

In addition, extra basis vectors can be added to generate higher dimensional vector spaces. These may not seem relevant to us, inhabiting as we do, a three dimensional space. However, it remains the case that it is possible to do calculations in higher dimensional spaces and these have produced many valuable results for applied mathematicians.

As was the case for vectors in two dimensions, we can represent vectors in three dimensions using column vectors as follows:

The position vector $\boldsymbol{a}$ = **OP** where P has coordinates (x, y, z) is given by

$$\boldsymbol{a} = \begin{pmatrix} x \\ y \\ z \end{pmatrix} = \begin{pmatrix} x \\ 0 \\ 0 \end{pmatrix} + \begin{pmatrix} 0 \\ y \\ 0 \end{pmatrix} + \begin{pmatrix} 0 \\ 0 \\ z \end{pmatrix} = x\begin{pmatrix} 1 \\ 0 \\ 0 \end{pmatrix} + y\begin{pmatrix} 0 \\ 1 \\ 0 \end{pmatrix} + z\begin{pmatrix} 0 \\ 0 \\ 1 \end{pmatrix}$$

$$= x\boldsymbol{i} + y\boldsymbol{j} + z\boldsymbol{k}$$

Where this time the base vectors are

$$\boldsymbol{i} = \begin{pmatrix} 1 \\ 0 \\ 0 \end{pmatrix}, \boldsymbol{j} = \begin{pmatrix} 0 \\ 1 \\ 0 \end{pmatrix} \text{ and } \boldsymbol{k} = \begin{pmatrix} 0 \\ 0 \\ 1 \end{pmatrix}$$

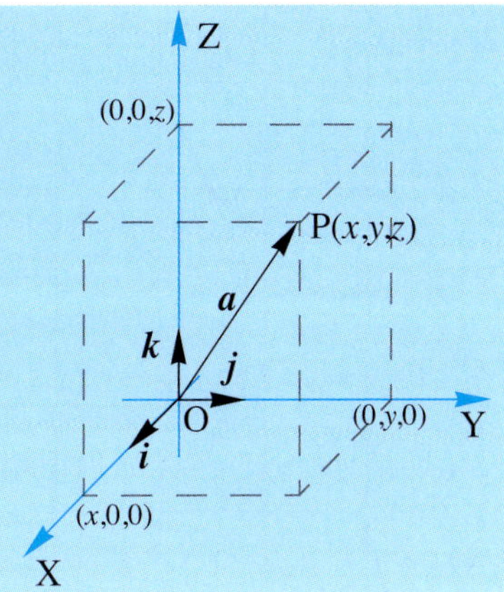

12.4.4 VECTOR OPERATIONS

Addition and Subtraction

If $\boldsymbol{a} = \begin{pmatrix} x_1 \\ y_1 \end{pmatrix} = x_1\boldsymbol{i} + y_1\boldsymbol{j}$ and $\boldsymbol{b} = \begin{pmatrix} x_2 \\ y_2 \end{pmatrix} = x_2\boldsymbol{i} + y_2\boldsymbol{j}$ then

$$\boldsymbol{a} \pm \boldsymbol{b} = \begin{pmatrix} x_1 \\ y_1 \end{pmatrix} \pm \begin{pmatrix} x_2 \\ y_2 \end{pmatrix} = \begin{pmatrix} x_1 \pm x_2 \\ y_1 \pm y_2 \end{pmatrix} = (x_1 \pm x_2)\boldsymbol{i} + (y_1 \pm y_2)\boldsymbol{j}$$

If $\boldsymbol{a} = \begin{pmatrix} x_1 \\ y_1 \\ z_1 \end{pmatrix} = x_1\boldsymbol{i} + y_1\boldsymbol{j} + z_1\boldsymbol{k}$ and $\boldsymbol{b} = \begin{pmatrix} x_2 \\ y_2 \\ z_2 \end{pmatrix} = x_2\boldsymbol{i} + y_2\boldsymbol{j} + z_2\boldsymbol{k}$ then

$$\boldsymbol{a} \pm \boldsymbol{b} = \begin{pmatrix} x_1 \\ y_1 \\ z_1 \end{pmatrix} \pm \begin{pmatrix} x_2 \\ y_2 \\ z_2 \end{pmatrix} = \begin{pmatrix} x_1 \pm x_2 \\ y_1 \pm y_2 \\ z_1 \pm z_2 \end{pmatrix} = (x_1 \pm x_2)\boldsymbol{i} + (y_1 \pm y_2)\boldsymbol{j} + (z_1 \pm z_2)\boldsymbol{k}$$

Scalar Multiplication

If $\boldsymbol{a} = \begin{pmatrix} x \\ y \end{pmatrix} = x\boldsymbol{i} + y\boldsymbol{j}$ then $k\boldsymbol{a} = k\begin{pmatrix} x \\ y \end{pmatrix} = \begin{pmatrix} kx \\ ky \end{pmatrix} = kx\boldsymbol{i} + ky\boldsymbol{j}$, $k \in \mathbb{R}$.

If $\boldsymbol{a} = \begin{pmatrix} x \\ y \\ z \end{pmatrix} = x\boldsymbol{i} + y\boldsymbol{j} + z\boldsymbol{k}$ then $k\boldsymbol{a} = k\begin{pmatrix} x \\ y \\ z \end{pmatrix} = \begin{pmatrix} kx \\ ky \\ kz \end{pmatrix} = kx\boldsymbol{i} + ky\boldsymbol{j} + kz\boldsymbol{k}$, $k \in \mathbb{R}$

EXAMPLE 12.10

If $\boldsymbol{a} = 2\boldsymbol{i} - \boldsymbol{j}$ and $\boldsymbol{b} = -\boldsymbol{i} + 3\boldsymbol{j}$, find:

(a) $\boldsymbol{a} + \boldsymbol{b}$ (b) $\boldsymbol{b} - \boldsymbol{a}$ (c) $3\boldsymbol{b} - 2\boldsymbol{a}$

Solution

Vectors are added 'nose to tail':

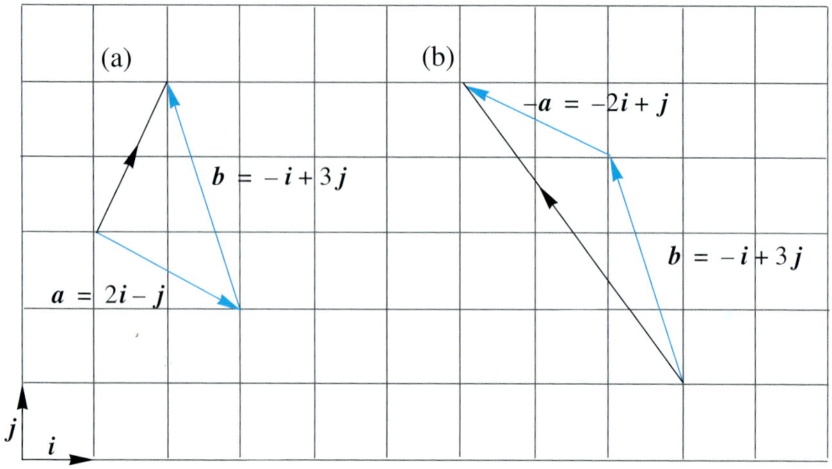

(a) Vectors are added in much the same way as are algebraic terms. Only like terms can be added or subtracted, so that

$$\begin{aligned} \boldsymbol{a} + \boldsymbol{b} &= (2\boldsymbol{i} - \boldsymbol{j}) + (-\boldsymbol{i} + 3\boldsymbol{j}) \\ &= (2-1)\boldsymbol{i} + (-1+3)\boldsymbol{j} \\ &= \boldsymbol{i} + 2\boldsymbol{j} \end{aligned}$$

(b) This problem is solved in a similar way:

$$\begin{aligned} \boldsymbol{b} - \boldsymbol{a} &= (-\boldsymbol{i} + 3\boldsymbol{j}) - (2\boldsymbol{i} - \boldsymbol{j}) \\ &= (-1-2)\boldsymbol{i} + (3-(-1))\boldsymbol{j} \\ &= -3\boldsymbol{i} + 4\boldsymbol{j} \end{aligned}$$

Note that we could also have expressed the sum as $\boldsymbol{b} - \boldsymbol{a} = \boldsymbol{b} + (-\boldsymbol{a}) = (-\boldsymbol{i} + 3\boldsymbol{j}) + (-2\boldsymbol{i} + \boldsymbol{j})$.

$$= -3\boldsymbol{i} + 4\boldsymbol{j}$$

(i.e., the negative of a vector is the same length as the original vector but points in the opposite direction.)

(c) Combining the properties of scalar multiplication with those of addition and subtraction we have

$$\begin{aligned} 3\boldsymbol{b} - 2\boldsymbol{a} &= 3(-\boldsymbol{i} + 3\boldsymbol{j}) - 2(2\boldsymbol{i} - \boldsymbol{j}) \\ &= -3\boldsymbol{i} + 9\boldsymbol{j} - 4\boldsymbol{i} + 2\boldsymbol{j} \\ &= -7\boldsymbol{i} + 11\boldsymbol{j} \end{aligned}$$

EXAMPLE 12.11

If $\boldsymbol{p} = \begin{pmatrix} 3 \\ -1 \\ 4 \end{pmatrix}$ and $\boldsymbol{q} = \begin{pmatrix} -2 \\ 0 \\ 3 \end{pmatrix}$, find:

(a) $\boldsymbol{p} + \boldsymbol{q}$ (b) $\boldsymbol{p} - \frac{\boldsymbol{q}}{2}$ (c) $\frac{3}{2}\boldsymbol{q} - \boldsymbol{p}$

Solution

(a)

$$p+q = \begin{pmatrix} 3 \\ -1 \\ 4 \end{pmatrix} + \begin{pmatrix} -2 \\ 0 \\ 3 \end{pmatrix} = \begin{pmatrix} 3-2 \\ -1+0 \\ 4+3 \end{pmatrix} = \begin{pmatrix} 1 \\ -1 \\ 7 \end{pmatrix}$$

(b)

$$p-\frac{q}{2} = \begin{pmatrix} 3 \\ -1 \\ 4 \end{pmatrix} - \frac{1}{2}\begin{pmatrix} -2 \\ 0 \\ 3 \end{pmatrix} = \begin{pmatrix} 3+1 \\ -1-0 \\ 4-1.5 \end{pmatrix} = \begin{pmatrix} 4 \\ -1 \\ 2.5 \end{pmatrix}$$

(c)

$$\frac{3}{2}q-p = 1.5\begin{pmatrix} -2 \\ 0 \\ 3 \end{pmatrix} - \begin{pmatrix} 3 \\ -1 \\ 4 \end{pmatrix} = \begin{pmatrix} -3-3 \\ 0+1 \\ 4.5-4 \end{pmatrix} = \begin{pmatrix} -6 \\ 1 \\ 0.5 \end{pmatrix}$$

EXAMPLE 12.12

A surveyor is standing at the top of a hill. Call this point 'the origin' (O). A lighthouse (L) is visible 4km to the West and 3km to the North. A town (T) is visible 5km to the South and 2km to the East. Using a vector basis in which $\boldsymbol{i}$ is a 1km vector running East and $\boldsymbol{j}$ is a 1km vector running North. Find the position vectors of the lighthouse, $\overrightarrow{OL}$ and the town $\overrightarrow{OT}$. Hence find the vector $\overrightarrow{LT}$ and the position of the town relative to the lighthouse.

Solution

The position vectors are: Lighthouse $\overrightarrow{OL} = -4\boldsymbol{i}+3\boldsymbol{j}$ and Town $\overrightarrow{OT} = 2\boldsymbol{i}-5\boldsymbol{j}$.

Then, to get from L to T we have

$$\begin{aligned} \overrightarrow{LT} &= \overrightarrow{LO}+\overrightarrow{OT} \\ &= -\overrightarrow{OL}+\overrightarrow{OT} \\ &= -(-4\boldsymbol{i}+3\boldsymbol{j})+(2\boldsymbol{i}-5\boldsymbol{j}) \\ &= 4\boldsymbol{i}-3\boldsymbol{j}+2\boldsymbol{i}-5\boldsymbol{j} \\ &= 6\boldsymbol{i}-8\boldsymbol{j} \end{aligned}$$

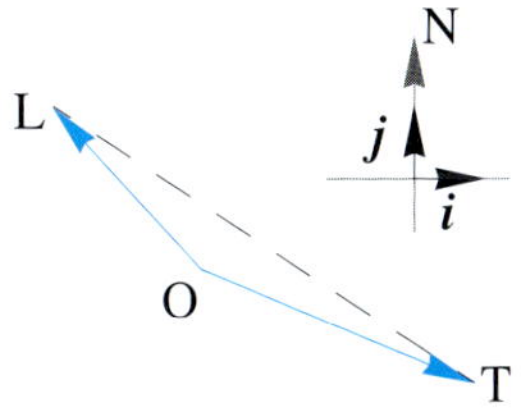

This means that the town is 6km East of the lighthouse and 8km South.

EXERCISES 12.4

1. If $\boldsymbol{a} = \boldsymbol{i}+7\boldsymbol{j}-\boldsymbol{k}$ and $\boldsymbol{b} = 4\boldsymbol{i}+7\boldsymbol{j}+5\boldsymbol{k}$ find:

(i) $4\boldsymbol{a}$ (ii) $3\boldsymbol{b}$ (iii) $2\boldsymbol{a}-\boldsymbol{b}$ (iv) $2(\boldsymbol{a}-\boldsymbol{b})$

2. The position vectors of A and B are $\overrightarrow{OA} = -3\boldsymbol{i}+4\boldsymbol{j}-2\boldsymbol{k}$ and $\overrightarrow{OB} = \boldsymbol{i}-4\boldsymbol{j}-3\boldsymbol{k}$. Find:

(i) $\overrightarrow{AO}$ (ii) $\overrightarrow{OA}-5\overrightarrow{OB}$ (iii) $-5\overrightarrow{OA}+3\overrightarrow{OB}$ (iv) $3\overrightarrow{OA}+6\overrightarrow{BO}$

3. The vectors $\boldsymbol{p}$ and $\boldsymbol{q}$ are defined by $\boldsymbol{p} = \begin{pmatrix} -1 \\ -2 \\ 4 \end{pmatrix}$ and $\boldsymbol{q} = \begin{pmatrix} 6 \\ 1 \\ 2 \end{pmatrix}$. Find:

(i) $\boldsymbol{p} + 2\boldsymbol{q}$ (ii) $-3\boldsymbol{p} - 5\boldsymbol{q}$ (iii) $3\boldsymbol{p}$ (iv) $2\boldsymbol{p} + 3\boldsymbol{q}$

4. Find the position vectors that join the origin to the points with coordinates A (2,–1) and B (–3,2). Express your answers as column vectors. Hence find $\overrightarrow{AB}$

5. A point on the Cartesian plane starts at the origin. The point then moves 4 units to the right, 5 units up, 6 units to the left and, finally 2 units down. Express these translations as a sum of four column vectors. Hence find the coordinates of the final position of the point.

6. Two vectors are defined as $\boldsymbol{a} = \boldsymbol{i} + \boldsymbol{j} + 4\boldsymbol{k}$ and $\boldsymbol{b} = -7\boldsymbol{i} - \boldsymbol{j} + 2\boldsymbol{k}$. Find:

(i) $-6\boldsymbol{a} - 2\boldsymbol{b}$ (ii) $-5\boldsymbol{a} + 2\boldsymbol{b}$ (iii) $4\boldsymbol{a} + 3\boldsymbol{b}$ (iv) $-2(\boldsymbol{a} + 3\boldsymbol{b})$

7. If $\boldsymbol{x} = \begin{pmatrix} 4 \\ -4 \\ 2 \end{pmatrix}$ and $\boldsymbol{y} = \begin{pmatrix} 4 \\ 3 \\ 7 \end{pmatrix}$, express the following as column vectors.

(i) $2\boldsymbol{x} + 3\boldsymbol{y}$ (ii) $\boldsymbol{x} + 2\boldsymbol{y}$ (iii) $5\boldsymbol{x} - 6\boldsymbol{y}$ (iv) $\boldsymbol{x} - 6\boldsymbol{y}$

8. Find the values of A and B if $A(7\boldsymbol{i} + 7\boldsymbol{j} + 4\boldsymbol{k}) - 3(3\boldsymbol{i} - \boldsymbol{j} + B\boldsymbol{k}) = -37\boldsymbol{i} - 25\boldsymbol{j} + 5\boldsymbol{k}$

9. Two vectors are defined as $\boldsymbol{a} = \begin{pmatrix} -3 \\ 1 \\ 4 \end{pmatrix}$ and $\boldsymbol{b} = \begin{pmatrix} 6 \\ -6 \\ -5 \end{pmatrix}$. Find values of the scalars X and Y if $X\boldsymbol{a} + Y\boldsymbol{b}$ is equal to: (i) $\begin{pmatrix} -36 \\ 32 \\ 33 \end{pmatrix}$ (ii) $\begin{pmatrix} 30 \\ -22 \\ -31 \end{pmatrix}$ (iii) $\begin{pmatrix} -12 \\ 24 \\ 1 \end{pmatrix}$

10. A submarine (which is considered the origin of the vector system) is 60 metres below the surface of the sea when it detects two surface ships. A destroyer (D) is 600 metres to the East and 800 metres to the South of the submarine. An aircraft carrier (A) is 1200 metres to the West and 300 metres to the South.

(a) Define a suitable vector basis for this problem.

(b) Using the submarine as the origin, state the position vectors of the destroyer and the aircraft carrier.

(c) A helicopter pilot, based on the aircraft carrier wants to make a supplies delivery to the destroyer. Find, in vector terms, the course along which the pilot should fly.

12.5 FURTHER PROPERTIES OF VECTORS IN 2-D AND 3-D

12.5.1 MAGNITUDE OF A VECTOR

If a vector is expressed in terms of the unit length basis of vectors, $\boldsymbol{i}$, $\boldsymbol{j}$, $\boldsymbol{k}$, the **length** of the vector can be found using the Theorem of Pythagoras.

The length of a vector is sometimes known as its **magnitude** or **absolute value**. The length of a vector $\boldsymbol{a}$ is often represented by $|\boldsymbol{a}|$ If the vector represents some physical quantity such as a force, the length of the vector is the size of the force, without its direction. The length of a vector is a scalar quantity.

The magnitude of a vector can be found by applying Pythagoras's Theorem in the case of vectors in two dimensions and repeated use of Pythagoras's Theorem in the case of vectors in three dimensions.

The magnitude of a vector

$$\boldsymbol{a} = \begin{pmatrix} x \\ y \end{pmatrix} = x\boldsymbol{i} + y\boldsymbol{j}$$

is given by $|\boldsymbol{a}| = \sqrt{x^2 + y^2}$.

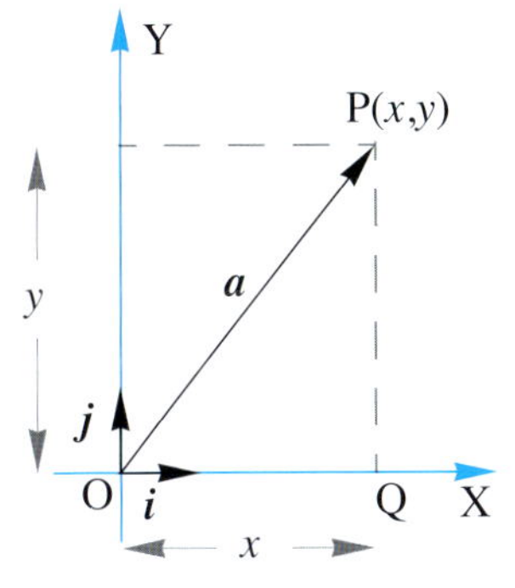

PROOF:

$$|\mathbf{OP}|^2 = |\mathbf{OQ}|^2 + |\mathbf{QP}|^2$$

$$\therefore |\boldsymbol{a}|^2 = x^2 + y^2$$

$$\therefore |\boldsymbol{a}| = \sqrt{x^2 + y^2}, \ (|\boldsymbol{a}| > 0)$$

The magnitude of a vector

$$\boldsymbol{a} = \begin{pmatrix} x \\ y \\ z \end{pmatrix} = x\boldsymbol{i} + y\boldsymbol{j} + z\boldsymbol{k}$$

is given by $|\boldsymbol{a}| = \sqrt{x^2 + y^2 + z^2}$.

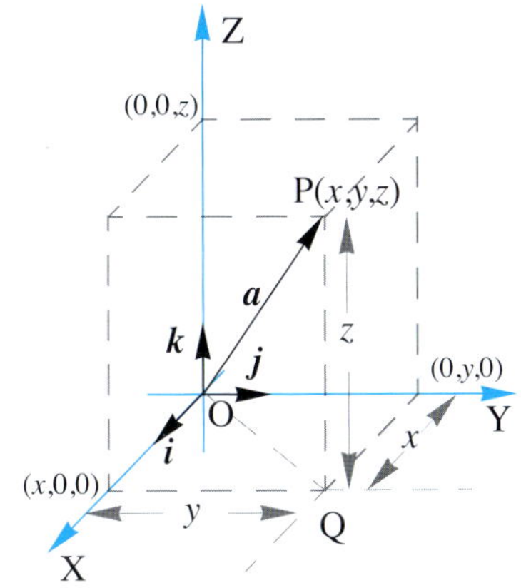

$$|\mathbf{OP}|^2 = |\mathbf{OQ}|^2 + |\mathbf{QP}|^2$$

$$\therefore |\boldsymbol{a}|^2 = (x^2 + y^2) + z^2$$

$$\therefore |\boldsymbol{a}| = \sqrt{x^2 + y^2 + z^2}, \ (|\boldsymbol{a}| > 0)$$

EXAMPLE 12.13

Find the lengths of the vectors:

(a) $3\boldsymbol{i} - 4\boldsymbol{j}$ (b) $-\boldsymbol{i} + 2\boldsymbol{j} - 5\boldsymbol{k}$ (c) $\begin{pmatrix} 3 \\ -1 \\ 2 \end{pmatrix}$

Solution

(a) The vector $3\boldsymbol{i} - 4\boldsymbol{j}$ is represented by the diagram:

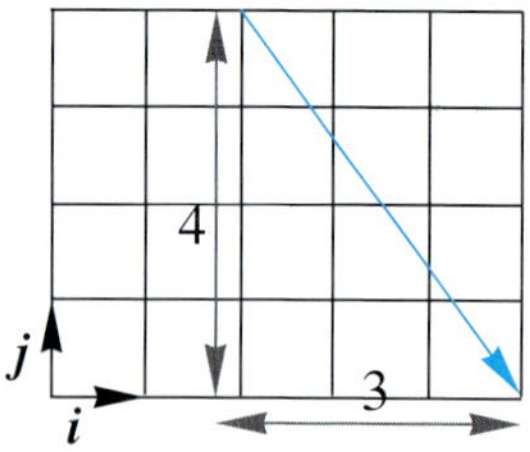

$$|3\boldsymbol{i} - 4\boldsymbol{j}| = \sqrt{3^2 + (-4)^2} = \sqrt{9 + 16} = \sqrt{25} = 5$$

(b) $|-\boldsymbol{i} + 2\boldsymbol{j} - 5\boldsymbol{k}| = \sqrt{(-1)^2 + 2^2 + (-5)^2} = \sqrt{30}$

(c) $\left|\begin{pmatrix} 3 \\ -1 \\ 2 \end{pmatrix}\right| = \sqrt{3^2 + (-1)^2 + 2^2} = \sqrt{14}$

12.5.2 UNIT VECTORS

The **unit vector**, as the name implies, is a vector having a magnitude (length) of one unit. We use the notation $\hat{\boldsymbol{a}}$, read as "*a hat*". Then, $\hat{\boldsymbol{a}}$ is a vector of length one unit in the same direction as $\boldsymbol{a}$.

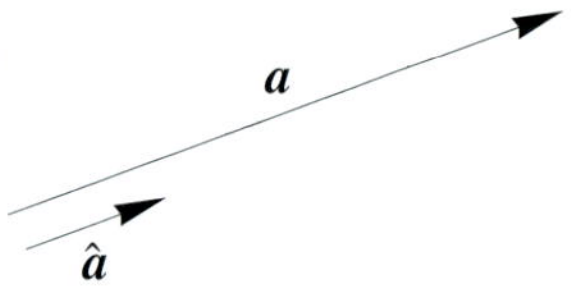

For example, if we had a vector $\boldsymbol{r}$ of length four units, to find the corresponding unit vector, $\hat{\boldsymbol{r}}$, we would need to divide the vector $\boldsymbol{r}$ by four, resulting in a vector parallel to $\boldsymbol{r}$ but of unit length.

We then have the definition that

$$\hat{\boldsymbol{a}} = \frac{\boldsymbol{a}}{|\boldsymbol{a}|}$$

EXAMPLE 12.14

Find unit vectors in the same directions as the vectors:

(a) $5\boldsymbol{i} - 2\boldsymbol{j}$ (b) $\begin{pmatrix} -3 \\ 6 \\ 4 \end{pmatrix}$

Solution

(i) $|5\boldsymbol{i} - 2\boldsymbol{j}| = \sqrt{5^2 + (-2)^2} = \sqrt{29}$. This is the length of the vector. In order to produce a vector of unit length, we can keep the original components of the vector, scaling them down to produce the required unit vector: $\frac{1}{\sqrt{29}}(5\boldsymbol{i} - 2\boldsymbol{j})$

(ii) $\left|\begin{pmatrix} -3 \\ 6 \\ 4 \end{pmatrix}\right| = \sqrt{(-3)^2 + 6^2 + 4^2} = \sqrt{61}$ The required unit vector is: $\frac{1}{\sqrt{61}}\begin{pmatrix} -3 \\ 6 \\ 4 \end{pmatrix}$.

EXAMPLE 12.15

Find a vector of length 10 units in the direction of $2\boldsymbol{i}-3\boldsymbol{j}+\boldsymbol{k}$.

Solution

Let $\boldsymbol{u} = 2\boldsymbol{i}-3\boldsymbol{j}+\boldsymbol{k}$, then $\hat{\boldsymbol{u}} = \frac{1}{|\boldsymbol{u}|}\boldsymbol{u} = \frac{1}{\sqrt{2^2+(-3)^2+1^2}}(2\boldsymbol{i}-3\boldsymbol{j}+\boldsymbol{k})$

$$= \frac{1}{\sqrt{14}}(2\boldsymbol{i}-3\boldsymbol{j}+\boldsymbol{k}).$$

That is, we have a vector that is parallel to $\boldsymbol{u}$ and of unit length. If we now multiply this unit vector by 10 we will produce a vector parallel to $\boldsymbol{u}$ but of length 10 units.

Required vector is $10\hat{\boldsymbol{u}} = \frac{10}{\sqrt{14}}(2\boldsymbol{i}-3\boldsymbol{j}+\boldsymbol{k})$.

EXERCISES 12.5

1. Find the lengths of these vectors, expressing your answers as surds. It is not necessary to simplify these surds.

(i) $\boldsymbol{i}+3\boldsymbol{j}$ (ii) $5(\boldsymbol{i}+\boldsymbol{j})$ (iii) $5\boldsymbol{i}-2\boldsymbol{j}+\boldsymbol{k}$ (iv) $-(2\boldsymbol{i}+\boldsymbol{j}+2\boldsymbol{k})$

(v) $4\boldsymbol{i}+6\boldsymbol{j}-\boldsymbol{k}$ (vi) $2\boldsymbol{i}+6\boldsymbol{j}+\boldsymbol{k}$ (vii) $\begin{pmatrix}2\\3\\1\end{pmatrix}$ (viii) $\begin{pmatrix}2\\-3\\-2\end{pmatrix}$

2. Find unit vectors in the same directions as these vectors:

(i) $4\boldsymbol{i}+4\boldsymbol{j}$ (ii) $4\boldsymbol{i}+5\boldsymbol{j}$ (iii) $-\boldsymbol{i}-2\boldsymbol{j}$ (iv) $\boldsymbol{i}+6\boldsymbol{j}-3\boldsymbol{k}$

(v) $2\boldsymbol{j}+4\boldsymbol{k}$ (vi) $2\boldsymbol{i}-2\boldsymbol{j}-3\boldsymbol{k}$ (vii) $\begin{pmatrix}2\\1\\2\end{pmatrix}$ (viii) $\begin{pmatrix}-1\\5\\1\end{pmatrix}$

3. A mass sitting on the ground is being pulled by a force of 4 Newtons in a Northerly direction, 3 Newtons in a Westerly direction and 1 Newton upwards.

(i) Express the forces acting on the mass in terms of an appropriate vector basis.
(ii) Find the total magnitude of the force acting on the mass

4. (a) Find a vector of length 3 units in the direction of $\boldsymbol{i}-\boldsymbol{j}+\boldsymbol{k}$.

(b) Find a vector of length $\sqrt{3}$ units in the direction of $3\boldsymbol{i}-\boldsymbol{j}+\sqrt{2}\boldsymbol{k}$.

5. The vectors $\boldsymbol{a} = 2\boldsymbol{j}+4\boldsymbol{k}$ and $\boldsymbol{b} = x\boldsymbol{i}+3\boldsymbol{k}$ are of equal length. Find x.

6. Find the maximum speed of a particle whose velocity, $\boldsymbol{v}$ m/s, at time t seconds is given by $\boldsymbol{v} = 2\sin t\boldsymbol{i}+\cos t\boldsymbol{j}+3\boldsymbol{k}$, $t \geq 0$.

12.6 SCALAR PRODUCT OF 2 VECTORS

12.6.1 DEFINITION OF THE SCALAR PRODUCT

The scalar product (or **dot product**) of two vectors is defined by:

$$\boldsymbol{a} \bullet \boldsymbol{b} = |\boldsymbol{a}||\boldsymbol{b}|\cos\theta$$

where θ is the angle between the two vectors which may be an obtuse angle. The angle must be measured between the directions of the vectors. That is, the angle between the two vectors once they are joined tail to tail.

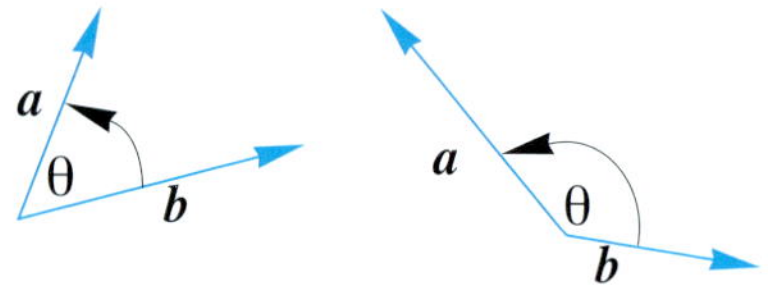

The three quantities on the right hand side of the equation are all scalars and it is important to realise that, when the **scalar** product of two **vectors** is calculated, the result is a **scalar**.

EXAMPLE 12.16

Find the scalar product of the vectors $2\boldsymbol{i}-3\boldsymbol{j}+\boldsymbol{k}$ and $\boldsymbol{i}+\boldsymbol{j}-\boldsymbol{k}$.

Solution

Let $\boldsymbol{a} = 2\boldsymbol{i}-3\boldsymbol{j}+\boldsymbol{k}$ and $\boldsymbol{b} = \boldsymbol{i}+\boldsymbol{j}-\boldsymbol{k}$, then to determine the scalar product, $\boldsymbol{a} \bullet \boldsymbol{b}$, we need to find

1. $|\boldsymbol{a}|$

2. $|\boldsymbol{b}|$

3. $\cos\theta$, where θ is the angle between $\boldsymbol{a}$ and $\boldsymbol{b}$.

Finding

1. $|\boldsymbol{a}| = \sqrt{2^2+(-3)^2+1^2} = \sqrt{14}$.

2. $|\boldsymbol{b}| = \sqrt{1^2+1^2+(-1)^2} = \sqrt{3}$.

3. $\cos\theta$:

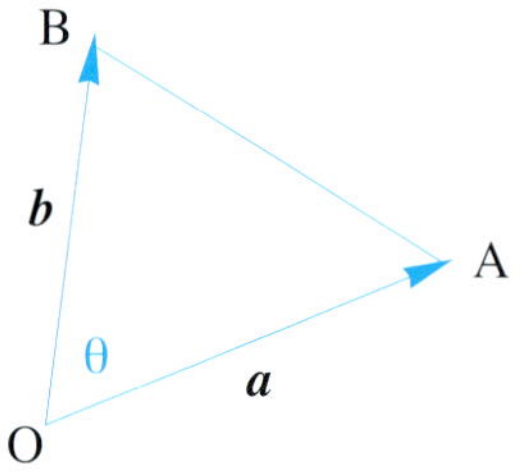

Finding $\cos\theta$ requires a little work. Relative to a common origin O, the points A(2, –3, 1) and B(1, 1, –1) have position vectors $\boldsymbol{a}$ and $\boldsymbol{b}$.

Before making use of the cosine rule we need to determine the length of AB. Using the distance formula between two points in space, we have

$$AB = \sqrt{(1-2)^2+(1-(-3))^2+(-1-1)^2} = \sqrt{1+16+4} = \sqrt{21}$$

Cosine rule:

$$AB^2 = OA^2+OB^2-2\cdot OA\cdot OB\cdot\cos\theta$$

$$(\sqrt{21})^2 = (\sqrt{14})^2+(\sqrt{3})^2-2\cdot\sqrt{14}\cdot\sqrt{3}\cdot\cos\theta$$

$$21 = 14+3-2\sqrt{42}\cos\theta$$

$$\therefore \cos\theta = -\frac{2}{\sqrt{42}}$$

Next, from the definition of the scalar product, $\boldsymbol{a} \bullet \boldsymbol{b} = |\boldsymbol{a}||\boldsymbol{b}|\cos\theta$, we have

$$\boldsymbol{a} \bullet \boldsymbol{b} = \sqrt{14}\times\sqrt{3}\times-\frac{2}{\sqrt{42}} = -2$$

The solution to Example 26.16 was rather lengthy, however, we now look at the scalar product from a slightly different viewpoint.

First consider the dot product $\boldsymbol{i} \bullet \boldsymbol{i}$: Using the definition, we have that

$$\boldsymbol{i} \bullet \boldsymbol{i} = |\boldsymbol{i}||\boldsymbol{i}|\cos 0 = 1 \times 1 \times 1 = 1$$

[the angle between the vectors $\boldsymbol{i}$ and $\boldsymbol{i}$ is 0 and so $\cos\theta = \cos 0 = 1$].

Next consider the product $\boldsymbol{i} \bullet \boldsymbol{j}$: Using the definition, we have that

$$\boldsymbol{i} \bullet \boldsymbol{j} = |\boldsymbol{i}||\boldsymbol{j}|\cos 90 = 1 \times 1 \times 0 = 0$$

[the angle between the vectors $\boldsymbol{i}$ and $\boldsymbol{j}$ is $90°$ and so $\cos\theta = \cos 90° = 0$].

Similarly, we end up with the following results for all possible combinations of the $\boldsymbol{i}$, $\boldsymbol{j}$ and $\boldsymbol{k}$ vectors:

$$\boldsymbol{i} \bullet \boldsymbol{i} = \boldsymbol{j} \bullet \boldsymbol{j} = \boldsymbol{k} \bullet \boldsymbol{k} = 1$$

and

$$\boldsymbol{i} \bullet \boldsymbol{j} = \boldsymbol{i} \bullet \boldsymbol{k} = \boldsymbol{j} \bullet \boldsymbol{k} = \boldsymbol{j} \bullet \boldsymbol{i} = \boldsymbol{k} \bullet \boldsymbol{i} = \boldsymbol{k} \bullet \boldsymbol{j} = 0$$

Armed with these results we can now work out the scalar product of the vectors $\boldsymbol{a} = x_1\boldsymbol{i} + y_1\boldsymbol{j} + z_1\boldsymbol{k}$ and $\boldsymbol{b} = x_2\boldsymbol{i} + y_2\boldsymbol{j} + z_2\boldsymbol{k}$ as follows,

$$\begin{aligned} a \bullet b &= (x_1\boldsymbol{i} + y_1\boldsymbol{j} + z_1\boldsymbol{k}) \bullet (x_2\boldsymbol{i} + y_2\boldsymbol{j} + z_2\boldsymbol{k}) \\ &= x_1x_2(\boldsymbol{i} \bullet \boldsymbol{i}) + x_1y_2(\boldsymbol{i} \bullet \boldsymbol{j}) + x_1z_2(\boldsymbol{i} \bullet \boldsymbol{k}) \\ &\quad + y_1x_2(\boldsymbol{j} \bullet \boldsymbol{i}) + y_1y_2(\boldsymbol{j} \bullet \boldsymbol{j}) + y_1z_2(\boldsymbol{j} \bullet \boldsymbol{k}) \\ &\quad + z_1x_2(\boldsymbol{k} \bullet \boldsymbol{i}) + z_1y_2(\boldsymbol{k} \bullet \boldsymbol{j}) + z_1z_2(\boldsymbol{k} \bullet \boldsymbol{k}) \\ \therefore a \bullet b &= x_1x_2 + y_1y_2 + z_1z_2 \end{aligned}$$

That is, if

$$\boldsymbol{a} = x_1\boldsymbol{i} + y_1\boldsymbol{j} + z_1\boldsymbol{k} \text{ and } \boldsymbol{b} = x_2\boldsymbol{i} + y_2\boldsymbol{j} + z_2\boldsymbol{k} \text{ then } a \bullet b = x_1x_2 + y_1y_2 + z_1z_2$$

Using this result with the vectors of Example 26.16, $2\boldsymbol{i} - 3\boldsymbol{j} + \boldsymbol{k}$ and $\boldsymbol{i} + \boldsymbol{j} - \boldsymbol{k}$ we have:

$$\begin{aligned} (2\boldsymbol{i} - 3\boldsymbol{j} + \boldsymbol{k}) \bullet (\boldsymbol{i} + \boldsymbol{j} - \boldsymbol{k}) &= 2 \times 1 + (-3) \times 1 + 1 \times (-1) \\ &= 2 - 3 - 1 \\ &= -2 \end{aligned}$$

A much faster process!

However, the most usual use of scalar product is to calculate the angle between vectors using a rearrangement of the definition of scalar product:

$$\cos\theta = \frac{\boldsymbol{a} \bullet \boldsymbol{b}}{|\boldsymbol{a}||\boldsymbol{b}|}$$

EXAMPLE 12.17

For the following pairs of vectors, find their magnitudes and scalar products. Hence find the angles between the vectors, correct to the nearest degree.

(a) $-\boldsymbol{i} + 3\boldsymbol{j}$ and $-\boldsymbol{i} + 2\boldsymbol{j}$ (b) $\begin{pmatrix} 0 \\ -5 \\ 4 \end{pmatrix}$ and $\begin{pmatrix} -5 \\ -1 \\ -3 \end{pmatrix}$

Solution

(a) In using the scalar product, it is necssary to calculate the magnitudes of the vectors.

$|-\boldsymbol{i}+3\boldsymbol{j}| = \sqrt{(-1)^2+3^2} = \sqrt{10}$ and $|-\boldsymbol{i}+2\boldsymbol{j}| = \sqrt{(-1)^2+2^2} = \sqrt{5}$

Next, calculate the scalar product: $-\boldsymbol{i}+3\boldsymbol{j} \qquad -\boldsymbol{i}+2\boldsymbol{j}$

$(-\boldsymbol{i}+3\boldsymbol{j}) \bullet (-\boldsymbol{i}+2\boldsymbol{j}) = -1 \times -1 + 3 \times 2 = 7$

Finally, the angle is: $\cos\theta = \dfrac{\boldsymbol{a} \bullet \boldsymbol{b}}{|\boldsymbol{a}||\boldsymbol{b}|} = \dfrac{7}{\sqrt{10} \times \sqrt{5}} \Rightarrow \theta \approx 8°$

(b) $\left|\begin{pmatrix} 0 \\ -5 \\ 4 \end{pmatrix}\right| = \sqrt{0^2+(-5)^2+4^2} = \sqrt{41}$ and $\left|\begin{pmatrix} -5 \\ -1 \\ -3 \end{pmatrix}\right| = \sqrt{(-5)^2+(-1)^2+(-3)^2} = \sqrt{35}$

Next, the scalar product:

$$\begin{pmatrix} 0 \\ -5 \\ 4 \end{pmatrix} \bullet \begin{pmatrix} -5 \\ -1 \\ -3 \end{pmatrix} = 0 \times (-5) + (-5) \times (-1) + 4 \times (-3) = -7$$

Finally, the angle can be calculated:

$$\cos\theta = \frac{\boldsymbol{a} \bullet \boldsymbol{b}}{|\boldsymbol{a}||\boldsymbol{b}|} = \frac{-7}{\sqrt{41} \times \sqrt{35}} \Rightarrow \theta \approx 101°$$

The use of cosine means that obtuse angles between vectors (which occur when the scalar product is negative) are calculated correctly when using the inverse cosine function on a calculator.

12.6.2 PROPERTIES OF THE SCALAR PRODUCT

Closure The scalar product of two vectors is a scalar (i.e., the result is not a vector). The operation is **not closed** and so closure does not apply.

Commutative Now, $\boldsymbol{a} \bullet \boldsymbol{b} = |a||b|\cos\theta = |b||a|\cos\theta = \boldsymbol{b} \bullet \boldsymbol{a}$

That is, $\boldsymbol{a} \bullet \boldsymbol{b} = \boldsymbol{b} \bullet \boldsymbol{a}$.

Therefore the operation of scalar product is commutative

Associative If the associative property were to hold it would take on the form $(\boldsymbol{a} \bullet \boldsymbol{b}) \bullet \boldsymbol{c} = \boldsymbol{a} \bullet (\boldsymbol{b} \bullet \boldsymbol{c})$. However, $\boldsymbol{a} \bullet \boldsymbol{b}$ is a real number and therefore the operation $(\boldsymbol{a} \bullet \boldsymbol{b}) \bullet \boldsymbol{c}$ has no meaning [you cannot 'dot' a scalar with a vector].

Distributive The **scalar product is distributive** (over addition)
We leave the proof of this result as an exercise.

Identity As the operation of scalar product is not closed, an identity cannot exist.

Inverse As the operation of scalar product is not closed, an inverse cannot exist.

Note that although the scalar product is non-associative, the following 'associative rule' holds for the scalar product:

$$\text{If } k \in \mathbb{R}, \text{ then, } \boldsymbol{a} \bullet (k\boldsymbol{b}) = k(\boldsymbol{a} \bullet \boldsymbol{b})$$

12.6.3 SPECIAL CASES OF THE SCALAR PRODUCT

1. **Perpendicular Vectors**

If the vectors $\boldsymbol{a}$ and $\boldsymbol{b}$ are perpendicular then, $\boldsymbol{a} \bullet \boldsymbol{b} = |\boldsymbol{a}||\boldsymbol{b}|\cos 90° = 0$.

[Note: We are assuming that $\boldsymbol{a}$ and $\boldsymbol{b}$ are non-zero vectors.]

2. **Zero vector**

For **any** vector $\boldsymbol{a}$, $\boldsymbol{a} \bullet \boldsymbol{0} = |\boldsymbol{a}||\boldsymbol{0}|\cos\theta = 0$.

3. **Parallel vectors**

If the vectors $\boldsymbol{a}$ and $\boldsymbol{b}$ are parallel then, $\boldsymbol{a} \bullet \boldsymbol{b} = |\boldsymbol{a}||\boldsymbol{b}|\cos 0 = |\boldsymbol{a}||\boldsymbol{b}|$.

If the vectors $\boldsymbol{a}$ and $\boldsymbol{b}$ are antiparallel then, $\boldsymbol{a} \bullet \boldsymbol{b} = |\boldsymbol{a}||\boldsymbol{b}|\cos\pi = -|\boldsymbol{a}||\boldsymbol{b}|$.

[Note: We are assuming that $\boldsymbol{a}$ and $\boldsymbol{b}$ are non-zero vectors.]

Combining the results of 1., and 2., above, we have the important observation that

If $\boldsymbol{a} \bullet \boldsymbol{b} = 0$ then either
1. $\boldsymbol{a}$ and/or $\boldsymbol{b}$ are both the zero vector, **0**.

Or

2. $\boldsymbol{a}$ and $\boldsymbol{b}$ are perpendicular with neither $\boldsymbol{a}$ nor $\boldsymbol{b}$ being the zero vector.

Notice how this result differs from the standard Null Factor Law when dealing with real numbers, where given $ab = 0$ then a or b or both are zero! That is, the **cancellation property** that holds for real numbers **does not hold for vectors**.

A nice application using the perpendicular property above can be seen in the next example.

EXAMPLE 12.18 Three towns are joined by straight roads. Oakham is the state capital and is considered as the 'origin'. Axthorp is 3km East and 9km North of Oakham and Bostock is 5km East and 5km South of Axthorp.

Considering $\boldsymbol{i}$ as a 1km vector pointing East and $\boldsymbol{j}$ a 1km vector pointing North:

(a) Find the position vector of Axthorp relative to Oakham.
(b) Find the position vector of Bostock relative to Oakham.

A bus stop (S) is situated two thirds of the way along the road from Oakham to Axthorp.

(c) Find the vectors $\overrightarrow{OS}$ and $\overrightarrow{BS}$.
(d) Prove that the bus stop is the closest point to Bostock on the Oakham to Axthorp road.

Solution

(a) Axthorp is 3km East and 9km North of Oakham so $\overrightarrow{OA} = 3\boldsymbol{i} + 9\boldsymbol{j}$

(b) $\overrightarrow{OB} = \overrightarrow{OA} + \overrightarrow{AB} = 3\boldsymbol{i} + 9\boldsymbol{j} + 5\boldsymbol{i} - 5\boldsymbol{j} = 8\boldsymbol{i} + 4\boldsymbol{j}$

(c) $\overrightarrow{OS} = \frac{2}{3}(\overrightarrow{OA}) = \frac{2}{3}(3\boldsymbol{i} + 9\boldsymbol{j}) = 2\boldsymbol{i} + 6\boldsymbol{j}$

$\overrightarrow{BS} = \overrightarrow{BO} + \overrightarrow{OS} = -(8\boldsymbol{i} + 4\boldsymbol{j}) + 2\boldsymbol{i} + 6\boldsymbol{j} = -6\boldsymbol{i} + 2\boldsymbol{j}$

(d) The next step is to calculate the angle between $\overrightarrow{OS}$ and $\overrightarrow{BS}$ by calculating the scalar product of the two vectors:

$\overrightarrow{OS} \bullet \overrightarrow{BS} = (2\boldsymbol{i} + 6\boldsymbol{j}) \bullet (-6\boldsymbol{i} + 2\boldsymbol{j}) = 2 \times (-6) + 6 \times 2 = 0$

This means that $\overrightarrow{OS}$ and $\overrightarrow{BS}$ are at right angles to each other. It follows that the bus stop is the closest point to Bostock on the Oakham to Axthorp road.

EXAMPLE 12.19

Find the value(s) of m for which the vectors $2m\boldsymbol{i} + m\boldsymbol{j} + 8\boldsymbol{k}$ and $\boldsymbol{i} + 3m\boldsymbol{j} - \boldsymbol{k}$ are perpendicular.

Solution

As the two vectors are perpendicular, then

$$(2m\boldsymbol{i} + m\boldsymbol{j} + 8\boldsymbol{k}) \bullet (\boldsymbol{i} + 3m\boldsymbol{j} - \boldsymbol{k}) = 0$$
$$\Rightarrow 2m + 3m^2 - 8 = 0$$
$$\Leftrightarrow 3m^2 + 2m - 8 = 0$$
$$\Leftrightarrow (3m - 4)(m + 2) = 0$$
$$\Leftrightarrow m = \frac{4}{3} \text{ or } m = -2$$

EXAMPLE 12.20

Find a vector perpendicular to $\boldsymbol{u} = 4\boldsymbol{i} - 3\boldsymbol{j}$.

Solution

Let the vector perpendicular to $\boldsymbol{u} = 4\boldsymbol{i} - 3\boldsymbol{j}$ be $\boldsymbol{v} = x\boldsymbol{i} + y\boldsymbol{j}$.

Then, as $\boldsymbol{u} \perp \boldsymbol{v} \Rightarrow \boldsymbol{u} \bullet \boldsymbol{v} = 0$ so that $(4\boldsymbol{i} - 3\boldsymbol{j}) \bullet (x\boldsymbol{i} + y\boldsymbol{j}) = 0$

$$\therefore 4x - 3y = 0 \quad - (1)$$

Unfortunately, at this stage we only have one equation for two unknowns! We need to obtain a second equation from somewhere. To do this we recognise the fact that if $\boldsymbol{v}$ is perpendicular to $\boldsymbol{u}$, then so too will the unit vector, $\hat{\boldsymbol{v}}$, be perpendicular to $\boldsymbol{u}$.

Then, as $|\hat{\boldsymbol{v}}| = 1 \Rightarrow \sqrt{x^2 + y^2} = 1 \therefore x^2 + y^2 = 1 \quad - (2)$

From (1) we have that $y = \frac{4}{3}x \quad - (3)$

Substituting (3) into (2) we have: $x^2 + \left(\frac{4}{3}x\right)^2 = 1 \Leftrightarrow 25x^2 = 9 \Leftrightarrow x = \pm\frac{3}{5}$

Substituting into (3) we have: $y = \pm\frac{4}{5}$

Therefore, both $\boldsymbol{v} = \frac{3}{5}\boldsymbol{i} + \frac{4}{5}\boldsymbol{j}$ and $\boldsymbol{v} = -\left(\frac{3}{5}\boldsymbol{i} + \frac{4}{5}\boldsymbol{j}\right)$ are perpendicular to $\boldsymbol{u}$.

EXAMPLE 12.21

Use a vector method to derive the cosine rule for the triangle shown.

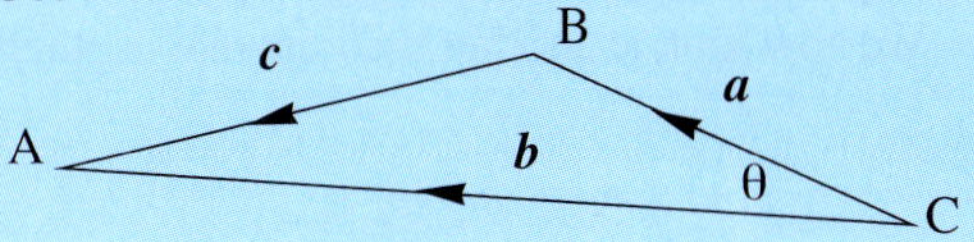

Solution

From the triangle rule for vector addition we have $\boldsymbol{a} + \boldsymbol{c} = \boldsymbol{b} \Leftrightarrow \boldsymbol{c} = \boldsymbol{b} - \boldsymbol{a}$.
Now, using the scalar product we have:

$$\begin{aligned} \boldsymbol{c} \bullet \boldsymbol{c} &= (\boldsymbol{b} - \boldsymbol{a}) \bullet (\boldsymbol{b} - \boldsymbol{a}) \\ &= \boldsymbol{b} \bullet \boldsymbol{b} - \boldsymbol{b} \bullet \boldsymbol{a} - \boldsymbol{a} \bullet \boldsymbol{b} + \boldsymbol{a} \bullet \boldsymbol{a} \\ &= |\boldsymbol{b}|^2 - 2\boldsymbol{a} \bullet \boldsymbol{b} + |\boldsymbol{a}|^2 \\ \therefore |\boldsymbol{c}|^2 &= |\boldsymbol{b}|^2 + |\boldsymbol{a}|^2 - 2|\boldsymbol{a}||\boldsymbol{b}|\cos\theta \end{aligned}$$

EXAMPLE 12.22

Find a vector perpendicular to both $\boldsymbol{a} = 2\boldsymbol{i} + \boldsymbol{j} - \boldsymbol{k}$ and $\boldsymbol{b} = \boldsymbol{i} + 3\boldsymbol{j} + \boldsymbol{k}$.

Solution

Let the vector $\boldsymbol{c} = x\boldsymbol{i} + y\boldsymbol{j} + z\boldsymbol{k}$ be perpendicular to both $\boldsymbol{a}$ and $\boldsymbol{b}$.
Then, we have that $\boldsymbol{a} \bullet \boldsymbol{c} = 0$ and $\boldsymbol{b} \bullet \boldsymbol{c} = 0$.
From $\boldsymbol{a} \bullet \boldsymbol{c} = 0$ we obtain: $(2\boldsymbol{i} + \boldsymbol{j} - \boldsymbol{k}) \bullet (x\boldsymbol{i} + y\boldsymbol{j} + z\boldsymbol{k}) = 2x + y - z = 0$ – (1)
From $\boldsymbol{b} \bullet \boldsymbol{c} = 0$ we obtain: $(\boldsymbol{i} + 3\boldsymbol{j} + \boldsymbol{k}) \bullet (x\boldsymbol{i} + y\boldsymbol{j} + z\boldsymbol{k}) = x + 3y + z = 0$ – (2)
In order to solve for the three unknowns we need one more equation. We note that if $\boldsymbol{c}$ is peprendicular to $\boldsymbol{a}$ and $\boldsymbol{b}$ then so too will the unit vector, $\hat{\boldsymbol{c}}$. So, without any loss in generality, we can assume that $\boldsymbol{c}$ is a unit vector. This will provide a third equation.
As we are assuming that $\boldsymbol{c}$ is a unit vector, we have: $|\boldsymbol{c}| = 1 \therefore x^2 + y^2 + x^2 = 1$ – (3)
We can now solve for x, y and z:

(1) + (2): $3x + 4y = 0$ – (4)

2×(1) – (2): $5y + 3z = 0$ – (5)

Substituting (4) and (5) into (3): $\left(-\frac{4}{3}y\right)^2 + y^2 + \left(-\frac{5}{3}y\right)^2 = 1$

$$\begin{aligned} &\Leftrightarrow 16y^2 + 9y^2 + 25y^2 = 9 \\ &\Leftrightarrow 50y^2 = 9 \\ &\Leftrightarrow y = \pm\frac{3}{5\sqrt{2}} \\ &\therefore y = \pm\frac{3\sqrt{2}}{10} \end{aligned}$$

Substituting into (4) and (5) we obtain $x = -\frac{4}{3} \times \pm\frac{3\sqrt{2}}{10} = \mp\frac{2\sqrt{2}}{5}$ and $z = -\frac{5}{3} \times \pm\frac{3\sqrt{2}}{10} = \mp\frac{\sqrt{2}}{2}$.

Therefore, $\mp\frac{2\sqrt{2}}{5}\boldsymbol{i} \pm \frac{3\sqrt{2}}{10}\boldsymbol{j} \mp \frac{\sqrt{2}}{2}\boldsymbol{k}$ or $\mp\left(\frac{2\sqrt{2}}{5}\boldsymbol{i} - \frac{3\sqrt{2}}{10}\boldsymbol{j} + \frac{\sqrt{2}}{2}\boldsymbol{k}\right)$ are two vectors perpendicular to $\boldsymbol{a}$ and $\boldsymbol{b}$. Of course any multiple of this vector will also be perpendicular to $\boldsymbol{a}$ and $\boldsymbol{b}$.

As we have already seen in Example 26.21, the scalar product is a very powerful tool when proving theorems in geometry. We now look at another theorem that is otherwise lengthy to prove by standard means.

EXAMPLE 12.23 Prove that the median to the base of an isosceles triangle is perpendicular to the base.

Solution

Consider the triangle ABC as shown, where M is the mid-point of the base $\overline{BC}$. Next, let $\boldsymbol{a} = \mathbf{AB}$ and $\boldsymbol{b} = \mathbf{AC}$. We then wish to show that $\mathbf{AM} \perp \mathbf{BC}$ (or $\mathbf{AM} \bullet \mathbf{BC} = 0$).

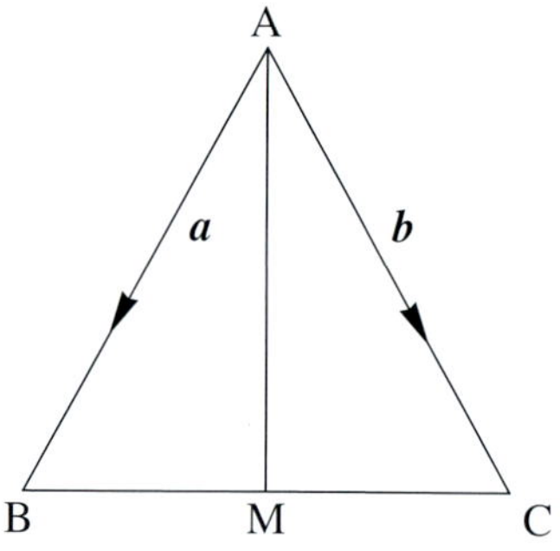

Now, $\mathbf{AM} = \mathbf{AB} + \mathbf{BM} = \mathbf{AB} + \frac{1}{2}\mathbf{BC}$

$= \boldsymbol{a} + \frac{1}{2}(\boldsymbol{b} - \boldsymbol{a})$

$= \frac{1}{2}(\boldsymbol{a} + \boldsymbol{b})$

Therefore, $\mathbf{AM} \bullet \mathbf{BC} = \frac{1}{2}(\boldsymbol{a} + \boldsymbol{b}) \bullet (\boldsymbol{b} - \boldsymbol{a})$

$= \frac{1}{2}(\boldsymbol{a} \bullet \boldsymbol{b} - \boldsymbol{a} \bullet \boldsymbol{a} + \boldsymbol{b} \bullet \boldsymbol{b} - \boldsymbol{b} \bullet \boldsymbol{a})$

$= \frac{1}{2}(-|\boldsymbol{a}|^2 + |\boldsymbol{b}|^2)$ [because $\boldsymbol{a} \bullet \boldsymbol{b} = \boldsymbol{b} \bullet \boldsymbol{a}$]

$= 0$ [because $|\boldsymbol{a}| = |\boldsymbol{b}|$]

Therefore, as $\mathbf{AM} \neq \mathbf{0}$ and $\mathbf{BC} \neq \mathbf{0}$, then $\mathbf{AM} \bullet \mathbf{BC} = 0 \Rightarrow \mathbf{AM} \perp \mathbf{BC}$.
i.e., the median is perpendicular to the base.

12.6.4 DIRECTION COSINES

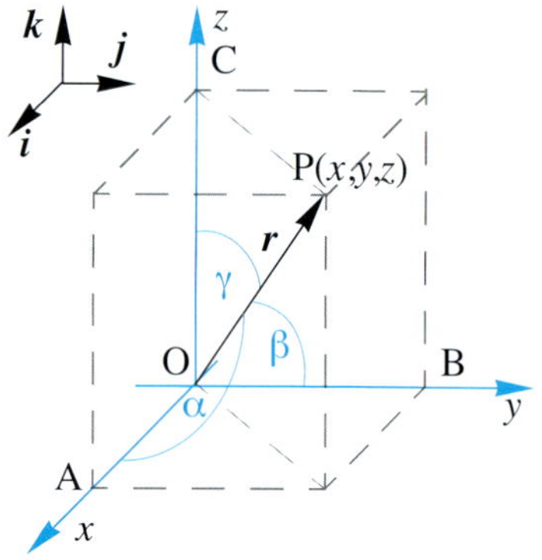

Let α, β and γ denote the angles that the position vector

$$\mathbf{OP} = \boldsymbol{r} = x\boldsymbol{i} + y\boldsymbol{j} + z\boldsymbol{k}$$

makes with the positive directions of the x-, y- and z- axes respectively.

These angles, are known as the direction angles of $\boldsymbol{r}$ and the cosines of these angles, i.e., $\cos\alpha$, $\cos\beta$ and $\cos\gamma$ are known as the **direction cosines** of $\boldsymbol{r}$.

We denote the direction cosines of $\boldsymbol{r}$ by the letters l, m and n respectively.
So that

$$l = \cos\alpha = \frac{x}{|\boldsymbol{r}|}, \quad m = \cos\beta = \frac{y}{|\boldsymbol{r}|}, \quad n = \cos\gamma = \frac{z}{|\boldsymbol{r}|}$$

Where $\cos\alpha = \frac{x}{|\boldsymbol{r}|}$ is obtained from ΔOAP

$\cos\beta = \frac{y}{|\boldsymbol{r}|}$ is obtained from ΔOBP

and $\cos\gamma = \frac{z}{|\boldsymbol{r}|}$ is obtained from ΔOCP

Also, realise then, that the unit vector, $\hat{\boldsymbol{r}}$, is such that

$$\hat{\boldsymbol{r}} = \frac{1}{|\boldsymbol{r}|}\boldsymbol{r} = \frac{1}{|\boldsymbol{r}|}(x\boldsymbol{i} + y\boldsymbol{j} + z\boldsymbol{k}) = \frac{x}{|\boldsymbol{r}|}\boldsymbol{i} + \frac{y}{|\boldsymbol{r}|}\boldsymbol{j} + \frac{z}{|\boldsymbol{r}|}\boldsymbol{k} = \cos\alpha\boldsymbol{i} + \cos\beta\boldsymbol{j} + \cos\gamma\boldsymbol{k}$$

This gives the results $\cos^2\alpha + \cos^2\beta + \cos^2\gamma = 1$ or $l^2 + m^2 + n^2 = 1$.

This means that if θ is the angle between two vectors, $\boldsymbol{a} = x_1\boldsymbol{i} + y_1\boldsymbol{j} + z_1\boldsymbol{k}$ and $\boldsymbol{b} = x_2\boldsymbol{i} + y_2\boldsymbol{j} + z_2\boldsymbol{k}$ with direction cosines l_1, m_1, n_1 and l_2, m_2, n_2 respectively, then

$$\cos\theta = l_1 l_2 + m_1 m_2 + n_1 n_2$$

We will come across this result at a later stage, when we look at the angle between two straight lines. At this stage however, you should realise that this result is a simple extension of using the scalar product to find the angle between two vectors. That is, the cosine of the angle between the vectors $\boldsymbol{a}$ and $\boldsymbol{b}$ is the same as the cosine of the angle between their unit vectors $\hat{\boldsymbol{a}}$ and $\hat{\boldsymbol{b}}$ – which is what we are saying when we use $\cos\theta = l_1 l_2 + m_1 m_2 + n_1 n_2$.

EXAMPLE 12.24

Consider the vectors $\boldsymbol{a} = 2\boldsymbol{i} - \sqrt{2}\boldsymbol{j} + \boldsymbol{k}$ and $\boldsymbol{b} = -\boldsymbol{i} + 2\boldsymbol{j} - \boldsymbol{k}$.

(a) Find the direction cosines of the vector $\boldsymbol{a}$.
(b) Find the direction angles of $\boldsymbol{b}$.
(c) Find the cosine of the angle between $\boldsymbol{a}$ and $\boldsymbol{b}$.

Solution

(a) We first need to find $|\boldsymbol{a}|$: $|\boldsymbol{a}| = \sqrt{2^2 + (-\sqrt{2})^2 + 1^2} = \sqrt{7}$.

Then, $\hat{\boldsymbol{a}} = \frac{1}{|\boldsymbol{a}|}(2\boldsymbol{i} - \sqrt{2}\boldsymbol{j} + \boldsymbol{k}) = \frac{1}{\sqrt{7}}(2\boldsymbol{i} - \sqrt{2}\boldsymbol{j} + \boldsymbol{k}) = \frac{2}{\sqrt{7}}\boldsymbol{i} - \frac{\sqrt{2}}{\sqrt{7}}\boldsymbol{j} + \frac{1}{\sqrt{7}}\boldsymbol{k}$

So that the direction cosines are $\cos\alpha = \frac{2}{\sqrt{7}}$, $\cos\beta = -\frac{\sqrt{2}}{\sqrt{7}}$ and $\cos\gamma = \frac{1}{\sqrt{7}}$.

(b) As in part (a) we find the unit vector of $\boldsymbol{b}$: $|\boldsymbol{b}| = \sqrt{1 + 4 + 1} = \sqrt{6}$.

$$\therefore \hat{\boldsymbol{b}} = -\frac{1}{\sqrt{6}}\boldsymbol{i} + \frac{2}{\sqrt{6}}\boldsymbol{j} - \frac{1}{\sqrt{6}}\boldsymbol{k}$$

To find the directional angles of $\boldsymbol{b}$ we have,

$$\cos\alpha = -\frac{1}{\sqrt{6}} \text{ so that } \alpha = \arccos\left(-\frac{1}{\sqrt{6}}\right) \approx 114°6'$$

$$\cos\beta = \frac{2}{\sqrt{6}} \text{ so that } \beta = \arccos\left(\frac{2}{\sqrt{6}}\right) \approx 35°16'$$

and

$$\cos\gamma = -\frac{1}{\sqrt{6}} \text{ so that } \gamma = \arccos\left(-\frac{1}{\sqrt{6}}\right) \approx 114°6'$$

(c) If θ is the angle between $\boldsymbol{a}$ and $\boldsymbol{b}$ then using the direction cosines of $\boldsymbol{a}$ and $\boldsymbol{b}$ we have

$$\cos\theta = l_1 l_2 + m_1 m_2 + n_1 n_2 = \frac{2}{\sqrt{7}} \times -\frac{1}{\sqrt{6}} + \left(-\frac{\sqrt{2}}{\sqrt{7}}\right) \times \frac{2}{\sqrt{6}} + \frac{1}{\sqrt{7}} \times \left(-\frac{1}{\sqrt{6}}\right)$$

$$= \frac{-2-2\sqrt{2}-1}{\sqrt{42}}$$

$$= -\frac{3+2\sqrt{2}}{\sqrt{42}}$$

EXERCISES 12.6

1. Find the scalar product, $\boldsymbol{a} \bullet \boldsymbol{b}$, for each of the following

(a)

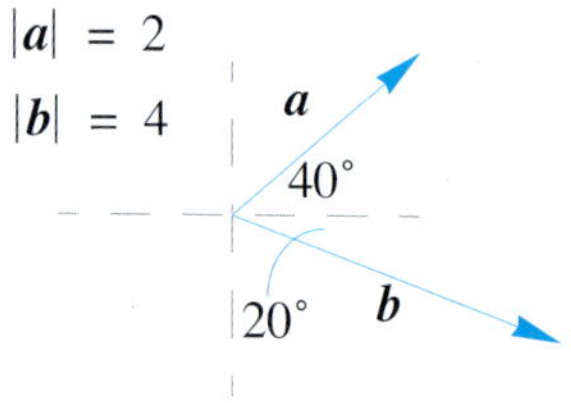

(b)

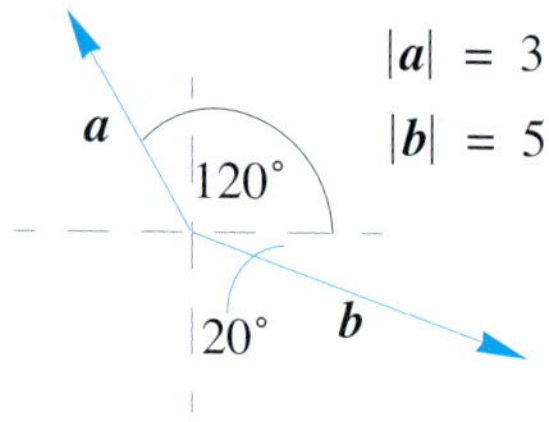

(c)

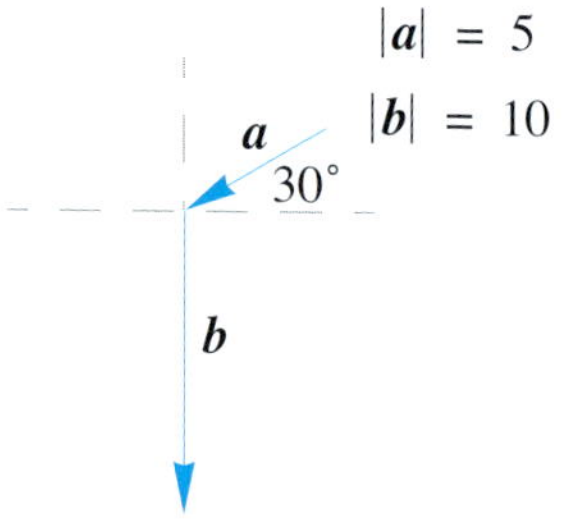

2. Find the scalar products of these pairs of vectors.

(i) $3\boldsymbol{i}+2\boldsymbol{j}$ and $2\boldsymbol{i}+3\boldsymbol{j}$ (ii) $3\boldsymbol{i}+7\boldsymbol{j}$ and $2\boldsymbol{i}+3\boldsymbol{j}$

(iii) $3\boldsymbol{i}-\boldsymbol{j}$ and $-2\boldsymbol{i}+2\boldsymbol{j}$ (iv) $6\boldsymbol{i}+\boldsymbol{j}-\boldsymbol{k}$ and $-7\boldsymbol{i}-4\boldsymbol{j}+3\boldsymbol{k}$

(v) $-\boldsymbol{j}+5\boldsymbol{k}$ and $-4\boldsymbol{i}+\boldsymbol{j}+\boldsymbol{k}$ (vi) $-\boldsymbol{i}+5\boldsymbol{j}+4\boldsymbol{k}$ and $5\boldsymbol{i}-4\boldsymbol{k}$

(vii) $\begin{pmatrix}0\\6\\1\end{pmatrix}$ and $\begin{pmatrix}7\\2\\-6\end{pmatrix}$ (viii) $\begin{pmatrix}-3\\-1\\7\end{pmatrix}$ and $\begin{pmatrix}3\\2\\1\end{pmatrix}$ (ix) $\begin{pmatrix}-6\\-1\\7\end{pmatrix}$ and $\begin{pmatrix}7\\3\\5\end{pmatrix}$

3. Find the angles between these pairs of vectors, giving the answers in degrees, correct to the nearest degree.

(i) $-4\boldsymbol{i}-4\boldsymbol{j}$ and $-3\boldsymbol{i}+2\boldsymbol{j}$ (ii) $\boldsymbol{i}-\boldsymbol{j}$ and $3\boldsymbol{i}+6\boldsymbol{j}$

(iii) $-4\boldsymbol{i}-2\boldsymbol{j}$ and $-\boldsymbol{i}-7\boldsymbol{j}$ (iv) $-7\boldsymbol{i}+3\boldsymbol{j}$ and $-2\boldsymbol{i}-\boldsymbol{j}$

(v) $i + 3j + 7k$ and $6i + 7j - k$ (vi) $j + 3k$ and $-j - 2k$

(vii) $\begin{pmatrix} -3 \\ -1 \\ -5 \end{pmatrix}$ and $\begin{pmatrix} 4 \\ 5 \\ -5 \end{pmatrix}$ (viii) $\begin{pmatrix} -2 \\ 7 \\ -7 \end{pmatrix}$ and $\begin{pmatrix} 5 \\ 2 \\ -5 \end{pmatrix}$

4. Two vectors are defined as $\boldsymbol{a} = 2\boldsymbol{i} + x\boldsymbol{j}$ and $\boldsymbol{b} = \boldsymbol{i} - 4\boldsymbol{j}$. Find the value of x if:
(i) the vectors are parallel.
(ii) the vectors are perpendicular.

5. If $\boldsymbol{a} = 2\boldsymbol{i} - 3\boldsymbol{j} + \boldsymbol{k}$, $\boldsymbol{b} = -\boldsymbol{i} + 2\boldsymbol{j} + 2\boldsymbol{k}$ and $\boldsymbol{c} = \boldsymbol{i} + \boldsymbol{k}$, find, where possible,
(a) $\boldsymbol{a} \bullet \boldsymbol{b}$ (b) $(\boldsymbol{a} - \boldsymbol{b}) \bullet \boldsymbol{c}$ (c) $\boldsymbol{a} \bullet \boldsymbol{b} \bullet \boldsymbol{c}$
(d) $(\boldsymbol{a} - \boldsymbol{b}) \bullet (\boldsymbol{a} + \boldsymbol{b})$ (e) $\dfrac{\boldsymbol{a}}{\boldsymbol{c}}$ (f) $\boldsymbol{b} \bullet \boldsymbol{0}$

6. If $\boldsymbol{a} = 2\boldsymbol{i} - \sqrt{3}\boldsymbol{j}$, $\boldsymbol{b} = \sqrt{3}\boldsymbol{i} - \boldsymbol{j}$ and $\boldsymbol{c} = \boldsymbol{i} + \boldsymbol{j}$ find, where possible,
(a) $\boldsymbol{a} \bullet (\boldsymbol{b} + \boldsymbol{c}) + \boldsymbol{b} \bullet (\boldsymbol{c} - \boldsymbol{a}) + \boldsymbol{c} \bullet (\boldsymbol{a} - \boldsymbol{b})$ (b) $(\boldsymbol{b} - \boldsymbol{c}) \bullet (\boldsymbol{c} - \boldsymbol{b}) + |\boldsymbol{b}|^2$
(c) $2|\boldsymbol{a}|^2 - \sqrt{3}\boldsymbol{c} \bullet \boldsymbol{c}$ (d) $\sqrt{\dfrac{\boldsymbol{a}}{|\boldsymbol{a}|} + \dfrac{\boldsymbol{b}}{|\boldsymbol{b}|} + \dfrac{\boldsymbol{c}}{|\boldsymbol{c}|}}$

7. Find the value(s) of x for which the vectors $x\boldsymbol{i} + \boldsymbol{j} - \boldsymbol{k}$ and $x\boldsymbol{i} - 2x\boldsymbol{j} - \boldsymbol{k}$ are perpendicular.

8. P, Q and R are three points in space with coordinates (2, –1, 4), (3, 1, 2) and (–1, 2, 5) respectively. Find angle Q in the triangle PQR.

9. Find the values of x and y if $\boldsymbol{u} = x\boldsymbol{i} + 2y\boldsymbol{j} - 8\boldsymbol{k}$ is perpendicular to both $\boldsymbol{v} = 2\boldsymbol{i} - \boldsymbol{j} + \boldsymbol{k}$ and $\boldsymbol{w} = 3\boldsymbol{i} + 2\boldsymbol{j} - 4\boldsymbol{k}$.

10. Find the unit vector that is perpendicular to both $\boldsymbol{a} = 3\boldsymbol{i} + 6\boldsymbol{j} - \boldsymbol{k}$ and $\boldsymbol{b} = 4\boldsymbol{i} + \boldsymbol{j} + \boldsymbol{k}$.

11. Show that if $\boldsymbol{u}$ is a vector in three dimensions, then $\boldsymbol{u} = (\boldsymbol{u} \bullet \boldsymbol{i})\boldsymbol{i} + (\boldsymbol{u} \bullet \boldsymbol{j})\boldsymbol{j} + (\boldsymbol{u} \bullet \boldsymbol{k})\boldsymbol{k}$.

12. (a) Find a vector perpendicular to both $\boldsymbol{a} = -\boldsymbol{i} + 2\boldsymbol{j} + 4\boldsymbol{k}$ and $\boldsymbol{b} = 2\boldsymbol{i} - 3\boldsymbol{j} + 2\boldsymbol{k}$.
(b) Find a vector perpendicular to $2\boldsymbol{i} + \boldsymbol{j} - 7\boldsymbol{k}$.

13. Show that if $|\boldsymbol{a} - \boldsymbol{b}| = |\boldsymbol{a} + \boldsymbol{b}|$, where $\boldsymbol{a} \neq \boldsymbol{0}$ and $\boldsymbol{b} \neq \boldsymbol{0}$, then $\boldsymbol{a}$ and $\boldsymbol{b}$ are perpendicular.

14. If $\boldsymbol{a} \bullet \boldsymbol{b} = \boldsymbol{a} \bullet \boldsymbol{c}$ where $\boldsymbol{a} \neq \boldsymbol{0} \neq \boldsymbol{b}$, what conclusion(s) can be made?

15. Find the direction cosine of the following vectors
(a) $3\boldsymbol{i} + 4\boldsymbol{j}$ (b) $\sqrt{2}\boldsymbol{i} + \boldsymbol{j} - \boldsymbol{k}$

16. (a) Find the direction cosines of the vector $\boldsymbol{r} = -2\boldsymbol{i} + 2\boldsymbol{j} + \boldsymbol{k}$.
(b) Hence, find the angles, α, β and γ that $\boldsymbol{r}$ makes with the positive x-, y- and z- axes respectively.

17. Using the scalar product for vectors prove that the cosine of the angle between two lines with direction cosines l_1, m_1, n_1 and l_2, m_2, n_2 is given by $\cos\theta = l_1 l_2 + m_1 m_2 + n_1 n_2$.

18. Find the cosine of the acute angle between
(a) two diagonals of a cube.
(b) the diagonal of a cube and one of its edges.

19. (a) On the same set of axes sketch the graphs of $x + 3y - 6 = 0$ and $2x - y + 6 = 0$, clearly labelling all intercepts with the axes.
(b) Find a unit vector along the line
i. $x + 3y - 6 = 0$.
ii. $2x - y + 6 = 0$.
(c) Hence find the acute angle between the two lines $x + 3y - 6 = 0$ and $2x - y + 6 = 0$.

20. Find a unit vector $\boldsymbol{a}$ such that $\boldsymbol{a}$ makes an angle of 45° with the z–axis and is such that the vector $\boldsymbol{i} - \boldsymbol{j} + \boldsymbol{a}$ is a unit vector.

21. Using the scalar product for vectors prove Pythagoras's Theorem for the triangle ABC shown.

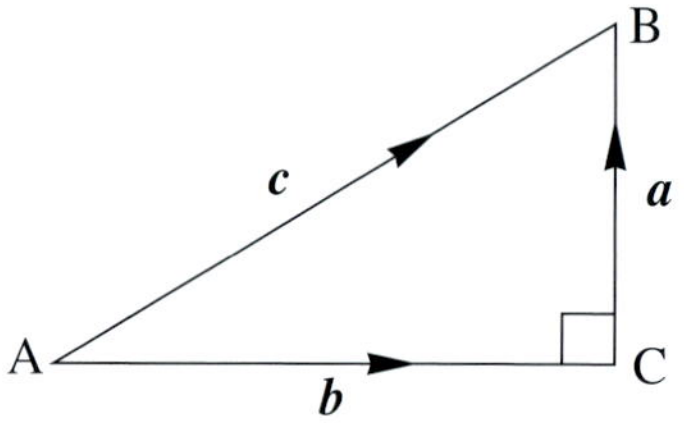

22. Prove that an angle inscribed in a semicircle is a right angle.

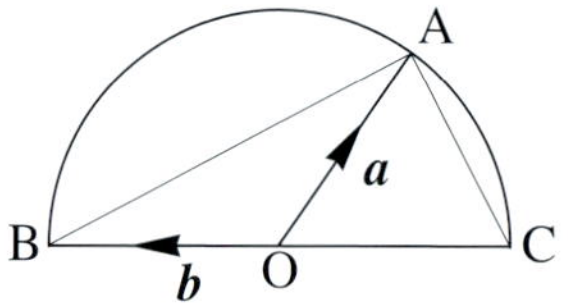

23. In the trapezium shown, BE:BC = 1:3.
Show that $3\mathbf{AC} \bullet \mathbf{DE} = 2(4m^2 - n^2)$
where $|\mathbf{AB}| = m$, $|\mathbf{DC}| = 2|\mathbf{AB}|$ and $|\mathbf{DA}| = n$.

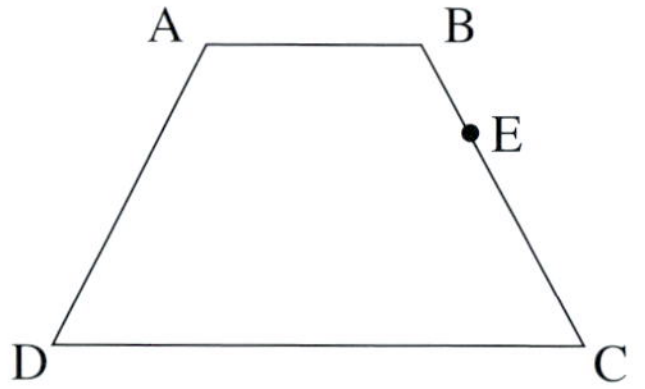

24. Prove that the altitudes of any triangle are concurrent.

25. An oil pipeline runs from a well (W) to a distribution point (D) which is 4km East and 8km North of the well. A second well (S) is drilled at a point 9km East and 7km South of the distribution point. It is desired to lay a new pipeline from the second well to a point (X) on the original pipeline where the two pipes will be joined. This new pipeline must be as short as possible.
(a) Set up a suitable vector basis using the first well as the origin.
(b) Express $\overrightarrow{WD}, \overrightarrow{WS}, \overrightarrow{DS}$ in terms of your basis.
(c) Write a unit vector in the direction of $\overrightarrow{WD}$.
(d) If the point X is d km along the pipeline from the first well, write a vector equal to $\overrightarrow{WX}$.
(e) Hence find the vector $\overrightarrow{WX}$ such that the new pipeline is as short as possible.

12.6.5 USING A GRAPHICS CALCULATOR

Many models of graphics calculator allow entry of vectors and limited calculations with them. The TI 82/3 models use the matrix menu system. After entering the vectors as column matrices, simple arithmetic calculations can be performed. The matrix menu structure will be considered in more detail later in this chapter. More advanced models such as the TI-85 can calculate absolute values and scalar products. These are under the **VECTOR**, **MATH** menus.

We consider two examples.

Example 1 (Addition/subtraction):

To enter and calculate a vector sum such as $3 \times \begin{pmatrix} 2 \\ -3 \\ 1 \end{pmatrix} - 4\begin{pmatrix} -5 \\ 2 \\ -4 \end{pmatrix}$ the steps are:

Step 1: Enter the two matrices using the **MATRIX** menu, **EDIT** option.

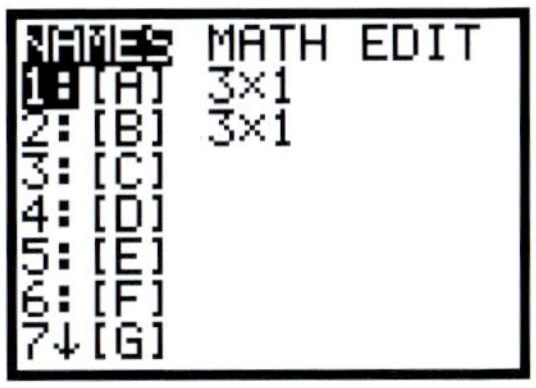

Step 2: Choose matrix A and define the correct dimensions for the vectors, i.e., 3×1 and enter the correct components. Remember to use the 'negative key' (–) rather than the subtraction key when entering the negative number. Repeat for matrix B.

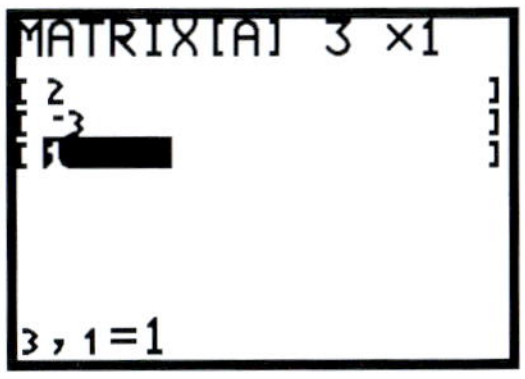

Step 3: The calculation is now completed by keying it as if it were a normal numerical calculation using the memories of the calculator. In this case the names of the two vectors are accessed using the **MATRIX NAMES** menu and highlighting the name required at each stage of the calculation.

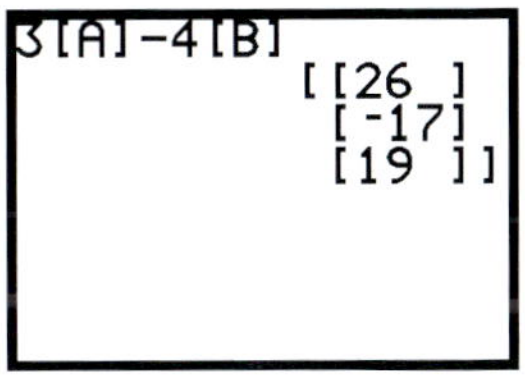

Example 2 (Scalar product):

To find the scalar product, $\begin{pmatrix} 2 \\ -3 \\ 1 \end{pmatrix} \bullet \begin{pmatrix} -5 \\ 2 \\ -4 \end{pmatrix}$ choose the Matrix A as having dimensions 1×3 and Matrix B as having dimension 3×1. Then multiply the matrices:

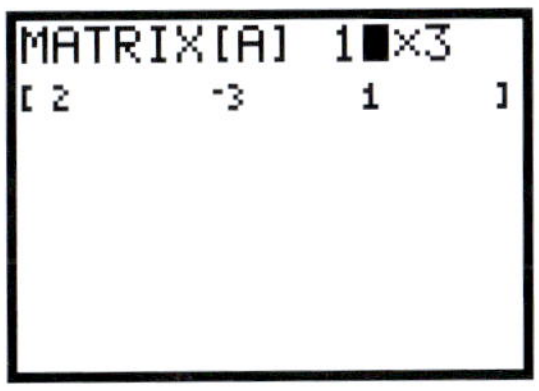

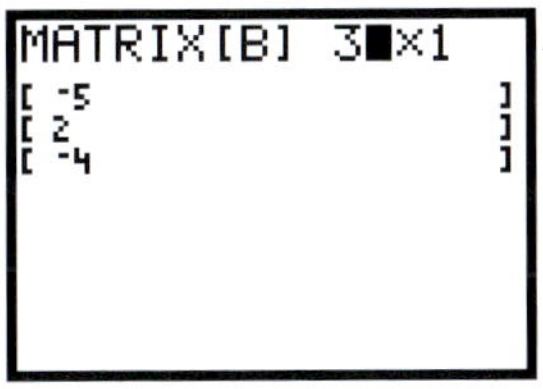

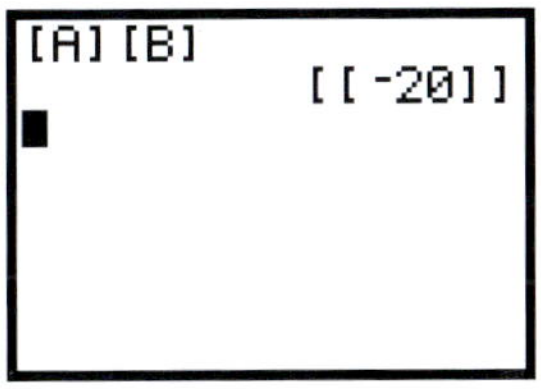

So that in this case, the dot product, is –20.

12.7 VECTOR EQUATION OF A LINE

12.7.1 VECTOR EQUATION OF A LINE IN 2-D

We start this section by considering the following problem:

Relative to an origin O, a house, situated 8 km North of O, stands next to a straight road. The road runs past a second house, located 4 km East of O. If a person is walking along the road from the house North of O to the house East of O, determine the position of the person while on the road relative to O.

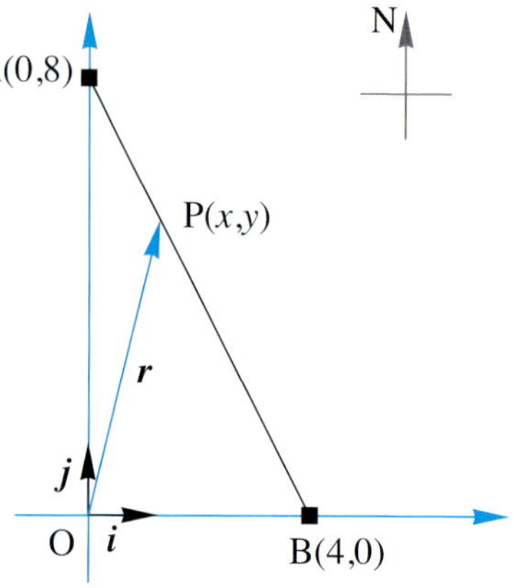

We start by drawing a diagram and place the person along the road at some point P.

We need to determine the position vector of point P.

We have $\boldsymbol{r} = \mathbf{OP} = \mathbf{OA} + \mathbf{AP}$

Now, as P lies somewhere along $\overline{AB}$, we can write $\mathbf{AP} = \lambda\mathbf{AB}$, where $0 \le \lambda \le 1$, so that when $\lambda = 0$ the person is at A and when $\lambda = 1$ the person is at B.

Next, $\mathbf{AB} = \mathbf{AO} + \mathbf{OB} = -8\boldsymbol{j} + 4\boldsymbol{i}$, and so we have that

$$\boldsymbol{r} = 8\boldsymbol{j} + \lambda(-8\boldsymbol{j} + 4\boldsymbol{i}).$$

This provides us with the position vector of the person while walking on the road.

We take this equation a little further. The position vector of P can be written as $\boldsymbol{r} = x\boldsymbol{i} + y\boldsymbol{j}$ and so we have that

$$x\boldsymbol{i} + y\boldsymbol{j} = 8\boldsymbol{j} + \lambda(-8\boldsymbol{j} + 4\boldsymbol{i})$$

That is, we have $x\boldsymbol{i} + y\boldsymbol{j} = 4\lambda\boldsymbol{i} + (8 - 8\lambda)\boldsymbol{j}$ meaning that

$$x = 4\lambda \text{ and } y = 8 - 8\lambda$$

The equations $x = 4\lambda$ – (1) and $y = 8 - 8\lambda$ – (2) are known as the **parametric form** of the equations of a straight line

Next, from these parametric equations we have $\lambda = \frac{x}{4}$ – (3) and $\lambda = \frac{y-8}{-8}$ – (4)

Then, equating (3) and (4) we have $\frac{x}{4} = \frac{y-8}{-8}$. This equation is known as the **Cartesian form** of the equation of a straight line. We can go one step further and simplify this last equation.

$$\frac{x}{4} = \frac{y-8}{-8} \Leftrightarrow -2x = y - 8 \Leftrightarrow y = -2x + 8$$

which corresponds to the straight line passing through A and B.

This approach to describe the position of an object (or person) is of great use when dealing with objects travelling in a straight line. When planes are coming in for landing, it is crucial that their positions along their flight paths are known, otherwise one plane could be heading for a collision with another plane in the air.

We now formalise the definition of the vector equation of a line in a plane:

The vector equation of a line L in the direction of the vector $\boldsymbol{b}$, passing through the point A with position vector **a** is given by

$$\boldsymbol{r} = \boldsymbol{a} + \lambda\boldsymbol{b}$$

where λ is a scalar parameter.

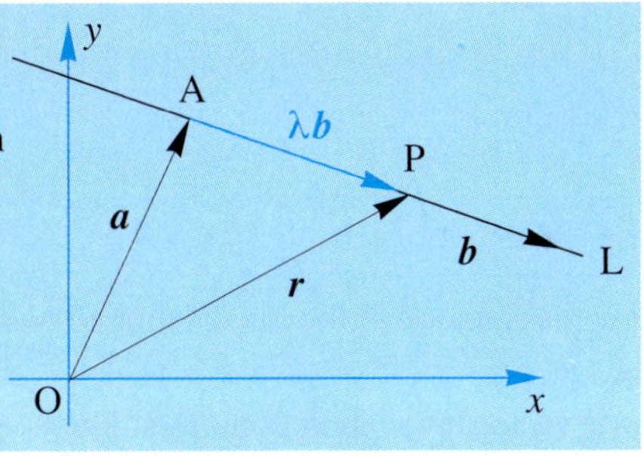

PROOF:

Let the point P(x, y) be any point on the line L, then the vector **AP** is parallel to the vector $\boldsymbol{b}$.

$$\boldsymbol{r} = \mathbf{OP}$$
$$= \mathbf{OA} + \mathbf{AP}$$
$$\therefore \boldsymbol{r} = \boldsymbol{a} + \lambda\boldsymbol{b}$$

So the equation of L is given by $\boldsymbol{r} = \boldsymbol{a} + \lambda\boldsymbol{b}$ as required.

We can now derive two other forms for equations of a line. We start by letting the coordinates of A be (a_1, a_2), the coordinates of P be (x, y) and the vector $\boldsymbol{b} = \begin{pmatrix} b_1 \\ b_2 \end{pmatrix}$.

From $\boldsymbol{r} = \boldsymbol{a} + \lambda\boldsymbol{b}$ we have, $\begin{pmatrix} x \\ y \end{pmatrix} = \begin{pmatrix} a_1 \\ a_2 \end{pmatrix} + \lambda\begin{pmatrix} b_1 \\ b_2 \end{pmatrix} \Leftrightarrow \begin{pmatrix} x \\ y \end{pmatrix} = \begin{pmatrix} a_1 + \lambda b_1 \\ a_2 + \lambda b_2 \end{pmatrix}$

This provides us with the

Parametric form for the equation of a straight line:

$$x = a_1 + \lambda b_1$$
$$y = a_2 + \lambda b_2$$

Next, from the parametric form we have $x = a_1 + \lambda b_1 \Leftrightarrow x - a_1 = \lambda b_1 \Leftrightarrow \lambda = \dfrac{x - a_1}{b_1}$ – (1)

and $y = a_2 + \lambda b_2 \Leftrightarrow y - a_2 = \lambda b_2 \Leftrightarrow \lambda = \dfrac{y - a_2}{b_2}$ – (2)

Equating (1) and (2) provides us with the

Cartesian form for the equation of a straight line:

$$\frac{x - a_1}{b_1} = \frac{y - a_2}{b_2}$$

EXAMPLE 12.25

Find the vector equation of the line L, as shown in the diagram.

Comment on the uniqueness of this equation.

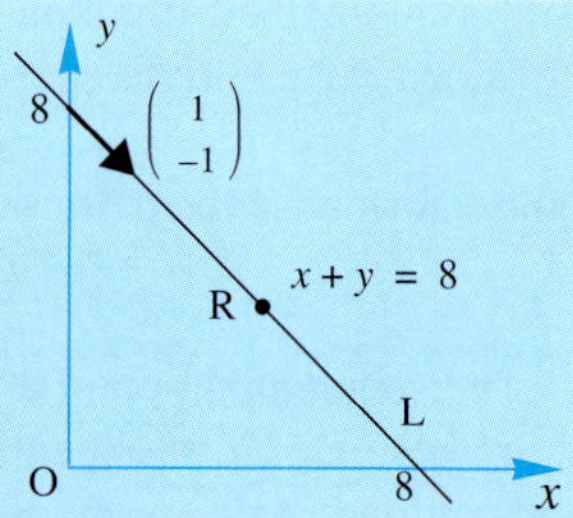

Solution

The vector equation of the line L is based on finding (or using) *any* point on the line, such as $\begin{pmatrix}0\\8\end{pmatrix}$, and *any* vector in the direction of the line L, such as $\begin{pmatrix}1\\-1\end{pmatrix}$.

The position vector of any point R on the line can then be written as $\boldsymbol{r} = \begin{pmatrix}0\\8\end{pmatrix} + \lambda\begin{pmatrix}1\\-1\end{pmatrix}$.

As λ varies, different points on the line are generated, and conversely any point on the line has a corresponding value of λ. For example, substituting $\lambda = 3$ gives the point $\begin{pmatrix}3\\5\end{pmatrix}$ and the point $\begin{pmatrix}8\\0\end{pmatrix}$ corresponds to $\lambda = 8$.

NB: the vector equation (in parametric form) is **not unique**. The equation $\boldsymbol{r} = \begin{pmatrix}4\\4\end{pmatrix} + \lambda\begin{pmatrix}-2\\2\end{pmatrix}$ is an equally valid description of the line, and in this case substituting $\lambda = 0.5$ generates the point $\begin{pmatrix}3\\5\end{pmatrix}$.

EXAMPLE 12.26

Find the vector equation of the line L, passing through the point A(2, 5) and parallel to the vector $3\boldsymbol{i} - 4\boldsymbol{j}$.

Solution

Rather than depend on a standard formula, it is always helpful to visualise problems such as these, in particular, when we move onto straight lines in space. We draw a general representation of this situation and work from there.

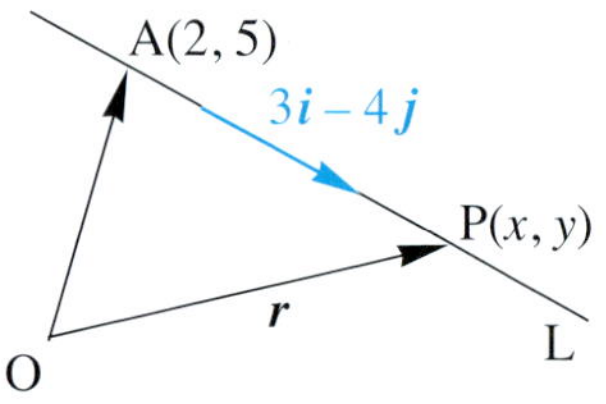

Let the point P be any point on the line L with position vector **r**,

then $\mathbf{OP} = \mathbf{OA} + \mathbf{AP}$

However, as A and P lie on the line L, then $\mathbf{AP} = \lambda(3\boldsymbol{i} - 4\boldsymbol{j})$.

Therefore, $\boldsymbol{r} = (2\boldsymbol{i} + 5\boldsymbol{j}) + \lambda(3\boldsymbol{i} - 4\boldsymbol{j})$

This represents the vector equation of the line L in terms of the parameter λ, where $\lambda \in \mathbb{R}$.

The equation could also be written as, $\boldsymbol{r} = (2 + 3\lambda)\boldsymbol{i} + (5 - 4\lambda)\boldsymbol{j}$.

EXAMPLE 12.27

Find the vector equation of the line L, passing through the points A(1, 4) and B(5, 8). Give both the parametric form and Cartesian form of L.

Solution

We start with a sketch of the situation described:

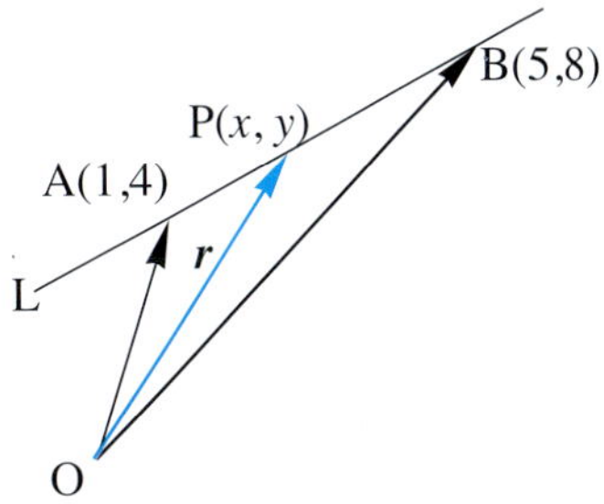

Let the point P be any point on the line L with position vector $\boldsymbol{r}$, then $\mathbf{OP} = \mathbf{OA} + \mathbf{AP}$

Then, as $\mathbf{AP} \parallel \mathbf{AB} \Rightarrow \mathbf{AP} = \lambda\mathbf{AB}$, where $\lambda \in \mathbb{R}$.

This means that we need to find the vector **AB** which will be the vector parallel to the line L. So, we have

$$\mathbf{AB} = \mathbf{AO} + \mathbf{OB} = -\begin{pmatrix}1\\4\end{pmatrix} + \begin{pmatrix}5\\8\end{pmatrix} = \begin{pmatrix}4\\4\end{pmatrix} = 4\begin{pmatrix}1\\1\end{pmatrix}$$

Therefore, from $\mathbf{OP} = \mathbf{OA} + \mathbf{AP}$ we have

$$\mathbf{OP} = \begin{pmatrix}1\\4\end{pmatrix} + \lambda \times 4\begin{pmatrix}1\\1\end{pmatrix}$$

That is, $\boldsymbol{r} = \begin{pmatrix}1\\4\end{pmatrix} + t\begin{pmatrix}1\\1\end{pmatrix}$ where $t = 4\lambda$

This represents the vector equation of the straight line L.

To find the **parametric form** of L we make use of the equation $\boldsymbol{r} = \begin{pmatrix}1\\4\end{pmatrix} + t\begin{pmatrix}1\\1\end{pmatrix}$.

As P(x, y) is any point on the line L, we write the vector equation as $\begin{pmatrix}x\\y\end{pmatrix} = \begin{pmatrix}1\\4\end{pmatrix} + t\begin{pmatrix}1\\1\end{pmatrix}$.

From where we obtain the parametric equations, $x = 1 + t$ and $y = 4 + t$.

To find the **Cartesian form** of L we now make use of the parametric equations.

From $x = 1 + t$ we have $t = x - 1$ – (1) and from $y = 4 + t$ we have $t = y - 4$ – (2)

Then, equating (1) and (2) we have $x - 1 = y - 4$ (or $y = x + 3$).

EXAMPLE 12.28

The vector equation of the line L, is given by $\boldsymbol{r} = \begin{pmatrix}3 + 2\lambda\\5 - 5\lambda\end{pmatrix}$.

(a) Express the vector equation in the standard form $\boldsymbol{r} = \boldsymbol{a} + \lambda\boldsymbol{b}$.

(b) Find a unit vector in the direction of L.

(c) Find the Cartesian form of the line L.

Solution

(a) $\boldsymbol{r} = \begin{pmatrix}3 + 2\lambda\\5 - 5\lambda\end{pmatrix} = \begin{pmatrix}3\\5\end{pmatrix} + \begin{pmatrix}2\lambda\\-5\lambda\end{pmatrix} = \begin{pmatrix}3\\5\end{pmatrix} + \lambda\begin{pmatrix}2\\-5\end{pmatrix}$ [which is in the form $\boldsymbol{r} = \boldsymbol{a} + \lambda\boldsymbol{b}$]

(b) The direction of the line L is provided by the vector $\boldsymbol{b}$, i.e., $\begin{pmatrix}2\\-5\end{pmatrix}$.

To find the unit vector we need $\left|\begin{pmatrix}2\\-5\end{pmatrix}\right| = \sqrt{4 + 25} = \sqrt{29}$. $\therefore \hat{\boldsymbol{b}} = \frac{1}{\sqrt{29}}\begin{pmatrix}2\\-5\end{pmatrix}$.

(c) Using the point P(x, y) as representing any point on the line L, we have that $\boldsymbol{r} = \begin{pmatrix} x \\ y \end{pmatrix}$.

Therefore, we can write the vector equation as $\begin{pmatrix} x \\ y \end{pmatrix} = \begin{pmatrix} 3 + 2\lambda \\ 5 - 5\lambda \end{pmatrix}$.

From this equation we then have that $x = 3 + 2\lambda$ – (1)

and $y = 5 - 5\lambda$ – (2)

We can now find the Cartesian equation by eliminating the parameter λ using (1) and (2).

From (1): $\lambda = \dfrac{x-3}{2}$. From (2): $\lambda = \dfrac{y-5}{-5}$

Therefore, $\dfrac{x-3}{2} = \dfrac{y-5}{-5}$.

EXAMPLE 12.29

Find the angle between the lines $\dfrac{x-2}{4} = \dfrac{y+1}{3}$ and $\dfrac{x+2}{-1} = \dfrac{y-4}{2}$.

Solution

We must first express the lines in their vector form. To do this we need to introduce a parameter for each line.

Let $\dfrac{x-2}{4} = \dfrac{y+1}{3} = \lambda$ giving the parametric equations $x = 2 + 4\lambda$ and $y = -1 + 3\lambda$.

We can now express these two parametric equations in the vector form:

$$\begin{pmatrix} x \\ y \end{pmatrix} = \begin{pmatrix} 2 + 4\lambda \\ -1 + 3\lambda \end{pmatrix} = \begin{pmatrix} 2 \\ -1 \end{pmatrix} + \lambda\begin{pmatrix} 4 \\ 3 \end{pmatrix}$$

This vector equation informs us that the line $\dfrac{x-2}{4} = \dfrac{y+1}{3}$ is parallel to the vector $\begin{pmatrix} 4 \\ 3 \end{pmatrix}$.

In the same way we can obtain the vector equation of the line $\dfrac{x+2}{-1} = \dfrac{y-4}{2}$.

Let $\dfrac{x+2}{-1} = \dfrac{y-4}{2} = t$ giving the parametric equations $x = -2 - t$ and $y = 4 + 2t$.

From where we obtain the vector equation $\begin{pmatrix} x \\ y \end{pmatrix} = \begin{pmatrix} -2 - t \\ 4 + 2t \end{pmatrix} = \begin{pmatrix} -2 \\ 4 \end{pmatrix} + t\begin{pmatrix} -1 \\ 2 \end{pmatrix}$.

This vector equation informs us that the line $\dfrac{x+2}{-1} = \dfrac{y-4}{2}$ is parallel to the vector $\begin{pmatrix} -1 \\ 2 \end{pmatrix}$.

To find the angle between the two lines we use their direction vectors, $\begin{pmatrix} 4 \\ 3 \end{pmatrix}$ and $\begin{pmatrix} -1 \\ 2 \end{pmatrix}$ along

with the scalar product: $\begin{pmatrix} 4 \\ 3 \end{pmatrix} \bullet \begin{pmatrix} -1 \\ 2 \end{pmatrix} = \left|\begin{pmatrix} 4 \\ 3 \end{pmatrix}\right| \times \left|\begin{pmatrix} -1 \\ 2 \end{pmatrix}\right| \cos\theta \Rightarrow -4 + 6 = \sqrt{16+9} \times \sqrt{1+4}\cos\theta$

$$\Leftrightarrow \cos\theta = \frac{2}{5\sqrt{5}}$$

$$\therefore \theta \approx 79°42'$$

Notice that when discusing the relationship between two straight lines in a plane there are a number of possible outcomes:

1. The lines are parallel and never meet

In this situation, the lines have the same direction vectors and no common points.

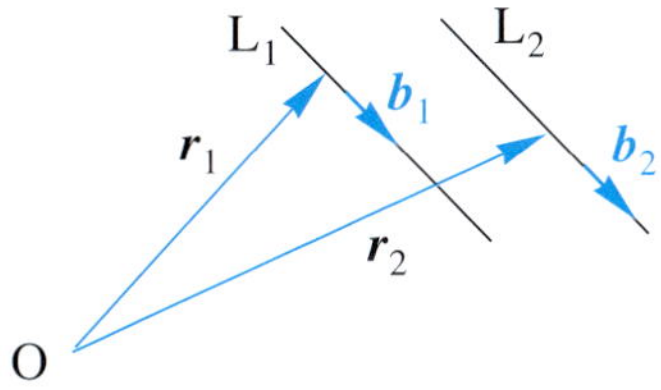

$L_1: \boldsymbol{r}_1 = \boldsymbol{a}_1 + \lambda \boldsymbol{b}_1$

$L_2: \boldsymbol{r}_1 = \boldsymbol{a}_2 + t\boldsymbol{b}_2$ and $\boldsymbol{b}_1 \| \boldsymbol{b}_2$

2. The lines are not parallel and therefore intersect at some point.

In this situation, the lines have non-parallel direction vectors and a common point.

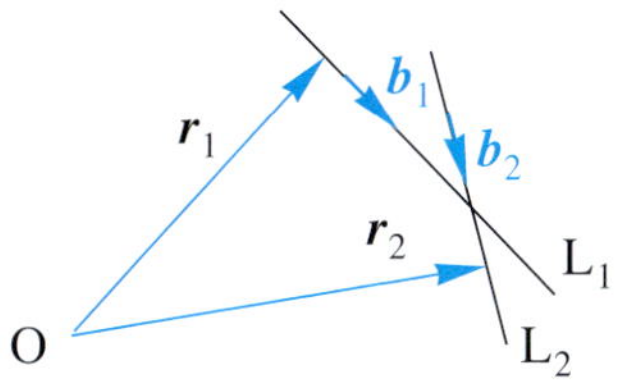

$L_1: \boldsymbol{r}_1 = \boldsymbol{a}_1 + \lambda \boldsymbol{b}_1$

$L_2: \boldsymbol{r}_1 = \boldsymbol{a}_2 + t\boldsymbol{b}_2$

3. The lines are parallel and have a common point, meaning that they are **coincident**.

In this situation, the lines have the same direction vectors and all their points are common.

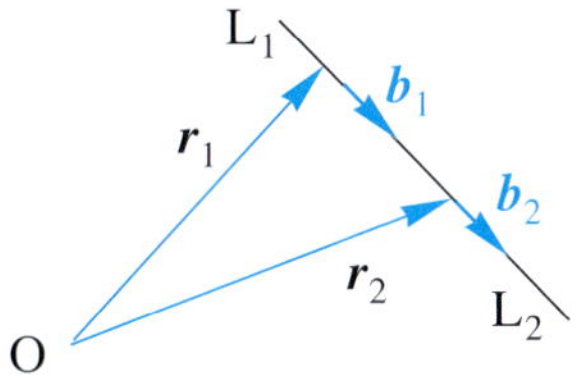

$L_1: \boldsymbol{r}_1 = \boldsymbol{a}_1 + \lambda \boldsymbol{b}_1$

$L_2: \boldsymbol{r}_1 = \boldsymbol{a}_2 + t\boldsymbol{b}_2$ and $\boldsymbol{b}_1 \| \boldsymbol{b}_2$

EXAMPLE 12.30

Find the point of intersection of the lines L_1 and L_2 whose vector equations are $\boldsymbol{r}_1 = 2\boldsymbol{i} + 3\boldsymbol{j} + \lambda(2\boldsymbol{i} + \boldsymbol{j})$ and $\boldsymbol{r}_2 = 5\boldsymbol{i} - 2\boldsymbol{j} + \mu(\boldsymbol{i} - 2\boldsymbol{j})$ respectively.

Solution

If the two lines intersect then there exists a point on L_1 and L_2 such that $\boldsymbol{r}_1 = \boldsymbol{r}_2$.

That is, $2\boldsymbol{i} + 3\boldsymbol{j} + \lambda(2\boldsymbol{i} + \boldsymbol{j}) = 5\boldsymbol{i} - 2\boldsymbol{j} + \mu(\boldsymbol{i} - 2\boldsymbol{j})$

$\therefore (2 + 2\lambda)\boldsymbol{i} + (3 + \lambda)\boldsymbol{j} = (5 + \mu)\boldsymbol{i} + (-2 - 2\mu)\boldsymbol{j}$

Giving the system of equations $2 + 2\lambda = 5 + \mu$ – (1)

and $3 + \lambda = -2 - 2\mu$ – (2)

Solving for λ: $2 \times (1) + (2)$: $7 + 5\lambda = 8 \Leftrightarrow \lambda = \frac{1}{5}$

Hence, using the vector equation $\boldsymbol{r}_1$, the point of intersection has the position vector

$\boldsymbol{r} = 2\boldsymbol{i} + 3\boldsymbol{j} + \frac{1}{5}(2\boldsymbol{i} + \boldsymbol{j}) = \frac{12}{5}\boldsymbol{i} + \frac{16}{5}\boldsymbol{j}$. That is, the lines intersect at the point $\left(\frac{12}{5}, \frac{16}{5}\right)$.

EXERCISES 12.7.1

1. For the straight line with equation $\boldsymbol{r} = \boldsymbol{a} + \lambda\boldsymbol{b}$ where $\boldsymbol{a} = \boldsymbol{i} + 2\boldsymbol{j}$ and $\boldsymbol{b} = -2\boldsymbol{i} + 3\boldsymbol{j}$ find the coordinates of the points on the line for which

(a) i. $\lambda = 0$ ii. $\lambda = 3$ iii. $\lambda = -2$

(b) Use part (a) to sketch the graph of $\boldsymbol{r} = \boldsymbol{i} + 2\boldsymbol{j} + \lambda(-2\boldsymbol{i} + 3\boldsymbol{j})$.

2. Find the vector equation of the line passing through the point A and parallel to the vector $\boldsymbol{b}$, where

(a) $A \equiv (2, 5)$, $\boldsymbol{b} = 3\boldsymbol{i} - 4\boldsymbol{j}$ (b) $A \equiv (-3, 4)$, $\boldsymbol{b} = -\boldsymbol{i} + 5\boldsymbol{j}$

(c) $A \equiv (0, 1)$, $\boldsymbol{b} = 7\boldsymbol{i} + 8\boldsymbol{j}$ (d) $A \equiv (1, -6)$, $\boldsymbol{b} = 2\boldsymbol{i} + 3\boldsymbol{j}$

(e) $A \equiv (-1, -1)$, $\boldsymbol{b} = \begin{pmatrix} -2 \\ 10 \end{pmatrix}$ (f) $A \equiv (1, 2)$, $\boldsymbol{b} = \begin{pmatrix} 5 \\ 1 \end{pmatrix}$

3. Find a vector equation of the line passing through the points A and B where

(a) $A(2, 3)$, $B(4, 8)$ (b) $A(1, 5)$, $B(-2, 1)$ (c) $A(4, -3)$, $B(-1, -2)$

4. Find the vector equation of the straight line defined by the parametric equations

(a) $x = 9 + \lambda, y = 5 - 3\lambda$ (b) $x = 6 - 4t, y = -6 - 2t$

(c) $x = -1 - 4\lambda, y = 3 + 8\lambda$ (d) $x = 1 + \frac{1}{2}\mu, y = 2 - \frac{1}{3}\mu$

5. Find the parametric form of the straight line having the vector equation

(a) $\boldsymbol{r} = \begin{pmatrix} -8 \\ 10 \end{pmatrix} + \mu\begin{pmatrix} 2 \\ 1 \end{pmatrix}$ (b) $\boldsymbol{r} = \begin{pmatrix} 7 \\ 4 \end{pmatrix} + \mu\begin{pmatrix} -3 \\ -2 \end{pmatrix}$

(c) $\boldsymbol{r} = \begin{pmatrix} 5 \\ 3 \end{pmatrix} + \frac{\mu}{2}\begin{pmatrix} 5 \\ 1 \end{pmatrix}$ (d) $\boldsymbol{r} = \begin{pmatrix} 0.5 - 0.1t \\ 0.4 + 0.2t \end{pmatrix}$

6. Find the Cartesian form of the straight line having the vector equation

(a) $\boldsymbol{r} = \begin{pmatrix} 1 \\ 3 \end{pmatrix} + \mu\begin{pmatrix} 3 \\ 1 \end{pmatrix}$ (b) $\boldsymbol{r} = \begin{pmatrix} 2 \\ 4 \end{pmatrix} - \lambda\begin{pmatrix} 7 \\ 5 \end{pmatrix}$

(c) $\boldsymbol{r} = -\begin{pmatrix} 2 \\ 4 \end{pmatrix} + \lambda\begin{pmatrix} 1 \\ 8 \end{pmatrix}$ (d) $\boldsymbol{r} = \begin{pmatrix} 0.5 \\ 0.2 \end{pmatrix} - t\begin{pmatrix} -1 \\ 11 \end{pmatrix}$

(e) $\boldsymbol{r} = \begin{pmatrix} 7 \\ 5 \end{pmatrix} + \lambda\begin{pmatrix} 0 \\ 1 \end{pmatrix}$ (f) $\boldsymbol{r} = \begin{pmatrix} 2 \\ 6 \end{pmatrix} + \lambda\begin{pmatrix} 5 \\ 0 \end{pmatrix}$

7. Write the following lines in vector form

(a) $y = \frac{1}{3}x + 2$ (b) $y = x - 5$ (c) $2y - x = 6$

8. Find the position vector of the point of intersection of each pair of lines.

(a) $\boldsymbol{r}_1 = \begin{pmatrix} 2 \\ 1 \end{pmatrix} + \begin{pmatrix} \lambda \\ 3\lambda \end{pmatrix}$ and $\boldsymbol{r}_2 = \begin{pmatrix} 1 \\ 3 \end{pmatrix} + \mu\begin{pmatrix} 1 \\ 2 \end{pmatrix}$

(b) $\boldsymbol{r}_1 = \begin{pmatrix} 0 \\ 4 \end{pmatrix} + \lambda\begin{pmatrix} 2 \\ 5 \end{pmatrix}$ and $\boldsymbol{r}_2 = \begin{pmatrix} 2 \\ -2 \end{pmatrix} + \mu\begin{pmatrix} 1 \\ 1 \end{pmatrix}$

9. Find the equation of the line that passes through the point A(2, 7) and is perpendicular to the line with equation $\boldsymbol{r} = -\boldsymbol{i} - 3\boldsymbol{j} + \lambda(3\boldsymbol{i} - 4\boldsymbol{j})$.

10. Let the position vectors of the points $P(x_1, y_1)$ and $Q(x_2, y_2)$ be $\boldsymbol{p}$ and $\boldsymbol{q}$ respectively. Show that the equation $\boldsymbol{r} = (1-\lambda)\boldsymbol{p} + \lambda\boldsymbol{q}$ represents a vector equation of the line through P and Q, where $\lambda \in \mathbb{R}$.

11. The line L is defined by the parametric equations $x = 4 - 5k$ and $y = -2 + 3k$.
(a) Find the coordinates of three points on L.
(b) Find the value of k that corresponds to the point (14, –8).
(c) Show that the point (–1, 4) does not lie on the line L.
(d) Find the vector form of the line L.
(e) A second line, M, is defined parametrically by $x = a + 10\lambda$ and $y = b - 6\lambda$. Describe the relationship between M and L for the case that
i. $a = 8$ and $b = 4$
ii. $a = 4$ and $b = -2$

12. Find the cartesian equation of the line that passes through the point A(2, 1) and such that it is perpendicular to the vector $4\boldsymbol{i} + 3\boldsymbol{j}$.

13. Find the direction cosines for each of the following lines
(a) $\boldsymbol{r} = \begin{pmatrix}3\\4\end{pmatrix} + \mu\begin{pmatrix}-3\\2\end{pmatrix}$ (b) $\boldsymbol{r} = \begin{pmatrix}5\\9\end{pmatrix} + \lambda\begin{pmatrix}4\\3\end{pmatrix}$

14. (a) Show that the line $ax + by + c = 0$ has a directional vector $\begin{pmatrix}b\\-a\end{pmatrix}$ and a normal vector $\begin{pmatrix}a\\b\end{pmatrix}$.
(b) By making use of directional vectors, which of the following lines are parallel to $L : 2x + 3y = 10$?
i. $5x - 2y = 10$
ii. $6x + 9y = 20$
iii. $4x + 6y = -10$

15. Find the point of intersection of the lines $\boldsymbol{r} = \begin{pmatrix}-2\\1\end{pmatrix} + \lambda\begin{pmatrix}3\\8\end{pmatrix}$ and $\frac{x-3}{2} = \frac{y}{5}$.

16. Find a vector equation of the line passing through the origin and through the point of intersection of the lines $\boldsymbol{u} = \begin{pmatrix}3\\2\end{pmatrix} + \lambda\begin{pmatrix}-1\\3\end{pmatrix}$ and $\boldsymbol{v} = \begin{pmatrix}-1\\1\end{pmatrix} + \mu\begin{pmatrix}2\\1\end{pmatrix}$.

17. Consider the line with vector equation $\boldsymbol{r} = (4\boldsymbol{i} - 3\boldsymbol{j}) + \lambda(3\boldsymbol{i} + 4\boldsymbol{j})$. Find the point(s) of intersection of this line with the line
(a) $\boldsymbol{u} = (4\boldsymbol{i} + 5\boldsymbol{j}) + \mu(2\boldsymbol{i} - \boldsymbol{j})$
(b) $\boldsymbol{v} = (-2\boldsymbol{i} + 3\boldsymbol{j}) + t(-6\boldsymbol{i} - 8\boldsymbol{j})$
(c) $\boldsymbol{w} = (13\boldsymbol{i} + 9\boldsymbol{j}) + s(3\boldsymbol{i} + 4\boldsymbol{j})$

12.7.2 APPLICATION OF THE VECTOR EQUATION $r = a + tb$

We now consider an application of the vector equation $\boldsymbol{r} = \boldsymbol{a} + t\boldsymbol{b}$ with t representing time and $\boldsymbol{b}$ representing velocity (so that $|\boldsymbol{b}|$ would then represent the speed).

An object moving in a plane has its position, $P(x, y)$, on that plane defined by the vector equation $\boldsymbol{r} = \begin{pmatrix} 2 \\ 0 \end{pmatrix} + t\begin{pmatrix} 1 \\ 2 \end{pmatrix}$, where t represents the time for which this object has been moving. As the vector $\boldsymbol{r}$, represents the position vector of P, we can write $\boldsymbol{r} = \begin{pmatrix} x \\ y \end{pmatrix}$ so that such an equation could also be written in the form $\begin{pmatrix} x \\ y \end{pmatrix} = \begin{pmatrix} 2 \\ 0 \end{pmatrix} + t\begin{pmatrix} 1 \\ 2 \end{pmatrix}$. This in turn, would mean that $x = 2 + t$ and $y = 2t$ and so, we could trace the path of the object by constructing a table of values:

t	$x = 2 + t$	$y = 2t$	$P(x,y)$
0	2	0	(2, 0)
1	3	2	(3, 2)
2	4	4	(4, 4)
3	5	6	(5, 6)
4	6	8	(6, 8)
5	7	10	(7, 10)
⋮	⋮	⋮	⋮

Plotting the resulting coordinates on a set of axes we have:

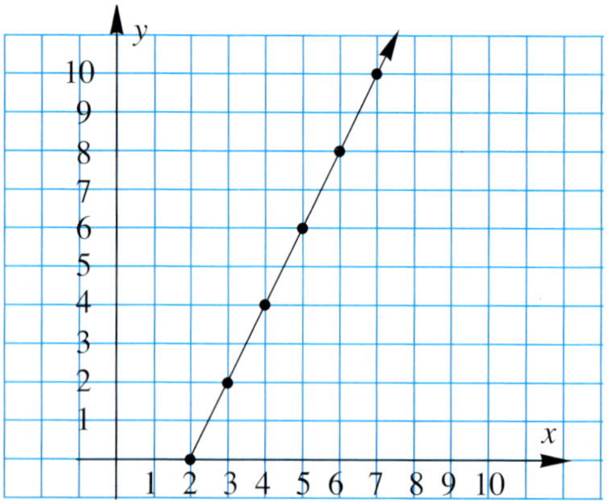

From these coordinates we see that the object moves in a straight line, in particular, we also observe that every second, the object moves 1 unit in the positive x-direction and 2 units in the positive y-direction.
That is, every second, the objectt has a displacement of $\begin{pmatrix} 1 \\ 2 \end{pmatrix}$ unit. If we let this unit of measure be the metre, we can then say that the object is displaced $\begin{pmatrix} 1 \\ 2 \end{pmatrix}$ m per second. Which is by definition, the **velocity**. We can then describe the motion of this object as follows:

The object with vector equation $\boldsymbol{r} = \begin{pmatrix} 2 \\ 0 \end{pmatrix} + t\begin{pmatrix} 1 \\ 2 \end{pmatrix}$, starts it motion at the point $(2, 0)$ and has a velocity vector $\begin{pmatrix} 1 \\ 2 \end{pmatrix}$. As speed is the magnitude of velocity, we can also say that the object has a **speed** of $\left|\begin{pmatrix} 1 \\ 2 \end{pmatrix}\right| = \sqrt{1^2 + 2^2} = \sqrt{5}$ m/s.

Such an application of the vector equation $\boldsymbol{r} = \boldsymbol{a} + t\boldsymbol{b}$ where the vector $\boldsymbol{b}$ represents the velocity vector and t the time is very useful when coordinating the paths of objects – i.e., making sure that two (or more) objects do not collide while moving.

EXAMPLE 12.31 Find the velocity vector and the speed of an object, with position vector defined by $\boldsymbol{r} = \begin{pmatrix} 1 \\ -1 \end{pmatrix} + t\begin{pmatrix} 3 \\ 4 \end{pmatrix}$, where the vectors $\begin{pmatrix} 1 \\ 0 \end{pmatrix}$ km and $\begin{pmatrix} 0 \\ 1 \end{pmatrix}$ km represent a displacement of 1 km in the easterly and northerly directions respectively and the parameter t represents the time in hours, $t \geq 0$. Describe the motion of the object.

Solution

From $\boldsymbol{r} = \begin{pmatrix} 1 \\ -1 \end{pmatrix} + t\begin{pmatrix} 3 \\ 4 \end{pmatrix}$ we have that the velocity vector is given by $\begin{pmatrix} 3 \\ 4 \end{pmatrix}$.

In turn, this means that the speed of the object is $\left|\begin{pmatrix} 3 \\ 4 \end{pmatrix}\right| = \sqrt{3^2 + 4^2} = 5$ km/hr.

The object moves in a straight line, starting at the point $(1, -1)$ and heading in a direction $\begin{pmatrix} 3 \\ 4 \end{pmatrix}$ with a speed of 5 km/hr.

Notice that we can also determine the equation of the straight line in the same way as we did in section 12.7.1. From $\boldsymbol{r} = \begin{pmatrix} 1 \\ -1 \end{pmatrix} + t\begin{pmatrix} 3 \\ 4 \end{pmatrix}$ we have, $x = 1 + 3t$ and $y = -1 + 4t$. Then,

eliminating t from the equations we have: $\frac{x-1}{3} = \frac{x+1}{4} \Leftrightarrow 4(x-1) = 3(y+1)$

$$\text{or, } 4x - 3y = 7$$

Also, as $t \geq 0$, then from $x = 1 + 3t$, we must have that $x \geq 1$, and so, the (Cartesian) equation of the object's path is given by $4x - 3y = 7, x \geq 1$.

Using the Graphics Calculator

On the TI–83 it is possible to enter the vector equation in its parametric form. To do so, we must first press **MODE** and then select **Par** and **Radian** from the mode screen. Then, when we select the equation editor screen we obtain the following screen:

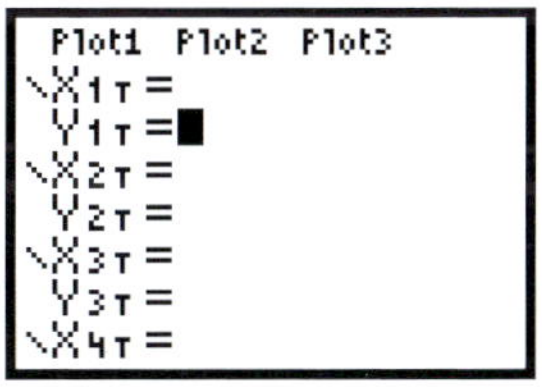

Unlike the case of functions, we have two equations to enter, X_{1T} and Y_{1T}. That is, the equation for x and y respectively.

For the case of Example 12.31, we have:

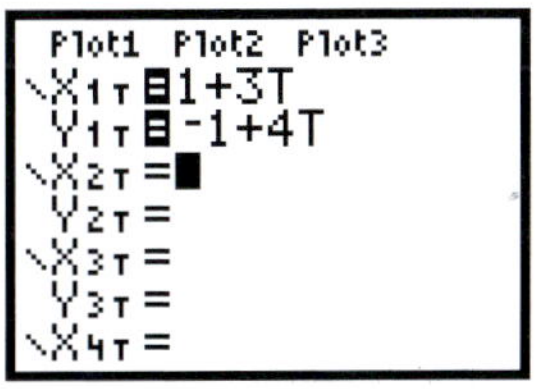

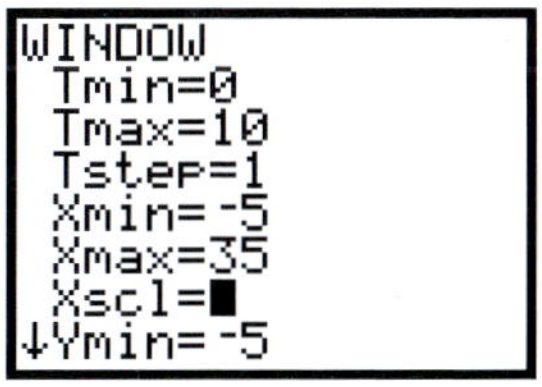

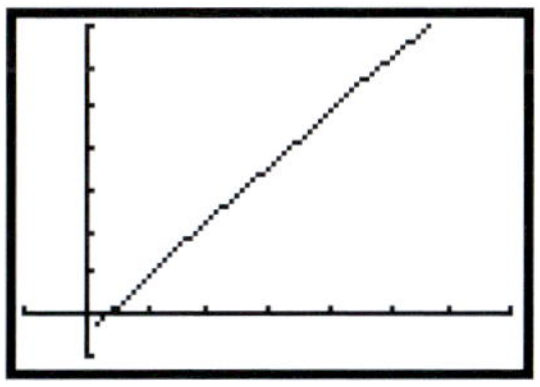

EXAMPLE 12.32

The parametric equations of a moving object, are given by $x = 4+5t$ and $y = 5-2t$, where $t \geq 0$, is measured in seconds and distances are measured metres.

(a) Find the vector equation defining the object's path.

(b) What is the object's speed?

(c) A pole is located at the point (39, –9). Will the object collide with the pole?

Solution

(a) From the parametric equations, $x = 4+5t$ and $y = 5-2t$ we have, $\boldsymbol{r} = \begin{pmatrix} 4+5t \\ 5-2t \end{pmatrix}$.

That is, $\boldsymbol{r} = \begin{pmatrix} 4 \\ 5 \end{pmatrix} + t\begin{pmatrix} 5 \\ -2 \end{pmatrix}, t \geq 0$.

(b) The velocity vector is $\begin{pmatrix} 5 \\ -2 \end{pmatrix}$, therefore, the speed $= \left|\begin{pmatrix} 5 \\ -2 \end{pmatrix}\right| = \sqrt{5^2+(-2)^2} = \sqrt{29}$ m/s.

(c) We need to find if the object's path passing through the point (39, –9).

To do this we equate $\boldsymbol{r} = \begin{pmatrix} 4+5t \\ 5-2t \end{pmatrix}$ to $\begin{pmatrix} 39 \\ -9 \end{pmatrix}$ and solve for t (if possible).

So, $\begin{pmatrix} 4+5t \\ 5-2t \end{pmatrix} = \begin{pmatrix} 39 \\ -9 \end{pmatrix} \Leftrightarrow \begin{cases} 4+5t = 39 \\ 5-2t = -9 \end{cases}$.

Solving $4+5t = 39$ we have $t = 7$ and solving $5-2t = -9$ we also have $t = 7$.
Therefore, as the equations are consistent, the object does collide with teh pole, after 7 sec.

Of course, we could have first obtained the equation of the straight line by removing the parameter t: $\frac{x-4}{5} = \frac{y-5}{-2}$ so that $-2(x-4) = 5(y-5)$ or $5y+2x-33 = 0$.

And then, checked to see if the point (39, –9) lies on the line:

$$\text{L.H.S} = 5 \times -9 + 2 \times 39 - 33 = 0 = \text{R.H.S}$$

Which agrees with our previous finding.

Determining when objects collide

The above example a simple version of what could be considered a more general situation. That is, determining if two moving objects collide. For example, if object A has its position defined by the vector equation $r_A = \begin{pmatrix} x_1 \\ y_1 \end{pmatrix} + t\begin{pmatrix} v_{xA} \\ v_{yA} \end{pmatrix}$ and object B has its position defined by the vector equation $r_B = \begin{pmatrix} x_2 \\ y_2 \end{pmatrix} + t\begin{pmatrix} v_{xB} \\ v_{yB} \end{pmatrix}$, then the objects would collide if we can find a consistent solution (for t) to the equation $\begin{pmatrix} x_1 \\ y_1 \end{pmatrix} + t\begin{pmatrix} v_{xA} \\ v_{yA} \end{pmatrix} = \begin{pmatrix} x_2 \\ y_2 \end{pmatrix} + t\begin{pmatrix} v_{xB} \\ v_{yB} \end{pmatrix}$.

EXAMPLE 12.33 Two objects, A and B have their positions defined by the vector equations $\boldsymbol{r}_A = \begin{pmatrix} -8 \\ 3 \end{pmatrix} + t\begin{pmatrix} 4 \\ 3 \end{pmatrix}$ and $\boldsymbol{r}_B = \begin{pmatrix} 2 \\ -7 \end{pmatrix} + t\begin{pmatrix} 2 \\ 5 \end{pmatrix}$, where $t \geq 0$ and is measured in minutes and all distances are in metres. Will the two particle collide?

Solution

To determine if they collide we must attempt to solve $\boldsymbol{r}_A = \boldsymbol{r}_B$:

So, we have $\begin{pmatrix} -8 \\ 3 \end{pmatrix} + t\begin{pmatrix} 4 \\ 3 \end{pmatrix} = \begin{pmatrix} 2 \\ -7 \end{pmatrix} + t\begin{pmatrix} 2 \\ 5 \end{pmatrix} \Leftrightarrow \begin{pmatrix} -8+4t \\ 3+3t \end{pmatrix} = \begin{pmatrix} 2+2t \\ -7+5t \end{pmatrix}$

Giving the simultaneous system of equations: $-8+4t = 2+2t \Leftrightarrow t = 5$

and $3+3t = -7+5t \Leftrightarrow t = 5$.

As the system of equations is consistent, then we conclude that the objects do collide. That is, they occupy the same spot at the same time.

EXERCISES 12.7.2

1. The position vectors, $\boldsymbol{r}$ km, and the velocity vectors, $\boldsymbol{v}$ km/h, of two boats at a certain time are given by

Boat A:	at 10:00 am	$\boldsymbol{r}_A = (6\boldsymbol{i} + 10\boldsymbol{j})$	$\boldsymbol{v}_A = 8\boldsymbol{i} - 2\boldsymbol{j}$
Boat B:	at 10:30 am	$\boldsymbol{r}_B = (8\boldsymbol{i} + 5\boldsymbol{j})$	$\boldsymbol{v}_B = 4\boldsymbol{i} + \boldsymbol{j}$

(a) If the boats continue to travel with these velocities will they collide?

(b) If they do collide, find the time when they collide as well as the position vectors of the point of collision.
If they do not collide, determine the time at which boat B should head out so that the two boats do collide.

2. The position vectors, $\boldsymbol{r}$ km, and the velocity vectors, $\boldsymbol{v}$ km/h, of two boats at a certain time are given by

Boat A:	at 12:00 noon	$\boldsymbol{r}_A = (5\boldsymbol{i} - \boldsymbol{j})$	$\boldsymbol{v}_A = 3\boldsymbol{i} + 4\boldsymbol{j}$
Boat B:	at 12:00 noon	$\boldsymbol{r}_B = (4\boldsymbol{i} + 5\boldsymbol{j})$	$\boldsymbol{v}_B = 2\boldsymbol{i} - \boldsymbol{j}$

(a) Find the position vectors of
i. boat A t hours after 12:00 noon.
ii. boat B t hours after 12:00 noon.

(b) If the boats continue to travel with these velocities will they collide?
If they do collide, find the time when they collide as well as the position vectors of the point of collision.

(c) The departure time of boat B is uncertain, except to say that it is sometime after 1:00 pm.
i. Find the position vector of boat B, t hours after boat A departs.
ii. Will the boats collide? Determine the time at which boat B should head out so that the two boats do collide.

12.7.3 LINES IN THREE DIMENSIONS

In three dimensional work always try to visualise situations very clearly. Because diagrams are never very satisfactory, it is useful to use the corner of a table with an imagined vertical line for axes; then pencils become lines and books or sheets of paper become planes.

It is tempting to generalise from a two dimensional line like $x + y = 8$ and think that the Cartesian equation of a three dimensional line will have the form $x + y + z = 8$. This is not correct - as we will see later **this represents a plane, not a line**.

We approach lines in three dimensions in exactly the same way that we did for lines in two dimensions. For any point P(x,y,z) on the line having the position vector $\boldsymbol{r}$, passing through the point A and parallel to a vector in the direction of the line, $\boldsymbol{b}$ say, we can write the equation of the the line as $\boldsymbol{r} = \boldsymbol{a} + \lambda\boldsymbol{b}$.

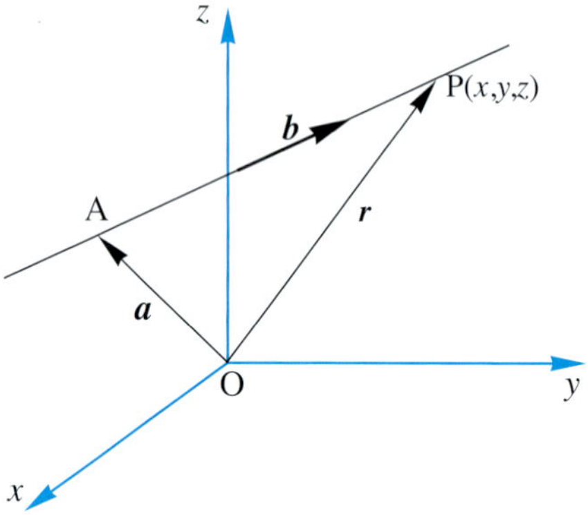

So, for example, the line passing through the point (4, 2, 5) and having the direction vector $\boldsymbol{i} - \boldsymbol{j} + 2\boldsymbol{k}$ can be written as:

$$\boldsymbol{r} = \begin{pmatrix} 4 \\ 2 \\ 5 \end{pmatrix} + \lambda\begin{pmatrix} 1 \\ -1 \\ 2 \end{pmatrix}.$$

Or, it could also have been written in $\boldsymbol{i}$, $\boldsymbol{j}$, $\boldsymbol{k}$ form as $\boldsymbol{r} = 4\boldsymbol{i} + 2\boldsymbol{j} + 5\boldsymbol{k} + \lambda(\boldsymbol{i} - \boldsymbol{j} + 2\boldsymbol{k})$.

As for the case in 2–D, the parametric form or Cartesian form of the equation is obtained by using a point P(x, y, z) on the line with position vector $\boldsymbol{r} = \begin{pmatrix} x \\ y \\ z \end{pmatrix}$ so that $\begin{pmatrix} x \\ y \\ z \end{pmatrix} = \begin{pmatrix} 4 \\ 2 \\ 5 \end{pmatrix} + \lambda\begin{pmatrix} 1 \\ -1 \\ 2 \end{pmatrix}$.

From where we first get the **parametric equations**:

$$x = 4 + \lambda,\ \ y = 2 - \lambda \ \text{ and } \ z = 5 + 2\lambda.$$

Solving each of these for λ we get: $\lambda = x - 4 = 2 - y = \dfrac{z-5}{2}$.

The parameter λ plays no part in the Cartesian equation, so we drop it and write the

Cartesian equation* as $$x - 4 = 2 - y = \frac{z-5}{2}.$$

It is important to be clear what this means: if we choose x, y and z satisfying the Cartesian equation, then the point P(x, y, z) will be on the line.

For example $x = 10$, $y = -4$ and $z = 17$ satisfies the Cartesian equation, and if we think back to our original parametric equation we can see that $\begin{pmatrix} 10 \\ -4 \\ 17 \end{pmatrix} = \begin{pmatrix} 4 \\ 2 \\ 5 \end{pmatrix} + 6\begin{pmatrix} 1 \\ -1 \\ 2 \end{pmatrix}$.

*The Cartesian equation for a line in 3-D is not examinable, but is included for completeness.

To convert a Cartesian equation into parametric form we reverse the process and introduce a parameter λ. For example if the Cartesian equation is $\frac{x-1}{3} = \frac{y+2}{2} = \frac{z-6}{4}$ we write:

$$\frac{x-1}{3} = \frac{y+2}{2} = \frac{z-6}{4} = \lambda$$

$$\Rightarrow x = 1 + 3\lambda$$

$$y = -2 + 2\lambda$$

$$z = 6 + 4\lambda$$

$$\text{and } \boldsymbol{r} = \begin{pmatrix} 1 \\ -2 \\ 6 \end{pmatrix} + \lambda \begin{pmatrix} 3 \\ 2 \\ 4 \end{pmatrix}$$

You will probably have noticed the strong connection between the numbers in the fractions in the Cartesian form and the numbers in the vectors in the parametric form.

Consider the **Cartesian form** of any straight L passing through the point $P(x_1, y_1, z_1)$:

$$\frac{x-x_1}{a} = \frac{y-y_1}{b} = \frac{z-z_1}{c}$$

From this equation we obtain the **parametric form** of the straight line:

$$\frac{x-x_1}{a} = \lambda \Leftrightarrow x = x_1 + \lambda a$$

$$\frac{y-y_1}{b} = \lambda \Leftrightarrow y = y_1 + \lambda b$$

$$\frac{z-z_1}{c} = \lambda \Leftrightarrow z = z_1 + \lambda c$$

which then leads to the **vector form** of the straight line:

$$\begin{pmatrix} x \\ y \\ z \end{pmatrix} = \begin{pmatrix} x_1 \\ y_1 \\ z_1 \end{pmatrix} + \lambda \begin{pmatrix} a \\ b \\ c \end{pmatrix}$$

That is, the denominators of the Cartesian form of a straight line provide the coefficients of the directional vector of the line. This is an important observation, especially when finding the angle between two lines when the equation of the line is provided in Cartesian form.

However, rather than simply commiting this observation to memory it is always a good idea to go through the (very short) working involved.

EXAMPLE 12.34

Find the Cartesian form of the straight line passing through the point (4,6,3) and having direction vector $3\boldsymbol{i} - 2\boldsymbol{j} + \boldsymbol{k}$. Draw a sketch of this line on a set of axes.

Solution

We start by sketching the line:

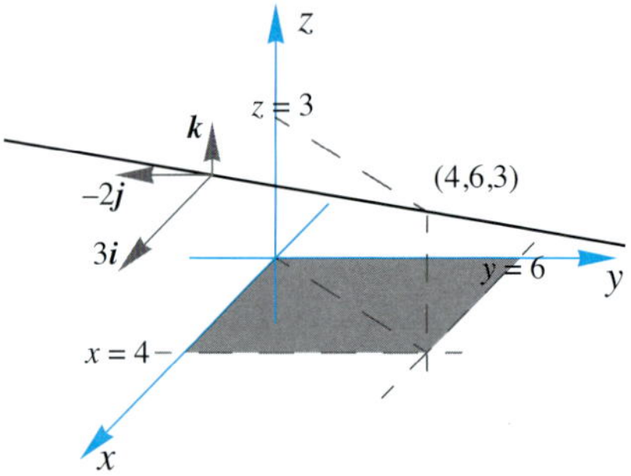

The direction vector of the line is $3\boldsymbol{i} - 2\boldsymbol{j} + \boldsymbol{k}$ and as the line passes through the point (4,6,3), the vector equation of the line is given by

$$\boldsymbol{r} = (4\boldsymbol{i} + 6\boldsymbol{j} + 3\boldsymbol{k}) + \lambda(3\boldsymbol{i} - 2\boldsymbol{j} + \boldsymbol{k}).$$

From the vector equation we obtain the parametric form of the line:

$$x = 4 + 3\lambda,\ y = 6 - 2\lambda \text{ and } z = 3 + \lambda.$$

From these equations we have, $\lambda = \frac{x-4}{3}$, $\lambda = \frac{y-6}{-2}$ and $\lambda = \frac{z-3}{1}$

Then, eliminating λ we have $\frac{x-4}{3} = \frac{y-6}{-2} = \frac{z-3}{1}$ or $\frac{x-4}{3} = \frac{y-6}{-2} = z-3$.

Which represents the Cartesian form of the line.

EXAMPLE 12.35

Find the vector form of the equation of the line through the point A(2,1,1) and the point B(4,0,3).

Solution

We make a very rough sketch - there is no point in trying to plot A and B accurately. Let the position vector of any point P on the line be $\boldsymbol{r}$.

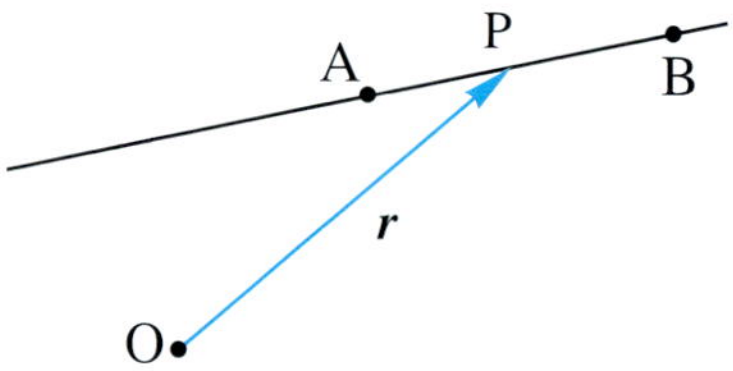

Then the vector form of the line is $\boldsymbol{r} = \mathbf{OA} + \lambda\mathbf{AB}$.

Now, $\mathbf{OP} = \boldsymbol{r} = \mathbf{OA} + \mathbf{AP}$.

But $\mathbf{AP} = \lambda\mathbf{AB} \therefore \boldsymbol{r} = \mathbf{OA} + \lambda\mathbf{AB}$ and $\mathbf{AB} = \mathbf{AO} + \mathbf{OB} = -\mathbf{OA} + \mathbf{OB}$

$$\therefore \mathbf{AB} = -\begin{pmatrix} 2 \\ 1 \\ 1 \end{pmatrix} + \begin{pmatrix} 4 \\ 0 \\ 3 \end{pmatrix} = \begin{pmatrix} 2 \\ -1 \\ 2 \end{pmatrix} \text{ and so, } \boldsymbol{r} = \begin{pmatrix} 2 \\ 1 \\ 1 \end{pmatrix} + \lambda\begin{pmatrix} 2 \\ -1 \\ 2 \end{pmatrix}.$$

EXAMPLE 12.36

Find the acute angle between the straight lines

$$L_1: \frac{x-3}{2} = \frac{y+2}{-1} = \frac{z}{\sqrt{3}} \text{ and } L_2: \frac{x+1}{1} = \frac{y-2}{1} = \frac{z-1}{\sqrt{3}}$$

Solution

Because the lines are given in their standard Cartesian form, we know that the denominators represent the coefficients of the direction vectors of these lines. As the angle between the lines is the same as the angle between their direction vectors we need only use the direction vectors of each line and then apply the dot product.

For L_1 the direction vector is $\boldsymbol{b}_1 = 2\boldsymbol{i} - \boldsymbol{j} + \sqrt{3}\boldsymbol{k}$ and for L_2 it is $\boldsymbol{b}_2 = \boldsymbol{i} + \boldsymbol{j} + \sqrt{3}\boldsymbol{k}$.

Using the dot product we have:

$$b_1 \bullet b_2 = |b_1||b_2|\cos\theta \therefore (2i - j + \sqrt{3}k) \bullet (i + j + \sqrt{3}k) = \sqrt{8} \times \sqrt{5}\cos\theta$$

$$2 - 1 + 3 = \sqrt{40}\cos\theta$$

$$\cos\theta = \frac{4}{\sqrt{40}}$$

$$\therefore \theta = 50°46'$$

EXAMPLE 12.37

Write the equation of the line $\frac{x+1}{3} = \frac{4-y}{2} = z$ in parametric form, and show that it is parallel to $-i + 5j + k + \mu(-6i + 4j - 2k)$.

Solution

From the Cartesian form of the line $\frac{x+1}{3} = \frac{4-y}{2} = z = \lambda$ (say) we obtain the parametric form

$$x = -1 + 3\lambda,\ y = 4 - 2\lambda \text{ and } z = \lambda.$$

We can then write this in the vector form $r = -i + 4j + \lambda(3i - 2j + k)$.

Comparing the direction vectors of the two lines we see that $-6i + 4j - 2k = -2(3i - 2j + k)$, and so the direction vectors (and hence the lines) are parallel.

It is worth emphasising, that lines will be parallel or perpendicular if their direction vectors are parallel or perpendicular.

Consider the two lines, $L_1 : r_1 = a_1 + \lambda b_1$ and $L_2 : r_2 = a_2 + \lambda b_2$, with directions vectors $b_1 = x_1 i + y_1 j + z_1 k$ and $b_2 = x_2 i + y_2 j + z_2 k$ respectively, then

1. If the two lines are perpendicular we have $b_1 \bullet b_2 = 0 \Rightarrow x_1 x_2 + y_1 y_2 + z_1 z_2 = 0$.

2. If the two lines are parallel we have $b_1 = m b_2, m \neq 0$ which in turn implies that $x_1 = m x_2, y_1 = m y_2$ and $z_1 = m z_2$. Then, eliminating the constant m, we have that

$$L_1 \parallel L_2 \Leftrightarrow \frac{x_1}{x_2} = \frac{y_1}{y_2} = \frac{z_1}{z_2}$$

EXAMPLE 12.38

Line L passes through the points (4, 3, 9) and (7, 8, 5), while line M passes through the points (12, 16, 4) and (k,26,–4), where $k \in \mathbb{R}$. Find the value(s) of k, if

(a) L is parallel to M.

(b) L is perpendicular to M.

Solution

We first need to determine direction vectors for both L and M.

For L: Let the points be A(4, 3, 9) and B(7, 8, 5), then a direction vector for L, $\boldsymbol{b}_1$ (say), is given by

$$\boldsymbol{b}_1 = \begin{pmatrix} 7-4 \\ 8-3 \\ 5-9 \end{pmatrix} = \begin{pmatrix} 3 \\ 5 \\ -4 \end{pmatrix}.$$

For M: Let the points be X(12, 16, 4) and Y(k,26,–4), then a direction vector for M, $\boldsymbol{b}_2$ (say), is given by

$$\boldsymbol{b}_2 = \begin{pmatrix} k-12 \\ 26-16 \\ -4-4 \end{pmatrix} = \begin{pmatrix} k-12 \\ 10 \\ -8 \end{pmatrix}.$$

(a) If L $\|$ M we must have that $\boldsymbol{b}_1 = c\boldsymbol{b}_2, c \in \mathbb{R}$.

i.e.,
$$\begin{pmatrix} 3 \\ 5 \\ -4 \end{pmatrix} = c\begin{pmatrix} k-12 \\ 10 \\ -8 \end{pmatrix} \Rightarrow \frac{3}{k-12} = \frac{5}{10} = -\frac{4}{-8}$$

So that
$$\frac{3}{k-12} = \frac{1}{2} \Leftrightarrow k-12 = 6 \Leftrightarrow k = 18.$$

(b) If L $\perp$ M we must have that $\boldsymbol{b}_1 \bullet \boldsymbol{b}_2 = 0$.

i.e.,
$$\begin{pmatrix} 3 \\ 5 \\ -4 \end{pmatrix} \bullet \begin{pmatrix} k-12 \\ 10 \\ -8 \end{pmatrix} = 0 \Rightarrow 3(k-12)+50+32 = 0$$

$$\Leftrightarrow 3k = -46 \Leftrightarrow k = -\frac{46}{3}$$

Intersection of 2 lines in 3–D

Two lines in space may

1. intersect at a point, or
2. be parallel and never intersect, or
3. be parallel and coincident (i.e., the same), or
4. be neither parallel nor intersect.

Of the above scenarios, the first three are consistent with our findings when dealing with lines in a plane (i.e., 2-D), however, the fourth scenario is new. We illustrate these now.

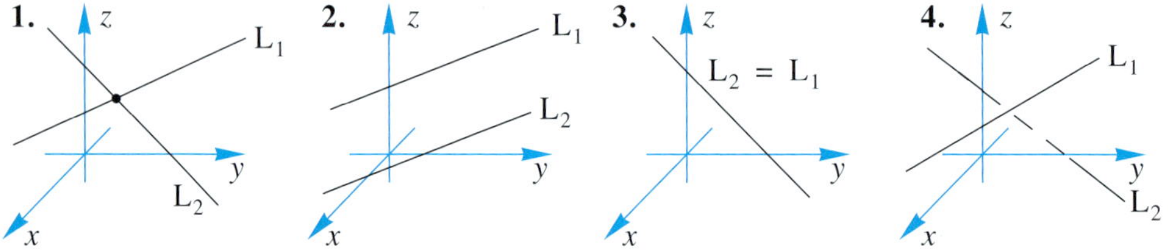

Two lines that meet at (at least) one point must lie in the same plane (cases **1.** and **3.**) Two intersecting lines or two parallel lines are said to be **coplanar** (cases **1.**, **2.**, and **3.**). Two lines which are not parallel and which do not intersect are said to be **skew** – skew lines do not lie on the same plane – i.e., they are not coplanar. (case **4.**)

Lines lying on the xy–, xz– and yz– planes

From the Cartesian form of the straight line, L: $\frac{x-x_1}{a} = \frac{y-y_1}{b} = \frac{z-z_1}{c}$ we can write:

$$\frac{x-x_1}{a} = \frac{y-y_1}{b} \Leftrightarrow b(x-x_1) = a(y-y_1) \text{ – (1)}$$

$$\frac{x-x_1}{a} = \frac{z-z_1}{c} \Leftrightarrow c(x-x_1) = a(z-z_1) \text{ – (2)}$$

$$\frac{y-y_1}{b} = \frac{z-z_1}{c} \Leftrightarrow c(y-y_1) = b(z-z_1) \text{ – (3)}$$

Equations (1), (2) and (3) represent the planes perpendicular to the *xy*–, *xz*– and *yz* planes respectively. Each of these equations is an equation of a plane containing L. The simultaneous solution of any pair of these planes will produce the same line. In fact, the three equations are not independent because any one of them can be derived from the other two.

If any one of the numbers a, b or c is zero we obtain a line lying in one of the *xy*–, *xz*– or *yz* planes. For example, consider the case that $c = 0$ and neither a nor b is zero.

In such a case we have, $\frac{x-x_1}{a} = \frac{y-y_1}{b}$ and $z = z_1$ meaning that the line lies on the plane containing the point $z = z_1$ and parallel to the *xy*–plane.

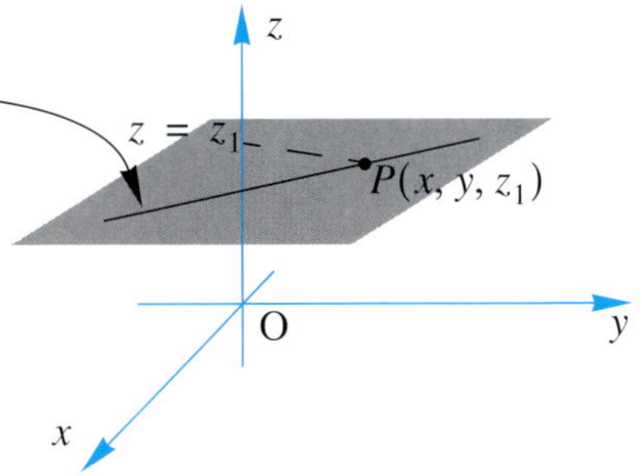

For 'convenience' we sometimes write the equation as $\frac{x-x_1}{a} = \frac{y-y_1}{b} = \frac{z-z_1}{0}$,

although clearly, $\frac{z-z_1}{0}$ has no meaning.

EXAMPLE 12.39

Line L passes through the points A(1,2,–1) and B(11,–2,–7) while line M passes though the points C(2,–1,–3) and D(9,–10,3). Show that L and M are skew lines.

Solution

We start by finding the vector equations of both lines.
For L we have a direction vector given by

$$\boldsymbol{b}_1 = (11-1)\boldsymbol{i} + (-2-2)\boldsymbol{j} + (-7-(-1))\boldsymbol{k} = 10\boldsymbol{i} - 4\boldsymbol{j} - 6\boldsymbol{k}$$

Then, as L passes through A(1,2,–1), it has a vector equation given by

$$\boldsymbol{r} = \boldsymbol{i} + 2\boldsymbol{j} - \boldsymbol{k} + \lambda(10\boldsymbol{i} - 4\boldsymbol{j} - 6\boldsymbol{k})$$

This gives the parametric form as, $x = 1 + 10\lambda$, $y = 2 - 4\lambda$ and $z = -1 - 6\lambda$ – (1)
Similarly, we can find the parametric form for M.

The vector form of M is given by $\quad \boldsymbol{r} = 2\boldsymbol{i} - \boldsymbol{j} - 3\boldsymbol{k} + \mu(7\boldsymbol{i} - 9\boldsymbol{j} + 6\boldsymbol{k})$

So that the parametric form is given by $\quad x = 2 + 7\mu,\ y = -1 - 9\mu$ and $z = -3 + 6\mu$ – (2)

Now, as the set of coefficients of the direction vector of M and L are not proportional,

$$\text{i.e., as } \frac{10}{7} \neq \frac{-4}{-9} \neq \frac{-6}{6}$$

the lines L and M are not parallel.

Then, for the lines to intersect there must be a value of λ and μ that will provide the same point (x_0, y_0, z_0) lying on both L and M. Using (1) and (2) we equate the coordinates and try to determine this point (x_0, y_0, z_0):

$$1 + 10\lambda = 2 + 7\mu \quad \text{– (3)}$$
$$2 - 4\lambda = -1 - 9\mu \quad \text{– (4)}$$
$$-1 - 6\lambda = -3 + 6\mu \quad \text{– (5)}$$

Solving for λ and μ using (4) and (5) we obtain: $\quad \mu = -\frac{5}{39}$ and $\lambda = \frac{18}{39}$.

Substituting these values into (1), we have L.H.S $= 1 + 10 \times \frac{18}{39} \neq 2 + 7 \times -\frac{5}{39} =$ R.H.S.

As the the first equation is not consistent with the other two, the lines do not intersect and as they are not parallel they must be skew.

EXERCISES 12.7.3

1. Find the vector form of the line passing through the point

(a) A(2, 1, 3) which is also parallel to the vector $\boldsymbol{i} - 2\boldsymbol{j} + 3\boldsymbol{k}$.

(b) A(2, –3, –1) which is also parallel to the vector $-2\boldsymbol{i} + \boldsymbol{k}$.

2. Find the vector form of the line passing through the points

(a) $A(2, 0, 5)$ and $B(3, 4, 8)$.

(b) $A(3, -4, 7)$ and $B(7, 5, 2)$.

(c) $A(-3, 4, -3)$ and $B(4, 4, 4)$.

3. Find the Cartesian form of the line having the vector form

(a) $\boldsymbol{r} = \begin{pmatrix} 0 \\ 2 \\ 3 \end{pmatrix} + s\begin{pmatrix} 3 \\ 4 \\ 5 \end{pmatrix}$ (b) $\boldsymbol{r} = \begin{pmatrix} -2 \\ 3 \\ -1 \end{pmatrix} + t\begin{pmatrix} 5 \\ 0 \\ -2 \end{pmatrix}$ (c) $\boldsymbol{r} = \begin{pmatrix} 0 \\ 0 \\ 0 \end{pmatrix} + s\begin{pmatrix} 1 \\ 1 \\ 1 \end{pmatrix}$

4. Find the Cartesian equation of the line passing through the points A(5, 2, 6) and B(–2, 4, 2). Also, provide the parametric form of this line.

5. For the line defined by the parametric equations $x = 3 + 2t$, $y = 4 - 3t$ and $z = 1 + 5t$ find the coordinates of where the line crosses the xy–plane.

6. Convert these lines to their parametric form.

(a) $\frac{x-2}{3} = y-5 = 2(z-4)$ (b) $\frac{2x-1}{3} = y = \frac{4-z}{2}$

(c) $\frac{x-3}{-1} = \frac{2-y}{3} = \frac{z-4}{2}$ (d) $\frac{2x-2}{4} = \frac{3-y}{-2} = \frac{2z-4}{1}$

7. Convert these lines to their Cartesian form:

(a) $\boldsymbol{r} = \begin{pmatrix} 4 \\ 1 \\ -2 \end{pmatrix} + t\begin{pmatrix} 3 \\ -4 \\ -2 \end{pmatrix}$ (b) $\boldsymbol{r} = 2\boldsymbol{i} + \boldsymbol{k} + \mu(\boldsymbol{j} - 3\boldsymbol{k})$

8. Show that the lines $\frac{x-1}{2} = 2-y = 5-z$ and $\frac{4-x}{4} = \frac{3+y}{2} = \frac{5+z}{2}$ are parallel.

9. Find the Cartesian equation of the lines joining the points
(a) $(-1, 3, 5)$ to $(1, 4, 4)$ (b) $(2, 1, 1)$ to $(4, 1, -1)$

10. (a) Find the coordinates of the point where the line $\boldsymbol{r} = \begin{pmatrix} -2 \\ 5 \\ 3 \end{pmatrix} + t\begin{pmatrix} -1 \\ 2 \\ 1 \end{pmatrix}$ intersects the x-y plane.

(b) The line $\frac{x-3}{4} = y+2 = \frac{4-z}{5}$ passes through the point $(a, 1, b)$. Find the values of a and b.

11. Find the Cartesian equation of the line having the vector form

(a) $\boldsymbol{r} = \begin{pmatrix} 1 \\ 4 \\ -2 \end{pmatrix} + t\begin{pmatrix} 1 \\ -1 \\ 0 \end{pmatrix}$ (b) $\boldsymbol{r} = \begin{pmatrix} 2 \\ 1 \\ 3 \end{pmatrix} + t\begin{pmatrix} 2 \\ 0 \\ 0 \end{pmatrix}$.

In each case, provide a diagram showing the lines.

12. Find the vector equation of the line represented by the Cartesian form

$$\frac{x-1}{2} = \frac{1-2y}{3} = z-2.$$

Clearly describe this line.

13. Find the acute angle between the following lines

(a) $\boldsymbol{r} = \begin{pmatrix} 0 \\ 2 \\ 3 \end{pmatrix} + s\begin{pmatrix} 3 \\ 4 \\ 5 \end{pmatrix}$ and $\boldsymbol{r} = \begin{pmatrix} -2 \\ 5 \\ 3 \end{pmatrix} + t\begin{pmatrix} -1 \\ 2 \\ 1 \end{pmatrix}$.

(b) $\boldsymbol{r} = \begin{pmatrix} 2 \\ 1 \\ 4 \end{pmatrix} + s\begin{pmatrix} -2 \\ 0 \\ 1 \end{pmatrix}$ and $\boldsymbol{r} = \begin{pmatrix} 1 \\ 1 \\ 1 \end{pmatrix} + s\begin{pmatrix} 1 \\ 1 \\ 3 \end{pmatrix}$

(c) $\frac{x-3}{-1} = \frac{2-y}{3} = \frac{z-4}{2}$ and $\frac{x-1}{2} = \frac{y-2}{-2} = z-2$

14. Find the point of intersection of the lines

(a) $\frac{x-5}{-2} = y - 10 = \frac{z-9}{12}$ and $x = 4, \frac{y-9}{-2} = \frac{z+9}{6}$

(b) $\frac{2x-1}{3} = \frac{y+5}{3} = \frac{z-1}{-2}$ and $\frac{2-x}{4} = \frac{y+3}{2} = \frac{4-2z}{1}$

15. (a) Find the Cartesian form of the lines with parametric equation given by

$$L : x = \lambda, y = 2\lambda + 2, z = 5\lambda \quad \text{and}$$
$$M : x = 2\mu - 1, y = -1 + 3\mu, z = 1 - 2\mu$$

(b) Find the point of intersection of these two lines.
(c) Find the acute angle between these two lines.
(d) Find the coordinates of the point where i. L cuts the x-y plane.
ii. M cuts the y-z plane.

16. Show that the lines $\frac{x-2}{3} = \frac{y-3}{-2} = \frac{z+1}{5}$ and $\frac{x-5}{-3} = \frac{y-1}{2} = \frac{z-4}{-5}$ are coincident.

17. Show that the lines $\frac{x-1}{-3} = y - 2 = \frac{7-z}{11}$ and $\frac{x-2}{3} = \frac{y+1}{8} = \frac{z-4}{-7}$ are skew.

18. Find the equation of the line passing through the origin and the point of intersection of the lines with equations $x - 2 = \frac{y-1}{4}, z = 3$ and $\frac{x-6}{2} = y - 10 = z - 4$.

19. The lines $\frac{x}{3} = \frac{y-2}{4} = 3 + z$ and $x = y = \frac{z-1}{2k}$, $k \in \mathbb{R} \setminus \{0\}$ meet at right angles. Find k.

20. Consider the lines $L : x = 0, \frac{y-3}{2} = z + 1$ and $M : \frac{x}{4} = \frac{y}{3} = \frac{z-10}{-1}$.
Find, correct to the nearest degree, the angle between the lines L and M.

21. Find the value(s) of k, such that the lines $\frac{x-2}{k} = \frac{y}{2} = \frac{3-z}{3}$ and $\frac{x}{k-1} = \frac{y+2}{3} = \frac{z}{4}$ are perpendicular.

22. Find a direction vector of the line that is perpendicular to both $\frac{x+1}{3} = \frac{y+1}{8} = \frac{z+1}{12}$ and $\frac{1-2x}{-4} = \frac{3y+1}{9} = \frac{z}{6}$.

23. Are the lines $\frac{x-1}{5} = \frac{y+2}{4} = \frac{4-z}{3}$ and $\frac{x+2}{3} = \frac{y+7}{2} = \frac{2-z}{3}$ parallel? Find the point of intersection of these lines. What do you conclude?

REVISION SET C - PAPER 1 & PAPER 2 STYLE QUESTIONS

1. (a) If $\overrightarrow{OA} = 3\boldsymbol{i} - 2\boldsymbol{j} + \boldsymbol{k}$ and $\overrightarrow{OB} = \boldsymbol{i} + 2\boldsymbol{k}$, find $2\overrightarrow{OB} + 3\overrightarrow{AO}$.

(b) The angle between the vectors $\boldsymbol{u} = 2\boldsymbol{i} - \boldsymbol{j} + 3\boldsymbol{k}$ and $\boldsymbol{v} = \boldsymbol{i} + 4\boldsymbol{j} - 2\boldsymbol{k}$ is θ.

Find a given that $\cos\theta = \dfrac{a}{\sqrt{14 \times 21}}$.

(c) Find a unit vector perpendicular to $2\boldsymbol{i} + \boldsymbol{j} - 3\boldsymbol{k}$.

2. Find, in i. parametric form ii. Cartesian form
the line of intersection of the planes $3x + 2y - z = 6$ and $x + 4y - z = -1$.

3. Find the cosine of the acute angle between the lines
$x = -1 + 2\lambda, y = 1 + 3\lambda, z = 2 - \lambda$ and $x = 7 + 5\mu, y = -8 - 3\mu, z = -2 + \mu$.

4. If $A = \begin{bmatrix} 2 & -1 \\ -1 & 2 \end{bmatrix}$, $X = \begin{bmatrix} x \\ y \end{bmatrix}$ where $x, y \neq 0$ and $AX = kX$ where k is a constant, find the values of k for which the equation is true.

5. The position vector of two particles, A and B is given by $\boldsymbol{r}_A = at\boldsymbol{i} + (bt - 5t^2)\boldsymbol{j} + t^2\boldsymbol{k}$ and $\boldsymbol{r}_B = 8t\boldsymbol{i} - 4t\boldsymbol{j} + 4\boldsymbol{k}$. If the two particles collide at a point in space find when and where they collide and the values of a and b.

6. Find A^2 if $A = \begin{bmatrix} 0 & 2 & 1 \\ 1 & 1 & 1 \\ -1 & -2 & -2 \end{bmatrix}$. **Hence** solve the system equations $\begin{aligned} x + 2y + 2z &= 1 \\ 2y + z &= -2 \\ x + y + z &= 0 \end{aligned}$.

7. The matrices A, B and X are such that $A^{-1} = \frac{1}{2}A$, $B^{-1} = \frac{1}{9}B$ and $ABX = I$, where I is the unit matrix, find BX, X and AX in terms of A and B.

8. Let the position vectors of points A and B be $\overrightarrow{OA} = -2\boldsymbol{i} - \boldsymbol{j} + \boldsymbol{k}$ and $\overrightarrow{OB} = \boldsymbol{i} - 2\boldsymbol{j} - \boldsymbol{k}$.

Find (a) $\overrightarrow{AB}$.
(b) the angle AOB to the nearest degree.
(c) the position vector of a point C if B is the mid-point of AC.

9. Let $A = \begin{bmatrix} 1 & 2i \\ -2i & 1 \end{bmatrix}$ where $i^2 = -1$.

(a) Show that $A^2 = 2A + 3I$.

(b) Find A^{-1} in terms of A and I.

10. The position vectors of particles A, B and C from a fixed point O, at any time t, are given by $\boldsymbol{r}_A = 2t^2\boldsymbol{i} + (t+1)\boldsymbol{j}$, $\boldsymbol{r}_B = 2t\boldsymbol{i} + (2t-4)\boldsymbol{j}$ and $\boldsymbol{r}_C = bt^3\boldsymbol{i} + (bt+4)\boldsymbol{j}$ where b is a constant.

(a) Find the minimum distance from O to the path of particle B.

(b) If particles A and C collide, find when they collide and the value of b.

11. If $\begin{bmatrix} a & 2 \\ 1 & 1 \end{bmatrix}\begin{bmatrix} x \\ y \end{bmatrix} = \begin{bmatrix} 1 \\ 2 \end{bmatrix}$, where $a \in \mathbb{R}$ (a) find $\begin{bmatrix} x \\ y \end{bmatrix}$ as a matrix in terms of a.

(b) state when $\begin{bmatrix} x \\ y \end{bmatrix}$ is not defined.

(c) find x and y if $a = 3$.

12. (a) The triangle ABC is such that $\boldsymbol{AB} = 3\boldsymbol{i} + 6\boldsymbol{j} - 2\boldsymbol{k}$ and $\boldsymbol{AC} = 4\boldsymbol{i} - \boldsymbol{j} + 3\boldsymbol{k}$.

Find i. $\angle BAC$ ii. the area of ΔABC

(b) PQRS is a trapezium with $\overrightarrow{PQ} = \boldsymbol{p}$, $\overrightarrow{PS} = \boldsymbol{s}$ and $\overrightarrow{SR} = 3\boldsymbol{p}$. T is the midpoint of [QR]. Express the following in terms of $\boldsymbol{p}$ and $\boldsymbol{s}$.

i. $\overrightarrow{PR}$ ii. $\overrightarrow{QR}$ iii. $\overrightarrow{PT}$ iv. $\overrightarrow{ST}$

13. (a) Given that $M = \begin{bmatrix} \cos\theta & -\sin\theta \\ \sin\theta & \cos\theta \end{bmatrix}$, find $det(M)$.

(b) If $A = \begin{bmatrix} 2 & a \\ 0 & 1 \end{bmatrix}$, deduce an expression for A^n, $n \geq 1$, .

(c) Find the value of a such that the matrix $M = \begin{bmatrix} 3 & 2 & 1 \\ 0 & 1 & 3 \\ -1 & a & 1 \end{bmatrix}$ is singular.

14. The position vectors of the points A, B and C are $\boldsymbol{i} - \boldsymbol{j} + 2\boldsymbol{k}$, $2\boldsymbol{i} + \boldsymbol{j} + 4\boldsymbol{k}$ and $3\boldsymbol{i} + 4\boldsymbol{k}$ respectively. Find

(a) the angle BAC to the nearest degree.

(b) the area of the triangle ABC.

15. (a) Find the i. vector ii. parametric and iii. Cartesian

equation of the line through the point with position vector $\begin{pmatrix} 2 \\ -3 \end{pmatrix}$ & parallel to $\begin{pmatrix} 3 \\ 7 \end{pmatrix}$.

(b) Find the position vector of the point of intersection of the lines

$$l_1 : \boldsymbol{r} = \begin{pmatrix} 14 \\ -1 \end{pmatrix} + \lambda\begin{pmatrix} 5 \\ -4 \end{pmatrix} \text{ and } l_2 : \boldsymbol{r} = \begin{pmatrix} 9 \\ -4 \end{pmatrix} + \mu\begin{pmatrix} -4 \\ 6 \end{pmatrix}$$

(c) i. Are the straight lines $x - 1 = \frac{y}{3} = \frac{z-1}{4}$ and $\frac{x-2}{4} = 3 - y = z$ parallel?

ii. Find the point(s) of intersection of the lines in i. What do you conclude?

16. Find the acute angle between the lines $\boldsymbol{r} = \begin{pmatrix}1\\2\\4\end{pmatrix} + \lambda\begin{pmatrix}4\\3\\1\end{pmatrix}$ and $\boldsymbol{r} = \begin{pmatrix}-1\\-4\\1\end{pmatrix} + \mu\begin{pmatrix}3\\2\\3\end{pmatrix}$.

17. The position vectors of the points A, B and C are given by $\overrightarrow{OA} = \boldsymbol{i} + 2\boldsymbol{j} + 2\boldsymbol{k}$, $\overrightarrow{OB} = \boldsymbol{i} + a\boldsymbol{j} - 2\boldsymbol{k}$ and $\overrightarrow{OC} = b\boldsymbol{i} + 3\boldsymbol{j} + c\boldsymbol{k}$ where a, b and c are constants. Find

(a) a if $\overrightarrow{OA}$ is perpendicular to $\overrightarrow{OB}$.

(b) b and c if O, A and C are collinear.

18. Find a unit vector perpendicular to both $2\boldsymbol{i} + \boldsymbol{j} + \boldsymbol{k}$ and $3\boldsymbol{j} - 2\boldsymbol{k}$.

19. Let $A = \begin{bmatrix}-2 & -3 & -1\\1 & 2 & 1\\3 & 3 & 2\end{bmatrix}$. Find all real numbers k such that the matrix $A - kI$ has no inverse.

20. Let $A = \begin{bmatrix}1 & 1 & -1\\3 & 4 & 6\end{bmatrix}$, $B = \begin{bmatrix}2 & 0\\-2 & 1\end{bmatrix}$, $C = \begin{bmatrix}1\\7\end{bmatrix}$ and $X = \begin{bmatrix}x\\y\\z\end{bmatrix}$.

(a) Evaluate i. BC ii. B^{-1}

(b) Find the equations of the planes produced from the matrix equation $AX = C$.

(c) Give the vector equation of the straight line where the two planes in (b) meet.

21. OABC, DEFG is a cuboid with $\overrightarrow{OA} = x\boldsymbol{i}, x > 0$, $\overrightarrow{OC} = 3\boldsymbol{j}$ and $\overrightarrow{OD} = 4\boldsymbol{k}$.

(a) Find x if the angle between the diagonals given by $\overrightarrow{OF}$ and $\overrightarrow{AG}$ is a right angle.

(b) Also find x if the same diagonals are 60° to each other.

22. State the position vectors, **OA** and **OB** of the points with coordinates A(2, –2, 1) and B(4, 0, –3) respectively. Find the angle between the vectors **OA** and **OB**.

23. If $A = \begin{bmatrix}1 & 0 & 2\\3 & 1 & 1\\-1 & 0 & -1\end{bmatrix}$ and $B = \begin{bmatrix}-1 & 0 & -2\\2 & 1 & 5\\1 & 0 & 1\end{bmatrix}$, evaluate AB and use your result to solve the simultaneous equations

$$\begin{aligned} x + 2z &= 1\\ 2x + y + 5z &= 1\\ x + z &= 2 \end{aligned}$$

24. Suppose $A = \begin{bmatrix}1 & 2\end{bmatrix}$, $B = \begin{bmatrix}3\\4\end{bmatrix}$, $C = \begin{bmatrix}1 & 2\\2 & 4\end{bmatrix}$, $D = \begin{bmatrix}1 & 2\\3 & 4\end{bmatrix}$. For each of the following, either find the matrix or give a reason why it cannot be found

(a) AB (b) BA (c) $C + D$ (d) $C^{-1}D$

25. Are the points P(3, 1, 0), Q(2, 2, 2) and R(0, 4, 6) collinear?

26. (a) Find the direction ratios and direction cosines of the lines

i. $l_1 : x = 2 + \lambda, y = 2 - \lambda, z = 2\lambda$

ii. $l_2 : x = 2 + 3t, y = 3 + 6t, z = 4 + 2t$

(b) Find the coordinates of the point of intersection of the lines l_1 and l_2.

27. Find the equation of the line throught A(1, 2, –3) and parallel to the line defined by the vector equation, $\boldsymbol{r} = \begin{pmatrix} 0 \\ 1 \\ -2 \end{pmatrix} + \lambda \begin{pmatrix} 3 \\ 2 \\ 1 \end{pmatrix}$

28. Find the point of intersection of the lines $r = \begin{pmatrix} 3 \\ 2 \end{pmatrix} + \lambda \begin{pmatrix} -1 \\ 1 \end{pmatrix}$ and $r = \begin{pmatrix} -2 \\ 1 \end{pmatrix} + \mu \begin{pmatrix} 2 \\ 3 \end{pmatrix}$

29. If the vectors $\begin{pmatrix} 3x \\ x+1 \end{pmatrix}$ and $\begin{pmatrix} x-4 \\ 4 \end{pmatrix}$ are perpendicular show that $3x^2 - 8x + 4 = 0$.

Hence find those values of x for which the vectors are perpendicular.

30. Particle A heads off on the path shown in the diagram.

Its velocity is given by the vector $\begin{pmatrix} 3 \\ -2 \end{pmatrix}$ ms^{-1}.

(a) Find the position vector of particle A relative to the origin O, t hours after its departure from point R(0, 80).

(b) Find particle A's position 2 hours after departure.

A second particle, B, is at rest at the point Q(0, –40).

(c) If particle B moves off with a velocity of $\begin{pmatrix} 4 \\ 1 \end{pmatrix}$ ms^{-1} at the same time that A departs, will the two particles collide?

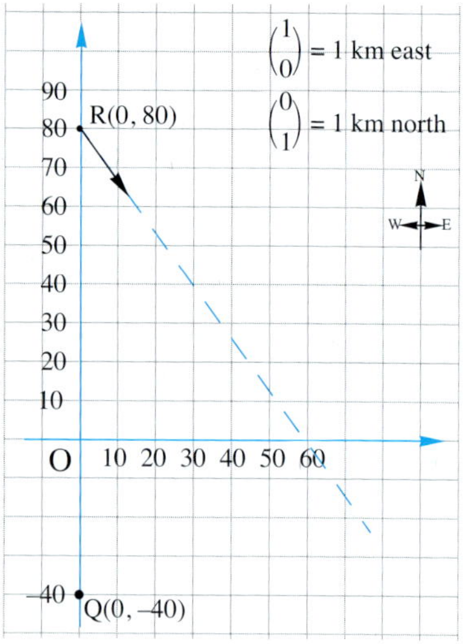

31. The ship, Excalibur, positioned 4 km east and 2 km north of a harbour sets off at a constant velocity of $\begin{pmatrix} 2 \\ 3 \end{pmatrix}$ kmh^{-1}. Fog has rendered visibility to a minimum.

(a) Find the position vector of Excalibur relative to the harbour, t hours after it sets off.

A lighthouse, no longer in operation, is located 25 km east and 13 km north of the harbour

(b) Find the position vector of Excalibur **relative to the lighthouse**.

(c) Hence, determine the closest that Excalibur will get to the lighthouse.

13.1 DESCRIBING DATA

13.1.1 DATA COLLECTION

Statistics is the science of getting 'facts from figures' and involves the collection, organisation and **analysis** of sets of observations called **data**. The data represents individual observations to which we can assign some numerical value or categorise in some way that identifies an **attribute**.

The set from which the data is collected is known as the **population**. The population can be finite or infinite, however, the data collected will be a subset from this population. Such a subset is called a **sample**. The process of gathering the data is known as **sampling**. Once we have our sample, we use characteristics of the sample to draw conclusions about the population. This is known as making **statistical inferences**. Note that statistical inference is quite different from simply collecting data and then displaying or summarising it as a 'diagram' – which is known as **descriptive statistics**.

The method that is used in collecting the sample affects the validity of the inferences that can or should be made. The aim then is to obtain a sample that is representative of the population.

This concept can be represented as follows:

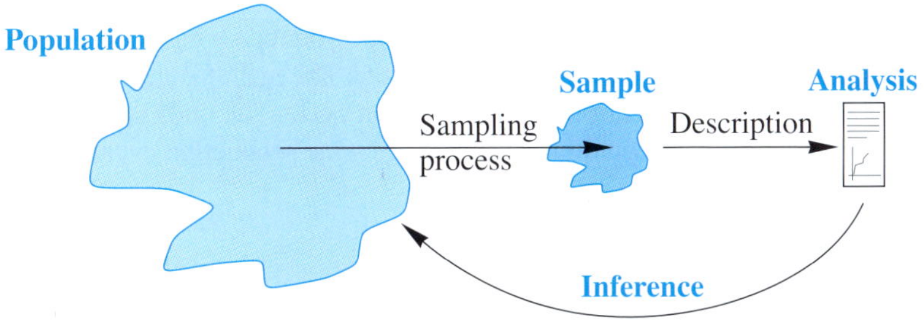

One approach to reduce bias in the sample we acquire is to use a **random sampling process**. By doing this we stand a better chance of obtaining samples that reflect the population as a whole.

13.1.2 TYPES OF DATA

Data can be classified as numerical or categorical –

1. **Numerical data**: These are made up of observations that are quantative and so have a numerical value associated with them.
For example, if the data is to represent heights, then the data would be collected as numerical values, e.g., 172 cm, 165 cm, etc.

2. **Categorical data**: These are made up of observations that are qualitative (which are sometimes also known as nominal data).
For example, if the data is to represent hair colour, then the data would be collected as qualitative data e.g., black, brown, blue, etc.

CHAPTER 13

13.1.3 DISCRETE AND CONTINUOUS DATA

As a rule of thumb, **discrete data** are sets of data that can be **counted**, which distinguishes it from **continuous data** that are sets of data that are **measured**.

EXERCISES 13.1

1. Micro Inc. produces 14,500 electrical components each month. Of these, 2000 are randomly selected and tested.
The test reveals 42 defective components.
(a) What is i. the population size.
ii. the sample size.
(b) Give an estimate of the number of defectives produced during that month.

2. A salmon farm is attempting to determine the number of salmon in its reservoir. On Monday 300 salmon were caught, tagged and then released back into the reservoir. The following Monday 200 salmon were caught and of these 12 were tagged.
(a) Comment on the sampling procedure. Is the sample size large enough? Is there a bias involved?
(b) Estimate the the number of salmon in the reservoir.

3. A manufacturer wishes to investigate the quality of his product – a measuring instrument that is calibrated to within a 0.01 mm reading accuracy. The manufacturer randomly selects 120 of these instruments during one production cycle. She finds that 8 of the instruments are outside the accepted measuring range. One production cycle produces 1500 of these measuring instruments.
(a) What is i. the population size.
ii. the sample size.
(b) Give an estimate of the number of unacceptable instruments produced during a complete production cycle.
(c) In any given week there are 10 production cycles. How many unacceptable instruments can the manufacturer expect at the end of a week. Comment on your result.

4. Classify the following as categorical or numerical data.

(a) The winning margin in a soccer game.
(b) The eye colour of a person.
(c) The number of diagrams in a magazine.
(d) The breed of a cat.
(e) The fire-hazard levels during summer.

5. Classify the following as discrete or continuous data.

(a) The number of cats in a town.
(b) The length of a piece of string.
(c) The time to run 100 metres.
(d) The number of flaws in a piece of glass.
(e) The volume of water in a one litre bottle.

13.2 FREQUENCY DIAGRAMS

13.2.1 PRODUCING A FREQUENCY DIAGRAM

The following figures are the heights (in centimetres) of a group of students:

156	172	168	153	170	160	170	156	160	160	172	174
150	160	163	152	157	158	162	154	159	163	157	160
153	154	152	155	150	150	152	152	154	151	151	154

These figures alone do not give us much information about the heights of this group of people. One of the first things that is usually done in undertaking an analysis is to make a frequency table. In this case as there are a large number of different heights, it is a good idea to group the height data into the categories (or classes) 148-150, 151-153, 154-156, etc. before making a tally.

Heights	Tally	Frequency
148-150	√√√	3
151-153	√√√√√√√√	8
154-156	√√√√√√√	7
157-159	√√√√	4
160-162	√√√√√√	6
163-165	√√	2
166-168	√	1
169-171	√√	2
172-174	√√√	3

Each height is recorded in the appropriate row of the tally column. Finally, the frequency is the number of tally marks in each row. As a check, the total of the frequency column should equal the count of the number of data items. In this case there are 36 heights.

The choice of class interval in making such a frequency table is generally made so that there are about ten classes. This is not inevitably the case and it is true to say that this choice is an art rather than a science. The objective is to show the distribution of the data as clearly as possible. This can best be seen when the data is shown graphically. There are a number of ways in which this can be done. In the present example, we are dealing with heights. Since heights vary continuously, we would most usually use a histogram to display the distribution.

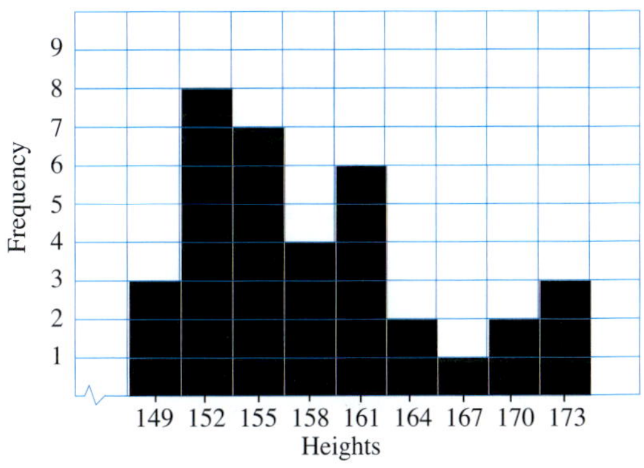

There are two details connected with the construction of histograms that you should not ignore. Firstly, as far as the horizontal scales are concerned, we are representing the continuous variable 'height'. The first class interval represents all the people with heights in the range 148 to 150cm. Since these have been rounded to the nearest whole centimetre, anyone with a height from 147.5 to 150.5cm, or [147.5,150.5), will have been placed in this class. Similarly, anyone with a height in the range [150.5,153.5) will be categorised in the class 151-153cm. If you want to label the divisions between the blocks on the histogram, technically these should be 147.5, 150.5 etc. Secondly, in a histogram, it is the area of the bars and not their height that represents the number of data items in each class. To be completely correct, we should give the area as a measure of the vertical scale. This definition allows us to draw histograms with class intervals of varying widths. This is sometimes done to smooth out the variations at the extremes of a distribution where the data is sparse. This aspect will not be considered in this chapter.

Once we have drawn a histogram, it should be possible to see any patterns that exist in the data. In this case, there is a big group of students with heights from about 150 to 160cm. There are also quite a few students with heights significantly larger than this and very few with heights below the main group. The distribution has a much larger 'tail' at the positive end than at the negative end and is said to be positively skewed. Such patterns can also be seen using other graphical devices. Such as a line graph:

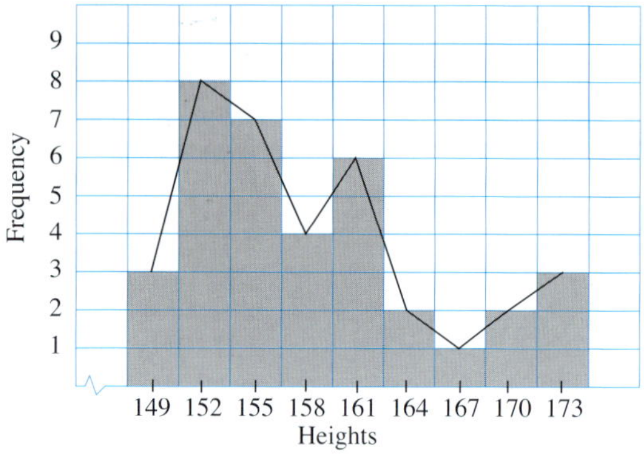

The same patterns are evident from this diagram as were seen from the histogram.

13.2.2 USING A GRAPHICS CALCULATOR

Data can be entered on the calculator either as separate figures from the original data or as a frequency table. In both cases, the data is entered as a list.

To enter the original data, press the **STAT** key and choose **EDIT** from the screen menus. If necessary, press 4 followed by the keys **L1** (2nd 1), **L2** (2nd 2) etc. **ENTER** to clear any previous lists. Next select **STAT EDIT**. The data can now be entered as a column under **L1**.

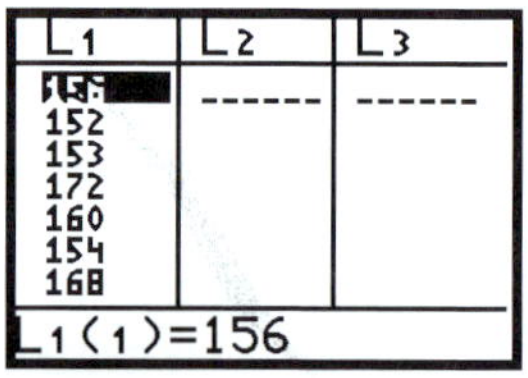

The data can now be displayed as a statistical graph. As with other types of graph, the appropriate window must be set. In the present case, the x data range should be set at 145 to 175. The **Xscl** setting determines the width of class interval that will be used.

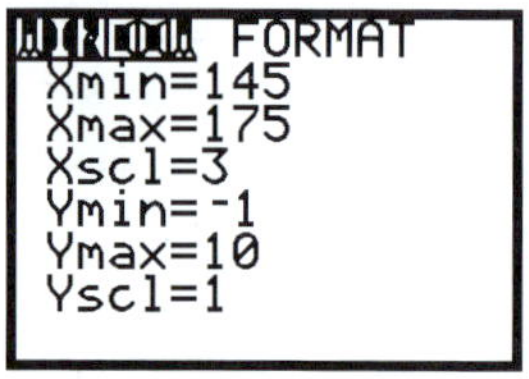

Next, any Cartesian graphs must be cleared. Press **Y=** and clear any rule that you see. The statistical plotting facility must now be activated. Press **2nd STAT PLOT**. Choose plot 1 and turn it on by using the arrows to the word **On** and then press **ENTER**. Also select the histogram symbol from the list of available plot types. We entered the data as **L1** so we must select this as the source of the data. Finally,

because each height was entered separately, the frequency must be defined as 1. Pressing **GRAPH** should now display the histogram. This should be similar to that produced earlier.

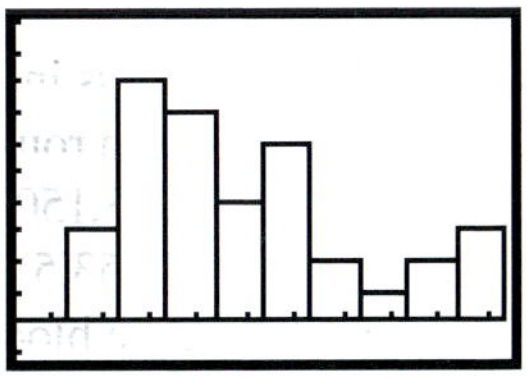

One advantage of using a calculator to produce such plots is that, once the data has been stored, the conditions of the plot can be varied rapidly. For example, if the **Xscl** is changed to 1, the class interval of the frequency table becomes 1 and the histogram is as shown.

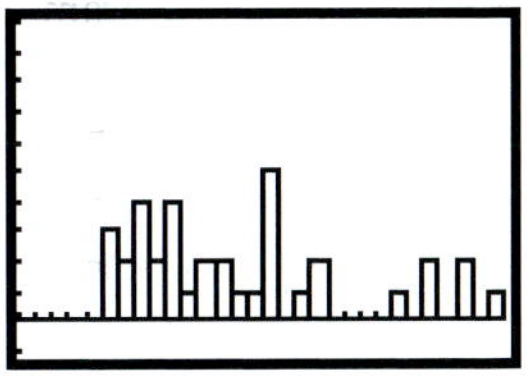

If the data is presented as a grouped list (frequency table), the *x* values should be entered as **L1** and the frequencies as **L2**. In the case of the height data, the mid-point of each class interval (149, 152, 155 etc.) should be entered in **L1**. The **STAT PLOT** instructions must also be set to record the fact that the frequencies are stored in **L2** before statistical plots or calculations will be successful.

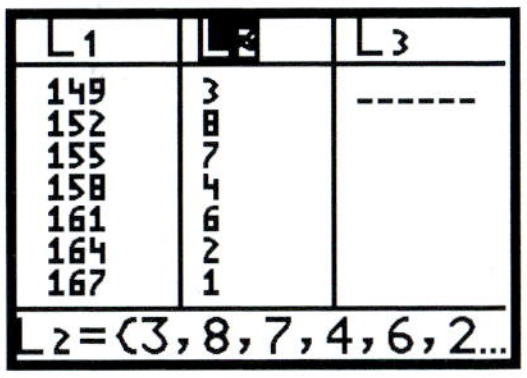

EXERCISES 13.2

1. The following figures are the weights (in grams) of a group of fish sampled from a reservoir:

226	233	233	244	224	235	238	244
222	239	233	243	221	230	237	240
225	230	236	242	222	235	237	240
220	235	238	243	222	232	232	242
229	231	234	241	228	237	237	245
229	231	237	244	225	236	235	240

(a) Construct a frequency table using the following class intervals: 218-220, 221-223, 224-226 etc. grams.

(b) Draw a histogram showing this distribution.

2. In a study of the weights of a sample of semi-precious gem-stones, the following results were obtained (grams):

1.33	1.59	1.82	1.92	1.46	1.57	1.82	2.06
1.59	1.70	1.81	2.02	1.24	1.53	1.69	2.01
1.57	1.62	1.61	1.93	1.11	1.90	1.79	1.91
1.19	1.53	1.90	1.90	1.17	1.97	1.92	2.06
1.41	1.64	1.83	1.90	1.11	1.81	1.83	1.90
1.15	1.68	1.82	1.98	1.39	1.54	1.92	2.04

(a) Construct a frequency table using the following class intervals: [1.1,1.2), [1.2,1.3), [1.3,1.4) etc. grams.

(b) Draw a histogram showing this distribution.

3. Make a frequency table and draw a histrogram for the following sets of data.
Set A:

21.1	28	26.9	31.9	23.7	28.8	27.9	31.3
21.5	26.8	27.4	31.2	21.4	29.9	29.4	31.5
20.4	25.1	25.8	33.6	23.7	25.6	29.1	30.3
21.5	28.2	28.2	31.3	22.4	25.7	25.1	30.3
21.9	29.1	28.7	30.1	21.8	27.8	29.1	34.3
22.5	25.2	25.5	32.9	22.3	29	27.2	33.3

Set B

7	6	5	70	9	9	25	72
7	7	4	72	8	9	28	73
9	9	9	72	6	7	27	71
7	7	9	70	6	8	27	73
8	5	26	73	5	6	26	70
9	9	28	73	5	8	26	71

Compare the two data sets.

13.3 STATISTICAL MEASURES 1

13.3.1 MEASURE OF CENTRAL TENDENCY

After using a graphical presentation of some sort to look at the general pattern of the data, we would usually calculate some representative 'statistics'. The aim of producing these is to reduce the amount of data to a small number of figures that represent the data as well as possible. In the case of the height data we have been studying, we have already observed that the heights group around the range 150-160 cm. This is sometimes known as a 'central tendency' and we have several ways in which we measure this:

13.3.2 MODE

This is the most frequent class of data. In the present case there were more students in the 151-153 cm class than any other so we would give this class as the mode. It is possible for some data to have more than one mode. We describe this as being bimodal, trimodal etc. The mode tends only to be used when there is no alternative such as when we are collecting data on the television stations that people like best.

13.3.3 MEAN

This is the measure commonly (and incorrectly) called average. Numeric data is added and the result is divided by the number of items of data that we have.

Notation: The notation used for the mean depends on whether or not we are claiming to have the mean of all (the **population**) or part (a **sample**) of the possible data set.

In the case of the students, we appear to have a small group of 36 selected from all the possible students in this age group and so we are looking at a sample. It is generally quite clear whether any set of data refers to a population (such as a **census**) or a sample (such as a **poll**).

The **population mean** is denoted by μ and a **sample mean** by $\bar{x}$.

For a data set x, with n items both means are calculated in the same way:

$$\text{Mean} = \frac{\sum x}{n} = \frac{1}{n}\sum x$$

The symbol Σ means 'add all the following'.

If the data is presented in the form of a frequency table in which each item of data x_i is present with a frequency of f_i, then the formula becomes:

$$\text{Mean} = \frac{\sum f_i x_i}{\sum f_i}$$

For the height data, we have two ways of approaching this calculation. One way is to return to the original data and add it all up.

The total is 5694. There are 36 measurements so: Mean $= \frac{5694}{36} = 158.16667$

Alternatively we can use the grouped data formula. There is a convenient way of doing this if we add an extra column to the orginal frequency table:

Heights	Mid-height	Frequency	$f \times h$
148-150	149	3	447
151-153	152	8	1216
154-156	155	7	1085
157-159	158	4	632
160-162	161	6	966
163-165	164	2	328
166-168	167	1	167
169-171	170	2	340
172-174	173	3	519
	Totals:	36	5700

From the table:

$$\sum f_i = 36 \text{ and } \sum f_i \times h_i = 5700 \text{ so Mean } = \frac{\sum f_i \times h_i}{\sum f_i} = \frac{5700}{36} = 158.33333.$$

This method of calculating the mean will not necessarily give exactly the same answer as the mean calculated from the original data as we have made the assumption that all the students with heights in the range 148-150 cm had a height of 149 cm. This will not generally be a seriously inaccurate assumption as the students with heights below this figure (148 cm) will be balanced by those with heights above this (150 cm). In this case, the difference is quite small.

13.3.4 MEDIAN

The median is found by arranging all the data in order of size and selecting the middle item. For the heights data, there is an even number of figures and so there is not a middle number. In this situation, we take the mean of the middle two data items.

Order	1	2	3	4	5	6	7	8	9
Ht:	150	150	150	151	151	152	152	152	152
	10	11	12	13	14	15	16	17	18
	153	153	154	154	154	154	155	156	156
	19	20	21	22	23	24	25	26	27
	157	157	158	159	160	160	160	160	160
	28	29	30	31	32	33	34	35	36
	162	163	163	168	170	170	172	172	174

The middle heights are the 18th and 19th (156 and 157cm) so the median is 156.5cm.

It is usual to take the mean of the two numbers to give an answer to represent the median, however, there are a number of interpolations that can be used. For our purposes however, we will continue to use the mean of the two observations.

When there are $2n + 1$ observation, i.e., there is an odd number of observations, the median corresponds to the $\frac{(2n+1)+1}{2}$ th observation (after they have been placed in order from lowest to highest).

e.g For the data set $\{2, 4, 12, 7, 9\}$ we first list the data from lowest to largest: 2, 4, 7, 9, 12.
Here $n = 5$ and so the middle observation is the $\frac{5+1}{2} = 3$ rd observation. i.e., 7.

EXERCISES 13.3

1. For the data set of Exercise 13.2 Q.1., find the mode, mean and median weights.

2. For the data set of Exercise 13.2 Q.2., find the mode, mean and median weights.

3. For the data sets of Exercise 13.2 Q.3., find the mode, mean and median weights.

4. The following numbers represent the annual salaries of the emplyoyees of a small company.

\$20910	\$20110	\$20390	\$20170	\$20060	\$20350
\$21410	\$21130	\$21340	\$21360	\$21360	\$21410
\$20350	\$20990	\$20690	\$20760	\$20880	\$20960
\$21240	\$21060	\$21190	\$21400	\$76000	\$125000

(a) Find the mean salary.
(b) Find the median salary.
(c) Which of the two figures is the better representative measure of salary?

5. The selling prices for the properties in a suburb over June 2004 were:

\$191000	\$152000	\$152000	\$181000
\$180000	\$163000	\$169000	\$189000
\$184000	\$169000	\$167000	\$172000
\$190000	\$169000	\$159000	\$172000
\$202000	\$162000	\$160000	\$154000
\$181000	\$166000	\$163000	\$196000
\$201000	\$154000	\$166000	\$154000
\$178000	\$164000	\$157000	\$185000
\$177000	\$169000	\$157000	\$172000
\$195000	\$150000	\$163000	\$1150000
\$186000	\$166000	\$151000	\$1155000
\$185000	\$151000	\$168000	\$1200000

(a) Find the mean selling price.
(b) Find the median selling price.
(c) Which of the two figures is the better representative measure of selling price?

6. For the figures given below, calculate the mean from the original data.

5	16	15	17	9	16	19	15
6	17	10	16	8	13	13	19
7	16	18	18	8	18	19	18
6	17	19	16	7	13	17	19
9	14	17	19	9	16	17	19
8	18	16	15	8	18	16	15

(a) Use the frequency table method with class intervals 4-6, 7-9 etc. to calculate the mean of the data.
(b) Use the frequency table method with class intervals 1-5, 6-10 etc. to calculate the mean of the data.

13.4 STATISTICAL MEASURES 2

13.4.1 MEASURES OF SPREAD

So far we have only looked at ways of measuring the central tendency of a set of data. This is not necessarily the only feature of a data set that may be important. The following sets of data are test results obtained by a group of students in two tests in which the maximum mark was 20.

Test 1.

4	12	11	10	5	10	12	12	6	8	19	13	3
7	11	13	4	9	12	10	6	13	19	11	3	12
14	11	6	13	16	11	5	10	12	13	7	8	13
14	6	10	12	10	7	10	12	10				

Test 2.

9	8	10	10	8	9	10	11	8	8	11	10	9
8	11	10	9	8	10	11	8	9	11	10	9	8
11	11	9	9	11	10	8	9	11	10	8	9	11
11	8	8	11	10	8	9	10	10				

The means of the two data sets are fairly close to one another (Test 1, 10.1, Test 2, 9.5). However, there is a substantial difference between the two sets which can be seen from the frequency tables.

Test 1

Mark	3	4	5	6	7	8	9	10	11	12	13	14	15	16	17	18	19
Freq.	2	2	2	4	3	2	1	8	5	8	6	2	0	1	0	0	2

Test 2

Mark	3	4	5	6	7	8	9	10	11	12	13	14	15	16	17	18	19
Freq.	0	0	0	0	0	13	11	12	12	0	0	0	0	0	0	0	0

The marks for Test 1 are quite well spread out across the available scores whereas those for Test 2 are concentrated around 9, 10 & 11. This may be important as the usual reason for setting tests is to rank the students in order of their performance. Test 2 is less effective at this than Test 1 because the marks have a very small spread. In fact, when teachers and examiners set a test, they are more interested in getting a good spread of marks than they are in getting a particular value for the mean. By contrast, manufacturers of precision engineering products want a small spread on the dimensions of the articles that they make. Either way, it is necessary to have a way of calculating a numerical measure of the spread of data. The most commonly used measures are variance, standard deviation and interquartile range.

13.4.2 VARIANCE AND STANDARD DEVIATION

To calculate the variance of a set of data, the frequency table can be extended as follows:
Test 1

Mark (M)	Frequency	$M - \mu$	$f(M - \mu)^2$
3	2	–7.10	100.82
4	2	–6.10	74.42
5	2	–5.10	52.02
6	4	–4.10	67.24
7	3	–3.10	28.83
8	2	–2.10	8.82
9	1	–1.10	1.21
10	8	–0.10	0.08
11	5	0.90	4.05
12	8	1.90	28.88
13	6	2.90	50.46
14	2	3.90	30.42
15	0	4.90	0.00
16	1	5.90	34.81
17	0	6.90	0.00
18	0	7.90	0.00
19	2	8.90	158.42
		Total:	640.48

The third column in this table measures the amount that each mark **deviates from the mean** mark of 10.10. Because some of these marks are larger than the mean and some are smaller, some of these deviations are positive and some are negative. If we try to calculate an average deviation using these results, the negative deviations will cancel out the positive deviations. To correct this problem, one method is to square the deviations. Finally, this result is multiplied by the frequency to produce the results in the fourth column.

The last row is calculated: $2 \times (3 - 10.10)^2 = 2 \times 50.41 = 100.82$.

The total of the fourth column is divided by the number of data items (48) to obtain the variance of the marks: $\text{Variance} = \dfrac{640.48}{48} = 13.34$

The measure most commonly used is the square root of the variance (remember that we squared the deviations). A measure that is known as the **standard deviation** of the marks. In the previous case: $\text{Standard deviation} = \sqrt{13.34} = 3.65$

Repeating this calculation for the second set of marks:

Mark (M)	Frequency	$M - m$	$f(M-m)^2$
8	13	–1.48	28.475
9	11	–0.48	2.534
10	12	0.52	3.245
11	12	1.52	27.725
		Total:	61.979

$$\text{Variance} = \frac{61.979}{48} = 1.291$$

$$\text{Standard deviation} = \sqrt{1.291} = 1.136$$

This figure is about one third of the figure calculated for Test 1. This reflects the fact that Test 2 has not spread the students very well.

In summary, the variance and population standard deviation are calculated using the formulas:

$$\text{Variance} = \frac{\sum f_i(x_i - \mu)^2}{\sum f_i} \qquad \text{Population Standard Deviation} = \sqrt{\frac{\sum f_i(x_i - \mu)^2}{\sum f_i}}$$

Of course, all of this type of mechanical work is very tedious, and seldom is it the case that we are required to work our way through these calculations. Statisticians (or anyone required to produce statistics such as these) would generally use a statistical package of some sort, where data could be easily transferred and then manipulated by any number of different commands. However, it is not necessary to have access to high powered statistical packages. Spreadsheets are widely used to produce the same results .

In many areas of this course we have seen (and will continue to see) how powerful the graphics calculator is. In the area of statistics, the TI-83 comes into its own. Once the data is entered as a list, it becomes a very powerful tool. And so, as we have said many a time before, make sure that you become familiar with all that your graphics calculator has to offer.

13.4.3 USING A GRAPHICS CALCULATOR

Standard deviation can be calculated directly by first entering the data as a list.

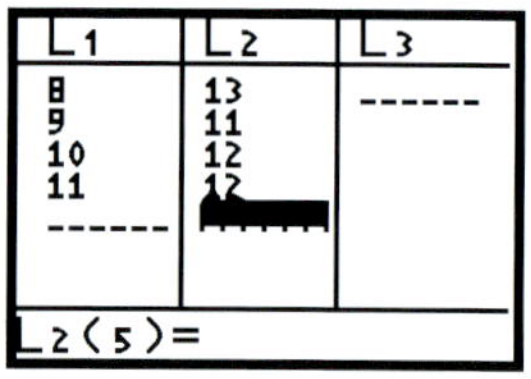

For Test 2, this is best done with the marks as list 1 and the frequencies as list 2.

Next, use **STAT CALC** to access the **1-VarStats** menu. Next, you must nominate the two lists that contain the data. List 1 is **2nd 1** and list 2 is **2nd 2**. The two list names must be separated by a comma.

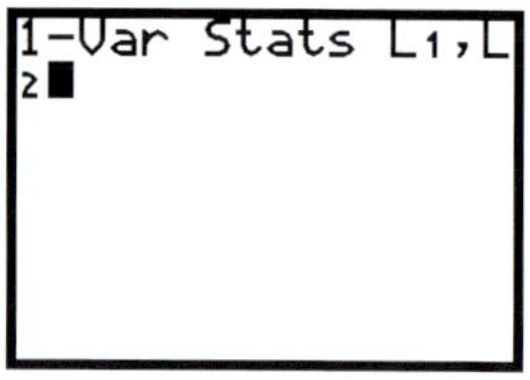

Finally, press **ENTER**. This screen gives the mean, $(\bar{x})$, the sum of the data, Σx, and the sum of the squares of the data, Σx^2.

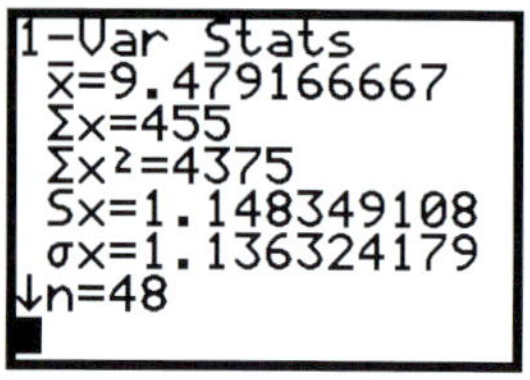

S*x* is known as the **sample standard deviation**. This is the same as the standard deviation discussed above but with one less than the number of data items in the denominator (47 in this case).
σ*x* is the **population standard deviation** discussed above.

Sample standard deviation? Population standard deviation? What's it all about? Unfortunately there are regional variations (as well as in textbooks) in the notation and the language that is used to define these terms.

When we refer to the *sample variance*, it suggests that we are finding the variance of a sample and by default, the sample is a subset of a population and so we are in fact finding an estimate of the population variance. This estimate is known as the *unbiased estimate* of the population variance.

The **unbiased estimate** of the population variance, σ^2, is given by $s_{n-1}^2 = \frac{1}{n-1}\sum_{i=1}^{n} f_i(x_i - \bar{x})^2$.

The standard deviation of the sample is given by the square root of s_{n-1}^2, i.e., $\sqrt{s_{n-1}^2}$, which corresponds to the value **S*x*** that is produced by the TI-83.

The **variance of a population**, σ^2, is given by $\sigma^2 = \frac{1}{n}\sum_{i=1}^{n} f_i(x_i - \bar{x})^2$. The standard deviation then is $\sigma = \sqrt{\sigma^2}$.

To differentiate between division by n and division by $n-1$ we use s_n^2 for division by n and s_{n-1}^2 for division by $n-1$. Giving the relationship $s_{n-1}^2 = \frac{n}{n-1}s_n^2$.

Then, as the population variance, σ^2, is generally unknown, s_{n-1}^2 serves as an **estimate** of σ^2.

On the TI–83 we have that **S*x*** $= s_{n-1}$ and **σ*x*** $= s_n$.

It is therefore important that you are familiar with the notation that your calculator uses for sample standard deviation (unbiased) and population standard deviation.

EXERCISES 13.4

1. The weights (kg) of two samples of bagged sugar taken from a production line

Sample from machine A

1.95	1.94	2.02	1.94	2.07	1.95	2.02	2.06
2.09	2.09	1.94	2.01	2.07	2.05	2.04	1.91
1.91	2.02	1.92	1.99	1.98	2.09	2.05	2.05
1.99	1.97	1.97	1.95	1.93	2.03	2.02	1.90
1.93	1.91	2.00	2.03	1.94	2.00	2.02	2.02
2.03	1.96	2.04	1.92	1.95	1.97	1.97	2.07

Sample from machine B

1.77	2.07	1.97	2.22	1.60	1.96	1.95	2.23
1.79	1.98	2.07	2.32	1.66	1.96	2.05	2.32
1.80	1.96	2.06	1.80	1.93	1.91	1.93	2.25
1.63	1.97	2.08	2.32	1.94	1.93	1.94	2.22
1.76	2.06	1.91	2.39	1.98	2.06	2.02	2.23
1.75	1.95	1.96	1.80	1.95	2.09	2.08	2.29

(a) Find the mean weights of the bags in each sample.

(b) Use the formula $S_x = \sqrt{\dfrac{\sum f_i(x_i - \bar{x})^2}{\sum f_i - 1}}$ to calculate the sample standard deviations of each sample.

(c) Use the formula $\sigma_x = \sqrt{\dfrac{\sum f_i(x_i - \mu)^2}{\sum f_i}}$ to calculate the population standard deviations of each sample.

2. The following frequency table gives the numbers of passengers using a bus service over a week long period.

Passengers	0-4	5-9	10-14	15-19	20-24	25-29
Frequency	3	5	11	15	10	7

(a) Find the mean number of passengers carried per trip.
(b) Find the population standard deviation of the number of passengers carried per trip.

3. The number of matches per box in a sample of boxes taken from a packing machine was:

Matches	47	48	49	50	51	52
Frequency	3	6	11	19	12	9

Find the mean and sample standard deviation of the number of matches per box.

13.5 STATISTICAL MEASURES 3

13.5.1 QUARTILES

We have already seen that there is more than one way of measuring central tendency and the same is true of measures of spread. The median is, in some circumstances, a good measure of central tendency. Following on from this, we can define quartiles, the data items that are one quarter and three quarters of the way through a list.

The following data represent the number of employees absent from work over a nine day period:

$$2, 6, 5, 4, 7, 1, 0, 5, 2.$$

Firstly, we order the data to get $0, 1, 2, 2, 4, 5, 5, 6, 7.$
The median is the middle figure:

0 1 2 2 4 5 5 6 7

Median

Lower quartile (Q1) Upper quartile (Q3)

The median divides the distribution into an upper and lower group. The lower quartile is the middle figure of the lower group and the upper quartile is the middle figure of the upper group. As with finding the median, it may be necessary when dealing with a group with an even number of data items to take the mid point between two numbers. This is the case with the current data set. The lower quartile is 1.5 and the upper quartile is 5.5.

When dealing with large data sets or grouped data, there is an alternative method of finding the median and quartiles based on **cumulative frequency**.
This is calculated as follows:

These figures represent the numbers of customers in a small cinema:

Customers	Frequency	Cumulative Frequency
0-9	1	1
10-19	4	5
20-29	9	14
30-39	11	25
40-49	32	57
50-59	23	80
60-69	10	90
70-79	9	99
80-89	1	100

The cumulative frequency is calculated by 'accumulating' the frequencies as we move down the table. Thus the figure 25 in the shaded box means that on 25 occasions there were fewer than 40 customers.

Cumulative frequencies can now be used to produce a **cumulative frequency curve**:

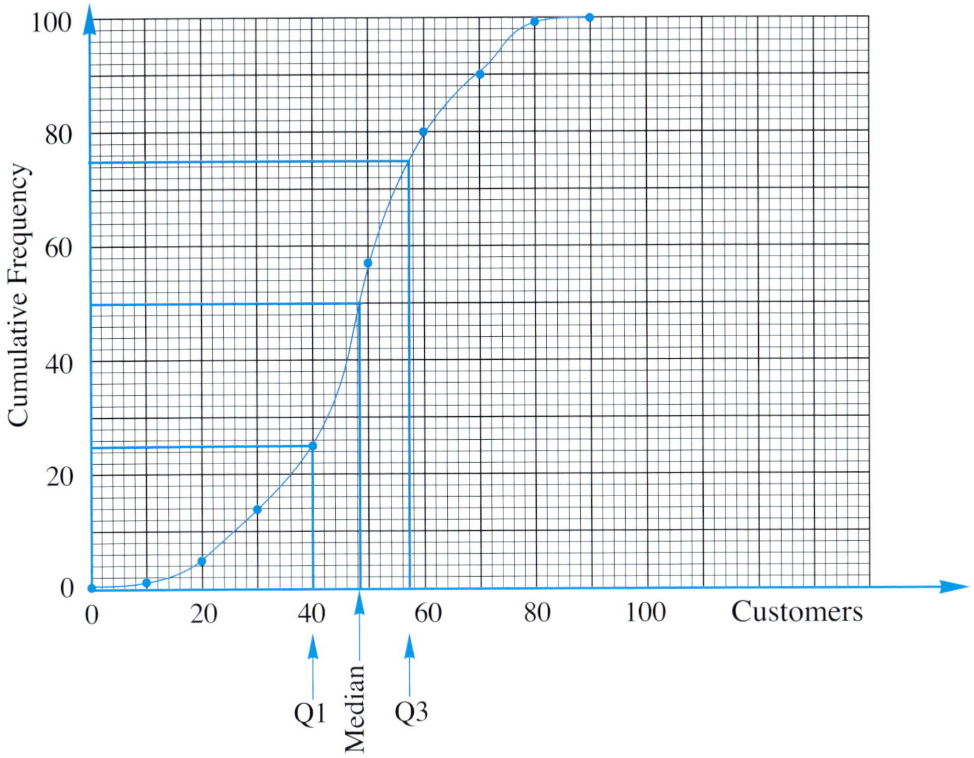

The cumulative frequency curve or **ogive** has effectively placed the data in order. This now enables us to read off estimates of the median and quartiles. The median is half way along the list of data. Since there are 100 figures, the median point is at 50. Technically this should be figure number 51, however, this method only produces an approximate figure and we seldom worry about this distinction. Reading across from 50 and down to the'customers' scale gives a figure of about 48 customers as the median. Similarly, the lower quartile can be found at a cumulative frequency of 25. Reading across from this figure to the graph and then to the horizontal axis gives a lower quartile of approximately 40 customers. Similarly, the upper quartile is about 57 customers.

The difference between the two quartiles is known as the **inter-quartile range**. In this case, the inter-quartile range is 57 – 40 = 17 customers. This is, like the standard deviation, a measure of the spread of the data. For these cinema attendance figures, the standard deviation is about 16 customers. It is not necessarily the case that these two measures of spread give similar answers. When comparing two data sets, choose which measure of spread you wish to use and use that measure throughout the analysis. Do not try to compare the inter-quartile range of one data set with the standard deviation of another.

In choosing which measure of spread to use, we generally use the quartiles and the median for a data set that contains a very few numbers that are very unusual. Such data are known as **outliers**.The data sets in Exercise 13.3 Questions 4 & 5 are examples of this type of data containing outliers. The standard deviation and mean are much more sensitive to outliers than are the median and inter-quartile range. Of course, you will need to look at a data set that has outliers and decide whether or not you want to minimise their effect on the representative statistics that you calculate.

13.5.2 BOX PLOT

Using a graphics calculator

The median and inter-quartile range of a data set can be found directly using a graphics calculator. The data can either be entered as a list or as a frequency table using two lists.

The data set: 5.7, 4.2, 7.9, 3.1, 9.4, 4.2, 7.7 & 8.0
can be entered as a single list.

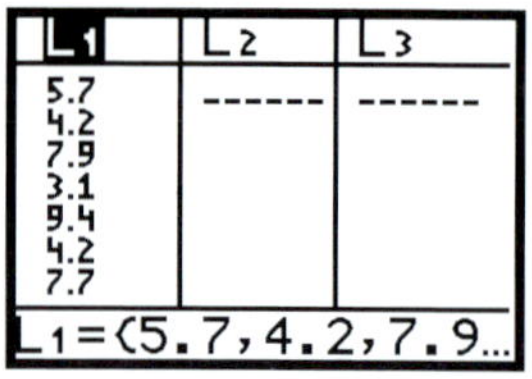

The statistics of this set can be calculated in the same way as the mean and standard deviation were calculated previously. The appropriate figures are found on the second screen (use the down arrow). This screen gives the **five figure summary** of the data: Minimum = 3.1, lower quartile = 4.2, median = 6.7, upper quartile = 7.95 and maximum = 9.4.

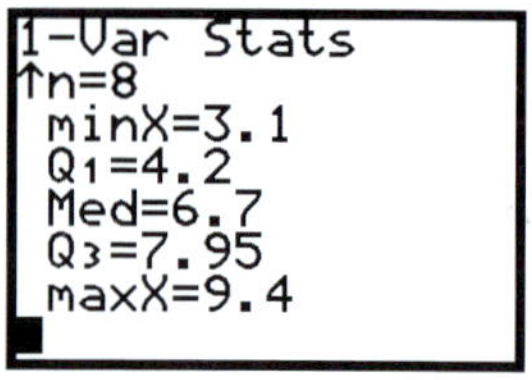

The five figure summary can be displayed using a graph known as a box-plot. This can be displayed on the TI-83 by choosing **2nd STAT PLOT**, turning on plot 1 and selecting the box-plot icon. An appropriate viewing window will also need to be set. The **TRACE** function can be used to identify the five figure summary. The diagram shows the median.

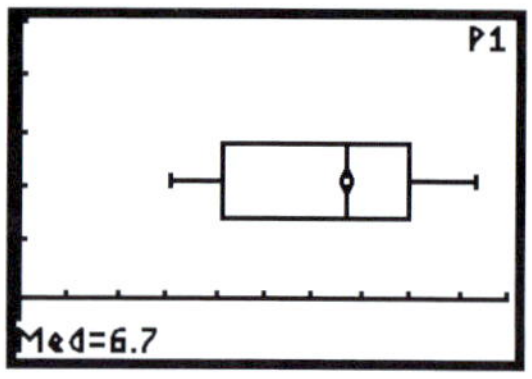

NB: The **first box-plot icon** produces a box plot showing any **outliers**.
The **second box-plot icon** produces a box plot with the **whiskers** running to the outliers.

To construct a **box-plot** (also known as a **box and whisker plot**), draw an appropriate horizontal scale. For this data set, a horizontal scale of 1 to 10 with marks every unit is appropriate. The box plot can now be drawn:

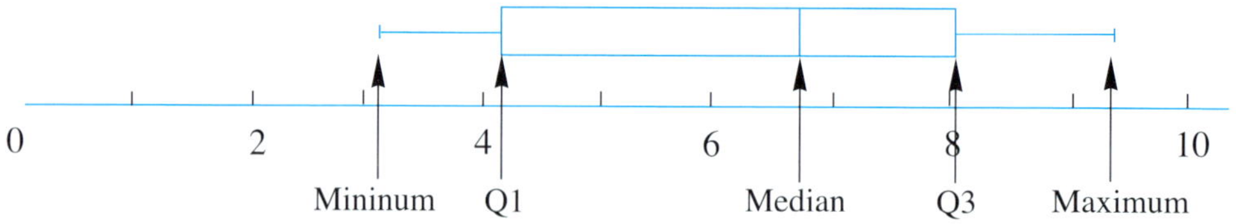

In the same way, the five figure summary for grouped data can be found by entering the data as list 1 and the frequencies as list 2. This frequency table gives the numbers of goals scored in 20 soccer matches:

Goals	0	1	2	3	4	5
Frequency	3	5	5	4	2	1

Enter the goals as **L1** and the frequencies as **L2**.

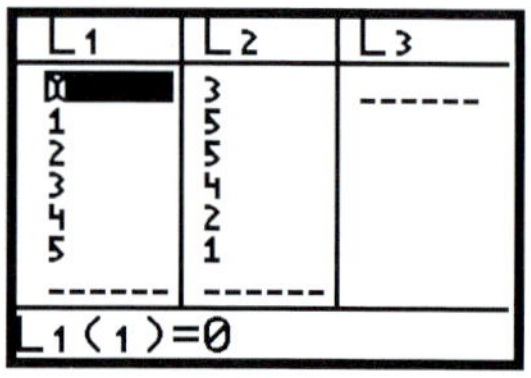

Next use the **STAT CALC** menu and identify the two lists that contain the data.

Press enter and use the down arrow to access the median part of the display.

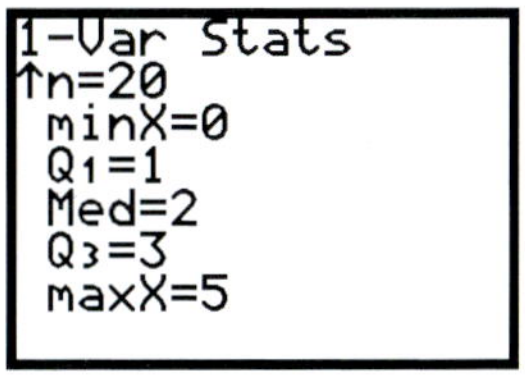

EXERCISES 13.5

1. Find the median, quartiles and inter-quartile range of these data sets:

(a) 5, 6, 2, 8, 9, 2, 7, 0, 5, 3

(b) 2.8, 4.9, 2.8, 0.9, 3.3, 5.8, 2.9, 3.7, 6.9, 3.3, 5.1

(c) 142, 167, 143, 126, 182, 199, 172, 164, 144, 163, 192, 101, 183, 153

(d) 0.02, 0.25, 1.72, 0.93, 0.99, 1.62, 0.67, 1.42, 1.75, 0.04, 1.12, 1.93

(e) 1200, 2046, 5035, 4512, 7242, 6252, 5252, 8352, 6242, 1232

2. Find the median, quartiles and inter-quartile range of these grouped data sets:

(a)

x	0	1	2	3	4	5
Frequency	1	3	6	6	7	1

(b)

x	10	11	12	13	14	15
Frequency	12	45	56	78	42	16

(c)

x	1.0	1.5	2.0	2.5	3.0	3.5
Frequency	2	4	9	9	2	1

(d)

x	10	20	30	40	50	60
Frequency	4	8	15	19	20	5

(e)

x	0	5	10	15	20	25
Frequency	0	3	0	6	7	5

3. The weekly expenses paid to a group of employees of a small company were

\$25	\$0	\$10	\$10
\$55	\$0	\$12	\$375
\$75	\$445	\$7	\$2

(a) Find the mean weekly expense.

(b) Find the population standard deviation of the expenses.

(c) Find the median weekly expense.

(d) Find the quartiles and the inter-quartile range of the expenses.

(e) Which of these statistics are the best representatives of the expenses data?

4. The table shows the numbers of cars per week sold by a dealership over a year.

Cars sold	0	1	2	3	4	5
Number of weeks	2	13	15	12	7	3

(a) Find the mean weekly sales.

(b) Find the population standard deviation of the sales.

(c) Find the median weekly sales.

(d) Find the quartiles and the inter-quartile range of the sales.

5. The table shows the weekly turnover of a small shop over a period during Spring and Summer.

Sales (\$)	\$0-\$99	\$100-\$199	\$200-\$299	\$300-\$399
Number of weeks	2	9	15	7

(a) Find the mean weekly sales.

(b) Find the population standard deviation of the sales.

(c) Construct a cumulative frequency table and draw the cumulative frequency curve.

(d) Find the median weekly sales from your graph.

(e) Find the quartiles and the inter-quartile range of the sales from your graph.

6. Plot the cumulative frequency curves for these data and hence estimate the median, quartiles and inter-quartile range of the data.

x	0-4	5-9	10-14	15-19	20-24	25-29
Frequency	2	5	11	9	7	2

EXERCISES 13.6 – MISCELLANEOUS EXERCISES

1. Identify the sample and the population in each of these situations.

(a) The University of Arkansas tests a new AIDS drug on 100 randomly chosen patients suffering from the disease. After a year of treatment, 42 patients had shown positive signs of recovery and 10 patients had died. The remainder appeared unchanged by the treatment.

(b) The Australian Government surveys 1000 working aged people in NSW to determine unemployment figures in that state.

(c) John asks all members of his IB higher maths class what theme they would like for the Nappa Valley High School senior dance.

2. Classify each of the following as discrete or continuous variables:

(a) The grade received by a student.
(b) The number of people in a passing car.
(c) The playing time of a movie.
(d) Cups of coffee drunk by your maths teacher in a day.
(e) Rainfall in Bali in a month.
(f) The length of a king cobra.
(g) The mass of silica in a sample of 1 kg of earth.

3. Two tetrahedral dice are rolled 60 times, and the sum of their scores recorded:

2, 3, 4, 5, 6, 7, 8, 4, 4, 6, 3, 5, 5, 6, 5, 8, 5, 5,4, 4, 2, 7, 6, 4, 5, 5, 3, 8, 7, 5, 5, 4, 4, 3, 7, 6, 6, 5, 5, 5, 6, 4, 5, 6, 3, 4, 5, 6, 3, 4, 5, 6, 2, 3, 4, 5, 6, 7, 8, 7

(a) Construct a frequency table from this data and draw the histogram.
(b) Complete the cumulative frequency distribution and draw a cumulative frequency graph

4. Choose suitable classes to group the data in each of these situations:

(a) About 15 classes are to be used to group 300 scores which range from 207 to 592.
(b) About 18 classes are to be used to group 900 scores which range from 112 to 418.
(c) SAT scores from 500 students are to be grouped. The lowest score is 458 and the highest score is 784. Use approximately 15 classes.

5. The weights of 50 year 9 students are measured to the nearest kg and recorded. Construct a frequency table using class intervals of 5kg and draw the histogram. Complete the cumulative frequency distribution and draw a cumulative frequency graph.

41	54	37	55	52	60	45	56	56	47
54	51	64	53	64	57	65	40	73	53
57	46	76	56	46	59	55	63	48	43
47	72	41	51	67	44	53	63	58	63
50	49	56	57	48	55	55	53	63	35

6. A group of 100 IB students was given a maths test that was graded out of 20. The following is the distribution of the marks obtained:

mark	9	10	11	12	13	14	15	16	17	18	19
number of students	1	1	3	5	8	13	19	24	14	10	2

(a) Write down the mode.
(b) Draw a cumulative frequency graph
(c) Calculate the mean.
(d) Find the median.
(e) Find the upper and lower quartiles.
(f) Another group of 50 students had a mean mark of 17.16 on the same test. Calculate the mean of the entire group of 150 students.

7. A biologist measures the lengths of 60 mature fern fronds to the nearest centimeter. The results are summarized in the table below:

Frond Length (cm)	Frequency
10 – 14	2
15 – 19	6
20 – 24	8
25 – 29	10
30 – 34	15
35 – 39	9
40 – 44	6
45 – 49	4

(a) Write down the modal class.
(b) Draw a cumulative frequency graph
(c) Calculate the mean.
(d) Find the median.

8. For a mathematics project, Eun-Kee timed the length of 30 popular Korean songs to the nearest second. His raw data is presented below:

185	230	205	215	217	206	192	187	207	245	205	181
216	227	239	214	242	248	193	222	217	219	204	234
227	236	234	217	186	236						

(a) Complete the frequency and cumulative frequency distribution table using class intervals of 10 seconds.
(b) Use your table to calculate the mean.
(c) Re-calculate the mean using the raw data.

9. The masses of a sample of new potatoes were measured to the nearest gram and are summarized below. Calculate the mean of the data.

Mass(g)	Frequency
10 – 19	2
20 – 29	14
30 – 39	21
40 – 49	73
50 – 59	42
60 – 69	13
70 – 79	9
80 – 89	4
90 – 99	2

10. Foodcity supermarket recorded the length of time, to the nearest minute, that a sample of 200 cars was parked in their car park. The results were:

Time (minutes)	Frequency
0 – 14	13
15 – 29	23
30 – 44	32
45 – 59	33
60 – 74	27
75 – 89	20
90 – 104	12
105 – 119	11
120 – 134	10
135 – 149	11
150 – 164	8

(a) Draw the cumulative frequency curve and use it to estimate the upper and lower quartiles.
(b) Estimate the 80th percentile.
(c) Estimate the percentage of cars parking for more than 50 minutes.
(d) Calculate the mean time parked for the sample of cars.

11. The heights of 10 students, to the nearest centimeter, are as follows. State the range and use a table to calculate the standard deviation.

172 169 163 175 182 170 165 176 177 169

12. The scores of 25 students in a quiz out of 10 are presented below. Use a table to calculate the mean and standard deviation.

7	6	7	4	5	6	4	8	7	6
8	4	5	5	6	7	5	7	4	3
6	4	9	5	7					

Verify your results by using the statistical functions found on your calculator.

13. A scientific researcher weighs a random sample of 30 lizards and records their weights to the nearest gram. Use your calculator to find the sample mean and standard deviation. Provide an unbiased estimate of the population standard deviation.

21	18	15	20	18	17	12	23	19	19
17	20	13	15	17	21	18	14	13	18
22	17	15	12	25	15	16	18	16	17

14. The scores of 100 students in a test out of 10 were:

score	3	4	5	6	7	8	9	10
frequency	5	11	19	24	21	12	6	2

(a) Find the median.
(b) State the range.
(c) Find the upper and lower quartiles.
(d) State the inter-quartile range.
(e) Calculate the mean.
(f) Calculate the standard deviation.

15. For a mathematics project, Eun-Kee timed the length of 30 popular Korean songs to the nearest second. His raw data is presented in Q.8 and below:

185	230	205	215	217	206	192	187	207	245	205	181
216	227	239	214	242	248	193	222	217	219	204	234
227	236	234	217	186	236						

Use your calculator to find the standard deviation, and provide an estimate of the standard deviation of all Korean popular songs.

16. The masses of a sample of new potatoes were measured to the nearest gram and are summarized in Q.9 and below. Estimate the standard deviation of weights of new potatoes.

Mass (g)	Frequency
10 – 19	2
20 – 29	14
30 – 39	21
40 – 49	73
50 – 59	42
60 – 69	13
70 – 79	9
80 – 89	4
90 – 99	2

14.1 MULTIPLICATION PRINCIPLE

14.1.1 DEFINITION

Permutations

Permutations represents a counting process where the **order must be taken into account**.

For example, the number of permutations of the letters A, B, C and D, if only two are taken at a time, can be enumerated as

AB, AC, AD, BA, BC, BD, CA, CB, CD, DA, DB, BC

That is, AC is a different permutation from CA (different order).

Instead of **permutation** the term **arrangement** is often used.

This definitions lead to a number of **Counting Principles** which we now look at.

14.1.2 MULTIPLICATION PRINCIPLE

Rule 1: If any one of n different mutually exclusive and exhaustive events can occur on each of k trials, the number of possible outcomes is equal to n^k.

For example, if a die is rolled twice, there are a total of $6^2 = 36$ possible outcomes.

Rule 2: If there are n_1 events on the first trial, n_2 events on the second trial, and so on, and finally, n_k events on the kth trial, then the number of possible outcomes is equal to $n_1 \times n_2 \times \ldots \times n_k$.

For example, if a person has three different coloured pairs of pants, four different shirts, five different ties and three different coloured pairs of socks, the total number of different ways that this person can dress is equal to $3 \times 4 \times 5 \times 3 = 180$ ways.

Rule 3: The total number of ways that n different objects can be arranged in order is equal to $n \times (n-1) \times (n-2) \times \ldots 3 \times 2 \times 1$.

Because of the common usage of this expression, we use the factorial notation. That is, we write

$$n! = n \times (n-1) \times (n-2) \times \ldots 3 \times 2 \times 1$$

which is read as ***n* factorial**. Notice also that 0! is defined as 1, i.e., 0! = 1.

For example, in how many ways can 4 boys and 3 girls be seated on a park bench? In this case any one of the seven children can be seated at one end, meaning that the adjacent position can be filled by any one of the remaining six children, similarly, the next adjacent seat can be occupied by any one of the remaining 5 children, and so on . . . .

Therefore, in total there are $7 \times 6 \times 5 \times 4 \times 3 \times 2 \times 1 = 7! = 5040$ possible arrangements.

Using the TI–83, we have:

Enter **7**:

Select **MATH** and **PRB**:

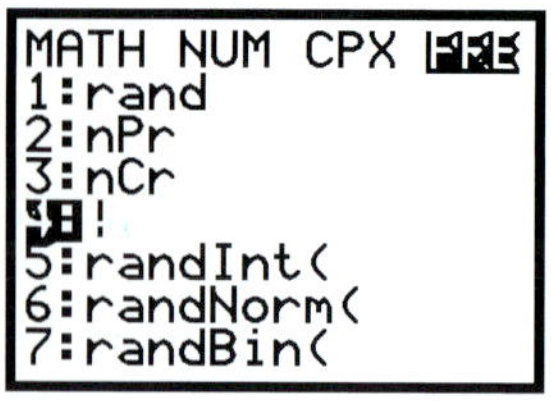

Select option **4: !** and press **ENTER:**

EXAMPLE 14.1

John wishes to get from town A to town C via town B. There are three roads connecting town A to town B and 4 roads connecting town B to town C. In how many different ways can John get from town A to town C?

Solution

We start by visualising this situation:

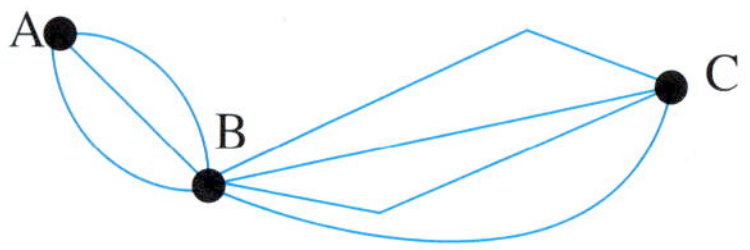

Consider the case where John uses Road 1 first.

The possibilities are:

Road 1 then *a*,
Road 1 then *b*,
Road 1 then *c*,
Road 1 then *d*.

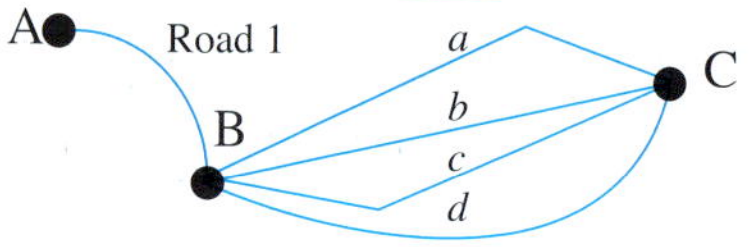

That is, there are 4 possible routes.

Then, for each possible road from A to B there are another 4 leading from B to C.

All in all, there are $4 + 4 + 4 = 3 \times 4 = 12$ different ways John can get from A to C via B.

EXAMPLE 14.2

Using the following street network, in how many different ways can a person get from point P to point Q if they can only move from left to right

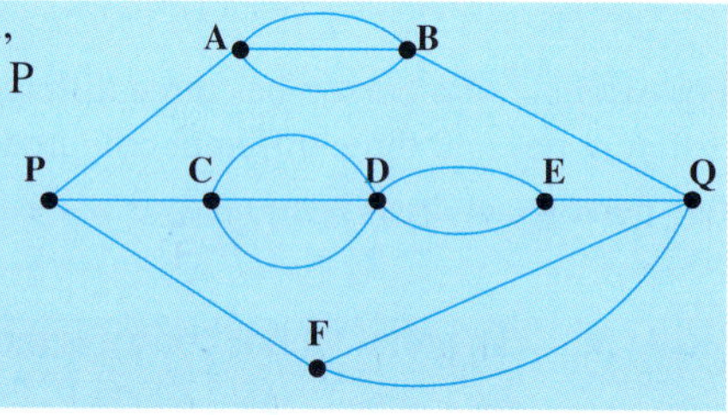

Solution

In travelling from P to Q there are:

$3 = 1 \times 3 \times 1$ paths (along P to A to B to Q)
$6 = 1 \times 3 \times 2 \times 1$ paths (along P to C to D to E to Q)
$2 = 1 \times 2$ paths (along P to F to Q)

In total there are $3 + 6 + 2 = 11$ paths

EXAMPLE 14.3

A golfer has 3 drivers, 4 tees and 5 golf balls. In how many ways can the golfer take his first hit.

Solution

We think of this problem as follows:

The golfer has 3 possible drivers to use and so the first task can be carried out in 3 ways.
The golfer has 4 possible tees to use and so the second task can be carried out in 4 ways.
The golfer has 5 golf balls to use and so the third task can be carried out in 5 ways.

Using the multiplication principle, there are a total of $3 \times 4 \times 5 = 60$ ways to take the first hit.

14.1.3 PERMUTATIONS

Based on the definition given in §14.1.1 we have the following rule:

Rule 4: **Permutations**:
The total number of ways of **arranging** n objects, taking r at a time is given by

$$\frac{n!}{(n-r)!}$$

Notation: We use the notation ${}^nP_r$ (read as "n–p–r') to denote $\frac{n!}{(n-r)!}$.

That is, ${}^nP_r = \frac{n!}{(n-r)!}$.

For example, the total number of arrangements of 8 books on a bookshelf if only 5 are used is given by ${}^8P_5 = \frac{8!}{(8-5)!} = \frac{8!}{3!} = 6720$.

When using the TI–83, we can either use the same approach as in the previous example or use the **nPr** function:
Type the first number, **8**, then select **MATH** and **PRB**, then select option **2:nPr**, enter the second number, **5**, and then press **ENTER**:

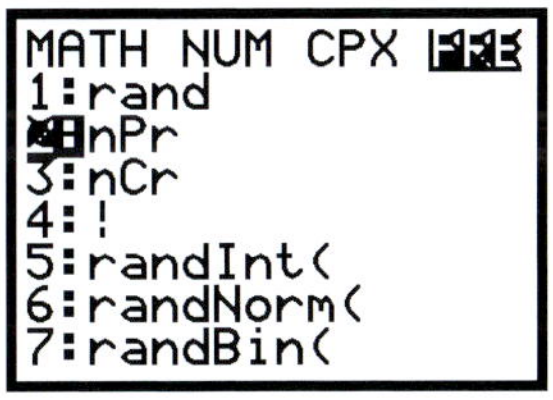

EXAMPLE 14.4

In how many ways can 5 boys be arranged in a row

(a) using three boys at a time?
(b) using 5 boys at a time?

Solution

We have 5 boys to be arranged in a row with certain constraints.

(a) The constraint is that we can only use 3 boys at a time. In other words, we want the number of arrangements (permutations) of 5 objects taken 3 at a time.
From rule 4: $n = 5, r = 3$,

$$\text{Therefore, number of arrangements} = {}^5P_3 = \frac{5!}{(5-3)!} = \frac{120}{2} = 60$$

(b) This time we want the number of arrangements of 5 boys taking all 5 at a time.
From rule 4: $n = 5, r = 5$,

$$\text{Therefore, number of arrangements} = {}^5P_5 = \frac{5!}{(5-5)!} = \frac{120}{0!} = 120$$

Box method

Problems like Example 14.4 can be solved using a method known as "the box method". In that particular example, part (a) can be considered as filling three boxes (with only one object per box) using 5 objects:

Box 1	Box 2	Box 3

The first box can be filled in 5 different ways (as there are 5 possibilities available). Therefore we 'place 5' in box 1:

Box 1	Box 2	Box 3
5		

Now, as we have used up one of the objects (occupying box 1), we have 4 objects left that can be used to fill the second box. So, we 'place 4' in box 2:

Box 1	Box 2	Box 3
5	4	

At this stage we are left with three objects (as two of them have been used). Meaning that there are 3 possible ways in which the third box can be filled. So, we 'place 3' in box 3:

Box 1	Box 2	Box 3
5	4	3

This is equivalent to saying, that we can carry out the first task in 5 different ways, the second task in 4 different ways and the third task in 3 different ways. Therefore, using the multiplication principle we have that the total number of arrangements is $5 \times 4 \times 3 = 60$.

Comparing this to the expression ${}^5P_3 = \frac{5!}{(5-3)!}$ we have $\frac{5!}{(5-3)!} = \frac{5!}{2!} = \frac{5 \times 4 \times 3 \times 2 \times 1}{2 \times 1}$

$$= 5 \times 4 \times 3$$
$$= 60$$

i.e., the last step in the evaluation process is the same as the step used in the 'box method'.

EXAMPLE 14.5

Vehicle licence plates consist of two letters from a 26–letter alphabet, followed by a three–digit number whose first digit cannot be zero. How many different licence plates can there be?

Solution

We have a situation where there are five positions to be filled:

Letter	Letter	Number	Number	Number

That is, the first position must be occupied by one of 26 letters, similarly, the second position must be occupied by one of 26 letters. The first number must be made up of one of nine different digits (as zero must be excluded), whilst the other two positions have 10 digits that can be used. Therefore, using Rule 2, we have:

Total number of arrangements = $26 \times 26 \times 9 \times 10 \times 10 = 608400$.

EXAMPLE 14.6

How many 5-digit numbers greater than 40 000 can be formed from the digits 0, 1, 2, 3, 4, and 5 if

(a) there is no repetition of digits allowed?

(b) repetition of digits is allowed?

Solution

(a) Consider the five boxes

Box 1	Box 2	Box 3	Box 4	Box 5

Only the digits 4 and 5 can occupy the first box (so as to obtain a number greater than 40 000). So there are 2 ways to fill box 1:

Box 1	Box 2	Box 3	Box 4	Box 5
2				

Box 2 can now be filled using any of the remaining 5 digits. So, there are 5 ways of filling box 2:

Box 1	Box 2	Box 3	Box 4	Box 5
2	5			

We now have 4 digits left to be used. So, there are 4 ways of filling box 3:

Box 1	Box 2	Box 3	Box 4	Box 5
2	5	4		

Continuing in this manner we have:

Box 1	Box 2	Box 3	Box 4	Box 5
2	5	4	3	2

Then, using the multiplication principle we have $2 \times 5 \times 4 \times 3 \times 2 = 240$ arrangements.

Otherwise, we could have relied on rule 4 and obtained $2 \times {}^5P_4 = 2 \times 120 = 240$

(b) As in part (a), only the digits 4 and 5 can occupy the first box.
If repetition is allowed, then boxes 2 to 5 can each be filled using any of the 6 digits:

Box 1	Box 2	Box 3	Box 4	Box 5
2	6	6	6	6

Using the multiplication principle there are $2 \times 6 \times 6 \times 6 \times 6 = 2592$ arrangements.

However, one of these arrangements will also include the number 40 000. Therefore, the number of 5 digit numbers greater than 40 000 (when repetition is allowed) is given by 2592 – 1 = 2591.

EXAMPLE 14.7

Find n if ${}^{n}P_3 = 60$.

Solution

$$ {}^{n}P_3 = 60 \Leftrightarrow \frac{n!}{(n-3)!} = 60 $$

$$ \Leftrightarrow \frac{n(n-1)(n-2)(n-3)!}{(n-3)!} = 60 $$

$$ \Leftrightarrow n(n-1)(n-2) = 60 $$

$$ \Leftrightarrow n^3 - 3n^2 + 2n = 60 $$

$$ \Leftrightarrow n^3 - 3n^2 + 2n - 60 = 0 $$

Using the TI–83 to solve this polynomial, we have:

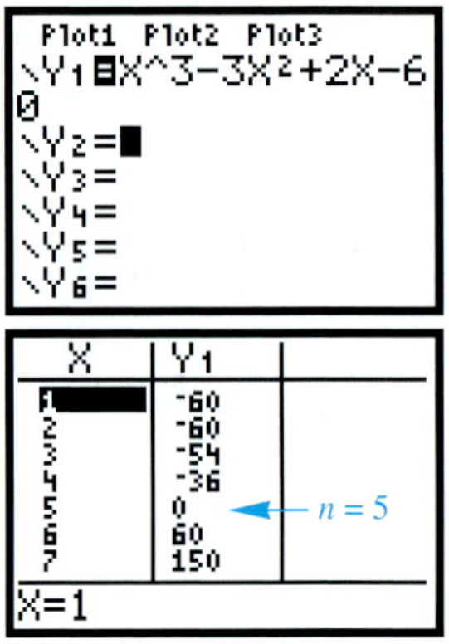

i.e., $n = 5$.

EXAMPLE 14.8

How many different arrangements of the letters of the word HIPPOPOTAMUS are there?

Solution

The word 'HIPPOPOTAMUS' is made up of 12 letters, unfortunately, they are not all different! Meaning that although we can swap the three P's with each other, the word will remain the same.

Now, the total number of times we can re–arrange the Ps (and not alter the word) is 3! = 6 times (as there are three Ps). Therefore, if we 'blindly' use Rule 2, we will have increased the number of arrangements 6 fold.

Therefore, we will need to divide the total number of ways of arranging 12 objects by 6.

That is, $\frac{12!}{3!} = 79833600$.

However, we also have 2 Os, and so, the same argument holds. So that in fact, we now have a total of $\frac{12!}{3! \times 2!} = 39916800$ arrangements.

This example is a special case of **permutations with repetitions**:

Rule 5: The number of permutations of n objects of which n_1 are identical, n_2 are identical, . . . , n_k are identical is given by $\frac{n!}{n_1! \times n_2! \times \ldots \times n_k!}$.

EXERCISES 14.1

1. A, B and C are three towns. There are 5 roads linking towns A and B and 3 roads linking towns B and C. How many different paths are there from town A to town C via town B?

2. In how many ways can 5 letters be mailed if there are
(a) 2 mail boxes available?
(b) 4 mail boxes available?

3. There are 4 letters to be placed in 4 letter boxes. In how many ways can these letters be mailed if
(a) only one letter per box is allowed?
(b) there are no restrictions on the number of letters per box?

4. Consider the cubic polynomial $p(x) = ax^3 + bx^2 - 5x + c$.
(a) If the coefficients, a, b and c come from the set $\{-3, -1, 1, 3\}$, find the number of possible cubics if no repetitions are allowed.
(b) Find the number of cubics if the coefficients now come from $\{-3, -1, 0, 1, 3\}$ (again without repetitions).

5. The diagram alongside shows the possible routes linking towns A, B, C and D.

A person leaves town A for town C. How many different routes can be taken if the person is always heading towards town C

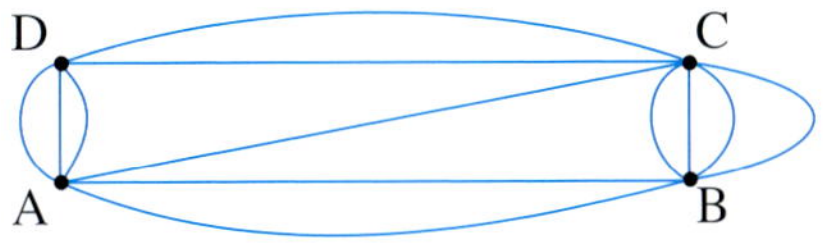

6. In how many different ways can Susan get dressed if she has 3 skirts, 5 blouses, 6 pairs of socks and 3 pairs of shoes to chose from?

7. In how many different ways can 5 different books be arranged on a shelf?

8. In how many ways can 8 different boxes be arranged taking 3 at a time?

9. How many different signals can be formed using 3 flags from 5 different flags.

10. Three Italian, two Chemistry and four Physics books are to be arranged on a shelf. In how many ways can this be done
(a) if there are no restrictions?
(b) if the Chemistry books must remain together?
(c) if the books must stay together by subject?

11. Find n if ${}^nP_2 = 380$.

12. 5 boys and 5 girls, which include a brother-sister, pair are to be arranged in a straight line. Find the number of possible arrangements if
(a) there are no restrictions.
(b) the tallest must be at one end and the shortest at the other end.
(c) the brother and sister must be i. together ii. separated

13. In how many ways can the letters of the word Mississippi be arranged?

14. In how many ways can three yellow balls, three red balls and four orange balls be arranged in a row if the balls are identical in everyway other than their colour?

15. In a set of 8 letters, m of them are the same and the rest different. If there are 1680 possible arrangements of these 8 letters, how many of them are the same?

14.2 COMBINATIONS

Combinations

On the otherhand, combinations represent a counting process where the order has no importance. For example, the number of combinations of the letters A, B, C and D, if only two are taken at a time, can be enumerated as:

AB, AC, AD, BC,BD, CD,

That is, the combination of the letters A and B, whether written as AB or BA, is considered as being the same.

Instead of **combination** the term **selection** is often used.

Rule 6: **Combinations**:
The total number of ways of **selecting** n objects, taking r at a time is given by

$$\frac{n!}{(n-r)!r!}$$

Notation: We use the notation $\binom{n}{r}$ (read as "n–c–r') to denote $\frac{n!}{(n-r)!r!}$.

That is, $\binom{n}{r} = \frac{n!}{(n-r)!r!}$. Note:Sometimes ${}^nC_r$ is used instead of $\binom{n}{r}$.

For example, in how many ways can 5 books be selected from 8 different books? In this instance, we are talking about selections and therefore, we are looking at combinations. Therefore we have, the selection of 8 books taking 5 at a time is equal to

$$\binom{8}{5} = \frac{8!}{(8-5)!5!} = \frac{8!}{3!5!} = 56 \quad .$$

Using the TI–83 we can make use of the **nCr** function:
Type the first number, **8**, then select **MATH** and **PRB**, then select option **3:nCr**, enter the second number, **5**, and then press **ENTER**:

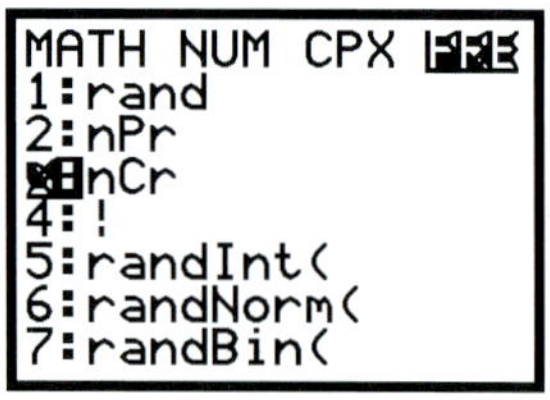

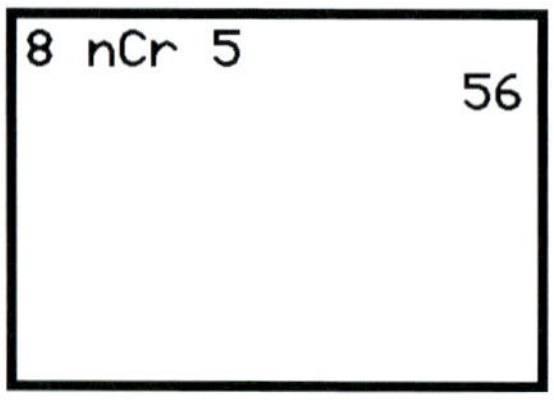

EXAMPLE 14.9

A sports committee at the local hospital consists of 5 members. A new committee is to be elected, of which 3 members must be women and 2 members must be men. How many different committees can be formed if there were originally 5 women and 4 men to select from?

Solution

First we look at the number of ways we can select the women members (using Rule 6):

We have to select 3 from a possible 5, therefore, this can be done in ${}^5C_3 = 10$ ways.

Similarly, the men can be selected in ${}^4C_2 = 6$ ways.

Using Rule 2, we have that the total number of possible committees $= {}^5C_3 \times {}^4C_2 = 60$.

EXAMPLE 14.10

A committee of 3 men and 2 women is to be chosen from 7 men and 5 women. Within the 12 people there is a husband and wife. In how many ways can the committee be chosen if it must contain either the wife or the husband but not both?

Solution

Case 1: Husband included

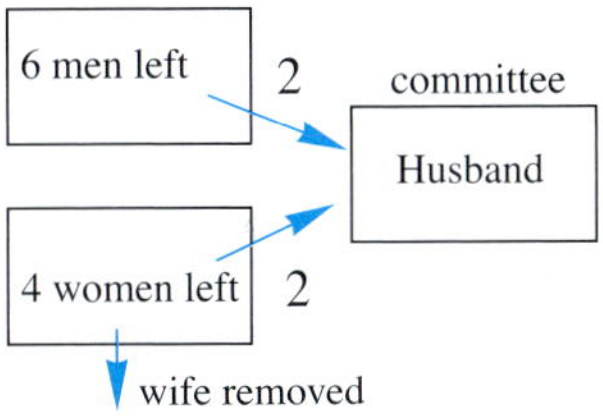

If the husband is included, the wife must be removed (so that she cannot be included). We then have to select 2 more men from the remaining 6 men and 2 women from the remaining 4 women.

This is done in ${}^6C_2 \times {}^4C_2 = 90$ ways

Case 2: Wife included

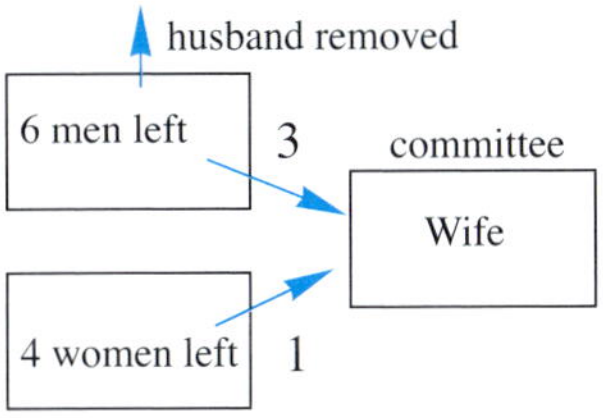

If the wife is included, the husband must be removed. We then have to select 3 men from the remaining 6 men and 1 woman from the remaining 4 women.

This is done is ${}^6C_3 \times {}^4C_1 = 80$ ways

Therefore there are a total of ${}^6C_2 \times {}^4C_2 + {}^6C_3 \times {}^4C_1 = 90 + 80 = 170$ possible committees.

EXERCISES 14.2

1. In how many ways can 5 basketball players be selected from 12 players?

2. A tennis club has 20 members.

(a) In how many ways can a committee of 3 be selected.

(b) In how many ways can this be done if the captain must be on the committee?

3. In how many ways can 3 red balls, 4 blue balls and 5 white balls be selected from 5 red balls, 5 blue balls and 7 white balls?

4. In how many ways can 8 objects be divided into 2 groups of 4 objects?

5. A cricket training squad consists of 4 bowlers, 8 batsmen, 2 wicket keepers and 4 fielders. From this squad a team of 11 players is to be selected. In how many ways can this be done if the team must consist of 3 bowlers, 5 batsmen, 1 wicket keeper and 2 fielders?

6. A class consists of 12 boys of whom 5 are prefects. How any committees of 8 can be formed if the committee is to have
(a) 3 prefects?
(b) at least 3 prefects?

7. In how many ways can 3 boys and 2 girls be arranged in a row if a selection is made from 6 boys and 5 girls?

8. If $\binom{n}{3} = 56$ show that $n^3 - 3n^2 + 2n - 336 = 0$. Hence find n.

9. In how many ways can a jury of 12 be selected from 9 men and 6 women so that there are at least 6 men and no more than 4 women on the jury.

10. Show that $\binom{n+1}{3} - \binom{n-1}{3} = (n-1)^2$. Hence find n if $\binom{n+1}{3} - \binom{n-1}{3} = 16$.

EXERCISES 14.3 – MISCELLANEOUS QUESTIONS

1. Five different coloured flags can be run up a mast.
(a) How many different signals can be produced if all five flags are used?
(b) How many different signals can be produced if any number of flags is used?

2. In how many different ways can 7 books be arranged in a row?

3. In how many different ways can three boys and four girls be seated in a row?
In how many ways can this be done if
(a) no two girls are sitting next to each other,
(b) the ends are occupied by girls?

4. In how many different ways can 7 books be arranged in a row if
(a) 3 specifed books must be together,
(b) two specified books must occupy the ends.

5. A school council consists of 12 members, 6 of whom are parents, 2 are students, the Principal and the remainder are teachers. The school captain and vice–captain must be on the council. If there are 10 parents and 8 teachers nominated for positions on the school council, how many different committees can there be?

6. A committee of 5 men and 5 women is to be selected from 9 men and 8 women. How many possible committees can be formed?
Amongst the 17 people, there is a married couple. If the couple cannot serve together, how many committees could there be?

7. A sports team consists of 5 bowlers (or pitchers), 9 batsmen and 2 keepers (or back–stops). How many different teams of 11 players can be chosen from the above squad if the team consists of
(a) 4 bowlers (pitchers), 6 batsmen and 1 keeper (back–stop)?
(b) 6 batsmen (pitchers) and at least 1 keeper (back–stop)?

8. Twenty people are to greet each other by shaking hands. How many hand shakes are there?

9. How many arrangements of the letters of the word "MARRIAGE" are possible?

10. How many arrangements of the letters of the word "COMMISSION" are possible?

11. A committee of 4 is to be selected from 7 men and 6 women. In how many ways can this be done if
(a) there are no restrictions?
(b) there must be an equal number of men and women on the committee?
(c) there must be at least one member of each sex on the committee?

12. Prove that (a) $\binom{n}{r} + \binom{n}{r+1} = \binom{n+1}{r+1}$. (b) ${}^{n+1}P_r = {}^{n}P_r + r \times {}^{n}P_{r-1}$.

13. A circle has n points on its circumference. How many chords joining pairs of points can be drawn?

14. A circle has n points on its circumference. What is the maximum number of points of intersection of chords inside the circle?

15. (a) Show that $2^n = \sum_{r=0}^{n} \binom{n}{r}$.
(b) In how many ways can 8 boys be divided into two unequal sets?

16. Whilst at the library, Patrick decides to select 5 books from a group of 10. In how many different ways can Patrick make the selection?

17. A fish tank contains 5 gold coloured tropical fish and 8 black coloured tropical fish.
(a) In how many ways can five fish be selected?
(b) If a total of 5 fish have been selected from the tank, how many of these contain two gold fish?

18. In how many ways can 4 people be accommodated if there are 4 rooms available?

19. A car can hold 3 people in the front seat and 4 in the back seat. In how many ways can 7 people be seated in the car if John and Samantha must sit in the back seat and there is only one driver?

20. In how many ways can six men and two boys be arranged in a row if:
(a) the two boys are together?
(b) the two boys are not together?
(c) there are at least three men separating the boys?

21. In how many ways can the letters of the word "TOGETHER" be arranged? In how many of these arrangements are all the vowels together?

22. In how many ways can 4 women and 3 men be arranged in a row, if there are 8 women and 5 men to select from?

23. In how many ways can 4 women and 3 men be arranged in a circle? In how many ways can this be done if the tallest woman and shortest man must be next to each other?

24. In how many ways can 5 maths books, 4 physics books and 3 biology books be arranged on a shelf if subjects are kept together?

25. How many even numbers of 4 digits can be formed using 5, 6, 7, 8 if
(a) no figure is repeated?
(b) repetition is allowed?

26. 5 men and 5 women are to be seated around a circular table. In how many ways can this be done if the men & women alternate?

27. A class of 20 students contains 5 student representatives. A committee of 8 is to be formed. How many different committees can be formed if there are
(a) only 3 student representatives?
(b) at least 3 student representatives?

28. How many possible juries of 12 can be selected from 12 women & 8 men so that there are at least 5 men and not more than 7 women?

29. In how many ways can 6 people be seated around a table if 2 friends are always
(a) together?
(b) separated?

15.1 PROBABILITY

We are often faced with statements that reflect an element of likelihood, For example, "It is likely to rain later in the day" or "What are the chances that I roll a six?". Such statements relate to a level of uncertainty (or indeed, a level of certainty). It is this element of likelihood in which we are interested. In particular, we need to find a measure of this likelihood — i.e., the associated probability of particular events.

15.1.1 PROBABILITY AS A LONG-TERM RELATIVE FREQUENCY

An experiment is repeated in such a way that a series of independent and identical trials are produced, so that a particular event A is observed to either occur or not occur. We let N be the total number of trials carried out and $n(A)$ (or $|A|$) be the number of times that the event A was observed.

We then call the ratio $\frac{n(A)}{N}$ $\left(\text{or } \frac{|A|}{N}\right)$ the **relative frequency** of the event A. This value provides some indication of the likelihood of the event A occuring.

In particular, for large values of N we find that the ratio $\frac{n(A)}{N}$ tends to a number called the **probability** of the event A, which we denote by $p(A)$ or $P(A)$.

As $0 \leq n(A) \leq N$, this number, $P(A)$, must lie between 0 and 1 (inclusive), i.e., $0 \leq P(A) \leq 1$.

A more formal definition is as follows:

If a random experiment is repeated N times, in such a way that each of the trials is identical and independent, where $n(A)$ is the number of times event A has occured after N trials, then

$$\text{As } N \to \infty, \frac{n(A)}{N} \to P(A).$$

It is possible to provide a graph of such a situation, which shows that as N increases, the ratio $\frac{n(A)}{N}$ tends towards some value p, where in fact, $p = P(A)$.

Such a graph is called a **relative frequency graph**.

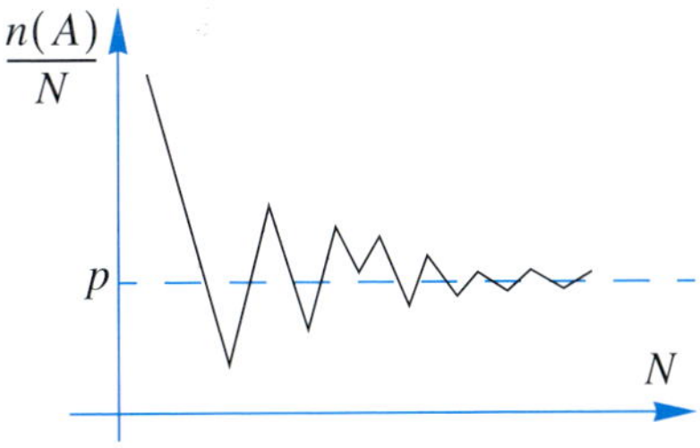

15.1.2 THEORETICAL PROBABILITY

When the circumstances of an experiment are always identical, we can arrive at a value for the probability of a particular event by using mathematical reasoning, often based on an argument reflecting some form of symmetry (i.e., without the need to repeatedly perform the experiment). This type of probability is called **theoretical probability**.

For example, when we roll a die, every possible outcome, known as the **sample space**, can be listed as $U = \{1, 2, 3, 4, 5, 6\}$ (sometimes the letter ε is used instead of U). The probability of obtaining a "four" (based on considerations of **symmetry of equal likelihood**) is given by $\frac{1}{6}$.

Such a probability seems obvious, as we would argue that;

"Given there are six possible outcomes and each outcome is equally likely to occur (assuming a fair die), then the chances that a "four" occurs must be one in six, i.e., $\frac{1}{6}$."

15.1.3 LAWS OF PROBABILITY

We will restrict our arguments to **finite sample spaces**. Recall, that a **sample space** is the set of every possible outcome of an experiment, and that an **event** is any subset of the sample space. This relationship is often represented with a Venn diagram:

The Venn diagram shows the sample space U, with the event A, as a subset.

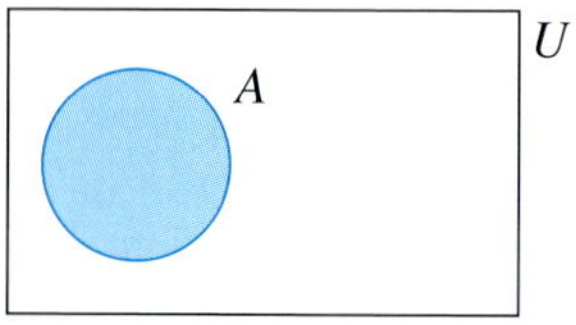

15.1.4 DEFINITION OF PROBABILITY

If an experiment has equally likely outcomes and of these the event A is defined, then the **theoretical probability of event A** occuring is given by

$$P(A) = \frac{n(A)}{n(U)} = \frac{\text{Number of outcomes in which A occurs}}{\text{Total number of outcomes in the sample space}}$$

Where $n(U)$ is the total number of possible outcomes in the sample space, U, (i.e., $n(U) = N$).

As a consequence of this definition we have what are known as the **axioms of probability**:

1. $0 \le P(A) \le 1$

2. $P(\varnothing) = 0$ and $P(\varepsilon) = 1$

That is, if $A = \varnothing$, then the event A can never occur.

$A = U$ implies that the event A is a certainty.

3. If A and B are both subsets of U and are mutually exclusive, then

$$P(A \cup B) = P(A) + P(B).$$

Note:

Two events A and B are said to be **mutually exclusive** (or disjoint) if they have no elements in common, i.e., if $A \cap B = \varnothing$.

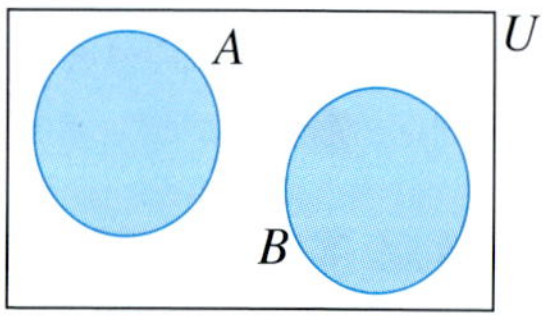

EXAMPLE 15.1

A fair die is thrown. List the sample space of the experiment and hence find the probability of observing: (a) a multiple of 3
(b) an odd number.

Are these events mutually exclusive?

Solution

(a) The sample space is $U = \{1, 2, 3, 4, 5, 6\}$.
Let A be the event 'obtaining a multiple of 3'.
We then have that $A = \{3, 6\}$. Therefore, $P(A) = \frac{n(A)}{n(U)} = \frac{2}{6} = \frac{1}{3}$.

(b) Let B be the event 'obtaining an odd number'.
Here $B = \{1, 3, 5\}$ and so $P(B) = \frac{n(B)}{n(U)} = \frac{3}{6} = \frac{1}{2}$.

In this case, $A = \{3, 6\}$ and $B = \{1, 3, 5\}$, so that $A \cap B = \{3\}$. Therefore, as $A \cap B \neq \varnothing$ A and B are not mutually exclusive.

EXAMPLE 15.2

Two coins are tossed. Find the probability that:
(a) two tails are showing (b) a tail is showing.

Solution

Let H denote the event a head is showing and T the event a tail is showing. This means that the sample space (with two coins) is given by $U = \{HH, HT, TH, TT\}$.

(a) The event that two tails are showing is given by the event $\{TT\}$, therefore, we have that $P(\{TT\}) = \frac{n(\{TT\})}{n(U)} = \frac{1}{4}$.

(b) The event that one tail is showing is given by $\{HT, TH\}$, therefore, we have that $P(\{HT, TH\}) = \frac{n(\{HT, TH\})}{n(U)} = \frac{2}{4} = \frac{1}{2}$.

EXAMPLE 15.3

A card is drawn from a standard deck of 52 playing cards. What is the probability that a diamond card is showing?

Solution

Let D denote the event "a diamond card is selected".
This means that $n(D) = 13$ as there are 13 diamond cards in a standard deck of cards.

Therefore, $P(D) = \frac{n(D)}{n(U)} = \frac{13}{52} = \frac{1}{4}$.

15.1.5 PROBLEM SOLVING STRATEGIES IN PROBABILITY

When dealing with probability problems it is often useful to use some form of diagram to help 'visualize' the situation. **Diagrams** can be in the form of

1. Venn diagrams.
2. Tree diagrams.
3. Lattice diagrams.
4. Karnaugh maps (probability tables).
5. As a last resort, any form of diagram that clearly displays the process under discussion (e.g. flow chart).

It is fair to say that some types of diagrams lend themselves well to particular types of problems. These will be considered in due course.

EXAMPLE 15.4

Find the probability of getting a sum of 7 on two throws of a die.

Solution

In this instance, we make use of a lattice diagram to display all possible outcomes. From the diagram, we can list the required event (and hence find the required probability):

Let S denote the event "A sum of seven is observed". From the lattice diagram, we see that there are 6 possibilities where a sum of seven occurs.
In this case we have

$S = \{(1, 6), (2, 5), (3, 4), (4, 3), (5, 2), (6, 1)\}$.

Therefore, we have that $P(S) = \frac{n(S)}{n(U)} = \frac{6}{36} = \frac{1}{6}$

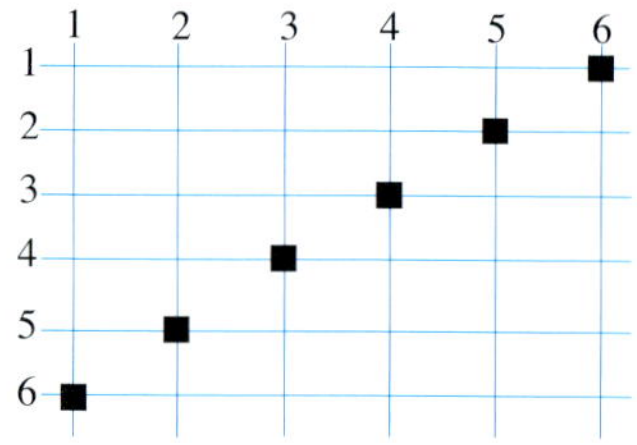

EXERCISES 15.1

1. From a bag containing 6 white and 4 red balls, a ball is drawn at random. What is the probability that the ball selected is
(a) red (b) white (c) not white

2. From an urn containing 14 marbles of which 4 are blue and 10 are red, a marble is selected at random. What is the probability that
(a) the marble is blue (b) the marble is red

3. A letter is chosen at random from the letters of the alphabet. What is the probability that
(a) the letter is a vowel (b) the letter is a consonant

4. A coin is tossed twice. List the sample space and find the probability of observing
(a) two heads.
(b) at least one head.

5. A coin is tossed three times. List the sample space and find the probability that
(a) two heads, show uppermost.
(b) at least two heads show uppermost.
(c) three heads or three tails are showing.

6. A letter is chosen at random from the word FERTILITY. Find the probability that the letter chosen is
(a) a T (b) an I (c) a consonant (d) a vowel.

7. A bag has 20 coins numbered from 1 to 20. A coin is drawn at random and its number is noted.
What is the probability that the coin drawn
(a) has an even number on it?
(b) has a number that is divisible by 3?
(c) has a number that is divisible by 3 or 5?

8. A die is rolled twice. Use a lattice diagram to illustrate the sample space. What is the probability of observing
(a) at least one five (b) a four and a three
(c) a pair (d) a sum of eight

9. A family has three children. List the sample space and hence find the probability that
(a) there are 3 boys.
(b) there are 2 boys and 1 girl.
(c) there are at least two girls.

10. A card is selected from a pack of 52 cards. Find the probability that the card is
(a) red (b) a heart (c) red and a heart

11. From an urn containing 16 cubes of which 6 are red, 4 are white and 6 are black, a cube is drawn at random.
Find the probability that the cube is
(a) red (b) white (c) black (e) red or black

12. A coin and a die are tossed simultaneously. Draw a lattice diagram to depict this situation.
(a) Using your lattice diagram, list the sample space.
(b) What is the probability of observing a tail and an even number?

13. A die is thrown three times. Find the probability of observing
(a) three sixes.
(b) three even numbers.
(c) two odd numbers.
Hint: You might need to draw a three dimensional lattice diagram.

15.2 PROBABILITY & VENN DIAGRAMS

From the axioms of probability we can develop further rules to help solve problems that involve chance. We illustrate these rules with the aid of Venn diagrams.

Event	Set language	Venn diagram	Probability result
The **complement** of A is denoted by A'.	A' is the complement to the set A, i.e., the set of elements that do not belong to the set A.	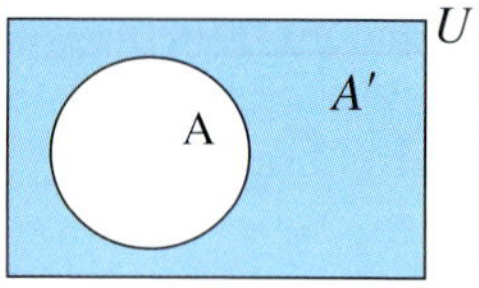	$P(A') = 1 - P(A)$ $P(A')$ is the probability that event A does not occur.
The **intersection** of A **and** B : $A \cap B$	$A \cap B$ is the intersection of the sets A and B, i.e., the set of elements that belong to **both** the set A **and** the set B.	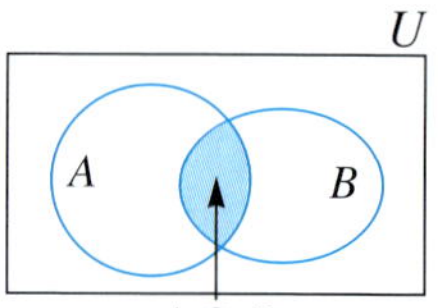 $A \cap B$	$P(A \cap B)$ is the probability that both A and B occur.
The **union** of events A and B : $A \cup B$	$A \cup B$ is the union of the sets A and B, i.e., the set of elements that belong to A **or** B or both A **and** B.	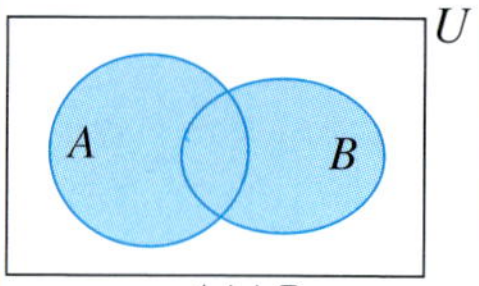 $A \cup B$	$P(A \cup B)$ is the probability that either event A or event B (or both) occur. From this we have what is known as the '**Addition rule**' for probability: $P(A \cup B) = P(A) + P(B) - P(A \cap B)$
If $A \cap B = \varnothing$ the events A and B are said to be **disjoint**. That is, they have no elements in common.	If $A \cap B = \varnothing$ the sets A and B are **mutually exclusive**.	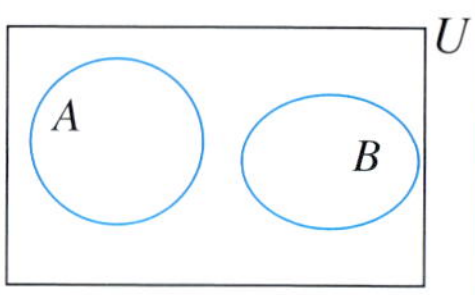 $A \cap B = \varnothing$	If A and B are mutually exclusive events then event A and event B cannot occur simultaneously, i.e., $n(A \cap B) = 0$ $\Rightarrow P(A \cap B) = 0$ Therefore: $P(A \cup B) = P(A) + P(B)$

Although we now have a number of 'formulas' to help us solve problems that involve probability, using other forms of diagrams to clarify situations and procedures should not be overlooked.

EXAMPLE 15.5

A card is randomly selected from an ordinary pack of 52 playing cards. Find the probability that it is either a 'black card' or a 'King'.

Solution

Let B be the event 'A black card is selected.' and K the event ' A King is selected'. We first note that event B has as its elements the Jack of spades(J♠), the Jack of clubs (J♣), the Queen of spades(Q♠), the Queen of clubs(Q♣) and so on.
This means that
B ={K♠,K♣,Q♠,Q♣,J♠,J♣,10♠,10♣,9♠,9♣,8♠,8♣,7♠,7♣,6♠,6♣,5♠,5♣,4♠,4♣,3♠, 3♣,2♠,2♣,A♠,A♣} and
K = {K♠, K♦, K♥, K♣}, so that $B \cap K$ = {K♠, K♣}.
Using the addition rule, $P(B \cup K) = P(B) + P(K) - P(B \cap K)$

we have $P(B \cup K) = \frac{26}{52} + \frac{4}{52} - \frac{2}{52} = \frac{7}{13}$.

Note the importance of subtracting $\frac{2}{52}$ as this represents the fact that we have included the event {K♠, K♣} twice when finding B and K.

We now consider one of the problems from Exercise 15.1, but this time we make use of the addition rule.

EXAMPLE 15.6

A bag has 20 coins numbered from 1 to 20. A coin is drawn at random and its number is noted. What is the probability that the coin has a number that is divisible by 3 or by 5?

Solution

Let T denote the event "The number is divisible by 3" and S, the event "The number is divisible by 5".
Using the addition rule we have $P(T \cup S) = P(T) + P(S) - P(T \cap S)$
Now, $T = \{3, 6, 9, 12, 15, 18\}$ and $S = \{5, 10, 15, 20\}$ so that $T \cap S = \{15\}$.

Therefore, we have $P(T) = \frac{6}{20}$ and $P(S) = \frac{4}{20}$ and $P(T \cap S) = \frac{1}{20}$.

This means that $P(T \cup S) = \frac{6}{20} + \frac{4}{20} - \frac{1}{20} = \frac{9}{20}$.

EXAMPLE 15.7

If $p(A) = 0.6$, $p(B) = 0.3$ and $p(A \cap B) = 0.2$, find

(a) $p(A \cup B)$ (b) $p(B')$ (c) $p(A \cap B')$

Solution

(a) Using the addition formula we have, $p(A \cup B) = p(A) + p(B) - p(A \cap B)$

$\Rightarrow p(A \cup B) = 0.6 + 0.3 - 0.2 = 0.7$

(b) Using the complementary formula, we have $p(B') = 1 - p(B) = 1 - 0.3 = 0.7$.

(c) To determine $p(A \cap B')$, we need to use a Venn diagram:

Using the second Venn diagram we are now in a position to form a new formula:

$p(A \cap B') = p(A) - p(A \cap B)$.

$\therefore\ p(A \cap B') = 0.6 - 0.2 = 0.4$.

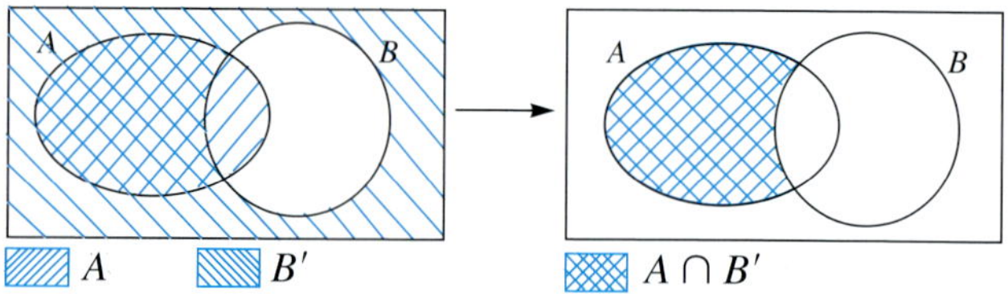

EXAMPLE 15.8

A coin is tossed three times. Find the probability of

(a) obtaining three tails (b) obtaining at least one head.

Solution

We begin by drawing a tree diagram to describe the situation:

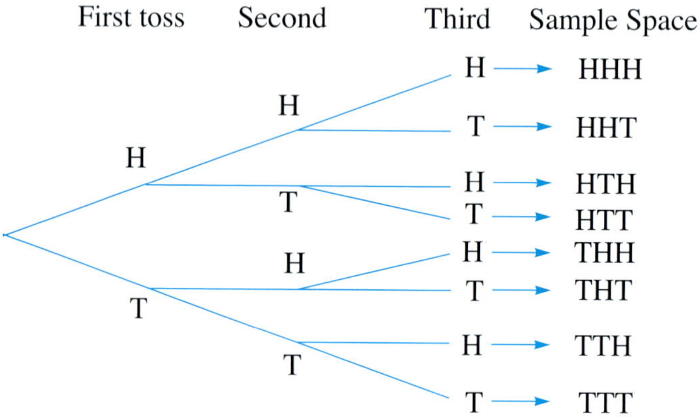

From the tree diagram we have a sample space made up of eight possible outcomes:

{HHH, HHT, HTH, HTT, THH, THT, TTH, TTT}

(a) Let X be the event "Obtaining three tails.", so $X = \{TTT\}$. Therefore $P(X) = \frac{1}{8}$.

(b) Although we can answer this question by using the tree diagram, we make use of complementary events to solve this problem.
Notice that "At least one head." is the complement of no heads.

Therefore, $p(\text{At least one head}) = P(X') = 1 - P(X) = 1 - \frac{1}{8} = \frac{7}{8}$.

EXERCISES 15.2

1. A letter is chosen at random from the letters of the word TOGETHER.

(a) Find the probability of selecting a T.

(b) Find the probability of selecting a consonant.

(c) Find the probability of not selecting an E.

2. A card is drawn at random from a standard deck.

(a) Find the probability that the card is an ace.

(b) Find the probability that the card is black.
(c) Find the probability that the card is an ace and black.
(d) Find the probability that the card is an ace or black.

3. A letter is selected at random from the alphabet. Find the probability that the letter is a vowel or comes from the word 'helpful'.

4. The events A and B are such that $P(A) = 0.5$, $P(B) = 0.7$ and $P(A \cap B) = 0.2$. Find

(a) $P(A \cup B)$ (b) $P(B')$ (c) $P(A' \cap B)$

5. The events A and B are such that $p(A) = 0.35$, $p(B) = 0.5$ and $p(A \cap B) = 0.15$. Using a Venn diagram (where appropriate) find:

(a) $p(A')$ (b) $p(A \cup B)$ (c) $p(A \cup B')$.

6. The events A and B are such that $p(A) = 0.45$, $p(B) = 0.7$ and $p(A \cap B) = 0.20$. Using a Venn diagram (where appropriate) find:

(a) $p(A \cup B)$ (b) $p(A' \cap B')$ (c) $p((A \cap B)')$.

7. A coin is tossed three times.

(a) Draw a tree diagram and from it write down the sample space.

Use the results from part (a) to find the probability of obtaining

(b) only one tail.
(c) at least 2 tails.
(d) 2 tails in succession.
(e) 2 tails.

8. In a class of 25 students it is found that 6 of the students play both tennis and chess, 10 play tennis only and 3 play neither. A student is selected at random from this group.

Using a Venn diagram, find the probability that the student.

(a) plays both tennis and chess.
(b) plays chess only.
(c) does not play chess.

9. A blue and a red die are rolled together (both numbered one to six).

(a) Draw a lattice diagram that best represents this experiment.
(b) Find the probability of observing an odd number.
(c) Find the probability of observing an even number with the red die.
(d) Find the probability of observing a sum of 7
(e) Find the probability of observing a sum of 7 or an odd number on the red die.

10. A card is drawn at random from a standard deck of 52 playing cards. Find the probability that the card drawn is

(a) a diamond.
(b) a club or spade.
(c) a black card or a picture card.
(d) a red card or a queen.

11. A and B are two events such that $P(A) = p$, $P(B) = 2p$ and $P(A \cap B) = p^2$.

(a) Given that $P(A \cup B) = 0.4$, find p.

Use a Venn Diagram to help you find the following:

(b) $P(A' \cup B)$

(c) $P(A' \cap B')$.

12. In a group of 30 students 20 hold an Australian passport, 10 hold a Malaysian passport and 8 hold both passports. The other students hold only one passport (that is neither Australian nor Malaysian). A student is selected at random.

(a) Draw a Venn diagram which describes this situation.
(b) Find the probability that the student has both passports.
(c) Find the probability that the student holds neither passport.
(d) Find the probability that the student holds only one passport.

15.3 CONDITIONAL PROBABILITY

15.3.1 INFORMAL DEFINITION OF CONDITIONAL PROBABILITY

Conditional probability works in the same way as simple probability. The only difference is that we are provided with some prior knowledge (or some extra condition about the outcome). So rather than considering the whole sample space, ε, given some extra information about the outcome of the experiment, we only need to concentrate on part of the whole sample space, ε^*. This means that the sample space is reduced from ε to ε^*. Before formalising this section, we use an example to highlight the basic idea.

EXAMPLE 15.9

(a) In the toss of a die, find the probability of obtaining a 'Two'.
(b) After tossing a die, it is noted that an even number appeared, what is the probability that it is a 'Two'?

Solution

(a) This part is straight forward: $U = \{1, 2, 3, 4, 5, 6\}$, and so P('Two') $= \frac{1}{6}$.

(b) This time, because *we know that an even number has occured*, we have a new sample space, namely $U^* = \{2, 4, 6\}$. The new sample size is $n(U^*) = 3$.

Therefore, P('Two' *given that an even number showed up*) $= \frac{1}{3}$.

15.3.2 FORMAL DEFINITION OF CONDITIONAL PROBABILITY

If A and B are two events, then **the conditional probability of event A given event B** is found using

$$P(A|B) = \frac{P(A \cap B)}{P(B)}, \quad P(B) \neq 0.$$

Note **1.** If A and B are mutually exclusive then $P(A|B) = 0$.

2. From the above rule, we also have the general **Multiplication rule**:

$$P(A \cap B) = P(A|B) \times P(B)$$

It should also be noted that usually $P(A|B) \neq P(B|A)$.

EXAMPLE 15.10

Two dice numbered one to six are rolled onto a table. Find the probability of obtaining a sum of five given that the sum is seven or less.

Solution

We first draw a lattice diagram:

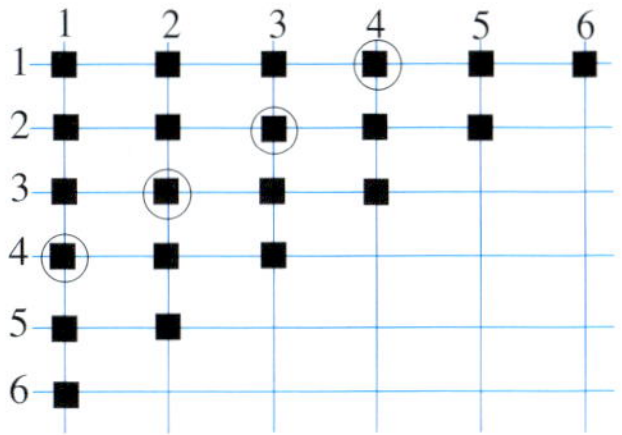

From the diagram we see that the new sample space is made up of 21 outcomes (black boxes) and the event we want (circled) consists of 4 outcomes.

Then, $P((X=5) \cap (X \leq 7)) = \frac{4}{36}$ and $P(X \leq 7) = \frac{21}{36}$.

Therefore, $P(X=5|X \leq 7) = \dfrac{\frac{4}{36}}{\frac{21}{36}} = \frac{4}{21}$.

EXAMPLE 15.11

A box contains 2 red cubes and 4 black cubes. If two cubes are chosen at random, find the probability that both cubes are red given that

(a) the first cube is not replaced before the second cube is selected.

(b) the first cube is replaced before the second cube is selected.

Solution

Let A be the event "the first cube is red" and B be the event "the second cube is red". This means that the event $A \cap B$ must be "both cubes are red".

Now, $p(A) = \frac{2}{6} = \frac{1}{3}$ (as there are 2 red cubes from a total of 6 cubes in the box).

The value of $P(B)$ depends on whether the selection is carried out with or without replacement.

(a) If the first cube selected is red and it is not replaced, then we only have 1 red cube left (in the box) out of a total of five cubes.

So, the probability that the second cube is red given that the first is red is $\frac{1}{5}$.

That is $p(B|A) = \frac{1}{5} \Rightarrow P(A \cap B) = P(B|A) \times P(A) = \frac{1}{5} \times \frac{1}{3} = \frac{1}{15}$.

(b) This time, because the cube is replaced, the probability that the second cube is red given that the first one is red is still $\frac{1}{3}$.

So that, $P(B|A) = \frac{1}{3} \Rightarrow P(A \cap B) = P(B|A) \times P(A) = \frac{1}{3} \times \frac{1}{3} = \frac{1}{9}$.

EXAMPLE 15.12

Two events A and B are such that $P(A) = 0.5$, $P(B) = 0.3$ and $P(A \cup B) = 0.6$.

Find (a) $P(A|B)$. (b) $P(B|A)$ (c) $P(A'|B)$

Solution

(a) $P(A|B) = \frac{P(A \cap B)}{P(B)}$, therefore we need to find $P(A \cap B)$.

Using the addition rule we have $P(A \cup B) = P(A) + P(B) - P(A \cap B)$

$$0.6 = 0.5 + 0.3 - P(A \cap B)$$

$$\therefore P(A \cap B) = 0.2$$

Therefore, $P(A|B) = \frac{P(A \cap B)}{P(B)} = \frac{0.2}{0.3} = \frac{2}{3}$

(b) $P(B|A) = \frac{P(B \cap A)}{P(A)} = \frac{0.2}{0.5} = 0.4$.

(c) $P(A'|B) = \frac{P(A' \cap B)}{P(B)} = \frac{P(B) - P(A \cap B)}{P(B)} = \frac{0.3 - 0.2}{0.3} = \frac{1}{3}$

15.3.3 INDEPENDENCE

The events *A* and *B* are said to be statistically independent if the probability of event *B* occuring is not influenced by event *A* occuring.

Therefore we have the mathematical definition:

Two events *A* and *B* are independent if, and only if,
$P(A|B) = P(A)$ and $P(B|A) = P(B)$

However, a more convenient definition for independence can be given as follows:

***A* and *B* are independent if, and only if**
$P(A \cap B) = P(A) \times P(B)$

This definition can be used as a test to decide if two events are independent. However, as a rule of thumb, if two events are 'physically independent' then they will also be statistically independent.

There are a few points that should always be considered when dealing with independence:

1. Never assume that two events are independent unless you are absolutely certain that they are independent.

2. How can you tell if two events are independent ?
A good rule of thumb is:
If they are physically independent, they are mathematically independent.

3. Make sure that you understand the difference between mutually exclusive events and independent events.
Mutually exclusive means that the events A and B have nothing in common and so there is no intersection. i.e., $A \cap B = \varnothing \Rightarrow \mathrm{P}(A \cap B) = 0$.
Independent means that the outcome of event A will not influence the outcome of event B i.e., $\mathrm{P}(A \cap B) = \mathrm{P}(A) \times \mathrm{P}(B)$.

4. Independence need not only be for two events, it can be extended, i.e., if the events A, B and C are each independent of each other then

$$\mathrm{P}(A \cap B \cap C) = \mathrm{P}(A) \times \mathrm{P}(B) \times \mathrm{P}(C)$$

5. Showing that two events, A and B are independent, requires three steps:
Step 1 Evaluate the product $\mathrm{P}(A) \times \mathrm{P}(B)$.
Step 2 Determine the value of $\mathrm{P}(A \cap B)$ using any means (other than step 1) i.e., use grids, tables, Venn diagrams, . . .
That is, you must not assume anything about A and B.
Step 3 If the answer using Step 1 is equal to the answer obtained in Step 2, then and only then will the events be independent. Otherwise, they are not independent.
Notice that not being independent does not therefore mean that they are mutually exclusive. They simply aren't independent. That's all.

6. Do not confuse the multiplication principle with the the rule for independence;
Multiplication principle is $\mathrm{P}(A \cap B) = \mathrm{P}(A|B) \times \mathrm{P}(B)$
Independence is given by $\mathrm{P}(A \cap B) = \mathrm{P}(A) \times \mathrm{P}(B)$

EXAMPLE 15.13

Two fair dice are rolled. Find the probability that two even numbers will show up.

Solution

Let the E_1 and E_2 denote the events "An even number on the first die." and "An even number on the second die." respectively.
In this case, the events are physically independent, i.e., the outcome on one die will not influence the outcome on the other die, and so we can confidently say that E_1 and E_2 are independent events.

Therefore, we have $\mathrm{P}(E_1 \text{ and } E_2) = \mathrm{P}(E_1 \cap E_2) = \mathrm{P}(E_1) \times \mathrm{P}(E_2) = \frac{1}{2} \times \frac{1}{2} = \frac{1}{4}$.

EXAMPLE 15.14

Debra has a chance of 0.7 of winning the 100 m race and a 60% chance of winning the 200 m race.

(a) Find the probability that she only wins one race.

(b) Find the probability that she wins both races.

Solution

Let W_1 denote the event "Debra wins the 100 m race." and W_2, the event "Debra wins the 200 m race.".

(a) If Debra wins only one race she must either

win the 100 m **and** lose the 200 m **or**

win the 200 m **and** lose the 100 m.

That is, we want $P(W_1 \cap W_2') = P(W_1) \times P(W_2') = 0.7 \times 0.4 = 0.28$ or

we can multiply the probabilities because the events are independent (why?):

$P(W_2 \cap W_1') = P(W_2) \times P(W_1') = 0.6 \times 0.3 = 0.18$.

Therefore, the required probability is $0.28 + 0.18 = 0.46$

Notice that if W_1 and W_2 are independent, then so too are their complements.

(b) Winning both races means that Debra will win the 100 m **and** 200 m race.

Therefore, we have $P(W_1 \cap W_2) = P(W_1) \times P(W_2) = 0.7 \times 0.6 = 0.42$.

Notice how we have made repeated use of the word '**and**', this emphasizes the fact that we are talking about the intersection of events.

EXAMPLE 15.15

Four seeds are planted, each one having an 80% chance of germinating. Find the probability that

(a) all four seeds will germinate.

(b) at least one seed will germinate.

Solution

(a) Let G_i denote the event that the i th seed germinates.

This means that $P(G_1) = P(G_2) = P(G_3) = P(G_4) = 0.8$

It is reasonable to assume that each seed will germinate independently of the other.

Therefore, P(All four seeds germinate) $= P(G_1 \cap G_2 \cap G_3 \cap G_4)$

$= P(G_1) \times P(G_2) \times P(G_3) \times P(G_4)$

$= (0.8)^4$

$= 0.4096$

(b) Now, p(At least one seed will germinate) $= 1 - p$(No seeds germinate).

P(Any **one** seed does not germinate) $= P(G_i') = 0.2$

Therefore, P(At least one seed will germinate) $= 1 - (P(G_i'))^4 = 1 - (0.2)^4 = 0.9984$.

EXAMPLE 15.16

A bag contains 5 white balls and 4 red balls. Two balls are selected in such a way that the first ball drawn is not replaced before the next ball is drawn.
Find the probability of selecting exactly one white ball.

Solution

We begin by drawing a diagram of the situation:

From our diagram we notice that there are two possible sample spaces for the second selection.

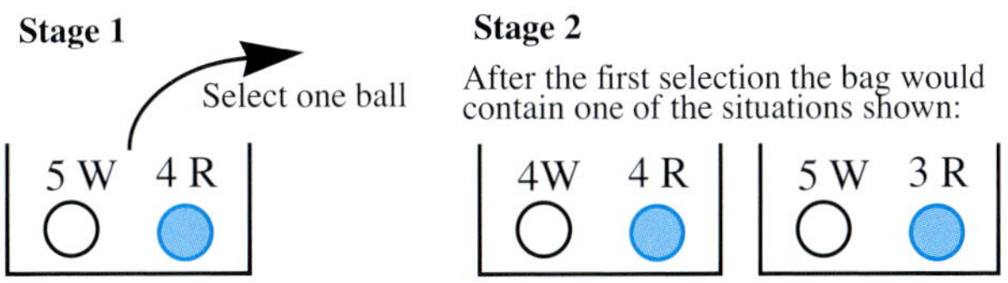

As an aid, we make use of a tree diagram, where W_i denotes the event "A white ball is selected on the ith trial" and R_i denotes the event "A red ball is selected on the ith trial".

The event "Only one white" occurs if the first ball is white **and** the second ball is red, **or** the first ball is red **and** the second ball is white.

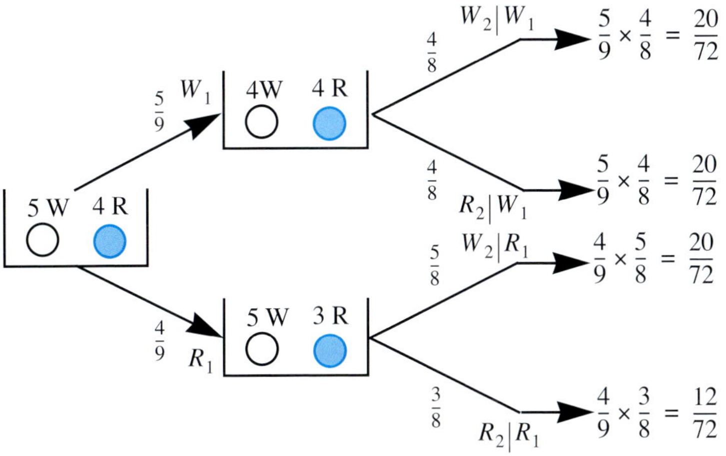

$$
\begin{aligned}
\text{P(One White ball)} &= P(W_1 \cap R_2) + P(R_1 \cap W_2) \\
&= P(R_2|W_1) \times P(W_1) + P(W_2|R_1) \times P(R_1) \\
&= \frac{4}{8}\times\frac{5}{9}+\frac{5}{8}\times\frac{4}{9} \\
&= \frac{5}{9}
\end{aligned}
$$

EXERCISES 15.3

1. Two events A and B are such that $p(A) = 0.6, p(B) = 0.4$ and $p(A \cap B) = 0.3$. Find the probability of the following events:

(a) $A \cup B$ (b) $A|B$ (c) $B|A$ (d) $A|B'$

2. A and B are two events such that $p(A) = 0.3$, $p(B) = 0.5$ and $p(A \cup B) = 0.55$
Find the probability of the following events:
(a) $A|B$ (b) $B|A$ (c) $A|B'$ (d) $A'|B'$

3. Urn A contains 9 cubes of which 4 are red. Urn B contains 5 cubes of which 2 are red. A cube is drawn at random and in succesion from each urn.
(a) Draw a tree diagram representing this process.
(b) Find the probability that both cubes are red.
(c) Find the probability that only 1 cube is red.
(d) If only 1 cube is red, find the probability that it came from urn A.

4. A box contains 5 red, 3 black, and 2 white cubes. A cube is randomly drawn and has its colour noted. The cube is then replaced, together with 2 more of the same colour. A second cube is then drawn.
(a) Find the probability that the first cube selected is red.
(b) Find the probability that the second cube selected is black.
(c) Given that the first cube selected was red, what is the probability that the second cube selected is black?

5. A fair coin, a double-headed coin and a double-tailed coin are placed in a bag. A coin is randomly selected. The coin is then tossed.
(a) Draw a tree diagram showing the possible outcomes.
(b) Find the probability that the coin lands with a tail showing uppermost.
(c) In fact, the coin falls "heads", find the probability that it is the "double-headed" coin.

6. Two unbiased coins are tossed together. Find the probability that they both display heads given that at least one is showing a head.

7. A money box contains 10 discs, 5 of which are yellow, 3 of which are black and 2 green. Two discs are selected in succession, with the first disc not replaced before the second is selected.
(a) Draw a tree diagram representing this process.
(b) Hence find the probability that the discs will be of a different colour.
(c) Given that the second disc was black, what is the probability that both were black?

8. Two dice are rolled. Find the probability that the faces are different given that the dice show a sum of 10.

9. Given that $p(A) = 0.6$, $p(B) = 0.7$ and that A and B are independent events.
Find the probability of the event
(a) $A \cup B$ (b) $A \cap B$ (c) $A|B'$ (d) $A' \cap B$

10. The probability that an animal will still be alive in 12 years is 0.55 and the probability that its mate will still be alive in 12 years is 0.60.
Find the probability that
(a) both will still be alive in 12 years.
(b) only the mate will still be alive in 12 years.
(c) at least one of them will still be alive in 12 years.
(d) the mate is still alive in 12 years given that only one is still alive in 12 years.

11. Tony has a 90% chance of passing his maths test, whilst Tanya has an 85% chance of passing the same test. If they both sit for the test, find the probability that
(a) only one of them passes.
(b) at least one passes the test.
(c) Tanya passed given that at least one passed.

12. The probability that Rory finishes a race is 0.55 and the probability that Millicent finishes the same race is 0.6. Because of team spirit, there is an 80% chance that Millicent will finish the race if Rory finishes the race.
Find the probability that
(a) both will finish the race.
(b) Rory finishes the race given that Millicent finishes.

13. If A and B are independent events, show that their complementary events are also independent events.

14. A student runs the 100 m, 200 m and 400 m races at the school athletics day. He has an 80% chance of winning any one given race. Find the probability that he will
(a) win all 3 races.
(b) win the first and last race only.
(c) win the second race given that he wins at least two races.

15. Dale and Kritt are trying to solve a physics problem. The chances of solving the problem are Dale—65% and Kritt— 75%. Find the probability that
(a) only Kritt solves the problem.
(b) Kritt solves the problem.
(c) both solve the problem.
(d) Dale solves the problem given that the problem was solved.

16. A coin which is weighted in such a way that there is a 70% chance of it landing heads. The coin is tossed three times in succession, find the probability of observing
(a) three tails.
(b) two heads.
(c) two heads given that at least one head showed up.

15.4 BAYES' THEOREM

15.4.1 LAW OF TOTAL PROBABILITY

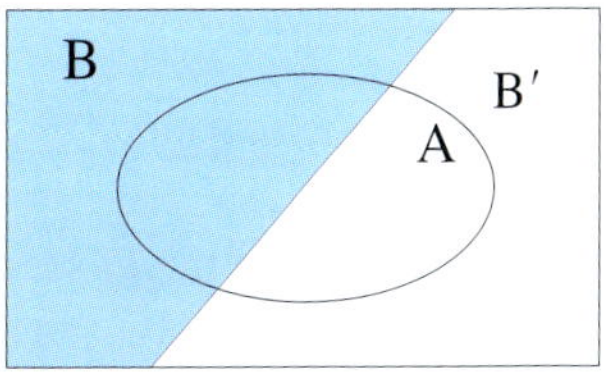

Using the Venn diagram, for any event A, we have that

$$A = A \cap \varepsilon = A \cap (B \cup B')$$
$$= (A \cap B) \cup (A \cap B')$$

As these two events are mutually exclusive, we have:

$$P(A) = P(A \cap B) + P(A \cap B')$$

However, $$P(A|B) = \frac{P(A \cap B)}{P(B)} \Rightarrow P(A \cap B) = P(B) \times P(A|B)$$

and $$P(A|B') = \frac{P(A \cap B')}{P(B')} \Rightarrow P(A \cap B') = P(B') \times P(A|B').$$

Which leads to the **Law of Total Probability**.

$$\therefore \ P(A) = P(B) \times P(A|B) + P(B') \times P(A|B')$$

Although this expression might look daunting, in fact, it represents the result that we would obtain if a tree diagram was used:

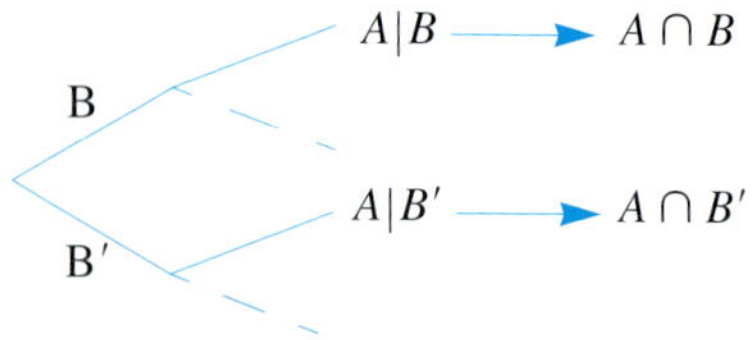

EXAMPLE 15.17

A box contains 3 black cubes and 7 white cubes. A cube is drawn from the box. Its colour noted and a cube of the other colour is then added to the box. A second cube is then drawn. What is the probability that the second cube selected is black?

Solution

We begin by setting up a tree diagram, where B_i denotes the event "A Black cube is observed on ith selection" and W_i denotes the event "A White cube is observed on ith selection".

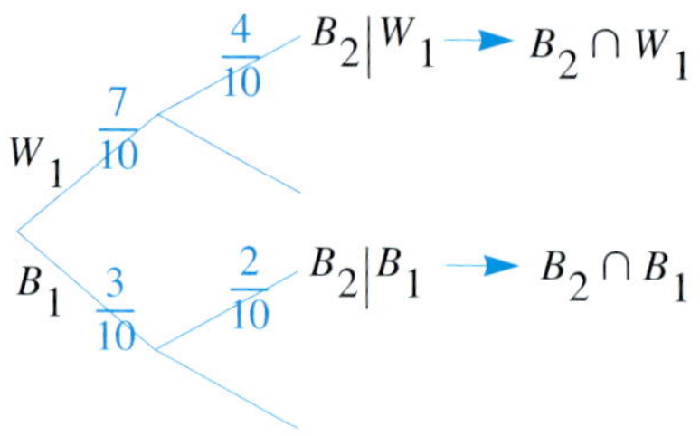

A black cube could have been observed on the second selection if

i. the first cube selected was white (i.e., $B_2|W_1$), or

ii. the first cube selected was black (i.e., $B_2|B_1$).

Therefore, $P(B_2) = P(B_2 \cap W_1) + P(B_2 \cap B_1) = \frac{7}{10} \times \frac{4}{10} + \frac{3}{10} \times \frac{2}{10} = \frac{17}{50}$.

15.4.2 BAYES' THEOREM FOR TWO EVENTS

As we saw earlier, conditional probability provides a mean by which we can adjust the probability of an event in light of new information. Bayes' Theorem, developed by Rev. Thomas Bayes (1702–1761), does the same thing, except this time it provides a means of adjusting a set of associated probabilities in the light of new information.

For two events, we have:

$$P(A|B) = \frac{P(A \cap B)}{P(B)} = \frac{P(A) \times P(B|A)}{P(A) \times P(B|A) + P(A') \times P(B|A')}$$

Again, the formula might seem daunting, however, it is only making use of a tree diagram.

EXAMPLE 15.18

A box contains 3 black cubes and 7 white cubes. A cube is drawn from the box. Its colour noted and a cube of the other colour is then added to the box. A second cube is then drawn. If both cubes are of the same colour, what is the probability that both cubes were in fact white?

Solution

Following on from the previous example, we have the same tree diagram:

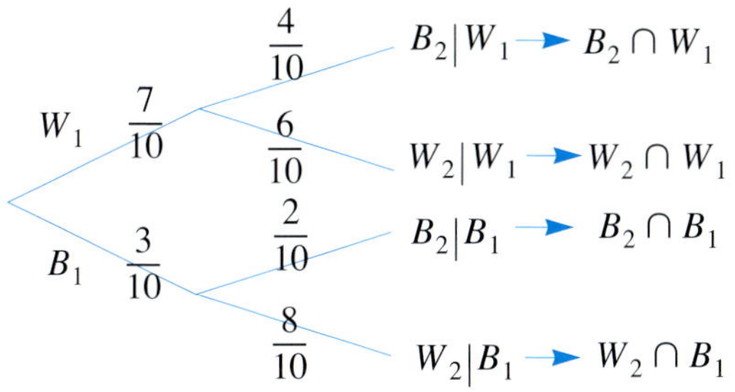

We require: P(Both White **given** that both are of the same colour)

Now, the probability that they are of the same colour is given by the probability that they are **both white or both black**, i.e., $P((W_2 \cap W_1) \cup (B_2 \cap B_1))$.

Next: P(Both White **given** both are the same colour)

$$= P(W_2 \cap W_1 | (W_2 \cap W_1) \cup (B_2 \cap B_1))$$

$$= \frac{P((W_2 \cap W_1) \cap ((W_2 \cap W_1) \cup (B_2 \cap B_1)))}{P((W_2 \cap W_1) \cup (B_2 \cap B_1))}$$

$$= \frac{P(W_2 \cap W_1)}{P(W_2 \cap W_1) + P(B_2 \cap B_1)}$$

$$= \frac{P(W_2|W_1) \times P(W_1)}{P(W_2|W_1)P(W_1) + P(B_2|B_1)P(B_1)}$$

$$= \frac{\frac{6}{10} \times \frac{7}{10}}{\frac{6}{10} \times \frac{7}{10} + \frac{2}{10} \times \frac{3}{10}}$$

$$= \frac{7}{8}$$

EXAMPLE 15.19

In a small country town, it was found that 90% of the drivers would always wear their seatbelts. On 60% of occasions, if a driver was not wearing a seatbelt they would be fined for speeding. Whereas if they were wearing a seatbelt, they would be fined for speeding 20% of the time. Find the probability that a driver who was fined for speeding was wearing a seatbelt.

Solution

Let the event A denote the event "driver wears a seatbelt' and B denote the event "Driver speeds". Using a tree diagram we have:

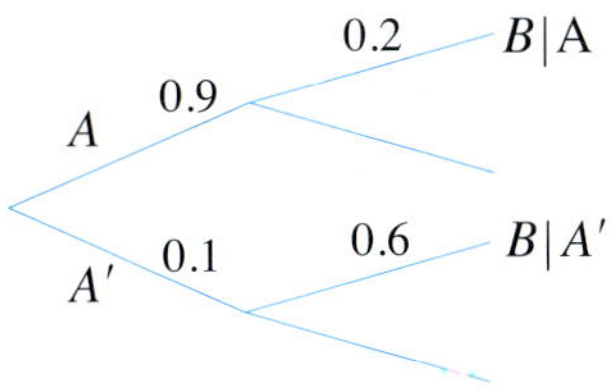

We need to find, *Pr*(Driver was wearing a seatbelt| driver was booked for speeding)

$$= P(A|B)$$

$$= \frac{P(A \cap B)}{P(B)}$$

$$= \frac{P(A) \times P(B|A)}{P(A) \times P(B|A) + P(A') \times P(B|A')}$$

Therefore, $P(A|B) = \frac{0.9 \times 0.2}{0.9 \times 0.2 + 0.1 \times 0.6} = \frac{18}{24}$

So, P(that a driver who was booked for speeding was in fact wearing a seatbelt) = 0.75

EXERCISES 15.4

1. Two machines, *A* and *B* produce 40% and 60% of the daily output respectively. Each machine also produces a total of 3% and 5% respectively, of items that are defective.

(a) An item is selected at random. Find the probability that it is defective.

(b) An item is selected and is found to be defective. Find the probability that it came from machine B.

2. At the Heights International School, it is found that 12% of the male students and 7% of the female students are taller than 1.8 m. Sixty percent of the school is made up of female students.

(a) A student selected at random is found to be taller than 1.8m. What is the probability that the student is a female?

(b) A second student selected at random is found to be shorter than 1.8m. What is the probability that the student is a male?

3. A box contains 4 black cubes and 6 white cubes. A cube is drawn from the box. Its colour noted and a cube of the other colour is then added to the box. A second cube is then drawn.

(a) If both cubes are of the same colour, what is the probability that both cubes were in fact white?

(b) The first ball is replaced before the second ball is added to the box.
What is the probability that both cubes were white given that both cubes were of the same colour?

4. An urn, labelled *A*, contains 8 cards numbered 1 through 8 whilst a second urn, labelled *B*, contains five cards numbered 1 through five. An urn is selected at random and from that urn a card is selected. Find the probability that the card came from urn A given that it is an even numbered card.

5. (a) An event *A* can occur only if one of the mutually exclusive events B_1, B_2 or B_3 occurs. Show that

$$P(A) = P(B_1) \times P(A|B_1) + P(B_2) \times P(A|B_2) + P(B_3) \times P(A|B_3)$$

(b) Of the daily output, machines *A* and *B* produce items of which 2% are defective, whilst machine C produces items of which 4% are defective. Machines B and C produce the same number of items, whilst machine A produces twice as many items as machine B.

i. An item is selected at random. Find the probability that it is defective.
ii. An item is selected and is found to be defective. Find the probability that it came from machine B.

6. A box contains N coins, of which m are fair coins whilst the rest are double–headed coins.
(a) A coin is selected at random and tossed.
i. What is the probability of observing a head?
ii. Given that a head was observed, what is the probability that the double–headed coin was selected?
(b) This time, a coin is selected at random and tossed n times. What is the probability that it was a fair coin, if it showed up heads on all n tosses?

7. A population of mice is made up of 75% that are classified as 'M+', of which, 30% have a condition classified as 'N–'. Otherwise, all other mice have the 'N–' condition. A mouse selected at random is classified as having the N–' condition. What is the probability that the mouse comes from the 'M+' classification group?

8. A survey of the adults in a town shows that 8% have liver problems. Of these, it is also found that 30% are heavy drinkers, 60% are social drinkers and 10% are non–drinkers. Of those that did not suffer from liver problems, 5% are heavy drinkers, 65% are social drinkers and 30% do not drink at all.
(a) An adult is selected at random. What is the probability that this person is a heavy drinker?
(b) If a person is found to be a heavy drinker, what is the probability that this person has liver problems?
(c) If a person is found to have liver problems, what is the probability that this person is a heavy drinker?
(d) If a person is found to be a non–drinker, what is the probability that this person has liver problems?

9. The probability that a person has a deadly virus is 5 in one thousand. A test will correctly diagnose this disease 95% of the time and incorrectly on 20% of occasions.
(a) Find the probability of this test giving a correct diagnosis.
(b) Given that the test diagnoses the patient as having the disease, what is the probability that the patient does not have the disease?
(c) Given that the test diagnoses the patient as not having the disease, what is the probability that the patient does have the disease?

10. The probability that a patient has a virus is 0.03. A medical diagnostic test will be able to determine whether the person in question actually has the virus. If the patient has the virus, the medical test will produce a positive result 90% of the time whilst if the patient does not have the virus, it will produce a negative result 98% of the time.
(a) What proportion of all tests provide a positive result?
(b) If the test shows a positive result, what is the probability that the patient actually has the virus?
(c) If the test shows a negative result, what is the probability that the patient does not have the virus?

11. The probability that a day of the week will be dry or wet is related to the state of the previous day in the following manner:
If it is dry one day, the chances of it being dry the next is 0.8.
If it is wet one day, the chances of it being dry the next is 0.4

(a) Given that Monday is dry, what is the probability that Tuesday is wet?
(b) Given that Monday is dry, what is the probability that Tuesday is wet and Wednesday is dry?
(c) Given that Monday is dry, what is the probability that Wednesday is dry?

Define the conditional probability matrix, P_i, for $i = 1, 2, 3, ...,$ as follows:

$$P_i = \begin{bmatrix} p(\text{Day } i \text{ is dry} \mid \text{Day } i-1 \text{ is dry}) & p(\text{Day } i \text{ is wet} \mid \text{Day } i-1 \text{ is dry}) \\ p(\text{Day } i \text{ is dry} \mid \text{Day } i-1 \text{ is wet}) & p(\text{Day } i \text{ is wet} \mid \text{Day } i-1 \text{ is wet}) \end{bmatrix}$$

(d) Find, P_1 and P_2.
(e) Verify that $P_2 = (P_1)^2$
(f) Show that $P_{i+1} = P_i P_1$.
(g) Use part f. to find the probability that Thursday will be dry, given that Monday was dry.

15.5 USING PERMUTATIONS AND COMBINATIONS IN PROBABILITY

Because enumeration is such an important part of finding probabilities, a sound knowledge of permutations and combinations can help to ease the workload involved.

EXAMPLE 15.20

Three maths books, three chemistry books and two physics books are to be arranged on a shelf. What is the probability that the three maths books are together?

Solution

The total number of arrangements of all 8 books is $8! = 40320$
To determine the number of arrangements that contain the three maths books together we make use of the box method:

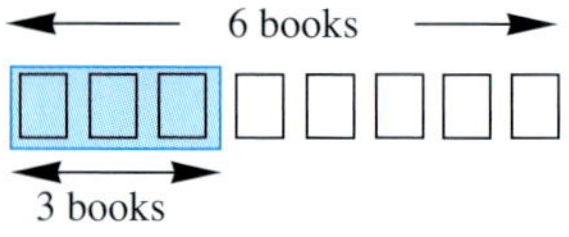

We now have 6 boxes to arrange, giving a total of 6! arrangements.
However, the three maths books (within the red box) can also be arranged in 3! ways.
Therefore, there are $6! \times 3! = 4320$ ways this can be done.

So, P(maths books are together) $= \dfrac{6! \times 3!}{8!} = \dfrac{6! \times 3!}{8 \times 7 \times 6!} = \dfrac{6}{8 \times 7}$

$= \dfrac{3}{28}$

EXAMPLE 15.21

A committee of 5 is randomly chosen from 8 boys and 6 girls. Find the probability that the committee consists of at least 3 boys.

Solution

The possibilities are:

Boys	Girls	No. of Selections
3	2	$\binom{8}{3} \times \binom{6}{2} = 840$
4	1	$\binom{8}{4} \times \binom{6}{1} = 420$
5	0	$\binom{8}{5} \times \binom{6}{0} = 56$

The total number of committees with at least 3 boys is 840 + 420 + 56 = 1316

However, the total number of committees of 5 from 14 is $\binom{14}{5} = 2002$.

If X denotes the number of boys on the committee, then $p(X \geq 3) = \frac{1316}{2002} = \frac{94}{143}$.

EXERCISES 15.5

1. Five red cubes and 4 blue cubes are placed at random in a row. Find the probability that
(a) the red cubes are together.
(b) both end cubes are red.
(c) the cubes alternate in colour.

2. Five books of different heights are arranged in a row. Find the probability that
(a) the tallest book is at the right end.
(b) the tallest and shortest books occupy the end positions.
(c) the tallest and shortest books are together.
(d) the tallest and shortest books are never next to each other.

3. Three cards are dealt from a pack of 52 playing cards. Find the probability that
(a) 2 of the cards are kings.
(b) all three cards are aces.
(c) all three cards are aces given that at least one card is an ace.

4. The letters of the word LOTTO are arranged in a row. What is the probability that the Ts are together?

5. A committee of 4 is to be selected from 7 men and 6 women. Find the probability that
(a) there are 2 women on the committee.
(b) there is at least one of each sex on the committee.

6. A basketball team of 5 is to be selected from 12 players. Find the probability that
(a) the tallest player is selected.
(b) the captain and vice-captain are selected.
(c) either one, but not both of the captain or vice-captain are selected.

7. Find the probability of selecting one orange, one apple and one pear from five oranges, four apples and three pears.

8. Three red cubes, four blue cubes and six yellow cubes are arranged in a row. Find the probability that
(a) the cubes at each end are the same colour.
(b) the cubes at each end are of a different colour.

9. A sample of three light bulbs is selected from a box containing 15 light bulbs. It is known that five of the light bulbs in the box are defective.
(a) Find the probability that the sample contains a defective.
(b) Find the probability that the sample contains at least two defectives.

10. Eight people of different heights are to be seated in a row. What is the probability that
(a) the tallest and shortest persons are sitting next to each other?
(b) the tallest and shortest occupy the end positions?
(c) there are at least three people sitting between the tallest and shortest?

11. Eight people of different heights are to be seated in a row. The shortest and tallest in this group are not seated at either end. What is the probability that
(a) the tallest and shortest persons are sitting next to each other?
(b) there is one person sitting between the tallest and shortest?

12. A committee of four is to be selected from a group of five boys and three girls. Find the probability that the committee consists of exactly two girls given that it contains at least one girl.

13. A bag contains 6 red marbles and 4 white marbles. Three marbles are randomly selected. Find the probability that
(a) all three marbles are red.
(b) all three marbles are red given that at least two of the marbles are red.

14. Four maths books, two chemistry books and three biology books are arranged in a row.
(a) What is the probability that the books are grouped together in their subjects?
(b) What is the probability that the chemistry books are not grouped?

15. A contestant on the game show "A Diamond for your Wife" gets to select 5 diamonds from a box. The box contains 20 diamonds of which 8 are fakes.

(a) Find the probability that the contestant will not bring a real diamond home for his wife.

Regardless of how many real diamonds the contestant has after his selection, he can only take one home to his wife. A second contestant then gets to select from the remaining 15 diamonds in the box. But only gets to select one diamond.

(b) What is the probability that this second contestant selects a real diamond?

16. Light bulbs are sold in packs of 10. A quality inspector selects two bulbs at random without replacement. If both bulbs are defective the pack is rejected. If neither are defective the pack is accepted. If one of the bulbs is defective the inspector selects two more from the bulbs remaining in the pack and rejects the pack if one or both are defective. What are the chances that a pack containing 4 defective bulbs will in fact be accepted?

16.1 DISCRETE RANDOM VARIABLES

16.1.1 CONCEPT OF A RANDOM VARIABLE

Consider the experiment of tossing a coin twice. The sample space, S, (i.e., the list of all possible outcomes) of this experiment can be written as S = {HH, HT, TH, TT}.

We can also assign a numerical value to these outcomes. For example, we can assign the number

0 to the outcome {HH},
1 to the outcomes {HT, TH} and
2 to the outcome {TT}.

These numerical values are used to represent the number of times that a tail was observed **after** the coin was tossed.

The numbers 0, 1 and 2 are **random** in nature, that is, until the coins are tossed we have no idea as to which one of the outcomes will occur. We define a random variable as follows:

A **random variable**, X (random variables are usually denoted by capital letters), which can take on exactly n numerical values, each of which corresponds to one and only one of the events in the sample space is called a **discrete random variable**.

Note that the values that correspond to the random variables X, Y, Z . . . are denoted by their corresponding lower case letters, x, y, z . . . . For the example above, $X = \{x: x = 0, x = 1, x = 2\}$.

16.1.2 DISCRETE RANDOM VARIABLE

A **discrete random** variable is one in which we can produce a **countable** number of outcomes. Because of this, discrete random variables are usually associated with a **counting process**. For example, the number of plants that will flower, the number of defective items in a box or the number of items purchased at a supermarket store.

We can display this concept using a simple diagram such as the one below:

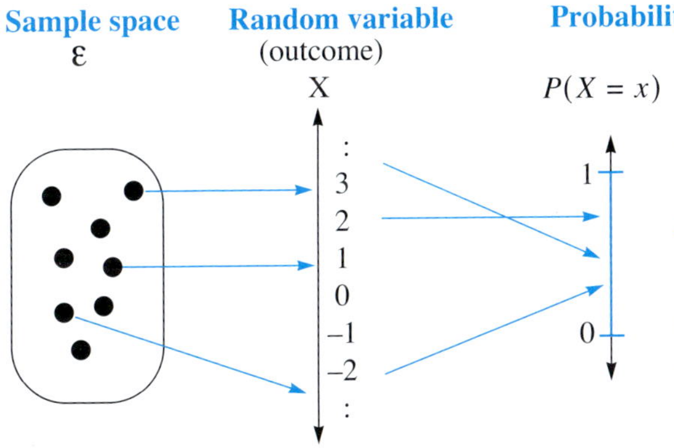

We write 'The **probability** that the random variable $X = x$ is p' as:

$$P(X = x) = p.$$

Note that the probability of any event must always lie between 0 and 1, inclusive.

To obtain the sample space, we may need to carry out an experiment. However, as we shall see, there are many types of random variables that already possess their own sample spaces, random outcomes and associated probability values. We will deal with these later.

EXAMPLE 16.1

Consider the experiment of tossing a coin three times in succession.

(a) List all possible outcomes.

(b) If the random variable X denotes the number of heads observed, list the values that X can have and find the corresponding probability values.

Solution

(a) In the experiment of tossing a coin three times the sample space is given by

$$S = \{HHH, HHT, HTH, THH, TTH, THT, HTT, TTT\},$$

where the event {HTH} represents the observation, Head, Tail, Head in that order.

(b) This means that on any one trial of this experiment, we could have obtained either

0 heads, 1 head, 2 heads or 3 heads.

Therefore, the random variable X has as its possible values the numbers 0, 1, 2, 3.

That is, $X = 0$ corresponds to the event {TTT}, that is, no heads.
$X = 1$ corresponds to the events {TTH, THT, HTT}, that is, one head.
$X = 2$ corresponds to the events {THH, HTH, HHT}, that is, two heads.
$X = 3$ corresponds to the event {HHH}, that is, three heads.

Once we have our sample space, we can look at the chances of each of the possible outcomes. In all there are 8 possible outcomes.

So that the chances of observing the event {HHH} would be $\frac{1}{8}$, i.e., $P(X = 3) = \frac{1}{8}$.

To find $P(X = 2)$, we observe that the outcome '$X = 2$' corresponds to {THH, HTH, HHT}.
So that in this case there is a chance of 3 in 8 of observing the event where '$X = 2$'.
Continuing in this manner we have

$$P(X = 0) = P(\{HHH\}) = \frac{1}{8}$$

$$P(X = 1) = P(\{TTH, THT, HTT\}) = \frac{3}{8}$$

$$P(X = 2) = P(\{HHT, HTH, THH\}) = \frac{3}{8}$$

$$P(X = 3) = P(\{HHH\}) = \frac{1}{8}$$

16.1.3 PROBABILITY DISTRIBUTIONS

We can describe a discrete random variable by making use of its probability distribution. That is, by showing the values of the random variable and the probabilities associated with each of its values.

A probability distribution can be displayed in any one of the following formats:

1. **Tabular form**
2. **Graphical representation**
   (*With the probability value on the vertical axis, and the values of the random variable on the horizontal axis.*)
3. **Function**
   (*A formula that can be used to determine the probability values.*)

EXAMPLE 16.2

Use each of the probability distribution representations discussed to display the results of the experiment where a coin is tossed three times in succession.

Solution

Let the random variable X denote the number of heads observed in three tosses of a coin.

1. **Tabular form:**

x	0	1	2	3
$p(X = x)$	$\frac{1}{8}$	$\frac{3}{8}$	$\frac{3}{8}$	$\frac{1}{8}$

2. **Graphical representation:**

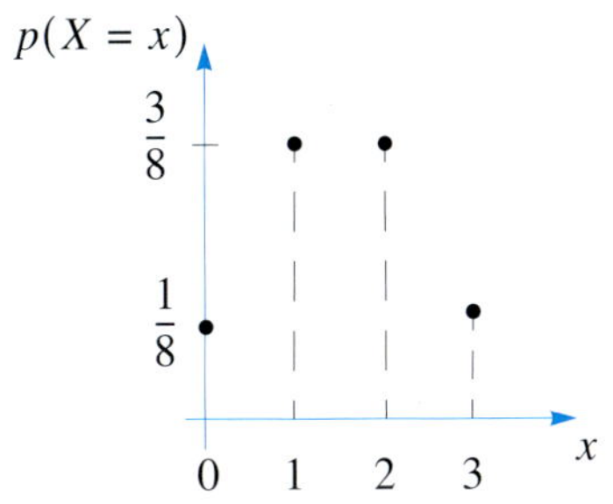

3. **Function:** $p(X = x) = \binom{3}{x}\left(\frac{1}{2}\right)^3, x = 0, 1, 2, 3$, where $\binom{3}{x} = \frac{3!}{(3-x)!x!}$.

16.1.4 PROPERTIES OF THE PROBABILITY FUNCTION

We can summarise the features of any discrete probability function as follows:

1. The probability for any value of X must **always lie between** 0 and 1 (inclusive).
That is, $0 \le P(X = x_i) \le 1$ for all values of x_i.

2. For the n mutually exclusive and exhaustive events, $A_1, A_2, \ldots, A_n$ that make up the sample space ε, then, the sum of the corresponding probabilities must be 1.

That is, $$\sum_{i=1}^{i=n} P(X = x_i) = P(X = x_1) + P(X = x_2) + \ldots + P(X = x_n) = 1$$

Where $P(X = x_i)$ is the probability of event A_i occuring.

Any function that does not abide by these two rules cannot be a probability function.

EXAMPLE 16.3

Consider the random variable X with probability function defined by
$$P(X = 0) = \alpha,\ P(X = 1) = 2\alpha \text{ and } P(X = 2) = 3\alpha.$$
Determine the value of α.

Solution

Because we are given that this is a probability function, then summing all the probabilities must give a result of 1.

Therefore we have that $P(X=0)+P(X=1)+P(X=2) = 1$

$$\therefore \alpha + 2\alpha + 3\alpha = 1$$

$$\Leftrightarrow 6\alpha = 1$$

$$\Leftrightarrow \alpha = \frac{1}{6}$$

EXAMPLE 16.4

The probability distribution of the random variable X is represented by the function $P(X=x) = \frac{k}{x}, x = 1, 2, 3, 4, 5, 6$

Find (a) the value of k (b) $P(3 \le X \le 5)$.

Solution

(a) Using the fact that the sum of all the probabilities must be 1, we have

$$P(X=1)+P(X=2)+\ldots+P(X=6) = \frac{k}{1}+\frac{k}{2}+\frac{k}{3}+\frac{k}{4}+\frac{k}{5}+\frac{k}{6} = 1$$

$$\Leftrightarrow \frac{147k}{60} = 1$$

Therefore, $k = \frac{60}{147} = \frac{20}{49}$.

(b) Now, $P(3 \le X \le 5) = P(X=3)+P(X=4)+P(X=5) = \frac{k}{3}+\frac{k}{4}+\frac{k}{5} = \frac{47k}{60}$

However, we know that $k = \frac{60}{147}$.

Therefore, $P(3 \le X \le 5) = \frac{47}{60} \times \frac{60}{147} = \frac{47}{147}$.

EXAMPLE 16.5

A discrete random variable X has a probability distribution defined by the function $P(X=x) = \binom{4}{x}\left(\frac{2}{5}\right)^x\left(\frac{3}{5}\right)^{4-x}$ where $x = 0, 1, 2, 3, 4$.

(a) Display this distribution using i. a table form ii. a graphical form

(b) Find i. $P(X=2)$ ii. $P(1 \le X \le 3)$

Solution

(a) i. We begin by evaluating the probability for each value of x:

$$p(X=0) = {}^4C_0\left(\frac{2}{5}\right)^0\left(\frac{3}{5}\right)^{4-0} = \frac{81}{625}, \text{ [Using the notation } \binom{4}{x} = {}^4C_x\text{]}.$$

$$p(X=1) = {}^4C_1\left(\frac{2}{5}\right)^1\left(\frac{3}{5}\right)^3 = \frac{216}{625}, \quad p(X=2) = {}^4C_2\left(\frac{2}{5}\right)^2\left(\frac{3}{5}\right)^2 = \frac{216}{625},$$

$$p(X=3) = {}^4C_3\left(\frac{2}{5}\right)^3\left(\frac{3}{5}\right)^1 = \frac{96}{625}, \quad p(X=4) = {}^4C_4\left(\frac{2}{5}\right)^4\left(\frac{3}{5}\right)^0 = \frac{16}{625}$$

We can now set up this information in a table:

x	0	1	2	3	4
$P(X = x)$	$\frac{81}{625}$	$\frac{216}{625}$	$\frac{216}{625}$	$\frac{96}{625}$	$\frac{16}{625}$

ii. Using the table found in part i., we can construct the following graph:

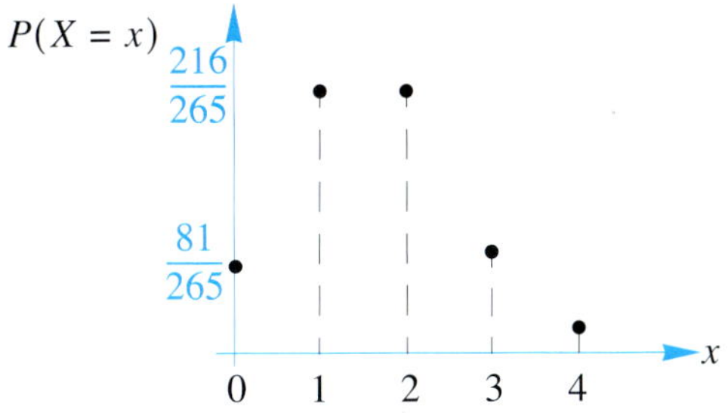

(b) i. From our probability table, we have that $P(X = 2) = \frac{216}{625}$.

ii. The statement $P(1 \le X \le 3)$ requires that we find the probability of the random variable X taking on the values 1, 2 or 3. This amounts to evaluating the sum of the corresponding probabilities.

Therefore, we have $P(1 \le X \le 3) = P(X = 1) + P(X = 2) + P(X = 3)$

$$= \frac{216}{625} + \frac{216}{625} + \frac{96}{625}$$

$$= \frac{528}{625}$$

16.1.5 CONSTRUCTING PROBABILITY FUNCTIONS

When we are given the probability distribution, we can determine the probabilities of events. However, there is still one issue that we must deal with:

How do we obtain the probabilities in the first place ?

Sometimes we recognise a particular problem and know of an existing model that can be used. However, resolving this question is not always an easy task, as this often requires the use of problem solving skills and modelling skills as well as interpretive skills.

EXAMPLE 16.6

A bag contains 5 white cubes and 4 red cubes. Two cubes are selected in such a way that the first cube drawn is not replaced before the next cube is drawn. Find the probability distribution of X, where X denotes the number of white cubes selected from the bag.

Solution

We start by drawing a diagram that will help us visualise the situation:

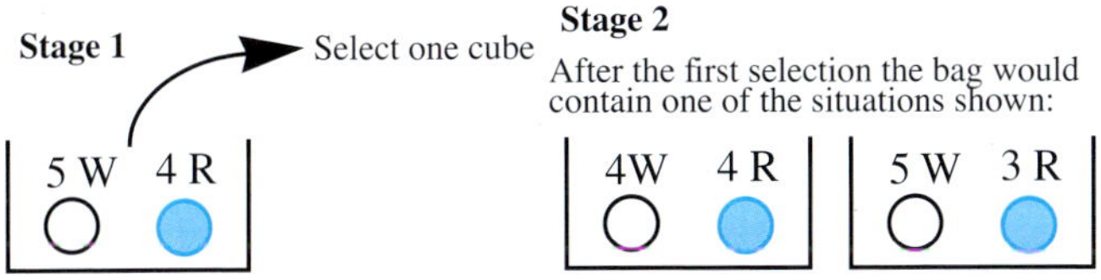

Next, we set up the corresponding tree diagram:

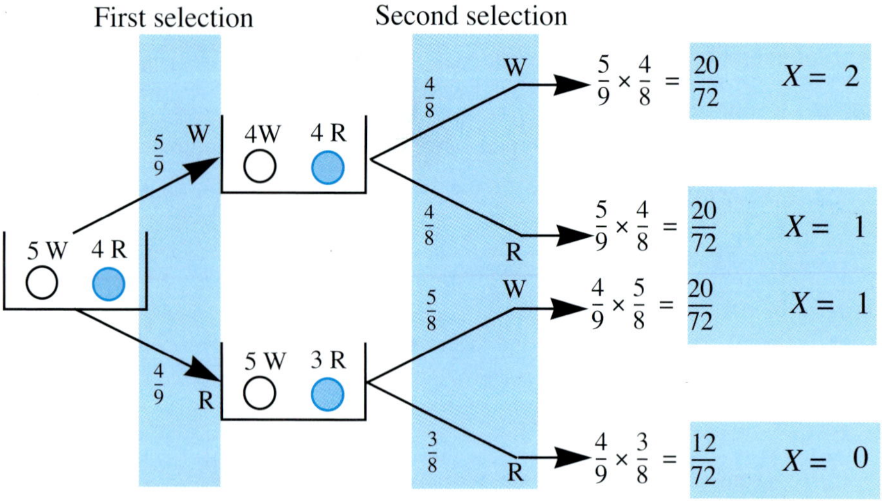

We are now in a position to complete the probability table.

x	0	1	2
$P(X = x)$	$\frac{12}{72}$	$\frac{40}{72}$	$\frac{20}{72}$

EXAMPLE 16.7

Two friends Kirsty and Bridget, independently applied for different jobs. The chances that Kirsty is successful is 0.8 and the chances that Bridget is successful is 0.75.

(a) If X denotes the number of successful applications between the two friends, find the probability distribution of X.

(b) Hence find the probability that i. both are successful
ii. that if one is successful, it is Kirsty.

Solution

a. Let K denote the event that Kirsty is successful, so that $P(K) = 0.8$ and let B denote the event that Bridget is successful, so that $P(B) = 0.75$.

Now, the event '$X = 0$' translates to 'noboby is successful':

That is, $P(X = 0) = P(K' \cap B') = P(K') \times P(B') = 0.2 \times 0.25 = 0.05$.

Similarly, the event '$X = 1$' translates to 'only one is successful':

That is, $P(X = 1) = P(K \cap B') + P(K' \cap B) = 0.8 \times 0.25 + 0.2 \times 0.75 = 0.35$.

Lastly, the event '$X = 2$' translates to 'both are successful':

That is, $P(X = 2) = P(K \cap B) = P(K) \times P(B) = 0.8 \times 0.75 = 0.6$

We can now construct a probability distribution for the random variable X:

x	0	1	2
$P(X = x)$	0.05	0.35	0.60

(b) i. $P(\text{Both successful}) = P(X = 2) = 0.60$

ii. $P(K|\text{Only one is successful}) = \dfrac{P(K \cap \{X=1\})}{P(\{X=1\})} = \dfrac{P(K \cap B')}{P(\{X=1\})}$

$= \dfrac{0.20}{0.35} (= 0.5714)$

EXERCISES 16.1

1. Find the value of k, so that the random variable X describes a probability distribution.

x	1	2	3	4	5
$P(X = x)$	0.25	0.20	0.15	k	0.10

2. The discrete random variable Y has the following probability distribution

y	1	2	3	4
$P(Y = y)$	β	2β	3β	4β

(a) Find the value of β.

(b) Find i. $P(Y = 2)$

ii. $P(Y > 2)$

3. A delivery of six television sets contains 2 defective sets. A customer makes a random purchase of two sets from this delivery. The random variable X denotes the number of defective sets purchased by the customer.

(a) Find the probability distribution table for X.

(b) Represent this distribution as a graph.

(c) Find $P(X \le 1)$.

4. A pair of dice are rolled. Let Y denote the sum showing uppermost.

(a) Determine the possible values that the random variable Y can have.

(b) Display the probability distribution of Y in tabular form.

(c) Find $P(Y = 8)$.

(d) Sketch the probability distribution of Y.

5. A fair coin is tossed 3 times.

(a) Draw a tree diagram representing this experiment.

(b) Display this information using both graphical and tabular representations.

(c) If the random variable Y denotes the number of heads that appear uppermost, find $P(Y \ge 2 | Y \ge 1)$.

6. The number of customers that enter a small corner newsagency between the hours of 8 p.m and 9 p.m can be modelled by a random variable X having a probability distribution given by $P(X = x) = k(3x + 1)$, where $x = 0, 1, 2, 3, 4$.

(a) Find the value of k.

(b) Represent this distribution in i. tabular form.
ii. graphical form.

(c) What are the chances that at least 2 people will enter the newsagency between 8 p.m and 9 p.m on any one given day?

7. The number of cars passing an intersection in the country during the hours of 4 p.m and 6 p.m follows a probability distribution that is modelled by the function

$$P(X = x) = \frac{(0.1)^x}{x!}e^{-0.1}, x = 0, 1, 2, 3, \ldots .$$

Where the random variable X denotes the number of cars that pass this intersection between 4 p.m and 6 p.m.

(a) Find i. $P(X = 0)$ ii. $P(X = 1)$

(b) Find the probability of observing at least three cars passing through this intersection during the hours of 4 p.m and 6 p.m.

8. The number of particles emitted during a one hour period is given by the random variable X, having a probability distribution

$$P(X = x) = \frac{(4)^x}{x!}e^{-4}, x = 0, 1, 2, 3, \ldots, \text{ find } P(X > 4).$$

9. A random variable X has the following probability distribution

x	0	1	2	3
$P(X = x)$	$\frac{1}{6}$	$\frac{1}{2}$	$\frac{1}{5}$	$\frac{2}{15}$

(a) Find the probability distribution of $Y = X^2 - 2X$.

(b) Find i. $P(Y = 0)$ ii. $P(Y < 3)$.

10. A bakery has six indistinguishable muffins on display. However, two of them have been filled with strawberry jam and the others with apricot jam. Claire, who hates strawberry jam, purchases two muffins at random. Let N denote the number of strawberry jam muffins Claire buys. Find the probability distribution of the random variable N.

11. A box contains four balls numbered 1 to 4. A ball is selected at random from the box and its number is noted.

(a) If the random variable X denotes the number on the ball, find the probability distribution of X.

After the ball is placed back into the box, a second ball is randomly selected.

(b) If the random variable S denotes the sum of the numbers shown on the balls after the second draw, find the probability distribution of S.

12. A probability distribution function for the random variable X is defined by

$$P(X = x) = k \times (0.9)^x, x = 0, 1, 2, \ldots$$

Find (a) $P(X \geq 2)$.

(b) $P(1 \leq X < 4)$

16.2 MEAN AND VARIANCE

16.2.1 CENTRAL TENDENCY AND EXPECTATION

For a discrete random variable X with a probability distribution defined by $P(X = x)$, we define the **expectation of the random variable X** as

$$E(X) = \sum_{i=1}^{i=n} x_i P(X = x_i)$$
$$= x_1 \times P(X = x_1) + x_2 \times P(X = x_2) + \ldots + x_n \times P(X = x_n)$$

Where $E(X)$ is read as "The expected value of X". $E(X)$ is interpreted as the **mean value** of X and is often written as μ_X (or simply μ). Often we write the expected value of X as $\sum xP(X = x)$. This is in contrast to the **mode** which is the most common value(s) and the **median** which is the value with half the probabilities below and half above the median value.

16.2.2 SO WHAT EXACTLY DOES $E(X)$ MEASURE?

The expected value of the random variable is a **measure of the central tendency** of X. That is, it is an indication of its 'central position'– based not only on the values of X, but also the **probability weighting** associated with each value of X. That is, it is the probability–weighted average of its possible values.

To find the value of $E(X)$ using the formula $E(X) = \sum_{i=1}^{i=n} x_i P(X = x_i)$ we take each possible value of x_i, multiply it by its associated probability $P(X = x_i)$ (i.e., its 'weight') and then add the results. The number that we obtain can be interpreted in two ways:

1. As a **probability–weighted average**, it is a summary number that takes into account the relative probabilities of each x_i value.

2. As a **long–run average**, it is a measure of what one could expect to observe if the experiment was repeated a large number of times.
   For example, when tossing a fair coin a large number of times (say 500 times) and the random variable X denotes the number of tails observed, we would expect to observe 250 tails, i.e., $E(X) = 250$.

Note: *Although we would expect 250 tails after tossing a coin 500 times, it might be that we do not observe this outcome! For example, if the average number of children per 'family' in Australia is 2.4, does this mean we expect to see 2.4 children per 'family'?*

In short, we might not be able to observe the value $E(X)$ that we obtain.

EXAMPLE 16.8

For the random variable X with probability distribution defined by

x	1	2	3	4
$P(X=x)$	$\frac{1}{10}$	$\frac{2}{10}$	$\frac{3}{10}$	$\frac{4}{10}$

Find the mode, median and mean values of X.

Solution

The mode is $X = 4$ (the most probable) and the median 3 (half are above and half below 3– this is probably best done by sketching the cdf of X).

To find the mean of X we use the formula $E(X) = \sum_{i=1}^{i=n} x_i P(X = x_i)$:

That is, $E(X) = \left(1 \times \frac{1}{10}\right) + \left(2 \times \frac{2}{10}\right) + \left(3 \times \frac{3}{10}\right) + \left(4 \times \frac{4}{10}\right) = \frac{1}{10} + \frac{4}{10} + \frac{9}{10} + \frac{16}{10} = 3$

Therefore, X has a mean value of 3.

EXAMPLE 16.9

A fair die is rolled once. If the random variable X denotes the number showing, find the expected value of X.

Solution

Because the die is fair we have the following probability distribution:

x	1	2	3	4	5	6
$P(X=x)$	$\frac{1}{6}$	$\frac{1}{6}$	$\frac{1}{6}$	$\frac{1}{6}$	$\frac{1}{6}$	$\frac{1}{6}$

Then, $E(X) = \left(1 \times \frac{1}{6}\right) + \left(2 \times \frac{1}{6}\right) + \left(3 \times \frac{1}{6}\right) + \left(4 \times \frac{1}{6}\right) + \left(5 \times \frac{1}{6}\right) + \left(6 \times \frac{1}{6}\right)$

$= \frac{21}{6} = 3.5$

Notice then, that in the case above we obtain an expected value of 3.5, even though the random variable X does not actually take on the value 3.5! This is the same as when a census might report that the average number of children per family is 2.4, where clearly, there cannot be 2.4 children in any family.

Do not forget to use your graphics calculator for such tasks. To do so, enter the data as lists, then multiply the lists and add the results in the resulting list:

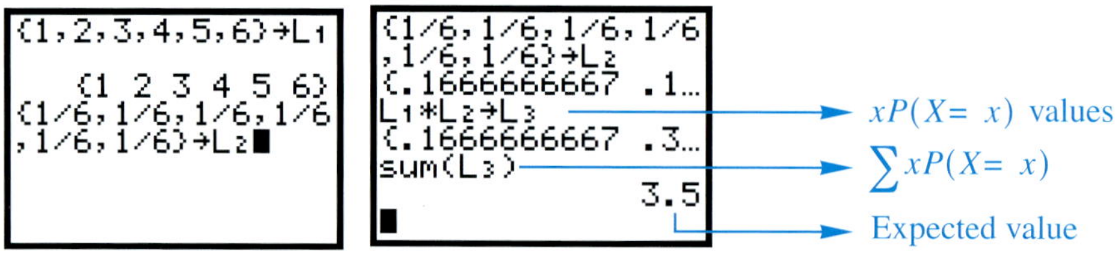

16.2.3 PROPERTIES OF THE EXPECTATION (FUNCTION)

1. $E(a) = a$, where a is a constant.
That is, the expected value of a constant is the constant itself.

2. $E(aX) = aE(X)$, where a is a constant.

3. $E(f(X)) = \sum_{i=i}^{i=n} f(x_i) \times P(X = x_i)$.
Where $f(X)$ is some real valued function of the random variable X.

4. $E(aX + b) = \sum_{i=i}^{i=n} (ax_i + b) \times P(X = x_i) = aE(X) + b$ where a and b are constants.
Note: This is a special case of 3., where in this case, $f(X) = aX + b$

NB: Be careful not to assume that $E(X^2)$ is the same as $(E(X))^2$, i.e., $E(X^2) \neq (E(X))^2$.

Nor does $E\left(\frac{1}{X}\right)$ equal $\frac{1}{E(X)}$, i.e., $E\left(\frac{1}{X}\right) \neq \frac{1}{E(X)}$.

In general, we have that $E(f(X)) \neq f(E(X))$ with $f(X) = aX + b$ being a notable exception.

EXAMPLE 16.10

For the probability distribution shown below

x	0	1	2	3
$P(X = x)$	$\frac{1}{6}$	$\frac{1}{2}$	$\frac{1}{5}$	$\frac{2}{15}$

Find (a) $E(X)$ (b) $E(X^2)$ (c) $E(X^2 + 3X - 1)$

Solution

(a) $E(X) = \sum xP(X = x) = 0 \times \frac{1}{6} + 1 \times \frac{1}{2} + 2 \times \frac{1}{5} + 3 \times \frac{2}{15} = 1.3$

(b) $E(X^2) = \sum x^2 P(X = x) = 0^2 \times \frac{1}{6} + 1^2 \times \frac{1}{2} + 2^2 \times \frac{1}{5} + 3^2 \times \frac{2}{15} = 2.5$

Notice that $(E(X))^2 = (1.3)^2 = 1.69 \neq E(X^2)$!

(c) First we simplify the expression by 'expanding' and using some of the properties:

$$
\begin{aligned}
E(X^2 + 3X - 1) &= E(X^2) + E(3X) - E(1) \\
&= E(X^2) + 3E(X) - 1 \\
&= 2.5 + 3 \times 1.3 - 1 \\
&= 5.4
\end{aligned}
$$

Again, making use of the graphics calculator can make the arithmetic in problems such as the one above much easier to deal with:

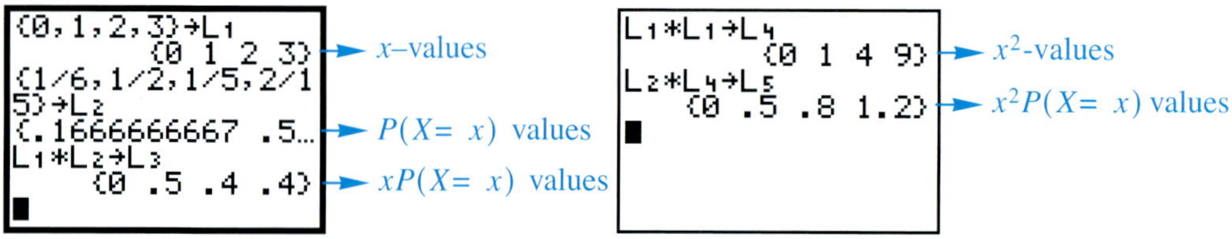

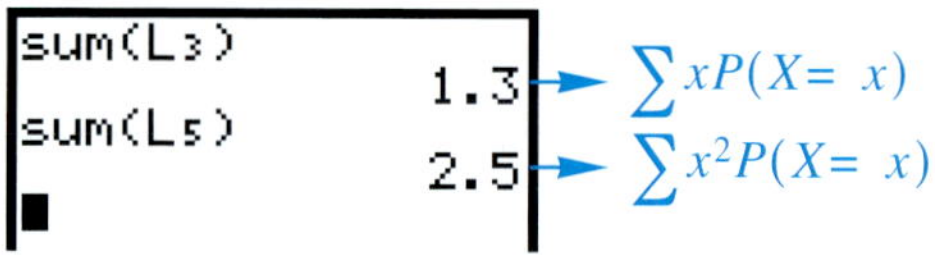

EXAMPLE 16.11

A dart board consisting of concentric circles of radius 1, 2 and 3 units is placed against a wall. Upon throwing a dart, which lands at some random location on the board, a player will receive \$8.00 if the smaller circle is hit, \$ 6.00 if the middle annular region is hit and \$ 4.00 if the outer annular region is hit. However, should the player miss the dart board altogther, they would lose \$ 7.00. The probability that the player misses the dart board altogether is 0.5.

(a) How will the player fare in this game?

(b) What can you say about this game?

Solution

(a) Let X denote the player's "winnings" on any one throw.
Therefore, $X = \{8, 6, 4, -7\}$.
Notice that a loss of \$7.00 is represented by –7.
The probabilities corresponding to the random numbers in the set X can be determined by finding the proportion of area for each of the values.

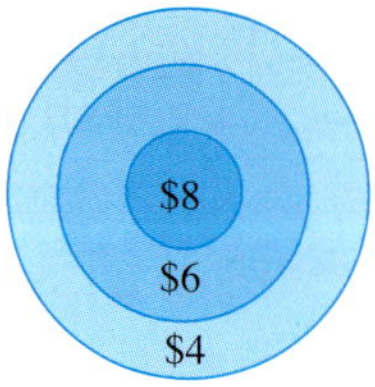

For example, to win \$8.00, the player must first hit the board, and then the dart must land in the inner circle (having a radius of 1 unit). Therefore, we have

$$P(X = 8) = \frac{1}{2} \times \frac{\pi \times 1^2}{\pi \times 3^2} = \frac{1}{18}.$$

Similarly,

$$P(X = 6) = \frac{1}{2} \times \frac{\pi \times 2^2 - \pi \times 1^2}{\pi \times 3^2} = \frac{1}{6},$$

$$P(X = 4) = \frac{1}{2} \times \frac{\pi \times 3^2 - \pi \times 2^2}{\pi \times 3^2} = \frac{5}{18}$$

and

$$P(X = -7) = 0.5$$

Therefore, we have that $E(X) = 8 \times \frac{1}{18} + 6 \times \frac{1}{6} + 4 \times \frac{5}{18} - 7 \times \frac{1}{2} = -\frac{17}{18} \approx -0.94$.

That is, the player can expect to lose \$0.94 (this is considered as a long–run average loss).

(b) Because $E(X) < 0$, the game is unfair.

A fair game would have $E(X) = 0$.

16.2.4 VARIANCE

Although we now have a means by which we can calculate a measure of the central tendency of a random variable, an equally important aspect of a random variable is its spread. For example, the mean of the three numbers, 100, 110 and 120 is 110. Similarly, the mean of the three numbers 10, 100, and 220 is also 110. Yet clearly, the values in the second set of data have a wider spread than those in the first set of data. The **variance** (or more so, the **standard deviation**) provides a better measure of this spread.

The **variance** of a discrete random variable may be considered as the average of the squared deviations about the mean. This provides a measure of the variability of the random variable or the probability dispersion. The variance associated with the random variable X is given by

$$Var(X) = E((X-\mu)^2) = \sum_{i=1}^{i=n}(x-\mu)^2 P(X=x)$$

However, for computational purposes, it is often better to use the alternative definition:

$$Var(X) = E(X^2) - (E(X))^2 = E(X^2) - \mu^2$$

The variance is also denoted by σ^2 (read as "**sigma squared**"), i.e., $Var(X) = \sigma^2$.

We also have the **standard deviation**, given by

$$Sd(X) = \sigma = \sqrt{Var(X)},$$

which also provides a measure of the spread of the distribution of X.

What is the difference between the Var(X) and the Sd(X)?

Because of the squared factor in the equation for $Var(X)$ (i.e., $Var(X) = E(X^2) - \mu^2$), the units of $Var(X)$ are not the same as those for X. However, because the $Sd(X)$ is the square root of the $Var(X)$, we have "adjusted" the units of $Var(X)$ so that they now have the same units as the random variable X.

The reason for using the $Sd(X)$ rather than the $Var(X)$ is that we can make clearer statistical statements about the random variable X (in particular, statements that relate to an overview of the distribution).

EXAMPLE 16.12

The probability distribution of the random variable X is shown below:

x	–2	–1	0	1	2
$P(X=x)$	$\frac{1}{64}$	$\frac{12}{64}$	$\frac{38}{64}$	$\frac{12}{64}$	$\frac{1}{64}$

Find (a) The variance of X. (b) The standard deviation of X.

Solution

(a) First we need to find $E(X)$ and $E(X^2)$:

$$E(X) = \sum xP(X=x) = -2\times\frac{1}{64}+(-1)\times\frac{12}{64}+0\times\frac{38}{64}+1\times\frac{12}{64}+2\times\frac{1}{64} = 0$$

and $$E(X^2) = \sum x^2P(X=x) = (-2)^2\times\frac{1}{64}+(-1)^2\times\frac{12}{64}+0^2\times\frac{38}{64}+1^2\times\frac{12}{64}+2^2\times\frac{1}{64} = \frac{1}{2}$$

Therefore, $Var(X) = E(X^2)-\mu^2 = \frac{1}{2}-0^2 = \frac{1}{2}$.

(b) Using $Sd(X) = \sigma = \sqrt{Var(X)}$, we have that, $Sd(X) = \sigma = \sqrt{\frac{1}{2}} \approx 0.707$.

16.2.5 PROPERTIES OF THE VARIANCE

1. If a is a constant, then $Var(a) = 0$.
That is, there is no variability in a constant!

2. $Var(aX) = a^2Var(X)$, where a is a constant.

3. $Var(aX+b) = a^2Var(X)$, where a and b are constants.

EXAMPLE 16.13

A random variable X has a probability distribution defined by

x	0	1	2	3	4
$P(X=x)$	$\frac{1}{16}$	$\frac{3}{16}$	$\frac{7}{16}$	$\frac{3}{16}$	$\frac{2}{16}$

Find (a) $E(X)$ (b) $Var(4X+2)$ (c) $Sd(3-X)$

Solution

(a) $$E(X) = 0\times\frac{1}{16}+1\times\frac{3}{16}+2\times\frac{7}{16}+3\times\frac{3}{16}+4\times\frac{2}{16} = \frac{17}{8}$$

(b) To evaluate $Var(4X+2)$, we need $E(X^2)$:

Now, $$E(X^2) = 0^2\times\frac{1}{16}+1^2\times\frac{3}{16}+2^2\times\frac{7}{16}+3^2\times\frac{3}{16}+4^2\times\frac{2}{16} = \frac{45}{8}$$

Therefore, $$Var(4X+2) = 4^2Var(X) = 16\left(\frac{45}{8}-\left(\frac{17}{8}\right)^2\right) = 16\times\frac{71}{64} = \frac{71}{4}.$$

(c) $Sd(3-X) = \sqrt{Var(3-X)}$

However, $Var(3-X) = (-1)^2Var(X) = \frac{71}{64}$ $\therefore Sd(3-X) = \sqrt{\frac{71}{64}} \approx 1.05$.

EXERCISES 16.2

1. A discrete random variable X has a probability distribution given by.

x	1	2	3	4	5
$P(X = x)$	0.25	0.20	0.15	0.3	0.10

(a) Find the mean value of X.
(b) Find the variance of X.

2. The discrete random variable Y has the following probability distribution

y	1	2	3	4
$P(Y = y)$	0.1	0.2	0.3	0.4

(a) Find the mean value of Y.
(b) Find i. $Var(Y)$ ii. $Sd(Y)$
(c) Find i. $E(2Y)$ ii. $E\left(\frac{1}{Y}\right)$

3. A random variable X has the following probability distribution

x	0	1	2	3
$P(X = x)$	$\frac{1}{6}$	$\frac{1}{2}$	$\frac{1}{5}$	$\frac{2}{15}$

(a) Find i. $E(X)$ ii. $E(X^2)$ iii. $E(X^2 - 2X)$.
(b) Find i. $Sd(X)$ ii. $Var(3X + 1)$
(c) If $Y = \frac{1}{X+1}$, find

i. $E(Y)$ ii. $E(Y^2)$

4. A delivery of six television sets contains 2 defective sets. A customer makes a random purchase of two sets from this delivery. The random variable X denotes the number of defective sets the customer purchased. Find the mean and variance of X.

5. A pair of dice are rolled. Let Y denote the sum showing uppermost.
(a) Find $E(Y)$
(b) Find $Var(Y)$

6. How many tails would you expect to observe when a fair coin is tossed 3 times.

7. The number of customers that enter a small corner newsagency between the hours of 8 p.m and 9 p.m can be modelled by a random variable X having a probability distribution given by $P(X = x) = k(2x + 1)$, where $x = 0, 1, 2, 3, 4$.

(a) Find the value of k.
(b) How many customers can be expected to enter the newsagency between 8 and 9 p.m?
(c) Find the standard deviation of X.

8. A discrete random variable Y has its probability distribution function defined as

y	-2	-1	0	1
$P(Y = y)$	k	0.2	$3k$	0.4

(a) Find k.
(b) Given that the function, F, is defined by $F(y) = P(Y \le y)$, find
i. $F(-1)$ ii. $F(1)$
(c) Find
i. the expected value of Y
ii. the variance of Y
iii. the expected value of $(Y + 1)^2$

9. A dart board consisting of consecutive circles of radius 1, 2 and 3 units is placed against a wall. A player throws darts at the board, each dart landing at some random location on the board. The player will receive \$9.00 if the smaller circle is hit, \$7.00 if the middle annular region is hit and \$4.00 if the outer annular region is hit. Should players miss the board altogether, they would lose \$$k$ each time. The probability that the player misses the dart board is 0.5. Find the value of k if the game is to be fair.

10. A box contains 7 black cubes and 3 red cubes. Debra selects three cubes from the box without replacement. Let the random variable N denote the number of red cubes selected by Debra.
(a) Find the probability distribution for N.
(b) Find i. $E(N)$ ii. $Var(N)$.
Debra will win \$2.00 for every red cube selected and lose \$1.00 for every black cube selected. Let the random variable W denote Debra's winnings.
(c) If $W = aN + b$, find a and b. Hence, find $E(W)$.

11. (a) A new gambling game has been introduced in a casino:
A player stakes \$8.00 in return for the throw of two dice, where the player wins as many dollars as the sum of the two numbers showing uppermost.
How much money can the player expect to walk away with?
(b) At a second casino, a different gambling game has been set up:
A player stakes \$8.00 in return for the throw of two dice, if two sixes come up, the player wins \$252.
Which game would be more profitable for the casino in the long run?

12. Given that $Var(X) = 2$, find
(a) $Var(5X)$ (b) $Var(-3X)$ (c) $Var(1 - X)$

13. Given that $Var(X) = 3$ and $\mu = 2$, find
(a) $E(2X^2 - 4X + 5)$ (b) $Sd\left(4 - \frac{1}{3}X\right)$ (c) $E(X^2) + 1 - E((X + 1)^2)$

14. A store has eight toasters left in its storeroom. Three of the toasters are defective and should not be sold. A salesperson, unaware of the defective toasters, selects two toasters for a customer. Let the random variable N denote the number of defective toasters the customer purchases. Find

(a) $E(N)$.

(b) $Sd(N)$.

15. (a) The random variable Y is defined by:

y	-1	1
$P(Y = y)$	p	$1 - p$

Find the mean and variance of Y.

(b) The random variable X is defined as $X = Y_1 + Y_2 + Y_3 + \ldots + Y_n$ where each Y_i, $i = 1, 2, 3, \ldots, n$ is independent and has the distribution defined in (a).

Find i. $E(X)$

ii. $Var(X)$

16. A game is played by selecting coloured discs from a box. The box initially contains two red and eight blue discs. Tom pays \$10.00 to participate in the game. Each time Tom participates he selects two discs. The winnings are governed by the probability distribution shown below, where the random variable N is the number of red discs selected.

n	0	1	2
Winnings	\$0	\$W	\$5W
$P(N = n)$			

(a) Complete the table.

(b) For what value of W will the game be fair?

17. A random variable X has the following probability distribution:

x	0	1	2
$P(X = x)$	a	$\frac{1}{3}(1 - b)$	$\frac{1}{3}b$

(a) What values may a and b take?

(b) Express, in terms of a and b i. $E(X)$ ii. $Var(X)$

18. (a) Find the mean and variance of the probability distribution defined by

$$P(Z = z) = k(0.8)^z, z = 0, 1, 2, \ldots$$

(b) i. Show $P(X = x) = p \times (1 - p)^x, x = 0, 1, 2, \ldots$ defines a probability distribution.

ii. Show that $E(X) = \frac{1 - p}{p}$

iii. Show that $Var(X) = \frac{1 - p}{p^2}$.

16.3 THE BINOMIAL DISTRIBUTION

16.3.1 THE BINOMIAL EXPERIMENT

The binomial distribution is a special type of discrete distribution which finds applications in many settings of everyday life. In this section we summarise the important features of this probability distribution.

16.3.2 BERNOULLI TRIALS

Certain experiments consist of repeated trials where each trial has only two mutually exclusive, possible outcomes. Such trials are referred to as **Bernoulli trials**. The outcomes of a Bernoulli trial are often referred to as "a success" or "a failure". The terms "success" and "failure" in this context do not necessarily refer to the everyday usage of the word success and failure. For example, a "success" could very well be referring to the outcome of selecting a defective transistor from a large batch of transistors.

We often denote P(Success) by p and P(Failure) by q, where $p + q = 1$ (or $q = 1 - p$).

16.3.3 PROPERTIES OF THE BINOMIAL EXPERIMENT

1. There are a fixed number of trials. We usually say that there are n trials.

2. On each one of the n trials there is only one of two possible outcomes, labelled "*success*" and "*failure*".

3. Each trial is identical and independent.

4. On each of the trials, the probability of a success, p, is always the same, and the probability of a failure, $q = 1 - p$, is also always the same.

16.3.4 THE BINOMIAL DISTRIBUTION

If a (discrete) random variable X has all of the above mentioned properties, we say that X has a **binomial distribution**. The probability distribution function is given by

$$p(X = x) = \binom{n}{x} p^x q^{n-x} = \binom{n}{x} p^x (1-p)^{n-x}, x = 0, 1, 2, \ldots n$$

Where X denotes the **number of successes** in n trials such that the **probability of a success on any one trial** is p, $0 \leq p \leq 1$ and $p + q = 1$ (or $q = 1 - p$)).

We can also express the binomial distribution in a compact form, written as $X \sim \mathrm{B}(n, p)$, read as "**X is distributed binomially with parameters n and p**",
where n is the number of trials and $p = P(success)$ [it is also common to use $X \sim \mathrm{Bin}(n, p)$].

For example, the probability function for $X \sim \mathrm{B}(6, 0.4)$ (i.e., 6 trials and $p = 0.4$) would be

$$P(X = x) = \binom{6}{x}(0.4)^x(0.6)^{6-x}, x = 0, 1, 2, \ldots n.$$

EXAMPLE 16.14

If $X \sim B(5, 0.6)$, find $P(X = 4)$.

Solution

$X \sim B(5, 0.6)$ means that $P(X = 4)$ is the probability of 4 successes in five trials, where each trial has a 0.6 chance of being a success, that is, $n = 5, p = 0.6$ and $x = 4$.

$$\therefore P(X = 4) = {}^5C_4(0.6)^4(0.4)^{5-4}$$
$$= \frac{5!}{1!4!}(0.6)^4(0.4)^1$$
$$= 0.2592$$

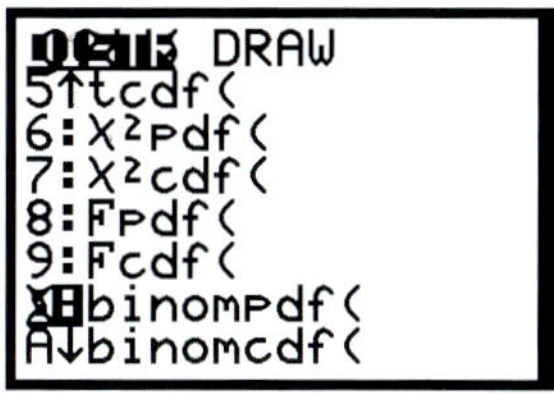

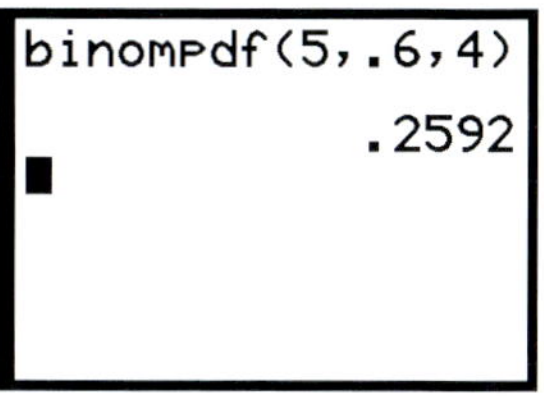

Again we see how useful the TI–83 is when dealing with probability. In particular, when dealing with known distributions.

Pressing **2nd VARS** brings up the **DIST** menu. From this menu we can then scroll down to the appropriate distribution and then, after we enter the appropriate parameters, we obtain the required probability.

When dealing with the binomial distribution there are two options:

1. the **binompdf** function – which evaluates individual probabilities and
2. the **binomcdf** function – which evaluates cumulative probabilities.

EXAMPLE 16.15

A manufacturer finds that 30% of the items produced from one of the assembly lines are defective. During a floor inspection, the manufacturer selects 6 items from this assembly line. Find the probability that the manufacturer finds

(a) two defectives. (b) at least two defectives.

Solution

(a) Let X denote the number of defectives in the sample of six. Therefore, we have that $n = 6, p(success) = p = 0.30$ ($\Rightarrow q = 1 - p = 0.70$), so that $X \sim B(6, 0.3)$.
Note that in this case, a 'success' refers to a defective.

i. $P(X = 2) = {}^6C_2(0.3)^2(0.7)^4 = 0.3241$.

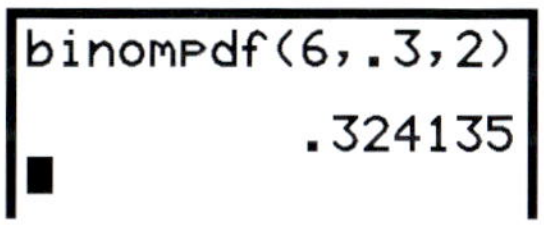

ii. $P(X \geq 2) = P(X = 2) + P(X = 3) + \ldots + P(X = 6)$

$$= {}^6C_2(0.3)^2(0.7)^4 + {}^6C_3(0.3)^3(0.7)^3 + \ldots + {}^6C_6(0.3)^6(0.7)^0$$

Evaluating the probabilities manually is rather time consuming as there are lots of calculations. However the TI–83 can provide a list of the individual probability values. We can then scroll across the screen to see the other values:

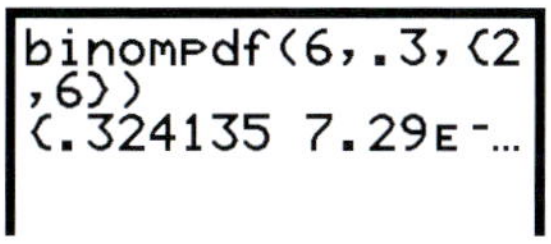

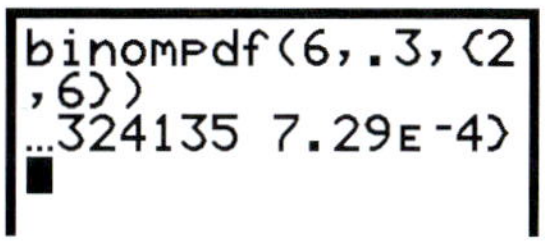

A second method makes use of the complementary event:

$$P(X \geq 2) = 1 - P(X < 2) = 1 - P(X \leq 1)$$
$$= 1 - [P(X = 1) + P(X = 0)]$$
$$= 1 - [0.1176 + 0.3025]$$
$$= 0.5798$$

```
1-binomcdf(6,.3,
1)
            .579825
```

Note: Using the cumulative binomial distribution on the TI–83, we have that
$P(X \geq 2) = 1 - P(X \leq 1) = 1 -$ **binomcdf**$(6,0.3,1) = 1 - 0.420175 = 0.5798$

EXAMPLE 16.16

Sophie has 10 pots labelled one to ten. Each pot, and its contents, is identical in every way. Sophie plants a seed in each pot such that each seed has a germinating probability of 0.8.

(a) What is the probability that all the seeds will germinate?
(b) What is the probability that only three seeds will not germinate?
(c) What is the probability that more than eight seeds do germinate?
(d) How many pots must Sophie use to be 99.99% sure that at least one seed germinates?

Solution

Let X denote the number of seeds germinating. Therefore we have that $X \sim B(10, 0.8)$, i.e., X is binomially distributed with parameters $n = 10$ and $p = 0.8$ (and $q = 1 - p = 0.20$)

(a) $P(X = 10) = \binom{10}{10}(0.8)^{10}(0.2)^{0} = 0.1074$.

(b) If only three seeds will **not** germinate, then only seven seeds must germinate!

We want, $P(X = 7) = \binom{10}{7}(0.8)^{7}(0.2)^{3} = 0.2013$.

```
binompdf(10,.8,1
0)
        .1073741824
binompdf(10,.8,7
)
         .201326592
```

(c) Now, $P(X > 8) = P(X = 9) + P(X = 10)$
$= 0.2684 + 0.1074$
$= 0.3758$

```
binompdf(10,.8,(
9,10))
(.268435456 .10…
```

```
binompdf(10,.8,(
9,10))
…456 .107374182…
```

(d) At least one flower means, $X \geq 1$, therefore we need to find a value of n such that $P(X \geq 1) \geq 0.9999$.

Now, $P(X \geq 1) = 1 - P(X = 0) = 1 - \binom{8}{0}(0.8)^{0}(0.2)^{n} = 1 - (0.2)^{n}$.

Solving for n we have: $1 - (0.2)^{n} \geq 0.9999 \Leftrightarrow (0.2)^{n} \leq 0.0001$

This inequality can be solved by trial and error, algebraically, or as we will solve it, graphically:

Enter relevant equations

```
Plot1 Plot2 Plot3
\Y1=(.2)^X
\Y2=.0001
\Y3=
\Y4=
\Y5=
\Y6=
\Y7=
```

Set domain and range

```
WINDOW
 Xmin=-1
 Xmax=8
 Xscl=1
 Ymin=0
 Ymax=.00015
 Yscl=.25
 Xres=1
```

Sketch graph

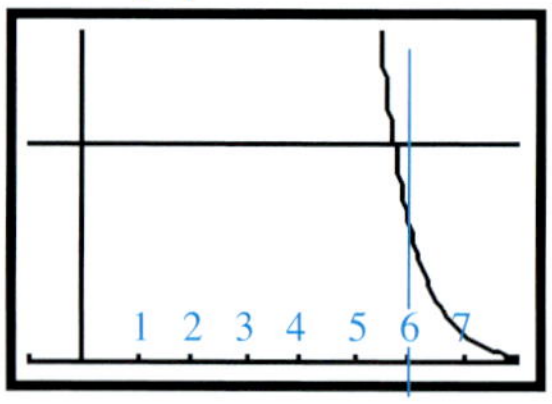

From the graph we see that for $(0.2)^{n} \leq 0.0001$, Sophie would need at least 6 pots, i.e., $n \geq 6$.

16.3.5 EXPECTATION, MODE AND VARIANCE FOR THE BINOMIAL DISTRIBUTION

If the random variable X, is such that $X \sim B(n, p)$, we have;

1. the **expected value of X** is $\mu = E(X) = np$

2. the **mode of X** is that value of x which has the largest probability

3. the **variance of X** is $\sigma^2 = Var(X) = npq = np(1-p)$.

Note 1. Although we can use our earlier definitions of the expected value and the variance of a random variable, the formulas above are a nice compact form and can only be used when dealing with the binomial distribution

2. The standard deviation, $Sd(X)$, is still given by $\sigma = \sqrt{Var(X)} = \sqrt{npq}$.

EXAMPLE 16.17

A fair die is rolled six times. If X denotes the number of fours obtained, find

(a) $E(X)$. (b) the mode of X. (c) $Sd(X)$.

Solution

In this case we have that $X \sim B\left(6, \frac{1}{6}\right)$, therefore $q = \frac{5}{6}$.

(a) $\mu = E(X) = 6 \times \frac{1}{6} = 1$.

(b) To find the mode, we need to know the probability of each outcome. We do this by constructing a table of values:

x	0	1	2	3	4	5	6
$P(X = x)$	$\frac{15625}{46656}$	$\frac{18750}{46656}$	$\frac{9375}{46656}$	$\frac{2500}{46656}$	$\frac{375}{46656}$	$\frac{30}{46656}$	$\frac{1}{46656}$

So that the mode of X is 1 (as it has the highest probability value).
Notice that in this case, the mode of X = expected value of X.
Will this always be the case?

(c) $\sigma = \sqrt{Var(X)} = \sqrt{npq} = \sqrt{6 \times \frac{1}{6} \times \frac{5}{6}} \approx 0.9129$.

EXAMPLE 16.18

An urn contains 7 marbles of which 2 are blue. A marble is selected, its colour noted and then replaced in the urn. This process is carried out 50 times. Find;

(a) The mean number of blue marbles selected.

(b) The standard deviation of the number of blue marbles selected.

Solution

Because we replace the marble before the next selection, each trial is identical and independent. Therefore, if we let X denote the number of blue marbles selected, we have that $p = \frac{2}{7}$, $n = 50$ and $q = \frac{5}{7}$.

(a) $E(X) = np = 50 \times \frac{2}{7} = 14.29$

(b) $Var(X) = npq = 50 \times \frac{2}{7} \times \frac{5}{7} = \frac{500}{49} \therefore \sigma = \sqrt{\frac{500}{49}} \approx 3.19$.

EXAMPLE 16.19

The random variable X is such that $E(X) = 8$ and $Var(X) = 4.8$. Find $p(X = 3)$.

Solution

This time we are given that $np = 8$ and $npq = np(1-p) = 4.8$.
Therefore, after substituting $\boldsymbol{np} = 8$ into $\boldsymbol{np}(1-p) = 4.8$, we have that

$$8(1-\text{p}) = 4.8$$

$$\therefore (1-p) = 0.6 \Rightarrow p = 0.4$$

Substituting $p = 0.4$ back into $np = 8$, we have that $n = 20$.

Therefore, $P(X = 3) = \binom{20}{3}(0.4)^3(0.6)^{17} = 0.0123$.

EXERCISES 16.3

1. At an election 40% of the voters favoured the Environment Party. Eight voters were interviewed at random. Find the probability that
(a) exactly 4 voters favoured the Environment Party.
(b) a majority of those interviewed favoured the Environment Party.
(c) at most 3 of the people interviewed favoured the Environment Party.

2. In the long run, Thomas wins 2 out of every 3 games. If Thomas plays 5 games, find the probability that he will win
(a) exactly 4 games.
(b) at most 4 games.
(c) no more than 2 games.
(d) all 5 games.

3. A bag consists of 6 white cubes and 10 black cubes. Cubes are withdrawn one at a time with replacement. Find the probability that after 4 draws
(a) all the cubes are black.
(b) there are at least 2 white cubes.
(c) there are at least 2 white cubes given that there was at least one white cube.

4. An X-ray has a probability of 0.95 of showing a fracture in the leg. If 5 different X-rays are taken of a particular leg, find the probability that
(a) all five X-rays identify the fracture.
(b) the fracture does not show up.
(c) at least 3 X-rays show the fracture
(d) only one X-ray shows the fracture.

5. A biased die, in which the probability of a '2' turning up is 0.4, is rolled 8 times. Find the probability that:
(a) a '2' turns up 3 times.
(b) a '2' turns up on at least 4 occasions.

6. During an election campaign, 66% of a population of voters are in favour of a food quality control proposal. A sample of 7 voters was chosen at random from this population. Find the probability that:
(a) there will be 4 voters that were in favour.
(b) there will be at least 2 voters who were in favour.

7. During an election 35% of the people in a town favoured the fishing restrictions at Lake Watanaki. Eight people were randomly selected from the town. Find the probability that:
(a) 3 people favoured fishing restrictions.
(b) at most 3 of the eight favoured fishing restrictions.
(c) there was a majority in favour of fishing restrictions.

8. A bag containing 3 white balls and 5 black balls has 4 balls withdrawn one at a time, in such a way that the first ball is replaced before the next one is drawn. Find the probability
(a) of selecting 3 white balls.
(b) of selecting at most 2 white balls.
(c) of selecting a white ball, two black balls and a white ball in that order.
(d) of selecting two white balls and two black balls.

9. A tennis player finds that he wins 3 out of 7 games he plays. If he plays 7 games straight, find the probability that he will win
(a) exactly 3 games.
(b) at most 3 games.
(c) all 7 games.
(d) no more than 5 games.
(e) After playing 30 games, how many of these would he expect to win?

10. A true–false test consists of 8 questions. A student will sit for the test, but will only be able to guess at each of the answers. Find the probability that the student answers
(a) all 8 questions correctly.
(b) 4 questions correctly.
(c) at most 4 of the questions correctly.
The following week, the same student will sit another true–false test, this time there will be 12 questions on the test, of which he knows the answer to 4.
(d) What are the chances of passing this test (assuming that 50% is a pass.)

11. The births of males and females are assumed to be equally likely, find the probability that in a family of 6 children:
(a) there are exactly 3 girls.
(b) there are no girls.

(c) the girls are in the majority.

(d) How many girls would you expect to see in a family of 6 children?

12. During any one production cycle it is found that 12% of items produced by a manufacturer are defective. A sample of 10 items are selected at random and inspected. Find the probability that:

(a) no defectives will be found.

(b) at least two defectives will be found.

A batch of 1000 such items are now inspected.

(c) How many of these items would you expect to be defective?

13. 10% of washers produced by a machine are considered to be either oversized or undersized. A sample of 8 washers is randomly selected for inspection.

(a) What is the probability that there are 3 defective washers?

(b) What is the probability that there is at least one defective washer?

14. Over a long period of time, an archer finds that she is successful on 90% of her attempts. In the final round of a competition she has 8 attempts at a target.

(a) Find the probability that she is successful on all 8 attempts.

(b) Find the probability that she is successful on at least 6 attempts.

The prize that is awarded is directly proportional to the number of times she is successful, earning 100 fold, in dollars, the number of times she is successful.

(c) What can she expect her winnings to be after one round?

She draws with another competitor, however, as there can be only one winner, a second challenge is put into place:

They must participate in another 3 rounds, with 5 attempts in each round.

(d) Find the probability that she manages 3 perfect rounds.

15. For each of the random variables, (a) $X \sim B(7, 0.2)$

(b) $X \sim B(8, 0.38)$

Find i. the mean,

ii. the mode and

iii. the standard deviation

iv. $P(X \geq 6 | X > 4)$

v. $P(X > 4 | X \leq 6)$

16. In a suburb, it is known that 40% of the population are blue collar workers. A delegation of one hundred volunteers are each asked to sample 10 people in order to determine if they are blue collar workers. The town has been broken up into 100 regions so that there is no possibility of doubling up (i.e., each worker is allocated one region). How many of these volunteers would you expect to report that there were fewer than 4 blue collar workers.

17. Show that if $X \sim B(n, p)$, then

$$P(X = x+1) = \left(\frac{n-x}{x+1}\right)\left(\frac{p}{1-p}\right)P(X = x),\ x = 0, 1, 2, \ldots, n-1$$

18. Show that if $X \sim B(n, p)$, then

(a) $E(X) = np$. (b) $Var(X) = np(1-p)$.

19. Mifumi has ten pots labelled one to ten. Each pot and its content can be considered to be identical in every way. Mifumi plants a seed in each pot, such that each seed has a germinating probability of 0.8.

(a) Find the probability that

i. all the seeds will germinate.

ii. exactly three seeds will germinate.

iii. more than eight seeds germinate.

(b) How many pots must Mifumi use to be 99.99% sure to obtain at least one flower?

20. A fair die is rolled eight times. If the random variable X denotes the number of fives observed, find

(a) $E(X)$ (b) $Var(X)$ (c) $E\left(\frac{1}{8}X\right)$ (d) $Var\left(\frac{1}{8}X\right)$

21. A bag contains 5 balls of which 2 are red. A ball is selected at random. Its colour noted and then replaced in the bag. This process is carried out 50 times. Find

(a) the mean number of red balls selected.

(b) the standard deviation of the number of red balls selected.

22. The random variable X is $B(n, p)$ distributed such that $\mu = 9$ and $\sigma^2 = 3.6$. Find

(a) $E(X^2 + 2X)$

(b) $P(X = 2)$

23. (a) If $X \sim Bin(10, 0.6)$, find i. $E(X)$ ii. $Var(X)$

(b) If $X \sim Bin(15, 0.4)$, find i. $E(X)$ ii. $Var(X)$

24. The random variable X has a binomial distribution with $E(X) = 12$ and $Var(X) = 4.8$. Find $P(X = 12)$.

25. Metallic parts produced by an automated machine have some variation in their size. If the size exceeds a set threshold, the part is labelled as defective. The probability that a part is defective is 0.08. A random sample of 20 parts is taken from the day's production. If X denotes the number of defective parts in the sample, find its mean and variance.

26. Quality control for the manufacturing of bolts is carried out by taking a random sample of 15 bolts from a batch of 10,000. Empirical data shows that 10% of bolts are found to be defective. If three or more defectives are found in the sample, that particular batch is rejected.

(a) Find the probability that a batch is rejected.

(b) The cost to process the batch of 10,000 bolts is \$20.00. Each batch is then sold for \$38.00, or it is sold as scrap for \$5.00 if the batch is rejected.
Find the expected profit per batch.

27. In a shooting competition, a competitor knows (that on average) she will hit the bullseye on three out of every five attempts. If the competitor hits the bullseye she receives \$10.00. However if the competitor misses the bullseye but still hits the target region she only receives \$5.00.

(a) What can the competitor expect in winnings on any one attempt at the target?

(b) How much can the competitor expect to win after 20 attempts?

28. A company manufactures bolts which are packed in batches of 10,000. The manufacturer operates a simple sampling scheme whereby a random sample of 10 is taken from each batch. If the manufacturer finds that there are fewer than 3 faulty bolts the batch is allowed to be shipped out. Otherwise, the whole batch is rejected and re–processed.

(a) If 10% of all bolts produced are known to be defective, find the proportion of batches that will be re–processed.

(b) Show that if $100p\%$ of bolts are known to be defective, then

$$P(\text{Batch is accepted}) = (1-p)^8(1+8p+36p^2),\ 0 \le p \le 1$$

(c) Using a graphics calculator, sketch the graph of P('Batch is accepted') versus p. Describe the behaviour of this curve.

29. Large batches of screws are produced by TWIST'N'TURN Manufacturers Ltd. Each batch consists of N screws and has a proportion p of defectives. It is decided to carry out an inspection of the product; by selecting 4 screws at random and accepting the batch if there is no more than one defective, otherwise the batch is rejected.

(a) Show that $P(\text{Accepting any batch}) = (1-p)^3(1+3p)$.

(b) Sketch a graph showing the relationship between the probability of accepting a batch and p (the proportion of defectives).

30. A quality control process for a particular electrical item is set up as follows:

A random sample of 20 items is selected. If there is no more than one faulty item the whole batch is accepted. If there are more than two faulty items the batch is rejected. If there are exactly two faulty items, a second sample of 20 items is selected from the same batch and is accepted only if this second sample contains no defective items.
Let p be the proportion of defectives in a batch.

(a) Show that the probability, $\Phi(p)$, that a batch is accepted is given by

$$\Phi(p) = (1-p)^{19}[1+19p+190p^2(1-p)^{19}],\ 0 \le p \le 1.$$

(b) Find the probability of accepting this batch if it is known that 5% of all items are defective.

(c) If 200 such batches are produced each day, find an estimate of the number of batches that can be expected to be rejected on any one day.

17.1 THE NORMAL DISTRIBUTION

17.1.1 WHY THE NORMAL DISTRIBUTION?

The examples considered in Chapter 16 dealt with data that was **discrete**. Discrete data is generally counted and can be found exactly. Discrete data is often made up of whole numbers. For example, we may have counted the number of occupants in each of the cars passing a particular point over a period of two hours. In this case the data is made up of whole numbers. If we collect information on the European standard shoe sizes of a group of people, we will also be collecting discrete data even though some of the data will be fractional: shoe size nine and a half.

Alternatively, sometimes we collect data using measurement. For example, we might collect the birth weights of all the babies delivered at a maternity hospital over a year. Because weight is a continuous quantity (all weights are possible, not just whole numbers or certain fractions), the data collected is **continuous**. This remains the case even though we usually round continuous data to certain values. In the case of weight, we might round the data to the nearest tenth of a kilogram. In this case, if a baby's weight is given as 3.7kg it means that the weight has been rounded to this figure and lies in the interval [3.65,3.75). If we are looking at data such as these weights it may seem as if the data is discrete even in cases when it is in fact continuous.

When dealing with continuous data, we use different methods. The most important distinction is that we can never give the number of babies that weigh *exactly* 3.7kg as there are *none* of these. All that we can give is the number of babies born that have weights in the range [3.65,3.75).

One of the ways in which we can handle continuous data is to use the **normal distribution**. This distribution is only a model for real data. This means that its predictions are only approximate. The normal distribution generally works best in a situation in which the data clusters about a particular mean and varies from this as a result of **random factors**. The birth weights of babies cluster about a mean with variations from this mean resulting from a range of chance factors such as genetics, nutrition etc. The variation from the mean is measured by the standard deviation of the data. In examples such as this, the normal distribution is often a fairly good model. The basis of all normal distribution studies is the **standard normal curve**.

CHAPTER 17

17.1.2 THE STANDARD NORMAL CURVE

The standard normal curve models data that has a mean of zero and a standard deviation of one. The equation of the standard normal curve is:

$$f(z) = \frac{1}{\sqrt{2\pi}}e^{\left(-\frac{z^2}{2}\right)}, -\infty < z < \infty$$

The equation of this distribution is complex and does not directly give us any information about the distribution. The shape of the curve, does, however, indicate the general shape of the distribution.

The shape of this curve is often referred to as the 'bell shaped curve'. On the next page we see how this function behaves.

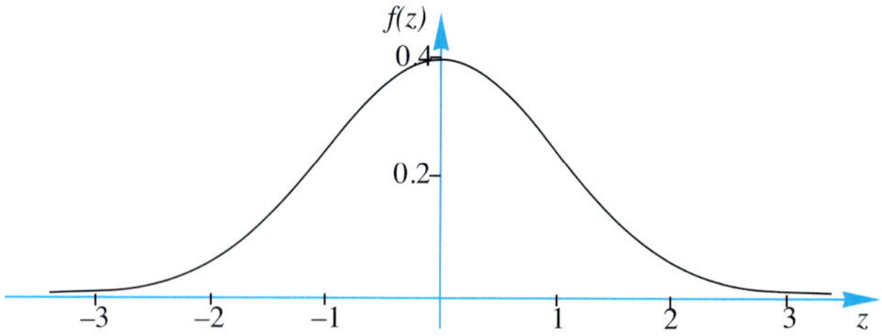

As a result of the fact that the variable z is continuous, it is not the height of the curve but the areas underneath the curve that represent the proportions of the variable that lie between various values. The total area under the curve is 1 (even though the curve extends to infinity in both directions without actually reaching the axis).

For example, the proportion of the standard normal data that lies between 1 and 2 is represented by the area shown.
Areas under curves are usually found using a method covered in Chapter 22. In the case of the normal curve, the complexity of the equation of the graph makes this impossible at least at this level. Instead, we rely on a table of values. The full table can be found at the end of this chapter.

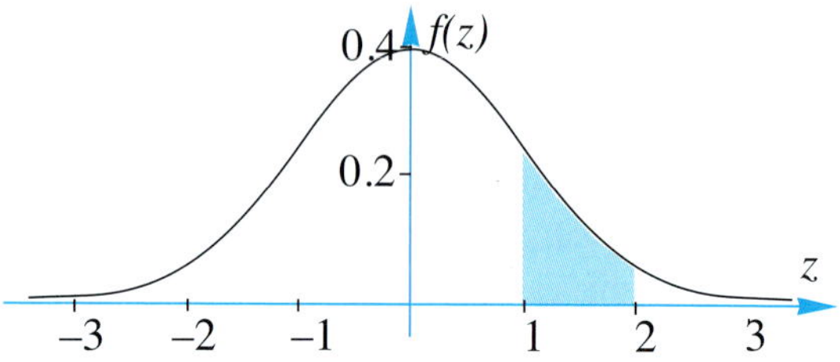

17.1.3 USING THE STANDARD NORMAL TABLE

The table tells us the proportion of values of the standard normal variable that are less than any given value. It is best to view this graphically.

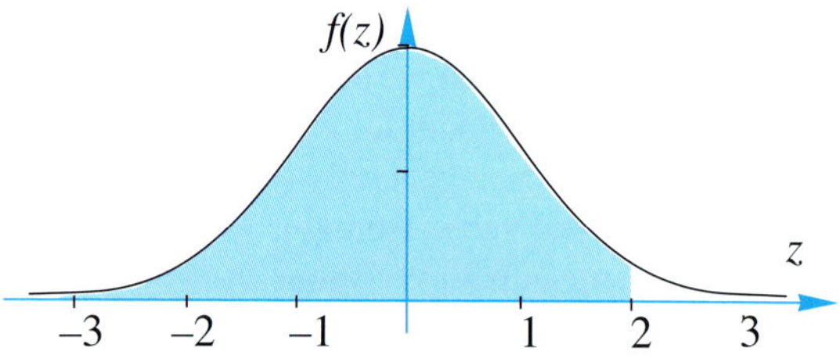

The diagram shows the area that represents the proportion of values for which $z < 2$.
This proportion can also be interpreted as the probability that a randomly chosen value of z will have a value of less than 2 or $p(Z < 2)$.
This value can be found from the row beginning with 2.0 in the table:

	0	**0.01**	**0.02**	**0.03**	**0.04**	**0.05**	**0.06**	**0.07**	**0.08**	**0.09**
0.0	0.5000	0.5040	0.5080	0.5120	0.5160	0.5199	0.5239	0.5279	0.5319	0.5359
1.8	0.9641	0.9649	0.9656	0.9664	0.9671	0.9678	0.9686	0.9693	0.9699	0.9706
1.9	0.9713	0.9719	0.9726	0.9732	0.9738	0.9744	0.9750	0.9756	0.9761	0.9767
2.0	0.9772	0.9778	0.9783	0.9788	0.9793	0.9798	0.9803	0.9808	0.9812	0.9817
2.1	0.9821	0.9826	0.9830	0.9834	0.9838	0.9842	0.9846	0.9850	0.9854	0.9857
2.2	0.9861	0.9864	0.9868	0.9871	0.9875	0.9878	0.9881	0.9884	0.9887	0.9890

The value in the 2.0 row and 0 column represents $p(Z < 2.0)$ and is 0.9772. This value can be interpreted as:
The proportion of values of z less than 2 is 0.9772
The percentage of values of z less than 2 is 97.72%
The probability that a randomly chosen value of the standard normal variable is less than 2 is 0.9772.

The following set of examples will illustrate how the table can be used to solve other standard normal distribution problems.

EXAMPLE 17.1

For the standard normal variable Z, find:

(a) $p(Z < 1)$ (b) $p(Z < 0.96)$ (c) $p(Z < 0.03)$

Solution

(a) All these examples can be solved by direct use of the tables

$p(Z < 1) = 0.8413$

	0	0.01	0.02	0.03	0.04	0.05	0.06	0.07	0.08	0.09
0.0	0.5000	0.5040	0.5080	0.5120	0.5160	0.5199	0.5239	0.5279	0.5319	0.5359
0.9	0.8159	0.8186	0.8212	0.8238	0.8264	0.8289	0.8315	0.8340	0.8365	0.8389
1.0	0.8413	0.8438	0.8461	0.8485	0.8508	0.8531	0.8554	0.8577	0.8599	0.8621
1.1	0.8643	0.8665	0.8686	0.8708	0.8729	0.8749	0.8770	0.8790	0.8810	0.8830

(b) $p(Z < 0.96)$ (= 0.8315) can be found by using the row for 0.9 and the column for 0.06. The required value can be found at the row and column intersection.

	0	0.01	0.02	0.03	0.04	0.05	0.06	0.07	0.08	0.09
0.0	0.5000	0.5040	0.5080	0.5120	0.5160	0.5199	0.5239	0.5279	0.5319	0.5359
0.8	0.7881	0.7910	0.7939	0.7967	0.7995	0.8023	0.8051	0.8078	0.8106	0.8133
0.9	0.8159	0.8186	0.8212	0.8238	0.8264	0.8289	0.8315	0.8340	0.8365	0.8389
1.0	0.8413	0.8438	0.8461	0.8485	0.8508	0.8531	0.8554	0.8577	0.8599	0.8621

(iii) $p(Z < 0.03)$ (= 0.5120) is found similarly

	0	0.01	0.02	0.03	0.04	0.05	0.06	0.07	0.08	0.09
0.0	0.5000	0.5040	0.5080	0.5120	0.5160	0.5199	0.5239	0.5279	0.5319	0.5359
0.1	0.5398	0.5438	0.5478	0.5517	0.5557	0.5596	0.5636	0.5675	0.5714	0.5753

Other problems are best solved using a combination of graphs and the table. Problems arise when we have 'greater than' problems or negative values of z.

EXAMPLE 17.2

For the standard normal variable Z, find:

(a) $p(Z > 1.7)$ (b) $p(Z < -0.88)$ (c) $p(Z > -1.53)$

Solution

(a) $p(Z > 1.7)$. Graphically this is the area shaded in the diagram. Since we can only look up 'less than' probabilities using the table, we must use the fact that the total area under the curve is 1. It follows that:

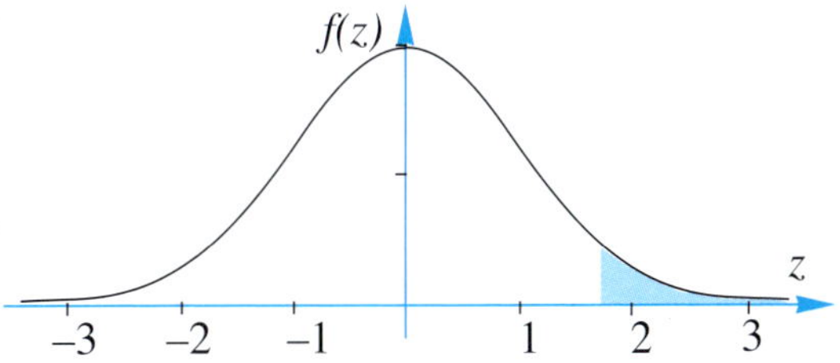

$$\begin{aligned} p(Z > 1.7) &= 1 - p(Z < 1.7) \\ &= 1 - 0.9554 \\ &= 0.0446 \end{aligned}$$

(ii) $p(Z < -0.88)$. The table does not give any negative z values. This question can be solved by looking at the diagram on the right. By the symmetry of the curve, the required area (shaded) is the same as the area shown with vertical stripes. It follows that:

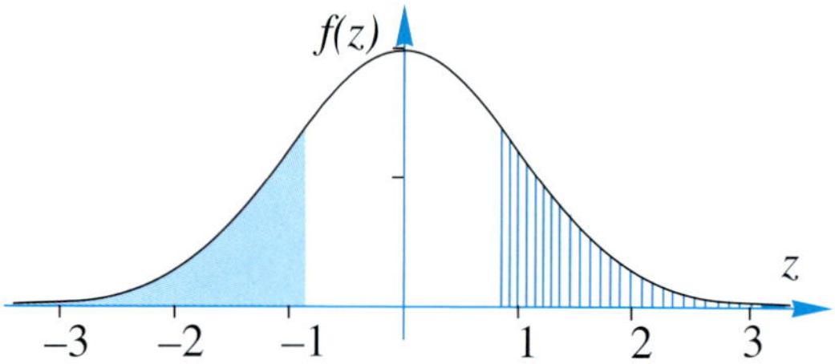

$$\begin{aligned} p(Z < -0.88) &= p(Z > 0.88) \\ &= 1 - p(Z < 0.88) \\ &= 1 - 0.8106 \\ &= 0.1894 \end{aligned}$$

(c) $p(Z > -1.53)$. Again, we cannot look up a negative z value, but we can use the symmetry of the graph.

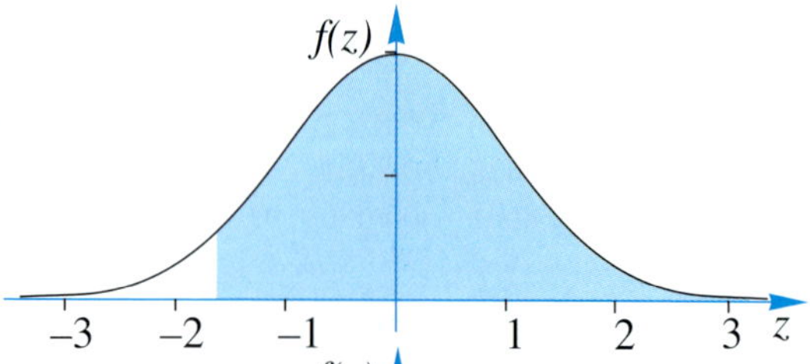

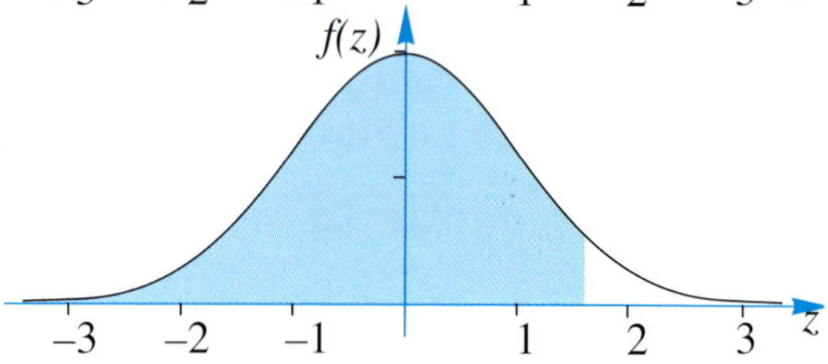

The shaded area in this diagram is the same as the required area in the diagram directly above, so:

$$\begin{aligned} p(Z > -1.53) &= p(Z < 1.53) \\ &= 0.9370 \end{aligned}$$

The final set of examples looks at some 'between values' type of problems.

EXAMPLE 17.3

For the standard normal variable Z, find:

(a) $p(1.7 < Z < 2.5)$ (b) $p(-1.12 < Z < 0.67)$ (c) $p(-2.45 < Z < -0.08)$

Solution

(a) $p(1.7 < Z < 2.5)$. This is found by using the tables to find $p(Z < 2.5)$ and $p(Z < 1.7)$.
The required answer is then the difference between these two values.

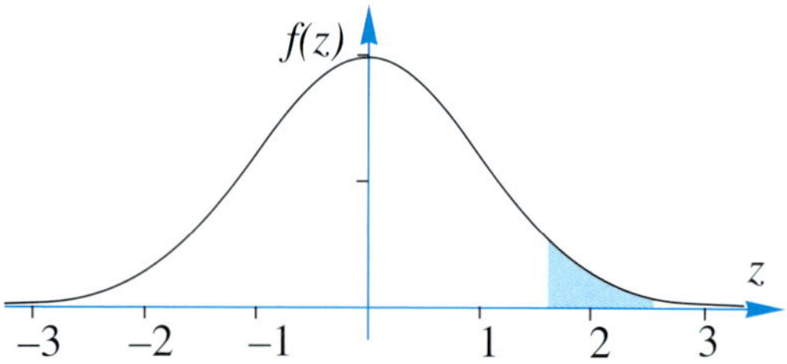

$$\begin{aligned} p(1.7 < Z < 2.5) &= p(Z < 2.5) - p(Z < 1.7) \\ &= 0.9938 - 0.9554 \\ &= 0.0384 \end{aligned}$$

(b) $p(-1.12 < Z < 0.67)$. The area is shown shaded. The same principal is used to solve this problem as the previous example.

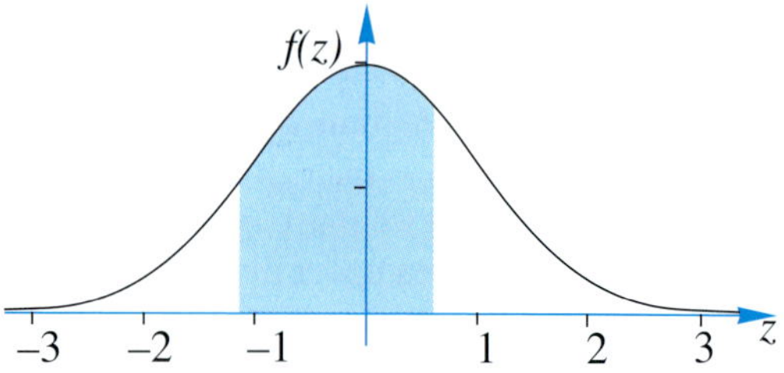

The additional difficulty is the negative z value.

i.e., $p(-1.12 < Z)$ is calculated as $p(Z > 1.12) = 1 - p(Z < 1.12)$.

$$\begin{aligned}\therefore p(-1.12 < Z < 0.67) &= p(Z < 0.67) - p(Z < -1.12)\\ &= p(Z < 0.67) - p(Z > 1.12)\\ &= p(Z < 0.67) - (1 - p(Z < 1.12))\\ &= 0.7486 - (1 - 0.8686)\\ &= 0.6172\end{aligned}$$

(c) $p(-2.45 < Z < -0.08)$ (shaded) is calculated using the symmetry of the curve as $p(0.08 < Z < 2.45)$ (vertical stripes). so,

$$\begin{aligned}p(-2.45 < Z < -0.08) &= p(0.08 < Z < 2.45)\\ &= 0.9928 - 0.5319\\ &= 0.4609\end{aligned}$$

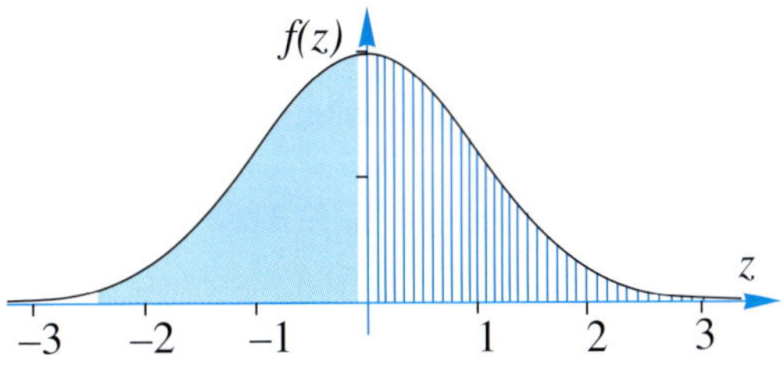

EXERCISES 17.1

1. For the standard normal variable Z, find:

(i)	$p(Z < 0.5)$	(ii)	$p(Z < 1.84)$	(iii)	$p(Z < 1.62)$	(iv)	$p(-2.7 < Z)$
(v)	$p(-1.97 < Z)$	(vi)	$p(Z < -2.55)$	(vii)	$p(1.9 < Z)$	(viii)	$p(Z < -1.56)$
(ix)	$p(2.44 < Z)$	(x)	$p(-0.95 < Z)$	(xi)	$p(Z < 0.37)$	(xii)	$p(1.39 < Z)$

2. For the standard normal variable z, find:

(i)	$p(1.75 < Z < 2.65)$	(ii)	$p(0.3 < Z < 2.5)$	(iii)	$p(1.35 < Z < 1.94)$
(iv)	$p(-1.92 < Z < -1.38)$	(v)	$p(2.23 < Z < 2.92)$	(vi)	$p(-1.51 < Z < -0.37)$
(vii)	$p(-2.17 < Z < 0.76)$	(viii)	$p(1.67 < Z < 2.22)$	(ix)	$p(-0.89 < Z < 0.8)$
(x)	$p(-2.64 < Z < -1.04)$	(xi)	$p(-1.43 < Z < 2.74)$	(xii)	$p(-1.59 < Z < -0.46)$
(xiii)	$p(-2.12 < Z < 0.58)$	(xiv)	$p(-2.61 < Z < 1.39)$	(xv)	$p(-1.86 < Z < 0.13)$
(xvi)	$p(-2.56 < Z < 0.92)$	(xvii)	$p(-1.75 < Z < 2.03)$	(xviii)	$p(-0.9 < Z < 1.34)$

17.2 FORMALISING THE DEFINITION OF THE NORMAL DISTRIBUTION

So far we have given a discussion of the notion of the standard normal distribution. We now provide a more mathematical approach in defining this important distribution.

17.2.1 THE NORMAL DISTRIBUTION

If the random variable X is normally distributed, then it has a **probability density function** given by

$$f(x) = \frac{1}{\sigma\sqrt{2\pi}}e^{-\frac{1}{2}\left(\frac{x-\mu}{\sigma}\right)^2}, -\infty < x < \infty$$

Where $\mu = E(X)$, the mean of X, and $\sigma = Sd(X)$, the standard deviation of X, are known as the **parameters of the distribution**.

If the random variable X is normally distributed with mean μ and variance σ^2, we write

$$X \sim N(\mu, \sigma^2)$$

The shape of this probability density function, known as the **normal curve**, has a characteristic '**bell shape**' that is symmetrical about its mean, μ.

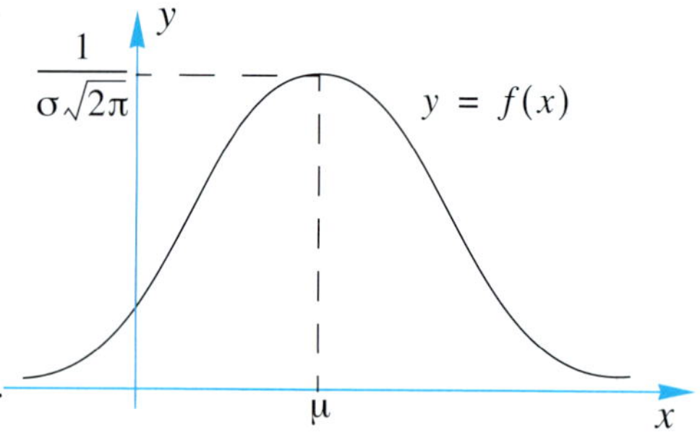

17.2.2 PROPERTIES OF THIS CURVE

1. The curve is **symmetrical** about the line $x = \mu$. In fact, the mode occurs at $x = \mu$.

2. The normal curve approaches the horizontal axis asymptotically as $x \to \pm\infty$.

3. The area under this curve is equal to one.

That is, $\int_{-\infty}^{\infty} f(x)dx = \int_{-\infty}^{\infty} \frac{1}{\sigma\sqrt{2\pi}} e^{-\frac{1}{2}\left(\frac{x-\mu}{\sigma}\right)^2} dx = 1$.

4. Approximately 95% of the observations lie in the region $\mu - 2\sigma \le x \le \mu + 2\sigma$.

17.2.3 FINDING PROBABILITIES USING THE NORMAL DISTRIBUTION

To find the probability, $p(a \le X \le b)$, where $X \sim N(\mu, \sigma^2)$ we would need to evaluate the integral:

$$p(a \le X \le b) = \int_a^b f(x)dx = \int_a^b \frac{1}{\sigma\sqrt{2\pi}} e^{-\frac{1}{2}\left(\frac{x-\mu}{\sigma}\right)^2} dx.$$

Evaluating such an integral is beyond the scope of this course. However, to help us calculate such probabilities we make use of a **transformation** that will convert any normal distribution with mean μ and variance σ^2 to a normal distribution with a **mean of 0 and a variance of 1**. This new curve is known as a **standard normal distribution**.

17.2.4 THE STANDARD NORMAL DISTRIBUTION

Making use of the transformation $Z = \frac{X - \mu}{\sigma}$, we obtain the standard normal distribution.

That is, we transform the distribution of $X \sim N(\mu, \sigma^2)$ to that of $Z \sim N(1, 0)$.
This can be shown as follows:

1. $E(Z) = E\left(\frac{X-\mu}{\sigma}\right) = \frac{1}{\sigma}E(X-\mu) = \frac{1}{\sigma}(E(X)-\mu) = \frac{1}{\sigma}(\mu-\mu) = 0$

2. $Var(Z) = Var\left(\frac{X-\mu}{\sigma}\right) = \frac{1}{\sigma^2}Var(X-\mu) = \frac{1}{\sigma^2}Var(X) = \frac{1}{\sigma^2} \times \sigma^2 = 1$

That is,

$$X \sim N(\mu, \sigma^2) \Rightarrow Z = \frac{X-\mu}{\sigma} \sim N(1, 0)$$

This means that the probability density function of the standard normal distribution, Z,

(with mean $\mu = 0$ and variance $\sigma^2 = 1$) is given by $f(z) = \frac{1}{\sqrt{2\pi}}e^{-\frac{1}{2}z^2}$

So that the probability is now given by $p(a \le Z \le b) = \int_a^b \frac{1}{\sqrt{2\pi}}e^{-\frac{1}{2}z^2}\,dz$

Although we still have the difficulty of evaluating this definite integral, the reason for doing this is that we have at our disposal tables that already have the probability (area) under the standard normal curve, as we have already seen in §17.1. It is from such tables that we can then calculate the required probabilities. These calculations can also be carried out using specialised calculators.

17.2.5 FINDING PROBABILITIES

Tables usually indicate the region (i.e., the area) that is being evaluated by displaying a graph of the standard normal curve together with the shaded region.

What follows are results based on tables that provide probabilities for which the shaded region is to the left of z. That is, the probability that z is less than (or equal to) a is given by

$p(Z < a)$ = shaded region to the left of $z = a, a > 0$.

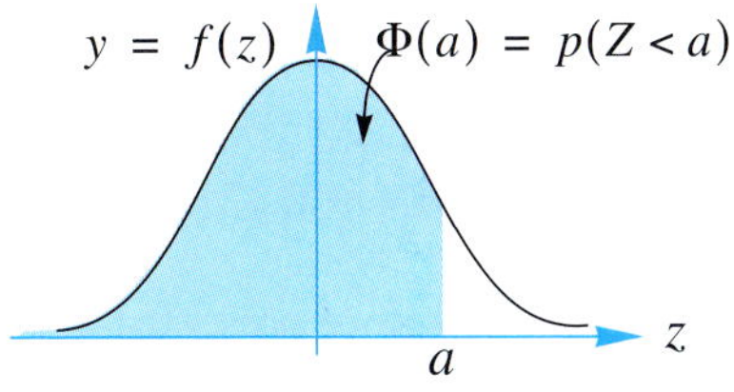

We usually denote this area by $\Phi(a)$.

That is, $\Phi(a) = p(Z \le a) = \int_{-\infty}^{a} \frac{1}{\sqrt{2\pi}}e^{-\frac{1}{2}z^2}\,dz$

The notation $\Phi(a)$ is very useful as it allows us to deal with probabilities in the same way as we would deal with functions. In §17.1 the table of the standard normal probabilities was introduced so that we could familiarise ourselves with how probabilities could be evaluated immediately or by making use of the symmetry properties of the normal curve. We summarise the symmetry properties that were seen to be useful in evaluating probabilities:

If $a \ge 0, b \ge 0$ and Z has a **standard normal distribution**, we have the standard results:

1. $p(Z < a) = \Phi(a)$

In this case we look up the tables
e.g $p(Z < 2.10) = \Phi(2.01) = 0.9778$

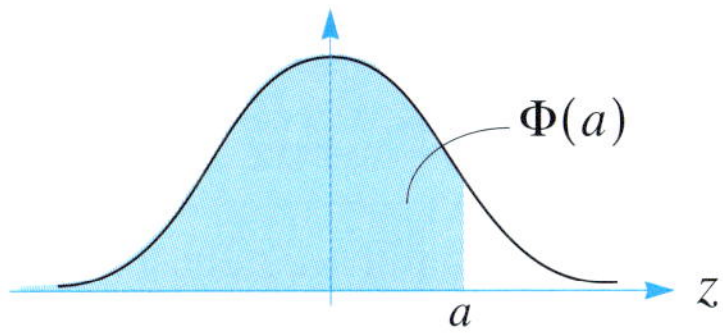

2. $p(Z > a) = 1 - p(Z < a) = 1 - \Phi(a)$

e.g $p(Z > 1.54) = 1 - p(Z < 1.54)$
$= 1 - \Phi(1.54)$
$= 1 - 0.9382$
$= 0.0618$

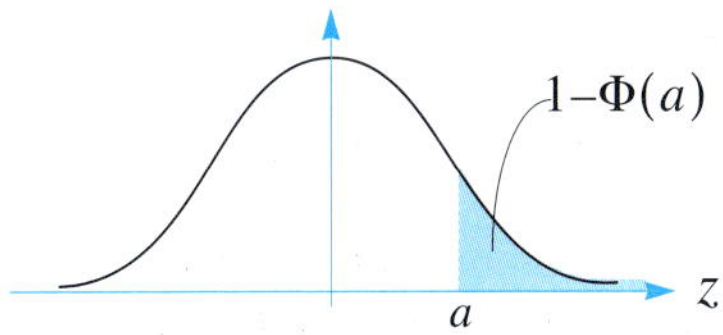

3. $p(Z < -a) = p(Z > a) = 1 - \Phi(a)$

e.g $p(Z < -1.26) = p(Z > 1.26)$
$= 1 - \Phi(1.26)$
$= 1 - 0.8962$
$= 0.1038$

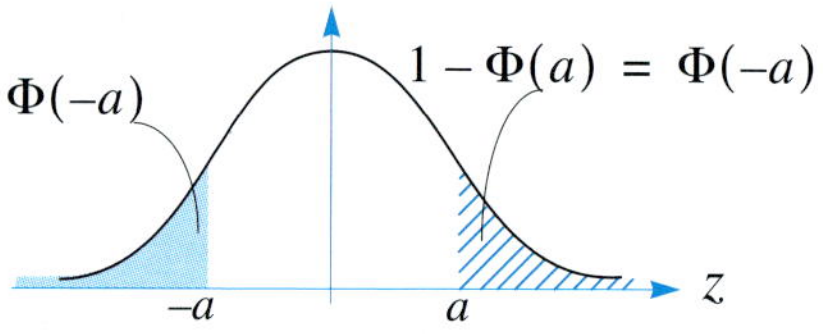

4. $p(Z > -a) = p(Z < a) = \Phi(a)$

e.g $p(Z > -0.1) = p(Z < 0.1) = 0.5398$

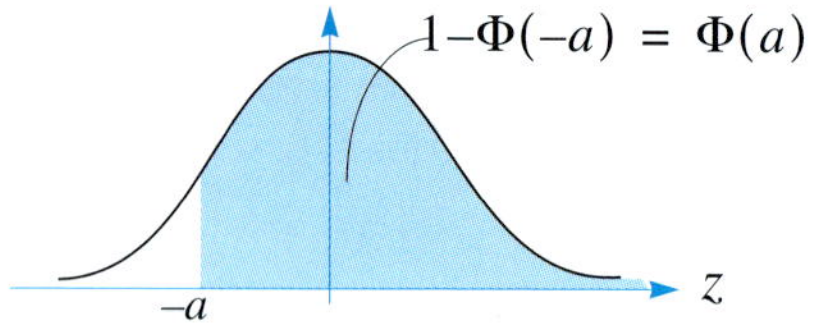

5. $p(a < Z < b) = p(Z < b) - p(Z < a)$
$= \Phi(b) - \Phi(a)$

e.g $p(0.40 < Z < 1.2) = p(Z < 1.2) - p(Z < 0.4)$
$= 0.8849 - 0.6554$
$= 0.2295$

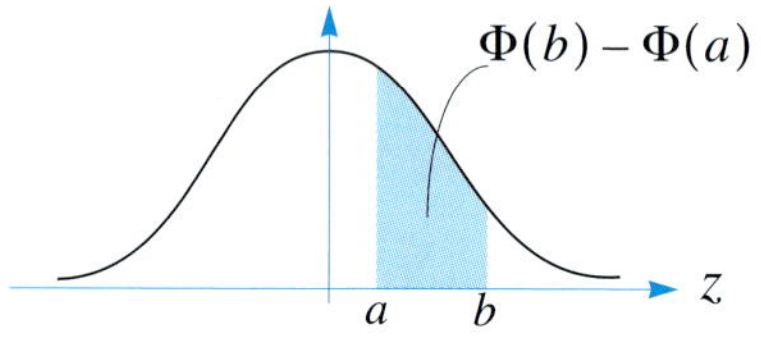

All of this is very good, if we are dealing with the standard normal distribution. However, most of the time (if not practically, all of the time) the data collected will be such that we do not have a mean of zero and a variance of 1. That is, the information gathered from a population (concerning some attribute) will be presented in such a way that the mean and the variance will not comply with those of the standard normal distribution. And so, it will be necessary to first carry out the transformation which we have already discussed, and then work out the required probabilities. That is, we will need to first standardise the statistics obtained from our data, using the *Z*–transformation, and then use the standard normal distribution table.

17.2.6 STANDARDISING ANY NORMAL DISTRIBUTION

Very few practical applications will have data whose mean is 0 and whose standard deviation is 1. The standard normal curve is, therefore, not directly usable in most cases. We get over this difficulty by relating every problem to the standard normal curve.

As we have already seen, a general variable, X, is related to the standard normal variable, Z, using the relation: $Z = \frac{X - \mu}{\sigma}$ where μ = the mean of the data and σ is the standard deviation.

We use an example to illustrate this.

EXAMPLE 17.4

A production line produces bags of sugar with a mean weight of 1.01kg and a standard deviation of 0.02kg.

(a) Find the proportion of the bags that weigh less than 1.03kg.
(b) Find the proportion of the bags that weigh more than 1.02kg.
(c) Find the percentage of the bags that weigh between 1.00kg and 1.05kg.

Solution

(a) The first step is to relate the x value of 1.03 to the z value using the values of the mean and standard deviation. So, we have that $z = \frac{x-\mu}{\sigma} = \frac{1.03-1.01}{0.02} = 1$

Graphically, this means that we have related the normal distribution that models the weights of the bags of sugar to the standard normal distribution.

Distribution of weights of sugar bags   Standard normal distribution

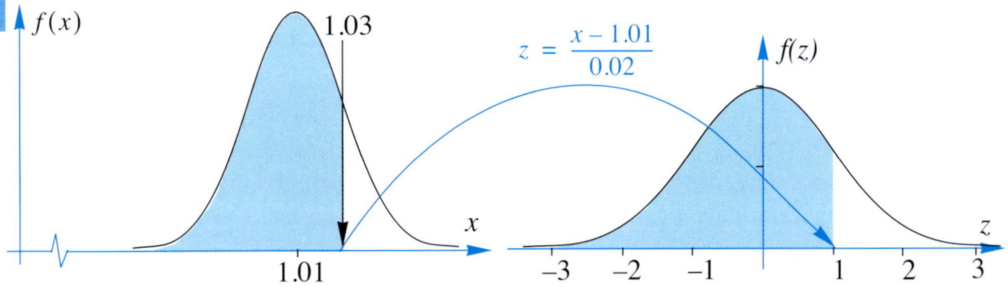

The problem that we are trying to solve: $p(X < 1.03)$ has the same solution as the standard problem $p(Z < 1)$. This can be solved directly from the table to get 0.8413.

(b) Again, transforming this into a standard problem with $x = 1.02$ gives:

$$z = \frac{x-\mu}{\sigma} = \frac{1.02-1.01}{0.02} = 0.5$$

Graphically, this is:

Distribution of weights of sugar bags   Standard normal distribution

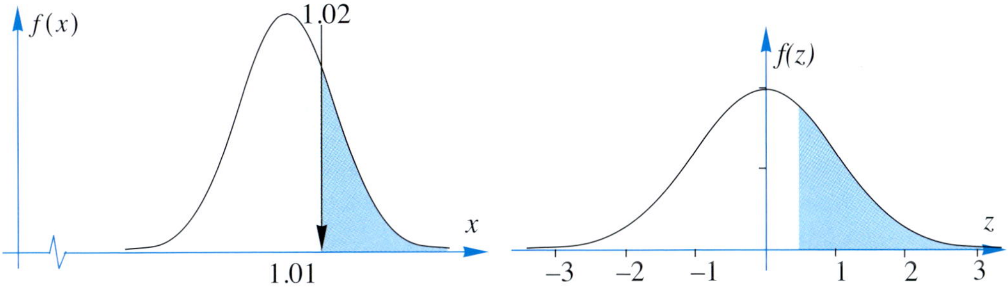

$$\begin{aligned} p(X > 1.02) = p(Z > 0.5) &= 1 - p(Z < 0.5) \\ &= 1 - 0.6915 \\ &= 0.3085 \end{aligned}$$

(c) Again, transforming both the x values to z values, we get:

$$z_1 = \frac{x_1-\mu}{\sigma} = \frac{1-1.01}{0.02} = -0.5 \text{ and } z_2 = \frac{x_2-\mu}{\sigma} = \frac{1.05-1.01}{0.02} = 2$$

The graphical interpretation of this is:

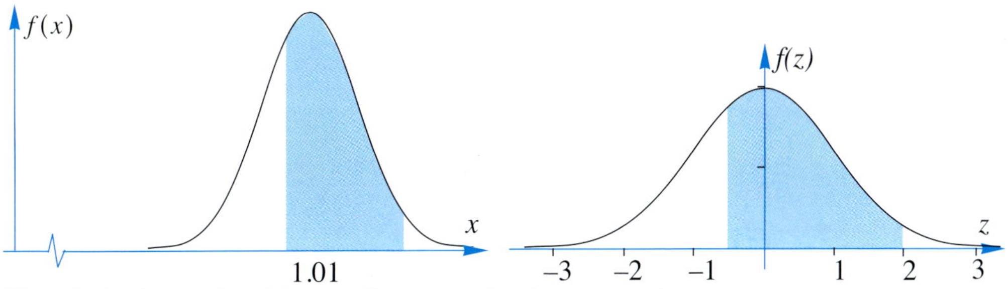

The solution is now found in a similar way to the above examples:

$$
\begin{aligned}
p(1 < X < 1.05) &= p(-0.5 < Z < 2) \\
&= p(Z < 2) - p(Z < -0.5) \\
&= p(Z < 2) - p(Z > 0.5) \\
&= p(Z < 2) - (1 - p(Z < 0.5)) \\
&= 0.9772 - (1 - 0.6915) \\
&= 0.6687
\end{aligned}
$$

SUMMARY

When evaluating probabilities for a random variable X which is **normally distributed** (as opposed to X having a **standard normal distribution**), the following steps should be carried out:

Step 1: Find the value of z which corresponds to the value of x
That is, transform the given random variable X, which is $X \sim N(\mu, \sigma^2)$ to that of $Z \sim N(1, 0)$, using the transformation $Z = \dfrac{X - \mu}{\sigma}$.

Step 2: Sketch a diagram of the standard normal curve with the required region shaded.

Step 3: Use the standard normal distribution tables to evaluate the required region.
NB: This last step often requires the use of the symmetrical properties of the curve to be able to evaluate the required region.

So far we have been making use of the probability tables for the standard normal distribution. However, a graphics calculator is extremely useful in cutting down the workload when determining these probabilities. It is highly recommended that you become familiar with these functions on your calculator. We illustrate this in this next example.

EXAMPLE 17.5

X is a normal random variable with mean $\mu = 80$ and variance $\sigma^2 = 16$, find (a) $p(X \leq 78)$ (b) $p(76 \leq X \leq 84)$ (c) $p(X \geq 86)$

Solution

(a)
$$
\begin{aligned}
p(X \leq 78) = p\left(Z \leq \frac{78 - \mu}{\sigma}\right) &= p\left(Z \leq \frac{78 - 80}{4}\right) \\
&= p(Z \leq -0.5) \\
&= 1 - p(Z \leq 0.5) \\
&= 1 - 0.6915 \\
&= 0.3085
\end{aligned}
$$

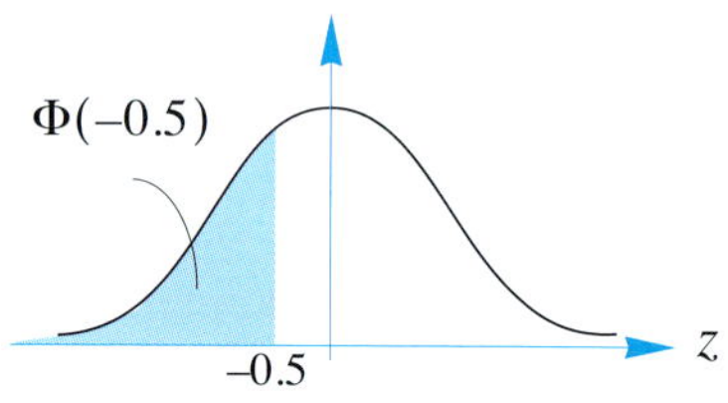

(b)
$$
\begin{aligned}
p(76 \leq X \leq 84) &= p\left(\frac{76 - 80}{4} \leq Z \leq \frac{84 - 80}{4}\right) \\
&= p(-1 \leq Z \leq 1) \\
&= 0.6826
\end{aligned}
$$

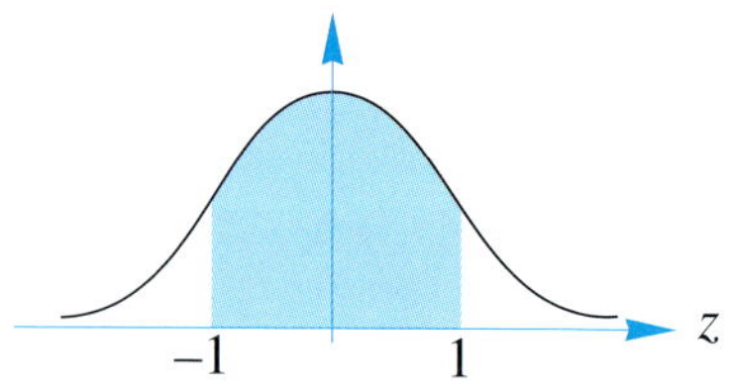

Notice that:
$$
\begin{aligned}
p(-1 \leq Z \leq 1) &= \Phi(1) - \Phi(-1) \\
&= \Phi(1) - (1 - \Phi(1)) \\
&= 2\Phi(1) - 1
\end{aligned}
$$

(c) $p(X \geq 86) = p\left(Z \geq \frac{86-80}{4}\right) = p(Z \geq 1.5)$

$= 1 - p(Z \leq 1.5)$

$= 1 - 0.9332$

$= 0.0668$

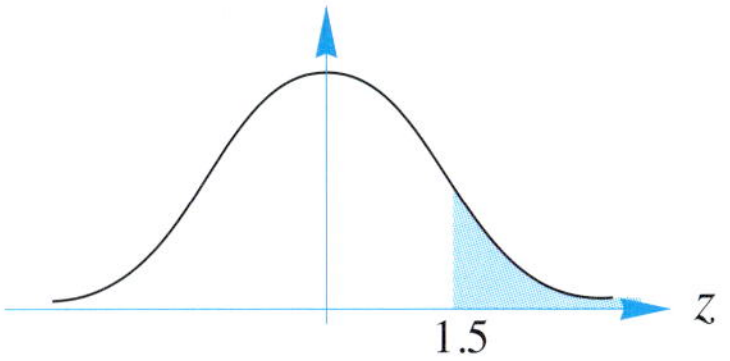

We now make use of the TI–83 to find these probabilities using the **DISTR** function:

(a) $p(Z \leq -0.5)$:

First, set up the window using an appropriate scale:

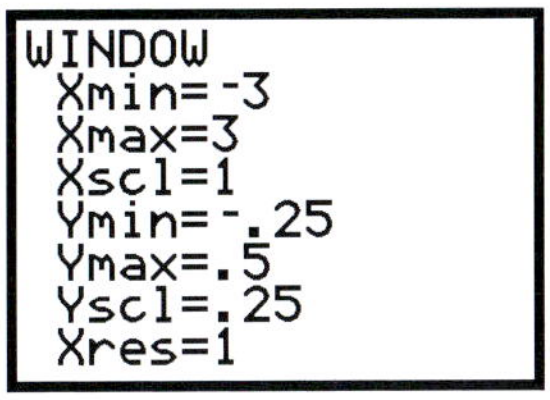

Using a domain of [–3,3] will be sufficient in most cases.

Next, follow through each of the displays shown:

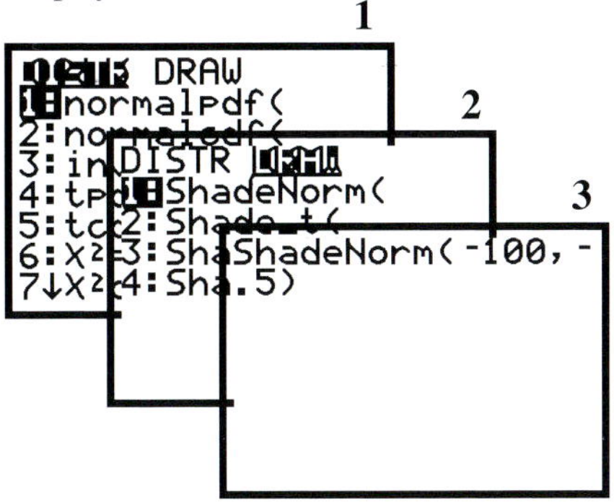

After using the **ShadeNorm(** function, use the **ENTER** function:

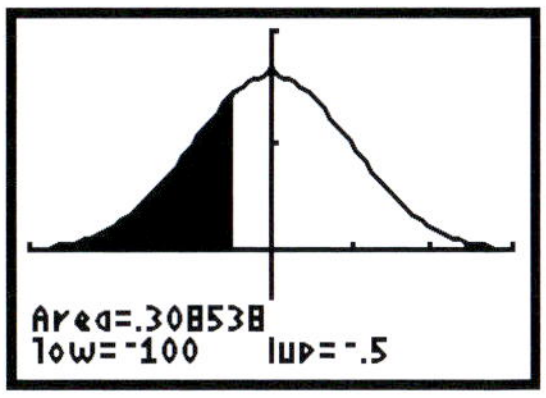

Notice: we have used a lower bound of –100, so that in fact, we are finding $p(-100 \leq Z \leq 0.5)$.

Having such a 'large' lower bound (of –100 say), provides an accurate (at least to 4 d.p) value for $p(Z \leq 0.5)$. That is, $p(Z \leq 0.5) = p(-100 \leq Z \leq -0.5) = 0.3085$.

It should be noted that although we are using values obtained after carrying out the standardization process, the TI–83 also enables us to find the required probability directly. We show this next:

This time the window has been set to reflect the information based on the data:

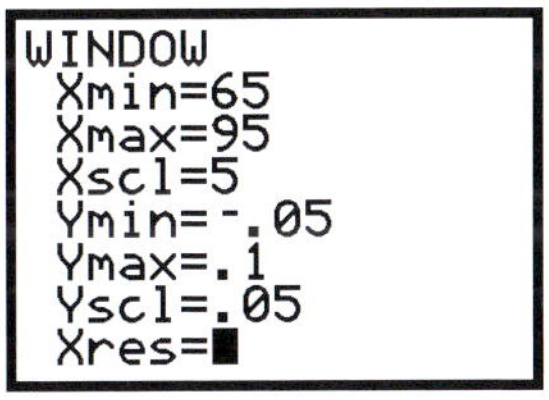

Again, we use a 'large' lower bound, the upper limit and then the parameters:

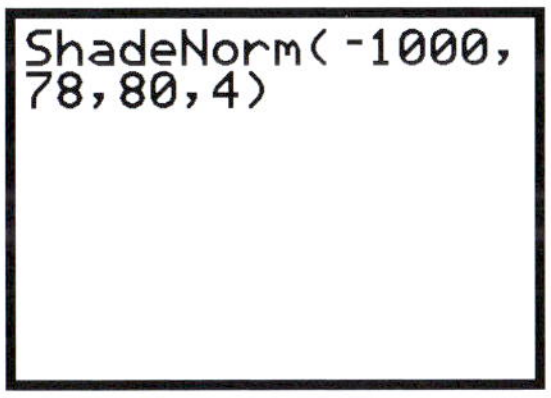

The graph shows the actual curve defined by its parameters:

$$f(x) = \frac{1}{4\sqrt{2\pi}}e^{-\frac{1}{2}\left(\frac{x-80}{4}\right)^2}$$

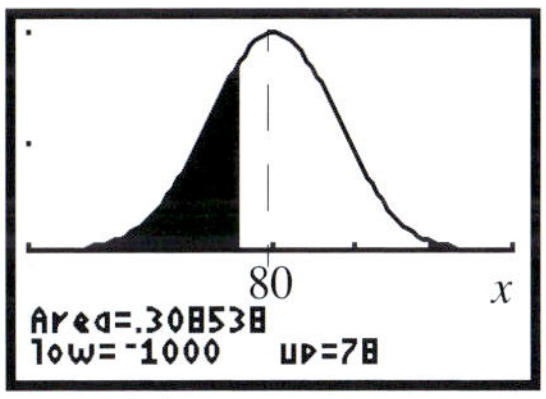

(b) Similarly, we have:

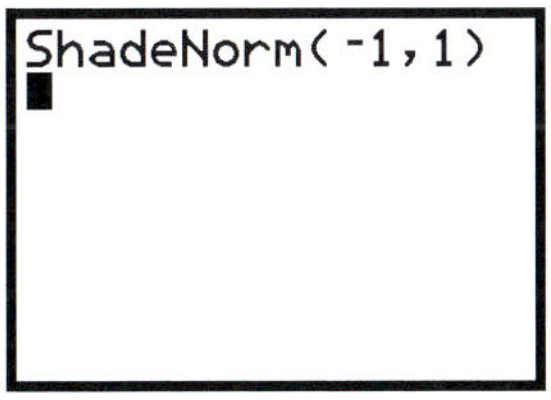

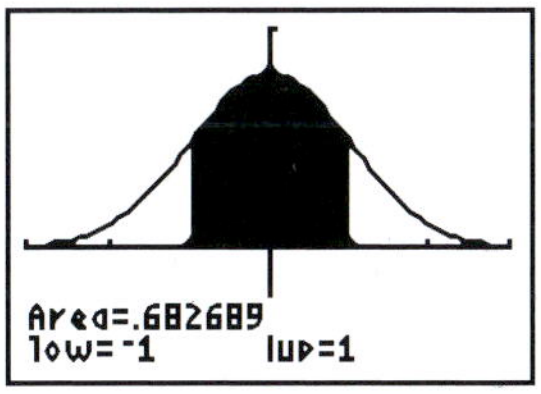

(c) This time we use a large upper bound:

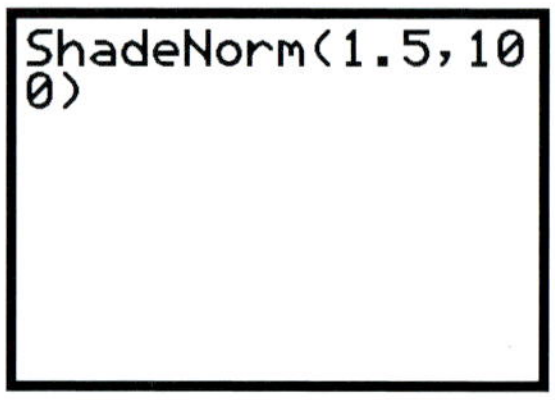

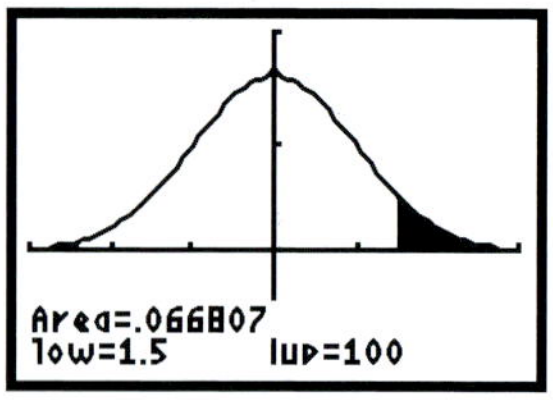

Note that it is not necessary to use the **ShadeNorm(** option on the TI–83. If you are not interested in seeing a sketch of the graph and its shaded region, but instead only want the probability, then make use of the **normalcdf(** option in the **DIST** menu.

We illustrate this for the previous example using **normalcdf(**lower value, upper value, μ , σ):

(a)

(b)

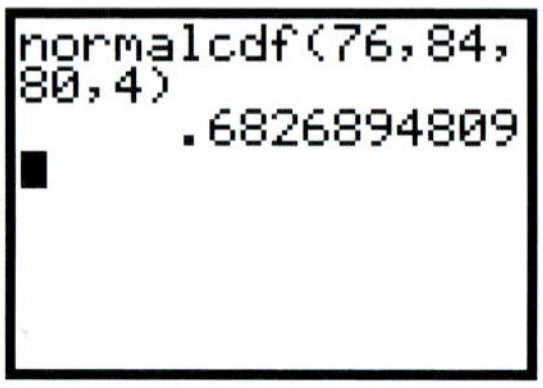

(c)

17.2.7 INVERSE PROBLEMS

There are occasions when we are told the proportion of the data that we are to consider and asked questions about the data conditions that are appropriate to these proportions.

EXAMPLE 17.6

Find the values of a in each of these statements that refer to the standard normal variable, z.

(a) $p(Z < a) = 0.5478$ (b) $p(Z > a) = 0.6$ (c) $p(Z < a) = 0.05$

Solution

(a) $p(Z < a) = 0.5478$. In this case, we are given the proportion and asked for the value of z which makes the condition true. Because we know the proportion, we must look for the figure 0.5478 in the body of the table.

	0	0.01	0.02	0.03	0.04	0.05	0.06	0.07	0.08	0.09
0.0	0.5000	0.5040	0.5080	0.5120	0.5160	0.5199	0.5239	0.5279	0.5319	0.5359
0.1	0.5398	0.5438	0.5478	0.5517	0.5557	0.5596	0.5636	0.5675	0.5714	0.5753
0.2	0.5793	0.5832	0.5871	0.5910	0.5948	0.5987	0.6026	0.6064	0.6103	0.6141

Once the figure has been found in the table, it is necessary to infer the value of z that fits the condition. In this case the value is in the row for 0.1 and the column for 0.02 and we can infer that $a = 0.12$. You should check that $p(Z < 0.12) = 0.5478$.

(b) $p(Z > a) = 0.6$. This is a 'greater than' problem and must be converted into the 'less than' problem $p(Z < a) = 0.4$

From the diagram, it is evident that a is negative. This gives us a problem as negative values are not present on the table.

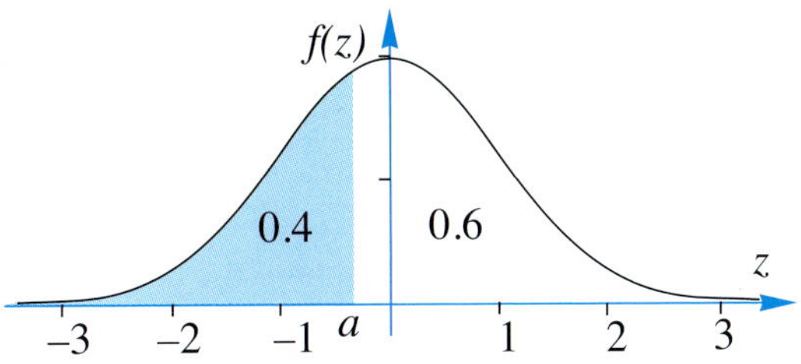

In this case, consider the associated problem $p(Z < b) = 0.6$.

By symmetry, if we can find the appropriate value of b, a will follow because $a = -b$.

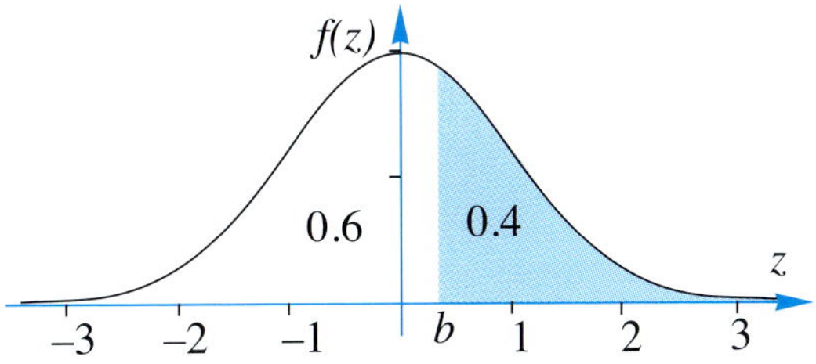

There is a second problem as 0.6 is not present (exactly) in the table. In this case, we find the entries that are as close to 0.6 as possible.

	0	**0.01**	**0.02**	**0.03**	**0.04**	**0.05**	**0.06**	**0.07**	**0.08**	**0.09**
0.0	0.5000	0.5040	0.5080	0.5120	0.5160	0.5199	0.5239	0.5279	0.5319	0.5359
0.1	0.5398	0.5438	0.5478	0.5517	0.5557	0.5596	0.5636	0.5675	0.5714	0.5753
0.2	0.5793	0.5832	0.5871	0.5910	0.5948	0.5987	0.6026	0.6064	0.6103	0.6141
0.3	0.6179	0.6217	0.6255	0.6293	0.6331	0.6368	0.6406	0.6443	0.6480	0.6517

From the table, $p(Z < 0.25) = 0.5987$ and $p(Z < 0.26) = 0.6026$ it is clear that the correct value of b is between 0.25 and 0.26 and closer to 0.25 than to 0.26 as 0.5987 is closer to 0.6 than is 0.6026. A reasonable value for b would seem to be about 0.253. There are several ways in which we could do better than this. Some texts provide 'difference values' in the main table and a separate inverse table. At the time of writing, neither of these were provided in IB exams. Also, there is a technique known as linear interpolation that can make the above argument more precise, but this is not strictly necessary in most applications. In the present case, $b \approx 0.253$ so the answer to our problem is that $a \approx -0.253$.

(iii) $p(Z < a) = 0.05$. Again, thinking graphically, there is a better associated problem:

$p(Z < a) = 0.05$ is the same as

$p(Z > b) = 0.05$ or $p(Z < b) = 0.95$

By symmetry $a = -b$.

Looking for the closest value to 0.95 in the table gives $b \approx 1.645$ and $a \approx -1.645$

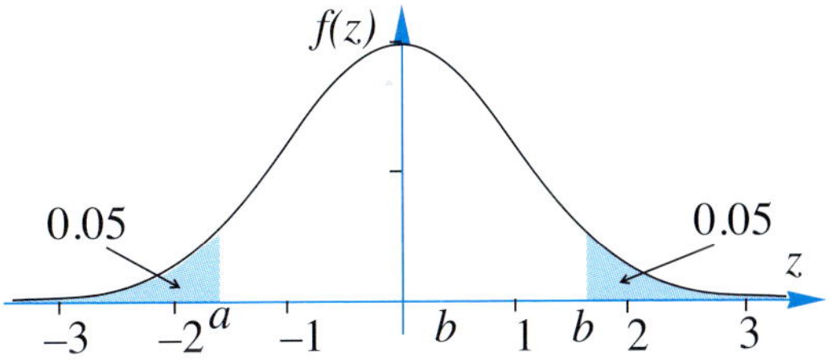

	0	**0.01**	**0.02**	**0.03**	**0.04**	**0.05**	**0.06**	**0.07**	**0.08**	**0.09**
1.6	0.9452	0.9463	0.9474	0.9484	0.9495	0.9505	0.9515	0.9525	0.9535	0.9545

As we saw in Example 17.5 (b) and (c), finding the inverse values can sometimes require an approximation. However, recall that we had defined the function $\Phi(a)$ as a means to represent the probability, $p(Z \le a)$. We will now make use of this function to help us deal with a more general approach to solving inverse problems.

17.2.8 FINDING QUANTILES

To find a quantile (or percentile) means to find the value of a, where $p(X \le a) = p$, where p is the pth percentile.

As we saw in the previous section, the process only requires that we read the normal distribution tables in reverse (or use the **Inverse Cumulative Normal Distribution Table** (if one is provided).

Dealing with this problem using mathematical notation we have,

$$p(X \le a) = p \Leftrightarrow p\left(Z \le \frac{a-\mu}{\sigma}\right) = p \text{ where } 0 \le p \le 1$$

$$\therefore \frac{a-\mu}{\sigma} = \Phi^{-1}(p)$$

For which we then solve for a.

To find the value of $\Phi^{-1}(p)$, we can look up the Normal tables in reverse, however, making use of a calculator reduces the workload significantly and increases the accuracy of the results.

On the TI–83, the inverse values are obtained by accessing the **invNorm(** option in the **DIST** menu. To use the **invNorm(** option we must have the z–values, i.e., the standardised values.

We illustrate this with the following examples.

EXAMPLE 17.7

If $X \sim N(100, 25)$ find the value of k, such that (a) $p(X \le k) = 0.90$
(b) $p(X \le k) = 0.20$.

Solution

(a) $p(X \le k) = 0.90 \Leftrightarrow p\left(Z \le \frac{k-100}{5}\right) = 0.90$

Therefore, $\frac{k-100}{5} = 1.2816$

$\Rightarrow k = 106.408$

(b) $p(X \le k) = 0.20 \Leftrightarrow p\left(Z \le \frac{k-100}{5}\right) = 0.20$

Therefore, $\frac{k-100}{5} = -0.8416$

$k = 100 + 5(-0.8416)$
$= 95.792$

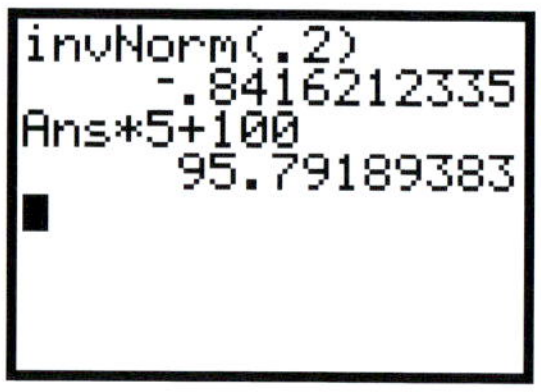

EXAMPLE 17.8 The Board of Examiners have decided that 85% of all candidates sitting Mathematical Methods will obtain a pass grade in the examination. The actual examination marks are found to be normally distributed with a mean of 55 and a variance of 16. What is the lowest score a student can get on the exam to be awarded a pass grade?

Solution

Let the random variable X denote the exam score. We then have that $X \sim N(55, 16)$.
We now need to find the score x, such that $p(X \geq x) = 0.85$ (or $p(X \leq x) = 0.15$).

$$\text{Now, } p(X \leq x) = 0.15 \Leftrightarrow p\left(Z \leq \frac{x-55}{4}\right) = 0.15$$

$$\Leftrightarrow \frac{x-55}{4} = -1.0364$$

$$\Leftrightarrow x = 55 + 4(-1.0363)$$

$$\therefore x = 50.8544$$

```
invNorm(.15)
        -1.03643338
4*Ans+55
        50.85426648
```

Therefore a student needs to score at least 51 marks to pass the exam.

EXAMPLE 17.9 The lifetime of a particular make of television tube is normally distributed with a mean of 8 years, and a standard deviation of σ years. The chances that the tube will not last 5 years is 0.05. What is the value of the standard deviation?

Solution

Let X denote the life-time of the television tubes, so that $X \sim N(8, \sigma^2)$.

Given that $p(X < 5) = 0.05 \Rightarrow p\left(Z < \frac{5-8}{\sigma}\right) = 0.05$.

That is, we have that $p\left(Z < -\frac{3}{\sigma}\right) = 0.05$

$$\Leftrightarrow -\frac{3}{\sigma} = -1.6449$$

$$\Leftrightarrow \sigma = 1.8238$$

```
invNorm(.05)
        -1.644853626
-3/Ans
        1.823870497
```

And so the standard deviation is approximately one year and 10 months.

EXAMPLE 17.10 The weight of a population of men is found to be normally distributed with mean 69.5 kg. 13% of the men weigh at least 72.1 kg, find the standard deviation of their weight.

Solution

Let the random variable X denote the weight of the men, so that $X \sim N(69.5, \sigma^2)$.
We then have that $p(X \geq 72.1) = 0.13$ or $p(X \leq 72.1) = 0.87$.

$$\therefore p\left(Z \leq \frac{72.1-69.5}{\sigma}\right) = 0.87 \Leftrightarrow \frac{72.1-69.5}{\sigma} = 1.1264$$

$$\therefore \sigma = 2.3083$$

```
(72.1-69.5)/invN
orm(.87)
        2.308256818
```

EXERCISES 17.2

1. If Z is a standard normal random variable, find

(a) $p(Z > 2)$ (b) $p(Z < 1.5)$ (c) $p(Z \geq 0.5)$

(d) $p(Z \leq 1.2)$ (e) $p(Z \geq 1.5)$ (f) $p(Z \leq 2)$

2. If Z is a standard normal random variable, find

(a) $p(Z > -2)$ (b) $p(Z < -1.5)$ (c) $p(Z \geq -0.5)$

(d) $p(Z \leq -1.2)$ (e) $p(Z \geq -1.5)$ (f) $p(Z \leq -2)$

3. If Z is a standard normal random variable, find

(a) $p(0 \leq Z \leq 1)$ (b) $p(1 \leq Z \leq 2)$ (c) $p(1.5 \leq Z < 2.1)$

4. If Z is a standard normal random variable, find

(a) $p(-1 \leq Z \leq 1)$ (b) $p(-2 \leq Z \leq -1)$ (c) $p(-1.5 \leq Z < -0.1)$

5. If X is a normal random variable with mean $\mu = 8$ and variance $\sigma^2 = 4$. Find:

(a) $p(X \geq 6)$ (b) $p(5 < X \leq 8)$ (c) $p(X < 9.5)$

6. If X is a normal random variable with mean $\mu = 100$ and variance $\sigma^2 = 25$. Find:

(a) $p(X \geq 106)$ (b) $p(105 < X \leq 108)$ (c) $p(X < 95)$

7. If X is a normal random variable with mean $\mu = 60$ and standard deviation $\sigma = 5$. Find:

(a) $p(X \geq 65)$ (b) $p(55 < X \leq 65)$ (c) $p(50 \leq X < 55)$

8. Scores on a test are normally distributed with a mean of 68 and a standard deviation of 8. Find the probability that a student scored:

(a) at least 75 on the test

(b) at least 75 on the test given that the student scored at least 70.

(c) In a group of 50 students, how many students would you expect to score between 65 and 72 on the test.

9. If X is a normally distributed variable with a mean of 24 and standard deviation of 2, find:

(a) $p(X > 28 | X \geq 26)$ (b) $p(26 < X < 28 | X \geq 27)$

10. The heights of men are normally distributed with a mean of 174 cm and a standard deviation of 6 cm. Find the probability that a man selected at random:

(a) is at least 170 cm tall (b) is no taller than 180 cm

(c) is at least 178 cm given that he is at least 174 cm.

11. If X is a normal random variable with a mean of 8 and a standard deviation of 1, find the value of c, such that

(a) $p(X > c) = 0.90$ (b) $p(X \leq c) = 0.60$

12. If X is a normal random variable with a mean of 50 and a standard deviation of 5, find the value of c, such that

(a) $p(X \leq c) = 0.95$ (b) $p(X \geq c) = 0.95$ (c) $p(-c \leq X \leq c) = 0.95$

13. The Board of Examiners have decided that 80% of all candidates sitting the Mathematical Methods Exam will obtain a pass grade. The actual examination marks are found to be normally distributed with a mean of 45 and a standard deviation of 7. What is the lowest score a student can get on the exam to be awarded a pass grade?

14. The weight of a population of women is found to be normally distributed with mean 62.5 kg. If 15% of the women weigh at least 72 kg, find the standard deviation of their weight.

15. The weights of a sample of a species of small fish are normally distributed with a mean of 37grams and a standard deviation of 3.8grams. Find the percentage of fish that weigh between 34.73 and 38.93grams. Give your answer to the nearest whole number.

16. The weights of the bars of chocolate produced by a machine are normally distributed with a mean of 232grams and a standard deviation of 3.6grams. Find the proportion of the bars that could be expected to weigh less than 233.91grams.

17. For a normal variable, X, $\mu = 196$ and $\sigma = 4.2$. Find:
(i) $p(X < 193.68)$ (ii) $p(X > 196.44)$ (iii) $p(193.68 < X < 196.44)$

18. The circumferences of a sample of drive belts produced by a machine are normally distributed with a mean of 292cm and a standard deviation of 3.3cm. Find the percentage of the belts that have diameters between 291.69cm and 293.67cm.

19. A normally distributed variable, X, has a mean of 52. $p(X < 51.15) = 0.0446$. Find the standard deviation of X.

20. The lengths of the drive rods produced by a small engineering company are normally distributed with a mean of 118cm and a standard deviation of 0.3cm. Rods that have a length of more than 118.37cm are rejected. Find the percentage of the rods that are rejected. Give your answer to the nearest whole number.

21. After their manufacture, the engines produced for a make of lawn mower are filled with oil by a machine that delivers an average of 270ml of oil with a standard deviation of 0.7ml. Assuming that the amounts of oil delivered are normally distributed, find the percentage of the engines that receive more than 271.12ml of oil. Give your answer to the nearest whole number.

22. A sample of detergent boxes have a mean contents of 234grams with a standard deviation of 4.6grams. Find the percentage of the boxes that could be expected to contain between 232.22 and 233.87grams. Give your answer to the nearest whole number.

23. A normally distributed variable, X, has a mean of 259. $p(X < 261.51) = 0.9184$. Find the standard deviation of X.

24. A normally distributed variable, X, has a standard deviation of 3.9. Also, 71.37% of the data are larger than 249.8. Find the mean of X.

25. The times taken by Maisie on her way to work are normally distributed with a mean of 26 minutes and a standard deviation of 2.3 minutes. Find the proportion of the days on which Maisie's trip takes longer than 28 minutes and 22 seconds.

26. In an experiment to determine the value of a physical constant, 100 measurements of the constant were made. The mean of these results was 138 and the standard deviation was 0.1. What is the probability that a final measurement of the constant will lie in the range 138.03 to 139.05?

27. In an experiment to determine the times that production workers take to assemble an electronic testing unit, the times had a mean of 322 minutes and a standard deviation of 2.6 minutes. Find the proportion of units that will take longer than 324 minutes to assemble. Answer to 2 significant figures.

28. A normally distributed variable, X, has a standard deviation of 2.6. $p(X < 322.68) = 0.6032$. Find the mean of X.

29. The errors in an experiment to determine the temperature at which a chemical catalyst is at its most effective, were normally distributed with a mean of 274˚C and a standard deviation of 1.2˚C. If the experiment is repeated what is the probability that the result will be between 275˚C and 276˚C?

30. The weights of ball bearings produced by an engineering process have a mean of 215g with a standard deviation of 0.1g. Any bearing with a weight of 215.32g or more is rejected. The bearings are shipped in crates of 10000. Find the number of bearings that might be expected to be rejected per crate.

31. If $X \sim N(\mu, 12.96)$ and $p(85.30 < X) = 0.6816$ find μ to the nearest integer.

32. At a Junior track and field meet it is found that the times taken for children aged 14 to sprint the 100 metres race are normally distributed with a mean of 15.6 seconds and standard deviation of 0.24 seconds. Find the probability that the time taken for a 14 year old at the meet to sprint the 100 metres is

(a) i. less than 15 seconds
ii. at least 16 seconds
iii. between 15 and 16 seconds.

(b) On one of the qualifying events, 8 children are racing. What is the probability that 6 of them will take between 15 and 16 seconds to sprint the 100 metres?

33. Rods are manufactured to measure 8 cm. Experience shows that these rods are normally distributed with a mean length of 8.02 cm and a standard deviation of 0.04 cm. Each rod costs \$5.00 to make and is sold immediately if its length lies between 8.00 cm and 8.04 cm. If its length exceeds 8.04 cm it costs an extra \$1.50 to reduce its length to 8.02 cm. If its length is less than 8.00 cm it is sold as scrap metal for \$1.00.

(a) What is the average cost per rod? (b) What is the average cost per usable rod?

34. The resistance of heating elements produced are normally distributed with mean 50 ohms and standard deviation 4 ohms.

(a) Find the probability that a randomly selected element has resistance less than 40 ohms.

(b) i. If specifications require that acceptable elements have a resistance between 45 and 55 ohms, find the probability that a randomly selected element satisfies these specifications.

ii. A batch containing 10 such elements is tested. What is the probability that exactly 5 of the elements satisfy the specifications?

(c) The profit on an acceptable element, i.e., one that satisfies the specifications, is \$2.00, while unacceptable elements result in a loss of \$0.50 per element. If \$ P is the profit on a randomly selected element, find the profit made after producing 1000 elements.

35. (a) Find the mean and standard deviation of the normal random variable X, given that $P(X < 50) = 0.05$ and $P(X > 80) = 0.1$.

(b) Electrical components are mass-produced and have a measure of 'durability' that is normally distributed with mean μ and standard deviation 0.5.
The value of μ can be adjusted at the control room. If the measure of durability of an item scores less than 5, it is classified as defective. Revenue from sales of non-defective items is \$ S per item, while revenue from defective items is set at \$ $\frac{1}{10}S$.
Production cost for these components is set at \$ $\frac{1}{10}\mu S$. What is the expected profit per item when μ is set at 6?

36. From one hundred first year students sitting the end-of-year Botanical Studies 101 exam, 46 of them passed while 9 were awarded a high distinction.

(a) Assuming that the students' scores were normally distributed, determine the mean and variance on this exam if the pass mark was 40 and the minimum score for a high distinction was 75.

Some of the students who failed this exam were allowed to sit a 'make-up' exam in early January of the following year. Of those who failed, the top 50% were allowed to sit the 'make-up' exam.

(b) What is the lowest possible score that a student can be awarded in order to qualify for the 'make-up' exam

37. The heights of men in a particular country are found to be normally distributed with mean 178 cm and a standard deviation of 5 cm. A man is selected at random from this population.

(a) Find the probability that this person is
i. at least 180 cm tall.
ii. between 177 cm and 180 cm tall.

(b) Given that the person is at least 180 cm, find the probability that he is
i. at least 184 cm.
ii. no taller than 182 cm.

(c) If ten such men are randomly selected, what are the chances that at least two of them are at least 176 cm?

Cumulative Standard Normal Distribution Table

$$F(x) = \frac{1}{\sqrt{2\pi}}\int_{-\infty}^{z} e^{\left(-\frac{1}{2}t^2\right)}\,dt$$

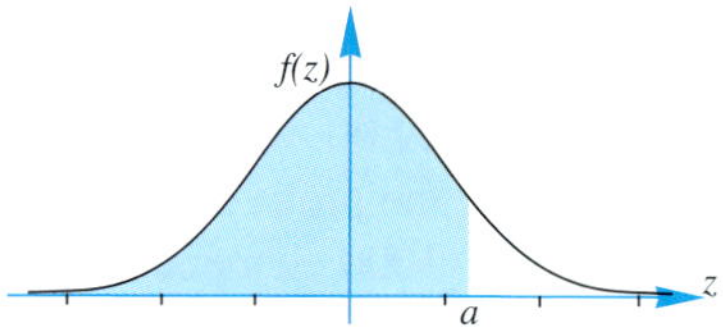

	0	0.01	0.02	0.03	0.04	0.05	0.06	0.07	0.08	0.09
0.0	0.5000	0.5040	0.5080	0.5120	0.5160	0.5199	0.5239	0.5279	0.5319	0.5359
0.1	0.5398	0.5438	0.5478	0.5517	0.5557	0.5596	0.5636	0.5675	0.5714	0.5753
0.2	0.5793	0.5832	0.5871	0.5910	0.5948	0.5987	0.6026	0.6064	0.6103	0.6141
0.3	0.6179	0.6217	0.6255	0.6293	0.6331	0.6368	0.6406	0.6443	0.6480	0.6517
0.4	0.6554	0.6591	0.6628	0.6664	0.6700	0.6736	0.6772	0.6808	0.6844	0.6879
0.5	0.6915	0.6950	0.6985	0.7019	0.7054	0.7088	0.7123	0.7157	0.7190	0.7224
0.6	0.7257	0.7291	0.7324	0.7357	0.7389	0.7422	0.7454	0.7486	0.7517	0.7549
0.7	0.7580	0.7611	0.7642	0.7673	0.7704	0.7734	0.7764	0.7794	0.7823	0.7852
0.8	0.7881	0.7910	0.7939	0.7967	0.7995	0.8023	0.8051	0.8078	0.8106	0.8133
0.9	0.8159	0.8186	0.8212	0.8238	0.8264	0.8289	0.8315	0.8340	0.8365	0.8389
1.0	0.8413	0.8438	0.8461	0.8485	0.8508	0.8531	0.8554	0.8577	0.8599	0.8621
1.1	0.8643	0.8665	0.8686	0.8708	0.8729	0.8749	0.8770	0.8790	0.8810	0.8830
1.2	0.8849	0.8869	0.8888	0.8907	0.8925	0.8944	0.8962	0.8980	0.8997	0.9015
1.3	0.9032	0.9049	0.9066	0.9082	0.9099	0.9115	0.9131	0.9147	0.9162	0.9177
1.4	0.9192	0.9207	0.9222	0.9236	0.9251	0.9265	0.9279	0.9292	0.9306	0.9319
1.5	0.9332	0.9345	0.9357	0.9370	0.9382	0.9394	0.9406	0.9418	0.9429	0.9441
1.6	0.9452	0.9463	0.9474	0.9484	0.9495	0.9505	0.9515	0.9525	0.9535	0.9545
1.7	0.9554	0.9564	0.9573	0.9582	0.9591	0.9599	0.9608	0.9616	0.9625	0.9633
1.8	0.9641	0.9649	0.9656	0.9664	0.9671	0.9678	0.9686	0.9693	0.9699	0.9706
1.9	0.9713	0.9719	0.9726	0.9732	0.9738	0.9744	0.9750	0.9756	0.9761	0.9767
2.0	0.9772	0.9778	0.9783	0.9788	0.9793	0.9798	0.9803	0.9808	0.9812	0.9817
2.1	0.9821	0.9826	0.9830	0.9834	0.9838	0.9842	0.9846	0.9850	0.9854	0.9857
2.2	0.9861	0.9864	0.9868	0.9871	0.9875	0.9878	0.9881	0.9884	0.9887	0.9890
2.3	0.9893	0.9896	0.9898	0.9901	0.9904	0.9906	0.9909	0.9911	0.9913	0.9916
2.4	0.9918	0.9920	0.9922	0.9925	0.9927	0.9929	0.9931	0.9932	0.9934	0.9936
2.5	0.9938	0.9940	0.9941	0.9943	0.9945	0.9946	0.9948	0.9949	0.9951	0.9952
2.6	0.9953	0.9955	0.9956	0.9957	0.9959	0.9960	0.9961	0.9962	0.9963	0.9964
2.7	0.9965	0.9966	0.9967	0.9968	0.9969	0.9970	0.9971	0.9972	0.9973	0.9974
2.8	0.9974	0.9975	0.9976	0.9977	0.9977	0.9978	0.9979	0.9979	0.9980	0.9981
2.9	0.9981	0.9982	0.9982	0.9983	0.9984	0.9984	0.9985	0.9985	0.9986	0.9986
3.0	0.9987	0.9987	0.9987	0.9988	0.9988	0.9989	0.9989	0.9989	0.9990	0.9990
3.1	0.9990	0.9991	0.9991	0.9991	0.9992	0.9992	0.9992	0.9992	0.9993	0.9993
3.2	0.9993	0.9993	0.9994	0.9994	0.9994	0.9994	0.9994	0.9995	0.9995	0.9995
3.3	0.9995	0.9995	0.9995	0.9996	0.9996	0.9996	0.9996	0.9996	0.9996	0.9997
3.4	0.9997	0.9997	0.9997	0.9997	0.9997	0.9997	0.9997	0.9997	0.9997	0.9998
3.5	0.9998	0.9998	0.9998	0.9998	0.9998	0.9998	0.9998	0.9998	0.9998	0.9998
3.6	0.9998	0.9998	0.9999	0.9999	0.9999	0.9999	0.9999	0.9999	0.9999	0.9999
3.7	0.9999	0.9999	0.9999	0.9999	0.9999	0.9999	0.9999	0.9999	0.9999	0.9999
3.8	0.9999	0.9999	0.9999	0.9999	0.9999	0.9999	0.9999	0.9999	0.9999	0.9999
3.9	1.0000	1.0000	1.0000	1.0000	1.0000	1.0000	1.0000	1.0000	1.0000	1.0000

REVISION SET D - PAPER 1 & PAPER 2 STYLE QUESTIONS

1. A and B are events with $P(A) = x$, $P(B) = 2x$ and $P(A \cup B) = \frac{3}{4}$. Using the expansion for $P(A \cup B)$, find x in each of the following cases:
(a) A and B are mutually exclusive.
(b) $A \subset B$.
(c) A and B are independent.

2. The annual rainfall at a certain locality is known to follow a normal distribution with mean value equal to 25 cm and standard deviation 5 cm. What is the probability that on any particular year, there will be an annual rainfall of at least 35 cm?

3. Twin sisters, Sue and Debbie, are both good maths students. Their Mathematics teacher calculated that Sue has a probability of 0.7 of getting an A at the end of the year and that Debbie has a probability of 0.6. Find the probability that:
(a) neither gets an A.
(b) if only one gets an A, it is Sue.

4. If $P(A) = 0.4$, $P(B) = 0.7$ and $P(A|B) = 0.3$. Find
(a) $P(A \cup B)$ (b) $P(B|A)$ (c) $P(A|A \cup B)$

5. In a certain town, 60% of the people are females and 40% are males. In an opinion poll taken, 30% of the females approved of the work done by the Town Mayor while 70% of the males approved of his work.
(a) What is the probability that a person chosen at random in the town will approve of the work done by the Town Mayor?
(b) Also, if the person chosen did approve of his work, what is the probability that person is a female?

6. (a) Write down the number of different arrangements of the letters of the word EQUILIBRIUM.
(b) One of these arrangements is chosen at random. Find the probability that:
i. the first two letters of the arrangements are consonants.
ii. all the vowels are together.

7. The weights of a large number of children at a Kindergarden are normally distributed with mean 20 kg and standard deviation of 2 kg. Find the percentage of children with weights
(a) less than 24 kg. (b) between 18 kg and 20 kg.

8. The probability that a certain darts player hits the bullseye with one dart is 0.4.
(a) Find the probability that the player scores at most two bullseyes with 3 darts.
(b) If the probability of scoring at least one bullseye with n darts is greater than 0.9, find the least possible value of n.

9. In a Football competition, 12 teams play a series of matches to determine the best 5 (which then play a further series of 'finals'). Before the start of this season, 'The Moon' newspaper ran a contest in which readers were invited to select the 5 teams which they expected to be the successful ones. The order was not important, and there was no restrictions on the number of entries per person.

(a) Wendy wanted to submit a sufficient number of entries to ensure that one of them must be correct. How many must she submit?

(b) In fact, nobody selected the correct 5 teams, so the price was divided among those who selected 4 correct. How many different selections of 5 teams could have qualified for this prize?

10. In the popular weekly 'Super 66' competition, contestants select any 6 digits in order, from the set $\{0, 1, 2, \ldots, 9\}$; any digit may be selected more than once.

(a) In how many ways may 6 different digits be arranged in order?

(b) Hence find the probability that the winning arrangement consists of 6 different digits.

11. It is known that exactly 10% of the students at Dexter High School are left-handed. If the Maths class contains 20 students, find the probability that exactly 2 of these are left-handed.

12. Three cards are drawn (together) at random from a full pack of 52. Find the probability that at least two of these are 'spades'.

13. As a result of a certain random experiment, the events A and/or B may occur. These events are independent, and $P(A) = 0.5$; $P(B) = 0.2$.

(a) Find the probability that both A and B occur.

(b) Find the probability that neither A nor B occurs.

(c) Let X denote the random variable which counts how many of the two events occur at a given time. Thus, for example, $X = 0$ if neither A nor B occur.
Find $P(X = x)$ for $x = 0, 1, 2$.

(d) Find the mean and variance of X.

14. A machine is set to manufacture circular metal discs of diameter 10 cm; in fact, the diameter (in cm) of the discs it produces is a random variable Y with normal distribution, having $\mu = 10.05$ and $\sigma = 0.10$.
The demands of quality control place 'acceptable limits' between 9.90 cm and 10.20 cm; all discs with diameters outside this range are rejected.

(a) Find the proportion of discs which are rejected.

(b) Given that a particular disc is accepted, find the probability that its diameter is more than 10 cm.

(c) Given that a particular disc has a diameter more than 10cm, find the probability that it is accepted.

(d) It is decided that too many discs are rejected by the standard. If the new rejection rate is to be 10%, find new 'acceptable limits', equally spaced on either side of the mean.

(e) Use the fact that $Var(Y) = E(Y^2) - [E(Y)]^2$ to estimate the mean area of the discs.

15. A certain cricket squad consists of 8 batsmen, 6 bowlers and a wicket keeper. A team of 11 is to be selected, consisting of 6 batsmen, 4 bowlers and the wicket keeper.

(a) How many different teams can be selected in this way, given that the captain (who is a batsman) must be selected in this team?

(b) In any given team, the 'batting order' is such that the six batsmen all bat before the wicket keeper, who bats at no.7, followed by the four bowlers. How many different 'batting order' are possible?

16. A fair coin is tossed 10 times. Find the probability that the number of 'heads' resulting is greater than the number of 'tails'.

17. A box contains three coins, one of which is known to be a 'double-headed'.
(a) If a coin is selected at random and tossed, find the probability that it falls 'heads'.
(b) If a coin selected at random falls 'heads' when tossed, what is the probability that it is the double-headed coin?

18. In the game of 'Monopoly', a player cannot start until she throws a 'double' with a set of two six-sided dice i.e. both dice must show the same number (from 1 to 6) uppermost when tossed.
(a) State the form of the probability distribution of X, the number of unsuccessful turns a player has before she manages to throw a 'double'.
(b) Find i. $P(X = 5)$ ii. $P(X \geq 5)$ iii. $P(X = 5 | X \geq 5)$

19. A 'Snooker' set consists of 22 balls of identical size; 15 are red, each of the others being a different colour. If all 22 are put into a bag, and 3 are selected at random, find the probability that:
(a) all are red; (b) exactly one is red.

20. A certain dart board consists of five concentric circles, with the respective radii r, $2r$, $3r$, $4r$, $5r$. Any dart thrown registers a score corresponding to the region in which it lands, if it misses the board altogether it does not count, and is returned to the contestant for another throw.
(a) Show that the circles divide the board into regions whose areas are in the ratio $1 : 3 : 5 : 7 : 9$.
(b) If Harry's aim is quite random (i.e., he is equally likely to hit any point on the target), and X represents his score on any one shot, specify the probability distribution of X. i.e. Find $P(X = 1)$, $P(X = 2)$, . . . ,$P(X = 20)$.
(c) Find the mean and the variance of X.
(d) If he throws three darts altogether in any one turn, find the probability that he scores a total of at least 50.

21. A certain brand of soft-drink is sold in so-called 'litre' bottles. In fact, the amount of liquid in each bottle (in litres) is a normally distributed random variable with mean 1.005 and standard deviation 0.01.
(a) Find the proportion of soft-drink bottles containing less than 1 litre.
(b) If I buy four bottles, find the probability that all four contain less than 1 litre.
(c) Find also the probability that the mean contents of the four bottles is less than 1 litre.

22. For the Diploma of Transcendental Studies, a student must take six subjects, one of which must be Maths. The other available subjects are classified into two groups, A and B, which contain 4 and 5 subjects respectively. If he must select at least two of his subjects from each group, how many different subject combinations are possible?

23. The letters of the word *GENERAL* are arranged at random in a row. Find the probability that: (a) *G* precedes *L*;
(b) *G* immediately precedes *L*;
(c) both *E*s occur together.

24. A bag contains 3 red and 2 black marbles. Let X be the number of marbles withdrawn (at random), one at a time without replacement, until the first black marble is drawn.
(a) Explain why X cannot take any value greater than 4.
(b) Specify the probability distribution of X. .
(c) Find i. $E(X)$ ii. $E(X^2)$ iii. $E(2X-1)$

25. The lengths of steel rods are normally distributed with a mean of 50 cm and a standard deviation of 0.5 cm. What is the probability of a rod having a length
i. between 49.5 cm and 51 cm if all measurements are considered to be accurate?
ii. of 50 cm if all measurements are correct to the nearest 0.1 cm?

26. (a) The independent probabilities of three students scoring more than 10 runs in a cricket match are 0.5, 0.5 and 0.25. If the random variable, X, denotes the number of students who score more than 10 runs, find $P(X = x)$ where $x = 0, 1, 2, 3$.

(b) In a sequence of independent trials, the probability of success is p for each trial. Let the random variable, X, denotes the number of failures before the first success occurs.
i. Show that $P(X = x) = (1-p)^x p, x = 0, 1, 2, \ldots$.
ii. Evaluate $P(X = 3)$ if $p = 0.8$.
iii. Evaluate $p(X \leq 10 | X \leq 20)$ if $p = 0.1$.

(c) A student is forced to guess on a multiple choice test 20% of the time. If he guesses, his probability of being correct is 0.2, while if he does not guess, his probability of being correct is 0.9.
i. In a 50 item multiple choice test, what would be his expected number of correct items?
ii. What is the probability that he guessed on a question, given that the answer he gave was correct?

27. The mean diameter of bolts from a machine can be adjusted so that the proportion of bolts greater than 1.00 cm is 0.05, and the proportion less than 0.90 cm is 0.01. Assuming the distribution of the bolt diameters to be Normal, find the mean and the standard deviation of the diameter.

28. (a) A pack of cards contains 4 red and 5 black cards. A hand of 5 cards is drawn without replacement. What is the probability of there being 2 red and 3 black cards in the hand?

(b) The distribution of errors in using an instrument to measure length in normally distributed with a mean of 1 cm and a standard deviation of 2 cm. What is the probability of a measurement
i. underestimating the true length?
ii. being correct to the nearest cm?

29. (a) A man who works in Melbourne drives home, either along Road A or Road B. He varies his route, choosing Road A $\frac{2}{3}$ of the time. If he drives along Road A, he arrives home before 6:00 pm 90% of the time, while by the more attractive route, along Road B, he gets home before 6:00 pm 60% of the time. What is the probability

i. that he gets home before 6:00 pm?

ii. that he travelled along Road B, if he gets home before 6:00 pm?

(b) A discrete random variable X may take the values 0, 1 or 2. The probability distribution of X is defined by $P(X = x) = \frac{k}{x!}$. Find

i. k. ii. the mean and variance of X.

30. (a) Two boxes each contain 9 balls. In box X there are four black and five white balls. In box Y there are three black and six white balls. A box is chosen at random and two balls are drawn from it without replacement. If B is the event of two black balls being drawn and C is the event of at least one black ball,

Find i. $P(B)$ ii. $P(C)$ iii. $P(B \cap C)$

iv. $P(B \cup C)$ v. $P(B|C)$

(b) Suppose that 25% of the Australian population who voted in the national anthem referendum voted for 'Waltzing Maltida'.
What is the probability that more than 85% of a random sample of 300 of this population voted for 'Waltzing Matilda''?

31. A mathematics competition consists of fifteen multiple choice questions each having four choices with only one choice correct. Williams works through this test and he knows the correct answer to the first seven questions, but he does not know anything at all about the remaining questions, so he guesses. Find the probability that he gets exactly 12 correct.

32. Yoghurt is sold to a customer in boxes containing 10 cartons. The customer selects two cartons at random from each box without replacement. If both cartons are defective he rejects the box of cartons. If none is defective he accepts the box. If one of the cartons is defective he selects two more at random from those remaining and rejects the batch if one or both of these is defective; otherwise he accepts the batch.
What is the probability that a batch containing 3 defectives will be accepted?

33. In a certain country 65% of the population support Ben's Party and 35% support Sam's Party. 90% of Ben's Party and 40% of those who support Sam's Party believe that Ben is a good leader. A member of the population is chosen at random. Given that the person believes Ben is a good leader use Bayes' Theorem to find the probability that the person supports Ben's Party.

34. From 3 coins, one of which is double headed, a coin is selected at random and tossed twice.

i. Let X denote the number of heads that appear. Find $P(X = x)$ for $x = 0, 1, 2$ and hence calculate the mean and variance of X.

ii. If two heads appear, calculate the probability of the double headed coin being chosen.

Using the same 3 coins, a coin is selected at random then tossed 5 times.
iii. What is the probability of tossing exactly 2 heads?

35. In a test match, the probability of New Zealand defeating the Australian cricket team is 0.2, of a draw is 0.3 and of an Australian win is 0.5. For a three math test series, what is the probability of
i. New Zealand winning the first two matches?
ii. New Zealand winning two matches?
iii. Australia winning two matches, the other being drawn?
iv. Australia winning the series (that is, winning more matches than New Zealand)?

36. Of 5 cards, 3 are labelled with a 1, the others with a 2. Three cards are drawn at random from the five cards, observed, then returned to the pack. This process is repeated a second tine. If X denotes the number of times two 1's and a 2 are drawn,
(a) Find the probability of two 1's and a 2 on the first draw.
(b) Find $P(X = x)$, for $x = 0, 1, 2$ for the two draws.
(c) Calculate the mean and variance of X.
(d) Calculate $P(X > 1|X > 0)$.

37. (a) Urn X contains 4 red and 2 green balls and urn Y contains 5 red and 1 green ball. An urn is chosen at random and 2 balls are drawn from it without replacement. If A is the event of two red balls being drawn and B is the event of at least one red ball being drawn from urn X, find

i. $P(A)$ ii. $P(B)$ iii. $P(A \cap B)$
iv. $P(A \cup B)$ v. $P(A'|B)$

(b) In x% of a day, the probability of a machine producing a defective article is p while for the rest of the day the probability of a defective article is q. If one item is chosen at random at the end of the day, what is the probability that it is defective?

38. The cross-sectional area of a rod produced by a machine is normally distributed with a standard deviation of σ cm^2 and a mean of 4.0 cm^2. If the proportion of rods with cross-sectional area of less than 3.0 cm^2 is 0.4, evaluate σ. If all rods with cross-sectional area of less than 3.0 cm^2 are rejected, what is the probability of an accepted rod having a cross-sectional area greater than 5 cm^2?

39. The I.Q. of a member of a population is determined from a scale which is normally distributed. The Stanford-Binet I.Q. scale has a mean of 100 and standard deviation of 16. Using this scale, find the

(a) probability that a person chosen at random has an I.Q. between 116 and 132.
(b) I.Q. of a person who is in the top 1% on the I.Q. scale.
(c) I.Q. of a person on the Stanford-Binet scale if their I.Q. on a scale which has a mean of 100 and standard deviation of 15 is given as 135.
(d) I.Q. of a person who is in the top 1% of those people who have an I.Q. greater than 100 on the Stanford-Binet scale.

40. Let A and B be events such that $P(A) = \frac{1}{3}$, $P(B) = \frac{1}{4}$ and $P(A \cup B) = \frac{5}{12}$.

i. Find $P(A|B)$ ii. $P(A|B')$ iii. Are A and B independent? Why?

41. The random variable X with sample space $\{1, 2, 3\}$ is such that

$$P(X = 1) = b, P(X = 2) = 3a - 3b, P(X = 3) = 2b$$

i. Express $E(X)$ in terms of a and b.
ii. What values may a and b takes?

42. A particular device is made by two manufacturers, ACME and EMCO. Any device produced by ACME is such that at any attempt the device operates with probability 0.9 independent of previous attempts.

i. Find the probability that a device made by ACME operates exactly twice before failing.

EMCO produces a cheaper version; any device produced by EMCO is such that at any attempt the device operates with a probability of 0.6 independent of previous attempts. A batch of five of these devices contains four made by ACME and one made by EMCO. One device is selected at random from the batch of five.

ii. If the selected device operates exatly twice before failing, find the probability that it is the device produced by EMCO.

43. A machine produces 800 items each day, which are packed 20 to a box. It is known that, on average, one per cent of items produced is defective.

(a) A box is selected at random. Find, correct to three decimal places, the probability that it contains more than one defective item.

From the day's production of 40 boxes, four boxes are selected at random. All items in these four boxes are inspected. If none of the four boxes is found to contain more than one defective item, the days production is passed as satisfactory. Otherwise, the entire day's production is inspected.

(b) i. Find, correct to three decimal places, the probability that the day's production is passed as satisfactory after the four boxes are inspected.
ii. Find, correct to the nearest integer, the mean number of items inspected.
iii. If all defective items found are replaced with good items, find, correct to three decimal places, the expected proportion of items which are defective after this inspection-and-replacement process.

44. A fruitgrower produces peaches whose weights are normally distributed with a mean of 180 g and a standard deviation of 20 g.

Peaches whose weights exceed 200 g are sold to canneries yielding a profit of \$0.40 per peach. Peaches whose weights are between 150 g and 200 g are sold to wholesale markets at a profit of \$0.20 per peach. Peaches whose weights are less than 150 g are sold for jam at a profit of \$0.10 per peach. Find the

i. percentage of peaches produced that are sold to canneries;
ii. percentage of peaches produced that are sold to wholesale markets;
iii. the mean profit per peach.

45. Consider the following set of data:

12, 4, 9, 10, 12, 13, 15, 11, 12, 15, 14, 8, 9, 10, 12, 9, 10, 16, 14, 13, 12, 15, 9, 10, 12

(a) Construct i. a frequency polygon.
ii. cummulative frequency polygon.

(b) Calculate i. the mean ii. the standard deviation.

(c) i. Construct a box-plot.
ii. Determine the median and the mode.
iii. Calculate the interquartile range

46. In order to fill unreserved seats in its planes, an airline sells tickets at half the normal economy class fare to people who purchase tickets at the airport within 30 minutes of the departure of a flight. However, experience shows that there is only a probability of 0.4 of getting such a ticket for any particular flight.

The airline has hourly flights to Northport and a man who wishes to do business in Northport has the choice of flying on the 10 am, 11 am or 12 noon flights. His travelling procedure is to try to purchase a half-price ticket on the 10 am flight and, if unsuccessful, to wait and try to purchase a half-price ticket on the 11 am flight. If he is again unsuccessful he travels on the 12 noon flight with a full fare first class ticket which he knows will always be available. The economy class fare is \$200 and the first class fare is \$300.

(a) Find the probability that the man catches
i. the 11 am flight;
ii. the 12 noon flight.

(b) If the man can earn $\$Q$ for every hour that he is in Northport, express as a function of Q the mean cost to him of his travelling procedure.

(c) For what values of Q would the man expect to earn more money, after allowing for his travelling costs, if he made no attempt to purchase a half-price ticket and travelled with a full fare economy class ticket on the 10 am flight?

47. Test scores, out of a possible 60 marks, for two classes were recorded as follows:

Class A: 23 24 30 31 32 33 34 34 40 41 43 43 50 51 25 26 28 35 35 37 39 39 46 48 48 56

Class B: 22 25 29 32 31 34 33 35 39 42 42 44 49 52 23 27 27 36 34 38 38 40 45 49 47 57

(a) For each set of data:
Construct i. a frequency polygon.
ii. cummulative frequency polygon.

(b) Calculate i. the mean ii. the standard deviation.

(c) i. Construct a box-plot.
ii. Determine the median and the mode.
iii. Calculate the interquartile range

(d) Compare the results of class A to those of Class B. What do you conclude?

18.1 QUANTITATIVE MEASURE

18.1.1 FUNCTIONAL DEPENDENCE

The notion of functional dependence of a function $f(x)$ on the variable x has been dealt with in Chapter 5. However, apart from this algebraic representation, sometimes it is desirable to describe a graphical representation using a qualitative rather than quantitative description. In doing so, there are a number of key words that are often used.

Words to be kept in mind are:

Rate of change (slow, fast, zero)	Increasing, decreasing
Positive, negative	Maximum, minimum
Average	Instantaneous
Stationary	Initial, final
Continuous, discontinuous	Range, domain

Such terms enable us to describe many situations that are presented in graphical form. There is one crucial point to be careful of when describing the graphical representation of a given situation. Graphs that look identical could very well be describing completely different scenarios. Not only must you consider the behaviour (shape) of the graph itself, but also take into account the variables involved.

Consider the two graphs below. Although identical in form, they tell two completely different stories. We describe what happens in the first five minutes of motion:

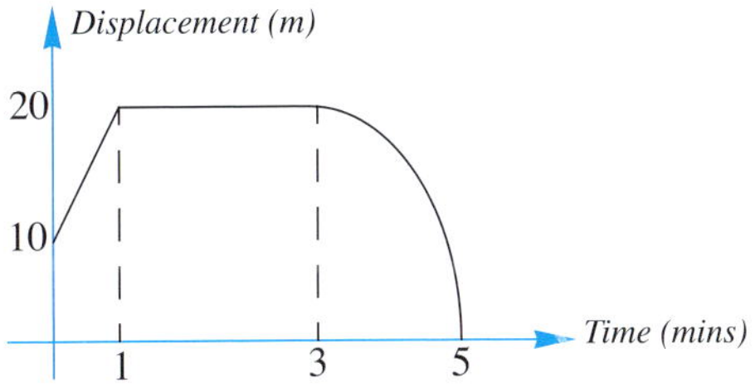

An object is moving in such a way that its displacement is increasing at a constant rate, that is, the object maintains a constant velocity (or zero acceleration) for the first minute. During the next two minutes the object remains stationary, that is, it maintains its displacement of 20 metres (meaning that it doesn't move any further from its starting position). Finally the particle returns to the origin.

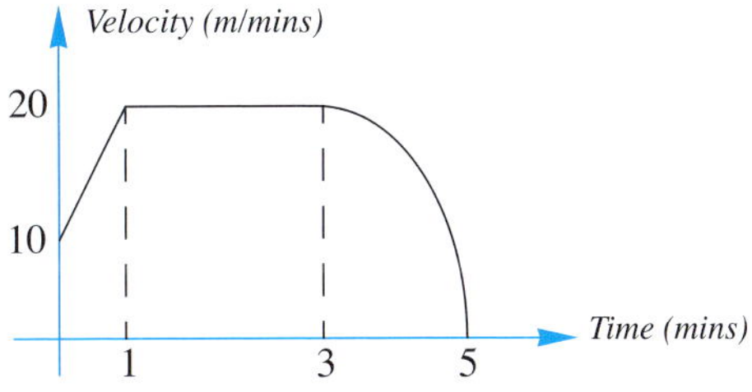

An object is moving at 10 m/min and keeps increasing its velocity at a constant rate until it reaches a velocity of 20 m/min, that is, it maintains a constant acceleration for the first minute. During the next two minutes the object is moving at a constant velocity of 20 m/mins (meaning that it is moving further away from its starting position). Finally, the particle slows to rest, far from the origin.

Although the shape of the graphs are identical, two completely different situations have been described!

18.1.2 QUANTITATIVE ASPECTS OF CHANGE

When dealing with the issue of rates of change, we need to consider two types of rates:

1. that of the **average rate** of change and
2. that of the **instantaneous rate** of change.

We start by considering the first of these terms, the average, and then see how the second, the instantaneous rate, is related to the first.

18.1.3 AVERAGE RATE OF CHANGE.

The average rate of change can be best described as an 'overview' of the effect that one variable (the independent variable) has on a second variable (the dependent variable). Consider the graph below, we can describe the change in the y-values (relative to the change in the x-values) as follows:

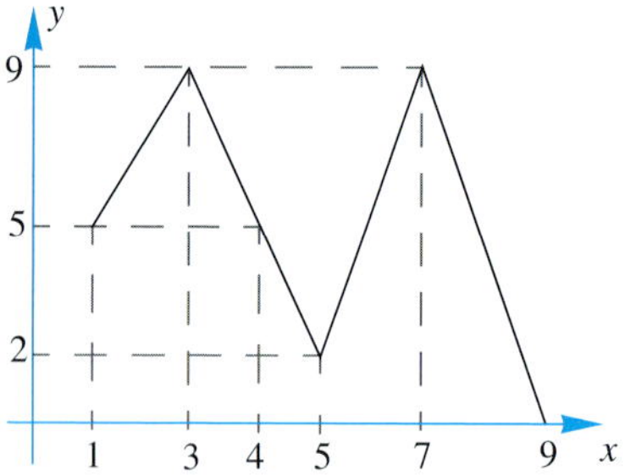

For $x \in [1, 3]$:

There is a **constant** increase from $y = 5$ to $y = 9$ as x increases from 1 to 3. So that an increase of 2 units in x has produced an increase of 4 units in y.
We say that the **average rate** of change of y with respect to x is $\frac{4}{2} = 2$

For $x \in [1, 4]$:

This time, the overall change in y is 0.
That is, although y increases from 5 to 9, it then decreases back to 5. So from its initial value of 5, because it is still at 5 as x increases from 1 to 4, the overall change in y is 0. This time the average rate of change is $\frac{0}{3} = 0$.

For $x \in [1, 5]$:

As x now increases from 1 to 5 we observe that there is an overall decrease in the value of y, i.e., there is an overall decrease of 3 units (y: $5 \to 9 \to 5 \to 2$). In this instance we say that the average rate of change is $-\frac{3}{4} = -0.75$.

Notice that we have included a negative sign to indicate that there was an overall decrease in the y values (as x has increased by 4).

Similarly for the rest of the graph. Note that we need not start at $x = 1$. We could just as easily have found the change in y for $x \in [3, 5]$. Here, the average rate of change is $-\frac{7}{2} = -3.5$.

The question then remains, is there a simple way to find these average rates of change and will it work for the case where we have non–linear sections? As we shall see in the next sections, the answer is 'yes'.

18.1.4 DETERMINING THE AVERAGE RATE OF CHANGE

To find the average rate of change in y it is necessary to have an initial point and an end point, as x increases from x_1 to x_2.

At A $x = x_1, y = y_1$ and at B $x = x_2, y = y_2$.

To obtain a numerical value, we find the gradient of the straight line joining these two points.

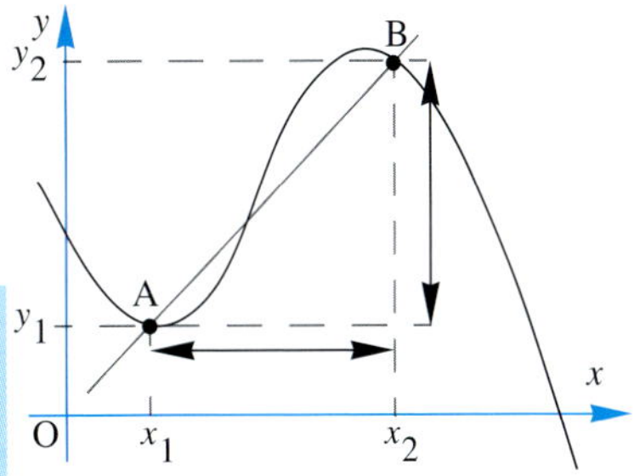

Average rate of change from A to B = gradient from A to B $= \dfrac{y_2 - y_1}{x_2 - x_1}$

EXAMPLE 18.1 For each of the graphs below, find the average rate of change of y with respect to x over the interval specified (i.e., over the domain L).

(a)

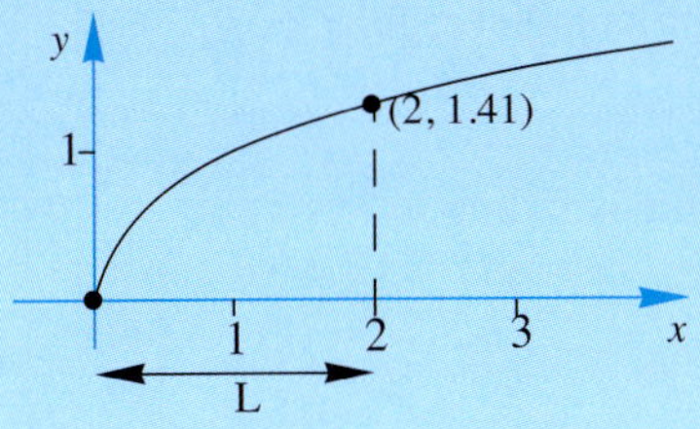

(b)

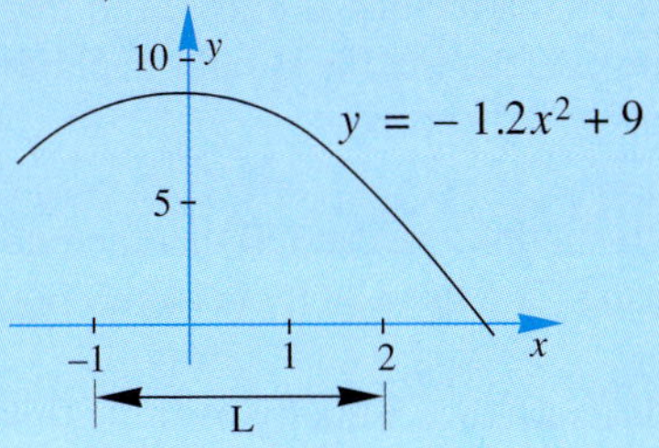

Solution

(a) For this case we have the 'starting point' at the origin (with coordinates (0,0)) and the 'end' point with coordinates (2,1.41).
This means that the average rate of change of y with respect to x, over the domain L is given by $\dfrac{y_2 - y_1}{x_2 - x_1} = \dfrac{1.41 - 0}{2 - 0} = 0.705$.

(b) This time we will need to first determine the coordinates of the extreme points:
For $x = -1$, $y = -1.2 \times (-1)^2 + 9 = 7.8$ and for $x = 2$, $y = -1.2 \times (2)^2 + 9 = 4.2$.

Therefore, the average rate of change is equal to $\dfrac{y_2 - y_1}{x_2 - x_1} = \dfrac{4.2 - 7.8}{2 - (-1)} = -1.2$.

It is not always necessary to have a graph in order to find the average rate of change. Often we are given information in the form of a table.

EXAMPLE 18.2 The table below shows the number of bacteria, N, present in an enclosed environment. Find the average growth rate of the population size over the first 4 hours

Time (hrs)	0	1	2	3	4	5	6	7	9
N	30	36	43	52	62	75	90	107	129

Solution

This time we need to consider the time interval $t = 0$ to $t = 4$. From the table we observe that the coordinates corresponding to these values are; (0,30) and (4,62).
Therefore, the average rate of growth of the number of bacteria over the first 4 hours is equal to $\frac{62-30}{4-0} = \frac{32}{4} = 8$.

This means that during the first 4 hours, the number of bacteria was increasing (on average) at a rate of 8 every hour.

Notice that in the 1st hour, the average rate was $\frac{36-30}{1-0} = \frac{6}{1} = 6$ (< 8), whereas in the 4th hour the average rate of increase was $\frac{62-52}{4-3} = \frac{10}{1} = 10$ (> 8).

18.1.5 VELOCITY AS A MEASURE OF THE RATE OF CHANGE OF DISPLACEMENT

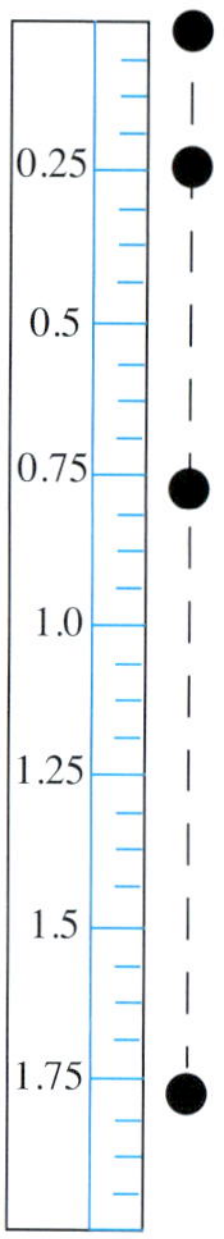

Consider a marble that is allowed to free fall from a height of 2 metres (see diagram). As the marble is falling, photographs are taken of its fall at regular intervals of 0.2 second.

From its motion, we can tell that the rate at which the marble is falling is increasing (i.e., its velocity is increasing).

What is its average velocity over the first 0.6 second?

Reading from the diagram, we see that the marble has fallen a total distance of 1.75 (approximately), therefore, the average velocity v_{ave} of the marble, given by the rate at which its displacement increases (or decreases), is given by

$$v_{ave} = \frac{1.75-0}{0.6-0} \approx 2.92 \text{ m/sec}$$

EXAMPLE 18.3

The displacement, x m, of an object, t seconds after it is dropped from the roof of a building is given by $x = 4.9t^2$ m.

(a) What is the object's displacement after 4 seconds?

(b) What is the average velocity of the object over the first 4 seconds of its motion?

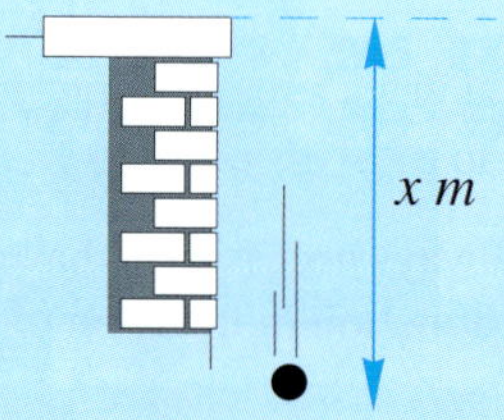

Solution

(a) After 4 seconds of free fall, the object's displacement will be $4.9(4)^2 = 78.4$ m. We obtained this result by substituting the value of $t = 4$ into the equation for the displacement $x = 4.9t^2$.

(b) The average velocity is given by the average rate of change of displacement, x m, with respect to the time t seconds.

Once we have the starting position and the end position we can determine the average velocity using: $v_{ave} = \frac{x_2 - x_1}{t_2 - t_1} = \frac{78.4 - 0}{4 - 0} = 19.6$.

That is, the object's average velocity over the first 4 seconds is 19.6 m/s.

EXAMPLE 18.4

The concentration of a drug, in milligrams per millilitre, in a patient's bloodstream, t hours after an injection is approximately modelled by the function

$$t \mapsto \frac{2t}{8 + t^3}, t \geq 0 .$$

Find the average rate of change in the concentration of the drug present in a patient's bloodstream; (a) during the first hour.
(b) during the first two hours.
(c) during the period $t = 2$ to $t = 4$.

Solution

To help us visualise the behaviour of this function we will make use of the TI–83. Begin by introducing the variable C, to denote the concentration of the drug in the patient's bloodstream t hours after it is administered. So that $C(t) = \frac{2t}{8 + t^3}, t \geq 0$.

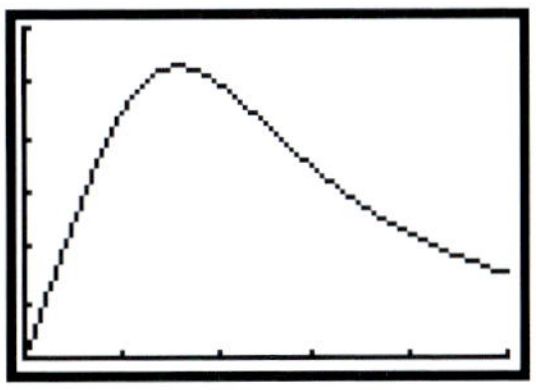

(a) Initially the concentration is 0 milligrams per millilitre, the concentration after 1hr is given by

$$C(1) = \frac{2 \times 1}{8 + 1^3} = \frac{2}{9} \approx 0.22 .$$

Therefore, the average rate of change in concentration (C_{ave}) during the first hour is given by $C_{ave} = \frac{0.22 - 0}{1 - 0} = 0.22$.

Note: the units are *mg/ml/hr.*

(b) The concentration 2 hours after the drug has been administered is $C(2) = \frac{2 \times 2}{8 + 2^3} = 0.25$.

That is, 0.25 *mg/ml*.
Therefore, the average rate of change in concentration with respect to time is

$$C_{ave} = \frac{0.25 - 0}{2 - 0} = 0.125 .$$

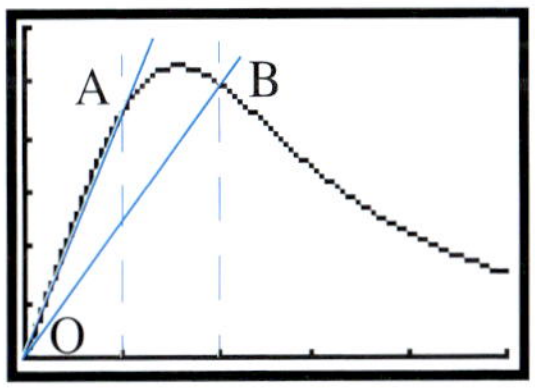

Notice that although the concentration has increased (compared to the concentration after 1 hour), the rate of change in the concentration has actually decreased!

This should be evident from the graph of $C(t)$ versus t.
The slope of the straight line from the origin to A(1, 0.22), m_{OA} , is greater than the slope from the origin O to the point B(2, 0.25), m_{OB} .
That is $m_{OA} > m_{OB}$.

(c) The average rate of change in concentration from $t = 2$ to $t = 4$ is given by $\frac{C(4) - C(2)}{4 - 2}$.

Now, $\frac{C(4)-C(2)}{4-2} = \frac{\frac{2\times 4}{8+4^3}-0.250}{4-2} \approx \frac{0.111-0.250}{2} = -0.0694$

Therefore, the average rate of change of concentration is –0.070 *mg/ml/hr*.
Meaning that the overall amount of drug in the patient's bloodstream is decreasing during the time interval $2 \le t \le 4$.

EXERCISES 18.1

1. For each of the following graphs determine the average rate of change over the specified domain.

(a) $x \in [4, 8]$

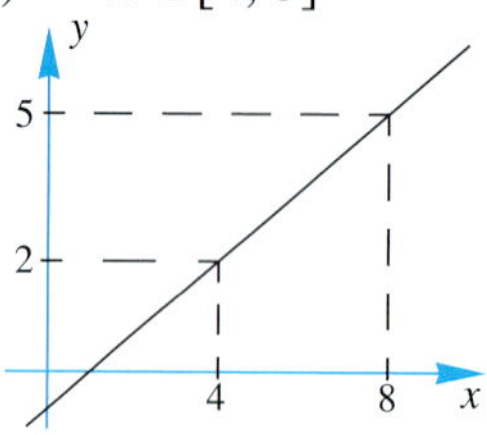

(b) $x \in [-b, 3b]$

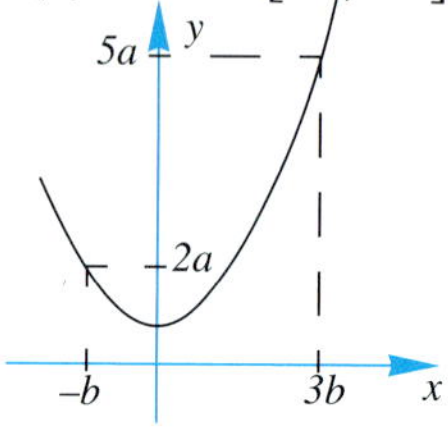

(c) $x \in [-2, 2]$

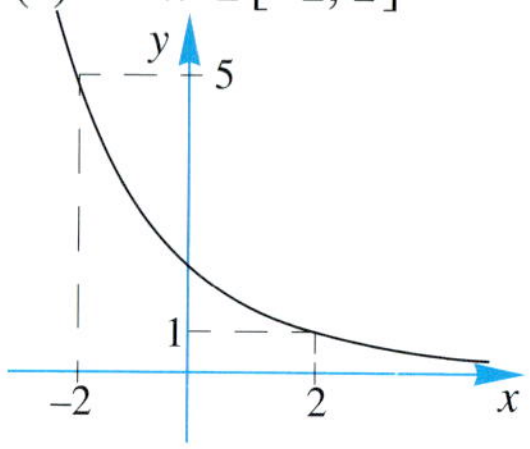

(d) $x \in [-3, 2]$

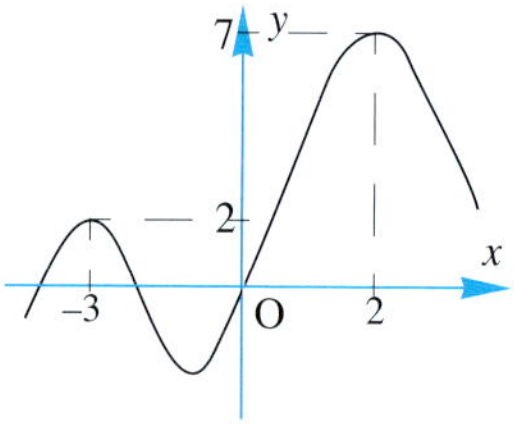

(e) $x \in [-2, 6]$

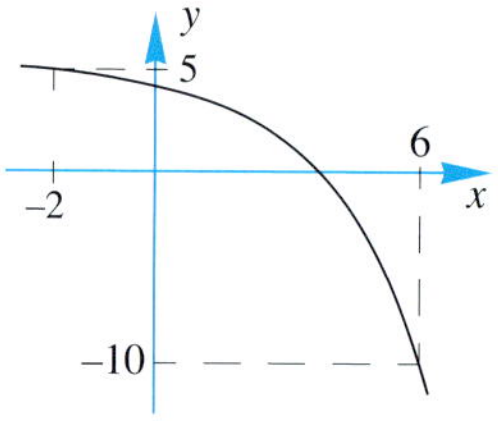

(f) $x \in [-1, 1]$

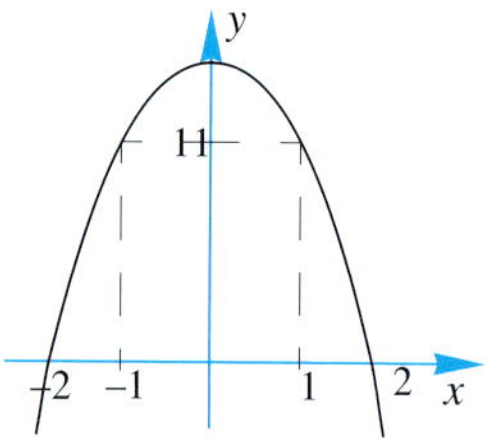

2. For each of the following functions, find the average rate of change over the given domain.

(a) $x \mapsto x^2 + 2x - 1, x \in [0, 2]$ (b) $x \mapsto \sqrt{x+1}, x \in [3, 8]$

(c) $x \mapsto 10 - \frac{1}{\sqrt{x}}, x \in [2, 20]$ (d) $x \mapsto \frac{x}{x+1}, x \in [0.1, 1.1]$

(e) $x \mapsto \frac{1}{1+x^2} - 1, x \in [0, 100]$ (f) $x \mapsto x\sqrt{400-x}, x \in [300, 400]$

(g) $x \mapsto 2^x, x \in [0, 5]$ (h) $x \mapsto (x-1)(x+3), x \in [-3, 2]$

3. The displacement of an object, t seconds into its motion is given by the equation,

$$s(t) = t^3 + 3t^2 + 2t, t \ge 0 .$$

Find the average rate of change of displacement during

(a) the first second.

(b) the first 4 seconds.

(c) the interval when $t = 1$ to $t = 1 + h$.

4. The distance s metres that a particle has moved in t seconds is given by the function $s = 4t + 2t^2, t \ge 0$. Find the particle's average speed over the first 4 seconds.

5. The distance s metres that a particle has moved in t seconds is given by the function $s = 4t + 2t^2, t \geq 0$. Find the particle's average speed during the time interval from when $t = 1$ to $t = 1 + h$.

6. The temperature T °C of food placed inside cold storage is modelled by the equation

$$T = \frac{720}{t^2 + 2t + 25} \text{ where } t \text{ is measured in hours.}$$

Find the average rate of change of the temperature, T°C, with respect to the time, t hours, during the first 2 hours that the food is placed in the cold storage.

7. The volume of water in a hemispherical bowl of radius r is given by $V = \frac{1}{3}\pi h^2(3r - h)$, where h is the height of the water surface inside the bowl. For the case where r is 20 cm,

(a) find the average rate of increase in the amount of water inside the bowl with respect to its height, h cm, as the water level rises from 2 cm to 5 cm.

(b) Find the average rate of increase in the amount of water inside the bowl with respect to its height, h cm, as the water level rises by

i. 1 cm ii. 0.1 cm iii. 0.01 cm

8. An amount of money is placed in a bank and is accumulating interest on a daily basis. The table below shows the amount of money in the savings account over a period of 600 days.

t (days)	100	200	300	400	500	600	700
\$$D$/day	1600	1709	1823	1942	2065	2194	2328

(a) Plot the graph of \$$D$ versus t (days).

(b) Find the average rate of change in the amount in the account during the period of 100 days to 300 days.

9. The temperature of coffee since it was poured into a cup was recorded and tabulated below.

t min	0	2	4	6	9
T °C	60	50	30	10	5

(a) Plot these points on a set of axes that show the relationship between the temperature of the coffee and the time it has been left in the cup.

(b) Find the average rate of change of temperature of the coffee over the first 4 minutes.

(c) Over what period of time is the coffee cooling the most rapidly?

10. The displacement, d metres, of an object, t seconds after it was set in motion is described by the equation

$$d = 4t + 5t^2, \text{ where } t \geq 0.$$

(a) Find the distance that the object travels in the first 2 seconds of its motion.

(b) Find the average rate of change of distance with respect to time undergone by the object over the first 2 seconds of its motion.

(c) What quantity is being measured when determining the average rate of change of distance with respect to time?

(d) How far does the object travel during the 5th second of motion?

(e) Find the object's average speed during the 5th second.

11. A person invested \$1000 and estimates that on average, the investment will increase each year by 16% of its value at the beginning of the year.
(a) Calculate the value of the investment at the end of each of the first 5 years.
(b) Find the average rate at which the investment has grown over the first 5 years.

18.2 QUALITATIVE MEASURE

18.2.1 QUALITATIVE ASPECTS OF CHANGE

Apart from quantitative measures (i.e., providing numerical values), it is also important to be able to provide qualitative descriptions of the behaviour of graphs. In doing so, many of the key words mentioned at the start of this chapter should be referred to.

18.2.2 DESCRIBING THE BEHAVIOUR OF A GRAPH

Consider the graph shown:

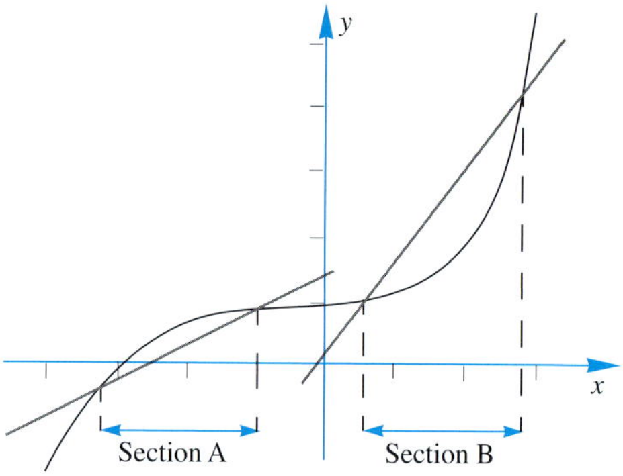

In both Section A and Section B, the gradients of the lines are positive. However, the gradient of the straight line in section B is steeper than that of the line in Section A. We can then say that over Section B the graph is increasing at a faster rate than it is over Section A.

In fact, if we were able to walk along this curve, from left to right, we could describe our 'journey' as follows:

As we walk from the left hand side and towards that part of the graph that lies above Section A, the **function is increasing**, i.e., as the values of x increase, so too do the values of y. As the values of x approach 0 (from the left side of the y–axis) the rate at which the function is increasing is slowing down. That is, I do not need to make as much effort to move as I get closer to the y–axis. Even though the function is still increasing (as we are getting closer to the y–axis), we then have that the rate of change of the function is in fact decreasing! Actually, by the stage when we have reached the y–axis we could almost say that the **function remains stationary**, i.e., it has stopped increasing. In this instance, we would say that the **rate of change of the function is zero**. As we pass the y–axis and keep moving along the curve we find it more difficult to walk along the curve. That is, the effort that we need to make to keep walking is increasing. In this instance the function is increasing but so too is the rate at which it is increasing.

18.2.3 PRODUCING A GRAPH FROM A PHYSICAL SITUATION

In this section we will concentrate on producing a graph to describe the behaviour of the flow of liquid into a container. The importance of such problems is that they enable us to describe how changing one variable will affect a second (related) variable. That is, the effect the independent variable has on the dependent variable. One way to do this is by increasing the independent variable (usually x) and observing the change in the dependent variable (usually y).

EXAMPLE 18.5 A cylindrical vase is placed under a tap and water is allowed to flow into it at a constant rate. Provide a graphical representation of the relationship between the volume of water in the vase and

(a) the time for which water flows into the vase.
(b) the level of water.

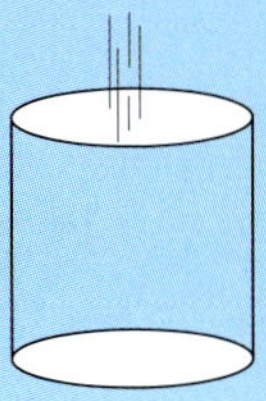

Solution

(a) The independent variable in this case is time, t seconds. Consider the volume of water, V cm^3, that flows into the vase in equal time intervals (of 2 seconds say). In these equal time intervals we have equal amounts of water flowing into the vase.
For example, if 10 cm^3 of water flows into the vase every 2 seconds, we could produce the following table of values:

t seconds	0	2	4	6	...
V cm^3	0	10	20	30	...

Based on the results of this table we can produce a graph of V cm^3 versus t seconds:

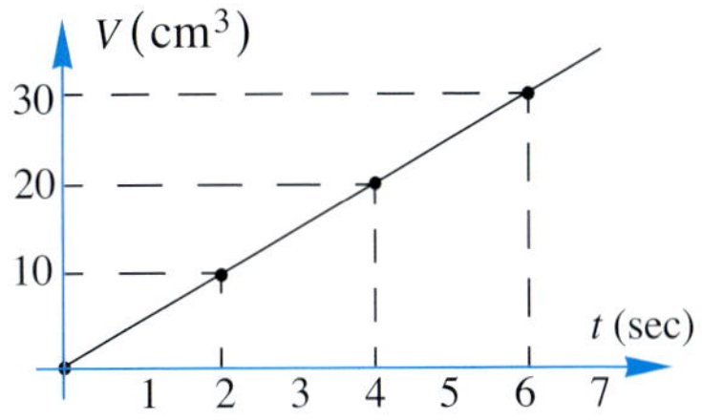

(b) To see how the volume changes with respect to the level of water, we use a different approach— this time we consider a 'frame–by–frame' sequence of the vase as it is filled.

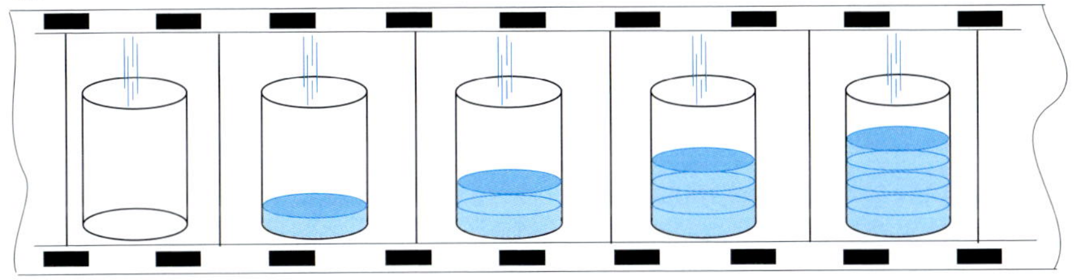

From our 'snap–shots', we see that for equal heights, equal amounts of water flow into the vase. So that every time the water level increases by 1cm, the volume increases by 8 cm^3.

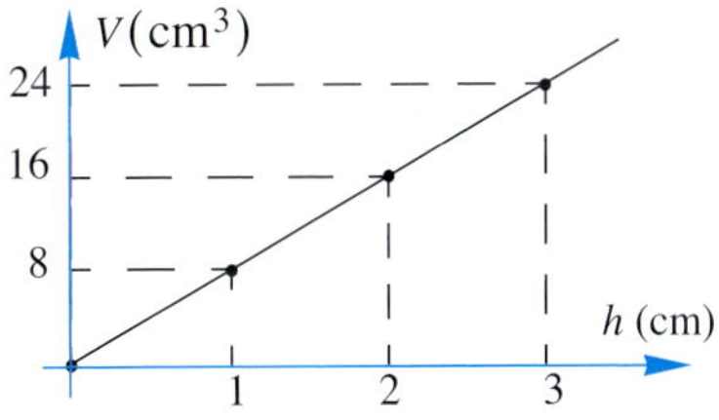

This would imply that the relationship between the volume, V cm^3 of water in the vase and the level of water, h *cm*, is linear.

EXAMPLE 18.6 Sketch a graph showing the relationship between the level of water in a flask and the time for which water has been flowing into the flask.

Solution

Let the level of water in the flask be denoted by h cm and the time for which water has been flowing be denoted by t seconds. Again, we use our 'frame–by frame' approach:

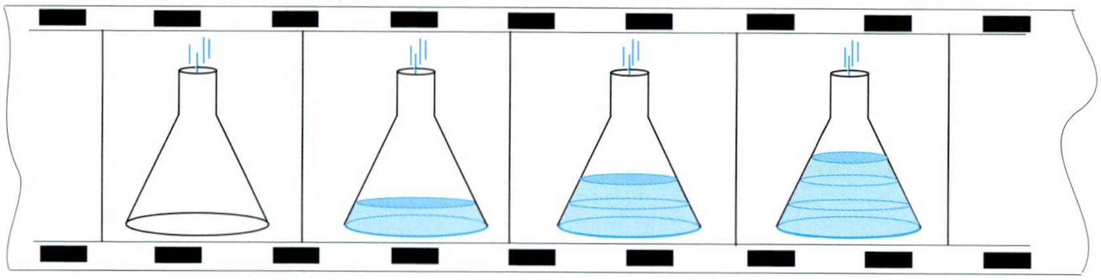

As we consider equal time intervals we see that the same amount of water will flow into the flask during each of these time intervals. However, because the flask becomes narrower as the level rises, then (because we still have the same volume of water flowing into the flask), the height of the space occupied by these equal volumes of water must increase at a faster rate than it had for the 'cone' section. A cross sectional view of the flask (shown above) shows this more clearly.

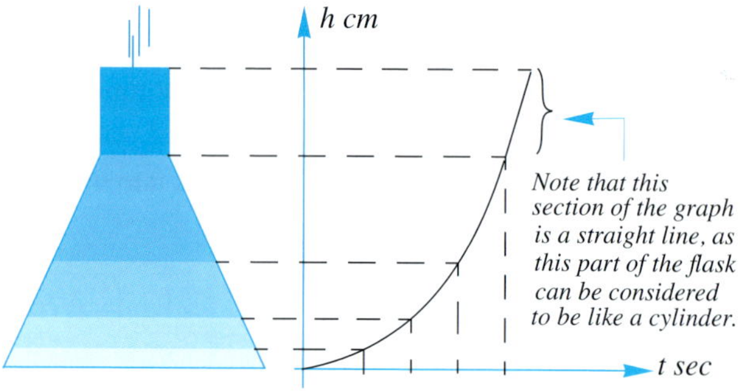

EXERCISES 18.2

1. (a) The cross section of a basin, shown in Figure A, is being filled by water flowing at a constant rate.

Sketch a graph of the relationship between the level of water, h cm, and the time, t sec, that water has been flowing.

Figure A

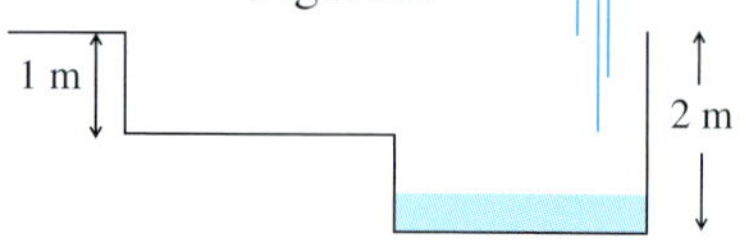

(b) The cross section of a second basin is shown in Figure B. Water is flowing into this basin at the same rate as in (a).

Sketch a graph of the relationship between the level of water, h cm, and the time, t sec, that water has been flowing for this basin.

Figure B

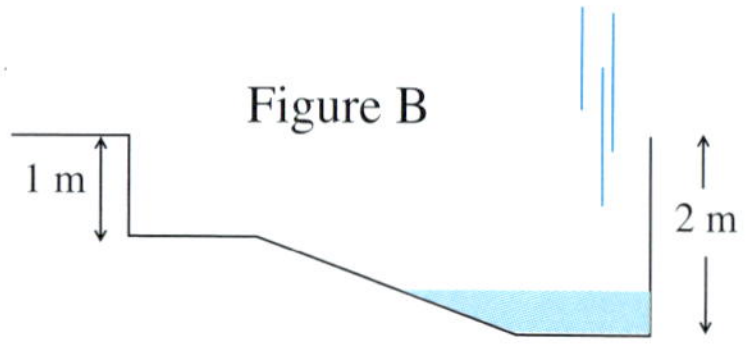

2. For each of the following bottles, sketch the graph that would show the relationship between the level of water, h cm, and the volume of water, V cm^3 in the bottle. That is, sketch a graph of h versus V.

You may assume that water is flowing into each bottle at a constant rate.

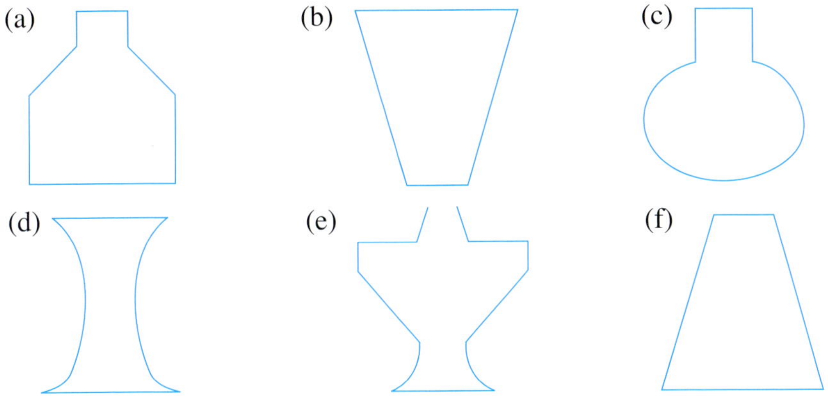

18.3 INSTANTANEOUS RATE OF CHANGE

18.3.1 INFORMAL IDEA OF LIMITS

As already discussed, the average rate of change between two points on a curve is determined by finding the gradient of the straight line joining these two points. However we often need to find the rate of change at a particular instant, and so the method used for finding the average rate of change is no longer appropriate. However, it does provide the foundation that leads to obtaining the instantaneous rate of change. We refine our definition of the average rate of change to incorporate the notion of the instantaneous rate of change. The basic argument revolves around the notion of magnifying near the point where we wish to find the instantaneous rate of change, that is, by repeatedly 'closing in' on a section of a curve. This will give the impression that over a very small section, the curve can be approximated by a straight line. Finding the gradient of that straight line will provide us with a very good approximation of the rate of change of the curve (over the small region under investigation). To obtain the exact rate of change at a particular point on the curve we will then need to use a **limiting** approach.

The process used to determine the rate of change at A is carried out as follows:

1. Start by drawing a secant from A to B, where B is chosen to be close to A. This will provide a reasonable first approximation for the rate of change at A. Then, to obtain a better approximation we move point B closer to point A.

2. Next, **zoom–in** towards point A, again. We move point B closer to point A, whereby a better measure for the rate of change at point A is now obtained. We then repeat step 2, i.e., move B closer to A and zoom in, move point B closer to A and zoom in, and so on.

3. Finally, the zooming–in process has reached the stage whereby the secant is now virtually lying on the curve at A. In fact the secant is now the tangent to the curve at the point A.

Using the process of repeatedly zooming–in to **converge** on a particular region lies at the heart of the **limiting process**. Once we have understood the concepts behind the limiting process, we can move on to the more formal aspect of limits. However, apart from an informal treatment of limits, work on limits is beyond the scope of the core work in HL mathematics.

We now provide a 'visual' representation of steps 1. to 3. described above.

As we magnify, and move point B closer to point A the secant from A to B becomes the tangent at A:

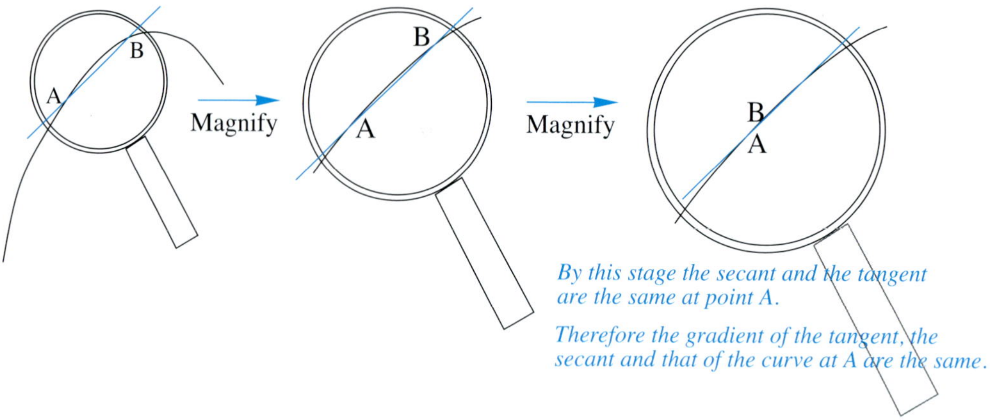

We now investigate this limiting process further through a number of examples. We will still maintain an informal approach to determining limits, but at the same time we will steer our work on limits towards the fundamentals behind the ideas in Chapter 19 – Differential calculus.

EXAMPLE 18.7

An object moves along a straight line. Its position, x metres (from a fixed point O), at time t seconds is given by $x(t) = t - \frac{1}{4}t^2, t \geq 0$. Determine

(a) its average velocity over the interval from $t = 1$ to $t = 2$
(b) its average velocity over the interval $t = 1$ to $t = 1.5$
(c) its average velocity over the interval $t = 1$ to $t = 1.1$
(d) its average velocity over the interval $t = 1$ to $t = 1 + h$, where h is small.

How can the last result help us determine the object's velocity at $t = 1$?

Solution

(a) The average velocity over the required second (from $t = 1$ to $t = 2$) is found by looking at the slope of the secant joining those two points on the graph of $x(t)$.

At $t = 2$, we have $x(2) = 2 - \frac{1}{4}(2)^2 = 1$, and at $t = 1$, $x(1) = 1 - \frac{1}{4}(1)^2 = \frac{3}{4}$.

Therefore, we have that
$$v_{ave} = \frac{x(2) - x(1)}{2 - 1} = \frac{1 - 0.75}{1} = 0.25$$

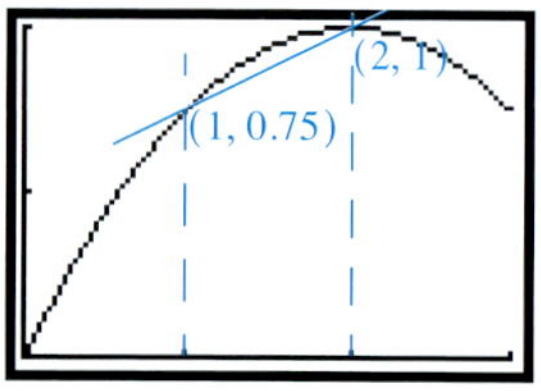

Therefore, the average velocity over the second is 0.75m/s.

(b) For $t = 1$ to $t = 1.5$ we have, $v_{ave} = \frac{x(1.5) - x(1)}{1.5 - 1} = \frac{(1.5 - 0.25 \times 1.5^2) - 0.75}{0.5} = 0.375$

(c) Similarly, for $t = 1$ to $t = 1.1$, we have $v_{ave} = \frac{x(1.1) - x(1)}{1.1 - 1} = 0.475$

(d) We are now in a position to determine the average rate over the interval $t = 1$ to $t = 1 + h$. The average velocity is given by $v_{ave} = \frac{x(1 + h) - x(1)}{1 + h - 1}$

Now, $x(1+h) = (1+h) - 0.25(1+h)^2 = 1 + h - 0.25(1 + 2h + h^2)$

$$= 0.75 + 0.5h - 0.25h^2$$

Therefore, $v_{ave} = \dfrac{0.75 + 0.5h - 0.25h^2 - 0.75}{1 + h - 1} = \dfrac{0.5h - 0.25h^2}{h}$

$$= \frac{h(0.5 - 0.25h)}{h}$$

$$= 0.5 - 0.25h,\ h \neq 0$$

Notice that for ii., (i.e., $t = 1$ to $t = 1.5$) $h = 0.5$, so that substituting $h = 0.5$ into this equation we have, $v_{ave} = 0.5 - 0.25(0.5) = 0.375$, providing the same result as before.

We can set up a table of values and from it determine what happens as we decrease the time difference.

h	v_{ave}
0.1	0.475
0.01	0.4975
0.001	0.4999

We notice that as h becomes very small, the average rate of change from $t = 1$ to $t = 1 + h$ becomes the instantaneous rate of change at t = 1! This is because we are zooming in onto the point where $t = 1$.

This means that the rate of change at $t = 1$ (h '= 0') would therefore be 0.5 m/s. So that the particle would have a velocity of 0.5 m/s after 1 second of motion.

EXAMPLE 18.8

For the graph with equation $f: x \mapsto (x+2)(x-1)(x-4)$,

(a) Find the average rate of change of f over the interval [–1,2].

(b) Find the rate of change of f, where $x = 4$.

Solution

(a) We first find the coordinates of the end points for the interval [–1,2]:

$x = -1, y = f(-1) = (-1+2)(-1-1)(-1-4) = 10$.

$x = 2, y = f(2) = (2+2)(2-1)(2-4) = -8$.

Therefore, the average rate of change in y with respect to x over the interval [–1,2] is given by

$$\frac{f(2) - f(-1)}{2 - (-1)} = \frac{-8 - 10}{3} = -6$$

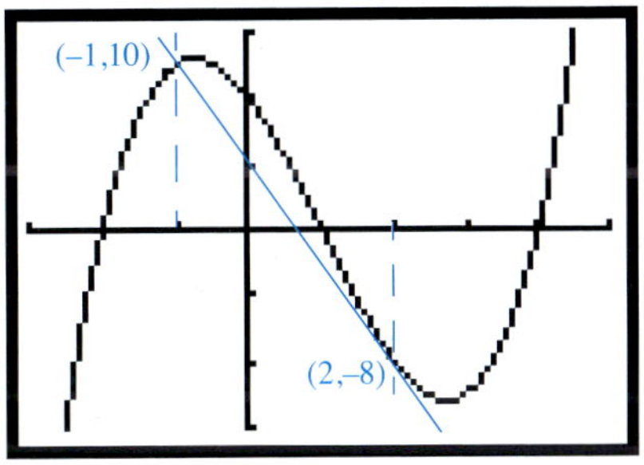

(b) To determine the rate of change at $x = 4$, we choose a second point close to $x = 4$. In this case, we use the point $x = 4 + h$, where h can be considered to be a very small number.

We will look at what happens to the gradient of the secant joining the points $(4, 0)$ and $(4 + h, f(4 + h))$ as h approaches zero.

The gradient of the secant is given by

$$\frac{f(4+h) - f(4)}{(4+h) - 4} = \frac{f(4+h) - f(4)}{h}$$

We now need to determine the value of $f(4+h)$ and $f(4)$. However, we already know that $f(4) = 0$.

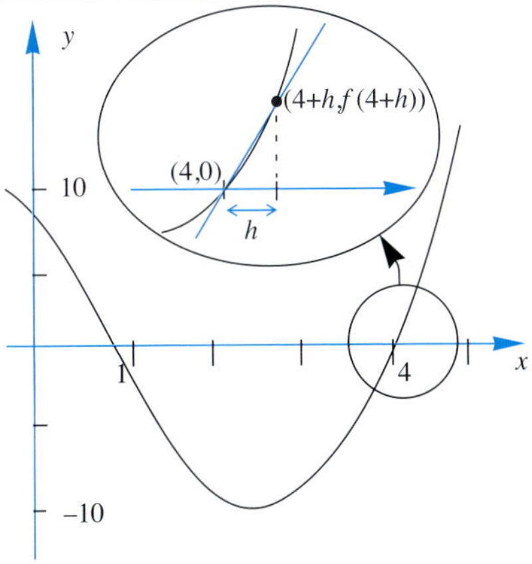

We can now find values for $f(4+h)$ as h approaches zero.

For $h = 0.1$, $f(4+0.1) = f(4.1) = (4.1+2)(4.1-1)(4.1-4) = 6.1 \times 3.1 \times 0.1 = 1.891$.

Therefore, $\frac{f(4+h)-f(4)}{h} = \frac{1.891-0}{0.1} = 18.91$, for $h = 0.1$.

We can continue in this same manner by making the value of h smaller still.
We do this by setting up a table of values:

h	$\frac{f(4+h)-f(4)}{h}$
0.01	18.09010000
0.001	18.00900100
0.0001	18.00090001

From the table, it appears that as h approaches zero, the gradient of the secant (which becomes the gradient of the tangent at (4,0)) approaches a value of 18.
Therefore, we have that the rate of change of f at (4,0) is 18.

More formally we write this result as $\lim_{h \to 0} \frac{f(4+h)-f(4)}{h} = 18$, which is read as

"The limit as h tends to zero of $\frac{f(4+h)-f(4)}{h}$ is equal to 18."

Which is saying that if we make h as small as we can, then $\frac{f(4+h)-f(4)}{h}$ equals 18.

EXAMPLE 18.9

The population of a city at the start of 2000 was 2.3 million, and its projected population, N million, is modelled by the equation $N(t) = 2.3e^{0.0142t}$, where $t \geq 0$ and is measured in years since the beginning of 2000. Find the rate of growth of the population in this city at the start of 2005.

Solution

Finding the rate of growth of the population at the start of 2005 as opposed to finding the rate over a period of time means that we are finding the instantaneous rate of change. To do this, we proceed as in the previous example, i.e., we use a limiting approach.

Consider the two points, $P(5, N(5))$ (start of 2005) and $A(5+h, N(5+h))$ on the curve representing the population size:

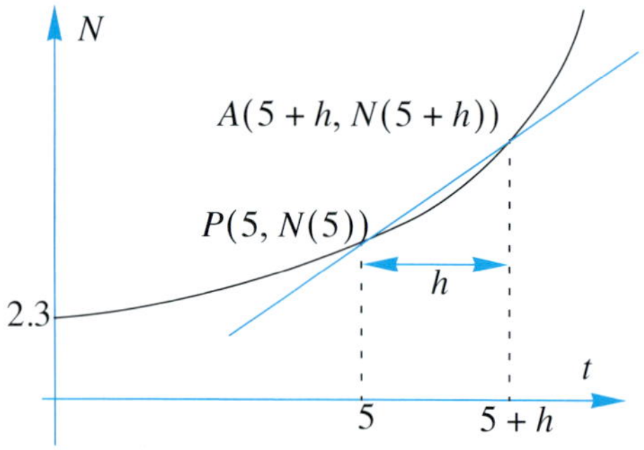

The gradient of the secant passing through P and A is given by $\frac{N(5+h)-N(5)}{(5+h)-5} = \frac{N(5+h)-N(5)}{h}$

Now, $N(5) = 2.3e^{0.0142 \times 5} = 2.3e^{0.071}$

and $N(5+h) = 2.3e^{0.0142(5+h)}$

Therefore, the gradient of the secant is given by

$$\frac{2.3e^{0.0142(5+h)} - 2.3e^{0.071}}{h} = \frac{2.3e^{0.071 + 0.0142h} - 2.3e^{0.071}}{h}$$

$$= \frac{2.3e^{0.071}(e^{0.0142h} - 1)}{h}$$

Again we set up a table of values:

h	$\frac{2.3e^{0.071}(e^{0.0142h} - 1)}{h}$
0.1	$\frac{2.3e^{0.071}(e^{0.0142 \times 0.1} - 1)}{0.1} = 0.035088$
0.01	$\frac{2.3e^{0.071}(e^{0.0142 \times 0.01} - 1)}{0.01} = 0.035066$
0.001	$\frac{2.3e^{0.071}(e^{0.0142 \times 0.001} - 1)}{0.001} = 0.035063$
0.0001	$\frac{2.3e^{0.071}(e^{0.0142 \times 0.0001} - 1)}{0.0001} = 0.035063$

Using limit notation we have, $\lim_{h \to 0} \frac{N(5+h) - N(5)}{h} = 0.035063$.

That is, the growth rate at the start of 2005 is 35,063 people per year.

EXERCISES 18.3

1. For each of the graphs shown, find the gradient of the secant joining the points P and Q.

(a)

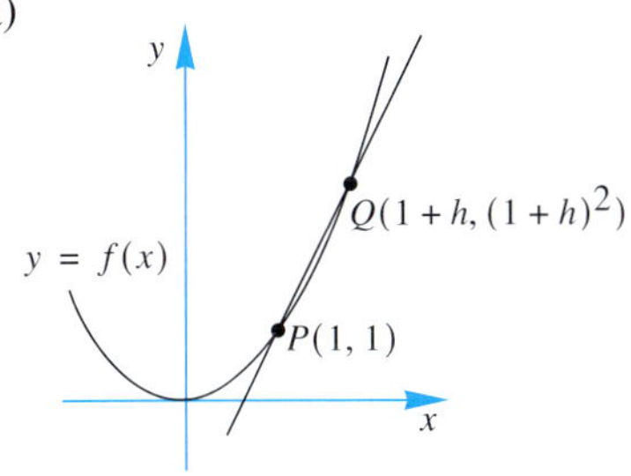

(b)

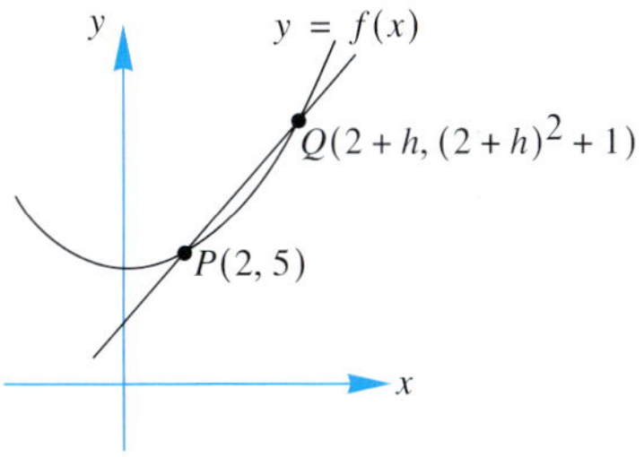

(c)

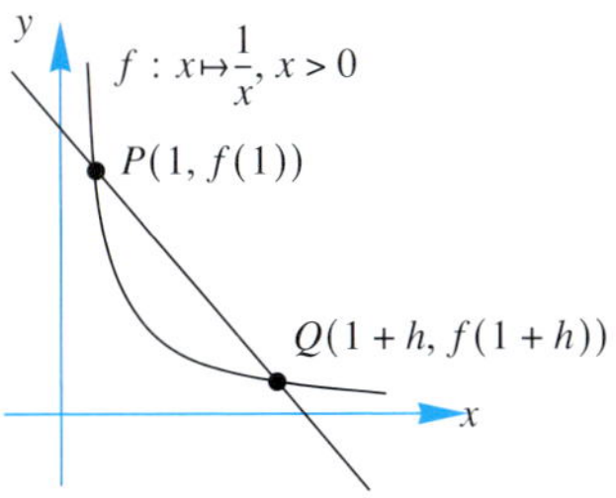

(d)

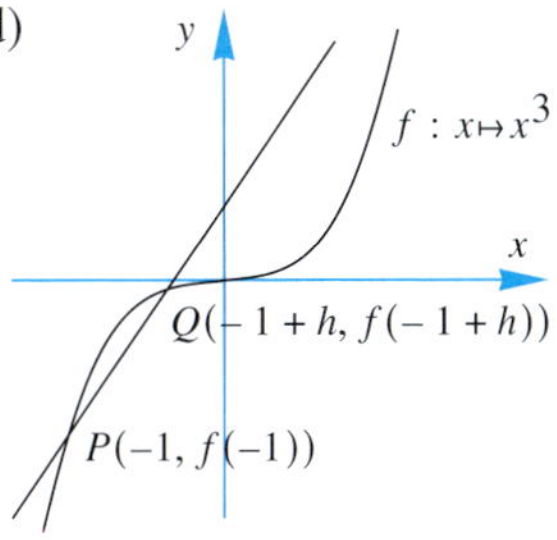

2. For each of the graphs in Q.1., use a limiting argument to deduce the instantaneous rate of change of the given function at the point *P*.

3. For each of the functions, *f*, given below, find the gradient of the secant joining the points $P(a, f(a))$ & $Q(a+h, f(a+h))$.

(a) $f(x) = 3 + x^2$ (b) $f(x) = 1 - x^2$ (c) $f(x) = (x+1)^2 - 2$

(d) $f(x) = x^3 + x$ (e) $f(x) = 2 - x^3$ (f) $f(x) = x^3 - x^2$

(g) $f(x) = \frac{2}{x}$ (h) $f(x) = \frac{1}{x-1}$ (i) $f(x) = \sqrt{x}$

4. For each of the functions, *f*, given below, find the gradient of the secant joining the points $P(a, f(a))$ & $Q(a+h, f(a+h))$ and hence deduce the gradient of the tangent drawn at the point *P*.

(a) $f(x) = x$ (b) $f(x) = x^2$ (c) $f(x) = x^3$ (d) $f(x) = x^4$.

Hence deduce the gradient of the tangent drawn at the point $P(a, f(a))$ for the function $f(x) = x^n, n \in N$.

5. An object moves along a straight line. Its position, *x* metres (from a fixed point O), at time *t* seconds is given by $x(t) = 2t^2 - 3t + 1, t \geq 0$.

(a) Sketch the graph of its displacement function.

(b) Determine
  i. its average velocity over the interval from $t = 1$ to $t = 2$.
  ii. its average velocity over the interval $t = 1$ to $t = 1.5$.
  iii. its average velocity over the interval $t = 1$ to $t = 1.1$.

(c) Show that its average velocity over the interval $t = 1$ to $t = 1 + h$, where *h* is small is given by $1 + 2h$.

(d) How can the last result help us determine the objects' velocity at $t = 1$?

(e) Show that its average velocity over any time interval of length *h* is given by $4t + 2h - 3$. Hence deduce the object's velocity at any time *t* during its motion.

6. The healing process of a certain type of wound is measured by the decrease in surface area that the wound occupies on the skin. A certain skin wound has its surface area modelled by the equation $S(t) = 20 \times 2^{-0.1t}$ where *S* sq. cm is the unhealed area *t* days after the skin received the wound.

(a) Sketch the graph of $S(t) = 20 \times 2^{-0.1t}, t \geq 0$.

(b) i. What area did the wound originally cover?
  ii. What area will the wound occupy after 2 days?
  iii. How much of the wound healed over the two day period?
  iv. Find the average rate at which the wound heals over the first two days.

(c) How much of the wound would heal over a period of *h* days?

(d) Find the rate at which the wound heals
  i. immediately after it occurs
  ii. one day after it occured.

18.4 DIFFERENTIATION PROCESS

18.4.1 THE DERIVATIVE AND THE GRADIENT FUNCTION

In the previous sections we concentrated our efforts on determining the average rate of change for a function over some fixed inteval. Then proceeded to find the instantaneous rate of change at a particular point (on the curve). We now consider the same process, with the exception that we will discuss the instantaneous rate at any point $P(x, f(x))$. The result will be an expression that will enable us to determine the instantaneous rate of change of the function at any point on the curve. Because the instantaneous rate of change at a point on a curve is simply a measure of the gradient of the curve at that point, our newly found result will be known as the **gradient function** (otherwise known as the derivative of the function).

For a continuous function, $y = f(x)$, we deduced that the instantaneous rate of change at the point $P(a, f(a))$ is given by $\dfrac{f(a+h)-f(a)}{h}$, where h is taken to be very small (in fact we say that h approaches or tends to zero).

So, to determine the rate at which a graph changes at a single point, we need to find the slope of the tangent line at that point.

This becomes obvious if we look back at our 'zooming in process—where the tangent line to the function at the point $P(a, f(a))$ is the line that best approximates the graph at that point.

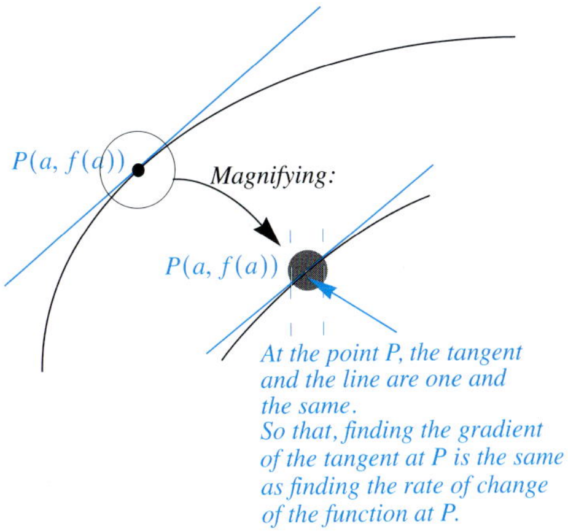

Rather than considering a fixed point $P(a, f(a))$, we now consider any point $P(x, f(x))$ on the curve with equation $y = f(x)$:

The rate of change of the function *f* at $P(x, f(x))$ is therefore given by the gradient of the tangent to the curve at *P*.

If point *Q* comes as close as possible to the point *P*, so that h approaches zero, then, the gradient of the tangent at *P* is given by the gradient of the secant joining the points $P(x, f(x))$ and $Q(x+h, f(x+h))$ as $h \to 0$.

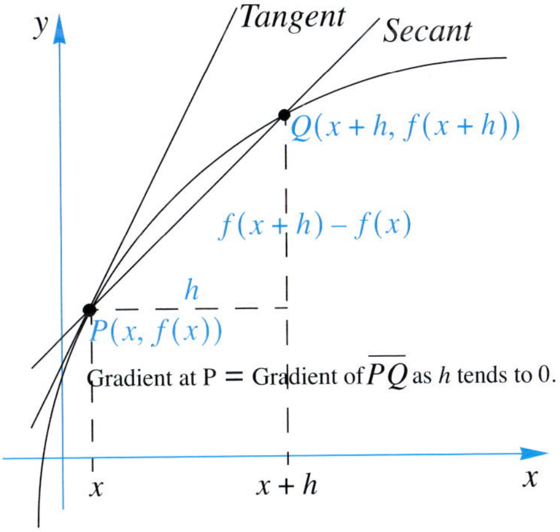

In mathematical notation we have:

$$\text{Rate of change at P} = \lim_{h \to 0} \frac{f(x+h)-f(x)}{h}$$

18.4.2 NOTATION AND LANGUAGE

We now introduce the term **derivative of a function**:

The **rate of change** of $f(x)$ at $P(x, f(x))$ = **Gradient function** of $f(x)$ at $P(x, f(x))$

= **The derivative** of $f(x)$

$$= \lim_{h \to 0} \frac{f(x+h) - f(x)}{h}$$

The derivative of a function $f(x)$ is denoted by $f'(x)$ and is read as "*f* dash of *x*".

That is, $$f'(x) = \lim_{h \to 0} \frac{f(x+h) - f(x)}{h}$$

Finding the derivative of a function using this approach is referred to as *finding the derivative of f from first principles*.

It is important to realise that in finding $f'(x)$ we have a new function—called the **gradient function**, because the expression $f'(x)$ will give the gradient anywhere on the curve of $f(x)$. So that if we want the gradient of the function $f(x)$ at $x = 5$, we first determine $f'(x)$ and then substitute the value of $x = 5$ into the equation of $f'(x)$.

EXAMPLE 18.10

Using the first principles method, find the derivative (*or the gradient function*) of the function $f(x) = 3x^2 + 4$. Hence, find the gradient of the function at $x = 3$.

Solution

Using the first principles method means that we must make use of the expression

$$f'(x) = \lim_{h \to 0} \frac{f(x+h) - f(x)}{h} \quad \text{– (1)}$$

We start by first evaluating the expression $f(x+h) - f(x)$:

That is:

$$\begin{aligned} f(x+h) - f(x) = 3(x+h)^2 + 4 - [3x^2 + 4] &= 3(x^2 + 2xh + h^2) + 4 - 3x^2 - 4 \\ &= 3x^2 + 6xh + 3h^2 - 3x^2 \\ &= 6xh + 3h^2 \end{aligned}$$

Substituting this result into (1):

$$\begin{aligned} \lim_{h \to 0} \frac{f(x+h) - f(x)}{h} &= \lim_{h \to 0} \frac{6xh + 3h^2}{h} \\ &= \lim_{h \to 0} \frac{h(6x + 3h)}{h} \\ &= \lim_{h \to 0} (6x + 3h),\ h \neq 0 \\ &= 6x \end{aligned}$$

That is, we now have the gradient function $f'(x) = 6x$.

To determine the gradient of the function at $x = 3$, we need to substitute the value $x = 3$ into the gradient function. That is, $f'(3) = 6 \times 3 = 18$.

Using the TI–83 we can determine the derivative at $x = 3$ by using the '**nDeriv(**' command from the **MATH** menu:

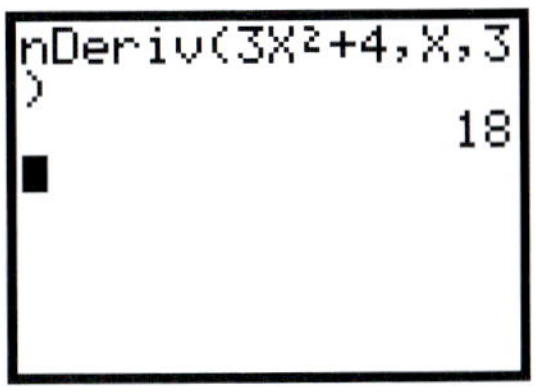

EXAMPLE 18.11

A particle moving along a straight line has its position at time t seconds governed by the equation $p(t) = t - 0.2t^2, t \geq 0$, where $p(t)$ is its position in metres from a fixed point O.

(a) Find the particle's velocity after it has been in motion for 1 second.

(b) Find the particle's velocity at time $t = a$, $a > 0$.

Solution

(a) This part is readily done by making use of the TI–83:
So, we have that the particle's velocity is 0.6 m/s

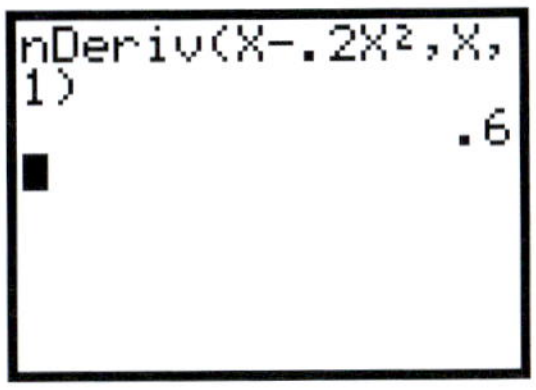

(b) Let us again make use of the TI–83:
We enter the function $p(t) = t - 0.2t^2, t \geq 0$ as $\mathbf{Y}_1$ and then enter $\mathbf{Y}_2$ as **nDeriv(**$\mathbf{Y}_1$**, X, X)**. Once this is done we sketch both $\mathbf{Y}_1$ and $\mathbf{Y}_2$.
The graph of $\mathbf{Y}_2$ represents the gradient function of $\mathbf{Y}_1$, i.e., the derivative of $\mathbf{Y}_1$.

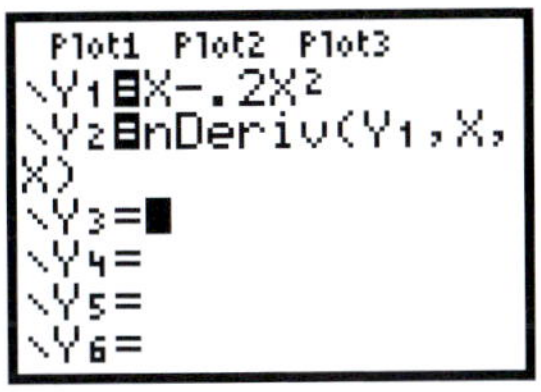

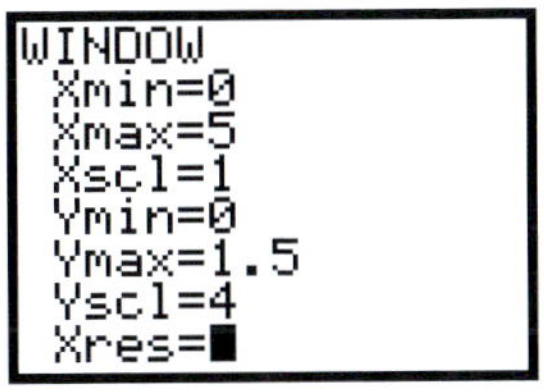

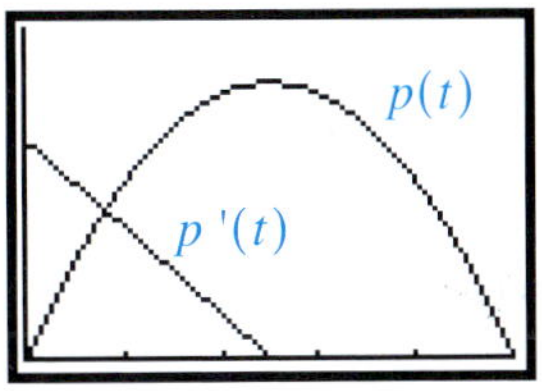

The graph of $p'(t)$ appears to be a straight line passing through the points $(0,1)$ and $(2.5, 0)$ and so we can obtain the equation of this straight line:

gradient $= m = \dfrac{0-1}{2.5-0} = -0.4$. Then, using $y - 1 = (-0.4)(x-0)$ we have, $y = -0.4x + 1$.

Changing to the appropriate variables we have, $v(t) = -0.4t + 1$.

Therefore at $t = a$, $v(a) = -0.4a + 1$.

Obviously, this method relied on our ability to spot the type of function **nDeriv(**$\mathbf{Y}_1$**, X, X)** produced. The fact that it could then be readily determined made it all fairly straight forward. However, more often than not, this will not be the case. So we make use of the definition of differentiation to obtain the velocity as the derivative of the position function.

$$v(t) = p'(t) = \lim_{h \to 0} \frac{p(t+h) - p(t)}{h}$$

We start by determining $p(t+h) - p(t) = (t+h) - 0.2(t+h)^2 - [t - 0.2t^2]$

$$= t + h - 0.2(t^2 + 2th + h^2) - t + 0.2t^2$$

$$= -0.4th + h - 0.2h^2$$

Therefore, $v(t) = \lim_{h \to 0} \dfrac{-0.4th + h - 0.2h^2}{h} = \lim_{h \to 0}(-0.4t + 1 - 0.2h) = -0.4t + 1$

And so, $v(a) = -0.4a + 1$.

EXERCISES 18.4

1. Use a limiting process to find the gradients of these curves at the points indicated:

(a) $x \mapsto x^3$ at $x = 1$ (b) $v = 2t^2 - 1$ at $t = 2$

(c) $f(x) = \dfrac{1}{x}$ at $x = 3$ (d) $x \mapsto 2^x$ at $x = 1$

(e) $f = t^2 - 2t + 3$ at $t = 0.5$ (f) $t \mapsto \dfrac{t^2 - 1}{t}$ at $t = 4$

2. An object is dropped from a high building. The distance, d metres, that the object has fallen, t seconds after it is released, is given by the formula $d = 4.9t^2, 0 \le t \le 3$.

(a) Find the distance fallen during the first second.
(b) Find the distance fallen between $t = 1$ and $t = h + 1$ seconds.
(c) Hence, find the speed of the object 1 second after it is released.

3. Find, from first principles, the gradient function, f' , of the following

(a) $f:x \mapsto 4x^2$ (b) $f:x \mapsto 5x^2$ (c) $f:x \mapsto 4x^3$
(d) $f:x \mapsto 5x^3$ (e) $f:x \mapsto 4x^4$ (f) $f:x \mapsto 5x^4$

Can you see a pattern in your results?

4. Find, from first principles, the derivative of the following functions

(a) $f(x) = 2x^2 - 5$ (b) $g(x) = 2 - x$ (c) $g(x) = 2 - x + x^3$

(d) $f(x) = \dfrac{1}{x}$ (e) $f(x) = \dfrac{2}{x + 1}$ (f) $f(x) = \sqrt{x}$

5. A particle moving along a straight line has its position at time t seconds governed by the equation $x(t) = 2t - 0.5t^2, t \ge 0$, where $x(t)$ is its position in metres from the origin O.

(a) Find the particle's velocity after it has been in motion for 1 second.
(b) Find the particle's velocity at time $t = a$, $a > 0$.

6. A particle moving along a straight line has its position at time t seconds governed by the equation $x(t) = 4t^2 - t^3, t \ge 0$, where $x(t)$ is its position in metres from the origin O.

(a) Sketch the displacement-time graph of the motion over the first five seconds
(b) Find the particle's velocity at time i. $t = 1$
ii. $t = 2$
(c) Find the particle's velocity at any time $t, t \ge 0$.
(d) When will the particle first come to rest?

19.1 DIFFERENTIATION

19.1.1 REVIEW

1. Rate of change, gradient and the derivative

The **rate of change** of a curve at a point gives a **measure of the gradient** of the curve at that point. When finding the **derivative** of the equation of a curve we obtain the **gradient function**. As the name suggests, the gradient function enables us to find the gradient at any point on the curve.

2. Differentiation

Differentiation is the process of finding the derivative of a function. The derivative of a function is often called its **derived function**.

3. Language and notation

The **derivative of $f(x)$ with respect to x** is usually written as $f'(x)$ (read as "*f* dash of *x*") or $\frac{d}{dx}(f(x))$ (read as "dee–dee–*x* of $f(x)$").

The **derivative of *y* with respect to *x*** is usually written as $\frac{dy}{dx}$ (read as "dee–*y*–dee–*x*") or *y*' (read as "*y*–dash of *x*").

4. Average rate → instantaneous rate

The **average rate** of change of the function $f(x)$ over the interval $x = x_1$ to $x = x_2$ is graphically represented by the **gradient of the secant** passing through the two points $(x_1, f(x_1))$ and $(x_2, f(x_2))$ on the curve $y = f(x)$.

Notice that when we talk about the average rate, we require two points on a curve.
i.e., ". . *over the interval* . ." or ". . *from* . . *to* . ."

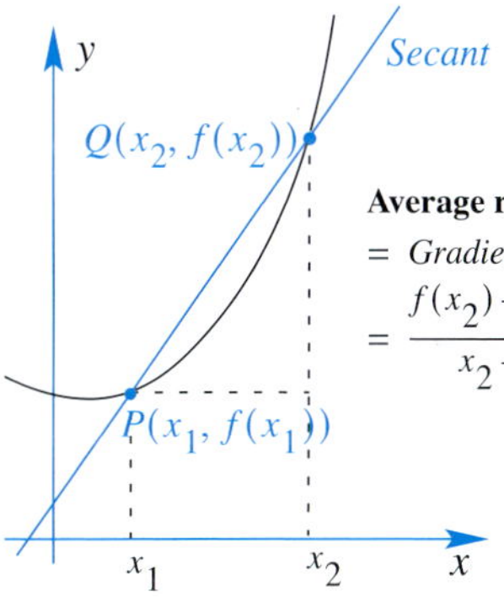

Average rate of change
= *Gradient of secant*
$= \frac{f(x_2) - f(x_1)}{x_2 - x_1}$

Whereas the **instantaneous rate** of change of $f(x)$ at the point $x = x_1$ is graphically represented by the **gradient of the tangent** at the point $(x_1, f(x_1))$ on the curve $y = f(x)$.

Notice that when we talk about the instantaneous rate, we refer to only one point,
i.e., ". . *at the point* . ."

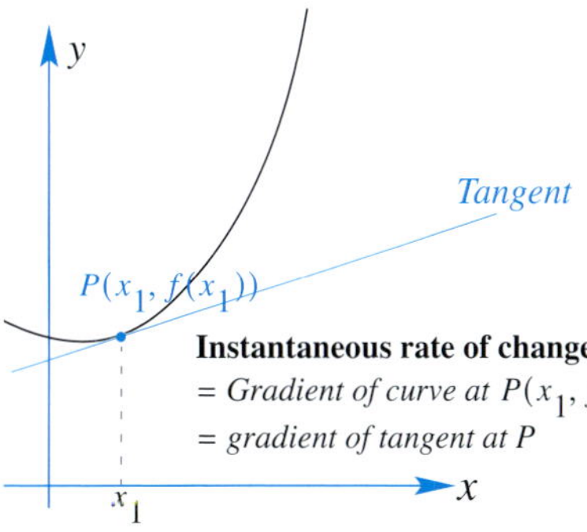

Instantaneous rate of change
= *Gradient of curve at* $P(x_1, f(x_1))$
= *gradient of tangent at P*

5. Different notation

The basic difference between the two can be seen in the diagram below:

Functional Notation

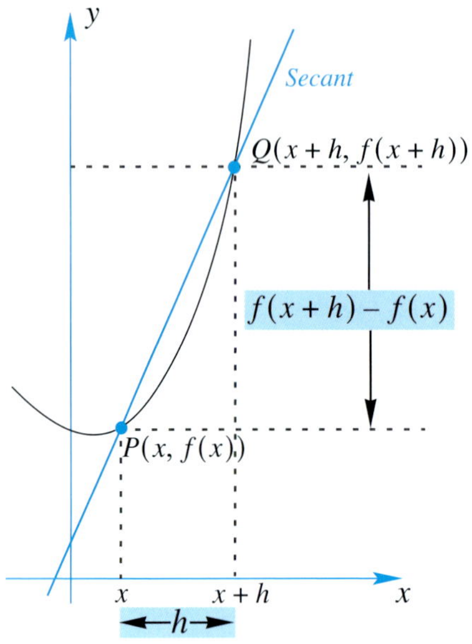

Gradient of secant $= \dfrac{f(x+h)-f(x)}{h}$

Delta Notation

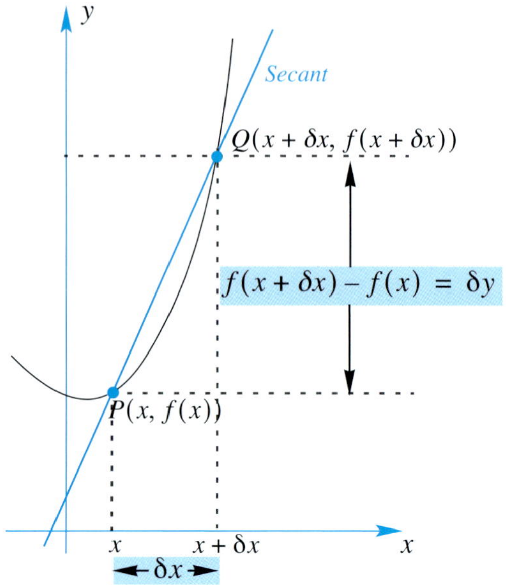

Gradient of secant $= \dfrac{f(x+\delta x)-f(x)}{\delta x} = \dfrac{\delta y}{\delta x}$

To find the gradient at some point *P*, that is, to find the derivative of the function at any point *P* on the curve defined by the equation $y = f(x)$, we use the method of first principles:

$$f'(x) = \lim_{h \to 0} \frac{f(x+h)-f(x)}{h}, h \neq 0$$

$$\frac{dy}{dx} = \lim_{\delta x \to 0} \frac{\delta y}{\delta x}, \delta x \neq 0$$

Using the delta notation, we read $\dfrac{dy}{dx} = \lim\limits_{\delta x \to 0} \dfrac{\delta y}{\delta x}, \delta x \neq 0$ as "dee–*y*–dee–*x* is equal to the limit as delta *x* (δx) tends to zero of delta *y* (δy) on delta *x* (δx)". This is in part where the expression "The derivative of *y* with respect to *x*" stems from. The notation $f'(x)$ is due to one of the greatest eighteenth century mathematicians, Joseph Louis Lagrange (1736–1813), whereas the notation $\dfrac{dy}{dx}$ is attributed to that other great mathematician, Gottfried Wilhelm Leibniz (1646–1716).

As we now have the definition of the derivative, given by the expression $\dfrac{dy}{dx} = \lim\limits_{\delta x \to 0} \dfrac{\delta y}{\delta x}, \delta x \neq 0$ or $f'(x) = \lim\limits_{h \to 0} \dfrac{f(x+h)-f(x)}{h}, h \neq 0$, then to differentiate a function $y = f(x)$ one of these expressions is used. Notice that the expression $f'(x)$ is itself a function and for this reason we also refer to the derivative as the **gradient function** of $y = f(x)$.

EXAMPLE 19.1

Find the derivative (the gradient function) of $f(x) = x^3 + 1$.

Solution

By definition, we have that $f'(x) = \lim_{h \to 0} \frac{f(x+h) - f(x)}{h}, h \neq 0$, and so, we start by simplifying the expression $f(x+h) - f(x)$:

$$\begin{aligned} f(x+h) - f(x) &= (x+h)^3 + 1 - (x^3 + 1) \\ &= x^3 + 3x^2h + 3xh^2 + h^3 + 1 - x^3 - 1 \\ &= 3x^2h + 3xh^2 + h^3 \\ &= h(3x^2 + 3xh + h^2) \end{aligned}$$

Therefore we have $f'(x) = \lim_{h \to 0} \frac{f(x+h) - f(x)}{h}, h \neq 0 = \lim_{h \to 0} \frac{h(3x^2 + 3xh + h^2)}{h}, h \neq 0$

$$\begin{aligned} &= \lim_{h \to 0} (3x^2 + 3xh + h^2), h \neq 0 \\ &= 3x^2 \end{aligned}$$

That is, $f(x) = x^3 + 1 \Rightarrow f'(x) = 3x^2$.

EXAMPLE 19.2

Differentiate the function $f(x) = \sqrt{x}$.

Solution

As in the previous example, we first simplify the expression $f(x+h) - f(x)$:

$$f(x+h) - f(x) = \sqrt{x+h} - \sqrt{x}.$$

However, in this case it appears that we can go no further. However;

$$f'(x) = \lim_{h \to 0} \frac{\sqrt{x+h} - \sqrt{x}}{h} = \lim_{h \to 0} \frac{\sqrt{x+h} - \sqrt{x}}{h} \times \frac{\sqrt{x+h} + \sqrt{x}}{\sqrt{x+h} + \sqrt{x}}$$

The difficulty is in realising that we need to multiply by $\frac{\sqrt{x+h} + \sqrt{x}}{\sqrt{x+h} + \sqrt{x}}$ (very tricky!). We do this so that we can rationalise the numerator. This will, hopefully, lead to an expression which will be more manageable. Carrying out the multiplication we have:

$$f'(x) = \lim_{h \to 0} \frac{(\sqrt{x+h} - \sqrt{x})(\sqrt{x+h} + \sqrt{x})}{h(\sqrt{x+h} + \sqrt{x})}, h \neq 0 = \lim_{h \to 0} \frac{((x+h) - x)}{h(\sqrt{x+h} + \sqrt{x})}, h \neq 0$$

$$\begin{aligned} &= \lim_{h \to 0} \frac{h}{h(\sqrt{x+h} + \sqrt{x})}, h \neq 0 \\ &= \lim_{h \to 0} \frac{1}{\sqrt{x+h} + \sqrt{x}}, h \neq 0 \\ &= \frac{1}{\sqrt{x} + \sqrt{x}} \\ &= \frac{1}{2\sqrt{x}} \end{aligned}$$

Notice that the function $f(x) = \sqrt{x}, x \geq 0$ has a domain defined by $x \geq 0$. Whereas its gradient function, $f'(x) = \frac{1}{2\sqrt{x}}, x > 0$ has a domain defined by $x > 0$.

So, in this instance, the domain of the function and that of its derivative (or gradient function) are not the same. We will investigate these matters in more depth later on.

19.1.2 POWER RULE FOR DIFFERENTIATION

Finding the derivative from first principles can be tedious. The previous two examples clearly show this. However, using the first principles approach, produces the results shown in the table below:

Function $y = f(x)$	x^4	x^3	x^2	x^1	x^{-1}	x^{-2}
Derivative $\frac{dy}{dx} = f'(x)$	$4x^3$	$3x^2$	$2x^1$	$1x^0$	$-1x^{-2}$	$-2x^{-3}$

Based on these results and following the general pattern, it is reasonable to assume the general result that if

$$y = x^n, n \in Z, \text{ then } \frac{dy}{dx} = nx^{n-1}.$$

In fact this rule is true for any exponent $n \in \mathbb{R}$, i.e., for any real number n.

For example, if we look at the square root function, then we have that $y = \sqrt{x} = x^{1/2}$. So in this case we have that $n = \frac{1}{2}$. Then, using our rule we have $y = \sqrt{x} = x^{\frac{1}{2}} \Rightarrow \frac{dy}{dx} = \frac{1}{2}x^{\frac{1}{2}-1} = \frac{1}{2}x^{-\frac{1}{2}}$

$$= \frac{1}{2\sqrt{x}}.$$

Which is the result we obtained in Example 19.2 when we used the first principle method. This result is known as the **power rule for differentiation**.

Notice that for the case $n = 0$, then $y = x^0$ and so we have that $\frac{dy}{dx} = 0x^{0-1} = 0$.

Note: The function $y = x^0$ represents the horizontal straight line $y = 1$, and so its gradient will always be 0. In fact, for the case where $y = k$ (a real constant) $y = kx^0 \Rightarrow \frac{dy}{dx} = k \times 0x^{0-1} = 0$.

As the function $y = k$ represents a horizontal straight line its gradient will always be 0. We therefore have the power rule:

> The function $f:x \mapsto x^n$, has a gradient function (or derivative) given by $f':x \mapsto nx^{n-1}$.
> This can also be written as:
>
> If $y = x^n$ then $\frac{dy}{dx} = nx^{n-1}$ or If $f(x) = x^n$ then $f'(x) = nx^{n-1}$

We prove the above for the case where n is a positive integer and leave further proofs till §19.8.

Let $f(x) = x^n$ where n is a positive integer. Using the definition of the derivative we have

$$f'(x) = \lim_{h \to 0} \frac{f(x+h) - f(x)}{h} = \lim_{h \to 0} \frac{(x+h)^n - x^n}{h}$$

Simplifying the numerator, we have:

$$(x+h)^n - x^n = \left[x^n + \binom{n}{1}x^{n-1}h + \binom{n}{2}x^{n-2}h^2 + \ldots + \binom{n}{n-1}xh^{n-1} + h^n\right] - x^n$$

$$= \binom{n}{1}x^{n-1}h + \binom{n}{2}x^{n-2}h^2 + \ldots + \binom{n}{n-1}xh^{n-1} + h^n$$

$$= nx^{n-1}h + \frac{n(n-1)}{2}x^{n-2}h^2 + \ldots + nxh^{n-1} + h^n$$

We now have that $$f'(x) = \lim_{h \to 0} \frac{nx^{n-1}h + \frac{n(n-1)}{2}x^{n-2}h^2 + \ldots + nxh^{n-1} + h^n}{h}$$

$$= \lim_{h \to 0}\left(nx^{n-1} + \frac{n(n-1)}{2}x^{n-2}h + \ldots + nxh^{n-2} + h^{n-1}\right)$$

$$= nx^{n-1}$$

EXAMPLE 19.3

Use the power rule to differentiate the following functions

(a) x^6 (b) $\frac{1}{\sqrt{x}}$ (c) $\sqrt[3]{x}$ (d) $\frac{1}{x^2}$

Solution

Before we differentiate these functions, each function must be rewritten in the form x^n before we can use the power rule.

(a) Let $f(x) = x^6 \Rightarrow f'(x) = 6x^{6-1} = 6x^5$

(b) Let $y = \frac{1}{\sqrt{x}}$. Remember, we first need to rewrite it in the form x^n:

Function →	Step 1 Rewrite: →	Step 2 Use power rule: →	Step 3 Simplify:
$y = \frac{1}{\sqrt{x}}$	$y = x^{-1/2}$	$\frac{dy}{dx} = -\frac{1}{2}x^{-\frac{1}{2}-1}$	$\frac{dy}{dx} = -\frac{1}{2}x^{-\frac{3}{2}} = -\frac{1}{2\sqrt{x^3}}$

(c) Let $y = \sqrt[3]{x}$. As in the previous example, we rewrite this function in the form x^n so that we can use the power rule:

Function →	Step 1 Rewrite: →	Step 2 Use power rule: →	Step 3 Simplify:
$y = \sqrt[3]{x}$	$y = x^{\frac{1}{3}}$	$\frac{dy}{dx} = \frac{1}{3}x^{\frac{1}{3}-1}$	$\frac{dy}{dx} = \frac{1}{3}x^{-\frac{2}{3}} = \frac{1}{3\sqrt[3]{x^2}}$

(d) Let $f(x) = \frac{1}{x^2}$ so that $f(x) = x^{-2} \therefore f'(x) = -2x^{-2-1} = -2x^{-3}$. That is, $f'(x) = -\frac{2}{x^3}$.

An extension of the power rule is the derivative of $f:x \mapsto ax^n$, where a is a real constant. In this case we have the general result:

$$\text{If } f:x \mapsto ax^n \text{ then } f':x \mapsto anx^{n-1}$$

EXAMPLE 19.4

Differentiate the following functions:

(a) $12x^3$ (b) $-\frac{4}{x}$ (c) $\frac{x^2}{7}$

Solution

(a) Let $y = 12x^3 \Rightarrow \frac{dy}{dx} = 12 \times 3x^{3-1} = 36x^2$.

(b) Let $f(x) = -\frac{4}{x}$, that is, $f(x) = -4x^{-1} \Rightarrow f'(x) = -4 \times -1x^{-1-1} = 4x^{-2} = \frac{4}{x^2}$.

(c) Let $f(x) = \frac{x^2}{7}$, that is, $f(x) = \frac{1}{7}x^2 \Rightarrow f'(x) = \frac{1}{7} \times 2x^{2-1} = \frac{2}{7}x$.

19.1.3 DERIVATIVE OF A SUM OR DIFFERENCE

This rule states that the derivative of a sum (or a difference) is equal to the sum (or the difference) of the derivatives. That is,

$$\text{If } y = f(x) \pm g(x) \text{ then } \frac{dy}{dx} = f'(x) \pm g'(x)$$

EXAMPLE 19.5

Differentiate the following functions

(a) $2x^3 + 5x - 9$ (b) $\sqrt{x} - \frac{5}{x^3} + x$ (c) $x^{1/3} + x^{5/4} - \sqrt{2x}$

Solution

(a) Let $y = 2x^3 + 5x - 9 \Rightarrow \frac{dy}{dx} = \frac{d}{dx}(2x^3 + 5x - 9) = \frac{d}{dx}(2x^3) + \frac{d}{dx}(5x) - \frac{d}{dx}(9)$

$$= 6x^2 + 5$$

Notice we have used a slightly different notation, namely that $f'(x) = \frac{d}{dx}(f(x))$. We can think of $\frac{d}{dx}$ as the differentiation operator, so that $\frac{d}{dx}(f(x))$ or $\frac{d}{dx}(y)$ is an operation of differentiation done on $f(x)$ or y respectively.

(b) Let $f(x) = \sqrt{x} - \frac{5}{x^3} + x \Rightarrow f'(x) = \frac{d}{dx}\left(\sqrt{x} - \frac{5}{x^3} + x\right) = \frac{d}{dx}(x^{1/2} - 5x^{-3} + x)$

$$= \frac{1}{2}x^{-1/2} - 5 \times -3x^{-3-1} + 1$$

$$= \frac{1}{2\sqrt{x}} + \frac{15}{x^4} + 1$$

(c) $$\frac{d}{dx}(x^{1/3} + x^{5/4} - \sqrt{2x}) = \frac{d}{dx}(x^{1/3}) + \frac{d}{dx}(x^{5/4}) - \frac{d}{dx}(\sqrt{2x})$$
$$= \frac{1}{3}x^{-2/3} + \frac{5}{4}x^{1/4} - \sqrt{2} \times \frac{1}{2}x^{-1/2}$$

Note, to find $\frac{d}{dx}(\sqrt{2x})$ we first express it as $\frac{d}{dx}(\sqrt{2} \times \sqrt{x}) = \sqrt{2} \times \frac{d}{dx}(\sqrt{x}) = \sqrt{2} \times \frac{d}{dx}(x^{1/2})$.

EXAMPLE 19.6 Miscellaneous Examples

Differentiate the following

(a) $5\sqrt{x} - 9, x \geq 0$ (b) $(x^3 + 2)^2$ (c) $\left(\sqrt{x} - \frac{2}{\sqrt{x}}\right)^2$

(d) $\frac{5x^2 + 4x - 3}{x}$ (e) $\sqrt{x}(7x^2 - 3x + 2)$ (f) $\frac{2x^4 - 4x + 3}{\sqrt{x}}$

Solution

(a) $$\frac{d}{dx}(5\sqrt{x} - 9) = \frac{d}{dx}(5x^{1/2} - 9) = 5 \times \frac{1}{2}x^{\frac{1}{2} - 1} - 0 = \frac{5}{2}x^{-1/2} = \frac{5}{2\sqrt{x}}.$$

(b) The first step is to expand the bracket so that each term is in the form of ax^n.

$$\frac{d}{dx}((x^3 + 2)^2) = \frac{d}{dx}(x^6 + 4x^3 + 4) = 6x^5 + 12x^2$$

(c) Again we expand first, and then differentiate:

$$\frac{d}{dx}\left(\left(\sqrt{x} - \frac{2}{\sqrt{x}}\right)^2\right) = \frac{d}{dx}\left((\sqrt{x})^2 - 2\sqrt{x} \times \frac{2}{\sqrt{x}} + \left(\frac{2}{\sqrt{x}}\right)^2\right) = \frac{d}{dx}\left(x - 4 + \frac{4}{x}\right)$$
$$= \frac{d}{dx}(x - 4 + 4x^{-1})$$
$$= 1 - \frac{4}{x^2}$$

(d) $$\frac{d}{dx}\left(\frac{5x^2 + 4x - 3}{x}\right) = \frac{d}{dx}\left(\frac{5x^2}{x} + \frac{4x}{x} - \frac{3}{x}\right) = \frac{d}{dx}(5x + 4 - 3x^{-1}) = 5 + \frac{3}{x^2}$$

(e) $$\frac{d}{dx}(\sqrt{x}(7x^2 - 3x + 2)) = \frac{d}{dx}(x^{1/2}(7x^2 - 3x + 2)) = \frac{d}{dx}(7x^{5/2} - 3x^{3/2} + 2x^{1/2})$$
$$= \frac{35}{2}x^{3/2} - \frac{9}{2}x^{1/2} + x^{-1/2}$$
$$= \frac{35}{2}\sqrt{x^3} - \frac{9}{2}\sqrt{x} + \frac{1}{\sqrt{x}}$$

(f) $$\frac{d}{dx}\left(\frac{2x^4 - 4x + 3}{\sqrt{x}}\right) = \frac{d}{dx}\left(\frac{2x^4}{\sqrt{x}} - \frac{4x}{\sqrt{x}} + \frac{3}{\sqrt{x}}\right) = \frac{d}{dx}\left(2x^{4 - \frac{1}{2}} - 4x^{1 - \frac{1}{2}} + 3x^{0 - \frac{1}{2}}\right)$$
$$= \frac{d}{dx}(2x^{7/2} - 4x^{1/2} + 3x^{-1/2})$$
$$= 7x^{5/2} - 2x^{-1/2} - \frac{3}{2}x^{-3/2}$$

$$= 7x^2\sqrt{x} - \frac{2}{\sqrt{x}} - \frac{3}{2x\sqrt{x}}$$

EXERCISES 19.1

1. Find the derivative of each of the following

(a)	x^5	(b)	x^9	(c)	x^{25}
(d)	$9x^3$	(e)	$-4x^7$	(f)	$\frac{1}{4}x^8$
(g)	$x^2 + 8$	(h)	$5x^4 + 2x - 1$	(i)	$-3x^5 + 6x^3 - x$
(j)	$20 - \frac{1}{3}x^4 + 10x$	(k)	$3x^3 - 6x^2 + 8$	(l)	$3x - 1 + \frac{x^2}{5} + x^4$

2. Find the derivative of each of the following

(a)	$\frac{1}{x^3}$	(b)	$\sqrt{x^3}$	(c)	$\sqrt{x^5}$
(d)	$\sqrt[3]{x}$	(e)	$4\sqrt{x}$	(f)	$6\sqrt{x^3}$
(g)	$2\sqrt{x} - \frac{3}{x} + 12$	(h)	$x\sqrt{x} + \frac{1}{\sqrt{x}} + 2$	(i)	$5\sqrt[3]{x^2} - 9x$
(j)	$5x - \frac{x}{\sqrt{x}} + \frac{4}{5x^2}$	(k)	$8\sqrt{x} + 3x^{-5} + \frac{x}{2}$	(l)	$\frac{x}{\sqrt{x^3}} - \frac{2}{x}\sqrt{x^3} + \frac{1}{3}x^3$

3. Find the derivative of each of the following

(a)	$\sqrt{x}(x+2)$	(b)	$(x+1)(x^3-1)$	(c)	$x\left(x^2 + 1 - \frac{1}{x}\right), x \neq 0$
(d)	$\frac{2x-1}{x}, x \neq 0$	(e)	$\frac{\sqrt{x}-2}{\sqrt{x}}, x > 0$	(f)	$\frac{x^2 - x + \sqrt{x}}{2x}, x \neq 0$
(g)	$\frac{3x^2 - 7x^3}{x^2}, x \neq 0$	(h)	$\left(x - \frac{2}{x}\right)^2, x \neq 0$	(i)	$\left(x + \frac{1}{x^2}\right)^2, x \neq 0$
(j)	$\sqrt{3x} - \frac{1}{3\sqrt{x}}, x > 0$	(k)	$(x - \sqrt[5]{x})^2, x \geq 0$	(l)	$\left(\frac{1}{\sqrt{x}} - \sqrt{x}\right)^3, x > 0$

4. (a) Show that if $f(x) = x^2 - x$, then $f'(x) = 1 + \frac{2f(x)}{x}$.

(b) Show that if $f(x) = \sqrt{2x} - 2\sqrt{x}, x \geq 0$, then $\sqrt{2x}f'(x) = 1 - \sqrt{2}, x > 0$.

(c) Show that if $y = ax^n$ where a is real and $n \in N$, then $\frac{dy}{dx} = \frac{ny}{x}, x \neq 0$.

(d) Show that if $y = \frac{1}{\sqrt{x}}, x > 0$, then $\frac{dy}{dx} + \frac{y}{2x} = 0$.

19.2 GRAPHICAL INTERPRETATION OF THE DERIVATIVE

19.2.1 THE VALUE OF THE DERIVATIVE AT A PARTICULAR POINT ON A CURVE

So far in this chapter we have looked at the **gradient function** of $f(x)$, namely $f'(x)$. The function $f'(x)$ represents the gradient at any point on the curve of the original function $f(x)$. In order to determine the gradient at a particular point $x = a$ (say), we revisit the definition of the derivative, $f'(x) = \lim_{h \to 0} \frac{f(x+h) - f(x)}{h}$. Rather than finding the derivative at 'any' point 'x', we determine the derivative at a particular point '$x = a$'. Note we are assuming that the derivative exists at the point '$x = a$'.

If the function $f(x)$ can be differentiated (i.e., is differentiable) at $x = a$, then

$$f'(a) = \lim_{h \to 0} \frac{f(a+h) - f(a)}{h}$$

Since $f'(a)$ gives the gradient of the tangent at $x = a$, it also gives the gradient of the graph of $y = f(x)$ and so the derivative may be used to determine gradients or to find where on a curve a particular gradient exists.

Another form of the derivative at $x = a$ is given by $f'(a) = \lim_{x \to a} \frac{f(x) - f(a)}{x - a}$. This result follows directly from our previous definition by replacing h with $x - a$. This can be seen by using the following 'replacements':

1. If $h = x - a$ then $h \to 0 \Rightarrow x - a \to 0 \Rightarrow x \to a$ and so we replace $\lim_{h \to 0}$ with $\lim_{x \to a}$.
2. If $h = x - a$ then $h + a = x$.

We first look at an example that uses a 'first principle' approach to determining the gradient of a curve at a particular point.

EXAMPLE 19.7

Let P and Q be points on the curve $y = x^2 - 2x$ for which $x = 2$ and $x = 2 + h$ respectively. Express the gradient of PQ in terms of h and hence find the gradient of the curve $y = x^2 - 2x$ at $x = 2$.

Solution

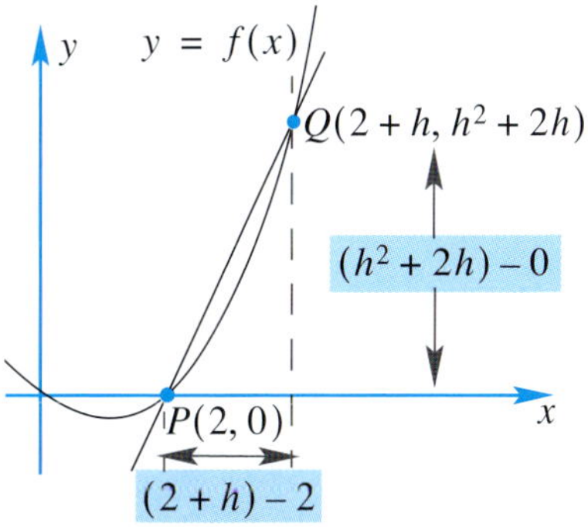

We begin by sketching the relevant curve and placing all the information on our diagram.

At P, $x = 2$ and so $y = 2^2 - 2(2) = 0$.

At Q, $x = 2 + h$ and so, $y = (2+h)^2 - 2(2+h) = h^2 + 2h$

Next, we find the gradient of PQ, $m_{PQ} = \frac{rise}{run} = \frac{y_2 - y_1}{x_2 - x_1}$.

So, $m_{PQ} = \frac{(h^2 + 2h) - 0}{(2+h) - 2} = \frac{h^2 + 2h}{h} = \frac{h(h+2)}{h}$

After cancelling the 'h' we have: $m_{PQ} = h + 2, h \neq 0$

Note the additional condition, $h \neq 0$. Why is this?

We are now in a position to determine the gradient of the curve at $x = 2$.

Gradient at $P = \lim_{h \to 0} m_{PQ} = \lim_{h \to 0} (h + 2) = 2$.

EXAMPLE 19.8

Find the gradient of the curve $y = \frac{1}{x-1}$ at the point (2, 1).

Solution

As in the previous example, we choose two points P and Q where $x = 2$ and $x = 2 + h$ respectively.

When $x = 2$, $y = \frac{1}{2-1} = 1$ and when $x = 2 + h$, $y = \frac{1}{2+h-1} = \frac{1}{h+1}$.

Meaning that P has coordinates $(2, 1)$ and

Q has coordinates $\left(2 + h, \frac{1}{1+h}\right)$.

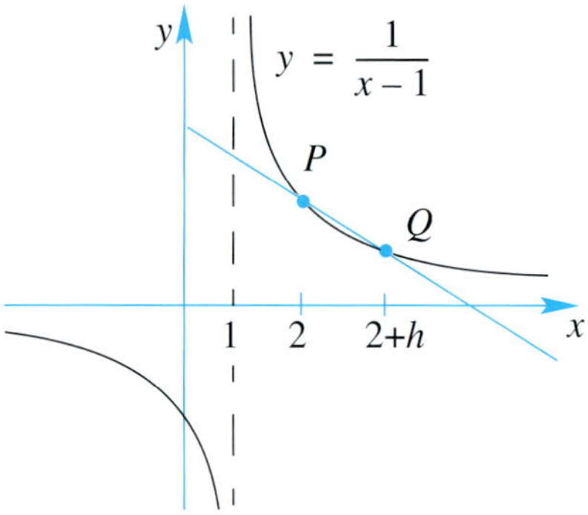

$$m_{PQ} = \frac{\frac{1}{1+h} - 1}{(2+h) - 2} = \frac{\frac{1-(1+h)}{1+h}}{h} = \frac{-\frac{h}{1+h}}{h} = -\frac{1}{1+h}$$

Then, as $h \to 0$, $m_{PQ} \to -\frac{1}{1}$.

That is, gradient at (2, 1) is given by $\lim_{h \to 0} -\frac{1}{1+h} = -1$.

As can be seen from the last two examples, finding the gradient of a curve at a particular point using a 'first principles approach' is rather lengthy. In fact, the process for finding the value of the derivative or the gradient at a particular point on a curve is rather straight forward. This process requires the use of two steps:

Step 1: Find the gradient function (i.e., the derivative). e.g., if $y = f(x)$ find $\frac{dy}{dx} = f'(x)$

Step 2: Substitute the x–value of the point in question into the equation of the derivative. i.e., if we want the gradient at $x = a$, determine $f'(a)$.

EXAMPLE 19.9

Find $f'(3)$ given that $f(x) = x^3 - 2x^2 + 10$.

Solution

Using the power rule for differentiation we have

$$f(x) = x^3 - 2x^2 + 10 \Rightarrow f'(x) = 3x^2 - 4x$$
$$\therefore f'(3) = 3 \times 3^2 - 4 \times 3$$
$$= 15$$

Notice how much more efficient this is compared to using $f'(3) = \lim_{h \to 0} \frac{f(3+h) - f(3)}{h}$

EXAMPLE 19.10 Find the gradient of the curve with equation $y = 9x - x^3$ at the point (2, 10).

Solution

First, find the equation that gives the gradient at any point on the curve, that is, find $\frac{dy}{dx}$.

Using the power rule we have, $y = 9x - x^3 \Rightarrow \frac{dy}{dx} = 9 - 3x^2$. Substituting $x = 2$ into the derivative equation, $\frac{dy}{dx} = 9 - 3(2)^2 = -3$. i.e., the gradient at the point (2, 10) is –3.

EXAMPLE 19.11 Determine the coordinate(s) on the curve $x \mapsto x^3 - x + 2$ where the gradient is 11.

Solution

Let $f(x) = x^3 - x + 2$, we need to find values of x for which $f'(x) = 11$:

We have that $f'(x) = 3x^2 - 1$, so that $3x^2 - 1 = 11 \Leftrightarrow 3x^2 - 12 = 0$

$$\Leftrightarrow 3(x^2 - 4) = 0$$

$$\Leftrightarrow 3(x-2)(x+2) = 0$$

$$x = 2 \text{ or } x = -2$$

For $x = 2$, $f(2) = 8$ and for $x = -2$, $f(-2) = -4$.

Therefore, the required coordinates are (2, 8) and (–2,–4).

Notice that it is possible for a curve to have the same gradient at different points.

EXAMPLE 19.12 Given that $f(x) = x^3 - x^2 - x + 1$, find the coordinates of all points for which the curve with equation $y = f(x)$ has a horizontal tangent.

Solution

If the tangent at a point on the curve is horizontal then the gradient of the curve at that point must be zero. So, to find those values of x where $f'(x) = 0$.

Now, $f'(x) = 3x^2 - 2x - 1 \therefore f'(x) = 0 \Leftrightarrow 3x^2 - 2x - 1 = 0$

$$\Leftrightarrow (3x+1)(x-1) = 0$$

$$\Leftrightarrow x = -\frac{1}{3} \text{ or } x = 1$$

For $x = -\frac{1}{3}$, $y = f\left(-\frac{1}{3}\right) = -\frac{1}{27} - \frac{1}{9} + \frac{1}{3} + 1 = \frac{32}{27}$ and for $x = 1$, $y = 1 - 1 - 1 + 1 = 0$.

So, the relevant points are $\left(-\frac{1}{3}, \frac{32}{27}\right)$ and $(1, 0)$.

We now make use of a graphics calculator to find the gradient at a particular point on a curve.

EXAMPLE 19.13

Use a graphics calculator to solve Example 19.10.

Solution

Using the TI–83 we first enter the equation, $y = 9x - x^3$ into the equation screen. Then we **QUIT** and have a blank screen. At this stage we select **MATH** and use the down arrow to select option **8:nDeriv(**, we now call up the **VARS** screen. Using the arrows, select **Y–VARS**. Next select option **1:Function** followed by option **1:Y1** and then press **ENTER**. At this stage we have a screen displaying **nDeriv(Y1**. Next we enter our parameters, **,X, 2)** – dont forget the commas. The '**X**' informs the calculator that the variable in question is '**X**' and the '**2**' informs the calculator that we wish to evaluate the derivative at '$x = 2$'. The screen sequence is now displayed:

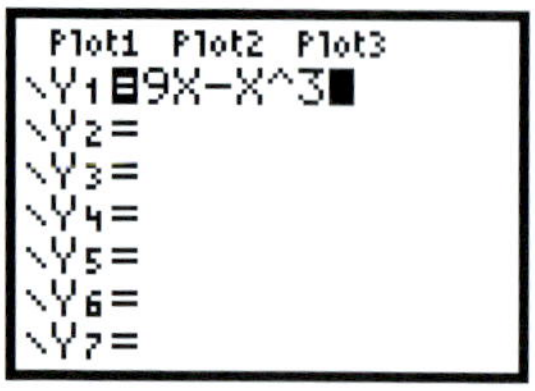

Notice that the answer we have obtained is –3.000001 (which, for all intended purposes is –3). The reason lies in our original 'accuracy' settings on the calculator – see your graphics calculator handbook for more details on this.

Note in this case it would have been easier to have entered the information on the one screen as follows:

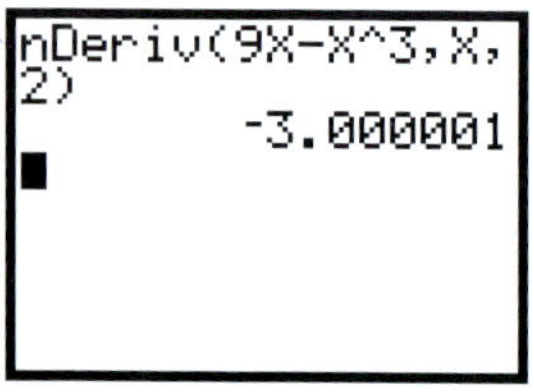

That is, type in the equation into the **nDeriv(** function as opposed to using the **VARS** approach. However, you should spend some time in considering when one approach is more beneficial than the other.

EXERCISES 19.2.1

1. Let P and Q be points on the curve $y = x^2 - 4x$ for which $x = 4$ and $x = 4 + h$ respectively. Express the gradient of PQ in terms of h and hence find the gradient of the curve $y = x^2 - 4x$ at $x = 4$.

2. For a curve with equation $y = \dfrac{2}{x+1}$ determine the coordinates of the points P and Q where $x = 1$ and $x = 1 + h$ respectively. Express the gradient of the line PQ in terms of h and hence find the gradient of the curve at $x = 1$.

3. Using a first principles method find the gradient of the curve with equation $y = 6 - x^3$ at the point where $x = 2$.

4. Find the gradient of the function at the indicated point.

(a) $f(x) = x^3 - 2$ at (1,–1) (b) $f(x) = \frac{1}{x}$ at (2, 0.5)

(c) $f(x) = (2x-1)^2$ at (2, 9) (d) $y = (2x+1)^2$ at (0,1)

(e) $y = x^2 - \frac{1}{x^2} + 2$ at (1, 2) (f) $y = \sqrt[3]{x^2} - \sqrt{x} + x$ at (1, 1)

(g) $f(x) = 1 - \sqrt[3]{x}$ at (8, –1) (h) $y = x\sqrt{x} + \frac{x}{\sqrt{x}} - \frac{\sqrt{x}}{x}$ at $(4, \frac{19}{2})$

5. Find the value(s) of x, so that $f'(x) = 0$ given that $f(x) = x^3 - 8x$.

6. For the curve with equation $y = x^2 - 12x$, find:

(a) $\frac{dy}{dx}$ (b) the gradient where $x = -3$

(c) the coordinates of the point where the gradient is 4.

7. For the curve with equation $y = -x^3 + 3x$, find:

(a) $\frac{dy}{dx}$ (b) the gradient where $x = 1$

(c) the coordinates of the point where the gradient is –3.

8. For the curve with equation $f(x) = \frac{1}{4}x^2(x^2 - 1)$, find:

(a) the coordinates where its gradient is zero.

(b) the set of values of x for which its gradient is positive.

9. Determine those values of x for which the curve with equation $y = 8 - x^2$ will have the same gradient as the curve with equation $y = x^3 - x$.

10. Find the gradient of the function $x : \mapsto x^3 + x^2 - 2x$ at the points where

(a) it crosses the x–axis,

(b) it cuts the y–axis.

11. The curve with equation $y = ax + \frac{b}{x^2}$ passes through the point (2, 0), where its gradient is found to be 3. Determine the values of a and b.

12. Given that $g(x) = x^2$, show that $f'(a+b) = f'(a) + f'(b)$.

13. Given that $h(x) = 4x - 2\sqrt{x}, x \geq 0$, find (a) $f(a^2)$ (b) $f'(a^2)$.

14. The function $f(x) = ax^3 - bx^2$ has a gradient of 2 at the point (1, 6). Find its gradient when $x = 2$.

15. Given that $f(x) = \frac{1}{x^n}$, show that $xf'(x) + nf(x) = 0$.

19.2.2 GRADIENT FUNCTION FROM A GRAPH

We now have a brief look at some geometrical properties of the gradient value $f'(a)$ at a point $(a, f(a))$ of the graph $y = f(x)$.

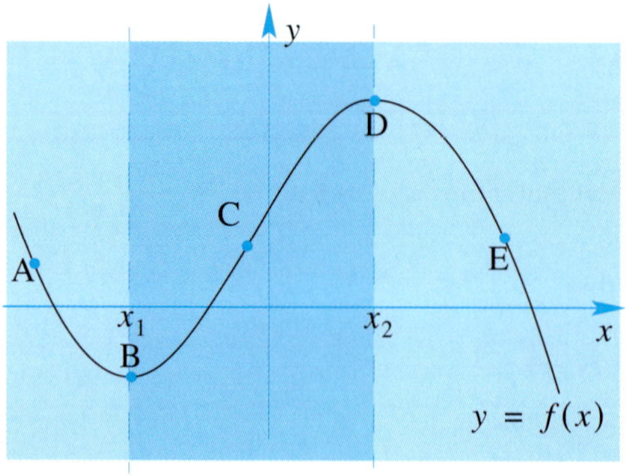

1. At A, tangent has a negative slope.
2. For $x < x_1$ gradient is < 0, i.e., $g'(x) < 0$.
3. At B, tangent is a horizontal line. Meaning that it has a gradient of zero.
4. At $x = x_1$ gradient is zero, i.e., $g'(x_1) = 0$.
5. At C, tangent has a positive slope.
6. For $x_1 < x < x_2$ gradient is positive.
7. At D, tangent is a horizontal line. Meaning that it has a gradient of zero.
8. At $x = x_2$ gradient is zero, i.e., $g'(x_2) = 0$.
9. For $x > x_2$ gradient is negative.
10. At E, tangent has a negative slope.

These properties are extremely useful in helping us 'find' and/or 'sketch' the gradient function of a graph when the equation of the function is not provided.

EXAMPLE 19.14

Sketch the graph of the gradient function of the curves shown below.

(a)

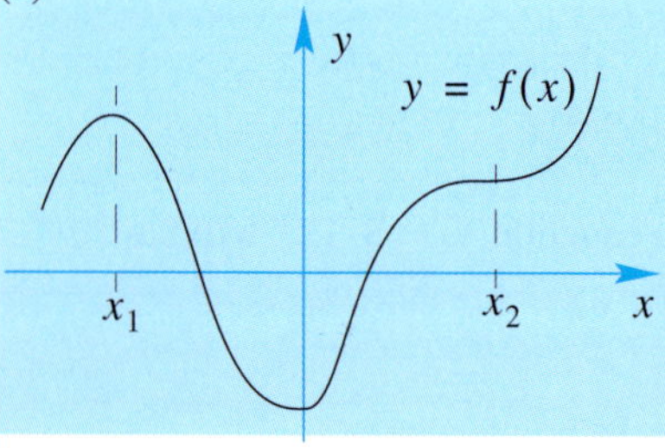

(b)

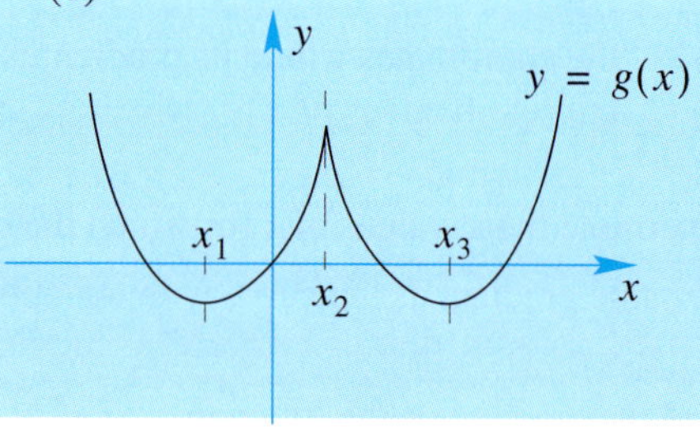

Solution

A general approach to these types of problems is to first locate where a horizontal tangent line would occur on the curve. Once the point(s) have been located we can then 'break up' the remainder of the curve into appropriate regions. e.g., regions of positive gradients, regions of negative gradients and so on.

(a) In this case, our 'key' points occur when $x = x_1$, $x = x_2$ and $x = 0$. At each of these points the gradient is zero. We observe that for $x < x_1$, $0 < x < x_2$ and $x > x_2$ the gradient is positive. Also, for $x_1 < x < 0$ the gradient is negative. We start by identifying these regions on the set of axes defining $f'(x)$ versus x – this will show us where we can sketch the gradient function.

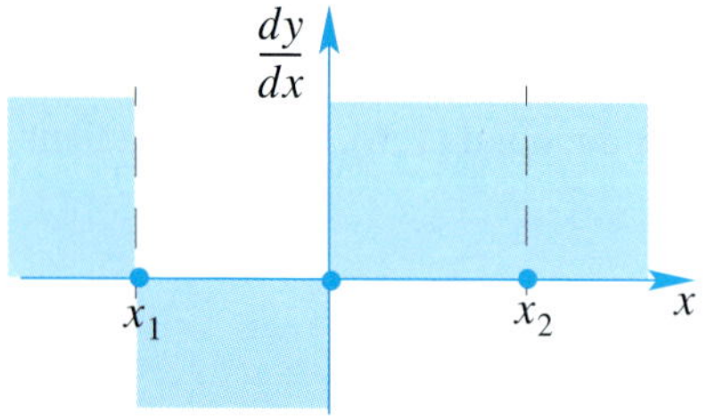

All that remains is to 'identify' the gradient values along the curve. By 'identify' we mean determine the relative gradient values. That is, the region where the gradient has been identified as being either positive or negative. As we move along the curve (from left to right) are the gradient values increasing or decreasing? As you move along the curve, keep asking yourself questions like, "Is the gradient value becoming more and more positive?", "Is the gradient value becoming more and more negative?", "Is the gradient becoming smaller and smaller?" and so on.

These questions can be more easily answered if you use a ruler or pencil and run it along the curve from left to right.

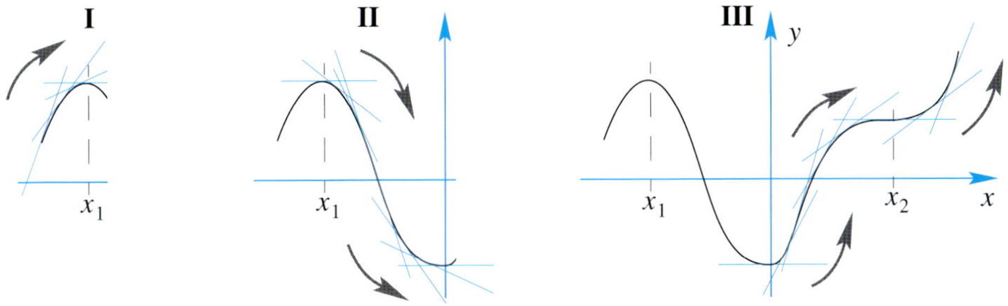

Section **I** ($x < x_1$):

Positive gradient that is decreasing in value as we get closer to x_1.

Section **II** ($x_1 < x < 0$):

Negative gradient becomes more and more negative, then, while still remaining negative, the gradient value becomes less and less negative until it reaches a value of zero at $x = 0$.

Section **III** ($x > 0$):

Positive gradient becomes more and more positive, then, while still remaining positive, the gradient value becomes less and less positive until it reaches a value of zero at $x = x_2$. As we continue along the curve, the gradient value remains positive and becomes more and more positive.

Combining all of our findings we can produce the following sketch of the gradient function:

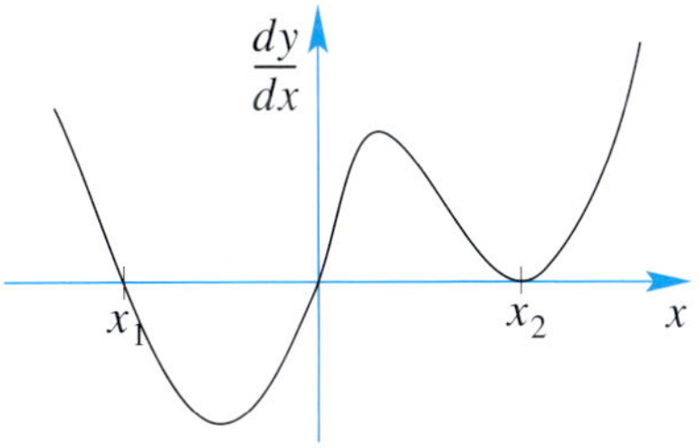

Realise that we are not so much interested in the numerical values of the gradient function but rather the general shape of the gradient function.

(b) Using the same approach we first determine the 'regions' where the gradient function lies (i.e., where it is positive and where it is negative).

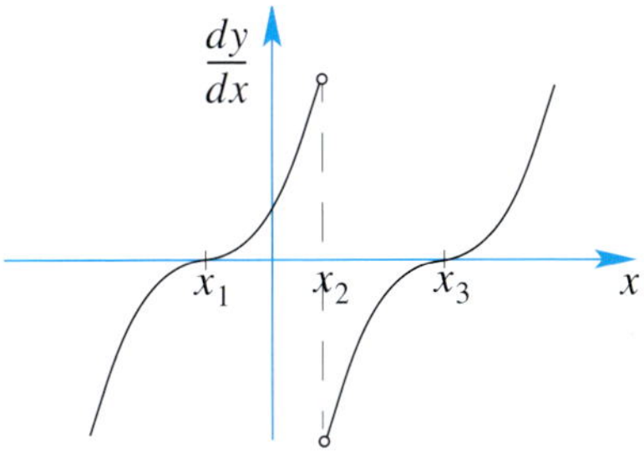

I. For $x < x_1$, $g'(x) < 0$.

II. At $x = x_1$, $g'(x) = 0$.

III. For $x_1 < x < x_2$, $g'(x) > 0$.

IV. At $x = x_2$ the gradient value cannot be found.

V. For $x_2 < x < x_3$, $g'(x) < 0$.

VI. At $x = x_3$, $g'(x) = 0$.

VI. For $x > x_3$, $g'(x) > 0$.

Notice that at $x = x_2$ the curve finishes in a peak. Trying to place a tangent at a point such as this is not possible – i.e., approaching $x = x_2$ from the left and from the right provides two different tangents that will not coincide as we get closer and closer to $x = x_2$. In order for a gradient-value to exist at $x = a$ (say) one of the conditions is that the tangents (as we approach $x = a$) from the left and from the right must coincide. This reinforces the need to understand that

$\lim\limits_{h \to 0} \dfrac{f(a+h) - f(a)}{h}$ exists if and only if $\lim\limits_{h \to 0^+} \dfrac{f(a+h) - f(a)}{h} = \lim\limits_{h \to 0^-} \dfrac{f(a+h) - f(a)}{h}$,

i.e., the right-hand side limit equals the left-hand side limit.

EXERCISES 19.2.2

1. For each of the following functions, sketch the corresponding gradient function.

(a)

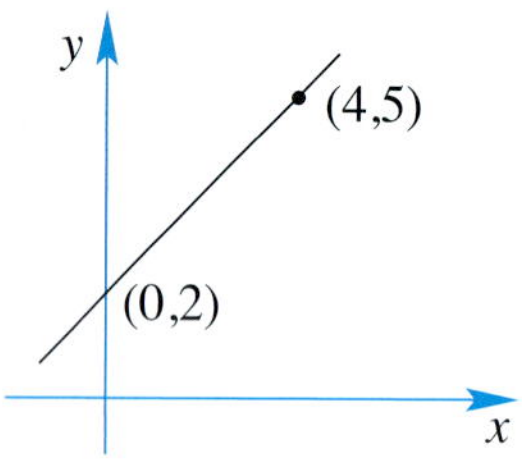

(b)

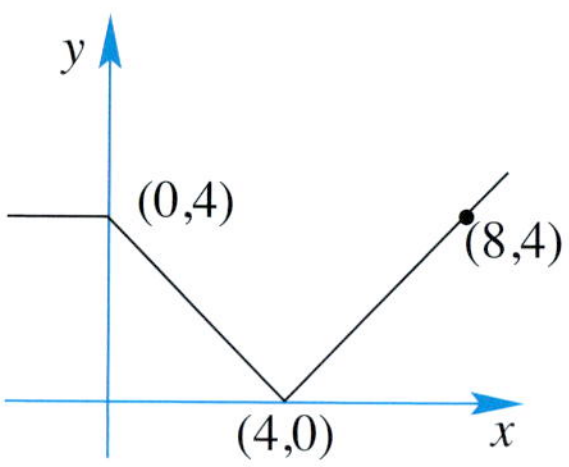

(c)

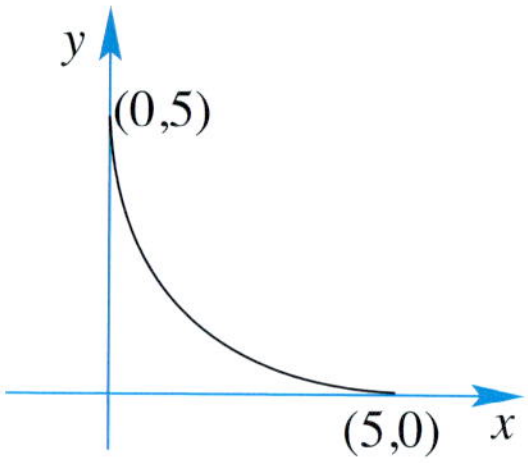

(d)

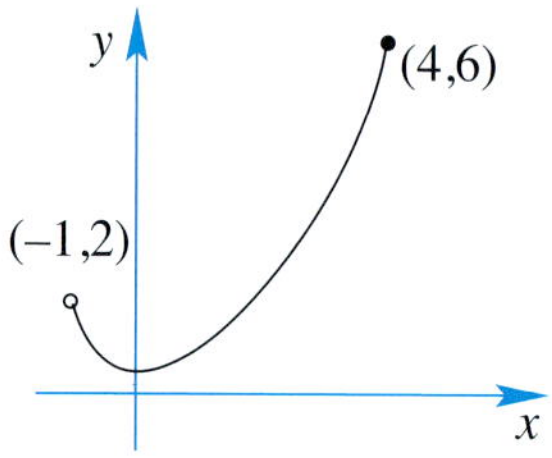

(e)

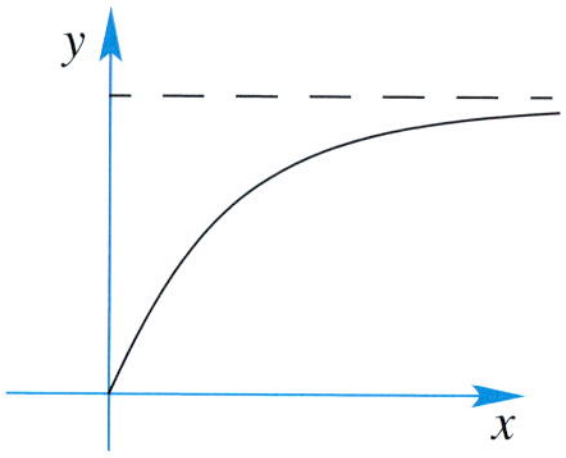

(f)

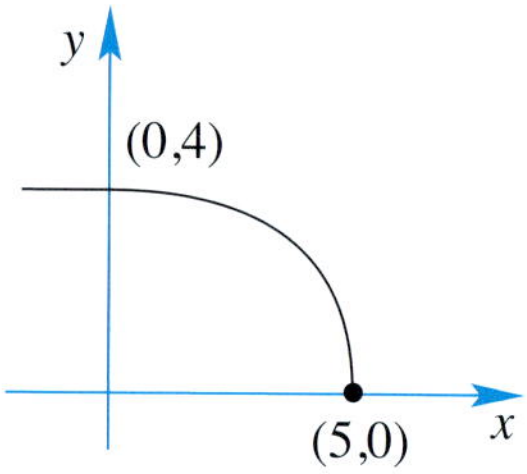

(g)

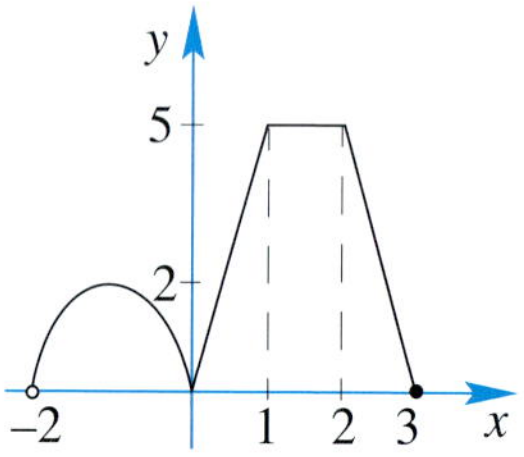

(h)

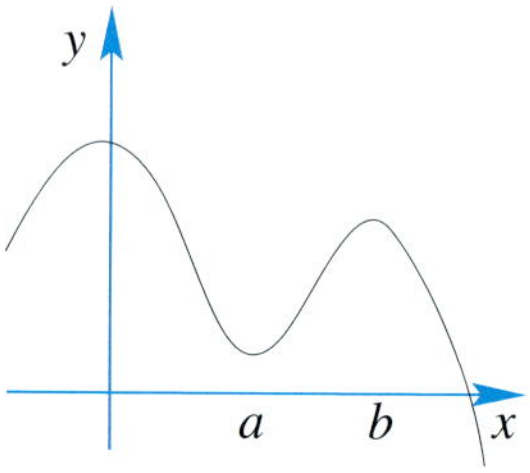

(i)

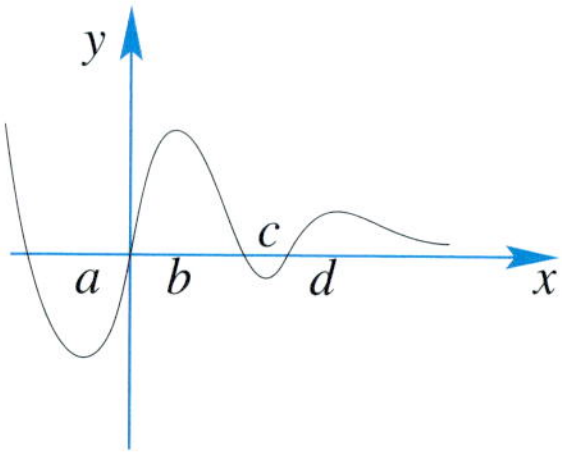

2. Sketch the graph of the function $f(x)$ given that $f(1) = 0$, $f'(x) = -1$ for all real x.

3. Sketch the graph of $f:\mathbb{R} \mapsto \mathbb{R}$ given that $f(2) = 2$, $f'(2) = 0$, $f(1) = 4$, $f'(1) = 0$, $f'(x) > 0$ for $x > 2$ and $x < 1$ and $f'(x) < 0$ for $1 < x < 2$.

19.2.3 DIFFERENTIATING WITH VARIABLES OTHER THAN x AND y

Although it was convenient to establish the underlying theory of differentiation based on the use of the variables x and y, it must be pointed out that not all expressions are written in terms of x and y. In fact, many of the formulae that we use are written in terms of variables other than y and x, e.g., volume, V, of a sphere is given by $V = \frac{4}{3}\pi r^3$, where r is its radius. The displacement of a particle moving with constant acceleration is given by $s = ut + \frac{1}{2}at^2$. However, it is reassuring to know that the rules are the same regardless of the variables involved. Thus, if we have that y is a function of x, we can differentiate y *with respect to* (w.r.t) x to find $\frac{dy}{dx}$. On the other hand, if we have that y is a function of t, we would differentiate y w.r.t. t and write $\frac{dy}{dt}$. Similarly, if W was a function of θ, we would differentiate W w.r.t. θ and write $\frac{dW}{d\theta}$.

EXAMPLE 19.15

Differentiate the following with respect to the appropriate variable

(a) $V = \frac{4}{3}\pi r^3$ (b) $p = 3w^3 - 2w + 20$ (c) $s = 10t + 4t^2$

Solution

(a) As V is a function of r, we need to differentiate V with respect to r:

$$V = \frac{4}{3}\pi r^3 \Rightarrow \frac{dV}{dr} = \frac{4}{3}\pi(3r^2) = 4\pi r^2.$$

(b) This time p is a function of w, and so we would differentiate p with respect to w:

$$p = 3w^3 - 2w + 20 \Rightarrow \frac{dp}{dw} = 9w^2 - 2.$$

(c) In this expression we have that s is a function of t and so we differentiate s w.r.t t:

$$s = 10t + 4t^2 \Rightarrow \frac{ds}{dt} = 10 + 8t.$$

EXERCISES 19.2.3

1. Differentiate the following functions with respect to the appropriate variable.

(a) $s = 12t^4 - \sqrt{t}$ (b) $Q = \left(n + \frac{1}{n^2}\right)^2$ (c) $P = \sqrt{r}(r + \sqrt[3]{r} - 2)$

(d) $T = \frac{(\theta - \sqrt{\theta})^3}{\theta}$ (e) $A = 40L - L^3$ (f) $F = \frac{50}{v^2} - v$

(g) $V = 2l^3 + 5l$ (h) $A = 2\pi h + 4h^2$ (i) $N = n^4 - \sqrt[3]{n} + \pi n$

2. Differentiate the following with respect to the independent variable.

(a) $v = \frac{2}{3}\left(5 - \frac{2}{t^2}\right)$ (b) $S = \pi r^2 + \frac{20}{r}$ (c) $q = \sqrt{s^5} - \frac{3}{s}$

(d) $h = \frac{2 - t + t^2}{t^3}$ (e) $L = \frac{4 - \sqrt{b}}{b}$ (f) $W = (m-2)^2(m+2)$

19.3 DERIVATIVE OF TRANSCENDENTAL FUNCTIONS

A **transcendental function** is a function that cannot be constructed in a finite number of steps from elementary functions and their inverses. Some examples of these functions are $\sin x$, $\cos x$, $\tan x$, the exponential function, e^x and the logarithmic function $\log_e x$ (or $\ln x$).

Also in this section we look at derivatives of expressions that involve the **product** of two functions, the **quotient** of two functions and the **composite** of two functions. Each of these types of expressions will lead to some standard rules of differentiation.

19.3.1 DERIVATIVE OF CIRCULAR TRIGONOMETRIC FUNCTIONS

We begin by considering the trigonometric functions, i.e., the sine, cosine and tangent functions. There are a number of approaches that can be taken to achieve our goal. In this instance we will use two different approaches to find the derivative of the function $x \mapsto \sin(x)$.

Method 1 (A limits approach):

Letting $f(x) = \sin(x)$ and using the definition from first principles we have

$$f'(x) = \lim_{h \to 0} \frac{f(x+h) - f(x)}{h}$$
$$= \lim_{h \to 0} \frac{\sin(x+h) - \sin(x)}{h}$$

It is not immediately obvious what this limit is, however, we can simplify this expression using the trigonometric identities that we have looked at in Chapter 10 and then make use of some algebra. We provide a full account of how to derive the result in §19.8. At this stage we simply quote the result.

$$f(x) = \sin(x) \text{ then } f'(x) = \cos(x)$$

Method 2 (Using the graph of the original function):

We will deduce the derivative of $\sin(x)$ by using its graph and the methods employed in Section 19.2.2.

We begin by sketching a graph of $f(x) = \sin(x)$ and by measuring its gradient (m) along different points on the curve. To do this we draw a tangent at each of these points and then measure the gradient of the tangent. Next we plot the values of the gradient that we obtain at each of these points. This in turn will provide us with the values of the gradient function. Plotting these values (as shown in the diagram alongside) will then produce a rough sketch of the gradient function (and hence the derivative). From the plot of the gradient values we can reasonably assume that the gradient function obtained corresponds to the cosine function.

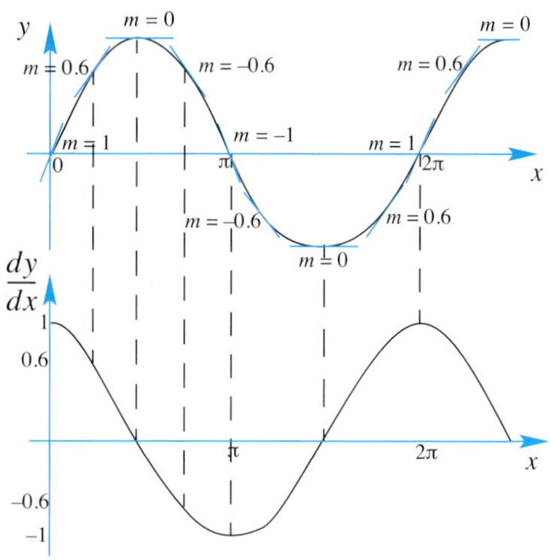

This is consistent with our earlier result (based on a limits approach). So, again we have that if $f(x) = \sin(x)$ then $f'(x) = \cos(x)$.

Similarly, using either approaches, we have that if

$$f(x) = \cos(x) \text{ then } f'(x) = -\sin(x)$$

and if

$$f(x) = \tan(x) \text{ then } f'(x) = \sec^2(x)$$

The derivative of other trigonometric functions can also be obtained in the same way, however, we will use other results that will make this process easier. Next, we consider the exponential function.

19.3.2 DERIVATIVE OF THE EXPONENTIAL FUNCTION

Consider the exponential function $x \mapsto e^x$. Although we could apply the same graphical method which was used to determine the derivative of the sine function, this time we shall make use of the definition of the derivative, i.e.,

$$\begin{aligned} f'(x) &= \lim_{h \to 0} \frac{f(x+h) - f(x)}{h} \\ &= \lim_{h \to 0} \frac{e^{x+h} - e^x}{h} \\ &= \lim_{h \to 0} \frac{e^x e^h - e^x}{h} \quad \text{using } e^x e^h = e^{x+h} \\ &= \lim_{h \to 0} \frac{e^x(e^h - 1)}{h} \\ &= e^x \lim_{h \to 0} \frac{(e^h - 1)}{h} \\ &= e^x \times 1 \\ &= e^x \end{aligned}$$

To determine the value of the limit we make use of a table of values. e.g., for $h = 0.01$, we have that

$$\frac{e^{0.01} - 1}{0.01} = \frac{0.01005}{0.01} = 1.005$$

As we use smaller values of h, this ratio will become closer to one.

Notice that this is the only function which has a derivative that is the same as itself. That is, if $f(x) = e^x$ then $f'(x) = e^x$

19.3.3 DERIVATIVE OF THE NATURAL LOG FUNCTION

Consider the function $x \mapsto \log_e x$. As in the previous case, we use the definition of the derivative to establish the gradient function of $x \mapsto \log_e x$.

$$f'(x) = \lim_{h \to 0} \frac{f(x+h) - f(x)}{h} = \lim_{h \to 0} \frac{\log_e(x+h) - \log_e(x)}{h}$$

$$= \lim_{h \to 0} \frac{\log_e\left(\frac{x+h}{x}\right)}{h} \qquad \text{(Using the log laws)}$$

The next step is a little tricky, so we write it down first and then see how we arrive at the result.

$$= \frac{1}{x} \lim_{h \to 0} \log_e\left(1 + \frac{h}{x}\right)^{\frac{x}{h}}$$

To get to this step we proceeded as follows:

$$\frac{\log_e\left(1 + \frac{h}{x}\right)}{h} = \frac{1}{x} \cdot x \frac{\log_e\left(1 + \frac{h}{x}\right)}{h} = \frac{1}{x} \cdot \left(\frac{x}{h}\right) \log_e\left(1 + \frac{h}{x}\right) = \frac{1}{x} \cdot \log_e\left(1 + \frac{h}{x}\right)^{\frac{x}{h}}$$

Then, as the argument in the limit is h (i.e., it is independent of x) we have

$$\lim_{h \to 0} \frac{1}{x} \cdot \log_e\left(1 + \frac{h}{x}\right)^{\frac{x}{h}} = \frac{1}{x} \lim_{h \to 0} \log_e\left(1 + \frac{h}{x}\right)^{\frac{x}{h}}$$

Then, as the log function is a continuous function, we have that the limit of the log is the same as the log of the limit so that

$$\lim_{h \to 0} \log_e\left(1 + \frac{h}{x}\right)^{\frac{x}{h}} = \log_e\left[\lim_{h \to 0}\left(1 + \frac{h}{x}\right)^{\frac{x}{h}}\right]$$

However, we also have that $\lim_{h \to 0}\left(1 + \frac{h}{x}\right)^{\frac{x}{h}} = e$ and so we end up with the result that

$$\frac{1}{x} \lim_{h \to 0} \log_e\left(1 + \frac{h}{x}\right)^{\frac{x}{h}} = \frac{1}{x} \log_e e = \frac{1}{x} \times 1 = \frac{1}{x}$$

And so, we have that if

$f(x) = \log_e x$ then $f'(x) = \frac{1}{x}$

19.3.4 DERIVATIVE OF A PRODUCT OF FUNCTIONS

Many functions can be written as the product of two (or more) functions. For example, the function $y = (x^3 - 2x)(x^2 + x - 3)$ is made up of the product of two simpler functions of x. In fact, expressions such as these take on the general form $y = u \times v$ or $y = f(x) \times g(x)$ where (in this case) we have $u = f(x) = (x^3 - 2x)$ and $v = g(x) = (x^2 + x - 3)$.

To differentiate such expressions we use the **product rule**, which can be written as:

Function	Derivative
If $y = u \times v$ then	$\frac{dy}{dx} = \frac{du}{dx} \times v + u \times \frac{dv}{dx}$
If $y = f(x) \times g(x)$ then	$\frac{dy}{dx} = f'(x) \times g(x) + f(x) \times g'(x)$

EXAMPLE 19.16

Differentiate the following

(a) $x^2 \sin(x)$ (b) $(x^3 - 2x + 1)e^x$ (c) $\frac{1}{x}\log_e(x)$

Solution

(a) Let $y = x^2 \sin(x)$ so that $u = x^2$ and $v = \sin(x)$. So that $\frac{du}{dx} = 2x$ and $\frac{dv}{dx} = \cos(x)$.

Using the product rule we have $\frac{dy}{dx} = \frac{du}{dx} \times v + u \times \frac{dv}{dx}$

$$= 2x \times \sin(x) + x^2 \times \cos(x)$$

$$= 2x\sin(x) + x^2\cos(x)$$

A useful method to find the derivative of a product makes use of the following table:

Function	Derivative
$u = x^2$	$\frac{du}{dx} = 2x$
$v = \sin(x)$	$\frac{dv}{dx} = \cos(x)$

$2x\sin(x)$

$x^2\cos(x)$

Adding: $2x\sin(x) + x^2\cos(x)$

(b) Let $y = (x^3 - 2x + 1)e^x$ so that $u = (x^3 - 2x + 1)$ and $v = e^x$.

Then, $\frac{du}{dx} = 3x^2 - 2$ and $\frac{dv}{dx} = e^x$.

$$\frac{dy}{dx} = \frac{du}{dx} \times v + u \times \frac{dv}{dx}$$

Using the product rule: $= (3x^2 - 2) \times e^x + (x^3 - 2x + 1) \times e^x$

$$= (3x^2 - 2 + x^3 - 2x + 1)e^x$$

$$= (x^3 + 3x^2 - 2x - 1)e^x$$

(c) Let $y = \frac{1}{x}\log_e x$ with $u = \frac{1}{x}$ and $v = \log_e x$. This time set up a table:

Function	Derivative
$u = \frac{1}{x}$	$\frac{du}{dx} = -\frac{1}{x^2}$
$v = \log_e x$	$\frac{dv}{dx} = \frac{1}{x}$

$-\frac{1}{x^2} \times \log_e x$

$\frac{1}{x} \times \frac{1}{x}$

Adding: $\frac{dy}{dx} = -\frac{1}{x^2} \times \log_e x + \frac{1}{x} \times \frac{1}{x}$

$$= -\frac{1}{x^2} \times \log_e x + \frac{1}{x^2}$$

$$= \frac{1}{x^2}(1 - \log_e x)$$

19.3.5 DERIVATIVE OF A QUOTIENT OF FUNCTIONS

In the same way as we have a rule for the product of functions, we also have a rule for the quotient of functions. For example, the function

$$y = \frac{x^2}{x^3 + x - 1}$$

is made up of two simpler functions of x. Expressions like this take on the general form

$$y = \frac{u}{v} \quad \text{or} \quad y = \frac{f(x)}{g(x)}.$$

For the example shown above, we have that $u = x^2$ and $v = x^3 + x - 1$.

As for the product rule, we state the result.

To differentiate such expressions we use the **quotient rule**, which can be written as:

Function	Derivative
If $y = \frac{u}{v}$ then	$\frac{dy}{dx} = \frac{\frac{du}{dx} \times v - u \times \frac{dv}{dx}}{v^2}$
If $y = \frac{f(x)}{g(x)}$ then	$\frac{dy}{dx} = \frac{f'(x) \times g(x) - f(x) \times g'(x)}{[g(x)]^2}$

EXAMPLE 19.17

Differentiate the following

(a) $\frac{x^2 + 1}{\sin(x)}$ (b) $\frac{e^x + x}{x + 1}$ (c) $\frac{\sin(x)}{1 - \cos(x)}$

Solution

(a) We express $\frac{x^2 + 1}{\sin(x)}$ in the form $y = \frac{u}{v}$, so that $u = x^2 + 1$ and $v = \sin(x)$.

Giving the following derivatives, $\frac{du}{dx} = 2x$ and $\frac{dv}{dx} = \cos(x)$.

Using the quotient rule we have,

$$\frac{dy}{dx} = \frac{\frac{du}{dx} \times v - u \times \frac{dv}{dx}}{v^2}$$

$$= \frac{2x \times \sin(x) - (x^2 + 1) \times \cos(x)}{[\sin(x)]^2}$$

$$= \frac{2x\sin(x) - (x^2 + 1)\cos(x)}{\sin^2(x)}$$

(b) First express $\frac{e^x + x}{x + 1}$ in the form $y = \frac{u}{v}$, so that $u = e^x + x$ and $v = x + 1$ and

$\frac{du}{dx} = e^x + 1$ and $\frac{dv}{dx} = 1$. Using the quotient rule, we have

$$\frac{dy}{dx} = \frac{\frac{du}{dx} \times v - u \times \frac{dv}{dx}}{v^2} = \frac{(e^x + 1) \times (x + 1) - (e^x + x) \times 1}{(x + 1)^2}$$

$$= \frac{xe^x + e^x + x + 1 - e^x - x}{(x+1)^2}$$

$$= \frac{xe^x + 1}{(x+1)^2}$$

(c) Express the quotient $\frac{\sin(x)}{1 - \cos(x)}$ in the form $y = \frac{u}{v}$, so that $u = \sin(x)$ and $v = 1 - \cos(x)$. Then $\frac{du}{dx} = \cos(x)$ and $\frac{dv}{dx} = \sin(x)$.

Using the quotient rule, we have

$$\frac{dy}{dx} = \frac{\frac{du}{dx} \times v - u \times \frac{dv}{dx}}{v^2} = \frac{\cos(x) \times (1 - \cos(x)) - \sin(x) \times \sin(x)}{(1 - \cos(x))^2}$$

$$= \frac{\cos(x) - \cos^2(x) - \sin^2(x)}{(1 - \cos(x))^2}$$

$$= \frac{\cos(x) - (\cos^2(x) + \sin^2(x))}{(1 - \cos(x))^2}$$

$$= \frac{\cos(x) - 1}{(1 - \cos(x))^2}$$

$$= -\frac{(1 - \cos(x))}{(1 - \cos(x))^2}$$

$$= -\frac{1}{(1 - \cos(x))}$$

19.3.6 THE CHAIN RULE

To find the derivative of $x^3 + 1$ we let $y = x^3 + 1$ so that $\frac{dy}{dx} = 3x^2$.

Next consider the derivative of the function $y = (x^3 + 1)^2$.

We first expand the brackets, $y = x^6 + 2x^3 + 1$, and obtain $\frac{dy}{dx} = 6x^5 + 6x^2$.

This expression can be simplified (i.e., factorised), giving $\frac{dy}{dx} = 6x^2(x^3 + 1)$.

In fact, it isn't too great a task to differentiate the function $y = (x^3 + 1)^3$.

As before, we expand; $y = x^9 + 3x^6 + 3x^3 + 1$ so that $\frac{dy}{dx} = 9x^8 + 18x^5 + 9x^2$.

Factorising this expression we now have $\frac{dy}{dx} = 9x^2(x^6 + 2x^3 + 1) = 9x^2(x^3 + 1)^2$.

But what happens if we need to differentiate the expression $y = (x^3 + 1)^8$? Of course, we could expand and obtain a polynomial with 9 terms (!), which we then proceed to differentiate and obtain a polynomial with 8 terms. . . and of course, we can then easily factorise that polynomial (not!). The question then arises, "Is there an easier way to do this?"

We can obtain some idea of how to do this by summarising the results found so far:

Function	Derivative	(Factored form)
$y = x^3 + 1$	$\frac{dy}{dx} = 3x^2$	$3x^2$
$y = (x^3+1)^2$	$\frac{dy}{dx} = 6x^5 + 6x^2$	$2 \times 3x^2(x^3+1)$
$y = (x^3+1)^3$	$\frac{dy}{dx} = 9x^8 + 18x^5 + 9x^2$	$3 \times 3x^2(x^3+1)^2$
$y = (x^3+1)^4$	$\frac{dy}{dx} = 12x^{11} + 36x^8 + 36x^5 + 12x^2$	$4 \times 3x^2(x^3+1)^3$

The pattern that is emerging is that if $y = (x^3+1)^n$ then $\frac{dy}{dx} = n \times 3x^2(x^3+1)^{n-1}$.

In fact, if we consider the term inside the brackets as one function, so that the expression is actually a composition of two functions, namely that of $x^3 + 1$ and the power function we can write $u = x^3 + 1$ and $y = u^n$.

So that $\frac{dy}{du} = nu^{n-1} = n(x^3+1)^{n-1}$ and $\frac{du}{dx} = 3x^2$, giving the result $\frac{dy}{dx} = \frac{dy}{du} \times \frac{du}{dx}$.

Is this a 'one–off' result, or can we determine a general result that will always work?
To explore this we use a graphical approach to see why it might be possible to obtain a general result.

We start by using the above example and then move onto a more general case. For the function $y = (x^3+1)^2$, we let $u = x^3 + 1\ (= g(x))$ and so $y = u^2\ (= f(g(x)))$. We need to find what effect a small change in x will have on the function y (via u). i.e., what effect will δx have on y?

We have a sort of **chain reaction**, that is, a small change in x, δx, will produce a change in u, δu, which in turn will produce a change in y, δy! It is the path from δx to δy that we are interested in.

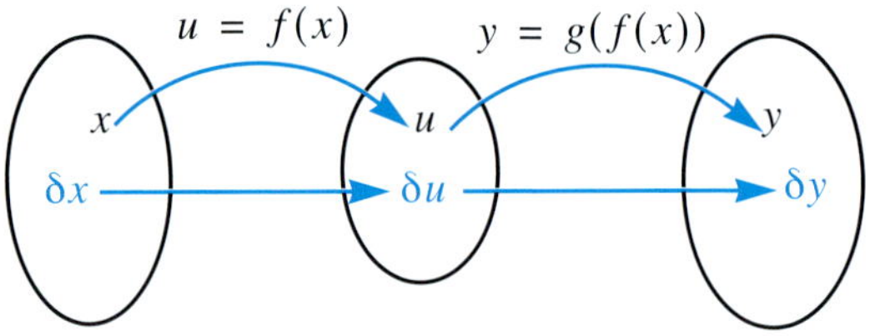

This can be seen when we produce a graphical representation of the discussion so far.

We start by looking at the effect that a change in x has on u:

Then we observe the effect that the change in u has on y:

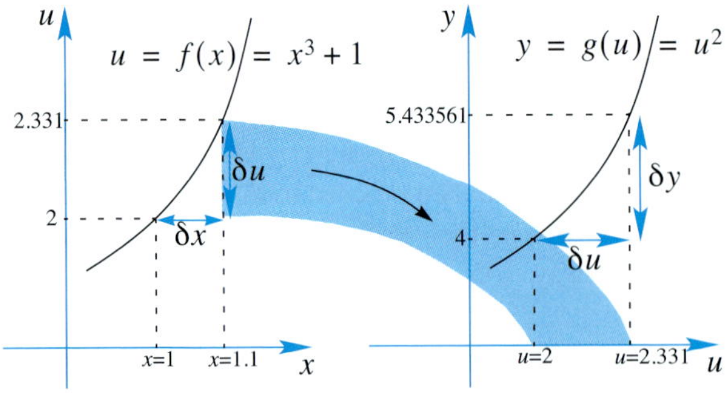

We then have $\delta x = 1.1 - 1 = 0.1$ *and* $\delta u = 2.331 - 2 = 0.331$.

Similarly, $\delta u = 2.331 - 2 = 0.331$ & $\delta y = 5.433561 - 4 = 1.433561$

Based on these results, the following relationship can be seen to hold:

$$\frac{\delta y}{\delta x} = \frac{\delta y}{\delta u} \times \frac{\delta u}{\delta x}$$

The basic outline in proving this result is shown in the following argument:

Let δx be a small increment in the variable x and let δu be the corresponding increment in the variable u. This change in u will in turn produce a corresponding change δy in y.
As δx tends to zero, so does δu. We will assume that $\delta u \neq 0$ when $\delta x \neq 0$. Hence we have that

$$\frac{\delta y}{\delta x} = \frac{\delta y}{\delta u} \cdot \frac{\delta u}{\delta x} \Rightarrow \lim_{\delta x \to 0} \frac{\delta y}{\delta x} = \lim_{\delta x \to 0} \frac{\delta y}{\delta u} \cdot \frac{\delta u}{\delta x}$$

$$= \left(\lim_{\delta x \to 0} \frac{\delta y}{\delta u}\right) \cdot \left(\lim_{\delta x \to 0} \frac{\delta u}{\delta x}\right)$$

$$= \left(\lim_{\delta u \to 0} \frac{\delta y}{\delta u}\right) \cdot \left(\lim_{\delta x \to 0} \frac{\delta u}{\delta x}\right)$$ *Given that:* $\delta x \to 0 \Rightarrow \delta u \to 0$

$$\therefore \frac{dy}{dx} = \frac{dy}{du} \cdot \frac{du}{dx}$$

We then have the result:

$$\frac{dy}{dx} = \frac{dy}{du} \cdot \frac{du}{dx}$$

Chain rule (composite function notation)

An alternative notation when using the chain rule occurs when the function is expressed in the form of a composite function, i.e., in the form $f \circ g$.

So, if $F = f \circ g$, then $F(x) = f(g(x))$ and $F'(x) = f'(g(x)) \cdot g'(x)$.

That is, the derivative of the composite function $f \circ g$ is $(f \circ g)' = (f' \circ g)g'$, or

$$\frac{d}{dx}(f \circ g) = \frac{df}{du}\frac{du}{dx}, \text{ where } u = g(x)$$.

In short, the chain rule provides a process whereby we can differentiate expressions that involve composite functions. For example, the function $y = \sin(x^2)$ is a composition of the *sine* function $\sin(\ )$ and the squared function, x^2. So that we would let u (or $g(x)$) equal x^2, giving $y = \sin(u)$, where $u = x^2$.

The key to differentiating such expressions is to recognise that the chain rule must be used, and to choose the appropriate function u (or $g(x)$).

Using the chain rule

We will work our way through an example, showing the critical steps involved when using the chain rule.

This is highlighted by finding the derivative of the function $y = \sin(x^2)$.

Step 1	**Recognition**
This is the most important step when deciding if using the chain rule is appropriate. In this case we *recognise* that the function $y = \sin(x^2)$ is a composite of the sine and the squared functions.	
Step 2	**Define u (or $g(x)$)**

Let the 'inside' function be u. In this case, we have that $u = x^2$

Step 3	**Differentiate u (with respect to x)**

$$\frac{du}{dx} = 2x$$

Step 4	**Express y in terms of u**

$$y = \sin(u)$$

Step 5	**Differentiate y (with respect to u)**

$$\frac{dy}{du} = \cos(u)$$

Step 6	**Use the chain rule**

$$\frac{dy}{dx} = \frac{dy}{du} \cdot \frac{du}{dx} = \cos(u) \times 2x = 2x\cos(x^2)$$

EXAMPLE 19.18

Differentiate the following functions (a) $y = \log_e(x + \cos x)$

(b) $f(x) = (1 - 3x^2)^4$

Solution

(a) Begin by letting $u = x + \cos(x) \Rightarrow \frac{du}{dx} = 1 - \sin(x)$.

Express y in terms of u, that is, $y = \log_e u \Rightarrow \frac{dy}{du} = \frac{1}{u}\left(= \frac{1}{x + \cos(x)}\right)$.

Using the chain rule we have $\frac{dy}{dx} = \frac{dy}{du} \cdot \frac{du}{dx} = \frac{1}{x + \cos(x)} \cdot (1 - \sin(x))$

$$= \frac{1 - \sin(x)}{x + \cos(x)}$$

(b) This time we let $g(x) = 1 - 3x^2$, so that $g'(x) = -6x$.

Now let $f(x) = (h \circ g)(x)$ so that $h(g(x)) = (g(x))^4$ and $h'(g(x)) = 4(g(x))^3$. Therefore, using the chain rule we have

$$f'(x) = (h \circ g)'(x) = h'(g(x)) \cdot g'(x)$$
$$= 4(g(x))^3 \times (-6x)$$
$$= -24x(1 - 3x^2)^3$$

Some standard derivatives

Often we wish to differentiate expressions of the form $y = \sin(2x)$ or $y = e^{5x}$ or other such functions, where the x term only differs by a constant factor from that of the basic function. That is, the only difference between $y = \sin(2x)$ and $y = \sin(x)$ is the factor '2'. We can use the chain rule to differentiate such expressions:

Let $u = 2x$, giving $y = \sin(u)$ and so $\frac{dy}{dx} = \frac{dy}{du} \cdot \frac{du}{dx} = \cos(u) \times 2 = 2\cos(2x)$

Similarly,

Let $u = 5x$, giving $y = e^u$ and so $\frac{dy}{dx} = \frac{dy}{du} \cdot \frac{du}{dx} = e^u \times 5 = 5e^{5x}$.

Because of the nature of such derivatives, functions such as these form part of a set of functions that can be considered as having derivatives that are often referred to as standard derivatives. Although we could make use of the chain rule to differentiate these functions, they should be viewed as standard derivatives.

These standard derivatives are shown in the table below (where k is some real constant):

y	$\frac{dy}{dx}$
$\sin(kx)$	$k\cos(kx)$
$\cos(kx)$	$-k\sin(kx)$
$\tan(kx)$	$k\sec^2(kx)$
e^{kx}	ke^{kx}
$\log_e(kx)$	$\frac{1}{x}$

Notice, the only derivative that does not involve the constant k is that of the logarithmic function. This is because letting $u = kx$, we have $y = \log(u)$ so $\frac{dy}{dx} = \frac{dy}{du} \cdot \frac{du}{dx} = \frac{1}{u} \times k = \frac{1}{\cancel{k}x} \times \cancel{k} = \frac{1}{x}$.

When should the chain rule be used?

A good rule of thumb:

A good **first** rule to follow is: If the expression is made up of a pair of brackets and a power, then, the chances are that you will need to use the chain rule.

As a start, the expressions in the table that follows would require the use of the chain rule. Notice then that in each case the expression can be (or already is) written in 'power form'. That is, of the form $y = [f(x)]^n$.

Expression	Express in power form	Decide on what u and y are
(a) $y = (2x+6)^5$	Already in power form.	Let $u = 2x+6$ and $y = u^5$
(b) $y = \sqrt{(2x^3+1)}$	$y = (2x^3+1)^{\frac{1}{2}}$	Let $u = 2x^3+1$ and $y = u^{\frac{1}{2}}$
(c) $y = \frac{3}{(x-1)^2}, x \neq 1$	$y = 3(x-1)^{-2}, x \neq 1$	Let $u = x-1$ and $y = 3u^{-2}$
(d) $f(x) = \sin^2 x$	$f(x) = (\sin x)^2$	Let $u = \sin x$ and $f(u) = u^2$
(e) $y = \frac{1}{\sqrt[3]{e^{-x}+e^x}}$	$y = (e^{-x}+e^x)^{-\frac{1}{3}}$	Let $u = e^{-x}+e^x$ and $y = u^{-\frac{1}{3}}$

However, this isn't always the case!

Although the above approach is very useful, often you have to recognise when the function of a function rule is more appropriate. By placing brackets in the appropriate places, we can recognise this feature more readily. The examples below illustrate this:

Expression	Express it with brackets	Decide on what u and y are
(a) $y = e^{x^2+1}$	$y = e^{(x^2+1)}$.	Let $u = x^2+1$ and $y = e^u$
(b) $y = e^{\sin 2x}$	$y = e^{(\sin 2x)}$	Let $u = \sin 2x$ and $y = e^u$
(c) $y = \sin(x^2-4)$	Already in bracket form.	Let $u = x^2-4$ and $y = \sin(u)$
(d) $f(x) = \log_e(\sin x)$	Already in bracket form.	Let $u = \sin x$ and $f(u) = \log_e(u)$

Completing the process for each of the above functions we have:

(a) $\frac{dy}{dx} = \frac{dy}{du}\frac{du}{dx} = e^u \times 2x = 2xe^{x^2+1}$.

(b) $\frac{dy}{dx} = \frac{dy}{du}\frac{du}{dx} = e^u \times 2\cos(2x) = 2\cos(2x)e^{\sin(2x)}$.

(c) $\frac{dy}{dx} = \frac{dy}{du}\frac{du}{dx} = \cos(u) \times 2x = 2x\cos(x^2-4)$.

(d) $\frac{dy}{dx} = \frac{dy}{du}\frac{du}{dx} = \frac{1}{u} \times \cos x = \frac{\cos x}{\sin x} = \cot x$.

We now look at some of the more demanding derivatives, i.e., derivatives which combine at least two rules of differentiation, for example, the need to use both the quotient rule and the chain rule, or the product rule and the chain rule.

EXAMPLE 19.19

Differentiate the following

(a) $y = \sqrt{1+\sin^2 x}$ (b) $y = e^{x^3}\sin(1-2x)$ (c) $x \mapsto \frac{x}{\sqrt{x^2+1}}$

Solution

(a) Let $y = \sqrt{(1+\sin^2 x)} = (1+\sin^2 x)^{1/2}$. Using the chain rule we have

$$\frac{dy}{dx} = \frac{1}{2} \times \frac{d}{dx}(1+\sin^2 x) \times (1+\sin^2 x)^{-1/2} = \frac{1}{2} \times (2\sin x \cos x) \times \frac{1}{\sqrt{(1+\sin^2 x)}}$$

$$= \frac{\sin x \cos x}{\sqrt{(1+\sin^2 x)}}$$

(b) Let $y = e^{x^3}\sin(1-2x)$. Using the product rule first, we have

$$\frac{dy}{dx} = \frac{d}{dx}(e^{x^3}) \times \sin(1-2x) + e^{x^3} \times \frac{d}{dx}(\sin(1-2x))$$

$$= 3x^2 e^{x^3}\sin(1-2x) + e^{x^3} \times -2\cos(1-2x)$$

$$= e^{x^3}(3x^2\sin(1-2x) - 2\cos(1-2x))$$

(c) Let $f(x) = \dfrac{x}{\sqrt{x^2+1}} \Rightarrow f'(x) = \dfrac{\frac{d}{dx}(x) \times \sqrt{x^2+1} - x \times \frac{d}{dx}(\sqrt{x^2+1})}{(\sqrt{x^2+1})^2}$ (*Quotient rule*).

$$= \frac{1 \times \sqrt{x^2+1} - x \times \frac{1}{2} \times 2x \times (x^2+1)^{-\frac{1}{2}}}{x^2+1}$$

$$= \frac{\sqrt{x^2+1} - \frac{x^2}{\sqrt{x^2+1}}}{(x^2+1)}$$

$$= \frac{\frac{(\sqrt{x^2+1})^2 - x^2}{\sqrt{x^2+1}}}{(x^2+1)}$$

$$= \frac{1}{(x^2+1)\sqrt{x^2+1}}$$

EXAMPLE 19.20

Differentiate the following

(a) $y = \ln\left(\dfrac{x}{x+1}\right), x > 0$ (b) $y = \sin(\ln t)$ (c) $y = x\ln(x^2)$

Solution

(a) $y = \ln\left(\dfrac{x}{x+1}\right) = \ln(x) - \ln(x+1) \therefore \dfrac{dy}{dx} = \dfrac{1}{x} - \dfrac{1}{x+1} = \dfrac{(x+1)-x}{x(x+1)} = \dfrac{1}{x(x+1)}$.

Notice that using the log laws to first simplify this expression made the differentiation process much easier.

The other approach, i.e., letting $u = \dfrac{x}{x+1}$, $y = \ln(u)$ and then using the chain rule would have meant more work – as not only would we need to use the chain rule but also the quotient rule to determine $\dfrac{du}{dx}$.

(b) Let $u = \ln t$ so that $y = \sin u$.

Using the chain rule we have $\dfrac{dy}{dt} = \dfrac{dy}{du} \cdot \dfrac{du}{dt} = \cos(u) \times \dfrac{1}{t} = \dfrac{\cos(\ln t)}{t}$

(c) Here we have a product $x \times \ln(x^2)$, so that the product rule needs to be used and then we need the chain rule to differentiate $\ln(x^2)$.

Notice that in this case we cannot simply rewrite $\ln(x^2)$ as $2\ln(x)$. Why?
Because the functions $\ln(x^2)$ and $2\ln(x)$ might have different domains. That is, the domain of $\ln(x^2)$ is all real values excluding zero (assuming an implied domain) whereas the domain of $2\ln(x)$ is only the positive real numbers. However, if it had been specified that $x > 0$, then we could have 'converted' $\ln(x^2)$ to $2\ln(x)$.

So, $\frac{dy}{dx} = \frac{d}{dx}(x) \times \ln(x^2) + x \times \frac{d}{dx}(\ln(x^2)) = 1 \times \ln(x^2) + x \times \frac{2x}{x^2} = \ln(x^2) + 2$

A short cut (?)

Once you have practiced the use of these rules and are confident in applying them, you can make use of the following table to speed up the use of the chain rule. Assuming that the function $f(x)$ is differentiable then we have:

y	$\frac{dy}{dx}$
$\sin[f(x)]$	$f'(x)\cos[f(x)]$
$\cos[f(x)]$	$-f'(x)\sin[f(x)]$
$\tan[f(x)]$	$f'(x)\sec^2[f(x)]$
$e^{f(x)}$	$f'(x)e^{f(x)}$
$\log_e[f(x)]$	$\frac{f'(x)}{f(x)}$
$[f(x)]^n$	$nf'(x)[f(x)]^{n-1}$

19.3.7 DERIVATIVE OF RECIPROCAL CIRCULAR FUNCTIONS

Dealing with the functions $\sec(x)$, $\cot(x)$ and $\text{cosec}(x)$ is a straight foward matter – simply rewrite them as their reciprocal counterparts. That is, $\sec(x) = \frac{1}{\cos(x)}$, $\cot(x) = \frac{1}{\tan(x)}$ and $\text{cosec}(x) = \frac{1}{\sin(x)}$. Once this is done, make use of the chain rule.

For example, $\frac{d}{dx}(\text{cosec}x) = \frac{d}{dx}\left(\frac{1}{\sin x}\right) = \frac{d}{dx}[(\sin x)^{-1}] = -1 \times \cos x \times (\sin x)^{-2} = -\frac{\cos x}{(\sin x)^2}$.

We could leave the answer as is or simplify it as follows; $-\frac{\cos x}{\sin x \sin x} = -\cot x \text{cosec} x$.
So, rather than providing a table of 'standard results' for the derivative of the reciprocal circular

trigonometric functions, we consider them as special cases of the circular trigonometric functions.

EXAMPLE 19.21

Differentiate the following

(a) $f(x) = \cot 2x, x > 0$ (b) $y = \sec^2 x$ (c) $y = \dfrac{\ln(\operatorname{cosec} x)}{x}$

Solution

(a) $f(x) = \cot 2x = \dfrac{1}{\tan 2x} = (\tan 2x)^{-1} \therefore f'(x) = -1 \times 2\sec^2 2x \times (\tan 2x)^{-2}$

$$= -\frac{2\sec^2 2x}{\tan^2 2x}$$

Now, $\dfrac{2\sec^2 2x}{\tan^2 2x} = 2 \times \dfrac{1}{\cos^2 2x} \times \dfrac{1}{\tan^2 2x} = 2 \times \dfrac{1}{\cos^2 2x} \times \dfrac{\cos^2 2x}{\sin^2 2x} = 2\operatorname{cosec}^2 2x$.

And so, $f'(x) = -2\operatorname{cosec}^2 2x$.

(b) $y = \sec^2 x = \dfrac{1}{(\cos x)^2} = (\cos x)^{-2} \therefore \dfrac{dy}{dx} = -2 \times -\sin x \times (\cos x)^{-3} = \dfrac{2\sin x}{(\cos x)^3}$.

Now, $\dfrac{2\sin x}{(\cos x)^3} = 2 \times \dfrac{\sin x}{\cos x} \times \dfrac{1}{(\cos x)^2} = 2\tan x \sec^2 x \therefore \dfrac{dy}{dx} = 2\tan x \sec^2 x$.

(c) $y = \dfrac{\ln(\operatorname{cosec} x)}{x} = \dfrac{\ln[(\sin x)^{-1}]}{x} = -\dfrac{\ln(\sin x)}{x} \therefore \dfrac{dy}{dx} = -\dfrac{\left(\dfrac{\cos x}{\sin x}\right) \times x - 1 \times \ln(\sin x)}{x^2}$

$$= -\frac{\dfrac{x\cos x - \sin x \ln(\sin x)}{\sin x}}{x^2}$$

$$= -\frac{x\cos x - \sin x \ln(\sin x)}{x^2 \sin x}$$

An interesting result

A special case of the chain rule involves the case $y = x$. By viewing this as an application of the chain rule $\dfrac{dy}{dx} = \dfrac{dy}{du} \cdot \dfrac{du}{dx}$ we have (after setting $y = x$):

$$\frac{d(x)}{dx} = \frac{dx}{du} \cdot \frac{du}{dx} \Rightarrow 1 = \frac{dx}{du} \cdot \frac{du}{dx} \text{ i.e., } \frac{dx}{du} = 1 \Big/ \frac{du}{dx}$$

This **important** result is often written in the form

$$\frac{dy}{dx} = \frac{1}{\left(\dfrac{dx}{dy}\right)}$$

We find that this result is very useful with problems that deal with related rates.

EXERCISES 19.3

1. Use the product rule to differentiate the following and then verify your answer by first expanding the brackets

(a) $(x^2+1)(2x-x^3+1)$ (b) $(x^3+x^2)(x^3+x^2-1)$

(c) $\left(\frac{1}{x^2}-1\right)\left(\frac{1}{x^2}+1\right)$ (d) $(x^3+x-1)(x^3+x+1)$

2. Use the quotient rule to differentiate the following

(a) $\frac{x+1}{x-1}$ (b) $\frac{x}{x+1}$ (c) $\frac{x+1}{x^2+1}$

(d) $\frac{x^2+1}{x^3-1}$ (e) $\frac{x^2}{2x+1}$ (f) $\frac{x}{1-2x}$

3. Differentiate the following

(a) $e^x \sin x$ (b) $x\log_e x$ (c) $e^x(2x^3+4x)$

(d) $x^4\cos x$ (e) $\sin x\cos x$ (f) $(1+x^2)\tan x$

(g) $\frac{4}{x^2}\times\sin x$ (h) $xe^x\sin x$ (i) $xe^x\log_e x$

4. Differentiate the following

(a) $\frac{x}{\sin x}$ (b) $\frac{\cos x}{x+1}$ (c) $\frac{e^x}{e^x+1}$

(d) $\frac{\sin x}{\sqrt{x}}$ (e) $\frac{x}{\log_e x}$ (f) $\frac{\log_e x}{x+1}$

(g) $\frac{e^x-1}{x+1}$ (h) $\frac{\sin x+\cos x}{\sin x-\cos x}$ (i) $\frac{x^2}{x+\log_e x}$

5. Differentiate the following

(a) $e^{-5x}+x$ (b) $\sin 4x-\frac{1}{2}\cos 6x$ (c) $e^{-\frac{1}{3}x}-\log_e(2x)+9x^2$

(d) $5\sin(5x)+3e^{2x}$ (e) $\tan(4x)+e^{2x}$ (f) $\cos(-4x)-e^{-3x}$

(g) $\log_e(4x+1)-x$ (h) $\log_e(e^{-x})+x$ (i) $\sin\left(\frac{x}{2}\right)+\cos(2x)$

(j) $\sin(7x-2)$ (k) $\sqrt{x}-\log_e(9x)$ (l) $\log_e(5x)-\cos(6x)$

6. Differentiate the following

(a) $\sin x^2+\sin^2 x$ (b) $\tan(2\theta)+\frac{1}{\sin\theta}$ (c) $\sin\sqrt{x}$

(d) $\cos\left(\frac{1}{x}\right)$ (e) $\cos^3\theta$ (f) $\sin(e^x)$

(g)	$\tan(\log_e x)$	(h)	$\sqrt{\cos(2x)}$	(i)	$\cos(\sin\theta)$
(j)	$4\sec\theta$	(k)	$\operatorname{cosec}(5x)$	(l)	$3\cot(2x)$

7. Differentiate the following

(a)	e^{2x+1}	(b)	$2e^{4-3x}$	(c)	$2e^{4-3x^2}$	(d)	$\sqrt{e^x}$
(e)	$e^{\sqrt{x}}$	(f)	$\frac{1}{2}e^{2x+4}$	(g)	$\frac{1}{2}e^{2x^2+4}$	(h)	$\frac{2}{e^{3x+1}}$
(i)	e^{3x^2-6x+1}	(j)	$e^{\sin(\theta)}$	(k)	$e^{-\cos(2\theta)}$	(l)	$e^{2\log_e(x)}$
(m)	$\frac{2}{e^{-x}+1}$	(n)	$(e^x - e^{-x})^3$	(o)	$\sqrt{e^{2x+4}}$	(p)	e^{-x^2+9x-2}

8. Differentiate the following

(a)	$\log_e(x^2+1)$	(b)	$\log_e(\sin\theta+\theta)$	(c)	$\log_e(e^x - e^{-x})$
(d)	$\log_e\left(\frac{1}{x+1}\right)$	(e)	$(\log_e x)^3$	(f)	$\sqrt{\log_e x}$
(g)	$\log_e(\sqrt{x-1})$	(h)	$\log_e(1-x^3)$	(i)	$\log_e\left(\frac{1}{\sqrt{x+2}}\right)$
(j)	$\log_e(\cos^2 x+1)$	(k)	$\log_e(x\sin x)$	(l)	$\log_e\left(\frac{x}{\cos x}\right)$

9. Differentiate the following

(a)	$x\log_e(x^3+2)$	(b)	$\sqrt{x}\sin^2 x$	(c)	$\cos^2\sqrt{\theta}$
(d)	$x^3e^{-2x^2+3}$	(e)	$\cos(x\log_e x)$	(f)	$\log_e(\log_e x)$
(g)	$\frac{x^2-4x}{\sin(x^2)}$	(h)	$\frac{10x+1}{\log_e(10x+1)}$	(i)	$\frac{\cos(2x)}{e^{1-x}}$
(j)	$x^2\log_e(\sin 4x)$	(k)	$e^{-\sqrt{x}}\sin\sqrt{x}$	(l)	$\cos(2x\sin x)$
(m)	$\frac{e^{5x+2}}{1-4x}$	(n)	$\frac{\log_e(\sin\theta)}{\cos\theta}$	(o)	$\frac{x}{\sqrt{x+1}}$
(p)	$x\sqrt{x^2+2}$	(q)	$(x^3+x)\sqrt[3]{x+1}$	(r)	$(x^3-1)\sqrt{x^3+1}$
(s)	$\frac{1}{x}\log_e(x^2+1)$	(t)	$\log_e\left(\frac{x^2}{x^2+2x}\right)$	(u)	$\frac{\sqrt{x-1}}{x}$
(v)	$e^{-x}\sqrt{x^2+9}$	(w)	$(8-x^3)\sqrt{2-x}$	(x)	$x^n\ln(x^n-1)$

10. Find the value of x where the function $x \mapsto xe^{-x}$ has a horizontal tangent.

11. Find the gradient of the function $x \mapsto \sin\left(\frac{1}{x}\right)$, where $x = \frac{2}{\pi}$.

12. Find the gradient of the function $x \mapsto \log_e(x^2 + 4)$ at the point where the function crosses the y–axis.

13. For what value(s) of x will the function $x \mapsto \ln(x^2 + 1)$ have a gradient of 1.

14. Find the rate of change of the function $x \mapsto e^{-x^2 + 2}$ at the point $(1, e)$.

15. Find (a) $\frac{d}{dx}(\sin x \cos x)$ (b) $\frac{d}{dx}(\sin x°)$ (c) $\frac{d}{dx}(\cos x°)$

16. (a) If y is the product of three functions, i.e., $y = f(x)g(x)h(x)$, show that $\frac{dy}{dx} = f'(x)g(x)h(x) + f(x)g'(x)h(x) + f(x)g(x)h'(x)$.

(b) Hence, differentiate the following i. $x^2 \sin x \cos x$

ii. $e^{-x^3} \sin(2x) \log_e(\cos x)$

17. (a) Given that $f(x) = 1 - x^3$ and $g(x) = \log_e x$, find i. $(f \circ g)'(x)$

ii. $(g \circ f)'(x)$

(b) Given that $f(x) = \sin(x^2)$ and $g(x) = e^{-x}$, find i. $(f \circ g)'(x)$

ii. $(g \circ f)'(x)$

18. Given that $T(\theta) = \frac{\cos k\theta}{2 + 3\sin k\theta}$, $k \neq 0$, determine $T'\left(\frac{\pi}{2k}\right)$.

19. If $f(x) = (x-a)^m (x-b)^n$, find x such that $f'(x) = 0$.

20. If $f(\theta) = \sin\theta^m \cos\theta^n$, find θ such that $f'(\theta) = 0$.

21. Differentiate the following

(a) $f(x) = \cot 4x$ (b) $g(x) = \sec 2x$ (c) $f(x) = \operatorname{cosec} 3x$

(d) $y = \sin\left(3x + \frac{\pi}{2}\right)$ (e) $y = \cot\left(\frac{\pi}{4} - x\right)$ (f) $y = \sec(2x - \pi)$

22. Differentiate the following

(a) $\sec x^2$ (b) $\sin x \sec x$ (c) $\ln(\sec x)$

(d) $\cot^3 x$ (e) $\frac{x}{\operatorname{cosec} x}$ (f) $\frac{\operatorname{cosec} x}{\sin x}$

(g) $x^4 \operatorname{cosec}(4x)$ (h) $\tan 2x \cot x$ (i) $\sqrt{\sec x + \cos x}$

23. Differentiate the following

(a) $e^{\sec x}$ (b) $\sec(e^x)$ (c) $e^x \sec x$

(d) $\cot(\ln x)$ (e) $\ln(\cot 5x)$ (f) $\cot x \ln x$

(g) $\operatorname{cosec}(\sin x)$ (h) $\sin(\operatorname{cosec} x)$ (i) $\sin x \operatorname{cosec} x$

19.4 SECOND DERIVATIVE

Since the derivative of a function f is another function, f', then it may well be that this derived function can itself be differentiated. If this is done, we obtain the **second derivative** of f which is denoted by f'' and read as "f double dash".

The following notation for $y = f(x)$ is used:

First derivative $\quad \frac{dy}{dx} = f'(x)$ [$= y'$]

Second derivative $\quad \frac{d}{dx}\left(\frac{dy}{dx}\right) = \frac{d^2y}{dx^2} = f''(x)$ [$= y''$]

So, for example, if $f(x) = x^3 - 5x^2 + 10$ then $f'(x) = 3x^2 - 10x$ and

$$f''(x) = 6x - 10.$$

The expression $\frac{d^2y}{dx^2}$ is read as "dee two y by dee x squared" and the expression y'' is read as "y double dash".

EXAMPLE 19.22

Find the second derivative of

(a) $x^4 - \sin 2x$ (b) $\ln(x^2 + 1)$ (c) xe^{-2x}

Solution

(a) Let $y = x^4 - \sin 2x$ then $y' = 4x^3 - 2\cos 2x$ and $y'' = 12x^2 + 4\sin 2x$.

(b) Let $f(x) = \ln(x^2 + 1)$ then $f'(x) = \frac{2x}{x^2 + 1}$ and $f''(x) = \frac{2(x^2 + 1) - 2x(2x)}{(x^2 + 1)^2}$

$$= \frac{2 - 2x^2}{(x^2 + 1)^2}$$

(c) Let $y = xe^{-2x}$ then $\frac{dy}{dx} = 1 \times e^{-2x} + x \times -2e^{-2x}$

$$= e^{-2x} - 2xe^{-2x}$$

So that $\frac{d^2y}{dx^2} = -2e^{-2x} - 2(e^{-2x} - 2xe^{-2x})$ [Notice that we have used the previous result]

$$= -4e^{-2x} + 4xe^{-2x}$$

EXERCISES 19.4

1. Find the second derivative of the following functions

(i) $f(x) = x^5$

(ii) $y = (1+2x)^4$

(iii) $f:x \mapsto \frac{1}{x}$ where $x \in \mathbb{R}$

(iv) $f(x) = \frac{1}{1+x}$

(v) $y = (x-7)(x+1)$

(vi) $f:x \mapsto \frac{x+1}{x-2}$ where $x \in \mathbb{R}\backslash\{2\}$

(vii) $f(x) = \frac{1}{x^6}$

(viii) $y = (1-2x)^3$

(ix) $y = \ln x$

(x) $f(x) = \ln(1-x^2)$

(xi) $y = \sin 4\theta$

(xii) $f(x) = x\sin x$

(xiii) $f(x) = x^3 \sin x$

(xiv) $y = x\ln x$

(xv) $f(x) = \frac{x^2-1}{2x+3}$

(xvi) $y = x^3 e^{2x}$

(xvii) $f(x) = \frac{\cos(4x)}{e^x}$

(xviii) $y = \sin(x^2)$

(xix) $f(x) = \frac{x}{1-4x^3}$

(xx) $y = \frac{x^2-4}{x-3}$

2. Find the second derivative of the function $f(x) = \frac{\log_e x}{x^2}$. Find a formula for the second derivative of the function $f(x) = x^n \cdot (\log_e x)$.

3. Find a formula for the second derivative of the family of functions $f(x) = \left(\frac{x+1}{x-1}\right)^n$ where n is a real number.

4. Find $f''(2)$ if $f(x) = x^2 - \sqrt{x}$.

5. Find the rate of change of the gradient of the function $g(x) = \frac{x^2-1}{x^2+1}$ where $x = 1$.

6. Find the the values of x where the rate of change of the gradient of the curve $y = x\sin x$ for $0 \le x \le 2\pi$ is positive.

19.5 PROOFS

For the sake of completeness and to allow a certain 'flow' to the chapter, we use this last section to include an assortment of proofs for results which we simply quoted in the body of this chapter.

19.5.1 DERIVATIVE OF $y = x^n$ WHERE n IS A NEGATIVE INTEGER

Consider the function with equation $y = x^n$ where n is a negative integer. Setting $n = -k$, where k is a positive integers we have

$$y = x^n = x^{-k} = \frac{1}{x^k}$$

Using the quotient rule we have

$$\frac{dy}{dx} = \frac{0 \times x^k - 1 \times kx^{k-1}}{(x^k)^2}$$

$$= -\frac{kx^{k-1}}{x^{2k}}$$

$$= -k \cdot x^{-k-1}$$

But remember, $n = -k$, so, $\frac{dy}{dx} = nx^{n-1}$.

That is, the rule is still true for negative integers.

19.5.2 DERIVATIVE OF $y = x^n$ WHERE n IS A FRACTION

Consider the function with equation $y = x^n$ where n is a fraction. Setting $n = \frac{k_1}{k_2}$, where k_1 and k_2 are positive integer we have

$$y = x^n = x^{\frac{k_1}{k_2}} \Leftrightarrow y^{k_2} = x^{k_1}.$$

Next, using implicit differentiation, we differentiate both sides with respect to x:

$$\frac{d}{dx}(y^{k_2}) = \frac{d}{dx}(x^{k_1}) \therefore k_2 y^{k_2-1} \cdot \frac{dy}{dx} = k_1 x^{k_1-1}$$

$$\frac{dy}{dx} = \frac{k_1}{k_2} \cdot \frac{x^{k_1-1}}{y^{k_2-1}}$$

As $y = x^{\frac{k_1}{k_2}}$ then $y^{k_2-1} = \left(x^{\frac{k_1}{k_2}}\right)^{k_2-1} = x^{\frac{k_1(k_2-1)}{k_2}}$ $\therefore \frac{x^{k_1-1}}{y^{k_2-1}} = \frac{x^{k_1-1}}{x^{\frac{k_1(k_2-1)}{k_2}}} = x^{k_1-1-\left[\frac{k_1(k_2-1)}{k_2}\right]}$

$$= x^{\frac{k_1}{k_2}-1}$$

So, we have that $\frac{dy}{dx} = \frac{k_1}{k_2}x^{\frac{k_1}{k_2}-1}$. i.e., $\frac{dy}{dx} = ny^{n-1}$.

19.5.3 PRODUCT RULE AND QUOTIENT RULE

Let $f(x)$ and $g(x)$ be two functions having derivatives $f'(x)$ and $g'(x)$, then the derivative of the product function $f(x)g(x)$ is given by $f'(x)g(x)+f(x)g'(x)$.

We will use the definition of the derivative, $(f(x)g(x))' = \lim_{h\to 0}\frac{f(x+h)g(x+h)-f(x)g(x)}{h}$.

The problem arises in trying to 'simplify' the numerator. The not so obvious solution to this problem is to add the expression $-f(x)g(x+h)+f(x)g(x+h)$ to the numerator.
This gives us

$$\begin{aligned}
(f(x)g(x))' &= \lim_{h\to 0}\frac{f(x+h)g(x+h)+[-f(x)g(x+h)+f(x)g(x+h)]-f(x)g(x)}{h}\\
&= \lim_{h\to 0}\frac{f(x+h)g(x+h)-f(x)g(x+h)}{h}+\lim_{h\to 0}\frac{f(x)g(x+h)-f(x)g(x)}{h}\\
&= \lim_{h\to 0}\frac{f(x+h)-f(x)}{h}\cdot g(x+h)+\lim_{h\to 0}f(x)\cdot\frac{g(x+h)-g(x)}{h}\\
&= \lim_{h\to 0}\frac{f(x+h)-f(x)}{h}\cdot\lim_{h\to 0}g(x+h)+\lim_{h\to 0}f(x)\cdot\lim_{h\to 0}\frac{g(x+h)-g(x)}{h}\\
&= f'(x)g(x)+f(x)g'(x)
\end{aligned}$$

We can prove the quotient rule by rewriting it as a product rule and then using the chain rule:

Let $y = \frac{f(x)}{g(x)} = f(x)\cdot\frac{1}{g(x)} = f(x)\cdot[g(x)]^{-1}$

$$\begin{aligned}
\therefore\frac{dy}{dx} &= f'(x)[g(x)]^{-1}+f(x)\times[-1\times g'(x)\times[g(x)]^{-2}]\\
&= \frac{f'(x)}{g(x)}-\frac{f(x)g'(x)}{[g(x)]^2}\\
&= \frac{f'(x)g(x)-f(x)g'(x)}{[g(x)]^2}
\end{aligned}$$

19.5.4 DERIVATIVE OF SOME TRIGONOMETRIC FUNCTIONS

The sine function

It must be pointed out at the very outset that the variable x is a real number and is given in radian measure – the proof, although carried out in the same way, would need to be altered slightly if we where considering x as being measured in degrees.

Letting $f(x) = \sin(x)$ and using the definition from first principle we have

$$f'(x) = \lim_{h\to 0}\frac{f(x+h)-f(x)}{h} = \lim_{h\to 0}\frac{\sin(x+h)-\sin(x)}{h}$$

Using the identity $\sin(A+B) = \sin A\cos B+\sin B\cos A$ we then have

$$= \lim_{h\to 0}\frac{\sin(x)\cos(h)+\sin(h)\cos(x)-\sin(x)}{h}$$

$$= \lim_{h \to 0}\left[\frac{\sin(x)\cos(h) - \sin(x)}{h} + \frac{\sin(h)\cos(x)}{h}\right]$$

$$= \sin(x)\lim_{h \to 0}\left[\frac{\cos(h) - 1}{h}\right] + \cos(x)\lim_{h \to 0}\frac{\sin(h)}{h} \quad \text{– (1)}$$

We now need to determine the value of the limits $\lim_{h \to 0}\left[\frac{\cos(h) - 1}{h}\right]$ and $\lim_{h \to 0}\frac{\sin(h)}{h}$.

We start with $\lim_{h \to 0}\left[\frac{\cos(h) - 1}{h}\right]$:

We have that $\lim_{h \to 0}\left[\frac{\cos(h) - 1}{h}\right] = \lim_{h \to 0}\left[\frac{\cos(h) - \cos(0)}{h - 0}\right]$ which, by definition, defines the gradient of the cosine function at $x = 0$. But, the cosine graph has a horizontal tangent at $x = 0$ and so the gradient of the cosine function at $x = 0$ must therefore be 0. Hence, the limit has a value of zero.

We are now left with determining $\lim_{h \to 0}\frac{\sin(h)}{h}$:

Using the graph of $f(h) = \frac{\sin(h)}{h}$ it is clear that the value of this limit is one.

i.e., $\lim_{h \to 0}\frac{\sin(h)}{h} = 1$.

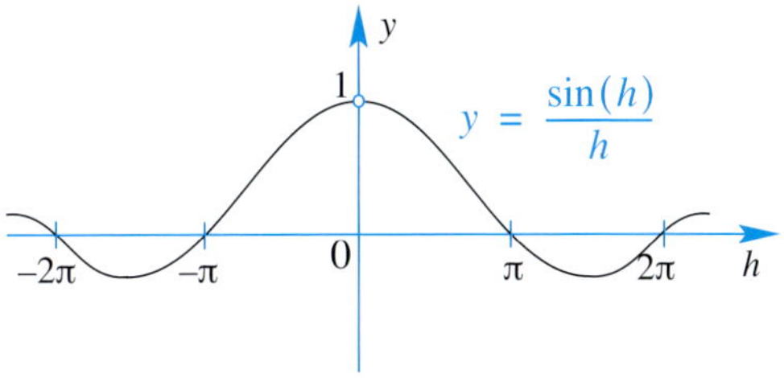

That is, although $f(0) = \frac{\sin(0)}{0}$ is undefined (hence the open circle at (0, 1)), the limit as h tends to zero exists!

We now continue from equation (1):

$$\sin(x)\lim_{h \to 0}\left[\frac{\cos(h) - 1}{h}\right] + \cos(x)\lim_{h \to 0}\frac{\sin(h)}{h} = \sin(x) \times 0 + \cos(x) \times 1$$

$$= \cos(x)$$

Therefore, if $f(x) = \sin(x)$ then $f'(x) = \cos(x)$.

Note that we have used a picture to provide very strong evidence that $\lim_{h \to 0}\frac{\sin(h)}{h} = 1$. However, for those who wish to have a more rigorous proof, we now provide one:

Consider the first quadrant of a circle of unit radius.

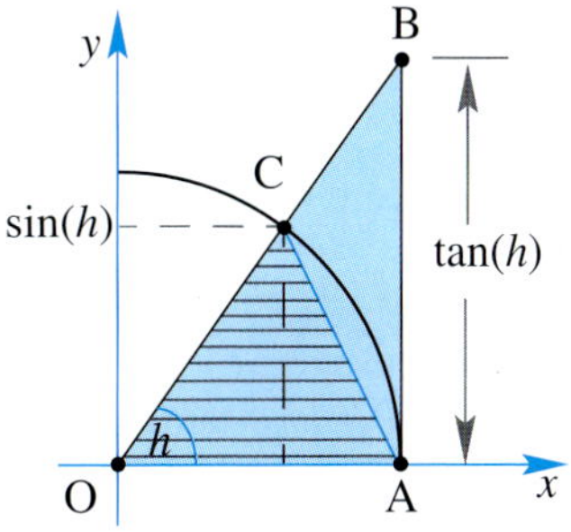

Area measure of $\Delta OAC = \frac{1}{2} \cdot 1 \cdot 1 \cdot \sin(h) = \frac{1}{2}\sin(h) = A_1$

Area measure of sector OAC $= \frac{1}{2} \cdot 1 \cdot 1 \cdot \theta = \frac{1}{2}h = A_2$ $(\theta = h)$

Area measure of $\Delta OAB = \frac{1}{2} \cdot 1 \cdot \tan(h) = \frac{1}{2}\tan(h) = A_3$

Now, $A_1 < A_2 < A_3 \Rightarrow \frac{1}{2}\sin(h) < \frac{1}{2}h < \frac{1}{2}\tan(h)$

$$\Leftrightarrow \sin(h) < h < \tan h \Leftrightarrow \frac{1}{\sin(h)} > \frac{1}{h} > \frac{1}{\tan(h)} \quad \left[\text{as } 0 < h < \frac{\pi}{2}\right]$$

Multiplying through by $\sin(h)$ we have: $\frac{1}{1} > \frac{\sin(h)}{h} > \cos(h)$.

Now, as $h \to 0$, $\cos(h) \to 1$ and seeing as $\frac{\sin(h)}{h}$ lies between 1 and $\cos(h)$, then, taking the limit we have: $\lim_{h \to 0} 1 > \lim_{h \to 0} \frac{\sin(h)}{h} > \lim_{h \to 0} \cos(h) \Leftrightarrow 1 > \lim_{h \to 0} \frac{\sin(h)}{h} > 1 \Rightarrow \lim_{h \to 0} \frac{\sin(h)}{h} = 1$.

Note: As the function $f(h) = \frac{\sin h}{h}$ is an even function, $f(h) = f(-h)$, it is only necessary to prove our result for the case where h was positive. The proof is complete because, as $f(h) = f(-h)$, then the result would hold true for h being negative.

What if x was in degrees?

We would need to convert from degrees to radians: i.e., $x^c = \frac{\pi}{180}x^\circ$.

So, $\frac{d}{dx}(\sin x^\circ) = \frac{d}{dx}\left(\sin \frac{\pi}{180}x\right) = \frac{\pi}{180}\cos\left(\frac{\pi}{180}x\right) = \frac{\pi}{180}\cos(x^\circ)$.

Notice that $\frac{d}{dx}(\sin x^\circ) \neq \cos(x^\circ)$

The tangent function

For this proof we use the ratio definition of the tangent, i.e., $\tan\theta = \frac{\sin\theta}{\cos\theta}$, $\cos\theta \neq 0$ and then make use of the quotient rule.

So, letting $f(\theta) = \tan\theta = \frac{\sin\theta}{\cos\theta}$ we have $f'(\theta) = \frac{\frac{d}{d\theta}(\sin\theta) \times \cos\theta - \sin\theta \times \frac{d}{d\theta}(\cos\theta)}{(\cos\theta)^2}$

$$= \frac{\cos\theta \times \cos\theta - \sin\theta \times (-\sin\theta)}{\cos^2\theta}$$

$$= \frac{\cos^2\theta + \sin^2\theta}{\cos^2\theta}$$

$$= \frac{1}{\cos^2\theta}$$

$$= \sec^2\theta$$

Again, a more formal proof can be obtained by using a first principle approach: i.e., using the definition $f'(\theta) = \lim_{h \to 0} \frac{\tan(\theta + h) - \tan\theta}{h}$.

Now, $\tan(\theta + h) - \tan\theta = \frac{\sin(\theta + h)}{\cos(\theta + h)} - \frac{\sin\theta}{\cos\theta} = \frac{\sin(\theta + h)\cos\theta - \sin\theta\cos(\theta + h)}{\cos\theta\cos(\theta + h)}$

$$= \frac{\sin[(\theta + h) - \theta]}{\cos\theta\cos(\theta + h)}$$

$$= \frac{\sin h}{\cos\theta\cos(\theta + h)}$$

$$\therefore f'(\theta) = \lim_{h\to 0}\frac{1}{h}\times\frac{\sin h}{\cos\theta\cos(\theta+h)} = \frac{1}{\cos\theta}\left[\lim_{h\to 0}\frac{\sin h}{h}\cdot\frac{1}{\cos(\theta+h)}\right] = \frac{1}{\cos\theta}\cdot 1\cdot\frac{1}{\cos\theta}$$
$$= \frac{1}{\cos^2\theta}$$

That is, if $f(\theta) = \tan\theta$ then $f'(\theta) = \sec^2\theta$.

19.5.5 EXPONENTIAL AND $y = x^n$ WHERE n IS REAL

The exponential function

This approach makes use of a result that will be looked at by students who study the Series and Differential Equations Option for this course. We only include it as a point of interest and so, suffice to state that a polynomial expression for the exponential function can be obtained. Such an expression is given by

$$e^x = 1 + x + \frac{x^2}{2!} + \frac{x^3}{3!} + \frac{x^4}{4!} + \ldots,\ x \in \mathbb{R}$$

Notice that this **series** continues indefinitely.

Then, using our definition of differentiation we have:

$$f'(x) = \lim_{h\to 0}\frac{f(x+h)-f(x)}{h} = \lim_{h\to 0}\frac{e^{x+h}-e^x}{h} = \lim_{h\to 0}\frac{e^x\cdot e^h - e^x}{h} = e^x\lim_{h\to 0}\frac{e^h-1}{h}$$

Now, $\dfrac{e^h-1}{h} = \dfrac{1}{h}(e^h-1) = \dfrac{1}{h}\left(\left[1+h+\dfrac{h^2}{2!}+\dfrac{h^3}{3!}+\dfrac{h^4}{4!}+\ldots\right]-1\right) = \dfrac{1}{h}\left(h+\dfrac{h^2}{2!}+\dfrac{h^3}{3!}+\dfrac{h^4}{4!}+\ldots\right)$

$$= 1 + \frac{h}{2!} + \frac{h^2}{3!} + \frac{h^3}{4!} + \ldots$$

Then, $\displaystyle\lim_{h\to 0}\frac{e^h-1}{h} = \lim_{h\to 0}\left(1+\frac{h}{2!}+\frac{h^2}{3!}+\frac{h^3}{4!}+\ldots\right) = 1$

And so, $\displaystyle f'(x) = e^x\lim_{h\to 0}\frac{e^h-1}{h} = e^x\times 1 = e^x$

The debate as to whether this is an actual proof or not can be argued both ways, however, it does provide us with another avenue of showing how a result can be confirmed via different and appropriate means.

The power function, $y = x^n$ where n is real

Let $y = x^n$ then $y = e^{\ln x^n} = e^{n\ln x}, x > 0$. Therefore, $\dfrac{dy}{dx} = \left(n\cdot\dfrac{1}{x}\right)\cdot e^{n\ln x} = \dfrac{n}{x}\cdot x^n$

$$= nx^{n-1}$$

A rather short and neat proof!

20.1 TANGENTS AND NORMALS

20.1.1 EQUATION OF TANGENT

The gradient of a curve $y = f(x)$ at any point (x_1, y_1) is equal to the gradient of the tangent to the curve at that point. To find the gradient at (x_1, y_1), you need to

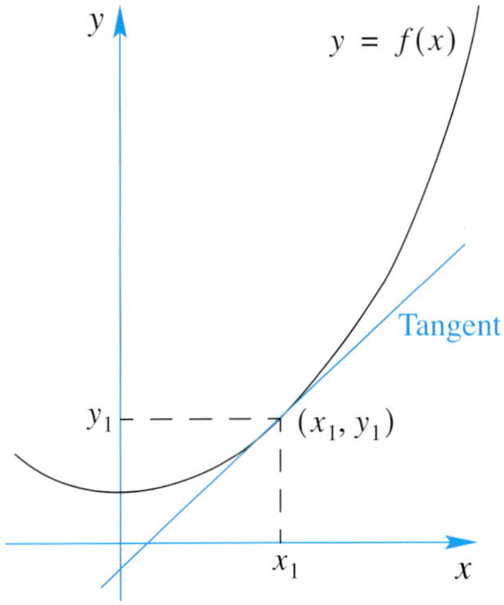

1. Find the gradient function of $y = f(x)$, that is, find the derivative $\frac{dy}{dx} = f'(x)$.

2. Find the gradient at (x_1, y_1), that is, find $f'(x_1)$.

This gives the gradient of the tangent, m, at the point (x_1, y_1).

Finally, to find the **equation of the tangent**, you need to use the straight line equation

$$y - y_1 = m(x - x_1), \text{ where } m = f'(x_1)$$

EXAMPLE 20.1

Find the equation of the tangent to the curve $y = 5 - x^2$ at the point (1,4).

Solution

1. Given that $y = 5 - x^2 \Rightarrow \frac{dy}{dx} = -2x$.

2. Then, for $x = 1$, we have $\frac{dy}{dx} = -2(1) = -2$.

Therefore, using $y - y_1 = m(x - x_1)$, with $m = -2$ and $(x_1, y_1) \equiv (1, 4)$, we have the equation of the tangent given by

$$y - 4 = (-2)(x - 1) \Leftrightarrow y - 4 = -2x + 2$$

That is, $y = -2x + 6$

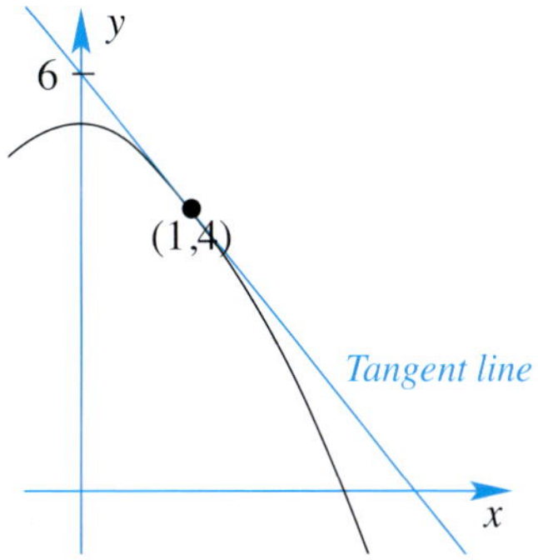

EXAMPLE 20.2

Find the equation of the tangent to the curve $y = x^3 - 8$ where $x = 2$.

Solution

Given that $y = x^3 - 8 \Rightarrow y' = 3x^2$. Then, for $x = 2$, $y' = 3 \times 2^2 = 12$. i.e., $m = 12$.
In order to use the equation $y - y_1 = m(x - x_1)$ we need both x- and y-values. As we are only given the x-value, we now determine the corresponding y-value, i.e., $x = 2 \Rightarrow y = 2^3 - 8 = 0$.
With $(x_1, y_1) \equiv (2, 0)$ the equation of the tangent is: $(y - 0) = 12(x - 2) \Leftrightarrow y = 12x - 24$.

EXAMPLE 20.3

For the curve $x \mapsto 3x^2e^{-0.2x}$, find the equation of the tangent at the point $(1, 3e^{-0.2})$.

Solution

Let $f(x) = 3x^2e^{-0.2x} \therefore f'(x) = 3\left(\frac{d}{dx}(x^2) \times e^{-0.2x} + x^2 \times \frac{d}{dx}(e^{-0.2x})\right)$ [Using the product rule]

$= 3(2xe^{-0.2x} - 0.2x^2e^{-0.2x})$

That is, $f'(x) = 3xe^{-0.2x}(2 - 0.2x)$.

Therefore, at $x = 1$, we have

$$f'(1) = 3(1)e^{-0.2}(2 - 0.2) = 5.4e^{-0.2}.$$

Using the general equation for a straight line,

$$y - y_1 = m(x - x_1),$$

where $m = 5.4e^{-0.2}$ and $(x_1, y_1) \equiv (1, 3e^{-0.2})$

we have:

$$y - 3e^{-0.2} = 5.4e^{-0.2}(x - 1)$$

$$\Leftrightarrow y - 3e^{-0.2} = 5.4e^{-0.2}x - 5.4e^{-0.2}$$

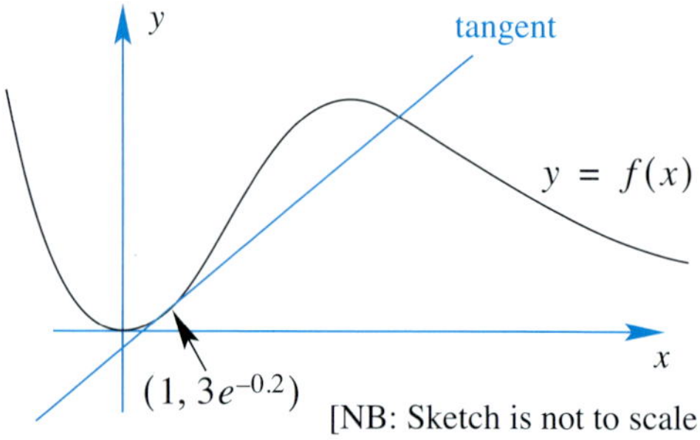

[NB: Sketch is not to scale.]

giving the equation of the tangent at $(1, 3e^{-0.2})$ as $y = 5.4e^{-0.2}x - 2.4e^{-0.2}$.

EXAMPLE 20.4

Find the equation of the tangent to the curve defined by $x^2y - y = x^2 - 4$ at the point where it crosses the positive x-axis.

Solution

We first determine the gradient function. To do this we obtain y in terms of x: $y = \frac{x^2 - 4}{x^2 - 1}$.

Using the quotient rule we have that $\frac{dy}{dx} = \frac{2x \times (x^2 - 1) - 2x \times (x^2 - 4)}{(x^2 - 1)^2}$

$$= \frac{6x}{(x^2 - 1)^2}$$

At the point where the curve crosses the x-axis we have that $y = 0$, so, substituting $y = 0$ into the equation of the curve we have: $x^2(0) - (0) = x^2 - 4 \Leftrightarrow 0 = x^2 - 4 \Leftrightarrow x = \pm 2$.

As we are only interested in the positive x-axis, we choose $x = 2$.

So, the gradient of the curve at the point (2, 0) is given by $\frac{dy}{dx} = \frac{6 \times 2}{(2^2 - 1)^2} = \frac{12}{9} = \frac{4}{3}$.

Using the equation of the straight line, $y - y_1 = m(x - x_1)$ with $m = \frac{4}{3}$ and $(x_1, y_1) \equiv (2, 0)$

we have the equation of the tangent given by $y - 0 = \frac{4}{3}(x - 2) \Leftrightarrow 3y - 4x + 8 = 0$.

20.1.2 EQUATION OF NORMAL

To find the equation of the normal at the point (x_1, y_1) we first need to determine the gradient of the tangent, m_t, and then use the relationship between the gradients of two perpendicular lines (given that the normal is perpendicular to the tangent).

To find the equation of the normal we need to

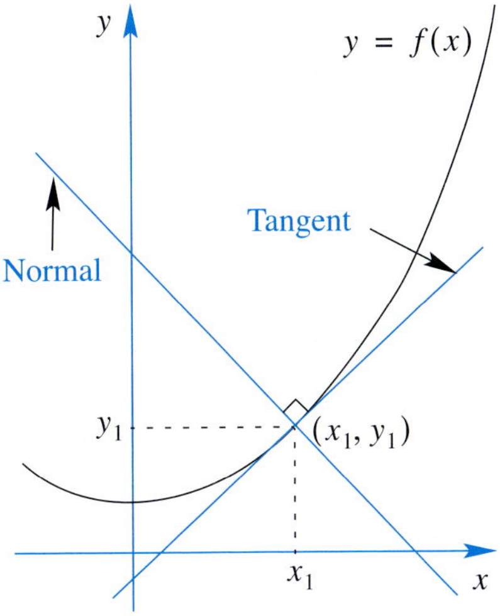

1. Find the gradient function of $y = f(x)$, that is, find the derivative $\frac{dy}{dx} = f'(x)$.

2. Find the gradient at (x_1, y_1), that is, find $f'(x_1)$.

This gives the gradient of the tangent, $m_t = f'(x_1)$, at the point (x_1, y_1).

3. *To find the gradient of the normal, m_N, we use the fact that the product of the gradients of two perpendicular lines is –1.*

i.e., $m_t \times m_N = -1$ $\left(\text{or } m_N = -\frac{1}{m_t}\right)$

Finally, to find the **equation of the normal**, we need to use the general equation for a straight line $y - y_1 = m(x - x_1)$, where this time $m = m_N = -\frac{1}{m_t} = -\frac{1}{f'(x_1)}$.

EXAMPLE 20.5

Find the equation of the normal to the curve $y = 2x^3 - x^2 + 1$ at the point $(1, 2)$.

Solution

First determine the gradient of the tangent: $\frac{dy}{dx} = 6x^2 - 2x$.

At $x = 1$, we have $\frac{dy}{dx} = 6(1)^2 - 2(1) = 4$. That is, $m_t = 4$.

We can now determine the gradient of the normal: using $m_N = -\frac{1}{m_t}$ we have $m_N = -\frac{1}{4}$.

Using the equation of a straight line, $y - y_1 = m(x - x_1)$ where $(x_1, y_1) \equiv (1, 2)$ and $m = -\frac{1}{4}$

we have that $$y - 2 = -\frac{1}{4}(x - 1) \Leftrightarrow 4y - 8 = -x + 1$$

Hence the equation of the normal is given by $4y + x = 9$.

EXAMPLE 20.6

Determine the equation of the normal to the curve $y = xe^{-2x} + 2$ at the point where the curve crosses the y–axis.

Solution

We first need to determine the y–intercept: $x = 0 \Rightarrow y = 0 \times e^0 + 2 = 2$.
That is, the curve passes through the point (0, 2).
Next, we need to determine the gradient of the tangent where $x = 0$.

From $y = xe^{-2x} + 2 \Rightarrow \frac{dy}{dx} = 1 \times e^{-2x} + x(-2e^{-2x})$ [Using the product rule]

$$= e^{-2x} - 2xe^{-2x}$$

Therefore, at $x = 0$, $\frac{dy}{dx} = e^{-2(0)} - 2(0)e^{-2(0)} = 1 - 0 = 1$.

That is, $m_t = 1 \Rightarrow m_N = -\frac{1}{1} = -1$. Then, using the general equation of a straight line we have the equation of the normal as $y - 2 = -1(x - 0)$, or $y = -x + 2$.

EXAMPLE 20.7

Find the equation of the normal to the curve $f(x) = \sqrt{x+2}$ at the point where $y = 3$.

Solution

This time, we need to first determine the x–value when $y = 3$:

$$3 = \sqrt{x+2} \Rightarrow 9 = x + 2 \Leftrightarrow x = 7$$

Therefore, we want the equation of the normal at the point (7, 3).

Now, $f(x) = \sqrt{x+2} = (x+2)^{1/2} \Rightarrow f'(x) = \frac{1}{2}(x+2)^{-1/2}$

Next, $f'(7) = \frac{1}{2\sqrt{7+2}} = \frac{1}{6}$ $(= m_t)$. Therefore $m_N = -6$ (= gradient of normal).

The equation of the normal is given by $y - 3 = -6(x - 7)$ or $y + 6x = 45$.

EXERCISES 20.1

1. Find the equations of the tangents to the following curves at the points indicated

(a) $y = x^3 - x^2 - x + 2$ at (2,4)

(b) $y = x^4 - 4x^2 + 3$ at $(1, 0)$

(c) $y = \sqrt{x+1}$ at (3, 2)

(d) $y = \frac{1}{\sqrt{x-1}} + \frac{1}{2}$ at $(5, 1)$

(e) $f(x) = \frac{x}{x+1}, x \neq -1$ at $\left(1, \frac{1}{2}\right)$

(f) $f(x) = \frac{2x}{x+2}, x \neq -2$ at (2, 1)

(g) $x \mapsto x(x^3 - 4)$ at (2, 8)

(h) $x \mapsto \frac{x^2}{x-1}, x \neq 1$ at (2, 4)

2. Find the equation of the normal for each of the curves in Question 1.

3. Find the equations of the tangents to the following curves at the points indicated

(a) $y = xe^x$ at $(1, e)$

(b) $x \mapsto \frac{e^x}{x}, x \neq 0$ at $(1, e)$

(c) $f(x) = x + \sin(x)$ at (π, π)

(d) $y = x\cos(x)$ at $(\pi, -\pi)$

(e) $y = \frac{x}{\sin(x)}$ at $\left(\frac{\pi}{2}, \frac{\pi}{2}\right)$

(f) $x \mapsto x\log_e(x+1)$ at $(e-1, e-1)$

(g) $x \mapsto xe^{x^2+1}$ at $(0, 0)$

(h) $f(x) = \sin(2x) + \cos(x)$ at $(0,1)$

4. Find the equation of the normal for each of the curves in Question 3.

5. Find the equation of the tangent to the curve $y = x^2(x^2 - 1)$ at the point A(2, 12).
The tangent at a second point, B(–2,12), intersects the tangent at A at the point C.
Determine the type of triangle enclosed by the points A, B and C.
Show that the tangents drawn at the points X and Y, where $x = a$ and $x = -a$ respectively will always meet at a third point Z which will lie on the y–axis.

6. Find the equation of the tangent and the normal to the curve $x \mapsto x + \frac{1}{x}, x \neq 0$ at the point $(1, 2)$.

Find the coordinates of the points where the tangent and the normal cross the x- and y-axes, and hence determine the area enclosed by the x–axis, the y–axis, the tangent and the normal.

7. Find the equation of the normal to the curve $y = \sqrt{25 - x^2}$ at the point (4, 3).

8. Show that every normal to the curve $y = \sqrt{a^2 - x^2}$ will always pass through the point (0,0).

9. Find the equation of the tangent to the curve $y = x^2 - 2x$ that is parallel to the line with equation $y = 4x + 2$.

10. Find the equation of the tangent to the curve $x \mapsto \log_e(x^2 + 4)$ at the point where the curve crosses the y–axis.

11. Find the equation of the tangent and the normal to the curve $x \mapsto x\tan(x)$ where $x = \frac{\pi}{4}$.

12. The straight line $y = -x + 4$ cuts the parabola with equation $y = 16 - x^2$ at the points A and B.

(a) Find the coordinates of A and B.
(b) Find the equation of the tangents at A and B, and hence determine where the two tangents meet.

13. For the curve defined by $x \mapsto \frac{x}{x^2+1}$ find the equation of the normal at the origin, and the equations of the tangents that are parallel to the x–axis. Find also the points where the tangents and the normal intersect.

14. The figure shows the curve whose equation is given by

$$y^2 = (x-1)^3 .$$

The tangent drawn at the point $P(5,8)$ meets the curve again at the point Q.

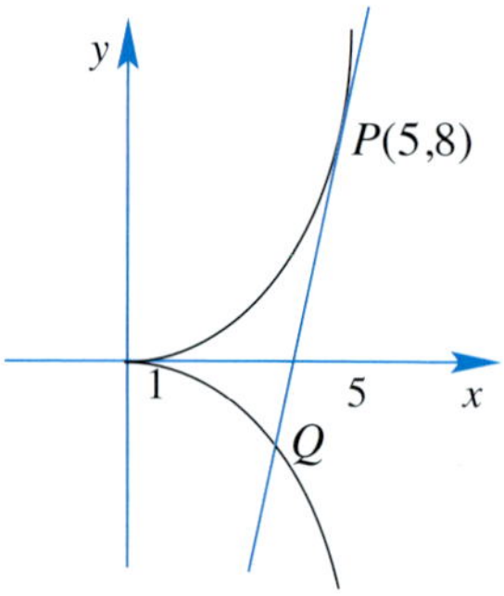

(a) Find the equation of the tangent at the point P.

(b) Find the coordinates of Q.

15. The line L and the curve C are defined as follows,

$$L: y = 4x - 2 \text{ and } C: y = mx^3 + nx^2 - 1$$

The line L is a tangent to the curve C at $x = 1$.

(a) Using the fact that L and C meet at $x = 1$, show that $m + n = 3$.

(b) Given that L is a tangent to C at $x = 1$, show that $3m + 2n = 4$.

(c) Hence, solve for m and n.

20.2 CURVE SKETCHING

20.2.1 INCREASING AND DECREASING FUNCTIONS

A function f is said to be **increasing** if its **graph rises** as it is sketched from left to right.

That is, if $x_2 > x_1 \Rightarrow f(x_2) > f(x_1)$

(i.e., the y–values increase as the x–values increase).

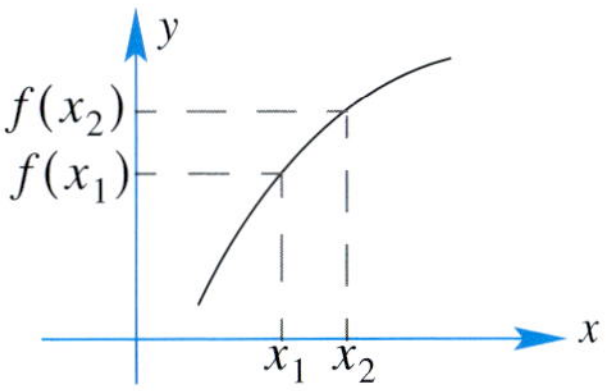

Similarly,

A function f is said to be **decreasing** if its **graph falls** as it is sketched from left to right.

That is, if $x_2 > x_1 \Rightarrow f(x_2) < f(x_1)$

(i.e., the y–values decrease as the x–values increase).

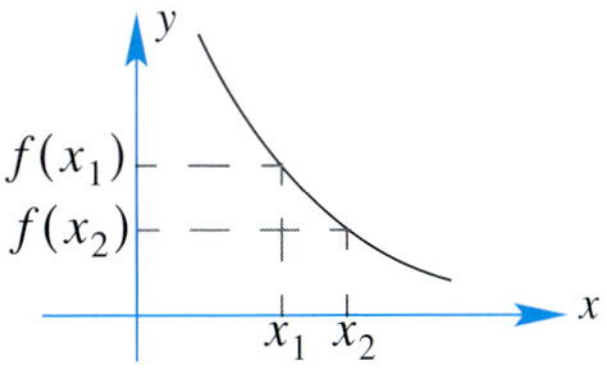

. . . A calculus point of view

The derivative can be used to determine whether a function is increasing or decreasing and so it can be used to help find those values of x for which the function is increasing or decreasing.

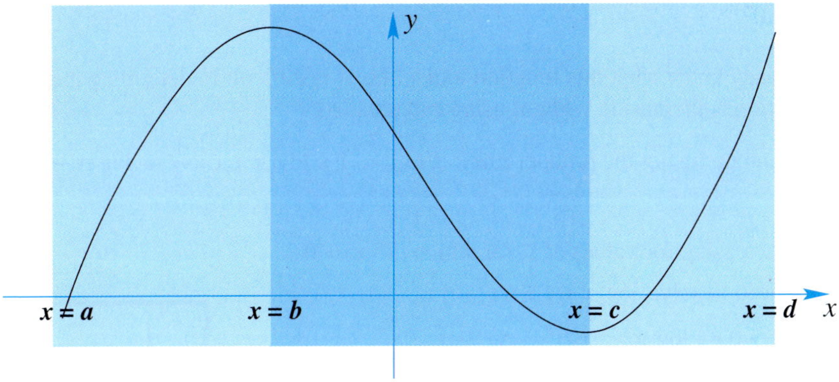

Any tangent line drawn in the interval $a < x < b$ will always have a positive gradient. This means that $\frac{dy}{dx} = f'(x) > 0$ in this region.

Any tangent line drawn in the interval $b < x < c$ will always have a negative gradient. This means that $\frac{dy}{dx} = f'(x) < 0$ in this region.

Any tangent line drawn in the interval $c < x < d$ will always have a positive gradient. This means that $\frac{dy}{dx} = f'(x) > 0$ in this region.

This means that to determine where a function is increasing or decreasing, the values of x for which $f'(x) > 0$ and $f'(x) < 0$ respectively need to be found.

EXAMPLE 20.8

Find the values of x for which the function $f(x) = 1 + 4x - x^2$ is increasing.

Solution

By definition, a function is increasing for those values of x for which $f'(x) > 0$.

Therefore: **1.** find $f'(x)$

2. find the values of x such that $f'(x) > 0$

Now, $f(x) = 1 + 4x - x^2 \Rightarrow f'(x) = 4 - 2x$

Then, $f'(x) > 0 \Leftrightarrow 4 - 2x > 0$

$\Leftrightarrow 4 > 2x$

$\Leftrightarrow x < 2$

We could also have determined this by sketching the graph of $f(x) = 1 + 4x - x^2$. The turning point can be determined by completing the square. i.e., $f(x) = -(x-2)^2 + 5$ giving the axis of symmetry as $x = 2$.

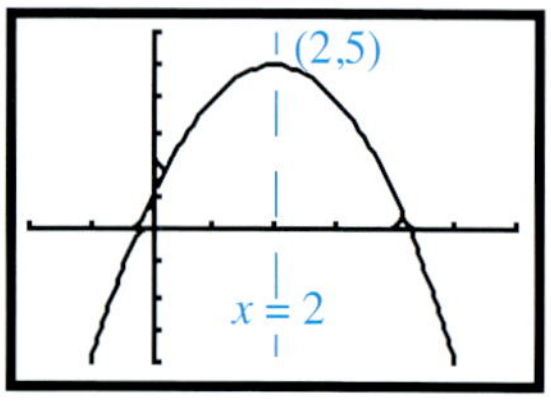

EXAMPLE 20.9

Find the values of x for which the function $f(x) = x\log_e x, x > 0$ is increasing.

Solution

Unless you already know what this function looks like, it is difficult to determine the interval for which the function is increasing without using calculus.

First we differentiate (using the product rule): $f'(x) = 1 \times \log_e x + x \times \frac{1}{x} = \log_e x + 1$

Now, $f(x)$ is increasing for values of x for which $f'(x) > 0$. Therefore we need to solve $\log_e x + 1 > 0$.

Now, $\log_e x + 1 > 0 \Leftrightarrow \log_e x > -1$

$\Leftrightarrow x > e^{-1}$

(The inequality can be determined by making use of a sketch of $f(x) = x\log_e x, x > 0$.)

$f(x) = x\log_e x, x > 0$:

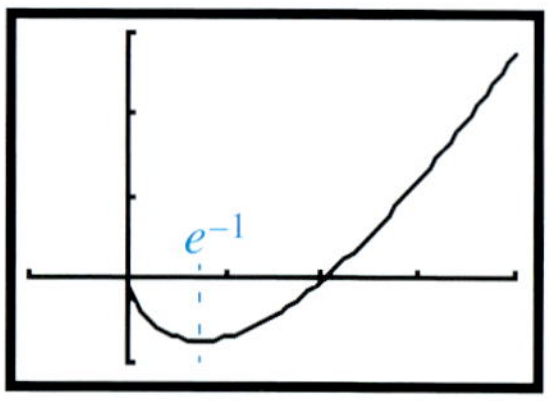

That is, $f(x) = x\log_e x, x > 0$ increases for values of x such that $x > e^{-1}$.

$f'(x) = \log_e x + 1$:

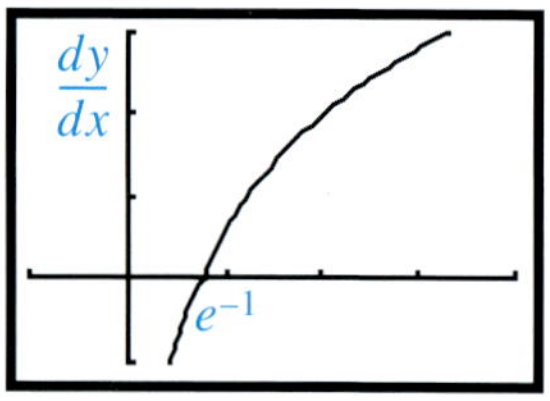

Note that we could have used the graph of the derivative function, $y = f'(x)$, and from it, determined those values of x for which the graph is above the x–axis.

20.2.2 STATIONARY POINTS

So far we have discussed the conditions for a function to be increasing $(f'(x) > 0)$ and for a function to be decreasing $(f'(x) < 0)$. What happens at the point where a function changes from an increasing state $((f'(x) > 0))$ to $(f'(x) = 0)$ and then to a decreasing state $((f'(x) < 0))$ or vice–versa?

Points where this happens are known as **stationary points.** At the point where the function is in a state where it is neither increasing nor decreasing, we have that $f'(x) = 0$. There are times when we can call these stationary points stationary points, but on such occassions, we prefer the terms **local maximum** and **local minimum** points.

At the point(s) where $\frac{dy}{dx} = f'(x) = 0$

we have a **stationary point**.

There are three types of stationary points, namely; **local maximum** point, **local minimum** point and **stationary point of inflection**.

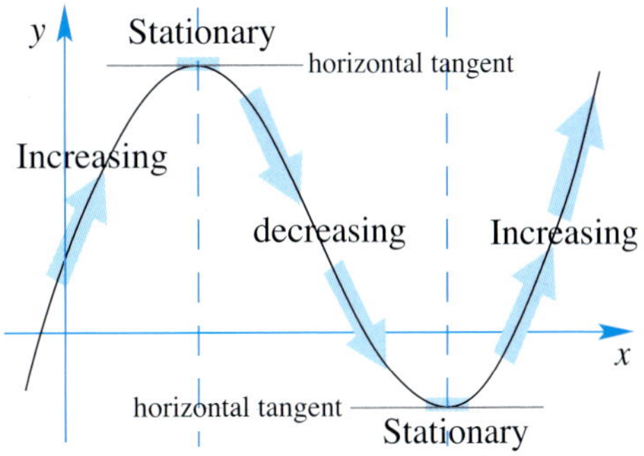

1. Local Maximum

When sketching a curve, if the following properties hold:

i. At $P(x_1, y_1)$, $\frac{dy}{dx} = f'(x) = 0$ that is $f'(x_1) = 0$.

ii. For $x < x_1$ then $\frac{dy}{dx} > 0$

$x > x_1$ then $\frac{dy}{dx} < 0$

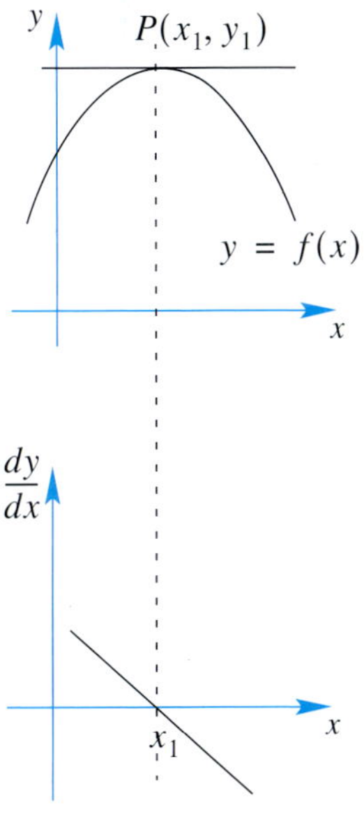

where the two chosen values of x are such that one is just slightly less than x_1 and the other is just slightly greater than x_1. Then,

$y = f(x)$ has a local maximum point (also known as a **relative maximum**) at the point $P(x_1, y_1)$.

iv. Graph of the gradient function:

Notice that the values of $\frac{dy}{dx}$ are changing from **positive** to **negative**. Sometimes this is referred to as the sign of the first derivative.

At this stage, it isn't so much the magnitude of the derivative that is important, but that there is a change in the sign of the derivative near $x = x_1$.

In this instance the sign of the derivative changes from positive to negative.

This change in sign is sometimes represented via the diagram alongside, which is referred to as a sign diagram of the first derivative. Such diagrams are used to confirm the nature of stationary points (in this case, that a local maximum occurs at $x = x_1$).

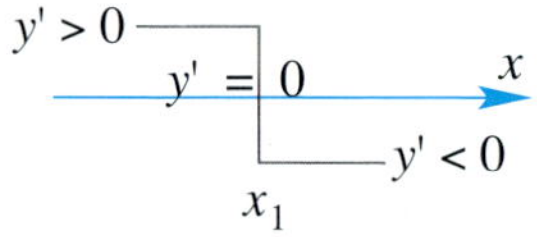

EXAMPLE 20.10 Find the local maximum value of the function whose equation is $f(x) = -3 + 4x - x^2$.

Solution

First we differentiate: $f(x) = -3 + 4x - x^2 \Rightarrow f'(x) = 4 - 2x$

Next, equate $f'(x)$ to 0 and solve for x: $0 = 4 - 2x$

$\Leftrightarrow x = 2$

To ensure that we have obtained a local maximum we choose values of x slightly less than 2 and slightly greater than 2, for example, choose $x = 1.9$ and $x = 2.1$.

For $x = 1.9$, we have that $f'(1.9) = 4 - 2(1.9) = 0.2$.

For $x = 2.1$ we have that $f'(2.1) = 4 - 2(2.1) = -0.2$.

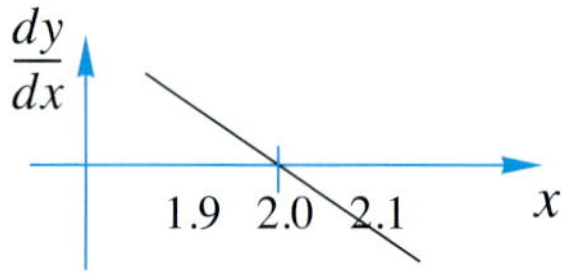

Using the graph of the gradient function, $\frac{dy}{dx}$, confirms that there is a local maximum at $x = 2.0$.

The local maximum value of $f(x)$, is found by substituting $x = 2$ into the given equation:

$f(2) = -3 + 4(2) - (2)^2 = 1$. That is, the local maximum occurs at the point (2,1).

This process can also be carried out using the TI–83:

1. Enter equation

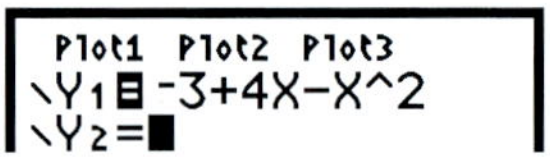

2. Use the **CALC** function, and then select **4:maximum**

3. Press **ENTER**. This will prompt you to choose a lower (left) bound then an upper (right) bound and finally to make a guess

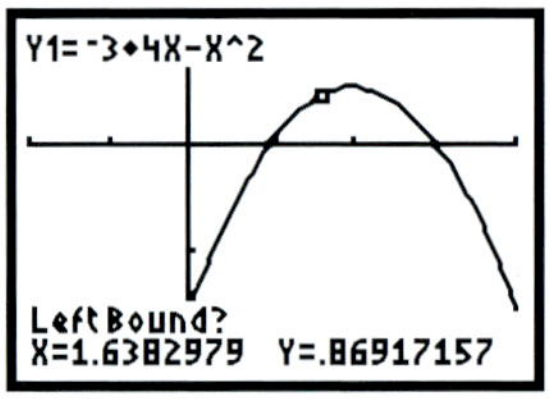

3.1

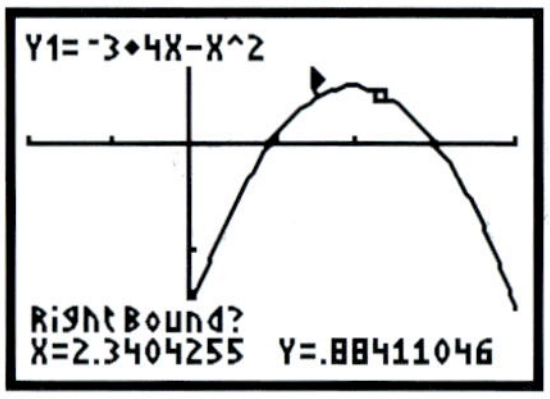

3.2

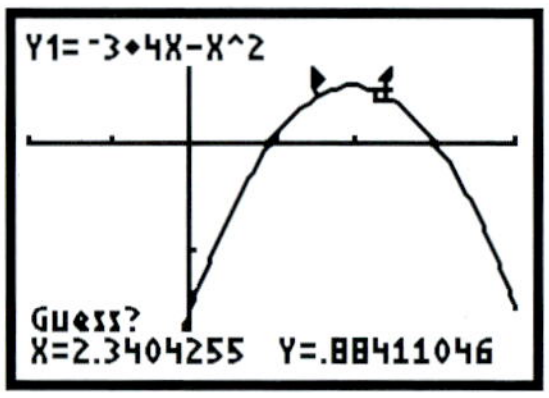

4. Press **ENTER** once more. This will provide both x– and y– values of the stationary point.

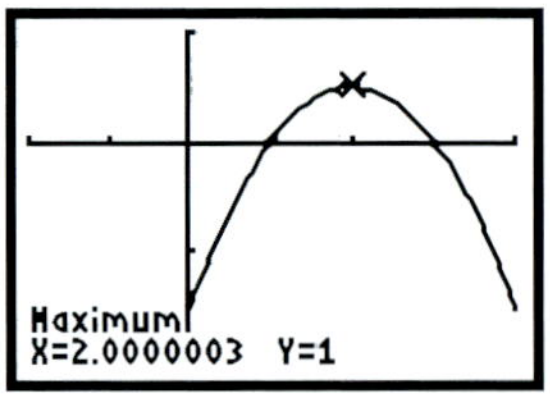

Move the cursor to a point between the two arrows and press **ENTER**.

The other option is to use the **fMax** command from the **MATH** screen. However, this will require some idea of the left and right bounds (which can be estimated from the graph of the function). We illustrate this in the next example.

EXAMPLE 20.11

Determine the coordinates of the local maximum for the function $f(x) = x^2 \sin(x), -1 \le x \le 4$

Solution

This time we start by sketching a graph of the function, setting the WINDOW parameters with the appropriate values (i.e., $-1 \le x \le 4$):

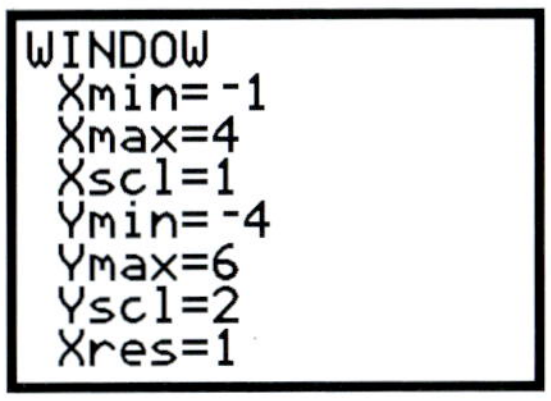

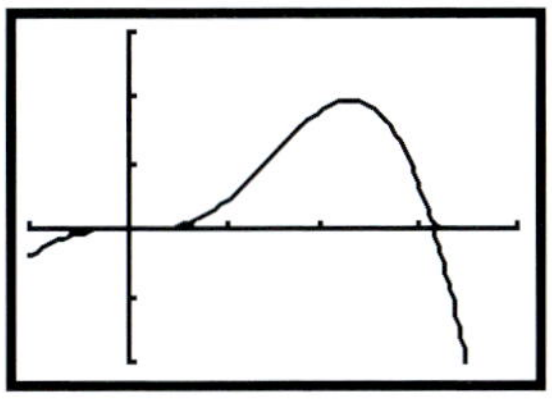

We can now use the **fMax** command from the **MATH** screen to determine the x–ordinate of the stationary point and then use the **CALC** function to find the y–ordinate:

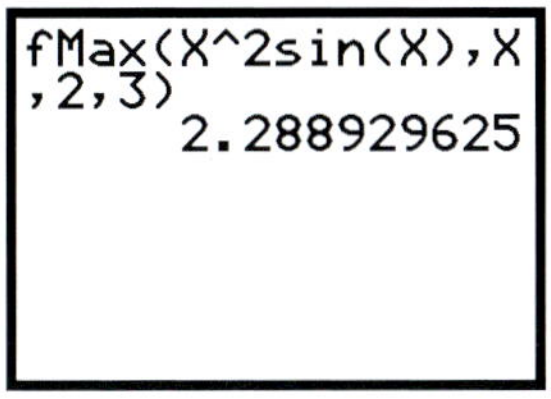

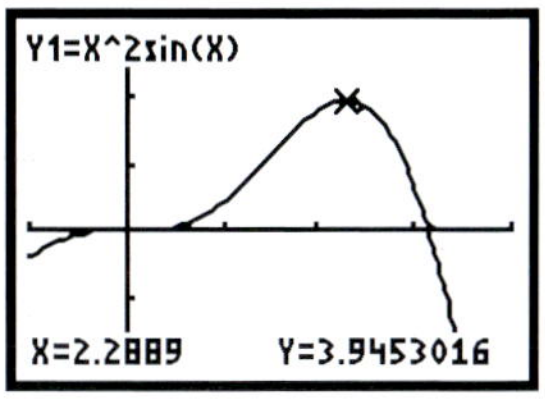

Therefore, the coordinate (to 2 d.p) is given by (2.29, 3.95).
Had we used an algebraic approach to this question, we would have had to solve for

$f'(x) = 0$ i.e., $2x\sin(x) + x^2\cos(x) = 0 \Leftrightarrow x[2\sin(x) + x\cos(x)] = 0$

$\Leftrightarrow x = 0$ or $2\sin(x) + x\cos(x) = 0$

For which, other than $x = 0$, no exact solution exist.

2. Local Minimum

When sketching a curve, if the following properties hold:

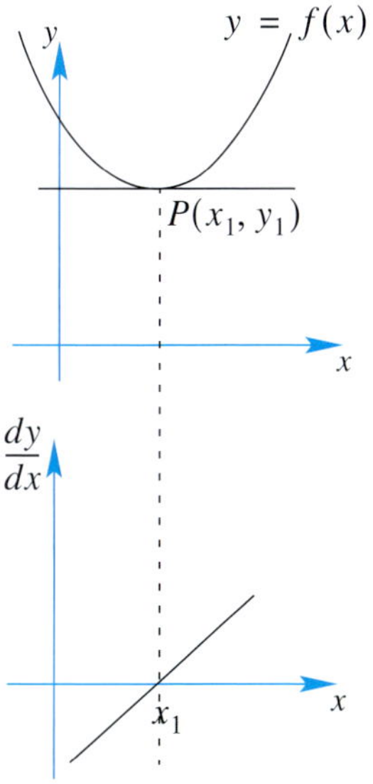

i. At $P(x_1, y_1)$, $\frac{dy}{dx} = f'(x) = 0$ that is $f'(x_1) = 0$.

ii. For $x > x_1$ then $\frac{dy}{dx} > 0$

$x < x_1$ then $\frac{dy}{dx} < 0$

where the two chosen values of x are such that one is just slightly greater than x_1 and the other is just slightly less than x_1. Then,

> $y = f(x)$ has a local minimum point (also known as a **relative minimum**) at the point $P(x_1, y_1)$.

iv. Graph of the gradient function:

Notice that the values of $\frac{dy}{dx}$ are changing from **negative** to **positive**. Sometimes this is referred to as the sign of the first derivative.

Again we can represent the change in the sign of the first derivative via the diagram alongside, which is referred to as a sign diagram of the first derivative. Such diagrams are used to confirm the nature of stationary points (in this case, that a local minimum occurs at $x = x_1$).

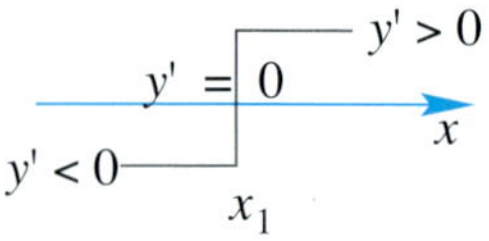

EXAMPLE 20.12

Find the minimum value of $x \mapsto \frac{e^x}{x}, x > 0$.

Solution

First differentiate (using the quotient rule):

$$\frac{d}{dx}\left(\frac{e^x}{x}\right) = \frac{\frac{d}{dx}(e^x) \times x - e^x \times \frac{d}{dx}(x)}{x^2} = \frac{e^x x - e^x}{x^2} = \frac{e^x(x-1)}{x^2}$$

We solve for $\frac{d}{dx}\left(\frac{e^x}{x}\right) = 0$, i.e., $\frac{e^x(x-1)}{x^2} = 0 \Leftrightarrow e^x(x-1) = 0$

However, $e^x \neq 0$ for all real values of x, therefore, the only possible solution occurs if $x = 1$. To verify that we have a local minimum we select a value of x slightly less than $x = 1$ and one slightly greater than $x = 1$:

For ($x < 1$): choose $x = 0.9$, we have that $\frac{d}{dx}\left(\frac{e^x}{x}\right) = -0.30$.

For ($x > 1$): choose $x = 1.1$, we have that $\frac{d}{dx}\left(\frac{e^x}{x}\right) = 0.25$.

Sign diagram of first derivative:

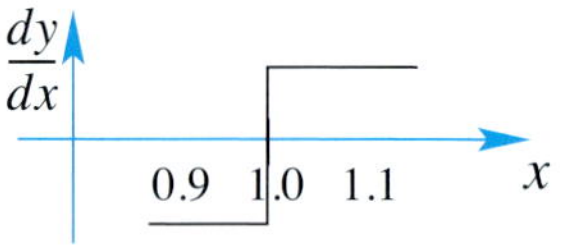

Graph of function:

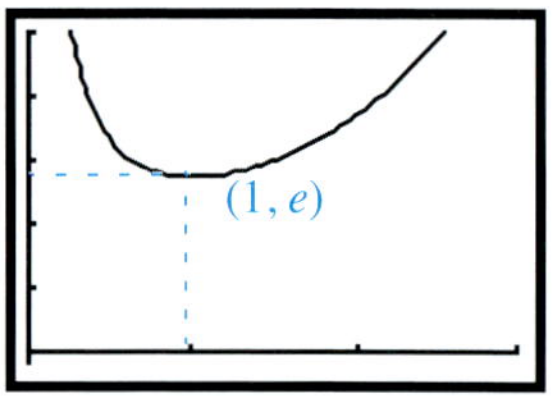

Therefore, for $x = 1$ we have a local minimum point.

The minimum value is therefore given by $y = \frac{e^1}{1} = e$, and occurs at the point $(1,e)$.

EXAMPLE 20.13

Find the local minimum of the function

$$y = \sin(x) + \frac{1}{2}\sin(2x), 0 \leq x \leq 2\pi .$$

Solution

We start by differentiating and finding the stationary points (i.e., solving for $\frac{dy}{dx} = 0$):

Now, $y = \sin(x) + \frac{1}{2}\sin(2x) \Rightarrow \frac{dy}{dx} = \cos(x) + \cos(2x)$.

Therefore, solving we have $\cos(x) + \cos(2x) = 0$

$$\cos(x) + (2\cos^2(x) - 1) = 0$$
$$(2\cos(x) - 1)(\cos(x) + 1) = 0$$

Therefore, $\cos(x) = \frac{1}{2}$ or $\cos(x) = -1$. i.e., $x = \frac{\pi}{3}, \frac{5\pi}{3}$ or $x = \pi$ for $x \in [0, 2\pi]$.

We can check the nature of the stationary points by making use of the sign of the first derivative:

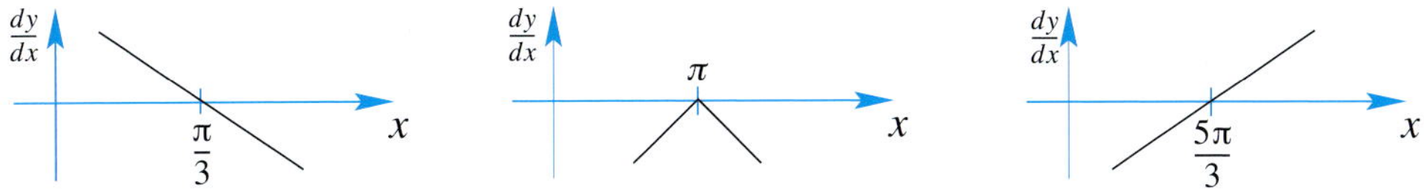

The graph of gradient function (near $x = \frac{5\pi}{3}$) indicates that a local minimum occurs at $x = \frac{5\pi}{3}$.

And so, the local minimum value is given by $y = \sin\left(\frac{5\pi}{3}\right) + \frac{1}{2}\sin\left(\frac{10\pi}{3}\right) = -\frac{3\sqrt{3}}{4}$.

NB: In the process we have come accross a new sign diagram (at $x = \pi$). This is dealt with in the next section.

3. Points of Inflection

There are two types: **A.** **Stationary points of inflection**
B. **Non–stationary points of inflection.**

A. Stationary point of inflection

The following properties hold at a stationary point of inflection.

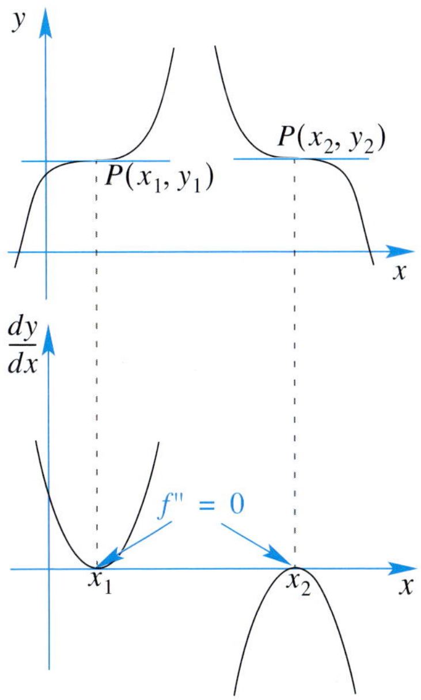

i. At $P(x_1, y_1)$, $f'(x) = 0$. That is $f'(x_1) = 0$.

ii. For $x < x_1$, $f'(x) > 0$ and for $x > x_1$, $f'(x) > 0$.

Similarly,

At $P(x_2, y_2)$, $f'(x) = 0$. That is $f'(x_2) = 0$

and for $x < x_2$, $f'(x) < 0$ and for $x > x_2$, $f'(x) < 0$.

That is, the gradient of the curve on either side of x_1 (or x_2) has the same sign.

iii. Graph of the **gradient function**, $y = f'(x)$:
Notice that the values of $f'(x)$ have the same sign on either side of $x = x_1$.

Notice that at $x = x_1$, the gradient of $f'(x)$ is also equal to zero. That is, the derivative of the derivative is equal to zero.

Therefore **if there is a** stationary point of inflection at $x = x_1$ then $f''(x_1) = 0$.

EXAMPLE 20.14

Find the stationary point of inflection for the graph with equation $y = (x-1)^3(x+2)$.

Solution

First we differentiate (using the product rule):

$$\text{Given } y = (x-1)^3(x+2) \Rightarrow \frac{dy}{dx} = 3(x-1)^2(x+2) + (x-1)^3(1)$$
$$= (x-1)^2[3(x+2) + (x-1)]$$
$$= (x-1)^2(4x+5)$$

Solving for $\frac{dy}{dx} = 0$, we have, $(x-1)^2(4x+5) = 0 \Leftrightarrow x = 1$ or $x = -\frac{5}{4}(= -1.25)$.

We can now check the sign of the derivative on either side of $x = 1$ and $x = -1.25$

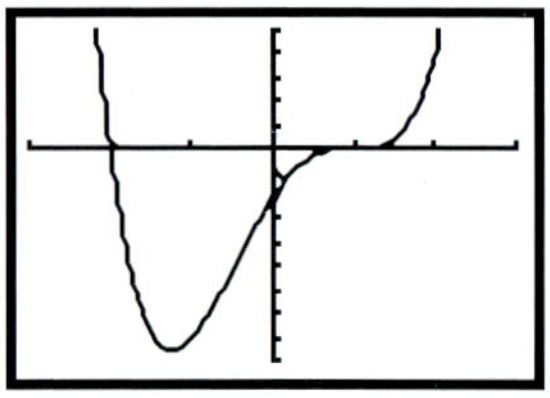

At $x = 1$: For $x = 0.9$, $\frac{dy}{dx} = (-0.1)^2(8.6) = 0.086 > 0$

For $x = 1.1$, $\frac{dy}{dx} = (0.1)^2(9.4) = 0.094 > 0$.

As the sign of the first derivative is the same on either side of $x = 1$, we have a stationary point of inflection at $x = 1$. i.e., at (1,0).

A sketch of the graph of $y = (x-1)^3(x+2)$ quickly confirms our result.
For $x = -1.25$, the graph shows a local minimum occurring at this point.

EXAMPLE 20.15

Locate the stationary points of inflection for the curve $f(x) = x^3e^{-x}$.

Solution

We begin by determining where stationary points occur:

$$f(x) = x^3e^{-x} \Rightarrow f'(x) = 3x^2e^{-x} - x^3e^{-x}$$

Setting $f'(x) = 0$, we have $3x^2e^{-x} - x^3e^{-x} = 0 \Leftrightarrow x^2e^{-x}(3-x) = 0$

$$\Leftrightarrow x = 0 \text{ or } x = 3$$

We can use the sign of the first derivative to help us determine the nature of the stationary point.

At $x = 3$:

For $x = 2.9$, $f'(2.9) = (2.9)^2e^{-2.9}(3-2.9) = 0.046 > 0$

For $x = 3.1$, $f'(3.1) = (3.1)^2e^{-3.1}(3-3.1) = -0.043 < 0$

Sign diagram:

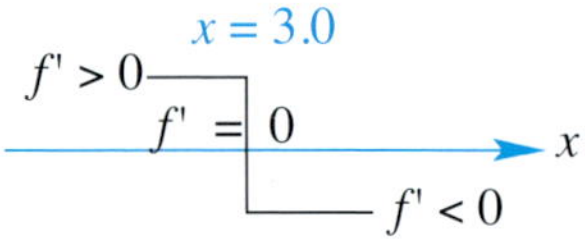

Therefore there exists a local maximum at $x = 3.1$.

At $x = 0$:

For $x = 0.1$, $f'(0.1) = (0.1)^2e^{-0.1}(3-0.1) = 0.026 > 0$

For $x = -0.1$, $f'(-0.1) = (-0.1)^2e^{0.1}(3+0.1) = 0.034 > 0$

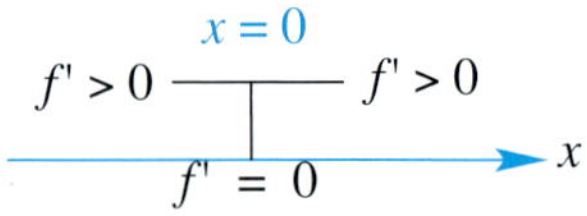

As there is no change in the sign of the first derivative there is a stationary point of inflection at $x = 0$.

Alternatively, we could sketch a graph of the function and use it to help us determine where the stationary point of inflection occurs:

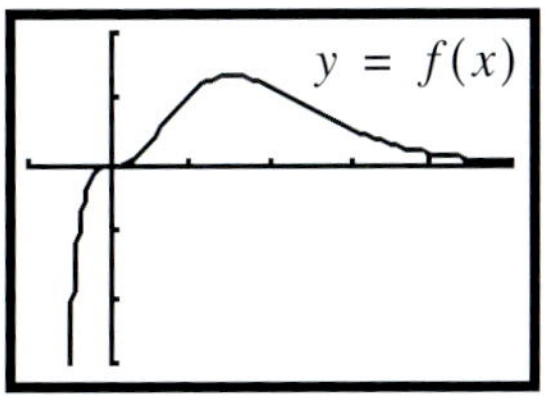

From the graph we can see that there is a local maximum at $x = 3$ and a stationary point of inflection at $x = 0$.

Therefore, the stationary point of inflection occurs at $(0, 0)$.

Notice that the sign diagrams of the first derivative in Examples 20.13, 20.14 and 20.16 all look slightly different. We have done this to emphasise that as long as the diagram provides a clear indication of the sign of the first derivative then its appearance can vary.

B. Non–stationary point of inflection

The following properties hold at a non-stationary point of inflection:

i. At $P(x_1, y_1)$, $f'(x_1) \neq 0$ and $f''(x_1) = 0$.

ii. For $x < x_1$, $f'(x) > 0$ and for $x > x_1$, $f'(x) > 0$

Similarly,

At $P(x_2, y_2)$, $f'(x_2) \neq 0$ and $f''(x_2) = 0$.

For $x < x_2$, $f'(x) < 0$ and for $x > x_2$, $f'(x) < 0$.

That is, the gradient of the curve on either side of x_1 (or x_2) has the same sign.

iii. Graph of the **gradient function**, $y = f'(x)$:
Notice that the values of $f'(x)$ have the same sign on either side of $x = x_1$.

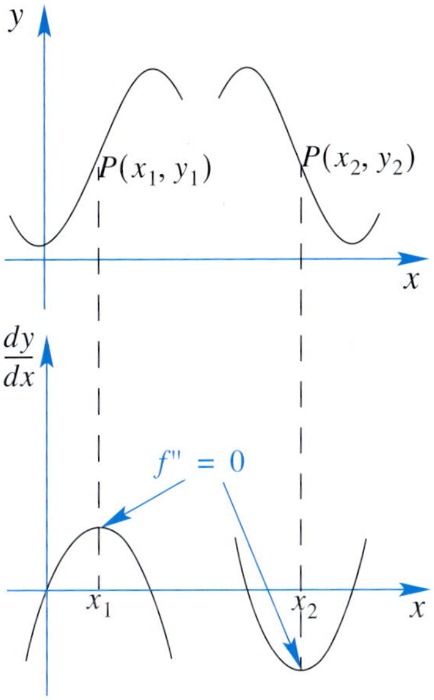

EXAMPLE 20.16

Show that the curve with equation $y = x^4 - 4x^3$ has a non-stationary point of inflection at $x = 2$.

Solution

For the curve to have a non–stationary point of inflection at $x = 2$ we need to show that:

1. $\frac{d^2y}{dx^2} = 0$ at $x = 2$ **and**

2. the sign of the gradient, $\frac{dy}{dx}$, is the same on both sides of $x = 2$ **and**

3. that $\frac{dy}{dx} \neq 0$ when $x = 2$.

A quick sketch of the function indicates that a point of inflection occurs at $x = 2$:

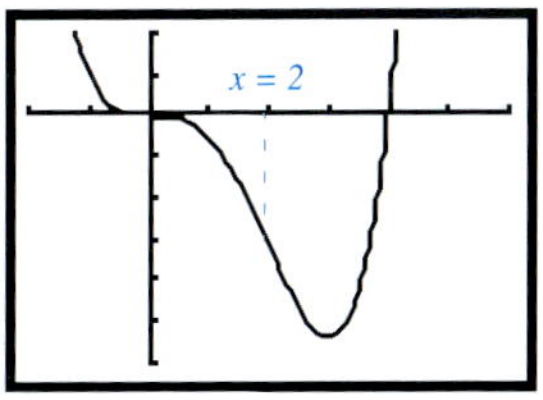

Now, $\frac{dy}{dx} = 4x^3 - 12x^2 \Rightarrow \frac{d^2y}{dx^2} = 12x^2 - 24x$.

For $x = 2$, $\frac{d^2y}{dx^2} = 12(2)^2 - 24(2) = 0$ and $\frac{dy}{dx} = -16 \neq 0$.

For $x = 2.1$, $\frac{dy}{dx} = -15.876$ and for $x = 1.9$, $\frac{dy}{dx} = -15.88$.

Sign diagram:

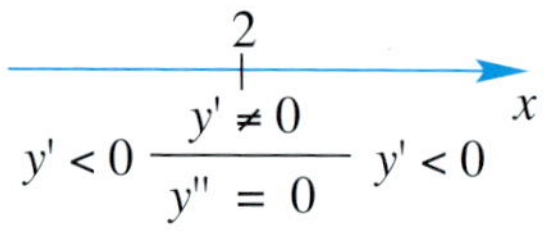

Therefore there is a non–stationary point of inflection at $x = 2$.

It is important to realise that it is **not sufficient** to say that "If $f''(x) = 0$ at $x = a$ then there must be a point of inflection at $x = a$."

Does $f''(a) = 0$ imply there is a point of inflection at $x = a$?

The answer is **NO**!

Although it is necessary for the second derivative to be zero at a point of inflection, the fact that the second derivative is zero at $x = a$ does **not** mean there must be a point of inflection at $x = a$.

That is:

> $f''(a) = 0$ is a necessary but not a sufficient reason for there to be an inflection point at $x = a$.

We use the following example to illustrate this.

Consider the case where $f(x) = x^4$.
Now, $f''(x) = 12x^2$, therefore solving for $f''(x) = 0$,
we have $12x^2 = 0 \Leftrightarrow x = 0$.

That is, $f''(0) = 0$. So, do we have a point of inflection at $x = 0$?

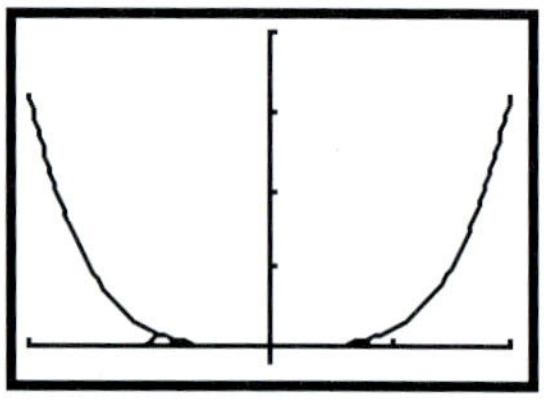

A sketch of f shows that although $f''(x) = 0$ at $x = 0$, there is in fact a local minimum and not a point of inflection at $x = 0$.

In other words, **finding where $f''(x) = 0$ is not enough** to indicate that there is an inflection point. To determine if there is a point of inflection you need to check the the sign of the first derivative on either side of the x-value in question. Or, check the **concavity** at the x-value in question. We will discuss the term concavity in §20.3.

EXAMPLE 20.17

Find and classify all stationary points (and inflection points) of $f(x) = x^3 - 3x^2 - 9x + 1$.

Solution

Now, $f(x) = x^3 - 3x^2 - 9x + 1 \Rightarrow f'(x) = 3x^2 - 6x - 9$

Solving for stationary points we have, $3x^2 - 6x - 9 = 0 \Leftrightarrow 3(x-3)(x+1) = 0$

$\Leftrightarrow x = 3 \text{ or } x = -1$

So that $f(3) = -26$ and $f(-1) = 6$.

Using the sign of the first derivative, we have:

At $x = 3$:

For $x < 3$ $(x = 2.9)$ $f'(2.9) < 0$ and

for $x > 3$ $(x = 3.1)$ $f'(3.1) > 0$.

Therefore, there is a local minimum at (3,–26).

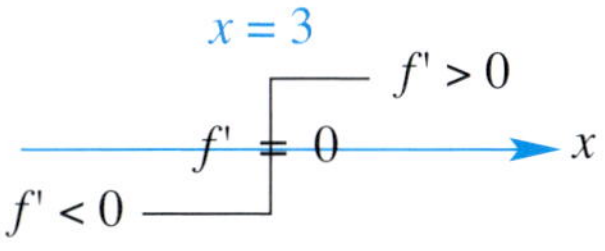

At $x = -1$:

For $x < -1$ $(x = -1.1)$ $f'(-1.1) > 0$ and

for $x > -1$ $(x = -0.9)$ $f'(-0.9) < 0$.

Therefore, there is a local maximum at (–1,6).

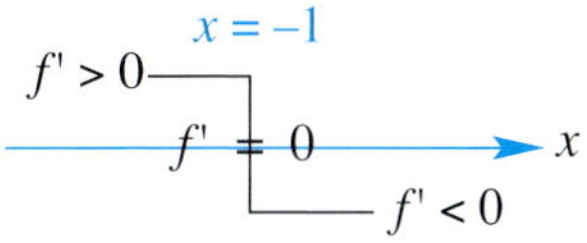

Checking for inflection points:

We have: $f''(x) = 0 \Leftrightarrow 6x - 6 = 0 \Leftrightarrow x = 1$.

For $x < 1$ $(x = 0.9)$ $f'(0.9) < 0$ and

for $x > 1$ $(x = 1.1)$ $f'(1.1) < 0$.

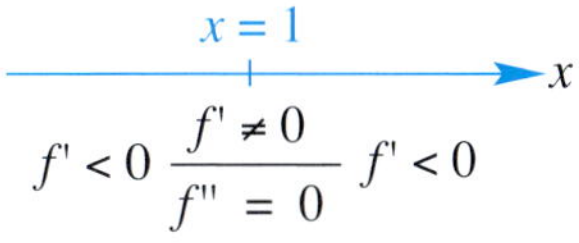

As the sign of the first derivative remains the same on either side of $x = 1$, there is a point of inflection at $(1, -10)$. Then, as the **first derivative** at $x = 1$ **is not zero**, we have a **non-stationary** point of inflection at $x = 1$.

EXAMPLE 20.18

Sketch the graph of the function $f(x) = x^4 - 2x^2$, clearly marking any stationary points and points of inflection.

Solution

We first find the stationary points (if any).

This means that we must solve for $f'(x) = 0$:

Now, $f'(x) = 4x^3 - 4x = 4x(x^2 - 1) \therefore 4x(x^2 - 1) = 0 \Leftrightarrow 4x(x+1)(x-1) = 0$

$\Leftrightarrow x = -1 \text{ or } x = 1 \text{ or } x = 0$

We now check for the nature of each point

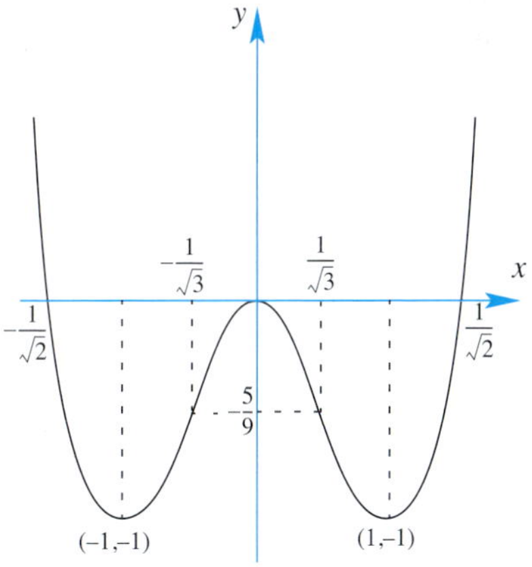

At $x = 0$: $x = -0.1$, $f'(-0.1) = 0.396 > 0$

$x = 0.1$, $f'(0.1) = -0.396 < 0$

Therefore, we have a **local maximum** point at (0,0).

At $x = 1$:

$x = 0.9$, $f'(0.9) = -0.684 < 0$

$x = 1.1$, $f'(1.1) = 0.924 > 0$

Therefore, we have a **local minimum** point at (1,–1).

At $x = -1$:

$x = -0.9$, $f'(-0.9) = 0.684 > 0$

$x = -1.1$, $f'(-1.1) = -0.924 < 0$

Hence, we have a **local minimum** point at (–1, –1).

We now look for possible points of inflection.

We need to solve for $f''(x) = 0$:

Now, $f''(x) = 0 \Rightarrow 12x^2 - 4 = 0$

$\Leftrightarrow 4(3x^2 - 1) = 0$

$\Leftrightarrow x = \frac{1}{\sqrt{3}}(\approx 0.58)$ or $x = -\frac{1}{\sqrt{3}}(\approx -0.58)$

At $x = \frac{1}{\sqrt{3}}, f\left(\frac{1}{\sqrt{3}}\right) = \left(\frac{1}{\sqrt{3}}\right)^4 - 2\left(\frac{1}{\sqrt{3}}\right)^2 = -\frac{5}{9}$ similarly $f\left(-\frac{1}{\sqrt{3}}\right) = -\frac{5}{9}$.

Next, we **check** to see if these are indeed points of inflection.

At $x = \frac{1}{\sqrt{3}}$ we have:

$x = 0.5,\ f'(0.5) \approx -1.5 < 0$

$x = 0.6,\ f'(0.6) \approx -1.54 < 0$

At $x = -\frac{1}{\sqrt{3}}$ we have:

$x = -0.5,\ f'(0.5) \approx 1.5 > 0$

$x = -0.6,\ f'(0.6) \approx 1.54 > 0$

From the signs of the first derivative, on either side of $x = \frac{1}{\sqrt{3}}$ and $x = -\frac{1}{\sqrt{3}}$ we have non–stationary points of inflection at $\left(\frac{1}{\sqrt{3}}, -\frac{5}{9}\right)$ and $\left(-\frac{1}{\sqrt{3}}, -\frac{5}{9}\right)$.

EXAMPLE 20.19

For the function $f:\left[0, \frac{5\pi}{2}\right] \mapsto \mathbb{R}$, where $f(x) = e^{-x}\cos x$, find

a. the x–intercepts,

b. the co-ordinates of the first two stationary points.

Use the above information to sketch the graph of f.

Solution

a. ***x*–intercepts:**

We need to solve for $e^{-x}\cos x = 0 \Leftrightarrow e^{-x} = 0$ or $\cos x = 0$.

As e^{-x} is always positive, we need only concern ourselves with $\cos x = 0$.

Given that $x \in \left[0, \frac{5\pi}{2}\right]$, we have $\cos x = 0 \Leftrightarrow x = \frac{\pi}{2}, \frac{3\pi}{2}, \frac{5\pi}{2}$.

That is, x–intercepts occur at $x = \frac{\pi}{2}, \frac{3\pi}{2}, \frac{5\pi}{2}$

b. **Stationary points**:

This time we need to solve for $f'(x) = 0$:

$f'(x) = \frac{d}{dx}(e^{-x}) \times \cos x + e^{-x} \times \frac{d}{dx}(\cos x)$ [product rule]

$= -e^{-x}\cos x - e^{-x}\sin x$

$= -e^{-x}(\cos x + \sin x)$

So, $f'(x) = 0 \Leftrightarrow -e^{-x}(\cos x + \sin x) = 0 \Leftrightarrow e^{-x} = 0$ or $\cos x + \sin x = 0$

Now, as e^{-x} is always positive we have $f'(x) = 0 \Leftrightarrow \cos x + \sin x = 0$.

Solving for $\cos x + \sin x = 0$ gives $\sin x = -\cos x \Leftrightarrow \dfrac{\sin x}{\cos x} = -1 \Leftrightarrow \tan x = -1$.

Therefore, $x = \dfrac{3\pi}{4}$ or $x = \dfrac{7\pi}{4}$.

Nature of stationary points:

For $x = \dfrac{3\pi}{4}(\approx 2.36)$: $\left.\begin{array}{l} x = 2.3 \text{ then } \dfrac{dy}{dx} = -0.008 \\ x = 2.4 \text{ then } \dfrac{dy}{dx} = 0.056 \end{array}\right\} \therefore$ local minimum at $x = \dfrac{3\pi}{4}$

For $x = \dfrac{7\pi}{4}(\approx 5.50)$: $\left.\begin{array}{l} x = 5.4 \text{ then } \dfrac{dy}{dx} = 0.0006 \\ x = 5.6 \text{ then } \dfrac{dy}{dx} = -0.0005 \end{array}\right\} \therefore$ local maximum at $x = \dfrac{7\pi}{4}$

Now, at $x = \dfrac{3\pi}{4}, y = e^{-\frac{3\pi}{4}}\cos\left(\dfrac{3\pi}{4}\right) = -\dfrac{\sqrt{2}}{2}e^{-\frac{3\pi}{4}}$, and at $x = \dfrac{7\pi}{4}, y = e^{-\frac{7\pi}{4}}\cos\left(\dfrac{7\pi}{4}\right) = \dfrac{\sqrt{2}}{2}e^{-\frac{7\pi}{4}}$

Note: The graph shown has been scaled so that the general shape can be seen more readily. In fact, the local maximum should be very close to the x-axis.

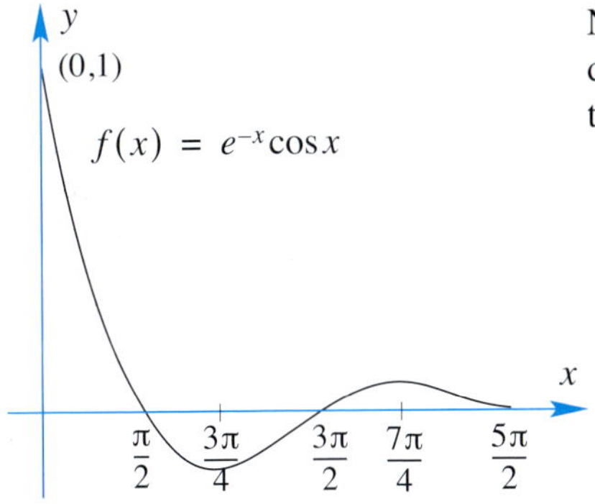

Notice that over the specified domain, the TI–83 produces the following graph:

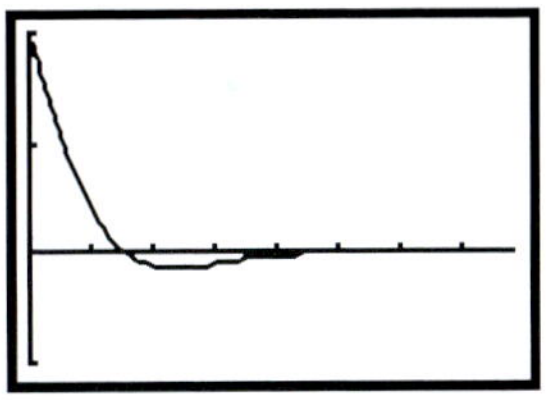

Also note that the behaviour of $f(x)$ takes on the combined behaviour of the decaying function, e^{-x}, and the oscillating function, $\cos(x)$.

20.2.3 GLOBAL MAXIMA AND MINIMA

Until now we have only considered locating the **local maxima** or the **local minima**. The process has been straightforward enough in the sense that there exists a procedure for locating these points, i.e., find the derivative, equating it to zero, solve and then use a sign diagram of the first derivative to identify the nature of the stationary point.

Consider the function $f(x) = 4x^3 + 9x^2 - 12x + 10$. Assuming that we have been able to find the stationary points we can sketch the graph of $f(x)$:

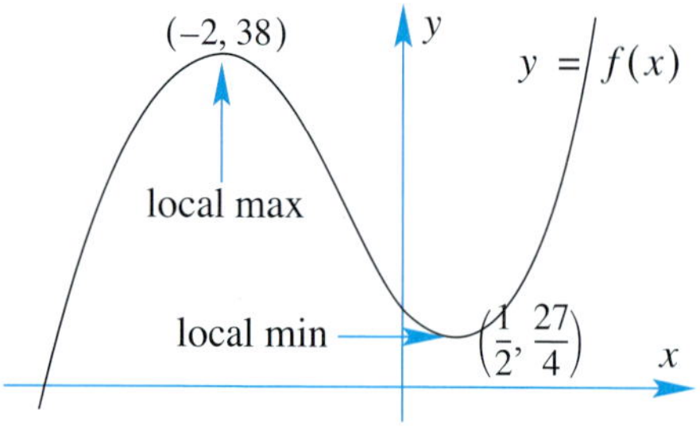

Going through our standard process we find that the curve has a

local maximum at $(-2, 38)$

and a

local minimum at $\left(\frac{1}{2}, \frac{27}{4}\right)$.

In sketching this curve we have (correctly) assumed that $x \in \mathbb{R}$ and as such the graph extends indefinitely and as such, no overall minimum or maximum can be given. However, what if we wished to find the maximum value of this function but this time have a restriction on the domain, e.g., $f(x) = 4x^3 + 9x^2 - 12x + 10, -3 \le x \le 2$.

In this instance, proceeding with our standard approach, i.e., finding the derivative and so forth gives the same results as above. However, we should now sketch the graph over the given domain. A sketch of this graph over the given domain is now produced:

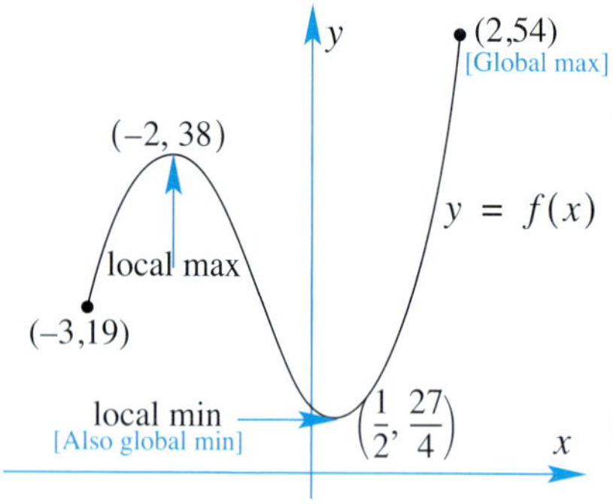

From the sketch we can see that although the point $(-2, 38)$ is still a stationary point, it only identifies the local maximum.

Over the new domain, the maximum value of $f(x)$ is in fact 54. We say that $f(x)$ has an endpoint maximum. This **endpoint** maximum occurs at $(2, 54)$. Then, as there is no other value greater than 54, we also say that the **global maximum** or the **absolute maximum** is 54.

Notice too that the point $\left(\frac{1}{2}, \frac{27}{4}\right)$ is both a local minimum **and** a **global minimum**, because for this domain the minimum value of $f(x)$ is $\frac{27}{4}$, which happens to coincide with the local miminum. Had the domain been $-4 \le x \le 2$, then the **absolute minimum** would have occured at $x = -4$ with a value of $f(-4) = -54$.

So, when using the term '**local**' we are in fact referring to points that are in the **immediate vicinity** (or **neighbourhood**) of the **critical point** (see definition below).

Notice also that the derivative, $f'(x)$, does not exist at $(2, 54)$ [or $(-3, 19)$ for that matter] – however, we still have a maximum at that point.
Why is it that the derivatives does not exist at these end point?

It should also be observed that there can be a local maximum or a local minimum for which the derivative is undefined at that point. We extend our definition of **local extrema** as follows:

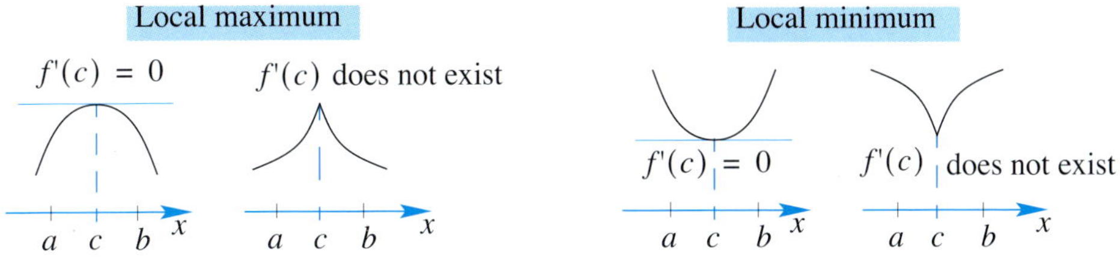

If f has a local maximum or a local miminum when $x = c$, where $a < c < b$, then c is a critical value of f. That is, either

$$f'(c) = 0 \text{ or } f'(c) \text{ does not exist.}$$

EXAMPLE 20.20

Find the critical value(s) of the function $f(x) = 2x - 3x^{2/3}$.

Solution

To determine the critical value(s) we must first find $f'(x)$:

$$f'(x) = 2 - 2x^{-1/3} = 2\left(1 - \frac{1}{x^{1/3}}\right) = 2\left(\frac{x^{1/3} - 1}{x^{1/3}}\right).$$

Next we must find value(s) of c such that $f'(c) = 0$ or $f'(c)$ does not exist.

So, $f'(x) = 0 \Leftrightarrow 2\left(\frac{x^{1/3} - 1}{x^{1/3}}\right) = 0 \Leftrightarrow x^{1/3} - 1 = 0 \Leftrightarrow x = 1$.

However, $f'(x)$ will be undefined if $x = 0$ (as division by 0 is not possible). So, we might also have a critical value at $x = 0$ (one for which the derivative is undefined).
In fact, a sketch of the function shows the critical points as well as the local **extrema**.
When $x = 1$, $f(1) = 2 - 3 = -1$ and when $x = 0$, $f(0) = 0$ and so we have local extrema at $(1, -1)$ and $(0, 0)$.

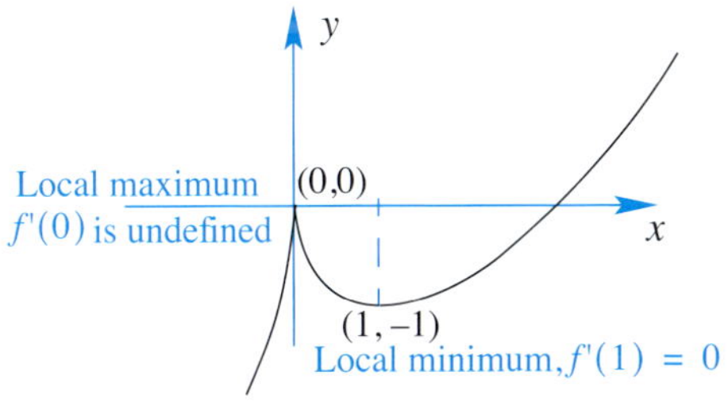

EXAMPLE 20.21

Given that $f(x) = \begin{cases} x^2 - 6x + 8 & \text{where } 1 \le x < 4 \\ x + 2 & \text{where } -5 \le x < 1 \end{cases}$

(a) Find all critical values of f.
(b) Find the absolute maximum of f.

Solution

We start by sketching the graph of f:

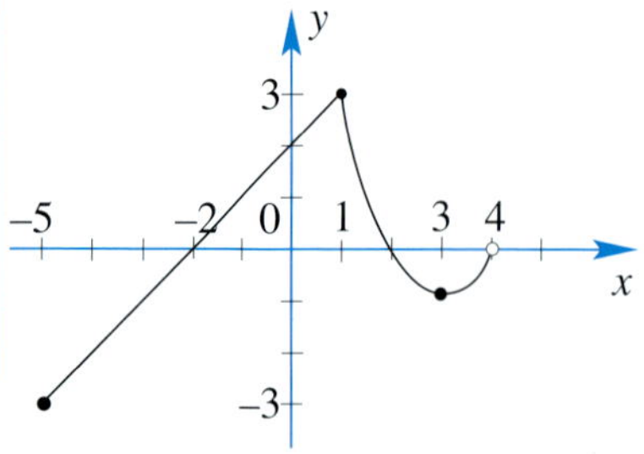

To determine the stationary point we solve

$\frac{d}{dx}(x^2 - 6x + 8) = 0 \Leftrightarrow 2x - 6 = 0 \Leftrightarrow x = 3$.

That is, $x = 3$ is a critical value.

However there is another critical value, where $f'(x)$ does not exist. Namely, $x = 1$.

Note that $x = -5$ provides an end point extremum.

We then have: An **absolute minimum** at $x = -5$.
An **absolute maximum** at the critical value $x = 1$.
A **local minimum** at the critical value $x = 3$.

Note: Even though $\lim_{x \to 4^-} f(x) = 0$, we cannot refer to this point as an end point extremum, as a value of $f(x)$ is not defined at this point.

EXAMPLE 20.22

For the function $h(x) = (x-1)x^{2/3}, -1 \le x \le 2$.

(a) Find its critical value(s).
(b) Find the coordinates of the i. local minimum ii. absolute minimum.
(c) Find the coordinates of the i. local maximum ii. absolute maximum.
(d) Find its point(s) of inflection.
(e) Sketch the graph of h.

Solution

We start by using the graphics calculator to get an idea of the general shape of the curve.

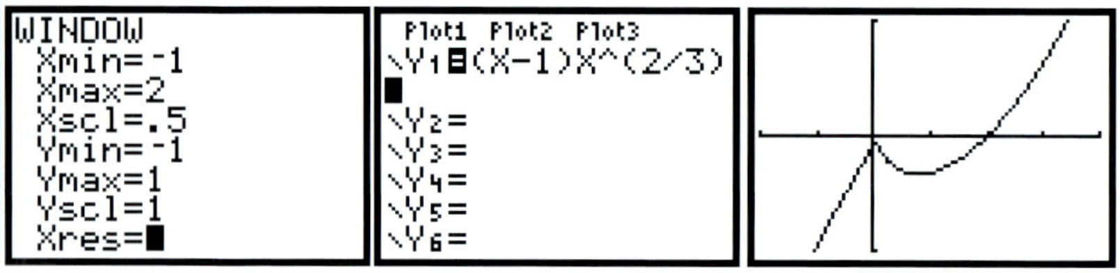

(a) Our first step in finding critical values is to solve for $h'(x) = 0$.

Now, $h(x) = (x-1)x^{2/3} \Rightarrow h'(x) = 1 \times x^{2/3} + (x-1) \times \frac{2}{3}x^{-1/3} = x^{2/3} + \frac{2(x-1)}{3x^{1/3}}$

$$= \frac{3x + 2(x-1)}{3x^{1/3}}$$

$$= \frac{5x-2}{3x^{1/3}}$$

Therefore, $h'(x) = 0 \Leftrightarrow \frac{5x-2}{3x^{1/3}} = 0 \Leftrightarrow x = \frac{2}{5} = 0.4$

And so, there is only one critical value that is a stationary point.

However, we also note that $h'(x)$ is undefined at $x = 0$, and from the graph above we see that this is also a critical value. In fact, we see that there is a **cusp** at $x = 0$.

(b) i. Using the above graph as an aid we have the local minimum occuring at $x = 0.4$, with coordinates $\left(0.4, -\frac{3}{5}\sqrt[3]{\left(\frac{2}{5}\right)^2}\right)$. i.e., $h\left(\frac{2}{5}\right) = \left(\frac{2}{5} - 1\right)\left(\frac{2}{5}\right)^{2/3} = -\frac{3}{5}\sqrt[3]{\left(\frac{2}{5}\right)^2}$.

ii. The absolute minimum occurs at the end point where $x = -1$

(c) i. The local maximum, even though $h'(x)$ is undefined there, occurs where $x = 0$. The coordinates of the local maximum are $(0, 0)$.

ii. The absolute maximum occurs at the endpoint where $x = 2$.
Now, $h(2) = (2-1) \times 2^{2/3} = 2^{2/3}$, therefore the coordinates of the absolute maximum are at $(2, 2^{2/3})$.

(d) For possible points of inflection we first solve $h''(x) = 0$.

$$\text{Now, } h'(x) = \frac{1}{3}(5x-2)x^{-1/3} \therefore h''(x) = \frac{1}{3}\left[5 \times x^{-1/3} + (5x-2) \times -\frac{1}{3}x^{-4/3}\right]$$

$$= \frac{1}{3}\left[\frac{15x - (5x-2)}{3x^{4/3}}\right]$$

$$= \frac{1}{9} \cdot \frac{10x+2}{x^{4/3}}$$

$$\text{Then, } h''(x) = 0 \Leftrightarrow \frac{10x+2}{9x^{4/3}} = 0 \Leftrightarrow x = -\frac{2}{10} = -0.2.$$

Again, making use of the graph from our graphics calculator. We see that as $h'(x) > 0$ for $x < -0.2$, $h'(x) > 0$ for $x > -0.2$ and $h''(x) = 0$ at $x = -0.2$, then there is a (non-stationary) point of inflection at $x = -0.2$.

(e) We are now in a position to sketch the graph of $h(x)$. Of course, we are also using the display from the graphics calculator as an aid – and it should be noted that in this case the graphics calculator has not clearly displayed the inflection behaviour at $x = -0.2$.

Using the available information we have gathered, we can now sketch the graph of h:

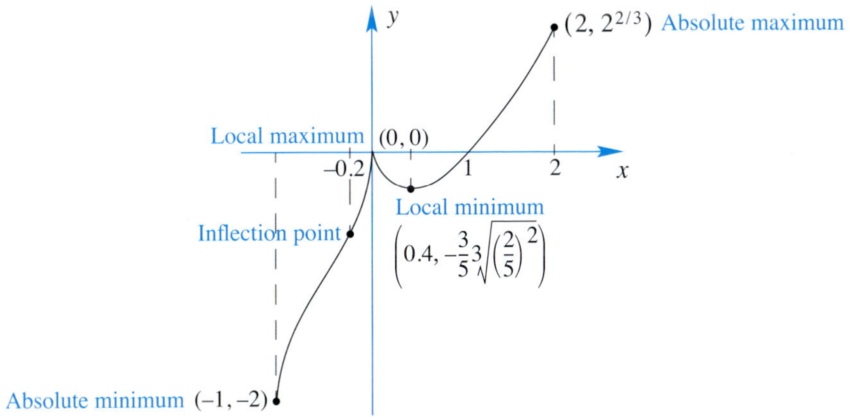

Note: The following section lies outside the scope of the course – we include it as an extension.

20.2.4 VERTICAL TANGENTS AND CUSPS

In the previous section we looked at critical points and how to find them. In this section we take a closer look at the properties of graphs that have critical points for which the derivative does not exist yet still provide a local (or **relative**) extrema. We also look at graphs for which neither a local maximum nor a local minimum exists at points where the first derivative does not exist.

We first display the graphs of such functions and then provide a definition. Note that in all cases we have that $f'(c)$ does not exist. In all four cases a **vertical tangent** exists at $x = c$.

> The graph of the function has a **vertical tangent line** at $x = c$ if f is continuous at $x = c$ and $|f'(x)| \to +\infty$ as $x \to c$.

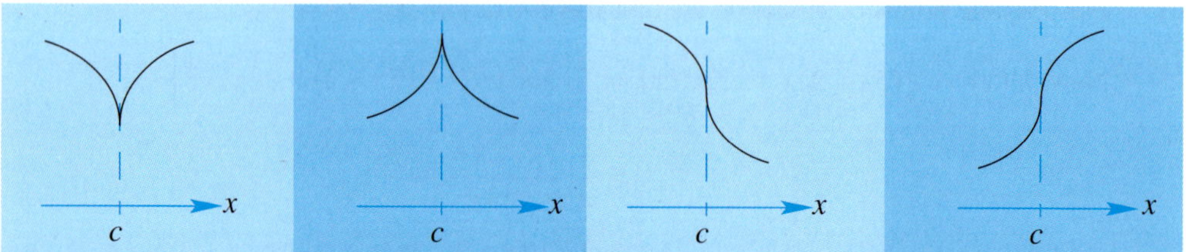

(a)	(b)	(c)	(d)
as:	as:	as:	as:
$x \to c^-, f'(x) \to -\infty$	$x \to c^-, f'(x) \to +\infty$	$x \to c^-, f'(x) \to -\infty$	$x \to c^-, f'(x) \to +\infty$
$x \to c^+, f'(x) \to +\infty$	$x \to c^+, f'(x) \to -\infty$	$x \to c^+, f'(x) \to -\infty$	$x \to c^+, f'(x) \to +\infty$

Figures (a) and (b):

> The graph of the function has a **cusp** at $x = c$ if f is continuous at c and $f'(x) \to +\infty$ as $x \to c$ from one side of $x = c$ and $f'(x) \to -\infty$ as $x \to c$ from the other side of $x = c$.

EXAMPLE 20.23

Sketch the graph of $f(x) = x^{1/3}$.

Solution

We approach this question in two ways.

The first is to realise that $f(x) = x^{1/3}$ is in fact the inverse of the x^3 function. That is, if we consider the function $g(x) = x^3$ we can determine its inverse as follows:

$$g(g^{-1}(x)) = x \Leftrightarrow [g^{-1}(x)]^3 = x \Leftrightarrow g^{-1}(x) = \sqrt[3]{x} \text{ i.e., } g^{-1}(x) = x^{1/3}.$$

Then, as $g(x) = x^3$ has a stationary point of inflection at $x = 0$, the tangent at $x = 0$ must be parallel to the x–axis, i.e., it has a **horizontal tangent** at $x = 0$. This then means that **its inverse**, i.e., $g^{-1}(x) = x^{1/3}$ would have a **vertical tangent** at $x = 0$.

Sketching the function $f(x) = x^{1/3}$ would then require us to reflect the graph of $g(x) = x^3$ about the line $y = x$.

The second method makes use of the definitions we have just looked at. We start by finding the derivative: $f(x) = x^{1/3} \Rightarrow f'(x) = \frac{1}{3}x^{-2/3} = \frac{1}{3x^{2/3}}$ then, at $x = 0$, the derivative is undefined.

However this **is not enough**[1] to imply that we have a vertical tangent. We need to look at **some more features** of the graph before we can **conclude** that we have **a vertical tangent** at $x = 0$. We produce a list of features that will allow us to conclude that we have a vertical tangent at $x = 0$:

1. $f'(x)$ is undefined at $x = 0$.
2. $f(x)$ is continuous at $x = 0$.
3. i. as $x \to 0^-$, $f'(x)\left(= \dfrac{1}{3x^{2/3}}\right) \to +\infty$.

   ii. as $x \to 0^+$, $f'(x)\left(= \dfrac{1}{3x^{2/3}}\right) \to +\infty$.

As all three conditions hold true, we must have a vertical tangent at $x = 0$.

We can now sketch the graph of the function $f(x) = x^{1/3}$:

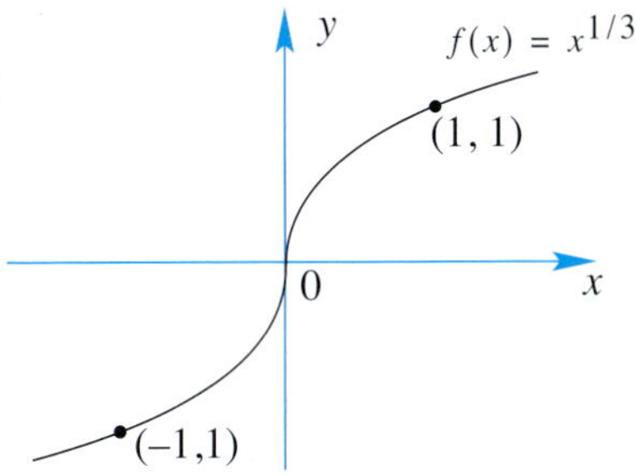

EXAMPLE 20.24

Sketch the graph of $f(x) = (x-2)^{2/3} + 1$.

Solution

A quick sketch of $f(x) = (x-2)^{2/3} + 1$ using the TI–83 shows us enough to suspect that there is a vertical tangent at $x = 2$, in fact, we suspect that it is a cusp.

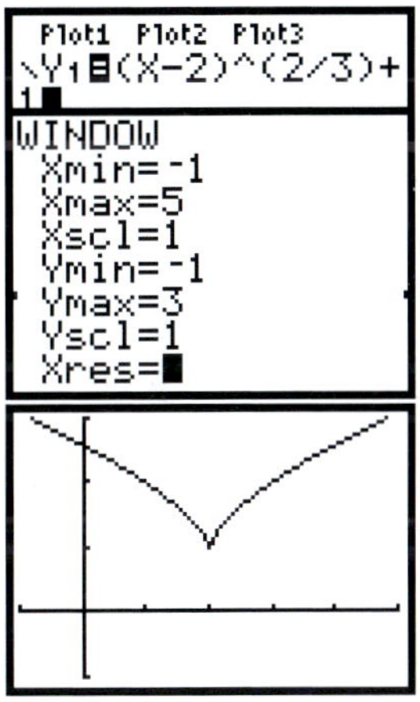

We now confirm this formally:

As $f(x) = (x-2)^{2/3} + 1 \Rightarrow f'(x) = \frac{2}{3}(x-2)^{-1/3} = \dfrac{2}{3(x-2)^{1/3}}$

Then, as

1. $f'(x)$ is undefined at $x = 2$.
2. $f(x)$ is continuous at $x = 2$.

and 3. i. as $x \to 2^-$, $f'(x)\left(= \dfrac{2}{3(x-2)^{1/3}}\right) \to -\infty$

ii. as $x \to 2^+$, $f'(x)\left(= \dfrac{2}{3(x-2)^{1/3}}\right) \to +\infty$

1. Note that the derivative of the function $f(x) = \dfrac{1}{x}$ is undefined at $x = 0$, but it does not have a vertical tangent at that point.

We have a vertical tangent at $x = 2$ and the curve forms a cusp at this point. This is shown in the sketch of $f(x) = (x-2)^{2/3} + 1$:

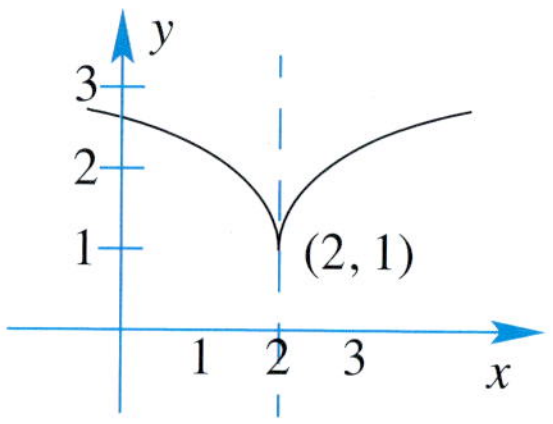

20.2.5 SUMMARY

A lot of ground has been covered with the many definitions encountered. So, below is a visual summary of the definitions we have covered to date.

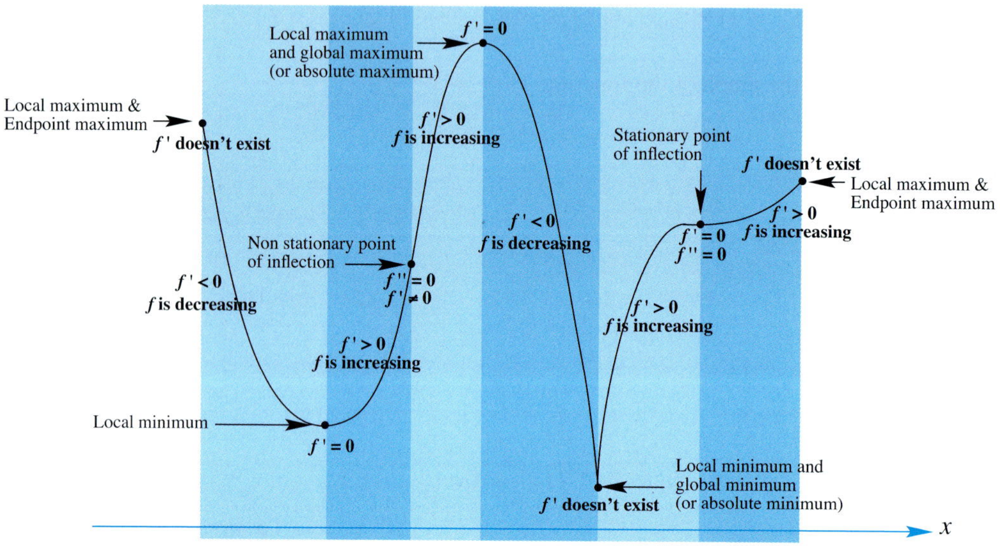

EXERCISES 20.2

1. Draw a sketch of the graph of the function $f(x), x \in \mathbb{R}$, where,

(a) $f(1) = 2, f'(1) = 0, f(3) = -2, f'(3) = 0, f'(x) < 0$ for $1 < x < 3$ and $f'(x) > 0$ for $x > 3$ and $x < 1$.

(b) $f'(2) = 0, f(2) = 0$, $f'(x) > 0$ for $0 < x < 2$ and $x > 2$, $f'(x) < 0$ for $x < 0$ and $f(0) = -4$.

(c) $f(4) = f(0) = 0$, $f'(0) = f'(3) = 0$, $f'(x) > 0$ for x > 3 and $f'(x) < 0$ for $x < 0$ and $0 < x < 3$.

(d) $f(4) = 4$, $f'(x) > 0$ for $x > 4$, $f'(x) < 0$ for $x < 4$, as $x \to 4^+$, $f'(x) \to +\infty$ and as $x \to 4^-$, $f'(x) \to -\infty$.

2. Find the coordinates and nature of the stationary points for the following:

(a) $y = 3 + 2x - x^2$ (b) $y = x^2 + 9x$

(c) $y = x^3 - 27x + 9$ (d) $f(x) = x^3 - 6x^2 + 8$

(e) $f(x) = 3 + 9x - 3x^2 - x^3$ (f) $y = (x-1)(x^2-4)$

(g) $f(x) = x - 2\sqrt{x}, x \ge 0$ (h) $g(x) = x^4 - 8x^2 + 16$

(i) $y = (x-1)^2(x+1)$ (j) $y = x\sqrt{x} - x, x \ge 0$

(k) $g(x) = x + \frac{4}{x}, x \ne 0$ (l) $f(x) = x^2 + \frac{1}{x^2}, x \ne 0$

3. Sketch the following functions:

(a) $y = 5 - 3x - x^2$ (b) $f(x) = x^2 + \frac{1}{2}x + \frac{3}{4}$

(c) $f(x) = x^3 + 6x^2 + 9x + 4$ (d) $f(x) = x^3 - 4x$

(e) $f(x) = \frac{1}{3}x^3 - x^2 + 4$ (f) $y = 4x^3 - x^4$

(g) $y = x^3 - 8$ (h) $y = x^4 - 16$

(i) $y = x - 4x\sqrt{x}, x \ge 0$ (j) $f(x) = x - 2\sqrt{x}, x \ge 0$

4. Find and describe the nature of all stationary points and points of inflection for the function $f(x) = x^3 + 3x^2 - 9x + 2$.

5. Sketch the graph of $x \mapsto x^4 - 4x^2$.

6. A function *f* is defined by $f:x \mapsto e^{-x}\sin x$, where $0 \le x \le 2\pi$.

(a) Find i. $f'(x)$ ii. $f''(x)$

(b) Find the values of x for which

i. $f'(x) = 0$ ii. $f''(x) = 0$

(c) Using parts (a), and (b), find the points of inflection and stationary points for *f*.

(d) Hence, sketch the graph of *f*.

7. A function *f* is defined by $f:x \mapsto e^{x}\sin x$, where $0 \le x \le 2\pi$.

(a) Find i. $f'(x)$ ii. $f''(x)$

(b) Find the values of x for which

i. $f'(x) = 0$ ii. $f''(x) = 0$

(c) Using parts (a) and (b), find the points of inflection and stationary points for *f*.

(d) Hence, sketch the graph of *f*.

8. A function *f* is defined by $f:x \mapsto e^{x}\cos x$, where $0 \le x \le 2\pi$.

(a) Find i. $f'(x)$ ii. $f''(x)$

(b) Find the values of x for which

i. $f'(x) = 0$ ii. $f''(x) = 0$

(c) Using parts (a) and (b), find the points of inflection and stationary points for *f*.

(d) Hence, sketch the graph of *f*.

9. A function *f* is defined by $f:x \mapsto xe^{-x}$, where $x > 0$.

(a) Find i. $f'(x)$ ii. $f''(x)$

(b). Find the values of x for which

i. $f'(x) = 0$ ii. $f''(x) = 0$

(c) Using parts (a) and (b), find the points of inflection and stationary points for *f*.

(d) Hence, sketch the graph of *f*.

10. (a) Find the maximum value of the function $y = 6x - x^2, 4 \le x \le 7$.

(b) Find the minimum value of the function $y = 6x - x^2, 2 \le x \le 6$.

(c) Find the maximum value of the function $y = 2x - x^3, -2 \le x \le 6$.

(d) Find the maximum value of the function $y = 36x - x^4, 2 \le x \le 3$.

11. For the function $f(x) = \frac{1}{3}x^3 - x^2 - 3x + 8, -6 \le x \le 6$ find;

i. its minimum value, ii. its maximum value.

12. For each of the labelled points on the following graphs state

i. whether the derivative exists at the point.

ii. the nature of the curve at the point.

(a)

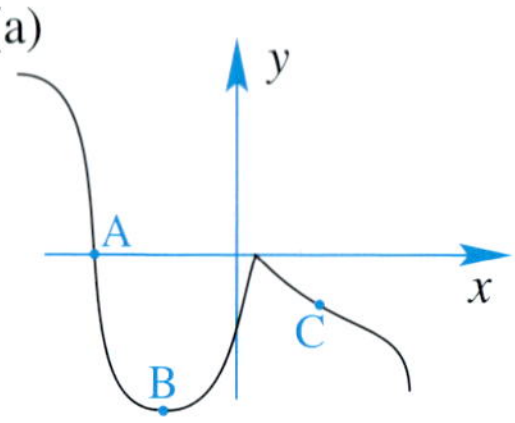

(b)

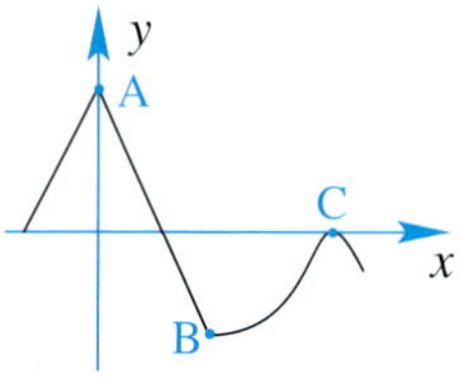

(c)

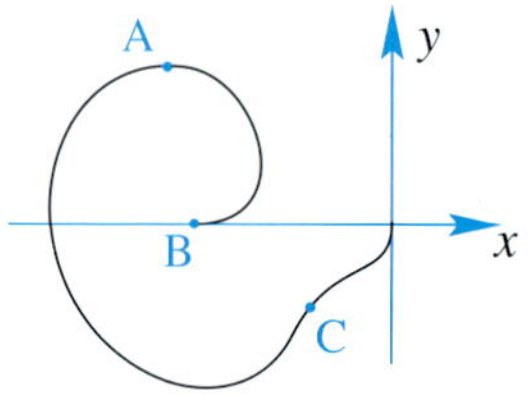

(d)

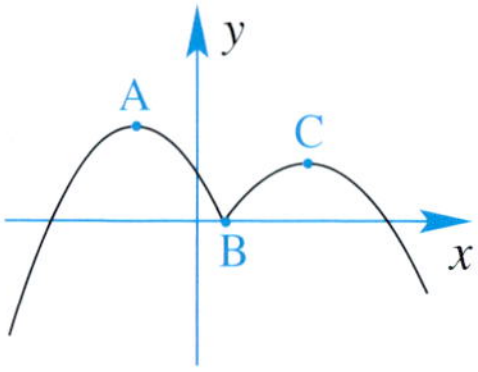

(e)

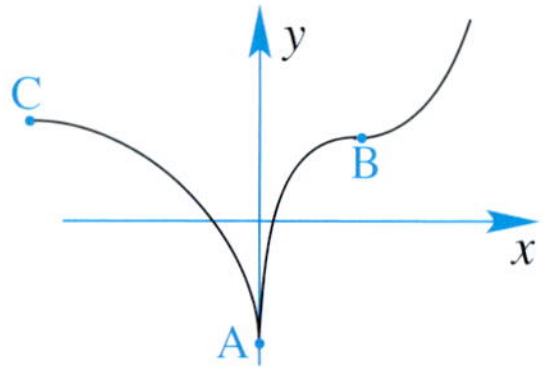

(f)

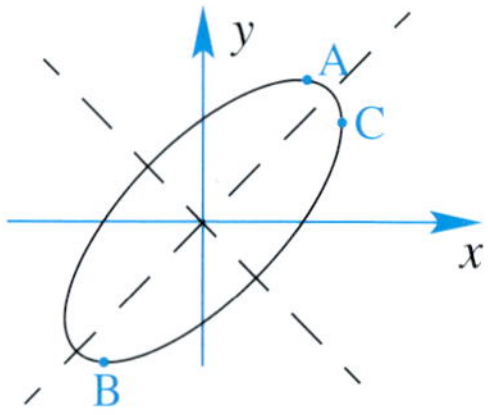

13. Identify which graph corresponds to

i. $f(x)$ ii. $f'(x)$ iii. $f''(x)$

(a)

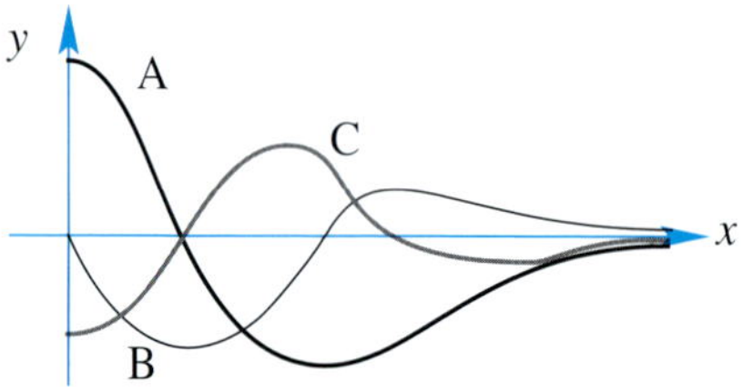

(b)

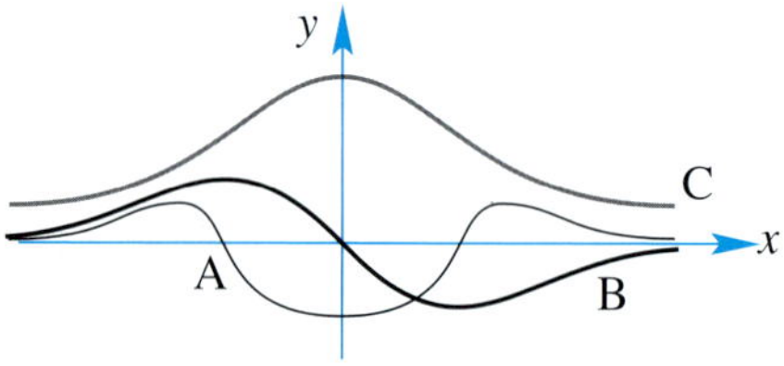

14. For each of the functions, $f(x)$, sketch (a) $f'(x)$ (b). $f''(x)$

i.

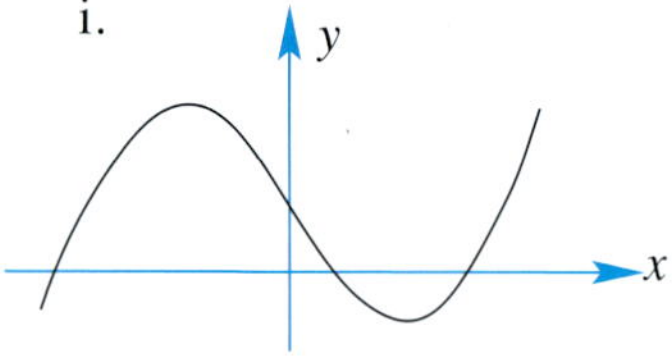

ii.

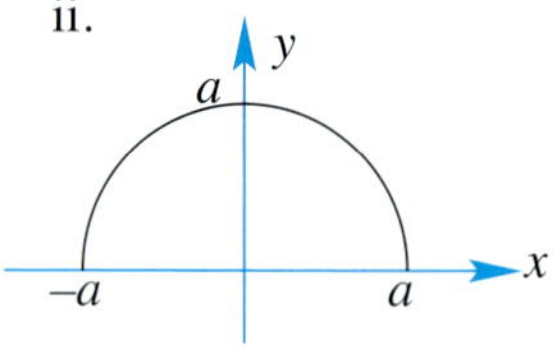

iii.

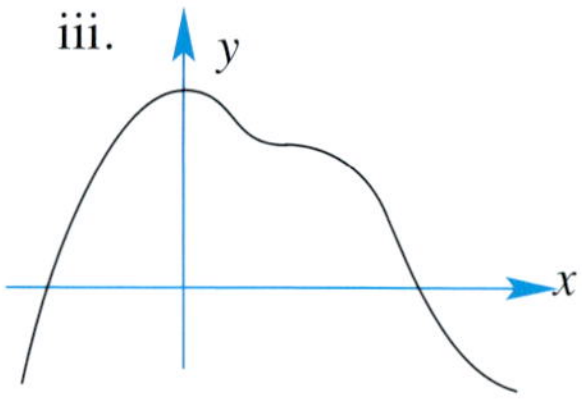

15. The curve with equation $y = ax^3 + bx^2 + cx + d$ has a local maximum where $x = -3$ and a local minimum where $x = -1$. If the curve passes through the points $(0,4)$ and $(1, 20)$ sketch the curve for $x \in \mathbb{R}$.

16. The function $f(x) = ax^3 + bx^2 + cx + d$ has turning points at $\left(-1, -\frac{13}{3}\right)$ and $(3, -15)$.

Sketch the graph of the curve $y = f(x)$.

17. The function $f: x \mapsto \mathbb{R}$, where $f(x) = ax^5 + bx^3 + cx$ has stationary points at $(-2, 64)$ $(2, -64)$ and $(0, 0)$. Find the values of a, b and c and hence sketch the graph of f.

18. Sketch the graph of the curve defined by the equation $y = x(10x - \ln x)$, $x > 0$ indentifying, where they exist, all stationary points and points of inflection.

19. Find m and n so that $f'(1)$ exists for the function $f(x) = \begin{cases} mx^2 + n & \text{if } x \leq 1 \\ \frac{1}{x} & \text{if } x > 1 \end{cases}$.

20. Consider the function $f(x) = (x - a)(x - 4)^{1/b}$.

(a) Sketch the graph of f for the case where

i. $a = 4$ and $b = 2$.

ii. $a = 2$ and $b = 2$.

(b) Find

i. $f'(x)$ if $a = 4$ and $b = 2$.

ii. $f'(x)$ if $a = 2$ and $b = 2$.

For each of i. and ii., in (b) find all stationary points and where $f'(x)$ is undefined.

21. The curve with equation $y = ax^3 + bx^2 + cx + 5$ has a stationary point at $(0, 5)$, an x–intercept at $x = -1$ and an inflection point where $x = 0.5$. Find the values of a, b and c.

22. Sketch the curve of the function with equation $y = (1 + x)^2 e^{-x}$ indentifying, if they exist, all stationary points and points of inflection.

23. The function $f(x) = (x^2 + bx + c)e^x$ passes through the point $(-2, 0)$ and the point $(1, 0)$. Sketch the graph of f identifying, if they exist, all stationary points and points of inflection.

24.* Sketch the graphs of the functions

(a) i. $g(x) = (x + 2)^{2/3}$ ii. $f(x) = (x + 2)^{1/3}$

(b) i. $h(x) = (x - 4)^{3/2} + 2$ ii. $h(x) = (x - 4)^{2/3} + 2$

25.* Sketch the graphs of the functions

(a) $f(x) = 5x^{4/5} - 4x$ (b) $h(x) = \frac{1}{2}x^2 - 3x^{5/3}$

(c) $g(x) = x^{2/3}(x - 3)$ (d) $g(x) = x^{3/2}(x - 3)$

26.* Determine the values of x for which the function $f(x) = (x^2 - 4)^{2/3}$ is increasing. Sketch the graph of f identifying, if they exist, all stationary points and points of inflection.

27.* (a) Find the local maximum, local minimum and inflection point for the function $f(x) = 5x^{2/3} - x^{5/3}$.

(b) Sketch the graph of $f(x) = 5x^{2/3} - x^{5/3}$,

(c) Find the maximum value of $f(x)$ over the interval $-2 \leq x \leq 6$.

28. The curve with equation $y = ax^3 + bx^2 + cx + d$ intersects the x-axis at $x = 1$ and cuts the y-axis at $(0, -34)$. Given that the curve has turning points at $x = 3$ and $x = 5$, determine the values of a, b, c and d. Sketch this curve.

29. (a) Given that $f(x) = axe^{-bx^2}$, show that $f''(x) = 2abx(2bx^2 - 3)e^{-bx^2}$. Hence show that there can be at most three points of inflection.

(b) If $f(x)$ has a stationary point where $x = \frac{1}{\sqrt{2}}$, find the value of b.

(c) For what value of a will the graph of the function pass through the point $(\sqrt{2}, e^{-2})$?

(d) Using the values of a and b found in parts (b) and (c), sketch the graph of the function.

30. (a) Find the smallest three positive critical values of $f(\theta) = \frac{\cos\theta}{\theta}, \theta \neq 0$.

(c) Sketch the graph of the function $f(\theta) = \frac{\cos\theta}{\theta}, 0 < \theta \leq 4\pi$.

* Functions such as these lie outside the scope of the syllabus. They have been included so that they can be used as part of classroom discussions.

20.3 THE SECOND DERIVATIVE AND ITS APPLICATIONS

20.3.1 DEFINITION

We have already encountered the second derivative, however, we have only done so as a process rather than as an in depth study. We now revisit the second derivative and look at its other uses.

Remember that if a function is differentiated twice, we obtain the **second derivative** of the function. The second derivative measures the rate at which the first derivative is changing.

If this **second derivative is small**, it means the first derivative is changing slowly and that the curve will look a bit like:  or  with a **small curvature**. If the **second derivative is a large positive** number, the gradient is increasing rapidly and the curve will look a bit like this:  , that is, a **large curvature**. For similar reasons, a **large negative value** for the second derivative means that the gradient of the curve is decreasing rapidly and the graph should look a bit like this:  . Again, producing a **large curvature**.

We now define the term **concavity** and look at its relationship to the second derivative.

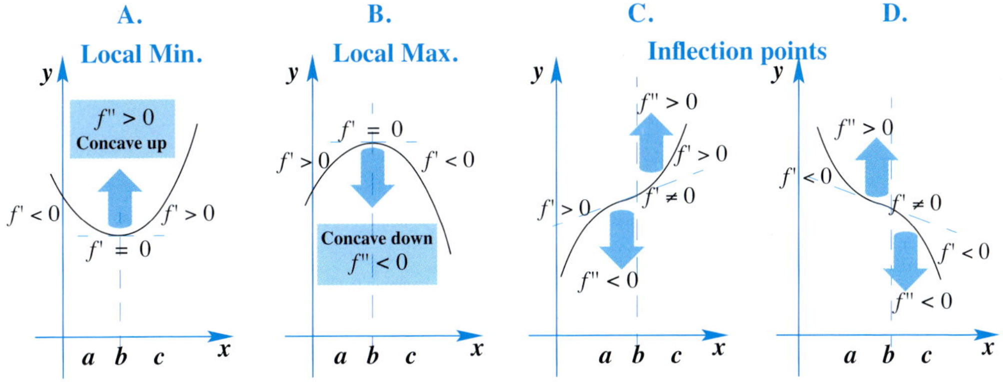

Graph A

As the x-values increase from a to b to c, i.e., from left to right,
then the values of f' increase from –ve to 0 to +ve $\Rightarrow f'' > 0 \Rightarrow$ a local minimum at $x = b$

Graph B

As the x-values increase from a to b to c, i.e., from left to right,
then the values of f' decrease from +ve to 0 to –ve $\Rightarrow f'' < 0 \Rightarrow$ a local maximum at $x = b$

Graph C

As the x-values increase from a to b to c, i.e., from left to right,
then the values of f' decrease from +ve to smaller +ve numbers and then increase to larger +ve numbers. However, this time f'' changes sign at $x = b$, from –ve values to +ve values
$\Rightarrow$ a point of inflection at $x = b$ (i.e., a non-stationary point of inflection).

Graph D

As the x-values increase from a to b to c, i.e., from left to right,
then the values of f' decrease from –ve to more –ve numbers and then increase to less –ve numbers. However, this time f'' changes sign at $x = b$, from +ve values to –ve values
$\Rightarrow$ a point of inflection at $x = b$ (i.e., a non-stationary point of inflection).

Note that a special case of Graphs C and D is where a stationary point occurs at $x = b$:

In such cases we say that a **stationary point of inflection** occurs at $x = b$.

That is at $x = b$ $f'' = 0$, $f' = 0$ and the sign of the second derivative changes (either from –ve to +ve or from +ve to –ve).

This situation has been addressed in detail earlier in this chapter

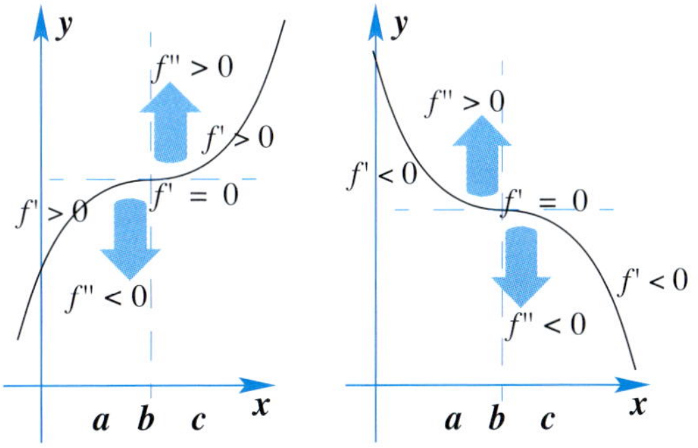

We can use the following sequence of derivatives to summarise our observations so far.

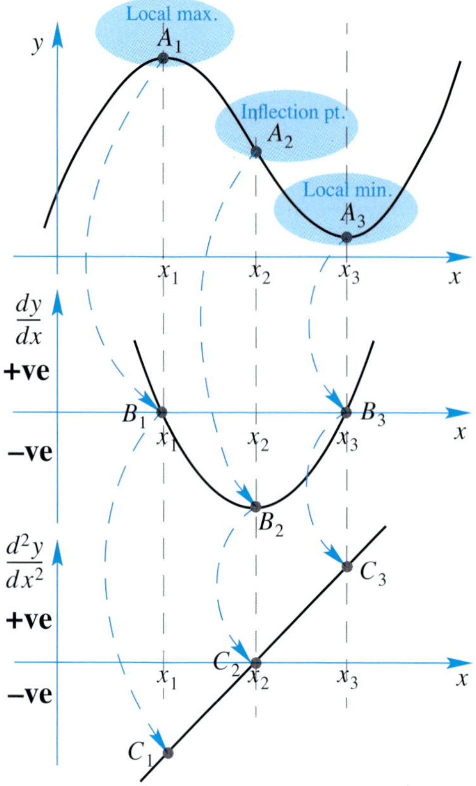

Local Maximum: $A_1 \to B_1 \to C_1$

So that at a local maximum we have that

1. $\frac{dy}{dx} = 0$ **2.** $\frac{d^2y}{dx^2} < 0$

Inflection point: $A_2 \to B_2 \to C_2$

So that at a point of inflection we have

Case A –

A non stationary point of inflection

1. $\frac{dy}{dx} \neq 0$ **2.** $\frac{d^2y}{dx^2} = 0$

and there is a change in the sign of the second derivative near x_2.

Case B –

A stationary point of inflection

1. $\frac{dy}{dx} = 0$ **2.** $\frac{d^2y}{dx^2} = 0$

and there is a change in the sign of the second derivative near x_2.

Local Minimum: $A_3 \to B_3 \to C_3$

So that at a local minimum we have that

1. $\frac{dy}{dx} = 0$ **2.** $\frac{d^2y}{dx^2} > 0$

EXAMPLE 20.25

Determine the values of x for which the function $f(x) = x^3 - 3x^2$ is concave up.

Solution

By definition, $f(x)$ will be concave up for those values of x where $f''(x) < 0$.

Now, $f'(x) = 3x^2 - 6x \therefore f''(x) = 6x - 6$.

Then, $f''(x) < 0 \Leftrightarrow 6x - 6 < 0 \Leftrightarrow x < 1$.

i.e., $f(x)$ is concave up for $x \in]-\infty, 1[$.

The use of a graphics calculator can confirm this result.

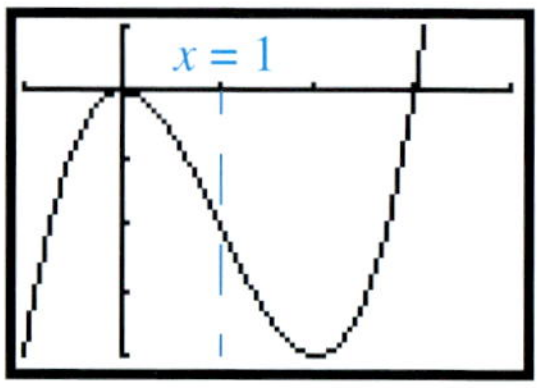

EXAMPLE 20.26

Find the first and second derivatives of the function:

$$f(x) = 2x^5 - 5x^4 - 10x^3, x \in [-5, 5]$$

Hence find the coordinates and nature of all the stationary points of the function.

Solution

$$f(x) = 2x^5 - 5x^4 - 10x^3 \Rightarrow f'(x) = 10x^4 - 20x^3 - 30x^2$$

$$\therefore f''(x) = 40x^3 - 60x^2 - 60x$$

We can locate the stationary points by equating the first derivative to zero:

$$10x^4 - 20x^3 - 30x^2 = 0 \Leftrightarrow 10x^2(x^2 - 2x - 3) = 0$$

$$\Leftrightarrow 10x^2(x-3)(x+1) = 0$$

$$\Leftrightarrow x = -1, 0, 3$$

Next, we must use a test of some sort to discriminate between the types of stationary point. Using the second derivative test, we have the following.

$x = -1$: $f''(x) = 40x^3 - 60x^2 - 60x \therefore f''(-1) = 40(-1)^3 - 60(-1)^2 - 60(-1) = -40$

From this we can infer that in the region of the stationary point, the gradient of the curve is decreasing. This means that the general shape must be:  and that we have a local maximum. The y-coordinate of this point is $f(-1) = 2(-1)^5 - 5(-1)^4 - 10(-1)^3 = 3$.

At the origin, both $f'(0) = 0$ and $f''(0) = 0$. However we can draw no conclusion as yet from this information as it could represent either a stationary point of inflection or a turning point. A second test is required which might be to look at the sign of the first derivative just to the left and just to the right of the stationary point:

$$f'(-0.01) = 10(-0.01)^4 - 20(-0.01)^3 - 30(-0.01)^2 = -0.0029799$$

$$f'(0.01) = 10(0.01)^4 - 20(0.01)^3 - 30(0.01)^2 = -0.0030199$$

Since the gradient is negative both to the left and right of $x = 0$, it follows that we have a descending stationary inflection point. The y-coordinate of this point is $f(0) = 0$.

Finally, $f''(3) = 40(3)^3 - 60(3)^2 - 60(3) = 360$ so that in the region of the stationary point at $x = 3$ the gradient is increasing so the graph must have the general shape:  and the stationary point is a local minimum. The y-coordinate of this point is $f(3) = -189$.

In summary, there is a local maximum at (–1,3), a descending inflection at the origin and a local minimum at (3,–189).

This graph is an example of one which is difficult to see using a graphics calculator. If you look at the graph using a window which will show the stationary points, you will probably miss the detail in the area of the origin.

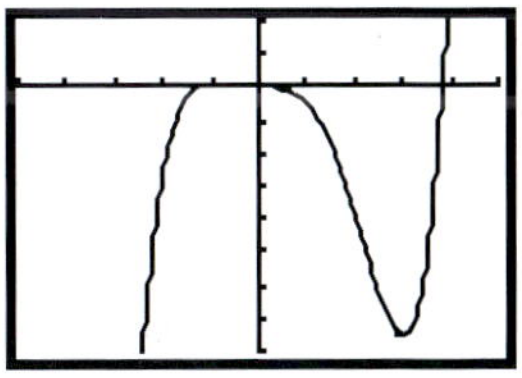

By contrast, if you only look at the region around the origin, you might well miss the minimum at (3,–189).

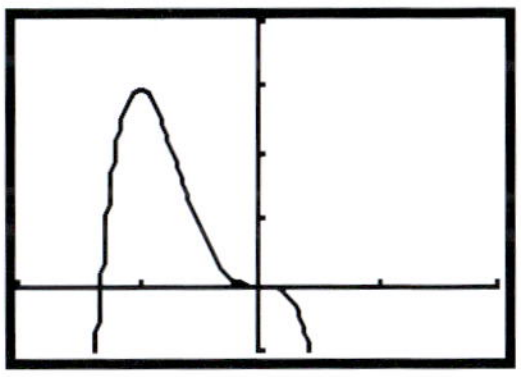

You should be careful when using a graphics calculator to show graphs and are advised to think carefully about the best way(s) of viewing them. In this example, the shape of the graph cannot be seen from a single window.

EXAMPLE 20.27

Determine the values of x for which the function $f(x) = \sin^2 x$ is concave down over the interval $[0, 2\pi]$.

Solution

$f(x) = \sin^2 x \Rightarrow f'(x) = 2(\cos x)(\sin x) = \sin 2x \therefore f''(x) = 2\cos 2x$

$f(x) = \sin^2 x$ is concave down where $f''(x) < 0 \Leftrightarrow 2\cos 2x < 0$

Consider the solution to $2\cos 2x = 0$, i.e., $\cos 2x = 0 \therefore 2x = \frac{\pi}{2}, \frac{3\pi}{2}, \frac{5\pi}{2}, \frac{7\pi}{2}$

$\therefore x = \frac{\pi}{4}, \frac{3\pi}{4}, \frac{5\pi}{4}, \frac{7\pi}{4}$

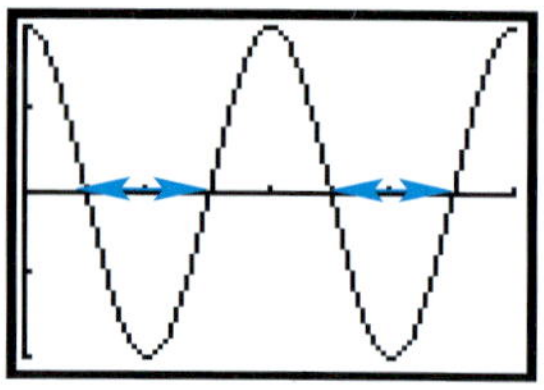

Then, using the graph of $f''(x) = 2\cos 2x$ as an aid, we have that $f''(x) < 0$ if $\frac{\pi}{4} < x < \frac{3\pi}{4}$ or $\frac{5\pi}{4} < x < \frac{7\pi}{4}$.

i.e., $x \in \left]\frac{\pi}{4}, \frac{3\pi}{4}\right[\cup \left]\frac{5\pi}{4}, \frac{7\pi}{4}\right[$.

EXAMPLE 20.28

The graph of the function $f(x) = \frac{\sin x}{x}, -2\pi < x < 0$ is concave up over the interval $a < x < b$. Find the values of a and b to two decimal places.

Solution

$f(x) = \frac{\sin x}{x} \therefore f'(x) = \frac{(\cos x)x - (\sin x) \times 1}{x^2} = \frac{x\cos x - \sin x}{x^2}$.

Then,

$$f''(x) = \frac{(\cos x - x\sin x - \cos x) \times x^2 - 2x \times (x\cos x - \sin x)}{x^4} = \frac{2x\sin x - 2x^2\cos x - x^3\sin x}{x^4}$$

$$= \frac{2\sin x - 2x\cos x - x^2\sin x}{x^3}$$

We now need to find those values of x for which $f''(x) < 0$.

We start by finding where $f''(x) = 0$ and then make use of a graph to identify the inequality.

The solution to $f''(x) = 0$ is provided by setting $\frac{2\sin x - 2x\cos x - x^2\sin x}{x^3} = 0$. That is, by solving $2\sin x - 2x\cos x - x^2\sin x = 0$ – which unfortunately cannot be readily solved. Seeing as we require answers to two decimal places, we make use of the graphics calculator:

Using the solve function we obtain the solutions $x = -2.08$ and $x = -5.94$

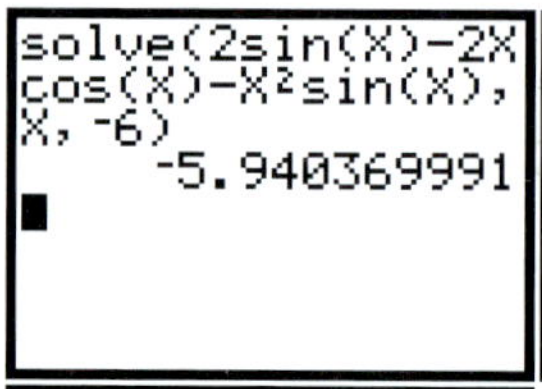

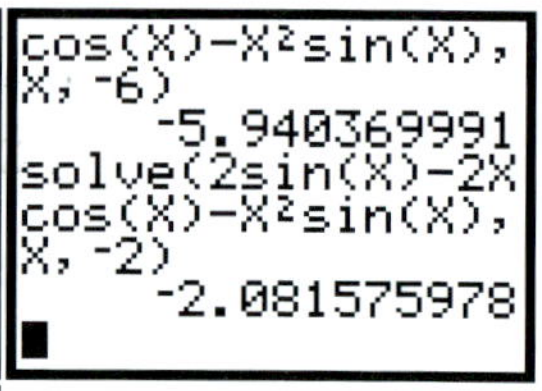

To determine the set of values for which the inequality $f''(x) < 0$ holds, we use the graph of $f(x)$ as an aid. From the graph of $f(x)$ we have:

$f''(x) < 0 \Leftrightarrow -5.95 < x < -2.08$

Therefore, $a = -5.94$ and $b = -2.08$

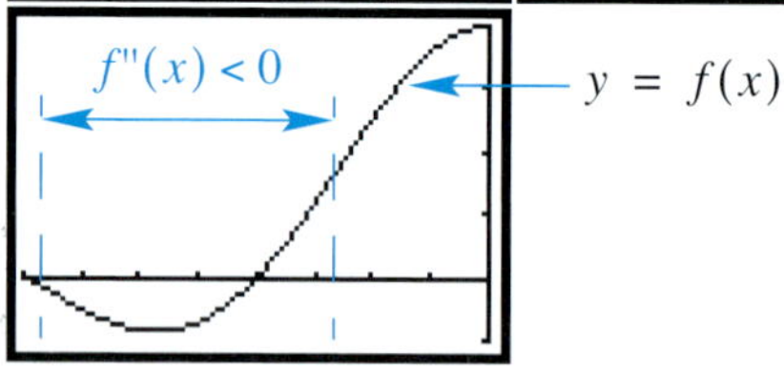

We now consider the previous problem and solve it by using the **nDerive** function on the TI–83:

We start by entering the equation of the function $f(x) = \frac{\sin x}{x}$ as Y_1 and then set Y_2 to be the first derivative of Y_1 and then set Y_3 to be the first derivative of Y_2 [which then becomes the second derivative of Y_1]. Once that is done, we use the solve function as shown.

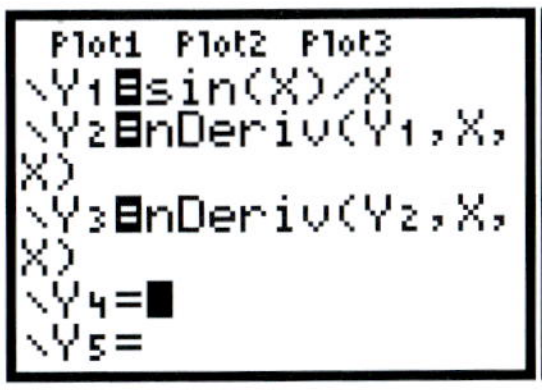

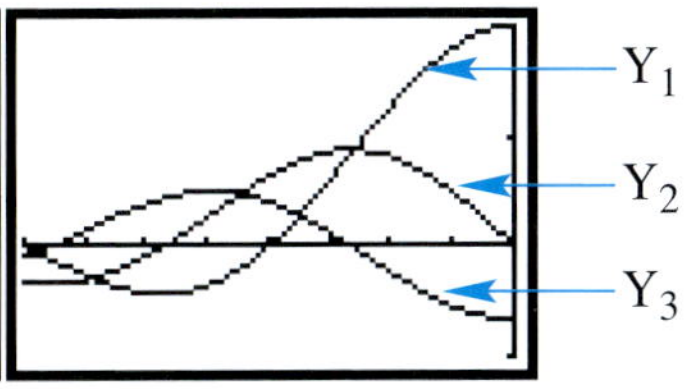

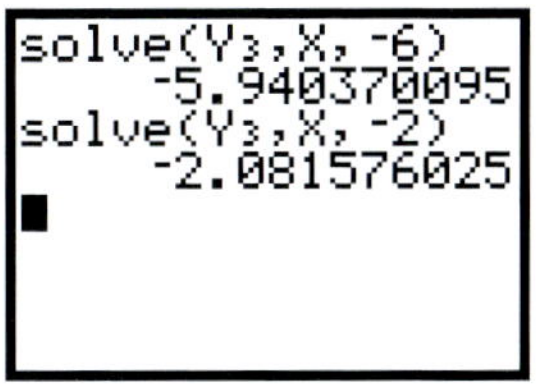

EXAMPLE 20.29

Determine the coordinates of all stationary points and their nature for the function with equation $f(x) = x^2e^{-x}$

Solution

First we differentiate the function $f(x) = x^2e^{-x}$: $f'(x) = 2xe^{-x} - x^2e^{-x} = xe^{-x}(2-x)$.

Stationary points occur where $f'(x) = 0 \Rightarrow xe^{-x}(2-x) = 0 \Leftrightarrow x = 0$ or $x = 2$.

Next we find the second derivative:

$$\begin{aligned} f''(x) &= (e^{-x} - xe^{-x}) \times (2-x) + xe^{-x} \times -1 \\ &= e^{-x}((1-x)(2-x) - x) \\ &= e^{-x}(x^2 - 4x + 2) \end{aligned}$$

We can now check for the nature of the stationary points:

At $x = 0$, $f''(0) = e^{-0}(0^2 - 4 \times 0 + 2) = 2 > 0$. Implying a local minimum at $x = 0$.

At $x = 2$, $f''(2) = e^{-2}(2^2 - 4 \times 2 + 2) = -2e^{-2} < 0$. Implying a local maximum at $x = 2$.

Therefore, we have a local minimum at (0,0) and a local maximum at $(2, 4e^{-2})$.

EXERCISES 20.3

1. Use the second derivative to determine the nature of the stationary points of:

(a) $f(x) = x^3 - 16x$ (b) $f(x) = x^4 - 2x^2$ (c) $f(x) = \sqrt{x} - x$

(d) $y = xe^{-x}$ (e) $y = (x+1)\log_e x$ (f) $y = x^2(x-1)^2$

(g) $y = \frac{x}{x^2+1}$ (h) $y = \frac{x}{x^2-1}$

2. Find the maximum and minimum values of the following:

(a) $f(x) = x^3 - 16x, 0 \le x \le 6$ (b) $f(x) = x^4 - 2x^2, -2 \le x \le 4$

(c) $y = \frac{x}{x^2+1}, 0 \le x \le 4$ (d) $y = \sqrt[3]{(x-1)^2}, 0 \le x \le 2$

3. Sketch the graph of $f(x) = \cos x + \frac{1}{2}\cos 2x, x \in [0, 2\pi]$ identifying all important features, including maximum, minimum and inflection points.

4. Sketch the graph of $f(x) = \sin x + \frac{1}{2}\sin 2x, x \in [0, 2\pi]$ identifying all important features, including maximum, minimum and inflection points.

5. Sketch the graph of $f(x) = x\sin x, x \in [-2\pi, 2\pi]$ identifying all important features, including maximum, minimum and inflection points.

6. Identify and justify the occurence of all maximum points, minimum points and points of inflection for the curves with equation

(a) $y = \sqrt{x} + \frac{1}{\sqrt{x}}$ (b) $y = \frac{x^2 + 3}{x + 1}$ (c) $y = \frac{\sqrt{x}}{x - 1}$

7. Sketch the graph of $f(x) = \sin^2 x - \cos x, x \in [0, 2\pi]$ identifying all important features, including maximum and minimum points.

8. Sketch the graph of (a) $f(x) = x^2 e^{-x}$ (b) $f(x) = e^{4 - x^2}$ (c) $g(x) = \frac{1}{x}e^{-\frac{1}{2}x}$
identifying all important features, including maximum, minimum and inflection points.

9. Sketch the graph of (a) $f(x) = \frac{1}{x}\ln(x)$ (b) $f(x) = x^2 - \ln x$

(c) $g(x) = \frac{2}{x} + \ln x$ (d) $h(x) = \ln[x(x - 1)]$

identifying all important features, including maximum, minimum and inflection points.

10. Consider the function $f(x) = (x - 2)^a (x + 2)^b$ where a and b are integers.

(a) Find $f'(x)$.

(b) Sketch the graph of $f(x)$, identifying all stationary points, for the case where
i. $a = 1, b = -1$ ii. $a = 2, b = 1$ iii. $a = 2, b = 2$

11. (a) For what values of c will $x\ln x - x + c \geq 1, x > 0$.

(b) Sketch the graph of the curve whose equation is $y = x\ln x - x + 2c, x > 0$, where c is the smallest value determined in part (a).

12. Making use of the second derivative, find the nature of
(a) all stationary points (b) all points of inflection

of the curve with equation $y = x^2 \ln\left(\frac{1}{x^3}\right), x > 0$.

(c) Sketch the graph of this curve.

13. For the function $f(x) = \frac{1}{x^2}\ln(x), x > 0$ determine, where they exist, the coordinates of

(a) stationary points (b) inflection points (c) Sketch the graph of $f(x)$.

20.4 RATIONAL FUNCTIONS

20.4.1 SKETCHING THE GRAPH OF $x \mapsto \frac{ax+b}{cx+d}, cx + d \neq 0$

Properties

Graphs of this nature possess two types of asymptotes, one vertical and the other horizontal.

1. The vertical asymptote

A vertical asymptote occurs when the denominator is zero, that is, where $cx + d = 0$. Where this occurs, we place a vertical line (usually dashed), indicating that the curve cannot cross this line under any circumstances. This must be the case, because the function is undefined for that value of x.

For example, the function $x \mapsto \frac{3x+1}{2x+4}$ is undefined for that value of x where $2x + 4 = 0$. That is, the function is undefined for $x = -2$. This means that we would need to draw a vertical asymptote at $x = -2$. In this case, we say that the asymptote is defined by the equation $x = -2$. Using limiting arguments provides a more formal approach to 'deriving' the equation of the vertical asymptote. The argument is based along the following lines:

as $x \to -2^-, \frac{3x+1}{2x+4} \to +\infty$

That is, as x tends to –2 from the left or 'below', (hence the minus sign next to the two) the function tends to positive infinity.

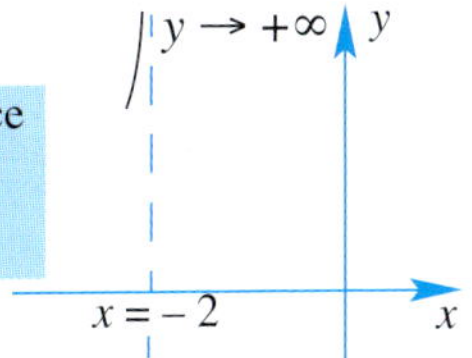

as $x \to -2^-, \frac{3x+1}{2x+4} \to +\infty$

That is, as x tends to –2 from the right or 'above', (hence the plus sign next to the two) the function tends to negative infinity.

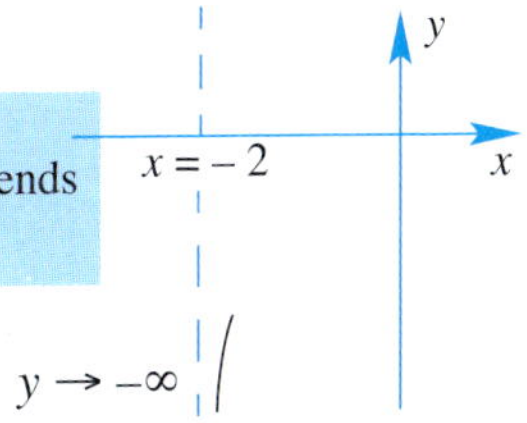

Therefore we write

$$\left.\begin{array}{ll}\text{As } x \to -2^+ & f(x) \to -\infty \\ \text{As } x \to -2^- & f(x) \to +\infty\end{array}\right\} \therefore x = -2 \text{ is a vertical asymptote of } f(x) = \frac{3x+1}{2x+4}, x \neq -2.$$

2. The horizontal asymptote

To determine the equation of the horizontal asymptote, again we use a limiting argument, however, this time we observe the behaviour of the function as $x \to \pm\infty$.

It will be easier to determine the behaviour of the function (as $x \to \pm\infty$) if we first 'simplify' the rational function (using long division): $f(x) = \frac{3x+1}{2x+4} = \frac{3}{2} - \frac{5}{2x+4}$.

Next we determine the behaviour for extreme values of x:

$$\left.\begin{array}{l} \text{As } x \to +\infty \quad f(x) \to \left(\frac{3}{2}\right)^{-} \\ \text{As } x \to -\infty \quad f(x) \to \left(\frac{3}{2}\right)^{+} \end{array}\right\} \text{Therefore, } y = \frac{3}{2} \text{ is the horizontal asymptote}$$

We can now add a few more features of the function:

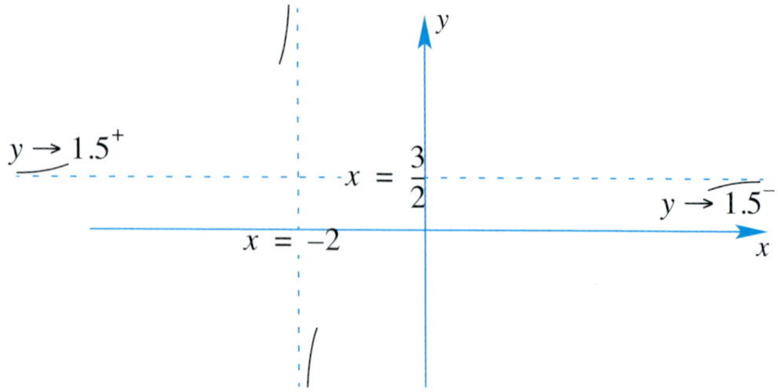

3. Axial intercepts

x–intercept

To determine the x–intercept(s) we need to solve for $f(x) = 0$.

In this case we have $f(x) = \frac{3x+1}{2x+4} = 0 \Leftrightarrow 3x + 1 = 0 \Leftrightarrow x = -\frac{1}{3}$.

That is, the curve passes through the point $\left(-\frac{1}{3}, 0\right)$.

y–intercept

To determine the y–intercept we find the value of $f(0)$ (if it exists, for it could be that the line $x = 0$ is a vertical asymptote).

In this case we have $f(0) = \frac{3 \times 0 + 1}{2 \times 0 + 4} = \frac{1}{4}$.

Therefore the curve passes through the point $\left(0, \frac{1}{4}\right)$.

Having determined the behaviour of the curve near its asymptotes (i.e., if the curve approached the asymptotes from above or below) and the axial–intercept, all that remains is to find the stationary points (if any).

4. Stationary points

To determine the coordinates of stationary points we need to first solve for $f'(x) = 0$ and then (if we do have a solution) substitute into the equation of $f(x)$ so that we can obtain the y coordinate.

In this case we have (*using the quotient rule*): $f'(x) = \frac{3(2x+4) - 2(3x+1)}{(2x+4)^2} = \frac{10}{(2x+4)^2}$.

Therefore, $f'(x) = 0 \Leftrightarrow \frac{10}{(2x+4)^2} = 0$, for which there are no real solutions.

This means that there are no stationary points for this curve. In fact, it can be shown that any function of the form $f(x) = \frac{ax+b}{cx+d}$ will never have a stationary point. We leave this as an exercise for you! We are now in a position to complete the sketch of our function. The information we have is

1. **Asymptotes**
   $x = -2$.
   $y = \frac{3}{2}$.
2. **Intercepts**
   x–$\left(-\frac{1}{3}, 0\right)$.
   y–$\left(0, \frac{1}{4}\right)$.
3. **Stationary points**
   None.

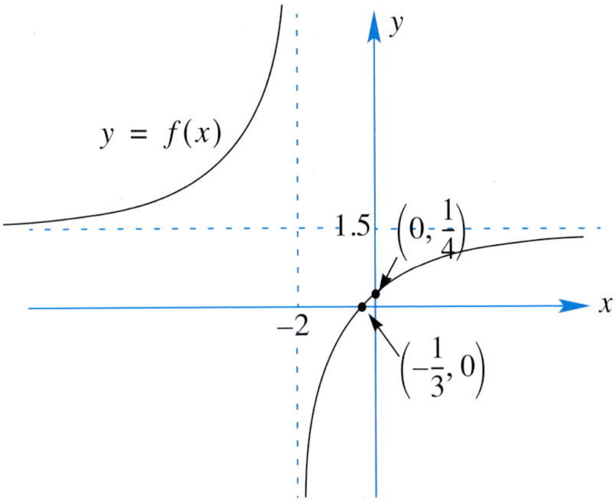

20.4.2 OTHER RATIONAL FUNCTIONS

EXAMPLE 20.30

Sketch the graph of $f(x) = 2 + \frac{x}{x^2-1}$, $x \neq \pm 1$.

Solution

The most obvious thing to do is— use your graphics calculator:

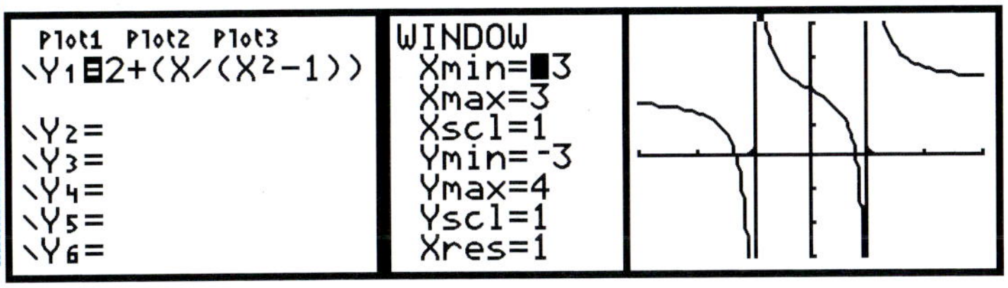

However, this only gives a general idea of the shape of the function, we still want to study some of its properties in more detail. We begin by investigating its asymptotic properties:

First we rewrite the function $f(x) = 2 + \frac{x}{x^2-1}$ as $f(x) = 2 + \frac{x}{(x-1)(x+1)}$. This will enable us to concentrate on x–values that will help us determine the asymptotes

Vertical Asymptotes:
These will correspond to values of x for which the denominator is zero.
That is, those values of x for which $(x-1)(x+1) = 0$, i.e., where $x = 1$ and $x = -1$.

Case 1 $x = 1$: As $x \to 1^+$ then $f(x) \to +\infty$
and as $x \to 1^-$ then $f(x) \to -\infty$.

Case 2 $x = -1$ As $x \to -1^+$ then $f(x) \to +\infty$
and as $x \to -1^-$ then $f(x) \to -\infty$.

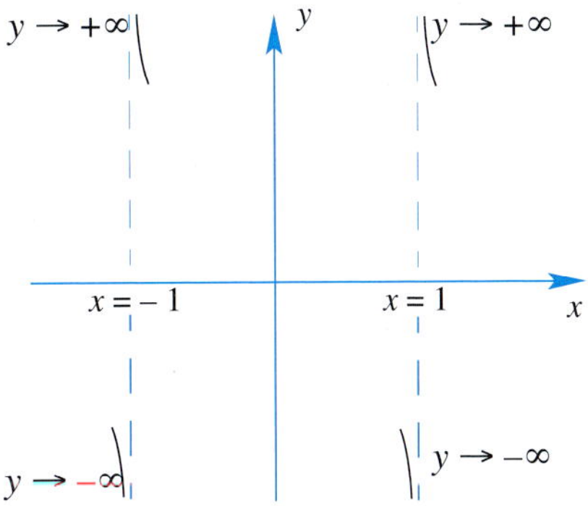

Horizontal asymptote:

That is, we look at the behaviour of $f(x)$ as $|x| \to \infty$

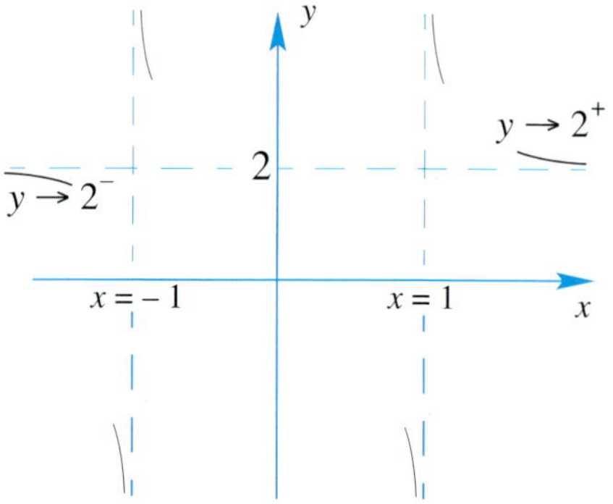

As $x \to +\infty$ then $\frac{x}{x^2-1} \to 0^+ \therefore f(x) \to 2^+$.

As $x \to -\infty$ then $\frac{x}{x^2-1} \to 0^- \therefore f(x) \to 2^-$.

We now determine the intercepts with the axes.

***x*–intercept(s):**

Setting $f(x) = 0$, we have, $2 + \frac{x}{x^2-1} = 0 \Leftrightarrow \frac{x}{x^2-1} = -2 \Leftrightarrow x = -2x^2+2$

$$\Leftrightarrow 0 = 2x^2 + x - 2$$

$$\Leftrightarrow x = \frac{-1 \pm \sqrt{17}}{4}$$

That is, the curve crosses the x–axis at $\left(\frac{-1+\sqrt{17}}{4}, 0\right)$ and $\left(\frac{-1-\sqrt{17}}{4}, 0\right)$.

***y*–intercept:**

We evaluate $f(0) = 2 + \frac{0}{0-1} = 2$. That is the curve passes through the point (0, 2).

Notice that we have in fact cut the horizontal asymptote at (0, 2)!

We can also look for possible stationary points and points of inflexion.

Stationary points:

We need to solve for $f'(x) = 0$.

Now $f(x) = 2 + \frac{x}{x^2-1} \Rightarrow f'(x) = \frac{1 \times (x^2-1) - x \times 2x}{(x^2-1)^2} = \frac{-(x^2+1)}{(x^2-1)^2}$.

Therefore, setting $\frac{-(x^2+1)}{(x^2-1)^2} = 0 \Rightarrow -(x^2+1) = 0$. For which there are no real solutions.

Therefore, there are no stationary points on this graph.

Points of inflection:

We first find the second derivative:

Now, $f'(x) = -\frac{(x^2+1)}{(x^2-1)^2} \Rightarrow f''(x) = -\frac{2x(x^2-1)^2 - (x^2+1)(2 \cdot 2x \cdot (x^2-1))}{(x^2-1)^2}$

$$= \frac{2x(x-1)(x^2+3)}{(x^2-1)^4} \text{ (After some simplification).}$$

Therefore, setting $f''(x) = 0 \Rightarrow \frac{2x(x-1)(x^2+3)}{(x^2-1)^4} = 0 \Leftrightarrow x = 0, \pm 1$.

But, $x \neq \pm 1$, therefore the only solution is $x = 0$. To check if in fact there is a point of inflection we need to check the sign of $f'(x)$ or $f''(x)$ on either side of $x = 0$.

We now show how this can be done using the TI–83. That is we sketch the graph of $y = f'(x)$.

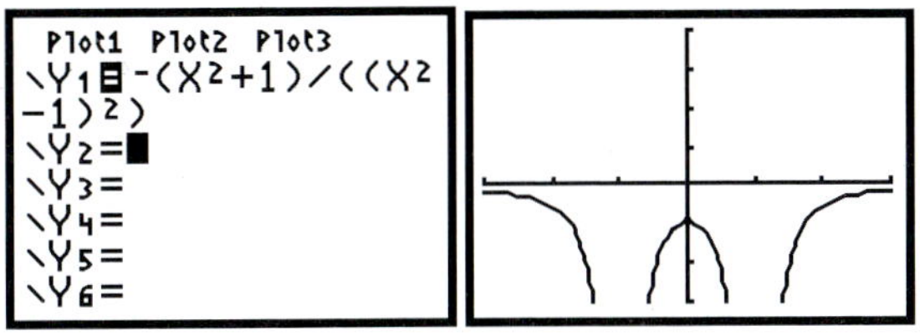

From the graph of $y = f'(x)$ we see that $f'(x) < 0$ for all values of x.

In particular, $f'(x) < 0$ for $x > 0$ and
$f'(x) < 0$ for $x < 0$

Therefore, at $x = 0$, i.e., at (0, 2) we have a non–stationary point of inflection.

Note that we could have sketched the graph of the second derivative using the TI-83 to verify this:

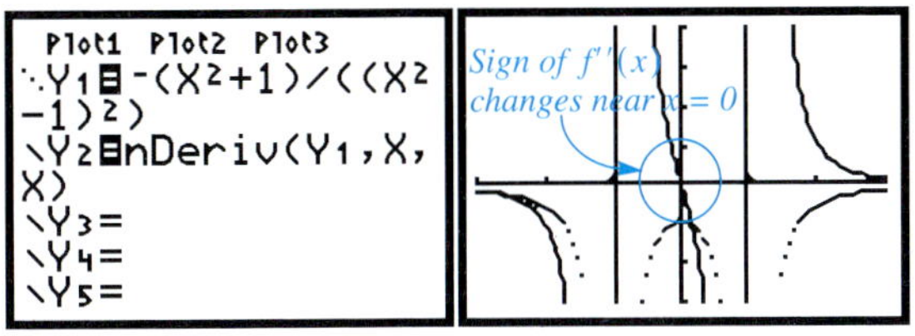

From this graph, we see that the sign of the second derivative changes from positive to negative as x increases near $x = 0$.
That is $f''(x) > 0$ just before $x = 0$
and $f''(x) < 0$ just after $x = 0$.

After all this, we can now sketch the graph of $f(x) = 2 + \frac{x}{x^2 - 1}$.

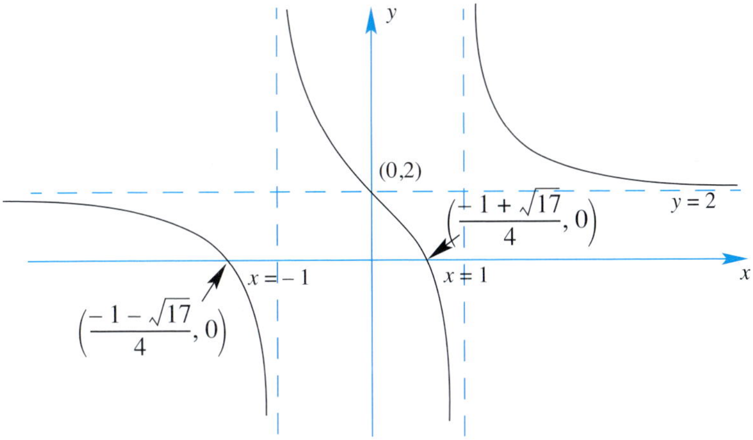

It should be noted that functions other than rational functions also have asymptotes. For example, graphs having the equation $f(x) = a + be^{kx}$ have a horizontal axis at $y = a$, and graphs whose equation is given by $f(x) = a\log_e(x - k) + c$ have a vertical asymptote at $x = k$. These functions have been dealt with in Chapter 7.

EXERCISES 20.4

1. Use a limiting argument to determine the equations of the vertical and horizontal asymptotes for the following:

(a) $f(x) = \dfrac{2x+1}{x+1}$ (b) $f(x) = \dfrac{3x+2}{3x+1}$

(c) $f(x) = \dfrac{2x-1}{4x+1}$ (d) $f(x) = \dfrac{4-x}{x+3}$

(e) $f(x) = 3 - \dfrac{1}{x}$ (f) $f(x) = 5 - \dfrac{1}{2-x}$

2. Make use of a graphics calculator to verify your results from Question 1 by sketching the graph of the given function.

3. Sketch the following curves, clearly labelling all intercepts, stating the equations of all asymptotes, and, in each case, showing that there are no stationary points:

(a) $x \mapsto \dfrac{3}{2x+1}$ (b) $x \mapsto \dfrac{x+1}{x+2}$

(c) $x \mapsto \dfrac{5-x}{2x-1}$ (d) $x \mapsto 3 + \dfrac{1}{x}$

(e) $x \mapsto \dfrac{1}{x-3} - 2$ (f) $x \mapsto 1 - \dfrac{2}{2x-3}$

4. The figure shows part of the graph of the function whose equation is $x \mapsto \dfrac{ax+2}{x-c}$.

Find the values of a and c.

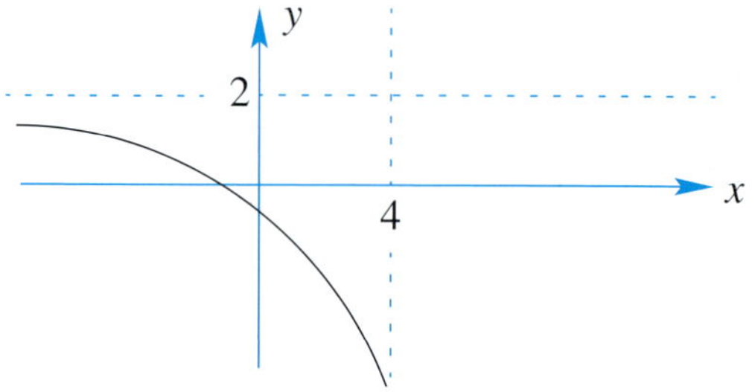

5. Given that $f: x \mapsto x+2$ and that $g: x \mapsto \dfrac{1}{x-1}$ sketch the graphs of

(a) $f \circ g$ (b) $g \circ f$

6. (a) Consider the function $f(x) = \dfrac{2-x}{2+x}$

i. Find the coordinates of the intercepts with the axes.

ii. Determine the equations of the asymptotes of f.

iii. Hence, sketch the graph of f.

iv. Determine the domain and range of f.

(b) Find f^{-1}, the inverse function of f.

(c) Deduce the graph of $(f(x))^2$.

7. (a) Express $\frac{8x-5}{x-3}$ in the form $A + \frac{B}{x-3}$, where A and B are integers.

Hence, state the equations of the vertical and horizontal asymptotes of the function $f(x) = \frac{8x-5}{x-3}$.

(b) Sketch the graph of $f(x) = \frac{8x-5}{x-3}$ and use it to determine its range.

21.1 RATES OF CHANGE

21.1.1 DEFINITIONS

We saw in Chapter 18 that $\frac{dy}{dx}$ measures the **rate of change of a quantity *y* with respect to another quantity *x***. In the same way, we have that

$\frac{dA}{dr}$ measures the rate of change of *A* with respect to *r*,

$\frac{dV}{dt}$ measures the rate of change of *V* with respect to *t* and

$\frac{dP}{dV}$ measures the rate of change of *P* with respect to *V*.

For example, if A m^2 measures the area of a circle of radius r m, then $\frac{dA}{dr}$ (or $A'(r)$) measures the rate of change of the area *A* with respect to its radius *r*. Then, as $A = \pi r^2 \Rightarrow \frac{dA}{dr} = 2\pi r$.

We note that a rate of change statement needs to have two quantities specified:

1. what quantity is changing, and
2. what it is changing with respect to.

However, often we use the expression '***the rate of change of ...***' with no reference to a second quantity. In such cases it can be assumed that we are referring to **the rate of change with respect to time**. So that "The rate of change of *N*", where *N* measures the population size of a herd of elephants, would be given by $\frac{dN}{dt}$, where *t* represents a unit of time.

CHAPTER 21

EXAMPLE 21.1

Find the rate of change of the volume of a sphere with respect to its radius.

Solution

If we consider a sphere as having a volume V units^3 for a corresponding radius r units, then we are looking for the expression $\frac{dV}{dr}$.

In order to determine this rate we need to have *V* as an expression in terms of *r*.

The volume of a sphere of radius *r* is given by $V = \frac{4}{3}\pi r^3$.

And so, we have that $\frac{dV}{dr} = \frac{4}{3}\pi \times 3r^2 = 4\pi r^2$.

Notice that a sphere of radius *r* has a surface area given by $S = 4\pi r^2$ and so, we have that $\frac{dV}{dr} = S$. That is, the rate of change of the volume of a sphere with respect to its radius is equal to its surface area! Are there other shapes for which this result is true?

21.1.2 RATES OF CHANGE AND THEIR SIGN

The following information relates to a quantity y as it varies with respect to the quantity x.

If we have a positive rate over some interval (a, b) then **y increases** over the interval (a, b).

i.e., if $\frac{dy}{dx} > 0$ on the interval (a, b), then the values of y increase as the values of x increase on the interval (a, b).

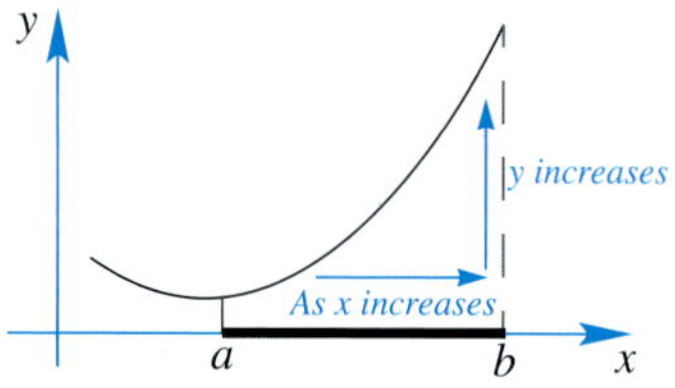

If we have a negative rate over some interval (a, b) then **y decreases** over the interval (a, b).

i.e., if $\frac{dy}{dx} < 0$ on the interval (a, b), then the values of y decrease as the values of x increase on the interval (a, b).

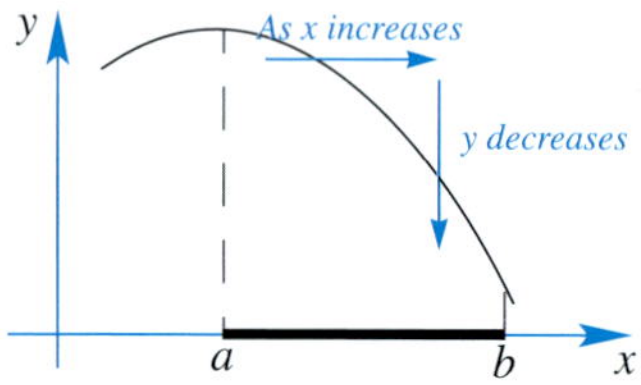

If we have a zero rate over some interval (a, b), then **y is constant** over the interval (a, b).

i.e., if $\frac{dy}{dx} = 0$ on the interval (a, b), then the values of y remain constant for all values of x on the interval (a, b).

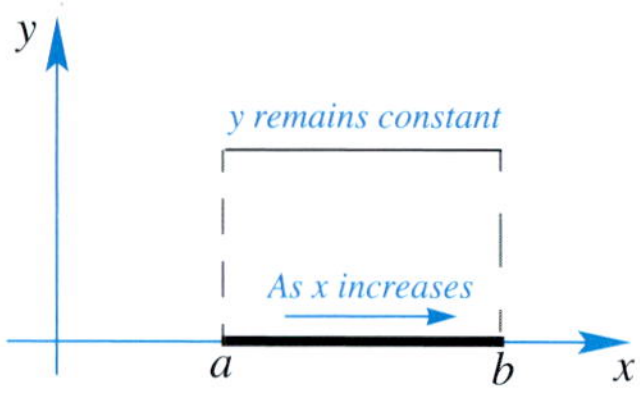

EXERCISES 21.1

1. Identify on the following graphs the values of x for which the function is
i. increasing ii. decreasing iii. constant

(a)

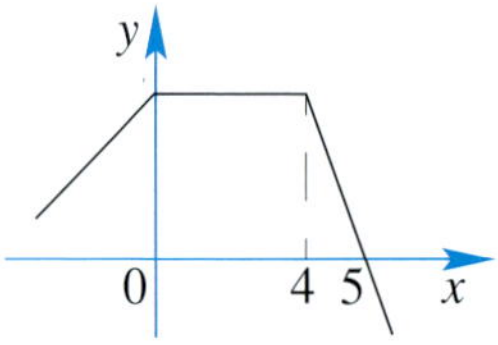

(b)

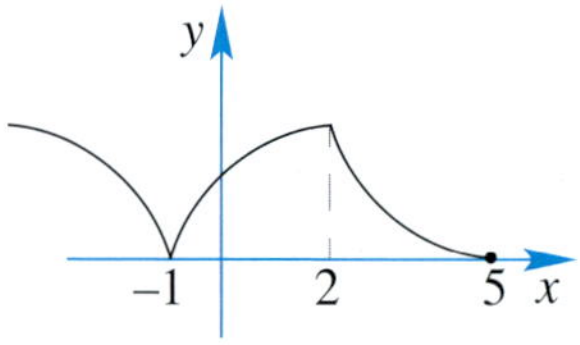

(c)

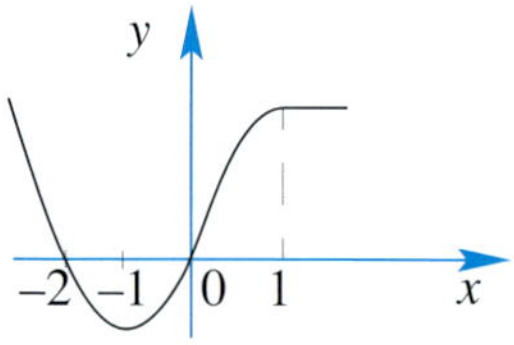

(d)

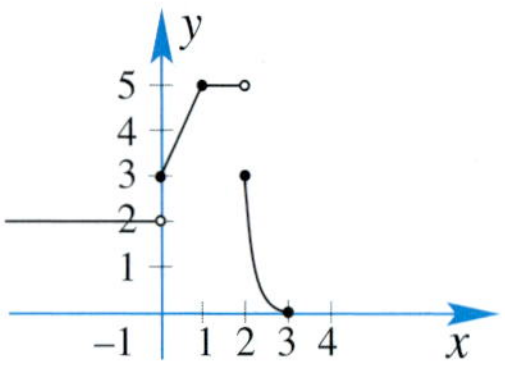

(e)

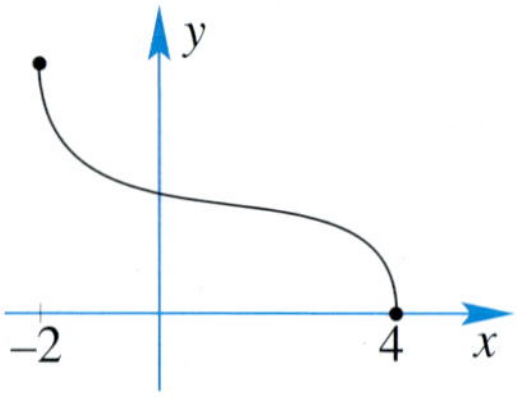

(f)

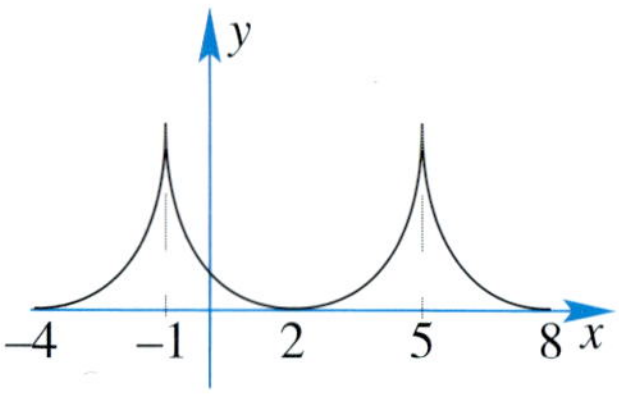

21.2 APPLIED RATES OF CHANGE

Applications of rates of change can be found in many situations. In this section we make use of a number of examples to highlight the diversity for which rates of change are useful.

EXAMPLE 21.2

The number, $N(t)$, of bacteria in a colony is given by the function

$t \mapsto 1.25t^2 + 20t + 980$, $t \geq 0$, where t is measured in hours.

(a) Find the rate of growth of the colony when $t = 4$.

(b) Show that $N(t)$ is an increasing function.

Solution

(a) Let $N(t) = 1.25t^2 + 20t + 980, t \geq 0$. The rate of growth is given by $\dfrac{dN}{dt}$ or $N'(t)$,

and so, $N'(t) = 2.5t + 20$. Then, when $t = 4$, $N'(4) = 2.5 \times 4 + 20 = 30$.

That is, the colony is growing at a rate of 30 bacteria per hour (when $t = 4$).

(b) To show that $N(t)$ is an increasing function, we need to show that its derivative is positive over its given domain. That is, we need to show that $N'(t) > 0$ (for $t \geq 0$).

Now, $N'(t) = 2.5t + 20$,

so that for $t \geq 0$, $N'(t) \geq 20$.

Therefore we have that

$N'(t) > 0$ for $t \geq 0$,

$\therefore$ $N(t)$ is an increasing function.

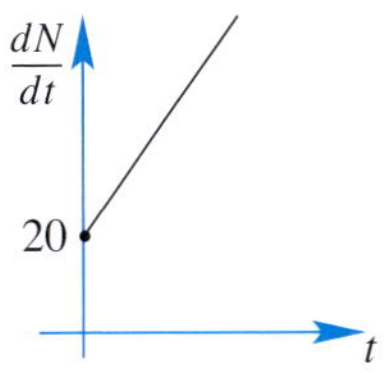

Graphical approach:

The graph of $N'(t)$ shows that $N'(t) > 0$ for $t \geq 0$, and so, $N(t)$ is an increasing function (for $t \geq 0$).

EXAMPLE 21.3

The population size of mosquitoes in a controlled laboratory experiment is modelled by the equation $P(t) = 250e^{0.09t}, t \geq 0$, where t is measured in days. Find the growth rate of the mosquito population after 5 days.

Solution

The rate of change of P with respect to t is given by $\dfrac{dP}{dt} = 250 \times 0.09e^{0.09t} = 22.5e^{0.09t}$

Therefore, when $t = 5$, $\dfrac{dP}{dt} = 22.5e^{0.09 \times 5} = 35.2870$.

That is, after five days, the number of mosquitoes is **increasing** at a rate of 35.29 mosquitoes per day (or 35 to the nearest whole number).

EXAMPLE 21.4

A hot metal bar is placed in an environment where the temperature remains constant at 34˚C. The temperature of the metal bar, T ˚C, is modelled by the equation

$$T = 34 + 90e^{-0.2t}, t \geq 0,$$

where t is measured in minutes.

Find the rate of change of the temperature of the bar 10 minutes after it is placed in this environment.

Solution

Given that $T = 34 + 90e^{-0.2t}$ then $\frac{dT}{dt} = 90 \times -0.2e^{-0.2t}$

$$= -18.0e^{-0.2t}.$$

Therefore, when $t = 10$, we have $\frac{dT}{dt} = -18e^{-2} \approx -2.436$.

That is, the rate of change of temperature is approximately –2.44˚C per minute.
This means that the temperature is **decreasing** at a rate of 2.44˚C per minute.

EXAMPLE 21.5

The concentration, C mol/litre, of a drug in the blood t minutes after being administered is given by the equation

$$C = bt2^{-kt}, 0 \le t \le 30.$$

How long will it be before the concentration level in the blood starts to decrease?

Solution

We need to determine $\frac{dC}{dt}$. Now, $C = bt2^{-kt}, 0 \le t \le 30$

$$\therefore \frac{dC}{dt} = \frac{d}{dt}(bt2^{-kt}) = b\left[\frac{d}{dt}(t) \times 2^{-kt} + t \times \frac{d}{dt}(2^{-kt})\right] \text{ [product rule]}$$

$$= b[2^{-kt} + t \times -k(\ln 2)2^{-kt}]$$

$$= b[1 - k(\ln 2)t]2^{-kt}$$

Then, $\frac{dC}{dt} = 0$ when $b[1 - k(\ln 2)t]2^{-kt} = 0 \Leftrightarrow [1 - k(\ln 2)t] = 0 \Leftrightarrow t = \frac{1}{k(\ln 2)}$.

Up until that time, the concentration of the drug in the blood is increasing and after that time the concentration of the drug in the blood is decreasing. A graph of this scenario best displays this.

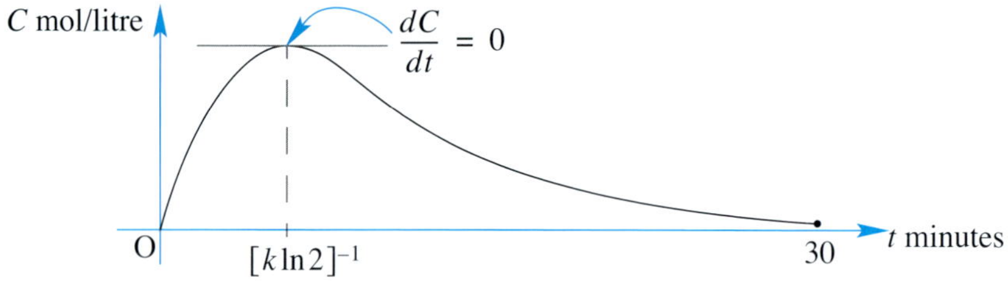

Application to economics

There are three important functions used in the area of manufacturing a commodity:

1. The **Cost** function, $C(x)$
2. The **Revenue** function, $R(x)$
3. The **Profit** function. $P(x)$,

where x is the number of items produced.

The rates of change with respect to x for each of these functions, $\frac{dC}{dx}$, $\frac{dR}{dx}$ and $\frac{dP}{dx}$ are referred to as the **Marginal Cost**, the **Marginal Revenue** and the **Marginal Profit** respectively.

EXAMPLE 21.6

The cost in dollars of manufacturing x units of a product is given by

$$C(x) = x^3 - 20x^2 + 300x + 1000\ ,\ x \geq 0.$$

If each item sells for \$300, find the marginal profit when 100 units are being manufactured.

Solution

First we need the profit equation: **Profit = Revenue – Cost**,

$$P(x) = R(x) - C(x)$$

where the revenue, $R(x) = 300x$ (as each item sells for \$300).

Therefore we have

$$P(x) = 300x - (x^3 - 20x^2 + 300x + 1000)$$
$$= -x^3 + 20x^2 - 1000$$

So the marginal profit,

$$\frac{dP}{dx} = -3x^2 + 40x\ .$$

Then, for $x = 100$, we have that

$$\frac{dP}{dx} = -3(100)^2 + 40(100)$$
$$= -26{,}000.$$

Application to kinematics

One important case of problems involving rates of change is kinematics. That is, the area concerned with the properties of motion, usually that of particles (or point masses). At this stage we will only provide two intial definitions and will leave a more detailed study of this topic until the next section.

Consider a particle having a **displacement**, $x(t)$ from a fixed point O at time t, then

its **instantaneous velocity**, $v(t) = \frac{dx}{dt} = x'(t)$ and

its **instantaneous acceleration**, $a(t) = \frac{dv}{dt} = x''(t)$

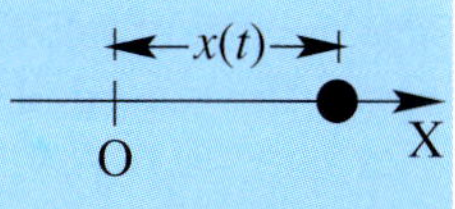

EXAMPLE 21.7

A body moves along the x-axis with displacement x cm where $x(t) = 10(t - \ln(t+1))\ ,\ t \geq 0$ and t is the time in seconds. Find the object's velocity and acceleration after being in motion for 1 second.

Solution

Using our definitions we have, $v(t) = \frac{dx}{dt} = 10\left(1 - \frac{1}{t+1}\right)\ \therefore v(1) = 10\left(1 - \frac{1}{2}\right) = 5$

The acceleration is given by $a = v'(t) = \frac{10}{(t+1)^2} \therefore a(1) = \frac{10}{4} = 2.5$

i.e., the object has a velocity of 5 m/s and an acceleration of 2.5 m/s^2 after 1 second.

EXERCISES 21.2

1. The number of deer, N, involved in a breeding program set up in a reserve has been modelled by the function $N = \frac{1}{10}t^2 + 4t + 50$, where t is measured in years since the program started. Find the rate at which the deer population is increasing 2 years after the program started.

2. The volume, $V\text{cm}^3$, of an object is given by the relation

$$V(t) = 0.5t^3 - 18t^2 + 216t + 200, t \geq 0 ,$$

where t is measured in days.

(a) Find the initial volume of the object.

(b) Find the rate of change of the volume when $t = 5$.

3. The number of organisms, N, present in a culture of bacteria, t hours from when observations were first made, is given by $N(t) = 3t^2 + 15t + 800, t \geq 0$.

(a) Find the rate of change of the number of organisms after 10 hours.

(b) Show that the number of organisms will always increase. Is this a realistic model?

4. The installation of a new electrical component into an existing product on an assembly line has an associated cost (per component), C dollars, that is closely approximated by the function

$$C = 8\sqrt[3]{(1.5t + 11.5)^4}, 0 \leq t \leq 7 ,$$

where t is measured in years.

(a) What will the initial cost per component be?

(b) Find the average rate at which the cost per component is changing over the first seven years.

(c) Find the rate of change in cost per component after 4 years of operation.

5. The number of sales, N thousand, made by a company is related to its advertising cost, x thousand dollars, by the equation

$$N = \frac{x^2}{9000}(300 - x), 0 \leq x \leq 200 .$$

Find the rate of change of the number of sales (*with respect to the advertising cost*) for (i) $x = 50$ (ii) $x = 100$ (iii) $x = 150$

6. The profit P made by an entertainment centre when selling x bags of lollies was modelled by the equation

$$P = 2.5x - \frac{1}{20000}x^2 - 3000, 0 \leq x \leq 50000 .$$

(a) For what values of x is the centre making a positive profit?

(b) For what values of x is the profit i. increasing?
ii. decreasing?

7. The revenue equation for a product is given by $R = 59xe^{-0.000015x}, x \geq 0$. Find the value of x for which its marginal revenue is zero. Hence, determine the maximum revenue.

8. The production strategy of a company manufacturing electrical components is based on the following models;

Demand equation: $x = 12000 - 30p$

where x is the number of components retailers are likely to buy per month at \$$p$ per component.

Cost equation: $C(x) = 50000 + 20x$

(a) Show that the revenue equation , $R(x)$, is given by $x \mapsto x\left(400 - \frac{x}{30}\right)$.

(b) Find the marginal revenue for a production level of 4000 units.

(c) Show that the profit, $P(x)$, is given by $x \mapsto 380x - \frac{1}{30}x^2 - 50000$.

(d) Determine the marginal profit when 5000 units are produced.

(e) For what values of x is the profit increasing?

9. Based on classical economic theory, the demand, $D(x)$, for a commodity in a free market decreases as the price x increases. The number of items, $D(x)$ of a particular product that people are willing to buy per week in a given town at a price \$$x$ is given by the function $$D(x) = \frac{40000}{x^2 + 12x + 20},\ 5 \le x \le 18 .$$

(a) Find the rate of change of demand with respect to price change.

(b) Find the rate of change of demand with respect to price change for $x = 10$.

10. During the early days of learning theory, a model to describe success based on the amount of practice undertaken by a person was given by

$$S(x) = 100\left(\frac{x+2}{x+32}\right)$$

where $S(x)$ measured the percentage of successful attempts after x practice attempts.

(a) Determine the rate at which learning occurs with respect to the percentage of practices undertaken.

(b) For what values of x is $S(x)$ i. increasing?
ii. decreasing?

11. The function $P(t) = 100 - 20\cos\left(\frac{5\pi}{3}t\right)$ approximates the blood pressure P millimetres of mercury at time t seconds for a person at rest.

(a) Find the rate of change in blood pressure when (i) $t = 0$ (ii) $t = \frac{1}{5}$

(b) Find the longest time period for which the blood pressure is increasing.

12. A particle moves in such a way that its displacement, x cm, t seconds after starting to move from a fixed point O is given by

$$x(t) = 10t + 20e^{-0.2t},\ t \ge 0 .$$

(a) Determine the rate of change of its displacement at time $t = 5$.

(b) Will this particle ever maintain a constant velocity for a period of time?

(c) Will this particle ever maintain a constant acceleration for a period of time?

13. Find the acceleration of a particle at $t = 1$, if its displacement at time t is $x = t^2e^{-t},\ t \ge 0$.

21.3 KINEMATICS

We have seen how rates can be used to define certain quantities, and briefly mentioned velocity and aceleration at the end of §21.2. We also saw in Chapter 18 how the average rate of change of displacement (with respect to time) provided a measure of the average velocity. In this section we provide a more in-depth study of kinematics. In particular, we concentrate on the relationship between displacement, velocity and acceleration.

21.3.1 MOTION ALONG A STRAIGHT LINE

We will be concentrating on motion along a straight line, where a body is free to move either forwards and backwards or up and down. When an object moves along a straight line OX, its position from a fixed point O, referred to as the origin, is determined by its displacement OP.

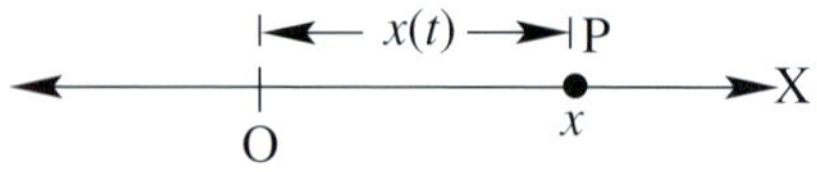

Usually the motion is along the x-axis and so its displacement is denoted by x, however, other representations, like s or h (usually for height) can be used. If the position of the body along the straight line can be specified at any instant, its displacement x can be written as a function of time, t units, so that $x = f(t)$ (or simply $x(t)$).

Representing a body's displacement–time relationship is often done by using a displacement–time graph. From such a graph, we can extend our initial definition of average velocity to that of instantaneous velocity:

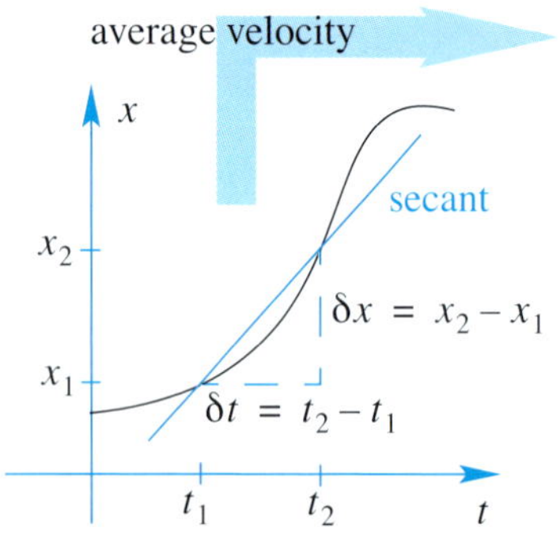

Average velocity, v_{ave}, is given by the gradient of the secant:

$$v_{ave} = \frac{\delta x}{\delta t} = \frac{x_2 - x_1}{t_2 - t_1}$$

Instantaneous velocity, v, occurs when $t_2 \to t_1$. i.e., $\delta t \to 0$ and is given by the gradient of the tangent:

$$v = \lim_{\delta t \to 0} \frac{\delta x}{\delta t} = \frac{dx}{dt}$$

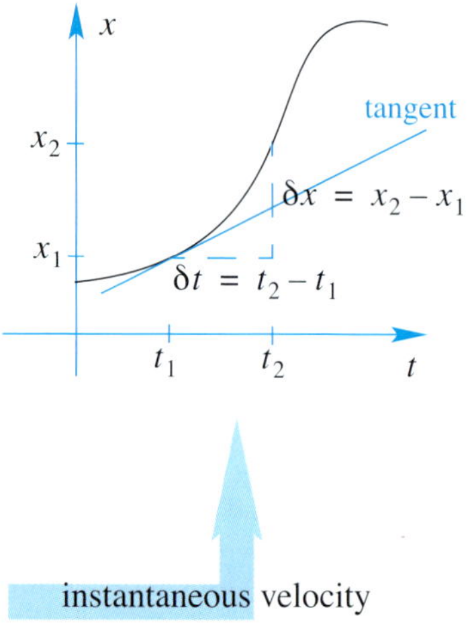

That is, the instantaneous velocity, v, at any time t is given by the first derivative of the displacement equation. Similarly, the average change in velocity gives the average acceleration, i.e., $a_{ave} = \frac{v_2 - v_1}{t_2 - t_1} = \frac{\delta v}{\delta t}$. Then, as $t_2 \to t_1$, i.e., $\delta t \to 0$, we have $a = \lim_{\delta t \to 0} \frac{\delta v}{\delta t} = \frac{dv}{dt}$ which gives the instantaneous acceleration at any time t. That is, the acceleration is given by the first derivative of the velocity or the second derivative of the displacement.

We must remember that we are considering motion of a particle along a straight line, even though the displacement-time graph is a two dimensional curve representing the motion. In short, the particle is not on what appears to be a 'roller-coaster ride' - it is always moving along a straight line. By projecting this curve onto the x-axis (as shown below or a parallel line) we can see where the particle is on this line at different times, which direction it is moving in, when and where its velocity is zero, and so on.

Projection of particle's position onto the vertical straight line

We follow the particle as it moves from A to B to C to D to E and finally to F by projecting the corresponding points on the displacement-time curve onto the vertical axis.

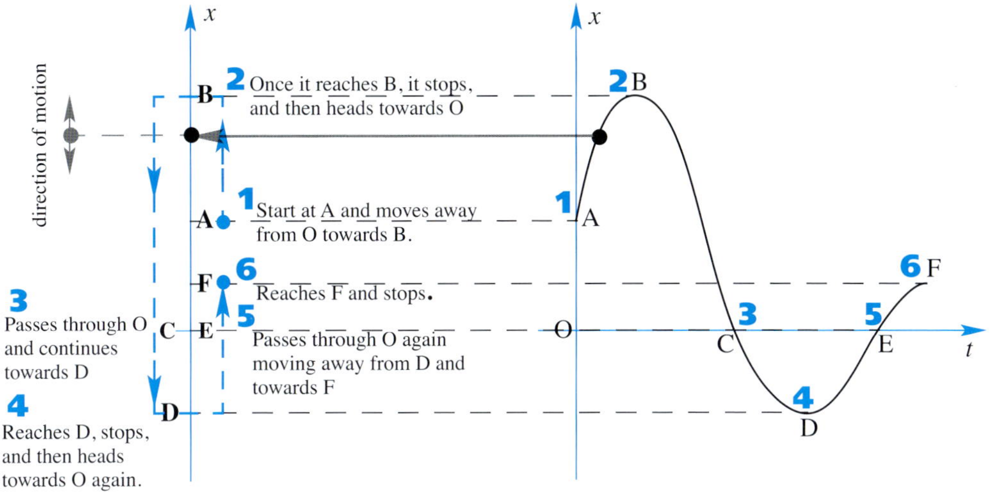

The relationships between displacement, velocity and acceleration are summarised as follows:

Velocity:

Velocity measures the rate of change of displacement.
So that if an object has a velocity v m/s and a displacement of x metres, then the relationship between v and x is given by

$$\frac{dx}{dt} = v$$

Acceleration:

Acceleration measures the rate of change of velocity.

So that if an object has an acceleration a m/s^2 and a velocity of v m/s, then the relationship between a and v is given by

$$\frac{dv}{dt} = a$$

Notice then that two other relationships can be derived:

1. From $a = \frac{dv}{dt}$, we have $a = \frac{dv}{dt} = \frac{d}{dt}(v) = \frac{d}{dt}\left(\frac{dx}{dt}\right) = \frac{d^2x}{dt^2}$.

2. From $a = \frac{dv}{dt}$, we have $a = \frac{dv}{dt} = \frac{dv}{dx}\frac{dx}{dt} = v\frac{dv}{dx}$.

EXAMPLE 21.8

A particle is projected vertically upwards in such a way that it experiences resistance so that its height, x metres above the ground after time t seconds, is given by

$$x = 220(1 - e^{-0.2t}) - 20t, t \geq 0 .$$

When will the particle reach its maximum height?

Solution

When the particle reaches its maximum height, it will momentarily stop, so that its velocity will be zero. That is $\frac{dx}{dt} = 0$. So, $x = 220(1 - e^{-0.2t}) - 20t \Rightarrow \frac{dx}{dt} = 220(0.2e^{-0.2t}) - 20$

$$= 44e^{-0.2t} - 20$$

So, $\frac{dx}{dt} = 0 \Rightarrow 44e^{-0.2t} - 20 = 0 \Leftrightarrow e^{-0.2t} = \frac{20}{44} \Leftrightarrow -0.2t = \log_e\left(\frac{5}{11}\right) \therefore t \approx 3.94$

Checking the nature of the stationary point (to verify the particle has in fact reached its maximum height):

For $t = 3.9$, $\frac{dx}{dt} = 0.17$ and for $t = 4.0$, $\frac{dx}{dt} = -0.23$.

Therefore there is a local maximum when $t = 3.94$. This means that the particle reaches its maximum height after approximately 3.94 seconds.

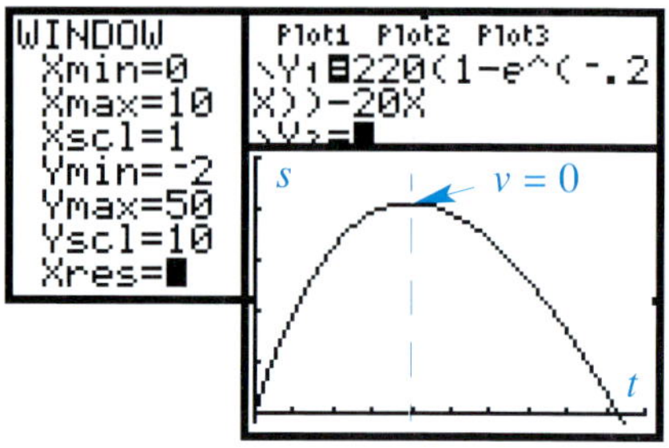

A sketch of the graph of the function x confirms this result.

When discussing velocity, acceleration and displacement as functions of t, it is also important to understand the significance of their signs. This can be summarised by the diagram below:

Displacement:

If $x > 0$, then P is to the right of O.
If $x < 0$, then P is to the left of O.
If $x = 0$, then P is at O.

$v = \frac{dx}{dt}$

Velocity:

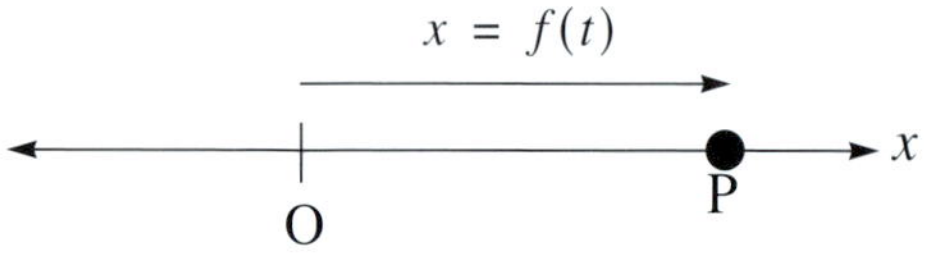

If $v > 0$, then P is moving to the right.
If $v < 0$, then P is moving to the left.
If $v = 0$, then P is stationary.

$a = \frac{dv}{dt}$

Acceleration:

If $a > 0$, then velocity of P is increasing.
If $a < 0$, then velocity of P is decreasing.
If $a = 0$, then velocity of P has a stationary. value

Note also that if both $v > 0$ and $a > 0$ (or $v < 0$ and $a < 0$) i.e., they have the **same sign**, then the

speed of the particle **is increasing**. Whereas, if they have **opposite signs**, then the **speed** of the particle **is decreasing**.

EXAMPLE 21.9

A particle moving in a straight line relative to some origin O, has its displacement, s metres, after being in motion for t seconds, governed by the equation

$$s = t^3 - 4t + 2 .$$

(a) Find the particle's velocity and acceleration at any time t seconds.
(b) Describe the particle's motion for $t \geq 0$.

Solution

(a) Given that $s = t^3 - 4t + 2$, then the velocity $v = \frac{ds}{dt} = 3t^2 - 4$.

Similarly, its acceleration $a = \frac{dv}{dt} = 6t$.

(b) We make use of the TI–83 to help us describe the motion of the particle. We first make use of the displacement–time graph:

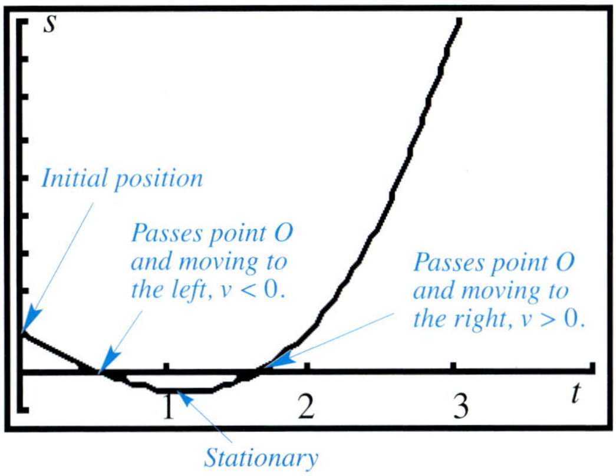

From the graph we see that the particle started 2 metres to the right of O and then started to move to the left, i.e., $v < 0$. After passing O, the particle kept moving left until it was stationary, at which point it started to move to the right. It passed the point O again and kept going.

Although this is a basic descriptive account of the particle's motion, more details can be included:

1. Solving for $s = 0$, will tell us when the particle reached the origin.
2. Solving for $v = 0$ will also tell us when the particle was stationary. Which we can then use to determine how far away from the origin it was before returning towards O.

1. Solving for $s = 0 \Rightarrow t^3 - 4t + 2 = 0$.
Making use of the **CALC** command on the TI–83, we have that $s = 0$ when $t = 0.54$ and $t = 1.68$.
That is, press **2nd CALC** then use the arrow keys to provide two suitable intervals over which a good guess can be made.

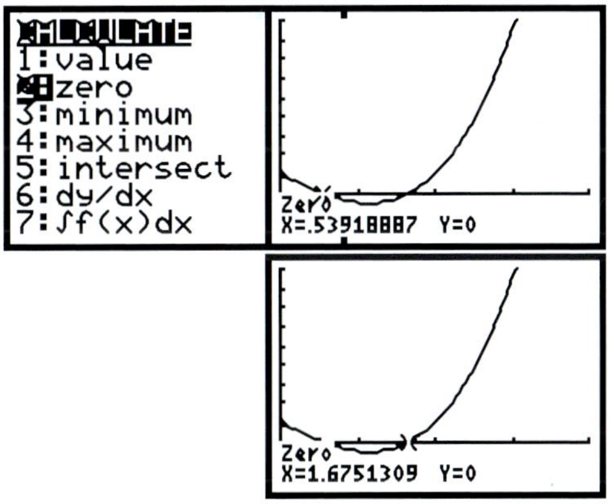

Also note that for $0 < t < 1$, the gradient of the displacement-time curve is negative, i.e., $v < 0$. Then, for $t > 1$, the gradient of the displacement-time curve is positive, i.e., $v > 0$.

Alternatively we could use the **Solve** option under the **MATH** menu: Either way, we can quote the result accurately to 2 decimal places.

```
solve(Y1,X,0)
        .5391888728
solve(Y1,X,1.5)
        1.675130871
```

2. Solving for $v = 0$, is slightly easier in this example as we end up solving a quadratic. That is, $v = 0 \Rightarrow 3t^2 - 4 = 0$, so that $t = \pm\sqrt{\frac{4}{3}} \approx \pm 1.15$.

However, $t \geq 0$, therefore we have that the particle becomes stationary after 1.15 seconds (and then moves to the right).

Note: For $0 \leq t < 1.15$, $v < 0$ but $a > 0$, meaning that the particle is slowing down.
However, for $t > 1.15$, $v > 0$ and $a > 0$, meaning that the particle is speeding up.

EXAMPLE 21.10

A particle's displacement, x metres, from an origin O is defined by the equation $x(t) = \sin 4t - \cos 4t$ at time t seconds.

(a) Show that the particle's acceleration is proportional to its displacement.
(b) What is the particle's maximum speed?
(c) Describe the particle's motion.

Solution

(a) We first find the particle's velocity: $v = \frac{dx}{dt} = 4\cos 4t + 4\sin 4t$

We then differentiate the velocity equation to obtain the acceleration equation:

$$a = \frac{dv}{dt} = -16\sin 4t + 16\cos 4t$$

Now, $a = -16\sin 4t + 16\cos 4t = -16(\sin 4t - \cos 4t) = -16x$. Then, as the acceleration is given in terms of x by the relation $a = -16x \Rightarrow a \propto x$, with a constant of proportionality of -16.

(b) The maximum speed will occur when the acceleration is zero.

That is, when $a = 0 \Leftrightarrow -16x = 0 \Leftrightarrow x = 0 \Leftrightarrow \sin 4t - \cos 4t = 0$.

Now, $\sin 4t - \cos 4t = 0 \Leftrightarrow \sin 4t = \cos 4t \Leftrightarrow \frac{\sin 4t}{\cos 4t} = 1 \Leftrightarrow \tan 4t = 1$.

$$\therefore 4t = \frac{\pi}{4}, \pi + \frac{\pi}{4}, 2\pi + \frac{\pi}{4}, \ldots$$

$$\Leftrightarrow t = \frac{\pi}{16}, \frac{5\pi}{16}, \frac{9\pi}{16}, \ldots$$

We can use a graph of the velocity function to help us determine which critical value(s) lead to a maximum and which ones lead to a minimum. However, it is important to understand the difference between speed and velocity. Velocity is a vector quantity and so has a direction associated with it, whereas speed is simply the magnitude of the velocity.

```
WINDOW
 Xmin=0
 Xmax=6.2831853...   (= 2π)
 Xscl=.78539816...   (= π/4)
 Ymin=-6
 Ymax=6
 Yscl=1
 Xres=■
```

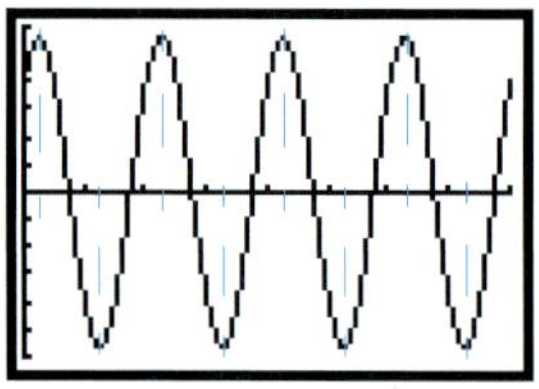

The maximum speed will then occur at all critical values (as we are only interested in the magnitude of the velocity).

So, at $t = \frac{\pi}{16}$, $v = 4\cos\left(\frac{4\pi}{16}\right) + 4\sin\left(\frac{4\pi}{16}\right) = 4 \times \frac{1}{\sqrt{2}} + 4 \times \frac{1}{\sqrt{2}} = \frac{8}{\sqrt{2}} = 4\sqrt{2}$ and so the maximum speed is $4\sqrt{2}$ m/s.

Had we selected $t = \frac{5\pi}{16}$, then $v = 4\cos\left(\frac{20\pi}{16}\right) + 4\sin\left(\frac{20\pi}{16}\right) = -4\sqrt{2}$ and so, although the velocity is $-4\sqrt{2}$ m/s, the speed is still $4\sqrt{2}$ m/s. Notice that in this case, the minimum speed is in fact 0 m/s!

(c) Using a displacement-time graph we see that the particle is moving back and forth about the origin O in a periodic manner. Particles describing this type of motion are said to undergo simple harmonic motion (S.H.M).

Its maximum distance from the origin is $\sqrt{2}$ m, which can be verified from the displacement-time equation as follows:

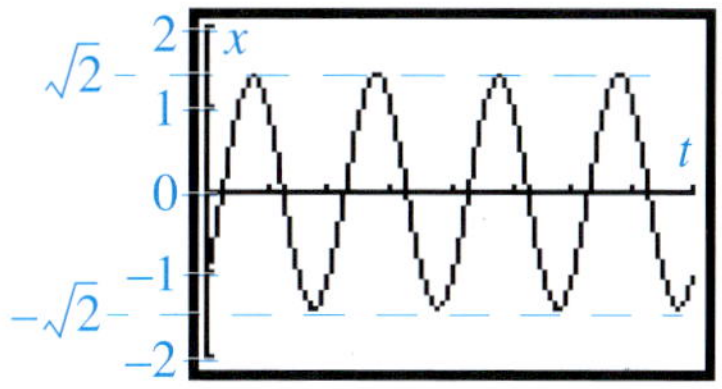

$$x(t) = \sin 4t - \cos 4t \Rightarrow \frac{dx}{dt} = 4\cos 4t + 4\sin 4t \therefore \frac{dx}{dt} = 0 \Leftrightarrow 4\cos 4t + 4\sin 4t = 0$$

$$\Leftrightarrow \sin 4t = -\cos 4t$$

$$\Leftrightarrow \tan 4t = -1$$

$$\text{i.e., } 4t = \frac{3\pi}{4}, \frac{7\pi}{4}, \ldots$$

$$\therefore t = \frac{3\pi}{16}, \frac{7\pi}{16}, \ldots$$

At $t = \frac{3\pi}{16}$, $x = \sin 4\left(\frac{3\pi}{16}\right) - \cos 4\left(\frac{3\pi}{16}\right) = \sin\left(\frac{3\pi}{4}\right) - \cos\left(\frac{3\pi}{4}\right) = \frac{1}{\sqrt{2}} - \left(-\frac{1}{\sqrt{2}}\right) = \frac{2}{\sqrt{2}} = \sqrt{2}$.

At $t = \frac{7\pi}{16}$, $x = \sin 4\left(\frac{7\pi}{16}\right) - \cos 4\left(\frac{7\pi}{16}\right) = \sin\left(\frac{7\pi}{4}\right) - \cos\left(\frac{7\pi}{4}\right) = -\frac{1}{\sqrt{2}} - \frac{1}{\sqrt{2}} = -\frac{2}{\sqrt{2}} = -\sqrt{2}$.

The particle's extreme displacements are $\pm\sqrt{2}$ m and it is undergoing S.H.M with a period of $\frac{\pi}{4}$.

EXERCISES 21.3

1. Find i. the velocity equation ii. the acceleration equation for the following displacement equations

(a) $x = \frac{t}{t-1}, t > 1$

(b) $x = (e^t - e^{-t})^2, t \geq 0$

(c) $x = at + be^{-t^2}, t \geq 0$

2. A particle moving in a straight line is such that its displacement in metres from some origin O at time t seconds is given by $s = t^2 + 4t - 5$, $t \geq 0$.

(a) What is the particle's velocity after travelling for 2 seconds?
(b) When is the particle at rest?
(c) i. Where is the particle (relative to O) when its motion begins?
ii. What is the particle's initial velocity?
(d) What is the furthest this particle gets from O during the first 5 seconds of motion?
(e) Sketch the displacement-time graph for this particle.

3. A particle moves in a straight line so that its displacement, s metres, from a fixed origin after t seconds, is given by the formula $s = t^3 - 2t^2 + t$, $t \geq 0$.

(a) What is the particle's initial velocity?
(b) How many times will the particle pass through the origin?
(c) When will the particle be stationary?
(d) What is the particle's acceleration after 4 seconds?

4. A particle moving in a straight line has its displacement governed by the equation

$$s = -2t^3 + 12t - 1 , t \geq 0,$$

where s is measured in metres and t in seconds.

(a) Find the particle's velocity and acceleration at time t.
(b) When will the particle come to rest?
(c) How often does the particle change its direction of motion?
(d) Sketch a displacement-time graph for this particle.

5. A particle has its displacement defined by $s = 4\sin t + 3\cos t$, $t \geq 0$.

(a) What is the particle's initial position?
(b) i. What is the particle's maximum displacement from the origin?
ii. What is the particle's maximum displacement from its initial position?
(c) What is the particle's maximum speed?
(d) Show that its accelaration, a, is given by $a = -s$.
(e) Describe the motion of this particle.

6. The displacement of an object from an origin O is given by $s = \frac{g}{k}\left(t + \frac{1}{k}e^{-kt} - \frac{1}{k}\right)$, $t \geq 0$.

(a) Show that its velocity is given by $v = \frac{g}{k}(1 - e^{-kt})$, $t \geq 0$.
(b) Show that its acceleration is given by $g - kv$.

7. A particle moving along a straight line has its displacement, x metres, from a fixed origin O, at time t seconds, governed by the equation

$$x = t^3 - 17t^2 + 80t - 100, t \geq 0$$

(a) Find the particle's initial position.
(b) How many times will the particle pass through the origin?
(c) Find the particle's initial
i. velocity.
ii. acceleration.
(d) Find the particle's maximum displacement from the origin for $0 \leq t \leq 10$.

8. A particle travelling in a straight line has its displacement from an origin O, given by the equation $x = 3\cos(2t - \pi) + 2, t \geq 0$.

(a) Find the maximum and minimum displacement.
(b) How long is it between successive times when the particle is at rest?
(c) Find the particle's acceleration i. in terms of t.
ii. in terms of its displacement.

9. A body is projected vertically upward so that its height, h metres, from ground level at any time t seconds is given by

$$h(t) = 15(1 - e^{-0.25t}) - 3t, t \geq 0.$$

(a) Find its position from ground level 1 second after being projected.
(b) Find the body's i. velocity, v ms^{-1} ii. acceleration, a ms^{-2}
at any time t seconds.
(c) What is the maximum height reached by the body?
(d) Find an expression for the acceleration in terms of its velocity.

10. A particle is moving along a straight line in such a way that its displacement, s metres, from a fixed point O at time t seconds is given by $s(t) = 2t^3 - 3t^2, t \geq 0$.
Find the times for which

(a) the particle's speed is increasing.
(b) the particle's velocity is increasing.
(c) the particle is at least 1 metre from O over the first 5 seconds.

11. Two particles, A and B, are moving along adjacent parallel straight tracks, so that their displacements are given by the equations $x_A = 4 - e^{0.3t}, t \geq 0$ and $x_B = 10te^{-t}, t \geq 0$.

(a) Describe the motion of each particle.
(b) How many times will the particles pass each other?
(c) At what times will the particles pass each other? Give your answer to two decimal places.
(d) Find an expression for the velocity of particle i. A ii. B
(e) Will the particles ever have the same speed at the same time?

12. A particle moving in a straight line has its displacement, x m, from a fixed point O given by $x(t) = \sqrt{t^2 + 4} - \sqrt{2}t, 0 \leq t \leq 10$, where t is measured in seconds.

(a) Find the particle's initial position.
(b) When will the particle first
i. reach the origin.
ii. come to rest.
(c) Find the particle's acceleration after 5 seconds.

13. A particle's position along a straight line is governed by the equation

$$x = 5\sin\left(2t - \frac{\pi}{2}\right) + 9, 0 \leq t \leq 3\pi$$

where x is measured in metres from a fixed origin O and time t is measured in seconds.

(a) Sketch the displacement-time graph for this particle.
(b) For what proportion of time will the particle be at least 12 metres from the origin?

21.4 APPLIED MAXIMA AND MINIMA PROBLEMS

21.4.1 MAXIMA–MINIMA PROBLEMS

The techniques and theories that have been developed in previous sections and chapters can be applied to practical problems in which the maximum or minimum value of a quantity is required.

Problems that require the use of this theory can be found in many real life situations: manufacturers wanting to minimize their costs, designers wanting to maximize the available space to be used (under specific constraints), farmers wanting to maximize the area of a paddock at a minimum cost, etc. These types of problems often require the construction of an appropriate function that models a particular situation, from which some optimum quantity can be derived or a critical value found for which this optimum quantity exists. We now consider a number of examples to highlight how differential calculus can be used to solve such problems.

EXAMPLE 21.11

Mirko had won at a raffle ticket draw. His prize was to be land. However he had to work hard for this plot of land. The rules specified that the winner was to be given 60 m of rope with which he could enclose a rectangular plot of land that ran along a straight river. Mirko thought for a while and then, with the rope in hand he mapped out his enclosure using the river bank as one side of his plot. What is the maximum land area that Mirko can enclose?

Solution

The object here is to determine the maximum possible area under the constraint that the total available rope measures 60 m. In order to solve problems such as these we need to introduce variables.

Let the dimensions of this plot of land be x m by y m and let the area enclosed by the rope be given by A m^2.
Then, we have that $A = xy$ – Eq. (1)
The constraint, that the rope is 60 m long, provides another equation: $2x + y = 60$ – Eq. (2)

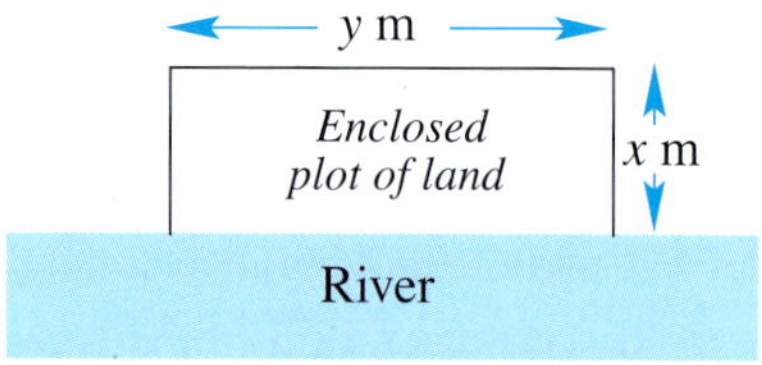

From (2) we have that $y = 60 - 2x$. Then, substituting into (1), we have $A = x(60 - 2x)$.

It is important to realise that we need to express the quantity we wish to optimise in terms of one variable and so, more often than not, we will need two equations: one that defines the quantity that we want to optimise and the second which provides a relationship between the variables that have been introduced.

We also need to determine the implied domain for our function. The physical restrictions are

$$1.\ x \geq 0 \text{ and } 2.\ y \geq 0 \Rightarrow 60 - 2x \geq 0 \Leftrightarrow x \leq 30$$

Combining these restrictions we have:

$$A(x) = x(60 - 2x), 0 \leq x \leq 30$$

We are now in a position to determine the stationary points of this function:

$$A(x) = 60x - 2x^2 \Rightarrow A'(x) = 60 - 4x \therefore A'(x) = 0 \Leftrightarrow 60 - 4x = 0 \Leftrightarrow x = 15$$

That is, we have a stationary point at $x = 15$.

As this value lies inside our domain we now check the nature of this stationary point, i.e., will this critical value provide a maximum or a minimum value of A?

Using the sign test (of the first derivative) we have:

$x = 15, A'(15) = 0$

$x > 15$ (say $x = 15.1$), $A'(15.1) = -0.4 < 0$.

$x < 15$ (say $x = 14.9$), $A'(14.9) = 0.4 > 0$.

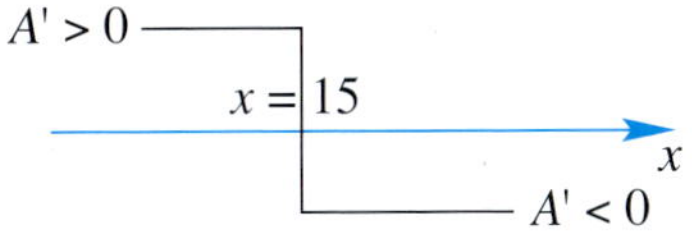

Using the sign of the first derivative we confirm that there exists a local maximum at $x = 15$.

So, when $x = 15$, $A = 15 \times (60 - 2 \times 15) = 450$.

That is, the maximum area that Mirko can enclose is 450 m^2.

A graph of the area function can verify this result.

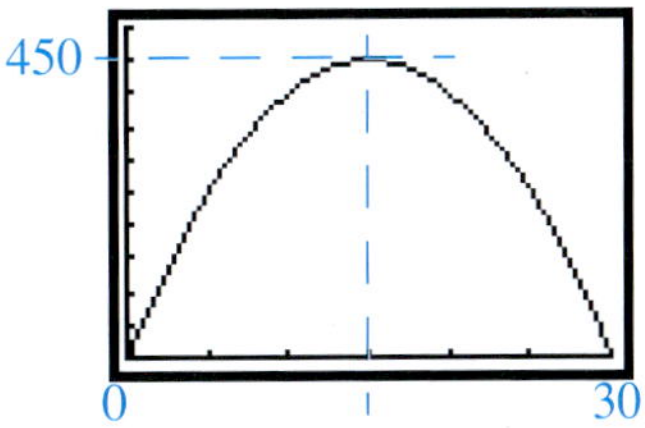

EXAMPLE 21.12

The points PQR form the corner of a house, where angle PQR is a right angle. Running parallel to these walls is a garden patch. There is only 20 metres of fencing available to create the enclosure PUTSRQ, where PU = RS $= x$ and PQ = QR = y.

a. Express ST in terms of x and y.

b. Find an expression for y in terms of x.

c. What area does this garden patch cover (give your answer in terms of x)?

d. Find the maximum area enclosed by this fence and the walls. Justify your answer.

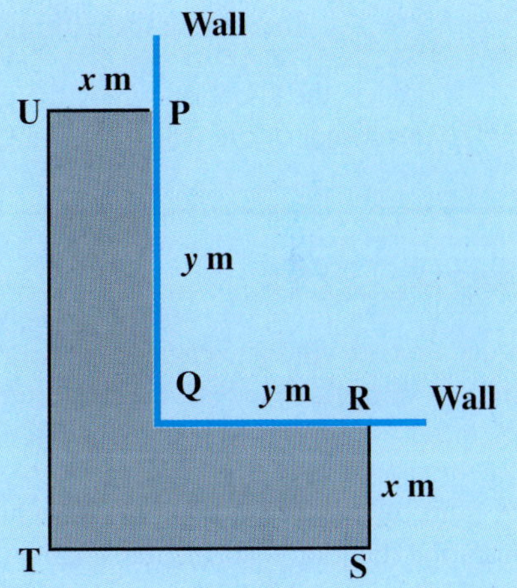

Solution

(a) ST = UP + QR $= x + y$.

(b) There is 20 m of fencing available, therefore, PU + UT + TS + SR = 20

That is, $2x + 2(x + y) = 20$

and so $y = 10 - 2x$ – Eq.1.

Note: As $y \geq 0 \Rightarrow 10 - 2x \geq 0 \Leftrightarrow x \leq 5$. We must also have that $x \geq 0$.

That is, there is a restriction on x, namely $0 \leq x \leq 5$.

(c) The required area, A m^2, is found by breaking the area into three sections. So that

$$A = xy + xy + x^2$$
$$= 2xy + x^2$$
$$= 2x(10 - 2x) + x^2, \text{ given that } y = 10 - 2x$$
$$= 20x - 3x^2$$

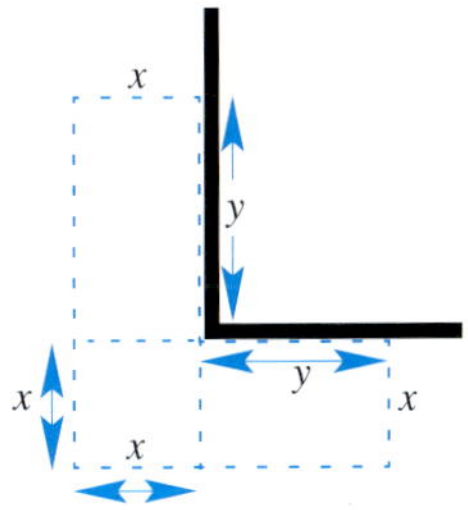

Therefore, we have that $A = 20x - 3x^2, 0 \leq x \leq 5$

(d) To find stationary points we first determine $\frac{dA}{dx} = 20 - 6x$

Then, solve $\frac{dA}{dx} = 0 \Leftrightarrow 20 - 6x = 0 \Leftrightarrow x = \frac{10}{3}$.

Using the sign test we see that a local maximum does occur at $x = \frac{10}{3}$.

(Note: For $x = 0, A = 0$ and for $x = 5, A = 25$.)

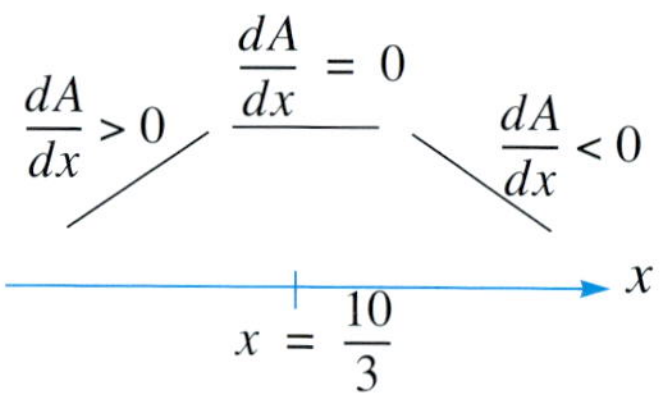

Substituting $x = \frac{10}{3}$ into the area equation, we have that the maximum area in square metres is $A = 20\left(\frac{10}{3}\right) - 3\left(\frac{10}{3}\right)^2 = \frac{100}{3}$.

Note that in this problem we did not make use of the function A to determine the domain. If we had only considered $A = 20x - 3x^2$ and then solved for $A \ge 0$, (i.e., make the assumption that an area is always positive) we would have obtained the inequalities $20x - 3x^2 \ge 0 \Leftrightarrow x(20 - 3x) \ge 0 \Leftrightarrow 0 \le x \le \frac{20}{3}$. However, it would not be possible for $x = \frac{20}{3}$, for even at the extreme where $y = 0$, the largest value x can have is 5. [Which depicts the situation where the garden patch is a square with a vertex making contact with the corner of the walls PQ and QR].

EXAMPLE 21.13

A rectangular sheet of cardboard measures 10 cm by 7 cm. Small squares of equal area are cut from each of the four corners of the sheet. The remaining sides are then folded to form an open box. Find the maximum volume that the box can have.

Solution

We start by drawing a diagram in the hope that it will help us introduce appropriate variables. From the diagram it seems reasonable that we should have variables to represent the dimensions of the box. This seems even more appropriate, given that we want to maximise the volume of the box, and in order to determine the volume we will need the width, length and height of the box.

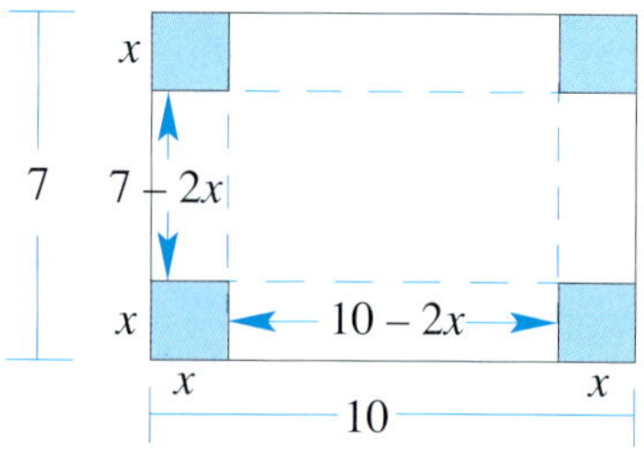

If we let the side length of the squares that are being cut out be x cm, then the length of the remaining sides will be of length $(7 - 2x)$ cm and $(10 - 2x)$ cm. We then complete our diagram by folding the sheet into an open box. Note that the length x cm on the original rectangular sheet becomes the height of the box.

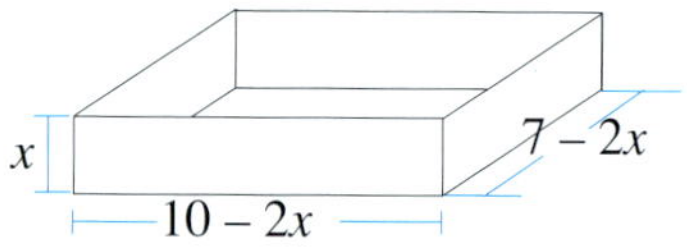

The next step is to decide what values of x we can use – i.e. we need to find the largest possible domain. This is usually done by considering the physical restrictions that are placed on the variables. In this case, all we know is that the lengths must be greater than (or equal to) zero.

Looking at each dimension we have:
1. $x \geq 0$
2. $10 - 2x \geq 0 \Leftrightarrow 10 \geq 2x \Leftrightarrow x \leq 5$
3. $7 - 2x \geq 0 \Leftrightarrow 7 \geq 2x \Leftrightarrow x \leq 3.5$

Therefore, the largest set of values for x that satisfy all three inequalities is $0 \leq x \leq 3.5$.

Now that we have the domain we need to find an expression for the volume, call it V cm^3.
As the volume of a box is given by length × width × height we have:

$$V(x) = (10 - 2x) \times (7 - 2x) \times x$$
$$= 4x^3 - 34x^2 + 70x$$

And so, our volume function is defined by

$$V(x) = 4x^3 - 34x^2 + 70x, \; 0 \leq x \leq 3.5$$

Next we search for turning points, i.e., we need to find those values of x for which $V'(x) = 0$.

So, $\quad V'(x) = 12x^2 - 68x + 70$

Then, $V'(x) = 0 \Leftrightarrow 12x^2 - 68x + 70 = 0 \Leftrightarrow x = \dfrac{68 \pm \sqrt{(-68)^2 - 4 \times 12 \times 70}}{2 \times 12}$

$$= \frac{68 \pm \sqrt{1264}}{24}$$

Then, as $0 \leq x \leq 3.5$ we have to choose $x = \dfrac{68 - 4\sqrt{79}}{24} \approx 1.3519$

Next, we determine the nature of this turning point. We do this by using the second derivative:

$$V''(x) = 24x - 68 \therefore V''(1.3519) = -35.553 < 0 \Rightarrow \text{ local maximum at } x = 1.3519$$

We also need to check the end points, i.e., we need to find $V(0) = 0$ and $V(3.5) = 0$. So, the local maximum is also the absolute maximum.
Therefore, the maximum volume is given by

$$V(1.3519) = 4(1.3519)^3 - 34(1.3519)^2 + 70(1.3519) = 42.3765 \approx 42.38 \text{ (to 2 d.p)}$$

Obviously, Example 21.24 required a fair amount of work, so we now look at how it could have been solved using a graphics calculator. Note however, the difficult part of the problem is finding the expression for the volume and the restrictions on x. The rest of the solution is fairly standard.

Using the TI-83, we first enter the equation $\mathbf{Y_1 = 4X\text{^}3 - 34X^2 + 70X}$, use the domain [0, 3.5] and then based on a sketch of the graph, we use the **fMax(** function from the **MATH** menu:

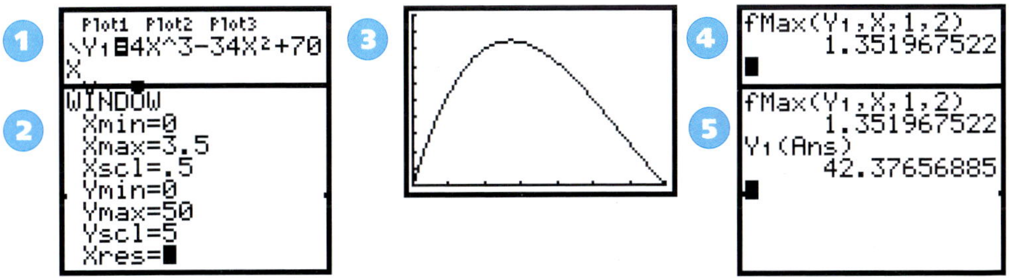

Five rather easy steps! However, often we are asked to find exact values, and a graphics calculator might not be able to provide such results, e.g., if the exact answer happens to be $\sqrt{20} - 4$, it might only be able to quote the answer as 0.47213... And so, to provide exact answers, we must develop our skills in solving such problems by classical means.

EXAMPLE 21.14

Find the point on the curve $y = x^2$ that is closest to the point (10,0).

Solution

We start by drawing a diagram:

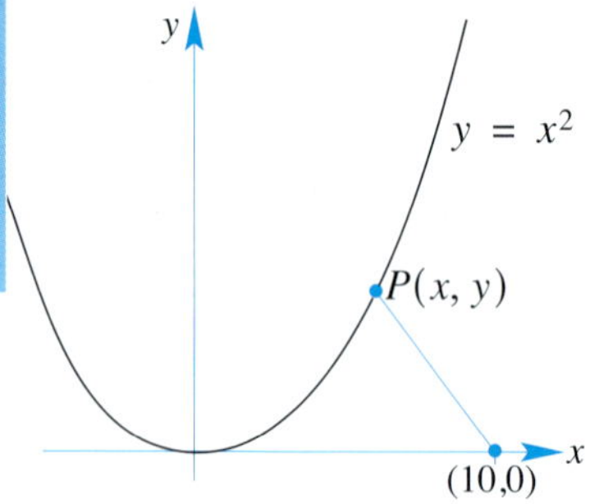

Consider some point $P(x, y)$ on the curve $y = x^2$.
By closest, we mean the shortest distance from the point P to the point (10,0).
In this case we use the formula for the distance between two points: $L = \sqrt{(x_2 - x_1)^2 + (y_2 - y_1)^2}$,
where L is the distance from (x_1, y_1) to (x_2, y_2).
Using the points $P(x, y)$ and (10,0) we have:

$$L = \sqrt{(x-10)^2 + (y-0)^2}$$
$$= \sqrt{(x-10)^2 + y^2}$$

However, we know that $y = x^2$ (as P lies on the curve). Therefore the distance from P to (10,0) (in terms of x) is given by

$$L = \sqrt{(x-10)^2 + (x^2)^2} = \sqrt{(x-10)^2 + x^4}$$
$$\therefore L = (x^2 - 20x + 100 + x^4)^{1/2}$$

Differentiating, we have: $\dfrac{dL}{dx} = \dfrac{1}{2}(2x - 20 + 4x^3)(x^2 - 20x + 100 + x^4)^{-1/2}$

$$= \frac{(x - 10 + 2x^3)}{\sqrt{x^2 - 20x + 100 + x^4}}$$

For stationary points we have $\dfrac{dL}{dx} = 0 \Rightarrow \dfrac{(x-10+2x^3)}{\sqrt{x^2 - 20x + 100 + x^4}} = 0 \Leftrightarrow x - 10 + 2x^3 = 0$.

We make use of a graphics calculator.
In this instance we sketch the graph of $y = x - 10 + 2x^3$ and then we can concentrate on where it crosses the x-axis:
Using the **TRACE** function on the TI–83,we see that the intercept occurs where $x \approx 1.6$.
Or, we could have simply used the **Solve** function on the TI–83

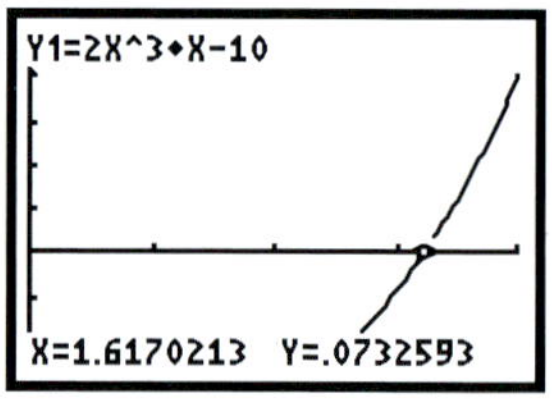

Notice:
Using the CALC function and Zero subcommand we obtain the value $x = 1.612$.

Now, use the sign of the first derivative to check the nature of the stationary point:

$$\left.\begin{array}{ll} x = 1.8 & \dfrac{dL}{dx} > 0 \\ x = 1.4 & \dfrac{dL}{dx} < 0 \end{array}\right\} \therefore \text{ we have a local maximum at } x = 1.6$$

Therefore, the point on the curve $y = x^2$ that is closest to the point (10,0) is (1.6, 2.56).
Note that the x–value is provided to only one decimal place. Using $x = 1.6126$, $y = 2.600$.

There is a very important observation to be made with Example 21.25. When we are looking to find the minimum value of $L = \sqrt{x^2 - 20x + 100 + x^4}$, rather than finding the critical value(s) of L, we could in fact find the critical values of L^2. The critical values of L and L^2 will be the same!
Note: The values of L and L^2 are not the same, only their critical values are.

We verify this now. $L = \sqrt{x^2 - 20x + 100 + x^4} \therefore L^2 = x^2 - 20x + 100 + x^4$.

Then, $$\frac{d}{dx}(L^2) = 2x - 20 + 4x^3$$

So that, $\frac{d}{dx}(L^2) = 0 \Leftrightarrow 2x - 20 + 4x^3 = 0 \Leftrightarrow x - 10x + 2x^3 = 0$, which will produce the same solutions as when we solved for $\frac{dL}{dx} = 0$. Using this aproach is much quicker (and neater).

We can show why this will always work for functions of the form $y = \sqrt{f(x)}$.

Using $y = \sqrt{f(x)}$:

$$\frac{dy}{dx} = \frac{1}{2} \times f'(x) \times \frac{1}{\sqrt{f(x)}} = \frac{f'(x)}{2\sqrt{f(x)}} \therefore \frac{dy}{dx} = 0 \Leftrightarrow \frac{f'(x)}{2\sqrt{f(x)}} \Leftrightarrow f'(x) = 0$$

Using $y^2 = f(x)$:

$$\frac{d}{dx}(y^2) = f'(x) \therefore 2y\frac{dy}{dx} = f'(x) \Leftrightarrow \frac{dy}{dx} = \frac{f'(x)}{2y} \therefore \frac{dy}{dx} = 0 \Leftrightarrow \frac{f'(x)}{2y} = 0 \Leftrightarrow f'(x) = 0.$$

That is, in both instances we solve the same equation, namely, $f'(x) = 0$.

EXAMPLE 21.15

In the lead up to the Christmas shopping period, a toy distributor has produced the following cost and revenue models for one of his toys.

Cost: $C(x) = 2.515x - 0.00015x^2, 0 \le x \le 6500$,

Revenue: $R(x) = 7.390x - 0.0009x^2, 0 \le x \le 6500$,

where x is the number of units produced.
What is the maximum profit that the distributor can hope for using these models?

Solution

The profit is found by determining the Cost – Revenue, so , letting \$ $P(x)$ denote the profit made for producing x units, we have

$$\begin{aligned} P(x) &= R(x) - C(x) \\ &= (7.390x - 0.0009x^2) - (2.515x - 0.00015x^2) \\ &= 4.875x - 0.00075x^2 \end{aligned}$$

To find the maximum value of $P(x)$, we first need to find the critical value(s) of $P(x)$:

Now, $P'(x) = 4.875 - 0.0015x \therefore P'(x) = 0 \Leftrightarrow 4.875 - 0.0015x = 0 \Leftrightarrow x = \frac{4.875}{0.0015} = 3250$.

So, when $x = 3250$, $P(3250) = 4.875(3250) - 0.00075(3250)^2 = 7921.875$.

Using the second derivative to check the nature of this turning point, we have:
$P''(x) = -0.0015$ and so, as $P''(x) < 0$ for all $x \in [0, 6500]$ we have a local maximum at (3250, 7921.875) and so, the maximum profit the distributor will make is \$7921.875 ≈ \$7922

Once again, we solve the previous example using a graphics calculator.
We enter the cost function as $\mathbf{Y_1}$, the revenue function as $\mathbf{Y_2}$ and the profit function as $\mathbf{Y_3 = Y_2 - Y_1}$. The rest follows:

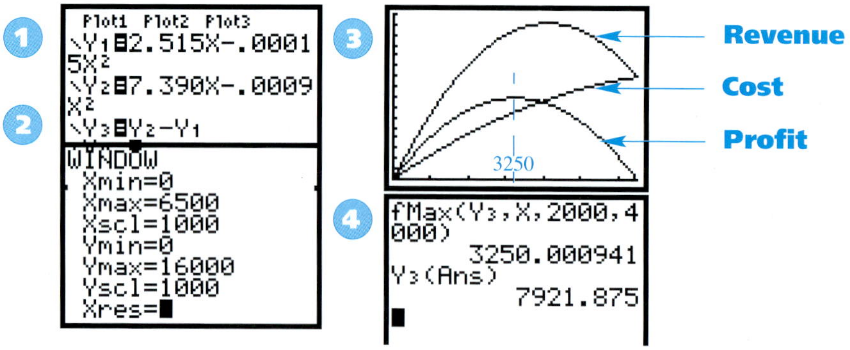

EXAMPLE 21.16

Two heavy industrial plants are located 12 kilometres apart. It is found that the concentration of particulate matter in parts per million in the polution created at a plant varies as the reciprocal of the square of the distance from the source. If plant 1 emits eight times the particulate matter of plant 2, the combined concentration, C, of particulate matter at any point between the two plants is found to be modelled by $C(x) = \frac{8}{x^2} + \frac{1}{(12-x)^2}$, $0.5 \le x \le 11.5$.

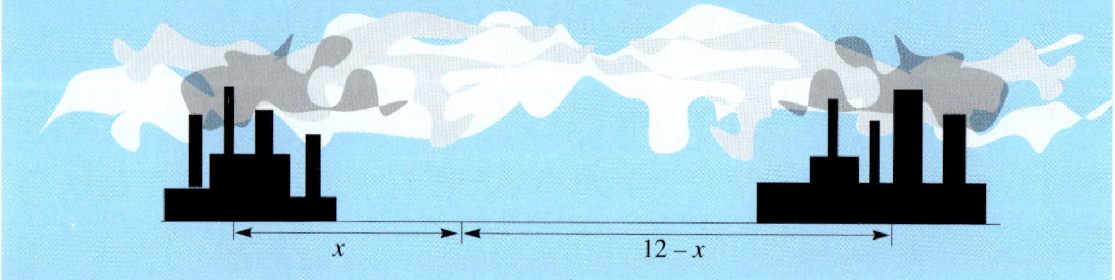

What is the minimum concentration of particulate matter that there can be between the two plants? How far from Plant 1 will this occur?

Solution

We need to determine where the stationary points occur, that is, we solve for $C'(x) = 0$.

Now, $C(x) = \frac{8}{x^2} + \frac{1}{(12-x)^2} = 8x^{-2} + (12-x)^{-2} \Rightarrow C'(x) = -16x^{-3} + 2(12-x)^{-3}$

Next, $C'(x) = 0 \Rightarrow -16x^{-3} + 2(12-x)^{-3} = 0 \Leftrightarrow -\frac{16}{x^3} + \frac{2}{(12-x)^3} = 0$

$$\Leftrightarrow \frac{2}{(12-x)^3} = \frac{16}{x^3}$$

$$\Leftrightarrow \frac{x^3}{(12-x)^3} = 8$$

Taking the cube root of both sides we have: $\frac{x}{(12-x)} = 2 \Leftrightarrow 24 - 2x = x$

$$\Leftrightarrow 3x = 24$$

$$\Leftrightarrow x = 8$$

That is, there is only one stationary point, at $x = 8$. Next we need to check the nature of this stationary point:

Sign of the first derivative:

$x = 7 \qquad C'(7) = -\dfrac{16}{343} + \dfrac{2}{125} = -0.03$

$x = 9 \qquad C'(9) = -\dfrac{16}{729} + \dfrac{2}{27} = 0.05$

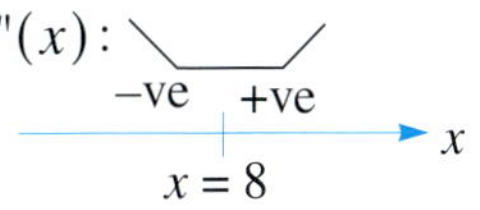

Therefore we have a local minimum at $x = 8$.
To find the minimum concentration we substitute $x = 8$ in the concentration equation.

Minimum concentration is $C(8) = \dfrac{8}{(8)^2} + \dfrac{1}{(12-8)^2} = \dfrac{8}{64} + \dfrac{1}{16} = \dfrac{3}{16}$.

So the minimum concentration is $\dfrac{3}{16}$ parts per million and occurs 8 km from Plant 1.

EXAMPLE 21.17

A non–uniform metal chain hangs between two walls. The height above ground level of this chain is given by the equation

$$h(x) = e^{-2x} + e^{x}, 0 \le x \le 2,$$

where x is the distance along the ground from the left wall.

How close to the ground will the chain get ?

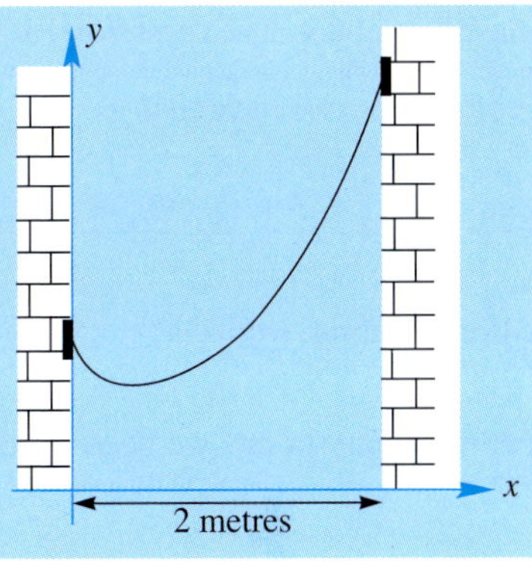

Solution

We start by finding the critical values: $h(x) = e^{-2x} + e^{x} \Rightarrow h'(x) = -2e^{-2x} + e^{x}$

Now,

$$\begin{aligned} h'(x) = 0 \Rightarrow -2e^{-2x} + e^{x} &= 0 \\ \Leftrightarrow 2e^{-2x} &= e^{x} \\ \Leftrightarrow \frac{2}{e^{2x}} &= e^{x} \\ \Leftrightarrow 2 &= e^{3x} \\ \Leftrightarrow 3x &= \log_e 2 \\ \Leftrightarrow x &= \frac{1}{3}\log_e 2 (\approx 0.23) \end{aligned}$$

Checking the nature of this stationary point (using the second derivative) we have:

$$h''(x) = 4e^{-2x} + e^{x} \therefore h''\left(\frac{1}{3}\log_e 2\right) = 4e^{-2 \times \frac{1}{3}\log_e 2} + e^{\frac{1}{3}\log_e 2} = 4 \times 2^{-2/3} + 2^{1/3} \approx 3.8 > 0$$

Therefore there is a local minimum point at $x = \frac{1}{3}\log_e 2$.

To find how close to the ground the chain gets we substitute this value into the equation:

$$h\left(\frac{1}{3}\log_e 2\right) = e^{-2 \times \frac{1}{3}\log_e 2} + e^{\frac{1}{3}\log_e 2} = 2^{-2/3} + 2^{1/3} \approx 1.89$$

Therefore the chain comes within 1.89 metres of the ground.

EXAMPLE 21.18

A box is to be constructed in such a way that it must have a fixed volume of 800 cm^3 and a square base. If the box is to be open ended at one end, find the dimensions of the box that will require the least amount of material.

Solution

Let the square base have side lengths x cm and the let the height be h cm.

Therefore the volume of the box is x^2h cm^3.

As the volume is 800 cm^3, we have $x^2h = 800$ – Eq. 1.

Next we denote the surface area of the box by S cm^2.

Therefore $S = x^2 + 4xh$ – Eq. 2.

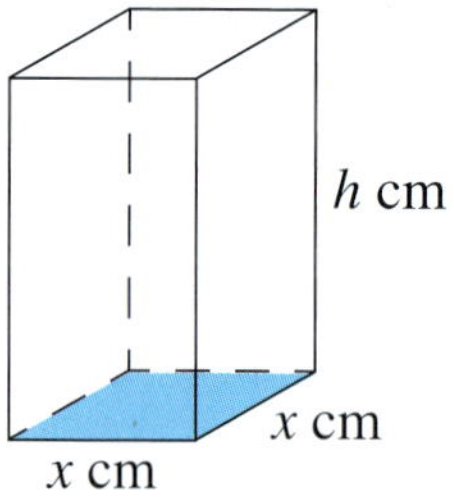

We wish to minimize S, therefore we need to find the critical point(s) of S. However, we must first obtain an expression for S in terms of x (exclusively).

From Eq. 1., we have that $h = \frac{800}{x^2}$ – Eq. 3.

Substituting Eq. 3. into Eq. 2. we have $S(x) = x^2 + 4x \times \frac{800}{x^2} = x^2 + \frac{3200}{x}$

Differentiating, we have, $S'(x) = 2x - \frac{3200}{x^2}$

For stationary points we need to solve $S'(x) = 0$, i.e., $2x - \frac{3200}{x^2} = 0 \Leftrightarrow 2x = \frac{3200}{x^2}$

$$\Leftrightarrow x^3 = 1600$$

$$\Leftrightarrow x = \sqrt[3]{1600}(\approx 11.70)$$

Next, we check the nature of the stationary point.

For $x = 12$, $S'(12) > 0$ and for $x = 11$, $S'(11) < 0$.

Therefore there is a local minimum at $x = 11.70$ and the amount of material required is least when $x = 11.70$ and $h = 5.85$.

EXAMPLE 21.19

A fisherman needs to go by the quickest route from pier A on one side of a strait to pier B on the other side of the strait. The fisherman can get across the strait at 8 km/hr and then run on land at 10 km/hr. Assuming that the banks of the strait are parallel straight lines, 12 km apart and pier B is 26 km along the bank from pier A, where should the fisherman aim to land on the other side of the strait to get to pier B as fast as possible?

Solution

In this problem we assume that the person is moving in a straight line. so, we start by drawing a diagram to represent this situation:

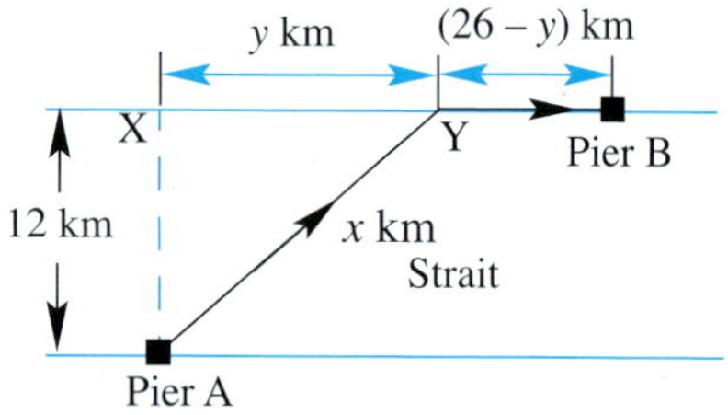

Next we introduce appropriate variables:

Let the fisherman land at Y, y km down the river after having travelled x km in the river from A to Y.

As we are looking for "*. . as fast as possible.*" we will need to derive an expression for the time taken to get from A to B. The journey is from A to Y and then from Y to B. Therefore, the total time taken is the time taken to get from A to Y plus the time taken to get from Y to B. Letting the total time taken be T hr, we have that $T = \frac{AY}{8} + \frac{YB}{10}$.

Now, AY = x and YB = 26 – y, therefore, $T = \frac{x}{8} + \frac{26-y}{10}$.

To proceed any further we will need to express y in terms of x. So, using Pythagoras's Theorem on ΔAXY we have: $x^2 = (AX)^2 + (XY)^2 \therefore x^2 = 12^2 + y^2 \therefore y = \sqrt{x^2 - 144}$ (as $y \geq 0$)

We now have a function for the time, T hr, in terms of one variable, x. Before we continue, we need to find the largest domain possible so that T is well defined. The smallest value of x occurs if the fisherman crosses straight to X, so that, $x \geq 12$. The largest value of x occurs if the fisherman heads straight to B and so, $x \leq \sqrt{12^2 + 26^2} = 2\sqrt{205}$.

We then have $$T(x) = \frac{x}{8} + \frac{26 - \sqrt{x^2 - 144}}{10}, \; 12 \leq x \leq 2\sqrt{205}.$$

To locate the critical values we solve $T'(x) = 0$:

$$T'(x) = \frac{1}{8} + \left[-\frac{1}{10} \times \frac{1}{2} \times 2x \times \frac{1}{\sqrt{x^2 - 144}}\right]$$

$$= \frac{1}{8} - \frac{x}{10\sqrt{x^2 - 144}}$$

Then, $T'(x) = 0 \Leftrightarrow \frac{1}{8} - \frac{x}{10\sqrt{x^2 - 144}} = 0 \Leftrightarrow \frac{x}{\sqrt{x^2 - 144}} = \frac{10}{8} \Leftrightarrow 4x = 5\sqrt{x^2 - 144}$

Solving for x we have:

$$16x^2 = 25(x^2 - 144) \Leftrightarrow 9x^2 = 25 \times 144 \Leftrightarrow x = \pm\sqrt{\frac{25 \times 144}{9}} = \pm\frac{5 \times 12}{3} = \pm 20$$

As $x \geq 12$ this means that the only valid solution is $x = 20$.

Using a sign diagram for $T'(x)$ we have:

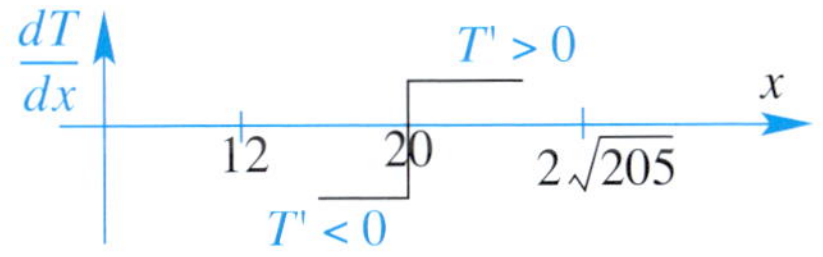

From the graph of $T(x)$ we can also see that the minimum time occurs when $x = 20$.

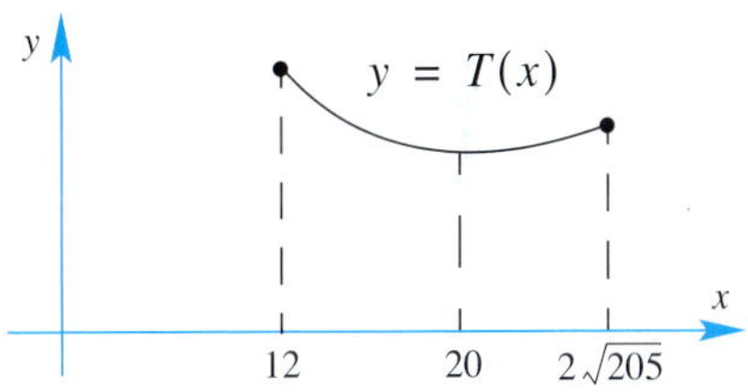

Therefore, $y = \sqrt{20^2 - 144} = 16$.

The fisherman should land 10 km from the pier.

EXAMPLE 21.20

A tourist decides to take a photo of a statue, 12 metres tall on a pillar that is also 12 metres tall. How far from the base of the pillar should the tourist stand so as to have the statue subtend the largest possible angle at their camera lens. You may assume that the camera is held at 1.70 metres above the horizontal and level ground.

Solution

We start by drawing a diagram and letting the statue subtend an angle θ at the lens of the camera. Let the camera be x m from the foot of the pillar.

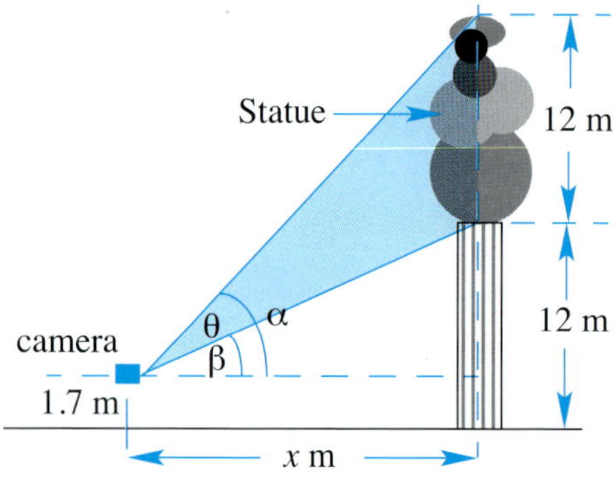

From the diagram we have that

$$\tan\alpha = \frac{24-1.7}{x} = \frac{22.3}{x} \quad \text{– Eq. (1)}$$

$$\tan\beta = \frac{12-1.7}{x} = \frac{10.3}{x} \quad \text{– Eq. (2)}$$

Then, to find an expression for θ we use the identity $\tan\theta = \tan(\alpha-\beta) = \dfrac{\tan\alpha - \tan\beta}{1+\tan\alpha\tan\beta}$.

Substituting the results of Eq. (1) and Eq. (2) we have:

$$\tan\theta = \frac{\frac{22.3}{x}-\frac{10.3}{x}}{1+\left(\frac{22.3}{x}\right)\left(\frac{10.3}{x}\right)} = \frac{\frac{12}{x}}{1+\frac{229.69}{x^2}} = \frac{\frac{12}{x}}{\frac{x^2+229.69}{x^2}} = \frac{12x}{x^2+229.69}$$

This provides us with an implicit function for θ in terms of x. So, differentiating both sides with respect to x, we have:

$$\frac{d}{dx}(\tan\theta) = \frac{d}{dx}\left[\frac{12x}{x^2+229.69}\right] \therefore \sec^2\theta \times \frac{d\theta}{dx} = \frac{12(x^2+229.69)-12x(2x)}{(x^2+229.69)^2}$$

$$= \frac{2756.28-12x^2}{(x^2+229.69)^2}$$

Solving for $\dfrac{d\theta}{dx} = 0$ we have $\dfrac{2756.28-12x^2}{(x^2+229.69)^2} = 0 \Leftrightarrow 12x^2 = 2756.28 \Leftrightarrow x^2 = 229.69$.

What happened to the "$\sec^2\theta$" term?

Now, as $x \geq 0$, $x = \sqrt{229.69} = 15.155\ldots \approx 15.16$.

That is, the tourist must stand 15.16 metres from the foot of the pillar in order to obtain the largest possible angle, which subtends the statue at their camera lens.

This time we will not use the sign test of the first or second derivative to check the nature of this stationary point but rather look at the relevant graph.

In this case, sketching the graph of $\tan\theta$ vs x as opposed to θ vs x will not make any difference in determining the critical value of θ. In this instance we deal with the graph of $\tan\theta$ vs x as this is the easier option. Otherwise we would need to sketch the graph of $\theta = \arctan\left(\dfrac{12x}{x^2+229.69}\right)$.

So, let's treat $\tan\theta$ as the variable T (say) and then sketch the graph of T vs x where

$$T = \frac{12x}{x^2+229.69},\ x \geq 0.$$

Why does this work? Would the same thing apply if we had been using the expression $\sin\theta = \dfrac{12x}{x^2+229.69}$ or $\cos\theta = \dfrac{12x}{x^2+229.69}$? What about $\ln(\theta) = \dfrac{12x}{x^2+229.69}$?

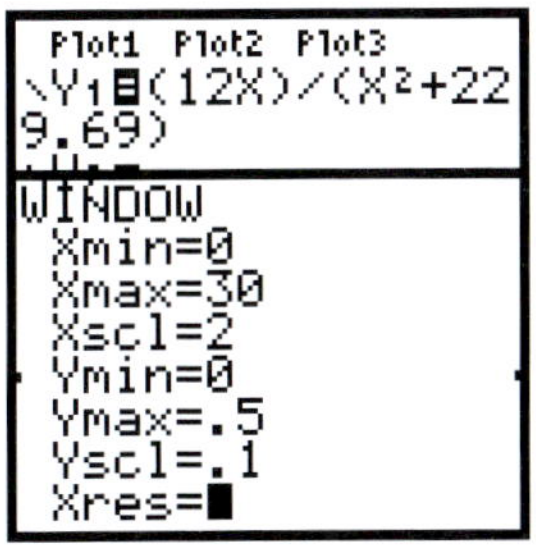

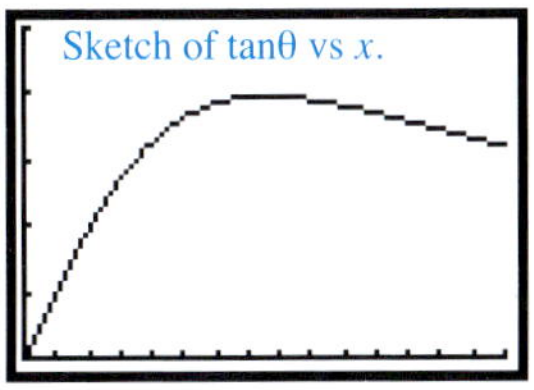

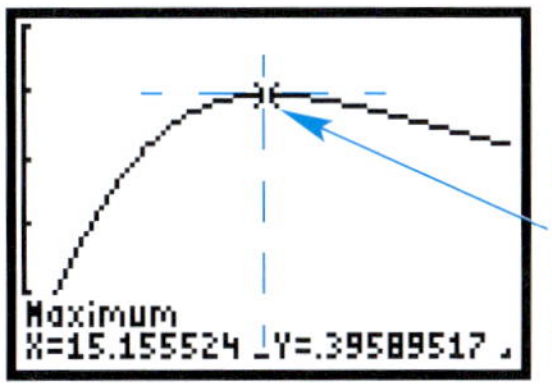

From the graph it is clear that the maximum occurs at x ≈ 15.16

Notice that when $x = 15.61$, $\tan\theta = 0.3958$, so that $\theta = \arctan(0.3958) = 21.59$. That is, the angle which the statue subtends at the camera lens when $x = 15.61$ is 21.59°. Had we sketched the graph of θ vs x, then the shape of the graph would be the same but the y-values would differ, in fact, at $x = 15.16$, we would have a θ-value of 0.3769 radians (which is equal to 21.59°).

21.4.2 END POINT MAXIMA/MINIMA PROBLEMS

So far we have dealt with problems that involved locating the stationary point via setting the first derivative to zero, solving and then verifying the nature of the stationary point (or turning point). We now consider, as we did in §20.4.3, problems where locating stationary points using our standard approach might only provide a local optimum whereas in fact, the optimum might exist at an endpoint. We now illustrate how to solve such questions.

EXAMPLE 21.21

A square sheet of cardboard of side length 12 cm has squares of side length x cm cut from each corner so that the remainder can be folded to form an open box. The four squares are then put together to form a hollow box (that is, open at both ends).

Show that the combined volume of the boxes, $V(x)$ cm^3, is given by

$$V(x) = 4x(6-x)^2 + x^3, 0 \le x \le 6 .$$

Hence find the maximum possible volume.

Solution

We start by drawing a diagram to help us visualise the situation.

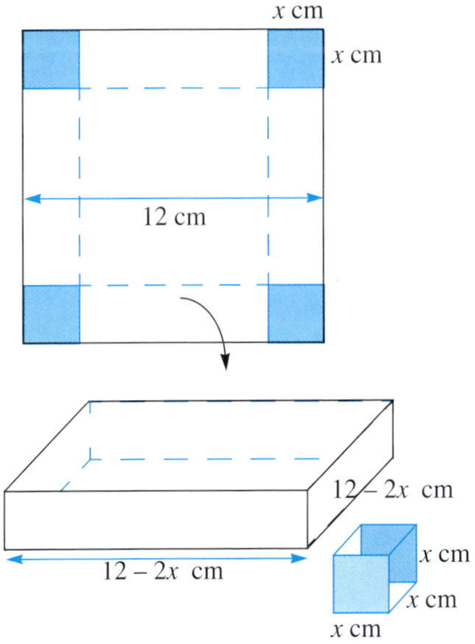

From the diagram, we have the combined volume

$$\begin{aligned} V(x) &= x(12-2x)(12-2x) + x^3 \\ &= x \times 2(6-x) \times 2(6-x) + x^3 \\ &= 4x(6-x)^2 + x^3 \end{aligned}$$

Now, as a length cannot be negative, we must have that $x \ge 0$ and $12 - 2x \ge 0$ so that $0 \le x \le 6$.

Therefore, $V(x) = 4x(6-x)^2 + x^3, 0 \le x \le 6$ as required.

Next, to determine the maximum volume, the normal course of action would be to find the stationary points, i.e., solve for $V'(x) = 0$.

Now, $V(x) = 4x(6-x)^2 + x^3 = 5x^3 - 48x^2 + 144x$

$$\therefore V'(x) = 15x^2 - 96x + 144$$

Setting $V'(x) = 0$ we have $15x^2 - 96x + 144 = 0 \Leftrightarrow x = 4$ or $x = 2.4$.

Using the sign of the first derivative we find that;

at $x = 4$ we have a local minimum,
at $x = 2.4$ we have a local maximum.

Now, $V(4) = 128$ and $V(2.4) = 138.24$, leading us to believe that the maximum volume would be 138.24 cm^3.

However, the graph of the function $V(x) = 4x(6-x)^2 + x^3, 0 \le x \le 6$, tells a different story:

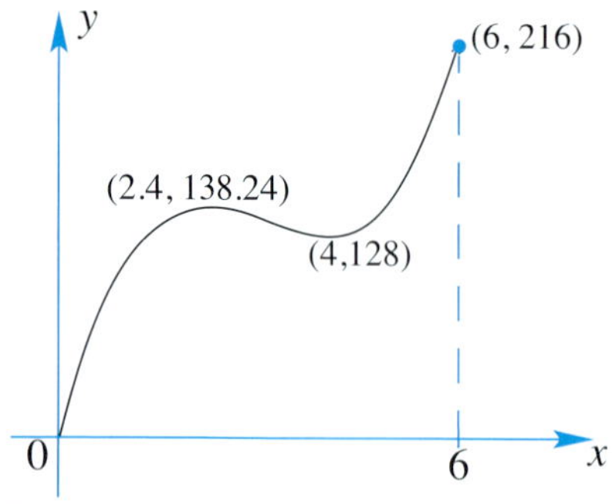

From the graph, it is clear that the maximum volume is in fact 216 cm^3 and not 138.24 cm^3 (as was thought). This is an example of an **end–point maximum**.

This highlights the importance of the restrictions placed on the domain. For example, if the problem had been to find the maximum value of $V(x)$, where

$$V(x) = 4x(6-x)^2 + x^3, 0 \le x \le 4,$$

then, the maximum value would have been 138.24 cm^3.

EXAMPLE 21.22

Two shops, A and B linked by a straight road, R1, are 4 kilometres apart. A second straight road, R2, bisects R1 at right angles. One kilometre along R2 from where it bisects R1 a third shop, C, can be found. A bus stop is to be placed on R2, somewhere between shop C and R1. Where should the bus stop be placed so that the sum of the direct distances from the shops to the bus stop is a minimum?

Solution

Let P be the location of the bus stop and x km the distance from R1 to P along R2. This means that $x \ge 0$ ($x = 0$ if on R1) but $x \le 1$ ($x = 1$ if at C)

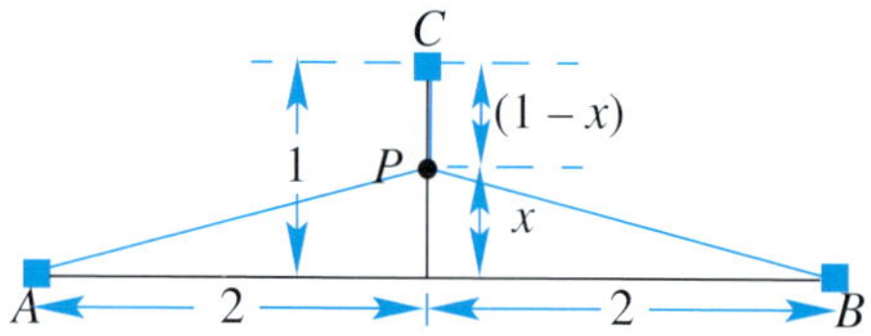

Let the sum of the distances be S km, then,

$$S = AP + PC + BP$$

Using Pythagoras's Theorem we have:

$$AP^2 = 2^2 + x^2 \therefore AP = \sqrt{4 + x^2}, \text{ as } AP \ge 0.$$

Similarly, $BP^2 = 2^2 + x^2 \therefore BP = \sqrt{4 + x^2}$, as $BP \ge 0$.

Therefore, $S = 2\sqrt{4 + x^2} + (1 - x), 0 \le x \le 1$.

Now, $\dfrac{dS}{dx} = 2 \times \dfrac{1}{2} \times 2x \times \dfrac{1}{\sqrt{4 + x^2}} + (-1) = \dfrac{2x}{\sqrt{4 + x^2}} - 1$

$$\therefore \frac{dS}{dx} = \frac{2x - \sqrt{4 + x^2}}{\sqrt{4 + x^2}}$$

And so, $\dfrac{dS}{dx} = 0 \Leftrightarrow \dfrac{2x - \sqrt{4 + x^2}}{\sqrt{4 + x^2}} = 0 \Leftrightarrow 2x - \sqrt{4 + x^2} = 0 \therefore 4x^2 = 4 + x^2$

That is, $3x^2 = 4 \therefore x = \pm\dfrac{2}{\sqrt{3}} = \pm\dfrac{2\sqrt{3}}{3} \approx \pm 1.1547$.

However, neither one of these critical values lies in the domain S (= [0, 1]). This means that we will need to look at an endpoint minimum.

When $x = 0$, $S = 2\sqrt{4} + (1) = 5$, giving us a maximum.

When $x = 1$, $S = 2\sqrt{4 + 1^2} + (1 - 1) = 2\sqrt{5}$.

Hence, the minimum value of S occurs at $x = 1$ and has a value of $2\sqrt{5}$ km (approx. 4.47 km). Therefore, bus stop should be placed at stop C.

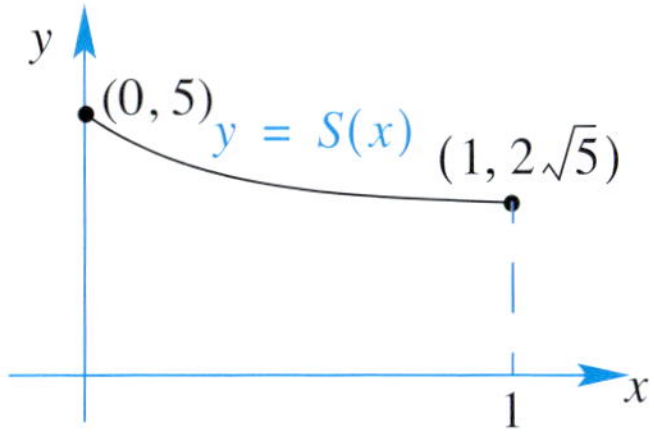

21.4.3 OPTIMISATION FOR INTEGER VALUED VARIABLES

When determining critical values, we do not always place importance on what they end up being. We simply quote an answer without further consideration to what the value is representing. For example, if we want to know how many globes need to be produced in order to maximise the profit, we might quote an answer of 2457.47... where in fact we know that the actual answer can only be an integer. We now consider such a problem – where the variables involved are integers.

EXAMPLE 21.23

Light and Co. purchase plastic components for their printing needs. They estimate that they will be using 800 such components every year. Each time they place an order for these plastic parts it costs them \$22. If they place an order x times per year, then, on average, they store $\frac{400}{x}$ units. Each stored unit costs \$8 to store. How often should they place an order so as to minimise their ordering and storage costs?

Solution

After placing x orders, the company's costs are made up of the ordering cost, \$ $22x$ and the storage cost \$ $\frac{400}{x} \times 8 = \$ \frac{3200}{x}$. We denote this cost by \$ $C(x)$, so that their expenses for ordering and storage are given by $C(x) = 22x + \frac{3200}{x}, x > 0$.

Next we determine the stationary point(s) for the cost function:

$$C'(x) = 22 - \frac{3200}{x^2}, x > 0.$$

Setting, $C'(x) = 0 \Rightarrow 22 - \frac{3200}{x^2} = 0 \Leftrightarrow 22 = \frac{3200}{x^2} \Leftrightarrow x^2 = \frac{3200}{22}$.

As $x > 0$, we choose the positive square root, so that the critical value occurs at $x = \sqrt{\frac{1600}{11}}$.

To check the nature of this stationary point, we use the second derivative: $C''(x) = \frac{6400}{x^3}$.

Then, at $x = \sqrt{\frac{1600}{11}}$, $C''(x) > 0$ and so we have a local minimum at $x = \sqrt{\frac{1600}{11}}$.

However, $x = \sqrt{\frac{1600}{11}} \approx 12.06\ldots$ does not provide us with an integer value, and the company can only make integer valued orders, for example, it cannot make 3.87 orders per year, it must make either 3 orders or 4 orders. So, we now need to find the minimum value of $C(x)$ for a positive integer – obviously, the positive integer we are looking for will lie in the vicinity of $x = 12.06$.

We now sketch the graph of $C(x)$ for integer values of x:

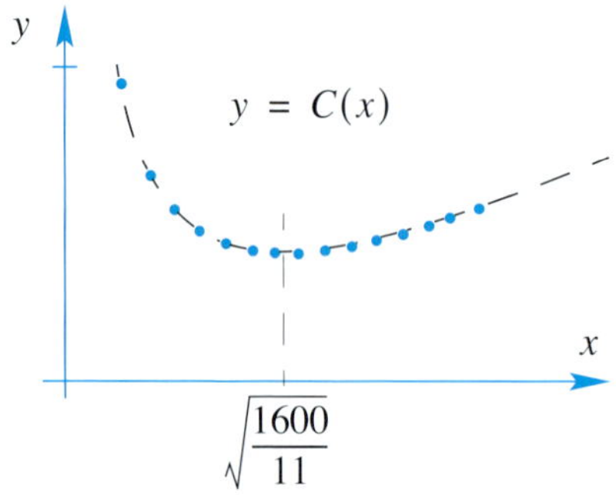

Using a table of values we have:

$C(11) = 532.91$

$C(12) = 530.67$

$C(13) = 532.15$,

and so, the minimum cost occurs when $x = 12$.

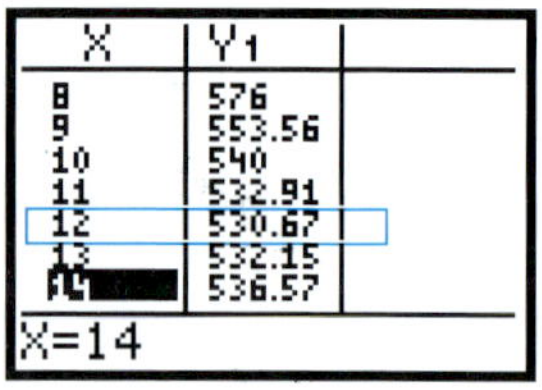

Therefore, they should place 12 orders per year in order to minimise their costs.

EXERCISES 21.4

1. A ball is thrown upwards and after t seconds reaches a height of h m above the ground. The height of the ball at time t is given by the equation $h = 19.6t - 4.9t^2 + 3$. What is the maximum height that the ball will reach from the ground?

2. The running cost, $\$C$ per kilometre for a new car is modelled by the equation $C = 20 + 0.2v^2 - 0.6v$, where v km/h is the average speed of the car during a trip.
(a) At what speed should the car be driven to minimise the running cost per km?
(b) What is the minimum running cost per km for this car?
(c) Comment on your answers.

3. The total revenue, $\$R$, that a company can expect after selling x units of its product – GIZMO – can be determined by the equation $R = -x^3 + 510x^2 + 72000x, x \geq 0$.
(a) How many units should the company produce to maximise their revenue?
(b) What is the maximum revenue to be made from the sales of GIZMOs?

4. A retailer has determined that the monthly costs, $\$C$, involved for ordering and storing x units of a product can be modelled by the function $C(x) = 2.5x + \frac{7500}{x}, 0 < x \leq 250$.
What is the minimum monthly cost that the retailer can expect? Note that x is an integer value.

5. The marketing department at DIBI Ltd. have determined that the demand, at $\$d$ per unit, for a product can be modelled by the equation, $d = \frac{80}{\sqrt{x}}$, where x is the number of units produced and sold. The total cost, $\$C$, of producing x items given by $C = 200 + 0.2x$.
What price will yield a maximum profit?

6. The cross-section of a small hill is modelled by the curve with equation

$$y = \frac{1}{8}x^2\sin\left(\frac{1}{2}x\right), 0 \le x \le 2\pi,$$

where x metres is the horizontal distance from the point O and y metres is the corresponding height. What is the maximum height of the hill?

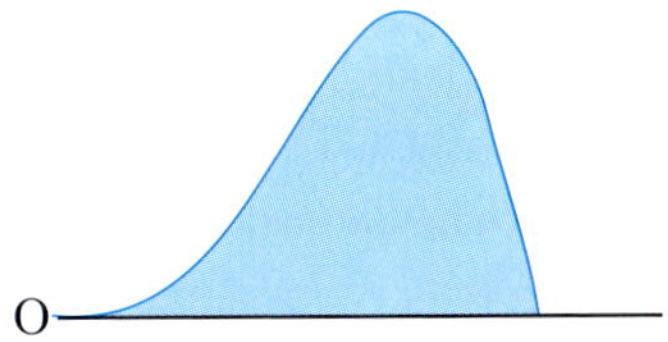

7. A 10 metres long sheet of tin of width 60 centimetres is to be bent to form an open gutter having a rectangular cross-section.
Find the maximum volume of water that this 10 metres stretch of guttering can carry.

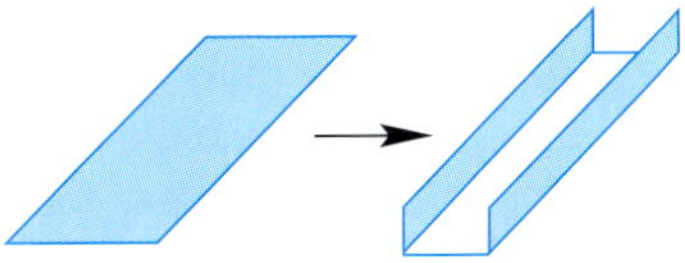

8. A 20 metre long piece of wire is bent into a rectangular shape. Find the dimensions of the rectangle that encloses the maximum area.

9. If $x + y = 8$ find the minimum value of $N = x^3 + y^3$.

10. A swimming pool is constructed as a rectangle and a semicircle of radius r m.

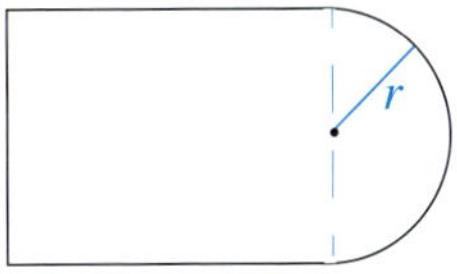

The perimeter of the pool is to be 50 metres. Find the value of r and the dimensions of the rectangular section of the pool if the surface area of the pool is to be a maximum.

11. A roof gutter is to be made from a long flat sheet of tin 21 cm wide by turning up sides of 7 cm so that it has a trapezoidal cross-section as shown in the diagram.
Find the value of θ that will maximise the carrying capacity of the gutter.

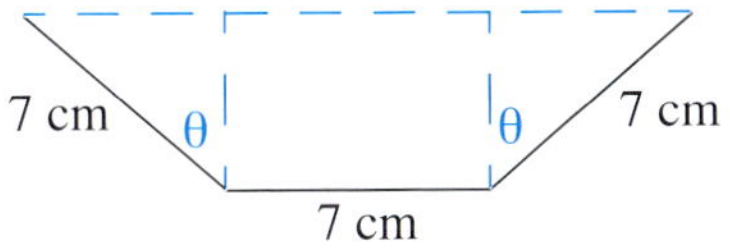

12. At the Happy Place amusement park, there is a roller coaster ride named 'The Not-So-Happy Ride'. A section of this ride has been created using a scaled version of the model given by the equation $y = \sin x + \frac{1}{2}\sin 2x, 0 \le x \le 2\pi$.

(a) Sketch the graph of this curve.
(b) What is the maximum drop that this ride provides?
(c) At what point(s) along the ride will a person come to the steepest part(s) of the ride?

13. A rectangle is cut from a circular disc of radius 18 metres. Find the maximum area of the rectangle.

14. Two real numbers x and y are such that $x + y = 21$. Find the value of x for which

(a) the product, xy, is a maximum.
(b) the product xy^3 is a maximum.

15. If $x + y = 12$, find the minimum value that $x^2 + y^2$ can have.

16. A farmer wishes to fence off a rectangular paddock using an existing stretch of a river as one side. The total length of wiring available is 100 m.
Let x m and y m denote the length and width of this rectangular paddock respectively, and let A m^2 denote its area.

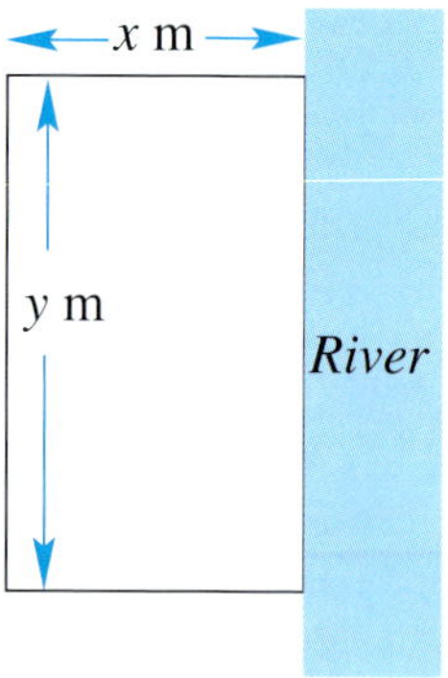

(a) Obtain an expression for y in terms of x.
(b) Find an expression for A in terms of x, stating any restrictions on x.
(c) Determine the dimensions which will maximize the area of the paddock.

17. A closed rectangular box with square ends is to be constructed in such a way that its total surface area is 400 cm^2. Let x cm be the side length of the ends and y cm its height.

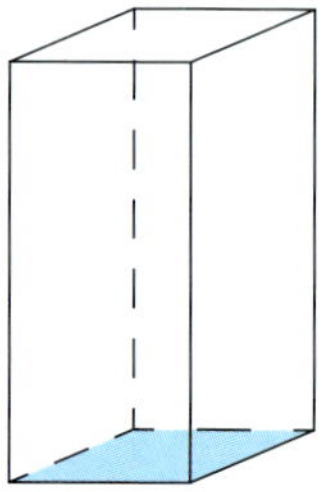

(a) Obtain an expression for y in terms of x, stating any restrictions on x.
(b) Find the largest possible volume of all such boxes.

18. A barrel is being filled with water in such a way that the volume of water, V ml, in the barrel after time t seconds is given by

$$V(t) = \frac{2}{3}\left(20t^2 - \frac{1}{6}t^3\right), 0 \le t \le 120 .$$

(a) Find the rate of flow into the barrel after 20 seconds.
(b) When will the rate of flow be greatest?
(c) Sketch the graph of $V(t), 0 \le t \le 120$.

19. The total cost, \$$C$, for the production of x items of a particular product is given by the linear relation $C = 600 + 20x, 0 \le x \le 100$, whilst its total revenue, \$$R$, is given by $R = x(100 - x), 0 \le x \le 100$.
(a) Sketch the graphs of the cost function and revenue function on the same set of axes.
(b) Determine the break–even points on your graph.
(c) For what values of x will the company be making a positive profit?
(d) Find an expression that gives the profit made in producing x items of the product and hence determine the maximum profit.

20. Find the points on the graph of $y = 9 - x^2$ that are closest to the point (0, 3).

21. A rectangle is bounded by the semi–circle with equation $y = \sqrt{25 - x^2}, -5 \le x \le 5$ and the x–axis. Find the dimensions of the rectangle having the largest area.

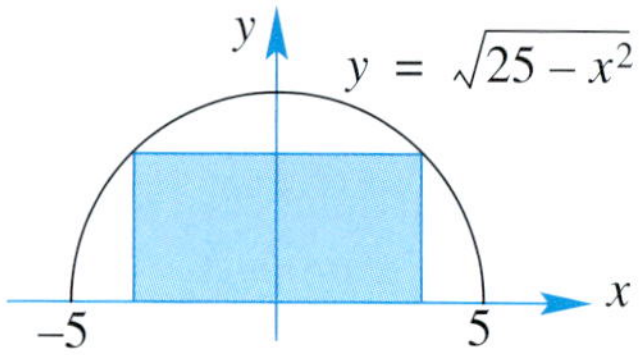

22. A rectangle is bounded by the positive x-axis the positive y-axis and the line with equation

$$y = \frac{2}{3}(8 - x)$$

Find the dimensions of the rectangle having the largest area.

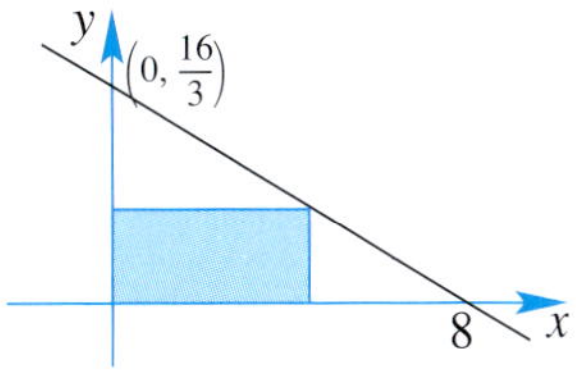

23. A certificate is to be printed on a page having an area of 340 cm^2. The margins at the top and bottom of the page are to be 2 cm and, on the sides, 1 cm.

(a) If the width of the page is x cm, show that the area, A cm^2 where printed material is to appear is given by $A = 348 - \frac{680}{x} - 4x$.

(b) Hence, determine the maximum area of print.

24. Find the minimum value of the sum of a positive integer and its reciprocal.

25. A closed circular cylinder is to have a surface area of 20π cm^2. Determine the dimensions of the cylinder which will have the largest volume.

26. A right circular cylinder of radius r cm and height h cm is to have a fixed volume of 30 cm^3.

(a) Show that the surface area, A cm^2 of such a cylinder is given by $A = 2\pi r\left(r + \frac{30}{\pi r^2}\right)$.

(b) Determine the value of r that will yield the minimum surface area.

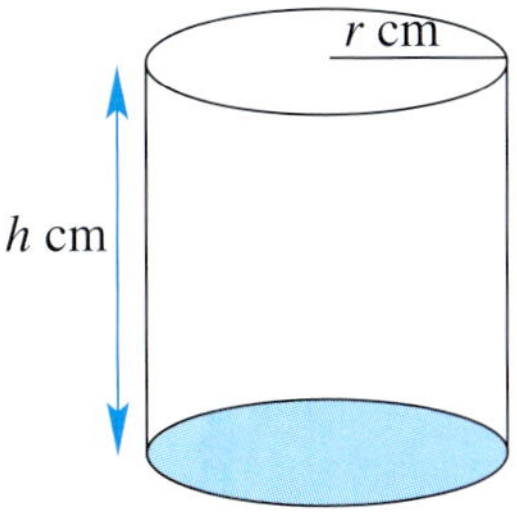

27. A rectangular container is made by cutting out squares from the corners of a 25 cm by 40 cm rectangular sheet of metal and folding the remaining sheet to form the container.

(a) If the squares that are cut out are x cm in length, show that the volume, V cm^3 of the container is given by

$$V = x(25 - 2x)(40 - 2x), 0 < x < \frac{25}{2}$$

(b) What size squares must be cut out in order to maximize the volume of the container?

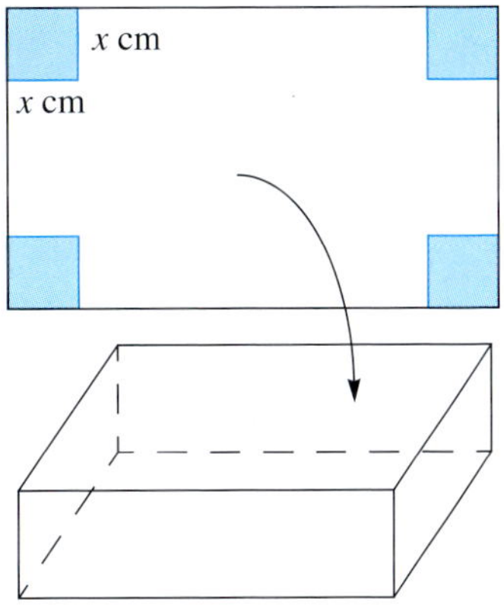

28. A right–circular cone of radius r cm contains a sphere of radius 12 cm.

(a) If the height of the cone is h cm, express h in terms of r.

(b) If V cm^3 denotes the volume of the cone, find an expression for V in terms of r.

(c) Find the dimensions of the cone with the smallest volume.

29. For a closed cylinder of radius r cm and height h cm, find the ratio $r : h$ which will produce the smallest surface area for a fixed volume.

30. Find the coordinates of the point on the curve with equation $y = x^2 + 1$ that is closest to the point $(2, 0)$.

31. A piece of wire is bent in the form of a sector of a circle of radius r metres containing an angle θ^c. The total length of the wire is 10 metres.

(a) Show that $\theta = \frac{2}{r}(5 - r)$.

(b) Find the value of r for which the area of this sector is a maximum.

32. Find the altitude of the cylinder with the largest volume that can be inscribed in a right circular cone.

33. A window is designed so that it has an equilateral triangle mounted on a rectangular base. If the perimeter totals 7 metres, what is the maximum area, that will allow the maximum amount of light to pass through the window?

34. A cone is formed by joining the two straight edges of a sector from a circle of radius r. If the angle contained by the two straight edges is α^c find the value of α^c which makes the volume of the cone a maximum.

35. Two houses in a new housing estate need to have cable connected to a distribution box, P, located somewhere along a straight path XY. House A is located a m from the path while house B is located b m from the path. The cost of installing the cable is \$ $6(AP + BP)$. Where should the box be placed so as to minimise the cost involved?

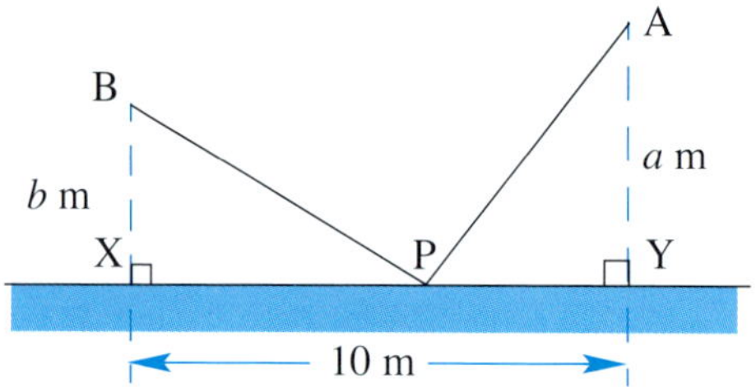

36. Felicity and Jane start walking at the same time towards an intersection of two roads that meet at right angles.

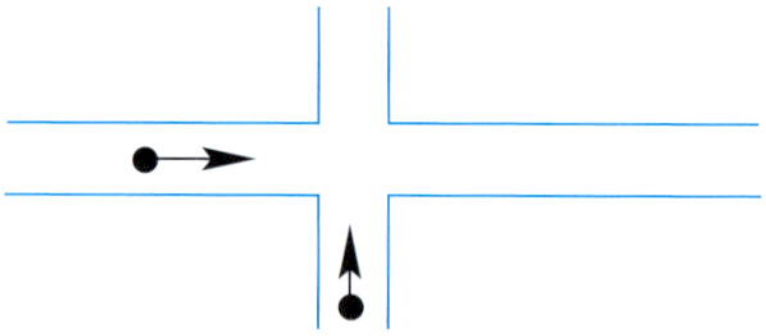

Felicity starts at 9 km from the intersection while Jane starts at 13 km from the intersection. Their speeds are 4 km/h and 3 km/h respectively. What is the closest that Felicity and Jane will get?

37. For an open cylinder of radius r cm and height h cm, find the ratio $r : h$ which will produce the smallest surface area for a fixed volume.

38. Find the height of a right–circular cone which can be inscribed in a sphere of radius 1 m, if this cone is to have the largest possible volume.

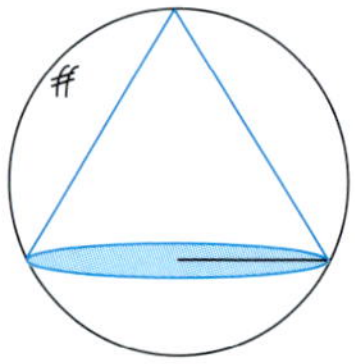

39. A piece of wire 30 cm long is cut into 2 pieces. One of the pieces is bent into a square while the other is bent into a circle. Find the ratio of the side length of the square to the radius of the circle which provides the smallest area sum.

40. A cylindrical tin with no lid is to be be made from a sheet of metal measuring 100 cm^2 . Given that the radius of the base of the tin is r cm show that its volume, $V \text{ cm}^3$, is given by

$$V = \frac{1}{2}(100r - \pi r^3)$$

Determine the value of r that will give the greatest volume.

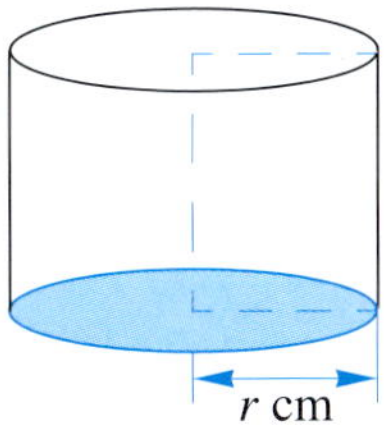

41. The last leg of a triathlon requires that you get from a point O, 2km from the nearest point P on a straight beach to a point Q, 3km down the coast.

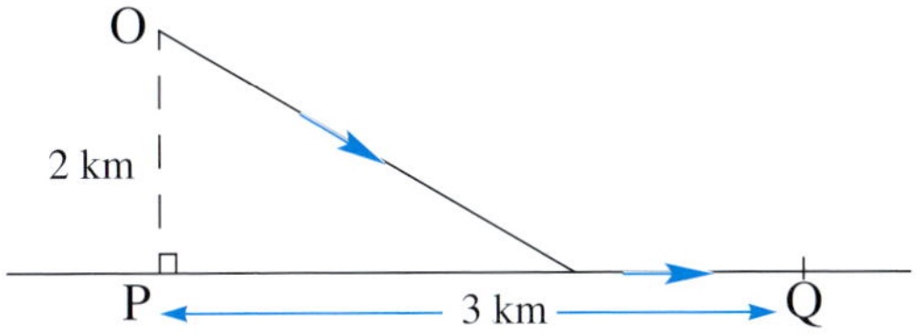

You may swim to any point on the beach and then run the rest of the way to point Q. If you can swim at a rate of 2 km/hr and run at 5 km/hr, where should you land on the beach so that you reach point Q in the least possible time?

42. A right circular cylinder is inscribed in a sphere of radius 6 cm. If α is the angle subtended at the centre of the sphere by the radius of the circular base of the cylinder,

(a) show that the curved surface area, $S \text{ cm}^2$, of the cylinder is given by $72\pi\sin 2\alpha$;

(b) find the radius and height of the cylinder having the largest curved surface area.

43. A person in a boat 4 km from the nearest point P on a straight beach, wishes to get to a point Q 8 km along the beach from P.

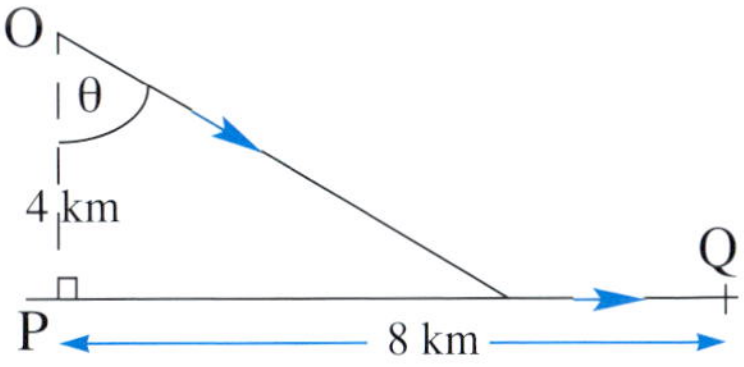

The person rows in a straight line to some point on the beach at a constant rate of 5 km/h.
Once on the beach the person walks towards Q at a steady rate of 6 km/hr.

(a) Show that the total time in hours taken for the trip is

$$T(\theta) = \frac{4\sec\theta}{5} + \frac{8 - 4\tan\theta}{6}, \theta \in \left[0, \frac{\pi}{2}\right) \text{ and } \tan\theta \le 2$$

(b) Where should the person land so that the trip takes the least amount of time?

44. A mast l metres tall erected on a building k metres tall, subtends an angle θ at a point on the ground x metres from the base.

(a) Find an expression for $\tan\theta$ in terms of x.

(b) Find the value of x that maximizes the value of θ .

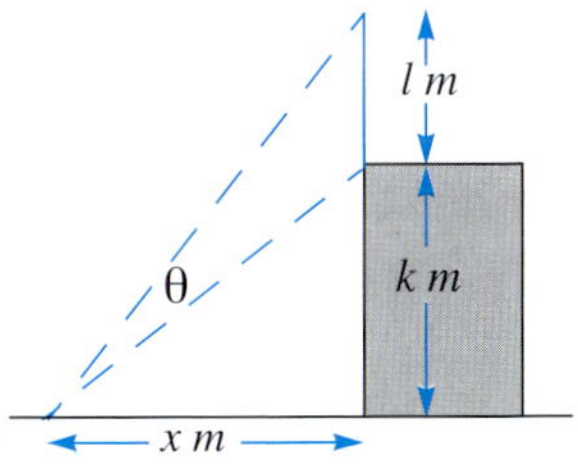

[Hint: As $\theta \in \left[0, \frac{\pi}{2}\right)$, θ is a maximum when $\tan\theta$ is a maximum.]

45. A person in a boat l km from the nearest point P on a straight beach, wishes to get to a point Q kl km along the beach from P.
The person rows in a straight line to some point on the shore at a constant rate of v km/h. Once on the beach the person walks towards Q at a steady rate of u km/hr where $(v < u)$.

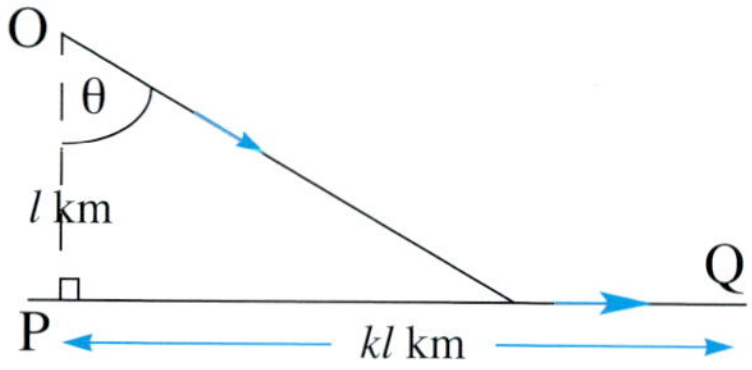

(a) Show that the total time in hours taken for the trip is

$$T(\theta) = \frac{l\sec\theta}{v} + \frac{kl - l\tan\theta}{u}, \theta \in \left[0, \frac{\pi}{2}\right) \text{ and } \tan\theta \le k .$$

(b) Show that if the person is to reach point Q in the least possible time, then

$\sin\theta = \frac{v}{u}$ where $k \ge c$, c being a particular constant.

(c) What would happen if $k < c$?

46. A closed tin is to be constructed as shown in the diagram. It is made up of a cylinder of height h cm and radius base r cm which is surmounted by a hemispherical cap.

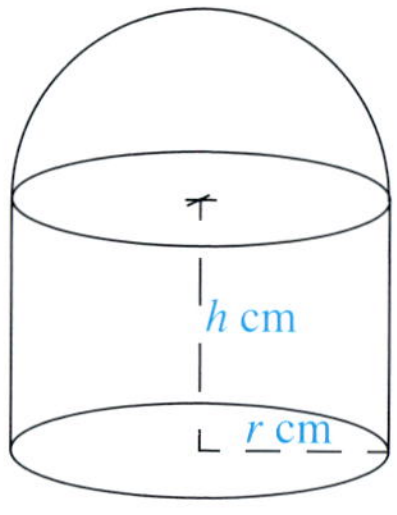

(a) Find an expression in terms of r and h for
  i. its volume, V cm^3 .
  ii. its surface area, A cm^2 .

(b) Given that $V = \pi k^3, k > 0$, show that its surface area is given by $A = 2\pi k^3\frac{1}{r} + \frac{5\pi}{3}r^2$.

(c) Find the ratio $r : h$ for A to be a minimum.

47. A ladder is to be carried horizontally around a corner from a corridor a m wide into a corridor b m wide.

What is the maximum length that the ladder can be?

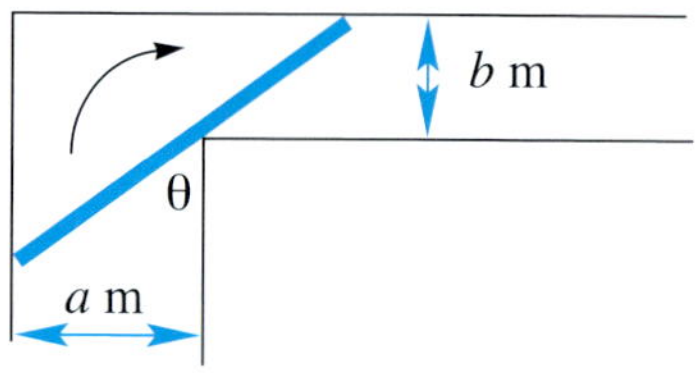

48. A man in a boat is 3 km from the nearest point of a straight beach. The man is to get to a point 6 km along the beach. The man can row at 4 km/hr and walk at 5 km/hr.

(a) Show that the time, T hours, taken to get to his destination is given by

$$T = \frac{\sqrt{x^2+9}}{4} + \frac{6-x}{5}, 0 \le x \le 6$$

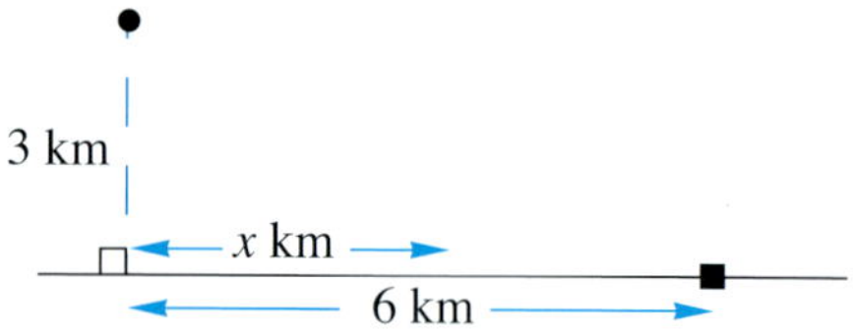

(b) Where should the man row to along the beach if he is to reach his destination in the least possible time?

(c) After some further training, the man finds that he can now row at 4.5 km/hr. Where should he now row to along the beach to minimise the time taken?

22.1 INTEGRATION

22.1.1 ANTIDIFFERENTIATION & THE INDEFINITE INTEGRAL

As the name suggests— ***Anti* differentiation** is the reverse process of differentiation. We are then searching for the answer to the following:

> Given an expression $f'(x)$ (i.e., the derivative of the function $f(x)$), what must the original function $f(x)$ have been?

For example, if $f'(x) = 2x$ then $f(x) = x^2$ is a *possible expression* for the original function. Why do we say ' *... is a possible expression for the original function.*'?

Consider the following results:
i. $f(x) = x^2 + 3 \Rightarrow f'(x) = 2x$
ii. $f(x) = x^2 - 5 \Rightarrow f'(x) = 2x$

From i., and ii., we see that given an expression $f'(x)$, there are a number of possible different original functions, $f(x)$. This is due to the fact that the derivative of a constant is zero and so when we are given an expression for $f'(x)$, there is no real way of knowing if there was a constant in the original function or what that constant might have been (unless we are given some extra information).

The best that we can do at this stage is to write the following:

Given that $f'(x) = 2x$, then $f(x) = x^2 + c$,

where c is some real number that is yet to be determined (it could very well be that $c = 0$).

The antidifferentiation process described above can be summarised as follows:

> Given that $\frac{dy}{dx} = f'(x)$, then (after antidifferentiating) $y = f(x) + c$, where $c \in \mathbb{R}$.
>
> We say that $y = f(x) + c$, where $c \in \mathbb{R}$ is the **antiderivative** of $f'(x)$.

22.1.2 LANGUAGE AND NOTATION

> The set of all antiderivatives of a function $h(x)$ is called the **indefinite integral** of $h(x)$, and is denoted by $\int h(x)dx$.
>
> The symbol $\int$ is called the **integral sign**,
> the function $h(x)$ is the **integrand** of the integral
> and x is the **variable of integration**.

CHAPTER 22

Once we have found an antiderivative (or indefinite integral) of $h(x)$, $H(x)$ (say) we can then write

$$\int h(x)dx = H(x) + c, \text{ where } c \in \mathbb{R}.$$

The constant c is called the **constant of integration**. The above result is read as:

'The **antiderivative** of $h(x)$ with respect to x is $H(x) + c$, where $c \in \mathbb{R}$'.
or
'The **indefinite integral** of $h(x)$ with respect to x is $H(x) + c$, where $c \in \mathbb{R}$'.

22.1.3 DETERMINING THE INDEFINITE INTEGRAL

So — how do we find the indefinite integral of $h(x)$?

We approach this problem by searching for a pattern (pretty much as we did when dealing with the derivative of a function). Recall the following results (when were searching for a rule to find the derivative of a function):

$h(x)$	x	x^2	x^3	x^4	x^5	. . .	x^n
$h'(x)$	1	$2x$	$3x^2$	$4x^3$	$5x^4$	. . .	nx^{n-1}

The differentiation process was then described using the following 'progress' diagram:

Differentiation process

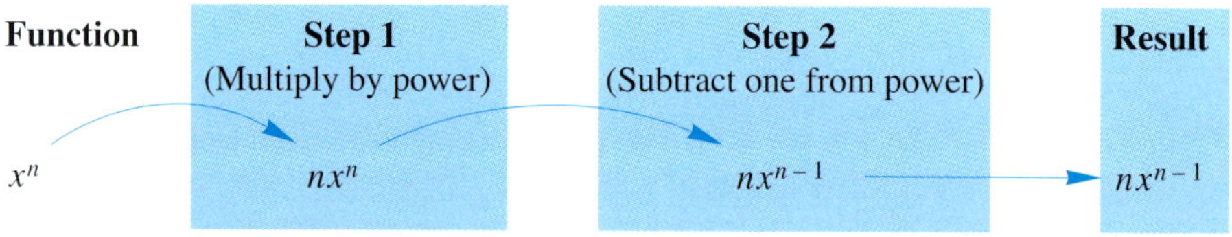

Finding the indefinite integral requires that we '*reverse the process*' (i.e., carry out the inverse operation). Again this can be illustrated using a 'progress' diagram:

Antidifferentiation process

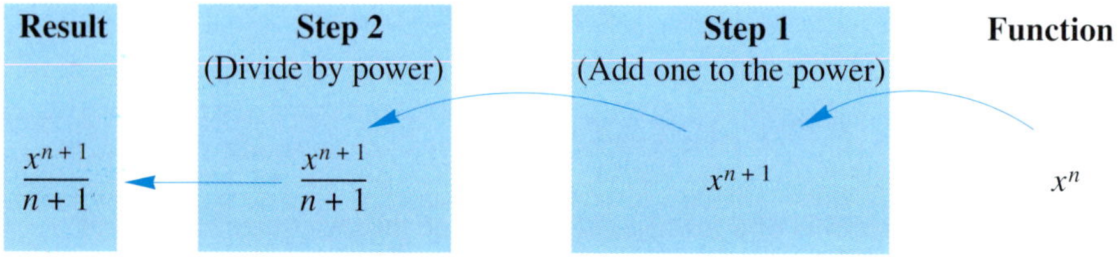

However, as we discussed previously, we must add a real constant to complete the process of antidifferentiation. We then have the general result:

$$\int x^n dx = \frac{x^{n+1}}{n+1} + c \quad \text{or} \quad \frac{1}{n+1}x^{n+1} + c \text{ as long as } n \neq -1.$$

A slightly more general result is one where we have ax^n rather than simply x^n. In this case we have that

$$\int ax^n dx = \frac{ax^{n+1}}{n+1} + c \quad \text{or} \quad \frac{a}{n+1}x^{n+1} + c \text{ as long as } n \neq -1.$$

EXAMPLE 22.1

Find the indefinite integral of the following

(a) $4x^2$ (b) x^{-3} (c) $5\sqrt{x}$ (d) $\sqrt[5]{x^3}$

Solution

In each case we use the 'progress' diagram on the previous page. That is, we first increase the power by one and then divide by the new power.

(a) $\int 4x^2 dx = \frac{4}{2+1}x^{2+1} + c, c \in \mathbb{R} = \frac{4}{3}x^3 + c, c \in \mathbb{R}$

(b) $\int x^{-3} dx = \frac{1}{-3+1}x^{-3+1} + c, c \in \mathbb{R} = -\frac{1}{2}x^{-2} + c, c \in \mathbb{R}$

$= -\frac{1}{2x^2} + c, c \in \mathbb{R}$

(c) $\int 5\sqrt{x}dx = \int 5x^{\frac{1}{2}} dx = \frac{5}{\frac{1}{2}+1}x^{\frac{1}{2}+1} + c, c \in \mathbb{R}$

$= \frac{5}{(3/2)}x^{3/2} + c, c \in \mathbb{R}$

$= \frac{10}{3}\sqrt{x^3} + c, c \in \mathbb{R}$

Notice that before we can start the antidifferentiation process we must express the integrand in the form ax^n, i.e., in power form.

(d) $\int \sqrt[5]{x^3} dx = \int x^{\frac{3}{5}} dx = \frac{1}{\frac{3}{5}+1}x^{\frac{3}{5}+1} + c, c \in \mathbb{R}$

$= \frac{1}{(8/5)}x^{\frac{8}{5}} + c, c \in \mathbb{R}$

$= \frac{5}{8}\sqrt[5]{x^8} + c, c \in \mathbb{R}$

Although we have been working through examples where the integrand has only one term, we can determine the indefinite integral of expressions that are made up of several terms.

EXAMPLE 22.2

Find the integral of

(a) $\int (2x^2 + x^3 - 4)dx$ (b) $\int (x-1)(x^4 + 3x)dx$ (c) $\int \frac{z^4 - 2z^2 + 3}{z^2} dz$

Solution

(a) $\int(2x^2+x^3-4)dx = \int 2x^2dx + \int x^3dx - \int 4dx = \frac{2}{2+1}x^{2+1} + \frac{1}{3+1}x^{3+1} - 4x + c$

$$= \frac{2}{3}x^3 + \frac{1}{4}x^4 - 4x + c,\ c \in \mathbb{R}$$

When determining the indefinite integral of 4, we have actually thought of '4' as '$4x^0$'. So that $\int 4dx = \int 4x^0dx = \frac{4}{0+1}x^{0+1} = 4x$.

(b) $\int(x-1)(x^4+3x)dx = \int(x^5-x^4+3x^2-3x)dx = \frac{1}{6}x^6 - \frac{1}{5}x^5 + x^3 - \frac{3}{2}x^2 + c,\ c \in \mathbb{R}$.

(c) $\int\frac{z^4-2z^2+3}{z^2}dz = \int\left(\frac{z^4}{z^2} - \frac{2z^2}{z^2} + \frac{3}{z^2}\right)dz = \int(z^2-2+3z^{-2})dz$

$$= \frac{1}{3}z^3 - 2z + \frac{3}{-1}z^{-1} + c$$

$$= \frac{1}{3}z^3 - 2z - \frac{3}{z} + c,\ c \in \mathbb{R}$$

Notice that in part (b) it was necessary to first multiply out the brackets **before** we could integrate. Similarly, for part (c) we had to first carry out the division **before** integrating.

22.1.4 PROPERTIES OF THE INDEFINITE INTEGRAL

In many of the above examples we made use of the following properties

Properties	Examples
1. $\int h'(x)dx = h(x) + c$	$\int \frac{d}{dx}(x^2)dx = \int(2x)dx = x^2 + c$
2. $\frac{d}{dx}\left(\int h(x)dx\right) = h(x)$	$\frac{d}{dx}\left(\int(x^3)dx\right) = \frac{d}{dx}\left(\frac{1}{4}x^4 + c\right) = x^3$
3. $\int kh(x)dx = k\int h(x)dx$	$\int 12x^3dx = 12\int x^3dx = 12 \times \left(\frac{1}{4}x^4 + c\right)$ $= 3x^4 + C$
4. $\int(h(x) \pm f(x))dx = \int h(x)dx \pm \int f(x)dx$	$\int(2x-3x^2)dx = \int 2xdx - \int 3x^2dx$ $= x^2 - x^3 + c$

EXERCISES 22.1

1. Find the indefinite integral of the following:

(a) x^3 (b) x^7 (c) x^5 (d) x^8

(e) $4x^2$ (f) $7x^5$ (g) $9x^8$ (h) $\frac{1}{2}x^3$

2. Find:

(a) $\int 5dx$ (b) $\int 3dx$ (c) $\int 10dx$ (d) $\int \frac{2}{3}dx$

(e) $\int -4dx$ (f) $\int -6dx$ (g) $\int -\frac{3}{2}dx$ (h) $\int -dx$

3. Find:

(a) $\int (1-x)dx$ (b) $\int (2+x^2)dx$ (c) $\int (x^3-9)dx$

(d) $\int \left(\frac{2}{5}+\frac{1}{3}x^2\right)dx$ (e) $\int \left(\frac{2}{4}\sqrt{x}-\frac{1}{x^2}\right)dx$ (f) $\int \left(\frac{5}{2}\sqrt{x^3}+8x\right)dx$

(g) $\int x(x+2)dx$ (h) $\int x^2\left(3-\frac{2}{x}\right)dx$ (i) $\int (x+1)(1-x)dx$

4. Find the antiderivative of the following:

(a) $(x+2)(x-3)$ (b) $(x^2-3x)(x+1)$ (c) $(x-3)^3$

(d) $(x+2x^3)(x+1)$ (e) $(1-\sqrt{x})(1+x)$ (f) $\sqrt{x}(x+1)^2-2$

5. Find:

(a) $\int \frac{x^2-3x}{x}dx$ (b) $\int \frac{4u^3+5u^2-1}{u^2}du$ (c) $\int \frac{(x+2)^2}{x^4}dx$

(d) $\int \frac{x^2+5x+6}{x+2}dx$ (e) $\int \frac{x^2-6x+8}{x-2}dx$ (f) $\int \left(\frac{t^2+1}{t}\right)^2 dt$

6. Find the indefinite integral of the following:

(a) $\sqrt[4]{x^3}+\frac{1}{\sqrt{x}}-5$ (b) $\sqrt{x}(\sqrt{x}-2x)(x+1)$ (c) $\frac{1}{z^3}-\frac{2}{z^2}+4z+1$

(d) $\left(2t+\frac{3}{t^2}\right)\left(t^2-\frac{1}{t}\right)+\frac{3}{t^3}$ (e) $\frac{(t-2)(t-1)}{\sqrt{t}}-\frac{2}{\sqrt{t}}$ (f) $\frac{u^3+6u^2+12u+8}{u+2}$

7. Given that $f(x) = ax^n$, $n \neq -1$ and $g(x) = bx^m$, $m \neq -1$ show that

(a) $\int [f(x)+g(x)]dx = \int f(x)dx + \int g(x)dx$

(b) $\int [f(x)-g(x)]dx = \int f(x)dx - \int g(x)dx$

(c) $\int kf(x)dx = k\int f(x)dx$

(d) $\int f(x)g(x)dx \neq (\int f(x)dx)(\int g(x)dx)$, $n+m \neq -1$.

(e) $\int \frac{f(x)}{g(x)}dx \neq \frac{\int f(x)dx}{\int g(x)dx}$, $n-m \neq -1$

8. (a) Show that $\frac{d}{dx}((2x+3)^4) = 8(2x+3)^3$. Hence find $\int (2x+3)^3 dx$.

(b) Show that $\frac{d}{dx}(\sqrt{x^2+4}) = \frac{x}{\sqrt{x^2+4}}$. Hence find $\int \frac{3x}{\sqrt{x^2+4}}dx$.

22.2 SOLVING FOR "C"

Although we have already discussed the reason for 'adding' a constant, c, when finding the indefinite integral, it is important that we can also determine the value of c.

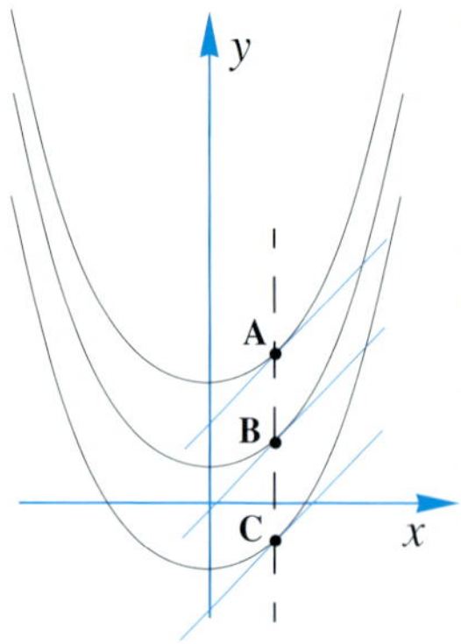

Given that $\frac{dy}{dx} = 2x$, upon antidifferentiating, we have $y = x^2 + c$.

This result is known as the **general solution**.

Some of the possible curves, $y = x^2 + c$, are shown, and we observe that at A, B and C, for a particular value of x, the gradients are equal.

To determine which of these curves is the one that we actually require, we must be provided with some extra information. In this case we would need to be given the coordinates of a point on the curve.

EXAMPLE 22.3

Find $f(x)$ given that $f'(x) = 2x$ and that the curve passes through the point (2,9).

Solution

As $f'(x) = 2x \Rightarrow f(x) = x^2 + c$.

Using the fact that at $x = 2$, $y = 9$, or that $f(2) = 9$, we have $9 = 2^2 + c \Leftrightarrow c = 5$.

Therefore, of all possible solutions of the form $y = x^2 + c$, the function satisfying the given information is $f(x) = x^2 + 5$.

EXAMPLE 22.4

Find $f(x)$ given that $f''(x) = 6x - 2$ and that the gradient at the point (1,5) is 2.

Solution

From the given information we have that $f'(1) = 2$ and $f(1) = 5$.

As $f''(x) = 6x - 2$ we have $f'(x) = 3x^2 - 2x + c_1$ – (1).

But, $f'(1) = 2 \therefore 2 = 3(1)^2 - 2(1) + c_1 \Leftrightarrow c_1 = 1$. [i.e., substituting into (1)]

Therefore, we have $f'(x) = 3x^2 - 2x + 1$.

Next, from $f'(x) = 3x^2 - 2x + 1$ we have $f(x) = x^3 - x^2 + x + c_2$ – (2).

But, $f(1) = 5 \therefore 5 = (1)^3 - (1)^2 + (1) + c_2 \Leftrightarrow c_2 = 4$. [i.e., substituting into (2)]

Therefore, $f(x) = x^3 - x^2 + x + 4$

Sometimes, information is not given in the form of a set of coordinates. Information can also be 'hidden' in the context of the problem.

EXAMPLE 22.5 The rate of change in pressure, p units, at a depth x cm from the surface of a liquid is given by $p'(x) = 0.03x^2$. If the pressure at the surface is 10 units, find the pressure at a depth of 5 cm.

Solution

Antidifferentiating both sides with respect to x, we have $\int p'(x)dx = \int 0.03x^2 dx$.

$$\therefore p(x) = 0.01x^3 + c \quad -(1)$$

At $x = 0, p = 10$. Substituting into (1) we have $10 = 0.01(0)^3 + c \Leftrightarrow c = 10$

Therefore, the equation for the pressure at a depth of x cm is $p(x) = 0.01x^3 + 10$.

At $x = 5$, we have $p(5) = 0.01(5)^3 + 10 = 11.25$. That is, the pressure is 11.25 units.

EXAMPLE 22.6 The growth rate of a city's population has been modelled by the equation $\frac{dN}{dt} = 400t^{1.05}, t \geq 0$ where t is the time in years after 1995 and N is the population size. In the year 2000 the population numbered 32,000. What will the population be in 2010?

Solution

We start by determining N:

$$\frac{dN}{dt} = 400t^{1.05} \Rightarrow \int\frac{dN}{dt}dt = \int 400t^{1.05}dt \quad \text{[i.e., antidiff. b.s. w.r.t. } t\text{]}$$

$$\therefore N = \frac{400}{2.05}t^{2.05} + c$$

$$= \frac{8000}{41}t^{2.05} + c$$

We are given that when $t = 5, N = 32000$.

$$\therefore 32000 = \frac{8000}{41}(5)^{2.05} + c \Leftrightarrow c = 26713.18$$

Therefore, we have that $N = \frac{8000}{41}t^{2.05} + 26713.18$.

Then, when $t = 15$, $N = \frac{8000}{41}(15)^{2.05} + 26713.18 = 76981.36$.

So, in 2010 the population would be 76981.

EXERCISES 22.2

1. Find the equation of the function in each of the following:

(a) $f'(x) = 2x + 1$, given that the curve passes through (1,5).

(b) $f'(x) = 2 - x^2$ and $f(2) = \frac{7}{3}$.

(c) $\frac{dy}{dx} = 4\sqrt{x} - x$, given that the curve passes through (4,0).

(d) $f'(x) = x - \frac{1}{x^2} + 2$, and $f(1) = 2$

(e) $\frac{dy}{dx} = 3(x+2)^2$, given that the curve passes through (0,8).

(f) $\frac{dy}{dx} = \sqrt[3]{x} + x^3 + 1$, given that the curve passes through (1,2).

(g) $f'(x) = (x+1)(x-1)+1$, and $f(0) = 1$

(h) $f'(x) = 4x^3 - 3x^2 + 2$, and $f(-1) = 3$

2. Find the equation of the function $f(x)$ given that it passes through the point (–1,2) and is such that $f'(x) = ax + \frac{b}{x^2}$, where $f(1) = 4$ and $f'(1) = 0$.

3. The marginal cost for producing x units of a product is modelled by the equation $C'(x) = 30 - 0.06x$. The cost per unit is \$40. How much will it cost to produce 150 units?

4. If $\frac{dA}{dr} = 6 - \frac{1}{r^2}$, and $A = 4$ when $r = 1$, find A when $r = 2$.

5. The rate, in cm^3/sec, at which the volume of a sphere is increasing is given by the relation $\frac{dV}{dt} = 4\pi(2t+1)^2, 0 \le t \le 10$. If initially the volume is $\pi\ \text{cm}^3$, find the volume of the sphere when $t = 2$.

6. The rate of change of the number of deer, N, in a controlled experiment, is modelled by the equation $\frac{dN}{dt} = 3\sqrt{t^3} + 2t, 0 \le t \le 5$. There are initially 200 deer in the experiment. How many deer will there be at the end of the experiment.

7. If $\frac{dy}{dx} \propto \sqrt{x}$, find an expression for y, given that $y = 4$ when $x = 1$ and $y = 9$ when $x = 4$.

8. A function with gradient defined by $\frac{dy}{dx} = 4x - m$ at any point P on its curve passes through the point (2,–6) with a gradient of 4. Find the coordinates of its turning point.

9. The marginal revenue is given by $\frac{dR}{dx} = 25 - 10x + x^2, x \ge 0$, where R is the total revenue and x is the number of units demanded. Find the equation for the price per unit, $P(x)$.

10. The rate of growth of a culture of bacteria is modelled by the equation $\frac{dN}{dt} = 200t^{1.01}$, where N is the number of bacteria present t hours after the culture begins to grow. Find the number of bacteria present in the culture at time t hours if initially there were 500 bacteria.

11. Sketch the graph of $y = f(x)$ for each of the following:

(a)

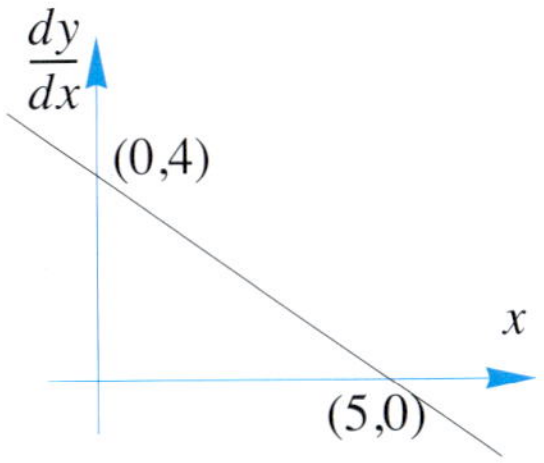

Where the curve passes through the point (5,10).

(b)

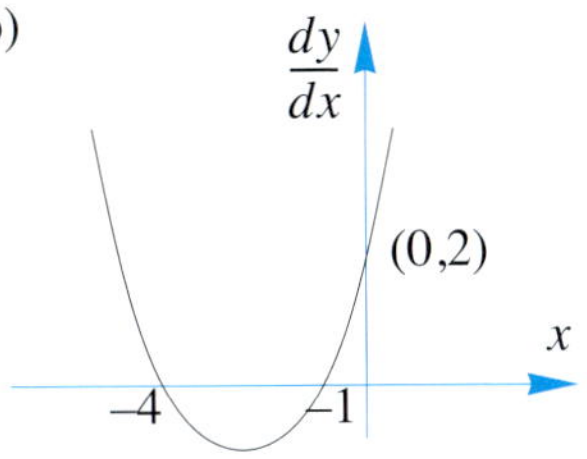

Where the curve passes through the point (0,0).

12. Find $f(x)$ given that $f''(x) = 12x + 4$ and that the gradient at the point (1,6) is 12.

13. Find $f(x)$ given that $f'(x) = ax^2 + b$, where the gradient at the point (1,2) is 4 and that the curve passes through the point (3,4).

14. The rate at which a balloon is expanding is given by $\frac{dV}{dt} = kt^{4.5}, t \geq 0$, where t is the time in minutes since the balloon started to be inflated and V cm^3 is its volume. Initially the balloon (which may be assumed to be spherical) has a radius of 5 cm.
If the balloon has a volume of 800 cm^3 after 2 minutes, find its volume after 5 minutes.

15. The area, A cm^2, of a healing wound caused by a fall on a particular surface decreases at a rate given by the equation $A'(t) = -\frac{35}{\sqrt{t}}$ where t is the time in days. Find the initial area of such a wound if after one day the area measures 40 cm^2.

22.3 STANDARD INTEGRALS

In the same way that there are rules for differentiating functions other than those of the form ax^n, we also have standard rules for integrating functions other than the ones that we have been dealing with so far. That is, there are standard rules for finding the indefinite integral of circular trigonometric functions, exponential functions and logarithmic functions.

We can deduce many such antiderivatives using the result $\int \frac{d}{dx}(h(x))dx = h(x) + c$ – (1).

For example, if we consider the derivative $\frac{d}{dx}(e^{2x}) = 2e^{2x}$,

We can write $\int \frac{d}{dx}(e^{2x})dx = \int 2e^{2x}dx$. But from (1) we have $\int \frac{d}{dx}(e^{2x})dx = e^{2x}$.

Therefore, $e^{2x} + c = \int 2e^{2x}dx$.

Or, we could write $e^{2x} + c = 2\int e^{2x}dx \Leftrightarrow \int e^{2x}dx = \frac{1}{2}(e^{2x} + c) = \frac{1}{2}e^{2x} + C$

Similarly, $\frac{d}{dx}(\ln x) = \frac{1}{x}$ and so, antidifferentiating both sides we have $\int \frac{d}{dx}(\ln x)dx = \int \frac{1}{x}dx$.

Then, as $\int \frac{d}{dx}(\ln x)dx = \ln x$, we can write, $\ln x + c = \int \frac{1}{x}dx$.

We summarise these rules in the table below:

$f(x)$	$\int f(x)dx$	**Examples**
$ax^n, n \neq -1$	$\frac{a}{n+1}x^{n+1} + c, n \neq -1$	$\int 2x^2 dx = \frac{2}{3}x^3 + c$
$\frac{1}{x}, x \neq 0$	$\log_e\lvert x\rvert + c, x \neq 0$ or $\log_e x + c, x > 0$	$\int \frac{6}{x}dx = 6\int \frac{1}{x}dx = 6\log_e x + c, x > 0$
$\sin(kx)$	$-\frac{1}{k}\cos(kx) + c$	$\int \sin(5x)dx = -\frac{1}{5}\cos(5x) + c$
$\cos(kx)$	$\frac{1}{k}\sin(kx) + c$	$\int \cos\left(\frac{x}{4}\right)dx = \int \cos\left(\frac{1}{4}x\right)dx = \frac{1}{\left(\frac{1}{4}\right)}\sin\left(\frac{x}{4}\right) + c$ $= 4\sin\left(\frac{x}{4}\right) + c$
$\sec^2(kx)$	$\frac{1}{k}\tan(kx) + c$	$\int \sec^2(2x)dx = \frac{1}{2}\tan(2x) + c$
e^{kx}	$\frac{1}{k}e^{kx} + c$	$\int e^{-3x}dx = -\frac{1}{3}e^{-3x} + c$

NB: In the above table (and from here on), the constant c is assumed to be a real number.

General power rule

The indefinite integral of $f(ax+b)$ where a and b are constants:

If $\int f(x)dx = F(x) + c$, then $\int f(ax+b)dx = \frac{1}{a}F(ax+b) + c$.

This means that all of the integrals in the table above can be generalised further.

In particular we consider the generalised power rule:

1. $\int (ax+b)^n dx = \frac{1}{a(n+1)}(ax+b)^{n+1} + c, n \neq -1$

2. $\int \frac{1}{ax+b}dx = \frac{1}{a}\log_e(ax+b) + c, ax+b > 0$

EXAMPLE 22.7

Find the antiderivative of

(a) $(3x+7)^4$ (b) $3(4-2x)^5$ (c) $\sqrt{5x+1}$ (d) $\dfrac{4}{(2x-3)^2}$

Solution

This question requires the use of the result: $\int (ax+b)^n dx = \frac{1}{a(n+1)}(ax+b)^{n+1} + c, n \neq -1$

(a) $\int (3x+7)^4 dx = \frac{1}{3 \times (4+1)}(3x+7)^{4+1} + c$

$= \frac{1}{15}(3x+7)^5 + c$

(b) $\int 3(4-2x)^5 dx = 3\int (4-2x)^5 dx = 3 \times \frac{1}{(-2) \times (5+1)}(4-2x)^{5+1} + c$

$= \frac{3}{-2 \times 6}(4-2x)^6 + c$

$= -\frac{1}{4}(4-2x)^6 + c$

(c) $\int \sqrt{5x+1}\,dx = \int (5x+1)^{1/2} dx$ [Note: we must first convert into power form!]

$= \frac{1}{5 \times \left(\frac{1}{2}+1\right)}(5x+1)^{\frac{1}{2}+1} + c$

$= \frac{2}{15}(5x+1)^{\frac{3}{2}} + c$

(d) $\int \frac{4}{(2x-3)^2} dx = 4\int (2x-3)^{-2} dx$

$= 4 \times \frac{1}{2 \times (-2+1)}(2x-3)^{-2+1} + c$

$= -2 \times (2x-3)^{-1} + c$

$= -\frac{2}{(2x-3)} + c$

EXAMPLE 22.8

Find the antiderivative of

(a) $\sin 4x$ (b) $\cos\left(\frac{1}{2}x\right)$ (c) $\sec^2(3x)$

Solution

(a) $\int \sin 4x dx = -\frac{1}{4}\cos 4x + c$. $\left[\text{Using } \int \sin(kx)dx = -\frac{1}{k}\cos(kx) + c\right]$

(b) $\int \cos\left(\frac{1}{2}x\right) dx = \frac{1}{\left(\frac{1}{2}\right)}\sin\left(\frac{1}{2}x\right) + c = 2\sin\left(\frac{1}{2}x\right) + c$. $\left[\text{Using } \int \cos(kx)dx = \frac{1}{k}\sin(kx) + c\right]$

(c) $\int \sec^2(3x)dx = \frac{1}{3}\tan(3x) + c$. $\left[\text{Using } \int \sec^2(kx)dx = \frac{1}{k}\tan(kx) + c\right]$

We can extend the results for the antiderivatives of circular trigonometric functions as follows:

$f(x)$	$\int f(x)dx$	**Examples**
$\sin(ax+b)$	$-\frac{1}{a}\cos(ax+b)+c$	$\int \sin(5x+2)dx = -\frac{1}{5}\cos(5x+2)+c$
$\cos(ax+b)$	$\frac{1}{a}\sin(ax+b)+c$	$\int \cos\left(\frac{x}{4}-1\right)dx = \int \cos\left(\frac{1}{4}x-1\right)dx$ $= 4\sin\left(\frac{x}{4}-1\right)+c$
$\sec^2(ax+b)$	$\frac{1}{a}\tan(ax+b)+c$	$\int \sec^2(2x+3)dx = \frac{1}{2}\tan(2x+3)+c$

EXAMPLE 22.9

Find the antiderivative of

(a) e^{2x} (b) $5e^{-3x}$ (c) $4x - 2e^{\frac{1}{3}x}$ (e) $\frac{5}{x}$ (f) $\frac{2}{3x+1}$

Solution

(a) $\int e^{2x}dx = \frac{1}{2}e^{2x} + c$. $\left[\text{Using } \int e^{kx}dx = \frac{1}{k}e^{kx} + c\right]$

(b) $\int 5e^{-3x}dx = 5\int e^{-3x}dx = 5 \times \frac{1}{-3}e^{-3x} + c = -\frac{5}{3}e^{-3x} + c$ $\left[\text{Using } \int e^{kx}dx = \frac{1}{k}e^{kx} + c\right]$

(c) $\int\left(4x - 2e^{\frac{1}{3}x}\right)dx = \int 4x\,dx - 2\int e^{\frac{1}{3}x}dx = 4 \times \frac{1}{2}x^2 - 2 \times \frac{1}{\left(\frac{1}{3}\right)}e^{\frac{1}{3}x} + c$ "

$= 2x^2 - 6e^{\frac{1}{3}x} + c$

(d) $\int \frac{5}{x}dx = 5\int \frac{1}{x}dx = 5\ln(x) + c, x > 0$

(e) $\int \frac{2}{3x+1}dx = 2\int \frac{1}{3x+1}dx = \frac{2}{3}\ln(3x+1) + c, x > -\frac{1}{3}$ $\left[\int \frac{1}{ax+b}dx = \frac{1}{a}\log_e(ax+b) + c\right]$

As we have just done for the circular trigonometric functions, we can extend the antiderivative of the exponential function to

$$\int e^{ax+b}dx = \frac{1}{a}e^{ax+b} + c$$

EXAMPLE 22.10

Find

(a) $\int 2(4x-1)^6 dx$ (b) $\int \frac{7}{4-3x}dx$ (c) $\int \frac{3}{\sqrt{2x+1}}dx$

(d) $\int e^{-2x+3}dx$ (d) $\int (e^{-4x} + \sqrt[5]{3x+2})dx$ (e) $\int \left(\cos\left(\frac{\pi}{2}x\right) - \frac{2}{6x+5}\right)dx$

Solution

(a) We have $n = 6$, $a = 4$ and $b = -1$, $\therefore \int 2(4x-1)^6 dx = \frac{2}{4(6+1)}(4x-1)^{6+1} + c$

$$= \frac{1}{14}(4x-1)^7 + c$$

(b) This time we use the case for $n = -1$, with $a = -3$ and $b = 4$, giving

$$\int \frac{7}{4-3x}dx = 7\int \frac{1}{4-3x}dx = -\frac{7}{3}\log_e(4-3x) + c,\ x < \frac{4}{3}$$

(c) We first rewrite the indefinite integral in power form: $\int \frac{3}{\sqrt{2x+1}}dx = \int 3(2x+1)^{-\frac{1}{2}}dx$

So that $a = 2$, $b = 1$ and $n = -\frac{1}{2}$. Therefore we have that

$$\int \frac{3}{\sqrt{2x+1}}dx = \int 3(2x+1)^{-\frac{1}{2}}dx = \frac{3}{2\left(-\frac{1}{2}+1\right)}(2x+1)^{-\frac{1}{2}+1} + c$$

$$= \frac{3}{2 \times \frac{1}{2}}(2x+1)^{\frac{1}{2}} + c$$

$$= 3\sqrt{2x+1} + c$$

(d) $\int e^{-2x+3}dx = -\frac{1}{2}e^{-2x+3} + c$

(e) $\int (e^{-4x} + \sqrt[5]{3x+2})dx = \int \left(e^{-4x} + (3x+2)^{\frac{1}{5}}\right)dx = -\frac{1}{4}e^{-4x} + \frac{1}{3\left(\frac{6}{5}\right)}(3x+2)^{\frac{6}{5}} + c$

$$= -\frac{1}{4}e^{-4x} + \frac{5}{18}\sqrt[5]{(3x+2)^6} + c$$

(f) $\int \left(\cos\left(\frac{\pi}{2}x\right) - \frac{2}{6x+5}\right)dx = \frac{1}{\frac{\pi}{2}}\sin\left(\frac{\pi}{2}x\right) - \frac{2}{6}\ln(6x+5) + c$

$$= \frac{2}{\pi}\sin\left(\frac{\pi}{2}x\right) - \frac{1}{3}\ln(6x+5) + c$$

EXAMPLE 22.11

The gradient at any point on the curve $y = f(x)$ is given by the equation $\frac{dy}{dx} = \frac{1}{\sqrt{x+2}}$. The curve passes through the point (2,3). Find the equation of this curve.

Solution

Integrating both sides of $\frac{dy}{dx} = \frac{1}{\sqrt{x+2}}$ with respect to x, we have $\int \frac{dy}{dx}dx = \int \frac{1}{\sqrt{x+2}}dx$.

But, $\int \frac{1}{\sqrt{x+2}}dx = \int (x+2)^{-1/2}dx = 2(x+1)^{1/2} + c$ [Using the general power rule.]

Therefore, we have $y = f(x) = 2\sqrt{x+2} + c$.

Now, $f(2) = 3 \Rightarrow 3 = 2\sqrt{4} + c \Leftrightarrow c = -1$.

Therefore, $f(x) = 2\sqrt{x+2} - 1$.

EXAMPLE 22.12

The gradient at any point on the curve $y = f(x)$ is given by $5\sin(2x+3)$. The curve passes through the point $\left(0, \frac{3}{2}\cos 3\right)$. Find the equation of this curve.

Solution

Given that $f'(x) = 5\sin(2x+3)$, then $f(x) = -\frac{5}{2}\cos(2x+3) + c$.

Now, we are given that $f(0) = \frac{3}{2}\cos 3 \therefore \frac{3}{2}\cos 3 = -\frac{5}{2}\cos(2 \times 0 + 3) + c$

$$\Leftrightarrow c = \frac{8}{2}\cos 3$$

Therefore, $f(x) = -\frac{5}{2}\cos(2x+3) + 4\cos 3$.

EXAMPLE 22.13

The rate of change of a radioactive substance, Q, is given by $-200ke^{-kt}$. Initially there were 200 milligrams of this substance. If the half-life of the substance is 1200 years, find the amount of the substance left after 2000 years.

Solution

We are given that $\frac{dQ}{dt} = -200ke^{-kt} \therefore Q = \frac{-200k}{-k}e^{-kt} + c$

$$Q = 200e^{-kt} + c$$

When $t = 0$, $Q = 200$, therefore $200 = 200 + c \Leftrightarrow c = 0$.

So that $Q(t) = 200e^{-kt}, t \geq 0$.

A half-life of 1200 years, this means that when $t = 1200$, $Q = 100$ (half its initial quantity).

That is, $100 = 200e^{-1200k} \Leftrightarrow e^{-1200k} = 0.5$

Therefore, $-1200k = \ln\left(\frac{1}{2}\right) \Leftrightarrow k = \frac{1}{1200}\ln 2$. That is, $Q(t) = 200e^{-\left(\frac{1}{1200}\ln 2\right)t}, t \geq 0$

Then, when t = 2000, $Q(t) = 200e^{-\left(\frac{1}{1200}\ln 2\right) \times 2000}$

$= 200e^{-\frac{5}{3}\ln 2}$

$= 62.996$

That is, there will be approximately 63 milligrams.

EXERCISES 22.3

1. Find the indefinite integral of

(a)	e^{5x}	(b)	e^{3x}	(c)	e^{2x}	(d)	$e^{0.1x}$
(e)	e^{-4x}	(f)	$4e^{-4x}$	(g)	$0.1e^{-0.5x}$	(h)	$2e^{1-x}$
(i)	$5e^{x+1}$	(j)	$-2e^{2-2x}$	(k)	$e^{\frac{1}{3}x}$	(l)	$\sqrt{e^x}$

2. Find the indefinite integral of

(a)	$\frac{4}{x}$	(b)	$-\frac{3}{x}$	(c)	$\frac{2}{5x}$	(d)	$\frac{1}{x+1}$
(e)	$\frac{1}{2x}$	(f)	$\left(1-\frac{1}{x}\right)^2$	(g)	$\left(\sqrt{x}-\frac{1}{\sqrt{x}}\right)^2$	(h)	$\frac{3}{x+2}$

3. Find the indefinite integral of

(a)	$\sin(3x)$	(b)	$\cos(2x)$	(c)	$\sec^2(5x)$	(d)	$\sin(-x)$

4. Find the indefinite integral of

(a)	$\sin(2x) + x$	(b)	$6x^2 - \cos(4x)$	(c)	e^{5x}
(d)	$4e^{-3x} + \sin\left(\frac{1}{2}x\right)$	(e)	$\cos\left(\frac{x}{3}\right) - \sin(3x)$	(f)	$e^{2x} + \frac{4}{x} - 1$
(g)	$(e^x + 1)^2$	(h)	$-5\sin(4x) + \frac{x-1}{x}$	(i)	$\sec^2(3x) - \frac{2}{x} + e^{\frac{1}{2}x}$
(j)	$(e^x - e^{-x})^2$	(k)	e^{2x+3}	(l)	$\sin(2x + \pi)$
(m)	$\cos(\pi - x)$	(n)	$\sin\left(\frac{1}{4}x + \frac{\pi}{2}\right)$	(o)	$\frac{e^x - 2}{\sqrt{e^x}}$

5. Using the general power rule, find the indefinite integral of

(a)	$(4x-1)^3$	(b)	$(3x+5)^6$	(c)	$(2-x)^4$

(d) $(2x+3)^5$ (e) $(7-3x)^8$ (f) $\left(\frac{1}{2}x-2\right)^9$

(g) $(5x+2)^{-6}$ (h) $(9-4x)^{-2}$ (i) $(x+3)^{-3}$

(j) $\frac{1}{x+1}$ (k) $\frac{2}{2x+1}$ (l) $\frac{4}{3-2x}$

(m) $\frac{-3}{5-x}$ (n) $\frac{9}{3-6x}$ (o) $\frac{5}{3x+2}$

6. Find the antiderivative of

(a) $\sin(2x-3)-2x$ (b) $3\cos\left(2+\frac{1}{2}x\right)+5$ (c) $\frac{1}{2}\cos\left(2-\frac{1}{3}x\right)-\frac{2}{2x+1}$

(d) $\sec^2(0.1x-5)-2$ (e) $\frac{4}{2x+3}-e^{-\frac{1}{2}x+2}$ (f) $\frac{4}{(2x+3)^2}-e^{2x-\frac{1}{2}}$

(g) $\frac{x+2}{x+1}-\frac{4}{x+2}$ (h) $\frac{2x+1}{x+2}+\frac{1}{2x+1}$ (i) $\frac{2}{(2x+1)^2}+\frac{2}{2x+1}$

7. Find $f(x)$ given that

(a) $f'(x) = \sqrt{4x+5}$ where $f(-1) = \frac{1}{6}$.

(b) $f'(x) = \frac{8}{4x-3}$ where $f(1) = 2$.

(c) $f'(x) = \cos(2x+3)$ where $f\left(\pi-\frac{3}{2}\right) = 1$.

(d) $f'(x) = 2-e^{-2x+1}$ where $f(0) = e$.

8. A bacteria population, N thousand, has a growth rate modelled by the equation

$$\frac{dN}{dt} = \frac{4000}{1+0.5t}, t \geq 0$$

where t is measured in days. Initially there are 250 bacteria in the population. Find the population size after 10 days.

9. The acceleration, in m/s^2 of a body in a medium is given by $\frac{dv}{dt} = \frac{3}{t+1}, t \geq 0$.

The particle has an initial speed of 6 m/s, find the speed (to 2 d.p) after 10 seconds.

10. The rate of change of the water level in a container, t seconds after it starts to be filled from a tap, is given by the relation $\frac{dh}{dt} = 0.2\sqrt[3]{t+8}, t \geq 0$, where h cm is the water level. Find the level of water after 6 seconds.

11. The gradient function of the curve $y = f(x)$ is given by $e^{0.5x}-\cos(2x)$. Find the equation of the function, given that it passes through the origin.

12. (a) Given that $\frac{d}{dx}(e^{ax}(p\sin bx+q\cos bx)) = e^{ax}\sin bx$, express p and q in terms of a and b.

(b) Hence find $\int e^{2x}\sin 3x\,dx$.

13. The rate of change of the charge, Q, in coulombs, retained by a capacitor t minutes after charging is given by $\frac{dQ}{dt} = -ake^{-kt}$. Using the graph shown, determine the charge remaining after

(a) one hour.
(b) 80 minutes.

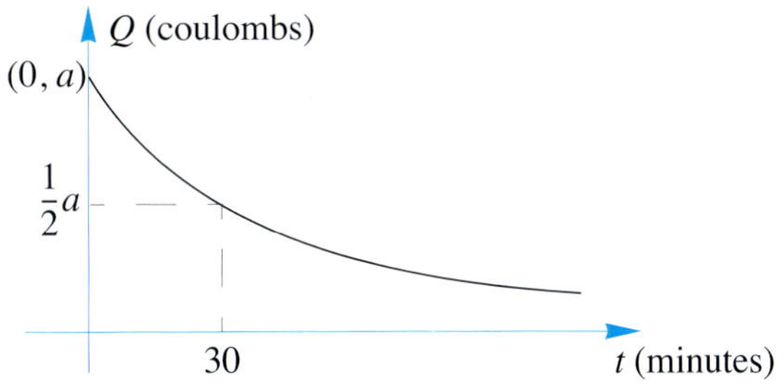

14. (a) Show that $\frac{d}{dx}(x\ln(x+k)) = \frac{x}{x+k} + \ln(x+k)$, where k is a real number.

(b) For a particular type of commercial fish, it is thought that a length-weight relationship exists such that their rate of change of weight, w kg, with respect to their length, x m, is modelled by the equation $\frac{dw}{dx} = 0.2\ln(x+2)$.

Given that a fish in this group averages a weight of 650 gm when it is 20 cm long, find the weight of a fish measuring 30 cm.

15. The rate of flow of water, $\frac{dV}{dt}$ litres/hour, pumped into a hot water system over a 24-hour period from 6:00 am, is modelled by the relation $\frac{dV}{dt} = 12 + \frac{3}{2}\cos\frac{\pi}{3}t$, $t \geq 0$.

(a) Sketch the graph of $\frac{dV}{dt}$ against t.

(b) For what percentage of the time will the rate of flow exceed 11 litres/hour.

(c) How much water has been pumped into the hot water system by 8:00 am?

16. The rates of change of the population size of two types of insect pests over a 4-day cycle, where t is measured in days, has been modelled by the equations

$$\frac{dA}{dt} = 2\pi\cos\pi t,\ t \geq 0$$

and

$$\frac{dB}{dt} = \frac{3}{4}e^{0.25t},\ t \geq 0$$

where A and B represent the number in thousands of pest A and B respectively. Initially there were 5000 insects of type A and 3000 insects of type B.

(a) On the same set of axes sketch the graphs, $A(t)$ and $B(t)$ for $0 \leq t \leq 4$.

(b) What is the maximum number of insects of type A that will occur?

(c) When will there first be equal numbers of insects of both types.

(d) For how long will the number of type B insects exceed the number of type A insects during the four days?

22.4 THE DEFINITE INTEGRAL

22.4.1 WHY THE DEFINITE INTEGRAL?

Unlike the previous section where the indefinite integral of an expression resulted in a new expression, when finding the **definite integral** we produce a **numerical value**.

Definite integrals are important because they can be used to find different types of measures, for example, areas, volumes, lengths and so on. It is, in essence, an extension of the work we have done in the previous sections.

22.4.2 LANGUAGE AND NOTATION

If the function $f(x)$ is continuous at every point on the interval [a,b] and $F(x)$ is any antiderivative of $f(x)$ on [a,b], then $\int_a^b f(x)dx$ is called the **definite integral** and is equal to $F(b)-F(a)$. That is, $\int_a^b f(x)dx = F(b)-F(a)$.

Which is read as:

"**the integral of $f(x)$ with respect to *x* from *a* to *b* is equal to $F(b)-F(a)$.**"

Usually we have an intermediate step to aid in the evaluation of the definite integral, this provides a somewhat 'compact recipe' for the evaluation process. This intermediate step is written as $[F(x)]_a^b$.

We therefore write

$$\int_a^b f(x)dx = [F(x)]_a^b = F(b)-F(a)$$

The process is carried out in 4 steps:

Step 1	Find an indefinite integral of $f(x)$, $F(x)$ (say).
Step 2	Write your result as $[F(x)]_a^b$.
Step 3	Substitute a and b into $F(x)$
Step 4	Subtract: $F(b)-F(a)$ to obtain the numerical value

Notice that the constant of integration *c* is omitted. This is because it would cancel itself out upon carrying out the subtraction: $(F(b)+c)-(F(a)+c) = F(b)-F(a)$.

In the expression $\int_a^b f(x)dx$, *x* is called the **variable of integration**, and *a* and *b* are called the **lower limit** and **upper limit** respectively. It should also be noted that there is no reason why the number *b* need be greater than the number *a* when finding the definite integral. That is, it is just as reasonable to write $\int_{2}^{-3} f(x)dx$ as it is to write $\int_{-3}^{2} f(x)dx$, both expressions are valid.

EXAMPLE 22.14

Evaluate the following

(a) $\int_{3}^{5} \frac{1}{x} dx$ (b) $\int_{2}^{4} \left(x + \frac{1}{x}\right)^2 dx$ (c) $\int_{0}^{1} \left(e^{2x} + \frac{3}{x+1}\right) dx$

(d) $\int_{\frac{\pi}{6}}^{\frac{\pi}{2}} \sin 3x dx$ (e) $\int_{5}^{2} (3x-4)^4 dx$ (f) $\int_{-2}^{0} (x - e^{-x}) dx$

Solution

(a) $\int_{3}^{5} \frac{1}{x} dx = [\log_e x]_3^5 = \log_e 5 - \log_e 3 = \log_e\left(\frac{5}{3}\right) \approx 0.511$

(b) $\int_{2}^{4} \left(x + \frac{1}{x}\right)^2 dx = \int_{2}^{4} \left(x^2 + 2 + \frac{1}{x^2}\right) dx = \left[\frac{1}{3}x^3 + 2x - \frac{1}{x}\right]_2^4$

$$= \left(\frac{1}{3}(4)^3 + 2(4) - \frac{1}{4}\right) - \left(\frac{1}{3}(2)^3 + 2(2) - \frac{1}{2}\right)$$

$$= \left(\frac{64}{3} + 8 - \frac{1}{4}\right) - \left(\frac{8}{3} + 4 - \frac{1}{2}\right)$$

$$= \frac{275}{12} (\approx 22.92)$$

(c) $\int_{0}^{1} \left(e^{2x} + \frac{3}{x+1}\right) dx = \left[\frac{1}{2}e^{2x} + 3\log_e(x+1)\right]_0^1 = \left(\frac{1}{2}e^2 + 3\log_e 2\right) - \left(\frac{1}{2}e^0 + 3\log_e 1\right)$

$$= \frac{1}{2}e^2 + 3\log_e 2 - \frac{1}{2}$$

$$\approx 5.27$$

(d) $\int_{\frac{\pi}{6}}^{\frac{\pi}{2}} \sin 3x dx = \left[-\frac{1}{3}\cos 3x\right]_{\frac{\pi}{6}}^{\frac{\pi}{2}} = -\frac{1}{3}[\cos 3x]_{\frac{\pi}{6}}^{\frac{\pi}{2}} = -\frac{1}{3}\left(\cos\left(\frac{3\pi}{2}\right) - \cos\left(\frac{3\pi}{6}\right)\right) = 0$

(e) $\int_{5}^{2} (3x-4)^4 dx = \left[\frac{1}{3(5)}(3x-4)^5\right]_5^2 = \frac{1}{15}((2)^5 - (11)^5) = -\frac{161019}{15} = -10734.6$

(f) $\int_{-2}^{0} (x - e^{-x}) dx = \left[\frac{1}{2}x^2 + e^{-x}\right]_{-2}^{0} = (0 + e^0) - \left(\frac{1}{2}(-2)^2 + e^2\right) = -1 - e^2 \approx -8.39$

EXAMPLE 22.15

Differentiate $y = e^{x^2+3}$. Hence find the exact value of $\int_0^1 5xe^{x^2+3}dx$.

Solution

Differentiating we have, $\frac{d}{dx}(e^{x^2+3}) = 2xe^{x^2+3}$.

Therefore, $\int_0^1 5xe^{x^2+3}dx = \frac{5}{2}\int_0^1 2xe^{x^2+3}dx = \frac{5}{2}\int_0^1 \frac{d}{dx}(e^{x^2+3})dx$

$$= \frac{5}{2}[e^{x^2+3}]_0^1 = \frac{5}{2}(e^4 - e^3)$$

Notice that in this case we made use of the fact that if $f'(x) = g(x)$ then $\int g(x)dx = f(x) + c$.

The above definite integrals can also be found using the **fnInt** function on the TI–83, although it would not provide an exact answer.

First, use the **MATH** menu and then select the **fnInt** command. Enter the equation of the integrand, the variable, and then the lower and upper limits.

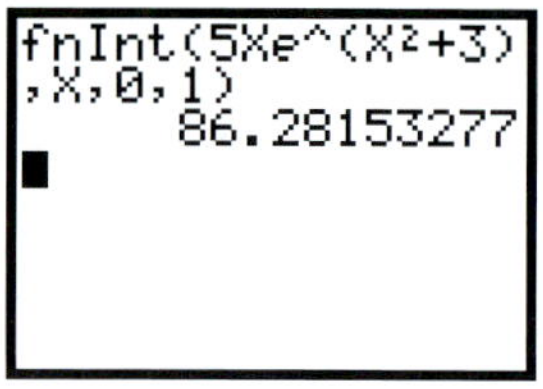

EXAMPLE 22.16

The production rate for radios by the average worker at Bat-Rad Pty Ltd t hours after starting work at 7:00 am, is given by $N'(t) = -2t^2 + 8t + 10, 0 \le t \le 4$. How many units can the average worker assemble in the second hour of production.

Solution

The second hour starts at $t = 1$ and ends at $t = 2$. Therefore, the number of radios assembled by the average worker in the second hour of production is given by $N = \int_{t=1}^{t=2} N'(t)dt$

That is, $N = \int_1^2 (-2t^2 + 8t + 10)dt = \left[-\frac{2}{3}t^3 + 4t^2 + 10t\right]_1^2$

$$= \left(-\frac{2}{3}(2)^3 + 4(2)^2 + 10(2)\right) - \left(-\frac{2}{3}(1)^3 + 4(1)^2 + 10(1)\right)$$

$$= \frac{52}{3}$$

Note that to evaluate the definite integral we made use of the graphics calculator:

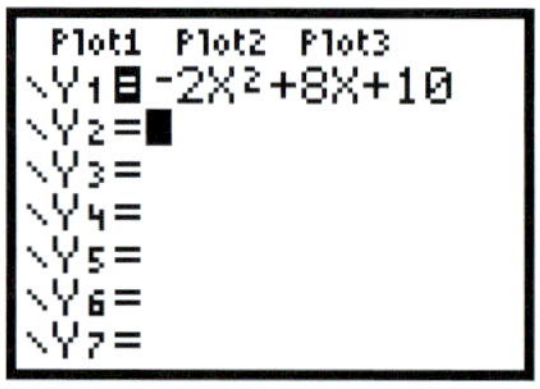

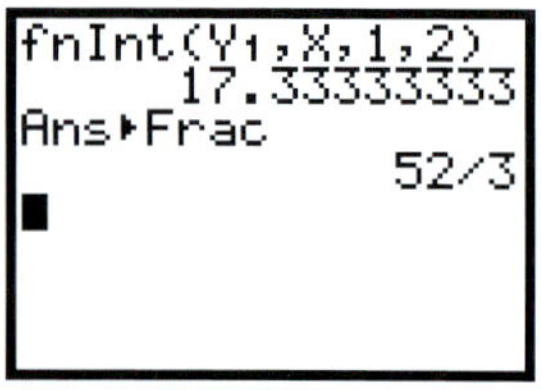

22.4.3 PROPERTIES OF THE DEFINITE INTEGRAL

There are a number of simple but very important properties of the definite integral (many of which we have already used). Some of these properties are summarised below (this list is not an exhaustive list) and it is assumed that the integrand, $f(x)$ is continuous on [a,b] and that the definite integral, $\int_a^b f(x)dx$ is finite.

Property	Examples
1. $\int_a^a f(x)dx = 0$	$\int_2^2 (x+1)dx = \left[\frac{x^2}{2}+x\right]_2^2 = \left(\frac{(2)^2}{2}+(2)\right)-\left(\frac{(2)^2}{2}+(2)\right)$ $= 0$
2. $\int_a^b f(x)dx = -\int_b^a f(x)dx$	$\int_1^2 e^{3x}dx = \left[\frac{1}{3}e^{3x}\right]_1^2 = \frac{1}{3}(e^6-e^3)$ $\int_2^1 e^{3x}dx = \left[\frac{1}{3}e^{3x}\right]_2^1 = \frac{1}{3}(e^3-e^6) = -\frac{1}{3}(e^6-e^3)$ $= -\int_1^2 e^{3x}dx$
3. $\int_a^b f(x)dx = \int_a^c f(x)dx + \int_c^b f(x)dx$ where $a \le c \le b$.	$\int_2^5 \frac{1}{x}dx = [\ln x]_2^5 = \ln 5 - \ln 2 = \ln 2.5$ $\int_2^3 \frac{1}{x}dx + \int_3^5 \frac{1}{x}dx = [\ln x]_2^3 + [\ln x]_3^5$ $= (\ln 3 - \ln 2) + (\ln 5 - \ln 3)$ $= \ln 5 - \ln 2 = \ln 2.5$
4. $\int_a^b kf(x)dx = k\int_a^b f(x)dx,\ k \in \mathbb{R}$	$\int_0^{\frac{\pi}{4}} 3\cos(2x)dx = 3\int_0^{\frac{\pi}{4}} \cos(2x)dx = 3\left[\frac{1}{2}\sin(2x)\right]_0^{\frac{\pi}{4}} = \frac{3}{2}$
5. $\int_a^b (f(x) \pm g(x))dx$ $= \int_a^b f(x)dx \pm \int_a^b g(x)dx$	$\int_1^3 (x+e^{-2x})dx = \left[\frac{1}{2}x^2-\frac{1}{2}e^{-2x}\right]_1^3 = \frac{1}{2}(8-e^{-6}+e^{-2})$ Similarly, $\int_1^3 xdx + \int_1^3 e^{-2x}dx = \left[\frac{1}{2}x^2\right]_1^3 + \left[-\frac{1}{2}e^{-2x}\right] = \frac{1}{2}(8-e^{-6}+e^{-2})$

The use of these properties relies on your ability to recognise their appropriateness in any given situation.

EXAMPLE 22.17

Suppose that $f(x)$ and $g(x)$ are continuous functions on the interval [1,5] and that $\int_1^3 f(x)dx = -2$, $\int_1^5 f(x)dx = 7$, $\int_1^3 g(x)dx = 3$ and $\int_1^5 g(x)dx = 5$.

Evaluate the following

(a) $\int_1^3 (3f(x) - g(x))dx$ (b) $\int_5^1 (f(x) - g(x))dx$ (c) $\int_1^3 (5 - 4f(x))dx$

Solution

(a) $\int_1^3 (3f(x) - g(x))dx = \int_1^3 3f(x)dx - \int_1^3 g(x)dx = 3\int_1^3 f(x)dx - 3 = 3 \times (-2) - 3 = -9$

(b) $\int_5^1 (f(x) - g(x))dx = \int_5^1 f(x)dx - \int_5^1 g(x)dx = -\int_1^5 f(x)dx - \left(-\int_1^5 g(x)dx\right) = -(7) + 5 = -2$

(c) $\int_1^3 (5 - 4f(x))dx = \int_1^3 5dx - 4\int_1^3 f(x)dx = [5x]_1^3 - 4(-2) = (15 - 5) + 8 = 18$

EXAMPLE 22.18

Find the value(s) of m, where

(a) $\int_0^m (2x - 4)dx = -4$ (b) $\int_0^m (2x - 4)dx = 5$

Solution

(a) $\int_0^m (2x - 4)dx = -4 \Rightarrow [x^2 - 4x]_0^m = -4 \Leftrightarrow m^2 - 4m = -4$

$\Leftrightarrow m^2 - 4m + 4 = 0$

$\Leftrightarrow (m - 2)^2 = 0 \Leftrightarrow m = 2$.

(b) $\int_0^m (2x - 4)dx = 5 \Leftrightarrow m^2 - 4m = 5 \Leftrightarrow m^2 - 4m - 5 = 0$

$\Leftrightarrow (m - 5)(m + 1) = 0$

$\Leftrightarrow m = 5$ or $m = -1$

Notice that **both** solutions must be given.

EXAMPLE 22.19

Evaluate

(a) $\int_1^3 \sqrt{2x + 3}dx$ (b) $\int_0^k \frac{2}{(3 - 2x)^2}dx$, $k < \frac{3}{2}$

Solution

(a) $\int_1^3 \sqrt{2x+3}\,dx = \int_1^3 (2x+3)^{1/2}dx = \left[\dfrac{1}{2\times\left(\frac{1}{2}+1\right)}(2x+3)^{\frac{1}{2}+1}\right]_1^3$ [Using power rule].

$$= \left[\frac{1}{3}(2x+3)^{\frac{3}{2}}\right]_1^3$$

$$= \frac{1}{3}[(9^{3/2}) - 5^{3/2}]$$

$$= \frac{1}{3}(27 - 5\sqrt{5})$$

(b) $\int_0^k \dfrac{2}{(3-2x)^2}dx = \int_0^k 2(3-2x)^{-2}dx = 2\left[\dfrac{1}{-2\times -1}(3-2x)^{-1}\right]_0^k = \left[\dfrac{1}{3-2x}\right]_0^k$

$$= \frac{1}{3-2k} - \frac{1}{3}$$

$$= \frac{2k}{3(3-2k)}$$

Exercises 22.4

1. Evaluate the following

(a) $\int_1^4 x\,dx$ (b) $\int_4^9 \sqrt{x}\,dx$ (c) $\int_2^3 \dfrac{2}{x^3}dx$ (d) $\int_{16}^9 \dfrac{4}{\sqrt{x}}dx$

2. Evaluate the following definite integrals (giving exact answers)

(a) $\int_1^2 \left(x^2 - \dfrac{3}{x^4}\right)dx$ (b) $\int_0^2 (x\sqrt{x} - x)dx$ (c) $\int_0^2 (1 + 2x - 3x^2)dx$

(d) $\int_{-2}^0 (x+1)dx$ (e) $\int_0^{-1} x^3(x+1)dx$ (f) $\int_{-1}^1 (x+1)(x^2-1)dx$

(g) $\int_1^4 (\sqrt{x} - 1)^2 dx$ (h) $\int_1^2 \left(x - \dfrac{1}{x}\right)^2 dx$ (i) $\int_1^3 \left(\dfrac{x^3 - x^2 + x}{x}\right)dx$

(j) $\int_{-1}^1 (x - x^3)dx$ (k) $\int_1^4 \dfrac{x+1}{\sqrt{x}}dx$ (l) $\int_1^4 \left(\sqrt{\dfrac{2}{x}} - \sqrt{\dfrac{x}{2}}\right)dx$

3. Use a graphics calculator to check your answers to Question 2.

4. Evaluate the following definite integrals (giving exact values)

(a) $\int_0^1 (e^x + 1)dx$ (b) $\int_1^2 \frac{4}{e^{2x}}dx$ (c) $\int_{-1}^1 (e^x - e^{-x})dx$

(d) $\int_{-1}^1 (e^x + e^{-x})dx$ (e) $\int_{-1}^1 (e^x + e^{-x})^2 dx$ (f) $\int_2^0 e^{2x+1}dx$

(g) $\int_0^1 (\sqrt{e^x} - 1)dx$ (h) $\int_0^1 \left(e^{\frac{1}{4}x} - e^{4x}\right)dx$ (i) $\int_1^{-1} e^{1-2x}dx$

5. Use a graphics calculator to check your answers to Question 4.

6. Evaluate the following definite integrals (giving exact values)

(a) $\int_1^2 \frac{3}{x}dx$ (b) $\int_0^4 \frac{2}{x+1}dx$ (c) $\int_2^6 \frac{x+4}{x}dx$

(d) $\int_4^5 \left(x^2 + \frac{1}{x}\right)^2 dx$ (e) $\int_{-1}^0 \frac{3}{1-2x}dx$ (f) $\int_0^1 \frac{2}{x+1}dx$

(g) $\int_0^1 \frac{2}{(x+1)^3}dx$ (h) $\int_2^4 \left(\sqrt{x} - \frac{2}{\sqrt{x}}\right)^2 dx$ (i) $\int_3^4 \frac{2x+1}{2x^2-3x-2}dx$

7. Use a graphics calculator to check your answers to Question 6.

8. Evaluate the following definite integrals (giving exact values)

(a) $\int_0^{\frac{\pi}{2}} \sin(2x)dx$ (b) $\int_{-\pi}^0 \cos\left(\frac{1}{3}x\right)dx$ (c) $\int_{\frac{\pi}{6}}^{\frac{\pi}{3}} \sec^2 4x dx$

(d) $\int_0^{\pi} \left(\cos x - \sin\left(\frac{x}{2}\right)\right)dx$ (e) $\int_0^{\frac{\pi}{4}} (x - \sec^2 x)dx$ (f) $\int_0^{\frac{\pi}{2}} 2\cos\left(4x + \frac{\pi}{2}\right)dx$

(g) $\int_{-\pi}^{\pi} \left(\sin\left(\frac{x}{2}\right) + 2\cos(x)\right)dx$ (h) $\int_0^{\frac{\pi}{12}} \sec^2\left(\frac{\pi}{4} - 2x\right)dx$ (i) $\int_0^{\pi} \cos(2x + \pi)dx$

9. Evaluate the following definite integrals (giving exact values)

(a) $\int_0^1 (x+1)^4 dx$ (b) $\int_1^3 \sqrt{2x+1}dx$ (c) $\int_{-1}^2 (1-2x)^3 dx$

(d) $\int_0^1 \frac{1}{(x+2)^3}dx$ (e) $\int_5^8 \frac{1}{\sqrt[3]{x-4}}dx$ (f) $\int_0^1 \frac{x}{(1+x)}dx$

10. Show that $\frac{2x+6}{x^2+6x+5} \equiv \frac{1}{x+1} + \frac{1}{x+5}$. Hence evaluate $\int_0^2 \frac{2x+6}{x^2+6x+5}dx$

11. Find $\frac{d}{dx}(x\sin 2x)$. Hence find the exact value of $\int_0^{\pi} x\cos 2x dx$.

12. Given that $\int_a^b f(x)dx = m$ and $\int_a^b g(x)dx = n$, find

(a) $\int_a^b 2f(x)dx - \int_a^b g(x)dx$ (b) $\int_a^b (f(x) - 1)dx$ (c) $\int_b^a 3g(x)dx$

(d) $\int_a^b (af(x) - m)dx$ (e) $\int_a^b (b^2 g(x) - 2nx)dx$

13. (a) Find $\frac{d}{dx}(xe^{0.1x})$. Hence, find $\int xe^{0.1x}dx$.

(b) Following an advertising initiative by the Traffic Authorities, preliminary results predict that the number of alcohol-related traffic accidents has been decreasing at a rate of $-12 - te^{0.1t}$ accidents per month, where t is the time in months since the advertising campaign started.

i. How many accidents were there over the first six months of the campaign.

ii. In the year prior to the advertising campaign there were 878 alcohol-related traffic accidents. Find an expression for the total number of accidents since the start of the previous year, t months after the campaign started.

14. The rate of Cable television subscribers in a city t years from 1995 has been modelled by the equation $\frac{2000}{\sqrt{(1 + 0.4t)^3}}$.

(a) How many subscribers were there between 1998 and 2002?

(b) If there were initially 40,000 subscribers, find the number of subscribers by 2010.

15. (a) Find $\frac{d}{dt}\left(\frac{800}{1 + 24e^{-0.02t}}\right)$.

(b) The rate at which the number of fruit flies appear when placed in an environment with limited food supply in an experiment was found to be approximated by the exponential model $\frac{384e^{-0.02t}}{(1 + 24e^{-0.02t})^2}$, $t \geq 0$, where t is the number of days since the experiment started. What was the increase in the number of flies after 200 days?

22.5 APPLICATIONS OF INTEGRATION

In Chapters 20 and 21 we saw that differentiation had a geometric meaning, that is, it provided a measure of the gradient of the curve at a particular point. We have also seen applications of the definite integral throughout the previous sections in this chapter. In this section we will investigate the geometric significance of the integral.

22.5.1 INTRODUCTION TO THE AREA BENEATH A CURVE

Consider the problem of finding the exact value of the shaded area, A sq. units, in the diagram shown.

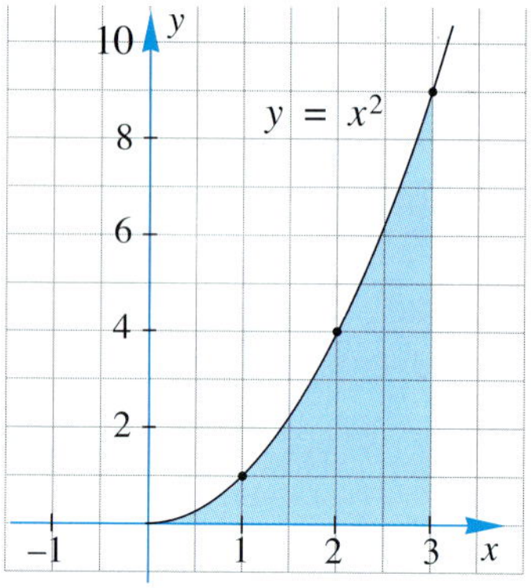

As a first step we make use of rectangular strips as shown below to obtain an approximation of the shaded area. We can set up a table of values, use it to find the area of each strip and then sum these areas.

In Figure 1, the rectangles lie below the curve, and so we call these the **lower rectangles**. In Figure 2, the rectangles lie above the curve, and so we call these the **upper rectangles**. Figure 3 shows that the true area (or exact area) lies somewhere between the sum of the areas of the lower rectangles, S_L, and the sum of the areas of the upper rectangles, S_U.

Lower rectangles, S_L Upper rectangles, S_U Exact area, A

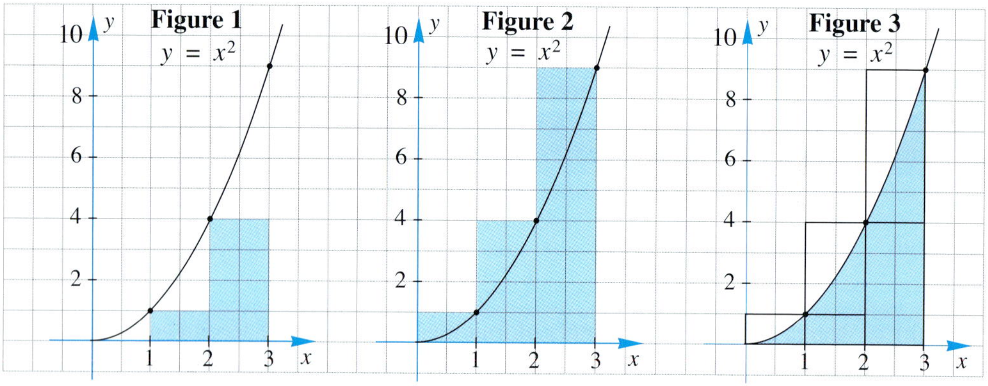

That is, we have that

Lower Sum = S_L < Exact Area, A < S_U = Upper Sum

In the case above we have that $S_L = 1 \times 1 + 1 \times 4 = 5$ and $S_U = 1 \times 1 + 1 \times 4 + 1 \times 9 = 14$.

Therefore, we can write $5 < A < 14$. However, this does seem to be a poor approximation as there is a difference of 9 sq. units between the lower approximation and the upper approximation. The problem lies in the fact that we have only used two rectangles for the lower sum and three rectangles for the upper sum. We can improve on our approximation by increasing the number of rectangles that are used. For example, we could used 5 lower rectangles and 6 upper rectangles, or 10 lower rectangles and 12 upper rectangles and so on.

22.5.2 IN SEARCH OF A BETTER APPROXIMATION

As shown in the diagrams below, as we increase the number of rectangular strips (or decrease the width of each strip) we obtain better approximations to the exact value of the area.

5 lower & 6 upper rectangles

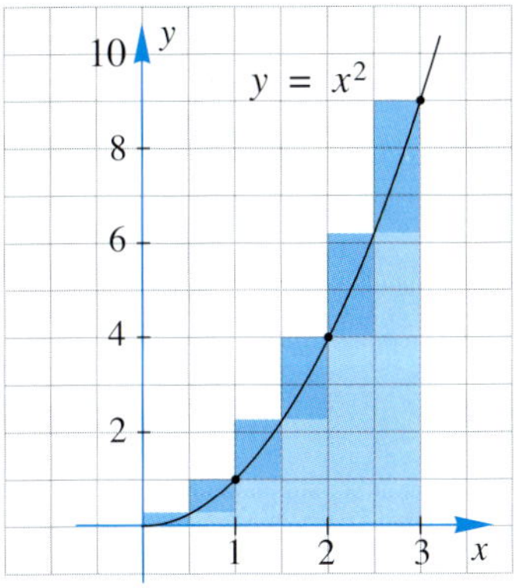

10 lower & 12 upper rectangles

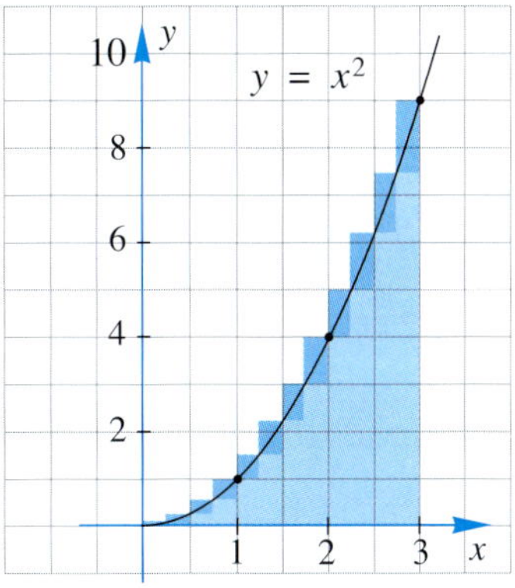

We can make use of a table of values to find the sum of the areas of the lower and upper rectangles:

Using intervals of width $\frac{1}{2}$ units.

X	Y1	
0	0	
.5	.25	
1	1	
1.5	2.25	
2	4	
2.5	6.25	
3	9	
X=0		

Using intervals of width $\frac{1}{4}$ units.

X	Y1	
0	0	
.25	.0625	
.5	.25	
.75	.5625	
1	1	
1.25	1.5625	
1.5	2.25	
X=0		

For intervals of width 0.5 we have:

$$\left.\begin{aligned} S_L &= \frac{1}{2}\times[0.25+1+2.25+4+6.25] = 6.875 \\ S_U &= \frac{1}{2}\times[0.25+1+2.25+4+6.25+9] = 11.375 \end{aligned}\right\} \text{ i.e., } 6.88 < A < 11.38$$

For intervals of width 0.25 we have:

$$\left.\begin{aligned} S_L &= \frac{1}{4}\times[0.25+0.5625+1+1.5625+2.25+\ldots+7.5625] \approx 7.89 \\ S_U &= \frac{1}{4}\times[0.0625+0.25+0.5625+\ldots+7.5625+9] \approx 10.16 \end{aligned}\right\} \text{ i.e., } 7.89 < A < 10.16$$

By continuing in this manner, the value of A will become *sandwiched* between a lower value and an upper value. Of course the more intervals we have the '*tighter*' the sandwich will be! What we can say is that if we partition the interval [0, 3] into n equal subintervals, then, as the number of rectangles increases, S_L increases towards the exact value A while S_U decreases towards the exact value A. That is,

$$\lim_{n\to\infty} S_L = A = \lim_{n\to\infty} S_U$$

As we have seen, even for a simple case such as $y = x^2$, this process is rather tedious. And as yet, we still have not found the exact area of the shaded region under the curve $y = x^2$ over the interval [0, 3].

22.5.3 TOWARDS AN EXACT AREA

We can produce an algebraic expression to determine the exact area enclosed by a curve. We shall also find that the definite integral plays a large part in determining the area enclosed by a curve.

As a starting point we consider a single rectangular strip.
Consider the function $y = f(x)$ as shown:

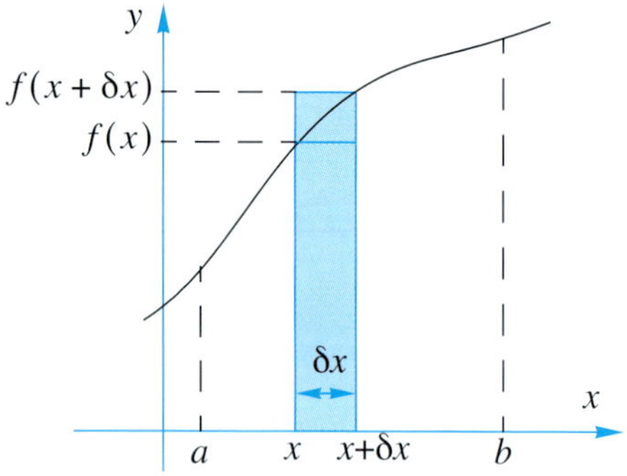

Divide the interval from $x = a$ to $x = b$ into n equal equal parts: $a = x_0, x_1, x_2, \ldots, x_n = b$.

This means that each strip is of width $\frac{b-a}{n}$.

We denote this width by δx so that $\delta x = \frac{b-a}{n}$.

The area of the lower rectangle is $f(x) \times \delta x$ and that of the upper rectangle is $f(x + \delta x) \times \delta x$.

Then, the sum of the areas of the lower rectangles for $a \le x \le b$ is $S_L = \sum_{x=a}^{b-\delta x} f(x)\delta x$

and the sum of the areas of the upper rectangles for $a \le x \le b$ is $S_U = \sum_{x=a}^{b} f(x+\delta x)\delta x$

Then, if A sq units is the area under the curve $y = f(x)$ over the interval $[a, b]$ we have that

$$\sum_{x=a}^{b-\delta x} f(x)\delta x < A < \sum_{x=a}^{b} f(x+\delta x)\delta x$$

As the number of strips increase, that is, as $n \to \infty$ and therefore $\delta x \to 0$ the area, A sq units, approaches a common limit, i.e., S_L from below, and S_U from above. We write this result as:

$$A = \lim_{\delta x \to 0} \sum_{x=a}^{b} f(x)\delta x$$

In fact this result leads to the use of the integral sign as a means whereby we can find the required area.

That is,

$$\lim_{\delta x \to 0} \sum_{x=a}^{b} f(x)\delta x = \int_a^b f(x)dx$$

Notice that we've only developed an appropriate notation and a 'recognition' that the definite integral provides a numerical value whose geometrical interpretation is connected to the area enclosed by a curve, the x–axis and the lines $x = a$ and $x = b$. We leave out a formal proof of this result in preference to having developed an intuitive idea behind the concept and relationship between area and the definite integral.

We can now combine our results of the definite integral with its geometrical significance in relation to curves on a Cartesian set of axes.

22.5.4 THE DEFINITE INTEGRAL AND AREAS

If $y = f(x)$ is **positive** and **continuous** on the interval [a,b], the area, A sq units, bounded by $y = f(x)$, the x-axis and the lines $x = a$ and $x = b$ is given by

$$\text{Area} = A = \int_a^b f(x)dx = \int_a^b y\,dx$$

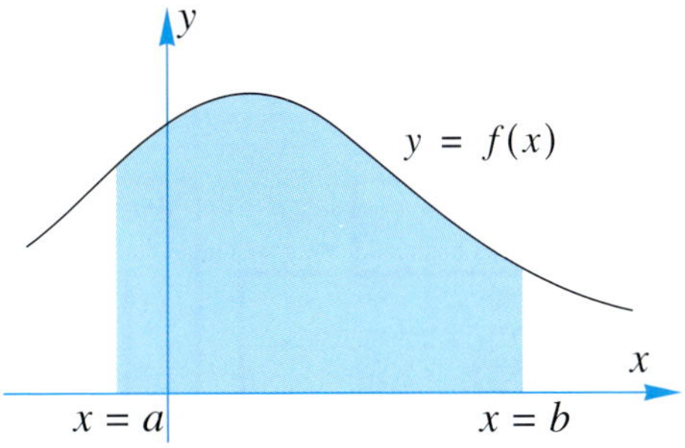

EXAMPLE 22.20

Find the area of the shaded region shown below.

(a)

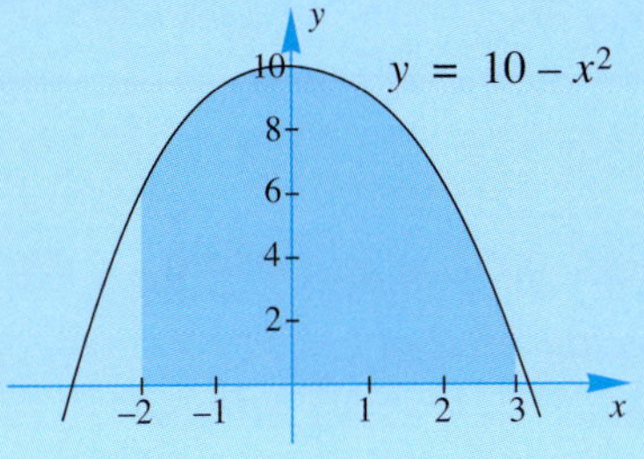

(b)

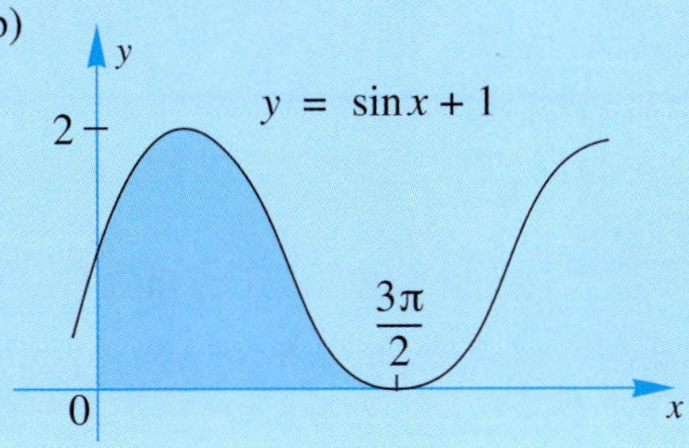

Solution

(a) $\text{Area} = \int_{-2}^{3}(10 - x^2)dx = \left[10x - \frac{1}{3}x^3\right]_{-2}^{3} = \left((30 - 9) - \left(-20 + \frac{8}{3}\right)\right) = \frac{115}{3}$

Therefore, the shaded area measures $\frac{115}{3}$ square units.

(b) $\text{Area} = \int_0^{\frac{3\pi}{2}}(\sin x + 1)dx = [-\cos x + x]_0^{\frac{3\pi}{2}} = \left(-\cos\left(\frac{3\pi}{2}\right) + \frac{3\pi}{2}\right) - (-\cos(0) + 0)$

$$= \frac{3\pi}{2} + 1$$

Therefore, the shaded area measures $\frac{3\pi}{2} + 1$ square units.

Apart from having the ability to 'find' an indefinite integral, or rather, sketch an indefinite integral, the TI–83 can also display the shaded regions required.

This is done by using the **Shade** command from the **DRAW** function.

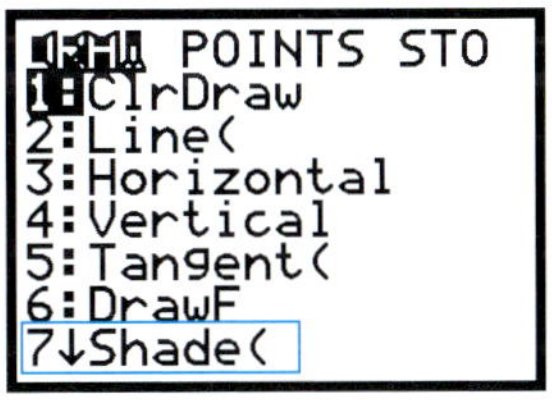

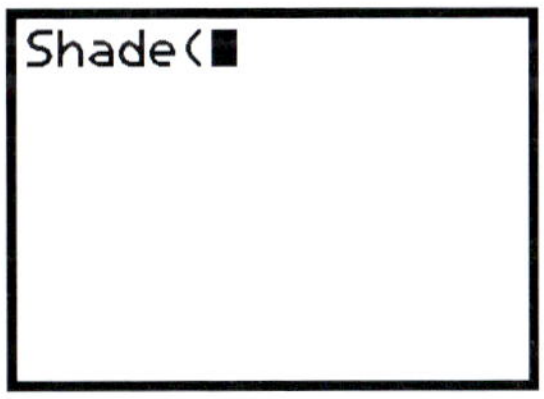

EXAMPLE 22.21

Find the area enclosed by the curve with equation $f(x) = 2 + x^2$, the x–axis and the lines $x = 1$ and $x = 3$.

Solution

Using the **DRAW** function, and then selecting **Shade**, we enter the required information, i.e., 0 (for $y = 0$. i.e., the x–axis), the equation of the function, the lower limit and then, the upper limit (do not forget to first enter the equation in the equation editor screen and to enter the appropriate window setting.) To find the measure of the area, we need to use the **fnInt** function:

1. Set up graph

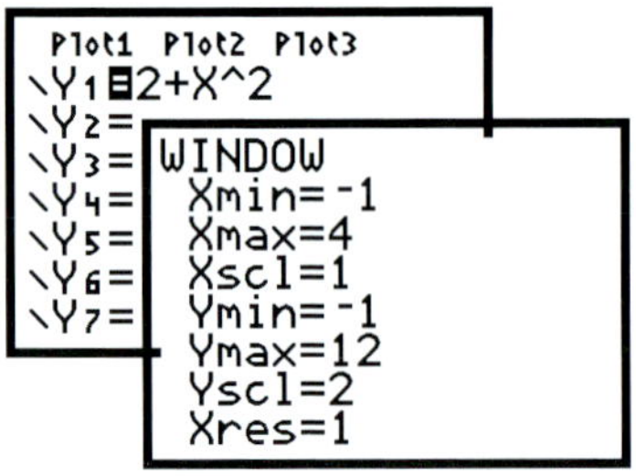

2. Display required region

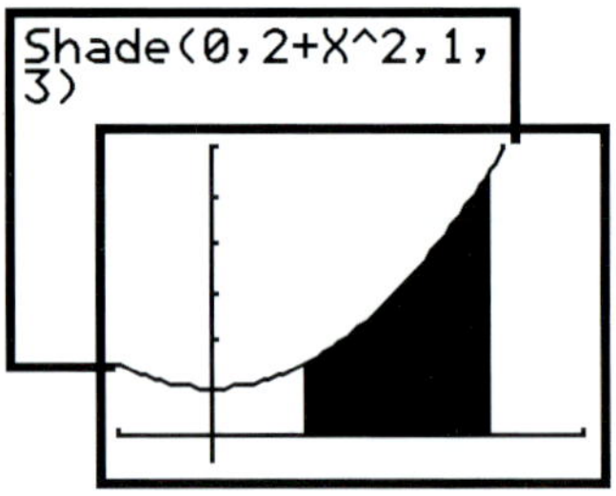

3. Find numerical value

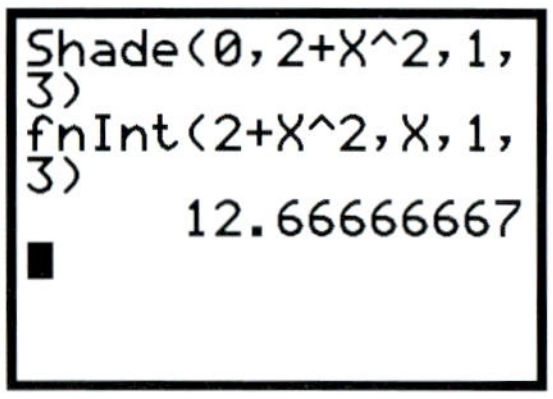

i.e., area is $12\frac{2}{3}$ sq units.

EXAMPLE 22.22

The area enclosed by the curve with equation $y = 4 - e^{-0.5x}$, the x–axis, the y–axis and the line $x = -2$, measures $k-2e$ sq units. Find the value of k.

Solution

Using the TI–83 to first visualise the situation we have:

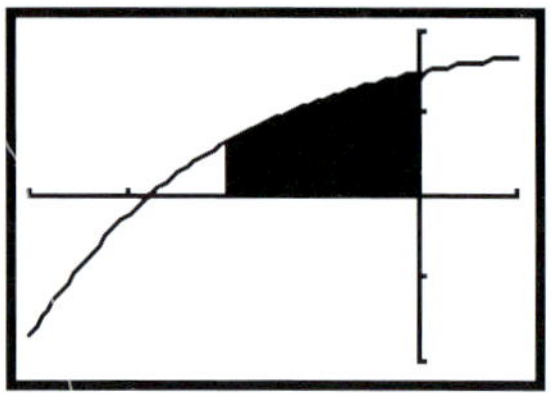

Therefore, we have $\int_{-2}^{0}(4 - e^{-0.5x})dx = k-2e$

$$[4x + 2e^{-0.5x}]_{-2}^{0} = k-2e$$

$$2 - (-8 + 2e^{1}) = k-2e$$

$$10 - 2e = k-2e$$

Therefore, $k = 10$.

22.5.5 FURTHER OBSERVATIONS ABOUT AREAS

1. To find the area bounded by $y = f(x)$, the y-axis and the lines $y = a$ and $y = b$ we carry out the following proces:

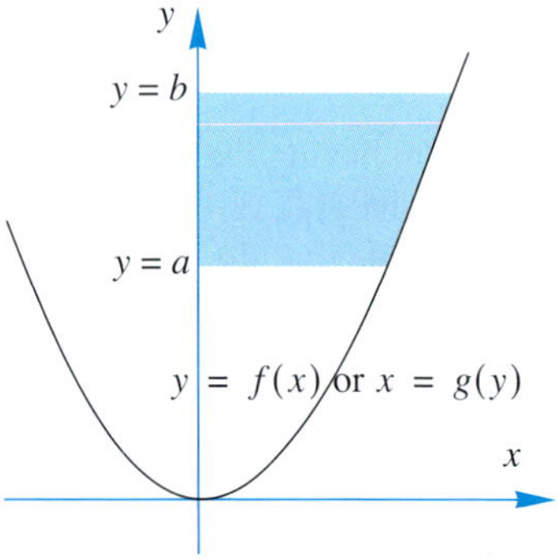

   1. First you need to make x the subject. i.e., from $y = f(x)$ obtain the new equation $x = g(y)$.
   2. Then find the definite integral,

   $$\int_a^b x\,dy = \int_a^b g(y)dy \text{ sq units.}$$

   which will give the shaded area.

2. If f is **negative** over the interval [a,b] (i.e $f(x) < 0$ for $a \leq x \leq b$), then the integral $\int_a^b f(x)dx$ is a negative number.

We therefore need to write the area, A, as

$$A = -\int_a^b f(x)dx$$

or, use the **absolute value** of the integral:

$$A = \left|\int_a^b f(x)dx\right|.$$

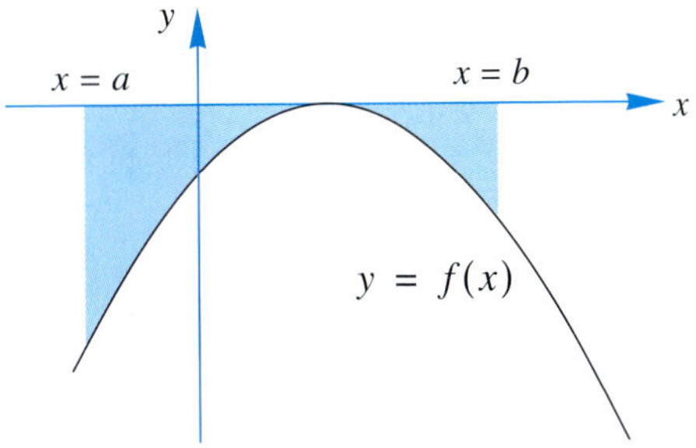

EXAMPLE 22.23

Find the area enclosed by the curve $y = \sqrt{x}$, the y–axis and the lines $y = 1$ and $y = 3$.

Solution

We begin by showing the required region on the cartesian plane:

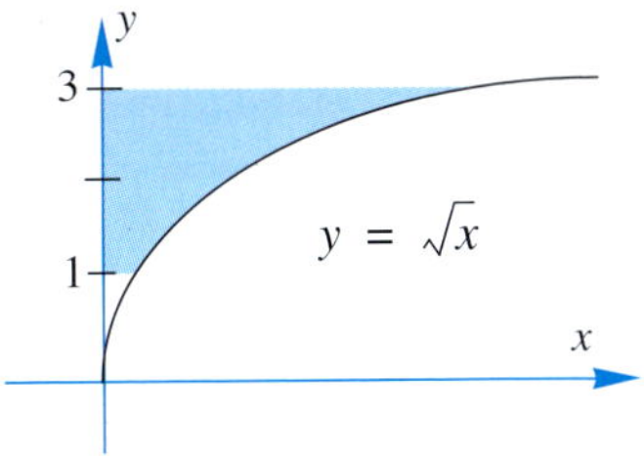

Next we need an expression for x in terms of y:

That is, $y = \sqrt{x} \Rightarrow y^2 = x, x > 0$

Therefore, the area of the shaded region is

$$A = \int_1^3 x dy = \int_1^3 y^2 dy = \left[\frac{1}{3}y^3\right]_1^3 = \frac{1}{3}(3^3 - 1^3) = \frac{26}{3}$$

So that the required area is $\frac{26}{3}$ sq units.

22.5.6 THE SIGNED AREA

It is possible for $y = f(x)$ to alternate between negative and positive values over the interval $x = a$ and $x = b$. That is there is at least one point $x = c$ where the graph crosses the x–axis, and so $y = f(x)$ changes sign when it crosses the point $x = c$.

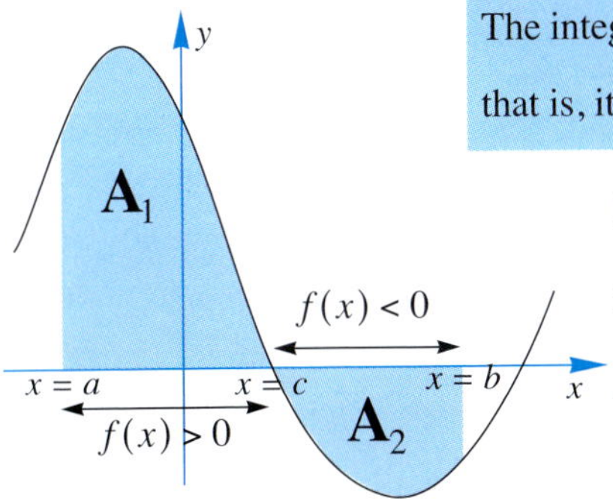

The integral $\int_a^b f(x)dx$ gives the **algebraic sum** of A_1 and A_2, that is, it gives the **signed area**.

For example, if $A_1 = 12$ and $A_2 = 4$, then the definite integral $\int_a^b f(x)dx = 12 - 4 = 8$.

This is because $\int_a^c f(x)dx = 12$, $\int_c^b f(x)dx = -4$ and so

$$\int_a^b f(x)dx = \int_a^c f(x)dx + \int_c^b f(x)dx = 12 + (-4) = 8$$

As $\int_c^b f(x)dx$ is a negative value, finding the negative of $\int_c^b f(x)dx$, that is $\left(-\int_c^b f(x)dx\right)$, would provide a positive value and therefore be a measure of the area of the region that is shaded below the x-axis.

The shaded area would then be given by $\int_a^c f(x)dx + \left(-\int_c^b f(x)dx\right)$.

This would provide the sum of two positive numbers.

22.5.7 STEPS FOR FINDING AREAS

It follows, that in order to find the area bounded by the curve $y = f(x)$, the x-axis and the lines $x = a$ and $x = b$, we first need to find where (and if) the curve crosses the x-axis at some point $x = c$ in the interval $a \le x \le b$. If it does, we must evaluate the area of the regions above and below the x-axis **separately**. Otherwise, evaluating $\int_a^b f(x)dx$ will provide the signed area (which only gives the correct area if the function lies above the x–axis over the interval $a \le x \le b$).

Therefore, we need to:

1. **Sketch** the graph of the curve $y = f(x)$ over the interval $a \le x \le b$. (In doing so you will also determine any x-intercepts).
2. **Integrate** $y = f(x)$ **over each region separately** (if necessary). (That is, regions above the x-axis and regions below the x-axis)
3. Add the required (positive terms).

EXAMPLE 22.24

Find the area of the region enclosed by the curve $y = x^3 - 1$, the x-axis and $0 \le x \le 2$.

Solution

First sketch the graph of the given curve: x-intercepts (when $y = 0$): $x^3 - 1 = 0 \Leftrightarrow x = 1$.
y-intercepts (when $x = 0$): $y = 0 - 1 = -1$.

From the graph we see that y is negative in the region [0,1] and positive in the region [1,2], therefore the area of the region enclosed is given by

$$A = \left(-\int_0^1 (x^3 - 1)dx\right) + \int_1^2 (x^3 - 1)dx = \left(-\left[\frac{x^4}{4} - x\right]_0^1\right) + \left[\frac{x^4}{4} - x\right]_1^2$$

$$= -\left(-\frac{3}{4}\right) + \left((2) - \left(-\frac{3}{4}\right)\right)$$

$$= 3.5$$

That is, the area measures 3.5 sq units.

Notice that $\int_0^2 (x^3 - 1)dx = \left[\frac{x^4}{4} - x\right]_0^2 = 2 \ (\ne 3.5)$.

Using the **CALC** option after sketching the curve $y = x^3 - 1$ on the TI–83, we can visualise these results more readily.

Signed area:

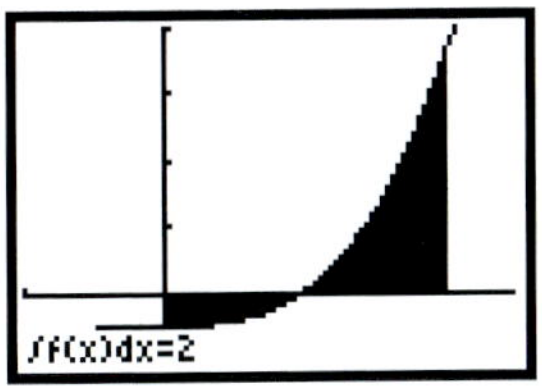

$\int_0^2 (x^3 - 1)dx = 2$

Actual area:

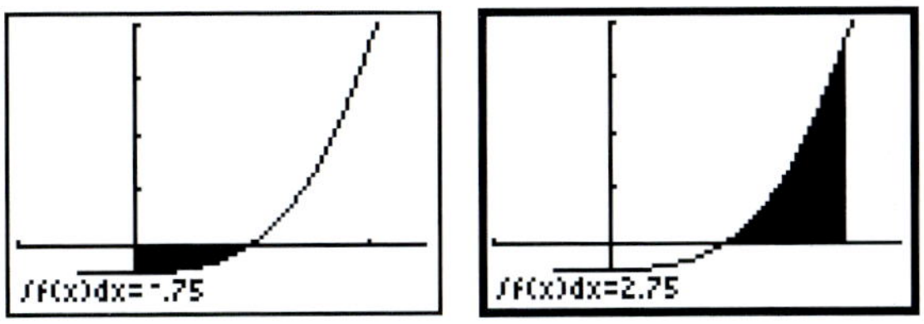

$A = \left(-\int_0^1 (x^3 - 1)dx\right) + \int_1^2 (x^3 - 1)dx = -(-0.75) + 2.75 = 3.5$

EXAMPLE 22.25

Find the area enclosed by the curve $y = x^3 - 3x^2 + 2x$, the x-axis and the lines $x = 0$ and $x = 3$.

Solution

First sketch the graph of the given curve:

x-intercepts (when $y = 0$): $x^3 - 3x^2 + 2x = 0 \Leftrightarrow x(x-2)(x-1) = 0 \therefore x = 0, 2, 1$

y-intercepts (when $x = 0$): $y = 0 - 0 + 0 = 0$.

From the diagram we have, Area = $A_1 - A_2 + A_3$.

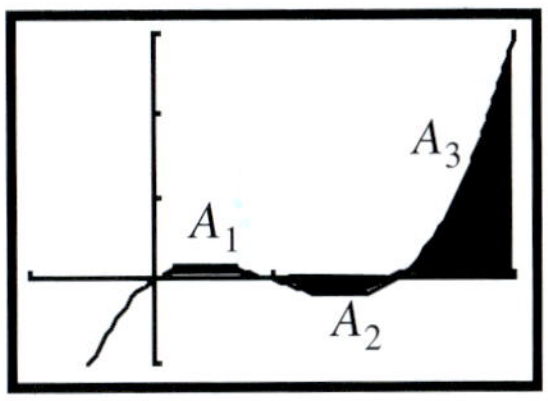

NB: $\int_0^3 (x^3 - 3x^2 + 2x)dx = \frac{9}{4}$

Now, $A_1 = \int_0^1 (x^3 - 3x^2 + 2x)dx = \left[\frac{x^4}{4} - x^3 + x^2\right]_0^1 = \frac{1}{4}$

$A_2 = \int_1^2 (x^3 - 3x^2 + 2x)dx = \left[\frac{x^4}{4} - x^3 + x^2\right]_1^2 = -\frac{1}{4}$

and $A_1 = \int_2^3 (x^3 - 3x^2 + 2x)dx = \left[\frac{x^4}{4} - x^3 + x^2\right]_2^3 = \frac{9}{4}$

Therefore, the required area is $\frac{1}{4} - \left(-\frac{1}{4}\right) + \frac{9}{4} = \frac{11}{4}$ sq units.

22.5.8 AREA BETWEEN TWO CURVES

The use of the definite integral in finding the area of a region enclosed by a single curve can be extended to finding the area enclosed between two curves. Although we do have a compact formula to find such areas, in reality it is a simple geometrical observation.

Consider two continuous functions, $f(x)$ and $g(x)$ on some interval $[a,b]$, such that over this interval, $g(x) \geq f(x)$. The area of the region enclosed by these two curves and the lines $x = a$ and $x = b$ is shown next.

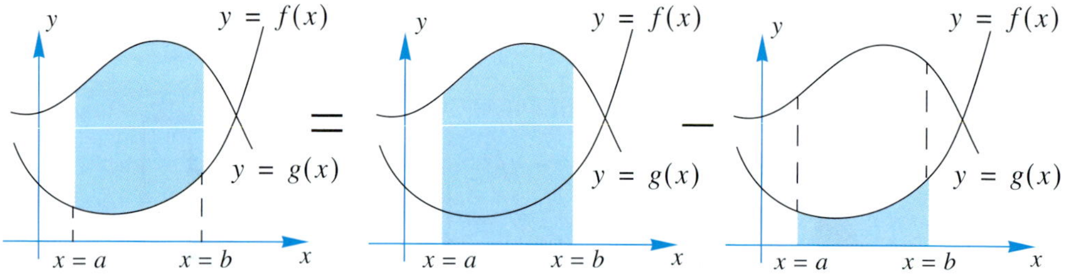

Area ***between*** *the curves* = *Area beneath y = g (x)* — *Area beneath y = f (x)*

That is,

> If $g(x) \geq f(x)$ on the interval $[a,b]$, then the area, A square units, enclosed by the two curves and the lines $x = a$ and $x = b$ is given by
>
> $$A = \int_a^b g(x)dx - \int_a^b f(x)dx = \int_a^b (g(x) - f(x))dx$$

EXAMPLE 22.26

Find the area of the region enclosed by the curves $g(x) = x + 2$, $f(x) = x^2 + x - 2$ and the lines $x = -1$ and $x = 1$.

Solution

The first step is to sketch both graphs so that it is clear which one lies above the other. In this case, as $g(x) \geq f(x)$ on $[-1,1]$, we can write the required area, A sq units, as

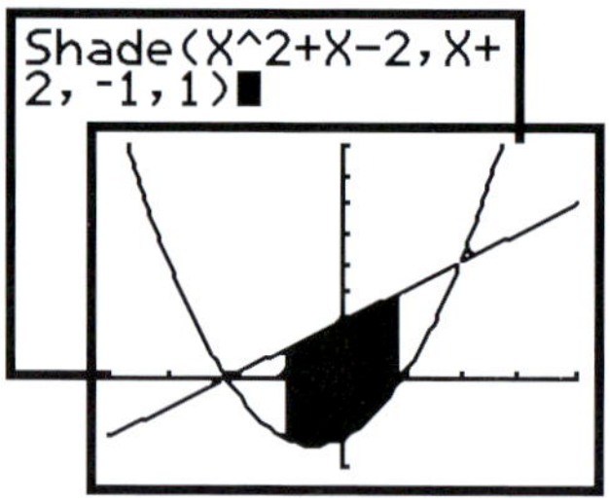

$$\int_{-1}^{1}((x+2)-(x^2+x-2))dx = \int_{-1}^{1}(4-x^2)dx$$

$$= \left[4x - \frac{x^3}{3}\right]_{-1}^{1}$$

$$= \left(4 - \frac{1}{3}\right) - \left(-4 + \frac{1}{3}\right)$$

$$= \frac{22}{3} \text{ sq units}$$

Note: If the question had been stated simply as:

"Find the area enclosed by the curves $g(x) = x + 2$ and $f(x) = x^2 + x - 2$."

This would indicate that we want the total area enclosed by the two curves, as shown:

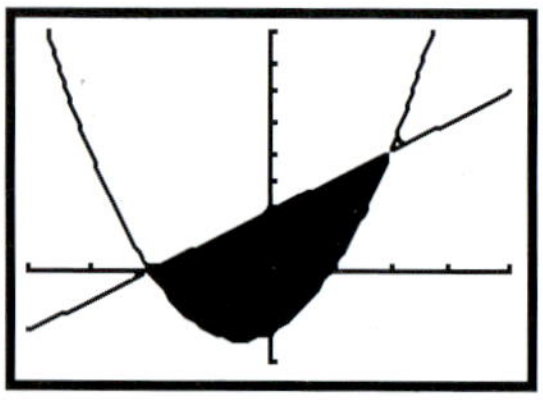

To find such an area we need first find the points of intersection:

$$x + 2 = x^2 + x - 2 \Leftrightarrow x^2 - 4 = 0$$

$$\therefore x = \pm 2$$

$\therefore$ required area $= \int_{-2}^{2}(4-x^2)dx = \left[4x - \frac{x^3}{3}\right]_{-2}^{2} = \frac{32}{3}$ sq units.

EXAMPLE 22.27

Find the area of the region enclosed by the curves $y = 4 - x$ and $y = \frac{2}{x-1}$.

Solution

Again we first sketch both graphs so that we can see which one lies above the other:

Next, we find the points of intersection:

$$4 - x = \frac{2}{x-1} \Leftrightarrow (4-x)(x-1) = 2$$

$$\Leftrightarrow (x-3)(x-2) = 0$$

Therefore, $x = 3$ or $x = 2$.

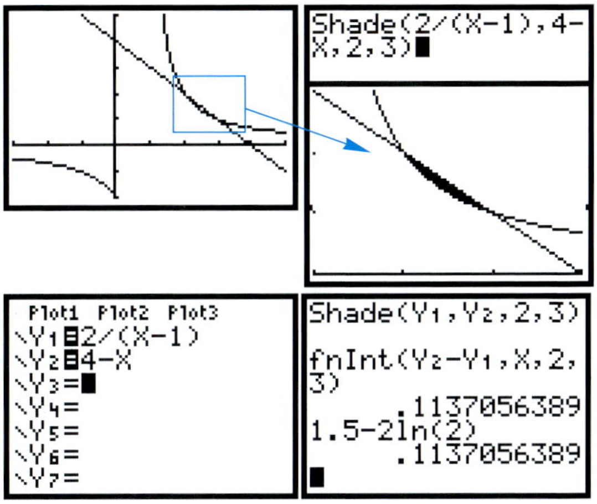

$$\text{Required area} = \int_2^3 \left((4-x) - \frac{2}{x-1} \right) dx$$

$$= \left[4x - \frac{x^2}{2} - 2\log_e(x-1) \right]_2^3$$

$$= \frac{3}{2} - 2\log_e 2 \text{ sq units}$$

EXAMPLE 22.28

Find the exact value of the area of the region enclosed by the curves $y = \sin 2x$ and $y = \cos^2 x$ over the region $0 \le x \le \pi$.

Solution

We start by sketching the graphs of $y = \sin 2x$ and $y = \cos^2 x$. Then determine the points of intersection.

$$\sin 2x = \cos^2 x \Leftrightarrow 2\sin x \cos x = \cos^2 x$$

$$\Leftrightarrow \cos x(2\sin x - \cos x) = 0$$

$$\Leftrightarrow \cos x = 0 \text{ or } \tan x = \frac{1}{2}$$

$$\text{i.e., } x = \frac{\pi}{2} \text{ or } x = \arctan\left(\frac{1}{2}\right)$$

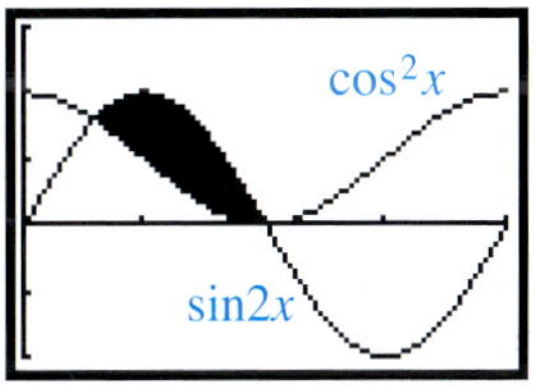

Therefore, area of shaded region, A sq units, is given by

$$A = \int_{\arctan\left(\frac{1}{2}\right)}^{\frac{\pi}{2}} [\sin 2x - \cos^2 x] dx = \int_{\arctan\left(\frac{1}{2}\right)}^{\frac{\pi}{2}} \left[\sin 2x - \frac{1}{2}(\cos 2x + 1) \right] dx$$

$$= \left[-\frac{1}{2}\cos 2x - \frac{1}{4}\sin 2x - \frac{x}{2} \right]_{\arctan\left(\frac{1}{2}\right)}^{\frac{\pi}{2}}$$

We will need to evaluate $\cos 2x$ and $\sin 2x$ when $x = \arctan\left(\frac{1}{2}\right)$.

We will need to construct a right-angled triangle for the given value of x:

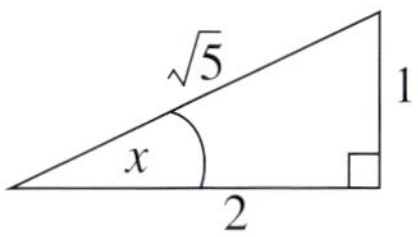

We first consider $\cos 2x$: Using the identity $\cos 2x = 2\cos^2 x - 1$ and the right-angled triangle we have that $\cos x = \frac{2}{\sqrt{5}}$. $\therefore \cos 2x = 2\left(\frac{2}{\sqrt{5}}\right)^2 - 1 = \frac{8}{5} - 1 = \frac{3}{5}$.

Next we consider $\sin 2x$: Using the identity $\sin 2x = 2\sin x \cos x$ we have that $\sin x = \frac{1}{\sqrt{5}}$. $\therefore \sin 2x = 2\left(\frac{1}{\sqrt{5}}\right)\left(\frac{2}{\sqrt{5}}\right) = \frac{4}{5}$

$$\therefore A = \left[-\frac{1}{2}\cdot\cos\pi - \frac{1}{4}\cdot\sin\pi - \frac{\pi}{4}\right] - \left[-\frac{1}{2}\left(\frac{3}{5}\right) - \frac{1}{4}\left(\frac{4}{5}\right) - \frac{1}{2}\left(\arctan\left(\frac{1}{2}\right)\right)\right]$$

$$= \left[\frac{1}{2} - \frac{\pi}{4}\right] - \left[-\frac{3}{10} - \frac{1}{5} - \frac{1}{2}\arctan\left(\frac{1}{2}\right)\right]$$

$$= 1 + \frac{1}{2}\arctan\left(\frac{1}{2}\right) - \frac{\pi}{4}$$

i.e., area of shaded region is $1 + \frac{1}{2}\arctan\left(\frac{1}{2}\right) - \frac{\pi}{4}$ sq. units

Checking with TI-83:

EXERCISES 22.5

1. Find the area of the region bounded by

(a) $y = x^3$, the x–axis, and the line $x = 2$.

(b) $y = 4 - x^2$, and the x–axis.

(c) $y = x^3 - 4x$, the x–axis, and the lines $x = -2$ and $x = 0$.

(d) $y = x^3 - 4x$, the x–axis, the line $x = 2$ and the line $x = 4$.

(e) $y = \sqrt{x} - x$, the x–axis, and the lines $x = 0$ and $x = 1$.

2. Find the area of the region bounded by

(a) $f(x) = e^x + 1$, the x–axis, and the lines $x = 0$ and $x = 1$.

(b) $f(x) = e^{2x} - 1$, the x–axis, the line $x = 1$ and the line $x = 2$.

(c) $f(x) = e^x - e^{-x}$, the x–axis, the line $x = -1$ and the line $x = 1$.

(d) $y = e^{\frac{1}{2}x + 1} - x$, the x–axis, the line $x = 0$ and the line $x = 2$.

3. Find the area of the region bounded by

(a) $y = \frac{1}{x}$, the x–axis, the line $x = 4$ and the line $x = 5$.

(b) $y = \frac{2}{x+1}$, the x–axis, the line $x = 0$ and the line $x = 4$.

(c) $f(x) = \frac{3}{2-x}$, the x–axis, the line $x = -1$ and the line $x = 1$.

(d) $f(x) = \frac{1}{x-1} + 1$, the x–axis, the line $x = -1$ and the line $x = \frac{1}{2}$.

4. Find the area of the region bounded by

(a) $f(x) = 2\sin x$, the x–axis, the line $x = 0$ and the line $x = \frac{\pi}{2}$.

(b) $y = \cos(2x) + 1$, the x–axis, the line $x = 0$ and the line $x = \frac{\pi}{2}$.

(c) $y = x - \cos\left(\frac{x}{2}\right)$, the x–axis, the line $x = \frac{\pi}{2}$ and the line $x = \pi$.

(d) $f(x) = \cos(2x) - \sin\left(\frac{x}{2}\right)$, the x–axis, the line $x = \frac{\pi}{2}$ and the line $x = \pi$.

(e) $y = 3\sec^2\left(\frac{x}{2}\right)$, the x–axis, the line $x = -\frac{\pi}{3}$ and the line $x = \frac{\pi}{3}$.

5. Verify your answers to Questions 1– 4, using a graphics calculator.

6. Find the area of the region enclosed by the curve $y = 8 - x^3$, the y–axis and the x–axis.

7. Find the area of the region enclosed by the curve $y = x^2 + 1$, and the lines $y = 2$ and $y = 4$.

8. Find the area of the region enclosed by the curve $f(x) = x + \frac{1}{x}$, the x–axis and the lines $x = -2$ and $x = -1$.

9. Find the area of the region enclosed by the curve $y = x^2 - 1$, the x–axis, the line $x = 0$ and $x = 2$.

10. Find the area of the region enclosed by the curve $y = x(x+1)(x-2)$ and the x–axis.

11. Find the area of the region enclosed by the curve $f(x) = 1 - \frac{1}{x^2}$

(a) the x–axis, the line $x = 1$ and $x = 2$

(b) the x–axis, the line $x = \frac{1}{2}$ and $x = 2$

(c) and the lines $y = -\frac{1}{2}$ and $y = \frac{1}{2}$.

12. The area of the region enclosed by the curve $y^2 = 4ax$ and the line $x = a$ is ka^2 sq units. Find the value of k.

13. Differentiate the function $y = \log_e(\cos 2x)$. Hence find the area of the region enclosed by the curve $f(x) = \tan(2x)$, the x–axis and the lines $x = 0$ and $x = \frac{\pi}{8}$.

14. (a) Find the area of the region enclosed by the curve $y = |2x - 1|$ the x–axis, the line $x = -1$ and the line $x = 2$.

(b) Find the area of the region enclosed by the curve $y = |2x| - 1$ the x–axis, the line $x = -1$ and the line $x = 2$.

15. Find the area of the region enclosed by the curve $f(x) = \frac{2}{(x-1)^2}$

(a) the x–axis, the lines $x = 2$ and $x = 3$,
(b) the y–axis, the lines $y = 2$ and $y = 8$.

16. (a) Differentiate the function $y = x\log_e x$, hence find $\int \log_e x dx$.

(b) Find the area of the region enclosed by the curve $y = e^x$, the y–axis and the lines $y = 1$ and $y = e$.

17. Find the area of the region bounded by the graphs of $y = x^2 + 2$ and $y = x$, over the the interval $0 \le x \le 2$.

18. (a) Find the area of the region bounded by the graphs of $y = 2 - x^2$ and $y = x$, over the interval $0 \le x \le 1$.

(b) Find the area of the region bounded by the graphs of $y = 2 - x^2$ and $y = x$.

19. (a) Find the area of the regions bounded by the following:

i. $y = x^3$, $x = 1$, $x = 2$ and $y = 0$.

ii. $y = x^3$, $y = 1$, $y = 8$ and $x = 0$.

(b) How could you deduce part ii. from part i.?

20. Find the area of the region bounded by the curves with equations $y = \sqrt{x}$, $y = 6 - x$ and the x–axis.

21. (a) Sketch the graph of the function $f(x) = |e^x - 1|$.

(b) Find the area of the region enclosed by the curve $y = f(x)$,

i. the x–axis and the lines $x = -1$ and $x = 1$.
ii. the y–axis and the line $y = e - 1$.
iii. and the line $y = 1$. Discuss your findings for this case.

22. (a) On the same set of axes, sketch the graphs of $f(x) = \sin\left(\frac{1}{2}x\right)$ and $g(x) = \sin 2x$ over the interval $0 \le x \le \pi$.

(b) Find the area of the region between by the curves $y = f(x)$ and $y = g(x)$ over the interval $0 \le x \le \pi$, giving your answer correct to 2 decimal places.

23. Consider the curve with equation $y^2 = x^3$ as shown in the diagram.
A tangent meets the curve at the point $A(a^2, a^3)$.

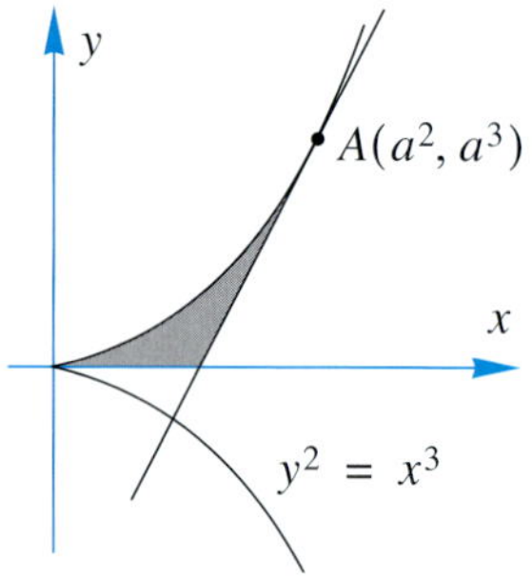

(a) Find the equation of the tangent at A.

(b) Find the area of the shaded region enclosed by the curve, the line $y = 0$ and the tangent.

24. (a) On a set of axes, sketch the graph of the curve $y = e^{x-1}$ and find the area of the region enclosed by the curve, the x-axis and the lines $x = 0$ and $x = 1$.

(b) Hence evaluate $\int_{e^{-1}}^{1} (\ln x + 1)dx$.

(c) Find the area of the region enclosed by the curves $y = e^{x-1}$ and $y = \ln x + 1$ over the $e^{-1} \le x \le 1$.

25. The area of the shaded region enclosed by the y-axis, the tangent to the curve at $x = a$ and the curve $y^2 = x$, $y \ge 0$, as shown in the diagram below, measures $\frac{16}{3}$ sq. units.

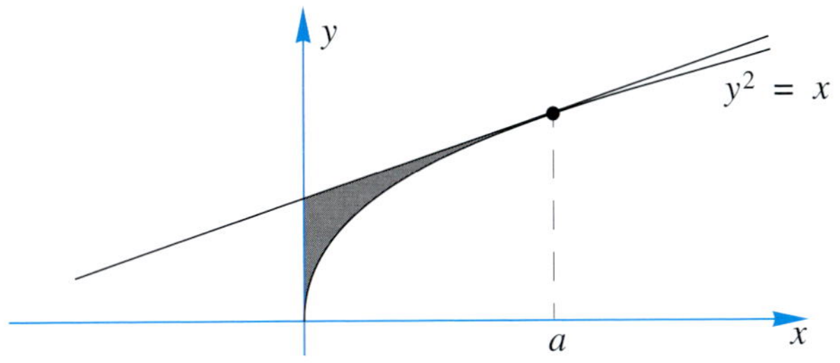

Find the exact value of a.

22.6 APPLICATIONS TO KINEMATICS

Another application of integration when relating it to areas is that of kinematics. Just as the gradient of the displacement–time graph produces the velocity–time graph, so too then, we have that the **area beneath the velocity–time graph** produces the **displacement–time** graph. Notice that the area provides the displacement (not necessarily the distance!).

Similarly with the acceleration–time graph, i.e., the **area under the acceleration–time graph** represents the **velocity**.

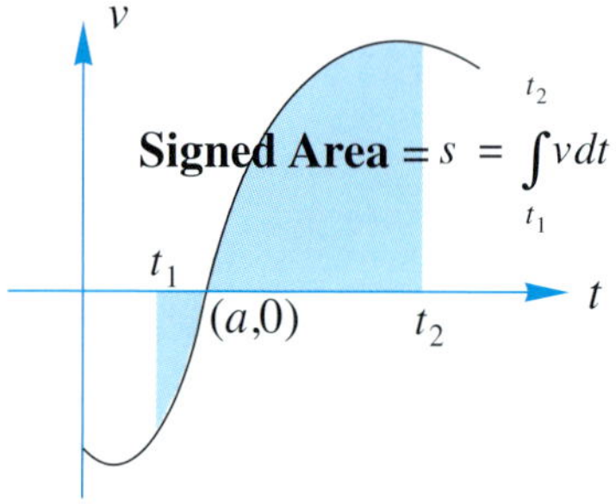

The displacement over the interval $[t_1, t_2]$ is given by

$$\textbf{Displacement} = s = \int_{t_1}^{t_2} v\,dt$$

However, the distance covered over the interval $[t_1, t_2]$ is given by

$$\textbf{Distance} = x = -\int_{t_1}^{a} v\,dt + \int_{a}^{t_2} v\,dt$$

EXAMPLE 22.29

The velocity of an object is v m/s after t seconds, where $v = 1 - 2\sin 2t$.

(a) Find the object's displacement over the first $\frac{3\pi}{4}$ seconds.

(b) Find the distance travelled by the object over the first $\frac{3\pi}{4}$ seconds.

Solution

(a) It is always a good idea to sketch the velocity–time graph:
The displacement is then given by

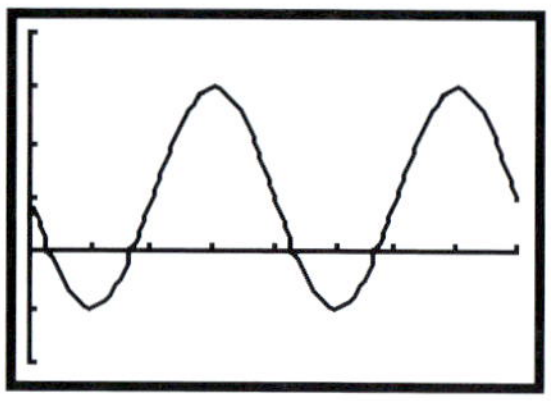

$$s = \int_0^{\frac{3\pi}{4}} (1 - 2\sin 2t)\,dt = \left[t + \cos 2t \right]_0^{\frac{3\pi}{4}}$$

$$= \left(\frac{3\pi}{4} + \cos\left(\frac{6\pi}{4}\right)\right) - (0 + \cos(0))$$

$$= \frac{3\pi}{4} - 1$$

$$\approx 1.36$$

That is, the object's displacement measures (approx.) 1.36 metres.

Once we have sketched the velocity-time graph using the TI–83 we can make use of the $\int f(x)dx$ option under the **CALC** menu.
When prompted for the Lower and Upper Limits, enter the required x–values.
However, it must be understood that the $\int f(x)dx$ option provides the value of the signed area. In this case it provides the displacement not the distance (although sometimes these are the same). We now display this result:

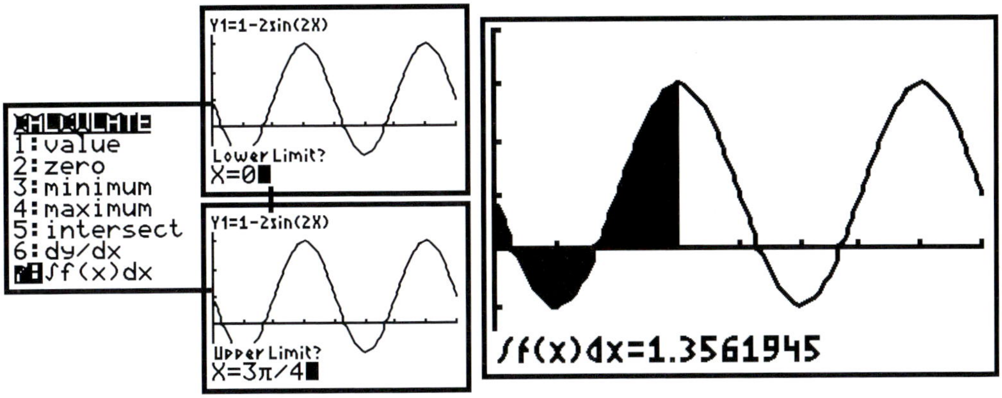

(b) As part of the graph lies below the t–axis, when determining the distance travelled we use the same principle as that which differentiates between the signed area and the actual area enclosed by a curve and the horizontal axis. In short:

Displacement = Signed area
Distance = Area.

The first step is to determine the t–intercepts:

Solving for $1 - 2\sin 2t = 0$ we have, $\sin 2t = \frac{1}{2} \Rightarrow 2t = \frac{\pi}{6}, \frac{5\pi}{6}, \ldots \quad \therefore t = \frac{\pi}{12}, \frac{5\pi}{12}, \ldots$

Note that we only require the first two intercepts.
Therefore, the distance is given by

$$x = \int_0^{\frac{\pi}{12}} (1 - 2\sin 2t)dt - \int_{\frac{\pi}{12}}^{\frac{5\pi}{12}} (1 - 2\sin 2t)dt + \int_{\frac{5\pi}{12}}^{\frac{3\pi}{4}} (1 - 2\sin 2t)dt$$

Evaluating this expression is rather lengthy, and—unless we require an exact value— we might as well make use of the graphics calculator.
There are a number of ways this can be done. Either we can make repeated use of the previous method i.e., using the $\int f(x)dx$ option in the **CALC** menu for each interval, or we could use the **9: fnInt(** option in the **MATH** menu.

Distance travelled
= 0.1278 – (–0.6849) + 1.9132 = 2.7259.

That is, object travelled (approx.) 2.73 m.

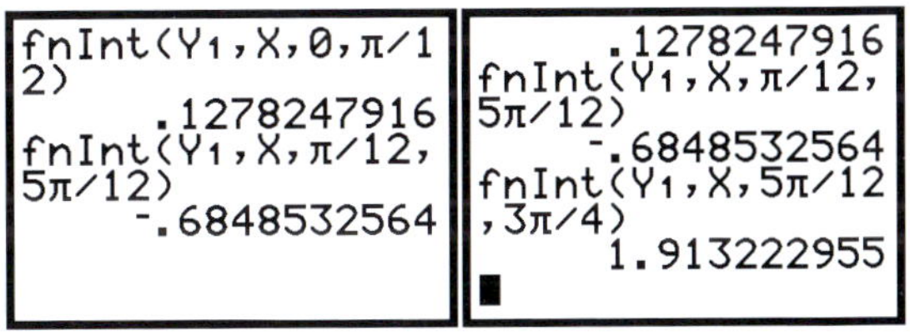

EXAMPLE 22.30

A rocket starts from rest and accelerates such that its acceleration is given by the formula $a(t) = 3t + t^2, 0 \le t \le 10$ where distances are measured in metres and time in seconds. Find the distance travelled in the first ten seconds of the rocket's motion.

Solution

The information is given as an acceleration. We must find the indefinite integral of this function to get a rule to give us the velocity.

$a(t) = 3t + t^2, 0 \le t \le 10 \Rightarrow v(t) = \int(3t + t^2)dt = \frac{3t^2}{2} + \frac{t^3}{3} + c$

Now, when $t = 0, v = 0 \therefore 0 = 0 + 0 + c \Rightarrow c = 0$

The constant is zero because we are told that the rocket starts from rest. The distance travelled is the area under this velocity time graph. Which must be found using definite integration. As the graph of $v(t)$ lies above the t–axis over the interval $0 \le t \le 10$, we have the distance D, given by

$$D = \int_0^{10}\left(\frac{3t^2}{2} + \frac{t^3}{3}\right)dt = \left[\frac{t^3}{2} + \frac{t^4}{12}\right]_0^{10}$$

$$= \frac{10^3}{2} + \frac{10^4}{12}$$

$$= 1333\frac{1}{3}\text{m}$$

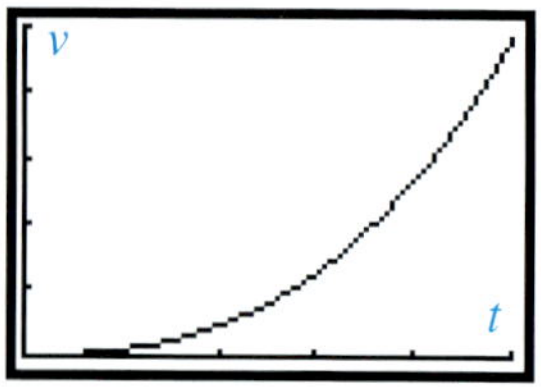

The technique described in this example is the basis of the inertial navigator. This senses acceleration and integrates it to infer velocity. The instrument then integrates a second time to calculate distance travelled. Of course, none of these quantities are generally expressed as exact mathematical formulae and the calculation has to be performed using numerical approximation.

EXERCISES 22.6

1. Find the displacement equation, $x(t)$, for each of the following

(a) $\frac{d^2x}{dt^2} = 6t$ where $\frac{dx}{dt} = 3$ and $x = 10$ when $t = 0$.

(b) $\frac{d^2x}{dt^2} = -(4\sin t + 3\cos t)$ where $\frac{dx}{dt} = 4$ and $x = 2$ when $t = 0$.

(c) $\frac{d^2x}{dt^2} = 2 - e^{-\frac{1}{2}t}$ where $\frac{dx}{dt} = 4$ and $x = 0$ when $t = 0$.

2. The acceleration, $a(t)$ ms^{-2}, of a body travelling in a straight line and having a displacement $x(t)$ m from an origin is governed by

$$a(t) = 6t - 2 \text{ where } \frac{dx}{dt} = 0 \text{ and } x = 0 \text{ when } t = 0.$$

(a) Find the displacement of the body at any time t.
(b) Find the displacement of the body after 5 seconds.
(c) Find the distance the body has travelled after 5 seconds.

3. A body moves along a straight line in such a way that its velocity, v ms^{-1}, is given by

$$v(t) = -\sqrt{t+4} + 2.$$

After 5 seconds of motion the body is at the origin O.
(a) Sketch the displacement-time graph for this body.
(b) How far will the body have travelled after another 5 seconds.

4. A particle starts from rest and moves with a velocity, v ms^{-1}, where $v = t(t-5)$. Find the distance travelled between the two occasions when the particle is at rest.

5. A stone is thrown vertically upwards from ground level with a velocity of 25 ms^{-1}. If the acceleration of the stone is 9.8 ms^{-2} directed downwards, find the time taken before the stone reaches its highest point and the total distance travelled when the stone falls back to the ground.

6. The velocity of a particle is given by $v(t) = 3 - 3\sin 3t$, which is measured in m/s.
(a) Find when the particle first comes to rest.
(b) Find the distance travelled by the particle from when it started to when it first comes to rest.

7. An object, starting from rest, moves in a straight line with an acceleration that is given by $a(t) = \dfrac{12}{(t+1)^2}$ ms^{-2}. Find the distance travelled during the first 9 seconds.

8. An object has its velocity governed by the equation $v(t) = 10\sin\left(\dfrac{\pi}{16}t\right)$ m/s.
(a) Given that $s(0) = 0$, find its displacement equation.
(b) Find its displacement after 20 seconds.
(c) Find its displacement during the 20th second.
(d) How far has it travelled in twenty seconds?

9. A particle moving in a straight line has its acceleration, a ms^{-2}, defined by the equation

$$a = \frac{2k}{t^3}\ \text{ms}^{-2}.$$

At the end of the first second of motion, the particle has a velocity measuring 4m/s.
(a) Find an expression for the velocity of the particle.
(b) Given that its velocity approaches a limiting value of 6 m/s, find k.
(c) Find the distance travelled by the particle after a further 9 seconds.

10. (a) Show that $\dfrac{d}{dx}\left[\dfrac{e^{ax}}{a^2+b^2}(a\cos bx + b\sin bx)\right] = e^{ax}\cos bx$.
(b) The velocity of a vibrating bridge component is modelled by the function $V = e^{-2t}\cos(3t)$. V ms^{-1} is the velocity of the component and t is the time in seconds after the observations begin.
Find the distance the component travels in the first tenth of a second.

22.7 VOLUMES (SOLID OF REVOLUTION)

A **solid of revolution** is formed by revolving a plane region about a line – called the **axis of revolution**. In this section we will only be using the x-axis or the y-axis.

For example, in the diagram alongside, if we revolve the triangular plane region about the vertical axis as shown, we obtain a cone.

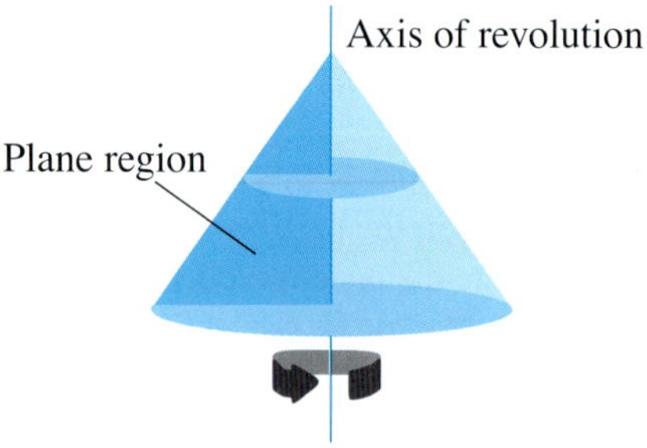

It is important to realise that depending on the axis of revolution, we can obtain very different shapes. For example, if a region bounded by the curve $y = x^2, x \geq 0$ is rotated about the x- and y–axes, two distinct solid shapes are formed:

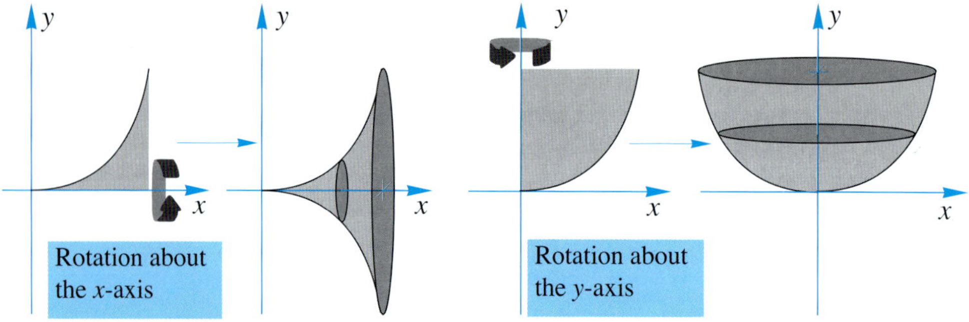

When the plane region (enclosed by the curve and the x-axis) is rotated about the x–axis, the solid object produced is rather like the bell of a trumpet (with a very narrow mouth piece!) or a Malay hat on its side. However, when the plane region (enclosed by the curve and the y-axis) is rotated about the y-axis, then the solid produced is like a bowl.

Using the same approach as that used when finding the area of a region enclosed by a curve, the x-axis and the lines $x = a$ and $x = b$ we have:

Then, the volume, V units^3, of such a solid can be cut up into a large number of slices (i.e., discs) each having a width δx and radius $f(x_i)$. The volume produced is then the sum of the volumes of these discs, i.e.,

$$V \approx \sum_{i=0}^{i=n-1} \pi[f(x_i)]^2\delta x, \text{ where } \delta x = \frac{b-a}{n}$$

So, as $n \to \infty$, $\delta x \to 0$ and so,

$$V = \lim_{\delta x \to 0} \sum_{i=0}^{i=n-1} \pi[f(x_i)]^2\delta x = \int_a^b \pi[f(x)]^2 dx$$

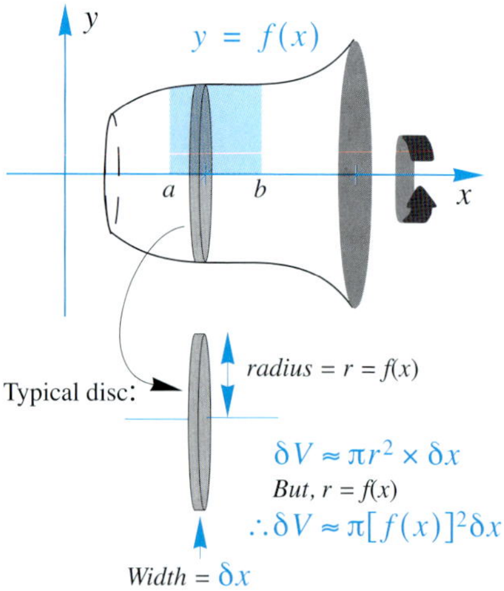

Therefore, we have:

The volume, V units^3, of a solid of revolution is given by

$$V = \pi \int_{x=a}^{x=b} [f(x)]^2 dx \quad \left[\text{or } V = \pi \int_a^b y^2 dx \right]$$

when a plane region enclosed by the curve, $y = f(x)$ and the lines $x = a$ and $x = b$ is **revolved about the x–axis**.

EXAMPLE 22.31

The curve $y = \sqrt{x-1}$, $1 \le x \le 5$ is rotated about the x-axis to form a solid of revolution. Sketch this solid and find its volume.

Solution

The curve has a restricted domain and is rotated about the x–axis, so, the solid formed has a volume given by

$$V = \pi \int_1^5 (\sqrt{x-1})^2 dx = \pi \int_1^5 (x-1) dx$$

$$= \pi \left[\frac{x^2}{2} - x \right]_1^5$$

$$= \pi \left(\frac{5^2}{2} - 5 - \left(\frac{1^2}{2} - 1 \right) \right)$$

$$= 8\pi$$

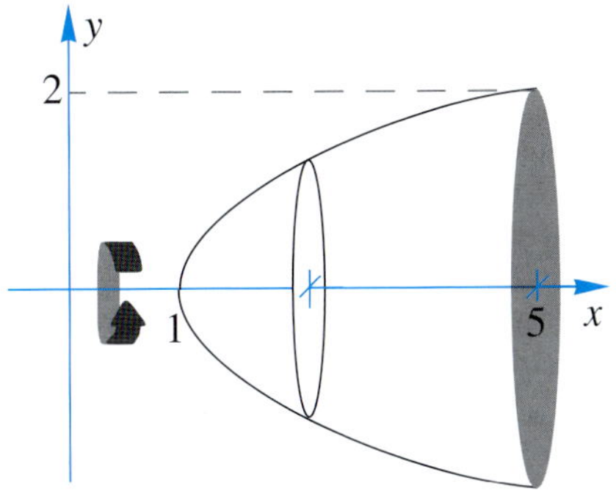

Therefore, the volume generated is 8π units^3.

If the curve is rotated about the y-axis, the solid formed looks like this:

EXAMPLE 22.32

Find the volume of the solid formed by revolving the region enclosed by the curve with equation $f(x) = \sqrt{25 - x^2}$ and the line $g(x) = 3$ about the x–axis.

Solution

We start by drawing a diagram of this situation: It is a bead.

Next we determine the points of intersection.

Setting $f(x) = g(x)$ we have

$$\sqrt{25 - x^2} = 3$$

$$\therefore 25 - x^2 = 9$$

$$\Leftrightarrow x^2 = 16$$

$$\therefore x = \pm 4$$

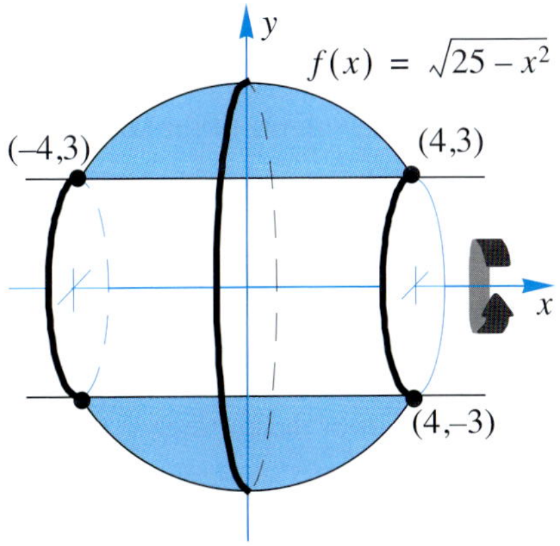

The solid formed is hollow inside, i.e., from $-3 \le y \le 3$.

Next, we find the difference between the two volumes generated (a little bit like finding the area between two curves)

$$V = V_{f(x)} - V_{g(x)} = \pi\int_{-4}^{4}[f(x)]^2dx - \pi\int_{-4}^{4}[g(x)]^2dx$$

$$= \pi\int_{-4}^{4}([f(x)]^2 - [g(x)]^2)dx$$

$$= 2\pi\int_{0}^{4}([f(x)]^2 - [g(x)]^2)dx \text{ [by symmetry]}$$

Therefore,

$$V = 2\pi\int_{0}^{4}([\sqrt{25-x^2}]^2 - [3]^2)dx$$

$$= 2\pi\int_{0}^{4}(16 - x^2)dx$$

$$= 2\pi\left[16x - \frac{1}{3}x^3\right]_0^4$$

$$= \frac{256}{3}\pi$$

i.e., required volume is $\frac{256}{3}\pi$ units^3.

EXERCISES 22.7

* Unless stated otherwise, all answers should be given as an exact value.

1. The part of the line $y = x + 1$ between $x = 0$ and $x = 3$ is rotated about the x-axis. Find the volume of this solid of revolution.

2. A curve is defined by $y = \frac{1}{\sqrt{x}}, x \in [1, 5]$. If this curve is rotated about the x-axis, find the volume of the solid of revolution formed.

3. Find the volume of the solid of revolution formed by rotating the part of the curve $y = e^x$ between $x = 1$ and $x = 5$ about the x-axis.

4. A solid is formed by rotating the curve $y = \sin x, x \in [0, 2\pi]$ about the x-axis. Find the volume of this solid.

5. The part of the curve $y = \frac{1}{1-x}$ between the x values 2 and 3 is rotated about the x-axis. Find the volume of this solid.

6. The part of the curve $y = \frac{x}{1+x}$ between the x values 0 and 2 is rotated about the x-axis. Find the volume of the solid formed in this way.

7. Find the equation of the straight line that passes through the origin and through the point (h,r). Hence use calculus to prove that the volume of a right circular cone with base radius r and height h is given by $V = \frac{1}{3}\pi r^2 h$.

8. Find the equation of a circle of radius r. Use calculus to prove that the volume of a sphere is given by the formula $V = \frac{4}{3}\pi r^3$.

9. The part of the curve $f(x) = \sin\frac{x}{10}$ between $x = 0$ and $x = 5$ is rotated about the x-axis. Find the volume of this solid of revolution.

10. The part of the curve $f(x) = x^2 - x + 2$ between $x = 1$ and $x = 2$ is rotated about the x-axis. Find the volume of this solid of revolution.

11. Find the volume generated by the region between by the x–axis and that part of the parabola $y = x^2$ from $x = 1$ to $x = 3$ when it is rotated about the x–axis.

12. Find the volume of the solid of revolution that is formed by rotating the region bounded by the curves $y = \sqrt{x}$ and $y = \sqrt{x^3}$ about the x–axis.

13. Find the volume of the solid of revolution that is formed when the region bounded by the curve with equation $y = 4 - x^2$ and the line $y = 1$ is rotated about the x–axis.

14. Find the volume of the solid generated by rotating the region bounded by the curves $y^2 = x^3$ and $y^2 = 2 - x$ about the x–axis.

15. The volume of the solid formed when the region bounded by the curve $y = e^x - k$, the x–axis and the line $x = \ln 3$ is rotated about the x–axis is $\pi \ln 3$ units^3. Find k.

16. Find the volume of the solid of revolution formed by rotating the region bounded by the axes and the curve $y = \sqrt{3}a\sin x + a\cos x$, $0 \le x \le 2\pi$, $a > 0$ about the x-axis.

17. If the curve of the function $f(\theta) = \sin k\theta$, $k > 0$, $\theta > 0$ is rotated about the θ–axis, a string of sausages is made. Find k such that the volume of each sausage is π units^3.

18. (a) On the same set of axes, sketch the curves $y = ax^2$ and $y = 1 - \frac{x^2}{a}$ where $a > 0$.

(b) Find the volume of the solid of revolution formed when the region enclosed by the curves in (a) is rotated about the x–axis.

19. On the same set of axes sketch the two sets of points $\{(x, y) : (x-2)^2 + y^2 \leq 4\}$ and $\{(x, y) : (x-a)^2 + y^2 \leq 4, a \in]-2,6[\}$. The intersection of these two sets is rotated about the x-axis to generate a solid. Find a if the volume of this solid is π units^3. Give your answer to three decimal places.

Extension Problems

20. Find the volume of the solid of revolution generated when the shaded region shown below is revolved about the line $y = 2$.

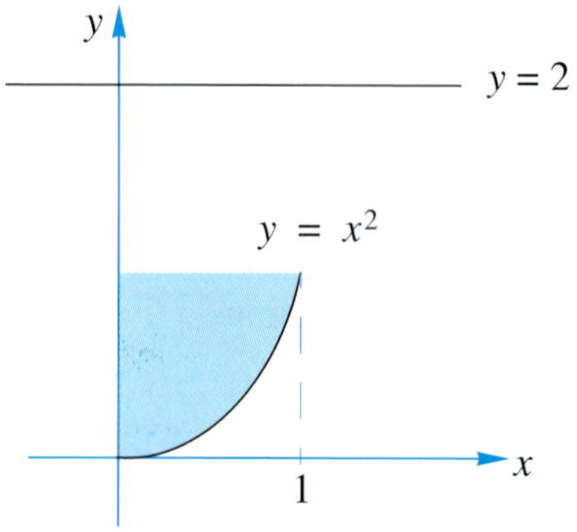

21. Find the volume of the solid of revolution generated by revolving the region enclosed by the curve $y = 4 - x^2$ and $y = 0$ about the line $x = 3$.
Draw the shape of the solid of revolution.

22. Using an argument similar to that found in Section 22.5 show why the **arc length**, L units, of a curve from $x = a$ to $x = b$ is given by $L = \int_a^b \sqrt{1 + [f'(x)]^2}dx$.

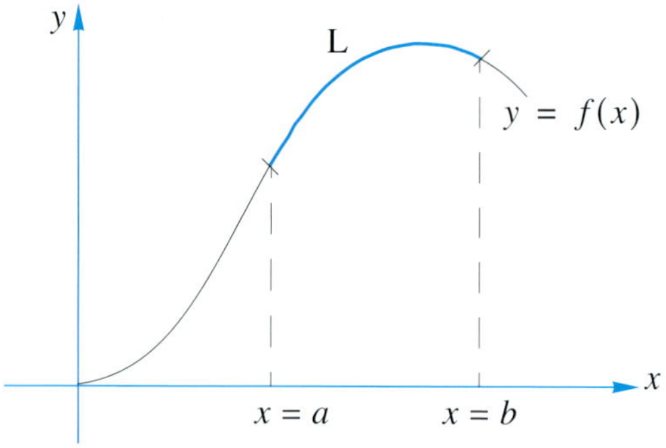

REVISION SET E - PAPER 1 & PAPER 2 STYLE QUESTIONS

1. Differentiate (a) $\sqrt{x^2+4}$ (b) $(2x-1)\cos 2x$

2. Evaluate $\int_1^4 \left(x + \frac{1}{\sqrt{x}}\right)^2 dx$.

3. (a) Find the values of x between 0 and 2π for which $\sin x = \sqrt{3}\cos x$.

(b) Sketch, on the same set of axes, the graphs of $y = \sin x$ and $y = \sqrt{3}\cos x$ for values of x from 0 to 2π. Hence find $\{x | \sin x > \sqrt{3}\cos x, 0 \le x \le 2\pi\}$.

(c) Find the area enclosed between the two curves in (b).

4. A cup of tea, initially at a temperature of 80˚C is left to cool. Its temperature, $\theta°C$, after t minutes is given by $\theta = 20 + 60e^{-0.04t}, t \ge 0$.

(a) How many degrees will its temperature fall in the first 10 minutes?

(b) At what rate, in degrees per minute, is its temperature falling at the end of 10 minutes?

(c) How many minutes will it take for its temperature to drop from 80˚C to 50˚C?
(Give all these answers correct to 1 decimal place.)

5. Let the function f be defined by $f(x) = x - 3 + \frac{4}{x^2}, x \ne 0$

(a) Find the values of x for which $f(x) \ge 0$.

(b) Find the values of x for which $f'(x) \ge 0$.

6. A tree trunk of length 12 metres is a truncated cone, the radii of the ends being 0.30 metres and 0.18 metres. If r metres is the radius of the cross-section at a distance h metres from the wider end, prove that

$$h = 100(0.3 - r)$$

A beam of uniform square cross-section is cut from the tree trunk, the centre line of the beam being along the axis of the cone.

If V m^3 is the volume of the widest beam of length h metres which can be cut from the trunk, prove that

$$V = 2r^2h.$$

Find the length of the beam of greatest volume which can be cut from the trunk.

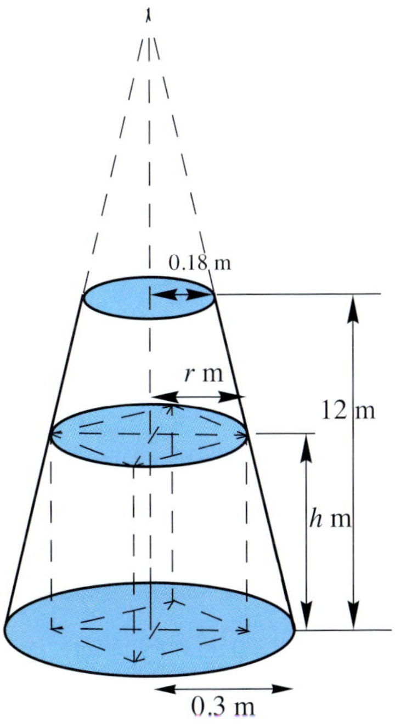

7. Find $f'(x)$ if (a) $f(x) = \frac{x^2-1}{x^2+1}$ (b) $f(x) = \cos^2 2x$

8. Let $f(x) = \sqrt{x}$ and $g(x) = 4 - x^2$.
(a) Find i. $f(g(2))$ ii. $g(f(2))$
(b) State all values of x for which $f(g(x))$ is defined.
(c) State all values of x for which $g(f(x))$ is defined.
(d) If $h(x) = f(g(x))$ find $h'(x)$.

9. Let P and Q be the points on the curve $y = \frac{1}{x^2}$ at which $x = 1$ and $x = 1 + h$ respectively.
(a) Express the gradient of the line [PQ] in terms of h.
(b) Hence find the gradient of the tangent to the curve at $x = 1$.

10. (a) Evaluate $\int_1^4 \frac{5x^2+3x-1}{\sqrt{x}} dx$.

(b) Find the derivative of $\log_e \cos x$, and your result to evaluate $\int_0^{\frac{\pi}{3}} \tan x dx$ correct to 2 decimal places.

11. A small object is oscillating up and down at the end of a vertical spring. The object is h metres above its starting point at time t seconds, where $h(t) = \frac{1}{2}\left(1 - e^{-0.05t}\cos\left(\frac{3\pi}{2}t\right)\right)$.
Find the rate at which the object is rising (or falling) 2.5 seconds after the motion starts.

12. (a) Sketch the graph of $y = 4(x^2 - x^4)$, locating the stationary points and giving their coordinates.

(b) Sketch on the same set of axes the graph of $y = \frac{1}{4(x^2 - x^4)}$
Label both graphs clearly.

13. Find $f'(x)$ if i. $f(x) = \sin^3 2x$. ii. $f(x) = \frac{x}{\sqrt{2x+3}}$.

14. Antidifferentiate $(e^x - e^{-x})^2$.

15. During a trench-digging operation, which takes 2 hours, the rate at which earth is being removed, after t minutes, is $(12 - 0.1t)$ cubic metres per minute, for $0 \le t \le 120$.
Find the total volume of earth removed.

16. (a) Sketch the graph of $y = \frac{1}{9}(4x^3 - x^4)$. Identify all intercepts with the axes and all stationary points, and state their coordinates.
(b) Find (correct to the nearest degree) the angle at which the graph cuts the positive x-axis.

17. A sheet of carboard measures 10 cm by 16 c. Squares x cm by x cm are cut from two corners, and strips x cm by $(x + y)$ cm are cut from two sides, as shown in the diagram. (The shaded regions are removed).

The carboard is then folded along the dotted lines to form a rectangular box with a lid.

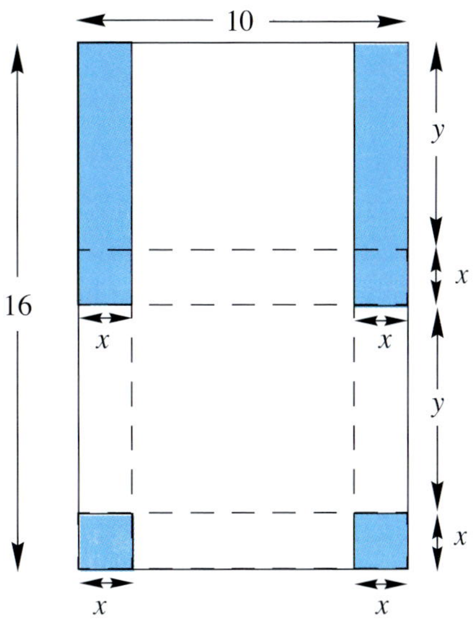

(a) Express y in terms of x, and hence show that the volume, V cm^3, of the box is given by

$$V = 2x^3 - 26x^2 + 80x.$$

(b) i. Find the value of x such that the volume of the box is a maximum. Justify that you have obtained a maximum by considering the sign of $V'(x)$.

ii. Find the maximum volume of the box.

18. A horizontal water trough, 0.9 metres long, has a parabolic vertical section, represented by the parabola $y = \frac{25}{4}x^2$ from $x = -0.2$ to $x = 0.2$.

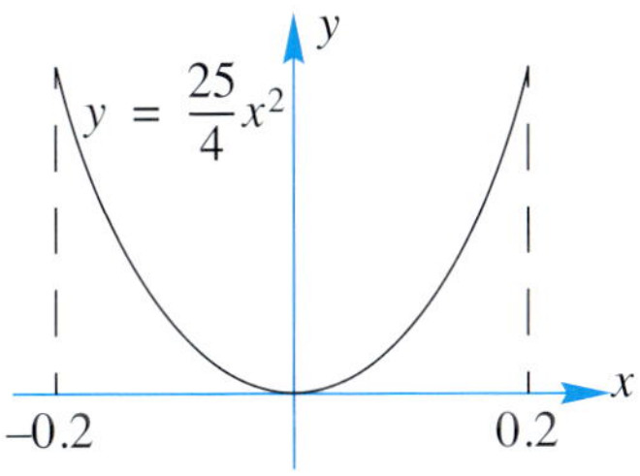

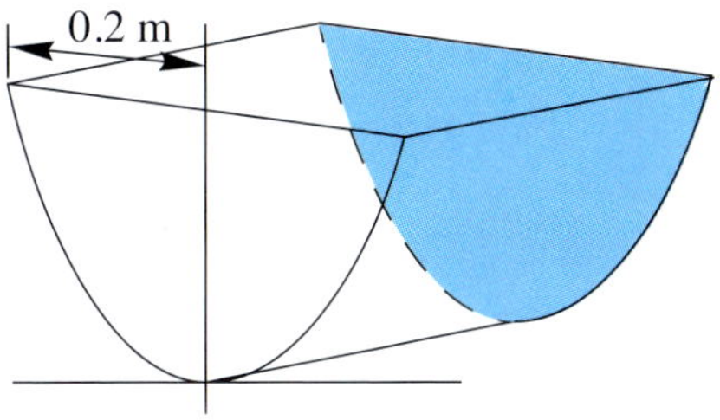

All measurements are in metres.
Water is flowing into the trough at the rate of 0.01 cubic metres / minute. At time t minutes, the depth of water in the trough is h metres and the volume of water in the trough is V cubic metres.

(a) Find, in terms of h, the area enclosed between the parabola $y = \frac{25}{4}x^2$ and the line $y = h$. Hence find V in terms of h.

(b) Given that $\frac{dV}{dt} = \frac{dV}{dh} \times \frac{dh}{dt}$, find the rate at which the water level in the trough is rising, when the depth is 0.16 metres.

19. Let $f(x) = 1 - \frac{3}{x}, x \neq 0$ and $g(x) = x^2 - 3x$

(a) Find the coordinates of the points of intersection of the graphs of f and g.
(b) Sketch on the same set of axes the graphs of f and g.
(c) Find the area of the region enclosed between these two graphs.

20. The graph of the function $y = f(x)$ is shown alongside. Sketch the graph of $y = f'(x)$.

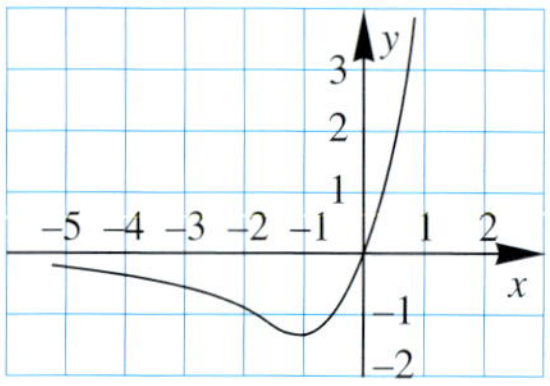

21. A closed cylindrical tin can is being designed. Its volume is to be 1000 cm^3. Its curved surface is to be shaped from a rectangular piece of tin sheet. Each of its circular ends (of radius r cm) is to cut from a square of tin sheet, as shown, with the shaded sections being wasted.

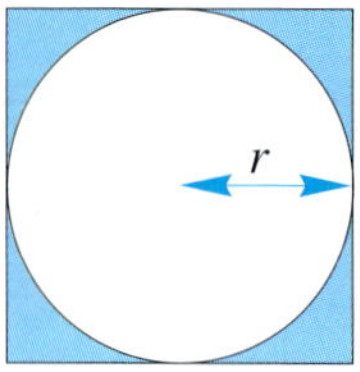

(a) If h cm is the height of the can, express h in terms of r.
Hence show that the area, $A \text{ cm}^2$, of tin sheet needed for a can is given by

$$A = \frac{2000}{r} + 8r^2$$

(b) Find the radius and height of the can for which the area of tin sheet needed is a minimum.
Justify that you have obtained a minimum by considering the sign of the first derivative.

22. Find $f'(x)$ if i. $f(x) = \sqrt{1 - 3x^2}$ ii. $f(x) = \dfrac{e^x}{1 + e^x}$

23. Given that $f(x) = x^3$

(a) Find $f(x + h) - f(x)$.

(b) Simplify $\dfrac{f(x+h) - f(x)}{h}$, $h \neq 0$.

(c) Find $\lim\limits_{h \to 0} \dfrac{f(x+h) - f(x)}{h}$ and hence state $f'(x)$.

24. (a) Evaluate $\displaystyle\int_0^{\frac{\pi}{2}} \left(\sin\frac{x}{2} + \cos 2x\right) dx$.

(b) Differentiate $\log_e(x^2 + 2)$. Hence evaluate $\displaystyle\int_0^2 \frac{2x}{x^2 + 2} dx$.

25. In a certain country which has a controlling birth rate and an ageing population, the predicted population P million, t years after 1980, is given by

$$P = 27.5 + 0.8te^{-0.02t}, t \geq 0.$$

(a) Find $\dfrac{dP}{dt}$.

(b) What is the predicted population for the year 2080?

(c) i. In the year 2080, will the population be increasing or decreasing?
ii. At what rate, in millions per year, will it be changing?

(d) In what year is it predicted that the population will reach a maximum, and what will this maximum population be?

26. A piece of airline baggage is in the shape of a cuboid of width x centimetres, length $3x$ centimetres and depth h centimetres.

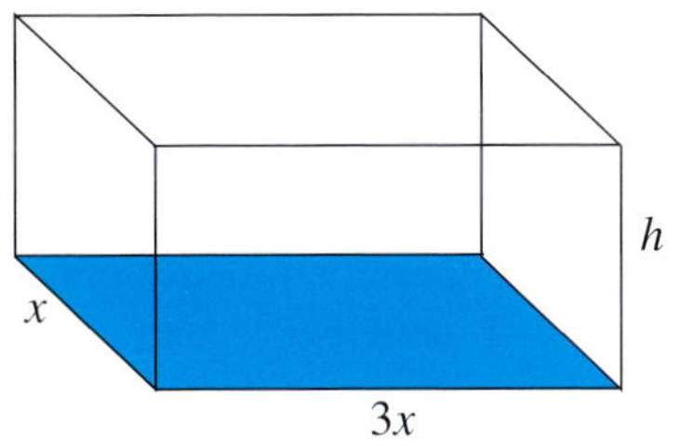

If the sum of its width and its length and its depth is 140 centimetres (the maximum allowed by the airline), find its greatest possible volume.

27. The curves with equations $y = 4x - x^2$ and $y = \frac{4}{x} - 1$ meet at 3 points A, B and C, where the x-coordinate of A is negative and the x-coordinates of B and C are positive.

(a) Find the coordinates of A, B and C.

(b) Sketch the two curves on the same set of axes.

(c) Find the area of the region enclosed by the two curves between B and C.

28. Find $f'(x)$ if i. $f(x) = \frac{x}{1 + \cos x}$ ii. $f(x) = \log_e\sqrt{x^2 + 1}$

29. Let P and Q be the points on the curve $y = x^3$ at which $x = 2$ and $x = 2 + h$, respectively.

(a) Express the gradient of the line [PQ] in terms of h.

(b) Hence find the gradient of the tangent to the curve $y = x^3$ at $x = 2$.

30. The graph of the function $f(x) = xe^{-x/2}$ is shown in the following diagram.

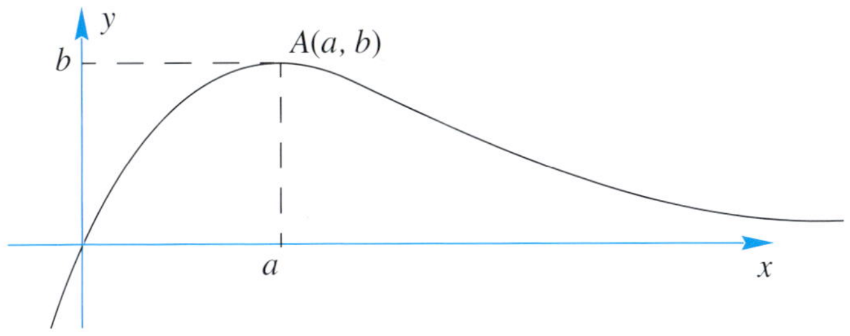

(a) Find the coordinates of the turning point, A, and justify that it is a maximum by considering the sign of the first derivative.

(b) i. Find the equation of the tangent at the origin O.

ii. Find $\frac{d}{dx}(xe^{-x/k})$ and hence show that $\int_0^a xe^{-x/2}dx = 4 - 2(2 + a)e^{-a/2}$.

iii. Find the area of the region enclosed by the curve $y = xe^{-x/2}$, the tangent to the curve $y = xe^{-x/2}$ at the origin and the straight line $x = a$.

(c) i. Show that $\frac{d}{dx}(x^2e^{-x}) = 2xe^{-x} - x^2e^{-x}$.

ii. Using the results of (b) and (c) i., find the volume of the solid generated by revolving the curve with equation $y = xe^{-x/2}$ about the x-axis from $x = 0$ to $x = 2$.

31. (a) Find the derivative of $x\ln x - x$

(b) Hence calculate $\int_1^2 \ln x dx$.

32. In a competition, a surf lifesaver has to race from a starting point S on the beach, at the edge of the water, to a buoy B which is 200 metres along the water's edge from S. He can run along the beach at an average speed of 4 metres per second and he can swim through the water at an average speed of 1 metre per second. Assume that he starts swimming as soon as he enters the water.

(a) How long will the lifesaver take

i. to swim in a straight line from S to B?

ii. to run along the beach to A and then swim straight out to B?

(b) Show that, if the lifesaver runs x metres along the beach to C and then swims in a straight line to B, the time taken will be T seconds, where

$$T = \frac{x}{4} + \sqrt{80000 - 400x + x^2}$$

(c) Hence find, to the nearest second, the shortest possible time for the lifesaver to go from S to B.

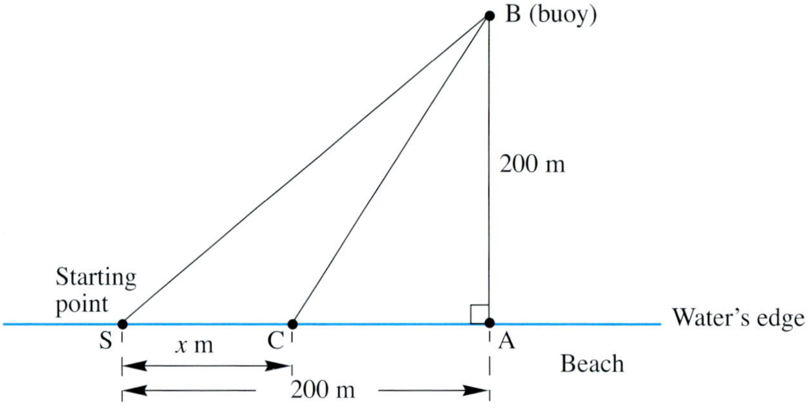

33. The graph of the function $y = e^{2x} - 4x$ is shown in the following diagram.

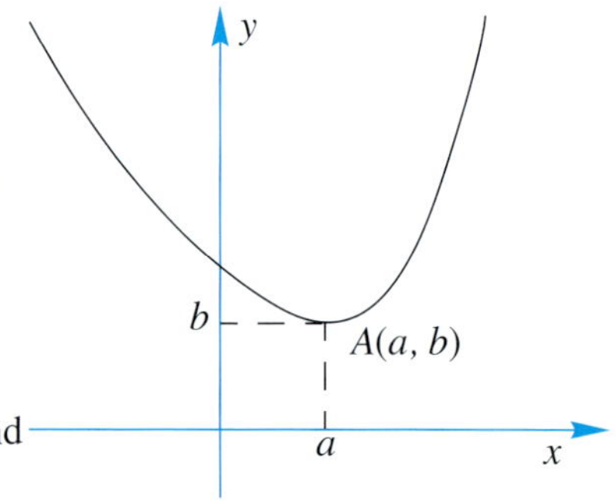

(a) Find the coordinates of the turning point, A, and justify that it is a minimum by considering the sign of the first derivative.

(b) Find the area of the region enclosed by the curve $y = e^{2x} - 4x$, the coordinates axes and the line $x = 1$.

(c) i. Find the equations of the tangents to the curve at the points $(0, 1)$ and $(1, e^2 - 4)$.

ii. Copy the diagram above and draw both tangents on the set of axes.

iii. The tangents intersect at the point (m, n). Show that $m = \frac{e^2 + 1}{2(e^2 - 1)}$ and $n = -\frac{2}{e^2 - 1}$.

(d) i. Given that $\frac{d}{dx}(xe^{2x}) = e^{2x} + 2xe^{2x}$, show that $\int_0^1 8xe^{2x}dx = 2e^2 + 2$.

ii. Find the volume of the solid of revolution formed when the curve with equation $y = e^{2x} - 4x$ is rotated about the x-axis from $x = 0$ to $x = 1$.

34. Consider the function $g:\mathbb{R}^+ \mapsto \mathbb{R}$, where $g(x) = 2(1-x) + \log_e x$

(a) Show that the graph of $y = g(x)$ has an intercept at $x = 1$.

(b) Find $g'(x)$ and specify the values of x for which $g'(x)$ is
i. positive ii. zero iii. negative.

(c) Find the coordinates of any stationary point of the graph of $y = g(x)$.

(d) Sketch the graph of $y = g(x)$, but do not make any attemp to determine any x-intercept other than $x = 1$.

(e) i. Find the equation of the tangent at the point where $x = 1$.
ii. Find the equation of the normal at the point where $x = 1$.

(f) i. Given that $\int(\log_e x)dx = x\log_e(x) - x + c$, find the area of the region enclosed by the curve $y = g(x)$, the tangent to the curve at $x = 1$ and the line $x = 0.5$.

ii. **Hence** find the area of the region enclosed by the curve $y = g(x)$, the normal to the curve at $x = 1$ and the line $x = 0.5$.

35. The volume, $V(t)$, of water in a reservoir at time t is given by $V(t) = 3 + 2\sin\frac{1}{4}t$.

(a) Find the volume of the water in the reservoir at time $t = 10$.

(b) i. Find $V'(t)$.
ii. Find the rate of change of the volume of water in the reservoir at time $t = 10$.
(Give your answers correct to two decimal places.)

36. The region bounded by the curve $y = e^x - 1$, the x-axis and the line $x = \log_e 3$ is rotated about the x-axis to form a solid of revolution. Find the volume of this solid.

37. The graph of $y = ax^3 + bx^2 + cx$ passes through the point $(-1, 16)$ and has a stationary point at $(1, -4)$. Find the values of a, b and c.

38. (a) Find $f'\left(\frac{\pi}{6}\right)$ if $f(x) = \log_e(\tan x)$.

(b) Find i. the derivative of $\cos^2 3x$.
ii. the antiderivative of $\cos^2 3x$.

39. Determine the constants A and B so that $x = t(A\cos 2t + B\sin 2t)$ will satisfy the equation $\frac{d^2x}{dt^2} + 4x = 2\cos 2t$ for all values of t.

40. (a) Find the area of the region enclosed by the curve with equation $y = 7x^2(1-x)$ and the x-axis.

(b) The region enclosed between the curve with equation $y = 7x^2(1-x)$ and the x–axis is rotated about the x-axis to form a solid of revolution. Find the volume of this solid.

41. A capsule is to be constructed by closing the ends of a circular cylinder, of height h and radius r, with hemispherical caps of radius r.

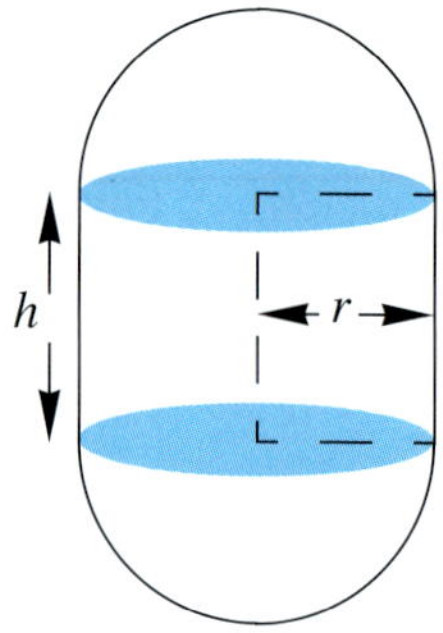

(a) Express the volume, V, of the capsule in terms of r and h.

(b) If the cylindrical part of the capsule can be made at a cost of k dollars per unit area of surface and the spherical parts at a cost of $\frac{3}{2}k$ dollars per unit area of surface, and the total cost of making the capsule is P dollars, express P in terms of k, r and h.

[The surface area of a sphere of radius r is $4\pi r^2$]

(c) If V is constant, express P as a function of r.

(d) What is the domain of this function?

(e) Find the value of r for which P is a minimum, and hence show that the least expensive capsule is one whose total length is twice its diameter.

42. A B C D is a square sheet of cardboard, with side a units. Creases WZ and XY are made parallel to AD as shown in diagram (i), to form flaps of width x units. These flaps are folded up to form a triangular prism with open ends. (see diagram (ii)).

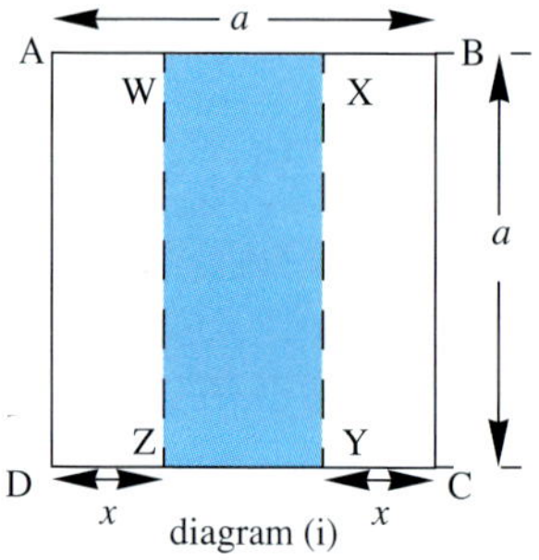

diagram (i)

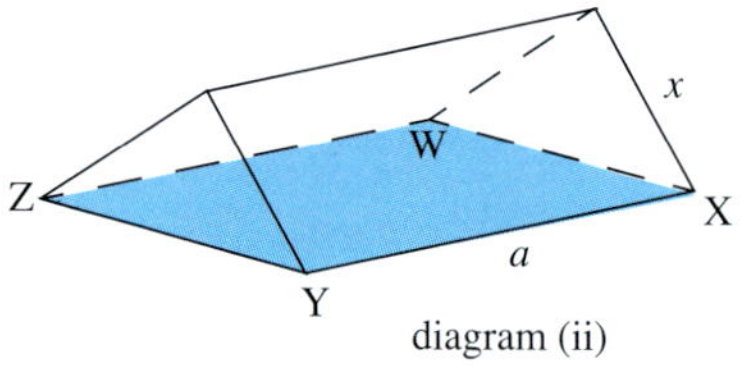

diagram (ii)

(a) Show that the volume, V cubic units, of the prism is given by

$$V = \frac{a}{4}(a-2x)\sqrt{4ax-a^2}\,.$$

(b) State the domain of this function.

(c) Find the maximum volume of the prism. Justify your answer.

43. (a) Find $\frac{dy}{dx}$ if $y = x^2\log_e x$. Hence, evaluate $\int_1^2 x\ln x\,dx$.

(b) P is any point on the curve $y = e^x$. If a is the x-coordinate of P, prove that the tangent to the curve at P may be drawn by joining P to the point $(a-1, 0)$.

(c) Hence, or otherwise, find the coordinates of the point on the curve $y = e^x$ at which the tangent passes through the origin and find the equation of this tangent.

(d) Find the area enclosed between the tangent, the curve and the y-axis.

44. A is the region bounded by the x-axis and the curve $y = \sin 2x$ between $x = 0$ and $x = \frac{\pi}{2}$. Lines $x = x_1, x = x_2$ and $x = x_3$ cut the region A into four parts of equal area. Find x_1, x_2 and x_3.

45. Let $f : \{x | x \geq a\} \mapsto \mathbb{R}$, where $f(x) = (x-2)^2$.

(a) If a has the least real value for which the inverse function, f^{-1} exists, find a and define f^{-1}.

(b) Find the volume of the solid of revolution formed by rotating about the x-axis the curve $y = f^{-1}(x)$ between $x = 0$ and $x = 4$.

46. Consider the functions $f(x) = \sqrt{e^x}$ and $g(x) = e^{\sqrt{x}}$.

(a) State the maximal domain of i. $f(x)$ ii. $g(x)$

(b) Find $f'(0)$ and show that $g'(0)$ does not exist.

(c) Find $\{x | f(x) = g(x)\}$.

(d) i. Sketch, on the same set of axes, the graphs of f and g.

ii. Hence, find $\{x | f(x) > g(x)\}$

(e) Given the following results

$$\int e^{\sqrt{x}} dx = 2e^{\sqrt{x}}(\sqrt{x} - 1) + c \text{ and } \int e^{2\sqrt{x}} dx = \frac{e^{2\sqrt{x}}(2\sqrt{x} - 1)}{2} + c$$

i. Find the area of the region enclosed by the curves $y = f(x)$ and $y = g(x)$.

ii. Find the volume of the solid of revolution generated, when the region enclosed by the curves in part i., is rotated about the x-axis.

47. A saucepan (without a lid) is in the shape of a cylinder with handle attached. The cylindrical part has an internal base radius r cm an internal height h cm and is to be lined with a non-stick coating whose cost is proportional to the internal surface area.

If the saucepan must be able to hold a fixed volume, V cm^3, show that the cost of lining it is a minimum when its height is equal to its base radius.

48. Differentiate with respect to x: i. $\cos^3 2x$ ii. $x\sqrt{1-x^2}$

49. (a) Differentiate $x \sin x$.

(b) Hence find an anti-derivative of $x \cos x$.

50. (a) Find the equation of the normal to the curve $y = e^{x-2}$ at the point where $x = 1$.

(b) Show the curve and the normal on a sketch graph.

(c) Find the area of the region enclosed by the normal, the curve and the y-axis.

51. (a) Find the area of the region enclosed by $y = x^2 + 1$ and $y = 2$

(b) If the region in (a) is now rotated about the x-axis. Find the volume of the solid of revolution generated.

52. When a bus travels along a straight road from one stop to the next, its velocity-time function for the trip is described by $v(t) = \frac{1}{10}t(5-t)$.

(a) State the domain of this function.
(b) Sketch the graph of the function.
(c) Find the maximum value of v.
(d) Find a rule for the acceleration of the bus between the stops.
(e) Sketch the acceleration-time graph for this bus.

53. A closed capsule is to be constructed as shown in the diagram. It consists of a cylinder of height h which has a flat base of radius r and is surmounted by a hemispherical cap.

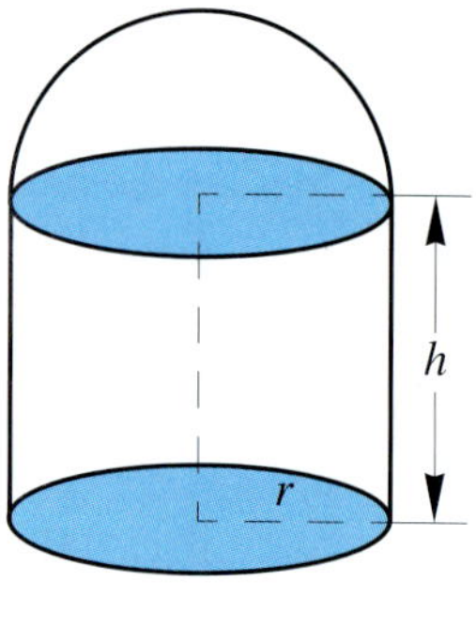

(a) Show that the volume, V, and the total surface area, A, are given by

i. $V = \pi r^2 h + \frac{2}{3}\pi r^3$.

ii. $A = 2\pi rh + 3\pi r^2$.

(b) If $V = \pi a^3$, where a is a positive constant, show that

$A = \frac{2\pi}{r}a^3 + \frac{5\pi}{3}r^3$.

(c) Hence find in terms of a the minimum and maximum surface area of the capsule, if its overall height (including the cap) must not be more than twice its diameter.

54. A car starts from the side of a road with its velocity v metres/second at any time t seconds given by $v = -\frac{1}{20}t^2 + 2t$. Five seconds later a second car starts from rest from the same position as the first car, accelerates uniformly to a speed of 20 metres / second in five seconds , then maintains this constant speed .

(a) On the same axes, draw a velocity-time graph for the two cars.
(b) Find when the second car overtakes the first.
(c) Find the distance between the two cars when the first car again comes to rest.

55. Find the derivatives with respect to x of each of the following:

(a) $e^{-x}\cos x$ (b) $\sqrt[3]{x^3 + 3x^2 + 3x + 1}$

56. $f(t) = \frac{(1+t)^2}{t}$ Find (a) the derivavtive (b) an antiderivative of f.

57. (a) Sketch the graph of $y = \cos x$ in the domain $\left\{x : -\frac{\pi}{2} \le x \le \frac{\pi}{2}\right\}$

(b) Using part of the x-axis as base, a rectangle is drawn under this loop of the curve, so that its upper two verticles lie on the curve. If one upper vertex is at the point (x, y), find an expression for the area of the rectangle, in terms of x only.

(c) Show that this rectangle will have maximum area when $\tan x = \frac{1}{x}$.

(d) Find the maximum area of the rectangle, giving your answer to 2 decimal places.

23.1 INTRODUCING THE PORTFOLIO

23.1.1 PURPOSE OF THE PORTFOLIO

During this course you will be required to submit two pieces of work for internal assessment. The purposes for this is to give you the opportunity to be rewarded for mathematics carried out without the time limitations set by examinations. In doing so, it will allow you to develop further insight into the nature of mathematics and develop your abilities to not only be inquisitive, but be judicious in the types of questions you ask about mathematics.

The benefits of carrying out these tasks will only be reciprocated by your efforts. These tasks aim to develop skills and techniques that will allow you to extract relevant information from, what might sometimes be, an overwhelming amount of information and data.

These tasks might require the use of technology, either in the form of a graphics calculator or a more sophisticated mathematical program. Along with the mathematical aspect of the task, you will be required to write up your findings and as such, develop your communication skills.

There will be two types of tasks that must be submitted, one of Type I and one of Type II. The differences between these two tasks are as follows:

Type I – mathematical investigation.
Type II – mathematical modelling

To define the exact dividing line between a mathematical investigation and mathematical modelling is not always the same as differentiating between black and white. There will always exist a shade of grey between the two types of tasks. However, we will try and provide a number of tasks that tend to belong more to one type of task than the other. Having said this, in very broad terms we could try and separate the two as follows:

Type I – mathematical investigation : tackling a problem solving task for which no solution, nor obvious means of finding a solution, is known

Type II – mathematical modelling : problem based on a 'real-life' situation, where problems are not presented in mathematical terms

However, we can provide the steps that should be considered when undertaking a mathematical investigation of either type.
The main steps in the process are:

1. Define the problem.
2. Identify the parameters that define a 'good decision'.
3. Research appropriate data.
4. Analyse the data.
5. Arrive at a decision and be prepared to justify it. This is known as evaluation.

Further, when dealing with tasks that are of a modelling specific type, the following approach can also be used as a guide

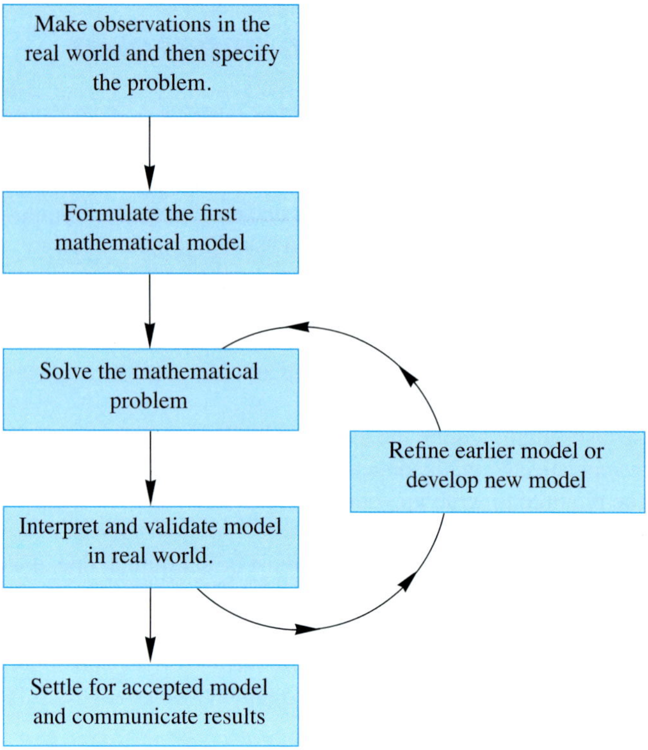

Of course, your portfolio work will also be assessed and graded. So, it is important that you are familiar with the '**Internal assessment criteria**' that will be used to grade your work. There are six criteria that will be used to assess your work. Notice the similarity between the assessment criteria for Type I and Type II tasks (hence the 'shade of grey' when trying to differentiate between the two). The criteria are based on:

TYPE I

A. **Use of notation and terminology**
B. **Communication**
C. **Mathematical process – searching for patterns**
D. **Results – generalisation**
E. **Use of technology**
F. **Structure and communication**

TYPE II

A. **Use of notation and terminology**
B. **Communication**
C. **Mathematical process – developing a model**
D. **Results – generalisation**
E. **Use of technology**
F. **Structure and communication**

Having said this, **it is your responsibility** to discuss your investigation with your teacher to ensure you are addressing the assessment criteria and internal assessment requirements.

23.1.2 PROBLEM SOLVING

We end this section with a set of challenging problems. These problems are known as 'extended-closed' problems. That is they require a combination of the skills required for Type I and Type II problems, but eventually, a 'solution' can be arrived at – how you go about getting there is a different matter. These Problem Solving Tasks could be looked upon as the begining of a Type I or Type II task. That is, they would require an extension of the question(s) posed. Should you wish to use one of these as a **starting point** for a Type I or Type II task, discuss it with your teacher first and make sure that you know how the problem could be extended so that it satisfies the Type I or Type II task requirements. Remember, **you must check with your teacher** to ensure that you will meet all the requirements as set out by the I.B.O Internal Assessment requirements for this task.

P Solving 1

PROBLEM: CIRCLES UPON CIRCLES

Here are two circles of radius 1 with centres A and B, which are 4 units apart.

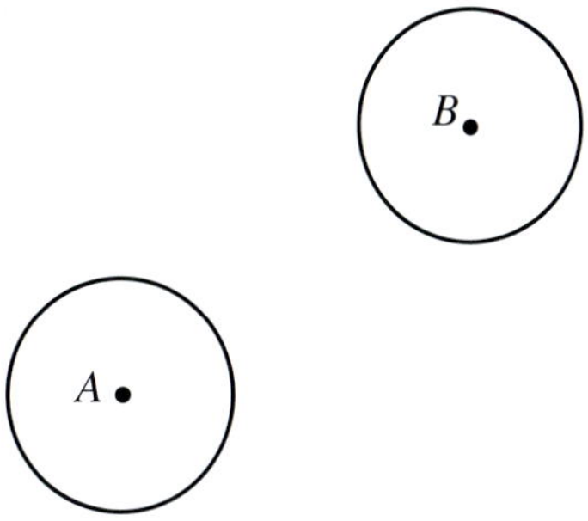

For a given pair of circles, any circle that intersects each of the original circles twice, with every intersection forming a right angle, we will call perpendicular to that pair.

For the pair of circles pictured,

a. i. give a precise description of a circle, perpendicular to the pair, with centre on the line joining A and B.

ii. give a precise description of any other circles perpendicular to the pair.

iii. comment on any noteworthy features of the family of all circles perpendicular to the pair.

b. Consider the circles, perpendicular to the pair, with centres above the line joining A and B. Each such circle has a point furthest from the line joining A and B, which we will call the top. Suppose that the spacing between two such circles is defined to be the distance between their tops. Find and describe a family of 'equispaced' circles, all of which are perpendicular to the original pair of circles.

c. Now consider the following. Each of the circles considered in part b. has a counterpart obtained by reflection about the line joining A and B. If we now take one of the circles from part b. and its reflected counterpart to form a pair of circles, describe the family of circles perpendicular to this pair and comment on any noteworthy features.

Generalise your results in part a. to the case where the original pair of circles have unequal radii.

PSolving 2

PROBLEM: CABLES

Part A

Engineers from a telephone company need to provide a telephone link between three towns A, B and C which form an isosceles triangle with AB = BC. The engineers wish to use the shortest possible cable length to join the three towns.

Instead of just solving the problem for the particular towns being considered, the engineers would like to obtain a more general result which could be used if a similar situation arose in the future. To do this they propose the following model: fix the distance between A and C at 2 units and change the isosceles triangle by moving B along a line perpendicular to AC and through the mid-point of AC.

The engineers have put forward three plans which they believe will be useful in solving the problem. These plans, showing the cable links, are shown in Figure 1.

Figure 1: Three plans for joining towns A, B and C

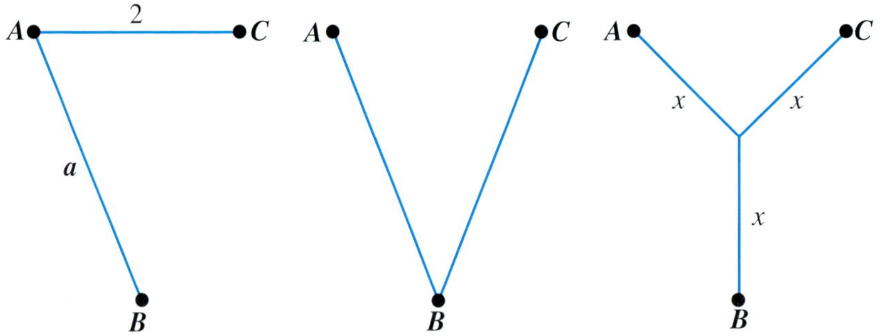

In the C–plan, cable is run from B to A to C. In the V–plan, cable is run from A to B to C. In the Y-plan, cables of equal length are run to A, B and C from a point between the towns.

For different allowable positions of B, when should a given plan be used to obtain the shortest cable length?

Part B

A consulting engineer suggests another option: to link the three towns using a modified Y-plan in which the cables are joined so that the angles between them are each 120°. This plan is shown in Figure 2.

Figure 2

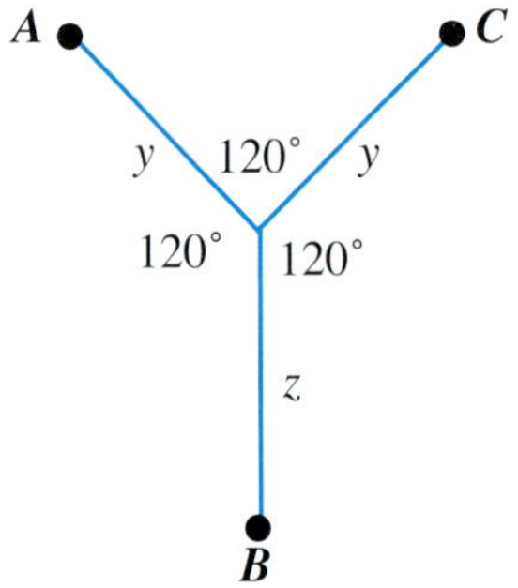

Is this plan better than any of the three plans in part A? Explain your answer fully.

Part C

The consulting engineer is working on the problem of connecting four towns A, B, C and D with telephone cable. She is considering four plans which are shown in Figure 3.

Figure 3: Four plans for joining towns A, B, C and D

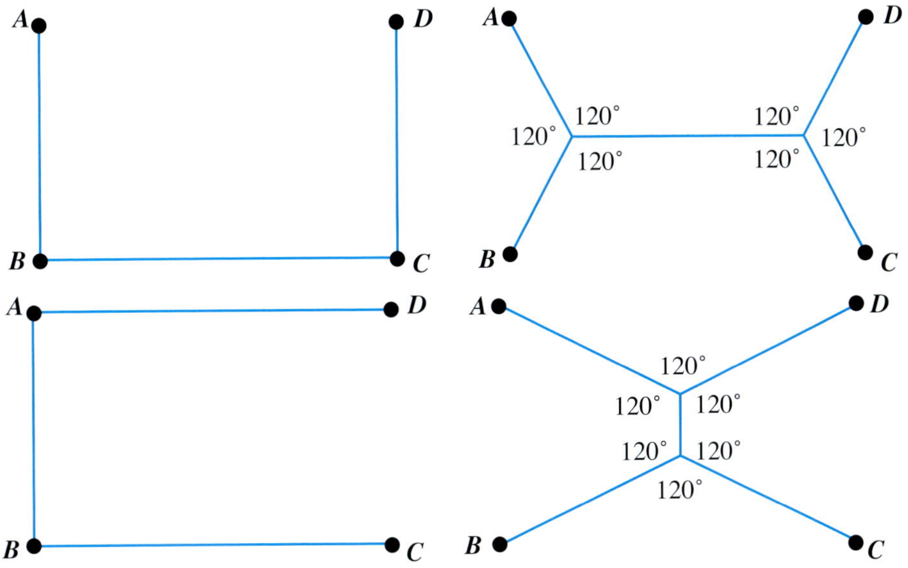

Suppose the distance between A and B is fixed at 2 units and we change the rectangle by adjusting the distance a between B and C.

If the engineers want to obtain the shortest cable length, which plan should be used (consider all possible values for a)?

P Solving 3

PROBLEM: LATTICE TRIANGLES

Consider the upper right hand quadrant of a square lattice with unit spacing between the lattice points. Suppose a point A is chosen such that the straight line OA from the origin O (0,0) to A does not pass through any lattice points between O and A. An example of a particular line OA is given below.

Figure 1

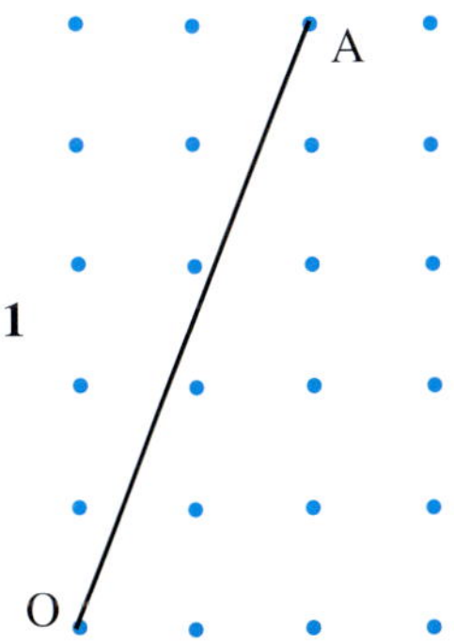

Figure 2

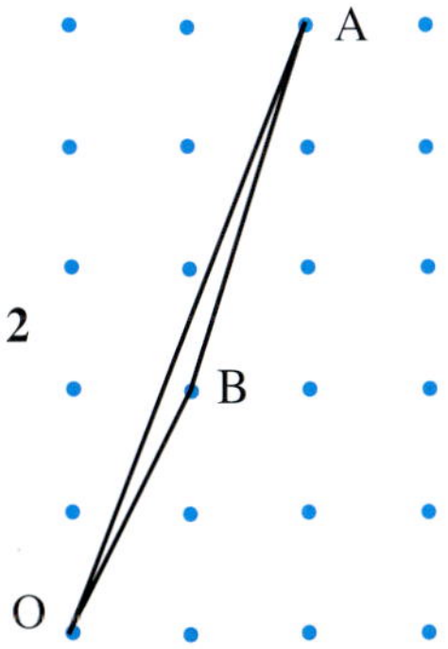

In Figure 1, what are the allowed coordinates of the point A?
For future reference, let N3, N4 and N5 be any positive integers such that (3, N3), (4, N4) and (5, N5) respectively are allowed points A.

Now suppose a triangle OAB is to be formed as in Figure 2, with the line OA as the longest side. The point B is to be chosen such that no lattice point other than the vertices OAB are contained within, or on the boundary of, the triangle.

> Find a formula for an allowed point B when A is the point (3, N3).
> For a general point A, give some criteria which the point B must satisfy and use these criteria to find a formula for an allowed point B when A is the point (4, N4) or (5, N5).

P Solving 4

PROBLEM: BURIED TREASURE

You have found a map and instructions to find treasure buried on an island on which are located a water tower, a large gum tree and a peppercorn tree. The instructions include the following directions.

'When you stand at the water tower you should be able to see the large gum tree ahead of you and somewhat to the left, and the peppercorn tree ahead of you and somewhat to the right.
Starting from the water tower, walk directly to the large gum tree, counting the number of steps taken. Turn left through an angle of 90° and walk the same distance away from the gum tree. Mark this spot.

Do this again by starting at the water tower, walking in a straight line to the peppercorn tree and counting the number of steps taken, turning right through an angle of 90°, walking the same distance away from the peppercorn tree and marking the spot.

Half-way along a straight line joining the two marked spots lies the treasure.'
Unfortunately, when you get to the island, the gum tree and the peppercorn tree are plainly visible, but the water tower is nowhere to be seen. It appears that the tower has been demolished and all traces removed. Where is the treasure buried?

You calculate the spot where you think the treasure is buried and have been digging unsuccessfully for some time when you remember that your compass was out of alignment and was not measuring angles accurately.

The angle at the gum tree through which you had turned left was only 85° and the angle at the peppercorn tree through which you had turned right was really 95°. Your compass has now broken down completely and you have no way of accurately determining a right-angle. What can you now do in your attempt to find the treasure?

In your concern to find the treasure you become exhausted and confused and you lose the map in a sudden gust of wind. You think that the instructions were as follows.

'Starting from the water tower, walk directly to the large gum tree, counting the number of steps taken. Turn right through an angle of 90° and walk twice the same distance away from the gum tree. Mark this spot. Do this again by starting at the water tower, walking in a straight line to the peppercorn tree and counting the number of steps taken, turning left through an angle of 90°, walking twice the same distance away from the peppercorn tree and marking the spot. Half-way along a straight line joining the two marked spots lies the treasure.'

If you were able to find this mid-point, how far from the treasure would you be if the gum tree and the peppercorn tree are 50 metres apart?

PSolving 5

PROBLEM: AREA AND PERIMETER

The following question was posed to a group of mathematics students.

> '*Are there shapes for which the numerical value of the length of the perimeter is the same as the numerical value of the area*?'

One student quickly saw that a square is a shape with this property because a square which has a side length of 4 units has a perimeter of 16 units and an area of 16 square units. The student could also easily show that there could only be one square with this property.

After looking at families of shapes like triangles, circles, rectangles and other polygons the students made the following conjecture.

> '*For every family of shapes there is at least one of these shapes for which the numerical value of the area and the numerical value of the perimeter are the same*.'

By a 'family of shapes' the students meant all shapes which are similar to a given shape. For example there is only one family of squares, but there is an infinite number of families of rectangles.
You are required to find the following.

> For which shapes does the conjecture hold?
>
> For each class of shapes for which the conjecture holds, give a method for finding an actual shape for which the numerical value of the area is the same as the numerical value of the perimeter.

PSolving 6

PROBLEM: CHOCOLATE WRAPPING

A triangular slice of chocolate has to be wrapped in a triangular piece of paper, which cannot be cut or torn. The chocolate is sliced so thinly that you can ignore its thickness.

Both the chocolate and paper are equilateral triangles. The paper can be folded along an edge of the chocolate. The edges of the wrapping do not have to overlap, they can just meet.

The chocolate can be positioned in various ways on the paper. Some examples appear in the diagram below:

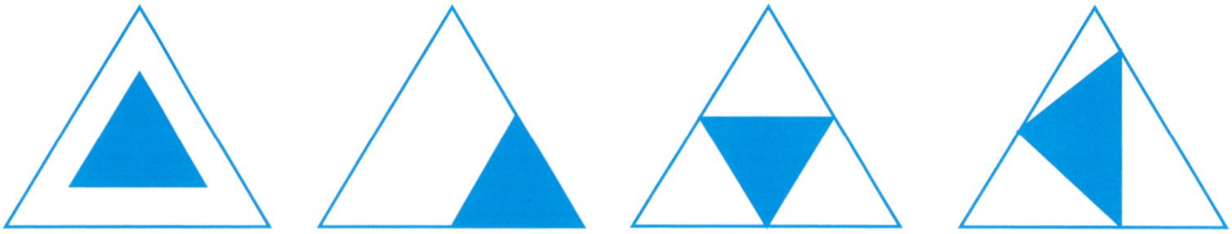

When the chocolate has a side length of 4 centimetres, what is the smallest side length of the paper that will allow the chocolate to be wrapped?

How does the solution differ if the chocolate and paper are both rectangular (not necessarily similar rectangles)?

What optimisation issues arise when dealing with such a situation?

PSolving 7

PROBLEM: SLIDING RECTANGLES

A rectangle ABCD is placed on the cartesian plane as shown in the diagram below.

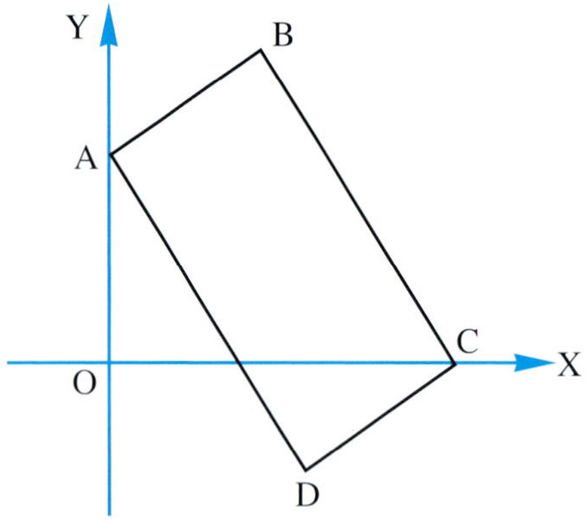

The rectangle is free to slide so that A is always somewhere on the positive Y-axis and C is always somewhere on the positive X-axis.

As the rectangle moves, what path will the point B follow? Explain your results.

What will be the path of the point D?

If the point A can move over both the positive and negative parts of the Y-axis, and the point C can move over both the positive and negative parts of the X-axis, what paths will the points B and D follow?

What effect will changing the dimensions of the rectangle have on the paths of points B and D?

P Solving 8

PROBLEM: GOAL SHOOTER

Hockey is played on a rectangular playing field with a goal area at each of the shorter ends of the rectangle. A player on the long boundary, as shown in the diagram below, wants to shoot a ball through the goal.

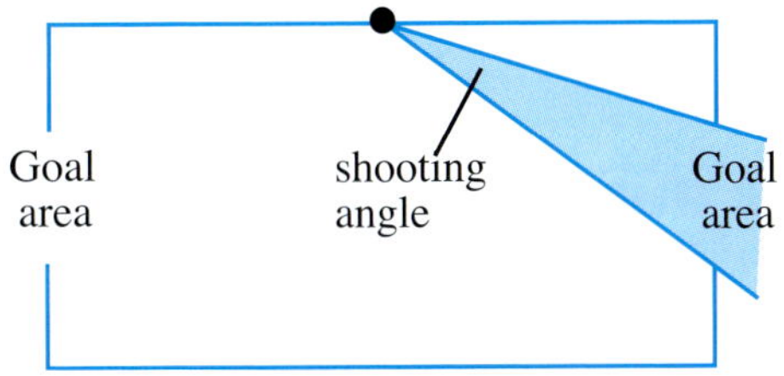

What position on the boundary will give the player the biggest shooting angle through the goal?

If the field was circular instead of rectangular, what position on the boundary would give the player the best shooting angle through the goal?

What is the best position on the boundary to shoot from if the field is elliptical?

P Solving 9

PROBLEM: JELLY BEANS

In the diagram below there is a grid that extends infinitely upwards and to the right.
At the bottom left corner there is a start dot (the square dot on the grid).

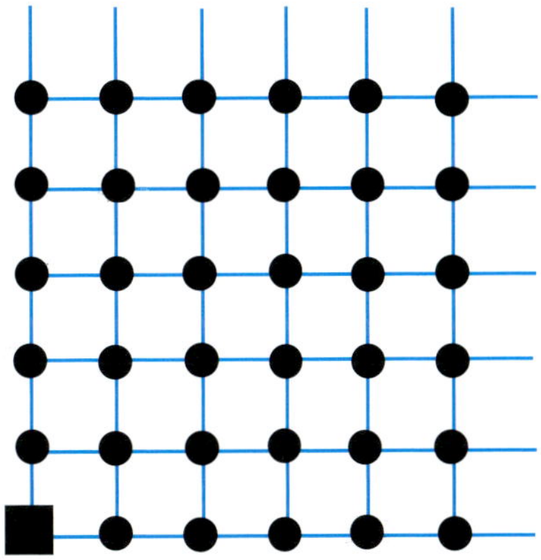

Christine is standing at the start and has a bag containing 3 black jelly beans and 2 purple jelly beans.

She takes a jelly bean from the bag and eats it. If it is a black jelly bean she moves up to the next dot. If it is a purple jelly bean she moves right to the next dot.

Which dots could lie on Christine's path and with what probabilities would she arrive at each of these dots?

What if the bag contained a different number of black and purple jelly beans?

What if Christine did not eat the jelly beans as she took them from the bag, but replaced them instead?

P Solving 10

PROBLEM: RUNS OF HEADS AND TAILS

When you toss a fair coin a number of times you get a sequence of heads (H) and tails (T). If you toss the coin a large number of times you might expect the number of heads and the number of tails to be about equal. However when you repeatedly toss a coin you will also notice that you get runs of heads and runs of tails of varying length.

One possible outcome of tossing a coin five times is HHHTH. This sequence contains a run of three heads, a run of one tail and a run of one head, a total of three runs.

Investigate the distribution of runs of different length for the 5 coin situation.

How does the number of coin tosses affect the distribution of runs of heads and tails?

What if there was a different number of equally likely outcomes (for example if you were rolling a die instead of tossing a coin) or if the outcomes were not equally likely (for example if you were using an unfair coin) ?

Investigate the distribution of runs in these situations..

P Solving 11

PROBLEM: TELEPHONE NUMBERS

For this problem you should begin by conducting the following experiment.

Obtain a copy of the white pages of a telephone directory and open it to any page in the main body of the directory. From the page you have chosen, take $2N$ consecutive telephone numbers (N must be fairly large, say 120 or more) and place these numbers into N pairs.

For each pair, note the pair of numbers formed by the last two digits of each telephone number. Square these two numbers and add them together.

Assign the pair to one of the four categories A, B, C or D according to the following rule.

A One or both of the numbers being added is equal to 0.
B The sum of the squares is equal to 1002.
C The sum of the squares is less than 1002, and neither number in the pair is zero.
D The sum of the squares is greater than 1002 and neither number in the pair is zero.

For example, a pair of telephone numbers is (3015403, 2133571). The last two digits of each number in the pair form another pair of numbers (03, 71). To obtain the sum of the squares of these numbers, proceed as follows.

$$\begin{aligned}(03)^2 + (71)^2 &= 3^2 + 71^2\\ &= 9 + 5041\\ &= 5050\end{aligned}$$

For this pair, the sum of the squares is less than 100^2. Therefore this pair should be assigned to category C.

For the $2N$ telephone numbers you have chosen, count the number of pairs a, b, and c in categories A, B and C respectively.

Now calculate the value of $4\left\{\dfrac{\frac{a}{2}+\frac{b}{2}+c}{N}\right\}$ for the $2N$ numbers you have chosen.

The value generated by the process followed in carrying out this experiment gives a good approximation to a particular irrational number. What is that number?

Give a mathematical argument which explains why the process gives a good approximation to this number.

Make sure your explanation accounts for the appearance of $\frac{a}{2}$ and $\frac{b}{2}$ in the above expression.

Why is it that any page in the telephone directory will give a similar result?

Suppose the sum of the squares of the last three digits of each pair of telephone numbers was computed instead of the last two digits.

How should the process be modified to give an approximation to the same irrational number?

How would this approximation compare with the previous one?

PSolving 12

PROBLEM: MAD MOUSE

The track of a section of a mad mouse railway at an amusement park is in disrepair and needs replacing. Viewed from the side, the track has the shape shown in the following diagram.

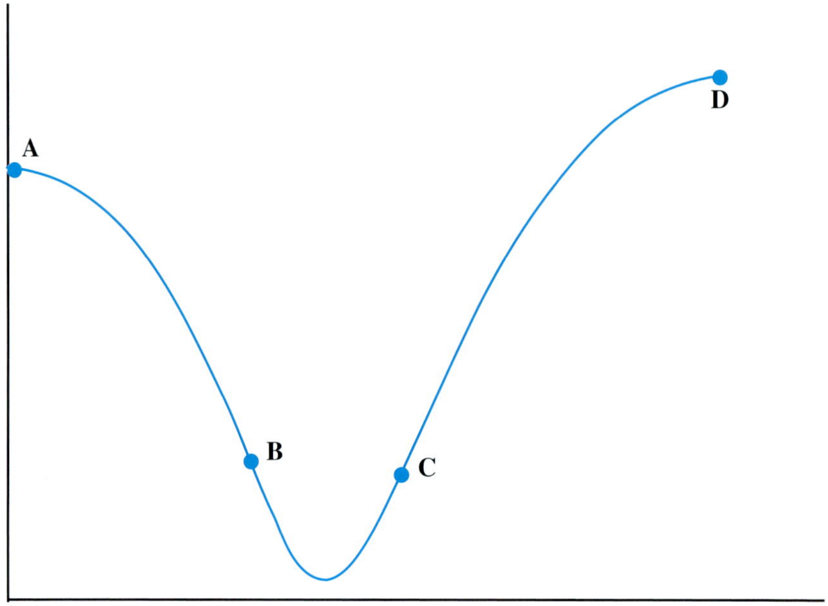

The profile of each of the segments AB, BC and CD is a parabola. The gradients of the profiles at the points B and C, where the segments meet, match so that the track is smooth.

The track at point A is 30 metres above ground level and inclined at an angle of 10° below the horizontal, while the track at point B is 10 metres above ground level and 8 metres horizontally across from A. The point C is at the same height as B and 11 metres across horizontally from A. The track at point D is 35 metres high and has zero slope.

The new track costs $100 per metre. Devise a method to estimate the cost of the new track.

Extension: Carry out an investigation addressing similar problems as those discussed so far, but this time visit a skateboard circuit.
Draw diagrams to scale.
Consider where the curved ramps meet the ground or other obstacles. How do the designers of this circuits ensure a 'smooth' ride for the skateboarder?

How would you go about designing your own skateboard ramp?

P Solving 13

PROBLEM: SPREAD OF DISEASE

One Sunday evening, five people with infectious influenza arrive in a large city with a population of about two million. They go to different locations in the city and thus the disease begins to spread throughout the population.

At first, when a person becomes infected he or she shows no sign of the disease and cannot spread it. About one week after first catching the disease this person becomes infective and can spread the disease to other people.

This infective phase also lasts for about one week. After this time the person is free from the influenza, although he or she may catch it again at some later time.

Epidemiologists are trying to model the spread of the influenza. They make a simplifying assumption that the infection progresses in one week units. That is, they assume that everybody who becomes infected does so on a Sunday evening, becomes infective exactly one week later, and is free of infection exactly one week after that. People free of the disease are called **susceptibles**.

The epidemiologists also assume that the city population is large enough so that the population size is constant for the duration of the disease. That is, they ignore births, deaths and other migrations into or out of the population.

Finally, they assume that each infective person infects a fixed fraction f of the number of susceptibles, so that

1. the **number of infectives** at week $n+1$ is equal to
   $f \times$ (the number of susceptibles at week n) $\times$ (the number of infectives at week n)

   and

2. the **number of susceptibles** at week $n+1$ is equal to
   (the number of susceptibles at week n) + (the number of infectives at week n)
   – (the number of infectives at week $n+1$).

Why must the number of infectives, plus the number of susceptibles, be constant from week to week?

Choose values of f between $\frac{1}{10^6}$ and $\frac{2}{10^6}$ and use this model to show how the number of infectives changes from week to week from the Sunday when the five infective people arrived.

What limiting values does the model predict for the number of infectives?
Will there always be a limiting value for the number of infectives? If so, how are the limiting values related to the population size and to f?

For what values of f will it be possible to have a situation where the number of infectives eventually oscillates between two values?

Extension: How does the work covered in this project relate to the concepts encountered in Project 2 – Fractals?

What other models for the spread of diseases exist?

P Solving 14

PROBLEM: SHORTEST PATH

For any triangle there exists a point X such that the sum of the distances from each vertex to X is a minimum.

Consider all isosceles triangles of perimeter 6 units. For which of these triangles is the sum of the distances from each vertex to X a minimum?

P Solving 15

PROBLEM: STRAIGHT UP AND DOWN

Suppose an object is projected vertically upwards from ground level with initial velocity 10 m/sec. Ignoring air resistance, plot the corresponding velocity - time and displacement - time graphs of the motion of the object.

On the same axes, sketch the respective graphs when the affect of air resistance is included. Be sure to give a careful account of your reasoning.

Does the affect of air resistance alter the time taken for the object to reach its maximum height?
Does the object return to ground level with speed 10 m/sec in both cases?
In what other ways is the motion of the object altered by the affect of air resistance?

P Solving 16

PROBLEM: TABLE-CLOTH

A square table top has sides of length 1 metre. A rectangular shaped table-cloth, 2 metres long and x metres wide, where $x < 1$, is to be placed on the table top.
When the table-cloth is placed on the table it does not cover the table completely. In addition, some of the table-cloth hangs down over the edge of the table.

For the case where both of the long sides of the tablecloth pass across opposite sides of the table, as shown in Figure 1, how can the table-cloth be placed on the table so that the maximum area of the table top is covered?

Figure 1

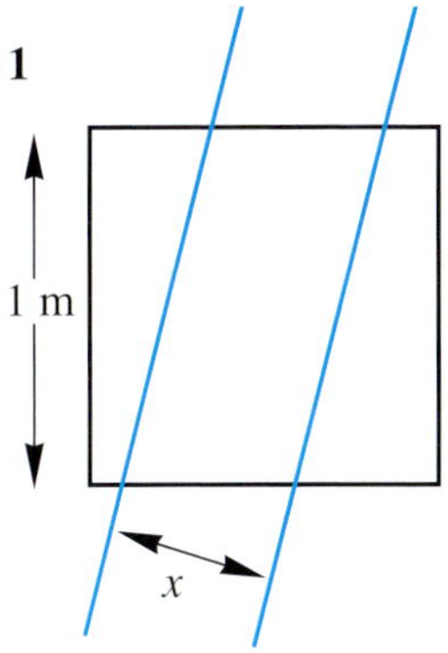

Figure 2

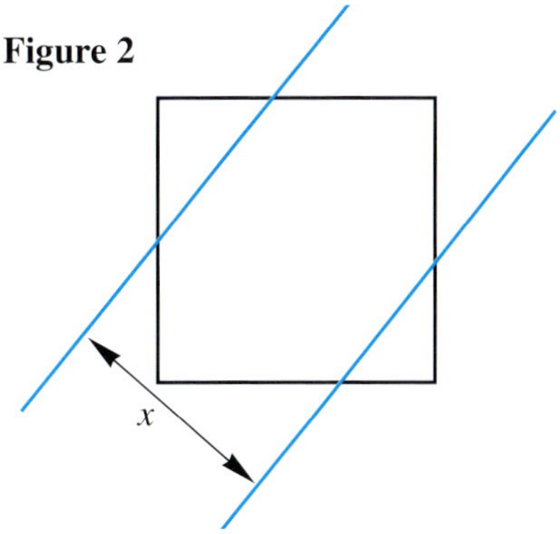

For $x = 0.9$ metre calculate the precise area covered by the table-cloth for this placement, accurate to at least four decimal places.

Now consider the case, where the long sides of the table-cloth pass across all sides of the table, as shown in Figure 2. For this case, with $x = 0.9$ metre, find a way to place the table-cloth so that more of the table top is covered than the maximum area found for the previous case.

P Solving 17

PROBLEM: THE REAL THING

Two Year 12 students were discussing equations they were expected to be able to solve in mathematics. One student argued that as most quadratic equations do not have real solutions, students could only be expected to learn a few examples off by heart. The other student said this must be wrong because most quadratic equations do have real solutions so it is important to learn the formula for the solution of a quadratic equation.

Determine the proportion of cases in which the equation $Ax^2 + Bx + C = 0$ has a real solution where the numbers A, B and C are restricted in various ways, as follows.

a. $A = 1$ and B and C are chosen from the list (0, 0.1, 0.2, 0.3, 0.4, 0.5, 0.6, 0.7, 0.8, 0.9).
b. $A = 1$ and B and C are chosen from real numbers between 0 and 1.
c. $A = 1$ and B and C are chosen from real numbers between 0 and 5.
d. $A = 1$ and B and C are chosen from real numbers greater than or equal to 0.

Explain how you interpret the phrase 'proportion of cases' in (b), (c) and (d).
Describe how you could go about finding the proportion of cases for which the equation $Ax^2 + Bx + C = 0$ has a real solution when A, B and C are chosen from real numbers between 0 and 1.

PSolving 18

PROBLEM: WALKING IN THE RAIN

On a rainy day a person with an umbrella is walking north at a speed v m/s. A wind is blowing from the South and rain is falling with speed 2 m/s at an angle a to the vertical.
At what angle b to the vertical should the person position the umbrella so that it blocks out the maximum amount of rain? Under what conditions will the umbrella need to be held vertically upwards?

Unfortunately this person has to walk for a long time in the rain. Eventually the umbrella becomes saturated and the rain begins to pass through it. However it is possible that by varying the speed of walking, the amount of rain falling on the umbrella can be kept to a minimum.
If the umbrella is positioned at angle b, as calculated previously, what is the desired walking speed which will minimise the amount of water falling on the umbrella?

PSolving 19

PROBLEM: MOOMBA FLOAT

A float in the Moomba parade is a rectangle 12 metres long and 2.5 metres wide travelling at a constant speed of u metres per second in a straight line along a road. A clown runs around the outside edge of the float at a speed of two metres per second continuously in a clockwise direction, always keeping within a negligible distance from the edge of the float. You may assume that the clown is at all times on the edge of the float.

a. Give sketches of the different types of paths which the clown traces out on the roadway for different values of u.

b. At one point the clown drops his hat. Exactly ten seconds later he is at the same spot again and can pick up his hat. Find the possible values of u.

PSolving 20

PROBLEM: WATER FOUNTAIN

A water fountain is designed to shoot water through its nozzle at a speed of 14 m/s. The water leaves the nozzle at all angles between $-\alpha$ and α to the vertical, where the particular value of α can be chosen and controlled by an adjustment on the nozzle. The nozzle of the fountain is at ground level, as is illustrated in the diagram.

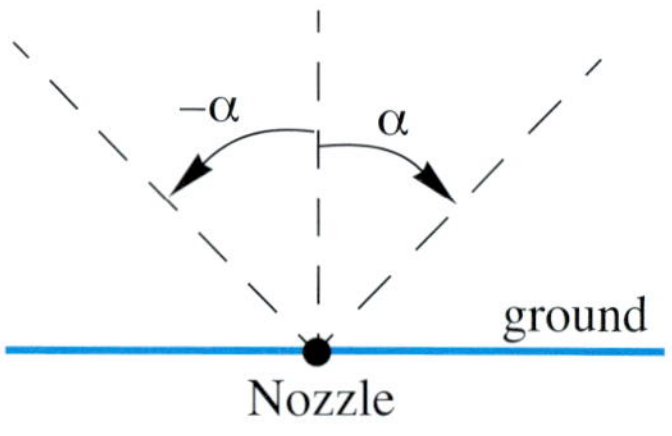

When viewed from a distance the trajectories of the water droplets form a shape. The boundary of the shape separates the region around the fountain which water can reach from the region which is too far away for water to reach. The boundary can be considered as a curve in a plane (the X–Y plane, say). This plane is vertical and passes through the fountain.

a. Neglecting air resistance, find the equation of the path of a typical water particle in the X–Y plane (take $g = 9.8\ ms^{-2}$).

b. What properties does the boundary curve have with relationship to the trajectories?

c. Find the equation of the boundary curve. Consider all cases $0 < \alpha < \frac{\pi}{2}$.

P Solving 21

PROBLEM: RECORD PLAYER

The diagram below shows a record being played on a record player. When the head is tangential to the groove there is said to be a zero tracking error and this is desirable for good reproduction of sound.

Can you design a record player which has zero tracking error while the record is being played or at least a very small tracking error?

You should limit your design to the tone-arm type design, that is, a pivoted arm similar to that shown in the diagram, but you may alter the shape of the arm and the position of the pivot point. The straight line distance between S the stylus tip and P the pivot point is fixed at 25 cm.

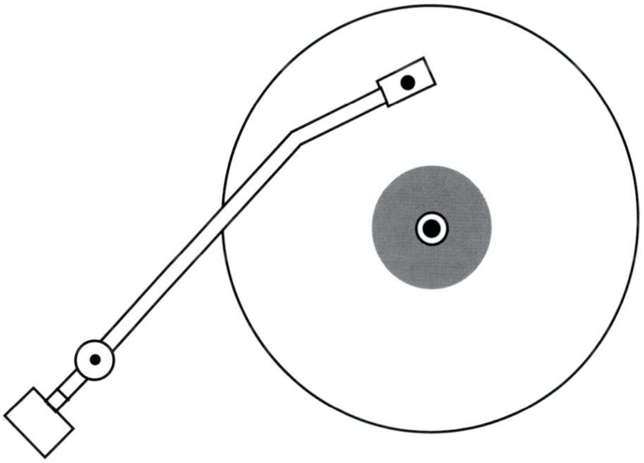

P Solving 22

PROBLEM: THE ART GALLERY

Question 1

A room in an art gallery contains a picture which you are interested in viewing. The picture is two metres high and is hanging so that the bottom of the picture is one metre above your eye level. How far from the wall on which the picture is hanging should you stand so that the angle of vision occupied by the picture is a maximum? What is this maximum angle?

Question 2

On the opposite wall there is another equally interesting picture which is only one metre high and which is also hanging with its base one metre above eye level, directly opposite the first picture. Where should you stand to maximise your angle of vision of this second picture?

Question 3

How much advantage would a person 20 centimetres taller than you have in viewing these two pictures?

Question 4

This particular art-gallery room is six metres wide. A gallery guide of the same height as you wishes to place a viewing stand one metre high in a fixed position which provides the best opportunity for viewing both pictures simply by turning around. The guide decides that this could best be done by finding the position where the sum of the two angles of vision is greatest. Show that the maximum value which can be obtained by this approach does not give a suitable position for the viewing stand.

Question 5

The gallery guide then decides to adopt an alternative approach which makes the difference between the angles of vision of the two pictures, when viewed from the viewing stand, as small as possible. Where should the viewing stand be placed using this approach? Comment on your answer.

P Solving 23

PROBLEM: OIL PIPELINES

An oil platform is situated at a point 0, at sea, 10 km from the nearest point P on a stretch of straight coastline. An oil refinery is at a point R, 16 km along the coast from the point P (see Figure 1 below). It is necessary to lay a pipeline from the platform to the refinery. This is to consist of straight line sections. Laying pipeline underwater is 4/3 times as expensive as laying pipeline on land. This ratio is known as the relative cost.

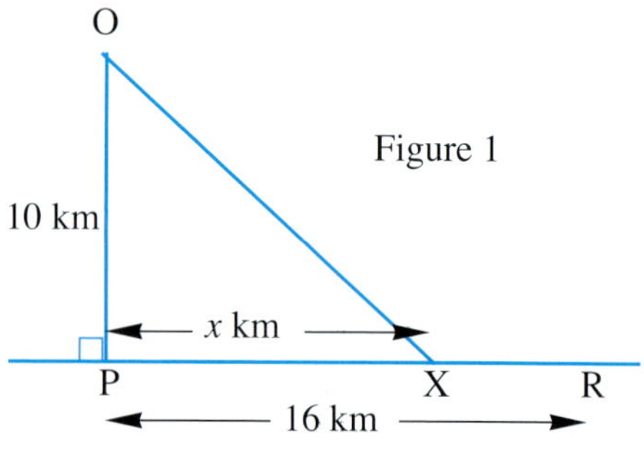

Figure 1

Question 1
If the pipeline reaches the coast at a point X, distance x km from P in the direction of R, find an expression for the cost C (in dollars) of laying the pipeline in terms of x, given that laying pipeline on land has a fixed cost of A dollars per kilometre.

Question 2
Find the least expensive route for laying the required pipeline.

Question 3
Compare the cost of the solution you have found with the cost of a pipeline directly from O to R and with a pipeline from O to P to R.

Question 4
If the cost of the underwater pipeline changes in relation to the cost of the coastal pipeline, explore how the least expensive route changes. In particular, see if there is a value of k (the relative cost) at which the direct OR route becomes the cheapest solution, or one for which the route from O to P to R is the least expensive.

Figure 2

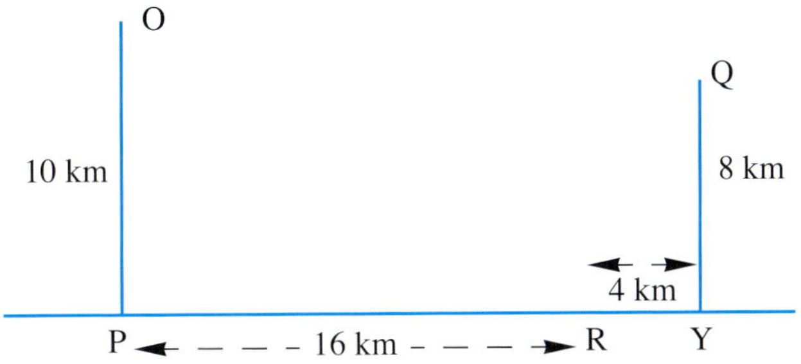

Question 5
Another platform is to be located at a point Q, 8 km from shore on the other side of R, opposite a point Y, 4 km from R (see Figure 2). With the relative cost of 4/3 as before, explore whether there are any cost advantages in combining pipelines, stating any assumptions you have made.

PSolving 24

PROBLEM: THROUGH THE FOG

Two boats approach each other in heavy fog. The fishing boat is travelling due east at 8.5 m/sec and is currently 400 metres west of the point marked X in the diagram. The ferry is going more cautiously, at 3 m/sec southwards. The distance between the ferry and point X is 120 metres. Suppose visibility in the fog is restricted to 50 metres, that is, only objects within 50 metres of the observer can be seen. For this problem–solving task, you may regard each boat as a moving point. For questions 1–5, assume that both boats maintain their present speed and direction.

Figure 1

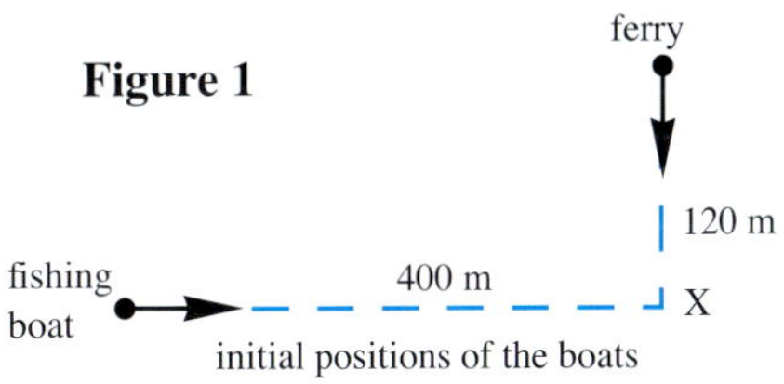

Question 1
Find when each boat reaches point X. Give the position of the other boat at each of these times. Show that they will be visible to each other at some point.

Question 2

Find the direct distance in metres between the boats as they are shown in the diagram.

Question 3

$d : [0, 60] \mapsto \mathbb{R}$ is a function which relates the time, t (in seconds), to the direct distance (in metres) between the two boats in the minute immediately following the situation shown in the diagram. Find the function $d(t)$.

Question 4

Find when the boats are closest to each other and describe their positions at this time with respect to the point X. How far apart are they at this time?

Question 5

For how many seconds are the boats visible to each other?

Question 6

Suppose the fog thickens and visibility is reduced to zero. The captain of the fishing boat hears the ferry through the fog and decides immediately to decrease his boat's speed uniformly at the rate of z m/s^2. Assume that this decision is made when the boats are at initial positions described in Figure 1. Show that, if the ferry's captain continues as before, a collision cannot occur. State any simplifying assumptions you have made.

Question 7

Show that if, instead, the captain of the ferry immediately decreases its speed uniformly at the rate of y m/s^2 while the fishing boat continues at constant speed, a collision can occur, and find the value of y for which this will happen.

Question 8

Assume now that, at their initial positions, both captains can choose either to increase or decrease the speeds of their boats. Find a combination of uniform speed changes which allows a collision to occur where neither y nor z is zero. Is the pair of values obtained for y and z unique for the circumstance you are exploring? If not, define and explain a function which relates the values of z and y which cause a collision to occur, including consideration of its domain and range.

PSolving 25

PROBLEM: RATIONAL POINTS ON CURVES

The aim of this task is to study the relationship between arithmetic, geometry and trigonometry. A well–known example of this relationship is the theorem of Pythagoras. This theorem states a property of right–angled triangles, hence it immediately involves geometry and trigonometry. But it also involves arithmetic, since right–angled triangles can be made from certain triples of integers, such as 3, 4, 5. This raises the question: how can all such triples be found?

Definitions

1. A Pythagorean triple is a triple of positive integers (a, b, c) such that $a^2 + b^2 = c^2$.

2. A rational number is a quotient $\frac{m}{n}$ of integers m and n with $n \neq 0$. Thus $\frac{1}{2}$ is rational, for example, but $\sqrt{2}$ is not rational.

3. A rational point in the plane is a point whose cartesian coordinates x, y are both rational numbers. For example, $\left(1, \frac{1}{2}\right)$ is a rational point, but $(1, \sqrt{2})$ is not a rational point.

4. An integer point in the plane is defined similarly.

Rational points on the unit circle

Since the members a, b, c of a Pythagorean triple satisfy $a^2 + b^2 = c^2$, the rational numbers $x = \frac{a}{c}$ and $y = \frac{b}{c}$ satisfy

$$x^2 + y^2 = \frac{a^2}{c^2} + \frac{b^2}{c^2} = \frac{a^2 + b^2}{c^2} = \frac{c^2}{c^2} = 1$$

But this says that (x, y) is a point on the unit circle (Figure 1), because (by Pythagoras' theorem) the distance of (x, y) from the origin is $\sqrt{x^2 + y^2} = 1$. Thus each Pythagorean triple (a, b, c) gives a rational point $\left(\frac{a}{c}, \frac{b}{c}\right)$ on the unit circle.

For example, the triple (3, 4, 5) gives the point $\left(\frac{3}{5}, \frac{4}{5}\right)$.

Figure 1

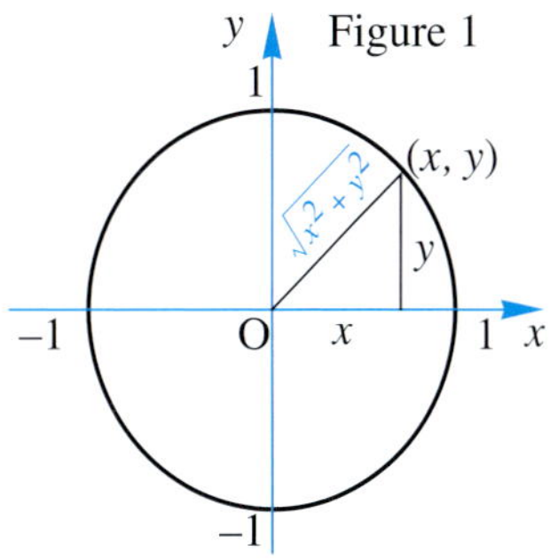

Question 1

The Pythagorean triple (a, b, c) corresponds to the rational point $\left(\frac{a}{c}, \frac{b}{c}\right)$ on the unit circle, where the fractions $x = \frac{a}{c}$ and $y = \frac{b}{c}$ are in lowest terms. What other Pythagorean triples also correspond to this point?

Question 2

By looking for rational points on the unit circle, instead of Pythagorean triples, we simplify the search for such triples a little. Now we have to solve the equation $x^2 + y^2 = 1$ which has two variables instead of three. Also geometry gives a hint on how to find solutions.

Start with an 'obvious' rational point on the unit circle, such as (–1,0). The line through this obvious rational point and any other rational point has rational slope, *t* say. Why is *t* rational?

Question 3

Conversely, the line $y = t(x+1)$ of rational slope t through the point (–1,0) meets the unit circle in a second rational point.

Why is the other intersection of $y = t(x+1)$ with $x^2 + y^2 = 1$ rational? Work out its *x* and *y* coordinates in terms of *t*.

Question 4

Now substitute $\frac{u}{v}$ for t in your formulas for x and y, where u and v are integers, and hence conclude that (a, b, c) is a Pythagorean triple where

$$a = v^2 - u^2, b = 2uv, c = u^2 + v^2 .$$

Question 5

But $a = 2uv, b = v^2 - u^2, c = u^2 + v^2$ is also a Pythagorean triple. Can your formulas for x and y also account for this?

Question 6

Do these formulas account for the Pythagorean triples $(3, 4, 5)$, $(5, 12, 13)$, $(8, 15, 17)$, $(9, 12, 15)$, $(7, 24, 25)$ and $(12, 16, 20)$?

Question 7

The same ideas apply to other curves such as $x^2 + 2y^2 = 1$ (an ellipse) and $x^2 - 2y^2 = 1$ (a hyperbola). It is possible to find formulas for x and y in terms of the slope t of a suitable line, and to use these formulas to find all triples of integers (a, b, c) such that

$$a^2 + 2b^2 = c^2 \text{ or } a^2 - 2b^2 = c^2 .$$

For each of these equations, find one integer triple which satisfies it.

Question 8

The hyperbola $x^2 - 2y^2 = 1$ is particularly interesting because it contains not only infinitely many rational points, but infinitely many integer points. A formula for the rational points is not likely to show this, but the following result gives a method for finding these integer points.

If (m_1, n_1) and (m_2, n_2) are integer points on $x^2 - 2y^2 = 1$, show that the point (m, n) is also on this curve, where m and n are the integers such that $m - n\sqrt{2} = (m_1 - n_1\sqrt{2})(m_2 - n_2\sqrt{2})$.

Question 9

Conclude from Question 8 that there are infinitely many integer points on the curve $x^2 - 2y^2 = 1$. Explain how these points give approximations to the irrational number $\sqrt{2}$.

Question 10

For the curve $x^2 - dy^2 = 1$, state and prove a generalisation of the result obtained in Question 8, including any limitations on the value of d. Also, comment on the possibility of using this generalised result to find rational approximations to $\sqrt{d}$.

PSolving 26

PROBLEM: STAYING DRY

A lawn watering sprinkler in a park revolves at a constant rate of one revolution every 32 seconds. It emits a stream of water which waters all points in a circle of radius 38.8 m as the sprinkler revolves. The sprinkler is set at a fixed point S, 26 m from a path which runs north-south through the park and so the path is watered between P and Q in the diagram. (Assume the path has negligible width.).

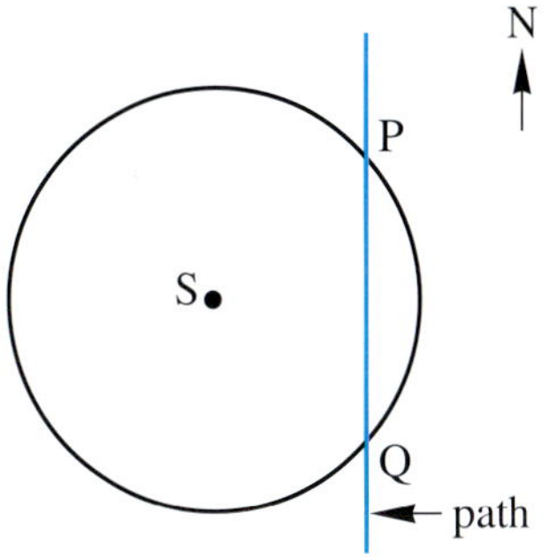

Question 1
What length of path is watered by the sprinkler?

Question 2
For how long on each revolution is the path watered?

Question 3
During one revolution of the sprinkler, the stream of water crosses the path at a distance *d* metres south of P after t seconds have elapsed. Take $t = 0$ when the water strikes the path at P.
Express *d* in terms of *t*, and state the implied domain and the range of the function with rule $d(t)$.

Question 4
An elderly gentleman is walking south on the path at the slow speed of 1.2 m/s. He arrives at P 10 seconds before the water strikes the path at P. When he hears the water strike the path at P, he immediately accelerates at the rate of 0.4 m/s^2 . But it is no good, and he is soaked as the stream of water catches him.

Suppose the distance between the elderly gentleman and the place where the stream of water is crossing the path is *w* metres, *t* seconds after the stream of water passes P. Express *w* in terms of *t*, and find when and where on the path the stream of water catches him.

Question 5
As soon as the stream of water hits him, the elderly gentleman stops. He gets out his handkerchief and dries his glasses. Fifteen seconds after stopping he starts walking again.

Find the constant speed at which he must walk in order to avoid getting wet again when the stream of water from the sprinkler next passes.

Question 6
The elderly gentleman would like to walk along the footpath at his normal walking speed of 1.2 m/s without getting wet. What is the minimum distance that the sprinkler would need to be moved back from its current position so that the elderly gentleman does not get wet. Assume that he plans his walk to give himself the maximum time.

Question 7
Later on, with the sprinkler still in its original position, a boy approaches P from the north. At the same time, a girl approaches Q from the south. Each waits for the water from the sprinkler to pass, hoping to avoid getting wet.

As soon as the water hits the path at P, the boy walks forward and reaches Q just before the water from the sprinkler does on its next pass so that he just escapes getting wet. The girl waits at Q until the water is just off the path and then jogs to P arriving there just in time to avoid the water from the sprinkler on its next revolution. At what speed was each person moving? When and where does boy meet girl? What is significant about this result? To what extent can this result be generalised mathematically? Justify your answer.

PSolving 27

PROBLEM: THE GREAT ESCAPE

Matthew is an escape artist who makes a living performing daring escapes from dangerous situations. He believes he has developed an escape trick that will prove he is the equal of the great escape artist Houdini.

In his escape his feet are chained to the top of a concrete block attached to the bottom of an enormous flask. He makes his escape from the chains as water is pumped into the flask. The flask has a circular transverse cross section. Its longitudinal cross section takes the shape of a curve with equation $y = \frac{a}{x^2} + b$, using the floor as the x-axis and the vertical centre of the flask as the y-axis (refer to Figure1).

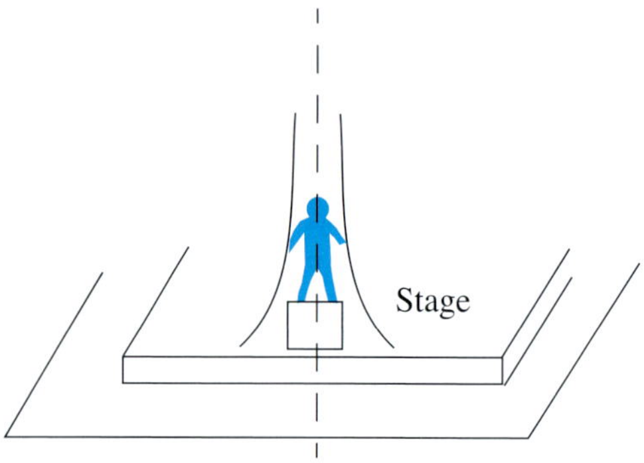

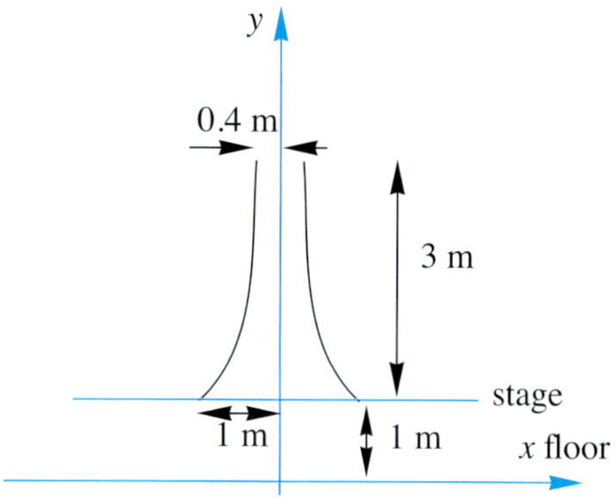

Question 1
Find the values of a and b.

Question 2
a. Find, in litres, the total volume of water that would be required to completely fill the flask, giving your answer to the nearest litre.

b. Express the volume of water (in litres) in the flask as a function of the height of the water above the floor.

When Matthew is placed in the flask, he plans to make his dramatic escape just as the level of water reaches the top of his head, when he is standing upright. Matthew is exactly 180 cm tall and it takes him 10 minutes to escape from the chains on the top of the concrete block. Assume that the volume occupied by Matthew and the concrete block is negligible for the purposes of these calculations.

Question 3
Find the height of a concrete block required which will allow Matthew to make his escape after exactly 10 minutes, when water is pumped into the flask at a constant rate of 300 litres per minute.

Matthew's manager plans to make the escape more exciting by increasing the rate at which the water enters the flask. He suggests that Matthew should time his escape from the chains so that he has been completely submerged in a standing position for a full minute before escaping from the chains. In addition, it is planned that the water should simultaneously start to overflow from the top of the flask just at the moment he makes his escape. Assume that the volume occupied by Matthew and the concrete block is negligible for the purposes of these calculations.

Question 4

a. Establish the constant rate in litres per minute at which water must be pumped into the flask so that it just begins to overflow after 10 minutes.

b. Hence, find the height of the concrete block required for this situation.

c. Express the height of the liquid in the flask, in terms of the time in minutes, after filling has commenced.

d. i. Find an expression that Matthew could use to find the height of a block for a given submersion time. [The submersion time is the time between the water reaching the top of Matthew's head (when he is in a standing position) and it reaching the top of the flask.]

ii. Sketch the graph of height versus submersion time.

Question 5

Matthew now realises that he has not considered his own volume or the volume of the block. Assume that the block is a cylinder of radius 13 cm, and that Matthew is a cylinder of radius 13 cm and height 180 cm.

a. If the block is of height 60 cm, find the number of litres of water to be pumped into the flask so that Matthew's head is just covered, when he is standing upright.

b. i. Suppose the block is of height H cm. Express, in terms of H, the number of litres of water to be pumped into the flask to just cover Matthew's head.

ii. If water is pumped in at 300 litres per minute, find the height (in cm) of the block so that Matthew's head is just covered in exactly 10 minutes.

Question 6

Matthew discovers that the pump can be adjusted to produce a constant flow rate of between 180 and 330 litres per minute and that he can change both the radius and the height of the cylindrical block.

a. Suppose that the block is of height H cm and radius r cm. Express, in terms of H and r, the number of litres of water to be pumped into the flask to just cover Matthew's head.

b. Establish 5 sets of values for the flow rate H and r which will allow Matthew's head to be just covered in exactly 10 minutes, commenting on any possible restrictions on the values of these variables.

PSolving 28

PROBLEM: SENSATIONAL SPINNING SANCHEZ SISTERS

The Sanchez sisters, Penny and Queenie, are circus performers. During their performance, Penny and Queenie run on the outside of separate cylinders, each of radius 1.5 metres. The cylinders are fixed to either end of a 10 metre long beam. The beam with the cylinders attached rotates in a vertical plane, around a horizontal axis at the centre of the beam.

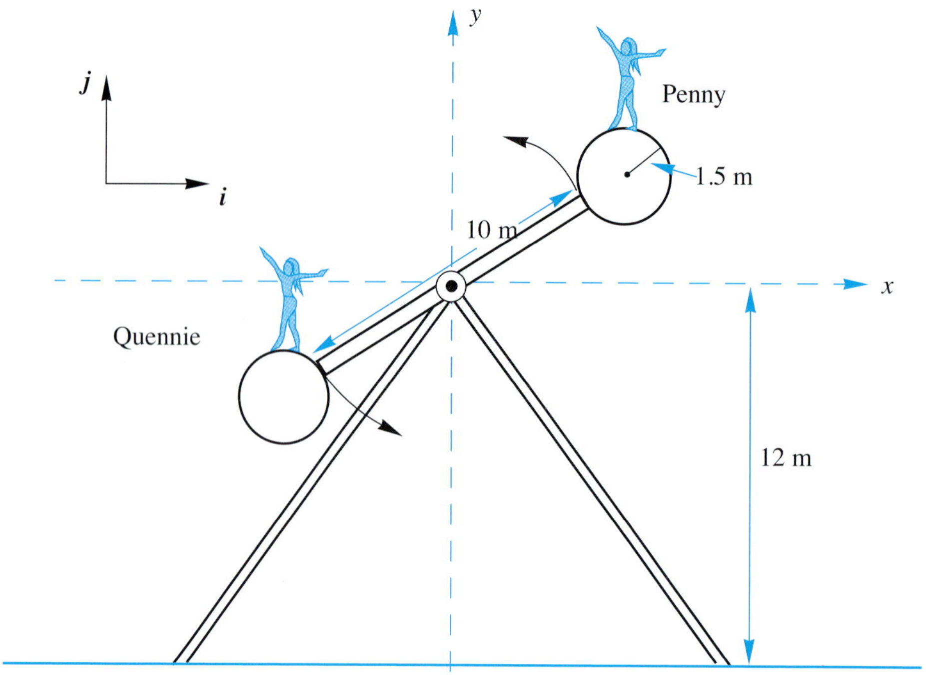

The axis is 12 metres above the ground. Each sister runs on the top of her cylinder as the beam rotates anticlockwise at a constant rate. We will assume that each sister is represented by a point located at her chest, which is 1.5 metres above her feet in each case. Penny's position can be

described by the equation $P(t) = \left(6.5\cos\left(\frac{\pi}{10}t\right) + 13\right)i + \left(6.5\sin\left(\frac{\pi}{10}t\right) + 15\right)j$.

where Penny's initial position is 6.5 metres to the right of the horizontal axis and 15 metres above the ground, t is the time in seconds after Penny has left her initial position, and n, in s^{-1}, is a constant related to the rate of rotation of the beam and attached cylinders.

Throughout this problem, assume air resistance is negligible, and take the acceleration due to gravity to have magnitude g, in ms^{-2}, where $g = 9.8$.

Question 1

a. How long does it take Penny to complete one revolution when

i. $n = 0.3$? ii. $n = \frac{\pi}{5}$?

b. What exact value of n allows Penny to take exactly 20 seconds to return to her starting point?

For all the remaining questions the beam rotates at a constant rate such that $n = \frac{\pi}{10}$.

Question 2

a. What is Penny's exact position at i. t = 5 ii. t = 7.5 iii. t = 10?

b. Derive the cartesian equation of Penny's path.

c. i. Write down Queenie's position as a vector function of time.
ii. Derive the cartesian equation of Queenie's path.

One of the tricks performed by Penny is juggling some clubs while she runs on her cylinder. At the end of the trick she releases the clubs, one at a time, in such a way that they fall freely to the ground. We will assume that the club is released from the point that represents Penny.

Question 3

a. Penny drops a club at $t = 7.5$.
   i. What is the horizontal velocity of the club at the moment it is dropped?
   ii. What is the vertical velocity of the club at the moment it is dropped?

Let T_c seconds be the time from the moment when Penny drops the club.

b. i. What is the horizontal velocity of the club, as it falls to the ground?
   ii. What is the vertical velocity of the club, as it falls to the ground?

c. i. Find an expression which describes the horizontal position of the club, as it falls to the ground.
   ii. Find an expression which describes the vertical position of the club, as it falls to the ground.

d. Write down the position of the club as a position vector as it falls to the ground.

e. Derive the cartesian equation of the path of the club as it falls to the ground. Sketch a graph of the path of the club.

Queenie's most exciting trick is catching balls shot from a cannon as she runs on her cylinder. The cannon propels the ball with a speed of u, in ms^{-1}. The initial angle of the ball's trajectory is θ degrees, $0 < \theta < 90°$, measured in an anticlockwise direction from the positive x-direction. When the cannon is fired, the initial position of the ball, as it leaves the cannon barrel, is $(-13, -10)$.

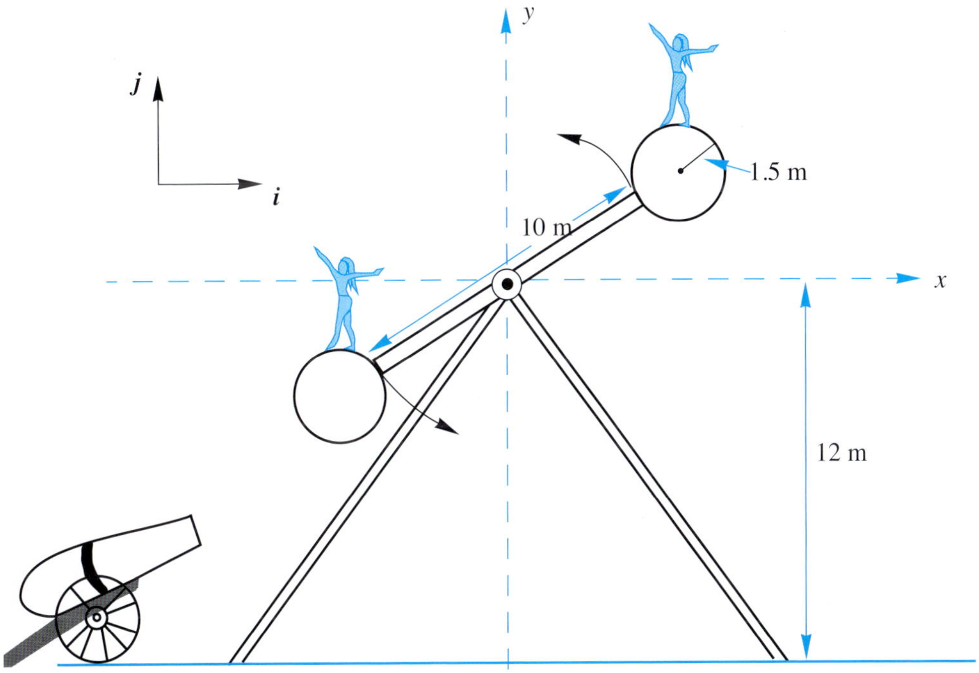

Question 4

For the first firing of the cannon, $\theta = 83°$, and $u = 20$.

a. i. What is the horizontal velocity of the ball at the moment the cannon is fired?

ii. What is the vertical velocity of the ball at the moment the cannon is fired?

b. i. Find an expression which describes the ball's horizontal position, in terms of T_b, where T_b represents the time in seconds from the moment the cannon is fired.

ii. Find an expression which describes the ball's vertical position, in terms of T_b .

c. Find the point at which the path of the ball first intersects Queenie's path and the time it takes the ball to reach this point.

d. Find where Queenie needs to be when the cannon is fired, so that she is able to catch the ball.

For subsequent firings of the cannon, θ and u can take other values.

Question 5

a. i. Write down the horizontal position of the ball in terms of T_b, u and θ.

ii. Write down the vertical position of the ball in terms of T_b, u and θ.

b. Derive the cartesian equation of the path of the ball.

To increase the thrill of Queenie's trick, the circus management suggest that Queenie should catch the ball just as she reaches the top of her path $(0, 9.5)$. Assume that she catches the ball when it reaches the highest point of its path.

Question 6

a. Find the values of u and θ.

b. Using these values for u and θ, find where Queenie needs to be, when the cannon is fired, so that she is able to catch the ball.

The circus management believe that Queenie's trick could be even more sensational if the ball takes a different path to Queenie. They still believe that she should catch the ball just as she reaches the top of her path.

Question 7

a. Using the equation found in 5b., find u as a function of θ so that the path of the ball passes through the top of Queenie's path. Sketch the graph of this function.

b. Find the minimum initial speed of the ball that allows it to pass through the top of Queenie's path.

c. Find the cartesian equations for two new paths for the ball that allow Queenie to catch it at the top of her path.

d. Sketch separate graphs for each of the new paths, which include a sketch of Queenie's path, from the instant the cannon is fired until Queenie catches the ball.

23.2 PORTFOLIO WORK TYPE I – MATHEMATICAL INVESTIGATION

We now provide a number of mathematical investigations that satisfy the requirements of the Type I – Mathematical investigation task, P work Type I . Each of the following investigations has a general theme and starting point(s). Although there are clear suggestions, these tasks will give you the opportunity to carry out your own investigation. However, as always, make sure that you discuss your intentions with your teacher. Once this has occurred and it is established that you will be satisfying the I.B.O requirements for the portfolio Type I task, then proceed. Do not proceed with any of the following tasks until your teacher has approved it.

23.2.1 All is fine and well . . . but where do I start?

The following flowchart highlights the important roles involved in carrying out your investigation (for both Type I and Type II tasks). If you are finding it difficult to start, use this as a guide.

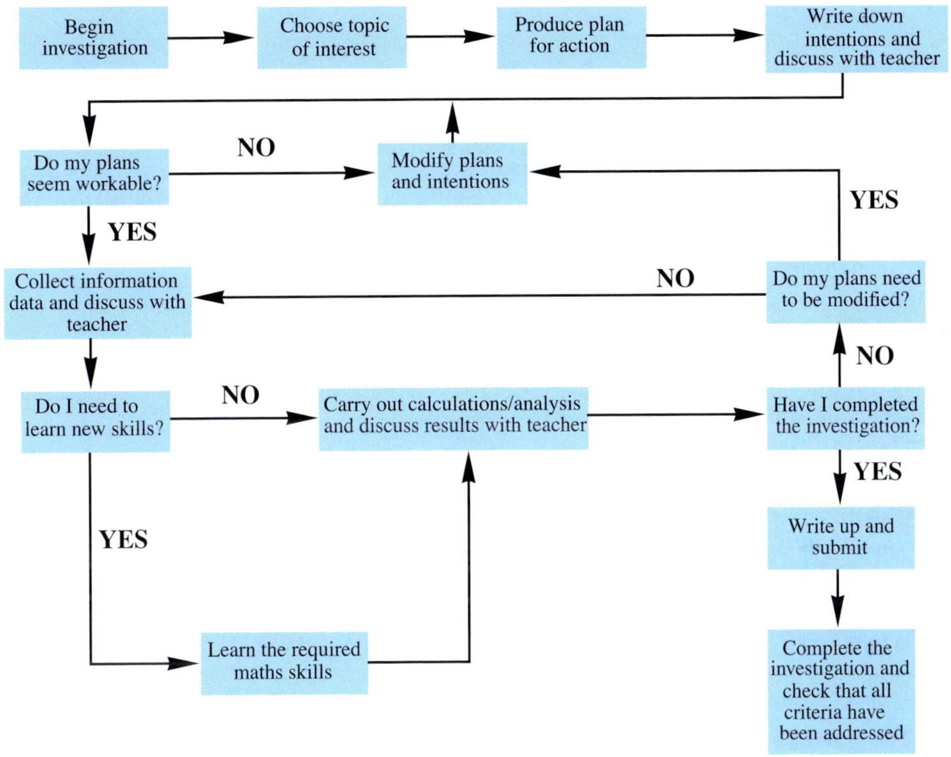

P work Type I 1

THEME: PERIODICITY

Your investigation must arise from or incorporate the study of periodicity. You are encouraged to show initiative and be independent in carrying out your project.

1. General Advice

Periodicity is a rich field for mathematical investigation. Many phenomena are periodic. Just consider, for example, the motion of planets, recurring decimals, calendars, gear mechanisms and waves. Such examples offer many opportunities to use equations (for the period), arithmetic (for finding the decimal period of a fraction), trigonometric functions (for modelling waves), geometry (for tilings and decorative patterns) and general mathematical judgement.

The general criteria for assessment should be borne in mind, but here are a few issues which arise specifically in the case of periodicity.

First, you could investigate whether periodicity is present in a specific situation and describe its properties mathematically. Alternatively, you could design and study a mechanism to produce a desired periodic phenomena, for example, a machine, a computer program or a mathematical rule. Second, what mathematics is appropriate ? Algebra, arithmetic, trigonometry, geometry? In some cases a full analysis will not be possible in terms of school mathematics. If so, you may have to approximate or quote appropriate advanced results without proof. Third (or perhaps first), does the problem have a history? Many problems have been solved correctly after a series of attempts, ranging from wildly wrong to approximately correct.

2. Starting points

You may investigate any topic related to the theme. You must discuss your choice of topic and how it relates to the theme with your teacher. The examples below show some starting points for projects. This list is by no means exhaustive.

2.1 **Calendar** - is the calendar exact ? Find the day of the week for a given date in a given year. What about alternative calendars, past and future?

2.2 **Modular arithmetic** - What is it? And how could you use it in investigating periodicity? What is the Chinese remainder Theorem? How does this relate to periodicity?

2.3 **Tides, length of the day, biorhythms.**

2.4 **Economics** - what, if any, periodicity can you find in, for example, the value of the Australian dollar, the stockmarket, the CPI?.

2.5 **Astronomy** - planets, moons, comets, distance of stars, epicycle theories.

2.6 **Waves** - sound, water, light.

2.7 **"Random" sequences** - investigate ways to create such sequences which are never periodic; create rules for repeating number sequences.

2.8 **Periodic patterns in wallpaper, tiling and crystals.**

2.9 **Mechanisms** - investigate mechanisms that produce periodic patterns in materials, carpets, zig-zag and other fancy stitches; in drawings (spirographs, tissajas figures); hammer drills, cogs in a gearing system, music boxes, pianolas.

2.10 **Electric currents and their trigonometric properties.**

2.11 **Sundials** - design a sundial; explain why it works. Where does the sun set?

P work Type I **2**

THEME: FRACTALS

Your investigation must be based on or incorporate the study of fractals. Some introductory reading is recommended in the general advice below, however you are encouraged to show initiative and be independent in carrying out your project.

1. General advice

The basic concepts of fractals have been summarised in this booklet. The following books are recommended for their general interest: Keith Devlin's Mathematics: The New Golden Age (Pelican 1988), Ian Stewart's The Problems of Mathematics (Oxford 1987) and James Gleick's Chaos (Viking 1987).

The important thing is to have some familiarity with the 'snowflake' curve and to understand the concepts of self-similarity and fractal dimension. Your report should include explanations of these concepts in your own words, and examples of your own which illustrate your explanations. The mathematical investigation of fractals is likely to involve geometry (length, area, volume), number and algebra (summing infinite series, exponents and logarithms). There is also the opportunity to use computers, particularly if you invent your own fractals.

A likely direction of investigation is in the modelling of nature (which is the reason for most of the current interest in fractals). Here it may also be convenient to use a computer to experiment with different models (though computer print-outs are generally less attractive than pictures carefully drawn by hand).

1. Families of fractals

How does a fractal vary when you vary the rule for making it? Invent examples. Investigate changes in appearance, dimension and so on when the rule is varied. Try including a random element.

2. Geometric properties

Find some properties of the snowflake curve; for example, its length and the area it encloses. Why does it have infinite length? Find out something about the Cantor set, the Sierpinski carpet and/or the Sierpinski sponge.

3. Models of nature

Is the snowflake curve a good model of a coast line? If not, how could it be unproved? Do all coast lines have the same fractal dimension? Investigate or create fractal models of other natural objects; for example, lightning flashes, ferns, the lung and circulatory systems. Can you generate them recursively by simple rules? (Being able to vary rules should be useful here.)

4. Rough and smooth objects

Find examples of smooth, self-similar objects. Do they occur in nature? How rough is the snowflake curve?
(What do tangents have to do with roughness?) How rough should an object be, to be regarded as fractal?

Some background

Fractals: the basic ideas

Fractals are geometric objects having fractional dimension, that is, a dimension between the usual dimensions 1, 2, 3 (for lines, surfaces and solids respectively). This seemingly strange property arises in quite a simple way from basic geometric ideas, which we now describe.

Similarity

Two objects are called similar if they are the same shape, in other words, if one becomes the same as the other under magnification. Figure 1 shows an example

Figure 1

Other examples: any two line segments are similar, any two squares are similar, but a square is not similar to a triangle. Since any segment of a line segment is also a line segment, a line segment is similar to parts of itself. Such 'self-similarity' also occurs in more interesting objects. Here is one.

The snowflake curve

The Koch snowflake (sometimes called the Koch island) is the union of infinitely many triangles, added in stages as indicated in Figure 2.

Figure 2

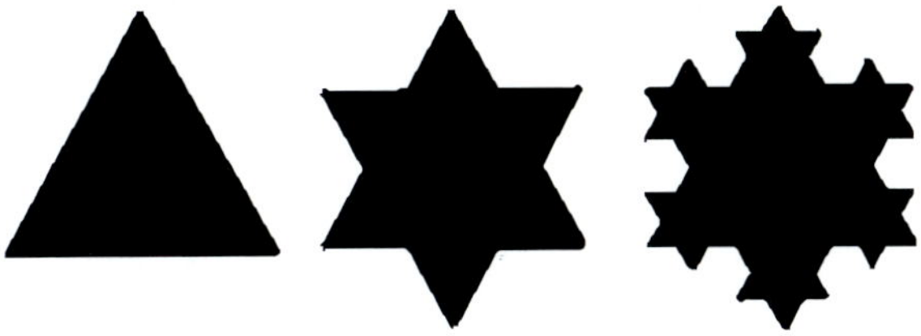

The rule for going from one stage to the next is: add an equilateral triangle to the middle third of each segment of the boundary. The Koch snowflake is interesting in itself (for example, can you find its area?) but we shall look only at its boundary, called the snowflake curve.

The snowflake curve is the limit of polygons obtained from an equilateral triangle by repeatedly replacing each line segment by a 'spiked line segment' as shown in Figure 3.

Figure 3

At the sixth stage a line segment becomes the polygon shown in Figure 4, which already gives a good impression of the snowflake curve.

Figure 4

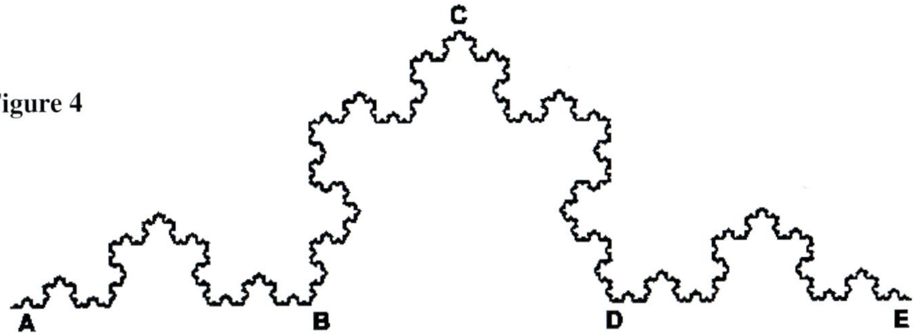

Self-similarity in the snowflake curve

Hold Figure 4 at arm's length so that it begins to blur. Now you can imagine that what you see is not just the sixth stage polygon, but a segment AE of the genuine snowflake curve. If you were to magnify part of it to get a better view - say by magnifying AB by 3, to the size of the whole segment AE - then you would see exactly the same thing! Indeed, it is clear, from the way the snowflake curve is constructed, that any part which begins as a line segment is similar to any other part which begins as a line segment.

Notice that assembling the four equal parts AB, BC, CD, DE, of the snowflake curve gives AE, which is three times the size of AB. In other words, assembling four copies of a segment of the snowflake curve magnifies the segment by 3. This is very strange: just how strange can be seen by looking again at ordinary geometric objects.

Dimension

The dimension of a geometric object is usually defined to be the number of coordinates needed to specify the position of a point in it. Thus the dimensions of a line segment, square and cube are 1,2,3, respectively (Figure 5).

Figure 5

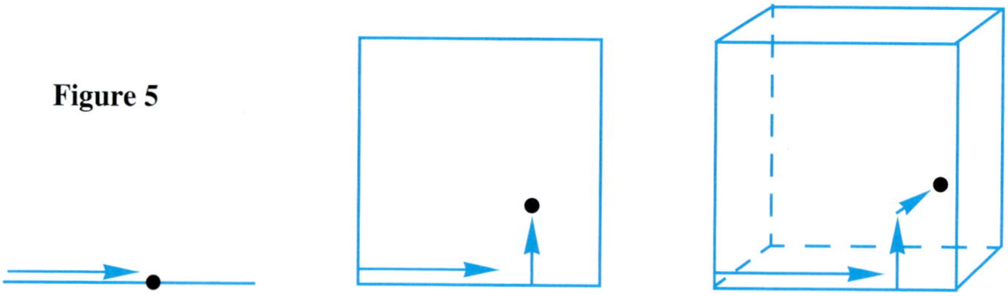

However, the numbers 1,2,3, also turn up as exponents when we calculate how many copies of an object have to be assembled to magnify the object by a given factor. This hints at a different definition of dimension, which we shall arrive at below. Consider, for example, how many copies of a line segment, square and cube have to be assembled to magnify by 3 (that is, to triple all lengths).

$3 = 3^1$ line segments

magnified by 3

$9 = 3^2$ squares

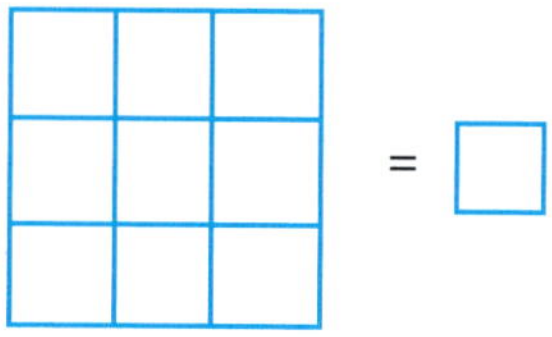

magnified by 3

$27 = 3^3$ cubes

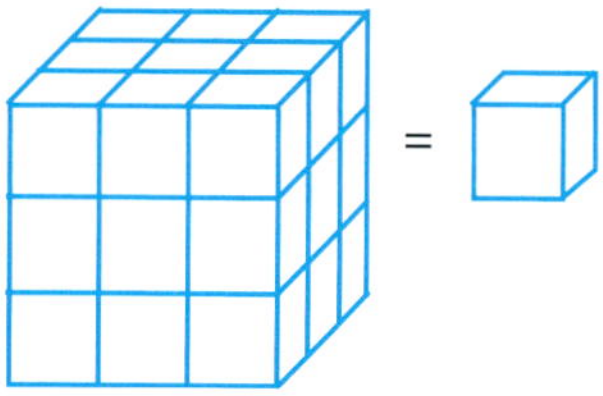

magnified by 3

In other words, another way to define dimension is to call an object d-dimensional if $3d$ copies have to be assembled to magnify the object by 3 (or, more generally, if md copies have to be assembled to magnify the object by m).

Fractal dimension

Now you can see what is strange about the snowflake curve: to magnify a segment of it by 3 one needs 4 copies, which is between the 3 copies needed to magnify a one dimensional object by 3 and the 9 copies needed to magnify a two-dimensional object by 3. Consequently, the snowflake curve has dimension between 1 and 2. In fact, according to the definition, its dimension d satisfies

$$3d = 4.$$

Taking logs gives $\quad d \log 3 = \log 4,$

and hence $\quad d = \log 4/\log 3$

$$= 1.2618\ldots$$

This number is called the fractal dimension of the snowflake curve. Similarly we can find the fractal dimension of other self-similar objects. If n copies of the object form the original, magnified by m, then fractal dimension = $\log n/\log m$

Example 1

Figure 6

Suppose we vary the snowflake rule as follows. Replace each line segment by a 'two-spiked segment' as shown in Figure 6. (Try drawing a few more stages.) Again, any part of the resulting limit curve which begins as a line segment is similar to any other part which begins as a line segment. But now we see that 7 copies of such a curve segment give the original, magnified by 5. (Because the double-spiked segment consists of 7 segments, and it is 5 times the length of each of them.) Consequently, fractal dimension = log 7/log 5.

Example 2 (The 'Sierpinski gasket')

Starting with an equilateral triangle, divide it into four equal equilateral triangles and remove the middle one (Figure 7.) Now repeat the process in each of the three triangles which remain, and continue indefinitely.

Figure 7

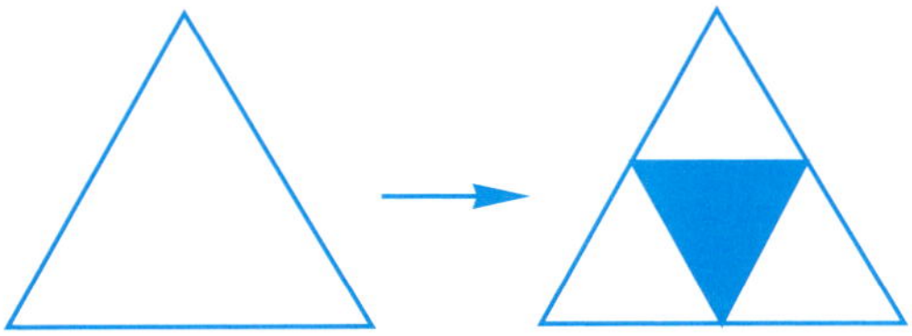

It follows that each small triangle becomes a set similar to the whole. We see from Figure 7 that 3 copies of the set yield the original, magnified by 2. Hence, fractal dimension = log3/log 2.

Objects such as these, with non-integer dimensions, are called fractals. Some books broaden the definition of fractal dimension (which is also called similarity dimensions) by allowing 'approximate or 'statistical' similarity. However, there is ample scope to invent fractal objects using strict similarity, along the lines of the examples above.

An example of a non-fractal, which nevertheless has some self-similarity, is the spiral shown in Figure 8.

Figure 8

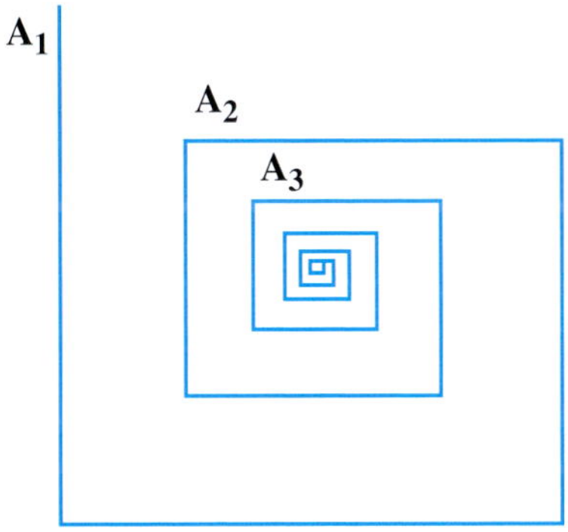

The whole spiral starting at A_1, is similar to the part starting at A_2, but twice as big. It follows that the whole is also similar to the part starting at A_3, but four times as big, and so on. However, one cannot assemble any number of copies of the spiral to make a similar spiral, so there is no way to obtain a fractal dimension. The spiral is in fact one-dimensional, like the line segments from which it is made.

P work Type I **3**

THEME: CIRCLES IN DESIGN

Your project must be based on, or incorporate the study of circles when used in design. You are encouraged to show initiative and be independent in carrying out your project.

1. General advice

Circles are involved in design in various ways ranging from the use of tools, such as a potter's wheel, to the visual effect shown in the design below.

The various characteristics of a circle, including radius and circumference, sector and tangent are well known and, where appropriate, you should refer to these in your project. Reference may be made to geometric properties of circles, their equations and the use of these properties in mensuration and trigonometry.

Acknowledge which area of mathematics you use: arithmetic, geometry, trigonometry, and/or algebra. You may choose to develop a computer program to assist you, or you may use a recognised computer package, but remember to include your own analysis of the problem.

2. Starting points

You may investigate any topic related to the theme: Circles in design. You must discuss your choice of topic and how it relates to the theme with your teacher. The examples listed below are possible starting points although it is not compulsory to use these ones. It is possible to combine related starting points but keep in mind that an investigative project is intended to include depth as well as breadth of research.

2.1 Investigate tools that use a circular motion; for example, lathes producing solids of revolution.

2.2 How does a tool convert circular motion to linear motion? What are the geometric properties involved?

2.3 How do gears and other circle-based mechanisms operate? What is their purpose?

2.4 Investigate patterns drawn by a spirograph. What are the equations of the curves produced?

2.5 How are circles used in sewing patterns? What properties of circles are used in pattern making, for example, altering sixes?

2.6 How are circles used in typefaces for printing or signwriting?

2.7 How are circles used in art? For example, how did Leonardo da Vinci and Albrecht Durer use circles in their art?

2.8 How are circles used in architecture? For example, the use of circles in the design of pneumatic structures, the shells of the Sydney Opera House (why was the original design of the shells changed to one based on circles?).

2.9 How were circles involved in the design of sundials and instruments used in navigation?

2.10 Investigate the use of geometric properties of circles in artistic designs. There are two examples below.

P work Type I **4**

THEME: INACCESSIBLE POINTS, MEASUREMENT AND ERRORS

Your investigation must be based on, or incorporate, the indirect measurement made necessary because of the inaccessibility of points needed for direct measurement. It must also incorporate an examination of the errors that result from such indirect measurement. You are encouraged to show initiative and be independent in carrying out your project.

1. General Advice

It is not always possible to measure distances or sizes of things outdoors directly For example, the height of Mt Everest, or of a flagpole on the top of a building, or the distance across the entrance to Port Phillip Bay or the area of Port Phillip Bay. These quantities cannot be directly measured with traditional measuring instruments.

Measuring these quantities has to be done indirectly, by combining measurements of directly accessible distances or sizes with measurements of angles. Problems like these led to the development of many parts of geometry (geo = earth, metry = measurement) and trigonometry (tri = three, gon = angle, metry = measurement).

Your task is to use trigonometry or geometry to find lengths or volumes in local outdoor situations where direct measurement is not practical. If necessary you can pretend that there are some obstacles in your local environment that make parts of the area inaccessible. These must be carefully specified when you describe your problem.

Nowadays, some modem instruments can measure the distance to an inaccessible point by using laser technology, but these are not to be used for this project. Methods such as using a barometer, or timing a stone as it falls to the bottom of a deep hole, are also not to be used in this project because they do not use trigonometry or geometry. Whilst the measurements of length and angles that you make should be done carefully, this project does not require you to make highly accurate measurements. NO extra credit will be given for very accurate direct measurement(s). One purpose of the project is for you to discuss the effect on the accuracy of your result that may arise from:

- assumptions you made in setting up your problem and its solutions.
- inaccuracies in each of the measurements you made.

Instructions for Starting Points 1,2,3 and 4

1. Carefully describe the length or volume you are going to measure indirectly and what factor(s) makes the target quantity inaccessible. (A diagram will be useful.)

2. Indirectly measure that length or volume which is not directly accessible for measurement using only one method. You may also wish to try out your indirect method on a length or volume where you already know the answer.
3. You must also make several indirect measurements using the same method. For example, if you were measuring the height of a tree using a directly measured distance from the tree and a directly measured angle of elevation, it would be a good idea to take measurements from several different points around the tree. This will enable you to practise making reasonably accurate measurements with your instruments and it will also help you judge how errors in the measurement of each quantity for a particular situation will effect the result.

4. Discuss the effect on your final answer of small errors in each of the direct measurements you made. For example, consider which measurements of length, angle etc need to be the most accurate for a satisfactory answer. This may vary for different parts of the domain.

5. Use one other method to measure indirectly the same length or volume.

6. Repeat Step 4 for your other method.

7. Compare the answers found using the methods and discuss which method is likely to give the best result.

8. Discuss the mathematical reasons why one method might be preferable and the conditions under which your recommendation is valid.

9 Extending your investigation.
Here are some suggestions for breadth or depth in your investigation. Because of time constraints, you should pursue only one of these suggestions.

- consider an analysis of errors based on data from step (4) or step (6) above.
- consider a mathematical justification of the valid conditions recommended in step (8).
- consider a third method to measure indirectly the same length or volume by repeating steps (5) to (8).

Measuring Instruments you may use

Only simple tools are allowed. These are

i. for distances: measuring tape, ruler, trundle wheel, bicycle, car odometer. Some measurements can be taken from a map (only for very large distances) to assist in calculating inaccessible distances for objects not shown on the map, for example, estimating the distance of a fire, or flare, from known reference points.

ii. for angles: an inclinometer for angles of elevation and depression constructed by yourself, a hiking compass or a protractor, a blackboard protractor with elongated arms.

2. Starting Points

You must investigate ONE of the following Starting Points. You must confirm with your teacher your choice of Starting Point (preferably by the end of the first week of the designated period).

2.1 Height

In the natural and/or the constructed environment, obstructions and conditions sometimes cause difficulty in the indirect measurement of the heights of objects such as trees and towers. Such obstructions include:

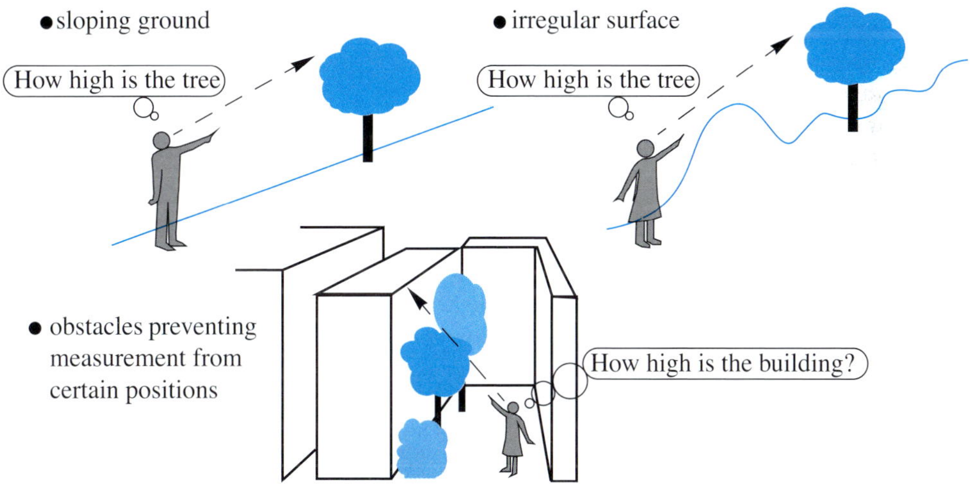

Follow the steps (1) to (9) in the General Advice to find the height of an object preferably in your local environment. where direct measurement is not possible. Provide a scale drawing of the situation, containing all relevant measurements.

2.2 Depth

In the natural and/or the constructed environment, obstructions and conditions sometimes cause difficulty in the indirect measurement of the depths of depressions such as quarries and valleys. Such obstructions include:

- sloping ground
- irregular surface
- obscured vision of the bottom of an object.
- obstacles preventing measurement from optimum positions.

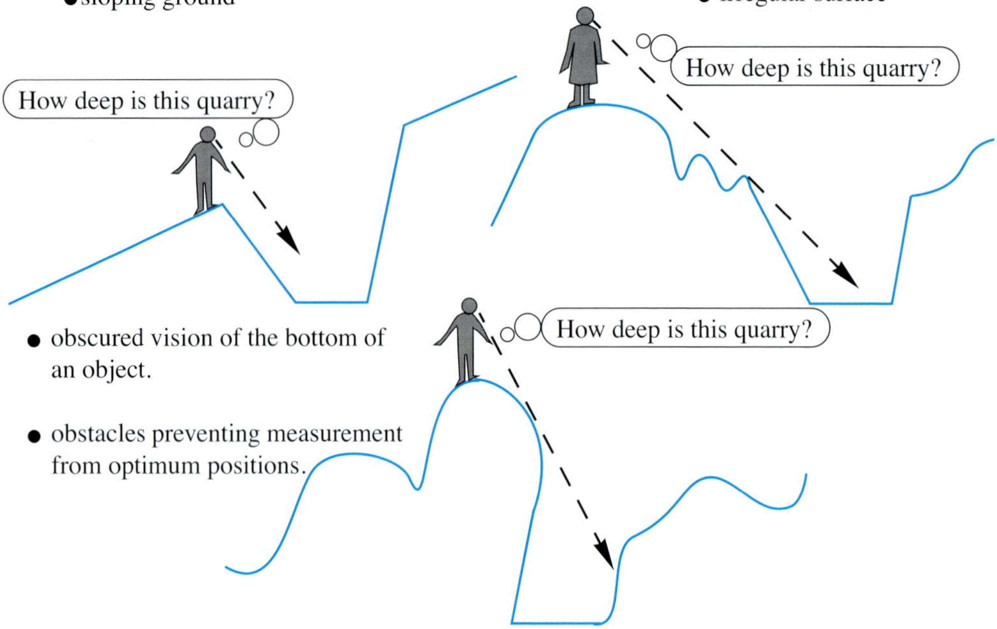

Follow the steps (1) to (9) in the General Advice to find the depth of a depression, preferably in your local environment, where direct measurement is not possible. Provide a scale drawing of the situation, containing all relevant measurements.

2.3 Distance between two points

In the natural and/or the constructed environment obstructions sometimes cause difficulty in measurement of the shortest distance between two points in situations such as those shown below.

- The shortest distance to an inaccessible point.
- The shortest distance between two inaccessible points as measured from a long way off.

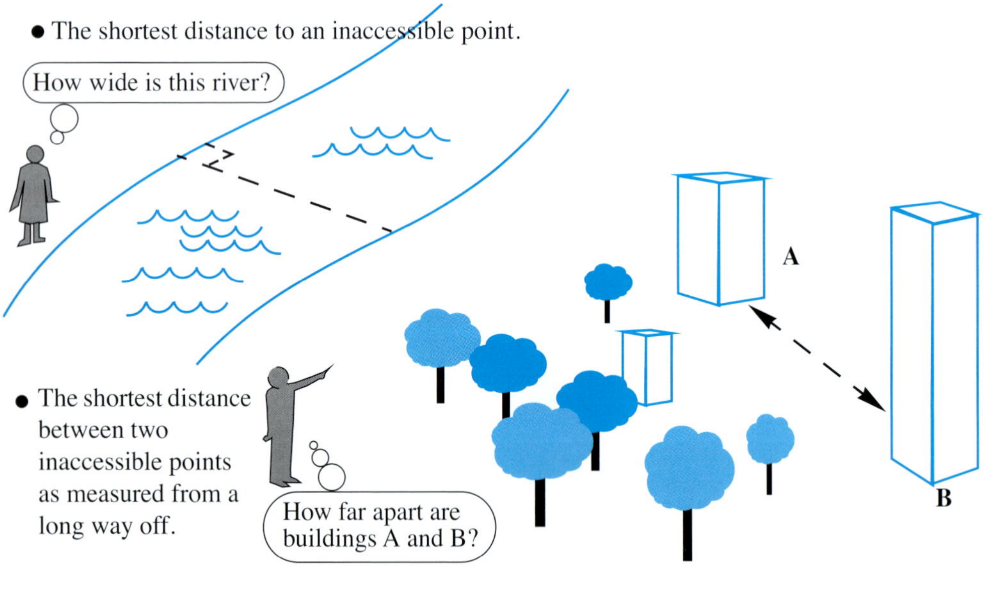

Investigate the indirect measurement of horizontal distances, preferably in your local environment, like those described above. In each case follow the steps (1) to (9) in the General Advice. Provide a scale drawing of the situation, containing all relevant measurements.

2.4 Volume

In the natural and/or the constructed environment, the volume of quantities needs to be found. Consider the situations of

– The need to estimate the volume of material removed from a quarry so that the number of earth carrying trucks required can be calculated.
– The need to estimate the volume of water to be added to a quarry when converting it into a lake.

Indirectly measure the volume of a depression such as a quarry or a valley. As you will need to make a number of indirect measurements, time may not allow a second method of calculation. Accordingly you will need to follow the steps (1) to (4) in the General Advice a number of times to enable indirect measurements of length to be used in the volume calculations. Note: steps (5) to (8) do not need to be followed at this stage. Drawings of the situation are required but need not necessarily be to scale.

Extending your investigation.

Here are some suggestions for breadth or depth in your investigation. You must pursue only one of these suggestions.

– For breadth you may wish to consider one other method to measure indirectly the volume by completing steps (5) to (8).
– for depth you may wish to consider an analysis of errors based on data from step (4).

P work Type I 5

THEME: IS NEAR ENOUGH GOOD ENOUGH? ERRORS AND APPROXIMATIONS

Your investigation must be based on or incorporate the study of errors and approximations. You are encouraged to show initiative and be independent in carrying out your project.

1. General advice

Errors arise in different ways and some are more significant than others. You should therefore distinguish between errors in data (due to human error, and limitations of measuring instruments), errors created by calculations with incorrect data, and methods which are inherently approximate; for example, truncation of series, replacement of curves by polygons.

The general criteria for assessment should be borne in mind, but here are some of the issues, which occur specifically when considering errors and approximations.

What is the effect of the error or approximation on the data and/or results?
Are the errors acceptable?

Acknowledge which areas of mathematics you use: coordinate geometry, calculus, and algebra. You may choose to develop a computer program to assist you or you may use a recognised computer package but remember to include your own analysis of the problem.

2. Starting points

You may investigate any topic related to the theme. You must discuss your choice of topic and how it relates to the theme with your teacher. The examples below show some starting points for projects. It is not compulsory to use the starting points.

2.1 Errors arising from inaccuracies in data

Some inaccuracies cause larger errors than others. How, for example, does the effect of a one percent error in setting a course for a ship compare with a one percent error in estimating the speed of the ocean?

- Compare the error of angle with error in speed for a tennis serve.
- How accurately does a baseball bat have to be swung compared to a cricket bat?
- Scientific experiments: how do errors in measurement affect results? Which data need to be more accurately measured?
- Compare the effects on predicted population due to errors in estimating birth rates, death rates, and so on.
- Make a mathematical model of a situation that interests you and investigate the effects of inaccuracies in the data.

2.2 Errors due to rounding off and calculations

- Calculators create an error due to rounding off. How does this error affect results when added, multiply, divide, find reciprocals and/or find square roots and so on?
- Devise calculations where the error is catastrophic and/or devise safeguards to avoid such catastrophes
- Find sets of linear equations for which the usual methods of solution by calculator give highly inaccurate results. Explain the inaccuracies.
- Find sets of equations that allow spreadsheets to give inaccurate answers. Explain the inaccuracies.
- How much money can a financial institution make by rounding off interest calculations?

2.3 Errors arising from approximate methods

- How are mathematical tables calculated for sine, cosine and tangent, exponential and square root tables?
- Finding π or e.

- Devise calculations for conversions in your head, for example, Fahrenheit to centigrade, kilograms to pounds, yards to metres.
- Accuracy of calculus methods; for example, approximating a curve by a tangent.
- Series truncation; for example, $\sin x = x - \frac{x^3}{6}$ for two terms. How does an increase in the number of terms increase accuracy?
- Approximating areas using geometry and/or calculus.
- Approximation of curves by polygons.
- Stopping after n iterations; for example, Newtons' method.
- Methods for finding square roots.

P work Type I 6

THEME: EXPONENTIAL AND LOGARITHMIC SCALES

Your project must be based on, or incorporate the study of exponential or logarithmic scales. You are encouraged to show initiative and be independent in carrying out your project.

1. General advice

In a relationship such as $y = e^x$, where y depends exponentially on x, the exponent x is called the logarithm of y. You should already be familiar with relationships involving exponents, such as the formula for compound interest. The goal of this project is to investigate relationships in which the exponent is used to measure some quantity. Many quantities in science and daily life are actually measured this way. For example, exponents (of energy) make up the scale used to measure earthquake strength. Your task is to explain the exponential relationship in one or more instances and investigate it in depth.

Exponential relationships offer the opportunity to use several parts of mathematics. Explaining the interplay between exponent and logarithm involves the concepts of function and inverse function. Exponential relationships can be illustrated by specific numerical calculations, and by graphing on log-log or semi-log paper. The precise meaning of logarithms can be given in terms of algebra and calculus. You are also invited to make and use actual logarithmic scales in connection with some of the topics suggested for investigation below.

Acknowledge which area of mathematics you use: coordinate geometry, calculus and/or algebra. You may choose to develop a computer program to assist you, or you may use a recognised computer package but remember to include your own analysis of the problem.

2. Starting points

You may investigate any topic related to the theme: Exponential and logarithmic scales. You must discuss your choice of topic and how it relates to the theme with your teacher. The examples listed below are possible starting points, although it is not compulsory to use these ones. You may investigate some or all of the points raised in one of the following topics.

2.1 Music

- How are exponents involved in musical scales?
- Describe mathematically the spacing of frets on a guitar neck.
- Describe the well tempered (or equally tempered) scale and how it departs from ideal harmony.
- Calculate the difference between the ideal fifth and the fifth on the well tempered scale.

2.2 Scientific scales

- Describe some logarithmic scales in science; for example, decibels, pH, the Richter scale, magnitude of stars, colour-coding on resistors.
- Collect and discuss information on these scales; for example, the loudness of typical sounds, the strength of earthquakes, the brightness of stars.
- Why are logarithmic scales used to measure such quantities?
- Suggest other quantities which might be appropriately measured on logarithmic scales.

2.3 The slide rule

A slide rule consists of two scales on which numbers are marked in such a way that numbers in a constant ratio are a constant distance apart. In particular, the powers of 2 appear on each scale as follows.

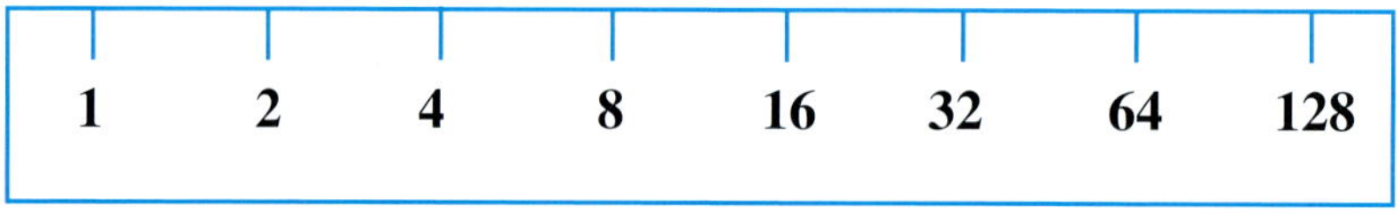

- Explain how a slide rule can be used lo do multiplication.
- Using the fact that $5^3 = 125$ is close to 128, estimate the positions of 5 and 25 on the above scale.
- Deduce approximate positions of 10, 20, 40, 50, 80 and then estimate the positions of 3
- Hence make a cardboard slide rule, accurate enough to multiply any two single digit numbers, using only a ruler and elementary arithmetic.
- Discuss different ways of estimating the positions of 3, 5, 7, 11, 13,. . . and ways of increasing accuracy.
- Is it feasible to make a slide rule by hand for multiplying numbers less than 20, or less than 100?

2.4 Logarithms and computation

- Find out about the contribution to the calculation of logarithms made by Napier, Briggs and other mathematicians.
- How were log tables originally calculated?
- Investigate the connection between integration and logarithms.
- Look up, or work out, some infinite series for exponentials, or logarithms.
- Compare the computational performance of different formulas.

2.5 Measurement of information

– The 'length' of a number can be measured by the number of its digits. For example, 27 has length 2, and 17 764 has length 5. If the number carries information (as does, say, a telephone number) then its length is a measure of the amount of information.
– Is this length scale logarithmic? If not, how does it differ from a logarithmic scale?
– How does the number of decimal digits compare with the number of binary digits (bits) in a number?
– If car licence plates were all letters, how long would they need to be? (Similarly for telephone numbers for Melbourne, for Australia and/or the world). How long do bar codes, credit card numbers, car licence plates and/or pin numbers need to be? If binary digits were used, how long would bar codes, credit card numbers, car licence plates, and/or pin numbers need to be?
– Investigate the role of redundancy in information - ways of reducing it, ways of making it useful, (for example, check digits in bar codes).

P work Type I 7

THEME: PATHS OF MOVING OBJECTS

Your investigation must arise from or incorporate the study of the paths of moving objects. You are encouraged to show initiative and be independent in carrying out your project.

1. General Advice

Moving objects are a rich field for mathematical investigation. Just consider, for example, the paths of a comet, a roller coaster, a cricket ball or a skateboard. Such examples offer many opportunities to use equations (for the description of curves), calculus (for finding direction and speed), approximation (neglect of complicating factors such as air resistance or friction) and general mathematical judgement.

The general criteria for assessment should be borne in mind, but here are a few issues, which arise specifically in the case of moving objects.

First, is it a question of "find a path" or "choose a path"? If the moving object is beyond our control, then the problem is to find its path. If the path is of our choosing, then the problem is to find how the object will behave on it. Second what mathematics is appropriate? Algebra, calculus, geometry? In many cases a full analysis will not be possible in terms of school mathematics. If so, you may have to quote approximate or quote appropriate advanced results without proof. Third (or perhaps first), does the problem have a history? Many problems have been solved correctly after a series of attempts, ranging from wildly wrong to approximately correct.

2. Starting points

You may investigate any topic related to the theme. You must discuss your choice of topic and how it relates to the theme with your teacher. The examples below show some starting points for projects. This list is by no means exhaustive.

2.1 Ballerinas, divers, gymnasts, skaters - motion of one part of the body about another and of the body as a whole. Eg. Compare paths of nose, finger, knee, toe of a skater performing a pirouette, develop a logo program to make the turtle do Olympic dives.

2.2 Balls - motion of bouncing balls, thrown balls, hit or kicked balls, motion of balls related to a specific sport. Eg. compare angle and direction to throw goals in netball and basketball.

2.3 Orbits of planets and satellites.

2.4 Design a skateboard ramp - compare the effectiveness, safety and fun of different curves.

2.5 Lifts - answer questions such as: How fast do lifts go? How does speed vary with distance between stops? How many floors should a building have before an express lift is required?

2.6 Rain drops on front or side windows of a moving car, train or aeroplane and on house windows. How does the path relate to the speed?

2.7 Side show rides - describe the path of a person in a side show ride. Eg. cup and saucer ride, rock and roll ride, octopus ride.

2.8 Constrained motion - paths of pens drawing spirographs, circles, ellipses and other curves, considering ancient and modern methods.

2.9 Path of a pendulum - some clocks use a pendulum swinging in a circular arc. How does the time of their swing relate to the amplitude? The cycloidal pendulum whose path is an inverted cycloid has swing time independent of amplitude. How did designers of clocks such as Huygens in the 17th, century make clocks using a cycloidal pendulum and why are they no longer in use? The cycloid also occurs as a curve of fastest descent called the brachistochrone.

2.10 Design a modification of a straight residential street that keeps traffic to a safe speed - such design features as speed bumps, traffic islands, roundabouts and so on are commonly used. Determine the spacing of such modifications taking into account their effect on the speed of passing cars.

P work Type I 8

THEME: DETERMINING CURVES

Your investigation must be based on, or incorporate, a study in determining curves. You are encouraged to show initiative and be independent in carrying out your project.

1. General Advice

Often, to solve a mathematical problem, it is necessary to find a particular line or curve which goes through a given set of points. For example, you have probably found the equation of a line which goes through two points. You will remember that only one straight line can be found that goes exactly through two given points. In this case we say that the two points determine the line. One given point and a given slope will also determine a particular line. However one point alone is not sufficient.

Sometimes, but not always, a line can be determined from a point on it and the area that is enclosed between the line and the positive parts of the co-ordinate axes.

For example, the line $x + y = 2$ (shown in Figure 1) is the only line which passes through (1, 1) and encloses an area of two square units with the positive parts of the co-ordinate axes. Therefore these two pieces of information determine that line. However for the point (2,2) and an enclosed area of nine square units, both $x + 2y = 6$ and $2x + y = 6$ can be drawn, so the given information does not determine a particular line (see Figure 2).

Figure 1

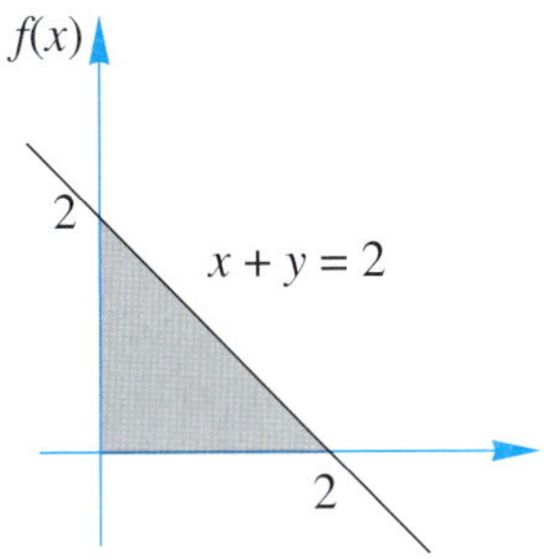

Figure 2

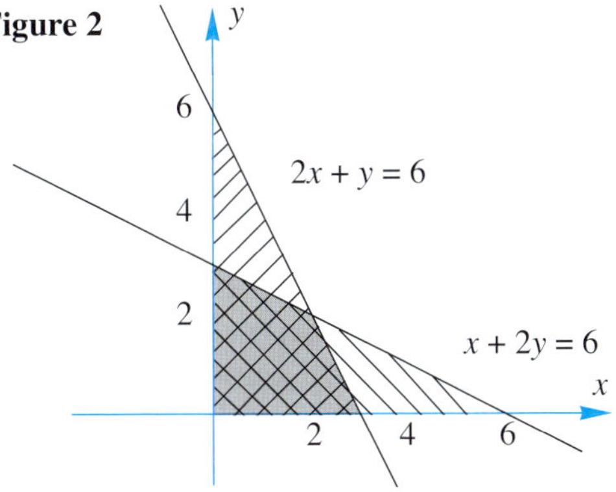

In this investigation you will explore what information can be used to determine curves and what curves can be found that fit particular sets of information. For example, you may investigate graphs like those shown in Figure 3, all of which pass through the points (–3,0), (0,0) and (5,0).

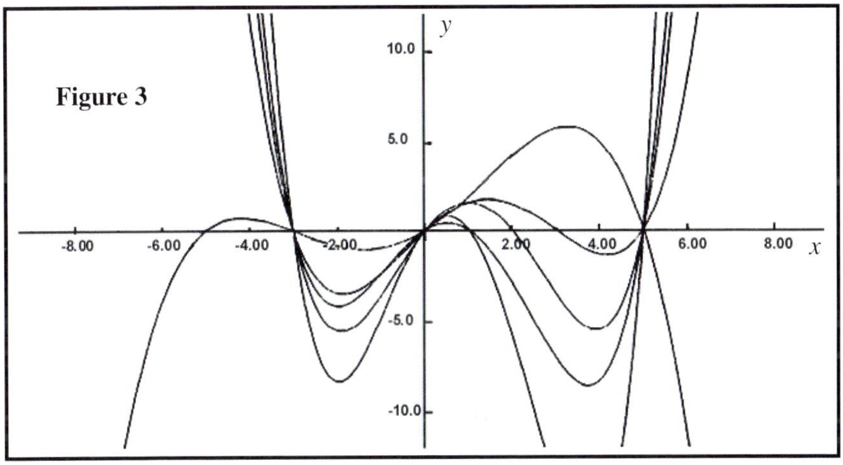

Figure 3

Types of information you may use to determine curves include: points, gradients, area beneath the curve, symmetry, asymptotes, maxima or turning points. However, you are advised to start by restricting your investigation to simple types of curves (say straight lines or parabolas) and to choose sets of points and other information that make your work easy. You can generalise your work after studying specific cases.

In this advice the word curve is used to include straight lines.
Examples of types of curve are straight lines, circles, graphs of quadratic equations of the form $y = ax^2 + bx + c$, cubic equations, sine functions, and exponential functions.

Remember, to test whether a given set of information determines a curve of a given type, you must show two things:

1. that there is a curve of that type which fits the information (this usually means finding its equation)

and

2. that no other curve of that type fits the information.

Acknowledge which area of mathematics you use: co-ordinate geometry, calculus or algebra. You may choose to develop a computer program to assist you, or you may use a recognised computer package but remember to include your own analysis of the problem.

1. Curves through sets of points.

i. Consider the three points (1,0), (3,1) and (5,6). Find some curves of different types that go through these three points. Which of these curves are determined by the three points?

ii. Identify some types of curves which cannot go through all of the above three points.

iii. Investigate the problems in i and ii above for other sets of three points and for sets with more than three points.

iv. Now select a type of curve. Identify the properties your sets of points must have so that curves of your selected type can be found to go through them. (For example, you will be able to find a circle through any three points that are not collinear.)

v. Investigate iv above for other types of curves.

2. How much information?

i. Imagine I am thinking of a curve and know its equation. You know that it is the graph of a quadratic equation of the form $y = ax^2 + bx + c$, but you do not know the values of a, b and c. What information would enable you to discover exactly which equation I am thinking of? Give a range of examples of sets of information that would identify the equation.

ii. How many pieces of information are required? You will have to make your own decision about what one piece of information is. Explain why you have chosen to count information in that way. (For example, if I told you the curve had a minimum value at (3, 3) is this one, two or more pieces of information? Why?)

iii. Extend this investigation to other types of curves.

3. Motion in a straight line.

If you are given the displacement-time function for a particle moving in a straight line you can calculate displacement, velocity and acceleration at an instant of time. Conversely, information about displacement, velocity and acceleration at a particular time may be used to determine the displacement-time curve for a particle if the type of curve is known.

i. Suppose you are given displacement, s; velocity, v; and acceleration, a; at an instant of time, t; say $s = 6$, $v = 2$, $a = 3$ when $t = 0$. Choose a type of curve for the displacement-time graph. (eg sine, parabola, exponential, part of a circle). Find all the curves of this type which satisfy the data. Does this information determine the curve? If it is not possible to find a curve which satisfies the information, explain why.

ii. Repeat the procedure in step i for other types of curves. Compare the motions predicted in each case by drawing the displacement-time graphs on the same axes. In each case describe what a person watching the particle would see, ie the way the particle moves in the line.

iii. Repeat steps i and ii above using different sets of given information. (For example acceleration at two different times and position at another time.)

iv. For each set of information, describe the common features of the graphs you have drawn and explain why they occur. (For example, all the curves that fitted the data in the first part went through a common point. Why?)

v. Possible extension. If a particle is moving in a plane and not just in a straight line, the motion can be described using a displacement-time graph in each dimension (e.g. horizontal and vertical). Choose sets of data as in part i and determine the displacement-time graph for each dimension. Describe what a person watching the particle would see, ie the way the particle moves in the plane.

4. Families of curves.

Choose a curve of a particular type (say a parabola of the form $y = ax^2 + bx + c$ or a circle). Imagine you had been given enough information to determine such a curve but then you lost one piece of information, say one of the points on the curve.

i. Draw graphs (on one set of axes) that show the curves that fit the information that is left. These curves belong to a "family of curves". Figure 4, for example, illustrates some of the members of the family of parabolas which pass through (0,6) and have gradient 2 at that point.

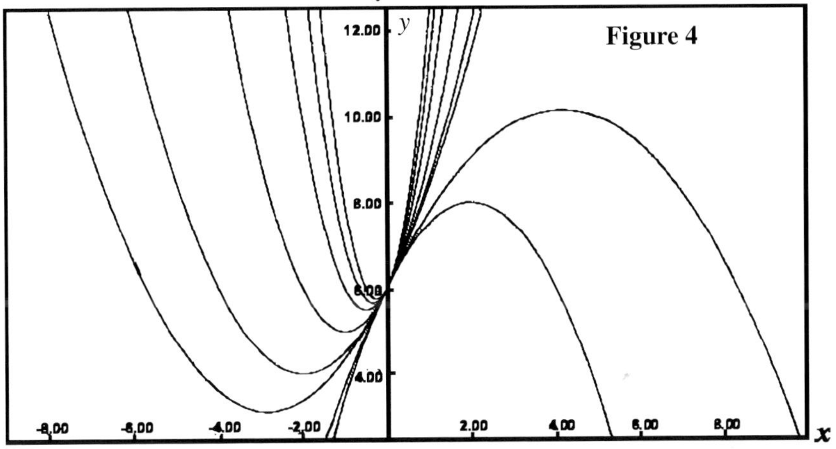

Figure 4

ii. Give some properties of the family of curves that you have sketched.

iii. What do the families look like if you lose information of different types, eg information about a point, a gradient, or an asymptote? Illustrate your answers by drawing some families of curves that would result if you lost these pieces of information.

iv. Repeat steps i, ii, and iii looking at other types of curves.

v. As an extension, you may like to draw families of curves which are solutions to differential equations and describe their properties. Alternatively you may investigate the questions in i to iv above for curves given in polar co-ordinates.

5. Approximating a complicated function with simpler functions.

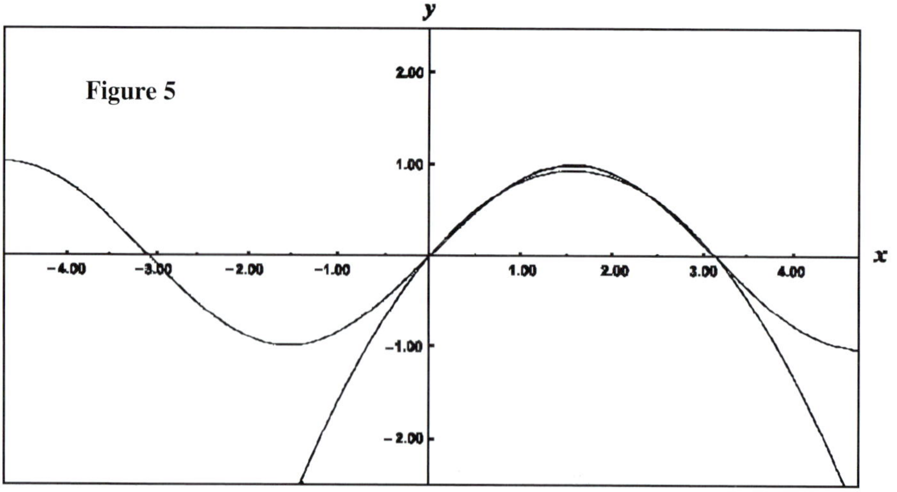

Figure 5 shows the graph of $y = \sin x$ and the graph of $y = -0.38x(x - p)$. The graph of the quadratic equation is close to the graph of $\sin x$ although only when x is between 0 and p. Calculating machines use simple functions like polynomials to calculate the values of complicated functions like $\sin x$ over a particular domain.

i. Choose a function and a domain for the function (like $\sin x$ or e^x or $\sqrt{x}$). Find a cubic curve that is as close as possible to the graph of your function, over your chosen domain.

ii. How did you measure as close as possible? You will have to invent your own method for this and describe it clearly. Is the cubic you have found the closest cubic there is?

iii. Will a different cubic curve be required to approximate your function over a different, perhaps larger, domain? Illustrate your answer by finding a cubic curve which is close to your function over a different domain. Is your curve the best approximating cubic over this domain?

iv. Can the approximation be improved by using a different type of curve (not a cubic)? For example, would it be bettter to use polynomials of degree 4 (quartics) or higher degree polynomials? Would it be better to use a combination of functions (eg a set of straight line segments, parts of two parabolas, etc)?

Pwork Type I 8

THEME: MAXIMA AND MINIMA

Your investigation must centre around the determination of maximum and minimum points for some given problems or set of functions. Whilst starting points are prescribed, there is room for student initiative. Independence and originality are encouraged.

1. General Advice

In this investigation, you are required to examine a problem or series of related problems, which involve the determination of maximum or minimum values of a function, or local maximum or minimum values with respect to one of the variables. Each starting point begins with a relatively straightforward problem, however you are advised to give careful consideration to domain limitations and other factors which might relate to the reasonableness of your solution. Following this inductory task, a more general version of the problem will be presented which requires a greater depth of analysis. Finally, there will be an extension which requires some individual initiative in placing the problem in a broader, or more realistic, context.

1. x^{α} and beyond

In this starting point you will explore the behaviour of functions of the form $y = x^{\alpha}$ defined on their maximum real domain and for different values of α where α is a rational number. In particular, you will be required to identify key points such as maxima or minima or points of inflection. You will then try to extend your investigation to the sums of such functions.

a. i. First consider positive integer values of α, $\alpha = 1, 2, 3, \ldots$ In general, for which values of α does the curve have a minimum value at $x = 0$?
Sketch the general shape of such a curve.

ii. For which values of a does the curve $y = x^{\alpha}$ have a point of inflection?
Sketch the general shape of such a curve.
(Note: A point of inflection is a point where a curve crosses its tangent without doubling back on itself. Alternatively, a point of inflection can be thought of as the point where a curve changes the direction of its concavity – see the diagram below. A stationary point of inflection is a special case of a point of inflection.)

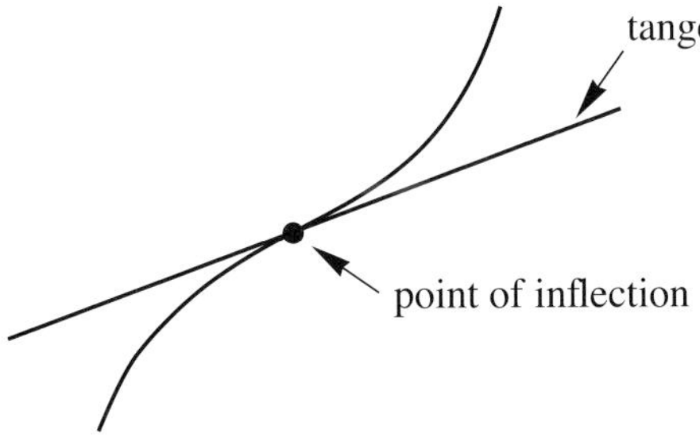

b. i. Now consider the reciprocals of the positive integer values:

$\alpha = \frac{1}{2}, \frac{1}{3}, \frac{1}{4}, \ldots$. For which values of α is the function $y = x^{\alpha}$ defined for all x? Describe the behaviour of the tangent to $y = x^{\alpha}$ at $x = 0$. For which values of α does $y = x^{\alpha}$ have a minimum at $x = 0$?

ii. Sketch graphs of $y = x^{\alpha}$ for other positive fractional values of α, for example $\alpha = \frac{2}{3}$ or $\alpha = \frac{3}{4}$. What is the domain of each of these functions?
(Note: Some computer packages may not graph these functions over their whole domain.)
How can you predict the type of graph or the domain of the function from the value of α? Does it matter whether $x^{2/3}$ is interpreted as $(x^{1/3})^2$ or $(x^2)^{1/3}$? Conside

this question for other values of α. There are curves $y = x^{\alpha}$ where a minimum at $x = 0$ is not detected by the condition $\frac{dy}{dx} = 0$. Identify and sketch several such curves. What is the value of $\frac{dy}{dx}$ at $x = 0$ in these cases?

c. Extension

i. Consider how the behaviour of $y = x^{\alpha}$ at a minimum or inflexion point is affected by adding or subtracting another function $y = x^{\beta}$, to get $y = x^{\alpha} \pm x^{\beta}$. Investigate integer values of α and β first.
Can maxima or minima or points of inflexion be created or destroyed by this process? Give several examples which justify your answer. Explain how a point of inflexion may be recognised from the graph of $\frac{dy}{dx}$.

ii. Some curves $y = x^{\alpha}$ have a **cusp** at $x = 0$, a sharp point like the spine on a holly leaf as shown in the diagram.

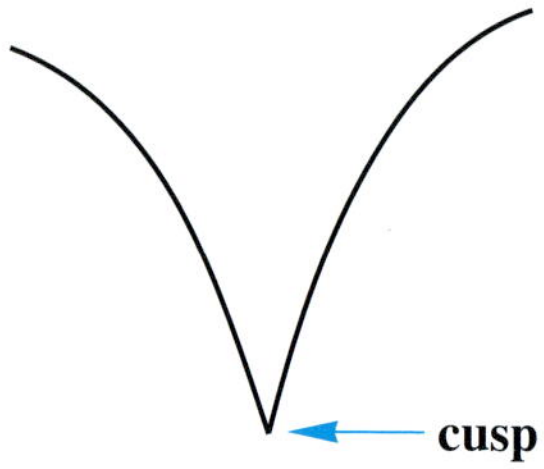

Give several examples of curves which have this property. If a function $y = x^{\alpha}$ has a cusp, what can you say about $y = x^{\alpha} + x$, $y = x^{\alpha} + 2$? Devise a curve that has cusps at $x = 0$ and $x = 1$, and sketch its graph.

iii. Devise and sketch another curve of the form $y = x^{\alpha} \pm x^{\beta}$ with another combination of features, for example, a cusp and a point of inflexion.

P work Type I 9

THEME: PRODUCTS OF FUNCTIONS

Your investigation must centre on the exploration of the properties of a particular function or class of functions which is itself the product of two more elementary functions. Whilst starting points are prescribed in this booklet, there is room for student initiative; independence and originality are encouraged.

1. General Advice

In this investigation, you are required to examine a problem or series of related problems which involve a function which is a product of two other functions. Each starting point begins with a relatively straightforward problem. However you are advised to give careful consideration to domain limitations and other factors which might relate to the reasonableness of your solution. Following this introductory task, a number of more complex variations of the original task are presented which require a greater depth of analysis. Finally there is an extension which requires some individual initiative in placing the problem in a broader or more realistic context and generalising earlier results.

2. Starting points

1. Productive polynomials

In this starting point you will explore the behaviour of functions of the general form

$$f(x) = x^n(A-x)^m$$

Consider first the case where A is a positive constant, and m and n are equal to 1.

a. Sketch this function on a suitable domain, showing any turning points.

Now consider the more general case where m and n are positive integers.

b. Systematically vary the values of A, m and n and determine the effect which each has on the behaviour of the function, paying particular attention to the number, location and nature of stationary points. Illustrate your findings with carefully selected graphs.

c. Show that there is a maximum number of stationary points for functions of this general form.

d. What combinations of stationary points are possible? What restrictions are there on the values of x for which these stationary points occur?

e. Sketch several well–chosen examples which illustrate your findings to d.

Extension

f. Extend your investigation in parts b., c., d., e. above by varying m and n in some further way.

2. Child's play

A child's swing is pulled back and then let go. Thereafter its displacement can be approximately modelled by a function of the form $x = Ae^{-t/b}\cos(ct)$, where x gives the horizontal displacement of the seat of the swing in metres from the equilibrium position (that is, the position at which the swing will eventually come to rest), t is the time in seconds and A, b and c are arbitrary positive constants.

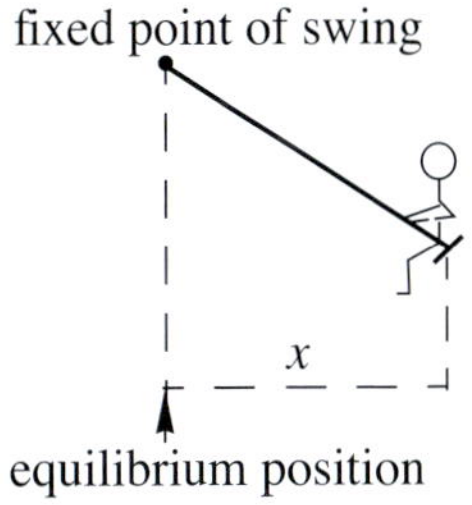

a. Consider first the function where $A = b = c = 1$. Sketch the graph of the displacement over a suitable domain and discuss its important features. Does this provide a good model of the swing's movement?

b. Systematically vary the values of A, b and c to determine the effect which each has on the behaviour of the displacement function, and how each is related to the physical features of the swing's movement.

c. Choose values which you think are realistic for the maximum horizontal displacement, the time between successive instants when the swing is vertical and any other physical features of the swing's movement which you have identified. Determine the values of A, b and c which allow a function to best model the swing's movement for these chosen values.

d. For any given set of values of A, b, c, the function has successive maxima and minima corresponding to the extreme positions on either side. Find the times and displacements of these extreme positions for a chosen set of values of A, b, c. Repeat for several other sets of values. Is there a pattern to the manner in which the magnitudes of these displacements decrease for each set of values?

e. Sketch several well–chosen examples which illustrate your findings to b., c. and d.

Extension.

f. Generalise your answers to part d. above in some way.

3 Drug dosages

The safe medicinal use of a drug requires knowledge both of its rate of absorption by the body (which may depend on how the drug is given) and its rate of clearance (principally through the kidneys). Typically, absorption is required to be swift, whilst clearance is more gradual. A class of functions which can be used to **model** this process are functions of the form $x = At^be^{-ct}$ where x represents the concentration of the drug in the blood in milligrams per litre (mg/l) and t represents the time, in minutes, since the administering of the drug. A, b and c are arbitrary positive constants.

a. Consider first the function where $A = b = c = 1$. Sketch the graph of the concentration over a suitable domain and discuss its important features. Would these values give a desirable model of the changes in concentration over time due to the effects of absorption and clearance?

b. Systematically vary the values of A, b and c to determine the effect which each has on the concentration function.

c. Determine the values of A, b and c which best model the drug concentration when both of the following conditions are satisfied:
- the drug reaches its maximum concentration of 0.2 mg/l after 30 minutes and
- the concentration after 2 hours is still half of the maximum value.

d. In many situations, drug concentration has to be maintained between two values by the use of repeat dosages which are possibly smaller than the initial dose.
Explore how a model could be developed to keep the concentration of a drug between 0.1 mg/l and 1 mg/l.
You may need to make some simplifying assumptions concerning the concentration of the drug remaining when the next dose is administered. It is obviously desirable to keep the time between successive dosages to a reasonable value.

e. Sketch several well–chosen examples which illustrate your findings to b., c. and d.

Extension.

f. Generalise your answers to part d. above in some way.

P work Type I **10**

THEME: FITTING FUNCTIONS

Your project must centre on the exploration of the 'closeness' of a particular function, class of functions or group of functions to a given curve, a given data set, or another function. You are encouraged to show initiative and originality.

1. General Advice

In this investigation, you are required to examine a problem or series of related problems which involve determination of a function which approximates a given function, or curve, or set of data. Each Starting Point begins with a relatively straightforward problem. Following this inductory task, a number of more complex variations of the original task are presented which require a greater depth of analysis. You are advised to give careful consideration to domain limitations and to how you measure 'closeness' or 'good fit', as well as other factors which might relate to the reasonableness of your solution. Finally, there is an extension which requires some individual initiative in extending earlier results or placing the problem in a broader, or more realistic, context.

2. Starting points

1. Approximations

Polynomial functions are often used to give an approximation to a more complex function. In this Starting Point you will compare polynomial functions of degree n of the form

$$p_n(x) = a_0 + a_1x + a_2x^2 + \ldots + a_nx^n \text{ with functions of the form } f(x) = A\log_e(kx + b),$$

where A, k and b are arbitrary real constants and the coefficients, $a_0, a_1, a_2, \ldots, a_n$, are arbitrary rational constants. The values of the coefficients for any particular logarithmic function can be determined by a process involving repeated differentiation and substitution of $x = 0$.

a. Let $A = k = b = 1$. The process for deriving a linear polynomial function, $p_1(x) = a_0 + a_1x$, which approximates $f(x)$ in a particular region, is as follows.

assume : $f(x) = \log_e(x + 1)$

substitute $x = 0$ into both sides : $\log_e 1 = a_0 \therefore a_0 = 0$

now differentiate both sides : $\dfrac{1}{x+1} = a_1$

and substitute $x = 0$ into both sides : $a_1 = 1$

therefore : $\log_e(x + 1) = 0 + 1 \times x \therefore p_1(x) = x$

Use a similar process to determine a quadratic function, $p_2(x)$, a cubic function, $p_3(x)$, and a quartic function, $p_4(x)$. Sketch the graphs of each of these functions and compare them with the graph of $f(x) = \log_e(x + 1)$.

Comment on the difference between $f(x)$ and $p(x)$ for each of these cases, paying particular attention to domains.

In parts b., c. and d., consider A and b to be positive arbitrary constants, with $k = 1$. Illustrate your findings throughout with carefully selected graphs.

b. Systematically vary A and b and determine the effect that each has on the behaviour of $f(x)$ and on the values of the coefficients for the approximating polynomials.

c. i. For a given set of values of A and b determine a quadratic, a cubic and a quartic function which will each give an approximation to this logarithmic function.

Determine how well each polynomial can be considered to be a 'close' approximation to $f(x)$ over the domain $[0, 1]$. You will need to explain how you have measured 'close'.

ii. Repeat c.i. for other sets of values and, using your measure of 'closeness', explore how well the polynomials approximate the logarithmic function as A and b vary. You may need to make some simplifying assumptions regarding domain.

d. Generalise the rule for $p_n(x)$. Can you determine the largest possible domain $[0, m]$ over which $p_n(x)$ could be used as an acceptable approximation to $f(x)$? Here, n can be as large as you like and m is a positive real number.

e. Now consider A, b and k to be positive arbitrary constants. Explore the effect of varying k on the behaviour of $f(x)$ and on the values of the coefficients for the approximating polynomials. Illustrate your findings with carefully selected graphs.

Extension

f. Extend your investigation in parts b., c. and e. in some way.

P work Type I **11**

TASK: BUSHWALKING WITH KIM

Bushwalkers travel through different types of country. The denseness of the bush and the ruggedness of the terrain influence the speed of travel. By planning a route to take such factors into consideration, the total time taken to travel from one point to another can be reduced. In calculating estimates of the time for a particular route, a walker uses his/her average speed for each different type of country.

For a walk through a particular type of country the travel time taken is given by $t = \frac{d}{v}$, where d is the distance in kilometres (km) and v is the average speed in kilometres per hour (km/h) for that particular type of country. A typical two-day walk for Kim covers a distance of up to 30 km with walking speeds of up to 5 km/h in cleared country.

a. For a typical two-day walk, choose several representative values for Kim's average speed and sketch a graph of the relationship between t and d for each of these values. Similarly choose several representative values for the distance to be travelled and draw a graph of the relationship between t and v for each of these values.
Discuss the key features of each of the two families of graphs, and the differences between the families.

Next weekend Kim is planning to walk from Ardale to Brushwood. The direct route, a distance of 14 km, will take her entirely through rugged bush country. However, there is a large square clearing situated as shown in the diagram below. This clearing has one diagonal along the perpendicular bisector of the direct route and one corner, C, at the midpoint of the direct route.

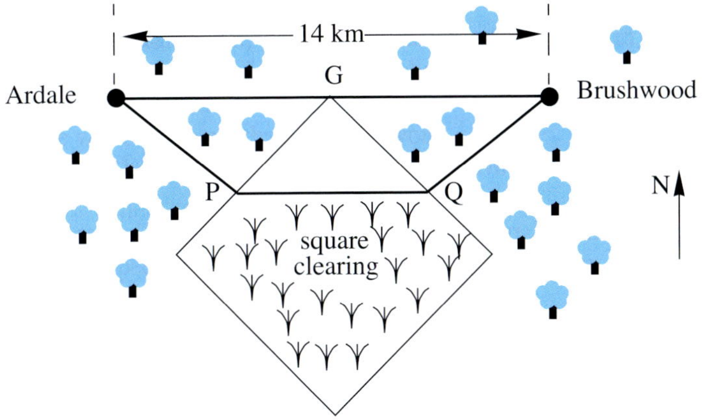

Kim believes that time will be saved if she travels from Ardale to Brushwood on a route similar to the one shown above passing through P and Q, where the section PQ is parallel to the direct route. Throughout this investigation assume that Kim's path across the square clearing is parallel to the direct route.

b. i. Kim's initial rough estimate of the side-length of the square clearing is 7 km. Select any route similar to that through P and Q and calculate the total time Kim would take if she travels at an average speed of 1 km/h in the bush and 5 km/h through the clearing. Compare this time with that for the direct route.

ii. Select an appropriate length variable and hence determine a mathematical relationship which will allow Kim to calculate the total time she would take for any route of this type. Sketch the graph of this relationship and discuss key features. Find and describe the route for which her travelling time will be least and compare it with the direct one.

iii. Investigate the effect that changes in the side-length of the square clearing would have on the route by which Kim can travel from Ardale to Brushwood in the shortest time possible.

c. Take the side-length of the square clearing to be 7 km. Consider now that Kim's speeds are given by v_b km/h and v_c km/h in bush and in cleared country respectively, where $v_c = kv_b$ and k is a real number such that $1 < k < 10$. For each of several carefully selected values of k, find the route for which Kim's travelling time will be least and the time taken. Discuss how the minimum time is related to k.

P work Type I **12**

TASK: BLOOD FLOW

The circulatory system of a human body comprises blood vessels of different radii and lengths with the radii generally decreasing as the distance from the heart increases. Blood flows from the heart through larger blood vessels to smaller blood vessels.

The efficiency of the flow of blood through a single blood vessel is related to the loss of energy due to friction in that blood vessel. This loss of energy in certain blood vessels can be approximately modelled by a function of the form $E = \frac{kl}{r^4}$ where E gives the energy loss of the flow of blood in joules, and r and l are the radius and length of the blood vessel, respectively, measured in millimetres. k is a positive constant whose value depends on different characteristics of the blood.

Throughout this investigation take $k = 9$ which is the normal value for a typical person. This function may be used for blood vessels with radii in the range of 2 mm to 15 mm and lengths of up to 30 mm.

a. Choose several representative values for the radius of a blood vessel. Sketch a graph of the relationship between E and l for each of these values. Similarly choose several representative values for the length of a blood vessel and sketch a graph of the relationship between E and r for each of these values.
Discuss the key features of each of the two families of graphs, and the differences between the families.

Throughout the circulatory system, blood vessels are constantly branching. At a branch point the total loss of energy can be shown to be related to the branch angle ABC, of size θ radians.

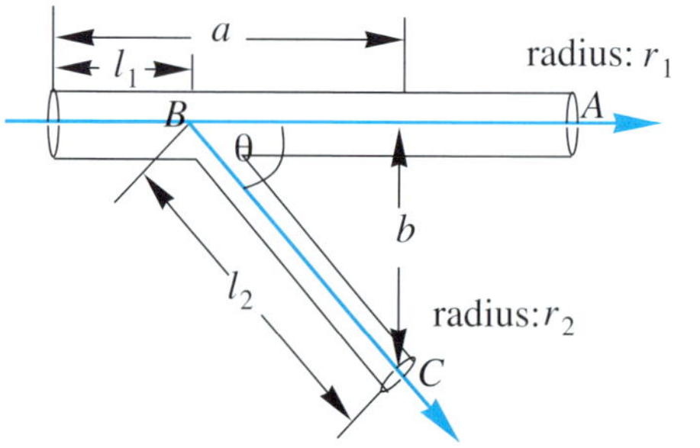

The total loss of energy, E_{Total}, as blood flows from one vessel into another is given by the sum of the energy losses

$$E_{\text{Total}} = E_1 + E_2 = \frac{kl_1}{r_1^4} + \frac{kl_2}{r_2^4}$$

where E_1 is the energy loss in the first (larger) blood vessel and E_2 is the energy loss in the second (smaller) blood vessel.

i. Choose a set of values for a, b, r_1 and r_2, where a and b may take any real value between 10 and 25 (inclusive) and $r_2 < r_1$. For this set, express l_1 and l_2 in terms of θ and hence express E_{Total} as a function of θ, clearly stating the domain.

ii. For the values chosen in i., determine the angle which would result in minimum total energy loss.

iii. Draw the graph of your function and discuss the key features.

c. i. Retain the values of r_1 and r_2 used in b.i. For each of several pairs of values for a and b, determine the angle for which the total energy loss is a minimum. Comment on your results.

ii. Investigate how the angle for minimum total energy loss is related to the radii of the two blood vessels.
Discuss whether your result is valid for all values of a and b.

P work Type I 13

TASK: CHRIS'S CUCKOO CLOCK

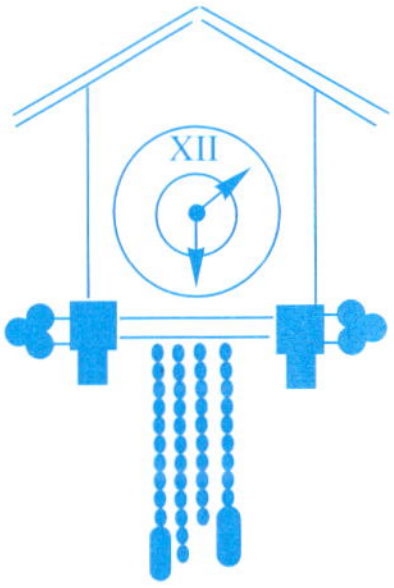

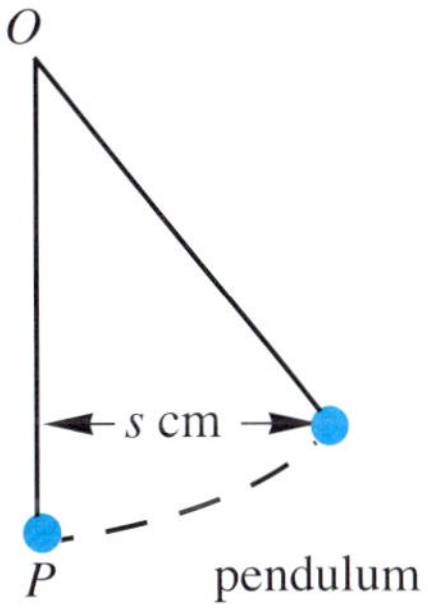

Chris has recently purchased a cuckoo clock. The rate at which the hands of the clock move is controlled by a pendulum which is kept in regular motion by slowly descending weighted chains. When the weights reach their lower point and stop moving, the pendulum swing begins to change, causing the hands of the clock to slow down and gradually stop. From the time when the swing begins to change, the horizontal displacement, s cm, of the point, P, at the end of the pendulum, from the vertical (see the diagram) can be modelled by functions with the rule

$$s(t) = 5e^{-kt}\sin(at),\ t \geq 0$$

where t is the time in seconds after the pendulum swing begins to change and k and a are real constants. Values of k and a vary from clock to clock but are always such that $0 < k < 1$ and $a > 0$.

a. Initially, consider the four functions with the rules $s_1(t) = 5e^{-kt}$, $s_2(t) = -5e^{-kt}$, $s_3(t) = 5\sin(at)$ and $s_4(t) = 5e^{-kt}\sin(at)$.
Choose one appropriate value for each of k and a.

i. Using your chosen values of k and a, sketch the graphs of rules
$s_1(t) = 5e^{-kt}$, $s_2(t) = -5e^{-kt}$, $s_3(t) = 5\sin(at)$ and $s_4(t) = 5e^{-kt}\sin(at)$.
over appropriate domains on separate sets of axes.
Discuss the key features of each of these graphs and any relationships between the graphs.

ii. Determine the turning points of your graph of $s(t)$ and the points of contact between the graph of $s(t)$ and $s_1(t)$ and between $s(t)$ and $s_2(t)$.
Discuss your findings.

Now consider the motion of the point P after the swing of the pendulum begins to change.

b. Continue to use your chosen values of k and a.

i. Consider the rate of change of the horizontal displacement, s cm, with time and investigate how this displacement changes over time.

ii. Investigate how the magnitudes of the successive local maxima of the horizontal displacement, s cm, change over time as the pendulum swings.

c. For several appropriate pairs of values of k and a, investigate the behaviour of $s(t)$.
Discuss how the values of these constants affect features of the motion of the point P after the swing of the pendulum begins to change.

P work Type I **14**

TASK: OF POWERS AND EXPONENTIALS

Jan has just started a course in computer science, and her class is discussing algorithms. An algorithm is a set of instructions designed to solve a given type of problem. A particular problem can often be solved by several different algorithms, some of which require more time or computer memory than others to carry out. The complexity of an algorithm is a measure of the time taken to apply the algorithm and depends on things like the number of computations and the amount of memory required to carry out the algorithm.

The time required for an algorithm to solve a problem can often be described either by a function of the form x^n or by one of the form n^x, where n is a positive integer which is constant for a particular algorithm and depends on the nature of the algorithm, and x is a variable which corresponds to the size of the input to which the algorithm is applied.

Jan becomes interested in this idea and agrees to carry out an investigation, the eventual aim of which is to try to determine whether one of the functions may be preferable to the other when consideration is given to the time required for an algorithm to solve a problem. Imagine that you are working with Jan on this problem.

a. Initially, Jan is asked to investigate and compare the behaviour of functions of both forms
$f(x) = n^x$ (exponential functions), and
$g(x) = x^n$ (power functions) where n is a positive integer constant and x is a real number.

Choose several representative values of n and consider the particular cases of the two types of functions for these values.

i. On one set of axes sketch the graphs of $y = f(x)$ for each of your chosen values of n. Repeat on another set of axes for $y = g(x)$. Discuss and compare key features of these families of graphs.

ii. Sketch the pair of graphs of $y = f(x)$ and $y = g(x)$ for each of your chosen values of n on separate axes with appropriate domains.
Determine the number of points of intersection of each pair of graphs.
Discuss the relative behaviour of each function within the intervals determined by the points of intersection.

b. Jan believes that she needs to further analyse the functions defined in a. and proceeds as follows.

i. Find an exact value of k such that $2^x = e^{kx}$ for all x. Hence find an exact expression for the derivative of 2^x.

ii. Consider the graphs of the derivative functions of $y = x^2$ and $y = 2^x$ and compare these functions over the intervals from a.ii. above.

c. To complete her investigation into the problem Jan now decides to take x to be a positive real number. Sketch the graph of the function over an appropriate domain.

Discuss key features of the graph. For $n > 2$, use this graph to investigate the number of intersection points and regions of the domain within which these points of intersection occur, for pairs of functions of the form $f(x) = n^x$ and $g(x) = x^n$.

d. Based on the results of her work in parts a., b. and c., Jan now concludes that an algorithm which solves a problem in time x^n, for some positive integer n, is preferred to an algorithm which solves the same problem in time n^x, for the same value of n.
Discuss the validity of Jan's conclusion.

P work Type I 15

TASK: IS FASTER BETTER?

Lee is the fleet manager of the Three C's Trading Company which maintains a fleet of cars for its executives. An important part of her job is to allocate the cars as efficiently as she can, so as to minimise the cost to the company.

The hourly cost, $C(v)$ dollars, of running a Three C's car at an average speed of v km/h is given by

$$C(v) = a + bv^n$$

where a, a positive integer, is a fixed cost related to registration and insurance while b, a positive real number, and n, a positive integer, are related to the operating costs such as the cost of petrol, oil, tyres and car maintenance. The constants a, b and n can take the following values:

$$1 \le a \le 5,\ 10^{-7} \le b \le 10^{-2} \text{ and } 2 \le n \le 4.$$

The cars in the Three C's fleet will be driven on roads where the maximum speed limit varies from 60 km/h for minor or suburban roads in the city or towns, to 110 km/h for roads outside the city or town areas or on highways or freeways. On some sections of road recommended speeds may be less than the maximum speed limit in that area due to particular driving conditions such as winding road, blind corners or poor road surface.

The total cost, in dollars, of running a Three C's car at an average speed of v km/h for a journey taking t hours is given by $T(v)$. The time taken, t hours, for a given journey depends on the average speed of travel and the distance covered. Initially consider a journey of 50 kilometres.

a. i. Obtain a general rule for the function $T(v)$ in terms of v only. Discuss domain constraints.

ii. Choose a particular set of values for a, b and n and sketch the graph of $T(v)$ as a function of v over an appropriate domain.
Discuss key features of the graph.

iii. Keeping your value of a fixed, systematically vary the values of b and n and sketch the resulting families of curves. Discuss the effects of b and n on the shape and key features of the graph of the function.

iv. For each possible value of n discuss the effect of the values of a and b on the shape and key features of the graph of the function and, in particular, on the location of any stationary points.

b. Investigate the effect of changing the length of the journey on the location of any stationary points and on other key features.

c. The advice that Lee provides to executives for the use of their cars includes the recommendation that travelling at an average speed of 55 km/h for city-type driving keeps the running costs to a minimum.
Use your findings from part a. to determine all sets of values of a, b and n which would allow this recommendation to be valid.

d. The Three C's Trading Company decides to buy two types of cars for the use of its executives. One model is a small 4-cylinder sedan while the other is a larger, more powerful 6-cylinder sedan. The known data for the two models are as follows:

model	Value of the constants		
	a	n	b
small	2	2	0.0016
large	5	3	0.00001

i. Investigate the relative total costs of running these two types of car, taking into account the possibility that these cars may be driven solely around the city, or on trips which are either essentially out of the city or on a combination of city and open-road driving.

ii. Draw up suggestions for Lee on how she could best allocate cars to her executives.

23.3 PORTFOLIO WORK TYPE II – MATHEMATICAL MODELLING

23.3.1 CARRYING OUT A MODELLING INVESTIGATION

You have already encountered elements of the modelling process as you have worked through some of the chapters in this book. However, before supplying you with a number of potential modelling investigations, P work Type II, we consider an example that highlights the differences between Type I investigations and Type II investigations – in essence, you will need to source the investigation, analyse the 'information', come up with a model and evaluate your results. Along with the work covered in Chapter 3, the following (brief) example only begins the investigation and provides some avenues to consider.

MODELLING EXAMPLE

TASK: LEAKING BOTTLE

Objective: To find and model the relationship between the velocity of the jet at a hole in the bottle and the height of the fluid surface above the hole.

Take a plastic bottle and make a small hole near the bottom of the bottle. Place the bottle on a flat, horizontal surface and place a ruler next to the bottle. You could also place a sheet of graph paper next to the bottle and use that instead of the ruler.

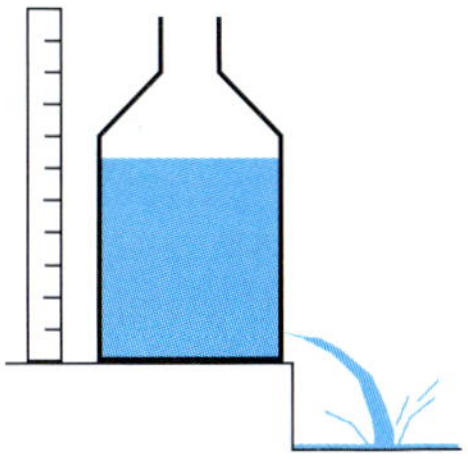

Fill the bottle with water and place it beside the sink

Make a record of the time it takes for the water level to drop every centimetre.

Draw up a table of values:

Height (cm)	10	9	8	7	6	5	4	3	2	1	0
Time (seconds)	0	3	10	20	35	50	62	83	100	120	145

The experiment should be repeated several times and then use the mean value (for example).

Plot the results with height, h cm, along the vertical axis and the time, t seconds, along the horizontal axis. This can be done using the TI–83:

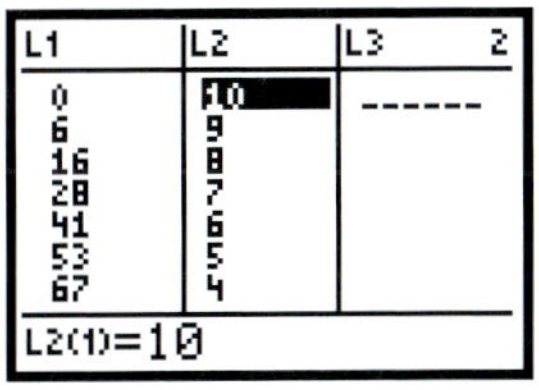

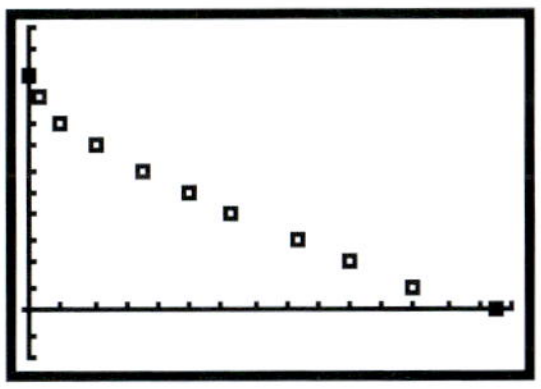

Is the relationship linear? That is, can we apply the model $h(t) = a + bt$?

Could the relationship be exponential, i.e., is $h(t) = a \times e^{-kt} + c$. Or, could there exist a better equation to model this data? What about a square root relationship? i.e., $h(t) = a\sqrt{t} + c$, or maybe, $\sqrt{h} = at + c$. Each of these should be considered (giving reasons as to why the others were rejected).

This is where the modelling aspect will come into play, and so, there will be a little trial and error and verification. If two models seem appropriate, then there might be the need to collect more data.

What about the size of the hole? Will that alter the model? Use identical bottles but with different hole sizes.

What about the position of the hole? What observations can you make if using identical bottles but having the hole placed at different heights on the bottle?

23.3.2 MODELLING INVESTIGATION TASKS

P work Type II 1

THEME: USE SIMULATION TO SOLVE A PROBLEM INVOLVING CHANCE

Your project must use simulation to solve a problem involving chance. You must-follow the instructions given below and report on each of the specified steps in the main text of your report.

1. General advice

Simulation is a useful method for investigating problems. When the problem involves chance, the simulation will involve a random experiment. The simulation process is as follows.

i. specify the problem carefully.
ii. identify the important mathematical relationships.
iii. find, use and test a model which represents the important features of the situation.

You need to check that your simulation is realistic, giving answers that agree with a real life situation. By working with a model, you should be able to investigate aspects of the real situation. It is important to evaluate the reliability of the answers obtained from your simulation. Acknowledge which area of mathematics you use: probability, statistics, logic and/or algebra. You may choose to develop a computer program to assist you, or you may use a recognised computer package, but remember to include your own analysis of the problem.

An Example

Imagine you wanted to start a shoe shop and needed to know how many pairs of shoes of each size to keep in stock. You can get data about the percentage of people who wear shoes of each size. This would give you a good idea about the percentage of pairs of shoes in each, size that are needed. If you know how many pairs of shoes you expect to sell, you would then be able to work out how many of each size to stock. However, to be confident that you do not run out of any particular size, you need to know when to order more pairs of shoes of each size. How many pairs should be left before you order more?

The question above can be answered by theoretical statistical methods that are not usually studied at school. However, simulation can be used to help answer the question. The purchase by a customer is simulated by the drawing of a marble from a bag (and afterwards replacing it). Use a different colour for each shoe size and be sure to allocate the right percentage of marbles for each size. Imagine that you had only six pairs left of a certain shoe size-what would be the chance that you would run out in a week if you had not ordered more? To answer this question, conduct some experiments with the marbles, keeping a careful tally of the results. From further experiments, you could also find out how many pairs of shoes of that size should be kept in stock so that you can be reasonably sure to meet the customer's demand. Is it the same quantity for all sizes of shoes? How do the answers vary if the assumptions you have made about the number of customers, frequency of delivery, for instance, are changed?

In the simulation above, physically choosing marbles from a bag is the way in which randomness is generated. The results of each experiment would also have to be tallied by hand. Using a computer to generate the random numbers and also to do the tallying makes the process much quicker, enabling many more trials to be carried out, so that a better idea of the reliability of your answer can be obtained.

Instructions

(Steps v., viii. and ix. are optional.)

i. Choose a real life situation involving chance or uncertainty and describe it carefully. Identify a small number of questions about it that you wish to investigate by simulation.

ii. Identify the factors that will be most important to simulate and describe the factors that you have chosen to disregard.

iii. Describe a way of investigating your questions using a physical simulation. List the materials required and specify the procedure to be carried out carefully.

iv. Clearly list the assumptions that have been made in setting up the simulation and explain why they are reasonable. (To do this, you may need to find some data from books or collect some data of your own, but do not let this take more than a couple of hours.)

v. If possible, describe a way of investigating your questions using a computer simulation instead of a physical simulation. (If you have access to a ready-made computer simulation suitable for answering your questions, you need to give full details of it and describe clearly how it works, including the assumptions that have been built into program and how, the chance element is generated.)

vi. Carry out simulation, either physically or by computer, (that is by writing your own program or using a ready-made program).

vii. Explain the answers clearly and discuss their validity.

viii. You may wish to investigate several different ways of simulating your situation or you may also wish to change some of the assumptions you have made and repeat the simulation. How and why do the answers vary? Which answers best reflect what might happen in the real world?

ix. You may also wish to compare the answers you have obtained with an answer obtained using theoretical probability or statistics, or with what has actually happened in practice.

2. Starting points

You may investigate any topic related to the theme: Use simulation to solve a problem involving chance. You must discuss your choice of topic and how it relates to the theme with your teacher and you must follow the instructions above. The examples listed below are possible starting points, although it is not compulsory to use these ones.

Use simulation to investigate

- How many cereal packets you would have to buy to get a complete set of cards
- The chances of winning a finals series given a particular position in a preliminary competition
- Stock control and inventory control (for example, the shoe shop)
- The chance of getting two identical birthday cards at a child's party

  The likelihood that two people, in a class have the same birthday
- Winners of horse races, or outcomes of bets for punters or bookmakers
- The variation in-time to travel from A to B by public transport
- How much time you should allow to drive from A to B to arrive by 9.00 am
- How much better for clients, in a bank or Medicare office, is a single queue compared with a multiple queuing system
- The likelihood of winning simple games using various strategies
- Winning prizes in a gambling game
- The chance of winning a tennis game after being

P work Type II 2

THEME: PREDICTION OF UNCERTAIN EVENTS

Your investigation must arise from or incorporate the prediction of uncertain events. You are encouraged to show initiative and be independent in carrying out your project.

1. General Advice

There can be significant benefits from being able to make accurate predictions, say for example, on the stockmarket, for trends in population, in games of chance, and for consumer prices. Mathematical techniques often enable us to make predictions and to analyse their limitations. Such techniques might involve finding equations to predict events or explain relationships; or the gathering and statistical analysis of data; or the counting of successful outcomes to determine probability.

The general criteria for assessment should be borne in mind, but here are a few issues, which arise specifically in the prediction of uncertain events.

First, consider in what sense, if any, is the event predictable. In some cases we understand how a system works and we can predict aspects of its behaviour via probability, for example in tattslotto and other games of chance. In other cases it may be necessary to collect data to mathematically model a system, for example, data on life expectancy or on traffic flow. Second what mathematics is appropriate? Algebra, statistics, probability, logical reasoning? Your investigation should include developing a mathematical method of prediction and discussion of the reliability of the prediction.

Please note that if gathering of data is appropriate for your project, you should not allow this to delay developing a detailed plan or model for the prediction of events using the data.

2. Starting points

You may investigate any topic related to the theme. You must discuss your choice of topic and how it relates to the theme with your teacher. The examples below show some starting points for projects. This list is by no means exhaustive.

2.1 **Trends** - employment trends in particular industries or local companies. How are they related to other economic indicators such as business investments, the value of the dollar, the CPI, et cetera?

2.2 **Trends** - population trends in countries and the world. How is rural population related to urban population?

2.3 **Queues** - compare waiting times in queues. When is it better to wait in an "8 items or less" queue (if you have less than 8 items)? In a bank, how long would you expect to wait if there were a single queue for many tellers, compared with separate queues for each teller?

2.4 **School enrolments** - local, state and national.

2.5 **Consumer prices** - predict the future cost of such items as a compact disk player, home computer, a dozen eggs, et cetera.

2.6 **Economic forecasting** - look for relationships among inflation, interest rates and the local currency which can be used to predict future levels.

2.7 **Games** - investigate games of chance (gambling). Is there a winning strategy? If you play a fair game and you play regularly what is your chance of "breaking even ?" Design a game of chance, modify the game's rules for differing numbers of players, for example, how should tattslotto be modified for half as many players or for twice as many players?

2.8 **Diseases** - predict the decline in the number of cases of measles or polio as a result of an immunization program.

2.9 **Diseases** - predict the spread of AIDS. Is the AIDS death rate exponential?

2.10 **Smoking and alcohol** - what are the long term trends? How are they related to health and life expectancy?

2.11 **Athletic records** - examine the long term trends in athletic performance. Could there be a 40 second 100 metre freestyle? Could there be a 10 metre long jump ? What will the 100 metre sprint record be in the year 2000? Will women's athletic perfomance in track events catch up to those of men?

2.12 **Extinction of species** - kangaroos, koalas, platypus, humans.

2.13 **Opinion polls** - how many people should you survey to gauge public opinion concerning, for example, political leaders, soap powders, perfumes, et cetera?

2.14 **Weather forecasting** - how accurate is your daily weather forecast? How accurate is the long range forecast?

2.15 **Life expectancy** - what are the long term trends? What is the relationship with life insurance premiums?

P work Type II 3

THEME: FITTING FUNCTIONS

Your investigation must centre on the exploration of the 'closeness' of a particular function, class of functions or group of functions to a given curve, a given data set, or another function. You are encouraged to show initiative and originality.

1. General Advice

In this investigation, you are required to examine a problem or series of related problems which involve determination of a function which approximates a given function, or curve, or set of data. The task begins with a relatively straightforward problem. Following this inductory task, a number of more complex variations of the original task are presented which require a greater depth of analysis. You are advised to give careful consideration to domain limitations and to how you measure 'closeness' or 'good fit', as well as other factors which might relate to the reasonableness of your solution. Finally, there is an extension which requires some individual initiative in extending earlier results or placing the problem in a broader, or more realistic, context.

You are at liberty to research a problem that is of interest to you. We provide the following either for you to use or to provide a guide of the type of question questions you should be considering (at the very least) when fitting a curve or curves to your own data.

Topic: Selling ice cream

Those who sell ice cream and gelati are well aware that the amount of ice cream/gelati sold depends both upon the price of the product and how hot the day is, that is, the temperature. Knowledge of both price and temperature and their relationship to consumption are important factors to economists and retailers for the analysis of market trends and sales. In this Starting Point, you will explore how a variety of functions can be used to form models of relationships, and discuss and evaluate their usefulness.

The following table summarises a projected sample of data for a popular retailer in a shopping strip.

price (cents per scoop)	ice cream consumption for different temperatures (litres per hour)			
	0°C	10°C	20°C	30°C
15	12	18	25	50
30	10	12	18	37
45	7	10	13	27
60	5	7	10	20
75	3	5	7	14
90	2	3	5	10
105	1	2	3	6

Data like this can be modelled by equations of two different forms:

1. $f(x) = \dfrac{A}{x-B} + C$, where A, B and C are arbitrary real constants; and

2. $g(x) = De^{kx} + E$, where k, D and E are arbitrary real constants.

a. i. For each temperature, plot consumption as a function of price on the same set of axes.

ii. First take $B = 0$ in $f(x)$ and $E = 0$ in $g(x)$. Select any two appropriate points for the 30°C curve and determine two equations, one of each type, which could model the consumption data.

Discuss how well each function models the data over the entire domain.

b. Now allow B and E to be non-zero and repeat a.ii. using three appropriate points for the 30°C curve. Do these curves fit the data more 'closely'?
You will need to explain how you have measured 'close'.

c. i. Systematically vary the values of the constants in each of the general functions to determine the effect of each constant on the behaviour of the function.

ii. Using your measure of 'closeness', decide which type of function is the better to use to model consumption for all temperatures.

The market cannot tolerate a price of more than 150 cents per scoop (that is, this is the maximum price) and at that price there are no sales. Likewise, no matter how cheap the bottom price is, consumption does not exceed 55 litres per hour on the hottest day shown.

d. Using the type of function that you have chosen to be 'better' in c.ii., find values of the constants which give a good model of the 30°C consumption data, over the domain [0,150]. Discuss any assumptions you have made.

e. It is believed that a better model for the 30°C data over the domain [0, 150] could be provided if it were to consist of a combination of a function of the type you have chosen to be better in c.ii. and two polynomial functions, one for prices over 105 cents and the other for prices below 15 cents.
Explore how such a model could be developed with the curves fitting smoothly together, and using quadratic functions of the form $q(x) = a_0 + a_1x + a_2x^2$, where a_0, a_1 and a_2

are real arbitrary constants, as the two polynomial functions.
Discuss why this model is better than the one developed in part d.

Extension

f. Extend your investigation in part e. in some way.

P work Type II 4

THEME: TRENDS AND RELATIONSHIPS

It will be your task to use statistics to investigate trends in the data and the relationships between variables in the data set provided, and to use these trends or relationships to make some predictions. You are required to carry out your project independently and are encouraged to show initiative and originality where this is possible.

In this investigation you should use only the data provided. Conclusions and interpretations should be made from the given data and the analysis performed in the investigation. You are not expected to carry out any additional data collection and/or research beyond that specified here. You are, however, encouraged to comment on any assumptions and limitations associated with the data provided.

The use of technology for dealing with the data is encouraged. You may choose to develop a computer program to assist you. You may use a statistical or graphics calculator or a recognised computer package, but remember to include your own analysis of the problem. You need to say what technology is being used; what is being calculated, giving the formula and defining the terms; what the purpose of the calculation is and its relevance to the problem.

In using a computer spreadsheet, for example, you must ensure that the appropriate formulation is included and all variables and units are correctly defined. All graphs, including computer generated graphs, must be appropriately scaled and correctly labelled.

Topic: Rabbits, the furry invaders

Rabbits were brought to Australia in 1859 and have since become a pest causing environmental damage and economic losses to farmers. One method of control used is a virus called myxomatosis which can be spread rapidly in various ways, one of which is transmission by mosquitoes and fleas. Many environmental, seasonal and climatic factors affect the number of fleas and/or mosquitoes and hence affect the spread of the virus.

The rabbit population over a five-year period in three different climatic regions is shown in the Data set for Rabbits. Rabbits were injected with the myxomatosis virus in the first year shown (Year 1). At Urana and Chidlow this was done in Spring, sometime during September, October or November. On Ile du Cimetière this was done in Autumn, sometime during March, April or May. Urana is situated inland near a lake in New South Wales. Urana has warm summers and mild winters and the rainfall is typically spread throughout the year. In wet years the region is flooded. Chidlow is situated near the coast close to Perth in Western Australia. Chidlow has warm wet winters and dry summers.

Ile du Cimetière is situated close to Antarctica, and there are no mosquitoes or fleas on this island because of the extreme climatic conditions.

a. i. Randomly choose a region and a three-year period. Explain your method of selection. Graph the rabbit population on a monthly basis for your selection. Describe any trends or patterns in the graph.

ii. Examine the data for the other two regions for the same period to decide if the trends or patterns are repeated. Comment on any similarities or differences between the three regions.

b. i. For either the region of Urana, or the region of Chidlow, use an appropriate method to determine a regression line which can be used to predict the rabbit population. Graph your regression line on the corresponding scatter plot to show the line of best fit. Discuss whether your selected three-year period is representative of the entire five years.

ii. Use your regression line formula to predict the following for your selected region.
- the rabbit population in July of the sixth year
- the rabbit population in December of the tenth year
- the month and year when the population will be half that at the end of the fifth year.

Discuss any assumptions you have made and comment on the reliability of your predictions. Do you believe that myxomatosis will ever cause rabbits to become extinct in your chosen region? Explain your answer using the results of your analyses in part a. or part b. or both.

iii. For your selected region, calculate an appropriate moving average, justifying your choice of length of the moving average. Investigate and discuss any long-term trends in the data. Compare your results with those from b.ii.

c. Randomly select two regions and two years, explaining your method of selection. For each of these years, draw a scatter plot and perform a correlation analysis between the populations for the two regions. Interpret and compare your results.

Rabbit population for the regions of Urana, Chidlow and Ile du Cimitière

URANA:

	YEAR 1	YEAR 2	YEAR 3	YEAR 4	YEAR 5
January	3000	100	25	15	10
February	3500	120	40	35	25
March	3900	140	55	45	35
April	4200	160	65	60	50
May	4500	160	70	65	60
June	4700	200	140	120	100
July	4900	300	250	230	200
Augus	4920	330	290	270	235
September	5000	460	350	330	290
October	5000	530	490	480	350
November	4900	510	480	465	390
December	4500	50	45	30	25

CHIDLOW:

	YEAR 1	YEAR 2	YEAR 3	YEAR 4	YEAR 5
January	1500	3640	1590	1290	200
February	1800	2940	1210	1000	190
March	1980	2720	1210	1000	190
April	2170	2240	800	700	210
May	2380	2000	750	790	240
June	2500	2300	920	830	380
July	3000	2500	1280	1000	490
Augus	3600	2000	1480	1120	500
September	4400	2400	1620	1000	520
October	5250	3000	1850	780	600
November	5210	2500	2000	500	610
December	4500	1860	1620	240	500

ILE DU CIMITIÈRE:

	YEAR 1	YEAR 2	YEAR 3	YEAR 4	YEAR 5
January	940	810	980	980	710
February	940	750	990	720	620
March	900	690	980	640	640
April	975	650	640	700	720
May	760	600	600	630	670
June	775	580	640	1000	680
July	700	590	800	990	800
Augus	1000	630	900	990	950
September	960	800	970	1000	960
October	1000	1000	980	910	1000
November	1000	1010	1010	900	950
December	920	1000	910	800	800

P work Type II 5

THEME: MATHEMATICAL MODELLING AND RATES OF CHANGE

Your project must be based on, or incorporate, mathematical modelling and rates of change. You are encouraged to show initiative and be independent in carrying out your project.

1. General Advice

In this investigation, you are required to examine the way in which different variables are connected in a mathematical model of a real situation. In particular you are asked to examine the connections between the rates of change of different variables in the model.

As a simple example, consider what happens when you pour liquid into a container. Obviously, the faster you pour, the faster the height of liquid will change. The height of the liquid is a function of the volume of liquid poured into the container. The rate of change of height depends upon the rate of pouring and the shape of the container. For example, in a cylindrical container, the rate of change of height is directly proportional to the rate of pouring, but if the container is of a more complex shape (eg a conical flask) this relationship will be more complicated.

In exploring these connections, it may be appropriate, in some cases, to perform simple experiments to approximate what happens, but it is essential that you develop a mathematical model which represents the situation and that you use mathematical techniques to establish results and connections. (It may, of course, be useful to compare mathematical with experimental results, but experiment alone is not sufficient.)

Where you use a mathematical function derived from an experiment, or data, you are not required to provide a theoretical justification for your choice of function.

Mathematical modelling

In developing a mathematical model to represent any of the situations given in the starting points, the following steps may be helpful.

1. Make assumptions which simplify a real situation. In the example above, you might ignore evaporation or assume a container to be a perfect cylinder. Consider the likely effects of these assumptions on your results.

2. Identify the key variables in your model. If there are too many variables to deal with initially, assume some of them to be constant so that you are left with a problem which you can work on mathematically. The choice of constants will, in general, be left up to you, but you should give some consideration to making realistic choices.

3. If you feel that you have fully explored a model with the assumptions you have made, it may be appropriate to change one, or more, of your assumptions. In most cases, the Starting Points suggest how this may be done.

4. At each stage, you should attempt to evaluate what your results are telling you. Look carefully at the domain in which your results might be valid and consider what might happen with extreme values of your variables. Consider the practical significance of your results.

2. Starting Points

You must investigate ONE of the following Starting Points. You must confirm with your teacher your choice of Starting Point.

2.1 Filter coffee-makers

In this starting point you should examine the connections between the flow of liquid in the different parts of a filter coffee-maker. You are to explore the relationships between the height and volume of liquid and how they vary with time.

A filter coffee system consists of three connected containers. Water is heated in a container, pumped through to the filter which contains the filter-paper and coffee-grounds and then drips into the coffee jug.

a. Working with a simple model

You should assume that water is initially drawn from a container which can be regarded as a rectangular prism and is pumped to a filter which has the shape of an inverted right circular cone. It then drips into a cylindrical coffee jug.

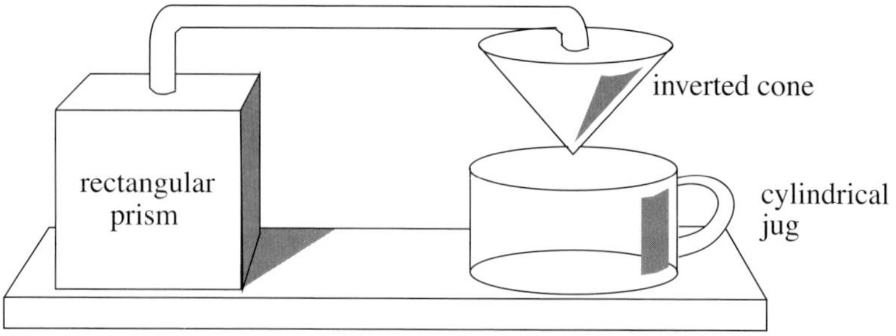

i. To begin, assume that water is pumped into the filter at a constant rate, and that the filter is blocked. That is, we are only looking at the first stage of transfer of liquid. Using a rate of flow which would allow a litre of water to pass into the filter in 15-20 minutes and by assuming that the filter is large enough to hold this amount of water, find the rate at which the volume and the height of water in the filter change with time. Sketch graphs of the rates of change which you have found.

ii. From these formulae for rates of change, you should now be able to find, by anti-differentiation or otherwise, formulae for volume and height in terms of time.

iii Now look at the second stage of transfer of liquid. Imagine that the filter is full of liquid. Allow this liquid from the filter to flow into the jug without any more water being pumped into the filter. Assume that the rate of flow is proportional to $\sqrt{h}$, where h is the height of liquid in the filter at time t. Find rate of change formulae, with respect to time for height and volume of both the liquid remaining in the filter and the liquid filling the jug.

iv From the various formulae for rates of change, you should now be able to find, by anti-differentiation, formulae for the volume and height of liquid in the jug in terms of time. Alternatively, sketch graphs of the rate of change functions to describe the behaviour of the system.

b. Extending the model

Here are a number of suggestions for extending the simple model given above. You are required to pursue only one of these or a similar alternative

i. Consider the whole system operating simultaneously. In particular this means looking at the height of liquid in the filter when water is pouring in and dripping out of the filter.

ii. You may wish to consider jugs of different shapes and/or the problem of where to place measurements showing cup gradations on the side of the jug. Which cup is filled the quickest?

iii. The assumption that water flows into the filter at a constant rate is questionable. In practice it starts slowly, builds up to a maximum and then decreases. Find a simple function with these properties over the domain under consideration and explore the effect of using this rate of flow on the rate of change of height and volume of water in the filter.

2.2 Blowing up balloons

A balloon when being blown up, may be considered, in the first instance, as a sphere of increasing radius from zero to a limiting value representing the maximum balloon size before bursting. If you assume that all parts of the surface of the balloon expand at the same rate, and that internal and external air pressure are the same, this simple model can be explored to find relationships between volume and surface area.

a. Working with a simple model

i If air is blown into the balloon at a constant rate (choose a realistic rate for an ordinary balloon), find the rates of change of volume, surface area and radius with respect to time. Sketch graphs of the rates of change which you have found.

ii Either by using your results from part i, or by other means, sketch separate graphs of radius against time, and surface area against time.

iii If the maximum size of the balloon is given in terms of its surface area, with a tolerance of plus or minus 10 per cent, how will this relate to maximum radius and maximum volume, and what will be the corresponding tolerances?

iv Develop a similar model for a 'sausage' shaped balloon, perhaps by assuming it to be an expanding cylinder. (You will have to decide on the relationship between radius and height as the volume increases.)

b. Extending the model

The assumption about internal and external air pressure is a weakness in the model. A simple experiment will enable you to develop an approximate relationship between radius of the balloon and mass of air blown in (measured, for example, by number of breaths). A function representing the situation more accurately can be developed from the data. You may wish to restrict the domain under consideration to help in choosing a suitable function. Develop such a function and repeat parts i and ii using this function instead of the earlier assumption. Interpret your results.

2.3. Water reservoirs

The main factors effecting the amount of volume of water in a reservoir are water usage (ie water drained out of the reservoir for consumption), evaporation and rainfall collected (which will depend not only on the size of the reservoir and the rate of rainfall, but also on the region from which water drains into the reservoir). In this Starting Point, you will examine the inter-relationship between some of these factors and the height and volume of water in the reservoir.

a. Working with a simple model

A reservoir can be modelled as a large frying-pan or bucket, that is, as a circular base, with sides which slope away from the base. Evaporation can be assumed to be proportional to the area of the water surface (ignore other factors, such as humidity). Rainfall and water usage can be initially regarded as constant.

i Develop some initial dimensions for such a reservoir of approximately 100 million cubic metres. Decide upon rates of water usage, evaporation and rainfall collected which seem reasonable. For these constants, determine the rates of change of height, surface area and volume with respect to time. If necessary, alter your initial values of these constants so that an approximate equilibrium is maintained, that is, the volume of water in the reservoir is maintained. Sketch graphs of the rates of change which you have found.

ii If there was no rainfall but the other constants remained the same, what would happen to the rates of change of height, surface area and volume with respect to time? How long before the reservoir would empty?

iii Suppose that because the reservoir has been allowed to empty completely, water usage is stopped to allow the reservoir to refill. Then it starts raining again at the same rate as before. Find the rates of change for the system at this time. How long will the reservoir take to fill?

b. Extending the model

This model could be extended in a number of ways. You are required to pursue only one of these or a similar alternative.

i The reservoir could be connected to another reservoir of a different capacity and of a different shape, for example, a reservoir with steeper sides. This second reservoir might be able to keep the first reservoir topped up in times of low rainfall. What then would be the effect on water levels in the second reservoir?

ii Develop a rate of change formula for non-constant rainfall and usage. Perhaps a simple function for changing rainfall could be obtained from looking at seasonal rainfall patterns in a likely reservoir location, and seasonal implications for water usage.

Note: data on seasonal rainfall patterns, for example, could be obtained from a school resource, such as a currently used geography text book or atlas or the internet. It is not necessary to seek information on this problem from public authorities.

EXERCISE 2.1

1.

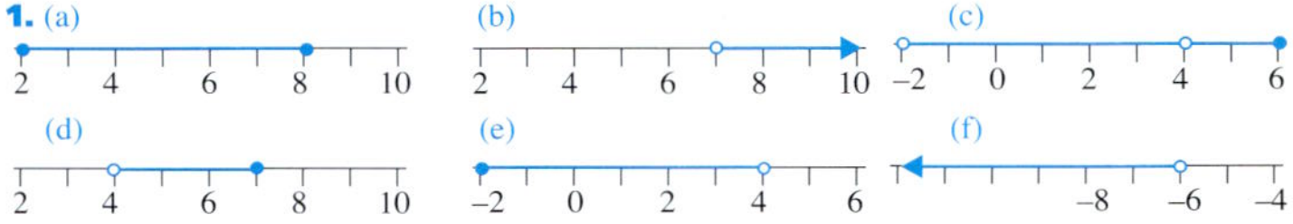

2. (a) $[-2,7]$ (b) $]9,\infty[$ (c) $]0,5]$ (d) $]-\infty,0]$ (e) $]-4,8[$ (f) $]-\infty,-1[\cup]2,\infty[$ **3.** (a) $5\sqrt{5}$ (b) $-\sqrt{3}$ (c) $\sqrt{3}$ **4.** (a) 4 (b) $4+\sqrt{6}$ (c) $6\sqrt{2}$ (d) $31+12\sqrt{3}$ **5.** (a) $2-\sqrt{3}$ (b) $\sqrt{7}+2$ (c) $2\sqrt{3}+\sqrt{15}$ (d) $-2-\sqrt{3}-4\sqrt{5}-2\sqrt{15}$ (e) $\frac{3+\sqrt{6}+\sqrt{10}+\sqrt{15}}{-2}$ (f) $3\sqrt{6}+2\sqrt{15}$ **6.** (a) i. $\frac{3\sqrt{5}+\sqrt{3}}{2}$ ii. $10+\frac{3\sqrt{15}}{2}$ (b) i. $\frac{14\sqrt{3}+48}{13}$ ii. $\frac{1344\sqrt{3}+3230}{169}$ **7.** (a) $\{\pm 3\}$ (b) $\{\pm 10\}$ (c) Ø (d) $\{-4,2\}$ (e) $\{-12,8\}$ (f) $\{0,4\}$

8.

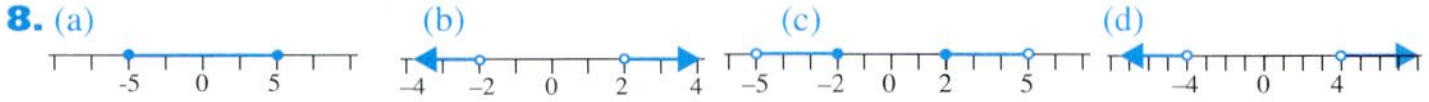

9. (a) $]1,\infty[$ (b) $]4,\infty[$ (c) $]4,6[$

EXERCISE 2.2.1

1. (a) 4 (b) 3 (c) –6 (d) $-\frac{11}{2}$ (e) $\frac{1}{10}$ (f) $\frac{3}{8}$ **2.** (a) $\frac{17}{5}$ (b) $\frac{4}{3}$ (c) $-\frac{3}{4}$ (d) $\frac{4}{3}$ (e) $\frac{35}{2}$ (f) $\frac{92}{41}$ **3.** (a) $-\frac{44}{5}$ (b) –39 (c) $-\frac{1}{7}$ (d) –3 (e) 2 (f) 4 **4.** (a) $2b-2$ (b) $b+1+\frac{b}{a}$ (c) $\frac{ab}{a+b}$ (d) $a(a+b)$ (e) ab (f) $\frac{ab}{a-b}$ (g) 0 (h) $\frac{a+b}{a^2+b^2}$ (i) $a+b$ **5.** (a) –4, 4 (b) $-\frac{9}{5}, 3$ (c) –6, 18 (d) $-\frac{11}{2}, \frac{17}{2}$ (e) $-\frac{7}{10}, \frac{1}{10}$ (f) $-\frac{5}{8}, \frac{3}{8}$ (g) $-\frac{7}{5}, \frac{17}{5}$ (h) $\frac{4}{3}, \frac{20}{3}$ (i) –3, 0 (j) $\frac{a-b}{2}, \frac{b-a}{2}, a \geq b$ (k) $\pm b(a-b), a \geq b$ (l) $-\frac{b}{a}, \frac{2b}{a}, b \geq 0$

EXERCISE 2.2.2

1. (a) $x<-4$ (b) $x \leq -\frac{1}{5}$ (c) $x>1$ (d) $x \leq -6$ (e) $x>\frac{18}{7}$ (f) $x>\frac{3}{8}$ **2.** (a) $x>\frac{52}{11}$ (b) $x \leq 1$ (c) $x \leq \frac{10}{3}$ **3.** (a) $x<1$ (b) $x<2-a$ (c) $x>\frac{2b}{3a}$ (d) $x \geq \frac{2}{(a+1)^2}$ **4.** (a) $-2 \leq x \leq 1$ (b) $-2 \leq x \leq 3$ (c) $-\frac{3}{2} \leq x \leq \frac{5}{2}$ (d) $x = -\frac{1}{2}$ (e) $-7 \leq x \leq 9$ (f) $-5 \leq x \leq 3$ (g) $-4 \leq x \leq 16$ (h) $-28 \leq x \leq 44$ (i) $-\frac{5}{12} \leq x \leq \frac{1}{12}$ **5.** (a) $x<-\frac{3}{2} \cup x>\frac{5}{2}$ (b) $x<\frac{3}{2} \cup x>\frac{7}{2}$ (c) $x \leq -12 \cup x \geq 16$ (d) $x \leq -24 \cup x \geq 6$ (e) $x<\frac{3}{4} \cup x>\frac{9}{4}$ (f) $-6<x<14$ (g) $x<-28 \cup x>44$ (h) $x<-\frac{5}{12} \cup x>\frac{1}{12}$ (i) $x \leq -4 \cup x \geq 16$ **6.** $p<3$

EXERCISE 2.3.1

1.

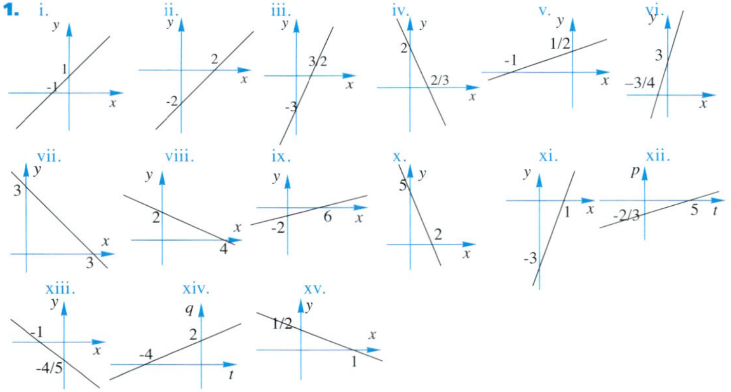

2. (a) 2 (b) 3 (c) $\frac{5}{3}$ **3.** (a) $y = 2x - 1$ (b) $y = 3x + 9$ (c) $y = -x - 1$ **4.** (a) $-\frac{1}{2}$ (b) $\frac{1}{3}$ (c) $\frac{3}{2}$ (d) $-\frac{4}{5}$ **5.** $y = 2x$ **6.** $y = -x + 1$ **7.** $y = \frac{x+2}{2}$ **8.** 2 **9.** (a) $y = \frac{5}{2}x$ (b) $y = -\frac{3}{2}x + 3$ (c) $y = \frac{5}{6}x - \frac{1}{2}$ (d) $y = -2x + 1$

10.

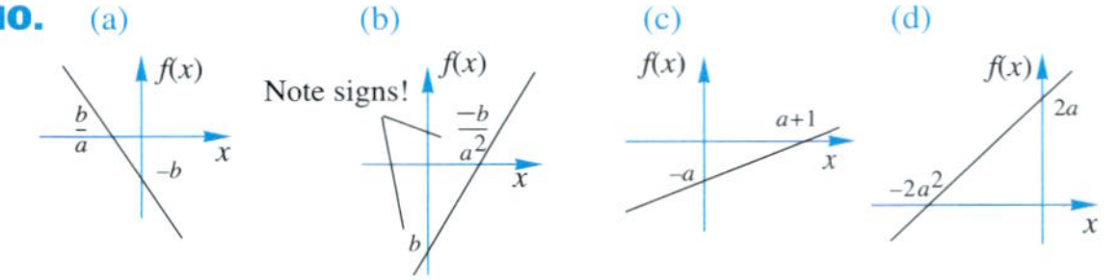

EXERCISE 2.3.2

1. (i) $x = 1, y = 2$ (ii) $x = 3, y = 5$ (iii) $x = -1, y = 2$ (iv) $x = 0, y = 1$ (v) $x = -2, y = -3$ (vi) $x = -5, y = 1$ **2.** (i) $x = \frac{13}{11}, y = \frac{17}{11}$ (ii) $x = \frac{9}{14}, y = \frac{3}{14}$ (iii) $x = 0, y = 0$ (iv) $x = \frac{4}{17}, y = -\frac{22}{17}$ (v) $x = -\frac{16}{7}, y = \frac{78}{7}$ (vi) $x = \frac{5}{42}, y = -\frac{3}{28}$ **3.** (i) –3 (ii) –5 (iii) –1.5

4. (i) $m = 2, a = 8$ (ii) $m = 10, a = 24$ (iii) $m = -6, a = 9$.

5. (a) $x = 1, y = a - b$ (b) $x = -1, y = a + b$ (c) $x = \frac{1}{a}, y = 0$ (d) $x = b, y = 0$ (e) $x = \frac{a-b}{a+b}, y = \frac{a-b}{a+b}$ (f) $x = a, y = b - a^2$

EXERCISE 2.3.3

1. (a) $x = 4, y = -5, z = 1$ (b) $x = 0, y = 4, z = -2$ (c) $x = 10, y = -7, z = 2$ (d) $x = 1, y = 2, z = -2$ (e) $\varnothing$ (f) $x = 2t - 1, y = t, z = t, t \in \mathbb{R}$

EXERCISE 2.4.1

1. (a) –5 (b) 4, 6 (c) –3, 0 (d) 1, 3 (e) –6, 3 (f) –2, 5/3 (g) 2 (h) –3, 6 (i) –6, 1 (j) 0, 1.5

2. (a) –1 (b) –7, 5 (c) –0.4, 3 (d) –2, 1 (e) –3, 1 (f) 4, 5

3. (a) $-1 \pm \sqrt{6}$ (b) $3 \pm \sqrt{5}$ (c) $1 \pm \sqrt{5}$ (d) $\frac{-1 \pm \sqrt{33}}{8}$ (e) $\frac{9 \pm \sqrt{73}}{4}$ (f) $\frac{1 \pm \sqrt{85}}{6}$

4. (a) $\frac{3 \pm \sqrt{37}}{2}$ (b) $\frac{5 \pm \sqrt{33}}{2}$ (c) $\frac{3 \pm \sqrt{33}}{2}$ (d) $\frac{7 \pm \sqrt{57}}{2}$ (e) $\frac{-7 \pm \sqrt{65}}{2}$ (f) –4, 2 (g) $-1 \pm 2\sqrt{2}$ (h) $\frac{-5 \pm \sqrt{53}}{2}$ (i) $\frac{3 \pm \sqrt{37}}{2}$ (j) no real solutions (k) $4 \pm \sqrt{7}$ (l) no real solutions (m) $\frac{2 \pm \sqrt{13}}{2}$ (n) $\frac{3 \pm 2\sqrt{11}}{5}$ (o) $\frac{6 \pm \sqrt{31}}{5}$ **5.** (a) $-2 < p < 2$ (b) $p = \pm 2$ (c) $p < -2$ or $p > 2$

6. (a) $m = 1$ (b) $m < 1$ (c) $m > 1$ **7.** (a) $m = \pm 2\sqrt{2}$ (b) $]-\infty, -2\sqrt{2}[\cup]2\sqrt{2}, \infty[$ (c) $]-2\sqrt{2}, 2\sqrt{2}[$ **8.** (a) $k = \pm 6\sqrt{2}$ (b) $]-\infty, -6\sqrt{2}[\cup]6\sqrt{2}, \infty[$ (c) $]-6\sqrt{2}, 6\sqrt{2}[$

EXERCISE 2.4.2

1. Graphs are shown using the ZOOM4 viewing window:

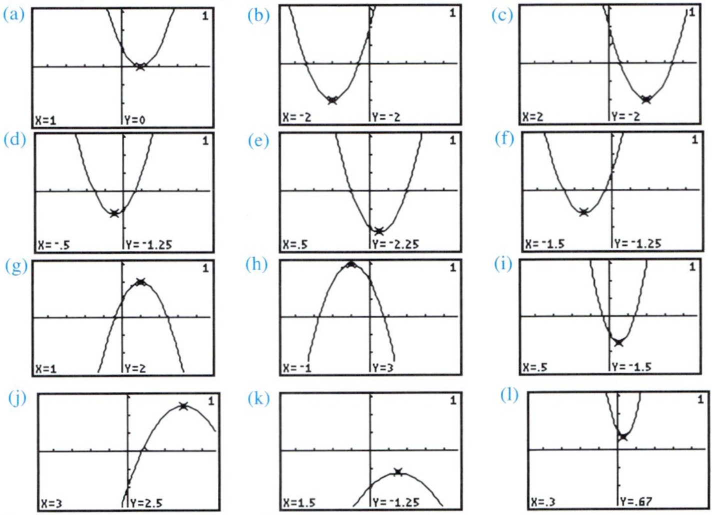

2. Graphs are shown using the ZOOM6 viewing window:

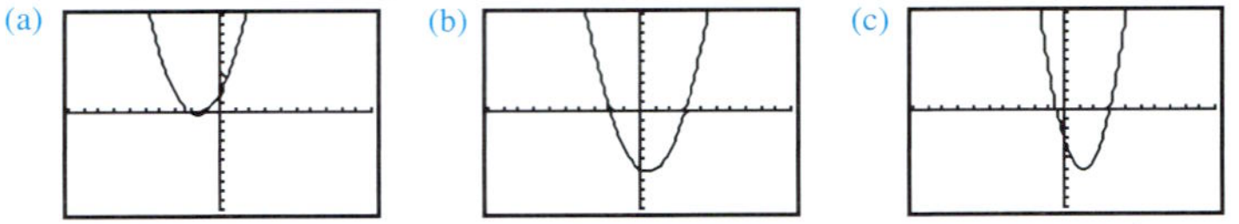

(a) (–2,0),(–1,0),(0,2) (b) (–2,0),(3,0),(0,–6) (c) (–0.5,0),(3,0),(0,–3)

(d)
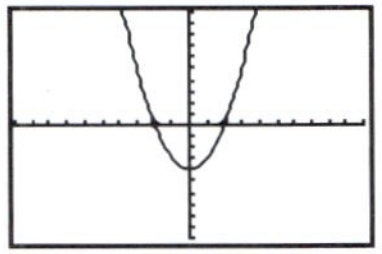
(–2,0),(2,0),(0,–4)

(e)
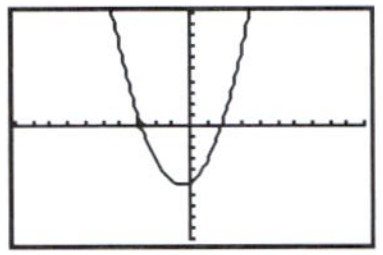
(–2.79,0),(1.79,0),(0,–5)

(f)
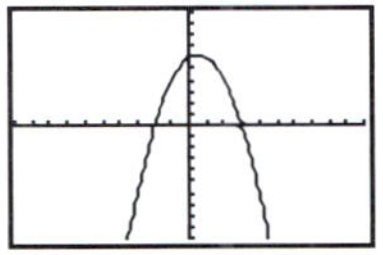
(–2,0),(3,0),(0,6)

(g)
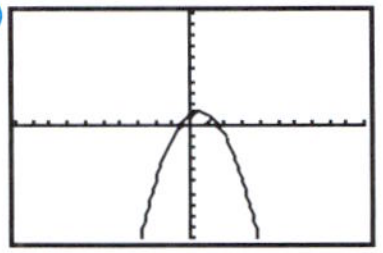
(–0.62,0),(1.62,0),(0,1)

(h)
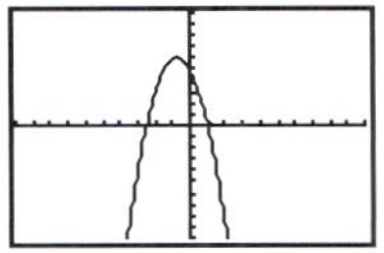
(–2.5,0),(1,0),(0,5)

(i)
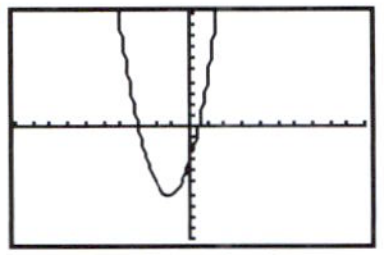
(–3,0),(0.5,0),(0,–3)

(j)
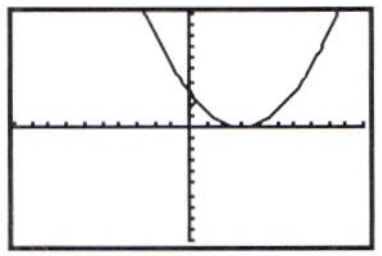
(3,0),(0,3)

(k)
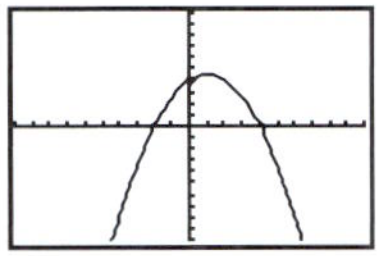
(–2,0),(4,0),(0,4)ß

(l)
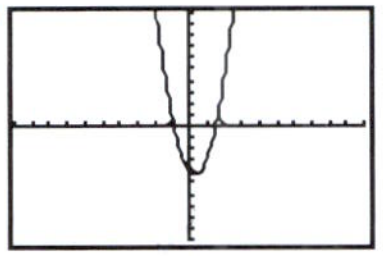
(–0.87,0),(1.54,0),(0,–4)

3. (a) $x = 1$ (b) (1,–1) (c) i. $\left(\frac{2 \pm \sqrt{2}}{2}, 0\right)$ ii. (0,1)
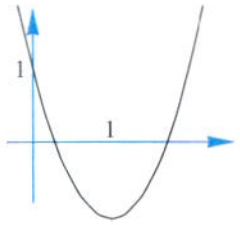

4. (a) $x = 1$ (b) (1,9)

(c) i. $\left(\frac{2 \pm 3\sqrt{2}}{2}, 0\right)$ ii. (0,7)
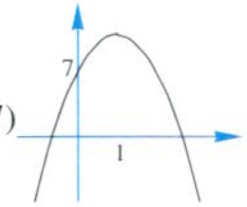

5. (a) $k = \frac{9}{4}$ (b) $k < \frac{9}{4}$ (c) $k > \frac{9}{4}$ **6.** (a) $k = \frac{25}{8}$

(b) $k < \frac{25}{8}$ (c) $k > \frac{25}{8}$ **7.** (a) $k = \pm 1$ (b) $-1 < k < 1$ (c) $k < -1 \cup k > 1$

8. (a) $y = \frac{5}{12}(x-2)(x-6)$ (b) $y = -\frac{3}{8}(x+4)^2$ (c) $y = \frac{3}{4}(x-2)^2 + 1$ (d) $y = 3x^2 - 6x + 7$

9. (a) $y = -\frac{2}{5}x(x-6)$ (b) $y = \frac{3}{4}(x-3)^2$ (c) $y = \frac{7}{9}(x+2)^2 + 3$ (d) $y = -\frac{7}{3}x^2 - 2x + \frac{40}{3}$

EXERCISE 2.4.3

1. (a)]–∞,–2[∪]1,∞[(b) [–3,2] (c)]–∞,0]∪[4,∞[(d) $]\frac{1}{3},3[$ (e)]–∞,–1.5]∪[–1,∞[(f)]0.75,2.5[

2. (a)]–∞,–2[∪]–1,∞[(b)]–2,3[(c)]–∞,–0.5]∪[3,∞[(d) [–2,2] (e) $]\frac{-1-\sqrt{21}}{2}, \frac{-1+\sqrt{21}}{2}[$

(f)]–∞,–2]∪[3,∞[(g) $[\frac{1-\sqrt{5}}{2}, \frac{1+\sqrt{5}}{2}]$ (h) [–2.5,1] (i)]–∞,–3[∪]0.5,∞[(j)]1,3[(k)]–1,0.5[(l) Ø

(m) Ø (n) [–1.5,5] (o) $]-\infty,-2[\cup]\frac{1}{3},\infty[$ **3.** (a) $-1 < k < 0$ (b) $-2\sqrt{2} < k < 2\sqrt{2}$ (c) $n \le -0.5$

4. (a) i.]–∞,–1[∪]2,∞[ii. [–1,2] (b) i.]–∞,2[∪]3,∞[ii. [2,3] (c) i.]1,3[ii.]–∞,1]∪[3,∞[

(d) i. $]-\frac{2}{3},1[$ ii. $]-\infty,-\frac{2}{3}]\cup[1,\infty[$ (e) i.]–∞,–2[∪]2,∞[ii. [–2,2] (f) i. $]2-\sqrt{3}, 2+\sqrt{3}[$

ii. $]-\infty, 2-\sqrt{3}]\cup[2+\sqrt{3},\infty[$ **5.**
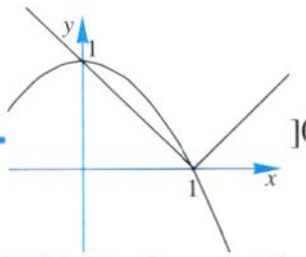

]0,1[**6.** [–2,0.5]

7. (a) $\{x: x < -3\}\cup\{x: x > 2\}$ (b) $\{x: -1 < x < 4\}$

EXERCISE 2.4.4

1. (a) (–2,–3) (2,5) (b) (–2,–1) (1,2) (c) $\left(-\frac{1}{3}, -2\right)$, (2, 5) (d) $\left(-\frac{3}{2}, -\frac{15}{4}\right)$, (1, 0)

(e) $\left(-\frac{9}{2}, -\frac{19}{4}\right)$, (1, –2) (f) $\left(\frac{3+\sqrt{73}}{4}, \frac{-3-\sqrt{73}}{8}\right), \left(\frac{3-\sqrt{73}}{4}, \frac{-3+\sqrt{73}}{8}\right)$

(g) $\left(\frac{1-\sqrt{13}}{2}, 1-\sqrt{13}\right), \left(\frac{1+\sqrt{13}}{2}, 1+\sqrt{13}\right)$ (h) no real solutions

(i) $\left(\frac{1-\sqrt{17}}{2}, \frac{5-3\sqrt{17}}{2}\right), \left(\frac{1+\sqrt{17}}{2}, \frac{5+3\sqrt{17}}{2}\right)$ (j) (–2,–3), (2,1) (k) no real solutions

2. (a) (1,4), (–7,84) (b) $\left(\frac{4}{3}, -\frac{56}{9}\right), \left(\frac{3}{4}, -\frac{7}{4}\right)$ (c) (0, 2), (3, 23) (d) $(-a, -a^2), \left(\frac{a}{2}, \frac{a^2}{2}\right)$ (e) Ø

(f) (2,8) (g) Ø (h) $\left(\frac{1}{2}, \frac{23}{4}\right)$ **3.** (a) $\pm 2\sqrt{6}$ (b) $m < -2\sqrt{6}, m > 2\sqrt{6}$ (c) $-2\sqrt{6} < m < 2\sqrt{6}$

4. $\sqrt{80}$ **5.** 1.75 **7.** $-\frac{23}{12}$ **8.** $c = \frac{a}{m}$ **10.** (a) i. (1,3), $\left(-\frac{14}{3}, \frac{196}{3}\right)$ ii. (–2, 12), $\left(\frac{7}{3}, \frac{49}{3}\right)$

(c) i. A(1,3), B(–2,2); $3y = x + 8$ ii. 4 sq. units

EXERCISE 3.1.1

1. 8 **2.** 4, 0.25 **3.** 8, 18 **4.** 8 and 11 or –8 and –11 **5.** 6, –10 **6.** 2 m **7.** 51 kmh^{-1} **8.** 11, 13; –11, –13 **9.** 25 days **10.** 30 **11.** (a) 30 (b) $50 each. **12.** 6 kmh^{-1} **13.** 16 **14.** 6 **15.** 3 hours **16.** 9 **17.** (a) 15 hrs (b) 10 hrs **18.** Chair-one: 20; Chair-two: 24 **19.** (a) 2 km (b) 2.5 km **20.** 7.5 hrs, 10.5hrs

EXERCISE 3.1.2

1. (a) i. $100 - 2x$ ii. $0 < x < 50$ [Nb: if $x = 0$ or 50, $A = 0$ and so there is no enclosure] (b) i. $A = 2x(50 - x)$, $0 < x < 50$ ii. 10m by 80m or 40m by 20m iii. 1250 m^2 iv. 25m by 50 m

2. (a) ii. $0 < x < 12$ (b) i. 20 m^2 ii. 32 m^2 iii. 32 m^2 (c)
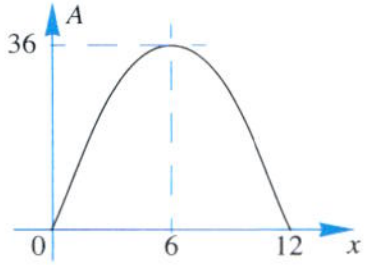

(d) 6 m by 6 m

3. (a)
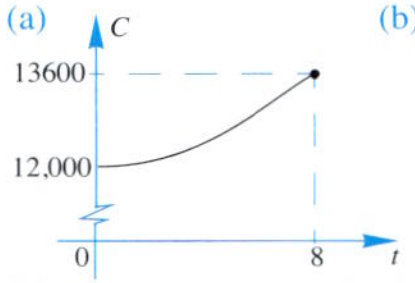

(b) $12,900

4. (a) $R(x) = xp = x(40 - 0.0004x)$, $0 \le x \le 100,000$ (b) i. $960,000 ii. 18377 or 81622 [as answer must be integer values] iii. $1,000,000

5. (a) $y = \frac{4}{3}(50 - x)$ (b) i. $A = \frac{8}{3}x(50 - x)$ ii. $0 < x < 50$ (c) i. $\frac{5000}{3}$ m^2

ii. $x = 25, y = \frac{200}{3}$; dimensions 50 m by $\frac{100}{3}$ m. **6.** 35.83 kmh^{-1} **7.** (a) 100%

(b) $t = 0.229$ (first time) then again at $t = 13.104$ (c) (d) i. 42.26% ii. 1.73 weeks

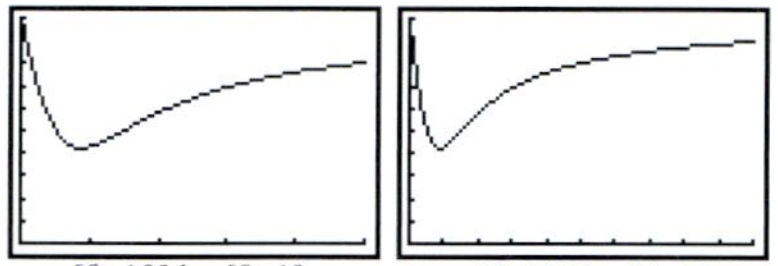

[0, 10] by [0, 1] [0, 20] by [0, 1]

(e) As time increases, oxygen level will be 100%

8. (a) (b) $95.26

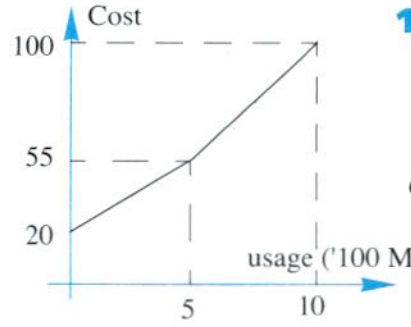

9. B(3254, 1953), C(6146,3687) units in metres

10. (a)

$$C = \begin{cases} 0.25, & 0 \le x < 50 \\ 0.65, & 50 \le x < 80 \\ 1.30, & 80 \le x < 160 \\ 1.95, & 160 \le x < 320 \\ 2.50, & 320 \le x < 500 \\ 3.40, & 500 \le x < 1000 \end{cases}$$

(b)

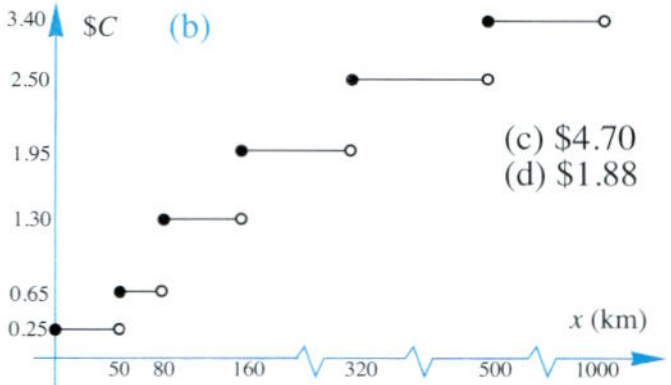

(c) $4.70
(d) $1.88

11. (a) $0 < x < 4$ (b) $A(x) = 3x + 0.25x^2, 0 < x < 4$ (c) i. ii. 3.25 iii. $-6 + \sqrt{48}$

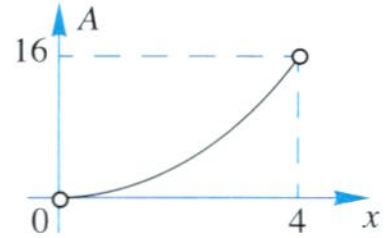

12. (a) i. 200 m ii. 320 m
(b) i. 0.34 sec and 11.66 sec ii. 11.31 sec (c) 12 sec
(d) 360 m

13. (a) 0.53 sec (on the way up) and 9.47 sec (on the way down)
(b) 10 sec (c) 500 m (d) 12.07 sec (e) 750 m

14. (a) $72500 (b) No. (Loss of $20000) (c) 2500

15. (a)

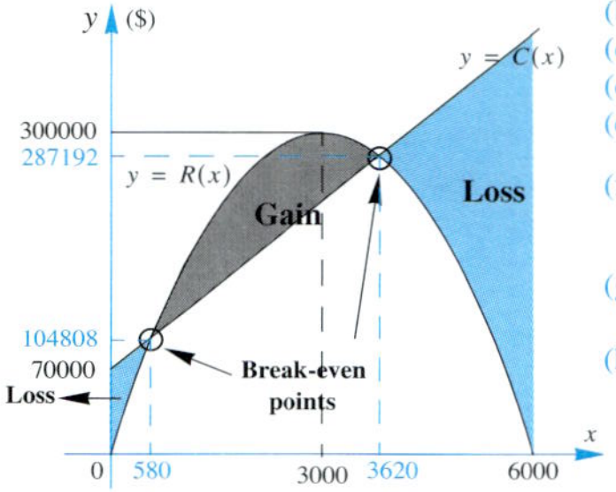

(b) i. $70,000 ii. $300,000
(c) Fixed cost (e.g., salary, electricity, ...)
(d) See graph in (a)
(e) $76667 (to nearest dollar)
(f) i. $P(x) = 140x - \frac{1}{30}x^2 - 70000, 0 \le x \le 6000$
ii. $77,000 iii. 2100
(g) i. $0 \le x \le 580$ or $3620 \le x \le 6000$
ii. $581 \le x \le 3619$
(h) See graph in (a)

16. (b) i. $P(x) = -\frac{1}{30}x^2 + 194x - 72000$ ii. $0 \le x \le 6000$

(d) The company will break-even at 399 radios and 5421 radios. Provided the company sells between 399 and 5421 radios they will make a profit. (e) 2910

10. (c)

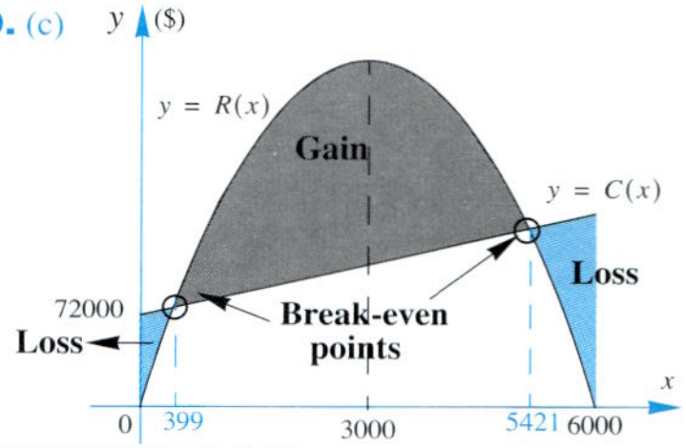

EXERCISE 3.1.3

1. i. (b) $y = 0.4x + 7.2$ ii. (b) $y = 6 - 2x$ iii. (b) $y = 0.5x + 3.2$ **2.** Second difference = 0.64

3. (b) $y = x^2 + 4x + 2$ **4.** (a) & (c)

(b) $y = 2x^2 - 3x + 7$ **5.** $y = 2x^2 - x + 3$ **6.**

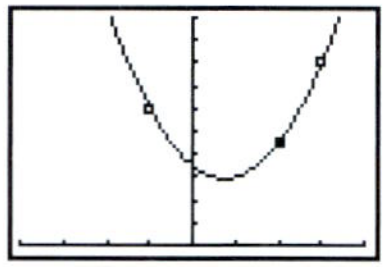

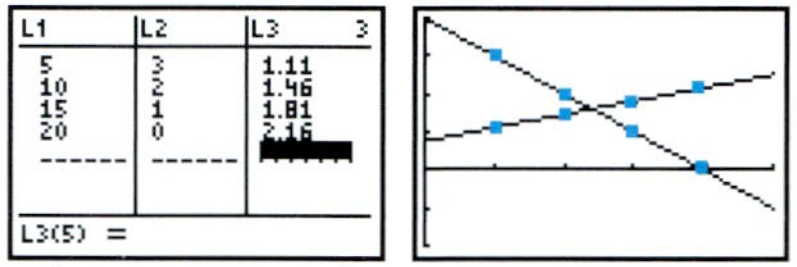

(c) i. $p = -0.2q + 4$ ii. $p = 0.07q + 0.76$
(d) Optimium scenario: demand = supply. This occurs when $p = 1.60, q = 12$.

7. (a) $y = -0.6333x^2 + 8.833x - 19.2$
(b) –40°C at 11 pm.
The model is not valid outside data range. Therefore extrapolation will not necessarily work.

8. Equation of path: $y = -\frac{31}{2400}x^2 + \frac{49}{48}x + 1$. Greatest height: 21.17 m.

9. (a) (b) Linear (c) i. ii. $M = 0, x = 1$. i.e., 1 m

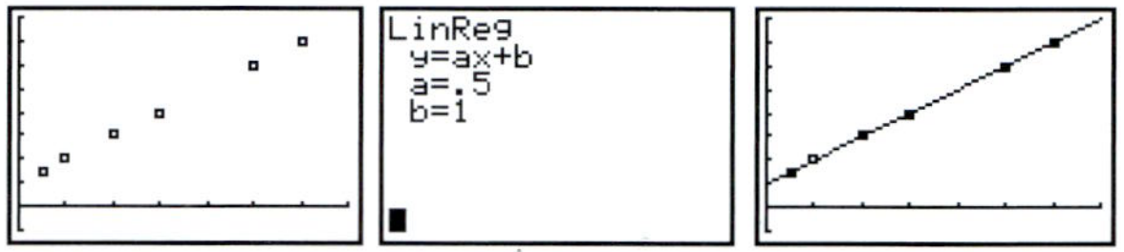

10. (a) (b) (c) i. parabola (d)

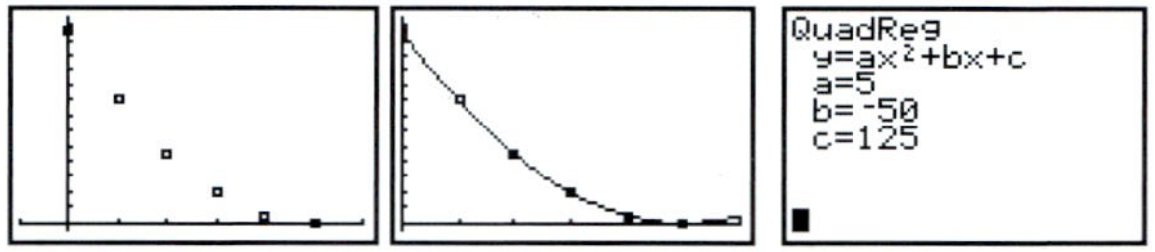

11. (a) $P(x) = -2x^2 + x + 3$ (b) $P(x) = (1-k)x^2 + x + k, x \in \mathbb{R}, k \neq 1$

12. (a) i. $a = 45000, b = 500000$ ii. $k_1 = 0, k_2 = 4$

(b)

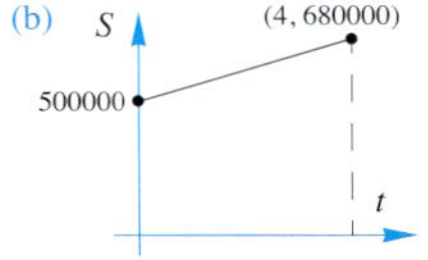

13. (a) (b) i. Parabolic

ii. $h(x) = -0.04694x^2 + 0.96518x + 1.7896$

(c) i. 6.75 m ii. 22.27 m

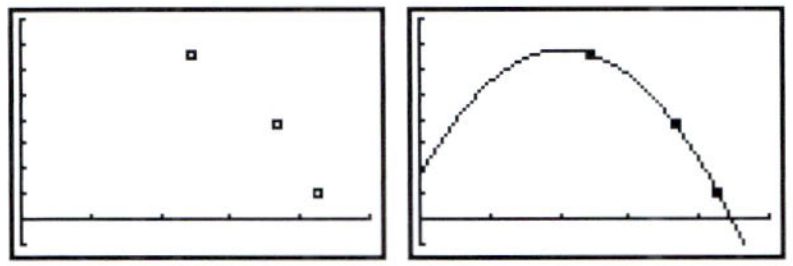

14. (a) $y = -\frac{2}{9}x^2 + \frac{11}{9}x + \frac{9}{2}, 0 \le x \le 5.5$ (b) [BE] : $y = -\frac{11}{9}x + \frac{121}{18}, 1 \le x \le 5.5$,

[BO] : $y = 5.5x, 0 \le x \le 1$ (c) $49°36'$

15. (a)

(b) second difference is constant = –50

(c) $y = -0.25x^2 + 25x + 580$ (d) \$22500 per car (e) i. \$824,750 ii. \$19750

16. (a) i. & ii. have a constant gradient. iii. results imply quadratic form (b) ii. $p = 10 - 0.001x$, $C(x) = 2x + 7000$,

$$R(x) = -0.001x^2 + 10x.$$

(c) $P(x) = -0.001x^2 + 8x - 7000$, max profit = $P(4000) = 9000$

EXERCISE 4.1.1

1. (a) $b^2 + 2bc + c^2$ (b) $a^3 + 3a^2g + 3ag^2 + g^3$ (c) $1 + 3y + 3y^2 + y^3$ (d) $16 + 32x + 24x^2 + 8x^3 + x^4$ (e) $8 + 24x + 24x^2 + 8x^3$ (f) $8x^3 - 48x^2 + 96x - 64$ (g) $16 + \frac{32}{7}x + \frac{24}{49}x^2 + \frac{8}{343}x^3 + \frac{1}{2401}x^4$ (h) $8x^3 - 60x^2 + 150x - 125$ (i) $27x^3 - 108x^2 + 144x - 64$ (j) $27x^3 - 243x^2 + 729x - 729$ (k) $8x^3 + 72x^2 + 216x + 216$ (l) $b^3 + 9b^2d + 27bd^2 + 27d^3$ (m) $81x^4 + 216x^3y + 216x^2y^2 + 96xy^3 + 16y^4$ (n) $x^5 + 15x^4y + 90x^3y^2 + 270x^2y^3 + 405xy^4 + 243y^5$ (o) $\frac{125}{p^3} + \frac{150}{p} + 60p + 8p^3$ (p) $\frac{16}{x^4} - \frac{32}{x} + 24x^2 - 8x^5 + x^8$ (q) $q^5 + \frac{10q^4}{p^3} + \frac{40q^3}{p^6} + \frac{80q^2}{p^9} + \frac{80q}{p^{12}} + \frac{32}{p^{15}}$ (r) $x^3 + 3x + \frac{3}{x} + \frac{1}{x^3}$

EXERCISE 4.1.2

1. (a) $160x^3$ (b) $21x^5y^2$ (c) $-448x^3$ (d) $-810x^4$ (e) $216p^4$ (f) $-20412p^2q^5$ (g) $-22680p$ **2.** (a) –1400000 (b) 6000 (c) 540 (d) –240 (e) 81648 (f) 40 **3.** 1.0406 0.0004% **4.** i. $64x^6 + 960x^5 + 6000x^4 + 20000x^3 + 37500x^2 + 37500x + 15625$ ii. 19750 iii. 20.6 iv. 0.1% **5.** 19 **6.** $-\frac{63}{8}$ **7.** $\frac{231}{16}$ **8.** $-\frac{130}{27}$ **9.** –20 **10.** $a = \pm 3$ **11.** $n = 5$ **12.** $n = 9$ **13.** (a) 0 (b) –59 **14.** $a = 3, n = 8$ **15.** $a = \pm 2, b = \pm 1$

EXERCISE 5.1

1. (a) dom = {2, 3, –2}, ran = {4, –9, 9} (b) dom = {1, 2, 3, 5, 7, 9}, ran = {2, 3, 4, 6, 8, 10} (c) dom = {0, 1}, ran = {1, 2} **2.** (a)]1, ∞[(b) [0, ∞[(c)]9, ∞[(d)]–∞, 1] (e) [–3, 3] (f)]–∞, ∞[(g)]–1, 0] (h) [0, 4] (i) [0, ∞[(j) [1, 5] (k)]0, 4[(l)]–∞, –1] ∪ [1, ∞[**3.** (a) $r = [-1, \infty[, d = [0, 2[$ (b) $r = \{y : y \ge 0\} \backslash \{4\}$, $\mathbb{R}$ (c) $r = [0, \infty[\backslash \{3\}, d = [-4, \infty[\backslash \{0\}$ (d) $r = [-2, 0[, d = [-1, 2[$ (e) $r =]-\infty, \infty[$ $d =]-\infty, -3] \cup [3, \infty[$ (f) $r = [-4, 4], d = [0, 8]$ **4.** (a) one to many (b) many to one (c) many to one (d) one to one (e) many to many (f) one to one **5.** (a) $\mathbb{R}\backslash\{-2\}$ (b)]–∞, 9[(c) [–4,4] (d)]–∞, –2] ∪ [2, ∞[(e) $\mathbb{R}\backslash\{0\}$ (f) $\mathbb{R}$ (g) $\mathbb{R}\backslash\{-1\}$ (h) [–a, ∞[(i) $[0, \infty[\backslash \{a^2\}$ (j)]–∞, –a] ∪ [a, ∞[(k) $\mathbb{R}$ (l) $\mathbb{R}\backslash\{-a^{-1}\}$ **6.** (a)]–∞,–a[(b)]0,ab] (c) $]-\infty, \frac{1}{4}a^3]$ (d) $[\frac{1}{4}a^3, \infty[$ (e) $\mathbb{R}\backslash\{a\}$ (f)]–∞,a[(g) [–a,∞[(h)]–∞, 0[

EXERCISE 5.2

Graphs with graphics calculator output have standard viewing window unless otherwise stated.

1. (a) 3, 5 (b) i. 2(x+a) + 3 ii. 2a (c) 3 **2.** (a) $0, \frac{10}{11}$ (b) $-\frac{5}{4}$ (c) $\left[0, \frac{10}{11}\right]$

3. (a) $-\frac{1}{2}x^2 - x + \frac{3}{2}, -\frac{1}{2}x^2 + x + \frac{3}{2}$ (b) $\pm\sqrt{2}$ (c) no solution

4. (a) $x = 0, 1$

(b)

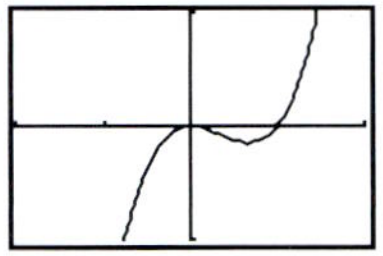

Window [–2,2], [–1,1]
Range: [–12, 4]

5. (a) i.

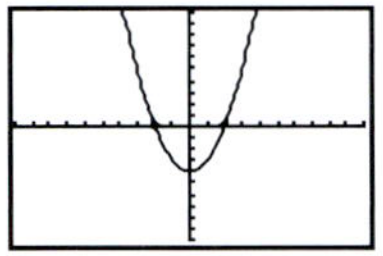

ii.

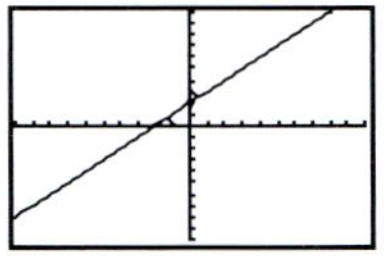

(b) i. $\{2\sqrt{2}, -2\sqrt{2}\}$ ii. {3, –2}

6. (b), (c), (d), (e) **8.** (a), (d), (e), (f)

9. (a)

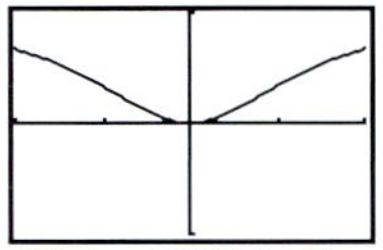

Window [-2,2], [-1,1]

(b) [0, 1[

10. (a) $\{y : y > 1\} \cup \{y : y \le -1.25\}$ (b) 10 **11.** (b) 1 **12.** (a) only - it is the only one with identical rules & domains.

13. (a) [–3, ∞[(b) [–3,0] (c) [3, ∞[(d) [1.5,3[∪]3,∞[

14. (a) i. $p(x) = 8 + 2\sqrt{16 - x^2}, 0 < x < 4$

ii. $A(x) = x\sqrt{16 - x^2}, 0 < x < 4$

(b) i.

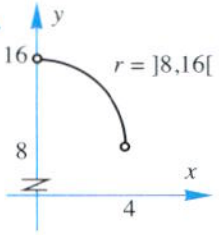

ii.

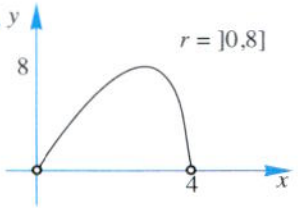

EXERCISE 5.3.1

1. (a)

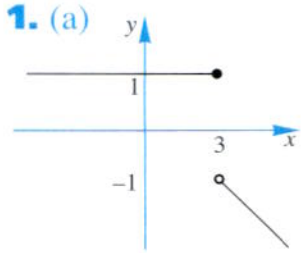

(b)

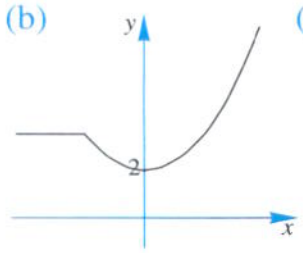

(c)

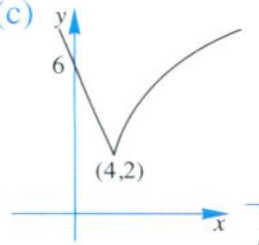

(d)

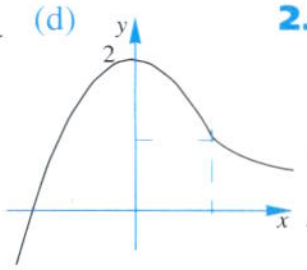

2. (a)

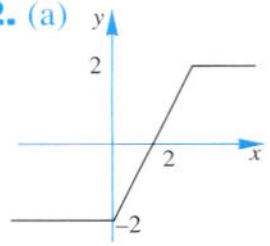

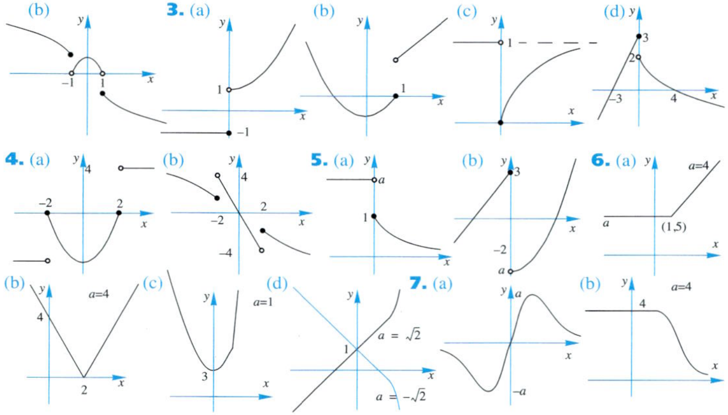

EXERCISE 5.3.2

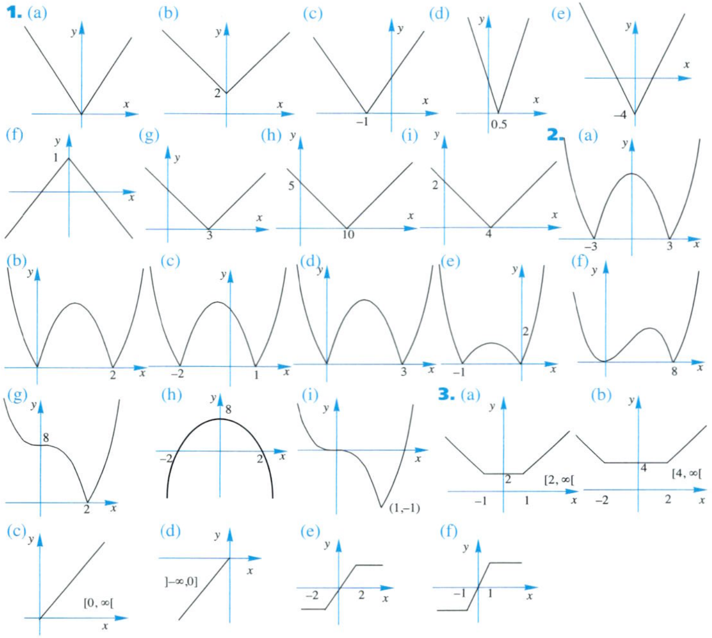

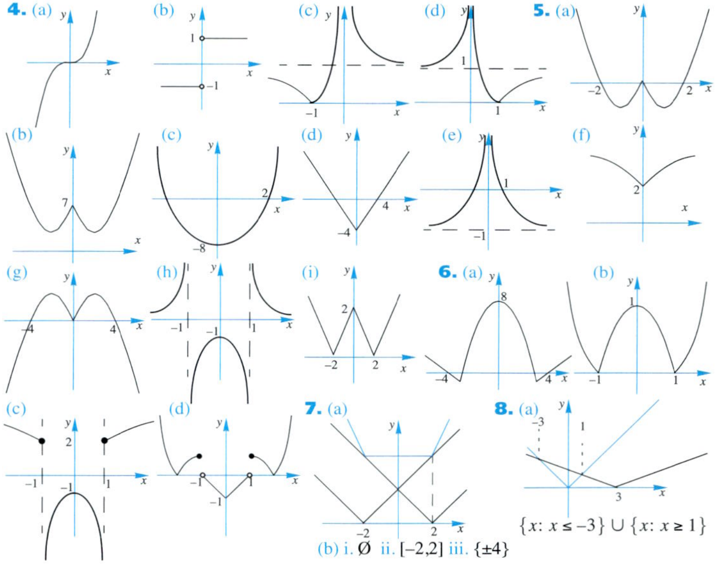

$\{x: x \leq -3\} \cup \{x: x \geq 1\}$

(b) i. Ø ii. [−2,2] iii. {±4}

EXERCISE 5.3.3

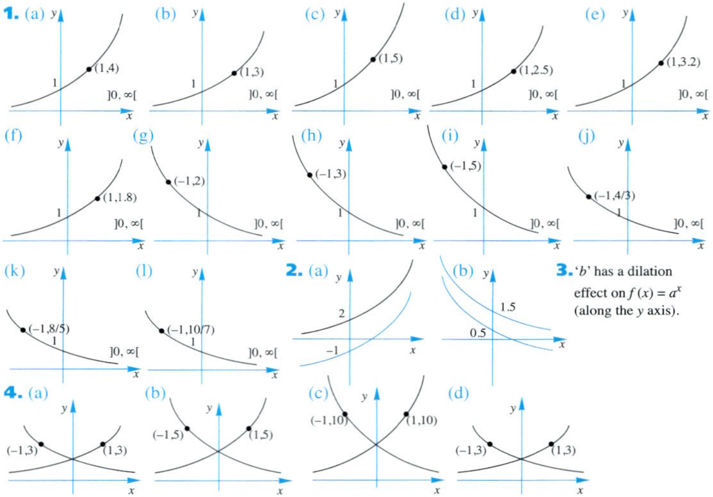

3. '*b*' has a dilation effect on $f(x) = a^x$ (along the *y* axis).

5. (a) [1, 16] (b) [3, 27] (c) [0.25, 16] (d) [0.5, 4] (e) [0.125, 0.25] (f) [0.1, 10]

6. (a) (b) (c) (d)
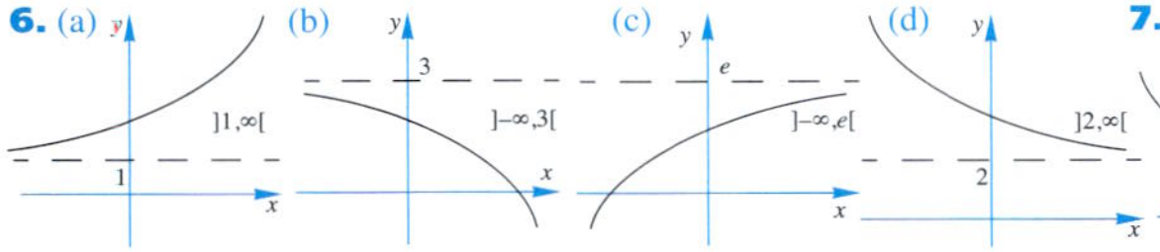

7. (a) −1.5 (b)
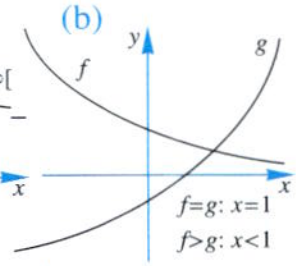

8. (a) $]2,2+e^{-1}[$ (b) $[-1,1[$ (c) $[1-e,1+e^{-1}]$

9. (a)
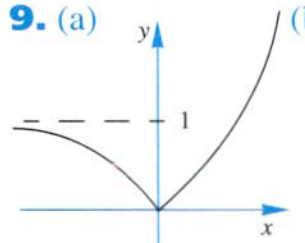
(b) 2

10. (a) (b) (c)
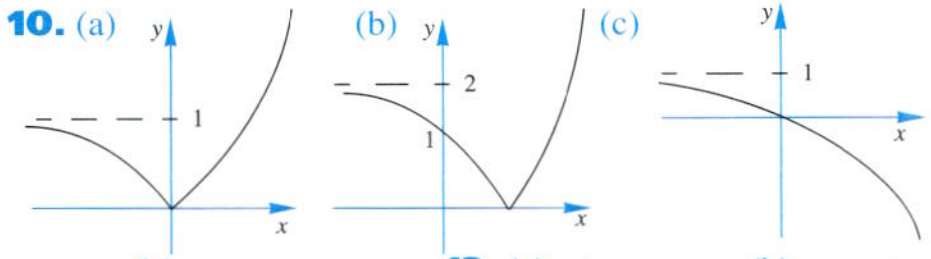

11. (a) (b) (c)
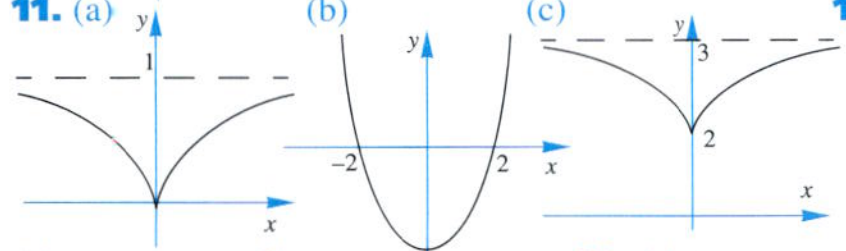

12. (a) (b)
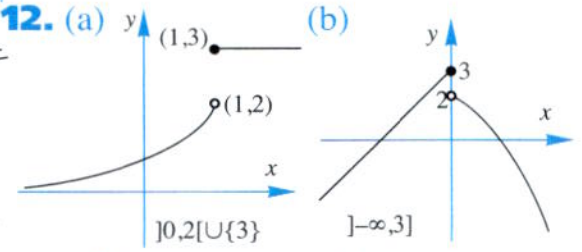

(c) (d)
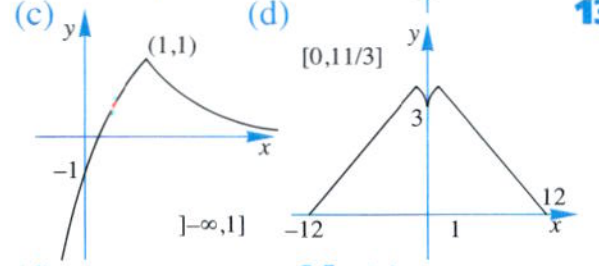

13. (a) (b) (c)
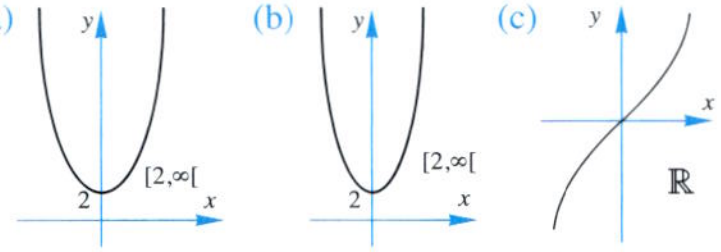

(d)
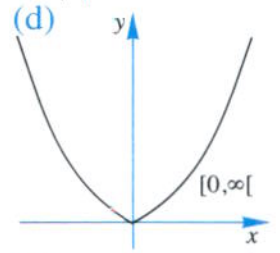

14. (a) (b) (c) (d)
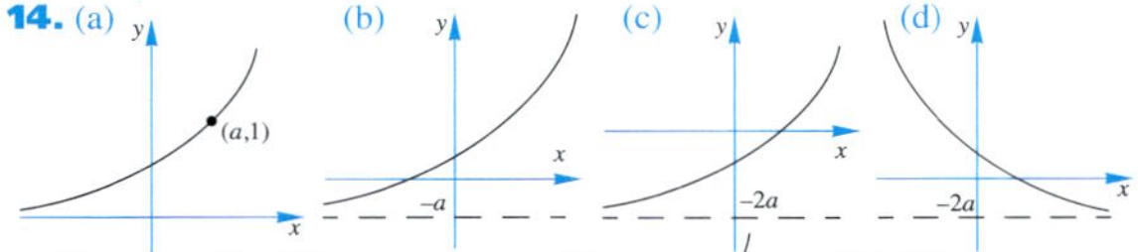

(e) (f)
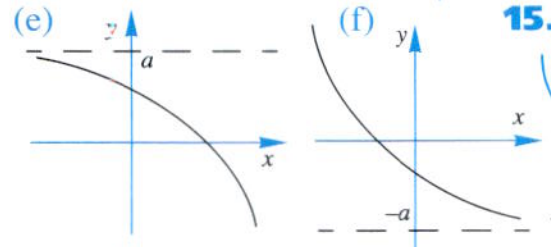

15. (a) (b)
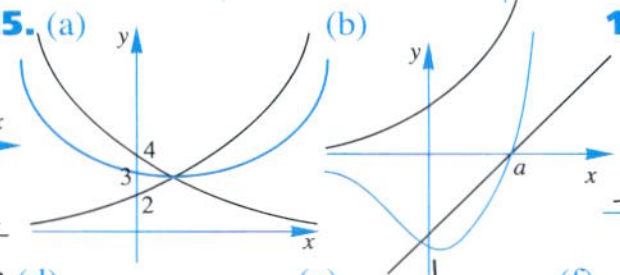

16. (a)
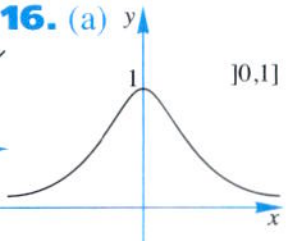

(b) (c) 1 < a < 2 (d) (e) (f)
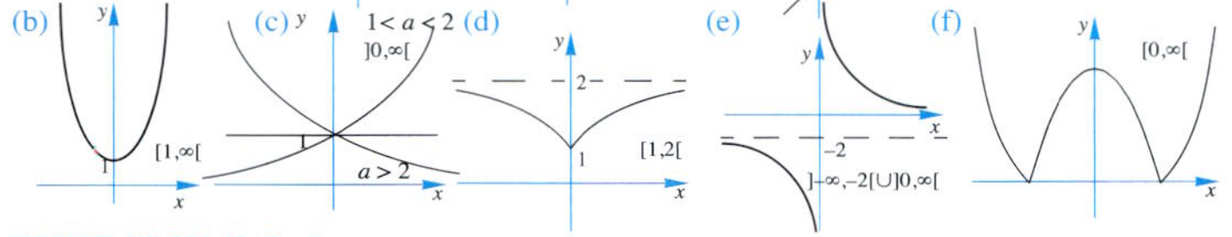

EXERCISE 5.3.4

1. (a) (b) (c) (d) (e)
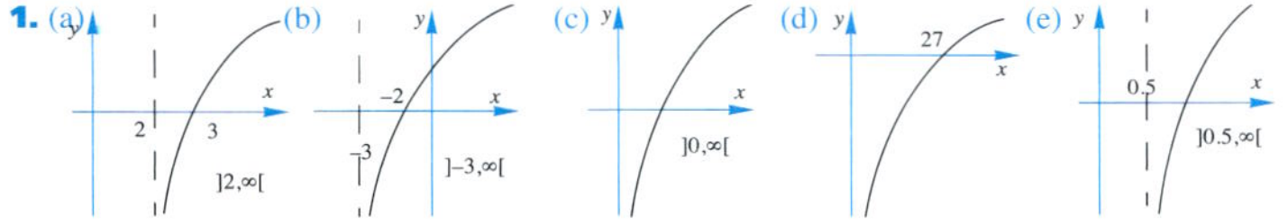

(f) (g) (h)
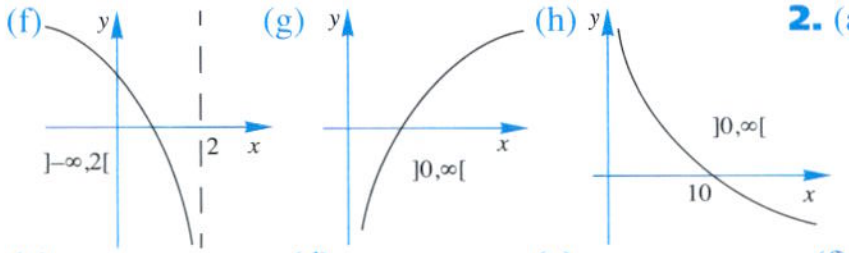

2. (a) (b)
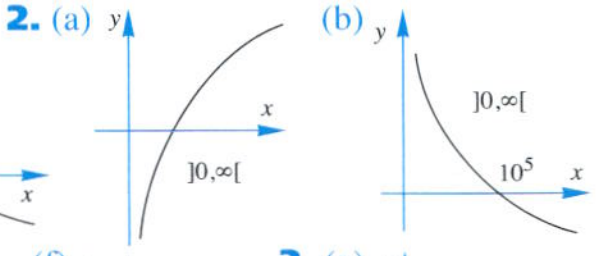

(c) (d) (e) (f)
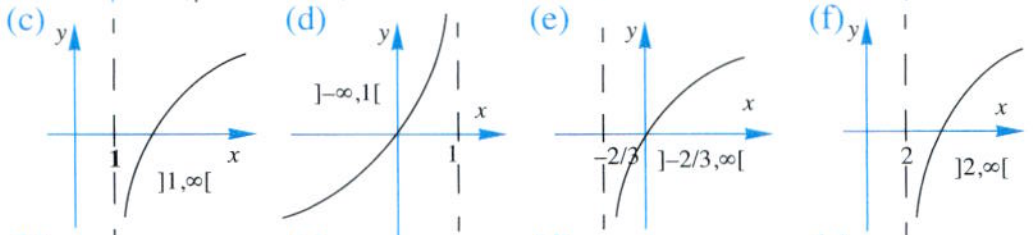

3. (a)
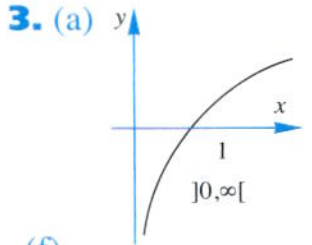

(b) (c) (d) (e) (f)
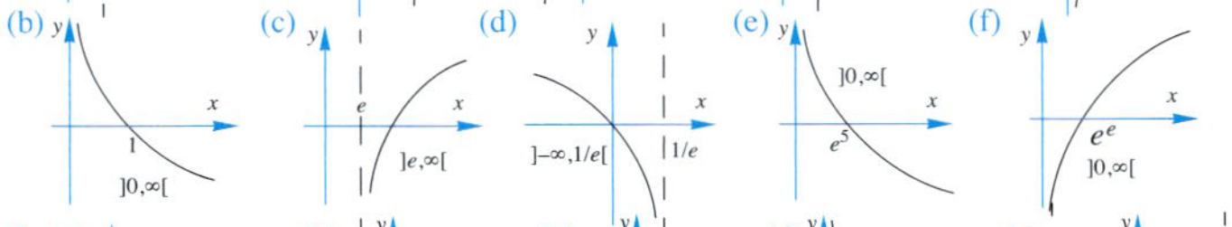

4. (a) (b) (c) (d) (e)
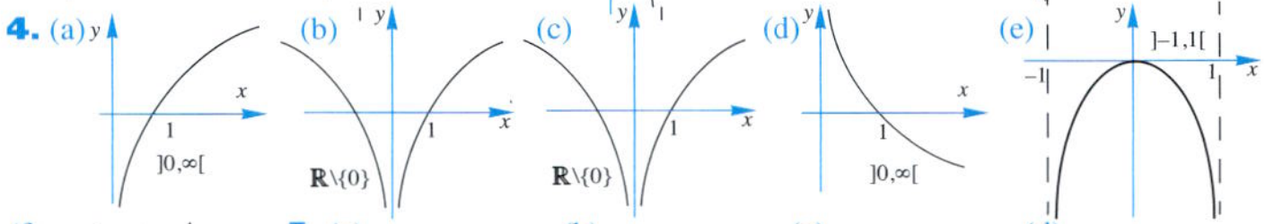

(f)
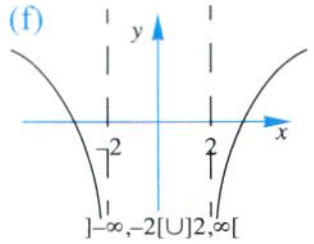

5. (a) (b) (c) (d)
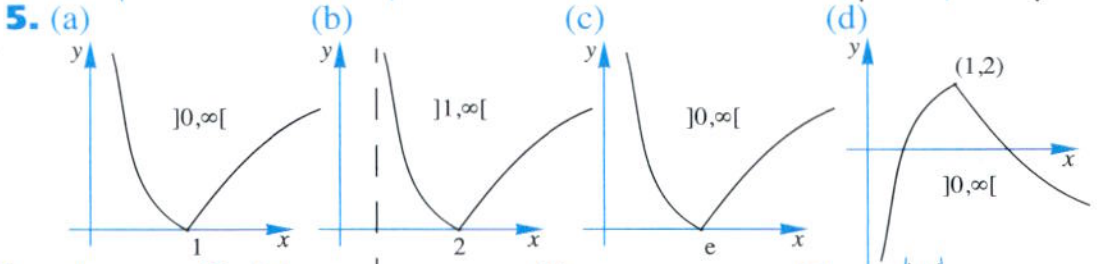

(e) (f)
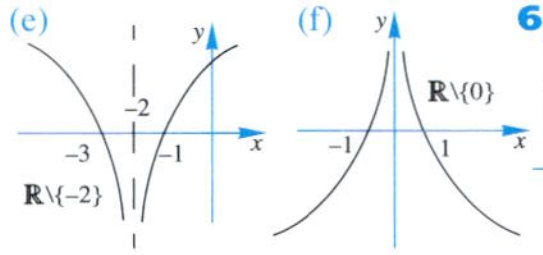

6. (a) (b) (c)
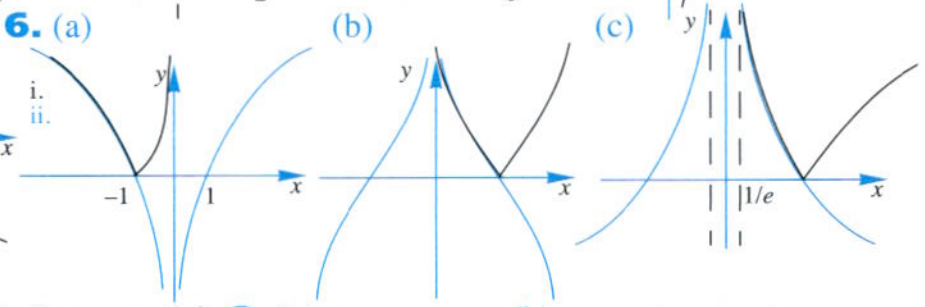

(d)
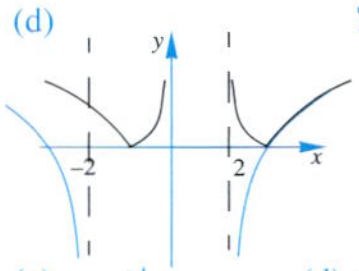

7. (a)
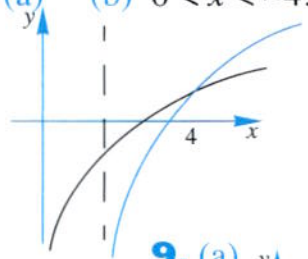
(b) 0 < x < ~4.3

8. (a) (b)
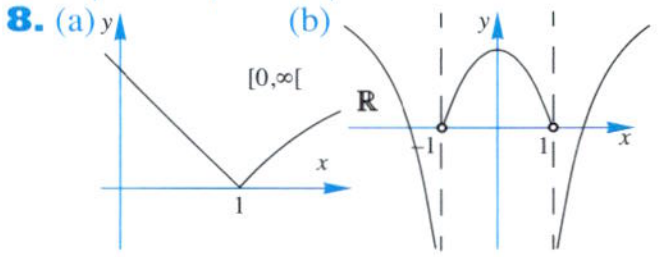

(c) (d)
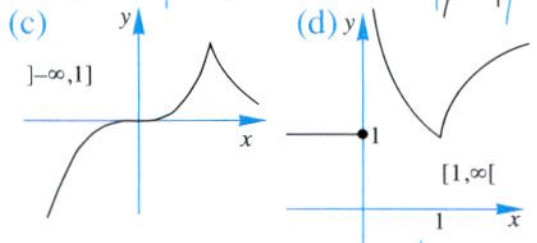

9. (a) (b) (c)
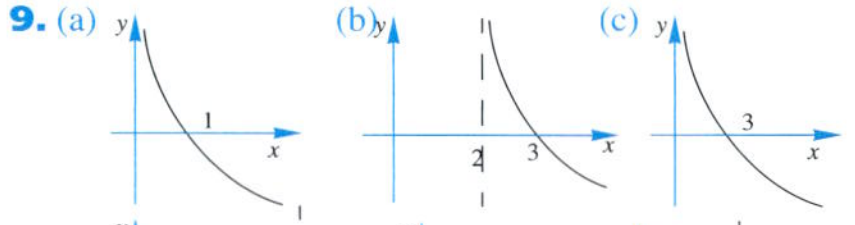

10. (a) (b) (c) (d) (e)
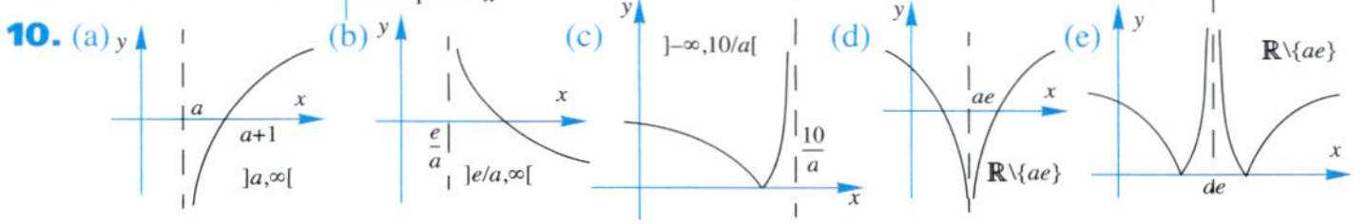

(f)

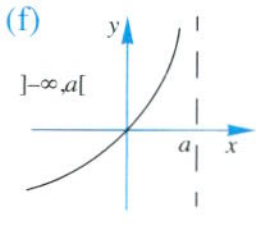

11. (a)

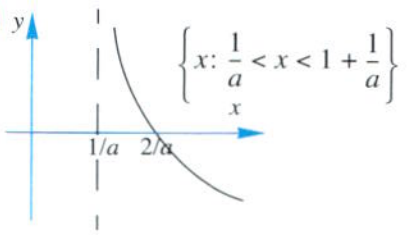

12.

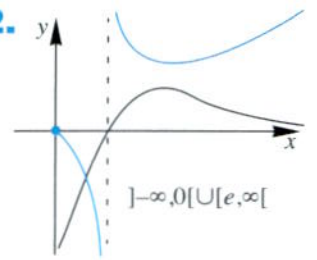

EXERCISE 5.3.5

1. (a) (b) (c) (d) (e)

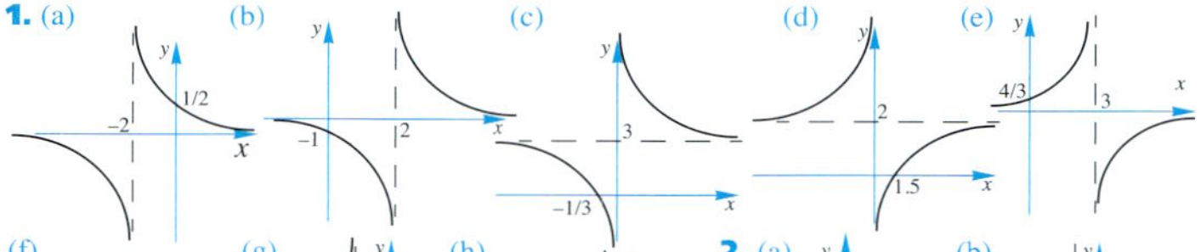

(f) (g) (h)

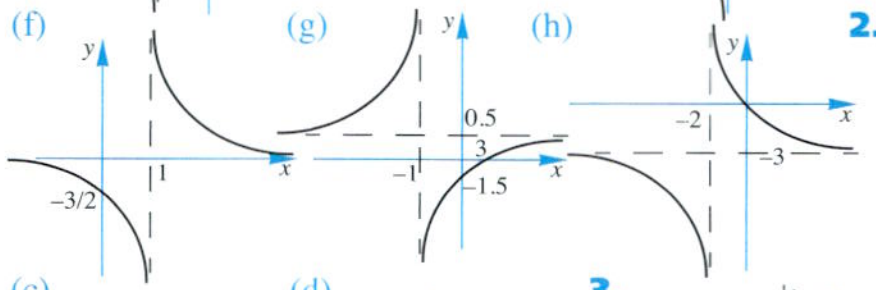

2. (a) (b)

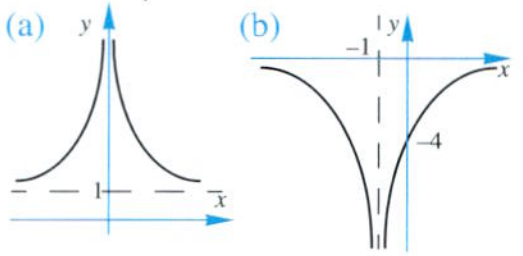

(c) (d)

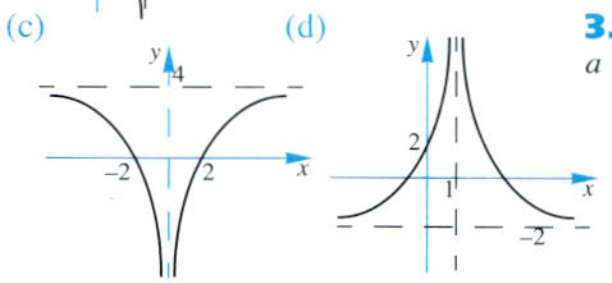

3. $a = 2, b = 1$

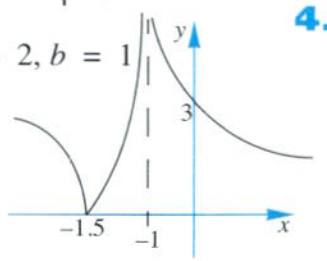

4. (a) i. ii.

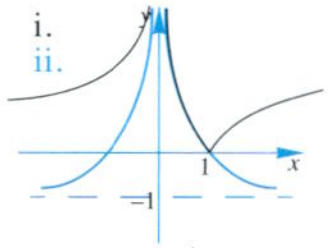

(b) (c) (d) (e) (f)

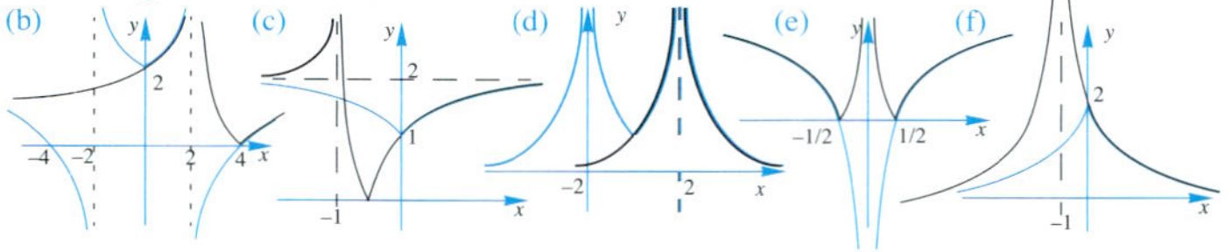

EXERCISE 5.4.1

1. (a) i. $f + g: [0, \infty[\mapsto \mathbb{R}$ where $(f + g)(x) = x^2 + \sqrt{x}$ $[0,\infty[$

ii. $f + g:]0, \infty[\mapsto \mathbb{R}$ where $(f + g)(x) = \frac{1}{x} + \ln(x)$ $]-\infty,\infty[$

iii. $f + g: [-3, -2] \cup [2, 3] \mapsto \mathbb{R}$ where $(f + g)(x) = \sqrt{9 - x^2} + \sqrt{x^2 - 4}$, $[\sqrt{5}, \sqrt{10}]$

(b) i. $fg: [0, \infty[\mapsto \mathbb{R}$ where $(fg)(x) = x^2\sqrt{x} = x^{5/2}$

ii. $fg:]0, \infty[\mapsto \mathbb{R}$ where $(fg)(x) = \frac{\ln(x)}{x}$

iii. $fg: [-3, -2] \cup [2, 3] \mapsto \mathbb{R}$ where $(fg)(x) = \sqrt{(9 - x^2)(x^2 - 4)}$

2. (a) i. $f - g:]-\infty, \infty[\mapsto \mathbb{R}$ where $(f - g)(x) = 2e^x - 1$, $]-1,\infty[$

ii. $f - g:]-1, \infty[\mapsto \mathbb{R}$ where $(f - g)(x) = (x + 1) - \sqrt{x + 1}$, $]-0.25,\infty[$

iii. $f - g:]-\infty, \infty[\mapsto \mathbb{R}$ where $(f - g)(x) = |x - 2| - |x + 2|$, $[-4,4]$

(b) i. $f/g: \mathbb{R}\setminus\{0\}, \mapsto \mathbb{R}$ where $(f/g)(x) = \frac{e^x}{1 - e^x}$

ii. $f/g:]-1, \infty[\mapsto \mathbb{R}$ where $(f/g)(x) = \sqrt{x + 1}$

iii. $f/g: \mathbb{R}\setminus\{-2\} \mapsto \mathbb{R}$ where $(f/g)(x) = \left|\frac{x - 2}{x + 2}\right|$

3. i. (a) $fog(x) = x^3 + 1$, $gof(x) = (x + 1)^3$ (b) $]-\infty, \infty[$, $]-\infty, \infty[$

ii. (a) $fog(x) = x + 1, x \ge 0$, $gof(x) = \sqrt{x^2 + 1}$ (b) $[1, \infty[$, $[1, \infty[$

iii. (a) $fog(x) = x^2$, $gof(x) = (x + 2)^2 - 2$ (b) $[0, \infty[$, $[-2, \infty[$

iv. (a) $fog(x) = x, x \ne 0$, $gof(x) = x, x \ne 0$ (b) $\mathbb{R}\setminus\{0\}$, $\mathbb{R}\setminus\{0\}$

v. (a) $fog(x) = x, x \ge 0$, $gof(x) = |x|$ (b) $[0, \infty[$, $[0, \infty[$

vi. (a) $fog(x) = \frac{1}{x^2} - 1, x \ne 0$, $gof(x)$ does not exist (b) $]-1, \infty[$

vii (a) $fog(x) = x^2, x \ne 0$, $gof(x) = x^2, x \ne 0$ (b) $]0, \infty[$, $]0, \infty[$

viii. (a) $fog(x) = |x| - 4$, $gof(x) = |x - 4|$ (b) $[-4, \infty[$, $[0, \infty[$

ix. (a) $fog(x) = |x + 2|^3 - 2$, $gof(x) = |x^3|$ (b) $[-2, \infty[$, $[0, \infty[$

x. (a) $fog(x)$ does not exist, $gof(x) = (4 - x), x \le 4$ (b) $[0, \infty[$

xi. (a) $fog(x) = \frac{x^2}{x^2 + 1}$, $gof(x) = \left(\frac{x}{x + 1}\right)^2, x \ne -1$ (b) $[0, 1[$, $[0, \infty[$

xii. (a) $fog(x) = x^2 + |x| + 1$, $gof(x) = |x^2 + x + 1|$ (b) $[1, \infty[$, $[0.75, \infty[$

xiii. (a) $fog(x) = 2^{x^2}$, $gof(x) = 2^{2x}$ (b) $[1, \infty[$, $]0, \infty[$

xiv. (a) $fog(x)$ does not exist, $gof(x) = \frac{1}{x + 1} - 1, x \ne -1$ (b) $\mathbb{R}\setminus\{-1\}$

xv. (a) $fog(x)$ does not exist, $gof(x) = \frac{4}{x - 1} + 1$ (b) $]1, \infty[$

xvi. (a) $fog(x) = 4^{\sqrt{x}}, x \ge 0$, $gof(x) = 4^{0.5x}$ (b) $[1, \infty[$, $]0, \infty[$

4. (a) $fog(x) = 2x + 3, x \in \mathbb{R}$ (b) $gof(x) = 2x + 2, x \in \mathbb{R}$ (c) $fof(x) = 4x + 3, x \in \mathbb{R}$

5. $g(x) = x^2 + 1, x \in \mathbb{R}$ **6.** (a) $fog(x) = \frac{1}{x} + x + 1, x \in \mathbb{R}\setminus\{0\}$,$]-\infty,-1] \cup [3,\infty[$

(b) $gof(x)$ does not exist. (c) $gog(x) = x + \frac{1}{x} + \frac{x}{x^2 + 1}, x \ne 0$, $]-\infty,-2.5] \cup [2.5,\infty[$

7. (a) 9 (b) 3 **9.** (a) $x = \pm 1$ (b) $x = 1, -3$ **10.** (a) $\frac{1}{x}$ (b) $\frac{-x}{2x + 1}$

11.

$$hof(x) = \begin{cases} (x - 1)^2 + 4, & x \ge 2 \\ 5 - x, & x < 2 \end{cases}$$

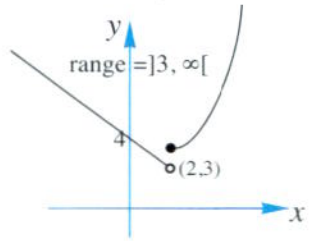

12. (a) $r_f \subseteq d_g$ and $r_{gof} \subseteq d_h$ (b) $g(x) = 4(x + 1)^2, x \in \mathbb{R}$

13. (a) $fog(x) = x, x \in]0,\infty[$ range $=]0,\infty[$

(b) $gof(x) = \frac{1}{2}(\ln(e^{2x - 1}) + 1), x \in \mathbb{R}$ $(= x)$ range $=]-\infty,\infty[$

(c) $fof(x) = e^{2(e^{2x - 1}) - 1}, x \in \mathbb{R}$ range $=]e^{-1},\infty[$

14. (a) hok does not exist.

(b) $koh(x) = 4\log(4x - 1) - 1, x > \frac{1}{4}$, $\mathbb{R}$

15. (a) $S = \mathbb{R}\setminus]-3,3[$; $T = \mathbb{R}$ (b) $T = \{x : |x| \ge 6, x = 0\}$; $S =]-\infty,-3] \cup [3,\infty[$

16. gof does not exist

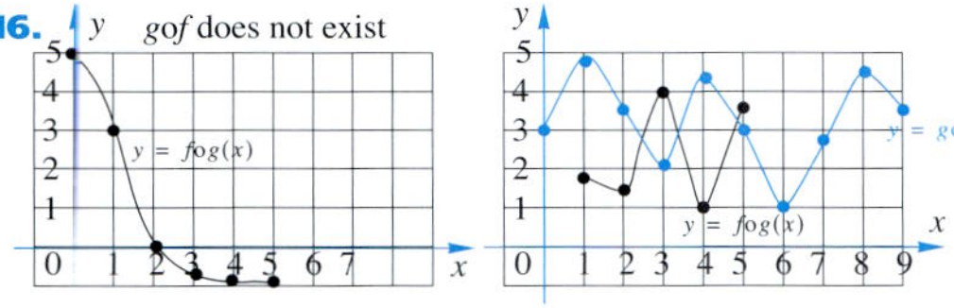

17. (a) Dom $f =]0,\infty[$, ran $f =]e,\infty[$, Dom $g =]0,\infty[$, ran $g = \mathbb{R}$

(b) fog does not exist: $r_g = \mathbb{R} \not\subset d_f =]0,\infty[$

gof exists as $r_f =]e, \infty[\subseteq d_g =]0,\infty[$

(c) gof: $]0,\infty[\mapsto \mathbb{R}$, where $gof(x) = (x+1) + \ln 2$

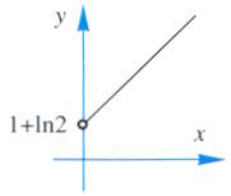

18. $(fog)(x) = |x|, x \in \mathbb{R}$; range $=[0, \infty[$

19. (a) (c)

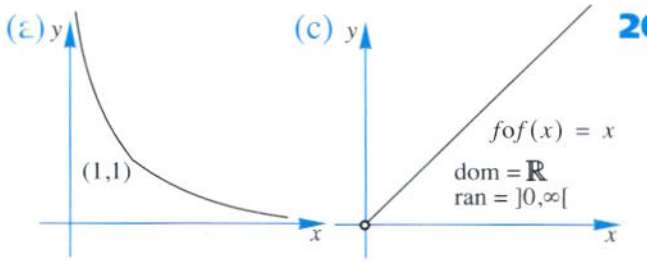

20. (a)

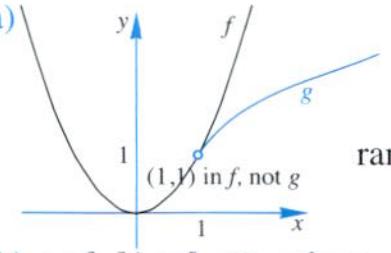

range $=]1,\infty[$

(b) gof: $]1,\infty[\mapsto \mathbb{R}$, where $gof(x) = x$

(d) fog^*: $]1,\infty[\mapsto \mathbb{R}$, where $gof(x) = x$

21. $d_f = \mathbb{R}\setminus\left\{\frac{a}{c}\right\}, r_f = \mathbb{R}\setminus\left\{\frac{a}{c}\right\}, r_f \subseteq d_f, fof(x) = x$

22. (a)

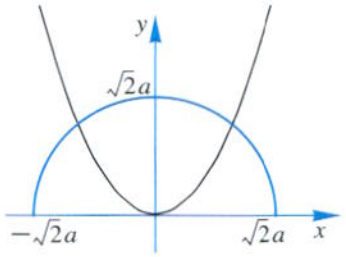

(b) $d_{fog} = [-\sqrt{2}a, \sqrt{2}a]$, $fog = 2a - \dfrac{x^2}{a}$

(c) $d_{gof} = [-2^{1/4}a, 2^{1/4}a]$, $fog = \dfrac{1}{a}\sqrt{2a^4 - x^4}$,

range $= [0, \sqrt{2}a]$

EXERCISE 5.4.2

1. (a) $\frac{1}{2}(x-1), x \in \mathbb{R}$ (b) $\sqrt[3]{x}, x \in \mathbb{R}$ (c) $3(x+3), x \in \mathbb{R}$ (d) $\frac{5}{2}(x-2), x \in \mathbb{R}$ (e) $x^2 - 1, x > 0$

(f) $(x-1)^2, x \geq 1$ (g) $\frac{1}{x} - 1, x > 0$ (h) $\dfrac{1}{(x+1)^2}, x > -1$

2. (a) (b) (c) (d) (e) (f) (g) (h)

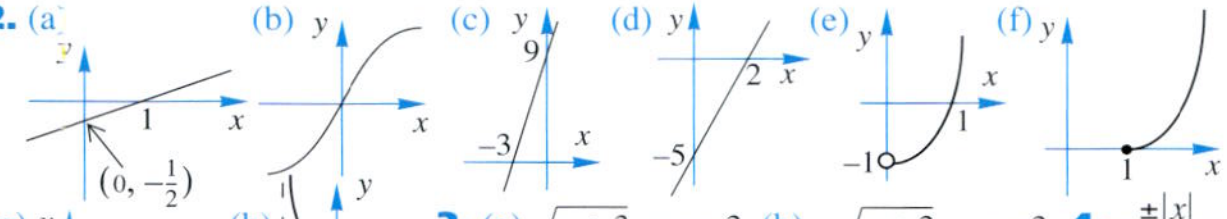

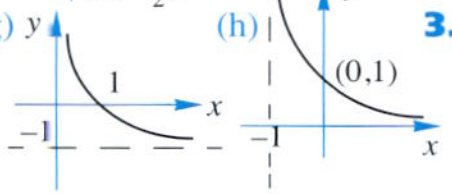

3. (a) $\sqrt{x+3}, x \geq -3$ (b) $-\sqrt{x+3}, x \geq -3$

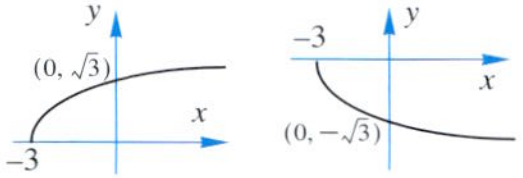

4. $\dfrac{\pm|x|}{\sqrt{1-x^2}}, -1 < x < 1$

5. (a) (b) (c) (d) (e) (f) (g) (h)

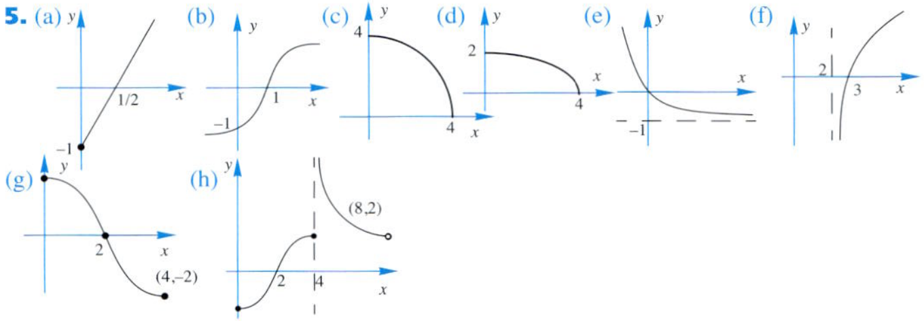

6. (a) $f^{-1}(x) = \log_3(x-1), x > 1$ (b) $f^{-1}(x) = \log_2(x+5), x > -5$

(c) $f^{-1}(x) = \frac{1}{2}(\log_3 x - 1), x > 0$ (d) $g^{-1}(x) = 1 + \log_{10}(3-x), x < 3$

(e) $h^{-1}(x) = \log_3\left(1 + \frac{2}{x}\right), x \in \mathbb{R}\setminus[-2,0]$ (f) $g^{-1}(x) = \log_2\left(\frac{1}{x+1}\right), x > -1$

7. (a) (b) (c) (d) (e) (f)

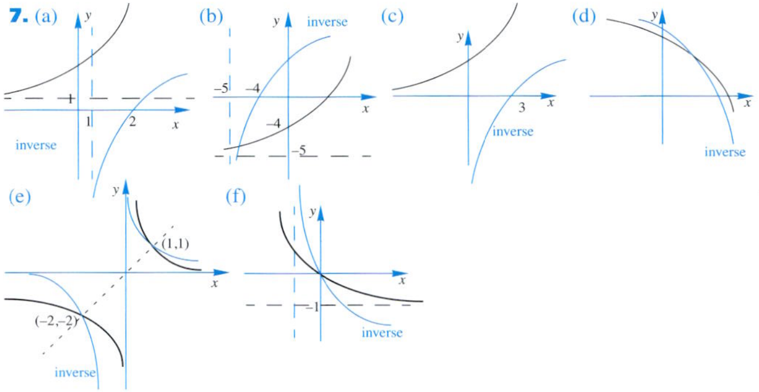

8. (a) $f^{-1}(x) = 2^x - 1, x \in \mathbb{R}$ (b) $f^{-1}(x) = \frac{1}{2} \cdot 10^x, x \in \mathbb{R}$ (c) $h^{-1}(x) = 2^{1-x}, x \in \mathbb{R}$

(d) $g^{-1}(x) = 3^{x+1} + 1, x \in \mathbb{R}$ (e) $h^{-1}(x) = 5^{x/2} + 5, x \in \mathbb{R}$ (f) $f^{-1}(x) = 1 - 10^{3(2-x)}, x \in \mathbb{R}$

9. $f^{-1}(x) = -1 + \sqrt{x+1}, x \geq -1$

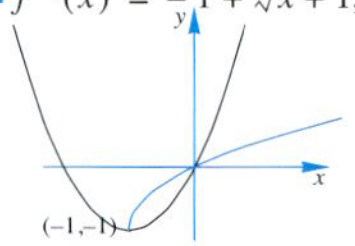

dom $= [-1, \infty[$, ran $= [-1, \infty[$

10. (a) $f^{-1}(x) = a - x$ (b) $h^{-1}(x) = \dfrac{2}{x-a} + a$

(c) $f^{-1}(x) = \sqrt{a^2 - x^2}$

11. $h^{-1}(x) = \sqrt[3]{2-x}$

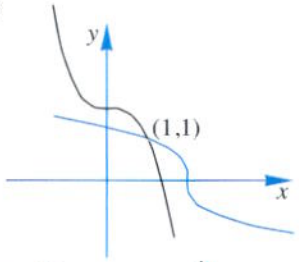

12. $[2, \infty[$ **13.** $\mathbb{R}^+\setminus\{1.5\}$

14. (a) Inverse exists as f is one:one

(b) **Case 1:** S = $]0, \infty[$

$g^{-1}(x) = \dfrac{x + \sqrt{x^2+4}}{2}$

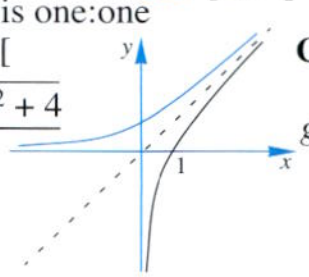

Case 2: S = $]-\infty, 0[$

$g^{-1}(x) = \dfrac{x - \sqrt{x^2+4}}{2}$

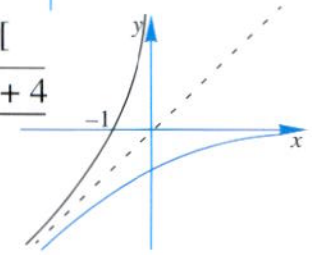

15. $f^{-1}(x) = a(x^2+1), x \geq 0$

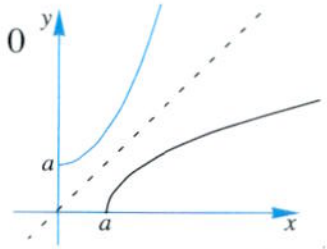

$\{x: f(x) = f^{-1}(x)\} = \varnothing$

16. (a)

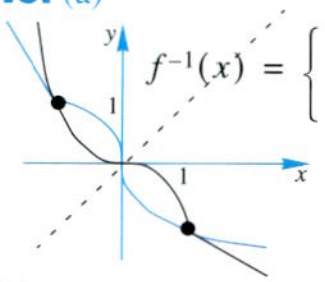

$f^{-1}(x) = \begin{cases} -(2x+1), & x < -1 \\ -\sqrt[3]{x}, & x \geq -1 \end{cases}$

(b)

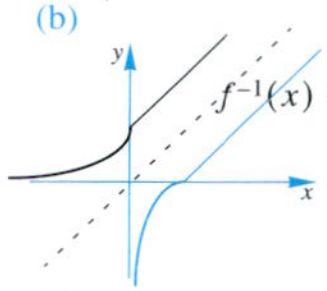

$f^{-1}(x) = \begin{cases} \ln(x) - 1, & 0 < x \leq e \\ x - e, & x > e \end{cases}$

(c)

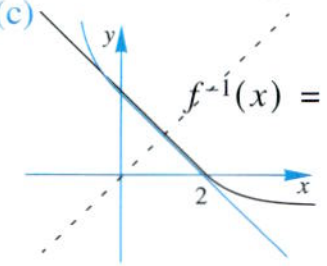

$f^{-1}(x) = \begin{cases} 1 + e^{-x}, & x < 0 \\ 2 - x, & x \geq 0 \end{cases}$

(d)

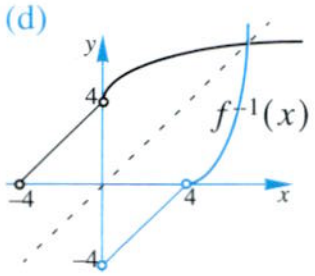

$f^{-1}(x) = \begin{cases} (x-4)^2, & x > 4 \\ x - 4, & 0 < x < 4 \end{cases}$

17. (a)

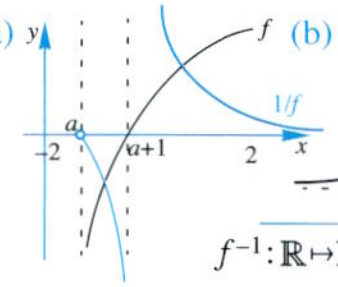

(b)

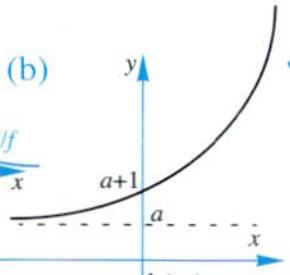

$f^{-1}: \mathbb{R} \mapsto \mathbb{R}$, $f^{-1}(x) = a + e^{ax}$

18. gof exists as $r_f \subseteq d_g$. It is one:one so the inverse exists:

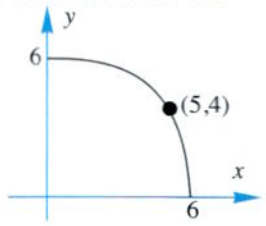

19. (a)

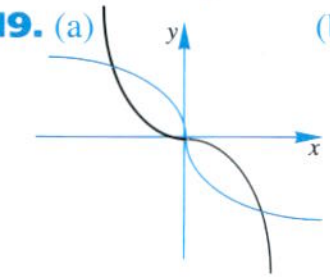

(b) i.

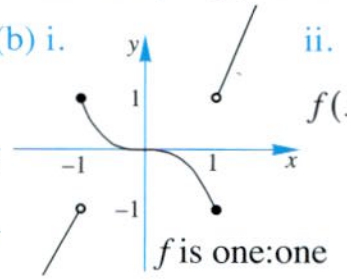

f is one:one

ii. $f(x) = \begin{cases} \frac{1}{2}(x-1) & x < -1 \\ -\sqrt[3]{x} & -1 \leq x \leq 1 \\ \frac{1}{2}(x+1) & x > 1 \end{cases}$

iii.

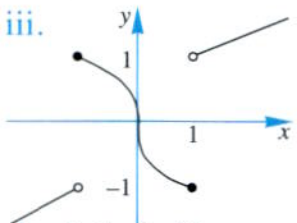

iv. $\{-1, 0, 1\}$

20. (a) i. $tom(x) = e^{\sqrt{x}}, x \geq 0$ ii. $mot(x) = \sqrt{e^x}, x \in \mathbb{R}$ (b) i. $(tom)^{-1}(x) = (\ln(x))^2, x > 1$
ii. $(mot)^{-1}(x) = \ln x^2, x > 0$ (c) i. & ii. neither exist
(d) Adjusting domains so that the functions in (c) exist, we have: $t^{-1}om^{-1}(x) = (mot)^{-1}(x)$ & $m^{-1}ot^{-1}(x) = (tom)^{-1}(x)$
(e) Yes as rules of composition OK.

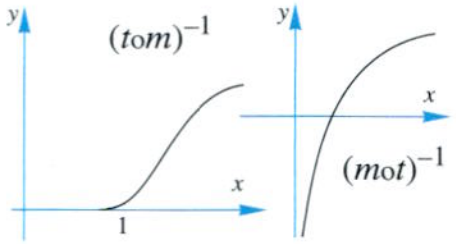

21. (a) 1 (b)

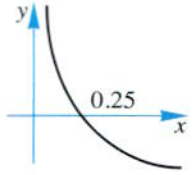

0.206 **22.** (a)

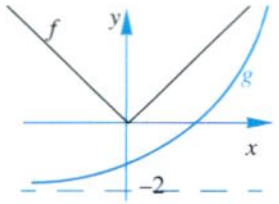

(b) fog exists but is not one:one
(c) i. B = [ln2, ∞[
ii. $(fog)^{-1}: [0, \infty[\mapsto \mathbb{R}$ where, $(fog)^{-1}(x) = \ln(x+2)$ iii.

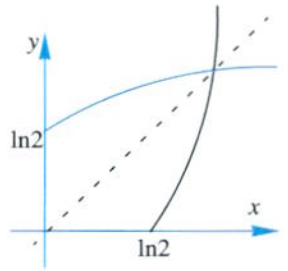

EXERCISE 6.1

1. (a) $y = (x-4)^2$ (b) $y = (x+2)^2$ (c) $y = x^2+5$ (d) $(x-2)^2 + y = 2$ (e) $x^2 + y = 4$
(f) $x^2 + y = 0$ (g) $y = \frac{8}{x-4}, x \neq 4$ (h) $y = \frac{8}{x} - 1, x \neq 0$ (i) $(x+1)^2 + y^2 = 4$
(j) $y^2 = \frac{9}{x-3}, x \neq 3$ (k) $(y+3)^2 = \frac{9}{x}, x \neq 0$ (l) $x + y^2 = 8$

2. (a) i (b) i (a) ii (b) ii (a) iii (b) iii (a) iv (b) iv

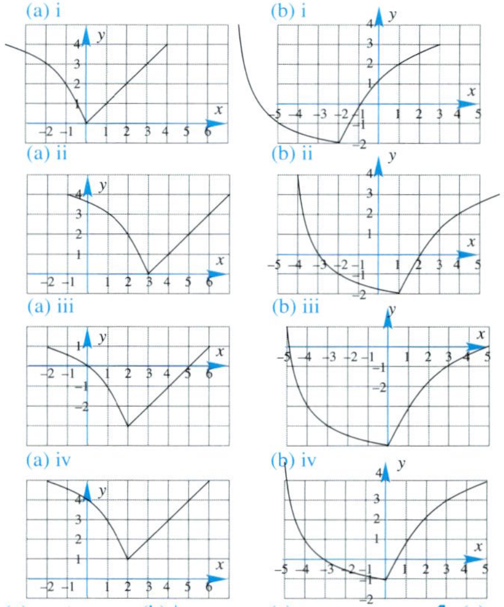

3. (a) (b) (c)

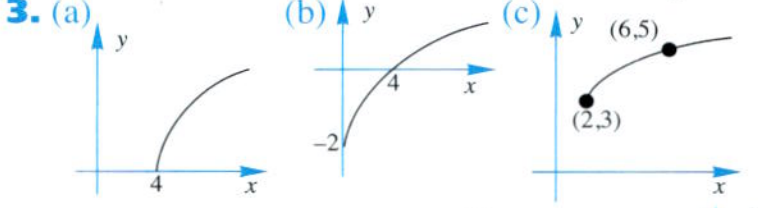

4. (a) (b)

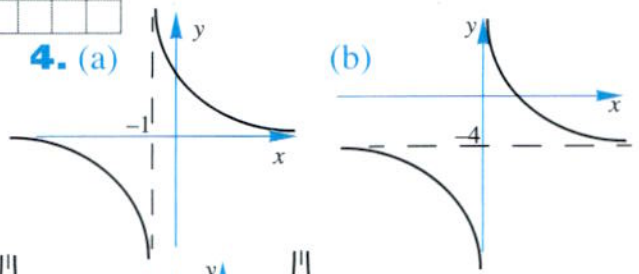

(c)

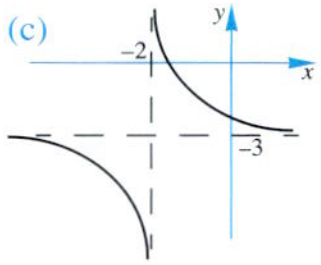

5. (a) (b) (c)

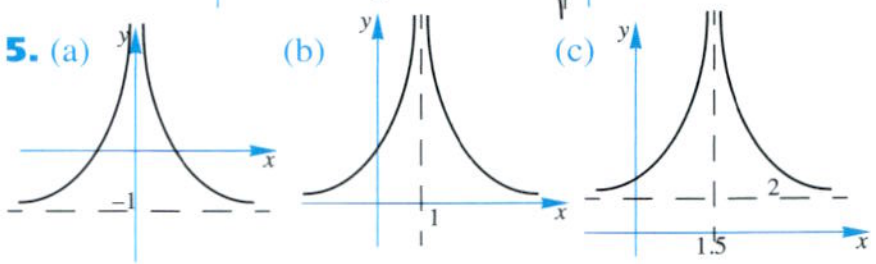

6. First function in black, second function in blue
(a) (b) (c) (d)

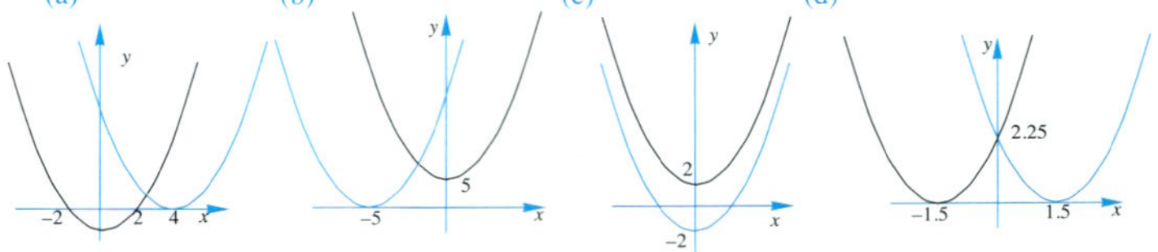

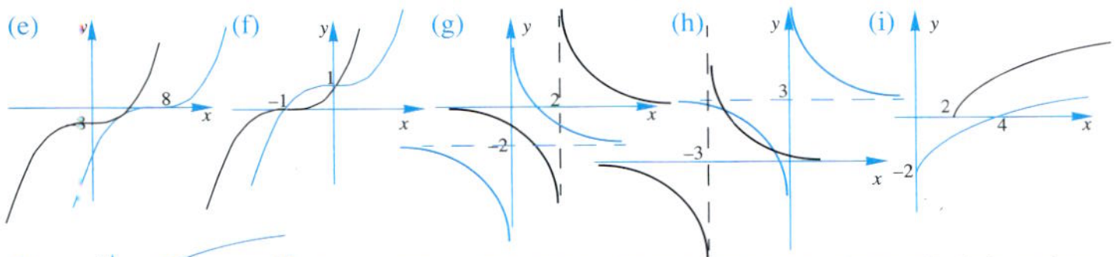

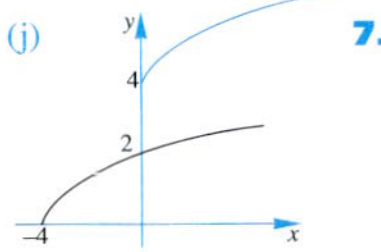

7. Note: coordinates were asked for. We have labelled most of these with single numbers.

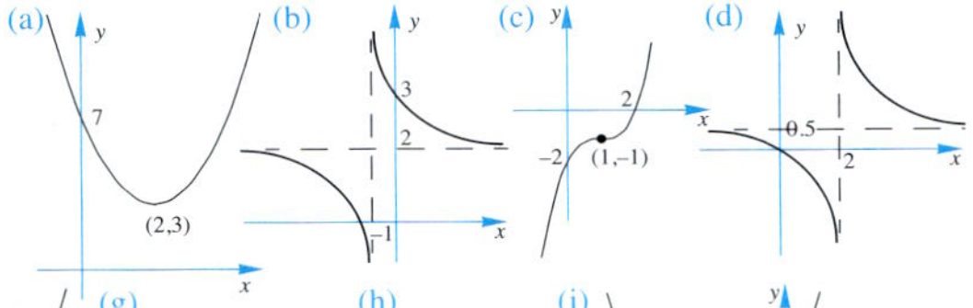

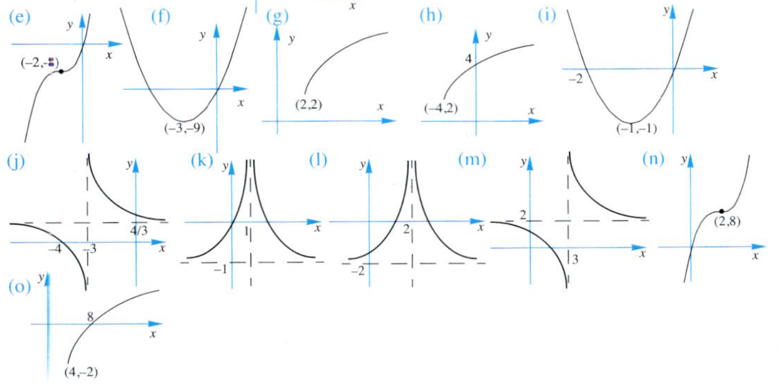

8. (a) $\begin{pmatrix} 0 \\ 4 \end{pmatrix}$ (b) $\begin{pmatrix} 0 \\ -2 \end{pmatrix}$ (c) $\begin{pmatrix} -1 \\ 0 \end{pmatrix}$ (d) $\begin{pmatrix} 2 \\ 0 \end{pmatrix}$ (e) $\begin{pmatrix} -2 \\ 0 \end{pmatrix}$ (f) $\begin{pmatrix} 0 \\ -4 \end{pmatrix}$ (g) $\begin{pmatrix} 2 \\ -2 \end{pmatrix}$ (h) $\begin{pmatrix} -2 \\ 3 \end{pmatrix}$

(i) $\begin{pmatrix} 4 \\ 2 \end{pmatrix}$ (j) $\begin{pmatrix} 2 \\ 3 \end{pmatrix}$ (k) $\begin{pmatrix} 3 \\ -1 \end{pmatrix}$ (l) $\begin{pmatrix} -k \\ h \end{pmatrix}$ (m) $\begin{pmatrix} -2 \\ 4 \end{pmatrix}$ (n) $\begin{pmatrix} 1 \\ -1 \end{pmatrix}$ (o) $\begin{pmatrix} -1 \\ 2 \end{pmatrix}$

9. (a) $g(x) = f(x-1)+1$ (b) $g(x) = f(x+2)-4$ (c) $g(x) = f(x-2)$
(d) $g(x) = f(x-1)+1$ (e) $g(x) = f(x-1)+3$

10.

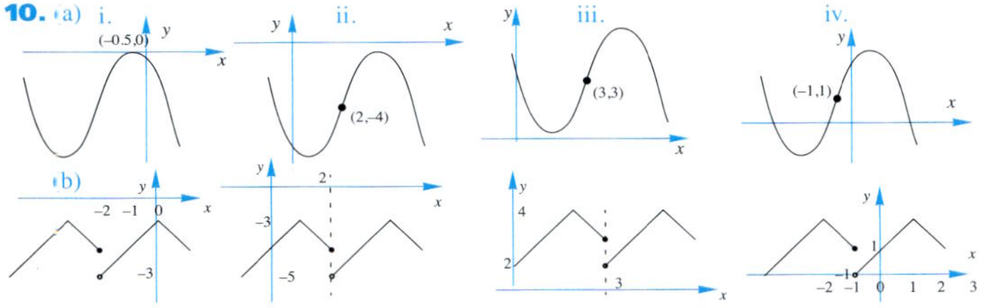

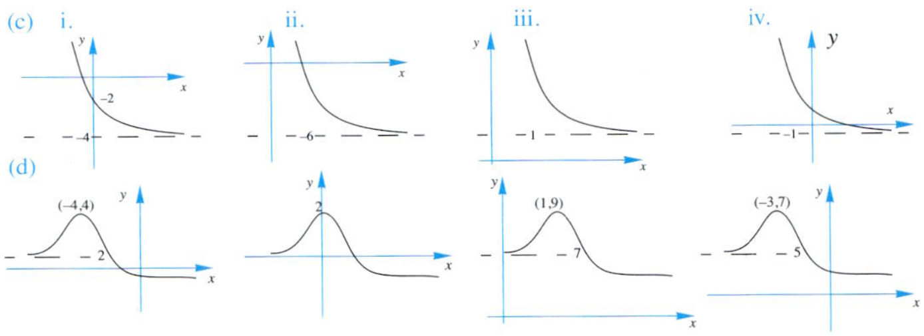

11. $y = \begin{cases} f(x+2)+2, & -3 \le x \le -1 \\ f(x+4)+2, & -5 \le x \le -3 \end{cases}$

EXERCISE 6.2

1.

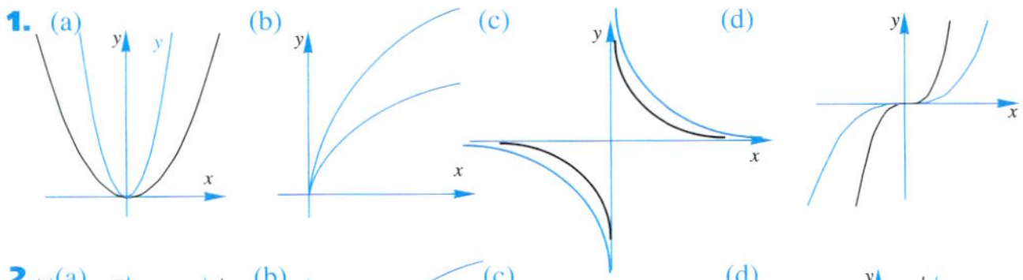

2.

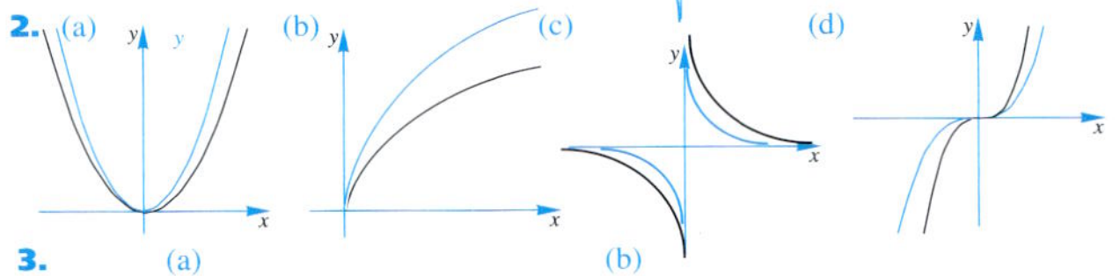

3.

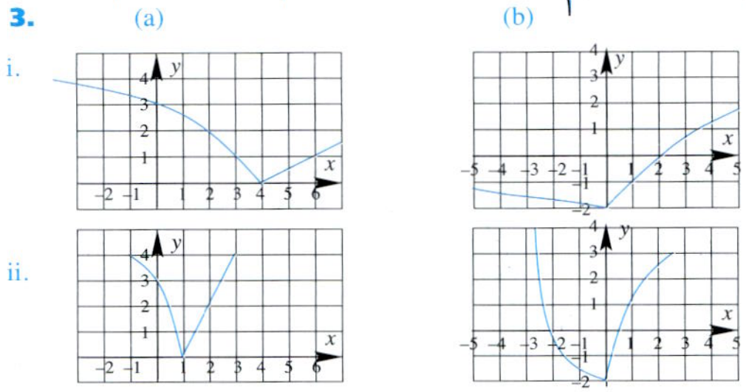

iii. (a) (b)

iv.

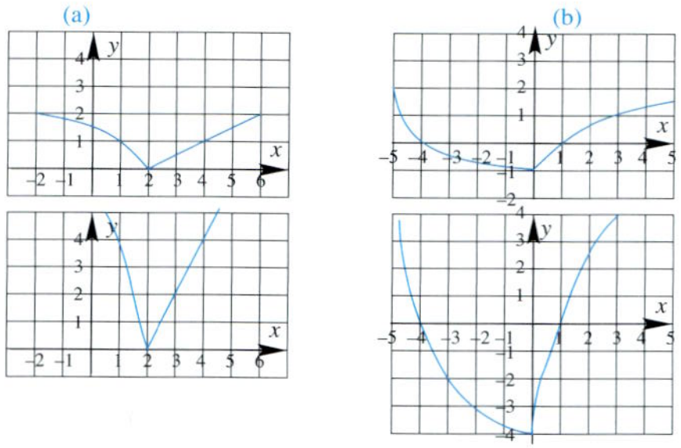

4. i.

$y = f\left(\frac{2}{3}x\right)$ (a) (b) (c) (d)

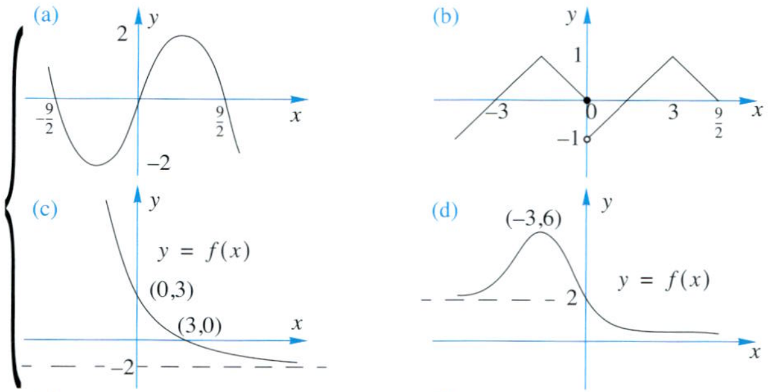

ii.

$y = 4f(x)$ (a) (b) (c) (d)

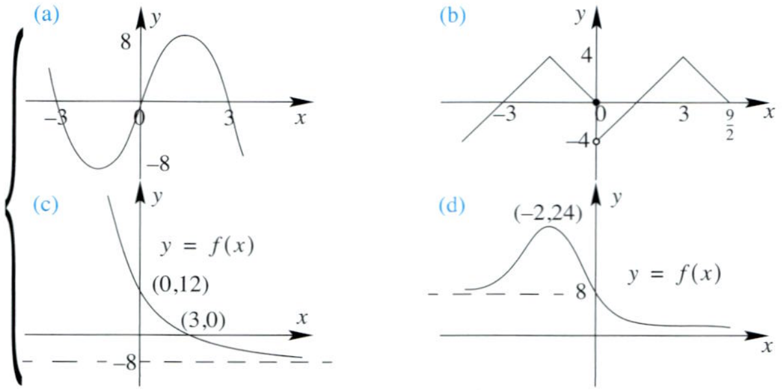

5. (a) $f(x) = |x| \quad y = f(2x) + 1$ (b) $f(x) = x^2 \quad y = \frac{1}{2}f(x-2) - 3$

(c) $f(x) = \frac{1}{x} \quad y = \frac{1}{2}f\left(x - \frac{1}{2}\right)$ (d) $f(x) = x^3 \quad y = 27f\left(x - \frac{2}{3}\right)$

(e) $f(x) = x^4 \quad y = 128f\left(x - \frac{1}{2}\right) - 2$ (f) $f(x) = \sqrt{x} \quad y = \sqrt{2}f(x) + 2$

6. (a) (b) (c)

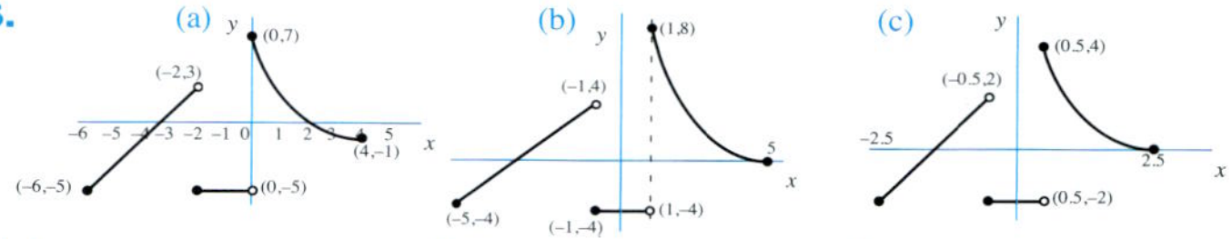

7. i. $f(x) = \begin{cases} (x+2)^2 & \text{if } x \geq 0 \\ 4 - x & \text{if } x < 0 \end{cases}$ ii. $h(x) = \begin{cases} x^2 - 3 & \text{if } x \geq 2 \\ 3 - x & \text{if } x < 2 \end{cases}$ iii. $h(x) = \begin{cases} 2x^2 & \text{if } x \geq 2 \\ 12 - 2x & \text{if } x < 2 \end{cases}$

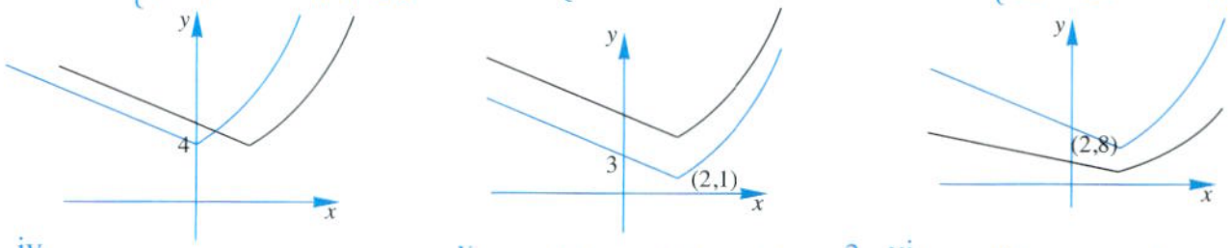

iv. $k(x) = \begin{cases} 4x^2 & \text{if } x \geq 1 \\ 6 - 2x & \text{if } x < 1 \end{cases}$ v. $k(x) = \begin{cases} (2x-1)^2 & \text{if } x \geq \frac{3}{2} \\ 7 - 2x & \text{if } x < \frac{3}{2} \end{cases}$ vi. $f(x) = \begin{cases} \frac{1}{2}(4x+2)^2 & \text{if } x \geq 0 \\ 2 - 2x & \text{if } x < 0 \end{cases}$

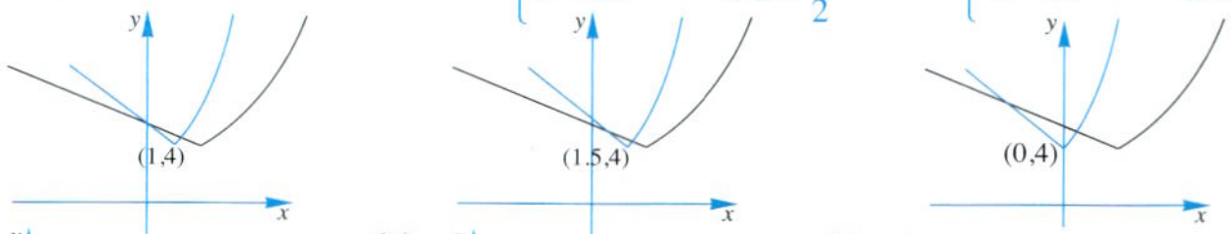

8. (a) (b)

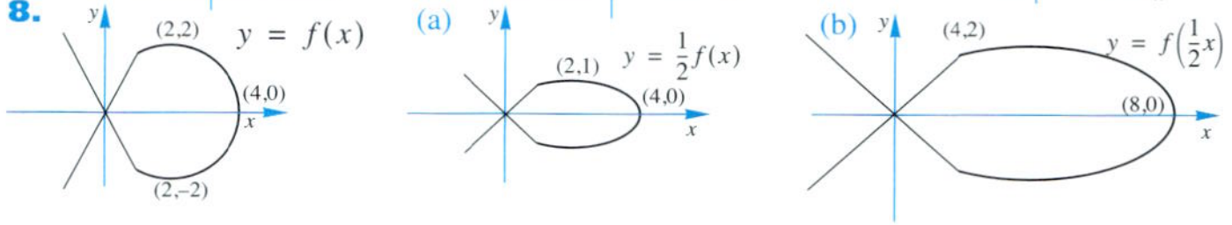

9. (a) (b) (c) (d)

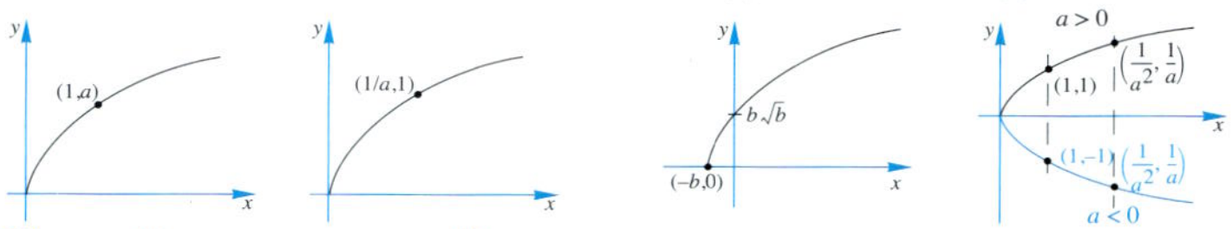

10. (a) (b)

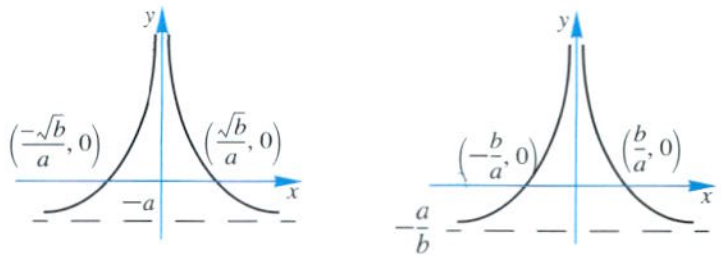

EXERCISE 6.3

1. i.

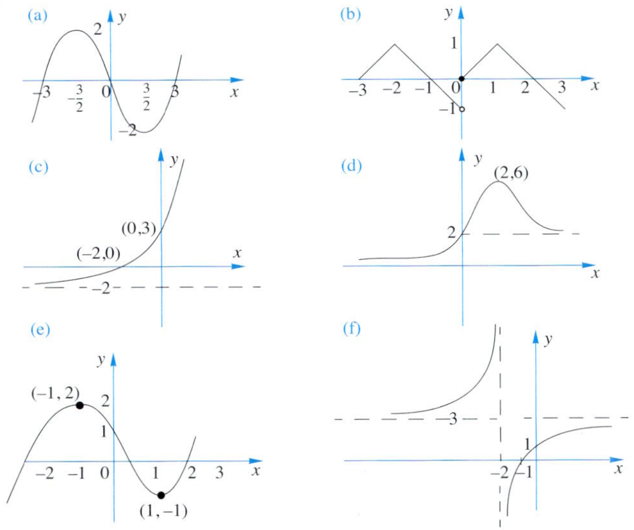

ii.

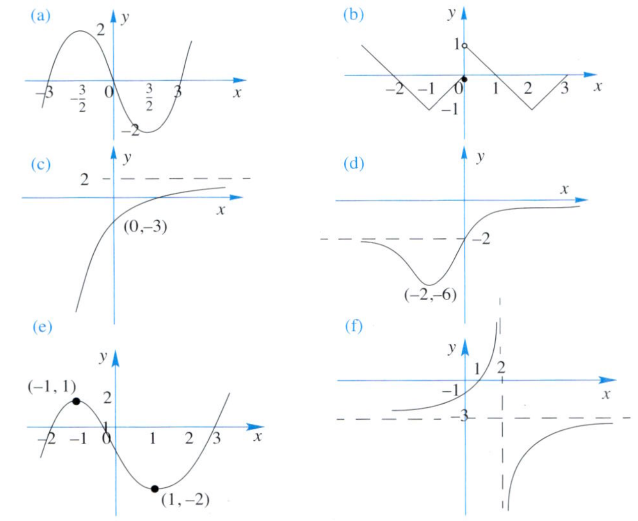

2. (a) $y = -f(x)$ (b) $y = f(-x)$ (c) $y = f(x+1)$ (d) $y = f(2x)$ (e) $y = 2f(x)$

3.

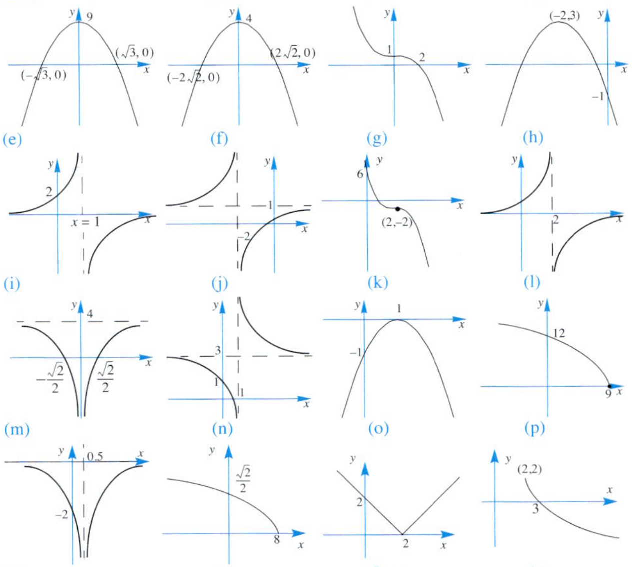

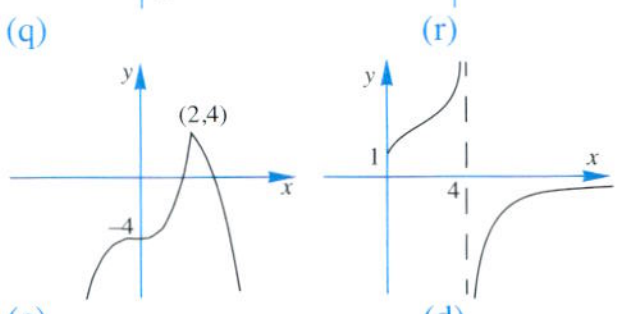

4.

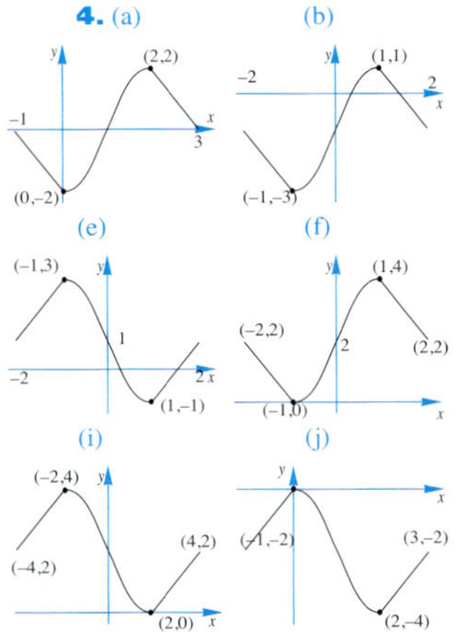

5. (a) i, (a) ii, (b) i, (b) ii, (c) i, (c) ii

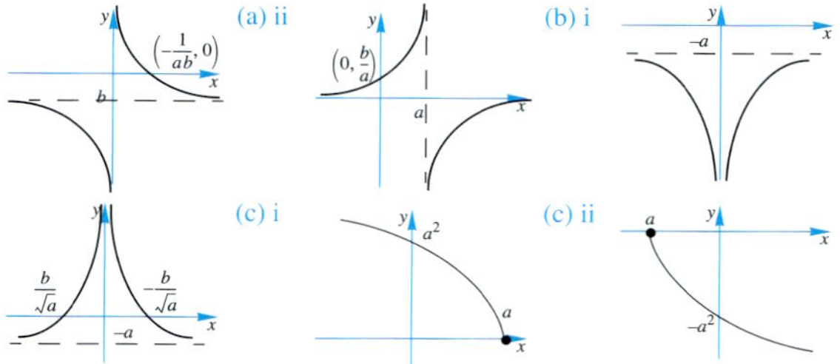

6. (a)

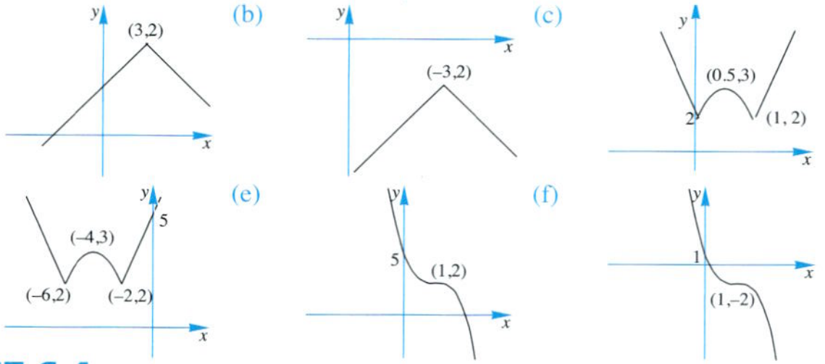

EXERCISE 6.4

1. (a)

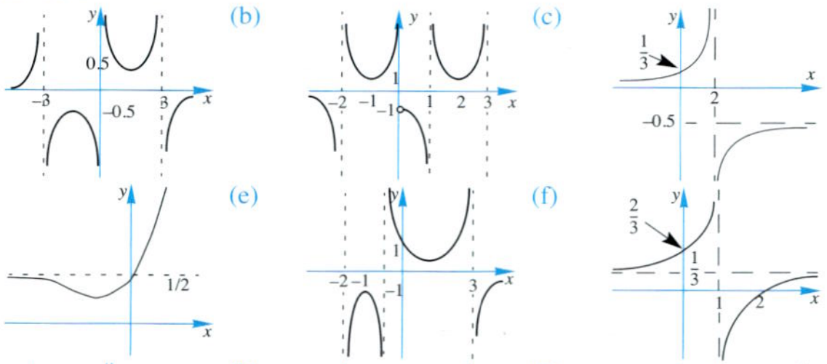

2. (a)

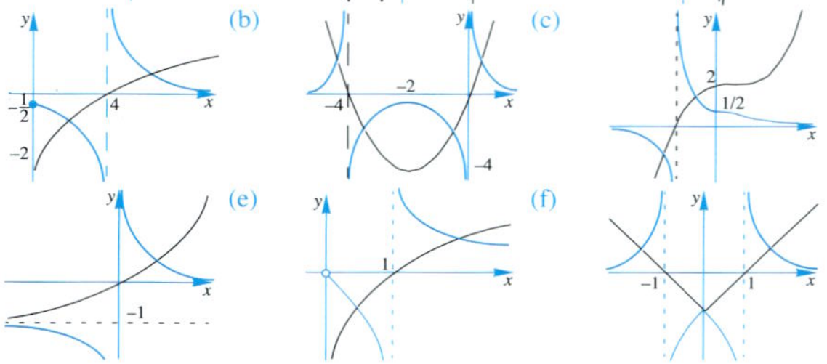

(g)

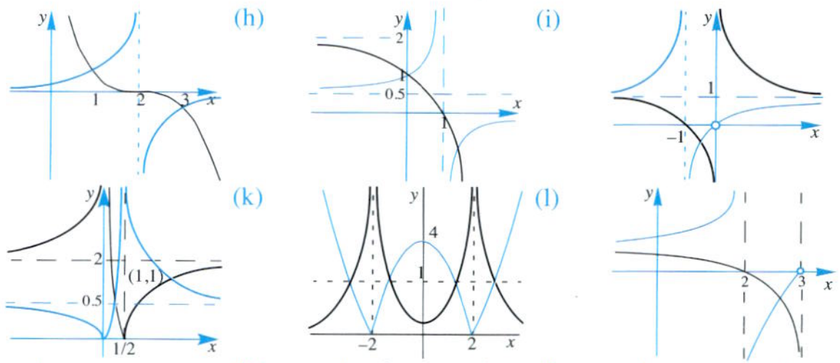

3. (a)

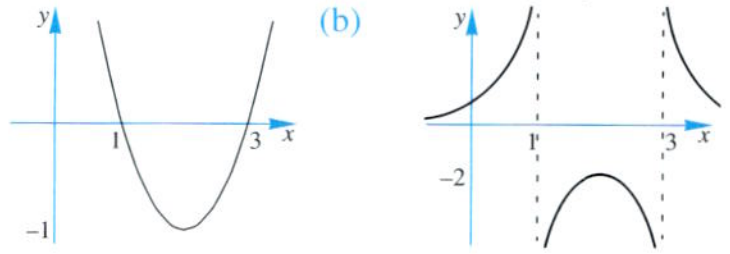

4.

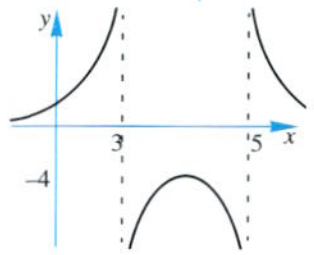

EXERCISE 6.5

1. (a) i

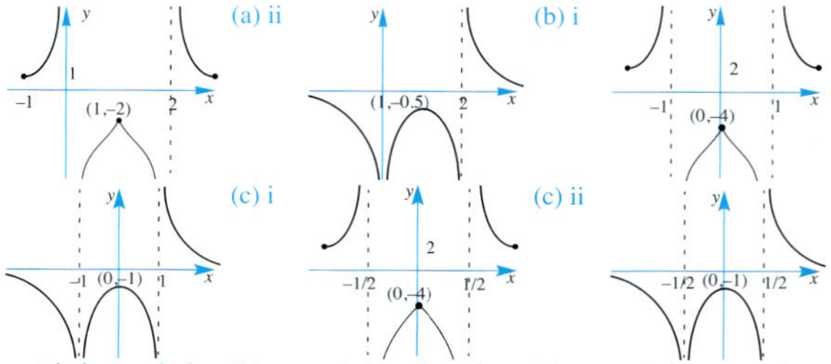

2. (a)

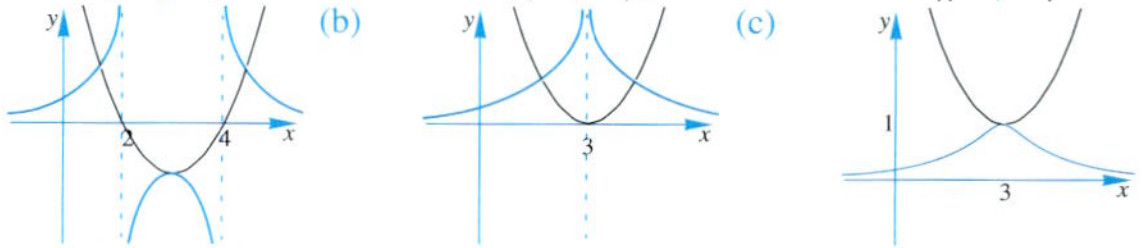

(d)

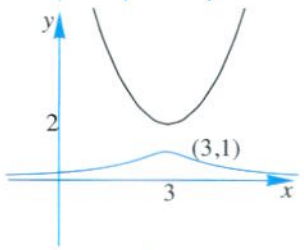

3. (a)

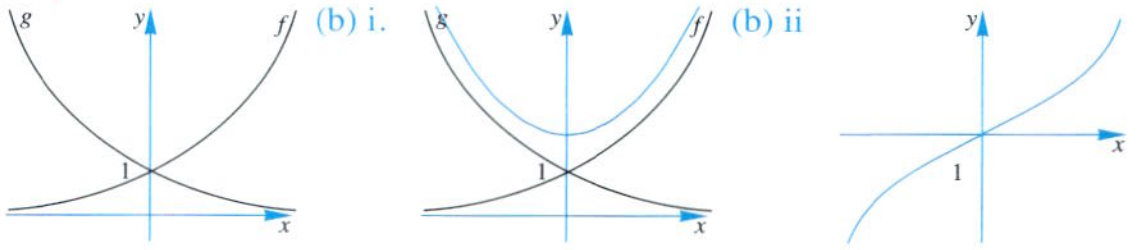

(c) i (c) ii (c) iii (c) iv (c) v (c) vi

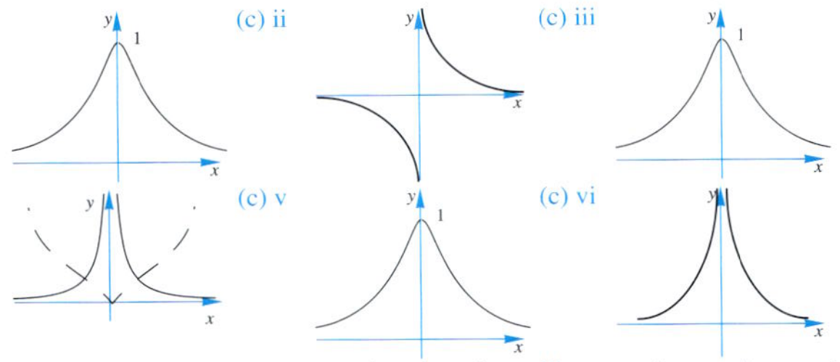

4. i ii iii

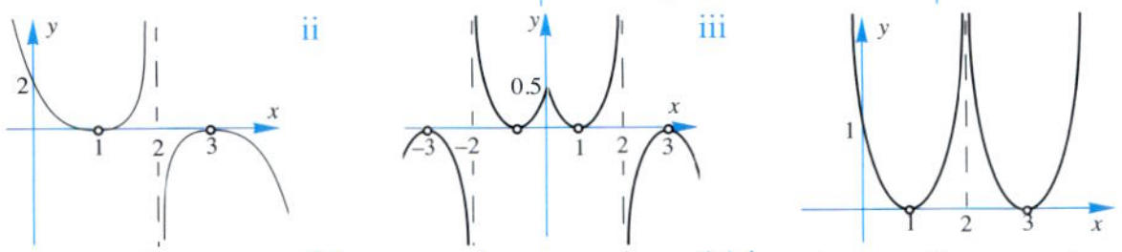

5. (a) (b) (c) i

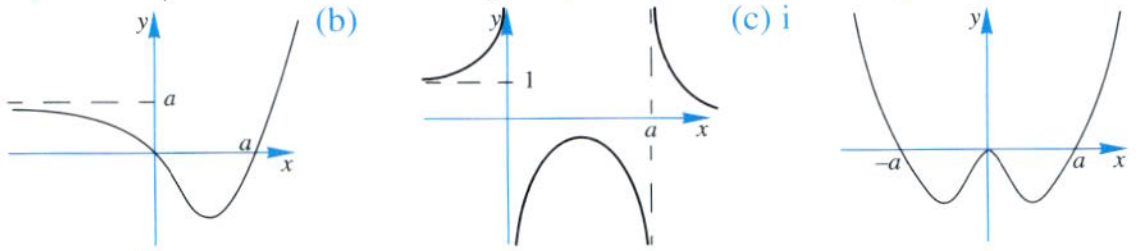

(c) ii

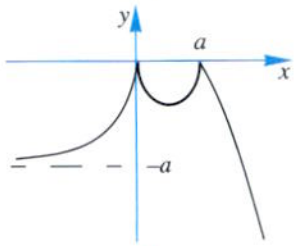

6. (a) (b) i (b) ii

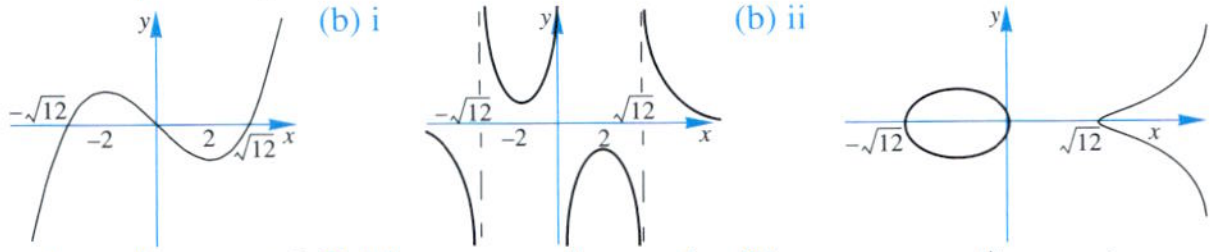

7.

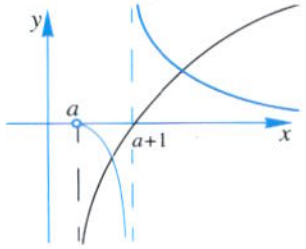

8.(a) (b)

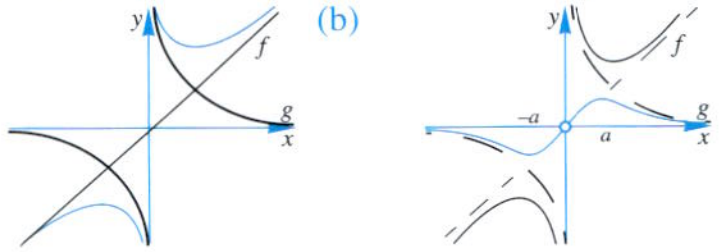

EXERCISE 7.1.1

1. (a) $\frac{27y^{15}}{8x^3}$ (b) $\frac{91}{216a^6}$ (c) 2^n+2 (d) $\frac{8x^{11}}{27y^2}$ (e) $\frac{3x^2y^2}{8}$ (f) $3^{n+1}+3$ (g) $4^{n+1}-4$ (h) $2(4^{n+1}-4)$ (i) $\frac{1-b^6}{16b^4}$ **2.** (a) 64 (b) $\left(\frac{2}{3}\right)^x$ (c) 2^{2y+1} (d) $\frac{1}{b^{2x}}$ (e) $\left(\frac{y}{2}\right)^6$ (f) $\left(\frac{9}{2}\right)^{n+2}$ **3.** (a) $\frac{z^2}{xy}$ (b) 3^{7n-2} (c) 5^{n+1} (d) 9 (e) 2^{6n+1} (f) 2^{1-3n} (g) x^{2+4n-n^2} (h) x^{3n^2+n+1} (i) 27 **4.** $\frac{y^{2m-2}}{x^m}$ **5.** (a) –81 (b) $-\frac{9x^8}{8y^4}$ (c) $y-x$ (d) $\frac{2x+1}{x+1}$ (e) –1 (f) $-b$ **6.** (a) $\frac{1}{x^2y^2}$ (b) $\frac{1}{x^4}$ (c) $-\frac{1}{x(x+h)}$

(d) $\frac{1}{x-1}$ (e) $\frac{1}{(x+1)(x-1)^5}$ (f) $\frac{1}{x^2}$ **7.** (a) $118\times5^{n-2}$ (b) 1 (c) $\frac{b^7}{a^4}$ (d) a^{mn} (e) $\frac{p+q}{pq}$ (f) $\frac{2\sqrt{a}}{a-1}$ (g) $\frac{7}{8}$ (h) $a^{7/8}$ **8.** (a) $x^{11/12}$ (b) $2a^{3n-2}b^{2n-2}$ (c) 2^n (d) $-\frac{7^{m-n}}{8}$ (e) $\frac{6\times5^n}{5^n+5}$ (f) $x+1$

EXERCISE 7.1.2

1. (a) 2 (b) –2 (c) $\frac{2}{3}$ (d) 5 (e) 6 (f) –2.5 (g) 2 (h) 1.25 (i) $\frac{1}{3}$ **2.** (a) –6 (b) $-\frac{2}{3}$ (c) –3 (d) 1.5 (e) 0.25 (f) 0.25 (g) $-\frac{1}{8}$ (h) $-\frac{11}{4}$ (i) –1.25

EXERCISE 7.1.3

1. (a) 3.5 (b) 3.5 (c) –3 (d) 1.5 (e) 3.5 (f) 1.5 (g) 1.8 (h) $-\frac{4}{7}$ (i) 0 **2.** (a) –0.75 (b) –1,4 (c) 0,1 (d) 3,4 (e) –1,4 (f) 0,2 **3.** (a) –1,1,2 (b) –3,1, 3,4 (c) $\frac{4}{3},\frac{5}{3}$, 2 (d) –1,1,2 (e) 3,7,$\frac{-1\pm\sqrt{233}}{2}$, $\frac{1}{3}$

EXERCISE 7.1.4

1. (a) i. 5.32 ii. 9.99 iii. 2.58 (b) i. 2.26 ii. 3.99 iii. 5.66 (c) i. 3.32 ii. –4.32 iii. –6.32 (d) i. –1.43 ii. 1.68 iii. –2.86 **2.** (a) 0 (b) 0.54 (c) –0.21 (d) –0.75, 0 (e) 1.13 (f) 0, 0.16

EXERCISE 7.1.5

1. (a) 2 (b) –1 (c) 0.5 (d) 0.5 **2.** (a) 1 (b) 0.6 (c) 0 **3.** (a) 0 (b) $\frac{2}{3}$ **4.** (a) –1,2 (b) –2,3 (c) –1 (d) –6,1 (e) 0,1 (f) 1 **5.** (a) 1.3863 (b) 2.1972 (c) 3.2189 (d) Ø **6.** (a) 0.4236 (b) 0.4055 (c) 0.3054 (d) –0.4176 **7.** (a) 0 (b) –0.6733 (c) 0 **9.** 36 **10.** $a=\sqrt{2}e, k=\ln(\sqrt{2})$

EXERCISE 7.2

1. (a) 1000 (b) 1516 (c) 2000 (d) 10 days **2.** (a) 0.0013 (b) 2.061 kg (c) 231.56 yrs
(d)

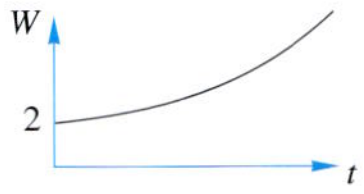

3. (a) 0.01398 (b) 52.53% (c) 51.53m (d) 21.53m

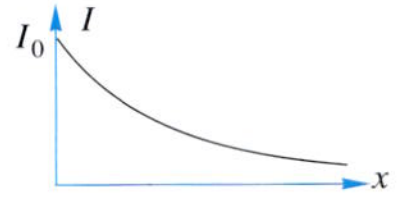

4. (a) i. 157 ii. 165 iii. 191 (b) 14.2 yrs (c) 20.1 yrs (d)

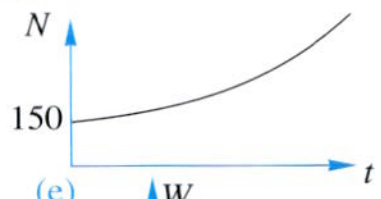

5. (a) 50 (b) 0.0222 (c) 17.99 kg (d)

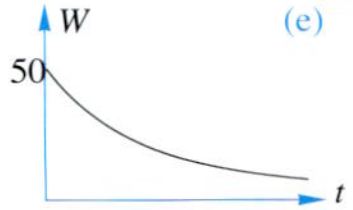

(e)

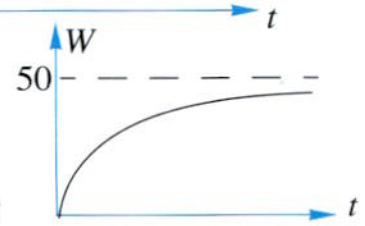

6. (a) 15 000°C (b) i. 11 900°C ii. 1500°C (c) 3.01 million yrs (d)

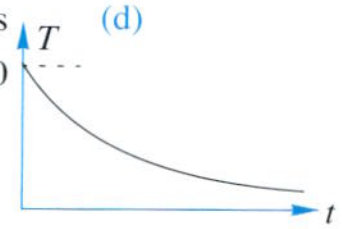

7. (a) 0.0151 (b) 12.50gm (c) 20 years (d)

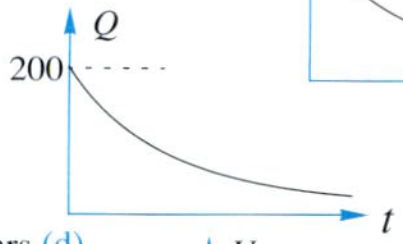

8. (a) \$2 million (b) \$1.589 mil (c) 30.1 years (d)

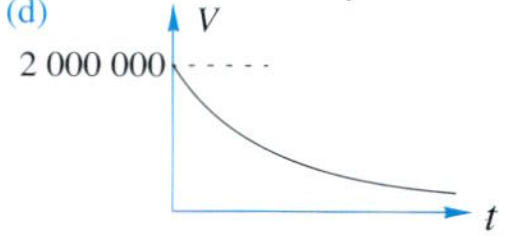

9. (b) 0.01761 (c) 199 230 (d) 22.6 years
10. (a) 20 cm^2 (b) 19.72 cm^2 (c) 100 days (d) 332 days
11. (a) 1 (b) i. 512170 ii. 517217 (c) 54.1 early 2014
12. (a) i. \$933.55 ii. \$935.50 (b) 11.95 years (c)

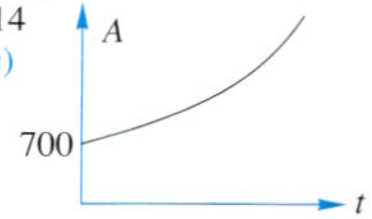

13. (a) 99 (b) $99 \times 2^{0.1394t}$ (c) 684

14. (a)

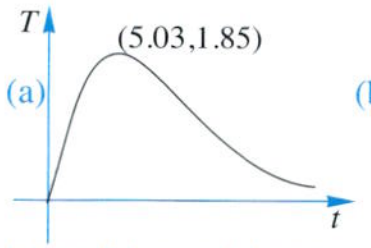

(b) 38.85°C at ~ midnight

15. (a) 19 (b) 2.63 (c) 100
16. (a) 18 cm (b) 4 cm (c) 1.28 m (d) 36 m (e) i. 21.7yr ii. 27.6yr iii. 34.5yr (f) 36 (g)
17. (a) 5 mg/min (b) 13.51 min (c) i. 2.1, ii. 13.9 iii. 68min
(d) 19.6 mg (e)

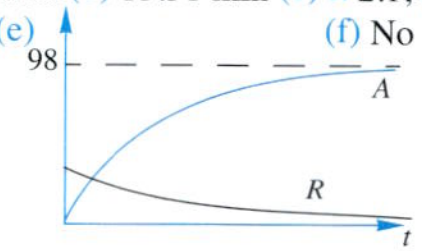

(f) No

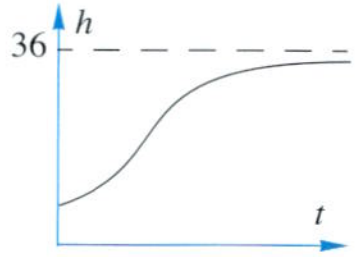

18. (a) i. \$499 ii. \$496 iii. \$467 (c) 15537 (d) i. \$499k ii. \$2.48mil iii. \$4.67mil (f) 12358
(g) \$5.14mil (b), (e)

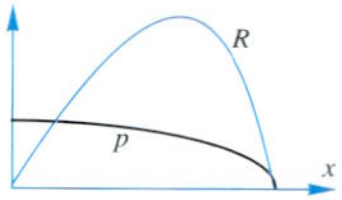

EXERCISE 7.3

1. (a) 2 (b) 2 (c) 5 (d) 3 (e) –3 (f) –2 (g) 0 (h) 0 (i) –1 (j) –2 (k) 0.5 (l) –2 **2.** (a) $\log_{10}10000 = 4$ (b) $\log_{10}0.001 = -3$ (c) $\log_{10}(x+1) = y$ (d) $\log_{10}P = 7$ (e) $\log_2(x-1) = y$ (f) $\log_2(y-2) = 4x$ **3.** (a) $2^9 = x$ (b) $b^x = y$ (c) $b^{ax} = t$ (d) $10^{x^2} = z$ (e) $10^{1-x} = y$ (f) $2^y = ax - b$ **4.** (a) 16 (b) 2 (c) 2 (d) 9 (e) $\sqrt[4]{2}$ (f) 125 (g) 4 (h) 9 (i) $\sqrt[3]{\frac{1}{3}}$ (j) 21 (k) 3 (l) 13
5. (a) 54.5982 (b) 1.3863 (c) 1.6487 (d) 7.3891 (e) 1.6487 (f) 0.3679 (g) 52.5982 (h) 4.7183 (i) 0.6065

EXERCISE 7.4

1. (a) 5 (b) 2 (c) 2 (d) 1 (e) 2 (f) 1 **2.** (a) $\log a = \log b + \log c$ (b) $\log a = 2\log b + \log c$ (c) $\log a = -2\log c$ (d) $\log a = \log b + 0.5\log c$ (e) $\log a = 3\log b + 4\log c$ (f) $\log a = 2\log b - 0.5\log c$ **3.** (a) 0.18 (b) 0.045 (c) –0.09 **4.** (a) $x = yz$ (b) $y = x^2$ (c) $y = \frac{x+1}{x}$ (d) $x = 2^{y+1}$ (e) $y = \sqrt{x}$ (f) $y^2 = (x+1)^3$ **5.** (a) $\frac{1}{2}$ (b) $\frac{1}{2}$ (c) $\frac{17}{15}$ (d) $\frac{3}{2}$ (e) $\frac{1}{3}$ (f) no real sol'n (g) 3,7 (h) $\frac{\sqrt{33}-1}{2}$ (i) 4 (j) $\sqrt{10}+3$ (k) $\frac{64}{63}$ (l) $\frac{2}{15}$ **6.** (a) $\log_3 2wx$ (b) $\log_4 \frac{x}{7y}$ (c) $\log_a[x^2(x+1)^3]$ (d) $\log_a\left[\frac{(x^5)(x+1)^3}{\sqrt{2x-3}}\right]$ (e) $\log_{10}\left(\frac{y^2}{x}\right)$ (f) $\log_2\left(\frac{y}{x}\right)$ **7.** (a) 1 (b) –2 (c) 3 (d) 9 (e) 2 (f) 9 **8.** (a) 1,4 (b) $1,3^{\pm\sqrt{3}}$ (c) $1,4^{3/4}$ (d) $1,5^{\pm 4/5}$ **9.** (a) $\frac{\log 14}{\log 2} = 3.81$ (b) $\frac{\log 8}{\log 10} = 0.90$ (c) $\frac{\log 125}{\log 3} = 4.39$ (d) $\frac{1}{\log 2} \times \log\left(\frac{11}{3}\right) - 2 = -0.13$ (e) $\frac{\log 10 - \log 3}{4\log 3} = 0.27$ (f) 5.11 (g) $\frac{-\log 2}{2\log 10} = -0.15$ (h) 7.37 (i) 0.93 (j) no real solution (k) $\frac{\log 3}{\log 2} - 2 = -0.42$ (l) $\frac{\log 1.5}{\log 3} = 0.37$ **10.** (a) 0.5,4 (b) 3 (c) –1,4 (d) $10,10^{10}$ (e) 5 (f) 3 **11.** (a) $(4, \log_4 11)$ (b) (100,10) (c) (2,1) **12.** (a) $y = xz$ (b) $y = x^3$ (c) $x = e^{y-1}$ **13.** (a) $\frac{1}{e^4-1}$ (b) $\frac{1}{3}$ (c) $\frac{\sqrt{5}-1}{2}$ (d) Ø **14.** (a) $\ln 21 = 3.0445$ (b) $\ln 10 = 2.3026$ (c) $-\ln 7 = -1.9459$ (d) $\ln 2 = 0.6931$ (e) $\ln 3 = 1.0986$ (f) $2\ln\left(\frac{14}{9}\right) = 0.8837$ (g) $e^3 = 20.0855$ (h) $\frac{1}{3}e^2 = 2.4630$ (i) $\pm\sqrt{e^9} = \pm 90.0171$ (j) Ø (k) $e^2 - 4 = 3.3891$ (l) $\sqrt[3]{e^9} = 20.0855$ **15.** (a) 0, ln2 (b) ln5 (c) ln2, ln3 (d) 0 (e) 0, ln5 (f) ln10 **16.** (a) 4.5222 (b) 0.2643 (c) 0,0.2619 (d) –1,0.3219 (e) –1.2925,0.6610 (f) 0,1.8928 (g) 0.25,2 (h) 1 (i) 121.5 (j) 2

EXERCISE 7.5

1. (a) 10 (b) 30 (c) 40 **2.** (a) 31.64 kg (b) 1.65 (c) $W = 2.4 \times 10^{0.8h}$

(d)

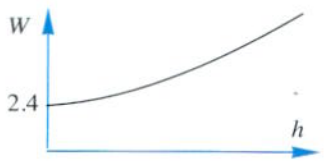

(e)

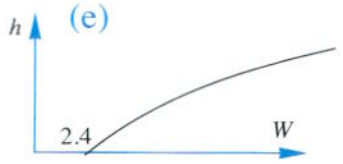

3. (a) 4.75 (b) $L = L_0 \times 10^{\left(\frac{6-m}{2.5}\right)}$ (c)

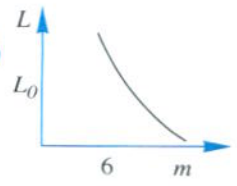

(d)

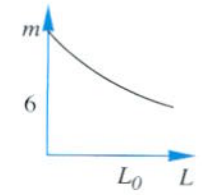

4. (a) [0,1[(b) i. 2.22 ii. 1.11 iii. 0.74 yrs (c) As c increases, reliability reduces.
(d) $x = 1 - 10^{-ct}$ (e)

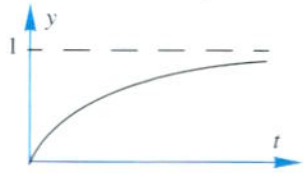

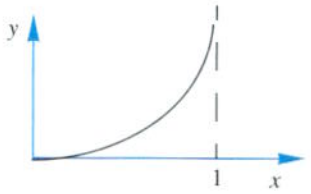

5. (a) $I = \frac{a}{n^k}$ **6.** (a) 0.10 (b) $\lambda = \lambda_0 \times 10^{-kx}$ (c) 16.82% (d) $k = -\frac{1}{x}\log\left(\frac{\lambda}{\lambda_0}\right)$

EXERCISE 8.1.1

1. i. (b) 4 (c) $t_n = 4n - 2$ ii. (b) –3 (c) $t_n = -3n + 23$ iii. (b) –5 (c) $t_n = -5n + 6$ iv. (b) 0.5 (c) $t_n = 0.5n$ v. (b) 2 (c) $t_n = y + 2n - 1$ vi. (b) –2 (c) $t_n = x - 2n + 4$ **2.** –28 **3.** 9,17 **4.** –43 **5.** 7 **6.** 7 **7.** –5 **8.** 0 **9.** (a) 41 (b) 31st **10.** 2, $\sqrt{3}$ **11.** (a) i. 2 ii. –3 (b) i. 4 ii. 11 **12.** $x - 8y$ **13.** $t_n = 5 + \frac{10}{3}(n-1)$ **14.** (a) –1 (b) 0

EXERCISE 8.1.2

1. (a) 145 (b) 300 (c) –170 **2.** (a) –18 (b) 690 (c) 70.4 **3.** (a) –105 (b) 507 (c) 224 **4.** (a) 126 (b) 3900 (c) 14th week **5.** 855 4. **6.** (a) 420 (b) –210 **7.** $a = 9, b = 7$

EXERCISE 8.1.3

1. 123 **2.** –3, –0.5, 2, 4.5, 7, 9.5, 12 **3.** 3.25 **4.** $a = 3$ $d = -0.05$ **5.** 10 000 **6.** 330 **7.** –20 **8.** 328 **9.** $725, 37wks **10.** i. $55 ii. 2750 **11.** (a) (i) 8m (ii) 40m (b) 84m (c) Dist = $2n^2 - 2n = 2n(n-1)$ (d) 8 (e) 26 players, 1300m **12.** (a) 5050 (b) 10200 (c) 4233 **13.** (a) 145 (b) 390 (c) –1845 **14.** (b) $3n - 2$

EXERCISE 8.2.1

1. (a) $r = 2, u_5 = 48, u_n = 3 \times 2^{n-1}$ (b) $r = \frac{1}{3}, u_5 = \frac{1}{27}, u_n = 3 \times \left(\frac{1}{3}\right)^{n-1}$

(c) $r = \frac{1}{5}, u_5 = \frac{2}{625}, u_n = 2 \times \left(\frac{1}{5}\right)^{n-1}$ (d) $r = -4, u_5 = -256, u_n = -1 \times (-4)^{n-1}$

(e) $r = \frac{1}{b}, u_5 = \frac{a}{b^3}, u_n = ab \times \left(\frac{1}{b}\right)^{n-1}$ (f) $r = \frac{b}{a}, u_5 = \frac{b^4}{a^2}, u_n = a^2 \times \left(\frac{b}{a}\right)^{n-1}$ **2.** (a) ±12

(b) $\frac{\pm\sqrt{5}}{2}$ **3.** (a) ±96 (b) 15th **4.** (a) $u_n = 10 \times \left(\frac{5}{6}\right)^{n-1}$ (b) $\frac{15625}{3888} \cong 4.02$ (c) $n = 5$ (4 times)

5. $-2, \frac{4}{3}$ **6.** (a) i. $4096 ii. $2097.15 (b) 6.2 yrs **7.** $\left(u_n = \frac{1000}{169} \times \left(\frac{12}{5}\right)^{n-1}\right), \frac{1990656}{4225} \cong 471.16$

8. 2.5,5,10 or 10,5,2.5 **9.** 53757 **10.** 108 952 **11.** (a) $56 156 (b) $299 284

EXERCISE 8.2.2

1. (a) 3 (b) $\frac{1}{3}$ (c) –1 (d) $-\frac{1}{3}$ (e) 1.25 (f) $-\frac{2}{3}$ **2.** (a) 216513 (b) 1.6384×10^{-10} (c) $\frac{256}{729}$

(d) $\frac{729}{2401}$ (e) $-\frac{81}{1024}$ **3.** (a) 11; 354292 (b) 7; 473 (c) 8; 90.90909 (d) 8; 172.778 (e) 5; 2.256

(f) 13; 111.1111111111 **4.** (a) $\frac{127}{128}$ (b) $\frac{63}{8}$ (c) $\frac{130}{81}$ (d) 60 (e) $\frac{63}{64}$ **5.** 4; 118096 **6.** $2109.50

7. 9.28cm **8.** (a) $V_n = V_0 \times 0.7^n$ (b) 7 **9.** 54 **10.** 53.5gms; 50 weeks. **11.** 7 **12.** 9

13. –0.5, –0.7797 **14.** $r = 5$, 1.8×10^{10} **15.** $8407.35

16. 1.8×10^{19} or about 200 billion tonnes.

EXERCISE 8.2.3

1. Term 9 AP = 180, GP = 256. Sum to 11 terms AP = 1650, GP = 2047. **2.** 18. **3.** 12 **4.** 12, 7 **5.** 8 weeks (Ken $220 & Bo-Youn $255) **6.** (a) week 8 (b) week 12 **7.** (a) 1.618 (b) 121379 [~121400, depends on rounding errors]

EXERCISE 8.2.4

1. (i) $\frac{81}{2}$ (ii) $\frac{10}{13}$ (iii) 5000 (iv) $\frac{30}{11}$ **2.** $23\frac{23}{99}$ **3.** 6667 fish. [Nb: $t_{43} < 1$. If we use $n = 43$ then ans is 6660 fish]; 20 000 fish. Overfishing means that fewer fish are caught in the long run. [*An alternate estimate for the total catch is 1665 fish.*] **4.** 27 **5.** 48,12,3 or 16,12,9 **6.** (a) $\frac{11}{30}$ (b) $\frac{37}{99}$

(c) $\frac{191}{90}$ **7.** 128 cm **8.** $\frac{121}{9}$ **9.** $2 + \frac{4}{3}\sqrt{3}$ **10.** $\frac{1-(-t)^n}{1+t}$ $\frac{1}{1+t}$ **11.** $\frac{1-(-t^2)^n}{1+t^2}$ $\frac{1}{1+t^2}$

EXERCISE 8.2.5

1. 3, –0.2 **2.** $\frac{2560}{93}$ **3.** $\frac{10}{3}$ **4.** (a) $\frac{43}{18}$ (b) $\frac{458}{99}$ (c) $\frac{413}{990}$ **5.** 9900 **6.** 3275 **7.** 3

8. $t_n = 6n - 14$ **9.** 6 **10.** $-\frac{1}{6}$ **11.** i. 12 ii. 26 **12.** 9, 12 **13.** ±2 **14.** (5, 5, 5), (5, –10, 20)

15. (a) 2, 7 (b) 2, 5, 8 (c) $3n - 1$ **16.** (a) 5 (b) 2 m

EXERCISE 8.3

1. $2773.08 **2.** $4377.63 **3.** $1781.94 **4.** $12216 **5.** $35816.95 **6.** $40349.37 **7.** $64006.80 **8.** $276971.93, $281325.41 **9.** $63762.25 **10.** $98.62, $9467.14, interest $4467.14. Flat interest = $6000 **11.** $134.41, $3790.44, 0.602% /month (or 7.22% p.a)

EXERCISE 9.1

1.	*a* cm	*b* cm	*c* cm	*A*	*B*	*C*
1	3.8	4.1	1.6	67°	90°	23°
2	81.5	98.3	55.0	56°	90°	34°
3	32.7	47.1	33.9	44°	90°	46°
4	1.61	30.7	30.7	3°	90°	87°
5	2.3	2.74	1.49	57°	90°	33°
6	48.5	77	59.8	39°	90°	51°
7	44.4	81.6	68.4	33°	90°	57°
8	2.93	13.0	12.7	13°	90°	77°
9	74.4	94.4	58.1	52°	90°	38°
10	71.8	96.5	64.6	48°	90°	42°
11	23.3	34.1	24.9	43°	90°	47°
12	43.1	43.2	2.3	87°	90°	3°
13	71.5	80.2	36.4	63°	90°	27°
14	33.5	34.1	6.5	79°	90°	11°
15	6.1	7.2	3.82	58°	90°	32°
16	29.1	30	7.3	76°	90°	14°
17	29.0	29.1	2.0	86°	90°	4°
18	34.5	88.2	81.2	23°	90°	67°
19	24.0	29.7	17.5	54°	90°	36°
20	41.2	46.2	21.0	63°	90°	27°
21	59.6	72.9	41.8	55°	90°	35°
22	5.43	6.8	4.09	53°	90°	37°

23	13.0	19.8	14.9	41°	90°	49°
24	14.0	21.3	16.1	41°	90°	49°
25	82.4	88.9	33.3	68°	90°	22°

2. (a) $2\sqrt{3}$ (b) $5(1+\sqrt{3})$ (c) 4 (d) $2(1+\sqrt{3})$ (e) $\frac{4}{3}(3+\sqrt{3})$ (f) $\sqrt{106}-5$ **4.** (a) $25(1+\sqrt{3})$ (b) $\frac{40\sqrt{3}}{3}$

EXERCISE 9.2

1. (a) i. 030°T ii. 330°T iii. 195°T iv. 200°T (b) i. N25°E ii. S iii. S40°W iv. N10°W **2.** 37.49m **3.** 18.94m **4.** 37° 18' **5.** $\frac{26}{9}$ m/s **6.** N58° 33'W, 37.23 km **7.** 199.82 m **8.** 10.58 m **9.** 72.25 m **10.** 25.39 km **11.** 15.76 m **12.** (a) 3.01km N, 3.99km E (b) 2.87km E 0.88km S (c) 6.86km E 2.13km N (d) 7.19km 253°T **13.** 524m

EXERCISE 9.3

1. (a) 39°48' (b) 64°46' **2.** (a) 12.81 cm (b) 61.35 cm (c) 77°57' (d) 60.83 cm (e) 80° 32' **3.** (a) 21°48' (b) 42°2' (c) 26°34' **4.** (a) 2274 (b) 12.7° **5.** 251.29 m **6.** (a) 103.5 m (b) 35.26° (c) 39.23° **7.** (b) 53.43 (c) 155.16 m (d) 145.68 m **8.** (b) 48.54 m **9.** (a) $\sqrt{(b-c)^2+h^2}$ (b) $\tan^{-1}\left(\frac{h}{a}\right)$ (c) $\tan^{-1}\left(\frac{h}{b-c}\right)$ (d) $2(b+c)\sqrt{h^2+a^2}+2a\sqrt{(b-c)^2+h^2}$ **10.** 82.80 m **11.** (a) 40.61 m (b) 49.46 m **12.** (a) 10.61 cm (b) 75° 58' (c) 93° 22' **13.** (a) 1.44 m (b) 73° 13' (c) 62° 11'

EXERCISE 9.4

1. (a) 1999.2 cm^2 (b) 756.8 cm^2 (c) 3854.8 cm^2 (d) 2704.9 cm^2 (e) 538.0 cm^2 (f) 417.5 cm^2 (g) 549.4 cm^2 (h) 14.2 cm^2 (i) 516.2 cm^2 (j) 281.5 cm^2 (k) 918.8 cm^2 (l) 387.2 cm^2 (m) 139.0 cm^2 (n) 853.7 cm^2 (o) 314.6 cm^2 **2.** 69345 m^2 **3.** $100\pi - 6\sqrt{91}$ cm^2 **4.** 17.34 cm **5.** (a) 36.77sq units (b) 14.70 sq units (c) 62.53 sq units **6.** 52.16 cm^2 **7.** 27° 2' **8.** $\frac{(b+a\times\tan\theta)^2}{2\tan\theta}$ **9.** Area of Δ ACD = 101.78 cm^2, area of Δ ABC = 61.38 cm^2

EXERCISE 9.5.1

	a cm	*b* cm	*c* cm	*A*	*B*	*C*
1	13.3	37.1	48.2	10°	29°	141°
2	2.7	1.2	2.8	74°	25°	81°
3	11.0	0.7	11.3	60°	3°	117°
4	31.9	39.1	51.7	38°	49°	93°
5	18.5	11.4	19.5	68°	35°	77°
6	14.6	15.0	5.3	75°	84°	21°
7	26.0	7.3	26.4	79°	16°	85°
8	21.6	10.1	28.5	39°	17°	124°
9	0.8	0.2	0.8	82°	16°	82°
10	27.7	7.4	33.3	36°	9°	135°
11	16.4	20.7	14.5	52°	84°	44°
12	21.4	45.6	64.3	11°	24°	145°
13	30.9	27.7	22.6	75°	60°	45°

14	29.3	45.6	59.1	29°	49°	102°
15	9.7	9.8	7.9	65°	67°	48°
16	21.5	36.6	54.2	16°	28°	136°
17	14.8	29.3	27.2	30°	83°	67°
18	10.5	0.7	10.9	52°	3°	125°
19	11.2	6.9	17.0	25°	15°	140°
20	25.8	18.5	40.1	30°	21°	129°

EXERCISE 9.5.2

	a	*b*	*c*	*A*°	*B*°	*C*°	*c**	*B**°	*C**°
1	7.40	18.10	21.06	20.00	56.78	103.22	12.95	123.22	36.78
2	13.30	19.50	31.36	14.00	20.77	145.23	6.49	159.23	6.77
3	13.50	17.00	25.90	28.00	36.24	115.76	4.12	143.76	8.24
4	10.20	17.00	25.62	15.00	25.55	139.45	7.22	154.45	10.55
5	7.40	15.20	19.55	20.00	44.63	115.37	9.02	135.37	24.63
6	10.70	14.10	21.41	26.00	35.29	118.71	3.94	144.71	9.29
7	11.50	12.60	22.94	17.00	18.68	144.32	1.16	161.32	1.68
8	8.30	13.70	18.67	24.00	42.17	113.83	6.36	137.83	18.17
9	13.70	17.80	30.28	14.00	18.32	147.68	4.27	161.68	4.32
10	13.40	17.80	26.19	28.00	38.58	113.42	5.24	141.42	10.58
11	12.10	16.80	25.63	23.00	32.85	124.15	5.30	147.15	9.85
12	12.00	14.50	24.35	21.00	25.66	133.34	2.72	154.34	4.66
13	12.10	19.20	29.34	16.00	25.94	138.06	7.57	154.06	9.94
14	7.20	13.10	19.01	15.00	28.09	136.91	6.30	151.91	13.09
15	12.20	17.70	23.73	30.00	46.50	103.50	6.93	133.50	16.50
16	9.20	20.90	27.97	14.00	33.34	132.66	12.59	146.66	19.34
17	10.50	13.30	21.96	20.00	25.67	134.33	3.03	154.33	5.67
18	9.20	19.20	26.29	15.00	32.69	132.31	10.80	147.31	17.69
19	7.20	13.30	18.33	19.00	36.97	124.03	6.82	143.03	17.97
20	13.50	20.40	25.96	31.00	51.10	97.90	9.01	128.90	20.10
21	10.80	20.80	24.48	26.00	57.59	96.41	12.91	122.41	31.59
22	13.00	12.20	23.91	19.00	17.79	143.21	0.84	162.21	1.21
23	13.60	20.40	22.92	36.00	61.85	82.15	10.09	118.15	25.85
24	11.40	12.50	22.88	16.00	17.59	146.41	1.15	162.41	1.59
25	8.00	16.80	23.99	10.00	21.39	148.61	9.10	158.61	11.39

26. (a-d) no triangles exist.

EXERCISE 9.5.3

1. 30.64 km **2.** 4.57 m **3.** 476.4 m **4.** 201°47'T **5.** 222.9 m **6.** (a) 3.40 m (b) 3.11 m **7.** (b) 1.000 m (c) 1.715 m **8.** (a) 51.19 min (b)1 hr 15.96 min (c) 14.08 km **9.** $4886 **10.**906 m

EXERCISE 9.5.4

	a cm	*b* cm	*c* cm	*A*	*B*	*C*
1	13.5	9.8	16.7	54°	36°	90°
2	8.9	10.8	15.2	35°	44°	101°
3	22.8	25.6	12.8	63°	87°	30°
4	21.1	4.4	21.0	85°	12°	83°
5	15.9	10.6	15.1	74°	40°	66°
6	8.8	13.6	20.3	20°	32°	128°
7	9.2	9.5	13.2	44°	46°	90°
8	23.4	62.5	58.4	22°	89°	69°

9	10.5	9.6	15.7	41°	37°	102°
10	21.7	36.0	36.2	35°	72°	73°
11	7.6	3.4	9.4	49°	20°	111°
12	7.2	15.2	14.3	28°	83°	69°
13	9.1	12.5	15.8	35°	52°	93°
14	14.9	11.2	16.2	63°	42°	75°
15	2.0	0.7	2.5	38°	13°	129°
16	7.6	3.7	9.0	56°	24°	100°
17	18.5	9.8	24.1	45°	22°	113°
18	20.7	16.3	13.6	87°	52°	41°
19	14.6	22.4	29.9	28°	46°	106°
20	7.0	6.6	9.9	45°	42°	93°
21	21.8	20.8	23.8	58°	54°	68°
22	1.1	1.7	1.3	41°	89°	50°
23	1.2	1.2	0.4	85°	76°	19°
24	23.7	27.2	29.7	49°	60°	71°
25	3.4	4.6	5.2	40°	60°	80°

EXERCISE 9.5.5

1. (a) 10.14 km (b) 121°T **2.** 7° 33' **3.** 4.12 cm **4.** 57.32 m **5.** 315.5 m **6.** (a) 124.3 km (b) W28° 47' S

EXERCISE 9.5.6

1. 39.60m 52.84m **2.** 30.2m **3.** 54°,42°, 84° **4.** 37° **5.** 028°T. **6.** 108.1cm **7.** (i) 135° (ii) 136cm **8.** 41°, 56°, 83° **9.** (i) 158° left (ii) 43.22km **10.** 264m **11.** 53.33cm **12.** 186m **13.** 50.12cm **14.** 5.17cm **15.** (a) 5950m (b) 13341m (c) 160° (d) 243° **17.** (a) 20.70° (b) 2.578 m (c) 1.994 m^3 **18.** (a) 4243 m^2 (b) 86 m (c) 101 m

EXERCISE 9.6

1. 5.36 cm **2.** 12.3 m **3.** 24 m **4.** 40.3 m, 48.2° **5.** 16.5 min, 8.9° **6.** ~10:49 am

7. (a) i. $\frac{d\sin\phi}{\sin(\phi-\theta)}$ ii. $\frac{d\sin\theta}{\sin(\phi-\theta)}$ (b) $\frac{d\sin\phi\tan\alpha}{\sin(\phi-\theta)}$ or $\frac{d\sin\theta\tan\beta}{\sin(\phi-\theta)}$ (c) $d\left(\frac{\sin\phi\cos\theta}{\sin(\phi-\theta)}-1\right)$

EXERCISE 9.7

1. (i) $\frac{169\pi}{150}\text{cm}^2$, $5.2+\frac{13\pi}{15}\text{cm}$ (ii) $\frac{529\pi}{32}\text{cm}^2$, $23+\frac{23\pi}{8}\text{cm}$ (iii) $242\pi\text{cm}^2$, $88+11\pi\text{cm}$ (iv) $\frac{1156\pi}{75}\text{m}^2$, $13.6+\frac{68\pi}{15}\text{m}$ (v) $\frac{96\pi}{625}\text{cm}^2$, $1.28+\frac{12\pi}{25}\text{cm}$ (vi) $\frac{361\pi}{15}\text{cm}^2$, $15.2+\frac{19\pi}{3}\text{cm}$ (vii) $5248.8\pi\text{m}^2$, $648+32.4\pi\text{cm}$ (viii) $\frac{12943\pi}{300}\text{cm}^2$, $17.2+\frac{301\pi}{30}\text{cm}$ (ix) $\frac{1922\pi}{75}\text{cm}^2$, $12.4+\frac{124\pi}{15}\text{cm}$ (x) $\frac{15884\pi}{3}\text{cm}^2$, $152+\frac{418\pi}{3}\text{cm}$ (xi) $12\pi\text{cm}^2$, $24+2\pi\text{cm}$ (xii) $\frac{98\pi}{3}\text{cm}^2$, $28+\frac{14\pi}{3}\text{cm}$ (xiii) $\frac{196\pi}{75}\text{cm}^2$, $5.6+\frac{28\pi}{15}\text{cm}$ (xiv) $\frac{11532\pi}{25}\text{cm}^2$, $49.6+\frac{186\pi}{5}\text{cm}$ (xv) $\frac{3\pi}{50}\text{cm}^2$, $2.4+\frac{\pi}{10}\text{cm}$ **2.** 0.63^c, 36° **3.** 0.0942m^3 **4.** 1.64^c **5.** 79cm. **6.** 5.25cm^2

7. (a) 31.83m (b) 406.28m (c) 11° **8.** 1.11^c **9.** 0.75^c **10.** (a) 1.85^c (b) i. 37.09 cm ii. 88.57 cm (c) 370.92cm^2 **11.** 26.57 cm^2 **12.** 193.5 cm **13.** (a) 105.22 cm (b) 118.83 cm **14.** (a) 9 cm

(b) 12 cm (c) 36°52′ **15.** (b)

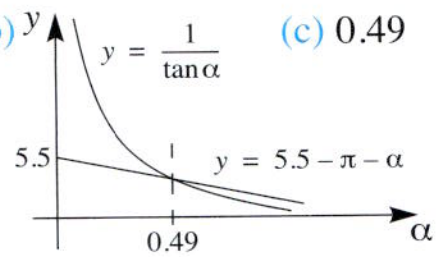

(c) 0.49 **16.** 1439.16 cm^2

EXERCISE 10.1

1. (a) 120° (b) 108° (c) 216° (d) 50° **2.** (a) π^c (b) $\frac{3\pi^c}{2}$ (c) $\frac{7\pi^c}{9}$ (d) $\frac{16\pi^c}{9}$ **3.** (a) $\frac{\sqrt{3}}{2}$ (b) $-\frac{1}{2}$ (c) $-\sqrt{3}$ (d) $-\frac{1}{2}$ (e) $-\frac{\sqrt{3}}{2}$ (f) $\frac{1}{\sqrt{3}}$ (g) $-\frac{1}{\sqrt{2}}$ (h) $-\frac{1}{\sqrt{2}}$ (i) 1 (j) $-\frac{1}{\sqrt{2}}$ (k) $\frac{1}{\sqrt{2}}$ (l) –1 (m) 0 (n) 1 (o) 0

4. (a) 0 (b) –1 (c) 0 (d) $\frac{1}{\sqrt{2}}$ (e) $-\frac{1}{\sqrt{2}}$ (f) –1 (g) $-\frac{1}{2}$ (h) $-\frac{\sqrt{3}}{2}$ (i) $\frac{1}{\sqrt{3}}$ (j) $-\frac{\sqrt{3}}{2}$ (k) $\frac{1}{2}$ (l) $-\sqrt{3}$ (m) $-\frac{1}{\sqrt{2}}$ (n) $\frac{1}{\sqrt{2}}$ (o) –1 **5.** (a) $\frac{1}{2}$ (b) $\frac{\sqrt{3}}{2}$ (c) 1 (d) $\frac{1}{2}$ **6.** (a) $-\frac{1}{2}$ (b) $-\frac{1}{\sqrt{2}}$ (c) $\sqrt{3}$ (d) $\frac{1}{2}$ (e) $-\frac{\sqrt{3}}{2}$ (f) $\frac{1}{\sqrt{3}}$ (g) $-\frac{\sqrt{3}}{2}$ **7.** (a) $\left(\frac{1}{2},\frac{\sqrt{3}}{2}\right)$ (b) $\left(-\frac{1}{2},\frac{\sqrt{3}}{2}\right)$ (c) $\left(-\frac{1}{\sqrt{2}},-\frac{1}{\sqrt{2}}\right)$ (d) $\left(\frac{\sqrt{3}}{2},-\frac{1}{2}\right)$ **8.** (a) 0 (b) $\frac{\sqrt{3}}{2}$ (c) $\frac{1}{\sqrt{3}}$ (d) $\frac{1+\sqrt{3}}{2\sqrt{2}}$ **10.** (a) $-\frac{2}{3}$ (b) $-\frac{2}{3}$ (c) $-\frac{2}{3}$ **11.** (a) $-\frac{2}{5}$ (b) $\frac{2}{5}$ **12.** (a) k (b) $-\frac{1}{k}$ (c) $-k$

13. (a) $\frac{\sqrt{5}}{3}$ (b) $-\frac{\sqrt{5}}{3}$ **14.** (a) $-\frac{3}{5}$ (b) $\frac{3}{4}$ (c) $\frac{4}{5}$ **15.** (a) $\frac{4}{5}$ (b) $\frac{3}{4}$ **16.** (a) $-k$ (b) $-\sqrt{1-k^2}$

17. (a) $-\sqrt{1-k^2}$ (b) $k/\sqrt{1-k^2}$ **18.** (a) $\sin\theta$ (b) $\cot\theta$ (c) 1

19. (a) $\frac{\pi}{3},\frac{2\pi}{3}$ (b) $\frac{\pi}{3},\frac{5\pi}{3}$ (c) $\frac{\pi}{3},\frac{4\pi}{3}$ (d) $\frac{5\pi}{6},\frac{7\pi}{6}$ (e) $\frac{5\pi}{6},\frac{11\pi}{6}$ (f) $\frac{7\pi}{6},\frac{11\pi}{6}$

EXERCISE 10.2.1

3. (a) $x^2+y^2=k^2, -k\le x\le k$ (b) $\frac{x^2}{b^2}+\frac{y^2}{a^2}=1, -b\le x\le b$ (c) $(x-1)^2+(2-y)^2=1, 0\le x\le 2$ (d) $\frac{(1-x)^2}{b^2}+\frac{(y-2)^2}{a^2}=1$ (e) $5x^2+5y^2+6xy=16$ **4.** (a) (i) $-\frac{4}{5}$ (ii) $-\frac{5}{3}$ (b) (i) $\frac{4}{\sqrt{7}}$ (ii) $-\frac{\sqrt{7}}{3}$

5. (a) $\frac{\pi}{3},\frac{2\pi}{3},\frac{4\pi}{3},\frac{5\pi}{3}$ (b) $\frac{\pi}{2},\frac{7\pi}{6},\frac{11\pi}{6}$ (c) $0,\frac{\pi}{6},\frac{5\pi}{6},\pi,2\pi$ (d) $\frac{\pi}{2},\frac{3\pi}{2}$ **9.** (a) $\frac{2a}{a^2+1}$ (b) $\frac{a^2-1}{a^2+1}$

10. (a) i. 1 ii. 1 (b) 1 **11.** (a) $\frac{1-\sqrt{x^2-1}}{x}$ (b) $\frac{1+\sqrt{x^2-1}}{x}$ (c) $\frac{2}{x^2}-1$ **12.** (a) $5\le\cos^2\theta+5\le 6$ (b) $1\le\frac{5}{3\sin^2\theta+2}\le\frac{5}{2}$ (c) $-2\le 2\cos^2\theta+\sin\theta-1\le\frac{9}{8}$ **13.** (a) ±2 (b) $\frac{\pi}{6}+2k\pi, k\in\mathbb{Z}$ or $\frac{7\pi}{6}+2k\pi, k\in\mathbb{Z}$ **14.** (a) $5^{-4}\le 5^{3\sin\theta-1}\le 25$ (b) $3^{-1}\le 3^{1-2\cos\theta}\le 27$ **15.** (a) $1+2k$ (b) $(1-k)\sqrt{1+2k}$ **16.** (a) $\frac{1-a}{2\sqrt{a}}$ (b) i. $2+\sqrt{2a-a^2}$ ii. $\frac{-\sqrt{2a-a^2}}{1-a}$ **17.** (a) $\frac{2}{3}$ (b) $0, \pm\frac{2\sqrt{2}}{3}$

18. $0,\frac{\pi}{3},\frac{2\pi}{3},\pi$

EXERCISE 10.2.2

1. (a) $-\frac{5\sqrt{11}}{18}$ (b) $-\frac{7}{18}$ (c) $\frac{5\sqrt{11}}{7}$ (d) $\frac{35\sqrt{11}}{162}$ **2.** (a) $-\frac{3}{5}$ (b) $-\frac{4}{5}$ (c) $\frac{3}{4}$ (d) $\frac{24}{7}$ **3.** $\frac{(\sqrt{3}-1)\sqrt{2}}{4}$

4. (a) $\frac{2ab}{a^2+b^2}$ (b) $\frac{a^2+b^2}{2ab}$ (c) $\frac{a^4-6a^2b^2+b^4}{(a^2+b^2)^2}$ (d) $\frac{2ab}{b^2-a^2}$ **7.** (a) $0, \frac{\pi}{3}, \pi, \frac{5\pi}{3}, 2\pi$ (b) $\frac{\pi}{6}, \frac{5\pi}{6}, \frac{3\pi}{2}$

(c) $0, \pi, 2\pi, \alpha, \pi \pm \alpha, 2\pi - \alpha, \alpha = \tan^{-1}\left(\frac{1}{\sqrt{2}}\right)$

EXERCISE 10.3

1. (a) 4π (b) $\frac{2\pi}{3}$ (c) 3π (d) 4π (e) 2 (f) $\frac{\pi}{2}$ **2.** (a) 5 (b) 3 (c) 5 (d) 0.5 **3.** (a) 2π, 2 (b) 6π, 3 (c) π

(d) π (e) π, 4 (f) π, 3 (g) 6π (h) $\frac{2\pi}{3}, \frac{1}{4}$ (i) 3π (j) $\frac{8\pi}{3}, \frac{2}{3}$

4.

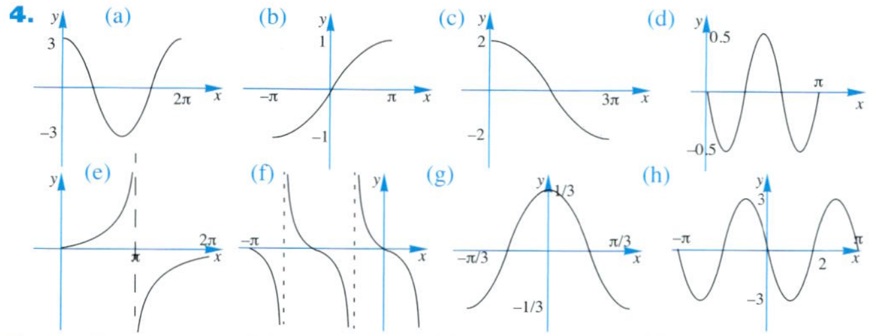

5.

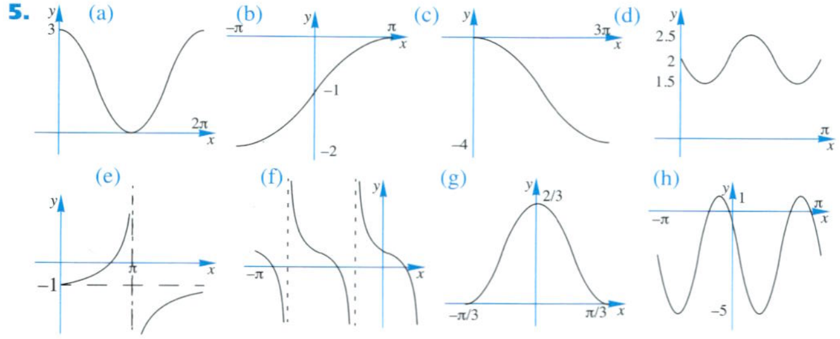

6.

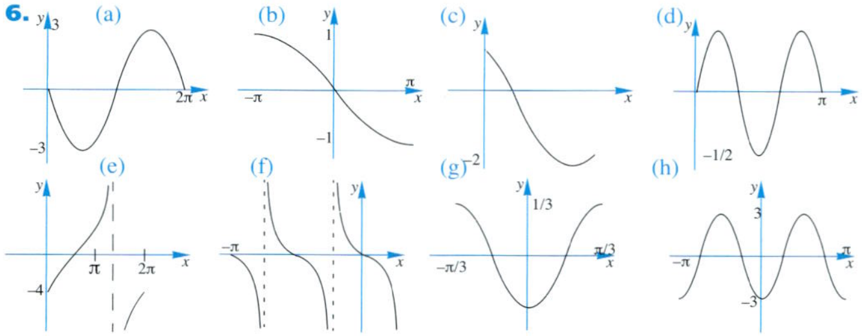

7.

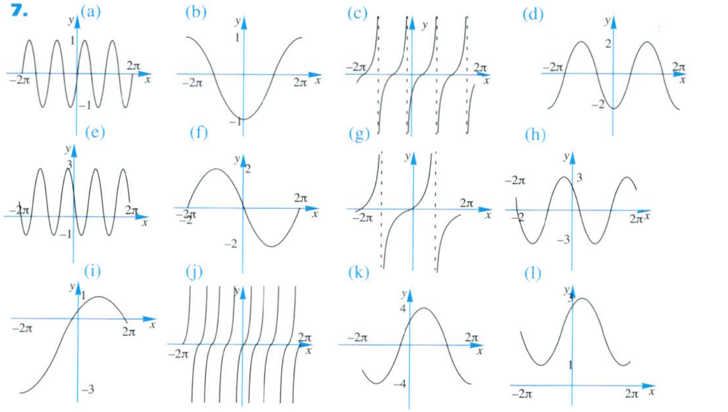

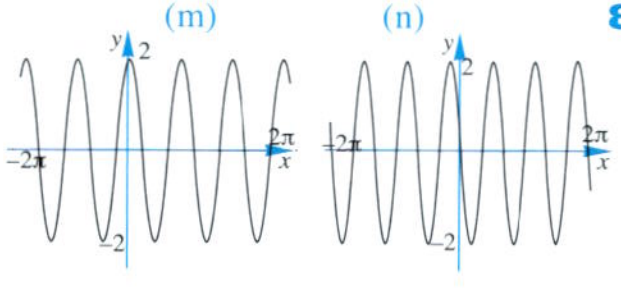

8.

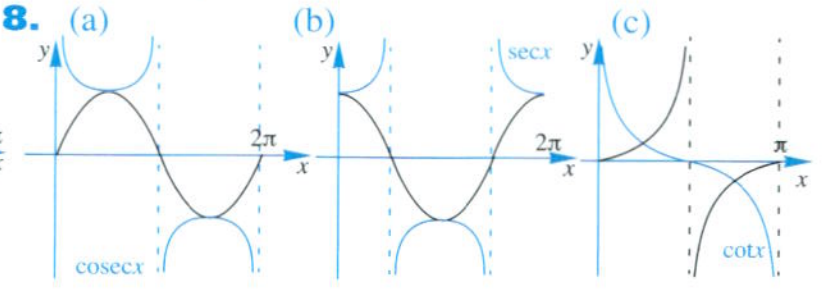

EXERCISE 10.4

1. (a) $\frac{\pi}{4}, \frac{3\pi}{4}$ (b) $\frac{7\pi}{6}, \frac{11\pi}{6}$ (c) $\frac{\pi}{3}, \frac{2\pi}{3}$ (d) $\frac{\pi}{18}, \frac{5\pi}{18}, \frac{13\pi}{18}, \frac{17\pi}{18}, \frac{25\pi}{18}, \frac{29\pi}{18}$ (e) $\frac{\pi}{3}, \frac{5\pi}{3}$ (f) $\frac{5}{4}, \frac{7}{4}, \frac{13}{4}, \frac{15}{4}, \frac{21}{4}, \frac{23}{4}$ **2.** (a) $\frac{\pi}{4}, \frac{7\pi}{4}$ (b) $\frac{2\pi}{3}, \frac{4\pi}{3}$ (c) $\frac{\pi}{6}, \frac{11\pi}{6}$ (d) π (e) $\frac{\pi}{6}, \frac{5\pi}{6}, \frac{7\pi}{6}, \frac{11\pi}{6}$ (f) $\frac{3}{2}, \frac{5}{2}, \frac{11}{2}$ **3.** (a) $\frac{\pi}{6}, \frac{7\pi}{6}$

(b) $\frac{3\pi}{4}, \frac{7\pi}{4}$ (c) $\frac{\pi}{3}, \frac{4\pi}{3}$ (d) $4\tan^{-1}2$ (e) $\frac{\pi}{3}, \frac{5\pi}{6}, \frac{4\pi}{3}, \frac{11\pi}{6}$ (f) 3 **4.** (a) 90°, 330° (b) 180°, 240°

(c) 90°, 270° (d) 65°, 155° (e) $\frac{\pi}{12}, \frac{5\pi}{12}, \frac{13\pi}{12}, \frac{17\pi}{12}$ (f) $0, \pi, 2\pi$ **5.** (a) 60°, 300° (b) $\frac{4\pi}{3}, \frac{5\pi}{3}$ (c) $\frac{\pi}{6}, \frac{7\pi}{6}$

(d) 23°35', 156°25' (e) $\frac{\pi}{3}, \frac{2\pi}{3}, \frac{4\pi}{3}, \frac{5\pi}{3}$ (f) $\frac{2\pi}{3}, \frac{5\pi}{3}$ (g) $\frac{5\pi}{6}, \frac{9\pi}{6}$ (h) $3.3559^c, 5.2105^c$ (i) $\frac{\pi}{3}, \frac{4\pi}{3}$

(j) $\frac{\pi}{3}, \frac{2\pi}{3}, \frac{4\pi}{3}, \frac{5\pi}{3}$ (k) $\frac{\pi}{6}, \frac{2\pi}{3}, \frac{7\pi}{6}, \frac{5\pi}{3}$ (l) 68°12', 248°12' **6.** (a) $-\frac{3\pi}{4}, \frac{\pi}{4}$ (b) $\pm\frac{\pi}{3}$

(c) $-\frac{7\pi}{8}, -\frac{3\pi}{8}, \frac{\pi}{8}, \frac{5\pi}{8}$ (d) $-\frac{\pi}{2}$ (e) $\pm\frac{\pi}{2}$ (f) $\frac{\pi}{8}, \frac{7\pi}{8}, \frac{9\pi}{8}, \frac{15\pi}{8}$ (g) $\frac{\pi}{2}, \frac{3\pi}{2}$ (h) $\frac{\pi}{2}, \frac{3\pi}{2}$

7. (a) $\frac{3\pi}{4}, \frac{7\pi}{4}, \tan^{-1}\left(\frac{2}{3}\right), \pi + \tan^{-1}\left(\frac{2}{3}\right)$ (b) $\frac{\pi}{3}, \frac{2\pi}{3}, \frac{3\pi}{4}, \frac{4\pi}{3}, \frac{5\pi}{3}, \frac{7\pi}{4}$

8. (a) $\frac{\pi}{12}, \frac{5\pi}{12}, \frac{7\pi}{12}, \frac{11\pi}{12}, \frac{13\pi}{12}, \frac{17\pi}{12}, \frac{19\pi}{12}, \frac{23\pi}{12}$ (b) $\frac{2\pi}{3}, \frac{4\pi}{3}$ (c) 0,1,2,3,4,5,6

9. i. $\left(\frac{\pi}{6}, \frac{5\pi}{6}\right) \cup \left(\frac{13\pi}{6}, \frac{17\pi}{6}\right)$ ii. $\left(\pi + \sin^{-1}\left(\frac{1}{\sqrt{3}}\right), 2\pi - \sin^{-1}\left(\frac{1}{\sqrt{3}}\right)\right) \cup \left(3\pi + \sin^{-1}\left(\frac{1}{\sqrt{3}}\right), 4\pi - \sin^{-1}\left(\frac{1}{\sqrt{3}}\right)\right)$

10. (a) ii. $\left[0, \frac{\pi}{4}\right) \cup \left(\frac{5\pi}{4}, 2\pi\right]$ (b) ii. $\left[0, \frac{\pi}{6}\right) \cup \left(\frac{\pi}{2}, \frac{5\pi}{6}\right) \cup \left(\frac{3\pi}{2}, 2\pi\right]$ **11.** (a) 90°,199°28',340°32'
(b)]199°28',340°32'[

EXERCISE 10.5

1. (a) 5, 24, 11, 19 (b) $T = 5\sin\left(\frac{\pi t}{12} - 3\right) + 19$ (c) 23.6° **2.** (a) 3. 4.2, 2, 7

(b) $L = 3\sin\left(\frac{\pi t}{2.1} - 3\right) + 7$ **3.** (a) 5, 11, 0, 7 (b) $V = 5\sin\left(\frac{2\pi t}{11}\right) + 7$ **4.** (a) 1, 11, 1, 12

(b) $P = \sin\left(\frac{2\pi}{11}(t-1)\right) + 12$ **5.** (a) 2.6, 7, 2, 6 (b) $S = 2.6\sin\frac{2\pi}{7}(t-2) + 6$ **6.** (a) 0.6, 3.5, 0, 11

(b) $P = 0.6\sin\left(\frac{4\pi t}{7}\right) + 11$ **7.** (a) 0.8, 4.6, 2.7, 11 (b) $D = 0.8\sin\frac{\pi}{2.3}(t-2.7) + 11$ **8.** (a) 3000

(b) 1000, 5000 (c) $\frac{4}{9}$ **9.** (a) 6.5 m, 7.5 m (b) 1.58 sec, 3.42 sec **10.** (a) 750, 1850 (b) 3.44

(c) Mid-April to End of August

11. (a) 15000 (b) 12 months (c) (d) 4 months

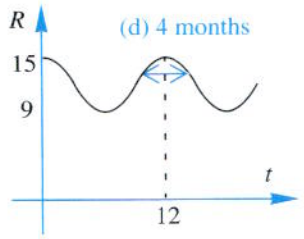

12. (a) π,–2,2 (b) $\frac{1}{3}$ m (c) $\frac{4}{3}$ m

13.(a)

t	0	0.5	1	1.5	2	2.5	3	3.5	4
$F(t)$	6	8	6	4	6	8	6	4	6
$G(t)$	4	4.0625	4.25	4.5625	5	5.5625	6.25	7.0625	8

(b)

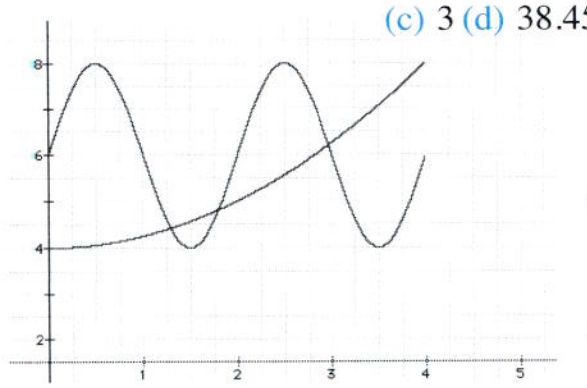

(c) 3 (d) 38.45%

14. (a)

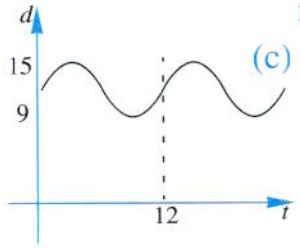

(b) i. 7,11,19,23
ii. $[0, 7] \cup [11, 19] \cup [23, 24]$
(c) 14.9m

EXERCISE 11.1.1

1. (i) $\begin{bmatrix} 2 & 8 \\ -2 & 16 \end{bmatrix}$ (ii) $\begin{bmatrix} -2 & 3 \\ 1 & -5 \end{bmatrix}$ (iii) $\begin{bmatrix} 6 & -9 \\ -3 & 15 \end{bmatrix}$ (iv) $\begin{bmatrix} 4 & 5 \\ -3 & 21 \end{bmatrix}$ (v) $\begin{bmatrix} 7 & 6 \\ -5 & 34 \end{bmatrix}$ (vi) $\begin{bmatrix} -2 & -19 \\ 3 & -27 \end{bmatrix}$

2. (i) $\begin{bmatrix} -1 & -1 & -2 \\ 0 & 6 & -9 \end{bmatrix}$ (ii) $\begin{bmatrix} -2 & -2 & -4 \\ 0 & 12 & -18 \end{bmatrix}$ (iii) $\begin{bmatrix} -2 & -5 \\ 2 & 2 \\ 3 & 1 \end{bmatrix}$ (iv) $\begin{bmatrix} 2 & 0 \\ 4 & 2 \\ 0 & -2 \end{bmatrix}$ (v) $\begin{bmatrix} 5 & 5 \\ 4 & 1 \\ -3 & -4 \end{bmatrix}$ (vi) $\begin{bmatrix} -6 & -15 \\ 6 & 6 \\ 9 & 3 \end{bmatrix}$

3. (i) $\begin{bmatrix} 3 & 3 & 6 \\ 0 & 3 & 12 \\ 0 & 6 & 3 \end{bmatrix}$ (ii) $\begin{bmatrix} -2 & -4 & 6 \\ 4 & 0 & 2 \\ 4 & 0 & -8 \end{bmatrix}$ (iii) $\begin{bmatrix} 0 & -1 & 5 \\ 2 & 1 & 5 \\ 2 & 2 & -3 \end{bmatrix}$ (iv) $\begin{bmatrix} 2 & 3 & -1 \\ -2 & 1 & 3 \\ -2 & 2 & 5 \end{bmatrix}$ (v) $\begin{bmatrix} 1 & -1 & 12 \\ 4 & 3 & 14 \\ 4 & 6 & -5 \end{bmatrix}$ (vi) $\begin{bmatrix} -3 & -5 & 4 \\ 4 & -1 & -2 \\ 4 & -2 & -9 \end{bmatrix}$

4. (i) 23 (ii) Nuts (iii) Taps (iv) Week 2 Wednesday - = B3 + I3, = B4 +I4 **5.** 7 by 5 **6.** $\begin{bmatrix} 3a & 3 \\ 0 & 0 \end{bmatrix}$

7. 3 **8.** 2 **9.** (a) $\begin{bmatrix} 1 & 0 \\ 0 & 1 \end{bmatrix}$ (b) $\begin{bmatrix} -\cos 2\theta & 0 \\ 0 & \cos 2\theta \end{bmatrix}$ (c) $\begin{bmatrix} -\cos 2\theta & 0 \\ 0 & \cos 2\theta \end{bmatrix}$ (d) $\begin{bmatrix} 0 & 0 \\ 0 & 0 \end{bmatrix}$ **10.** (a) $\begin{bmatrix} 1 & 0 \\ 0 & 1 \end{bmatrix}$

(b) $\begin{bmatrix} -\cos 2\theta & 2\sin\theta \\ -2\cos\theta & \cos 2\theta \end{bmatrix}$ **11.** $a = 3, b = 4$ **12.** $x = 1$

EXERCISE 11.1.2

1. (i) $\begin{bmatrix} 5 & -4 \\ 2 & -1 \end{bmatrix}$ (ii) $\begin{bmatrix} 6 & -8 \\ -15 & 26 \end{bmatrix}$ (iii) $\begin{bmatrix} -7 & 17 \\ 3 & 3 \end{bmatrix}$ (iv) $\begin{bmatrix} -15 & -10 \\ -10 & -36 \end{bmatrix}$ (v) $\begin{bmatrix} \frac{19}{6} & \frac{89}{3} \\ -\frac{1}{8} & \frac{11}{2} \end{bmatrix}$ (vi) $\begin{bmatrix} 2.6 & 5.5 \\ -8.55 & 0 \end{bmatrix}$

(vii) $\begin{bmatrix} 12 & -4 & 2 \\ 6 & 0 & -3 \end{bmatrix}$ (viii) $\begin{bmatrix} -9 & 3 & 2 \\ 4 & -2 & -2 \end{bmatrix}$ (ix) $\begin{bmatrix} \frac{7}{6} & \frac{3}{2} & \frac{7}{3} \\ \frac{13}{3} & 5 & \frac{29}{3} \end{bmatrix}$ (x) $\begin{bmatrix} 1.6 & -0.7 & 3.8 \\ -3.9 & -4.2 & -5.7 \end{bmatrix}$ (xi) $\begin{bmatrix} 11 & -2 \\ 11 & 3 \\ 1 & -7 \end{bmatrix}$

(xii) $\begin{bmatrix} 7 & 5 \\ 21 & -26 \\ 6 & 4 \end{bmatrix}$ (xiii) $\begin{bmatrix} 12 & 6 & 2 \\ 12 & 1 & 4 \\ 21 & 3 & 6 \end{bmatrix}$ (xiv) $\begin{bmatrix} -5 & 6 & 1 \\ 4 & -4 & -12 \\ -1 & 4 & 2 \end{bmatrix}$ (xv) $\begin{bmatrix} -7 & 2 & 13 \\ 8 & -8 & -17 \\ -4 & -11 & 4 \end{bmatrix}$ (xvi) $\begin{bmatrix} -\frac{25}{2} & -\frac{50}{3} & 4 \\ -\frac{79}{4} & -\frac{83}{6} & \frac{27}{4} \\ -2 & \frac{11}{6} & \frac{3}{4} \end{bmatrix}$

(xvii) $\begin{bmatrix} x + 3x^2 & -x^2 + 1 \\ 2x + 5x^2 & -x^2 + 2 \\ -x^2 + 5x & -2x \end{bmatrix}$ (xviii) $\begin{bmatrix} a + 2x^2 - a^2 & 2a + 4x & -2x^2 + 2a \\ a - 2ax & -2a & 3a + 2ax \\ 0 & 2a & -ax - x \end{bmatrix}$

2. (i) $\begin{bmatrix} 1 & 0 \\ 0 & 1 \end{bmatrix}$ (ii) $\begin{bmatrix} -1 & 0 \\ 0 & -1 \end{bmatrix}$, $\begin{bmatrix} 1 & 0 \\ 0 & 1 \end{bmatrix}$, $\begin{bmatrix} -1 & 0 \\ 0 & -1 \end{bmatrix}$ (iii) $\begin{bmatrix} -1 & 0 \\ 0 & -1 \end{bmatrix}$ if n is odd and $\begin{bmatrix} 1 & 0 \\ 0 & 1 \end{bmatrix}$ if n is even.

3. (i) $\begin{bmatrix}0&3&4&2\\1&0&0&6\\1&2&0&2\\4&3&0&0\end{bmatrix}\begin{bmatrix}120\\105\\110\\100\end{bmatrix}$ (ii) $\begin{bmatrix}955\\720\\530\\795\end{bmatrix}$ (iii) Loading costs for each depot. **4.** $\begin{bmatrix}0&-1\\1&0\end{bmatrix}$ $n=4$.

5. (a) 2×1 (b) – (c) 3×2 (d) – (e) – (f) 1×3 **6.** (a) No (b) Yes (c) $AB = BA$ **7.** $\begin{bmatrix}n\\ \frac{12-5n}{4}\end{bmatrix}, n \in \mathbb{R}$

8. $a = -0.4, b = 2$ **9.** B **10.** (a) $\begin{bmatrix}1&0\\0&1\end{bmatrix}, S^3 = S$ (b) $\begin{bmatrix}1&0\\0&1\end{bmatrix}, S^{2n+1} = S$ (c) $0, \pm 2\pi, \pm 4\pi, \ldots$

11. $x = 0 = y, w = z$ **12.** 2, 6 **13.** $A = \begin{bmatrix}1&0&1\\1&-1&p^2\\p&1&-(2p+1)\end{bmatrix}, a = 2p, b = p-1, c = p$

15. $x = 3, y = -2, a = -9, b = -4$

17. (a) $A^2 = \begin{bmatrix}2&1\\1&1\end{bmatrix}, A^3 = \begin{bmatrix}3&2\\2&1\end{bmatrix}, A^4 = \begin{bmatrix}5&3\\3&2\end{bmatrix}$ (b) $A^n = \begin{bmatrix}F_{n+1}&F_n\\F_n&F_{n-1}\end{bmatrix}$

EXERCISE 11.2

1. (i) –4 (ii) 9 (iii) 12 (iv) 2 (v) 15 (vi) $\frac{4}{9}$ (vii) 0.5 (viii) 4 (ix) 1 (x) $x^2 - 2$ (xi) 0 (xii) 1

2. (i) $\begin{bmatrix}0&1\\\frac{1}{4}&\frac{1}{2}\end{bmatrix}$ (ii) $\begin{bmatrix}-1&-1\\2&\frac{3}{2}\end{bmatrix}$ (iii) $\begin{bmatrix}-\frac{3}{14}&\frac{1}{7}\\-\frac{2}{7}&-\frac{1}{7}\end{bmatrix}$ (iv) $\begin{bmatrix}1&0\\-\frac{1}{2}&-\frac{1}{4}\end{bmatrix}$ (v) $\begin{bmatrix}0&1\\-1&-4\end{bmatrix}$ (vi) no inverse

(vii) $\begin{bmatrix}-\frac{1}{5}&\frac{1}{5}\\-\frac{4}{15}&-\frac{1}{15}\end{bmatrix}$ (viii) $\begin{bmatrix}1&-\frac{1}{2}\\1&-\frac{3}{4}\end{bmatrix}$ (ix) $\begin{bmatrix}-4&-1\\1&-1\end{bmatrix}$ (x) $\begin{bmatrix}1&3\\1&0\end{bmatrix}$ (xi) $\begin{bmatrix}2&3\\2&-3\end{bmatrix}$ (xii) $\begin{bmatrix}3&3\\1&-2\end{bmatrix}$ (xiii) $\begin{bmatrix}3&-4\\-2&3\end{bmatrix}$

(xiv) $\begin{bmatrix}\frac{3}{x}&-\frac{1}{x}\\-2&1\end{bmatrix}$ (xv) $\begin{bmatrix}-1&1\\\frac{3}{x}&-\frac{2}{x}\end{bmatrix}$ (xvi) $\frac{x-1}{-x^3+x^2+2x-1}\begin{bmatrix}\frac{1}{x-1}&1-x\\-x-1&x\end{bmatrix}$ **3.** –2, 2 **4.** –24 **5.** $-\frac{1}{2}$

6. (a) –29 (b) 80 (c) 0 (d) –11 **7.** (a) $\frac{1}{29}\begin{bmatrix}5&-7&13\\-7&4&5\\4&6&-7\end{bmatrix}$ (b) $\frac{1}{80}\begin{bmatrix}9&10&7\\-1&-10&17\\14&-20&2\end{bmatrix}$ (c) does not exist.

(d) $\frac{1}{11}\begin{bmatrix}2&3&-1\\5&2&3\\4&6&9\end{bmatrix}$ **8.** $-5\frac{1}{5}\begin{bmatrix}1&-2&4\\-1&2&1\\2&1&-2\end{bmatrix}$ **9.** $p = 2\ \frac{1}{35}\begin{bmatrix}5&4&7\\10&-6&7\\5&11&-7\end{bmatrix}$ **11.** $-\frac{23}{4}\ \frac{1}{23}\begin{bmatrix}-15&6&-8\\26&8&20\\-4&-3&4\end{bmatrix}$

12. $x = 1$

13. $-2\begin{bmatrix}11&-9&-10\\-1&1&1\\8&-\frac{13}{2}&-\frac{15}{2}\end{bmatrix}$ **15.** (a) $AB = \begin{bmatrix}5&0&0\\0&5&0\\0&0&5\end{bmatrix}, A^{-1} = \frac{1}{5}\begin{bmatrix}1&2&0\\0&1&2\\1&0&1\end{bmatrix}$

(b) $AB = \begin{bmatrix}1&0&0\\0&1&0\\0&0&1\end{bmatrix}, A^{-1} = \begin{bmatrix}2&3&5\\1&2&3\\1&-1&1\end{bmatrix}$ (c) $AB = \begin{bmatrix}3&0&0\\0&3&0\\0&0&3\end{bmatrix}, A^{-1} = \frac{1}{3}\begin{bmatrix}1&1&1\\2&-1&2\\1&3&2\end{bmatrix}$

16. (a) 0,1,3 (b) –2,0 **18.** (d) $\alpha = 14, \beta = 3$ (e) 1 (or 0, but we will assume that A is not singular). **21.** (a) –2 (b) $k = -1$

EXERCISE 11.3

1. (i) (1,–1) (ii) (4,–3) (iii) (–3,–3) (iv) (–2,–1) (v) (2,5) (vi) (1,0) (vii) (0,3) (viii) (–2,4)

(ix) (0.5,2) (x) (2,0.6) (xi) (–1,–3) (xii) (–2,2) (xiii) $\left(3, \frac{1}{3}\right)$ (xiv) (5,–2) (xv) (5,4)

2. (i) $a = -2\ b = -4$ (ii) $a = -2\ b \neq -4$ **3.** ±2 **4.** (a) (1,–3,2) (b) (5, –3, –1) (c) (–0.6, –4.2, –1.4)

5. (78,29,–47) **6.** $\left(\frac{4}{11}, \frac{9}{11}, \frac{7}{11}\right)$ **7.** (a) $m \neq -1, 2$ (b) $m = -1$ (c) $m = 2$ **8.** (a) $p \neq 1.2$

(b) $p = 1.2, k \neq 6$ (c) $p = 1.2, k = 6$ **9.** (a) (4,0) (b) Ø (c) (2,–3) (d) (2, 3, –2)
(e) $(4+2\lambda, 3-\lambda, \lambda)$ (f) (1, 2, –2) (g) Ø (h) $(2\lambda - 1, \lambda, \lambda)$ (i) Ø

10. $y = x^2 + 6x - 14$ **11.** $y = -\left(\frac{\lambda}{2}+6\right)x^2 + \left(\frac{\lambda}{2}+13\right)x + \lambda$ **12.** (a) i. $\left(-\frac{7\lambda}{11}, -\frac{\lambda}{11}, \lambda\right)$

ii. $\left(\frac{7-7\lambda}{11}, \frac{1-\lambda}{11}, \lambda\right)$ (b) (0,0,0) **13.** (a) 1,–2 (b) i. 1,–2 ii. neither 1 nor –2 iii. Ø

EXERCISE 12.1

1. vector **2.** scalar **3.** scalar **4.** vector **5.** vector **6.** vector **7.** scalar **8.** scalar

EXERCISE 12.2

1. (a) (b) (c) (d)

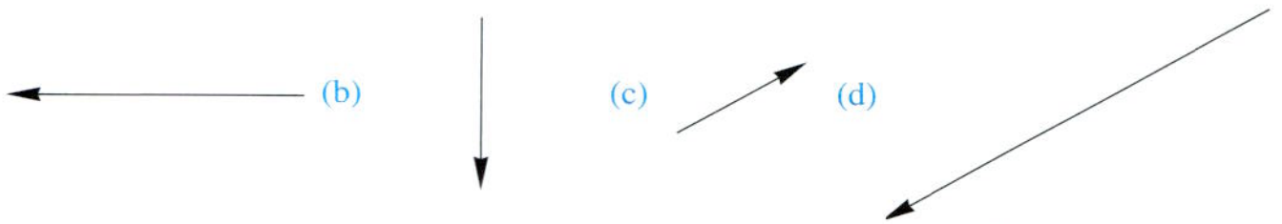

2. (a) (b) (c) (d)

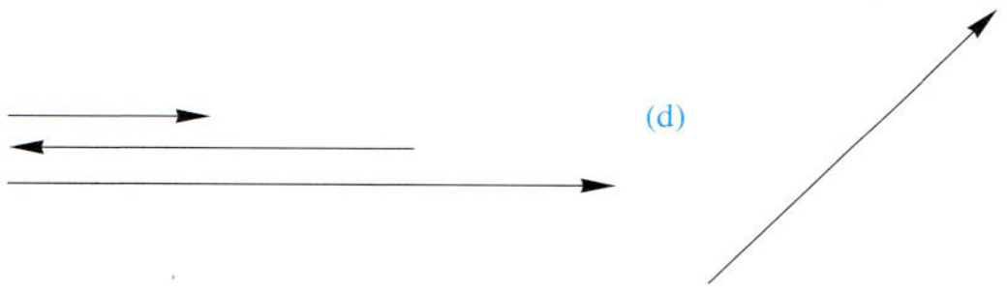

3. (a) {a, b, e, g, u} (b) {a, c}, {d, f}, {b, e} (c) {a,c}, {c, g} (d) {d, f}, {b, e}
(e) {d, f},{b, e},{a, c}

4.

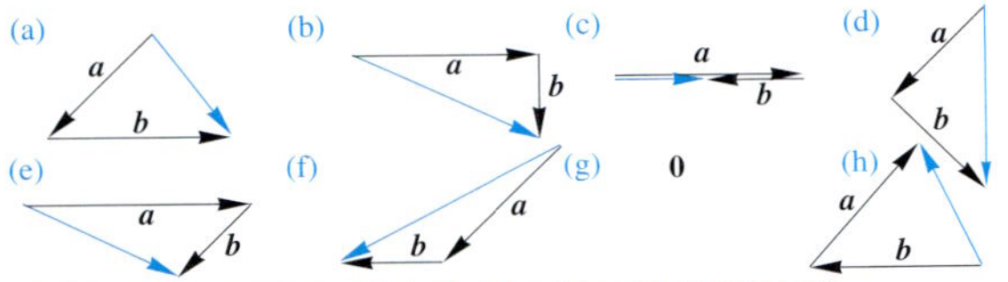

5. (a) **AC** (b) **AB** (c) **AD** (d) **BA** (e) **0** **6.** (a) Y (b) N (c) Y (d) Y (e) N

7. (a) i.

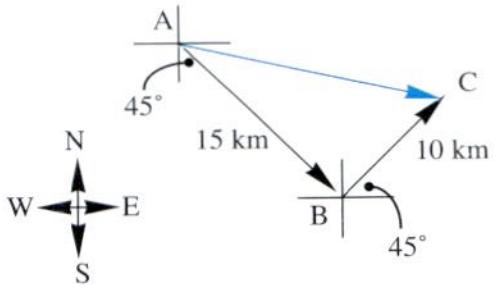

ii. $\sqrt{325}$

(b) i.

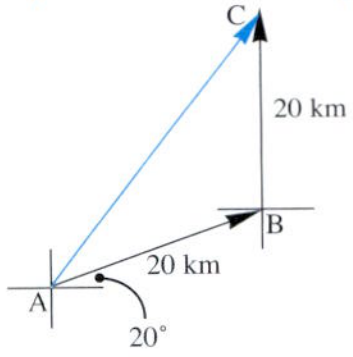

ii. $20\sqrt{2(1-\cos 110°)}$

(c) i.

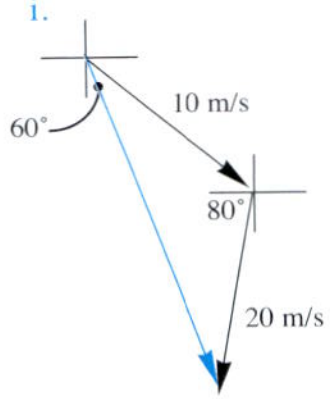

ii. $10\sqrt{5-4\cos 110°}$

8. 72.11 N, E 33°41′ N **9.** 2719 N along river **10.** (b) i. 200 kph N ii. 213.6 kph, N 7°37′ W
11. i. 200 ii. 369.32

EXERCISE 12.3

1. (a) $\boldsymbol{c}-\boldsymbol{a}$ (b) $\boldsymbol{b}-\boldsymbol{c}$ (c) $\frac{1}{2}(\boldsymbol{b}+\boldsymbol{a})$ **2.** (a) $\boldsymbol{b}-\boldsymbol{a}$ (b) $\boldsymbol{b}-2\boldsymbol{a}$ (c) $2\boldsymbol{b}-3\boldsymbol{a}$ (d) $\frac{1}{2}(\boldsymbol{b}+2\boldsymbol{a})$

3. (a) **0** (b) **PS** (c) **AY** (d) 6**OC** **4.** (a) $\frac{1}{2}(\boldsymbol{b}+\boldsymbol{a})$ (b) $\frac{1}{3}(2\boldsymbol{b}+\boldsymbol{a})$ (c) $\frac{1}{4}(\boldsymbol{a}+\boldsymbol{b}+2\boldsymbol{c})$

7. (a) $\boldsymbol{c}-\boldsymbol{b}$ (b) $\boldsymbol{c}+\boldsymbol{a}$ (c) $\boldsymbol{a}+\boldsymbol{c}-2\boldsymbol{b}$ **8.** (a) $2\sqrt{21}$ (b) $2\sqrt{26}$ **15.** $m=\frac{13}{23}, n=\frac{50}{23}$ **16.** $m=\frac{4}{3}$

EXERCISE 12.4

1. (i) $4\boldsymbol{i}+28\boldsymbol{j}-4\boldsymbol{k}$ (ii) $12\boldsymbol{i}+21\boldsymbol{j}+15\boldsymbol{k}$ (iii) $-2\boldsymbol{i}+7\boldsymbol{j}-7\boldsymbol{k}$ (iv) $-6\boldsymbol{i}-12\boldsymbol{k}$
2. (i) $3\boldsymbol{i}-4\boldsymbol{j}+2\boldsymbol{k}$ (ii) $-8\boldsymbol{i}+24\boldsymbol{j}+13\boldsymbol{k}$ (iii) $18\boldsymbol{i}-32\boldsymbol{j}+\boldsymbol{k}$ (iv) $-15\boldsymbol{i}+36\boldsymbol{j}+12\boldsymbol{k}$

3. (i) $\begin{pmatrix}11\\0\\8\end{pmatrix}$ (ii) $\begin{pmatrix}-27\\1\\-22\end{pmatrix}$ (iii) $\begin{pmatrix}-3\\-6\\12\end{pmatrix}$ (iv) $\begin{pmatrix}16\\-1\\14\end{pmatrix}$ **4.** $\begin{pmatrix}-5\\3\end{pmatrix}$ **5.** $\begin{pmatrix}-2\\3\end{pmatrix}$, (–2,3)

6. (i) $8\boldsymbol{i}-4\boldsymbol{j}-28\boldsymbol{k}$ (ii) $-19\boldsymbol{i}-7\boldsymbol{j}-16\boldsymbol{k}$ (iii) $-17\boldsymbol{i}+\boldsymbol{j}+22\boldsymbol{k}$ (iv) $40\boldsymbol{i}+4\boldsymbol{j}-20\boldsymbol{k}$

7. (i) $\begin{pmatrix}20\\1\\25\end{pmatrix}$ (ii) $\begin{pmatrix}12\\2\\16\end{pmatrix}$ (iii) $\begin{pmatrix}-4\\-38\\-32\end{pmatrix}$ (iv) $\begin{pmatrix}-20\\-22\\-40\end{pmatrix}$ **8.** A = –4, B = –7

9. i. (2,–5) ii. (–4,3) iii. (–6,–5) **10.** Depends on basis used. Here we used: East as $\boldsymbol{i}$, North $\boldsymbol{j}$ and vertically up $\boldsymbol{k}$ (b) $D=600\boldsymbol{i}-800\boldsymbol{j}+60\boldsymbol{k}$, $A=-1200\boldsymbol{i}-300\boldsymbol{j}+60\boldsymbol{k}$ (c) $1800\boldsymbol{i}-500\boldsymbol{j}$

EXERCISE 12.5

1. (i) $\sqrt{10}$ (ii) $5\sqrt{2}$ (iii) $\sqrt{30}$ (iv) 3 (v) $\sqrt{53}$ (vi) $\sqrt{41}$ (vii) $\sqrt{14}$ (viii) $\sqrt{17}$

2. (i) $\frac{1}{\sqrt{2}}(\boldsymbol{i}+\boldsymbol{j})$ (ii) $\frac{1}{\sqrt{41}}(4\boldsymbol{i}+5\boldsymbol{j})$ (iii) $\frac{1}{\sqrt{5}}(-\boldsymbol{i}-2\boldsymbol{j})$ (iv) $\frac{1}{\sqrt{46}}(\boldsymbol{i}+6\boldsymbol{j}-3\boldsymbol{k})$ (v) $\frac{1}{\sqrt{5}}(\boldsymbol{i}+2\boldsymbol{k})$

(vi) $\frac{1}{\sqrt{17}}(2\boldsymbol{i}-2\boldsymbol{j}-3\boldsymbol{k})$ (vii) $\frac{1}{3}\begin{pmatrix}2\\1\\2\end{pmatrix}$ (viii) $\frac{1}{3\sqrt{3}}\begin{pmatrix}-1\\5\\1\end{pmatrix}$ **3.** i. Depends on the basis: $-3\boldsymbol{i}+4\boldsymbol{j}+\boldsymbol{k}$

or $-4\boldsymbol{i}-3\boldsymbol{j}+\boldsymbol{k}$ ii. $\sqrt{26}$ **4.** (a) $\sqrt{3}(\boldsymbol{i}-\boldsymbol{j}+\boldsymbol{k})$ (b) $\frac{1}{4}(3\boldsymbol{i}-\boldsymbol{j}+\sqrt{2}\boldsymbol{k})$ **5.** $\pm\sqrt{11}$ **6.** $\sqrt{13}$

EXERCISE 12.6

1. (a) 4 (b) –11.49 (c) 25 **2.** (i) 12 (ii) 27 (iii) –8 (iv) –49 (v) 4 (vi) –21 (vii) 6 (viii) –4 (ix) –10 **3.** (i) 79° (ii) 108° (iii) 55° (iv) 50° (v) 74° (vi) 172° (vii) 80° (viii) 58° **4.** (i) –8 (ii) 0.5
5. (a) –6 (b) 2 (c) not possible (d) 5 (e) not possible (f) 0 **6.** (a) $4-2\sqrt{3}$ (b) $2\sqrt{3}-4$
(c) $14-2\sqrt{3}$ (d) not possible **7.** 1 **8.** 105.2° **9.** $x=-\frac{16}{7}, y=-\frac{44}{7}$ **10.** $\pm\frac{1}{\sqrt{11}}(-\boldsymbol{i}+\boldsymbol{j}+3\boldsymbol{k})$

12. (a) $\lambda(-16\boldsymbol{i}-10\boldsymbol{j}+\boldsymbol{k})$ (b) e.g. $\boldsymbol{i}+\boldsymbol{j}+\frac{3}{7}\boldsymbol{k}$ **14.** $\boldsymbol{a}\perp\boldsymbol{b}-\boldsymbol{c}$ if $\boldsymbol{b}\neq\boldsymbol{c}$ or $\boldsymbol{b}=\boldsymbol{c}$ **15.** (a) $\left(\frac{3}{5},\frac{4}{5}\right)$

(b) $\left(\frac{\sqrt{2}}{2},\frac{1}{2},-\frac{1}{2}\right)$ **16.** (a) $\left(-\frac{2}{3},\frac{2}{3},\frac{1}{3}\right)$ (b) 131.8°, 48.2°, 70.5° **18.** (a) $\frac{1}{3}$ (b) $\frac{1}{\sqrt{3}}$

19. (a)

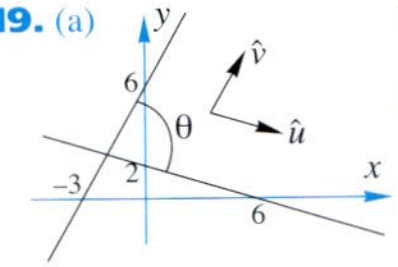

(b) $\hat{u}=\frac{1}{\sqrt{10}}(3\boldsymbol{i}-\boldsymbol{j})$, $\hat{v}=\frac{1}{\sqrt{5}}(\boldsymbol{i}+2\boldsymbol{j})$ **20.** $\frac{1}{2}(-\boldsymbol{i}+\boldsymbol{j}+\sqrt{2}\boldsymbol{k})$
(c) 81.87°

25. (a) Use $\boldsymbol{i}$ as a 1 km eastward vector and $\boldsymbol{j}$ as a 1 km northwards vector (b) $\overrightarrow{WD}=4\boldsymbol{i}+8\boldsymbol{j}$, $\overrightarrow{WS}=13\boldsymbol{i}+\boldsymbol{j}$ and $\overrightarrow{DS}=9\boldsymbol{i}-7\boldsymbol{j}$ (c) $\frac{1}{\sqrt{80}}(4\boldsymbol{i}+8\boldsymbol{j})$ (d) $\frac{d}{\sqrt{80}}(4\boldsymbol{i}+8\boldsymbol{j})$ (e) $3\boldsymbol{i}+6\boldsymbol{j}$

EXERCISE 12.7.1

1. (a) i. $\boldsymbol{r}=\boldsymbol{i}+2\boldsymbol{j}$ ii. $\boldsymbol{r}=-5\boldsymbol{i}+11\boldsymbol{j}$ iii. $\boldsymbol{r}=5\boldsymbol{i}-4\boldsymbol{j}$ (b) line joins (1,2) & (5,–4)
2. (a) $\boldsymbol{r}=2\boldsymbol{i}+5\boldsymbol{j}+\lambda(3\boldsymbol{i}-4\boldsymbol{j})$ (b) $\boldsymbol{r}=-3\boldsymbol{i}+4\boldsymbol{j}+\lambda(-\boldsymbol{i}+5\boldsymbol{j})$ (c) $\boldsymbol{r}=\boldsymbol{j}+\lambda(7\boldsymbol{i}+8\boldsymbol{j})$

(d) $\boldsymbol{r}=\boldsymbol{i}-6\boldsymbol{j}+\lambda(2\boldsymbol{i}+3\boldsymbol{j})$ (e) $\boldsymbol{r}=\begin{pmatrix}-1\\-1\end{pmatrix}+\lambda\begin{pmatrix}-2\\10\end{pmatrix}$ or $\boldsymbol{r}=-\boldsymbol{i}-\boldsymbol{j}+\lambda(-2\boldsymbol{i}+10\boldsymbol{j})$

(f) $\boldsymbol{r}=\begin{pmatrix}1\\2\end{pmatrix}+\lambda\begin{pmatrix}5\\1\end{pmatrix}$ or $\boldsymbol{r}=\boldsymbol{i}+2\boldsymbol{j}+\lambda(5\boldsymbol{i}+\boldsymbol{j})$

3. (a) $\boldsymbol{r}=2\boldsymbol{i}+3\boldsymbol{j}+\lambda(2\boldsymbol{i}+5\boldsymbol{j})$ (b) $\boldsymbol{r}=\boldsymbol{i}+5\boldsymbol{j}+\lambda(-3\boldsymbol{i}-4\boldsymbol{j})$ (c) $\boldsymbol{r}=4\boldsymbol{i}-3\boldsymbol{j}+\lambda(-5\boldsymbol{i}+\boldsymbol{j})$
4. (a) $\boldsymbol{r}=9\boldsymbol{i}+5\boldsymbol{j}+\lambda(\boldsymbol{i}-3\boldsymbol{j})$ (b) $\boldsymbol{r}=6\boldsymbol{i}-6\boldsymbol{j}+t(-4\boldsymbol{i}-2\boldsymbol{j})$ (c) $\boldsymbol{r}=-\boldsymbol{i}+3\boldsymbol{j}+\lambda(-4\boldsymbol{i}+8\boldsymbol{j})$
(d) $\boldsymbol{r}=\boldsymbol{i}+2\boldsymbol{j}+\mu\left(\frac{1}{2}\boldsymbol{i}-\frac{1}{3}\boldsymbol{j}\right)$ **5.** (a) $\begin{matrix}x=-8+2\mu\\y=10+\mu\end{matrix}$ (b) $\begin{matrix}x=7-3\mu\\y=4-2\mu\end{matrix}$ (c) $\begin{matrix}x=5+2.5\mu\\y=3+0.5\mu\end{matrix}$

(d) $\begin{array}{l} x = 0.5 - 0.1t \\ y = 0.4 + 0.2t \end{array}$ **6.** (a) $\frac{x-1}{3} = y - 3$ (b) $\frac{x-2}{-7} = \frac{y-4}{-5}$ (c) $x + 2 = \frac{y+4}{8}$

(d) $x - 0.5 = \frac{y-0.2}{-11}$ (e) $x = 7$ (f) $y = 6$ **7.** (a) $r = 2j + t(3i + j)$ (b) $r = 5i + t(i + j)$

(c) $\boldsymbol{r} = -6i + t(2i + j)$ **8.** (a) $6i + 13j$ (b) $-\frac{16}{3}i - \frac{28}{3}j$ **9.** $r = 2i + 7j + t(4i + 3j)$

11. (a) (4,–2), (–1,1), (9,–5) (b) –2 (d) $r = 4i - 2j + \lambda(-5i + 3j)$ (e) i. $M \parallel L$ ii. $M = L$

12. $4x + 3y = 11$ **13.** (a) $\frac{-3}{\sqrt{13}}, \frac{2}{\sqrt{13}}$ (b) $\frac{4}{5}, \frac{3}{5}$ **14.** (b) ii. & iii. **15.** (–83,–215)

16. $r = \frac{k}{7}(19i + 20j)$ **17.** (a) $\left(\frac{92}{11}, \frac{31}{11}\right)$ (b) Ø (c) Lines are coincident, all points are common.

EXERCISE 12.7.2

1. (a) No (b) 52.5 mins after A **2.** (a) i. $\boldsymbol{r}_A = \begin{pmatrix}5\\-1\end{pmatrix} + t\begin{pmatrix}3\\4\end{pmatrix}$ ii. $\boldsymbol{r}_B = \begin{pmatrix}4\\5\end{pmatrix} + t\begin{pmatrix}2\\-1\end{pmatrix}$ (b) No

(c) i. $\begin{pmatrix}4\\5\end{pmatrix} + (t-1)\begin{pmatrix}2\\-1\end{pmatrix}$ ii. 11 am

EXERCISE 12.7.3

1. (a) $r = 2i + j + 3k + t(i - 2j + 3k)$ (b) $r = 2i - 3j - k + t(-2i + k)$

2. (a) $r = 2i + 5k + t(i + 4j + 3k)$ (b) $r = 3i - 4j + 7k + t(4i + 9j - 5k)$

(c) $r = 4i + 4j + 4k + t(7i + 7k)$ **3.** (a) $\frac{x}{3} = \frac{y-2}{4} = \frac{z-3}{5}$ (b) $\frac{x+2}{5} = \frac{z+1}{-2}, y = 3$

(c) $x = y = z$ **4.** $\begin{array}{l} x = 5 - 7t \\ y = 2 + 2t \\ z = 6 - 4t \end{array}$ $r = \begin{pmatrix}5\\2\\6\end{pmatrix} + t\begin{pmatrix}-7\\2\\-4\end{pmatrix}$ $\frac{x-5}{-7} = \frac{y-2}{2} = \frac{z-6}{-4}$ **5.** $\left(\frac{13}{5}, \frac{23}{5}, 0\right)$

6. (a) $\begin{array}{l} x = 2 + 3t \\ y = 5 + t \\ z = 4 + 0.5t \end{array}$ (b) $\begin{array}{l} x = 1 + 1.5t \\ y = t \\ z = 4 - 2t \end{array}$ (c) $\begin{array}{l} x = 3 - t \\ y = 2 - 3t \\ z = 4 + 2t \end{array}$ (d) $\begin{array}{l} x = 1 + 2t \\ y = 3 + 2t \\ z = 2 + 0.5t \end{array}$

7. (a) $\frac{x-4}{3} = \frac{y-1}{-4} = \frac{z+2}{-2}$ (b) $x = 2, y = \frac{z-1}{-3}$ **9.** (a) $\frac{x+1}{2} = y - 3 = \frac{z-5}{-1}$

(b) $\frac{x-2}{2} = \frac{z-1}{-2}, y = -1$ **10.** (a) (1, –1, 0) (b) $a = 15, b = -11$

11. (a) $\begin{array}{l} x = 1 + t \\ y = 4 - t \\ z = -2 \end{array}$ (b) $\begin{array}{l} x = 2 + 2t \\ y = 1 \\ z = 3 \end{array}$

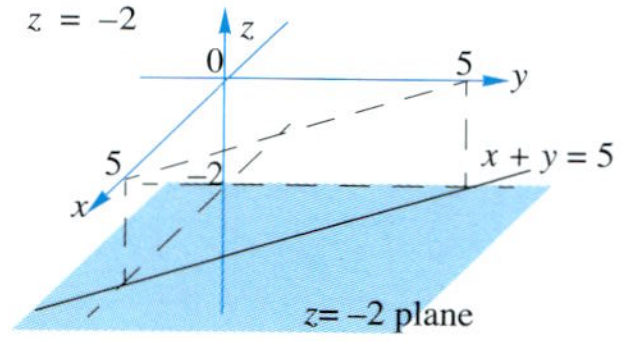

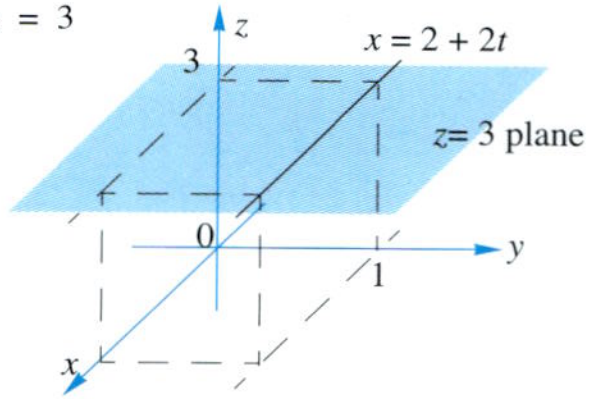

12. $\boldsymbol{r} = \begin{pmatrix}1\\0.5\\2\end{pmatrix} + t\begin{pmatrix}2\\-1.5\\1\end{pmatrix}$. Line passes through (1, 0.5, 2) and is parallel to the vector $2i - \frac{3}{2}j + k$

13. (a) 54.74° (b) 82.25° (c) 57.69° **14.** (a) (4, 10.5, 15) (b) Does not intersect.

15. (a) L: $x = \frac{y-2}{2} = \frac{z}{5}$, M: $\frac{x+1}{2} = \frac{y+1}{3} = \frac{z-1}{-2}$ (b) Ø (c) 84.92° (d) i. (0, 2, 0) ii. $\left(0, \frac{1}{2}, 0\right)$

18. $\frac{x}{4} = \frac{y}{9} = \frac{z}{3}$ **19.** $k = -\frac{7}{2}$ **20.** 64° **21.** 3 or –2 **22.** $12i + 6j - 7k$ (or any multiple thereof) **23.** Not parallel. Do not intersect. Lines are skew.

EXERCISE 13.1

1. (a) i. 14 500 ii. 2 000 (b) 305 (304.5) **2.** Sample size is large but may be biassed by factors such as the location of the catch. Population estimate is 5 000. **3.** (a) i. 1500 ii. 120 (b) 100 (c) 1 000 **4.** (a), (c) numerical, (b), (d), (e) categorical **5.** (a), (d) discrete, (b), (c), (e) continuous

EXERCISE 13.2

1.

218-220	221-223	224-226	227-229	230-232	233-235	236-238	239-241	242-244	245-247
1	4	4	3	6	8	9	5	7	1

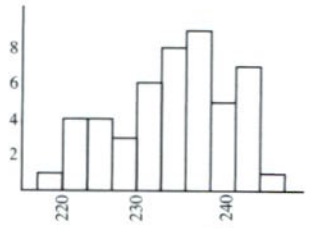

2.

1.1-1.2	1.2-1.3	1.3-1.4	1.4-1.5	1.5-1.6	1.6-1.7	1.7-1.8	1.8-1.9	1.9-2.0	2.0-2.1
5	1	2	2	7	6	1	12	7	5

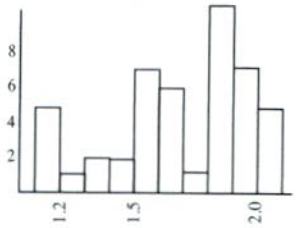

3. Set A Mode = 29.1 Mean = 27.2 Median = 27.85
Set B Mode = 9 Mean = 26.6 Median = 9. Set B is much more spread out than set A and although the two sets have a similar mean, they have very different mode and median.

EXERCISE 13.3

1. Mode = 236-238gms Mean = 234gms Median = 235gms **2.** Mode = 1.8-1.9gms Mean = 1.69gms Median = 1.80gms **3.** Set A Mode = 29.1 Mean = 27.2 Median = 27.85 Set B Mode = 9 Mean = 26.6 Median = 9. **4.** (a) \$27522 (b) \$21025 (c) Median **5.** i. \$233 300 ii. \$169 000 iii. Median **6.** (a) 14.375 (b) 14.354

EXERCISE 13.4

1. (a) Sample A Mean = 1.99kg Sample B Mean = 2.00kg (b) Sample A Sample std = 0.0552kg Sample B Sample std = 0.1877kg (c) Sample A Population std = 0.0547kg Sample B Population std = 0.1858kg **2.** (a) 16.41 (b) 6.84 **3.** Mean = 49.97 Std = 1.365

EXERCISE 13.5

1. (a) Med = 5, Q1 = 2, Q3 = 7, IQR = 5 (b) Med = 3.3, Q1 = 2.8, Q3 = 5.1, IQR = 2.3 (c) Med = 163.5, Q1 = 143, Q3 =182, IQR = 39 (d) Med = 1.055, Q1 = 0.46, Q3 = 1.67, IQR = 1.21 (e) Med = 5143.5, Q1 = 2046, Q3 = 6252, IQR = 4206
2. (a) Med = 3, Q1 = 2, Q3 = 4, IQR = 2 (b) Med = 13, Q1 = 12, Q3 = 13, IQR = 1 (c) Med = 2, Q1 = 2, Q3 = 2.5, IQR = 0.5 (d) Med = 40, Q1 = 30, Q3 = 50, IQR = 20

(e) Med = 20, Q1 = 15, Q3 = 22.5, IQR = 7.5
3. (a) \$84.67 (b) \$147.85 (c) \$11 (d) Q1 = \$4.50, Q3 = \$65 IQR = \$60.50 (e) Median & IQR.
4. (a) 2.35 (b) 1.25 (c) 2 (d) Q1 = 1, Q3 = 3, IQR = 2 **5.** (a) \$232 (b) \$83

(c-e)
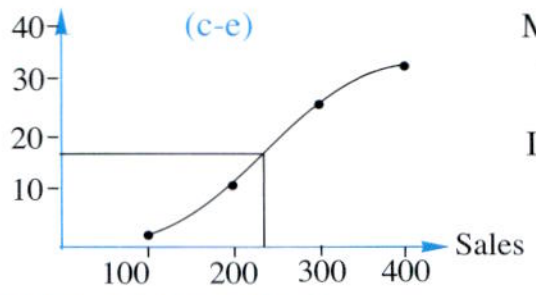

Med = \$220
Q1 = \$160
Q3 = \$310
IQR = \$150

6.
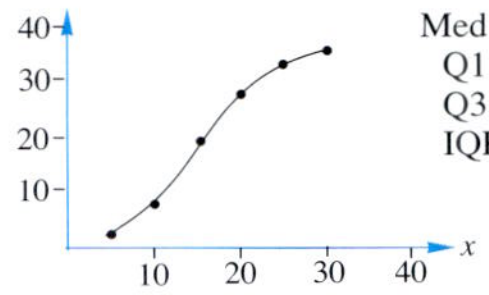

Med = 14
Q1 = 10
Q3 = 19
IQR = 9

EXERCISE 13.6

1. (a) Sample–100 randomly selected patients, population – all suffering from AIDS
(b) Sample–1000 working aged people in N.S.W, population – all working aged people in N.S.W
(c) Sample – John's I.B Higher Maths class, population – all seniors at Nappa Valley High School.

2. Discrete: a, b, d; Continuous: c, e, f, g.

3. (b)
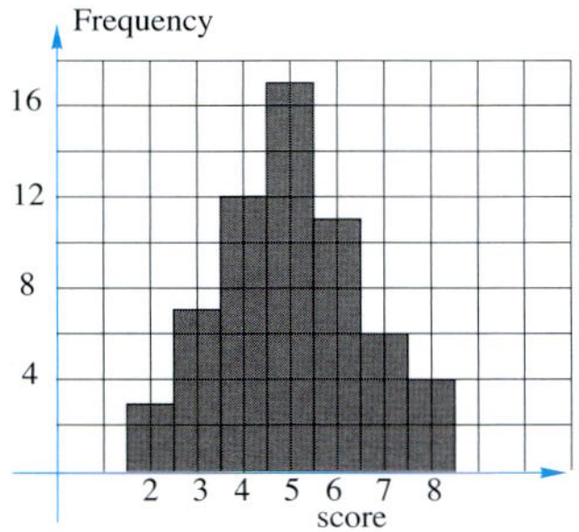

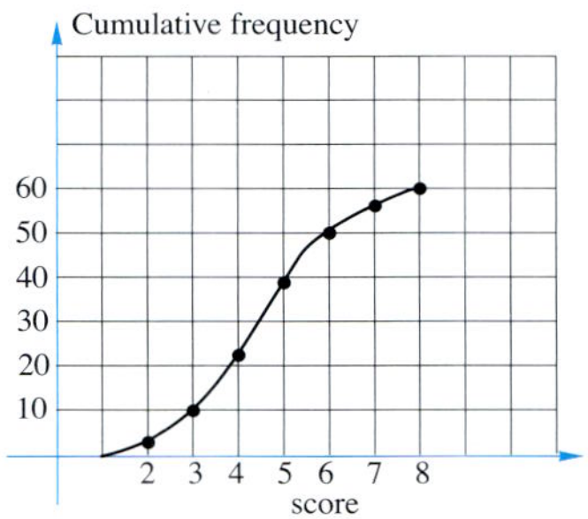

4. suggested answers only: (a) 200–224; 225–249; 250–274; . . . 575–599
(b) 100–119; 120–139; . . . 400–419 (c) 440–459; 460–479; . . . 780–799.

5. Make use of your graphics calculator

6. (a) 16 (b) graphics calculator (c) 15.23
(d) 15.5 (e) Q1 = 14, Q3 = 17 (f) 15.87 (2dec.pl.)

7. (a) 30–34 (b) graphics calculator (c) 30.4
(d) 32 (approx)

8. (b) 215.5 (c) 216.2 **9.** 48.17

10. (a) Q1~ 35, Q3~ 95 (b) ~ 104 (c) 60% (d) 67.15

11. range = 19, s = 5.49 **12.** 5.8; 1.50 **13.** 17.4; $s_n = 3.12$ $s_{n-1} = 3.18$

14. (a) 6 (b) 7 (c) Q1 = 5, Q3 = 7
(d) 2 (e) 6.15 (f) 1.61

15. $s_n = 18.8$, $s_{n-1} = 19.1$ **16.** 14.18

EXERCISE 14.1

1. 15 **2.** (a) 25 (b) 625 **3.** (a) 24 (b) 256 **4.** (a) 24 (b) 48 **5.** 15 **6.** 270 **7.** 120 **8.** 336 **9.** 60
10. (a) 362880 (b) 80640 (c) 1728 **11.** 20 **12.** (a) 10! (b) 2×8! (c) i. 2×9! ii. 8×9! **13.** 34650
14. 4200 **15.** 4

EXERCISE 14.2

1. 792 **2.** (a) 1140 (b) 171 **3.** 1050 **4.** 70 **5.** 2688 **6.** (a) 210 (b) 420 **7.** 24000 **8.** 8 **9.** 155
10. 5

EXERCISE 14.3

1. (a) 120 (b) 325 **2.** 5040 **3.** 5040; (a)144 (b) 1440 **4.** (a) 720 (b) 240 **5.** 11760 **6.** 7056; 4606
7. (a) 840 (b) 1680 **8.** 190 **9.** 10080 **10.** 226800 **11.** (a) 715 (b) 315 (c) 665 **13.** ${}^nC_2$ **14.** ${}^nC_4$
15. (b) 92 **16.** 252 **17.** (a) 1287 (b) 560 **18.** 256 **19.** 288 **20.** (a) 10080 (b) 30240 (c) 14400
21. 10080, 1080 **22.** 3528000 **23.** 720; 240 **24.** 103680 **25.** (a) 12 (b) 128 **26.** 2880
27. (a) 30030 (b) 37310 **28.** 77055 **29.** (a) 48 (b) 72

EXERCISE 15.1

1. (a) $\frac{2}{5}$ (b) $\frac{3}{5}$ (c) $\frac{2}{5}$ **2.** (a) $\frac{2}{7}$ (b) $\frac{5}{7}$ **3.** (a) $\frac{5}{26}$ (b) $\frac{21}{26}$ **4.** {HH, HT, TH, TT} (a) $\frac{1}{4}$ (b) $\frac{3}{4}$

5. {HHH,HHT,HTH,THH,TTT,TTH,THT,HTT} (a) $\frac{3}{8}$ (b) $\frac{1}{2}$ (c) $\frac{1}{4}$ **6.** (a) $\frac{2}{9}$ (b) $\frac{2}{9}$ (c) $\frac{2}{3}$ (d) $\frac{1}{3}$

7. (a) $\frac{1}{2}$ (b) $\frac{3}{10}$ (c) $\frac{9}{20}$ **8.** (a) $\frac{11}{36}$ (b) $\frac{1}{18}$ (c) $\frac{1}{6}$ (d) $\frac{5}{36}$ **9.** {GGG, GGB. GBG, BGG, BBB, BBG,
BGB, GBB} (a) $\frac{1}{8}$ (b) $\frac{3}{8}$ (c) $\frac{1}{2}$ **10.** (a) $\frac{1}{2}$ (b) $\frac{1}{4}$ (c) $\frac{1}{4}$ **11.** (a) $\frac{3}{8}$ (b) $\frac{1}{4}$ (c) $\frac{3}{8}$ (d) $\frac{3}{4}$

12. (a){(1, H),(2, H),(3, H),(4,H),(5, H),(6, H),(1, T),(2, T),(3, T),(4, T),(5, T),(6,T)}
(b) $\frac{1}{4}$ **13.** (a) $\frac{1}{216}$ (b) $\frac{1}{8}$ (c) $\frac{3}{8}$

EXERCISE 15.2

1. (a) $\frac{1}{4}$ (b) $\frac{5}{8}$ (c) $\frac{3}{4}$ **2.** (a) $\frac{1}{13}$ (b) $\frac{1}{2}$ (c) $\frac{1}{26}$ (d) $\frac{7}{13}$ **3.** $\frac{9}{26}$ **4.** (a) 1.0 (b) 0.3 (c) 0.5 **5.** (a) 0.65
(b) 0.70 (c) 0.65 **6.** (a) 0.95 (b) 0.05 (c) 0.80 **7.** (a) {TTT,TTH,THT,HTT,HHH,HHT,HTH,
THH} (b) $\frac{3}{8}$ (c) $\frac{1}{2}$ (d) $\frac{1}{4}$ (e) $\frac{3}{8}$ **8.** (a) $\frac{6}{25}$ (b) $\frac{6}{25}$ (c) $\frac{13}{25}$ **9.** (b) $\frac{3}{4}$ (c) $\frac{1}{2}$ (d) $\frac{1}{6}$ (e) $\frac{7}{13}$

10. (a) $\frac{1}{4}$ (b) $\frac{1}{2}$ (c) $\frac{8}{13}$ (d) $\frac{7}{13}$ **11.** (a) 0.1399 (b) 0.8797 (c) 0.6 **12.** (b) $\frac{4}{15}$ (c) $\frac{4}{15}$ (d) $\frac{11}{15}$

EXERCISE 15.3

1. (a) 0.7 (b) 0.75 (c) 0.50 (d) 0.5 **2.** (a) 0.5 (b) 0.83 (c) 0.10 (d) 0.90

3. (a)
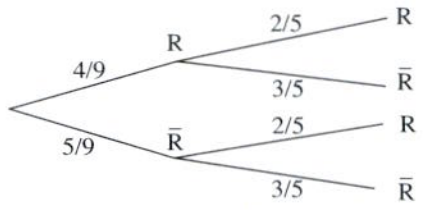

(b) $\frac{8}{45}$ (c) $\frac{22}{45}$ (d) $\frac{6}{11}$ **4.** (a) 0.5 (b) 0.30 (c) 0.25

5. (a)
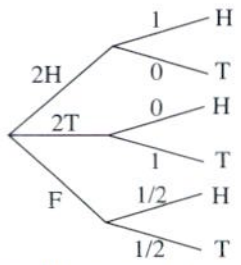

(b) $\frac{1}{2}$ (c) $\frac{2}{3}$ **6.** $\frac{1}{3}$ **7.** (a)
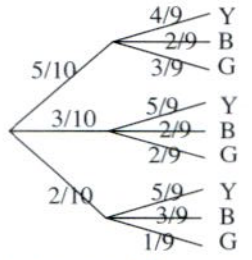

(b) $\frac{31}{45}$ (c) $\frac{2}{9}$ **8.** $\frac{2}{3}$

9. (a) 0.88 (b) 0.42 (c) 0.6 (d) 0.28 **10.** (a) 0.33 (b) 0.49 (c) 0.82 (d) 0.551 **11.** (a) 0.22 (b) 0.985

(c) 0.8629 **12.** (a) 0.44 (b) 0.733 **14.** (a) 0.512 (b) 0.128 (c) 0.8571 **15.** (a) 0.2625 (b) 0.75 (c) 0.4875 (d) 0.7123 **16.** (a) 0.027 (b) 0.441 (c) 0.453

EXERCISE 15.4

1. (a) 0.042 (b) 0.7143 **2.** (a) 0.4667 (b) 0.3868 **3.** (a) $\frac{5}{7}$ (b) $\frac{9}{13}$ **4.** $\frac{5}{9}$ **5.** (b) i. $\frac{1}{40}$ ii. 0.2

6. (a) i. $\frac{2N-m}{2N}$ ii. $\frac{2(N-m)}{2N-m}$ (b) $\frac{m}{m+(N-m)2^n}$ **7.** $\frac{9}{19}$ **8.** a. 0.07 b. 0.3429 c. 0.30 d. 0.0282 **9.** a. 0.8008 b. 0.9767 c. 0.0003 **10.** a. 0.0464 b. 0.5819 c. 0.9969

11. a. 0.2 b. 0.08 c. 0.72 d. $\begin{bmatrix}0.8 & 0.2\\0.4 & 0.6\end{bmatrix}$; $\begin{bmatrix}0.72 & 0.28\\0.56 & 0.44\end{bmatrix}$ g. 0.688

EXERCISE 15.5

1. (a) $\frac{5}{126}$ (b) $\frac{5}{18}$ (c) $\frac{1}{126}$ **2.** (a) $\frac{1}{5}$ (b) $\frac{1}{10}$ (c) $\frac{2}{5}$ (d) $\frac{3}{5}$ **3.** (a) $\frac{72}{5525}$ (b) $\frac{1}{5525}$ (c) $\frac{1}{1201}$ **4.** $\frac{2}{5}$

5. (a) $\frac{63}{143}$ (b) $\frac{133}{143}$ **6.** (a) $\frac{5}{12}$ (b) $\frac{5}{33}$ (c) $\frac{5}{6}$ **7.** $\frac{3}{11}$ **8.** (a) $\frac{4}{13}$ (b) $\frac{9}{13}$ **9.** (a) $\frac{67}{91}$ (b) $\frac{22}{91}$

10. (a) $\frac{1}{4}$ (b) $\frac{1}{28}$ (c) $\frac{5}{14}$ **11.** (a) $\frac{5}{28}$ (b) $\frac{1}{28}$ **12.** $\frac{6}{13}$ **13.** (a) $\frac{1}{6}$ (b) $\frac{1}{4}$ **14.** (a) $\frac{1}{210}$ (b) $\frac{7}{9}$

15. (a) $\frac{7}{1938}$ (b) 0.6 **16.** $\frac{11}{21}$

EXERCISE 16.1

1. 0.3 **2.** (a) 0.1 (b) i. 0.2 ii. 0.7

3. (a) $p(0) = \frac{6}{15}, p(1) = \frac{8}{15}, p(2) = \frac{1}{15}$ (b)

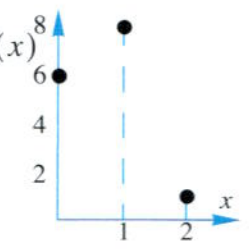

(c) $\frac{14}{15}$

4. (a) {2, 3, 4, 5, 6, 7, 8, 9, 10, 11, 12}

(b)

x	2	3	4	5	6	7	8	9	10	11	12
$p(x)$	$\frac{1}{36}$	$\frac{2}{36}$	$\frac{3}{36}$	$\frac{4}{36}$	$\frac{5}{36}$	$\frac{6}{36}$	$\frac{5}{36}$	$\frac{4}{36}$	$\frac{3}{36}$	$\frac{2}{36}$	$\frac{1}{36}$

(c) $\frac{5}{36}$

(d)

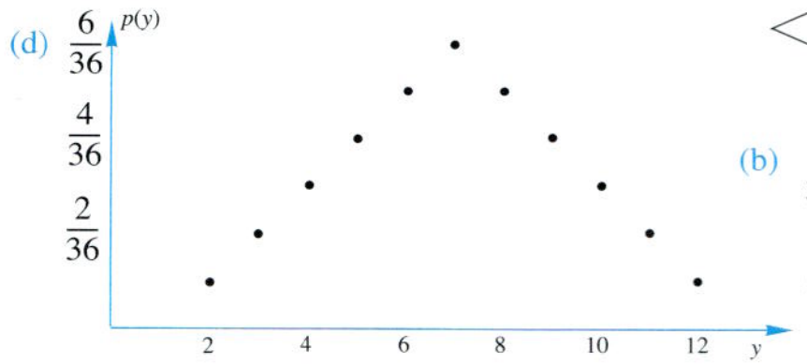

5. (a)

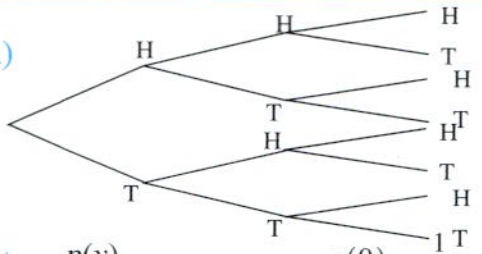

(b)

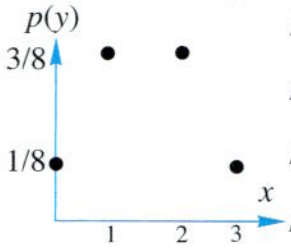

$p(0) = \frac{1}{8}$, $p(1) = \frac{3}{8}$, $p(2) = \frac{3}{8}$, $p(3) = \frac{1}{8}$ (c) $\frac{4}{7}$

6. (a) $\frac{1}{35}$ (b) i. $p(0) = \frac{1}{35}, p(1) = \frac{4}{35}, p(2) = \frac{7}{35}$, $p(3) = \frac{10}{35}, p(4) = \frac{13}{35}$ ii.

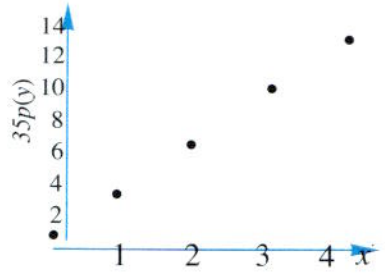

(c) $\frac{6}{7}$ **7.** (a) i. 0.9048 ii. 0.09048

(b) 0.0002 **8.** 0.3712 **9.** $p(0) = \frac{11}{30}, p(-1) = \frac{1}{2}, p(3) = \frac{2}{15}$ (b) i. $\frac{11}{30}$ ii. $\frac{13}{15}$

10.

n	0	1	2
$P(N = n)$	$\frac{6}{15}$	$\frac{8}{15}$	$\frac{1}{15}$

11. (a)

n	1	2	3	4
$P(N = n)$	$\frac{1}{4}$	$\frac{1}{4}$	$\frac{1}{4}$	$\frac{1}{4}$

(b)

s	2	3	4	5	6	7	8
$P(S = s)$	$\frac{1}{16}$	$\frac{2}{16}$	$\frac{3}{16}$	$\frac{4}{16}$	$\frac{3}{16}$	$\frac{2}{16}$	$\frac{1}{16}$

12. (a) 0.81 (b) 0.2439

EXERCISE 16.2

1. (a) 2.8 (b) 1.86 **2.** (a) 3 (b) i. 1 ii. 1 (c) i. 6 ii. 0.4 **3.** (a) i. 1.3 ii. 2.5 iii. –0.1 (b) i. 0.9 ii. 7.29 (c) i. $\frac{31}{60}$ ii. 0.3222 **4.** $\mu = \frac{2}{3}, \sigma^2 = 0.3556$ **5.** (a) 7 (b) 5.8333 **6.** $np = 3 \times \frac{1}{2} = 1.5$

7. (a) $\frac{1}{25}$ (b) 2.8 (c) 1.166 **8.** (a) 0.1 (b) i. 0.3 ii. 1 (c) i. 0 ii. 1 iii. 2 **9.** 5.56

10. $p(0) = \frac{35}{120}, p(1) = \frac{63}{120}, p(2) = \frac{21}{120}, p(3) = \frac{1}{120}$ (b) i. 0.9 ii. 0.49 (c) $W = 3N - 3$

$E(W) = -0.3$ **11.** (a) \$–1.00 (b) both the same **12.** (a) 50 (b) 18 (c) 2 **13.** (a) 11 (b) $\frac{\sqrt{3}}{3}$ (c) –4

14. (a) 0.75 (b) 0.6339 **15.** (a) $E(X) = 1 - 2p$, $Var(X) = 4p(1-p)$ (b) i. $n(1-2p)$ ii. $4np(1-p)$

16.

n	0	1	2
$P(N = n)$	$\frac{28}{45}$	$\frac{16}{45}$	$\frac{1}{45}$

W = 21.43 **17.** (a) $a = \frac{2}{3}, 0 \le b \le 1$

(b) $E(X) = \frac{b+1}{3}$, $Var(X) = \frac{1}{9}(2 + 7b - b^2)$ **18.** (a) $E(X) = 4$, $Var(X) = 20$

EXERCISE 16.3

1. (a) 0.2322 (b) 0.1737 (c) 0.5941 **2.** (a) 0.3292 (b) 0.8683 (c) 0.2099 (d) 0.1317 **3.** (a) 0.1526 (b) 0.4812 (c) 0.5678 **4.** (a) 0.7738 (b) 3.125×10^{-7} (c) 0.9988 (d) 3×10^{-5} **5.** (a) 0.2787 (b) 0.4059 **6.** (a) 0.2610 (b) 0.9923 **7.** (a) 0.2786 (b) 0.7064 (c) 0.1061 **8.** (a) 0.1318 (b) 0.8484 (c) 0.0549 (d) 0.3296 **9.** (a) 0.2938 (b) 0.6531 (c) 0.0027 (d) 0.9726 (e) 12.86 **10.** (a) 0.0039 (b) 0.2734 (c) 0.6367 (d) 0.9648 **11.** (a) 0.3125 (b) 0.0156 (c) 0.3438 (d) 3 **12.** (a) 0.2785 (b) 0.3417 (c) 120 **13.** (a) 0.0331 (b) 0.5695 **14.** (a) 0.4305 (b) 0.9619 (c) \$720 (d) 0.2059 **15.** (a) i. 1.4 ii. 1 iii. 1.058 iv. 0.0795 v. 0.0047 (b) i. 3.04 ii. 3 iii. 1.373 iv. 0.2670 v. 0.1390 **16.** 38.23 **19.** (a) i. 0.1074 ii. 7.9×10^{-4} iii. 0.3758 (b) at least 6

20. (a) $\frac{4}{3}$ (b) $\frac{10}{9}$ (c) $\frac{1}{6}$ (d) $\frac{5}{288}$ **21.** (a) 20 (b) 3.4641 **22.** (a) 102.6 (b) 0.000254 **23.** (a) i. 6 ii. 2.4 (b) i. 6 ii. 3.6 **24.** 0.1797 **25.** 1.6, 1.472 **26.** (a) 0.1841 (b) \$11.93 **27.** (a) \$8 (b) \$160

28. (a) 0.0702 (c)

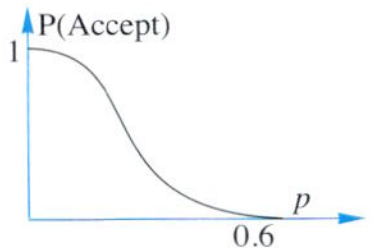

29. (b)

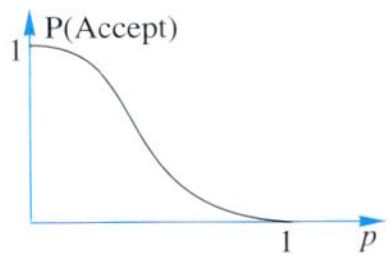

30. (b) 0.8035 (c) 39.3

EXERCISE 17.1

1. (i) 0.6915 (ii) 0.9671 (iii) 0.9474 (iv) 0.9965 (v) 0.9756 (vi) 0.0054 (vii) 0.0287 (viii) 0.0594 (ix) 0.0073 (x) 0.8289 (xi) 0.6443 (xii) 0.0823 **2.** (i) 0.0360 (ii) 0.3759 (iii) 0.0623 (iv) 0.0564 (v) 0.0111 (vi) 0.2902 (vii) 0.7614 (viii) 0.0343 (ix) 0.6014 (x) 0.1450 (xi) 0.9206 (xii) 0.2668 (xiii) 0.7020 (xiv) 0.9132 (xv) 0.5203 (xvi) 0.8160 (xvii) 0.9388 (xviii) 0.7258

EXERCISE 17.2

1. (a) 0.0228 (b) 0.9332 (c) 0.3085 (d) 0.8849 (e) 0.0668 (f) 0.9772 **2.** (a) 0.9772 (b) 0.0668 (c) 0.6915 (d) 0.1151 (e) 0.9332 (f) 0.0228 **3.** (a) 0.3413 (b) 0.1359 (c) 0.0489 **4.** (a) 0.6827 (b) 0.1359 (c) 0.3934 **5.** (a) 0.8413 (b) 0.4332 (c) 0.7734 **6.** (a) 0.1151 (b) 0.1039 (c) 0.1587 **7.** (a) 0.1587 (b) 0.6827 (c) 0.1359 **8.** (a) 0.1908 (b) 0.4754 (c) 16.88 **9.** (a) 0.1434 (b) 0.6595 **10.** (a) 0.2425 (b) 0.8413 (c) 0.5050 **11.** (a) –1.2816 (b) 0.2533 **12.** (a) 58.2243 (b) 41.7757 (c) 59.80 **13.** 39.11 **14.** 9.1660 **15.** 42% **16.** 0.7021 **17.** (i) 0.2903 (ii) 0.4583 (iii) 0.2514 **18.** 23% **19.** 0.5 **20.** 11% **21.** 5% **22.** 14% **23.** 1.8 **24.** 252 **25.** 0.1517 **26.** 0.3821 **27.** 0.22 **28.** 322 **29.** 0.1545 **30.** 7 **31.** 87 **32.** (a) i. 0.0062 ii. 0.0478 iii. 0.9460 (b) 0.0585 **33.** (a) \$5.11 (b) \$7.39 **34.** (a) 0.0062 (b) i. 0.7887 ii. 0.0324 (c) \$1472 **35.** (a) $\mu = 66.86, \sigma = 10.25$ (b) \$0.38S **36.** (a) $\mu = 37.2, \sigma = 28.2$ (b) 20 (19.9) **37.** (a) i. 0.3446 ii. 0.2347 (b) i. 0.3339 ii. 0.3852 (c) 0.9995

EXERCISE 18.1

1. (a) $\frac{3}{4}$ (b) $\frac{3a}{4b}$ (c) –1 (d) 1 (e) $-\frac{15}{8}$ (f) 0 **2.** (a) 4 (b) 0.2 (c) 0.027 (d) 0.433 (e) –0.01 (f) –30 (g) 6.2 (h) 1 **3.** (a) 6 m/s (b) 30 m/s (c) $11 + 6h + h^2$ m/s **4.** 12 m/s **5.** $8 + 2h$ **6.** –3.49 °C/sec **7.** (a) 127π cm^3/cm (b) i. 19.6667π cm^3/cm ii. 1.9967π cm^3/cm iii. 0.2000π cm^3/cm **8.** 1.115 **9.** (a) –7.5 °C/min (b) $t = 2$ to $t = 6$ **10.** (a) 28 m (b) 14 m/s (c) average speed (d) 49 m (e) 49 m/s **11.** (a) \$1160, \$1345.6, \$1560.90, \$1810.64,\$ 2100.34 (b) \$220.07 per year

EXERCISE 18.2

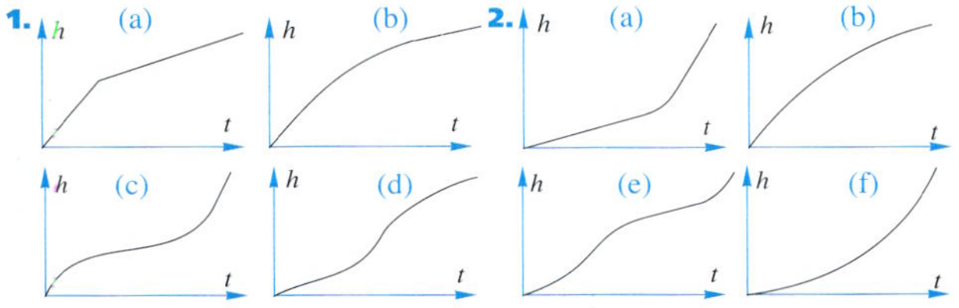

EXERCISE 18.3

1. (a) $h + 2$ (b) $4 + h$ (c) $\frac{-1}{1+h}$ (d) $3 - 3h + h^2$ **2.** (a) 2 (b) 4 (c) –1 (d) 3 **3.** (a) $2a + h$ (b) $-(2a + h)$ (c) $(2a + 2) + h$ (d) $3a^2 + 1 + 3ah + h^2$ (e) $-(3a^2 + 3ah + h^2)$ (f) $3a^2 - 2a + (3a - 1)h + h^2$ (g) $\frac{-2}{a(a+h)}$ (h) $-\frac{1}{(a-1)(a-1+h)}$ (i) $\frac{1}{\sqrt{a+h}+\sqrt{a}}$

4. (a) 1 ; 1 (b) $2a + h$; $2a$ (c) $3a^2 + 3ah + h^2$; $3a^2$ (d) $4a^3 + 6a^2h + 4ah^2 + h^3$; $4a^3$

5. (a)

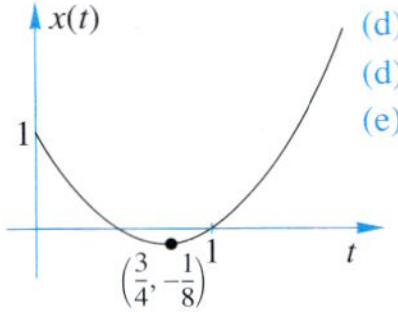

(b) i. 3 ms^{-1} ii. 2 ms^{-1} iii. 1.2 ms^{-1} (d) $1 + 2h$ (d) Find (limit) as $h \to 0$ (e) $4t - 3$

6. (a)

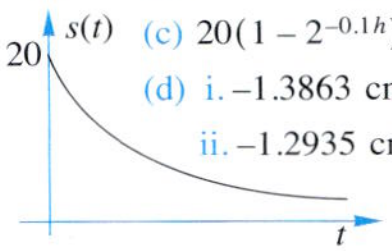

(b) i. 20 cm^2 ii. 17.41 cm^2 iii. 2.59 cm^2 iv. –1.29 cm^{2}day^{-1} (c) $20(1 - 2^{-0.1h})$ cm^{2}day^{-1} (d) i. –1.3863 cm^{2}day^{-1} ii. –1.2935 cm^{2}day^{-1}

EXERCISE 18.4

1. (a) 3 (b) 8 (c) $-\frac{1}{9}$ (d) 1.39 (e) –1 (f) $\frac{17}{16}$ **2.** (a) 4.9 m (b) $4.9(h^2 + 2h)$ m (c) 9.8 m/s

3. (a) $8x$ (b) $10x$ (c) $12x^2$ (d) $15x^2$ (e) $16x^3$ (f) $20x^3$ **4.** (a) $4x$ (b) –1 (c) $-1 + 3x^2$ (d) $-x^{-2}$ (e) $-2(x + 1)^{-2}$ (f) $0.5x^{-1/2}$ **5.** (a) 1 ms^{-1} (b) $(2 - a)$ ms^{-1}

6. (a)

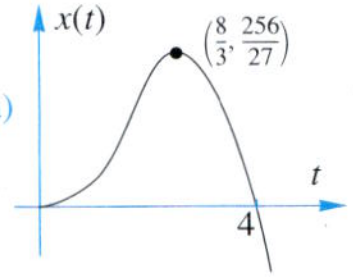

(b) i. 5 ms^{-1} ii. 4 ms^{-1} (c) $8t - 3t^2$ ms^{-1} (d) $\frac{8}{3}$ sec

EXERCISE 19.1

1. (a) $5x^4$ (b) $9x^8$ (c) $25x^{24}$ (d) $27x^2$ (e) $-28x^6$ (f) $2x^7$ (g) $2x$ (h) $20x^3 + 2$ (i) $-15x^4 + 18x^2 - 1$ (j) $-\frac{4}{3}x^3 + 10$ (k) $9x^2 - 12x$ (l) $3 + \frac{2}{5}x + 4x^3$ **2.** (a) $-\frac{3}{x^4}$ (b) $\frac{3}{2}\sqrt{x}$ (c) $\frac{5}{2}\sqrt{x^3}$ (d) $\frac{1}{3\sqrt[3]{x^2}}$ (e) $\frac{2}{\sqrt{x}}$

(f) $9\sqrt{x}$ (g) $\frac{1}{\sqrt{x}} + \frac{3}{x^2}$ (h) $\frac{3}{2}\sqrt{x} - \frac{1}{2\sqrt{x^3}}$ (i) $\frac{10}{3\sqrt[3]{x}} - 9$ (j) $5 - \frac{1}{2\sqrt{x}} - \frac{8}{5x^3}$ (k) $\frac{4}{\sqrt{x}} - \frac{15}{x^6} + \frac{1}{2}$

(l) $-\frac{1}{2\sqrt{x^3}} - \frac{1}{\sqrt{x}} + x^2$ **3.** (a) $\frac{3}{2}\sqrt{x} + \frac{1}{\sqrt{x}}$ (b) $4x^3 + 3x^2 - 1$ (c) $3x^2 + 1$ (d) $\frac{1}{x^2}$ (e) $\frac{1}{\sqrt{x^3}}$

(f) $\frac{1}{2} - \frac{1}{4\sqrt{x^3}}$ (g) –7 (h) $2x - \frac{8}{x^3}$ (i) $2x - \frac{2}{x^2} - \frac{4}{x^5}$ (j) $\frac{1}{2}\sqrt{\frac{3}{x}} + \frac{1}{6\sqrt{x^3}}$ (k) $2x - \frac{12}{5}\sqrt[5]{x} + \frac{2}{5\sqrt[5]{x^3}}$

(l) $-\frac{3}{2\sqrt{x}}\left(\frac{1}{x} + 1\right)\left(\frac{1}{\sqrt{x}} - \sqrt{x}\right)^2$

EXERCISE 19.2.1

1. $m_{PQ} = 4 + h$; $\lim_{h \to 0} m_{PQ} = 4$ **2.** $P(1, 1)$, $Q\left(1 + h, \frac{2}{2+h}\right)$; $m_{PQ} = -\frac{1}{2+h}$; $\lim_{h \to 0} m_{PQ} = -\frac{1}{2}$

3. –12 **4.** (a) 3 (b) $-\frac{1}{4}$ (c) 12 (d) 4 (e) 4 (f) $\frac{7}{6}$ (g) $-\frac{1}{12}$ (h) $\frac{53}{16}$ **5.** $\pm\sqrt{\frac{8}{3}}$ **6.** (a) $2x - 12$ (b) –18

(c) $(8, -32)$ **7.** (a) $-3x^2+3$ (b) 0 (c) $(\sqrt{2}, \sqrt{2}), (-\sqrt{2}, -\sqrt{2})$ **8.** (a) $\left(\pm\frac{\sqrt{2}}{2}, -\frac{1}{16}\right)$, (0,0)

(b) $\left\{x: \frac{-1}{\sqrt{2}} < x < 0\right\} \cup \left\{x: x > \frac{1}{\sqrt{2}}\right\}$ **9.** $x = \frac{1}{3}, -1$ **10.** (a) $-2, 6, 3$ (b) -2 **11.** $a = 1$ $b = -8$

12. $f'(a+b) = 2(a+b) = 2a+2b$ **13.** (a) $4a^2-2a, a \geq 0$ (b) $4-\frac{1}{a}, a > 0$ **14.** -56

EXERCISE 19.2.2

1.

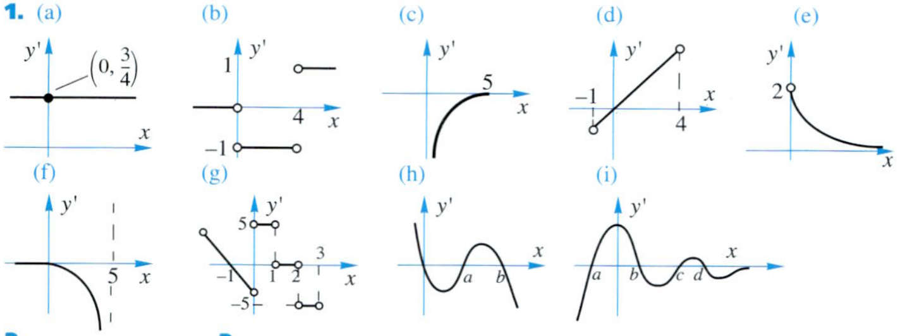

2.

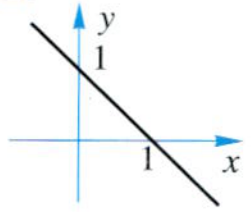

3.

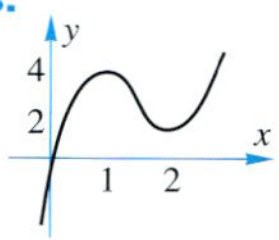

EXERCISE 19.2.3

1. (a) $48t^3 - \frac{1}{2\sqrt{t}}$ (b) $2n - \frac{2}{n^2} - \frac{4}{n^5}$ (c) $\frac{3}{2}\sqrt{r} + \frac{5}{6\sqrt[6]{r}} - \frac{1}{\sqrt{r}}$ (d) $2\theta - \frac{9}{2}\sqrt{\theta} + 3 - \frac{1}{2\sqrt{\theta}}$ (e) $40 - 3L^2$

(f) $-\frac{100}{v^3} - 1$ (g) $6l^2+5$ (h) $2\pi + 8h$ (i) $4n^3 - \frac{1}{3\sqrt[3]{n^2}} + \pi$

2. (a) $\frac{8}{3t^3}$ (b) $2\pi r - \frac{20}{r^2}$ (c) $\frac{5}{2}s^{3/2} + \frac{3}{s^2}$ (d) $-\frac{6}{t^4} + \frac{2}{t^3} - \frac{1}{t^2}$ (e) $-\frac{4}{b^2} + \frac{1}{2b^{3/2}}$ (f) $3m^2 - 4m - 4$

EXERCISE 19.3

1. (a) $3x^2 - 5x^4 + 2x + 2$ (b) $6x^5 + 10x^4 + 4x^3 - 3x^2 - 2x$ (c) $-\frac{4}{x^5}$ (d) $6x^5 + 8x^3 + 2x$

2. (a) $-\frac{2}{(x-1)^2}$ (b) $\frac{1}{(x+1)^2}$ (c) $\frac{1-x^2-2x}{(x^2+1)^2}$ (d) $\frac{-(x^4+3x^2+2x)}{(x^3-1)^2}$ (e) $\frac{2x^2+2x}{(2x+1)^2}$ (f) $\frac{1}{(1-2x)^2}$

3. (a) $(\sin x + \cos x)e^x$ (b) $\ln x + 1$ (c) $e^x(2x^3+6x^2+4x+4)$ (d) $4x^3\cos x - x^4 \sin x$

(e) $-\sin^2 x + \cos^2 x$ (f) $2x\tan x + (1+x^2)\sec^2 x$ (g) $\frac{4}{x^3}(x\cos x - 2\sin x)$

(h) $e^x(x\cos x + x\sin x + \sin x)$ (i) $(\ln x + 1 + x\ln x)e^x$

4. (a) $\frac{\sin x - x\cos x}{\sin^2 x}$ (b) $\frac{-[\sin x(x+1) + \cos x]}{(x+1)^2}$ (c) $\frac{e^x}{(e^x+1)^2}$ (d) $\frac{2x\cos x - \sin x}{2x\sqrt{x}}$ (e) $\frac{\ln x - 1}{(\ln x)^2}$

(f) $\frac{(x+1) - x\ln x}{x(x+1)^2}$ (g) $\frac{xe^x+1}{(x+1)^2}$ (h) $\frac{-2}{(\sin x - \cos x)^2}$ (i) $\frac{x^2 - x + 2x\ln x}{(x+\ln x)^2}$ **5.** (a) $-5e^{-5x}+1$

(b) $4\cos 4x + 3\sin 6x$ (c) $-\frac{1}{3}e^{-\frac{1}{3}x} - \frac{1}{x} + 18x$ (d) $25\cos 5x + 6e^{2x}$ (e) $4\sec^2 4x + 2e^{2x}$

(f) $-4\sin(4x) + 3e^{-3x}$ (g) $\frac{4}{4x+1} - 1$ (h) 0 (i) $\frac{1}{2}\cos\left(\frac{x}{2}\right) - 2\sin 2x$ (j) $7\cos(7x-2)$ (k) $\frac{1}{2\sqrt{x}} - \frac{1}{x}$

(l) $\frac{1}{x} + 6\sin 6x$ **6.** (a) $2x\cos x^2 + 2\sin x\cos x$ (b) $2\sec^2 2\theta - \frac{\cos\theta}{\sin^2\theta}$ (c) $\frac{1}{2\sqrt{x}}\cos\sqrt{x}$ (d) $\frac{1}{x^2}\sin\left(\frac{1}{x}\right)$

(e) $-3\sin\theta \cdot \cos^2\theta$ (f) $e^x\cos(e^x)$ (g) $\frac{1}{x}\sec^2(\log_e x)$ (h) $\frac{-\sin 2x}{\sqrt{\cos 2x}}$ (i) $-\cos\theta \cdot \sin(\sin\theta)$

(j) $4\sin\theta \cdot \sec^2\theta$ (k) $-5\cos 5x \cdot \csc^2(5x)$ (l) $-6\csc^2(2x)$

7. (a) $2e^{2x+1}$ (b) $-6e^{4-3x}$ (c) $-12xe^{4-3x^2}$ (d) $\frac{1}{2}\sqrt{e^x}$ (e) $\frac{1}{2\sqrt{x}}e^{\sqrt{x}}$ (f) e^{2x+4} (g) $2xe^{2x^2+4}$

(h) $-\frac{6}{e^{3x+1}}$ (i) $(6x-6)e^{3x^2-6x+1}$ (j) $\cos(\theta)e^{\sin\theta}$ (k) $2\sin(2\theta)e^{-\cos 2\theta}$ (l) $2x$ (m) $\frac{2e^{-x}}{(e^{-x}+1)^2}$

(n) $3(e^x + e^{-x})(e^x - e^{-x})^2$ (o) e^{x+2} (p) $(-2x+9)e^{-x^2+9x-2}$ **8.** (a) $\frac{2x}{x^2+1}$ (b) $\frac{\cos\theta + 1}{\sin\theta + \theta}$

(c) $\frac{e^x + e^{-x}}{e^x - e^{-x}}$ (d) $-\frac{1}{x+1}$ (e) $\frac{3}{x}(\ln x)^2$ (f) $\frac{1}{2x\sqrt{\ln x}}$ (g) $\frac{1}{2(x-1)}$ (h) $\frac{-3x^2}{1-x^3}$ (i) $-\frac{1}{2(x+2)}$

(j) $\frac{-2\sin x\cos x}{\cos^2 x + 1}$ (k) $\frac{1}{x} + \cot x$ (l) $\frac{1}{x} + \tan x$

9. (a) $\ln(x^3+2) + \frac{3x^3}{x^3+2}$ (b) $\frac{\sin^2 x}{2\sqrt{x}} + 2\sqrt{x}\sin x\cos x$ (c) $-\frac{1}{\sqrt{\theta}}\sin\sqrt{\theta} \cdot \cos\sqrt{\theta}$

(d) $(3x^2 - 4x^4)e^{-2x^2+3}$ (e) $-(\ln x + 1)\sin(x\ln x)$ (f) $\frac{1}{x\ln x}$

(g) $\frac{(2x-4)\sin(x^2) - 2x\cos(x^2)(x^2-4x)}{(\sin x^2)^2}$ (h) $\frac{10[\ln(10x+1) - 1]}{[\ln(10x+1)]^2}$ (i) $(\cos 2x - 2\sin 2x)e^{x-1}$

(j) $2x\ln(\sin 4x) + 4x^2\cot 4x$ (k) $(\cos\sqrt{x} - \sin\sqrt{x})\frac{1}{2\sqrt{x}}e^{-\sqrt{x}}$

(l) $-(2\sin x + 2x\cos x) \cdot \sin(2x\sin x)$ (m) $\frac{e^{5x+2}(9-20x)}{(1-4x)^2}$ (n) $\frac{\cos^2\theta + \sin^2\theta\ln(\sin\theta)}{\sin\theta\cos^2\theta}$

(o) $\frac{x+2}{2(x+1)\sqrt{x+1}}$ (p) $\frac{2x^2+2}{\sqrt{x^2+2}}$ (q) $\frac{10x^3+9x^2+4x+3}{3(x+1)^{2/3}}$ (r) $\frac{3x^2(3x^3+1)}{2\sqrt{x^3+1}}$

(s) $\frac{2}{x^2+1} - \frac{1}{x^2}\ln(x^2+1)$ (t) $\frac{2}{x(x+2)}$ (u) $\frac{2-x}{2x^2\sqrt{x-1}}$ (v) $\frac{-x^2+x-9}{\sqrt{x^2+9}} \cdot e^{-x}$ (w) $\frac{7x^3 - 12x^2 - 8}{2\sqrt{2-x}}$

(x) $nx^{n-1}\ln(x^n - 1) + \frac{nx^{2n-1}}{x^n - 1}$ **10.** $x = 1$ **11.** 0 **12.** 0 **13.** 1 **14.** $-2e$ **15.** (a) $\cos^2 x - \sin^2 x$

(b) $\frac{\pi}{180}\cos x°$ (c) $-\frac{\pi}{180}\sin x°$ **16.** (b) i. $2x\sin x\cos x + x^2\cos^2 x - x^2\sin^2 x$

ii. $e^{-x^3}(2\cos 2x\ln\cos x - 3x^2\sin 2x\ln\cos x - \sin 2x\tan x)$ **17.** (a) i. $-\frac{3}{x}(\ln x)^2$ ii. $-\frac{3x^2}{1-x^3}$

(b) i. $-2e^{-2x}\cdot\cos(e^{-2x})$ ii. $-2x\cos x^2\cdot e^{-\sin x^2}$ **18.** $-\frac{1}{5}k$ **19.** $x = a, b, \frac{mb+na}{m+n}$

20. $\{\theta : n\tan\theta^m\cdot\tan\theta^n = m\theta^{m-n}\}$ **21.** (a) $-4\csc(4x)$ (b) $2\sec(2x)\tan(2x)$

(c) $3\cot(3x)\csc(3x)$ (d) $-3\sin(3x)$ (e) $\csc^2\left(\frac{\pi}{4}-x\right)$ (f) $-2\sec(2x)\tan(2x)$

22. (a) $2x\sec(x^2)\tan(x^2)$ (b) $\sec^2 x$ (c) $\tan x$ (d) $-3\cot^2 x\csc^2 x$ (e) $x\cos x + \sin x$
(f) $-2\cot x\csc^2 x$ (g) $4x^3\csc(4x) - 4x^4\cot(4x)\csc(4x)$ (h) $2\cot x\sec^2(2x) - \csc^2 x\tan(2x)$
(i) $\frac{\sec x\tan x - \sin x}{2\sqrt{\cos x + \sec x}}$ **23.** (a) $e^{\sec x}\sec x\tan x$ (b) $e^x\sec(e^x)\tan(e^x)$

(c) $e^x\sec(x) + e^x\sec(x)\tan(x)$ (d) $\frac{-\csc^2(\log x)}{x}$ (e) $-5\csc(5x)\sec(5x)$

(f) $\frac{\cot(x)}{x} - \csc^2(x)\log x$ (g) $-\cos x\cot(\sin x)\csc(\sin x)$ (h) $-\cos(\csc x)\cot x\csc x$ (i) 0

EXERCISE 19.4

1. i. $20x^3$ ii. $48(1+2x)^2$ iii. $\frac{2}{x^3}$ iv. $\frac{2}{(1+x)^3}$ v. 2 vi. $\frac{6}{(x-2)^3}$ vii. $\frac{42}{x^8}$

viii. $24(1-2x)$ ix. $-\frac{1}{x^2}$ x. $\frac{-2(x^2+1)}{(1-x^2)^2}$ xi. $-16\sin 4\theta$ xii. $2\cos x - x\sin x$

xiii. $6x^2\cos x + 6x\sin x - x^3\sin x$ xiv. $\frac{1}{x}$ xv. $\frac{10}{(2x+3)^3}$ xvi. $6xe^{2x} + 12x^2e^{2x} + 4x^3e^{2x}$

xvii. $\frac{8\sin 4x - 15\cos 4x}{e^x}$ xviii. $2\cos x^2 - 4x^2\sin x^2$ xix. $\frac{-48(x^2+2x^5)}{(4x^3-1)^3}$ xx. $\frac{10}{(x-3)^3}$

2. $\frac{6\ln x - 5}{x^4}, \frac{n^2\ln x + n\ln x - 2n - 1}{x^{n+2}}$ **3.** $f(x) = \left(\frac{x+1}{x-1}\right)^n \Rightarrow f''(x) = \frac{4n(n+x)}{(x^2-1)^2}\left(\frac{x+1}{x-1}\right)^n$

4. $2 + \frac{1}{8\sqrt{2}}$ **5.** -1 **6.** $[0, 1.0768[\, \cup\,]3.6436, 2\pi]$

EXERCISE 20.1

1. (a) $y = 7x - 10$ (b) $y = -4x + 4$ (c) $4y = x + 5$ (d) $16y = -x + 21$ (e) $4y = x + 1$ (f) $4y = x + 2$ (g) $y = 28x - 48$ (h) $y = 4$ **2.** (a) $7y = -x + 30$ (b) $4y = x - 1$ (c) $y = -4x + 14$ (d) $y = 16x - 79$ (e) $2y = 9 - 8x$ (f) $y = -4x + 9$ (g) $28y = -x + 226$ (h) $x = 2$ **3.** (a) $y = 2ex - e$ (b) $y = e$ (c) $y = \pi$ (d) $y = -x$ (e) $y = x$ (f) $ey = (2e-1)x - e^2 + 2e - 1$ (g) $y = ex$ (h) $y = 2x + 1$
4. (a) $2ey = -x + 2e^2 + 1$ (b) $x = 1$ (c) $x = \pi$ (d) $y = x - 2\pi$ (e) $y = -x + \pi$
(f) $(2e-1)y = -ex + 3e^2 - 4e + 1$ (g) $ey = -x$ (h) $2y = -x + 2$ **5.** A: $y = 28x - 44$,
B: $y = -28x - 44$, Isosceles. $z \equiv (0, a^2 - 3a^4)$ **6.** 2 sq. units, $y = 2$ $x = 1$ **7.** $4y = 3x$
8. $by = \sqrt{a^2 - b^2}x$ **9.** $y = 4x - 9$ **10.** $y = \log_e 4$ **11.** $8y = 4(\pi+2)x - \pi^2$;

$4(\pi+2)y = -8x + 4\pi + \pi^2$ **12.** A: $y = -8x + 32$, B: $y = 6x + 25$, $\left(\frac{1}{2}, 28\right)$ **13.** $y = -x$, Tangents:

$y = \frac{1}{2}, y = -\frac{1}{2}$ $\left(-\frac{1}{2}, \frac{1}{2}\right), \left(\frac{1}{2}, -\frac{1}{2}\right)$ tangent and normal meet at (0.5, –0.5)

14. (a) $y = 3x - 7$ (b) $Q \equiv (2, -1)$ **15.** $m = -2, n = 5$

EXERCISE 20.2

1. (a) (b) (c) (d)

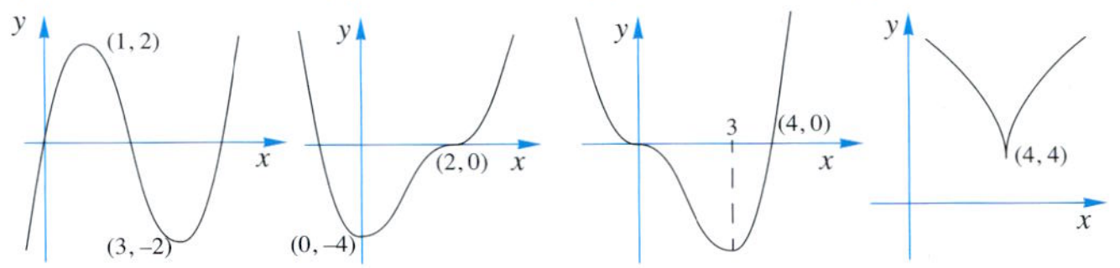

2. (a) max at (1, 4) (b) min at $\left(-\frac{9}{2}, -\frac{81}{4}\right)$ (c) min at (3, –45) max (–3, 63) (d) max at (0, 8),

min at (4, -24) (e) max at (1, 8), min at (–3, –24) (f) min at $\left(\frac{1+\sqrt{13}}{3}, \frac{70-26\sqrt{13}}{27}\right)$,

max at $\left(\frac{1-\sqrt{13}}{3}, \frac{70+26\sqrt{13}}{27}\right)$ (g) min at (1, –1) (h) max at (0, 16), min at (2, 0),

min at (–2, 0) (i) min at (1, 0) max at $\left(-\frac{1}{3}, \frac{32}{27}\right)$ (j) min at $\left(\frac{4}{9}, -\frac{4}{27}\right)$ (k) min at (2, 4),

max at (–2, –4) (l) min at (1, 2), min at (–1, 2)

3. (a) (b) (c) (d) (e) (f) (g) (h) (i) (j)

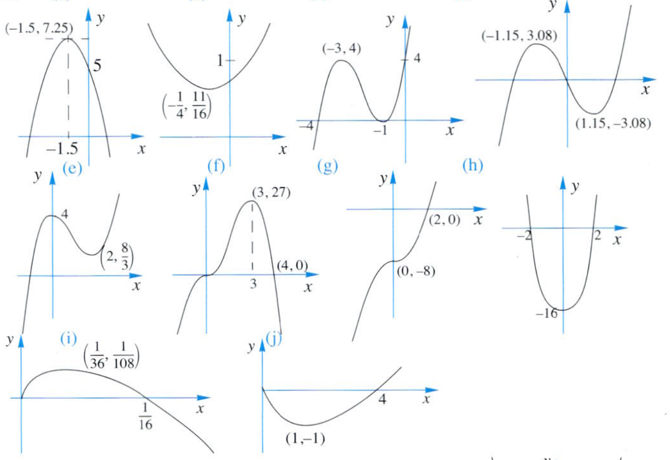

4. min at (1, –3), max at (–3, 29), non-stationary infl (–1, 13) **5.**

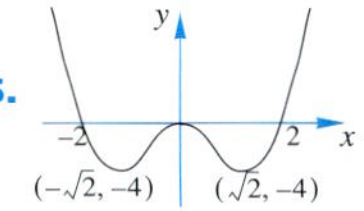

6. (a) i. $(\cos x - \sin x)e^{-x}$ ii. $-2\cos x . e^{-x}$ (b) i. $\frac{\pi}{4}, \frac{5\pi}{4}$ ii. $\frac{\pi}{2}, \frac{3\pi}{2}$ (c) Inf. $\left(\frac{\pi}{2}, e^{-\frac{\pi}{2}}\right)$ $\left(\frac{3\pi}{2}, -e^{-\frac{3\pi}{2}}\right)$

(d)
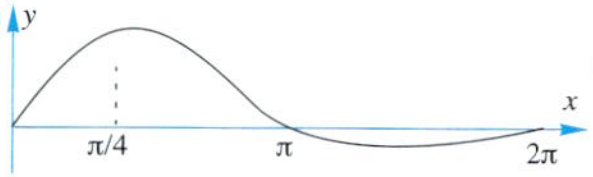

7. (a) i. $e^x(\sin x + \cos x)$ ii. $2e^x\cos x$ (b) i. $x = \frac{3\pi}{4}, \frac{7\pi}{4}$

ii. $x = \frac{\pi}{2}, \frac{3\pi}{2}$ (c) st.pts. $\left(\frac{3\pi}{4}, \frac{1}{\sqrt{2}}e^{\frac{3\pi}{4}}\right)$, $\left(\frac{7\pi}{4}, -\frac{1}{\sqrt{2}}e^{\frac{7\pi}{4}}\right)$ Infl. pts. $\left(\frac{\pi}{2}, e^{\frac{\pi}{2}}\right)$, $\left(\frac{3\pi}{2}, -e^{\frac{3\pi}{2}}\right)$

(d)
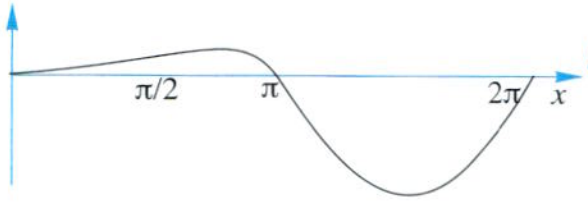

8. (a) i. $e^x(\cos x - \sin x)$ ii. $-2\sin x.e^x$ (b) i. $\frac{\pi}{4}, \frac{5\pi}{4}$

ii. $0, \pi, 2\pi$ (c) st.pts. $\left(\frac{\pi}{4}, \frac{1}{\sqrt{2}}e^{\frac{\pi}{4}}\right)$, $\left(\frac{5\pi}{4}, -\frac{1}{\sqrt{2}}e^{\frac{5\pi}{4}}\right)$ Inf. pts. $(0,1), (\pi, -e^{\pi}), (2\pi, e^{2\pi})$

(d)
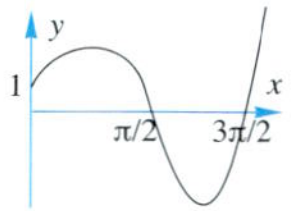

9. (a) i. $(1-x)e^{-x}$ ii. $(x-2)e^{-x}$
(b) i. $x = 1$ ii. $x = 2$
(c) st. pt. $(1, e^{-1})$ Inf. pt. $(2, 2e^{-2})$

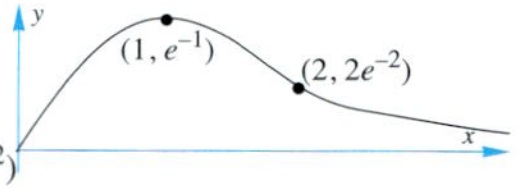

10. (a) 8 (b) 0 (c) 4 (d) $27\sqrt[3]{9} \approx 56.16$ **11.** i. min value –82. ii. max value 26

12. (a) A i. Yes ii. non-stationary pt of inflect. B. i. Yes ii. Stationary point (local/global min) C. i. Yes ii. non-stationary pt of inflect.
(b) A. i. No ii. Local/global max B. i. No. ii. Local/global min C. i. Yes ii. Stationary point (local max)
(c) A. i. Yes ii. Stationary point (local/global max) B. i. Yes ii. Stationary point (local min) C. i. Yes ii. non-stationary pt of inflect.
(d) A. i. Yes ii. Stationary pt (local/global max) B. i. No ii. Local min C. i. Yes ii. Stationary point (local max)
(e) A. i. No ii. Cusp (local min) B. i. Yes ii. Stationary pt of inflect C. i. Yes ii. Stationary point (local max)
(f) A. i. Yes ii. Stationary point (local/global max) B. i. Yes ii. Stationary point (local/global min) C. i. No ii. Tangent parallel to y–axis.

13. (a) i. A ii. B iii. C (b) i. C ii. B iii. A

14. i.
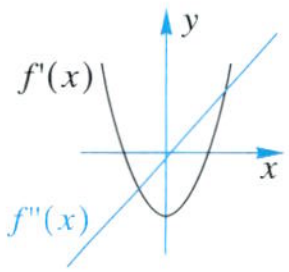

ii.
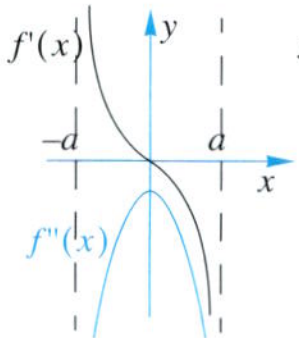

iii.
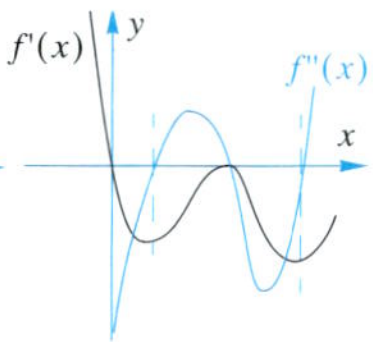

15. $y = x^3 + 6x^2 + 9x + 4$
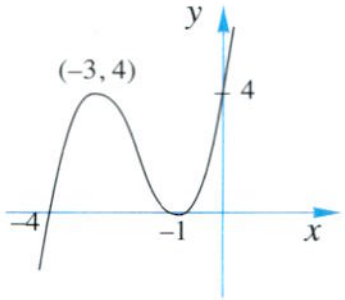

16. $f(x) = \frac{1}{3}x^3 - x^2 - 3x - 6$ **17.** $f(x) = 3x^5 - 20x^3$

18.
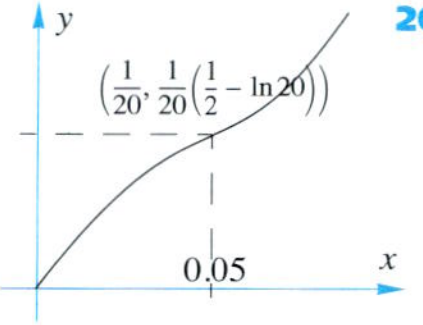

19. $m = -0.5, n = 1.5$

20. (a) i.
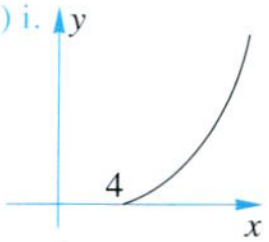

ii.
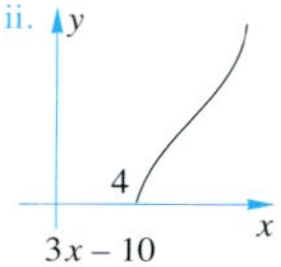

(b) i. $\frac{3}{2}\sqrt{x-4}$ ii. $\frac{3x-10}{2\sqrt{x-4}}$

21. $a = 2, b = -3, c = 0$ **22.** Stationary points: local min at $(-1, 0)$ and local max at $(1, 4e^{-1})$. inflection. pts are: $(1+\sqrt{2}, (6+4\sqrt{2})e^{-(1+\sqrt{2})})$ and $(1-\sqrt{2}, (6-4\sqrt{2})e^{-(1-\sqrt{2})})$

23. Absolute min at ~ $\left(\frac{-3+\sqrt{13}}{2}, -2.1733\right)$, Local max at ~ $\left(\frac{-3-\sqrt{13}}{2}, 0.2062\right)$

Inflection pts at ~ $(-0.4384, -1.4489)$ and $(-4.5615, 0.1488)$

24. – 27. are left as questions for classroom discussion.

28. $a = 1, b = -12,$
$c = 45, d = -34$
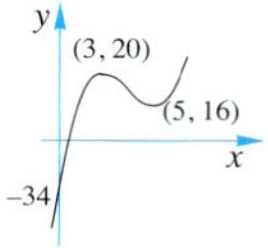

29. (b) $b = 1$
(c) $a = \frac{1}{\sqrt{2}}$
(d) $f(x) = \frac{1}{\sqrt{2}}xe^{-x^2}$
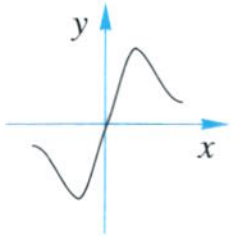

30. (a) 2.7983, 6.1212, 9.3179 (b) Make use of Graphics calculator to verify your sketch.

EXERCISE 20.3

1. (a) L.min at $x = \frac{4}{\sqrt{3}}$. L.max at $x = -\frac{4}{\sqrt{3}}$ (b) L.max at $x = 0$, L.min at $x = \pm 1$
(c) L.max at $x = 0.25$ (d) L.max at $x = 1$ (e) none (f) L.max at $x = 0.5$, L.min at $x = 1$ and $x = 0$
(g) L.max at $x = 1$, L.min at $x = -1$ (h) none **2.** (a) max = 120, min = $-\frac{128}{3\sqrt{3}}$ (b) max = 224, min = –1 (c) max = 0.5, min = 0 (d) max = 1, min = 0.

3.
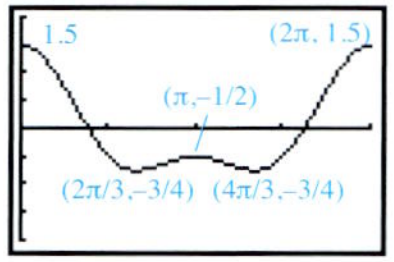

4.
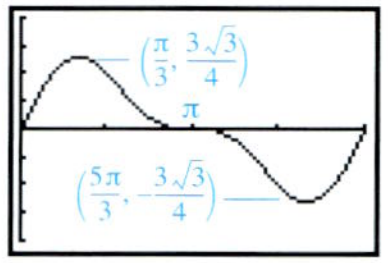

5. stationary points occur where $\tan x = x$
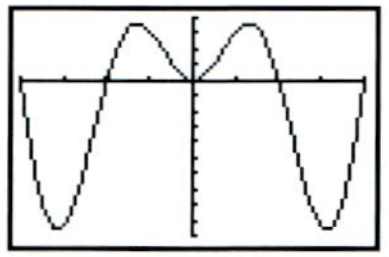

6. (a) L.min at (1, 2); Infl. pt at $\left(3, \sqrt{3} + \frac{1}{3}\sqrt{3}\right)$ (b) L.min at (1, 2); L.max at (–3, –6) (c) none

7.
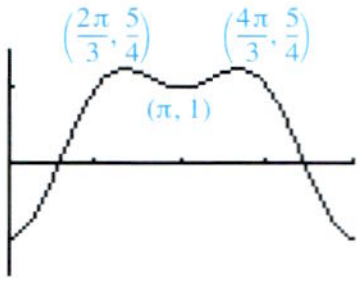

For **Q.8. – Q.11** verify your graphs with graphics calculator

8. (a) Global. min at (0, 0); local max at $(2, 4e^{-2})$
Infl. pts $(2-\sqrt{2}, (6-4\sqrt{2})e^{-(2-\sqrt{2})})$
$(2+\sqrt{2}, (6+4\sqrt{2})e^{-(2+\sqrt{2})})$
(b) Global. max: $(0, e^4)$. Infl. pt: $\left(\pm\frac{1}{\sqrt{2}}, e^{3.5}\right)$ (c) L.max: $\left(-2, -\frac{1}{2}e\right)$

9. (a) Global max at (e, e^{-1}). Infl. pt at $(e^{1.5}, 1.5e^{-1.5})$ (b) Global min at $\left(\frac{1}{\sqrt{2}}, 2+\frac{1}{2}\ln 2\right)$

(c) Global min at $(2, 1+\ln 2)$; Infl. pt at $(4, 2+\ln 4)$ (d) none

10. (a) $f'(x) = (x-2)^{a-1}(x+2)^{b-1}((a+b)x+2(a-b))$

(b) i. $f(x) = \frac{x-2}{x+2}$; none ii. $f(x) = (x-2)^2(x+2)$ local max at $\left(-\frac{2}{3}, \frac{256}{27}\right)$; local min at $(2, 0)$

iii. $f(x) = (x-2)^2(x+2)^2$; local min at $(\pm 2, 0)$, local max at $(0, 16)$.

11. (a) Global. min at $(1, c-1)$ **12.**

(b)

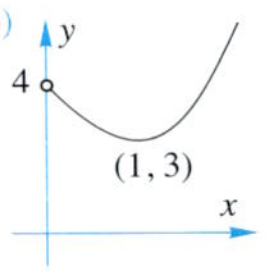

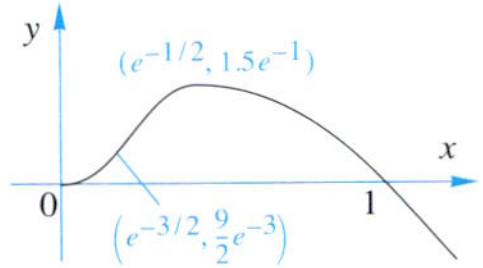

13. Global max at $(e^{0.5}, 0.5e^{-1})$; Infl. pt at $(e^{5/6}, (5/6)e^{-5/3})$.

EXERCISE 20.4

1. (a) $y = 2, x = -1$ (b) $y = 1, x = -\frac{1}{3}$ (c) $y = \frac{1}{2}, x = -\frac{1}{4}$ (d) $y = -1, x = -3$ (e) $y = 3, x = 0$

(f) $y = 5, x = 2$

3. (a)

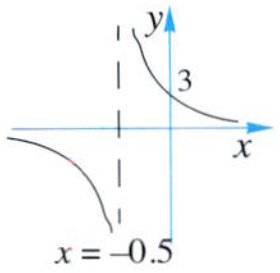

(b)

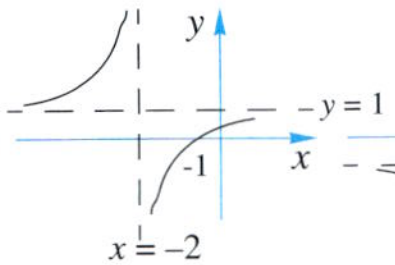

(c)

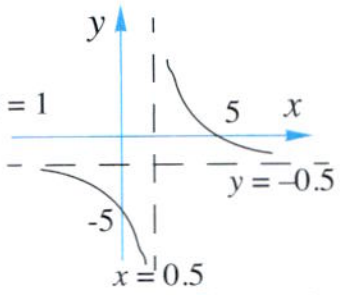

(d)

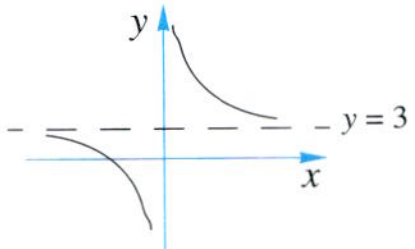

(e)

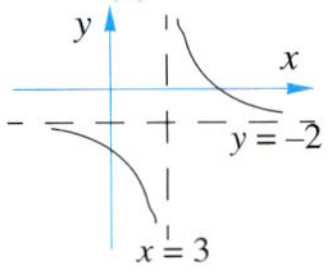

(f)

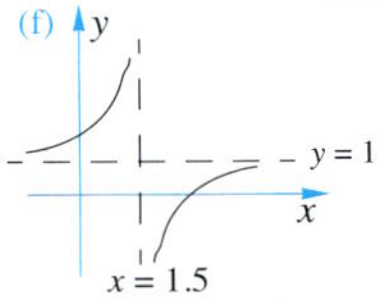

4. $a = 2, c = 4$

5. (a)

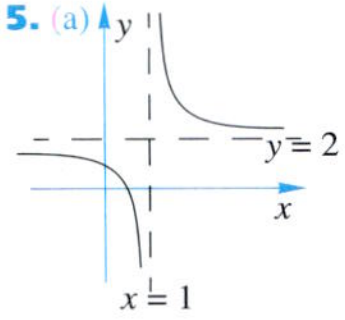

(b)

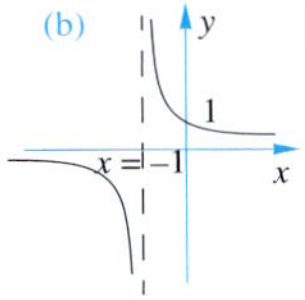

6. (a) i. $(0, 1), (2, 0)$ ii. $y = -1, x = -2$

iii.

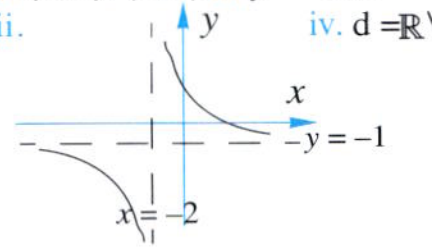

iv. d = $\mathbb{R}\setminus\{-2\}$

(b) $f^{-1}: \mathbb{R}\setminus\{-1\} \mapsto \mathbb{R}$, where $f^{-1}(x) = \frac{2(1-x)}{(1+x)}$

7. (a) $y = 8, x = 3$

(b)

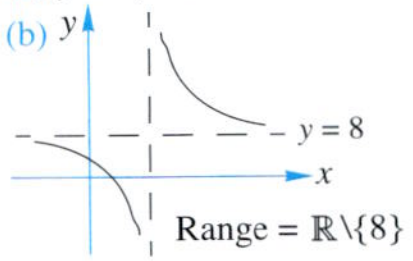

6. (c)

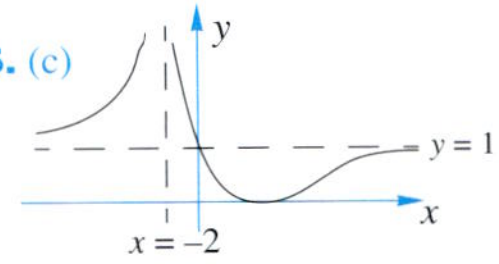

EXERCISE 21.1

1.(a) i. $x < 0$ ii. $x > 4$ iii. $0 \le x \le 4$ (b) i. $-1 < x < 2$ ii. $x < -1, 2 < x < 5$ iii. $\varnothing$
(c) i. $-1 < x < 1$ ii. $x < -1$ iii. $x \ge 1$ (d) i. $0 < x < 1$ ii. $2 < x < 3$ iii. $x < 0, 1 \le x < 2$
(e) i. $\varnothing$ ii. $-2 < x < 4$ iii. $\varnothing$ (f) i. $-4 < x < -1, 2 < x < 5$ ii. $-1 < x < 2, 5 < x < 8$ iii. $\varnothing$

EXERCISE 21.2

1. 4.4 (4 deer per year, to nearest integer) **2.** (a) 200 cm^3 (b) 73.5 $cm^3 day^{-1}$ **3.** (a) 75 (b) No
4. (a) \$207.66 (b) \$ 40.79 per year (c) \$41.54 per year **5.** i. 2.50 ii. 3.33 iii. 2.50
6. (a) $1230 < x < 48770$ (approx) (b) i. $0 \le x < 25000$ ii. $25000 < x \le 50000$ **7.** 66667 (to nearest integer), 1446992 (to nearest integer) **8.** (b) 133.33 (d) 46.67 (e) $0 < x < 5700$

9. (a) $D'(x) = \frac{-40000(2x+12)}{(x^2+12x+20)^2}$ $5 \le x \le 18$ (b) 22.22 (22 items/dollar) **10.** (a) $\frac{3000}{(x+32)^2}$

(b) i. $x \ge 0$ ii. $x \in \varnothing$ **11.** (a) i. 0 mm/s ii. ~ 90.69 mm/s (b) 0.6 sec **12.** (a) 8.53 cm/s (b) (c) never

13. $-e^{-1}$ ms^{-2}

EXERCISE 21.3

1. (a) i. $v = -\frac{1}{(t-1)^2}, t > 1$ ii. $a = \frac{2}{(t-1)^3}, t > 1$ (b) i. $v = 2(e^{2t} - e^{-2t}), t \ge 0$

ii. $a = 4(e^{2t} + e^{-2t}), t \ge 0$ (c) i. $v = a - 2bte^{-t^2}, t \ge 0$ ii. $a = 2be^{-t^2}(2t^2 - 1), t \ge 0$

2. (a) 8 ms^{-1} (b) never at rest (c) i. 5m from O in negative direction ii. 4 ms^{-1} (d) 40 m

(e)

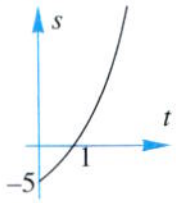

3. (a) 1 ms^{-1} (b) never (c) $t = \frac{1}{3}$ or $t = 1$ (d) 20 ms^{-2}

4. (a) $v = -6t^2 + 12$; $a = -12t$ (b) ~ 1.41 sec (c) once (d) use graphics calculator

5. (a) 3 m in positive direction (b) i. 5 m ii. 2 m (c) 5 ms^{-1} (e) oscillation about origin with amplitude 5 m and period 2π sec **7.** (a) 100 m, in negative direction (b) 3 times (c) i. 80 ms^{-1}

ii. –34 ms^{-2} (d) 14.81m **8.** (a) max = 5 units, min = –1 unit (b) $\frac{\pi}{2}$ sec (c) i. $a = -12\cos(2t - \pi)$

ii. $a = -4(x-2)$ **9.** (a) 0.318 m above (b) i. $v = 3.75e^{-0.25t} - 3$ ii. $a = -0.9375e^{-0.25t}$

(c) 0.322 m (d) $a = -0.25(v+3)$ **10.** (a) $0 < t < 0.5$ or $t > 1$ (b) $t > 0.5$ (c) $t = 1$ or $1.68 \le t \le 5$.

11. (a) This question is best done using a graphics calculator:
(b) From the graph the particles pass each other three times.
(c) 0.45 sec; 2.85 sec; 3.87 sec
(d) i. $v_A = -0.3e^{0.3t}$ ms^{-1} ii. $v_B = 10e^{-t}(1-t)$ ms^{-1}
(e) Yes. On two occassions

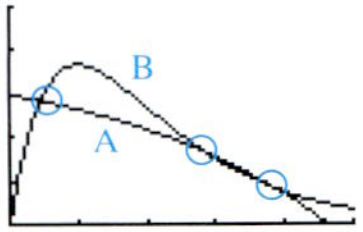

12. (a) 2m in positive direction (b) i. 2 sec ii. never (c) 0.026 ms^{-2}

13. (a) (b) 0.295

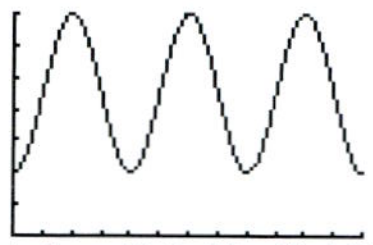

window: $[0, 3\pi]$ by $[0, 14]$

EXERCISES 21.4

1. 22.6 m **2.** (a) 1.5 kmh^{-1} (b) \$19.55 per km **3.** (a) 400 (b) \$46,400,000 **4.** \$273.86 **5.** \$0.40 **6.** 1.97 m **7.** 0.45 m^3 **8.** 5m by 5m **9.** 128 **10.** $r = \frac{50}{4+\pi} \approx 7.00$, dim of rect. $\frac{50}{4+\pi} \times \frac{50}{4+\pi}$ i.e., aprrox 7.00 m by 7.00 m **11.** $\theta = \frac{\pi}{6}$

12. (a)

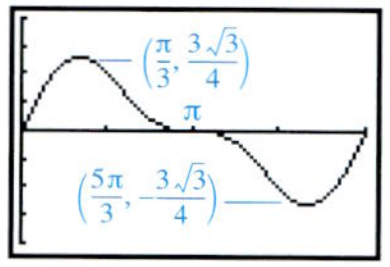

(b) $\frac{3\sqrt{3}}{2}$ units (c) At points of inflections, when $\cos x = -\frac{1}{4}$.

13. 648 m^2 **14.** (a) 10.5 (b) 5.25 **15.** 72 **16.** (a) $y = 100 - 2x$ (b) $A = x(100 - 2x), 0 < x < 50$ (c) $x = 25, y = 50$ **17.** (a) $\frac{100}{x} - \frac{1}{2}x, 0 < x < 10\sqrt{2}$ (b) $\frac{2000}{9}\sqrt{6} \approx 544.3$ cm^3

18. (a) 400 $ml\ s^{-1}$ (b) 40 sec (c)

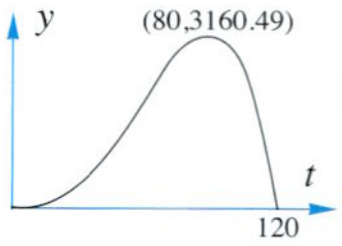

19. (a)

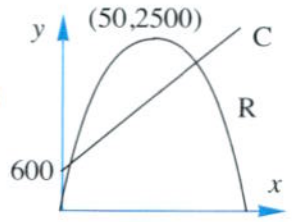

(b) 8.38, 71.62 (c) $9 \le x \le 71$ (d) $80x - x^2 - 600$, \$1000 **20.** $\left(\sqrt{\frac{11}{2}}, \frac{7}{2}\right)$ & $\left(-\sqrt{\frac{11}{2}}, \frac{7}{2}\right)$

21. $5\sqrt{2}$ by $\frac{5}{2}\sqrt{2}$ **22.** 4 by $\frac{8}{3}$ **23.** $348 - 8\sqrt{170} \sim 243.7$ cm^2 **24.** 2 **25.** radius $= \sqrt{\frac{10}{3}}$ cm, height $= 2\sqrt{\frac{10}{3}}$ cm **26.** $\sqrt[3]{\frac{15}{\pi}}$ **27.** 5 cm **28.** (a) $h = \frac{24r^2}{r^2 - 144}$ (b) $\frac{8\pi r^4}{r^2 - 144}$ (c) $r = 12\sqrt{2}$, $h = 48$ **29.** $r : h = 1 : 2$ **30.** ~ (0.55, 1.31) **31.** (b) 2.5 m **32.** altitude $= \frac{1}{3}$ height of cone

33. ~ 1.640 m wide and 1.040 m high **34.** $\frac{2\sqrt{2}}{\sqrt{3}}\pi$ **35.** where XP : PY = $b : a$ **36.** 5 km

37. $r : h = 1 : 1$ **38.** $\frac{4}{3}$ cm **39.** 2 : 1 **40.** $\frac{10}{\sqrt{3\pi}}$ **41.** 0.873 km from P

42. (b) $r = 3\sqrt{2}$, $h = 6\sqrt{2}$ **43.** (b) when $\theta = \arcsin\left(\frac{5}{6}\right)$, i.e., approx. 6.030 km from P.

44. (a) $\tan\theta = \frac{xl}{x^2 + k(l+k)}$ (b) $x = \sqrt{k^2 + kl}$

45. (c) if $k < c$, swimmer should row directly to Q. **46.** (a) i. $\pi r^2 h + \frac{2}{3}\pi r^3$ ii. $3\pi r^2 + 2\pi rh$ (c) $r : h = 1:1$ **47.** $(a^{2/3} + b^{2/3})^{3/2}$ **48.** (b) 4 km along the beach (c) row directly to destination

EXERCISE 22.1

1. (a) $\frac{1}{4}x^4 + c$ (b) $\frac{1}{8}x^8 + c$ (c) $\frac{1}{6}x^6 + c$ (d) $\frac{1}{9}x^9 + c$ (e) $\frac{4}{3}x^3 + c$ (f) $\frac{7}{6}x^6 + c$ (g) $x^9 + c$ (h) $\frac{1}{8}x^4 + c$ **2.** (a) $5x + c$ (b) $3x + c$ (c) $10x + c$ (d) $\frac{2}{3}x + c$ (e) $-4x + c$ (f) $-6x + c$ (g) $-\frac{3}{2}x + c$ (h) $-x + c$ **3.** (a) $x - \frac{1}{2}x^2 + c$ (b) $2x + \frac{1}{3}x^3 + c$ (c) $\frac{1}{4}x^4 - 9x + c$ (d) $\frac{2}{5}x + \frac{1}{9}x^3 + c$ (e) $\frac{1}{3}x^{3/2} + \frac{1}{x} + c$ (f) $x^{5/2} + 4x^2 + c$ (g) $\frac{1}{3}x^3 + x^2 + c$ (h) $x^3 - x^2 + c$ (i) $x - \frac{1}{3}x^3 + c$ **4.** (a) $\frac{1}{3}x^3 - \frac{1}{2}x^2 - 6x + c$ (b) $\frac{1}{4}x^4 - \frac{2}{3}x^3 - \frac{3}{2}x^2 + c$ (c) $\frac{1}{4}(x-3)^4 + c$ (d) $\frac{2}{5}x^5 + \frac{1}{2}x^4 + \frac{1}{3}x^3 + \frac{1}{2}x^2 + c$ (e) $x + \frac{1}{2}x^2 - \frac{2}{3}x^{3/2} - \frac{2}{5}x^{5/2} + c$ (f) $\frac{2}{7}x^{7/2} + \frac{4}{5}x^{5/2} + \frac{2}{3}x^{3/2} - 2x + c$ **5.** (a) $\frac{1}{2}x^2 - 3x + c$ (b) $2u^2 + 5u + \frac{1}{u} + c$ (c) $-\frac{1}{x} - \frac{2}{x^2} - \frac{4}{3x^3} + c$ (d) $\frac{1}{2}x^2 + 3x + c$ (e) $\frac{1}{2}x^2 - 4x + c$ (f) $\frac{1}{3}t^3 + 2t - \frac{1}{t} + c$ **6.** (a) $\frac{4}{7}\sqrt[4]{x^7} + 2\sqrt{x} - 5x + c$ (b) $\frac{1}{3}x^3 + \frac{1}{2}x^2 - \frac{4}{7}x^{7/2} - \frac{4}{5}x^{5/2} + c$ (c) $-\frac{1}{2z^2} + \frac{2}{z} + 2z^2 + z + c$ (d) $\frac{1}{2}t^4 + t + c$ (e) $\frac{2}{5}\sqrt{t^5} - 2\sqrt{t^3} + c$ (f) $\frac{1}{3}u^3 + 2u^2 + 4u + c$ **8.** (a) $\frac{1}{8}(2x+3)^4 + c$ (b) $3\sqrt{x^2 + 4} + c$

EXERCISE 22.2

1. (a) $x^2 + x + 3$ (b) $2x - \frac{1}{3}x^3 + 1$ (c) $\frac{8}{3}\sqrt{x^3} - \frac{1}{2}x^2 - \frac{40}{3}$ (d) $\frac{1}{2}x^2 + \frac{1}{x} + 2x - \frac{3}{2}$ (e) $(x+2)^3$ (f) $\frac{3}{4}\sqrt[3]{x^4} + \frac{1}{4}x^4 + x$ (g) $\frac{1}{3}x^3 + 1$ (h) $x^4 - x^3 + 2x + 3$ **2.** $\frac{1}{2}x^2 + \frac{1}{x} + \frac{5}{2}$ **3.** \$3835.03

4. 9.5 **5.** $\frac{251\pi}{3}$ cm^3 **6.** 292 **7.** $\frac{5}{7}\sqrt{x^3} + \frac{23}{7}$ **8.** (1, –8) **9.** $P(x) = 25 - 5x + \frac{1}{3}x^2$

10. $N = \frac{20000}{201}t^{2.01} + 500, t \ge 0$ **11.** (a) $y = -\frac{2}{5}x^2 + 4x$ (b) $y = \frac{1}{6}x^3 + \frac{5}{4}x^2 + 2x$

12. $y = 2(x^3 + x^2 + x)$ **13.** $f(x) = -\frac{3}{10}x^3 + \frac{49}{10}x - \frac{13}{5}$ **14.** Vol ~ 43202 cm^3 **15.** 110 cm^2

EXERCISE 22.3

1. (a) $\frac{1}{5}e^{5x}+c$ (b) $\frac{1}{3}e^{3x}+c$ (c) $\frac{1}{2}e^{2x}+c$ (d) $10e^{0.1x}+c$ (e) $-\frac{1}{4}e^{-4x}+c$ (f) $-e^{-x}+c$ (g) $-0.2e^{-0.5x}+c$ (h) $-2e^{1-x}+c$ (i) $5e^{x+1}+c$ (j) $e^{2-2x}+c$ (k) $3e^{x/3}+c$ (l) $2\sqrt{e^x}+c$

2. (a) $4\log_e x+c, x>0$ (b) $-3\log_e x+c, x>0$ (c) $\frac{2}{5}\log_e x+c, x>0$ (d) $\log_e(x+1)+c, x>-1$ (e) $\frac{1}{2}\log_e x+c, x>0$ (f) $x-2\log_e x-\frac{1}{x}+c, x>0$ (g) $\frac{1}{2}x^2-2x+\log_e x+c, x>0$ (h) $3\ln(x+2)+c$ **3.** (a) $-\frac{1}{3}\cos(3x)+c$ (b) $\frac{1}{2}\sin(2x)+c$ (c) $\frac{1}{5}\tan(5x)+c$ (d) $\cos(x)+c$

4. (a) $-\frac{1}{2}\cos(2x)+\frac{1}{2}x^2+c$ (b) $2x^3-\frac{1}{4}\sin(4x)+c$ (c) $\frac{1}{5}e^{5x}+c$ (d) $-\frac{4}{3}e^{-3x}-2\cos\left(\frac{1}{2}x\right)+c$ (e) $3\sin\left(\frac{x}{3}\right)+\frac{1}{3}\cos(3x)+c$ (f) $\frac{1}{2}e^{2x}+4\log_e x-x+c, x>0$ (g) $\frac{1}{2}e^{2x}+2e^{x}+x+c$ (h) $\frac{5}{4}\cos(4x)+x-\log_e x+c, x>0$ (i) $\frac{1}{3}\tan(3x)-2\log_e x+2e^{x/2}+c, x>0$ (j) $\frac{1}{2}e^{2x}-2x-\frac{1}{2}e^{-2x}+c$ (k) $\frac{1}{2}e^{2x+3}+c$ (l) $-\frac{1}{2}\cos(2x+\pi)+c$ (m) $\sin(x-\pi)+c$ (n) $-4\cos\left(\frac{1}{4}x+\frac{\pi}{2}\right)+c$ (o) $2e^{x/2}+4e^{-x/2}+c$ **5.** (a) $\frac{1}{16}(4x-1)^4+c$ (b) $\frac{1}{21}(3x+5)^7+c$ (c) $-\frac{1}{5}(2-x)^5+c$ (d) $\frac{1}{12}(2x+3)^6+c$ (e) $-\frac{1}{27}(7-3x)^9+c$ (f) $\frac{1}{5}\left(\frac{1}{2}x-2\right)^{10}+c$ (g) $-\frac{1}{25}(5x+2)^{-5}+c$ (h) $\frac{1}{4}(9-4x)^{-1}+c$ (i) $-\frac{1}{2}(x+3)^{-2}+c$ (j) $\ln(x+1)+c, x>-1$ (k) $\ln(2x+1)+c, x>-\frac{1}{2}$ (l) $-2\ln(3-2x)+c, x<\frac{3}{2}$ (m) $3\ln(5-x)+c, x<5$ (n) $-\frac{3}{2}\ln(3-6x)+c, x<\frac{1}{2}$ (o) $\frac{5}{3}\ln(3x+2)+c, x>-\frac{2}{3}$ **6.** (a) $-\frac{1}{2}\cos(2x-3)-x^2+c$ (b) $6\sin\left(2+\frac{1}{2}x\right)+5x+c$ (c) $\frac{3}{2}\sin\left(\frac{1}{3}x-2\right)-\ln(2x+1)+c$ (d) $10\tan(0.1x-5)-2x+c$ (e) $2\ln(2x+3)+2e^{-\frac{1}{2}x+2}+c$ (f) $-\frac{2}{2x+3}-\frac{1}{2}e^{2x-\frac{1}{2}}+c$ (g) $x+\ln(x+1)-4\ln(x+2)+c$ (h) $2x-\ln(x+2)+\frac{1}{2}\ln(2x+1)+c$ (i) $-\frac{1}{2x+1}+\ln(2x+1)+c$ **7.** (a) $f(x)=\frac{1}{6}(4x+5)^{3/2}$ (b) $f(x)=2\ln(4x-3)+2$ (c) $f(x)=\frac{1}{2}\sin(2x+3)+1$ (d) $f(x)=2x+\frac{1}{2}e^{-2x+1}+\frac{1}{2}e$

8. 14334 **9.** 13.19m/sec or 1.19m/sec **10.** 2.66cm **11.** $2e^{x/2}-\frac{1}{2}\sin(2x)-2$

12. (a) $p=\frac{a}{a^2+b^2}$, $q=-\frac{b}{a^2+b^2}$ (b) $\frac{1}{13}e^{2x}(2\sin 3x-3\cos 3x)+c$ **13.** (a) $0.25a$ (b) $a\times\left(\frac{1}{2}\right)^{8/3}\approx 0.1575a$ **14.** (b) 666 gm

15. (a) (b) 73.23% (c) ~25.24 litres **16.** (a) (b) 7000 (c) 1.16 day (d) 2 days

15. (a)

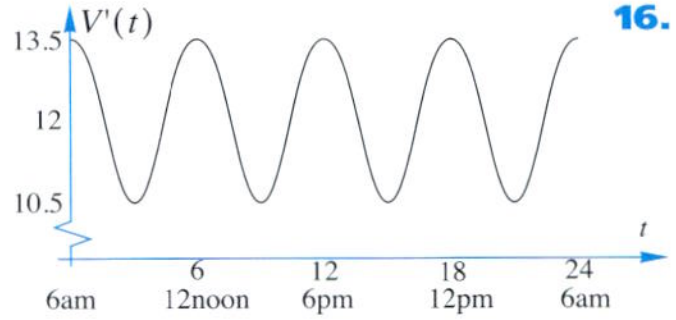

16. (a)

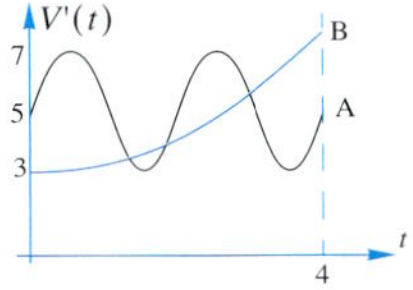

EXERCISES 22.4

1. (a) $\frac{15}{2}$ (b) $\frac{38}{3}$ (c) $\frac{5}{36}$ (d) -8 **2.** (a) $\frac{35}{24}$ (b) $\frac{8}{5}\sqrt{2}-2$ (c) -2 (d) 0 (e) $\frac{1}{20}$ (f) $-\frac{4}{3}$ (g) $\frac{7}{6}$ (h) $\frac{5}{6}$ (i) $\frac{20}{3}$ (j) 0 (k) $\frac{20}{3}$ (l) $-\frac{\sqrt{2}}{3}$ **4.** (a) e (b) $2(e^{-2}-e^{-4})$ (c) 0 (d) $2(e-e^{-1})$ (e) e^2+4-e^{-2} (f) $\frac{1}{2}(e-e^5)$ (g) $2\sqrt{e}-3$ (h) $\frac{1}{4}(16e^{1/4}-e^4-15)$ (i) $\frac{1}{2}(e^{-1}-e^3)$

6. (a) $3\ln 2$ (b) $2\ln 5$ (c) $4+4\ln 3$ (d) $\frac{1717}{4}$ (e) $\frac{3}{2}\ln 3$ (f) $2\ln 2$ (g) $\frac{3}{4}$ (h) $4\ln 2-2$ (i) $\ln 2$

8. (a) 1 (b) $\frac{3\sqrt{3}}{2}$ (c) $\frac{\sqrt{3}}{2}$ (d) -2 (e) $\frac{\pi^2}{32}-1$ (f) 0 (g) 0 (h) $\frac{\sqrt{3}}{2}-\frac{1}{2}$ (i) 0 **9.** (a) $\frac{31}{5}$ (b) $\frac{7\sqrt{7}}{3}-\sqrt{3}$ (c) 0 (d) $\frac{5}{72}$ (e) $3\sqrt[3]{2}-\frac{3}{2}$ (f) $1-\ln 2$

10. $\ln\left(\frac{21}{5}\right)$ **11.** $\sin 2x+2x\cos 2x$; 0 **12.** (a) $2m-n$ (b) $m+a-b$ (c) $-3n$ (d) $m(2a-b)$ (e) na^2 **13.** (a) $e^{0.1x}+0.1xe^{0.1x}$; $10xe^{0.1x}-100e^{0.1x}+c$ (b) i. 99 accidents ii. $N=12t+10te^{0.1t}-100e^{0.1t}+978$ **14.** (a) 1612 subscribers (b) 46220 **15.** (b) ~524 flies

EXERCISES 22.5

1. (a) 4 sq.units. (b) $\frac{32}{3}$ sq.units. (c) 4 sq.units. (d) 36 sq.units. (e) $\frac{1}{6}$ sq.units. **2.** (a) e sq.units. (b) $\frac{1}{2}(e^4-2-e^2)$ sq.units. (c) $2(e+e^{-1}-2)$ sq.units. (d) $2(e^2-2-e)$ sq.units.

3. (a) $\ln\left(\frac{5}{4}\right)$ sq.units. (b) $2\ln 5$ sq.units. (c) $3\ln 3$ sq.units. (d) 0.5 sq.units. **4.** (a) 2 sq.units. (b) $\frac{\pi}{2}$ sq.units. (c) $\frac{3}{8}\pi^2+\sqrt{2}-2$ sq.units. (d) $\sqrt{2}$ sq.units. (e) $4\sqrt{3}$ sq.units. **6.** 12 sq.units.

7. $4\left(\sqrt{3}-\frac{1}{3}\right)$ sq.units. **8.** $\ln 2+1.5$ sq.units. **9.** 2 sq.units. **10.** $\frac{37}{12}$ sq.units. **11.** (a) 0.5 sq.units. (b) 1 sq.unit. (c) $2(\sqrt{6}-\sqrt{2})$ sq.units. **12.** $\frac{8}{3}$ **13.** $-2\tan 2x$; $\frac{1}{4}\ln 2$ sq.units. **14.** (a) $\frac{9}{2}$ sq.units. (b) 3 sq.units. **15.** (a) 1 sq.unit. (b) 10 sq.units or 2 sq.units **16.** (a) $x\ln x-x+c$ (b) 1 sq.unit. **17.** $\frac{14}{3}$ sq.units **18.** (a) $\frac{7}{6}$ sq.units (b) $\frac{9}{2}$ sq.units **19.** i. $\frac{15}{4}$ sq.units ii. $\frac{45}{4}$ sq.units

20. $\frac{22}{3}$ sq.units. **21.** (b) i. $e^{-1}+e-2$ sq. units ii. 1 sq. unit iii. 2ln(2) sq. units

22. (b) 3.05 sq. units **23.** (a) $2y = 3ax - a^3$ (b) $\frac{1}{15}a^5$ sq. units **24.** (a) $1 - e^{-1}$ sq. units (b) e^{-1} sq. units (c) $1 - e^{-e^{-1}-1} - e^{-1} \approx 0.10066$ sq. units **25.** $a = 16$

EXERCISES 22.6

1. (a) $x = t^3 + 3t + 10, t \geq 0$ (b) $x = 4\sin t + 3\cos t - 1, t \geq 0$ (c) $x = t^2 - 4e^{-\frac{1}{2}t} + 2t + 4, t \geq 0$ **2.** (a) $x = t^3 - t^2, t \geq 0$ (b) 100 (c) $100\frac{8}{27}$ m

3. (a) $x = -\frac{2}{3}(4+t)^{3/2} + 2t + 8$ (b) 6.92 m **4.** $\frac{125}{6}$ m **5.** $\frac{125}{49}$ sec; 63.8 m **6.** (a) $\frac{\pi}{6}$ sec (b) $\frac{\pi}{2} - 1$ m **7.** 80.37 m **8.** (a) $s(t) = \frac{160}{\pi}\left[1 - \cos\left(\frac{\pi}{16}t\right)\right], t \geq 0$ (b) 86.94 m (c) –6.33 m (d) 116.78 m **9.** (a) $v = 4 + k - \frac{k}{t^2}, t > 0$ (b) $k = 2$ (c) 52.2 m **10.** (b) 0.0893 m

EXERCISES 22.7

All values are in cubic units

1. 21π **2.** $\pi\ln 5$ **3.** $\frac{\pi}{2}(e^{10} - e^2)$ **4.** π^2 **5.** $\frac{\pi}{2}$ **6.** $\pi\left(\frac{8}{3} - 2\ln 3\right)$ **9.** $\frac{\pi}{5}(5 - 5\sin 1)$

10. $\frac{251}{30}\pi$ **11.** $\frac{242}{5}\pi$ **12.** $\frac{\pi}{4}$ **13.** $\frac{88}{5}\sqrt{3}\pi$ **14.** $\frac{3\pi}{4}$ **15.** $k = 1$

16. $4\pi^2a^2$ **17.** $k = \frac{\pi}{2}$ **18.** $\frac{8\pi}{15}\sqrt{\frac{a}{1+a^2}}\left(\frac{3a^2+2}{1+a^2}\right)$ **19.** (a) two possible solutions: solving $a^3 - 6a^2 - 36a + 204 = 0$, $a = 4.95331$; solving $a^3 - 6a^2 - 36a - 28 = 0$, then $a = -0.95331$.

20. $\frac{28}{15}\pi$ **21.** 64π

REVISION EXERCISES – SET A

1. –84 **2.** (a)

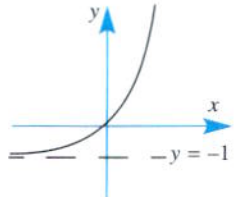

(b) i. $]-1, \infty[$ ii. $f^{-1}(x) = \ln(x+1)$ (c)

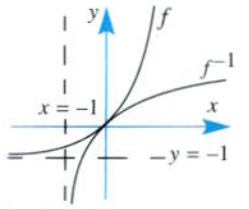

3. 840 **4.** (a) i. 0 ii. 2 (b) $-2 \leq x \leq 2$ (c) $x \geq 0$ **5.** (a) $(1,-2), (-1,4)$ and $(3,0)$

(b)

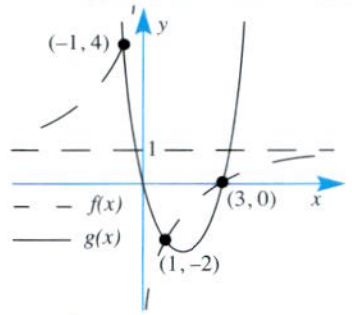

6. (a) 2 (b) $S = [0, \infty[$, range $= [1, \infty[$ (c) $f^{-1} : [1, \infty[\mapsto \mathbb{R}$, $f^{-1}(x) = (\ln x)^2$

7. (a)

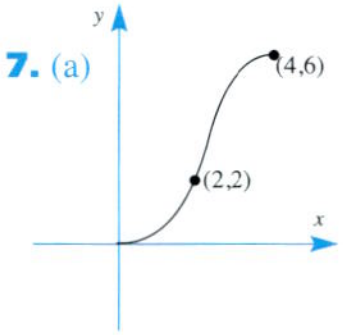

(b)

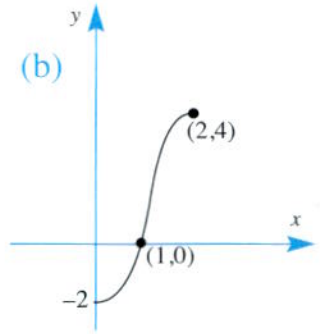

(c)

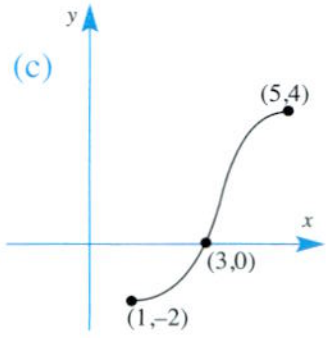

(d)

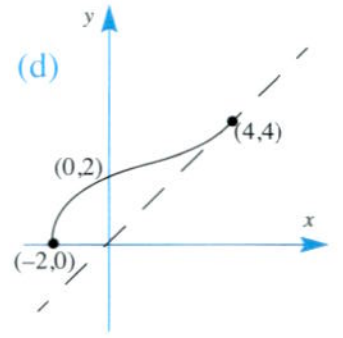

8. (a) i. 512 ii. 2 (b) i. $3x^2h + 3xh^2 + h^3$ ii. $3x^2 + 3xh + h^2$

9. (a) i. –1 or 6 ii. $\frac{3}{e-1}$ (b) i. $\mathbb{R}\backslash\{3\}$ ii. 0.2 iii. 0

10. (a) i. 2 or 6 ii. $\frac{1}{3}(e^2 - 4)$ (b) i. $0 < x < 1$ ii. $\mathbb{R}$ iii. $\log_e 4 \approx 0.72$ iv. $\frac{e^{0.8}}{1 + e^{0.8}} \approx 0.69$

11. (a) $g(f(x)) = -\frac{2x}{1-x}, x \in \mathbb{R}\backslash\{\pm 1\}$ (b) $P \equiv (2, 4)$

12. (a) i. $x = \frac{\ln 6}{\ln 3}$ ii. $\frac{15}{7}$ (b) $1 + \sqrt{3}$ (c) i. $f(g(x)) = \sqrt{\frac{1}{x^2} - 1}$, $g(f(x)) = \frac{1}{x-1}$ ii. $[-1,1]\backslash\{0\}$

13. 0.5 **14.** (a)

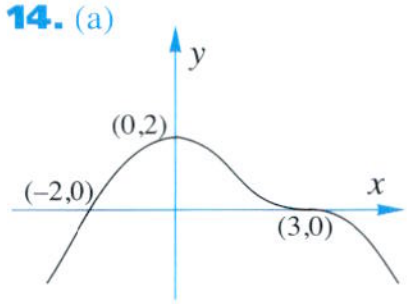

(b)

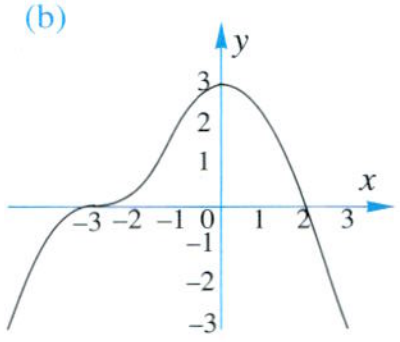

(c)

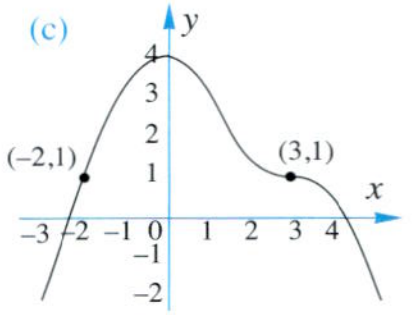

15. (a) $k = 0$ or 1 (b) $0 < x < 3$ **16.** (a) $0 < x < 5$ (b) 70 **17** i. 9 ii. –4 **18.** ±3

19. (a) $y = -2x$ (b) $\frac{x-y}{x+y}$ **20.** (b) $x = -\frac{4}{9}, y = \frac{1}{9}$ **21.** (b) ii. $p^5 = 3 + 5p$, $p^4 = 2 + 3p$

22. (a) $\frac{2}{9}$ (b) 59136 **23.** $a = -\frac{3}{5}, b = -\frac{648}{25}, n = 10$ **24.** (a) ii. $\{\pm 1\}$ (b) i. $y = \sqrt{6(x-3)}$ ii. $x = 9, y = 6$ **25.** $1792x^5$ **26.** (a) $\frac{5}{2}, -\frac{3}{2}$ (b) $\frac{3}{2}, -\frac{1}{2}$ (c) $\frac{17}{2}$ **27.** (a) $\mathbb{R}$ (b) $]-\infty,4]$ (c) $]-\infty,4[$

28. (b) $\frac{5}{8}$ sq. units

29. i.

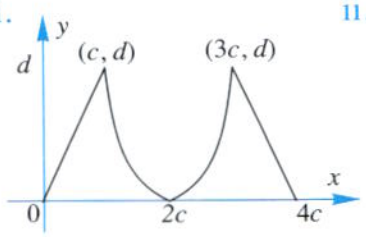

ii.

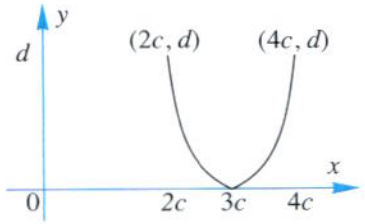

30. (a) i. & ii.

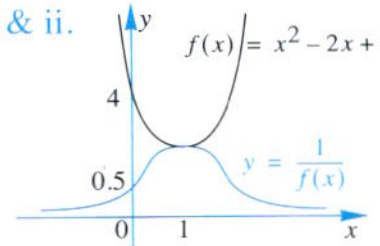

(b) i.

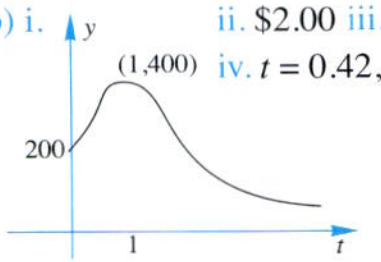

ii. $2.00 iii. $4 iv. $t = 0.42, 1.57$

31. (a)

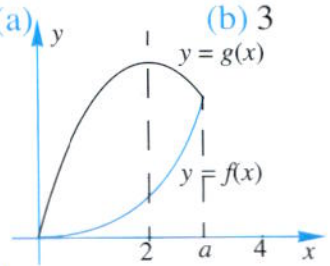

(b) 3

32. (a) $g^{-1}(x) = e^{2x} + e, x \in \mathbb{R}$ (b)

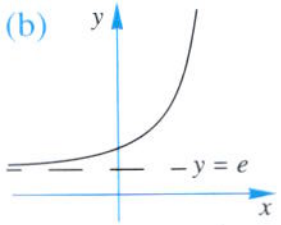

33. (a) i. $]0, \infty[$ ii. $\left]\left(-\frac{a}{eb}\right), \infty\right[$ (b)

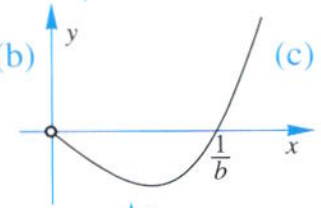

(c) $(1, a\log_e b)$ (d) $x = b^{\frac{1}{x} - 1}$

34. (a) $a = -36, b = 900$
(b) (c) 20 (d) 1000 (e) $t > 1.95$ (f)

t	1	2	3.5	5
$B(t)$	131.04	527.02	957.23	997.78

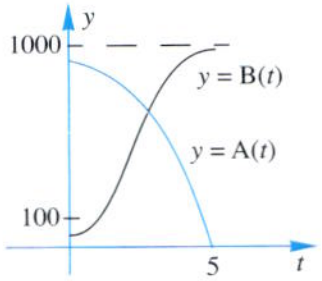

35. (a) 150 cm (b) 138 cm (c) 94 hrs (d) [0, 94] (e) $h^{-1}(x) = \frac{12.5 - \sqrt{t}}{0.13}$ (f) Use g.c (g) 17.3 hrs

36. (a)

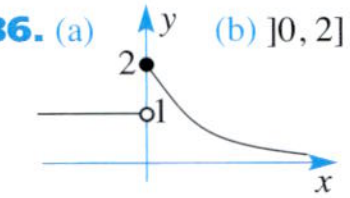

(b) $]0, 2]$ (c) No $(x = 0)$ **37.** 78 **38.** (a) 0 (b) $-\sqrt{2}$ (c) $r_f \not\subseteq d_g$, doesn't exist

39. $-\frac{63}{8}x^5$ **40.** (a) $g^{-1}(x) = -1 + \sqrt{x-2}, x \geq 2$ (b)

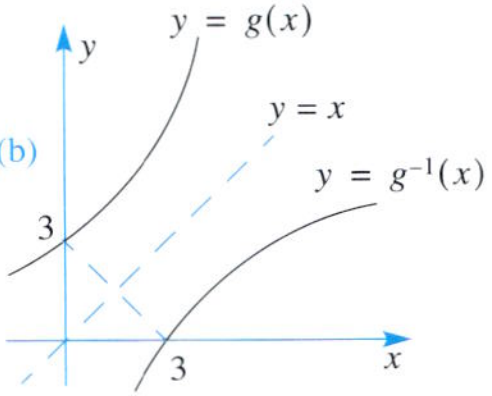

(c) no

41. ii. $h(x) = 4 - x, x \geq 0$, range $=]-\infty, 4]$ **42.** i. use g.c ii. $f^{-1}(x) = -\log_e(1-x), x < 1$
iii. use g.c **43.** –10 **44.** (a) $r_g \subseteq d_f \Rightarrow f \circ g$ exists; $r_f \not\subseteq d_g \Rightarrow g \circ f$ doesn't exist
(b) $x < -2$ or $x > 2$ (c) $\mathbf{S} =]-3, 2[$ **45.** (a) $f^{-1}(x) = (2-x)^2, x < 2$ (b) $r_g \not\subseteq d_{f^{-1}} \Rightarrow f^{-1} \circ g$ does not exist; $r_{f^{-1}} \subseteq d_g \Rightarrow g \circ f^{-1}$ exists (c) $F(x) = x - 2, x \leq 2$ **46.** (a) i. 50 ii. $50e \approx 135.9$

(c)

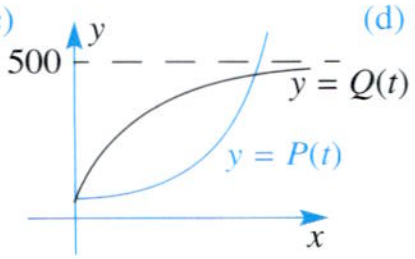

(d) i. 50 ii. 334.5 (f) Increasing at a decreasing rate
(g) ~ 460 wasps
(h) ii. $t = 0$ and $t = 10\log_e 9$

REVISION EXERCISES – SET B

1. (a) 189 (b) 99 (c) –96 (d) 36 **2.** (b) –65 **3.** (b) 23.9 km (c) 26.4° (d) 15.3 km (e) 107.5
4. (a) i. A: $49000; B: $52400; C: $19200 ii. A: $502400; B: $506100; C: $379400 (b) 4.6%
(c) i. 14 months ii. C never reaches its target **5.** (a) $r = 0.5$ (b) 62.5 cm **6.** (b) 26°34′ or 135°
7. (b) $\frac{7\pi}{6}, \frac{11\pi}{6}, \frac{\pi}{2}$ **8.** 28 **9.** (a) $\frac{\pi}{3}, \frac{2\pi}{3}, \frac{4\pi}{3}, \frac{5\pi}{3}$ (b) $0, \frac{\pi}{2}, \pi, 2\pi$ **10.** (a) max value is $\frac{17}{2}$ for $x = \frac{\pi}{2} + 2k\pi$ or $x = \frac{3\pi}{2} + 2k\pi$, where k is an integer; min value is $\frac{17}{5}$ for $x = k\pi$, where k is an integer (b) $\frac{\pi}{3}, \frac{5\pi}{3}$ **11.** (a) $u_n = 74 - 6n$ (b) $n = \frac{1}{6}(74 - p)$ (c) $\frac{1}{12}(74-p)(68+p)$, 420

12. $\frac{24(4\sqrt{3}-3)}{39}$ **14.** (a) 60°, 109°28′, 250°32′, 300° (b) i. 2cosecθ ii. $\frac{\pi}{3}, \frac{2\pi}{3}$ **15.** (a) ~342
(b) 20 terms (c) $0 < x < 2$ (d) $\{1, 3, 8, 18, \ldots\}$ (e) $u_n = 23 - 3n$ (f) $4131.45 **16.** (a) $-\frac{1}{2}$
(b) 4 **17.** (a) 120° (b) $14\sqrt{3}$ cm² **18.** (a) i. $0.3\sqrt{3}$ m ii. $0.2\sqrt{3}$ m (b) ~1.15 m (c) 73°13′
19. (a) $\frac{\pi}{3}, \frac{4\pi}{3}$ (b) $\left\{x \mid \frac{\pi}{3} < x < \frac{4\pi}{3}\right\}$ **20.** (a) 8 cm (b) 28°4′ **21.** 3 **22.** (a) $\left\{\frac{\pi}{12}, \frac{7\pi}{12}, \frac{13\pi}{12}, \frac{19\pi}{12}\right\}$
(b) $\frac{\pi}{3}$ **23.** (a) $77156.10 (b) $u_1 = -\sqrt{3}, u_3 = -3\sqrt{3}$ **24.** (a) $f(x) = 3\cos(2x)$ (b) $\left\{\frac{7\pi}{6}\right\}$
(c) 3 **25.** (b) i. BP = 660 m, PQ = 688 m **26.** 216° **27.** (b) 906 m **28.** (a) 38°40′
(b) 0.08004 m² (c) $493.71 **29.** (a) $\tan\alpha = -\frac{1+\sqrt{5}}{2}$ (b) range = [3, 3.5] (c) i. 3 ii. 2
30. (a) i. $W(4) = 19.38, P(4) = 14.82$ ii. $W(20) = 10.95, P(20) = 27.02$
iii. $W(35) = 13.45, P(35) = 23.25$ (b) Amp = 5, period = 50 weeks
(c)

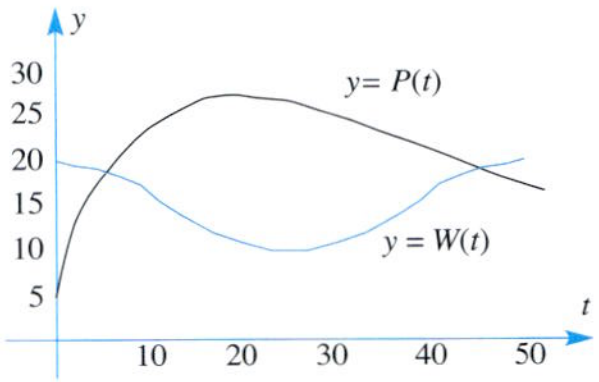

(d) $27.07 (e) during 7th & 46th week
31. (a) $49000, $47900, $46690 (b) $34062.58
(c) 18.8 yrs (d) ~$248564
32. (a) ii. 26 cards (b) 26, 40, 57, 77
(c) $a = 3, b = -1$ (d) 155 cards
(e) $t_n = \frac{n}{2}(3n+1)$
33. (a) ~2.77 m (b) i. 3.0 m ii. 2.0 m (c) 4.15 pm
(d) use g.c. (e) $2\frac{1}{6} < t < 6\frac{1}{3}$

34. 1.262 ha 35. $\left\{\frac{\pi}{4}, \frac{3\pi}{4}\right\}$ **36.** (a) $x = -\frac{2\pi}{3}, \frac{2\pi}{3}$ (b)

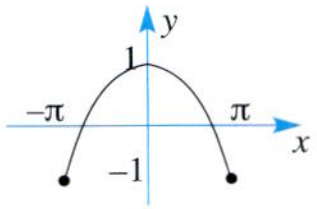

(c) $-\frac{2\pi}{3} < x < \frac{2\pi}{3}$

37. 1623 m **38.** (a) 19.5°C (b) $D(t) = -1 + 2\cos\left(\frac{\pi}{12}t\right)$ (d) use g.c (e) 8 am to midnight

39. 1939 m **40.** (a) ii. $N_0 = 2000, \alpha = 10$ (b) 2000, 2200, 2420, 2662, 2988.2 (c) 52 hrs
(d) 176995

41. (a) $(4-\pi)$ cm^2 (b) $\frac{(4-\pi)}{2}$ cm^2 (c) ii. $r=\frac{1}{2}$ iii. $A_n = (4-\pi)\times\left(\frac{1}{2}\right)^{n-1}, n = 1, 2\ldots$

(d) i. $\frac{31}{16}(4-\pi)$ cm^2 ii. $2(4-\pi)$ cm^2 (e) geometric

REVISION EXERCISES – SET C

1. (a) $-7\boldsymbol{i}+6\boldsymbol{j}+\boldsymbol{k}$ (b) -8 (c) $a = \frac{1}{\sqrt{3}}(\boldsymbol{i}+\boldsymbol{j}+\boldsymbol{k})$ **2.** i. $x = \frac{7}{2}+\lambda, y = \lambda, z = \frac{9}{2}+5\lambda$

ii. $\frac{x-3.5}{1} = \frac{y}{1} = \frac{z-4.5}{5}$ **3.** 0 **4.** $k=1$ or 3 **5.** $t=2, (16,-8,4)$ **6.** $A^2 = I_{3\times3}$; $(-1,-3,4)$

7. $BX = \frac{1}{2}A$, $X = \frac{1}{18}BA$, $AX = \frac{1}{18}ABA$ **8.** (a) $3\boldsymbol{i}-\boldsymbol{j}-2\boldsymbol{k}$ (b) 100° (c) $4\boldsymbol{i}-3\boldsymbol{j}-3\boldsymbol{k}$

9. (b) $A^{-1} = \frac{1}{3}(A-2I)$ **10.** (a) $|\boldsymbol{r}_B|_{min} = 2\sqrt{2}$ (b) $t = 5, b = \frac{2}{5}$ **11.** (a) $\frac{1}{a-2}\begin{bmatrix}-3\\2a-1\end{bmatrix}$

(b) $a=2$ (c) $x=-3, y=5$ **12.** (a) i. 90° ii. $\frac{7}{2}\sqrt{26}$ unit^2 (b) i. $\boldsymbol{s}+3\boldsymbol{p}$ ii. $\boldsymbol{s}+2\boldsymbol{p}$ iii. $\frac{1}{2}\boldsymbol{s}+2\boldsymbol{p}$

iv. $-\frac{1}{2}\boldsymbol{s}+2\boldsymbol{p}$ **13.** (a) 1 (b) $A^n = \begin{bmatrix}2^n & (2^n-1)a\\0 & 1\end{bmatrix}$ (c) $a = -\frac{2}{9}$ **14.** (a) 27° (b) $\frac{1}{2}\sqrt{17}$ unit^2

15. (a) i. $\binom{x}{y} = \binom{2}{-3}+\lambda\binom{3}{7}, \lambda\in\mathbb{R}$ ii. $x = 2+3\lambda, y = -3+7\lambda, \lambda\in\mathbb{R}$ iii. $\frac{x-2}{3} = \frac{y+3}{7}$

(b) $-\boldsymbol{i}+11\boldsymbol{j}$ (c) i. no ii. lines are skew **16.** 28°35′ **17.** (a) $a = \frac{3}{2}$ (b) $b = \frac{3}{2}, c = \frac{1}{3}$

18. $\frac{4}{\sqrt{77}}\left(-\frac{5}{4}\boldsymbol{i}+\boldsymbol{j}+\frac{3}{2}\boldsymbol{k}\right)$ or $-\frac{4}{\sqrt{77}}\left(-\frac{5}{4}\boldsymbol{i}+\boldsymbol{j}+\frac{3}{2}\boldsymbol{k}\right)$ **19.** $k=-1$, 1 or 2 **20.** (a) i. $\begin{bmatrix}2\\5\end{bmatrix}$ ii. $\frac{1}{2}\begin{bmatrix}1 & 0\\2 & 2\end{bmatrix}$

(b) $x+y-z = 1, 3x+4y+6z = 7$ (c) $\begin{pmatrix}x\\y\\z\end{pmatrix} = \begin{pmatrix}-3\\4\\0\end{pmatrix}+\lambda\begin{pmatrix}10\\-9\\1\end{pmatrix}$ **21.** (a) 5 (b) $\frac{5}{3}\sqrt{5}$

22. $\boldsymbol{OA} = 2\boldsymbol{i}-2\boldsymbol{j}+\boldsymbol{k}$, $\boldsymbol{OB} = 4\boldsymbol{i}-3\boldsymbol{k}$; 70°32′ **23.** $AB = I$; $x=3, y=0, z=-1$

24. (a) [11] (b) $\begin{bmatrix}3 & 6\\4 & 8\end{bmatrix}$ (c) $\begin{bmatrix}2 & 4\\5 & 8\end{bmatrix}$ (d) does not exist **25.** Yes **26.** (a) i. $(1,-1,2)$; $\left(\frac{1}{\sqrt{6}}, -\frac{1}{\sqrt{6}}, \frac{2}{\sqrt{6}}\right)$

ii. $(3,6,2)$; $\left(\frac{3}{7},\frac{6}{7},\frac{2}{7}\right)$ (b) lines do not meet **27.** $\frac{x-1}{3} = \frac{y-2}{2} = \frac{z+3}{1}$ **28.** $\left(\frac{2}{5},\frac{23}{5}\right)$

29. $\frac{2}{3}$ or 2 **30.** (a) $\boldsymbol{r}_A = \binom{0}{80000}+t\binom{3}{-2}$ (b) $\binom{21600}{65600}$ (units in metres) (c) They do not collide **31.** (a) $\boldsymbol{r} = \binom{4}{2}+t\binom{2}{3}$ (b) $\boldsymbol{LP} = \binom{-21}{-11}+t\binom{2}{3}$ (c) ~ 129.31 km

REVISION EXERCISES – SET D

1. (a) $\frac{1}{4}$ (b) $\frac{3}{8}$ (c) 0.3169 **2.** 0.0228 **3.** (a) 0.12 (b) 0.6087 **4.** (a) 0.89 (b) $\frac{21}{40}$ (c) $\frac{40}{89}$

5. (a) 0.46 (b) $\frac{9}{23}$ **6.** (a) 3326400 (b) i. $\frac{2}{11}$ ii. $\frac{2}{77}$ **7.** (a) 0.9772 (b) 0.3413 **8.** (a) 0.936 (b) 5

9. (a) 792 (b) 35 **10.** (a) 151200 (b) 0.1512 **11.** 0.2852 **12.** $\frac{128}{850}\approx 0.1506$ **13.** (a) 0.10 (b) 0.40

(c) $(x, P(X=x))$ values are: (0, 0.40), (1, 0.50), (2, 0.10) (d) $E(X) = 0.70$, $var(X) = 0.41$

14. (a) 0.8664 (b) 0.7210 (c) 0.9034 (d) $9.8855 < Y < 10.2145$ (e) 79.3350

15. (a) 315 (b) 17280 **16.** $\frac{193}{512}$ **17.** (a) $\frac{2}{3}$ (b) $\frac{1}{2}$ **18.** (a) $P(X=x) = \frac{1}{6}\times\left(\frac{5}{6}\right)^x, x = 0, 1, \ldots$

(i.e., Geometric) (b) i. 0.0670 ii. 0.4019 iii. $\frac{1}{6}$ **19.** (a) $\frac{13}{44}$ (b) $\frac{9}{44}$

20. (b) $(x, P(X=x))$ values are: $\left(1,\frac{9}{25}\right), \left(3,\frac{7}{25}\right), \left(5,\frac{5}{25}\right), \left(10,\frac{3}{25}\right), \left(20,\frac{1}{25}\right)$

(c) $E(X) = \frac{105}{25}\approx 4.2$, $var(X) = \frac{11400}{625}\approx 18.24$ (d) 0.00064 **21.** (a) 0.3085 (b) 0.0091 (c) 0.1587

22. 100 **23.** (a) $\frac{1}{2}$ (b) $\frac{1}{7}$ (c) $\frac{2}{7}$ **24.** (b) $(x, P(X=x))$ values are: (1,0.4), (2,0.3), (3,0.2),(4,0.1)

(c) i. 2 ii. 5 iii. 3 **25.** i. 0.8186 ii. 0.1585

26. (a) $(x, P(X=x))$ values are: $\left(0,\frac{3}{16}\right), \left(1,\frac{7}{16}\right), \left(2,\frac{5}{16}\right), \left(3,\frac{1}{16}\right)$ (b) ii. 0.0064 iii. 0.7705

27. $\mu = 0.9586, \sigma = 0.0252$ **28.** (a) $\frac{10}{21}$ (b) i. 0.3085 ii. 0.1747 **29.** (a) i. 0.8 ii. 0.25

(b) i. 0.4 ii. $E(X) = 0.8$, $var(X) = \frac{14}{25}$ **30.** (a) i. $\frac{1}{8}$ ii. $\frac{47}{72}$ iii. $\frac{1}{8}$ iv. $\frac{47}{72}$ v. $\frac{9}{47}$ **31.** $\frac{189}{8192}$

32. $\frac{43}{60}\approx 0.7167$ **33.** $\frac{117}{145}\approx 0.8069$ **34.** i. $(x, P(X=x))$ values are: $\left(0,\frac{1}{6}\right), \left(1,\frac{1}{3}\right), \left(2,\frac{1}{2}\right)$;

$E(X) = \frac{4}{3}$, $var(X) = \frac{5}{9}$ ii. $\frac{2}{3}$ iii. $\frac{5}{24}$ **35.** i. 0.4 ii. 0.096 iii. 0.225 iv. 0.635 **36.** (a) $\frac{3}{5}$

(b) $(x, P(X=x))$ values are: $\left(0,\frac{4}{25}\right), \left(1,\frac{12}{25}\right), \left(2,\frac{9}{25}\right)$ (c) $E(X) = 1.2$, $var(X) = 0.48$

(d) $\frac{3}{7}$ **37.** (a) i. $\frac{8}{15}$ ii. $\frac{7}{15}$ iii. $\frac{1}{5}$ iv. $\frac{4}{5}$ v. $\frac{4}{7}$ (b) $\frac{x(p-q)+100q}{100}$ **38.** $\frac{2}{3}$ **39.** (a) 0.1359

(b) 137.22 (c) $137\frac{1}{3}$ (d) $a = 141.21$ **40.** i. $\frac{2}{3}$ ii. $\frac{2}{9}$ iii. not independent

41. i. $b+6a$ ii. $0\le b\le\frac{1}{3}$ **42.** i. 0.081 ii. $\frac{4}{13}$ **43.** (a) 0.0169 (b) i. 0.9342 ii. 127 iii. 0.008

44. i. 0.1587 ii. 0.7745 iii. \$0.23 **45.** (a)

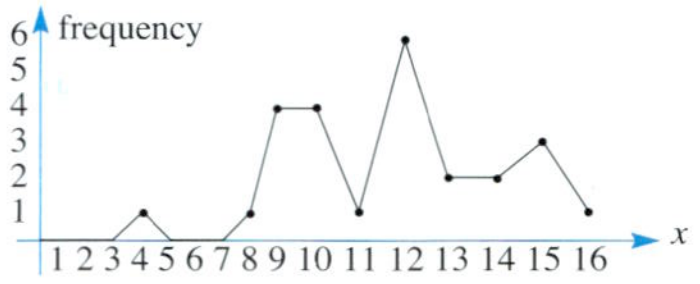

ii. 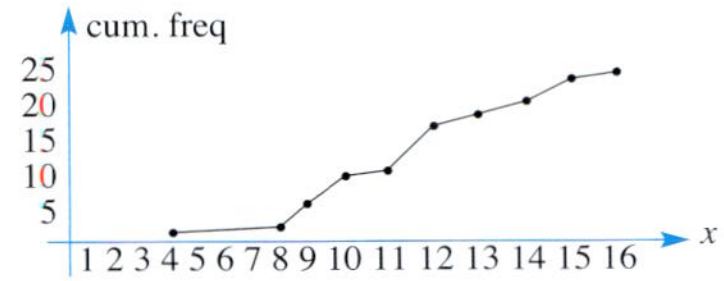

(b) i. 11.44 ii. 2.6695

(c) i. $x_{min} = 4$, $x_{max} = 16$, med = 12, $Q_3 = 13.5$, $Q_1 = 9.5$

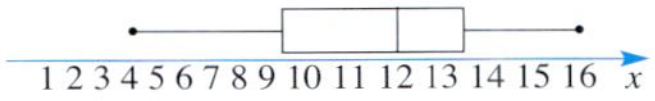

ii. med = 12, mode = 12 iii. 4

46. (a) i. 0.24 ii. 0.36 (b) $172 + 0.96Q$ (c) $Q > 29.17$

47. (a) use g.c (b) i. $\bar{x}_A = 37.35$, $\bar{x}_B = 37.31$ ii. A: $s_n = 8.801$, B: $s_n = 9.025$

(c) i. Class A: $x_{min} = 23$, $x_{max} = 56$, med = 36, $Q_3 = 43$, $Q_1 = 31$

Class B: $x_{min} = 22$, $x_{max} = 57$, med = 37, $Q_3 = 44$, $Q_1 = 31$

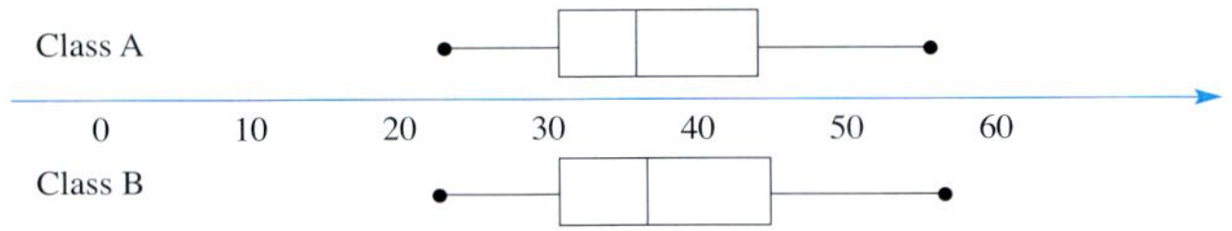

ii. Class A: med = 36; multimodal – 34, 35, 39, 43, 48
Class B: med = 37; multimodal – 27, 34, 38, 42, 49

iii. Class A: IQR = 12, Class B: IQR = 13 (d) Results from both classes are very close, however, Class B does slightly better as it has a larger median as well as the larger maximum value.

REVISION EXERCISES – SET E

1. (a) $\frac{x}{\sqrt{x^2+4}}$ (b) $2\cos 2x - 2(2x-1)\sin 2x$ **2.** $30\frac{1}{3} + \log_e 4$ **3.** (a) $\frac{\pi}{3}, \frac{4\pi}{3}$

(b) $\left\{x \mid \frac{\pi}{3} < x < \frac{4\pi}{3}\right\}$ (c) 4 sq. units **4.** (a) 19.8°C (b) 1.6°C per minute (c) 17.3 min

5. (a) $x \in [-1, 0[\cup]0, \infty[$ (b) $x \in]-\infty, 0[\cup [2, \infty[$ **6.** 10 m **7.** (a) $\frac{4x}{(x^2+1)^2}$

(b) $-4\sin 2x\cos 2x$ (or $-2\sin 4x$) **8.** (a) i. 0 ii. 2 (b) $x \in [-2, 2]$ (c) $x \geq 0$

(d) $-\frac{x}{\sqrt{4-x^2}}, -2 < x < 2$ **9.** (a) $-\frac{2+h}{(1+h)^2}, h \neq 0$ (b) –2 **10.** (a) 74 (b) 0.69 **11.** 1.455 ms^{-1}

12. (a) Absolute maximum at $\left(\pm\frac{1}{\sqrt{2}}, 1\right)$; local min at (0, 0); x-intercept at (±1, 0)

(b) Local min at $\left(\pm\frac{1}{\sqrt{2}}, 1\right)$; asymptotes at $x = \pm 1$, $y = 0$. **13.** i. $6\cos 2x\sin^2 2x$ ii. $\frac{x+3}{(2x+3)^{3/2}}$

14. $\frac{1}{2}(e^{2x} - 4x + e^{-2x}) + c$ **15.** 720 m^3 **16.** (a)

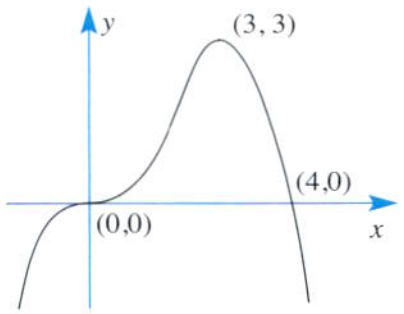

(b) 98°

17. (b) i. 2 ii. 72 cm^3 **18.** (a) Area = $A = \frac{8}{15}h^{3/2}$, Volume = $V = 0.48h^{3/2}$ (b) $\frac{5}{144}$ m/min

19. (a) (–1, 4), (1, –2), (3, 0) (b) use g.c (c) $\frac{16}{3} - 3\log_e 3$ sq. units **20.**

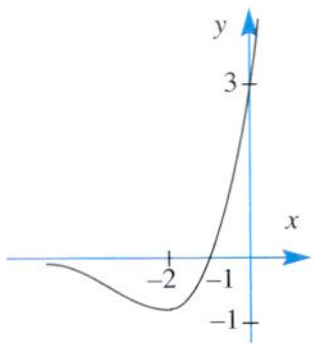

21. (a) $h = \frac{1000}{\pi r^2}$ (b) radius = 5 cm, height = 12.7 cm **22.** i. $-\frac{3x}{\sqrt{1-3x^2}}$ ii. $\frac{e^x}{(1+e^x)^2}$

23. (a) $3x^2h + 3xh^2 + h^3$ (b) $3x^2 + 3xh + h^2$ (c) $3x^2$ **24.** (a) $2 - \frac{3}{4}\sqrt{3}$ (b) $\log_e 3$

25. (a) $p'(t) = 0.8(1 - 0.02t)e^{-0.02t}$ (b) ~ 38.3 million (c) i. decreasing ii. `0.1 million/year (d) 50 years time, i.e., 2030; 42.2 million **26.** 76222 cm^3

27. (a) $A \equiv (-1, 5), B \equiv (1, 3), C \equiv (4, 0)$ (b) use g.c (c) $12 - 4\log_e 4$ sq. units

28. i. $\frac{1 + \cos x + x\sin x}{(1+\cos x)^2}$ ii. $\frac{x}{x^2+1}$ **29.** (a) $12 + 6h + h^2, h \neq 0$ (b) 12 **30.** (a) $A \equiv (2, 2e^{-1})$

(b) i. $y = x$ ii. $\frac{d}{dx}(xe^{-x/k}) = \left(1 - \frac{x}{2}\right)e^{-x/2}$ iii. $4 - 2(2+a)e^{-a/2} - \frac{1}{2}a^2$ (c) i. $(2x - x^2)e^{-x}$

ii. $\pi(2 - 10e^{-2})$ cubic units **31.** (a) $\log_e x$ (b) $2(\log_e 2 - 1)$ **32.** (a) i. 283 sec ii. 250 sec

(c) 244 sec **33.** (a) $A \equiv \left(\frac{1}{2}ln2, 2(1 - ln2)\right)$ (b) $\frac{1}{2}(e^2 - 5)$ sq. units (c) i. At (0,1): $y = -2x + 1$

At $(1, e^2 - 4)$: $y = (2e^2 - 4)x - e^2$ ii.

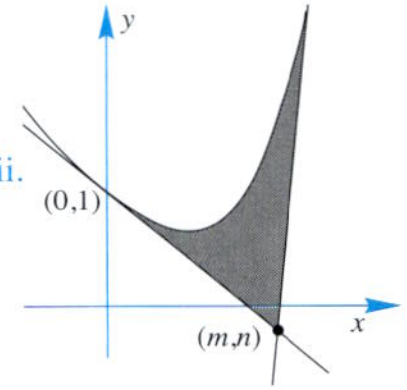

iii. $\frac{1}{2}(e^2 - 5)$ sq. units

(d) ii. $\frac{\pi}{12}(3e^4 - 24e^2 + 37)$ cubic units **34.** (b) i. $0 < x < 0.5$ ii. $x = 0.5$ iii. $x < 0$ or $x > 0.5$

(c) $\left(\frac{1}{2}, 1 - ln2\right)$ (d)

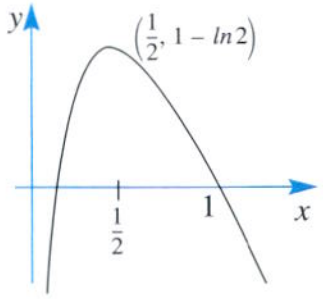

(e) i. $y = -x + 1$ ii. $y = x - 1$ (f) i. $\frac{3}{8} - \frac{1}{2}ln2$ sq. units

ii. $\frac{1}{8} + \frac{1}{2}ln2$ sq. units **35.** (a) 4.20 (b) i. $\frac{1}{2}\cos\frac{1}{4}t$ ii. –0.40 **36.** $\pi\log_e 3$ cubic units

37. $a = -1, b = 6, c = -9$ **38.** (a) $\frac{4\sqrt{3}}{3}$ (b) i. $-6\sin 3x\cos 3x$ ii. $\frac{x}{2} + \frac{1}{12}\sin 6x + c$

39. $A = 0, B = 0.5$ **40.** (a) $\frac{7}{12}$ sq. units (b) $\frac{7}{15}\pi$ cubic units **41.** (a) $V = \pi r^2 h + \frac{4}{3}\pi r^3$

(b) $P = 2\pi krh + 6\pi kr^2$ (c) $P = \frac{2kV}{r} + \frac{10\pi}{3}kr^2$ (d) $0 < r < \left(\frac{3V}{4\pi}\right)^{1/3}$ (e) $r = \left(\frac{3V}{10\pi}\right)^{1/3}$

42. (b) $\left[\frac{a}{4}, \frac{a}{2}\right]$ (c) $\frac{\sqrt{3}}{36}a^3$ cubic units **43.** (a) $2x\log_e x + x$; $2\log_e 2 - \frac{3}{4}$ (c) $(1, e)$; $y = ex$

(d) $\left(\frac{1}{2}e - 1\right)$ sq. units **44.** $x_1 = \frac{\pi}{6}, x_2 = \frac{\pi}{4}, x_3 = \frac{\pi}{3}$ **45.** (a) $a = 2$; $f^{-1}(x) = 2 + \sqrt{x}, x \geq 0$

(b) $\frac{136}{3}\pi$ cubic units **46.** (a) i. $\mathbb{R}$ ii. $[0, \infty[$ (b) 0.5 (c) $\{0, 4\}$ (d) i.

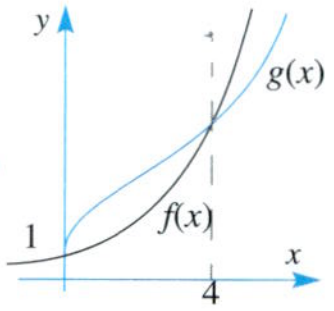

ii. $\{x \mid x > 4\}$ (e) i. 4 sq. units ii. $\frac{\pi}{2}(e^4 + 3)$ cubic units **48.** i. $-6\sin 2x\cos^2 2x$ ii. $\frac{1 - 2x^2}{\sqrt{1 - x^2}}$

49. (a) $\sin x + x\cos x$ (b) $x\sin x + \cos x$ **50.** (a) $y = -ex + e + e^{-1}$ (b)

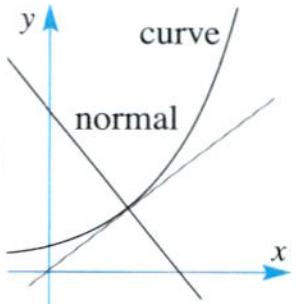

(c) $\frac{1}{2}e + e^{-2}$ sq. units **51.** (a) $\frac{4}{3}$ sq. units (b) $\frac{64}{15}\pi$ cubic units **52.** (a) $[0, 5]$ (b) use g.c (c) 0.625

(d) $a = \frac{1}{2} - \frac{1}{5}t, 0 \leq t \leq 5$ **53.** (c) Minimum, $3\pi a^2\left(\frac{5}{3}\right)^{1/3}$; Maximum $3\pi a^2\left(\frac{9}{4}\right)^{1/3}$

54. (a)

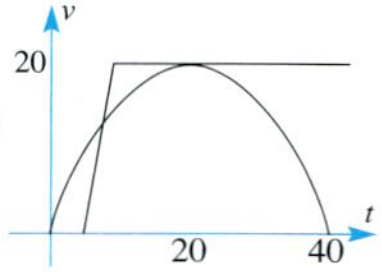

(b) 30 seconds (c) $116\frac{2}{3}$ metres **55.** (a) $-e^{-x}(\cos x + \sin x)$

(b) 1 **56.** (a) $-\frac{1}{t^2} + 1$ (b) $\ln t + 2t + \frac{1}{2}t^2 + c$ **57.** (a) use g.c (b) $A(x) = 2x\cos x, 0 < x < \frac{\pi}{2}$

(d) 1.12 sq. units

Note: A Solutions Manual – which includes fully worked solutions to almost every question in the textbook is available. Check the ibid press website for details as well as further updates on answers and solutions.

www.ibid.com.au

or contact fabio@ibid.com.au